SECOND EDITION

ELECTRICAL CONTACTS

PRINCIPLES AND APPLICATIONS

SECOND EDITION

ELECTRICAL CONTACTS

PRINCIPLES AND APPLICATIONS

Edited by

Paul G. Slade

CRC Press
Taylor & Francis Group
Boca Raton London New York

CRC Press is an imprint of the
Taylor & Francis Group, an **informa** business

CRC Press
Taylor & Francis Group
6000 Broken Sound Parkway NW, Suite 300
Boca Raton, FL 33487-2742

First issued in paperback 2017

Version Date: 20140130

ISBN 13: 978-1-138-07710-2 (pbk)
ISBN 13: 978-1-4398-8130-9 (hbk)

Library of Congress Cataloging-in-Publication Data

Electrical contacts : principles and applications / editor, Paul G. Slade. -- Second edition.
     pages cm
  Includes bibliographical references and index.
  ISBN 978-1-4398-8130-9 (hardback)
  1. Electric contacts. I. Slade, Paul G., 1941-

TK2821.E34 2013
621.31'042--dc23   2013045571

Visit the Taylor & Francis Web site at
http://www.taylorandfrancis.com

and the CRC Press Web site at
http://www.crcpress.com

Contents

Preface to the Second Edition.. xxix
Preface to the First Edition.. xxxi
Introduction xxxiii
Editor.. xli
Contributors..xliii

Part I   Contact Interface Conduction

1. Electrical Contact Resistance: Fundamental Principles 3
Roland S. Timsit

1.1   Introduction 4
1.2   Electrical Constriction Resistance..5
    1.2.1   Circular *a*-spots 5
    1.2.2   Non-Circular and Ring *a*-Spots...7
    1.2.3   Multiple Contact Spots..11
    1.2.4   Effect of the Shape of Contact Asperity on Constriction
            Resistance...17
1.3   Effect of Surface Films on Constriction and Contact Resistance..........18
    1.3.1   Electrically Conductive Layers on an Insulated Substrate 18
        1.3.1.1   Calculation of Spreading Resistance in a Thin Film.................18
    1.3.2   Electrically Conducting Layers on a Conducting Substrate....22
        1.3.2.1   Electrically Conducting Layers and Thin Contaminant
                  Films 23
    1.3.3   Growth of Intermetallic Layers..28
    1.3.4   Possible Effect of Electromigration on Intermetallic Growth
            Rates..42
    1.3.5   Electrically Insulating or Weakly Conducting Films...............45
        1.3.5.1   Growth Rate and Electrical Resistivity of Oxides of
                  Selected Contact Materials...47
    1.3.6   Fritting of Electrically Insulating Surface Films.......................56
1.4   Temperature of an Electrically Heated *a*-Spot...57
    1.4.1   Voltage–Temperature Relation..58
    1.4.2   Voltage–Temperature Relation with Temperature-Dependent
            Electrical Resistivity and Thermal Conductivity 60
    1.4.3   The Wiedemann–Franz Law...62
    1.4.4   Temperature Distribution in the Vicinity of an *a*-Spot............63
    1.4.5   Deviation of the Voltage–Temperature Relation in an Assymetric
            Contact...64
        1.4.5.1   Case I: Two Metals in Contact 64
        1.4.5.2   Case II: A Metal in Contact with a Non-metal.....................65
    1.4.6   Special Considerations on the "Melting" Voltage in Electrical
            Contacts...65

1.5   Mechanics of *a*-Spot Formation..69
        1.5.1   Smooth Interfaces..69
        1.5.2   Rough Interfaces 75
1.6   Breakdown of Classical Electrical Contact Theory in
        Small Contact Spots 79
        1.6.1   Electrical Conduction in Small *a*-Spots....................................79
                1.6.1.1   Contact Resistance..79
                1.6.1.2   Joule Heat Flow through *a*-Spots 82
        1.6.2   Observations of Breakdown of Classical Electrical Contact Theory
                in Aluminum Contacts 84
                1.6.2.1   Experimental Data on Aluminum...............................84
        1.6.3   Observations of Breakdown of Classical Electrical Contact Theory
                in Gold Contacts..87
        1.6.4   Observations of Breakdown of Classical Electrical Contact Theory
                in Tin Contacts 88
1.7   Constriction Resistance at High Frequencies 89
        1.7.1   Skin Depth and Constriction Resistance 89
        1.7.2   Evaluation of Constriction Resistance at High Frequencies....................91
        1.7.3   Constriction versus Connection Resistance at
                High Frequencies 94
1.8   Summary..95
Acknowledgments..96
References 104

2. **Introduction to Contact Tarnishing and Corrosion**....................................113
    Paul G. Slade

2.1   Introduction 114
2.2   Corrosion Rates 114
2.3   Corrosive Gases.. 117
2.4   Types of Corrosion.. 117
        2.4.1   Dry Corrosion.. 117
        2.4.2   Galvanic Corrosion... 118
        2.4.3   Pore Corrosion... 120
        2.4.4   Creep Corrosion 122
        2.4.5   Metallic Electromigration 123
        2.4.6   Stress Corrosion Cracking 124
        2.4.7   Contacts under Mineral Oil.. 125
2.5   Gas Concentrations in the Atmosphere.. 126
2.6   Measurements 129
        2.6.1   Weight Gain Measurement... 129
        2.6.2   Visual Inspection... 129
        2.6.3   Cathodic Reduction 129
        2.6.4   Scanning Electron Microscopy with Energy-Dispersive X-Ray
                Spectroscopy (SEM/EDAX) 129
        2.6.5   X-Ray Photoelectron Spectroscopy (XPS)..................................... 129
        2.6.6   Other Techniques 130
        2.6.7   Contact Resistance Measurements... 130
2.7   Mixed Flow Gas Laboratory Testing... 131

2.8   Electronic Connectors 131
      2.8.1   Background.. 131
      2.8.2   MFG Test Results 132
2.9   Power Connectors 133
2.10 Other Considerations 134
Acknowledgments 135
References 135

3. Gas Corrosion.. 139
William H. Abbott and Paul G. Slade

3.1   Introduction 140
      3.1.1   Scope.. 140
      3.1.2   Background.. 140
3.2   The Field Environments for Electrical Contacts................................... 143
      3.2.1   Environmental Variables 144
      3.2.2   Corrosion Rates 145
              3.2.2.1   Copper and Silver.. 145
              3.2.2.2   Other Metals 145
              3.2.2.3   Film Effects... 146
              3.2.2.4   Shielding Effects.. 147
      3.2.3   Reactivity Distributions.. 148
              3.2.3.1   Severity versus Performance 149
              3.2.3.2   Environmental Classes 150
              3.2.3.3   Specifications 150
      3.2.4   Corrosion Mechanisms 151
              3.2.4.1   Silver.. 151
              3.2.4.2   Copper... 151
              3.2.4.3   Nickel 152
              3.2.4.4   Tin.. 152
              3.2.4.5   Porous Gold Coatings 153
              3.2.4.6   Pore Corrosion 153
              3.2.4.7   Corrosion Product Creep 154
3.3   Laboratory Accelerated Testing 157
      3.3.1   Objectives.. 157
      3.3.2   Definition of Acceleration Factor.. 158
      3.3.3   Historical Background 159
      3.3.4   Single-Gas Corrosion Effects.. 159
              3.3.4.1   Hydrogen Sulfide 159
              3.3.4.2   Sulfur Dioxide (SO_2)... 163
              3.3.4.3   Nitrogen Dioxide (NO_2)... 163
              3.3.4.4   Chlorine.. 163
              3.3.4.5   Mixed-Gas Sulfur Environments................................... 164
              3.3.4.6   Humidity... 164
              3.3.4.7   Temperature... 164
              3.3.4.8   Gas Flow Effects... 165
      3.3.5   Mixed-Gas Environments.. 167
              3.3.5.1   Test Systems 169
              3.3.5.2   Monitoring Reactivity... 169

3.3.6　Test Applications..171
　　　3.3.6.1　Electronic Connectors.....................................171
　　　3.3.6.2　Mated versus Unmated Exposures.................172
　　　3.3.6.3　Other Considerations.......................................173
3.4　Lubrication and Inhibition of Corrosion 174
Acknowledgment..181
References 181

4. Effect of Dust Contamination on Electrical Contacts.................................185
Ji Gao Zhang

4.1　Introduction 186
　　4.1.1　Background..186
　　4.1.2　The Importance of the Dust Problem................................187
　　4.1.3　The Complexity of the Problem...189
　　4.1.4　The Purpose of the Studies..189
4.2　Dusty Environment and Dust Composition..............................190
　　4.2.1　The Source of Dust..190
　　4.2.2　The Collection of the Dust Particles for Testing 191
　　4.2.3　The Shape of the Dust Particles 191
　　4.2.4　The Identification of the Inorganic Materials 192
　　4.2.5　The Organic Materials in Dust...192
　　4.2.6　The Water Soluble Salts in Dust..192
4.3　The Characteristics of Dust Particles 193
　　4.3.1　The Electrical Behavior 193
　　　　4.3.1.1　Measurement of the Electric Charge 193
　　　　4.3.1.2　The Electrostatic Attracting Force on the Particle.................195
　　4.3.2　Mechanical Behavior...198
　　　　4.3.2.1　Load Effect 198
　　　　4.3.2.2　For Stationary Contacts 199
　　　　4.3.2.3　For Sliding Contacts..201
　　　　4.3.2.4　The Effect of Lubricants Coated on Contact Surface..............201
　　　　4.3.2.5　Sliding Contacts on Lubricated and Dusty Contacts..............203
　　　　4.3.2.6　Fretting (Micro Motion) on Lubricated and Dusty Contacts 204
　　4.3.3　Chemical Behavior...205
　　　　4.3.3.1　Dust Particles Create Pores...................................205
　　　　4.3.3.2　Corrosion Appears as a Result of Dusty Water Solutions......205
　　　　4.3.3.3　Indoor Exposure Results..206
　　　　4.3.3.4　Construction of the Corrosion Stain....................208
　　　　4.3.3.5　Fretting Experiments on Dust Corroded Coupon Surfaces...209
4.4　Application Conditions in Dusty Environment 211
　　4.4.1　Explanation of the Special Features.................................211
　　　　4.4.1.1　Covered by Accumulated Small Particles 211
　　　　4.4.1.2　Accumulative Particles Caused by Micro Motion.................211
　　　　4.4.1.3　High and Erratic Contact Resistance....................212
　　　　4.4.1.4　The Element of Si Causes High Contact Resistance................213
　　　　4.4.1.5　Organics Act as Adhesives 214
　　　　4.4.1.6　Corrosion Products Trap the Dust Particles............................216

4.4.1.7 Difference Between Short Life and
Longer Life Contacts.. 218
4.4.1.8 Large Pieces of "Stepping Stones" 218
4.4.1.9 The Performance of Failed Mobile Phones............... 218
4.4.2 Other Examples... 218
4.5 Theoretical Analysis of Connector Contact Failure due to the Dust................ 219
4.5.1 Two Micro Worlds in Contact 219
4.5.1.1 Particles Get into the Contact Interface................... 219
4.5.2 "Preliminary Attachment"... 220
4.5.2.1 Adhesive Effect.. 220
4.5.2.2 Trapping Effect of Corrosion Products 221
4.5.3 Contact Failure Mechanism 222
4.5.3.1 Single Particle and Ideal Model 222
4.5.3.2 Complicated Model – Number of Particles and
Morphology of Contact Pairs.................................... 223
4.5.4 Micro Movement... 223
4.5.4.1 Contact Failure... 223
4.6 Future Work.. 224
4.6.1 Dust Test for Connectors.. 224
4.6.2 Suggestion of the Dust Test 224
4.6.3 Minimizing the Dust Problem.. 225
4.6.3.1 Cleaning the Samples 225
References.. 226

Part II Nonarcing Contacts

5. Power Connectors 231
Milenko Braunović

5.1 Introduction... 233
5.2 Types of Power Connectors 234
5.2.1 Plug-and-Socket Connectors 236
5.2.2 Wire Connectors... 236
5.2.3 Bolted Connectors.. 236
5.2.4 Insulation Piercing Connectors 236
5.3 Properties of Conductor and Connector Materials.......................... 237
5.3.1 Definition of Conductor and Connector Systems 237
5.3.2 Factors Affecting Conductivity.. 237
5.3.2.1 Effect of Temperature 238
5.3.2.2 Effect of Lattice Imperfections 238
5.3.2.3 Magnetoresistance 239
5.3.2.4 Skin Effect... 240
5.3.3 Conductor Materials... 240
5.3.3.1 Copper and Copper Alloys... 240
5.3.3.2 Aluminum and Its Alloys 243
5.3.4 Materials for Connector Systems... 244
5.3.4.1 Pure Metals and Alloys 244
5.3.5 Electroplating and Cladding... 246

5.4   Parameters Affecting Performance of Power Connections 247
       5.4.1   Factors Affecting Reliability of Power Connections 247
       5.4.2   Contact Area 249
       5.4.3   Plastic Deformation 252
       5.4.4   Elastic Deformation 253
       5.4.5   Plated Contacts..253
       5.4.6   Oxidation..256
       5.4.7   Corrosion..260
                5.4.7.1   Atmospheric Corrosion 260
                5.4.7.2   Localized Corrosion 261
                5.4.7.3   Crevice Corrosion 261
                5.4.7.4   Pitting Corrosion 262
                5.4.7.5   Pore Corrosion 262
                5.4.7.6   Creep Corrosion 262
       5.4.8   Dust Corrosion...263
       5.4.9   Galvanic Corrosion...263
       5.4.10   Thermal Expansion 264
       5.4.11   Fretting...267
                5.4.11.1   Factors Affecting Fretting...............................268
                5.4.11.2   Mechanisms of Fretting 270
                5.4.11.3   Examples of Fretting Damage in Power Connections 274
                5.4.11.4   Compression Connectors 274
                5.4.11.5   Bus-Stab Contacts..276
                5.4.11.6   Plug-In Connectors.......................................278
                5.4.11.7   Bolted Connections 278
                5.4.11.8   Fretting in Aluminum Connections.........................283
                5.4.11.9   Effect of Electrical Current.............................285
                5.4.11.10 Fretting in Coatings (Platings) 286
                5.4.11.11 Fretting in Circuit Breaker Contact
                           Materials..287
       5.4.12   Intermetallic Compounds..288
       5.4.13   Intermetallics in Copper–Tin Systems...............................296
                5.4.13.1   Example of Intermetallics Formation in
                           Power Connections 298
       5.4.14   Stress Relaxation and Creep..299
       5.4.15   Nature of the Effect of Electric Current...........................301
       5.4.16   Effect of Electric Current on Stress Relaxation....................302
       5.4.17   Creep..305
5.5   Palliative Measures..306
       5.5.1   Contact Area...307
       5.5.2   Contact Pressure 313
       5.5.3   Mechanical Contact Device..320
       5.5.4   Disc-Spring (Belleville) Washers 322
       5.5.5   Wedge Connectors...325
       5.5.6   Automatic Splices..326
       5.5.7   Dead-end Connectors..327
       5.5.8   Shape-Memory Alloy Connector Devices...............................327
       5.5.9   Coating (Plating)..331

        5.5.10   Lubrication—Contact Aid Compounds 334
        5.5.11   Bimetallic Inserts 336
        5.5.12   Transition Washers 337
        5.5.13   Multilam Contact Elements..338
        5.5.14   Welded Connections 339
                    5.5.14.1   Thermite (Exothermic) Welding...339
                    5.5.14.2   Friction Welding..339
                    5.5.14.3   Explosion Welding 340
                    5.5.14.4   Resistance Welding 341
                    5.5.14.5   Resistance Brazing 341
        5.5.15   Connector Design 342
                    5.5.15.1   Fired Wedge-Connectors 343
                    5.5.15.2   Stepped Deep Indentation Connectors 344
   5.6   Connector Degradation...345
        5.6.1   Economical Consequences of Contact Deterioration............................346
        5.6.2   Power Quality..346
   5.7   Prognostic Models 349
        5.7.1   Prognostic Model 1 for Contact Remaining Life...............................349
        5.7.2   Prognostic Model 2 for Contact Remaining Life...............................350
        5.7.3   Physical Model 351
   5.8   Shape-Memory Alloys (SMA) 351
        5.8.1   Origin of Shape-Memory Effect..351
                    5.8.1.1   One-Way Memory Effect..352
                    5.8.1.2   Two-Way Memory Effect..353
        5.8.2   Applications of SMA in Power Connections 353
        5.8.3   Electrical Connections...353
        5.8.4   Temperature Indicators..355
   5.9   Metal Foam Materials..355
        5.9.1   Aluminum Foam Materials..355
                    5.9.1.1   Electrical and Thermal Properties of Foam Materials............357
                    5.9.1.2   Power Connection Applications...357
        5.9.2   Copper Foam Materials 358
                    5.9.2.1   Applications of Copper Foam Materials...............................358
        5.9.3   Silver Foam Materials..360
   5.10   Installation of Power Connections 361
        5.10.1   Examples of Improper Installations...361
   5.11   Accelerated Current-Cycling Tests (Standards)..363
        5.11.1   Present Current-Cycling Tests 363
   References 366

6. Low-Power Commercial, Automotive, and Appliance Connections 375
   Anthony Lee and George Drew

   6.1   Introduction...376
   6.2   Connectors 377
        6.2.1   Functional Requirements...377
        6.2.2   Types of Connectors 378
        6.2.3   Mechanical Considerations 381

6.3   Contact Terminals 384
       6.3.1   Contact Physics 384
       6.3.2   Terminal Types 384
       6.3.3   Other Electrical Contact Parameters...392
6.4   Degradation of Connector Contact 393
       6.4.1   Surface Films 393
       6.4.2   Fretting Corrosion of Tin-Plated Contacts 396
       6.4.3   Examples of Contact Failures..402
                6.4.3.1   Automotive Position Sensor Connector 402
                6.4.3.2   Fuel Injector Connector 403
                6.4.3.3   Glowing Contacts 403
                6.4.3.4   Electrolytic Corrosion 404
                6.4.3.5   Incompatible Plating and Low Contact Force.............404
6.5   Automotive Connector Contacts...405
       6.5.1   Vehicle Conditions...405
       6.5.2   High Power Connectors for Electric and
                Hybrid Vehicles...405
       6.5.3   Aluminum Wiring Connections 406
       6.5.4   Connections for High-Vibration Environment 408
6.6   Summary..408
References 409

7. Tribology of Electronic Connectors: Contact Sliding Wear, Fretting, and
   Lubrication...413
   Roland S. Timsit and Morton Antler

7.1   Introduction 414
7.2   Sliding Wear...415
       7.2.1   Early Studies...415
       7.2.2   Adhesion 416
                7.2.2.1   "Wiping" Contaminant from Contact Surfaces.........417
                7.2.2.2   Mild and Severe...418
                7.2.2.3   Prow Formation...419
                7.2.2.4   Rider Wear..421
                7.2.2.5   Gold Platings: Intrinsic Polymers and
                           Junction Growth 422
                7.2.2.6   Electroless Gold Plating 423
       7.2.3   Abrasion...424
       7.2.4   Brittle Fracture...426
       7.2.5   Delamination and Subsurface Wear 427
       7.2.6   Effect of Underplate and Substrate...428
                7.2.6.1   Hardness...428
                7.2.6.2   Roughness 431
       7.2.7   Electrodeposited Gold: Relationship of Wear to Underplate
                Hardness 431
                7.2.7.1   Hardener Metal Content 432
       7.2.8   Clad Metals..435
       7.2.9   Tin and Tin–Lead Alloys 437
       7.2.10   Silver 439

7.3 Fretting 441
   7.3.1 Background 441
   7.3.2 Fretting Regimes 442
   7.3.3 Static versus Dynamic Contact Resistance 445
   7.3.4 Field and Laboratory Testing Methodologies 447
      7.3.4.1 Generation of Fretting Displacement 447
      7.3.4.2 Determination of Contact Resistance 449
   7.3.5 Materials Studies 449
      7.3.5.1 Apparatus 449
      7.3.5.2 Metals Having Little or No Film-Forming Tendency 450
      7.3.5.3 Non-Noble Metals/Fretting Corrosion 452
      7.3.5.4 Frictional Polymer-Forming Metals 457
      7.3.5.5 Dissimilar Metals on Mating Contacts 460
   7.3.6 Wear-Out Phenomena 464
      7.3.6.1 Gold-Based Systems 464
      7.3.6.2 Palladium-Based Systems 465
      7.3.6.3 Tin and Tin–Lead Alloy Systems 466
      7.3.6.4 Role of Underplate and Substrate 467
   7.3.7 Parametric Studies 468
      7.3.7.1 Cycle Rate 468
      7.3.7.2 Wipe Distance 470
      7.3.7.3 Force 472
   7.3.8 Environmental Effects 474
   7.3.9 Thermal 475
   7.3.10 Effect of Current 476
   7.3.11 Surface Finish and Contact Geometry 479
   7.3.12 Material Transfer, Wear, Film Formation, and Contact
       Resistance 479
      7.3.12.1 Summary of Physical Processes 479
7.4 Lubrication 481
   7.4.1 Introduction 481
   7.4.2 Metallic Films 481
      7.4.2.1 Principles of Metallic Film Lubrication 481
      7.4.2.2 Sliding and Wiping Contacts 482
      7.4.2.3 Fretting Contacts 487
   7.4.3 Fluid Lubricants 488
      7.4.3.1 Background 488
      7.4.3.2 Some Fundamental Properties of Lubricants 488
      7.4.3.3 Requirements 491
      7.4.3.4 Types of Fluid Lubricants: A Sliding Contact
           Investigation 492
      7.4.3.5 Control of Fretting Degradation 496
   7.4.4 Grafted and Self-Assembled Lubricant Layers 500
   7.4.5 Greases and Solid Lubricants 503
      7.4.5.1 Greases 503
      7.4.5.2 Solids 504
   7.4.6 Lubricant Durability 504
   7.4.7 Other Considerations 505
References 506

8. Materials, Coatings, and Platings 519
Morton Antler and Paul G. Slade

  8.1   Introduction 520
        8.1.1   Scope 520
        8.1.2   Requirements of Contact Finishes and Coatings 520
        8.1.3   Terminology 520
  8.2   Metallic Finishes 520
        8.2.1   Wrought Metals 521
        8.2.2   Electrodeposits and Electroless Deposits 522
               8.2.2.1   Thickness of Platings 522
               8.2.2.2   Plating Hardness 523
               8.2.2.3   Classification of Platings 523
        8.2.3   Contact Finishes Produced by Non-Chemical Methods 525
        8.2.4   Metal-in-Elastomer Materials 525
        8.2.5   Overview 526
  8.3   Properties Related to Porosity 526
        8.3.1   Origins of Porosity 527
        8.3.2   Tests of Porosity 527
        8.3.3   Relationships between Porosity, Thickness of Finish, and
               Substrate Roughness 529
        8.3.4   Effect of Underplatings, Flash Coatings, and Strikes on the
               Porosity of Electrodeposits 530
        8.3.5   Reduction in the Chemical Reactivity of Finishes by the Use of
               Underplates 531
  8.4   Metallurgical and Structural Properties 532
        8.4.1   Thermal Diffusion 532
        8.4.2   Intermetallics 539
        8.4.3   Tin Whiskers 541
        8.4.4   Silver Whiskers 542
  8.5   Physical and Mechanical Properties 543
        8.5.1   Characteristics of Layered Systems 543
               8.5.1.1   Hardness 543
               8.5.1.2   Contact Resistance 543
        8.5.2   Topography 545
  Acknowledgement 546
  References 546

Part III   The Electric Arc and Switching Device Technology

9. The Arc and Interruption 553
Paul G. Slade

  9.1   Introduction 554
  9.2   The Fourth State of Matter 554
  9.3   Establishing an Arc 558
        9.3.1   Long-Gap Gas Breakdown 558
        9.3.2   Vacuum Breakdown and Short-Gap Breakdown 566
        9.3.3   The Volt–Current Characteristics of Separated Contacts 569

9.4 The Formation of the Electric Arc 570
    9.4.1 The Formation of the Electric Arc during Contact Closing................... 570
    9.4.2 The Formation of the Electric Arc during Contact Opening................ 571
9.5 The Arc in Air at Atmospheric Pressure 578
    9.5.1 The Arc Column... 578
    9.5.2 The Cathode Region 581
    9.5.3 The Anode Region 584
    9.5.4 The Minimum Arc Current and the Minimum
           Arc Voltage... 585
    9.5.5 Arc Volt–Ampere Characteristics 588
9.6 The Arc in Vacuum 592
    9.6.1 The Diffuse Vacuum Arc 592
    9.6.2 The Columnar Vacuum Arc 595
    9.6.3 The Vacuum Arc in the Presence of a Transverse
           Magnetic Field.. 596
    9.6.4 The Vacuum Arc in the Presence of an Axial Magnetic Field.............. 596
9.7 Arc Interruption 597
    9.7.1 Arc Interruption in Alternating Current Circuits............................... 597
          9.7.1.1 Stage 1 – Instantaneous Dielectric Recovery............................ 600
          9.7.1.2 Stage 2 – Decay of the Arc Plasma and
                  Dielectric Reignition 602
          9.7.1.3 Thermal Reignition 603
    9.7.2 Arc Interruption in Direct Current Circuits 604
    9.7.3 Vacuum Arc Interruption in Alternating Circuits 607
    9.7.4 Arc interruption of Alternating Circuits: Current Limiting 608
    9.7.5 Interruption of Low Frequency and High Frequency
           Power Circuits.. 609
    9.7.6 Interruption of Megahertz and Gigahertz
           Electronic Circuits 609
Acknowledgments.. 612
References 612

10. The Consequences of Arcing 617
Paul G. Slade

10.1 Introduction 618
10.2 Arcing Time 618
    10.2.1 Arcing Time in an AC Circuit.. 618
    10.2.2 Arcing Time in a DC Circuit 618
    10.2.3 Activation of the Contact 622
    10.2.4 Arcing Time in Very Low-Current DC Circuits:
            Showering Arcs... 624
10.3 Arc Erosion of Electrical Contacts 628
    10.3.1 Erosion on Make and Erosion on Break 631
    10.3.2 The Effect of Arc Current 631
    10.3.3 The Effect of Contact Size 633
    10.3.4 Determination of Contact Size in AC Operation................................ 635
    10.3.5 Erosion of Contacts in Low-Current DC Circuits 636
    10.3.6 Erosion of Contacts in Low-Current AC Circuits 644

10.4 Blow-Off Force...646
    10.4.1 Butt Contacts...647
10.5 Contact Welding...651
    10.5.1 Welding of Closed Contacts 651
    10.5.2 Welding during Contact Closure...654
    10.5.3 Welding as Contacts Open 657
10.6 Changes in the Contact Surface as a Result of Arcing 657
    10.6.1 Silver-Based Contacts 659
    10.6.2 Silver-Refractory Metal Contacts..659
    10.6.3 Other Ambient Effects on the Arcing Contact Surface: Formation
        of Silica and Carbon and Contact Activation 665
Acknowledgments...667
References 667

11. Reed Switches

11. Reed Switches..673
Kunio Hinohara

11.1 Principles and Design of the Reed Switch.. 674
    11.1.1 Pull-In Characteristics of a Reed Switch 674
    11.1.2 Drop-Out Characteristics of a Reed Switch.......................................682
    11.1.3 Magnet Drive Characteristics of a Reed Switch................................683
        11.1.3.1 X–Y Characteristic H (Horizontal) 684
        11.1.3.2 X–Z Characteristic H (Horizontal)..684
        11.1.3.3 X–Y Characteristic V (Vertical) 685
11.2 Recommended Contact Plating..686
    11.2.1 Materials for Contact Plating 686
    11.2.2 Ground Plating..686
    11.2.3 Rhodium Plating...687
    11.2.4 Ruthenium Plating...687
    11.2.5 Other Platings...688
        11.2.5.1 Copper Plating...688
        11.2.5.2 Tungsten Plating..689
        11.2.5.3 Rhenium Plating..689
        11.2.5.4 Iridium Plating 689
        11.2.5.5 Nitriding the Permalloy (Ni-Fe [48 wt%])
             Blade Material..689
11.3 Contact Surface Degradation and Countermeasures 689
    11.3.1 Surface Deactivation Treatment...690
        11.3.1.1 Life Test of Samples Left for 24 Hours after Sealing..............691
        11.3.1.2 Life Test of Samples Left for One Week after Sealing............691
        11.3.1.3 Life Test of Samples Left for One Month after Sealing...........693
        11.3.1.4 Life Test of Samples Left for Three Months, Six Months,
             and One Year after Sealing 693
    11.3.2 Prevention of Contact Adhesion..695
11.4 Applications of Reed Switches 696
    11.4.1 Reed Relays..697
    11.4.2 Applications of Magnetic-Driven Reed Switches...............................698
References 701

12. Low Current and High Frequency Miniature Switches: Microelectromechanical Systems (MEMS), Metal Contact Switches 703
Benjamin F. Toler, Ronald A. Coutu, Jr., and John W. McBride

12.1 Introduction 703
    12.1.1 Common MEMS Actuation Methods 704
12.2 Micro-Contact Resistance Modeling 705
12.3 Contact Materials for Performance and Reliability 713
12.4 Failure Modes and Reliability 720
12.5 Conclusion 725
References 725

13. Low Current Switching 731
John W. McBride

13.1 Introduction and Device Classification 733
13.2 Device Types 734
    13.2.1 Hand-Operated Switches 734
        13.2.1.1 The Rocker Switch Mechanism 734
        13.2.1.2 Lever Switches 737
        13.2.1.3 Slide Switches 738
        13.2.1.4 Rotary Switches 738
        13.2.1.5 Push-Button Switches 738
        13.2.1.6 Switching Devices Used below 0.5 A 739
    13.2.2 Actuated Switches 739
        13.2.2.1 Limit Switches 739
        13.2.2.2 Thermostatic Controls 740
        13.2.2.3 Electro-Mechanical Relay 741
13.3 Design Parameters for Static Switching Contacts 741
    13.3.1 Small-Amplitude Sliding Motion 742
    13.3.2 Contact Force and Contact Materials 742
        13.3.2.1 Contacts at Current Levels below 1 A 742
        13.3.2.2 Contacts at Current Levels between 1 and 30 A 742
        13.3.2.3 Contact Force 743
13.4 Mechanical Design Parameters 743
    13.4.1 Case Study (1): Hand-Operated Rocker-Switch Mechanism 744
        13.4.1.1 Moving-Contact Dynamics of a Rocker-Switch Mechanism 745
        13.4.1.2 Design Optimization of a Rocker-Switch Mechanism 746
    13.4.2 The Opening Characteristics of Switching Devices 746
        13.4.2.1 Moving Contact Dynamics at Opening 747
    13.4.3 The Make Operation 747
        13.4.3.1 Impact Mechanics 748
        13.4.3.2 The Coefficient of Restitution 749
        13.4.3.3 Impact Mechanics for a Pivoting Mechanism 751
        13.4.3.4 The Velocity of Impact 752
        13.4.3.5 Bounce Times 752
        13.4.3.6 Total Bounce Times 754

          13.4.3.7 Impact Times...755
          13.4.3.8 Design Parameters for the Reduction of
                    Contact Bounce...755
13.5 The Measurement of Contact Wear and Contact Dynamics 755
      13.5.1 The Measurement of Contact Surfaces 756
      13.5.2 Three Dimensional (3-D) Surface Measurement
              Systems 757
              13.5.2.1 Contact Systems...757
              13.5.2.2 Non-Contact Systems..758
      13.5.3 Case Study (2): Example of Volumetric Erosion 758
      13.5.4 The Measurement of Arc Motion and Contact
              Dynamics 759
13.6 Electrical Characteristics of Low-Current Switching
      Devices at Opening..761
      13.6.1 Low-Current DC Arcs 761
              13.6.1.1 Arc Voltage Characteristics.....................................762
              13.6.1.2 Voltage Steps below 7 A 762
              13.6.1.3 Case Study (3): Arc Voltage, Current and Length under
                        Quasi-Static Conditions for Ag/CdO Contacts.........763
              13.6.1.4 Opening Speed and Arc Length................................764
              13.6.1.5 Case Study (4): Automotive Systems 765
      13.6.2 DC Erosion...766
              13.6.2.1 Ag and Ag/MeO Contact Erosion/Deposition.......766
      13.6.3 Low-Current AC Arcs 768
              13.6.3.1 Typical Waveforms and Arc Energy........................769
      13.6.4 AC Erosion...770
              13.6.4.1 Point-on-Wave (POW) Studies Using Ag/CdO
                        Contact Materials 771
13.7 Electrical Characteristics of Low Current Switching
      Devices at Closure...773
      13.7.1 Contact Welding on Make 774
      13.7.2 Reducing Contact Bounce...775
      13.7.3 Pre-Impact Arcing 775
      13.7.4 Influence of Velocity during the First Bounce 777
              13.7.4.1 The First Bounce...778
      13.7.5 Bounces after the First...778
      13.7.6 Summary of Contact Bounce 779
13.8 Summary...779
      13.8.1 Switch Design...779
      13.8.2 Break Operation 779
              13.8.2.1 DC Operation..780
              13.8.2.2 AC Operation..780
      13.8.3 Make Operation 780
              13.8.3.1 Design Parameters 780
              13.8.3.2 Reducing Contact Bounce 781
              13.8.3.3 Arcing during the Bounce Process..........................781
Acknowledgments...781
References 781

14. Medium to High Current Switching: Low Voltage Contactors and Circuit Breakers, and Vacuum Interrupters..785
Manfred Lindmayer

    14.1   General Aspects of Switching in Air..787
          14.1.1 Arc Chutes 787
          14.1.2 Magnetic Blast Field 788
          14.1.3 Arc Dwell Time on the Contacts...789
          14.1.4 Sticking and Back-Commutation of the Arc 790
    14.2   Contacts for Switching in Air...791
    14.3   Low-Voltage Contactors 792
          14.3.1 Principle/Requirements...794
          14.3.2 Mechanical Arrangement..794
          14.3.3 Quenching Principle and Contact and Arc Chute Design.....................798
          14.3.4 Contact Materials...802
          14.3.5 Trends..803
                14.3.5.1 Contactors versus Electronics.................................804
                14.3.5.2 Vacuum Contactors 805
                14.3.5.3 Hybrid Contactors..805
                14.3.5.4 Integration with Electronic Systems.......................805
    14.4   Low-Voltage Circuit-Breakers and Miniature Circuit-Breakers..806
          14.4.1 Principle/Requirements...806
          14.4.2 General Arrangement 807
          14.4.3 Quenching Principle and Design of Arc Chute and Contact System ..808
                14.4.3.1 Quenching Principles 808
                14.4.3.2 Arc Chute and Contact Arrangement 810
          14.4.4 Trip System 814
          14.4.5 Examples of Miniature Circuit-Breakers...................................815
          14.4.6 Contact Materials...816
          14.4.7 Special Requirements for DC Switching 820
          14.4.8 Current Limitation by Principles Other than Deion Arc Chutes..........820
                14.4.8.1 Arcs Squeezed in Narrow Insulating Slots 821
                14.4.8.2 Reversible Phase Changes of Liquid or Low-Melting Metal...821
                14.4.8.3 Temperature-Dependent Ceramics or Polymers 821
                14.4.8.4 Contact Resistance between Powder Grains.............821
                14.4.8.5 Superconductors..821
    14.5   Simulations of Low-Voltage Switching Devices 822
          14.5.1 Simulation of Low-Voltage Arcs 822
                14.5.1.1 General Principle of Simulation 822
                14.5.1.2 Arc Roots on Cathode and Anode.............................824
                14.5.1.3 Radiation 825
                14.5.1.4 Interaction between Arc and Electrode or Wall Material (Ablation)...826
                14.5.1.5 Plasma Properties..826
                14.5.1.6 Simplification by Porous Media 827
           14.5.2 Further Simulations of Contact and Switching Device Behavior.........827

14.6 Vacuum Interrupters 828
      14.6.1 Principle/Applications 828
      14.6.2 Design..829
      14.6.3 Recovery and the Influence of the Design 831
      14.6.4 Contact Materials for Vacuum Interrupters and Their Influence on
              Switching...835
              14.6.4.1 Requirements..835
              14.6.4.2 Arc Interruption 836
              14.6.4.3 Interruption of High Frequency Transients 837
              14.6.4.4 Current Chopping...838
      14.6.5 Simulation of Arcs in Vacuum Interrupters....................................839
References 840

15. Arc Faults and Electrical Safety...849
    John J. Shea

    15.1 Introduction 849
    15.2 Arc Fault Circuit Interrupters (AFCIs)..850
    15.3 Arcing Faults..853
          15.3.1 Short-Circuit Arcing..854
          15.3.2 Series Arcing 855
    15.4 Glowing Connections..857
    15.5 Arcing Fault Properties...865
          15.5.1 Frequency..865
          15.5.2 Electrode Materials...866
          15.5.3 Arc Fault Current...867
          15.5.4 Cable Impedance and Cable Length Effects...................................868
    15.6 Other Types of Arcing Faults 871
    15.7 Conclusions..872
    References 872

Part IV   Arcing Contact Materials

16. Arcing Contact Materials...879
    Gerald J. Witter

    16.1 Introduction 880
    16.2 Silver Metal Oxides...883
          16.2.1 Types..883
          16.2.2 Manufacturing Technology...884
                  16.2.2.1 Internal Oxidation...884
                  16.2.2.2 Post-Oxidized Internally Oxidized Parts (Process B 1.0) 885
                  16.2.2.3 One-Sided Internally Oxidized Parts (Process B 2.01)...........887
                  16.2.2.4 Preoxidized Internally Oxidized Parts (Process B.2.02) 887
                  16.2.2.5 Powder Metallurgical (PM) Silver Metal Oxides
                            (Processes C and D) 887
          16.2.3 Electrical Performance Factors...890
                  16.2.3.1 AC versus DC Testing...890
                  16.2.3.2 High Current Inrush DC Automotive and AC Loads.............890

16.2.3.3 Inductive Loads 890
16.2.3.4 Silver–Tin Oxide Type Materials and Additives....................891
16.2.3.5 Material Factor 895
16.2.3.6 Interpreting Material Research, Example from Old Silver
Cadmium Oxide Research...895
16.2.4 Material Considerations Based on Electrical Switching
Characteristics 899
16.2.4.1 Erosion/Materials Transfer/Welding....................................899
16.2.5 Transfer/Welding..902
16.2.6 Erosion/Mechanisms/Cracking 904
16.2.7 Erosion/Arc Mobility 906
16.2.8 Interruption Characteristics 906
16.2.9 Contact Resistance 906
16.2.9.1 Summary Metal Oxides 908
16.3 Silver Refractory Metals...908
16.3.1 Manufacturing Technology...909
16.3.1.1 Manufacturing Technology/Press Sinter Repress
(Process D 1.0) 909
16.3.2 Material Technology/Extruded Material 910
16.3.2.1 Material Technology/Liquid Phase Sintering
(Process D 2.0)...910
16.3.2.2 Material Technology/Press Sinter Infiltration
(Process D 3.0) 911
16.3.3 Metallurgical/Metallographic Methods...912
16.3.3.1 Metallurgical/Metallographic Methods/Preparation 912
16.3.3.2 Metallurgical/Metallography/Quantitative Analysis 912
16.3.4 Metallurgical/Structure/Strength and Toughness.............................914
16.3.5 Electrical Properties (EP) 918
16.3.5.1 EP/Arc Erosion/Microstructure and Properties.....................919
16.3.5.2 EP/Arc Erosion/Silver Refractory 921
16.3.5.3 EP/Graphite Additions to Silver Tungsten and Silver
Tungsten Carbide 922
16.3.5.4 EP/Copper Refractory Metals..922
16.3.5.5 EP/Erosion/Summary 927
16.3.5.6 EP/Composite Refractory Materials/Contact
Resistance 928
16.4 Vacuum Interrupter Materials 935
16.5 Tungsten Contacts...936
16.6 Non-Noble Silver Alloys 937
16.6.1 Fine Silver..937
16.6.2 Hard Silver and Silver–Copper Alloys..938
16.7 Silver–Nickel Contact Materials...939
16.8 Silver Alloys and Noble Metals..940
16.8.1 Palladium and Silver–Palladium Alloys 940
16.8.2 Platinum...942
16.9 Silver–Graphite Contact Materials 943
16.10 Conclusion..945
Acknowledgements..946
References 946

17. Contact Design and Attachment..953
Gerald J. Witter and Guenther Horn

   17.1   Introduction 954
       17.1.1   Arc-Induced Contact Stresses and Interface Bond Quality 954
   17.2   Staked Contact Assembly Designs 955
       17.2.1   Contact Rivets...955
           17.2.1.1 Solid Rivets..955
           17.2.1.2 Machine-Made Composite Rivets....................................956
           17.2.1.3 Brazed Composite Rivets 959
           17.2.1.4 Rivet Staking 960
   17.3   Welded Contact Assembly Designs..960
       17.3.1   Resistance Welding...962
           17.3.1.1 Button Welding..962
           17.3.1.2 Wire-Welding..963
           17.3.1.3 Contact Tape Welding..963
       17.3.2   Special Welding Methods...965
           17.3.2.1 Percussion Welding..965
           17.3.2.2 Ultrasonic Welding of Contacts 965
           17.3.2.3 Friction Welding of Contacts 966
   17.4   Brazed Contact Assembly Designs..966
       17.4.1   Methods for Brazing Individual Parts..966
           17.4.1.1 Torch Brazing..967
           17.4.1.2 Induction Brazing..967
           17.4.1.3 Direct and Indirect Resistance Brazing 967
           17.4.1.4 Furnace Brazing..968
           17.4.1.5 Continuous Laminated Strip Brazing, "Toplay" 968
           17.4.1.6 Brazed Assembly Quality Control Methods...........................968
   17.5   Clad Metals, Inlay, and Edge Lay 969
   17.6   Contact Alloys for Non-Arcing Separable Contacts 970
       17.6.1   Gold and Gold Alloys..970
       17.6.2   Manufacturing Technology...970
       17.6.3   Physical and Chemical Properties..970
       17.6.4   Metallurgical Properties 971
       17.6.5   Contact Applications and Performance...972
   Acknowledgments..972
   References 972

18. Electrical Contact Material Testing Design and Measurement...............................975
Gerald J. Witter and Werner Rieder

   18.1   Objectives 976
   18.2   Device Testing and Model Switch Testing 976
       18.2.1   Device Testing 976
       18.2.2   Model Switch Testing 977
   18.3   Electrical Contact Testing Variables 978
       18.3.1   AC versus DC Testing 978
       18.3.2   Switching Load Type..979
       18.3.3   Opening and Closing Velocity Effects..980
       18.3.4   Contact Bounce...980

18.3.5 Contact Carrier Mass and Conductivity 981
18.3.6 Contact Closing Force and Over Travel 981
18.3.7 Enclosed and Open Contact Devices 981
18.3.8 Testing at Different Ambient Temperatures 982
18.3.9 Erosion Measurement 982
18.3.10 Summary Electrical Contact Testing Variables 982
18.4 Electrical Testing Result Types and Measurement Methods 982
18.4.1 Contact Resistance 982
18.4.1.1 Model Testing 983
18.4.1.2 Evaluation and Presentation of Results 985
18.4.2 Contact Bounce Measurement 986
18.4.2.1 Model Testing 988
18.4.2.2 Evaluation 988
18.4.3 Contact Welding Measurement 989
18.4.3.1 Weld Strength Measured 991
18.4.4 Contact Erosion Measurements 991
18.4.4.1 Accelerated and Model Testing 993
18.4.4.2 Extrapolation at Rated Stress 993
18.4.4.3 Increase of the Switching Frequency 993
18.4.4.4 Testing at Increased Electrical Load 994
18.4.4.5 Fixed-Gap Models 994
18.4.4.6 Moving Contact Models 994
18.4.4.7 Evaluation and Presentation of Results 995
18.4.5 AC Arc Reignition Measurement 996
18.4.6 Arc Motion Measurements 996
18.4.6.1 Measurement 996
18.4.6.2 Electronic Optical 996
18.4.6.3 Model Switch Arc Motion Control 998
18.4.6.4 Evaluation and Presentation of Results 999
18.4.7 Arc-Wall Interaction Measurements 999
References 999

19. Arc Interactions with Contaminants 1005
Gerald J. Witter and Werner Rieder

19.1 Introduction 1005
19.2 Organic Contamination and Activation 1006
19.2.1 The Phenomena 1006
19.2.2 Sources of Organic Vapors 1007
19.2.3 Processes of Contact Activation 1007
19.2.4 Activation Effects 1011
19.2.5 Activation and Contact Resistance Problems 1012
19.2.6 Methods for Detecting Carbon Contamination 1014
19.3 Mineral Particulate Contamination of Arcing Contacts 1015
19.4 Silicone Contamination of Arcing Contacts 1018
19.4.1 Contamination from Silicone Vapors 1019
19.4.2 Contamination from Silicone Migration 1023
19.4.3 Summary of Silicone Contamination Mechanisms 1029
19.5 Lubricants with Refractory Fillers 1030

19.6 Oxidation of Contact Materials 1031
19.7 Resistance Effects from Long Arcs 1032
Acknowledgments 1034
References 1034

Part V   Sliding Electrical Contacts

20. **Sliding Electrical Contacts (Graphitic Type Lubrication)** 1041
Kiochiro Sawa and Erle I. Shobert II

20.1 Introduction 1042
20.2 Mechanical Aspects 1044
      20.2.1 Hardness 1045
      20.2.2 Friction and Wear 1047
      20.2.3 Tunnel Resistance and Vibration 1048
20.3 Chemical Aspects 1051
      20.3.1 Oxidation 1051
      20.3.2 Moisture Film 1051
20.4 Electrical Effects 1055
      20.4.1 Constriction Resistance 1055
      20.4.2 Film Resistance 1055
      20.4.3 Fundamental Aspects of Commutation 1056
      20.4.4 Equivalent Commutation Circuit and DC Motor Driving
              Automotive Fuel Pump 1061
      20.4.5 Arc Duration and Residual Current 1061
20.5 Thermal Effects 1064
      20.5.1 Steady State 1064
      20.5.2 Actual Temperature 1067
      20.5.3 Thermal Mound 1068
20.6 Brush Wear 1068
      20.6.1 Holm's Wear Equation 1071
      20.6.2 Flashes and Smutting 1072
      20.6.3 Polarities and Other Aspects 1073
20.7 Brush Materials and Abrasion 1073
      20.7.1 Electro- and Natural Graphite Brushes 1074
      20.7.2 Metal Graphite Brush and Others 1075
20.8 Summary 1076
References 1077

21. **Illustrative Modern Brush Applications** 1081
Wilferd E. Yohe and William A. Nystrom

21.1 Introduction 1081
21.2 Brush Materials 1082
      21.2.1 Electrographite 1082
      21.2.2 Carbon-Graphite 1083
      21.2.3 Graphite 1083
      21.2.4 Resin-Bonded 1083

21.2.5 Metal-Graphite 1084
21.2.6 Altitude-Treated Brushes 1084
21.3 Brush Applications 1085
21.3.1 Minature Motors 1085
21.3.2 Fractional Horsepower Motors 1085
21.3.2.1 Wound Field/Permanent Magnet-Motor
Characteristics 1086
21.3.3 Automotive Brush Applications 1086
21.3.3.1 Auxiliary Motors 1087
21.3.3.2 Alternators 1088
21.3.3.3 Starter Motors 1088
21.3.4 Industrial Brushes 1089
21.3.5 Diesel Electric Locomotive Brushes 1090
21.3.6 Aircraft and Space Brushes 1090
21.3.7 Brush Design 1091

22. **Sliding Contacts for Instrumentation and Control** 1093
Glenn Dorsey and Jax Glossbrenner

22.1 Introduction 1094
22.2 Sliding Contact—The Micro Perspective 1097
22.2.1 Mechanical Aspects 1098
22.2.2 Motion Initiation (Pre-Sliding) 1100
22.2.3 Friction Forces 1100
22.2.4 Motion Continuation 1102
22.2.5 Adhesion 1102
22.2.6 Adhesive Transfer 1103
22.2.7 Plowing, or "Two-Body," Abrasion 1104
22.2.8 Hard Particle, or "Three-Body," Abrasion 1105
22.2.9 Motion Over Time 1105
22.3 Electrical Performance 1107
22.3.1 Contact Resistance Variation (Noise) 1108
22.3.2 Non-Ohmic Noise 1109
22.3.3 Non-Linear Noise (Frequency Dependent) 1111
22.3.4 Contact Impedance 1112
22.3.5 Data Integrity 1114
22.4 Micro-Environment of Contact Region 1114
22.4.1 Film Forming on the a-Spots 1115
22.4.2 Unintentional Contamination 1117
22.4.2.1 Particulates 1117
22.4.2.2 Contamination or "Air Pollution" 1118
22.4.2.3 Organic Off-Gasses 1118
22.4.2.4 Friction Polymers 1119
22.4.3 Lubrication (Intentional Contamination) 1119
22.4.4 Lubrication Modes (Anaerobic and Aerobic) 1121
22.4.4.1 Anaerobically Lubricated Contacts 1122
22.4.4.2 Aerobically Lubricated Contacts 1123
22.4.4.3 Temperature Extremes 1124
22.4.4.4 Submerged in Flammable Fuels 1125

                22.4.4.5 Low-Pressure/Vacuum Operation... 1125
                22.4.4.6 Vapor and Gas Lubrication 1125
        22.5 Macro Sliding Contact.. 1126
                22.5.1 Counterface Configuration.. 1126
                        22.5.1.1 Flat Surfaces 1127
                        22.5.1.2 Cylindrical Surfaces... 1127
                        22.5.1.3 Counterface Contact Shapes 1127
                22.5.2 Real versus Apparent Area of Contact 1128
                22.5.3 Brush Configurations.. 1128
                        22.5.3.1 Cartridge Brush 1128
                        22.5.3.2 Cantilever Composite Brush.. 1128
                        22.5.3.3 Cantilever Metallic Finger 1129
                        22.5.3.4 Cantilever Wire Brush 1129
                        22.5.3.5 Multifilament or Fiber Brush.. 1130
                        22.5.3.6 Benefits of Multiple Brushes 1130
                22.5.4 Forces on the Brush 1131
        22.6 Materials for Sliding Contacts.. 1132
                22.6.1 Materials for Counterfaces 1132
                22.6.2 Solid Lubricated Composite Materials for Brushes 1133
                22.6.3 Wire Brush Materials Criteria.. 1133
        22.7 Friction and Wear Characteristics 1135
                22.7.1 Friction.. 1135
                22.7.2 Wear... 1137
        22.8 Contact Parameters and Sliding-Contact Assemblies 1138
                22.8.1 Contact Noise 1138
                22.8.2 Slip Rings as Transmission Lines.. 1139
                22.8.3 Results of Normal Operation 1140
        22.9 Future.. 1141
        22.10 Summary... 1142
        Acknowledgments.. 1143
        References 1143

23. Metal Fiber Brushes.. 1151
Glenn Dorsey and Doris Kuhlmann-Wilsdorf

        23.1 Introduction 1152
                23.1.1 Fiber Brushes for Power.. 1152
                23.1.2 Diversification of Applications 1154
                23.1.3 Outline of Chapter 1156
        23.2 Sliding Wear of Multi-Fiber Brushes.. 1156
                23.2.1 Adhesive Wear 1157
                23.2.2 Holm-Archard Wear Equation.. 1158
                23.2.3 Low Wear Equilibrium 1159
                23.2.4 High Wear Regime 1160
                23.2.5 Plastic and Elastic Contact..1161
                23.2.6 Critical or Transition Brush Pressure.. 1163
                23.2.7 Wear of Fiber Brushes 1164
                23.2.8 Effects of Sliding Speed 1166
                23.2.9 Effect of Arcing and Bridge Transfer 1169

23.3 Surface Films, Friction, and Materials Properties 1170
    23.3.1 Thin Film Behavior 1170
    23.3.2 Water Molecules 1170
    23.3.3 Film Disruption 1172
    23.3.3 Lubrication 1172
23.4 Electrical Contact 1174
    23.4.1 Dependence of Electrical Resistance on Fiber Brush
           Construction 1174
23.5 Brush Dynamics 1177
    23.5.1 Speed Effect 1180
23.6 Future 1180
23.7 Summary 1181
Acknowledgments 1181
References 1187

Part VI   Contact Data

24. Useful Electric Contact Information 1195
Paul G. Slade

24.1 Introduction 1195
24.2 Notes to the Tables 1196
References 1210

Author Index 1211
Subject Index 1237

Preface to the Second Edition

Since the publication of the first edition of this book there have been some very costly system failures, which could have been prevented with a better knowledge of electrical contact phenomena. I will give two examples. The first is an electrical connector that supplied power to the "Main Fuel Shut-off Valve" in the F-16 fighter airplane. This connector used tin plated pins plugged into gold plated sockets. As will be briefly discussed in Chapter 3, the failure of this combination from fretting corrosion in the aircraft's vibration environment caused the fuel to stop flowing to the jet engines. Several F-16 crashes have been attributed to this connector failure with a subsequent cost of over $100 M. In hindsight it is probable that this pin socket combination used extensively in the earlier F-111 airplane resulted in it cancellation. Failure of the connectors most probably resulted in this plane's performance changing from a "terrain following" aircraft to a "terrain impacting" one. The second example occurred in the Large Hadron Collider (LHC), which began its initial testing in September 2008. Soon after it began to operate, a connection to a 12 MVA transformer failed. This cut power to the main compressors that operated the cryogenic system for cooling the super conducting magnets in two sections. This failure caused extensive wiring damage that cost more than $20 M to repair and set back the initial operation of this expensive experimental system by about nine months.

Electrical contact theory and practice does not have the hectic pace that we have become used to in recent years with the computer and communications technology, where obsolescence occurs not in one's lifetime, but within one's recent memory! In the generation since the first edition of this book was published, however, there have been significant advances in the use and understanding of electrical contacts. Therefore, we have developed this second edition in order to bring the subject up to date. Sadly, in the 15 years since the development of the first edition, five of the original authors have died: Morton Antler, Jax Glossbrenner, Doris Kulmann-Wilsdorf, Erle Shobert, and Werner Rieder. While this book continues to retain the essence of their original material, we are fortunate that a new generation of researchers has stepped in to contribute to this new edition: nine of the present contributors have received the Ragnor Holm Scientific Achievement Award for their contributions to the science of electrical contacts. The contributing authors continue to reflect the great diversity and international research in this subject. There are chapter authors from China, Japan, Europe, and the Americas. I wish again to extend a personal note of thanks to all of these contributors. The inclusion of their chapters has distilled the knowledge amassed from their years of research and of their practical experience.

Paul G. Slade, 2013

Preface to the First Edition

I began my studies in the fascinating world of electrical contacts as a graduate student at the University of Wales in the mid-1960s. Since that time, I have been involved with electrical contacts both as a research scientist and as a developer of switching components. Even though the subject is as old as electricity, it has continued to evolve. To those of us who have continued working in this field, it has provided a stimulating and ever-expanding subject for research. It has, however, always been—and still is—on the periphery of other major technology achievements. For example, plug-in connectors are a vital, but ignored, component of an integrated circuit—ignored, that is, until the connector fails. Another example that comes to mind is the development of sliding contacts that work in a space environment, which have been essential for the successful development of the communication satellite. Experts in electrical contact technology can cite many such examples. In fact, electrical contacts of one type or another are found in every electrical component, and the proper operation of the contacts is always vital to the reliable operation of that component.

The subject of electrical contacts is by necessity multidisciplinary. No matter which academic career you initially begin with, as soon as you start working in this area, you soon develop a good general knowledge of many others. The study of electrical contacts requires knowledge of physics, chemistry, mechanical engineering, electrical engineering, materials science, and environmental science. It also requires a broad applications knowledge; contacts can be found in electronic circuits that may carry currents of less than 10^{-6} A and also in power circuits that may carry currents of greater than 10^6 A.

The aim of this book is to provide information on the current state of electrical contact science and engineering to practicing scientists and engineers, as well as to provide a comprehensive introduction to the subject for technology graduate students. To do this we use the knowledge and experience of 17 contributors who are actively involved in the research, development, manufacture, and application of electrical contacts. Nine of these contributors have received the Ragnar Holm Scientific Achievement Award for their contributions to the science of electrical contacts, which is given by the IEEE Holm Conference on Electrical Contacts. The contributors reflect the great diversity that characterizes electrical contact applications.

This type of book is long overdue. The latest books published on this subject in English have been long out of print. Many times it would have been extremely helpful to have had readily available, in one source book, a thorough, up-to-date overview of the subject, application information, pitfalls to avoid, and design tips. There has also been a need for application guidance that is independent of that given by commercial contact manufacturers. This book provides a practical approach to the subject and gives the reader insights into electrical contact application that have been derived from the contributors' extensive experience. For those readers who would like to investigate a particular aspect of the subject in more detail, each chapter includes an extensive list of references where such information can be found.

The book is divided into six parts. Part I has three chapters that introduce the contact interface and the subject of contact tarnishing and corrosion. In Part II, static contacts are presented. The four chapters in this part discuss materials, design, and applications covering the range from electronic connectors to high-power electrical utility connectors.

The five chapters in Part III develop the subject of arcing contacts. The part begins with a discussion of the electric arc and continues with a review of practical design considerations for switching devices that interrupt currents in the range 0.1 to 100,000 A. Part IV continues the discussion of arcing contacts, covering the choice of contact materials, their attachment, testing, and contamination effects. Part V deals entirely with sliding contacts. Its four chapters describe the fundamentals of sliding contacts and their application over a wide range of currents and conditions. Finally, in Part VI tables of useful data are provided.

I wish to extend a personal note of thanks to all the contributors. The inclusion of their chapters, which distill the knowledge amassed from their years of research and experience, has made this book the comprehensive and definitive volume that we originally envisioned. Special mention is due to Erle Shobert, Gerry Witter, Guenther Horn, and Tony Lee, who were on the initial committee that began this project. I would especially like to thank Gerry Witter, who developed Part Four. I also wish to acknowledge the valuable help that I have received from Mort Antler and Werner Rieder. They have always been prepared to read manuscripts, offer positive and constructive criticism, and make helpful suggestions as the book progressed.

Paul G. Slade, 1999

Introduction

Those who cannot remember the past are condemned to repeat it.

The Life of Reason **(Vol. 1), George Santayana**

The electrical contact has always been an essential part of the electric circuit. The reliability of the electrical contact has also been an essential, but often ignored, factor—ignored, that is, until it fails. Thus, it sometimes seems as if problems that should have been solved many years ago keep recurring. This perception is, of course, only partially true. In their earliest forms, electric circuits usually carried only a limited range of currents: perhaps up to a few hundred amperes. Since then, this current range has increased considerably. Now electrical contacts of one type or another are found in very high power circuits passing currents in excess of 10^6 A (mega-amperes) [1] and in electronic circuits where currents can be as small as 10^{-6} A (microamperes) [2]. In past 60 years, there have been major advances in electrical contact science, including the recognition and the understanding of fretting corrosion; the development of flowing mixed gas laboratory systems for studying electrical contact corrosion; the study of the effects of dust on connector performance; the miniaturization of electro-magnetic relays; the introduction of MEMS (microelectromechanical systems) switches; the ease of use and the availability of a broad range of surface analysis systems; the advent of computers for recording and analyzing data, especially statistical data; the advent of user friendly computer software to design and develop switching systems; the continued development of silver-metal oxide contact materials; the development of contacts for use in new operating ambients such as SF_6 and vacuum; a whole new class of electrical connectors, especially for the electronics and automobile industries; and new types of sliding contacts. The reader should, however, be aware of the considerable body of knowledge that has been accumulated since this subject was first studied. Published research on electrical contact phenomena and the switching of electric circuits can be found dating back to 1835. Reference to this early work is given in the bibliography and abstracts on electric contacts from 1835 to 1951 that were published in 1952 by the ASTM (American Society for Testing Materials) [3]. A large part of this work is still relevant today. In fact, one paper from the end of the ninteenth century that is still frequently referenced by electric contact researchers is that by Kohlrausch and Diesselhorst [4]. They first developed the relationship between the voltage drop across a conductor and its temperature (see Chapter 1). The ASTM continued to publish abstracts of electrical contact research until 1965 [5]. For those interested in the history of electrical contact studies as a subject in its own right, this series of books provides an invaluable source of reference. From 1965 to 1977, the recording of electric contact abstracts was continued by the Holm Conference Organization [6]. The complete papers from the various embodiments of the Holm Conference (now called the IEEE Holm Conference on Electrical Contacts) can be found on a DVD [7] edited by Schoepf.

In the United States, the study of electrical contacts as a distinct discipline can be traced back to the first Holm Seminar, which was held at the Pennsylvania State University in 1953. Here Dr. Ragnar Holm presented a series of lectures on the state of electrical contact research and knowledge up to that time. This seminar was organized by Dr. Erle Shobert (then at Stackpole Carbon, Inc.) and Professor Ralph Armington (then on the faculty of the Pennsylvania State University's Electrical Engineering Department). The Holm Seminars developed into the

IEEE Holm Conferences on Electrical Contacts [7]; 2013 being the 59th conference. The first International Conference on Electric Contacts (ICEC) was held in Orono, Maine, in 1961 and was followed by a second in Graz, Austria, in 1964. Since that time, the International Contact Conference has been held every two years in the Americas, Europe, and Asia [8]; 2012 being the 26th conference. The Albert Keil Kontacktseminar has also been held in Germany since 1972. The proceedings of these conferences contain all the important advances in the subject for the past half century and can also be found on the DVD edited by Schoepf [7,8]. There has been an annual Japanese conference (International Session on Electromechanical Devices) since 2001. There has also been and ongoing committee in Japan that discusses electrical contacts and electromechanical devices. They hold meetings about ten times a year. In China the International Conference on Reliability of Electrical Products and Electrical Contacts was begun in 2004. Further Conferences were held in 2007, 2009, and 2012.

Although a reliable electrical contact is essential to the successful operation of every electric circuit, remarkably few books have been written on the subject. In 1940, Windred [9] published the first comprehensive book, *Electric Contacts*, in English. This book treated the subject in great detail and must have been of great practical value in analyzing and designing current switching devices. In 1941, Ragnar Holm published his first book on the subject in German [10]. He continued to update and revise this work until the publication of his seminal work, *Electric Contacts: Theory and Application*, in 1967, which was reprinted in 2000 [11]. This book contained a comprehensive review of all aspects of electrical contact phenomena known up to 1966, and is still a frequently referenced source book for many active researchers in the field. In 1957, F. Llewellyn Jones published *The Physics of Electrical Contacts* [12], which had a more limited scope. It concentrated on arcing and the erosion of relay-type contacts. Unfortunately, this book has long been out of print. Two books, one in German [13] and one in French [14], have been published in 1983 and 1996, and the first edition of this book filled an important niche in 1999. Braunovic et al. published their book *Electrical Contacts, Fundamentals, Applications and Technology*, on static contacts in 2007 [15]. In this new edition of our book, Electrical Contact, Principles and Applications, we continue to present, in one volume, the basic background to this subject as well as a comprehensive review of the present state of electrical contact research, development, and application. Because the subject continues to grow it is impossible for one author to have detailed knowledge of its every aspect. We therefore present individual specialized chapters written by recognized experts in that particular field of electrical contact research and application, making it possible to cover the whole range of electrical contact phenomena. There is, of course, a strong link between the study of electrical contacts, the design of electrical connection systems, and the design of circuit interruption devices. We therefore show many examples of the use of electrical contacts and component design criteria. It would be impossible, however, to include a complete discussion of all the design intricacies for every component in which electrical contacts are used. The reader should therefore use this book in conjunction with other specialized books that cover the particular aspects of electrical component design [16–28]. This book is divided into six parts, as follows.

Part I: Contact Interface Conduction

The four chapters in Part I discuss the basic principles of making contact and the effects of the ambient atmosphere on the passage of current from one conductor to the other. Chapter 1 presents the theory of two metal surfaces coming into contact. It discusses the true area of contact and develops the concept of contact resistance, a subject that has been investigated since the earliest studies on electrical contacts [29,30]. The effect of an electric current passing through the true area of contact, its effect on contact heating, the effect of very high frequency currents and the formation of intermetallic compounds are also explained. Chapter 2 introduces contact corrosion and tarnishing, another subject that has a long history of study [31]. It discusses the variety of ambient atmospheres that contacts may be exposed to, gives a general outline of the important corrosion mechanisms, presents methods of measuring the effects of corrosion, and introduces Chapters 3 and 4. In Chapter 3, the work on developing laboratory test atmospheres is discussed in detail, indicating the importance of the flowing gas test chamber that has revolutionized laboratory corrosion studies. It also discusses how the mixture and concentration of the test gases play a critical role in the corrosion effects. The corosion of thin noble metal plating on electronic connections is covered in detail and the palliative effects of lubricants. Chapter 4 concludes this part with a discussion of the the deleterious effects of dust on electronic connectors.

Part II: Nonarcing Contacts

The four chapters in Part II present the practical requirements of making permanent contact. The discussion also includes the periodic breaking of such a contact under the condition that no current is passed through the contact interface. Chapter 5 discusses the important aspects of producing high-power contact joints such as may be found in power transmission and distribution circuits; this type of joint has been studied for many years [32]. Here, currents in excess of 10^5 A may be carried. The discussion is continued in Chapter 6, where connections that carry currents of less than 1 ampere up to tens of amperes are covered. Here also the effects of loose connections are reviewed. This topic has gained considerable interest in recent years. Chapter 7 discusses the effects of contact wear when plug-in connections are made and broken. It also extends the discussion begun in Chapters 5 and 6 on contact fretting, especially as applied to electronic connectors. The final chapter in this part reviews the subject of plating a thin layer of metal that has good connector contact characteristics onto a substrate metal.

Part III: The Electric Arc and Switching Device Technology

Part III begins the discussion of opening contacts that have a current passing through them. The electric circuits being switched have currents from about 10 mA to 2×10^5 A and, in this book, a few volts to about 1000 V. Chapter 9 introduces the formation of the electric arc from both the breakdown of an open gap and the opening of conductors in contact; the discussion of these phenomena dates back to the 19th century [33,34]. Here, a new interpretation is developed for the minimum current required for an arc to form. It also covers arcing in ac and dc circuits and how the arc is extinguished and the circuit is interrupted. In Chapter 10, the effects of the arc on the contacts—contact erosion [35], welding, and contamination—are addressed. Chapters 11–15 present aspects of contact and arc chamber design for efficient operation of contacts interrupting electric circuits. The low current reed relay is discussed in Chapter 11 and Chapter 12 introduces contact requirements for the MEMS switch. Chapter 13 includes design aspects for other types of relay [36] and low current switches. In Chapter 14, the control of the high current arc in contactors [37], molded case circuit breakers used in circuits up to 1000 V, and vacuum interrupters are described together with the computer friendly software that is increasingly being used for switch design. Finally, Chapter 15 concludes this part with a discussion of the detection and elimination of low current arcing faults.

Part IV: Arcing Contact Materials

Part IV was developed by Gerry Witter, who authored or coauthored the four chapters. These chapters expand on the design of arcing contacts and the on how the electrical arc can affect their performance. This part also discusses the compromises that have to be made when choosing contacts that will be exposed to the electric arc [38]. These contacts not only have to conduct current reliably when closed, but also must resist severe surface damage when they are opening. The choice of materials is presented for a number of voltage and current ranges. The development of specialized contact compositions also has a long history [38–42]. Surface-arc-contamination reactions that involve changes in contact resistance during the life of the contact are also discussed. Chapter 16 describes arcing contact materials relating properties to performance. Chapter 17 provides information on arcing contact construction and attachment technology options. Chapter 18 gives theoretical guidelines to help in selecting contact materials based on contact properties and also provides guidelines for development of test methods to compare the performance of arcing contact materials. Chapter 19 discusses the effects on arcing contact performance from arc interaction with various contaminants on contact surfaces.

Part V: Sliding Electrical Contacts

Part V deals with sliding contacts, one of the earliest subjects of electrical contact research [43,44]. Chapter 20 presents the fundamentals of sliding contacts and the aspects of testing and evaluating the test results on sliding contacts. Although the principles addressed in this chapter are general, the authors are mostly concentrating on currents of a few amperes up to hundreds of amperes that can occur large in electric motors. Chapter 21 gives examples of the application of such sliding contacts. In Chapter 22, low current sliding contacts are discussed, in particular the low current contacts that would be used in slip rings and would be employed in harsh environments, such as outer space. Chapter 23 presents the work on high current density brushes.

Part VI: Contact Data

The last part of this book presents updated tables of physical data for a wide range of contact materials. It also gives some common conversion factors as well as a table of physical constants used frequently by those performing research on electrical contacts.

Future Developments

Based on my own experience of electrical contact research over five decades, I would anticipate that continuing research and the use of computer-aided recording and analysis equipment will continue to develop the science of electrical contacts. I would also expect to see advances in material selection and application specifications for power connections. Electronics connector technology will continue to be driven by cost and reliability, two requirements that are not always compatible. For circuit interrupter applications, the control of the electric arc and contact material development will lead to more efficient and reliable component design. The marriage of miniaturized electronics for detection and control of switches will expand the use of switching systems that use electrical contacts and electric arcs for making and interrupting electric circuits. Also, the increasingly user-friendly computer software will greatly facilitate the future design and development of these switching systems. For sliding contacts, the continued work on brush contacts promises a new generation of contact that will complement the existing styles manufactured from solid materials. Finally the three environmental initiatives begun by the European Union will continue to have a significant effect in the use of materials that are considered risky, these initiatives are: (1) RoHS (Restriction of Hazard Materials), (2) REACH (Regulation, Evaluation, Authorization and Restriction of Chemicals) and (3) SVHC (Substances of very high Concern) [45].

References

1. HA Calvin, JJ Anderson, PG Slade. The development of mega-ampere contact structures for an electromagnetic launcher. *IEEE Trans Comp Hybrids Manuf Technol* 16(2):203–210, March 1993.
2. HC Slade. A study of electroplated and evaporated low resistance ohmic contacts to n type GaAs. Southeaster, 1994 IEEE Students Paper Book, part of MS Thesis, Department Electrical Engineering, University of Virginia, Charlottesville, VA, 1994.
3. Bibliography *and Abstracts on Electrical Contacts 1835–1951*. Philadelphia, PA: American Society for Testing Materials, 1952.
4. F Kohlrausch, H Diesselhorst. Über den Stationaren Temperaturzustand Eines. Elektrisch Geheizten Leiters. *Ann Phys* 1:132–138, 1900.
5. Bibliography *and Abstracts on Electrical Contacts*. Philadelphia, PA: American Society for Testing Materials, in 11 volumes, one for each year, 1954–1964.
6. (a) Bibliography and Abstracts on Electrical Contacts. Circuit Breakers and Arc Phenomena 1965–1969. (IEEE No. 70M65 PMP), Holm Conference, 1970. (b) Bibliography and Abstracts on Electrical Contacts. Circuit Breakers and Arc Phenomena 1970–1971; Proceedings of the 1971 Holm Seminar on Electric Contact Phenomena, IIT. (c) Bibliography and Abstracts on Electrical Contacts. Circuit Breakers and Arc Phenomena, 1972; 1972 Holm Seminar on Electric Contact Phenomena, IIT. (d) Bibliography and Abstracts on Electrical Contacts. Circuit Breakers and Arc Phenomena 1975; Proceedings of the 1975 Holm Seminar on Electric Contact Phenomena, IIT. (e) Bibliography and Abstracts on Electrical Contacts, Circuit Breakers and Arc Phenomena 1976–1977; Proceedings of the 1977 Holm Conference on Electric Contact Phenomena, IIT. (f) Title Word Index for the 1st 24 Holm Conference and the 1st, 3rd, 6th and 9th International Contact Conferences; Proceedings of the 25th Holm Conference on Electric Contacts, IIT, 1979. (g) Title Word Index for the 1st 29 Holm Conferences and the 1st 11 International Contact Conferences; Proceedings of the 29th Holm Confernce on Electric Contacts, IIT, 1983. (h) Cumulative Index of Holm Conferences and International Contact Conferences 1953–1988 (IEEE Catalog No. JH 9412–8), 1989.

7. Proceedings of the Holm Seminars on Electrical Contact Phenomena 1953–1976, of the Holm Conference on Electrical Contacts 1977–1984, and of the IEEE Holm Conference on Electrical Contacts 1985–2012 are now available on the DVD (ISBN 978-3-8007-3522-4) from VDE Verlag GMBH. An updated version of this compendium is given each year at the IEEE Holm Conference on Electrical Contacts. Selections of papers from these conferences have been published in IEEE Transactions Parts Materials and Packaging; Components, Hybrids and Manufacturing Technology; and Components Packaging and Manufacturing Technology since 1968.

8. Proceedings of the International Conference on Electrical Contacts: 1st, Orono, Maine, 1961 to the 26th, Beijing, China, and the complete proceedings of the Albert Keil Kontaktseminars are also available on the DVD (ISBN 978-3-8007-3522-4) from VDE Verlag GMBH.

9. G Windred. *Electric Contacts*. London: Macmillan and Co., and New York: D. Van Nostrand Co. Inc., 1940.

10. R Holm. *Technical Physics of Electrical Contacts*. Berlin: Julius Springer, 1941.

11. R Holm. *Electric Contacts: Theory and Application*. New York: Springer, 2000.

12. F. Llewellyn Jones. *The Physics of Electrical Contacts*. Oxford: Clarendon Press, 1957.

13. A Keil, WA Merl, E Vinaricky. *Elektrische Kontakte und ihre Werkstoffe*. Berlin: Springer Verlag, 1984.

14. L Féchant (ed.). *Le Contact Electrique*, Vol. 1 and 2. Paris: Hermes, 1996.

15. M Braunovic, V Konchits, N Myshkin. *Electrical Contacts: Fundamentals, Applications and Technology*. Boca Raton, FL: CRC Press, 2007.

16. TE Browne (ed.). *Circuit Interruption: Theory and Techniques*. New York: Marcel Dekker, 1984.

17. RD Garzon. *High Voltage Circuit Breakers: Design and Application*. New York: Marcel Dekker, 1997.

18. GL Ginsberg (ed.). *Connections and Interconnections Handbook*, Vol. 2, Connector Types. Philadelphia, PA: The Electronic Connector Study Group, Inc. 1979.

19. RS Mroczkowski. *Electronic Connectors*. New York: McGraw Hill, Inc., 1998.

20. CH Flurscheim. *Power Circuit Breaker Theory and Design*. London: IEE, 1985.

21. A Greenwood. *Vacuum Switchgear*. London: IEE, 1994.

22. RT Lythall. *JSP Switchgear Book*. London: Newnes-Butterworths, 1972.

23. RL Peck, HN Wagar. *Switching Relay Design*. Princeton, NJ: Van Nostrand, 1955

24. S Dzierbicki, E Walcuk. *Wylaczniki Orgraniczajace pradu Przemiennego*. Warsaw: Wydawnictwa Naukowo-Techniczne, 1976.

25. TH Lee. *Physics and Engineering of High Power Switching Devices*. Cambridge, MA: MIT Press, 1975.

26. WT Shugg. *Handbook of Electrical and Electronic Insulating Materials*, 2nd Edition. New York: IEEE Press, 1998.

27. PG Slade. *The Vacuum Interrupter: Theory, Design and Application*. Boca Raton, FL: CRC Press, 2007.

28. RS Mroczkowski. *Electronic Handbook*. New York: McGraw-Hill Press, 1998.

29. L Binder. Contact resistance. *E und M* 30:781–788, 1912.

30. R Holm. Contact resistance especially at carbon contact. *Zeit fur Tech Phys* 3(9):290–294; 3(1):320–327; 3(11):349–357, 1922.

31. GE Luke. Resistance of connections. *Elec J* 21:66–69, 1924.

32. SW Melsom, HC Booth. Efficiency of overlapping joints in copper and aluminium bus-bars. *IIEE* 60:889–899, 1922.

33. OE Gunther. Energy and resistance of sparks caused by closing and breaking electric circuits. *Ann der Phys* 42(11):94–132, 1913.

34. W. Höpp. Arc formation in switchgear. *ETZ* 34:33–38; 55–58, 1913.

35. R Holm. Vaporization of electrodes in butt contacts. *Zeits fur Tech Phys* 15(11):483–487, 1934.

36. S Keilien. Testing snap switches. *Elec J* 33:521, 1936.

37. EW Seeger. Electromagnets and contacts for magnetic contactors. *Prod Eng* 5(5):181–182, 1934.

38. G Windred. Electrical contacts. *Engineer* 160:222, 1935.

39. H. von Fleischbein. Improved contact design and material. *ETZ* 39:445–446, 1918.

40. EF Kingsbury. *The Use of Noble Metals for Electrical Contacts*. Reprint-298. Murray Hill, NJ: Bell Telephone Laboratory, April 1928.
41. Sintered silver alloys for electrical contacts. *Met Ind (Lond)* 51:461, 1937.
42. G Windred. Compound metals for electrical contacts. *Elec Eng* 56:1045, 1937.
43. FG Baily, WSH Cleghorne. Communication. *J IEE* 38:150–182, 1907.
44. HFT Erben, AH Freeman. Brush friction and contact losses. *J AIEE* 32:559, 1913.
45. Full details of these initiatives can be found at: en.wikipedia.org

Editor

Paul G. Slade earned his BSc in physics and PhD in applied physics from the University of Wales. He also has an MBA from the University of Pittsburgh. He has more than 45 years of experience in electrical contact technology. He began his career at the Westinghouse Science and Technological Center where he became chief scientist responsible for the Industrial part of the Corporation. After the Westinghouse switching business was sold to the Eaton Corporation, he became the technical director for Vacuum Interrupter Research, Development, and Design for Eaton's Vacuum Interrupter business. In this position, he was responsible for the R&D into vacuum arc and vacuum breakdown phenomena. He also introduced over 100 new vacuum interrupter designs into the market. He has also developed an intensive course on vacuum interrupter technology and application for presentation to vacuum interrupter customers and users. After retirement from Eaton he has been working as a Consultant in circuit interruption, electrical contacts and vacuum interrupter technology.

During his career, Dr. Slade has become recognized as a leader in electrical contact research and application. Dr. Slade has maintained world-class R&D efforts in the science of electrical contacts. He has helped explain the complex resistance changes in arced Ag-W contacts and his work has led to design rules for use of this material in molded case breakers. He is the course director and lecturer for the IEEE Intensive Course on Electrical Contacts. During his career he has also led and administered the development of advanced circuit interruption devices, the maintenance of state-of-the-art plasma deposition facilities, and the introduction of unique nuclear instrumentation.

Dr. Slade received the 1985 Ragnar Holm Scientific Achievement Award, IEEE Holm Conference, the 1995 Armington Recognition Award, IEEE Holm Conference, and the 2014 Albert Keil Priese, from the German Förderverein Kontakte und Schalter for outstanding accomplishments in the field of electrical contacts. In 1989, he was elected as the Fellow of IEEE.

Dr. Slade has written more than 100 papers in archival journals and technical conference presentations. He has 23 U.S. patents. He has contributed three chapters to the book *Circuit Interruption* (T. E. Browne, editor) including the chapters on vacuum circuit breakers, molded case breakers and electrical contacts. He has edited, and is the primary author, for this widely referenced book, *Electrical Contacts Principles and Application* (this being the second edition, the first edition was published in 1999). He is also the author of the book, *The Vacuum Interrupter: Theory, Design and Application* (published in 2008).

Contributors to the Second Edition

Milenko Braunović*
Principal Consultant
MB Interfaces, Scientific
    Consultants
Montréal, Québec, Canada
milenkobraunovic@yahoo.ca

Ronald A. Coutu, Jr
Assistant Professor, Department of
    Electrical Engineering
Air Force Institute of Technology
    (AFIT/ENG)
Wright-Patterson AFB, Ohio
Ronald.Coutu@afit.edu

Glenn Dorsey
Moog Inc.
Blacksburg, Virginia
GDorsey@moog.com

George Drew
Delphi, CSE Technical Center
Warren, Ohio
george.drew@delphi.com

Guenther Horn
ElConMat Consulting Associates
Richmond, Virginia
ghorn.elconmat@hotmail.com

Anthony Lee
Retired former Director, Advanced
    Engineering
Delphi Packard Electric Systems
Warren, Ohio
tony11208@yahoo.com

Manfred Lindmayer*
Professor
Institut für Hochspannunsgtechnik und
    Elektrische Energieanlagen
Technische Universität
    Braunschweig
Braunschweig, Germany
M.Lindmayer@tu-braunschweig.de

John W. McBride*
Faculty of Engineering and the
    Environment
University of Southampton
Southampton, England
J.W.Mcbride@soton.ac.uk

Kiochiro Sawa*
Professor Emeritus
Keio University
Saitama, Japan
sawa@sd.keio.ac.jp

John J. Shea*
Eaton Corporation
Pittsburgh, Pennsylvania
johnjshea@eaton.com

Paul G. Slade*
Consultant: Electrical Contacts and Circuit
    Interruption
Ithaca, New York
paulgslade@verizon.net

Roland S. Timsit*
Timron Advanced Connector Technologies,
    Timron Scientific Consulting Inc.
Toronto, Ontario, Canada
rtimsit@timron-inc.com

Benjamin F. Toler
USAF

Gerald J. Witter*
CEO & President
Electrical Contacts Plus LLC
Waukegan, Illinois
Gwitter@electricalcontactsplus.com

Ji Gao Zhang*
Professor Emeritus
Beijing University of Posts and
    Telecommunications
Beijing, China
jigao33@hotmail.com

Contributors to the First Edition Whose Essential Input Is Retained

William H. Abbott*
Retired Program Manager
Department of Advanced Manufacturing
    Technology
Battelle
Columbus, Ohio

Morton Antler*d
Principal Consultant
Contact Consultants, Inc.
Columbus, Ohio

Jax Glossbrenner*d
Chief Slip Ring Engineer
Litton Poly-Scientific
Blacksburg, Virginia

Kunio Hinohara
Senior Manager
Components Plant, Components
    Division
Oki Electric Industry Company, Ltd.
Tokyo, Japan

Doris Kulhmann-Wilsdorf*d
University Professor of Applied Science
Department of Materials Science and
    Engineering
University of Virginia
Charlottesville, Virginia

William A. Nystrom
Carbone of America
St. Marys, Pennsylvania

Werner Rieder*d
Professor
Institute of Switching Devices and High-
    Voltage Technology, University of
    Technology
Vienna, Austria

Erle I. Shobert II*d
Technical Advisor
Contact Technologies, Inc.
St. Marys, Pennsylvania

Wilferd E. Yohe
Carbone of America
St. Marys, Pennsylvania

---

* Recipients of the *Ragnor Holm Scientific Achievement Award* for their contribution to the science of Electrical
   Contacts
d Deceased

Part I

Contact Interface Conduction

1

Electrical Contact Resistance: Fundamental Principles

Roland S. Timsit

Only connect!

Howards End, **E M Forster**

CONTENTS

1.1 Introduction 4
1.2 Electrical Constriction Resistance 5
    1.2.1 Circular *a*-spots 5
    1.2.2 Non-Circular and Ring *a*-Spots 7
    1.2.3 Multiple Contact Spots 11
    1.2.4 Effect of the Shape of Contact Asperity on Constriction Resistance 17
1.3 Effect of Surface Films on Constriction and Contact Resistance 18
    1.3.1 Electrically Conductive Layers on an Insulated Substrate 18
        1.3.1.1 Calculation of Spreading Resistance in a Thin Film 18
    1.3.2 Electrically Conducting Layers on a Conducting Substrate 22
        1.3.2.1 Electrically Conducting Layers and Thin Contaminant Films 23
    1.3.3 Growth of Intermetallic Layers 28
    1.3.4 Possible Effect of Electromigration on Intermetallic Growth Rates 42
    1.3.5 Electrically Insulating or Weakly Conducting Films 45
        1.3.5.1 Growth Rate and Electrical Resistivity of Oxides of Selected
                Contact Materials 47
    1.3.6 Fritting of Electrically Insulating Surface Films 56
1.4 Temperature of an Electrically Heated *a*-Spot 57
    1.4.1 Voltage–Temperature Relation 58
    1.4.2 Voltage–Temperature Relation with Temperature-Dependent Electrical
        Resistivity and Thermal Conductivity 60
    1.4.3 The Wiedemann–Franz Law 62
    1.4.4 Temperature Distribution in the Vicinity of an *a*-Spot 63
    1.4.5 Deviation of the Voltage–Temperature Relation in an Assymetric Contact 64
        1.4.5.1 Case I: Two Metals in Contact 64
        1.4.5.2 Case II: A Metal in Contact with a Non-metal 65
    1.4.6 Special Considerations on the "Melting" Voltage in Electrical Contacts 65
1.5 Mechanics of *a*-Spot Formation 69
    1.5.1 Smooth Interfaces 69
    1.5.2 Rough Interfaces 75

1.6   Breakdown of Classical Electrical Contact Theory in Small Contact Spots.................79
    1.6.1   Electrical Conduction in Small *a*-Spots...79
        1.6.1.1   Contact Resistance 79
        1.6.1.2   Joule Heat Flow Through *a*-Spots.................................82
    1.6.2   Observations of Breakdown of Classical Electrical Contact Theory in
        Aluminum Contacts...84
        1.6.2.1   Experimental Data on Aluminum..................................84
    1.6.3   Observations of Breakdown of Classical Electrical Contact Theory in
        Gold Contacts 87
    1.6.4   Observations of Breakdown of Classical Electrical Contact Theory in
        Tin Contacts...88
1.7   Constriction Resistance at High Frequencies 89
    1.7.1   Skin Depth and Constriction Resistance..................................89
    1.7.2   Evaluation of Constriction Resistance at High Frequencies...........91
    1.7.3   Constriction versus Connection Resistance at High Frequencies 94
1.8   Summary...95
Acknowledgments 96
References...104

1.1 Introduction

All solid surfaces are rough on the microscale. Surface microroughness consists of peaks and troughs whose shape, variations in height, average separation, and other geometrical characteristics depend on fine details of the surface generation process [1]. Contact between two engineering bodies, thus, occurs at discrete spots produced by the mechanical contact of asperities on the two surfaces, as illustrated in Figure 1.1. For all solid materials, the area of true contact is, thus, a small fraction of the nominal contact area, for a wide range of contact loads [1,2]. The mode of deformation of contacting asperities is either elastic, plastic, or mixed elastic–plastic depending on local mechanical contact stresses and on properties of the materials, such as elastic modulus and hardness. In a bulk electrical interface where the mating components are metals, the contacting surfaces are often covered with oxide or other electrically insulating layers. Generally, the interface becomes electrically conductive only when metal-to-metal contact spots are produced, that is, where electrically insulating films are ruptured or displaced at the asperities of the contacting surfaces. In a typical bulk electrical junction, the area of electrical contact is, thus, appreciably smaller than the area of true mechanical contact.

In a bulk electrical junction, the electric current lines become increasingly distorted as the contact interface is approached and the flow lines bundle together to pass through the separate contact spots (or "*a*-spots"), as illustrated in Figure 1.1. Constriction of the electric current by *a*-spots reduces the volume of material used for electrical conduction and thus increases electrical resistance. This increase in resistance is defined as the *constriction resistance* of the interface. Often, the presence of contaminant films of relatively large electrical resistivity on the contacting surfaces increases the resistance of *a*-spots beyond the value given by constriction resistance. The total interfacing resistance

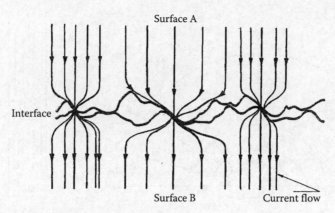

Surface A

Interface

Surface B　　　　Current flow

FIGURE 1.1
Schematic diagram of a bulk electrical interface.

provided by the constriction and film resistances determines the *contact resistance* of the interface.

The present chapter reviews some fundamental properties of electrical contacts and updates the reader on the results of recent research. The review focuses on the effect of constriction of current flow on electrical resistance, interdiffusion processes at electrical interfaces, the relationship between the drop in electrical potential and temperature in an electrical contact (the so-called *voltage–temperature* relation), sintering, softening, and melting in contact spots, the effect of deformation of asperity on contact resistance and so on. The chapter thus attempts to update the reader on similar topics covered in Holm's classic text [3].

1.2 Electrical Constriction Resistance

For the sake of simplicity, the evaluation of constriction resistance generally assumes *a*-spots to be circular. This assumption provides an acceptable geometrical description of electrical contact spots "on the average" where the roughness topographies of the mating surfaces are isotropic. The assumption becomes invalid where the mating surfaces are characterized by a directional roughness, as in rolled metal sheets or extruded rods, where the shape of the *a*-spot would be characterized by a large aspect ratio. Although most of the properties of electrical contacts are usually discussed in terms of circular *a*-spots, the constrictive properties of contact spots of other shapes are addressed below for the sake of completeness. Unless otherwise stated, all descriptions given below assume a DC electric current. The evaluation of constriction resistance under AC conditions will be discussed later in the chapter.

1.2.1 Circular *a*-Spots

The mathematical problem of constricted current flow is treated in several textbooks, for example, [4]. For a circular constriction located between two semi-infinite solids (one of

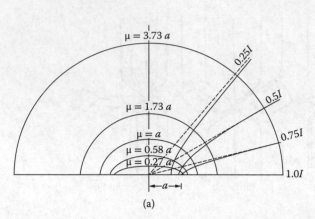
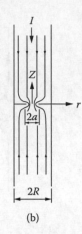

(a)   (b)

FIGURE 1.2
(a) Equipotential surfaces and current flow lines near an electrical constriction; the parameter μ is the vertical axis of the vertical ellipsoidal surface. The curves corresponding to current flow identify the boundaries enclosing the current fraction indicated, (b) Electrically conducting cylinder of radius R carrying a circular constriction of radius a.

such surfaces shown in Figure 1.2a), it is found that the equipotential surfaces in the contact members consist of ellipsoids defined by the equation

$$\frac{r^2}{a^2+\mu^2}+\frac{z^2}{\mu^2}=1$$

where μ is the length of the vertical semi-axis of the ellipsoid and (r, z) are cylindrical coordinates. The resistance between the equipotential surface with semi-axis μ and the constriction is given as [3,4]

$$R_\mu = \frac{\rho}{2\pi}\int_0^\mu \frac{d\mu}{(a^2+\mu^2)} = \frac{\rho}{2\pi a}\tan^{-1}\left(\frac{\mu}{a}\right) \tag{1.1}$$

where ρ is the resistivity of the conductor. Sufficiently far away from the constriction where μ is very large, the constriction resistance between the equipotential surface and the constriction, that is, the spreading resistance, is given as

$$R_s = \frac{\rho}{4a} \tag{1.2}$$

The total constriction resistance for the entire contact is, thus, twice the spreading resistance or

$$R_c = \frac{\rho}{2a} \tag{1.3}$$

Equations 1.2 and 1.3 are widely used in the literature and in problems relating to the design of electrical contacts. Later in this chapter, we shall see that the general form of Equation 1.3 is true for monometallic contacts even when they are an agglomeration of a-spots that are not necessarily circular. If the upper and lower half of a contact consist respectively of materials with resistivity ρ_1 and ρ_2, the spreading resistance associated with each half of the contact is then $\rho_i/4a$ where $i = 1,2$. The electrical constriction resistance, then, becomes

$$R_c = \frac{\{\rho_1+\rho_2\}}{4a}$$

TABLE 1.1

Electrical Resistance of a Circular Constriction in a Copper–Copper Interface

a-Spot Radius (µm)	Constriction Resistance (Ω)
0.01	0.88
0.1	8.8×10^{-2}
1	8.8×10^{-3}
10	8.8×10^{-4}

It is instructive to evaluate the magnitude of contact resistance as a function of the radius of the a-spot for a circular constriction in a copper–copper interface ($\rho = 1.75 \times 10^{-8} \, \Omega$ m). This is illustrated in Table 1.1.

Note that the resistance of a constriction 10 µm in radius is approximately 1 mΩ. This is very small! As will be shown later in this chapter, the passage of an electric current of 20 A through such a constriction will not cause appreciable heating within this constriction. Similarly, a constriction 100 µm will be able to pass a current of 200 A without significant heating. Thus, the area of electrical contact between two surfaces need not be large to "short out" the electrical interface, that is, to generate an electrical contact of acceptably low resistance.

The electrical resistance R_c presented by a circular constriction in a cylindrical conductor of radius R, as illustrated Figure 1.2b, can be calculated from a solution of Laplace's equation using appropriate boundary conditions [5,6]. It may be shown that the electrical contact resistance is accurately given as

$$R_c = \left\{\frac{\rho}{2a}\right\}\left(1 - 1.41581\left\{\frac{a}{R}\right\} + 0.06322\left\{\frac{a}{R}\right\}^2 + 0.15261\left\{\frac{a}{R}\right\}^2 + 0.19998\left\{\frac{a}{R}\right\}^4\right) \quad (1.4)$$

Equation 1.4 reduces to Equation 1.3 where the constriction radius becomes small in comparison with the cylinder diameter. Figure 1.3 shows the measured and calculated dependence of R_c on the ratio a/R [7]. The agreement between the measured and calculated values reveals the reliability of Equation 1.4 over the entire range of values of a/R. This reliability is often important in applications where a/R is close to unity, as in cylindrical bus bar junctions carrying large electric currents. In these cases, a small change in constriction radius may translate into a significant change in constriction resistance and into a correspondingly significant change in power dissipation. The current density distribution in the constriction is given by the expression [5,6]

$$j(r) = \frac{mI}{2\pi a^2 \left\{1 - \frac{r^2}{a^2}\right\}^{1/2}} \quad (1.5)$$

where I is the electric current, r is the radial location within the constriction and m is a factor that deviates significantly from unity only for values of a/R greater than approximately 0.5 [6].

1.2.2 Non-Circular and Ring a-Spots

As mentioned earlier, under conditions where the microtopography of the surfaces of the mating bodies is not isotropic, the assumption that a-spots may be treated as circular "on the average" may not be valid and could lead to erroneous conclusions. This section

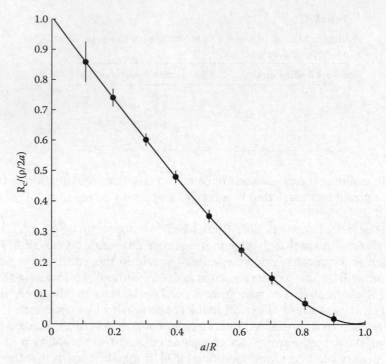

FIGURE 1.3
Measured and calculated (full curve) constriction resistance in a constricted cylinder of radius R, as a function of a/R. (From RS Timsit, *Proc 14th Int Conf Elect Cont*, Paris, 21, 1988 [7].)

examines the effect of a departure from circular symmetry on the constriction resistance of a single a-spot.

It is shown by Holm [3] that the spreading resistance $R_c (a, b)$ associated with an elliptical a-spot with semi-axes a and b is given as

$$R_c(a,b) = \frac{\rho}{2\pi} \int_0^\infty \frac{d\mu}{\left[\left(a^2 + \mu^2\right)\left(b^2 + \mu^2\right)\right]^{1/2}}$$

and may be expressed as

$$R_c = \frac{\rho}{4a_c} f(\gamma) \tag{1.6}$$

where $\gamma = \sqrt{a/b}$ is the square root of the aspect ratio of the constriction, the function $f(\gamma)$ is a form factor and the quantity a_c is the radius of a circular spot with area identical to that of the elliptical a-spot. The form factor is shown graphically in Figure 1.4 and decreases in value from 1 to 0 as the aspect ratio increases from 1 to ∞, that is, as the contact spot becomes increasingly elongated. Note that the constriction resistance decreases slowly as the aspect ratio increases from 1 to approximately 10. The constriction resistance of an elliptical a-spot located between two semi-infinite solids is given as $2R_s (a, b)$. Aichi and Tahara [8] measured the spreading resistance of rectangular a-spots using an electrolytic

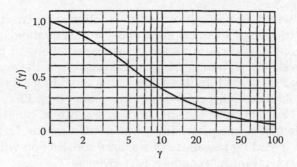

FIGURE 1.4
Dependence of the form factor $f(\gamma)$ on the aspect ratio γ. (With kind permission from Springer Science+Business Media: *Electric Contacts, Theory and Applications*, 2000, R Holm [3].)

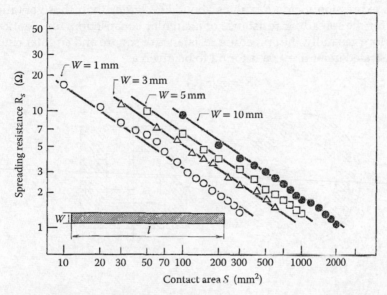

FIGURE 1.5
Dependence of spreading resistance on the area of the rectangular a-spot; the area is given as wl where w and l are, respectively, the width and length of the spot. The resistivity of the electrolytic bath is 175 Ω mm. (From H Aichi and N Tahara, *Proc 20th Int Conf Elect Cont*, Nagoya, Japan, 1, 1994 [8]. With permission.)

bath. Their data is illustrated in Figure 1.5. A regression analysis of the data of Figure 1.5 indicates that the spreading resistance R_s is given empirically as

$$R_s = k \frac{\rho}{S^{0.63}} \tag{1.7}$$

where S is the area of the rectangular constriction, when the aspect ratio of the constriction is 10 or larger. The quantity k is a parameter that depends on the width of the constriction. It varies from 0.36 to approximately 1 (when S and ρ are expressed respectively in mm^2 and Ω mm) as the width of the constriction increases from 1 mm to 10 mm. The constriction resistance of a rectangular a-spot located between two semi-infinite solids is, thus, given as $2R_s$. Note that Equation 1.7 may be expressed as

$$R_s = \frac{\rho}{4L} \left\{ \frac{k'}{w^{0.26}} \right\} \left\{ \frac{w}{l} \right\}^{0.13} \tag{1.8}$$

where w and l are the width and length of the constriction respectively, $L = \sqrt{wl}$ is the width of a square constriction of identical area, and $k' = 4k$. Equation 1.8 is of the same general form as Equation 1.6 and confirms the general predictions of Equation 1.6 that the resistance decreases slowly with increasing aspect ratio l/w. If the aspect ratio is 10 or greater, Equations 1.6 and 1.8 predict similar values of the spreading resistance. Recall that Equation 1.8 does not hold true for aspect ratios much smaller than 10.

A rigorous numerical evaluation of the spreading resistance of square a-spots, and of a-spots consisting of circular and square rings, was carried out by Nakamura [9]. This work indicates that the spreading resistance of a square constriction with side length $2L$ and located between two semi-infinite conductors is given as

$$R_s = 0.434 \frac{\rho}{L}$$

Note that this expression yields R_s values about 70% larger than those obtained from relation $0.25\rho/a$ for the spreading resistance of a circular constriction, for identical values of ρ/L and ρ/a. More generally, the spreading resistance of square and circular ring-shaped constriction illustrated in Figure 1.6a is found to be given as

$$R_s = R_0 F(\zeta)^{-1}$$

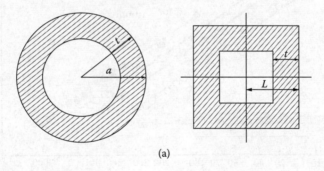

(a)

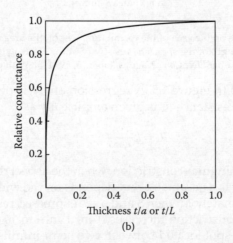

(b)

FIGURE 1.6
(a) Square and circular ring constrictions (b) Form factor F(ζ) for the ring constrictions shown in (a); the relative spreading conductance is given as $(\rho/4a)/R_s$ for the circular ring, and as $(0.434\rho/L)/R_s$ for the square ring. The difference between the two curves is too small to be seen in the plot. (From M Nakamura, *IEEE Trans Comp Hyb Manuf Tech*, CHMT-16: 339, 1993 [9]. With permission.)

TABLE 1.2

Resistance of Constrictions of the Same Area (100 μm²) and Different Shapes at a Copper–Copper Interface

Constriction Type	Radius (μm)	Length (μm)	Width (μm)	Ring Thickness (μm)	Resistance (Ω)
Circular disk	5.64				1.55×10^{-3}
Square		10	10		3.04×10^{-3}
Rectangular		50	2		0.43×10^{-3}
Ring	16.41			1	0.71×10^{-3}

where R_o is the spreading resistance of the full circular or square a-spot, and $F(\zeta)$ is a conductance form factor. In the case of the circular ring constriction, $\zeta = t/a$ where t and a are the thickness and outer radius respectively. In the case of the square constriction, $\zeta = t/L$. The form functions $F(\zeta)$ for the square and circular rings are essentially identical and are shown in Figure 1.6b. Note that the difference in the relative spreading resistances $R_s/[\rho/4a]$ and $R_s/[0.434\rho/L]$ for the circular and square ring is too small to be evident in the plot (Figure 1.6b).

The expressions for constriction resistance of a-spots that are rectangular, circular ring or rectangular ring shaped find application where contacting surfaces are specially textured, as for example, in many types of power utility connectors where surfaces carry pyramidal knurls that penetrate conductors to increase friction forces. The a-spots are, then, square or rectangular. Ring-shaped constrictions are rarer but occur where one contacting surface carries a highly conductive cladded layer on a knurled surface. In this instance, passage of electric current through the constriction occurs largely through the cladding material of the knurl in contact with the mating surface. Table 1.2 compares the resistance of constrictions of identical areas (100 μm²), but having different shapes in a copper–copper interface ($\rho = 1.75 \times 10^{-8}$ Ωm). Note from Table 1.2 that the constriction resistance is significantly affected by the shape of the constriction.

1.2.3 Multiple Contact Spots

In practice, an electrical junction comprises a multitude of contact a-spots through which electric current passes from one connector component to another. The a-spots are formed from the contact of asperities on the mating surfaces as illustrated in Figure 1.1. The number of asperity contacts increases with normal load as illustrated in the series of micrographs of Figure 1.7 reproduced from the work of Thomas and Probert [10]. In practice, contact between nominally flat surfaces occurs at clusters of a-spots. The positions of the clusters are determined by the large-scale waviness of the contact surfaces, and the a-spots by the small-scale surface roughness. Contact resistance is, then, determined by the number and dimensions of the a-spots and by the grouping and dimensions of the clusters. Although mechanical contact occurs at many contact spots, electrically conducting a-spots are generated only if surface insulating layers such as oxide films are fractured or dispersed. Because the fracture mode of oxide films may depend on the deformation mode of the contacting asperities, that is, elastic or plastic deformation, the number of metal-to-metal a-spots is generally difficult to predict and may be appreciably smaller than the number of asperities that made initial contact with the surface insulating films.

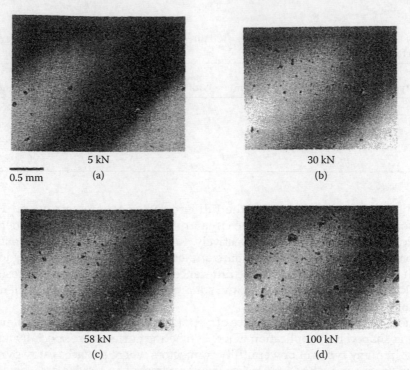

0.5 mm

5 kN
(a)

30 kN
(b)

58 kN
(c)

100 kN
(d)

FIGURE 1.7
Optical micrographs showing the increase in the number of contact spots on a soft but smooth steel optical flat following contact with a sand-blasted tool steel surface, with increasing load; the contact spots represent imprints of contacting asperities on the soft but smooth optical flat. (From TR Thomas and SD Probert, *J Phys D: Appl Phys* 3: 277, 1970 [10]. With permission.)

For the sake of ease of calculation, the collective action of metal-to-metal a-spots is generally treated in terms of the properties of circular spots. In the simplest case of a large number n of circular a-spots situated within a single cluster, it was shown by Greenwood [11] that, to a good approximation, the contact resistance is given as

$$R_c = \rho\left\{\frac{1}{2na} + \frac{1}{2\alpha}\right\} \tag{1.9}$$

where a is the mean a-spot radius defined as $\Sigma a_i/n$ (a_i is the radius of the ith spot) and α is the radius of the cluster sometimes defined as the Holm radius.

Table 1.3 shows the relative magnitudes of the two terms in Equation 1.9 calculated by Greenwood [11] for the regular array of 76 identical a-spots illustrated in Figure 1.8, as the a-spot radius is increased. In the calculation, the spot spacings are taken as one and the maximum a-spot radius is 0.5. Note that the cluster resistance $1/2\alpha$ exceeds the a-spot resistance when the a-spot radius increases beyond approximately 0.05. The contact radius of a single spot of the same resistance is, then, given to a reasonable approximation by the Holm radius, α. The radius of equivalent single contacts and the Holm radius for a variety of a-spot distributions, also calculated by Greenwood [11], are illustrated in Figure 1.9. Note in each case that the circular area defined by the Holm radius (i.e., the Holm circle) provides an excellent representation of the area over which electrical contact occurs. These results are significant because they suggest that the details of the number and spatial distribution of the a-spots are not important to the evaluation of contact resistance in many practical applications where

TABLE 1.3

Effect of a-Spot Radius on Constriction Resistance $1/2na$ and Holm Radius α

a-Spot Radius	a-Spot Resistance $1/2na$	Holm Radius α	Cluster Resistance $1/2\alpha$	Radius of Single Spot of Same Resistance
0.02	0.3289	5.34	0.0937	1.18
0.04	0.1645	5.36	0.0932	1.94
0.1	0.0658	5.42	0.0923	3.16
0.2	0.0329	5.50	0.0909	4.04
0.5	0.0132	5.68	0.0880	4.94

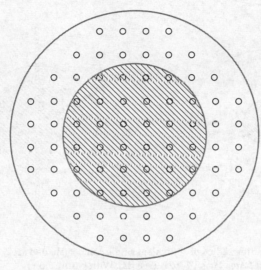

FIGURE 1.8
Regular array of a-spots; the shaded area is the single continuous contact with the same resistance; the outer circle is the Holm radius of the cluster. (From JA Greenwood, *Brit J Appl Phys* 17: 1621, 1966 [11]. With permission.)

electrical contact occurs reasonably uniformly over the nominal contact area, that is, in the absence of electrically insulating surface films. This conclusion is supported by Nakamura and Minowa [12] and Minowa et al. [13] who used finite element analysis and Monte Carlo techniques to examine the effect of a-spot distribution on electrical resistance. They found that the resistance of an interface characterized by a *fixed* electrical area was not significantly affected by the a-spot locations within the selected nominal contact area. Only if the a-spot distribution was limited to areas close to the periphery of the nominal geometrical interface was contact resistance affected appreciably. Even under these circumstances, resistance increased only by a factor of approximately two over the resistance produced by concentrating all the a-spots at the centre of the nominal area of electrical contact. The results of the investigations mentioned above suggest that for many engineering purposes, knowledge of the Holm radius may be sufficient to estimate contact resistance. As a first approximation, the Holm radius may be estimated from the true area of contact, A, as $[A/\pi]^{1/2}$.

Because the true area of contact is smaller than the apparent area of contact, a-spots must support local pressures that are comparable with the strengths of the materials of the contacting bodies. It is generally accepted that the true contact area is controlled by the plastic deformation of the asperities projecting from the surface. Although Archard [14]

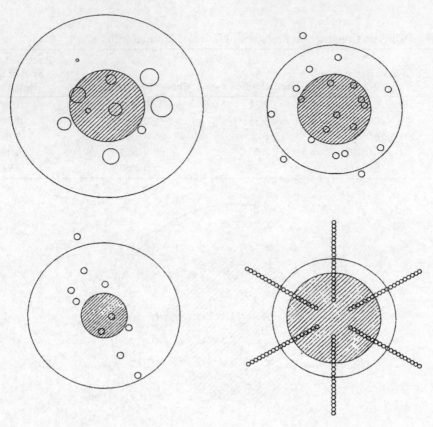

FIGURE 1.9
Clusters of *a*-spots with corresponding radius of equivalent single contact (shaded area) and Holm radius (outer radius). (From JA Greenwood, *Brit J Appl Phys* 17: 1621, 1966 [11]. With permission.)

proposed an elastic theory of contact, Greenwood and Williamson [2] have shown that deformation of the asperity is generally plastic in most practical applications. Bowden and Tabor [15] proposed that the contact pressure on contact asperities is equal to the flow pressure of the softer of the two contacting materials and the normal load is then supported by plastic flow of the softer asperities. Under this assumption, the area of mechanical contact, A_c, is related to the load F applied to the electrical interface and to the plastic flow stress (or hardness) H of the softer material as

$$F = A_c H \tag{1.10}$$

Expression 1.10 is extremely important and relevant for the interpretation of measurements of electrical contact resistance, as will be evident shortly. It states that the true area of mechanical contact between two surfaces is independent of the area of nominal contact of the surfaces; that is, $A_c = F/H$ depends only the contact force and the hardness of the contacting bodies, and is independent of the dimensions of the contacting objects. This is a remarkable statement. The physical origin of Equation 1.10 may be elucidated through the following simple argument: consider two sets of coupons of identical materials but of different dimensions, say 1 cm² and 10 cm² respectively, subjected to the same contact load, F. If the materials have the same surface finish, they will carry the same surface density of asperities. Thus, if the true contact area between the smaller coupons is generated through n asperities, the true contact area between the larger coupons is necessarily generated through $10n$ asperities. The average mechanical load developed at each asperity in the smaller interface is, then, F/n, whereas the same quantity in the larger interface is only $F/10n$. If deformation of the asperity is fully plastic, the area of contact at each asperity in the smaller interface will be 10 times larger on the average than in the larger interface, but the total contact area is identical in the two interfaces. Hence, Equation 1.10 is valid under conditions of plastic deformation.

If the electrical interface does not carry electrically insulating films and is characterized by a sufficiently large number of a-spots distributed within a Holm radius α, the data of Table 1.3 suggest that the contact resistance can be approximated as

$$R_c = \frac{\rho}{2\alpha}$$

and $A_c = \eta\pi\alpha^2$, where η is an empirical coefficient of order unity for clean interfaces. Using Equation 1.10, the contact resistance may be expressed as

$$R_c = \left\{ \frac{\rho^2\eta\pi H}{4F} \right\}^{1/2}$$

(1.11)

Expressions of the form as in Equation 1.11 have been used by several workers [3,15,16,17]. The general validity of this expression over a wide range of mechanical loads appears to be consistent with much published experimental data for a variety of contact materials (see [18–23]) as illustrated in Figures 1.10 through 1.12. The presence of interfacing contaminant films modifies the right-hand side of Equation1.11 by the addition of a new term, as will be shown later.

The relative simplicity of Equation 1.11 and the broad agreement between its predictions and the experimental data mask the underlying complexity of the physical phenomena involved in the generation of a bulk electrical interface. The decrease in contact resistance with increasing mechanical load depicted in Figures 1.10 through 1.12 stems from a combination of several factors, the most important ones being:

(1) An increase in the number of contacting surface asperities as the nominal surfaces are brought closer together under the influence of an increasing load, (2) A permanent

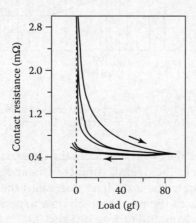

FIGURE 1.10
Contact resistance as a function of load for nominally clean gold electric contacts in air. The arrows show the direction of load application. (From RE Cuthrell and DW Tipping, *J Appl Phys* 44: 4360, 1973 [20].)

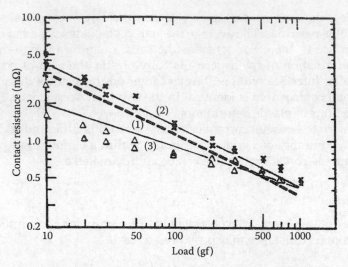

FIGURE 1.11
Contact resistance as a function of load for surfaces consisting of Ag90Pd10. Curve (1), results calculated using Equation 1.11; curves (2) and (3), experimental results (x, first test load increased; Δ, first test, load decreased). (From Y Watanabe, *Wear* 112: 1, 1986 [22].)

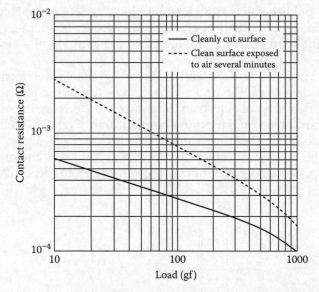

FIGURE 1.12
Contact resistance as a function of load for freshly cut copper and crossed rods and for identical rods after long exposure to air. (From LP Solos, *Elect Cont-1962: Eng Sem Elect Cont*, University of Maine, June 1962 [23].)

flattening of the contacting asperities, which reduces the constriction resistance associated with each a-spot and thus reduces the overall contact resistance, and (3) Work-hardening of the deformed contact asperities. The last effect decreases the rate at which contacting asperities flatten, and thus, reduces the rate at which new asperities are brought into play as the load is increased further. Note in Figures 1.10 and 1.11, contact resistance increases relatively slowly with decreasing load, after application of the initial load. This stems from permanent flattening and adhesion of the asperity following contact.

Mathematical models that attempt to explain the behaviors depicted in Figures 1.10 through 1.12 in terms of mechanical deformation properties of single asperities [1,2,14,24] predict a relationship between contact resistance and mechanical load that does not differ appreciably from that given in Equation 1.11. This remarkable result suggests that the details of deformation of single asperities are relatively unimportant and that assumptions used in the derivation of Equation 1.11 are not overly simplistic. Equation 1.11 is widely used by design engineers to estimate the expected constriction resistance for selected values of hardness and contact force in an electrical contact. This estimate generally agrees to 20% or better with actual measured values.

1.2.4 Effect of the Shape of Contact Asperity on Constriction Resistance

The surface asperities of solid bodies exhibit a wide variety of geometrical shapes. In general, the surface in the immediate vicinity of an a-spot is not parallel to the average plane of the electrical interface, as illustrated in Figure 1.13. It was shown by Sano [25] that the spreading resistance R_θ associated with an a-spot generated by an asperity making an angle θ with the mating surface, as shown in Figure 1.13, is given as

$$R_\theta = \frac{\rho}{2\pi a} \tan\left\{\frac{\pi + 2\theta}{4}\right\} \tan^{-1}\left\{\frac{z_e \cos\theta}{a}\right\} \tag{1.12}$$

where a is the radius of the contact spot and z_e is the normal distance from the contact interface. Equation 1.12 was derived under the assumption that the current distribution within the constriction is given by Equation 1.5 with $m = 1$, for all values of θ. Since this cannot be valid for large θ (for example, for $\theta \sim 90°$ where the current distribution is constant across the constriction), Equation 1.12 is valid only for relatively small values of θ. For values of z_e large in comparison with a, Expression 1.12 becomes

$$R_\theta = \frac{\rho}{4a} \tan\left\{\frac{\pi + 2\theta}{4}\right\}$$

The right-hand side of this relation reduces to the well-known expression for the spreading resistance of a circular constriction, $R_o = \rho/4a$, for $\theta = 0$. Figure 1.14 shows the variation of the ratio R_θ/R_o with increasing θ, as evaluated from Equation 1.12. The data indicate that the effect of the slope of the asperity for values of θ as large as 10° is negligible. This is an important result since it indicates that the presence of knurls, which are often embossed on connector surfaces to increase friction with the conductor, has a negligible effect on contact resistance unless the knurl slope θ is appreciably larger than 10°.

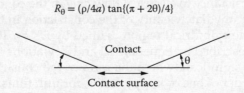

$$R_\theta = (\rho/4a) \tan\{(\pi + 2\theta)/4\}$$

Contact

Contact surface

FIGURE 1.13
The a-spot produced by the contact of a conical asperity making an angle θ with the mating surface.

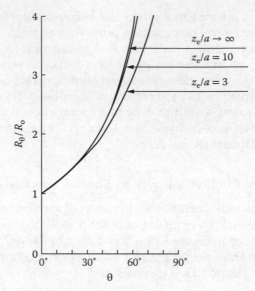

FIGURE 1.14
Variation of the ratio R_θ/R_o with increasing θ. (From Y Sano, *J Appl Phys* 58: 2651, 1985 [25]).

1.3 Effect of Surface Films on Constriction and Contact Resistance

1.3.1 Electrically Conductive Layers on an Insulated Substrate

Because thin conducting films have becomes ubiquitous as contact platforms in semi-conductor-based devices, it has become increasingly important to understand electrical constriction resistance in thin films deposited on an electrically insulated substrate or a substrate of substantially larger resistivity than that of the thin film. Such contacts would occur, for example, in a contact between a microwire and a thin film on a semiconductor. In these cases, optimization of configurations of the electrical contact with thin films would minimize contact resistance and associated factors such as Joule heating at the contacts. It is for this reason that constriction resistance in thin films is addressed in some detail in this section.

1.3.1.1 Calculation of Spreading Resistance in a Thin Film

Expression 1.2 for the electrical spreading resistance of a circular constriction in a bulk interface stems from spreading of electric current flow lines from the constriction towards the bulk of the conductor as illustrated in Figure 1.15a. In a conducting thin film, spreading resistance stems from the resistance to electrical flow *only* in the conducting region where the current spreads in the immediate vicinity of the constriction in the film as illustrated in Figure 1.15b, that is, in the cylindrical region defined by $r \leq r_A$. Thus, spreading resistance in a thin film is not described by Equation 1.2 and deviations from this relation are expected to be particularly significant where the ratio of the constriction radius a to the film thickness L_F is of the order of one or larger, that is, for thin films. In this situation, the current streamlines bend sharply away from the constriction edge and flow parallel to the film boundaries after a short distance $r \sim r_A$. In contrast, spreading occurs over a much

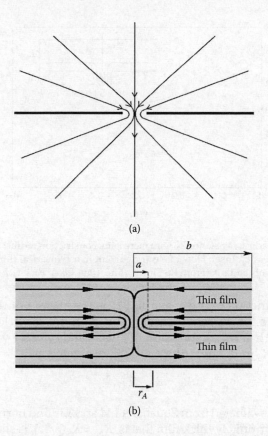

(a)

(b)

FIGURE 1.15
(a) Spreading of current streamlines in two "bulk" conductors in contact over a circular spot of radius a. The spreading resistance in each conductor is given by Equation 1.2. (b) Spreading of current streamlines near a constriction between two thin films.

larger region in the case of the bulk (i.e., two semi-infinite solids in contact) in Figure 1.15a. In Figure 1.15b, the resistance presented by the thin film in the region $r > r_A$ where the current streamlines in the radial direction are parallel to the horizontal boundaries of the films, does not contribute to constriction resistance. The resistance to radial flow in a hollow cylinder with an inner radius a, an outer radius b, and a length L_F is given as [3]

$$R_B = \frac{\rho}{2\pi L_F} \ln\left(\frac{b}{a}\right) \tag{1.13}$$

and will be designated as "bulk" resistance. Timsit [26] provided an analytical expression for the resistance R_T between the center of a circular constriction of radius a located on one flat side of a cylinder of height L_F and the outer circular surface of radius b of the cylinder as illustrated in the inset of Figure 1.16. The expression is

$$R_T = \frac{\rho}{\pi a} \sum_1^\infty \coth\left\{\frac{\lambda_n L_F}{b}\right\} \frac{\sin\left\{\frac{\lambda_n a}{b}\right\}}{J_1^2\{\lambda_n\}\lambda_n^2} \tag{1.14}$$

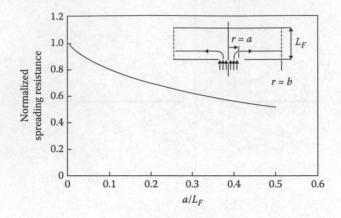

FIGURE 1.16
Variation of the normalized spreading resistance R_N with increasing constriction radius; the constriction radius is normalized to the film thickness L_F. Inset: Flow of electric current in a cylindrical film of outer radius b and thickness L_F, from a constriction of radius a. (From RS Timsit, *IEEE Trans Comp Pack Tech* 33: 636, 2010 [26].)

where $J_1(x)$ is the Bessel function of order one and λ_n is the nth root of the Bessel function of zero order, $J_0(x)$. Because the current distribution within the constriction was approximated by Equation 1.5 (with $m = 1$), Expression 1.14 was valid only for values of $a/L_F < 0.5$. The spreading resistance R_S in the thin film was then evaluated as

$$R_S = R_T - R_B \tag{1.15}$$

The spreading resistance evaluated from Equations 1.14 and 1.15 and normalized as R_N to the spreading resistance in an infinitely thick film that is, $R_N = R_S/(\rho/4a)$, is shown in Figure 1.16. The graph illustrates variations of the normalized spreading resistance with the normalized constriction radius a/L_F and indicates that the spreading resistance is identical with the value given by the classical expression, $\rho/4a$, that is, $R_N \approx 1$, only for small values of a/L_F (≤ 0.01).This result was found to be independent of the actual values of a and L_F and was verified for a wide range of values of L/b. These results show trends similar to those published by Norberg et al. [27] who evaluated contact resistance in thin films on the basis of approximations based on empirical modifications of Equation 1.3, which precludes a direct comparison with the results in Figure 1.16. The data of Figure 1.16 indicate that the spreading resistance decreases to about one-half the classical value at $a/L_F \sim 0.5$. The result that spreading resistance in a radially conducting film *decreases* with *decreasing* film thickness appears counter-intuitive since the resistance of a solid conductor increases with decreasing thickness. An explanation for this will be given later in relation to the behavior of constriction resistance at high signal frequency.

1.3.1.1.1 *Remarks on the Calculation of Spreading Resistance*

In the evaluation of spreading resistance from Equation 1.15, it is important to note that the choice of the bulk resistance R_B is somewhat *arbitrary*, because the current flow lines do not bend sharply and become parallel to the thin-film surface exactly at a radial distance $r = a$. As illustrated schematically in Figure 1.15b, the transition in current flow direction is completed at a radius $r_A > a$. Because the radius r_A is unknown *a priori*, the definition of the spreading resistance R_S in Equation 1.15 can only be considered as approximate. In a series of papers, Zhang et al. [28–31] evaluated numerically the spreading resistance of various 2D and 3D thin-film contact geometries for a wide range of values of both a/L_F and the ratio of resistivity of the contacting materials. The evaluations were performed to great accuracy using the MAXWELL 2D and 3D simulation codes [28]. For rectangular contact geometries with 2D symmetry, the results agreed exactly with the spreading resistance expressions of Hall [32,33] based on conformal mapping. For the cylindrical thin film illustrated in Figure 1.15, the authors summarized the dependence of the normalized spreading resistance R_N on $y = a/L_F$ by the following best-fit expressions

$$R_F\{y\} = 1 - 2.2968y + 4.9412y^2 - 6.1773y^3 + 3.811y^4 - 0.8836y^5, 0.001 \leq y \leq 1 \quad (1.16)$$

and

$$R_N\{y\} = 0.295 + 0.037y + 0.0595y^2, 1 < y < 10 \quad (1.17)$$

Equations 1.16 and 1.17 are independent of the outer film radius b since the bulk film resistance given by Equation 1.13 is subtracted out. The results of these computer simulations also confirmed the validity of Equation 1.14 to values of $a/L_F \sim 0.5$. Additional work by these authors and Timsit [31] extended the evaluation of R_N for values of a/L_F approaching 100. The results of all numerical evaluations are shown in Figure 1.17. They indicate that the normalized spreading resistance, based on the spreading resistance calculated using Equation 1.14, decreases from the value of 1 to reach a limiting value of 0.28 as L_F approaches zero that is, as a/L_F becomes large. This limiting value is independent of the film outer radius b. The result of a non-vanishing R_N for a vanishing film thickness was unexpected, but may be understood from an examination of current flow lines into the thin film, as shown in Figure 1.18.

The current flow lines of Figure 1.18 were calculated for the case of $a/L_F = 10.1$ [31] and show unambiguously that the streamlines are concentrated near the edge of the constriction and curve inwards significantly at radial distances smaller than a. The calculations in [31] show that the current streamlines in the thin film are concentrated in a radial region defined by a' and a with

$$a' \approx a\left\{1 - 0.44\frac{L_F}{a}\right\} \quad (1.18)$$

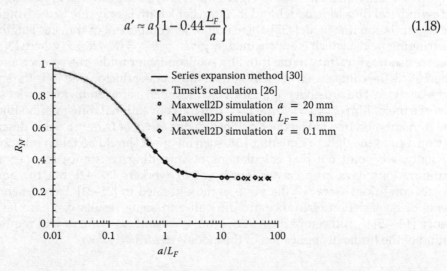

FIGURE 1.17
Normalized thin-film spreading resistance R_N as a function of a/L_F, for the cylindrical structure in Figure 1.15. The solid line is calculated from Equation 1.15, synthesized from the results of series expansion calculations [30]. The dashed line and the symbols describe respectively the results of Timsit's calculations [26] and the data from the MAXWELL 2D simulation. Three sets of simulation were performed. The first set was fixed at $a = 20$ mm (circles), and varying L_F from 20 mm to 1 mm; the second set was fixed at $L_F = 1$ mm (crosses), and varying a from 30 mm to 70 mm; the third set was fixed at $a = 0.1$ mm (diamonds), and varying L_F from 0.25 mm to 0.0015 mm.

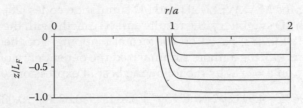

FIGURE 1.18

Field lines in the radial direction in the cylindrical thin film in the inset of Figure 1.16, calculated from the series expansion method [31], for the case of $a/L_F = 10.1$. (From P Zhang et al., *IEEE Trans Comp Pack Tech* 39: 1936, 2012 [31].)

Crowding of the current and the ensuing spreading inward in the region $r \leq a$ occurs independently of the value of film thickness, L_F and describes the effect responsible for the normalized spreading resistance of 0.28 for any non-zero value of a even as L_F tends to 0. Thus, a non-zero value of normalized spreading resistance would be obtained even if the film is very thin, since in this case spreading of the current line would occur within the volume defined by $r \leq a$.

There is an additional important result from the analyses in [28–31]. The crowding of the current streamlines near the edge between a' defined in Equation 1.18 and a could lead to significant ohmic heating in this region. Such enhanced ohmic heating is well-known in bulk electrical contacts [34], on the basis of the distribution of current density in the constriction as given by Equation 1.5. Current crowding in metal–semiconductor contacts can lead to significant ohmic heating and deleterious effects on contact resistance [35]. There are differences between metal–metal and metal–semiconductor contacts. For example, in the transmission line model of a metal–semiconductor contact [35,36], the length over which most of the current from a contact into a semiconductor thin film flows is called the transfer length, L_T. From Equation 1.18 and Figure 1.18, it may be argued that $L_T \sim 0.44 L_F$ for the present cylindrical thin-film model, and this transfer length is only due to the fringing fields. In the transmission line model [35], there is another component of transfer length, neglecting the fringing fields, which is approximately given by $L_{T2} = (R'_C/R'_S)^{1/2}$, where $R'_S = \rho/L_F$ is the sheet resistance (in Ωm^{-2}) in the thin film semiconductor under the contact, and $R'_C = A_C \rho_C$, where A_C is the contact area of the film with the semiconductor, and ρ_C is the so-called contact resistivity. The resistivity ρ_C arises from the metal–semiconductor barrier so that in the case of a metal film $\rho_C = 0$, yielding $L_{T2} = 0$ in the conventional transmission line model [33]. The transmission line model does not include the effect of fringing fields described in this section, but in the light of Equation 1.18, such an effect should be taken into account.

Finally, we point out that calculations of spreading resistance for various thin-film configurations have been carried out by other workers [37–42], but the boundary and contact conditions were not the same as those treated in [28–31]. Experimental measurements of constriction resistance are difficult and some results compare favorably with theory [43–45]. A full comparison between theory and experiment will require an appreciation of the limits of application of the models described above.

1.3.2 Electrically Conducting Layers on a Conducting Substrate

The presence of a film at the interface of an electrical junction affects contact resistance in a variety of ways. If the film is present initially on one of the contact surfaces and is electrically conducting, the constriction resistance of an a-spot is either decreased or increased relative to the resistance produced by the identical a-spot on the uncoated surface, depending on the electrical resistivity of the film material relative to that of the substrate. A change in mechanical hardness due to the presence of the film may also affect contact resistance as suggested by Equation 1.11. If the interfacing layer is formed by the interdiffusion of dissimilar materials across the contact, contact resistance will often increase through the formation of electrically resistive intermetallic compounds. If the film is electrically insulating or only weakly conducting, a good electrical contact is established only if the film is mechanically disrupted to allow the formation of metal-to-metal junctions. This section reviews the effect of electrically conducting layers such as electroplated layers, electrically resistive layers such as contaminant surface films and electrically insulating interfacing layers on constriction and contact resistance.

1.3.2.1 Electrically Conducting Layers and Thin Contaminant Films

Electrically conducting coatings (electroplates) are often used to minimize electrical contact resistance. Contact resistance may also be reduced through the action of several mechanisms such as a decrease in surface hardness, large electrical conductivity of the electroplate relative to that of the substrate, elimination of electrically insulating surface oxide films, and so on. Conducting coatings are also used to protect contact surfaces against tarnishing and oxidation, corrosion, mechanical wear, and the like. Because of its resistance to oxidation and mechanical wear, gold is the plating material of choice for producing reliable electrical contacts in copper–base electrical connectors. However, environmental testing involving exposure to high humidity and severely polluted laboratory or out-of-door environments [6,47] show that even gold coatings do not protect against corrosion if they are porous (see Chapters 2 and 3). These tests also show that the gold layer must be sufficiently thick to be pore-free and hence to perform satisfactorily in electrical connectors. Because this is generally not cost-effective, other plating materials have been examined as a replacement for gold. This replacement is not straightforward. For example, substitution by palladium has been hampered by the tendency of this metal to tarnish and form frictional polymers [48]. Many alloys have been evaluated, such as palladium–silver, tin–lead, tin–nickel, cobalt–gold, and so on. (see for example [49–53]). For aluminum–base connectors, the use of tin and nickel platings has been considered [54] to mitigate the effects of corrosion and surface oxidation on the electrical connectibility of aluminum. The introduction of plated layers in a connector system is discussed in detail in Chapter 7. In this section, we focus on the models used in the interpretation of contact resistance properties of conducting plated layers and surface contaminant films on metal surfaces.

A rigorous evaluation of the effect of electrically conducting films on contact resistance requires the use of methods of numerical analysis. It would be expected that electrical contact resistance depends on the electrical resistivity of the plating relative to that of the substrate, and on the ratio of the radius of the a-spot to the thickness of the plating. Where the resistivity of the plating material is larger than that of the substrate material and radius of the a-spot is of the same order of magnitude as the thickness of the film, the electric current emanating from the a-spot spreads out significantly more into the substrate than into the plating, as illustrated in Figure 1.19a. In this case, the potential drop in the immediate vicinity of the a-spot in the substrate is negligible in comparison with the potential drop across the film in a direction normal to the film–substrate interface [3]. Therefore, the film–metal interface defines a nearly equipotential surface, the current density in the film is, thus, approximately uniform across the a-spot as illustrated in Figure 1.19a, and the spreading resistance is still nearly given by [3]

$$R_S = \frac{\rho}{4a}$$

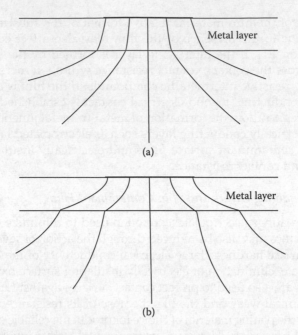

FIGURE 1.19
Current distribution in a metal surface film under conditions where: (a) The film resistivity is larger than that of the substrate material and the *a*-spot radius is of the same order of magnitude as the film thickness. (b) The film resistivity is smaller than that of the substrate and the current lines spread out more appreciably within the plating than within the substrate.

where ρ is the resistivity of the substrate material. Since the current also passes through the resistive film of area πa^2, thickness d and resistivity ρ_f, the additional film resistance is approximately $\rho_f d/\pi a^2$. To a first approximation and for the case where the film is sufficiently thin, the total resistance R_t then becomes

$$R_t = R_S + \rho_f \frac{d}{\pi a^2} \tag{1.19a}$$

This expression reduces to

$$R_t = \frac{\rho}{4a} \left\{ 1 + \frac{4}{\pi} \frac{\rho_f}{\rho} \frac{d}{a} \right\} \tag{1.19b}$$

We recall that Equation 1.19b represents the *resistance* only of the coated surface. An evaluation of contact resistance as measured by a contact probe requires the addition of the spreading resistance from the probe to this equation. Under conditions where the *a*-spot radius and the plated layer thickness do not differ greatly, the spreading resistance thus increases approximately linearly with plating thickness. For a sufficiently thick film, the spreading resistance will of course deviate from the above expression and approach the value $\rho_f/4a$. Equation 1.19b is useful in pointing out that the effect of constriction resistance is overshadowed by the film resistance whenever the ratio $(\rho_f/\rho)(d/a)$ is much larger than unity. The validity of this conclusion was verified by computer simulations reported by Nakamura and Minowa [55].

Where the resistivity of the plating material is smaller than that of the substrate, the current lines spread out more appreciably within the plating than within the substrate, as illustrated in Figure 1.19b, and the spreading resistance decreases with increasing film thickness. Here again, the spreading resistance approaches the value $\rho_f/4a$, where the a-spot is much smaller than the film thickness. Whether ρ_f is smaller or larger than the substrate resistivity, the effect of the plating on contact resistance is often evaluated via the ratio

$$P_f\{d/a, \rho_f, \rho\} = \left\{\frac{\rho_{eff}}{4a}\right\} \Big/ \left\{\frac{\rho}{4a}\right\} \tag{1.20}$$

where P_f (d/a, ρ_f, ρ) is the "plating factor" and ρ_{eff} is the effective resistivity of the plated substrate; and $\rho_{eff} = \rho$ for $d = 0$. The plating factor may be evaluated easily using the algorithm described by Williamson and Greenwood [56]. Figures 1.20 and 1.21 illustrate the calculated dependence of the plating factor on the ratio $d/2a$ respectively for typical cases for which $\rho_f/\rho > 1$ and $\rho_f/\rho < 1$ [56]. Note that P_f reaches a limiting value for $d/2a \approx 1$ for all the cases illustrated. The spreading resistance of a plated surface is, thus, given as $\rho_{eff}/4a$.

In practice and as indicated earlier, the effect of surface deformation on contact resistance must also be taken into account. If the resistivities of the two rough contacting materials (e.g. the plated surface and a measuring probe) are respectively ρ and ρ_p, and the effective resistivity of the plated material is ρP_f, Equation 1.11 immediately yields the contact resistance as

$$R_t = \frac{\rho_p + \rho P_f}{2}\left\{\frac{\eta\pi H}{4F}\right\}^{1/2} \tag{1.21}$$

where again H is the hardness of the softer metal in the interface. Recall also that P_f becomes a function of the load F since the dimensions of a-spots in the interface are affected by F. Thus, in practice, the contact resistance of an electrical interface in which the surfaces are rough, and in which one of the surfaces carries a conductive metallic film, may decrease with increasing load *somewhat differently* than as $F^{-1/2}$.

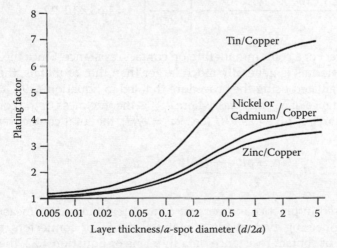

FIGURE 1.20
Dependence of the plating factor on the ratio $d/2a$ for the case of $\rho_f/\rho > 1$. (From JBP Williamson and JA Greenwood, *Proc Int Conf Elect Cont Electromech Comp*, Appendix, Beijing, China, Oxford: Pergamon Press, 1989 [56].)

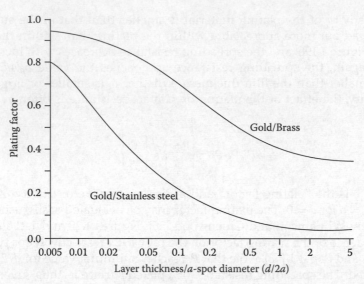

FIGURE 1.21
Dependence of the plating factor on the ratio $d/2a$ for the case of $\rho_f/\rho < 1$. (From JBP Williamson and JA Greenwood. *Proc Int Conf Elect Cont Electromech Comp*, Appendix, Beijing, China, Oxford: Pergamon Press, 1989 [56].)

It is instructive to consider one example of the use of Equation 1.21. Consider a gold probe at room temperature in contact with a load of 0.1 kgf on a copper surface plated with a tin layer of thickness 10 μm. From Table 24.1 (see Chapter 24), the hardness of gold (30 kg mm^{-2}) is much larger than that of tin (4 kg mm^{-2}), so that the hardness value H used in Equation 1.21 is that of tin. From Equation 1.10, the area of metal-to-metal contact is calculated as F/H or $0.1/4 = 0.025$ mm^2, thus yielding an average contact radius of $\sqrt{0.025/\pi} = 0.089$ mm or 89 μm. The ratio of layer thickness to average diameter of contact spot is, thus, $10/178 = 0.06$, which yields a plating factor P_f of 2 from Figure 1.20. Using $\rho_p = 2.3 \times 10^{-5}$ Ωmm for gold and $\rho = 1.75 \times 10^{-5}$ Ωmm for copper, the contact resistance is given as

$$R_t = \frac{2.3 \times 10^{-5} + 1.75 \times 10^{-5} \times 2}{2}\left\{\frac{\pi 4}{4 \times 0.1}\right\}^{1/2} = 1.63 \times 10^{-4}\,\Omega$$

Consider now the effect of a contaminant film on contact resistance. Since the resistivity ρ_{cont} of contaminant materials is generally much larger than that of metals, the effect on contact resistance is evaluated using the procedure that led to Equation 1.19a. This procedure yields a film resistance of $\rho_{cont}d_{cont}/\pi a^2$, where d_{cont} is the thickness of the contaminant film. Since the contact area is given as F/H (i.e., $\pi a^2 = F/H$), the total contact resistance is now given as

$$R_t = \frac{\rho_p + \rho P_f}{2}\left(\frac{\eta\pi H}{4F}\right)^{1/2} + \frac{\rho_{cont}d_{cont}H}{F} \tag{1.22}$$

Expression 1.22 is used in practice to interpret contact resistance data measured from plated surfaces. In engineering evaluations of metal coatings for connectors, and in the ensuing interpretation of contact resistance data in terms of Equation 1.22, the electrical contact properties are generally measured by loading a metal probe onto the coated surface and recording the electrical resistance as a function of the applied force. The probe material is often pure gold with a smooth hemispherical end with a radius of 1.6 mm. A dependence of contact resistance on load as F^{-1} is usually taken as evidence (from Equation 1.22) of the presence of contaminant material on the surface [57,58]. Details of the metal probe and measuring procedure are given in the ASTM Standard Practice [59]. We now provide examples of such measurements of contact resistance.

Figure 1.22 shows a plot of contact resistance as a function of contact force for tips of five different gold probes pressed against a gold target [60]. Four of the probes were character-ized by a spherical contact surface with a radius respectively of 3.2 mm, 1.6 mm, 1.2 mm, 0.9 mm; the fifth was prepared by machining the end of a gold rod 3.18 mm in diameter to a 60° cone with a sharply pointed tip. The measured contact resistance is very close to the values predicted on the basis of Equation 1.22 in the absence of a contaminant film (i.e., $d_{cont} = 0$). The one exception to the predicted behavior is for the conical tip. In this case, the investigators found that the larger resistance measured with this tip stemmed from a larger bulk resistance [60]. When this resistance component is subtracted from the mea-sured contact resistance, the curve obtained using the conical tip coincides with the curves yielded by the others. Note that the curves of Figure 1.22 are essentially independent of the radius of the gold probe and that contact resistance depends only on the applied force. This is indeed as predicted by Equation 1.22 with $d_{cont} = 0$. Thus, the area of metal-to-metal contact in the electrical interface is independent of the details of probe geometry and depends only on hardness.

Figure 1.23 shows the results of contact resistance measurements on freshly polished copper using the five gold tips referred to in Figure 1.22. There is a significant difference between these results and those of Figure 1.22 for gold-gold contacts. The slope of the ini-tial parts of the curves, to a load of 20 gf, is approximately −1 and the contact resistance is generally larger than obtained from gold–gold. The initial inverse proportionality to load F suggests that the contact behavior is described by Equation 1.22 where $d_{cont} \neq 0$, that is, in the presence of a contaminant film of relatively large resistivity such as an oxide film. The investigators found a good fit to Equation 1.22 using a value of $2.75 \times 10^{-14}\Omega$ m² for $\rho_{cont}d_{cont}$. Note again the curves of Figure 1.23 are essentially independent of probe geom-etry as predicted by Equation 1.22. The observations of Figures 1.22 and 1.23 support the

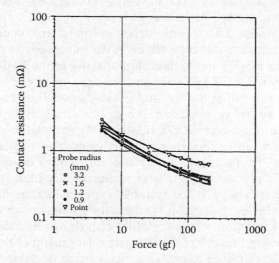

FIGURE 1.22
Contact resistance versus applied force for gold probes of various tip radius: Clean gold-to-gold. (From MR Pinnel and KF Bradford, *Proc 28th Elect Comp Conf*, Anaheim, CA 129, 1978 [60].)

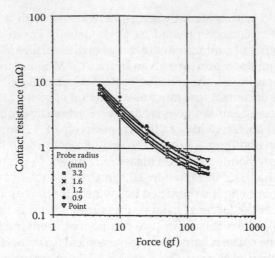

FIGURE 1.23
Contact resistance versus applied force: Clean gold on freshly polished copper. (From MR Pinnel and KF Bradford, *Proc 28th Elect Comp Conf*, Anaheim, CA 129, 1978 [60].)

premise that the area of metal-to-metal contact is determined only by plastic deformation (hardness) of the contacting surfaces. Additional examples of the type of contact resistance measurements depicted in Figures 1.22 and 1.23 are easily accessible in the literature on contact-resistance. These are described in [61,62].

1.3.3 Growth of Intermetallic Layers

The formation of intermetallic compounds arises from the interdiffusion of materials across a bimetallic interface. In electrical contacts, this interdiffusion occurs when the electrical interface is operated in a high-temperature environment or when sufficient electric current is passed through the contact to raise the temperature of the a-spot to well above the ambient temperature. The relationship between a-spot temperature and voltage drop across the contact will be addressed in Section 1.4.

Recent decades have witnessed a surge of interest in interdiffusion phenomena at bulk and thin-film bimetallic interfaces due respectively to the increased use of bimetallic welds in a variety of applications [63–72] and to the ubiquitous use of multi–thin-film structures in microelectronic devices [73]. Bulk joints are made using a variety of techniques ranging from pressure and friction welding to flash and explosive welding. In electrical applications, frequent electrical surges or operation of a device at relatively elevated temperatures may generate conditions favorable to formation of intermetallic layers in bimetallic electrical contacts, since interdiffusion is thermally activated. The formation of these layers is generally deleterious to the electrical stability and mechanical integrity of a bimetallic joint because intermetallic phases are usually characterized by high electrical resistivity and low mechanical strength [63–72]. For example, experimental evidence indicates the growth of an intermetallic layer considerably weakens the strength of an aluminum–copper joint [74]. The subject of formation of intermetallics also draws considerable attention in the microelectronics industry because of the intimate relationship of this phenomenon to the reliability of bimetallic thin-film contacts in microelectronic devices (see for example references [75–77]). This section reviews the intermetallic growth at interfaces of bulk electrical contacts of common interest such as aluminum–copper, aluminum–brass and plated layers on brass and phosphor bronze. For illustrative purposes, data obtained from aluminum–brass joints [70] will be used.

Figure 1.24 shows micrographs of two aluminum–brass interfaces obtained by scanning electron microscopy (SEM) after the joints had been heated in a furnace respectively for 765 hours at 250°C and for 6 hours at 400°C [70]. The brass consisted of wt% Cu70Zn30. The intermetallic layers consist of four distinctive bands and the total interdiffusion layer thicknesses differ significantly due largely to the large difference in the temperature of exposure. Figure 1.25 shows the growth of the layers measured at temperatures of 250°C–240°C. For each case, note that the thickness x of the separate layers grows according to the relation

$$x^2 = kt \tag{1.23}$$

where k is the interdiffusion rate constant at the selected temperature and t is the interdiffusion time. There is a clear proportionality of layer thickness to $t^{1/2}$ for all temperatures. At first glance, this observation is surprising since atoms that diffuse from one end of the bimetallic joint to the other do not cross the same number of interlayer boundaries and travel ever increasing distances as the thickness x increases. However, the work of Gosele and Tu [78] indicates that this observation is consistent with expected growth of the interdiffusion layer if diffusion of atoms through intermetallic layers is considerably slower than diffusion across interlayer boundaries. The activation energy characterizing the

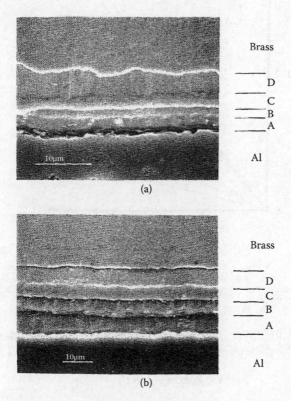

FIGURE 1.24
Micrographs of two aluminum–brass interfaces obtained by scanning electron microscopy (SEM) after the joints had been heated in a furnace respectively for (a) 765 hours at 250°C and (b) 6 hours at 400°C. (From RS Timsit, *Acta Met* 33: 97, 1985 [70].)

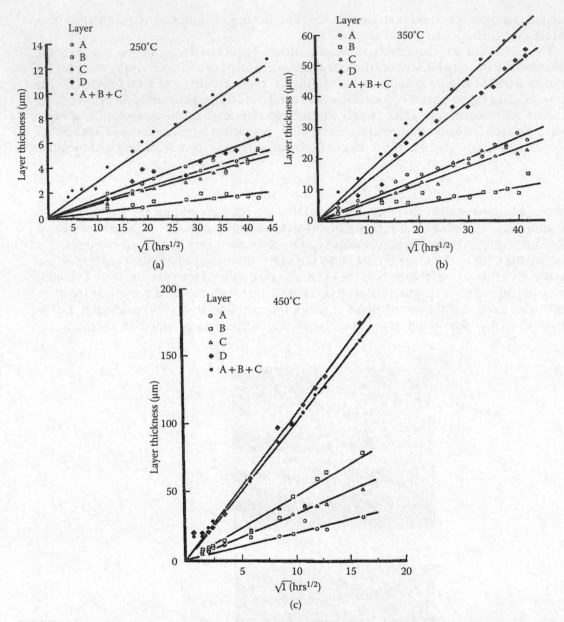

FIGURE 1.25
Growth rate of intermetallic layers at temperatures of 250°C–450°C at aluminum–brass interfaces. (From RS Timsit, *Acta Met* 33: 97, 1985 [70].)

growth of each intermetallic layer in aluminum–brass interfaces was determined in [70] on the assumption that the dependence of k on temperature is expressed as

$$k = k_0 \exp(-Q/RT) \tag{1.24}$$

where k_0 is a constant, Q is the activation energy, R is the universal gas constant (1.987 cal K^{-1}) and T is the absolute temperature. The results of the analysis are shown in Figure 1.26

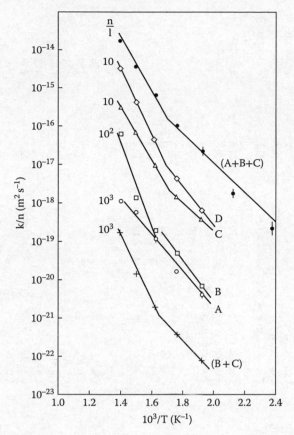

FIGURE 1.26

Arrhenius plots of the rate constants for growth of the various interdiffusion bands at an aluminum–brass interface. (From RS Timsit, *Acta Met* 33: 97, 1985 [70].)

and are summarized in Table 1.4. It was found that intermetallic growth could generally not be described satisfactorily by a single activation energy over the entire temperature range but that an adequate description was possible by considering the data from the temperature regions 150°C–300°C and 350°C–450°C separately. As shown in Table 1.4, the activation energies determined for temperatures lower than approximately 300°C are considerably smaller than those obtained for the more elevated temperatures. The lower activation energies may be indicative of intermetallic growth by mechanisms such as grain boundary diffusion or dissociated dislocation [79]. The activation energies above 300°C are characteristic of bulk diffusion. Similar observations of changes in activation energies over the two temperature regions were made [72] in an investigation of interdiffusion in bulk aluminum–copper diffusion couples similar to that described in [70]. As indicated in Table 1.4, the values of Q in the temperature ranges 150°C–300°C and 350°C–450°C for the diffusion bands $(A + B + C)$ at aluminum–brass interfaces are nearly identical to those measured in aluminum–copper bulk diffusion couples [72]. The chemical compositions of the layers formed at various temperatures at aluminum–brass interfaces are shown in Figure 1.27. Note the compositions do not vary appreciably as the temperature increases. A summary of the composition and crystal structure of the interdiffusion layers is also given in Table 1.4.

TABLE 1.4

Summary of Composition, Crystal Structure, and Growth Rates of Intermetallic Compounds Formed at Aluminum–Brass, Aluminum–Copper, Plating–Brass, and Plating–Phosphor-Bronze Interfaces

Diffusion Couple	Layer	Temperature Range (°C)	Composition Range (%wt)			Crystal Structure	k_0 (m² s⁻¹)	Activation Energy (kcal mol⁻¹)	Micro-Hardness (kg mm⁻²)
			Al	Cu	Zn				
Al-brass	A	250–450	46–43	53–56	~1	$CuAl_2$	5.6×10^{-9}	21.9	500–700
	B	250–300	34–27	64–69	2–4	$Cu_{0.61}Al_{0.39}$	4.2×10^{-9}	23.4	850–1000
		350–450	23–20	69–70	8–10	$Cu_{0.61}Al_{0.39}$	8.6	50.7	850–1000
	C	250–300	21–15	77–75	2–10	$Cu_{0.61}Al_{0.39}$	3.9×10^{-11}	16.8	850–1000
		350–450	19–15	76	5–9	$Cu_{0.61}Al_{0.39}$ Traces of Cu_3Al_2	1.14×10^{-5}	31.5	850–1000
	D	250–300	3–2	60–57	37–41	Largely $Cu_{0.6}Zn_{0.4}$	8.5×10^{-9}	21.8	230–330
		350–450	3–2	60–57	37–41	$Cu_{0.6}Zn_{0.4}$ (CuZn component at 450°C)	1.2×10^{-2}	38.4	230–330
	A + B + C	150–300					3.4×10^{-10}	17.3	
		350–450					4.0×10^{-4}	33.7	
	B + C	250–300					6.8×10^{-10}	19.0	
		350–450					2.0×10^{-3}	40.1	
Al–Cu		250–380				Cu_2Al	2.2×10^{-10}	17.2	35
		425–520				Cu_3Al_2	0.8	32.7	180
		250–520				Cu_4Al_3			624
						$CuAl$			648
						$CuAl_2$			413
Zn-brass		80–150		16	84	mainly Cu_5Zn_8	4.5×10^{-4}	22.5	650
				31	69				
In-brass		80–150		34	9 (57 In)		5.75×10^{-11}	10.9	200–300
Sn-brass		80–150		13–16	(86–83 Sn)	Cu_3Sn, Cu_6Sn_5	8.34×10^{-14}	6.37	180
Sn–Cu		100–200						6.0	180
Zn-phosphor-bronze		80–150		36	61 (2 Sn)	mainly Cu_3Zn_8	1.58×10^{-4}	20.0	700
In-phosphor-bronze		80–150					1.22×10^{-9}	12.9	200–300
Sn-phosphor-bronze		80–150					3.17×10^{-13}	7.2	200–300

Source: RS Timsit, Acta Met 33: 97, 1985 [70]; RS Timsit, IEEE Trans Comp Hyb Manuf Tech CHMT-9: 106, 1986 [71]; M. Braunovic and N. Alexandrov, IEEE Trans CompPack Manuf Tech CHMT-A17: 78, 1994 [72]. With permission.

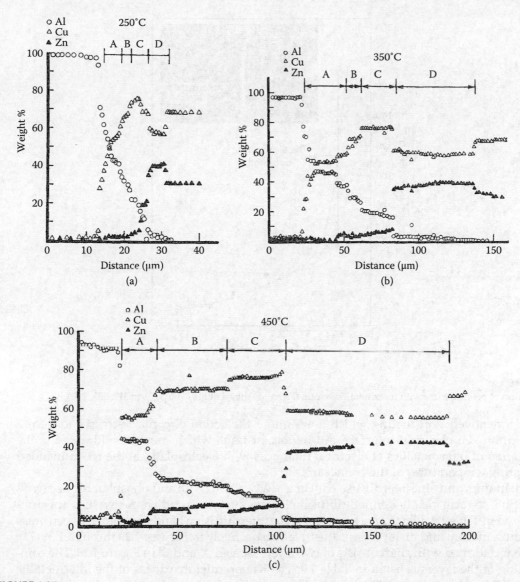

FIGURE 1.27
Chemical composition of intermetallic layers formed at aluminum–brass interfaces at various temperatures. Note that the compositions do not vary appreciably as the temperature increases. (From RS Timsit, *Acta Met* 33: 97, 1985 [70].)

It was mentioned that the brittleness and related properties of intermetallic layers can lead to a weakening of the interface in which the layers grow. The data of Figure 1.28 illustrate this effect for an aluminum–copper interface where the peeling force between the two mating surfaces was measured after various time intervals for interdiffusion [74]. The strength of the interface decreased precipitously with increasing thickness of the intermetallic layer once the width of the layer exceeded approximately 2 μm.

Many electrical connectors produced from copper–base alloys use electroplated layers such as tin, indium or even zinc to improve stability of the contact or mitigate attack by corrosion. The choice of plating material is often based on the ability of the material to

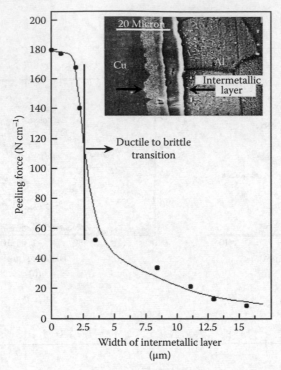

FIGURE 1.28
Variation of peeling force with intermetallic width. (From M Abbasi et al., *J Alloy Comp* 319: 233, 2001 [74].)

form a relatively soft coating which flows under the action of contact stresses to form a long unbroken electrical interface. An important issue which must be addressed is the formation of intermetallics at electrical interfaces with electroplates at the recommended operating temperatures of the connectors.

Tin, indium and zinc interdiffuse with brass (wt% Cu70Zn30) and phosphor bronze (wt% Cu95Sn5) to form clearly defined diffusion layers. The measured characteristics are summarized in Table 1.4. Illustrative examples of the interdiffusion bands formed between brass and zinc, indium and tin layers are illustrated respectively in Figures 1.29 through 1.31 [71]. The band formed with zinc consists of two layers labeled A' and B' in Figure 1.29. The composition of the layers is listed in Table 1.4. The Knoop microhardness of the intermetallic layer B' was measured as 650 kg mm^{-2}. The interdiffusion layer formed with indium also consists of two layers labeled A'' and B'' in Figure 1.30. Layer B'' consists of an intermetallic compound whose composition is given in Table 1.4. The Knoop micro-hardness is of the order of 300 kg mm^{-2}. The interdiffusion band produced with tin exhibited widely spaced nucleation regions as illustrated in Figure 1.31. The Knoop microhardness of the layer whose composition is given in Table 1.4 is ~ 180 kg mm^{-2}. Examples of growth curves for interdiffusion bands produced on brass and phosphor bronze are shown respectively in Figures 1.32 and 1.33. Note that the growth rates in brass and phosphor bronze are similar. The width of the interdiffusion bands is found to increase with time in accordance with Equations 1.23 and 1.24. The activation energies for the two substrates are listed in Table 1.4.

The growth of intermetallic layers in an electrical interface produces material gradients characterized by large changes in microhardness. These abrupt variations in hardness cause mechanical strains that may lead to mechanical fracture of the interface, particularly during thermal excursions in the service life of the connection. For these reasons and

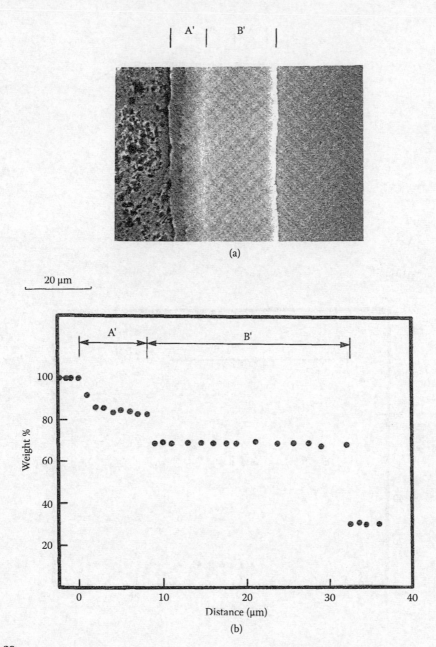

(a)

(b)

FIGURE 1.29
(a) Example of interdiffusion band formed at a brass–zinc interface (b) Zn concentration in interdiffusion bands shown in (a); the composition balance consists of Cu. (From RS Timsit, *IEEE Trans Comp Hyb Manuf Tech* CHMT-9: 106, 1986 [71].)

as already illustrated in Figure 1.28 [63–65,74], intermetallic growth may, thus, be highly deleterious to the mechanical and electrical integrity of an interface. There is an additional effect due to intermetallics. Since intermetallics are characterized by relatively large electrical resistivities, their growth within *a*-spots increases the electrical contact resistance. The experimental data presented below represent the few direct illustrations of this effect in the published literature [80] and is worth describing in detail.

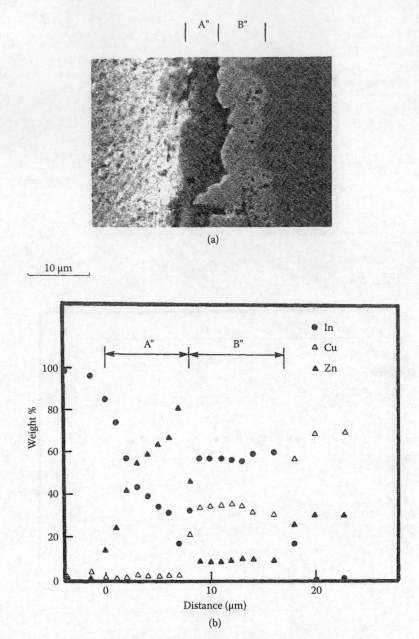

(a)

(b)

FIGURE 1.30
(a) Example of interdiffusion band formed at a brass–indium interface (b) Chemical composition of interdiffusion layer in (a). (From RS Timsit, *IEEE Trans Comp Hyb Manuf Tech* CHMT-9: 106, 1986 [71].)

Using Equation 1.1, it may be shown that the cold resistance R_{cl} of a layered contact spot,
such as the aluminum–brass a-spot illustrated in the inset of Figure 1.34, is given as [80]

$$R_{cl} = \frac{\rho_{Al} + \rho_{Br}}{2} \left(\frac{1}{2a_0} - \frac{1}{\pi a_0} \tan^{-1} \left\{ \frac{\delta}{2a_0} \right\} \right) + \frac{\rho_{int}}{\pi a_0} \tan^{-1} \left\{ \frac{\delta}{2a_0} \right\} \qquad (1.25)$$

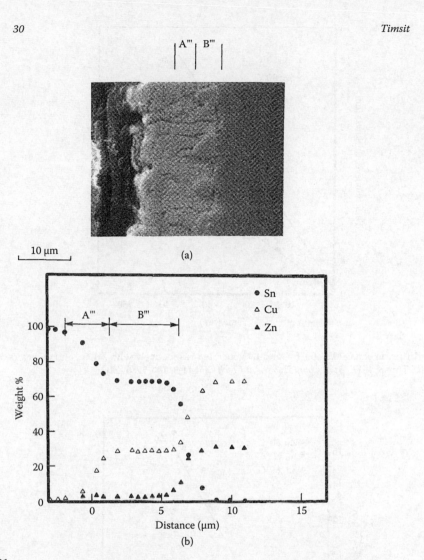

FIGURE 1.31
(a) Example of interdiffusion band formed at a brass–tin interface (b) Chemical composition of interdiffusion layer in (a). (From RS Timsit, *IEEE Trans Comp Hyb Manuf Tech* CHMT-9: 106, 1986 [71].)

where ρ_{Al}, ρ_{Br} and ρ_{int} are respectively the resistivities of aluminum, brass and an intermetallic layer formed in the contact; a_0 is the true contact radius. In the derivation of Equation 1.25, the temperature is assumed to be uniform within the contact. Since $\rho_{int} > \rho_{Al}$, ρ_{Br} [81], the contact resistance given by Equation 1.25 is larger than the resistance without the intermetallic layer, that is, when $\delta = 0$. An "equivalent" radius of the contact may then be defined as

$$a_{eq} = \frac{\rho_{Al} + \rho_{Br}}{4R_{cl}}$$

representing the radius of a circular a-spot formed between aluminum and brass in which the constriction resistance has the value given by R_{cl} from Equation 1.25, but without

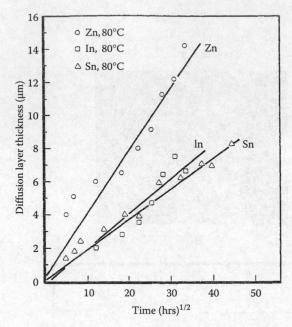

FIGURE 1.32
Growth of interdiffusion bands (A' + B'), B'' , and B''' generated respectively with Zn, ln, and Sn layers on brass at 80°C. (From RS Timsit, *IEEE Trans Comp Hyb Manuf Tech* CHMT-9: 106, 1986 [71].)

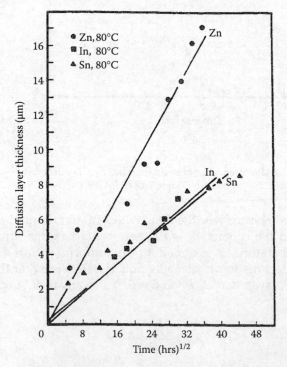

FIGURE 1.33
Growth of interdiffusion bands (A' + B'), B'', and B''' generated respectively with Zn, ln, and Sn layers on phosphor bronze at 80°C. (From RS Timsit, *IEEE Trans Comp Hyb Manuf Tech* CHMT-9: 106, 1986 [71].)

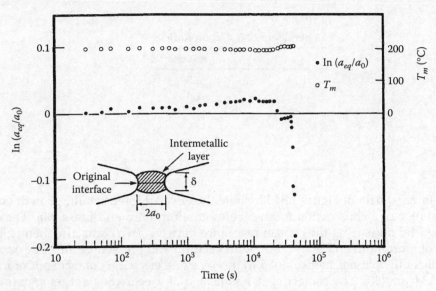

FIGURE 1.34
Effect of intermetallic growth on the equivalent contact radius a_{eq}; $a_0 = 0.34$ μm in an aluminum–brass contact. (From RS Timsit, *Can Elect Assoc Rep 76-19*, 1976 [80].)

intermetallics. The equivalent radius a_{eq} is clearly *smaller* than the true radius a_0. From Equation 1.25, it follows that

$$a_{eq} = \frac{a_0}{1+\dfrac{2}{\pi}(r-1)\tan^{-1}\left\{\dfrac{\delta}{2a_0}\right\}} \tag{1.26}$$

where

$$r = \frac{2\rho_{int}}{\rho_{Al}+\rho_{Br}}$$

From Equation 1.26, the thickness $\delta_{1/2}$ of intermetallics required to decrease the equivalent radius a_{eq} from the initial value a_0 to $a_0/2$ is found to be

$$\frac{\delta_{1/2}}{a_0} = 2\tan\left\{\frac{\pi}{2(r-1)}\right\}$$

Table 1.5 lists the values of $\delta_{1/2}/a_0$ for various values of the ratio r. Since $\rho_{Al} = 2.6 \times 10^{-8}$ Ωm, $\rho_{Br} = 6.3 \times 10^{-8}$ Ωm and $\rho_{int} \sim 2 \times 10^{-7}$ Ωm [79] at room temperature, r has the value of approximately 4 and should be reasonably independent of temperature. For aluminum–brass contacts, Table 1.5 indicates that $\delta_{1/2}/a_0$ takes the value of 1.2 for $r = 4$. Thus, the equivalent radius a_{eq} should drop to one-half the value of a_0 when

$$\delta_{1/2} \sim a_0$$

that is, when the thickness of the intermetallic layer is approximately equal to the *a*-spot radius.

TABLE 1.5

Dependence of $\delta_{1/2}/a_0$ on Ratio r

r	$\delta_{1/2}/a_0$
2.5	3.5
3	2.0
4	1.2
5	0.8
6	0.7
7	0.5

The experimental data of Figure 1.34 illustrate the effect of intermetallic growth on the contact resistance of a constriction formed between aluminum and brass [80]. The data were obtained by measuring the contact resistance between two contacting hemispherical surfaces of aluminum and brass, cleaned in vacuum and pressed against one another under a sufficiently low mechanical load to produce one electrical contact spot of initial radius $a_0 = 0.34$ μm [80]. The contact spot was heated at a near-constant temperature of 198°C for a number of hours to grow intermetallics within the a-spot. The "cold" resistance was evaluated as a function of time from measurements of drop in voltage across the contact and of bulk temperature (this evaluation procedure is described in Section 1.4). The "equivalent" radius a_{eq} plotted in Figure 1.34 was found to remain constant for approximately 4×10^4 s, after which it decreased very rapidly. We now show that this time interval for a significant increase in contact resistance (or decrease in a_{eq}) due to intermetallics is consistent with the growth rates of intermetallics expected on the basis of the information in Table 1.4.

From Table 1.4, the intermetallic layer of high resistivity in the aluminum–brass interface is the (A + B + C) band. Since this band grows in a parabola with the rate constant

$$k = 3.4 \times 10^{-10} \exp(-17{,}300/RT) \, \text{m}^2\text{s}^{-1} \tag{1.27}$$

according to Table 1.4, at the contact temperature of 198°C (471 K) corresponding to Figure 1.34, the rate constant k has the value of $3.4 \times 10^{-10}\exp(-17{,}300/(1.987 \times 471))$ or 3.19×10^{-18} m² s⁻¹. From Equation 1.23 the time required for the intermetallic layer to attain the thickness $\delta_{1/2}$ at 198°C is given as $[\delta_{1/2}]^2/k$. Since $\delta_{1/2} = a_0 = 0.34$ μm, this time interval has the value $[0.34 \times 10^{-6}]^2/3.19 \times 10^{-18}$ or 3.7×10^4 s. This is in excellent agreement with the time interval $[4.2 \times 10^4$ s] shown in Figure 1.34 when the equivalent radius a_{eq} decreased by a factor of two. Figure 1.34 illustrates vividly the effect of intermetallic growth on contact resistance. In practice, the operating temperature of a bimetallic electrical contact must be maintained low to preclude this type of growth. However, because the activation energies associated with interdiffusion of plated layers with brass are relatively small, interdiffusion occurs relatively rapidly even at room temperature as illustrated in Table 1.6.

Table 1.6 lists the thickness of the intermetallics formed by Zn, ln, and Sn layers with brass both at room temperature and at 55°C (the maximum recommended operating temperature of some switching devices) for various time intervals. Clearly, all the platings interdiffuse relatively rapidly to generate a surface layer harder than the original brass even at room temperature. Similar conclusions have been made regarding the effect of intermetallic growth on the contact resistance properties of tin-plated copper [82]. As shown in Table 1.4, the activation energy for intermetallic growth in tin–copper interfaces is almost identical to that measured in the tin–brass system. These results suggest that

TABLE 1.6

Thickness of Interdiffusion Layers Produced by Zinc, Indium, and Tin Deposits on Brass

Time Interval	T (°C)	Interdiffusion Layer Thickness (µm)		
		Zn	In	Sn
1 month	20	0.14	1.0	1.9
1 year	20	0.48	3.5	6.7
1 month	55	3.4	2.8	3.5
1 year	55	11.8	9.8	12.1

any electrical junction made with tin-plated copper or brass is potentially unstable *even* if the contact is relatively cool. The shelf storage lifetime of devices coated with any of the plating materials listed in Table 1.6 is clearly severely limited. Of greatest concern must be devices coated with zinc since the hardness of the interdiffusion layer is very large in this case. The generation of a hard surface layer would not only act to reduce the area of electrical contact but would also produce a brittle interface, which is deleterious to the mechanical stability of the junction. The growth of a hard surface layer may explain in part the unreliability of zinc-plated connector plates in household electrical wiring devices [83].

It is clear from the above considerations that intermetallic growth may be highly deleterious to the reliability of electrical contacts. Although the illustrative examples of this section have focused on interfaces involving copper-based alloys, the effects are not restricted to these types of interfaces and apply generally to a wide number of bimetallic couples. The literature on the effects of intermetallic layers on the mechanical integrity of an electrical interface is vast and has expanded significantly in recent years. This is due, in large part, to the adoption of lead-free solders and lead-free tin coatings on electrical contact surfaces in the electronics industry and the need to characterize the effect of intermetallic growth on the mechanical strength of soldered electrical joints. Intermetallic growth also has deleterious effect on the strength of ultrasonic bonded electrical interfaces in microchips. A great deal of the literature on intermetallic growth focuses on electrical contact materials such as copper, copper-based alloys, tin, nickel, aluminum, and gold. This is addressed in detail in references [84–86].

Interdiffusion in electrical interfaces comprising copper-based alloys is usually mitigated through the use of a diffusion barrier such as nickel. The effectiveness of the barrier depends on the microstructure as well as the thickness of the layer. A layer of nickel about 1 µm thick inhibits interdiffusion in a copper–tin contact. Interdiffusion between the tin and nickel still occurs but is characterized by slow intermetallic growth over a wide temperature range, generally to form Ni_3Sn_4 [87–89]. The use of a nickel–iron-plated layer to inhibit interdiffusion of Sn with copper has also been recommended [90]. Although electroplated and electroless nickel have been the interdiffusion barrier materials of choice in connector applications, efforts are being made to develop other materials such as nanocrystalline nickel–tungsten alloys to further enhance resistance to not only interdiffusion but also corrosion and mechanical wear of overplates [91]. Some of the intrinsically materials-related requirements for thin films to act as effective diffusion barriers have been reviewed by Nicolet [92].These requirements are likely valid for thick coatings such as electroplates.

The selection of an appropriate diffusion barrier for surfaces in an electrical contact subjected to a mechanical contact load must satisfy another major criterion in addition to that of mitigating interdiffusion. The barrier layer must also be characterized by mechanical properties that are compatible with those of the substrate material to preserve physical integrity.

For example, although a nickel layer mitigates tin diffusion into aluminum [93], the softness of aluminum relative to nickel may lead to fracturing of the latter under the action of a sufficiently large contact load. Cracking of the nickel leads to penetration of tin into the aluminum and may also lead to severe galvanic corrosion between tin, nickel and aluminum.

Because diffusion barriers add to fabrication cost, they are often not used in connectors if it is assessed that the operating temperature of the connector shall always remain relatively low.

1.3.4 Possible Effect of Electromigration on Intermetallic Growth Rates

The data on interdiffusion described in Section 1.3.3 were obtained from thermally annealed specimens. If these data are to be extrapolated reliably to predict intermetallic growth in practical electrical contacts, it must be established that this growth is unaffected by the presence of an electric current of high density, that is, by the effects of electromigration. Because the open literature on electromigration in bulk electrical contacts deals largely with aluminum–brass, aluminum–copper and aluminum–aluminum interfaces [72,80,94,95], in this section, we will focus on these types of electrical contacts in examining electromigration.

Consider some fundamental principles of electromigration. Theoretically, the passage of a direct electric current in a metal specimen in which thermal diffusion occurs does not affect either impurity or self-diffusion rates in the metal. The action of the electric current is *only* to shift the zone of material gradients which results from diffusion [96]. Although the effect of this shift on the mechanical integrity of the interface is unclear, the effect would be expected to be small if the shift were much smaller than the width of the diffusion bands. The direction of this shift depends on the direction of the electric current and its magnitude y is given as [96]

$$y = vt$$

where v is the velocity with which the shift occurs and t is the time, with

$$v = \frac{eZ^* D\rho j}{kT}$$

In the above expression, e is the magnitude of the electronic charge, Z^* and D are respectively the effective charge number and the diffusion coefficient of the diffusing species, ρ is the electrical resistivity of the diffusive host, j is the electric current density, k is Boltzmann's constant and T is the absolute temperature. Consider now the magnitude of the shift expected in the location of the diffusion band at an aluminum–brass interface (discussed in Section 1.3.3 above) due to the action of an electric current. Whether the diffusing species is aluminum, copper, or zinc, the diffusion coefficient D in the above expression cannot be larger than the rate constant for the growth of the intermetallic layer (A + B + C) given in Equation 1.27. Taking D = rate constant as an upper bound, $\rho \sim 2 \times 10^{-7} \Omega\text{m}$ (for the intermetallic) [81] and $Z^* \sim 10$ [96] yields

$$v = 1.0 \times 10^{-22} j \text{ ms}^{-1}$$

at a temperature of 200°C (say an upper bound for an operating contact). Since the current density in a functional bulk contact seldom exceeds 10^8 Am^{-2}, the shift cannot exceed approximately 36 nm after 1000 hours of operation. This is totally negligible in comparison with the thickness of the intermetallic band grown thermally under these conditions.

There is, thus, no theoretical basis for expecting an effect of electromigration on the performance of aluminum–brass electrical interfaces operating within this limit.

In order to determine (1) Whether this conjecture regarding the effect of electric current is indeed correct, and (2) Whether the values of the parameters used to evaluate the velocity v defined above indeed provide an upper bound for the shift, experiments were performed to measure both the interdiffusion bandwidth and the band shift in diffusion couples heated by passing an electric current [80]. In the experiments, aluminium–brass couples were Joule-heated with a current density of $0.1 \times 10^8 – 1 \times 10^8$ Am^{-2} at the electrical interface, to a temperature of the order of 300°C to grow intermetallic layers reasonably quickly. The experimental set-up is shown schematically in Figure 1.35. The diffusion couples were formed by roll-bonding sheets of aluminum and brass to obtain a composite material consisting of aluminium–brass–aluminum [80]. A direct electric current was passed through the specimen as shown in the diagram so that two diffusion interfaces could be exposed to the electrical flow simultaneously. A thermocouple was located on one side of the specimen to measure the temperature resulting from Joule-heating.

Because of the difference in the direction of electrical flow across each aluminum–brass interface in the specimen, any electrically induced shift in the diffusion band should occur into the aluminum in one interface and into the brass in the other. The shift was, thus, measured by determining the average difference between (1) The mean distance between one edge of the diffusion band and the original aluminum–brass interface on one side and (2) The mean distance between the corresponding boundaries on the opposite side. Typical results on bandwidths and bandshifts are shown in Table 1.7.

The measured bandwidths were always a little larger than those calculated from the rate constant defined in Table 1.4 on the basis of the mean temperature recorded on the electrical couple during the experiment. However, since the measured temperature was likely to be lower than the temperature of the interface, the differences between measured and calculated band thickness cannot be unequivocally attributed to the presence of electrical flow. This point is illustrated in column 7 of Table 1.7, which lists the temperature (T_{calc}) required to bring the calculated and measured band thickness into agreement in each run of the experiment. The deviation between the required interface temperature and that measured on the specimen skin is, indeed, not large with the exception of the last entry. It is emphasized, however, that the large deviation in this last case is not inconsistent with the large temperature gradient expected within the specimen under the action of the large electric current (200 A) used in the experiment. Hence, within the uncertainties of these experiments, the rate of intermetallic growth is consistent with that predicted from Table 1.4 and appears unaffected by the presence of electric current. This is in keeping with the predictions of conventional theory of electromigration . It is the present author's opinion that other investigations claiming observations of significant electromigration components to interdiffusion suffer similar uncertainties [72].

The bandshift arising from possible electromigration (column 9 of Table 1.7) is negligible and consistent with the small shifts predicted theoretically (column 10). These observations,

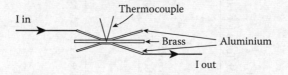

FIGURE 1.35
Experimental setup for the investigation of electromigration at aluminium–brass interfaces. (From RS Timsit, *Can Elect Assoc Rep 76-19*, 1976 [80].)

TABLE 1.7

Results of Tests for Effects of Electromigration on Intermetallic Growth in Aluminum–Brass Contacts

Mean Recorded Temperature (°C)	Mean Current (A)	Mean Current Density at Interfaces (Am^{-2})	Mean Thickness of Diffusion Band (A + B + C)(μm)			$T_{calc}{}^a$(°C)	Run Time (s)	Experimental Electromigration Shift (μm)	Calculated ElectroMigration shift (μm)
			Anode Side	Cathode Side	From Table 1.4				
318	42.8	7.1×10^6	9.6	9.3	5.9	338	2.6×10^5	None	0.01
342	138.2	0.35×10^8	17.3	15.2	14.4	347	5.0×10^5	None	0.24
310	200	0.13×10^8	35.0	32.1	17.0	369	2.6×10^6	None	1.6

[a] T_{calc} is the temperature required to bring the calculated band thickness into agreement with the experimental value.

along with the insensitivity of growth kinetics to electric current, lend credence to the estimates obtained above that electromigration in aluminum–brass contacts is negligible under conditions of interest in practice ($T - 200°C$, $j \sim 10^8$ Am^{-2}).

In summary, intermetallic growth at aluminum–brass interfaces under conditions of interest in bulk electrical contacts is sensitive to temperature and is not expected to be affected significantly by the presence of an electric current. Although the illustrative examples described above focused on aluminum/brass interfaces, the author's experience indicates that this conclusion applies to a wide range of bulk electrical interfaces of practical interest such as copper–brass, copper–tin-plated brass, and so on at current densities on the order of 0.1×10^8–1×10^8 Am^{-2}. However, investigations of interdiffusion at copper–tin interfaces at temperatures approaching the melting point of tin have pointed to an effect of electromigration on Cu–Sn intermetallic growth rates at current densities of $\sim 4 \times 10^8$ A m^{-2} [97]. Thus, it is possible that electromigration effects on intermetallic growth were not detected in aluminum–brass interfaces due to the insufficient current density and low temperature (i.e., much lower than the melting point of aluminum or brass).

1.3.5 Electrically Insulating or Weakly Conducting Films

Electrically insulating or weakly conducting surface films usually stem from the formation of oxide or corrosion products on contact surfaces. Often, these types of films are mechanically brittle, for example, Al$_2$O$_3$ films on aluminum, and electrical contact is established only after the films are fractured and metal-to-metal contact spots are formed by metal extrusion through cracks in the insulating layers. Under these conditions, the shapes of a-spots, and thus the constriction resistance, are determined by the shapes of cracks and the fracture mode of the insulating films [98]. Figure 1.36 shows the transmission electron micrograph of an Al$_2$O$_3$

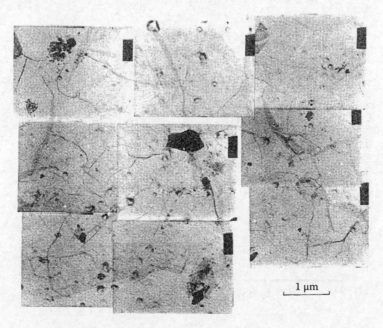

1 µm

FIGURE 1.36
Transmission electron micrograph of an Al$_2$O$_3$ film removed from an aluminum surface in a compression electrical contact. The dark areas correspond to fracture regions in the film. (From RS Timsit, *Can Elect Assoc Rep 76-19*, 1976 [80].)

film removed from an AA1350 aluminum surface after formation of an electrical contact. The contact consisted of two smooth hemispherical aluminum surfaces [80]. The electrical contact spots were formed by metal flow through the narrow fissures evident in the film.

The contact resistance in the electrical interface corresponding to Figure 1.36, measured in air as a function of the applied load, is illustrated in Figure 1.37a [80,99]. Note the threshold load of approximately 1 N to fracture the insulative oxide layers in the interface. Figure 1.37b

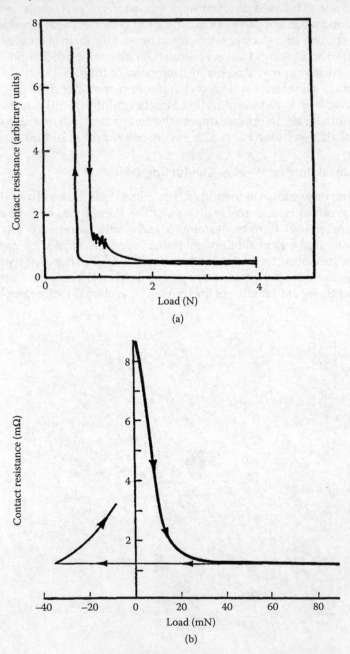

(a)

(b)

FIGURE 1.37
Contact resistance in an aluminum/aluminum contact with: (a) Oxide films present on the contact surfaces (b) Oxide films removed from the contact surfaces. (From RS Timsit, *IEEE Trans Comp Hyb Manuf* Tech 3: 71, 1980 [99].)

shows the behavior of the identical contact interface formed in ultra-high vacuum where the oxide films on the two contacting aluminum surfaces had first been removed by ion-beam etching [80,99]. In contrast to the data of Figure 1.37a, the absence of oxide not only leads to instant electrical contact when the surfaces touch but also leads to cold welding of the clean metal surfaces, evidenced by the generation of a significant adhesion force on removal of the contact load. A comparison of the curves of Figure 1.37 provides clear evidence of the deleterious effect that oxide or other surface films with high resistivity can have on the generation of a reliable electrical contact. In this section, we shall review the electrical properties of a few metal oxides, such as copper and aluminum oxides, before discussing the full effect of surface oxide films on electrical contact resistance.

1.3.5.1 Growth Rate and Electrical Resistivity of Oxides of Selected Contact Materials

1.3.5.1.1 Copper Oxide

The data available on the oxidation of copper at temperatures between 100°C and 200°C [100,101] suggests that the law of oxidation rate is relatively independent of the oxygen pressure and is either inverse logarithmic, direct logarithmic or of the form

$$d^n = k''t \tag{1.28}$$

where d is the oxide film thickness, $n \leq 3$, k'' is a constant, and t is the time. Assuming a cubic rate law, the data of Campbell and Thomas [102] suggest that at a temperature of 100°C, Equation 1.28 should take the form

$$d^3 = 0.86t$$

where d is in nanometers and t is in seconds. Thus, the growth rate is relatively rapid and it requires approximately 9 seconds for a copper oxide film to grow to a thickness of 2 nm at a temperature of 100°C. The electrical conductivity of copper oxide is highly dependent on the nature of the oxide formed that is, cupric (CuO), cuprous (Cu_2O) or the other ($CuO_{0.67}$) oxide phase. Cupric oxide is a relatively better electrical conductor than the other two oxides, with a specific resistance of 1–10 Ωm [103]. Cuprous oxide (and presumably $CuO_{0.67}$), on the other hand, is a p-type semiconductor with a specific resistance that is highly sensitive to temperature. The experimental data of Anderson and Greenwood [104] indicate that the specific resistance of cuprous oxide may be written as

$$\rho_{Cu_2O} = 0.38 \exp \left\{ \frac{0.3}{kT} \right\} \tag{1.29}$$

where ρ_{Cu_2O} and kT are expressed in Ωm and eV respectively. At a temperature of 100°C, it is found that $\rho_{Cu_2O} = 4.5 \times 10^3$ Ωm, a value considerably larger than for cupric oxide.

The dependence of the composition of a copper oxide film on oxygen pressure and temperature may be understood from Figure 1.38 reproduced from the work of Honjo [105,106]. The curves delineate the regions in which electron diffraction studies of oxide films revealed either Cu or Cu_2O as the outermost oxide in the film. Thus, below approximately 200°C, no cupric oxide is formed on the oxide surface in open air. Other work [100,107,108] has indicated that little, if any, CuO is present within a copper oxide film

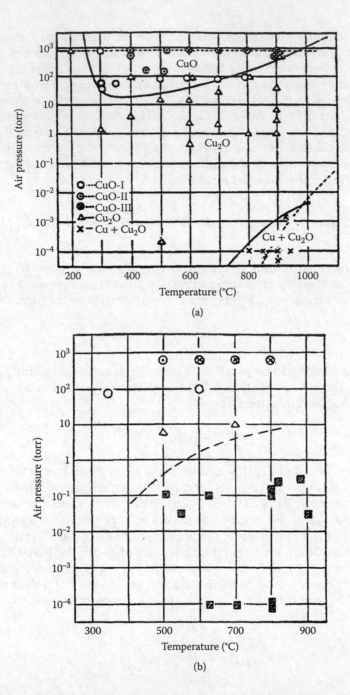

FIGURE 1.38

Pressure–temperature diagram for copper oxides formed by oxidation in air: (a) Pure copper, where three different CuO phases were identified as I, II and III by electron diffraction (b) wt% Cu93Ni7 alloy, where the points are labeled according to the oxide phases identified in (a). The shaded squares correspond to unidentified phases of oxides of copper or nickel. (From G. Honjo, *J Phy Soc Japan* 4: 330, 1949 [105]; G Honjo, *J Phy Soc Japan* 8: 113, 1953 [106].)

grown below 200°C either in air or oxygen at low-to-atmospheric pressures. The electrical resistance, R_f of a copper oxide film at 100°C may thus be calculated as

$$R_f = \frac{\rho_{Cu_2O}d}{A}$$

(1.30)

where d is the film thickness and A is the surface area of the oxide film.

1.3.5.1.2 Tin Oxide

There are two main oxides of tin: Stannic oxide (SnO_2) and Stannous oxide (SnO). The existence of these two oxides reflects the dual valency of tin, with oxidation states of 2^+ and 4^+. Figure 1.39 shows the Sn–O phase diagram for atmospheric pressure conditions [109,110]. This diagram indicates the presence of an intermediate tin oxide phase between SnO and SnO_2 at elevated temperatures. Sn_3O_4 is often given for its composition but Sn_2O_3 has also been considered. Stannic oxide is the thermodynamically most stable form of tin oxide. Stannic oxide is an n-type semiconductor and has been used widely in thin-film form as transparent electrode. The electrical resistivity of SnO_2 at room temperature varies greatly, from about 10^2 Ωm for well-prepared bulk oxide material [111] to about 4×10^4 Ωm for oxide layers grown on tin. Because SnO_2 is a semiconductor, the conductivity of the oxide rises rapidly with increasing temperature, especially at temperatures higher than approximately 750°C [111].

In a series of investigations, Tamai and coworkers characterized the growth rate and the composition of tin oxide surface films grown in air after electrodeposition of tin layers on a copper–nickel substrate [112–114]. Oxide growth rates were measured at temperatures ranging from 25°C to 150°C. Contact resistance properties were also measured. Figure 1.40 shows the oxide growth data at 25°C, indicating a relatively rapid linear oxide growth for the first five hours of exposure of the tin to laboratory air [113]. Following this initial phase of oxidation, the rate eventually decreased to a growth law characterized by $n \sim 4$ in Equation 1.28. Similar observations were made at the higher temperatures, thus indicating only a small effect of temperature on oxide growth rates up to 150°C [113]. Chemical analyses of the oxide using surface analytical techniques showed that the initial oxide film formed at a temperature lower than 120°C consisted of a mixture of crystalline and possibly amorphous SnO, as well as a mixture of SnO_2 and Sn_3O_4. The presence of these

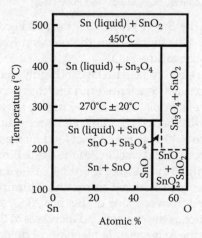

FIGURE 1.39
Sn–O phase diagram. (From M Batzill and U Diebold, *Prog Surf Sci* 79: 47, 2005 [109]; U Kuxmann and R Dobner, *Metallwissenschaft und Technik* 34: 821, 1980 [110].)

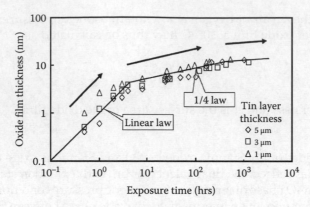

FIGURE 1.40
Growth rate of tin oxide in an indoor environment at 25°C and 75% RH. (From Y Nabeta et al., *Proc 55th Holm Conf Elect Cont*, Vancouver, BC, Canada, 174, 2009 [113].)

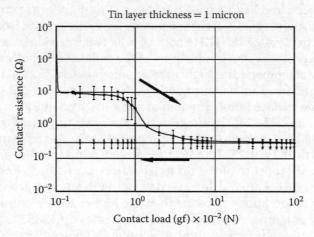

FIGURE 1.41
Variations of contact resistance with contact load for an interface between an air-oxidized tin layer plated on a Cu–Zn alloy substrate, and a platinum hemisphere; the arrows indicate uploading and downloading. (From T Tamai et al., *Proc 55th Holm Conf.* Vancouver, BC, Canada, 160, 2009 [112].)

oxide forms is consistent with the Sn–O phase diagram of Figure 1.39. Oxide grown at temperatures higher than 120°C (i.e., higher than the recrystallization temperature of tin) was found to consist largely of crystalline SnO_2.

The tin oxide film formed on an electroplated tin layer is highly electrically insulating. Figure 1.41 illustrates the variations in contact resistance with increasing contact load between a tin layer (thickness of a few microns) electroplated on a Cu–Zn alloy flat and exposed to air at room temperature, and a platinum hemisphere [112]. The contact resistance dropped sharply from a large value of about 10^3 Ω to about 10 Ω as the contact load increased to slightly over 10^{-3} N. This fall in resistance stemmed from initial cracking of the oxide film which had a thickness of between 5 nm and 10 nm according to the data of Figure 1.40. This initial decrease in resistance is reminiscent of the sharp resistance in oxidized aluminum–aluminum contacts as illustrated in Figure 1.37, which also stemmed from cracking of the thin Al_2O_3 surface film. In Figure 1.41, the continued decrease in resistance as the contact load increases to 0.1 N stemmed from extensive fracturing of the oxide film due to large-scale mechanical deformation of the underlying tin metal as tin grains piled up at the contact [112]. The electrical resistivity of tin oxide films is addressed with resistivity properties of other metal oxides in Figure 1.46.

1.3.5.1.3 Nickel Oxide

Nickel oxidizes to NiO at room temperature to form a crystalline rather than amorphous material. Other phases of nickel oxide may form but only at temperatures of a few to several hundred degrees Celsius [115].

Figure 1.42 shows the growth rate of nickel oxide in air at a temperature of 40°C and a relative humidity of 95% [116]. The oxide was grown on a nickel layer electroplated on a copper–alloy substrate from a sulfamate solution to a thickness of 5 µm. The growth rate is essentially linear up to a self-limiting thickness of about 2 nm. The limiting thickness of oxide film is not greatly affected by temperature up to 200°C, but is affected significantly by relative humidity, as shown in Figure 1.43 [116]. The native nickel oxide layer on nickel metal thus remains thin. The slow oxidation of nickel in dry oxygen has been confirmed [117]. The effect of moisture in promoting oxidation was mentioned by Holm [3].

NiO thin films usually exhibit *p*-type conductivity due to holes generated by Ni vacancies in the lattice. Therefore, the electrical resistivity depends sensitively on the preparation technique, grain structure, thermal history, and other factors (see for example [103,118]). The resistivity of vacuum-sputtered NiO thin films ranges from about 0.5 Ωm at room temperature to about 6 Ωm at temperature of 400°C, but the resistivity of the oxide in bulk form may be much larger [103]. Over this temperature range, the microhardness of NiO varies from

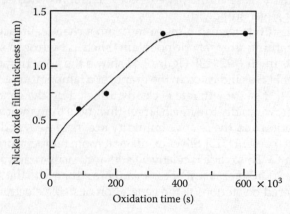

FIGURE 1.42
Growth rate of nickel oxide film at 200°C. (From SH Kulpa and RP Frankenthal, *J Electrochem Soc* 124: 1588, 1977 [116].)

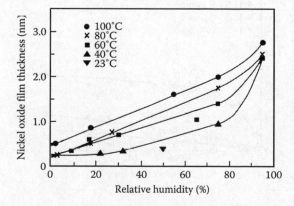

FIGURE 1.43
Thickness of nickel oxide film with increasing relative humidity, at various temperatures. (From SH Kulpa and RP Frankenthal, *J Electrochem Soc* 124: 1588, 1977 [116].)

about 400 kg mm^{-2} (~ 4 GPa) to approximately 900 kg mm^{-2} [118]. As is the case with other contact materials that oxidize, the nickel oxide film present on a nickel plating in an electrical contact must be fractured to achieve a low contact resistance. However, NiO is more difficult to fracture than other oxides, such as tin oxide for example, because the underlying nickel substrate is harder and mechanically stiffer than many other contact materials. It, thus, requires a relatively large contact load to deform the oxide film to its fracture limit. The variation of contact resistance with increasing contact force is, thus, similar to that illustrated in Figure 1.41 for oxidized tin, or Figure 1.37 for oxidized aluminum. Such a contact resistance–force curve was reported for a gold electroless nickel contact [119].

1.3.5.1.4 *Silver Sulfide*

Silver is widely used as a connector finish in high-power permanent electrical connections such as bolted joints where a relatively large contact force (e.g. tens of N) is required and durability requirements are not stringent. The susceptibility of silver to form electrically insulating tarnish film, such as silver sulfide and/or silver chloride, impedes use of the metal as a finish on electronic connectors since these connectors rely on a relatively low contact force with high durability requirements [120,121]. Nevertheless, the exploration of silver finish as a replacement for gold in separable electronic connectors is actively being pursued due to the large increase in the price of gold. The following description focuses on selected properties of silver sulfide films.

The formation of silver sulfide tarnish films stems from a chemical reaction of silver with free sulfur or sulfur-containing atmospheric pollutants such as hydrogen sulfide (H$_2$S) present even in trace amounts in air [122,123]. Figure 1.44 shows the growth rate of silver sulfide in air containing traces of H$_2$S as indicated in the graph, at a temperature of 25°C, and a relative humidity of 75% [124]. The growth rate is clearly affected by the concentration of H$_2$S. Silver tarnish films can grow rapidly to a much larger thickness than indicated in Figure 1.44 if both the H$_2$S concentration and the relative humidity are increased, and if traces of nitrogen dioxide (NO$_2$) are also present [125]. Silver oxidizes at room temperature only in the presence of ozone to form Ag$_2$O. This oxide is relatively soft and easily removable mechanically, and decomposes at about 200°C. Because it does not adhere strongly to the metallic substrate, it is generally not considered deleterious to the generation of stable electrical contacts.

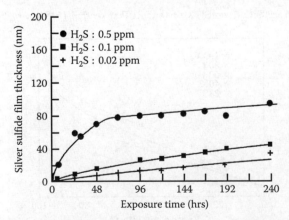

FIGURE 1.44

Growth rate of silver sulfide in air containing various trace amounts of H$_2$S, at a temperature of 25°C and a relative humidity of 75% . (From J Guinement and C Fiaud, *Proc 13th Int Conf Elect Cont*, Lausanne, Switzerland 383, 1986 [124].)

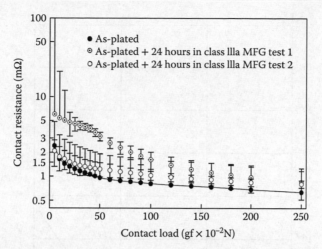

FIGURE 1.45
Variation of contact resistance with increasing contact force between a gold hemisphere and as-plated and tarnished silver. The tarnished samples (test 1 and test 2) were obtained by exposing as-plated silver to a Class IIIa Mixed Flow Gas (MFG) environment containing 20 ppb Cl_2, 200 ppb NO_2 and 100 ppb H_2S in air at 30°C and 75% RH. (From M Myers, Technical Paper 503–1016, Tyco Electronics, 2009 [127].)

The electrical conductivity of silver sulfide stems largely from ionic motion at temperatures lower than 177°C and largely from electronic motion above this temperature, and increases rapidly with increasing temperature [126]. At 100°C, the conductivity of silver sulfide ranges from about 5×10^{-2} to about $8\ \Omega^{-1}m^{-1}$ depending on the amount of excess sulfur [126]. At room temperature, the resistivity ranges approximately from 1 Ωm to 20 Ωm. Because of this high resistivity at room temperature, and as is the case with contact materials that oxidize, a silver sulfide film present on a silver surface in an electrical contact must be fractured to achieve a low contact resistance. A typical variation of contact resistance with increasing contact load on a tarnished silver surface is shown in Figure 1.45 [127]. It is important to note that contact resistance does not decrease initially as sharply as was the case in Figures 1.37 and 1.41. This suggests that the tarnish film is partially "displaceable" from the silver surface at low loads, and possibly susceptible to "tearing" rather than cracking in brittle fracture as suddenly as tin oxide, nickel oxide or aluminum oxide.

Figure 1.46 shows the expected values of film resistivity per unit area versus film thickness, that is, ρd versus d, for a number of common oxide or contaminant layers often found on electrical contact surfaces [128]. The curves were calculated on the basis of data quoted by Holm [3]. Note the rapid increase of film resistance with film thickness. This increase stems in part from an increase in bulk resistivity due to changes in the defect structure of the films as thickness increases.

1.3.5.1.5 Aluminum Oxide

The information available on the oxidation rate of aluminum (and its alloys) in air at temperatures ranging from room temperature to about 400°C indicates that the growth rate of Al_2O_3 is extremely rapid [129,130]. In air or dry oxygen, the growth rate is self-limiting with a maximum film thickness of 2–4 nm over a wide temperature range [131,132]. At temperatures as high as 300°C, the Al_2O_3 film seldom exceeds 20 nm in thickness in practical oxidation times even on surfaces originally subjected to extensive mechanical deformation [133]. At about 400°C and above, there is a change in the morphology of the thermally grown

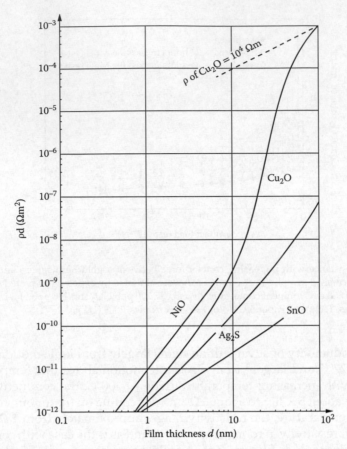

FIGURE 1.46
Resistivity per unit area (ρd) of various oxide and sulfide films versus film thickness d. (From P Johannet. Report from Electricité de France, 1995 [128].)

oxide. Instead of the oxide continuing to grow uniformly, discrete crystals of oxide appear and grow laterally until the whole film appears to consist of crystalline material [134].

The electrical resistivity of Al_2O_3 is so large (~ 10^{14}–10^{16} Ωcm) at 100°C [103] that the film is a perfect electrical insulator for all intensive purposes. Under these conditions, film resistance arises only through quantum tunneling effects. It may easily be shown that the tunneling resistance of an electrically insulating film sandwiched between two identical metallic conductors is as given in [135]

$$R_f = \frac{d}{A}\left\{\frac{h}{e}\right\}^2 \frac{2}{3\sqrt{2m\varphi}}\exp\left\{\frac{4\pi d}{h}\sqrt{2m\varphi}\right\} \tag{1.31}$$

where d and A are respectively the thickness and surface area of the film, the quantities h, e and m are respectively Planck's constant, the electronic charge and mass of the electron, and φ is the work function of the metal conductors. For an Al–Al_2O_3–Al sandwich, it is known that $\varphi \approx 2eV$ [136]. Equation 1.31 immediately yields $R_f \approx 11.6$ Ω for $d \approx 2$ nm and $A = 10^{-4}$ m^2.

It may now be shown that the electrical resistance offered by oxide films at an interface at ordinary contact temperatures can be expected to be much larger than that presented

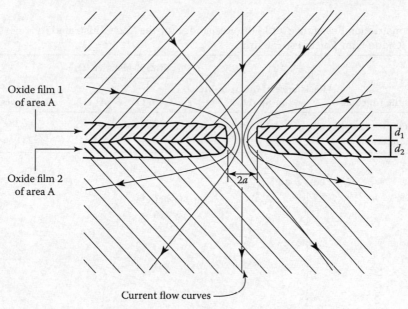

FIGURE 1.47
The *a*-spot surrounded by oxide films.

by metal-to-metal contact spots produced through cracks in the surface films. For this purpose, it will be sufficient to restrict the discussion to the behavior of a single contact spot surrounded by oxide layers as illustrated in Figure 1.47. If the radius of the *a*-spot is small in comparison with the dimensions of the surrounding oxide films, the resistance presented by the film is approximately given by

$$R_f = \frac{\rho_1 d_1 + \rho_2 d_2}{A} \qquad (1.32)$$

where ρ_1, ρ_2 and d_1, d_2 are respectively the resistivities and mean oxide film thicknesses in the electrical interface, and A is the total area of contact. The relative magnitudes of the constriction resistance of the *a*-spot and R_f may now be estimated for copper-to-copper and aluminum-to-aluminum contacts. Consider an electrical contact in which the area of nominal contact is given as A and has a value of 1 cm². If the radius of the metal-to-metal *a*-spot is 10 μm, the area of metal-to-metal contact represents only about 0.0003% of the total contact area, which would represent a highly unfavorable situation in practice. The values of the constriction resistance and R_f calculated respectively from Equation 1.3, using $\rho = (\rho_1 + \rho_2)/2$, and Equations 1.29 through 1.32 are listed in Table 1.8. In these calculations, the film thicknesses were taken as 2 nm and the junction temperature assumed to be 100°C. It is clear from the data in Table 1.8 that the resistance presented by the metallic bridge is considerably smaller than that of the surface films, although the contact area provided by the bridge is a small fraction of the total area of contact. In practical electrical contacts, the area of metal-to-metal contact is much larger than 0.0003%, so that current flow through the oxide film can be ignored in comparison to current flow through the metal-to-metal contacts. From Table 1.8, it is clear that contact resistance can be reduced drastically if surface oxide layers are removed or dispersed from the electrical interface.

TABLE 1.8

Values of Constriction Resistance and Film Resistance for the Case Illustrated in
Figure 1.47 (Oxide Film Thickness = 2 nm)

Type of Metallic Junction	Resistance of Metal-to-Metal a-Spot (Ω)	Film Resistance (Ω)		Total Film Resistance (Ω)
		Cu_2O	Al_2O_3	
Copper–copper	1.4×10^{-3}	0.09	—	0.18
Aluminum–aluminum	2.9×10^{-3}	—	11.6	23.2

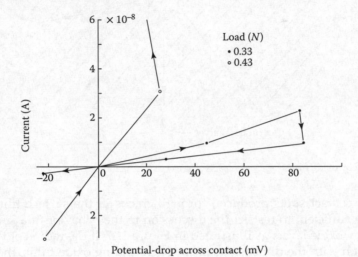

FIGURE 1.48

Example of current–voltage characteristics from aluminum–aluminum contact couples in which the surface oxide films were initially unfractured. (From RS Timsit, *Can Elect Assoc Rep 76-19*, 1976 [80].)

1.3.6 Fritting of Electrically Insulating Surface Films

In the presence of electrically insulating films on metal surfaces, the formation of metal-to-metal a-spots in a bulk electrical interface requires that the films be mechanically disrupted first, as shown earlier in Figures 1.36 and 1.37a. In the absence of the action of mechanical forces to fracture the films, disruption may be achieved by electrically induced breakdown if a sufficiently large electric field is applied across the contact interface. This dielectric breakdown, simultaneously yielding a metal-to-metal contact, has been labeled by Holm as *fritting* [3]. Film breakdown, leading to immediate melting of metallic bridges formed through the ruptured insulating layers, is usually referred to as *A-fritting*. It occurs when the voltage across the metallic bridges established through the fractured insulating film is larger than the so-called *melting voltage*. This melting voltage is usually on the order of 1 V and will be addressed in detail in the following section. If the voltage drop across the contact is sufficient to soften *metallic a*-spots but not to melt them, the film breakdown is labeled as *B-fritting* [3]. The two types of fritting are generally observed when metal surfaces covered with tarnish films or oxide layers are brought into contact and the potential difference across the interface is slowly increased. When the electric field strength across the insulating layers reaches a magnitude of the order of 10^8 Vm^{-1}, fritting occurs and is characterized by a sudden rise in the current caused by the abrupt decrease in the contact resistance when metallic a-spots are formed.

Figure 1.48 shows an example of B-fritting in an electrical interface consisting of two hemispherical oxide-covered aluminum surfaces in vacuum under a mechanical load lesser than 1 N [80]. The curve obtained under a load of 0.33 N gives the resistance of the contact as ~ 7.3×10^6 Ω. This curve indicates some initial electrical instability in the oxide films as the potential is increased because there are obvious signs of hysteresis as the voltage is lowered. The progressive degeneration of the oxide films as the contact load is increased is illustrated by the data obtained at 0.43 N. Here, the current surges rapidly to large values as the voltage drop across the contact is increased to approximately 25 mV. This fritting was probably assisted by mechanical weakening of the oxide surface layers under the larger mechanical load. Several other interesting examples of A-fritting and B-fritting are described by Holm in his classic book [3].

1.4 Temperature of an Electrically Heated *a*-Spot

The definition of constriction resistance as $\rho/2a$ is valid so long as the electric current through the constriction is sufficiently small and the effect on contact resistance due to heat generated is negligible. Under these conditions, the voltage drop across the constriction produced by an electric current I is simply given as $\rho I/2a$. Under conditions where significant Joule-heating is produced within the constriction, the crowding of current lines within the *a*-spot generates a thermal gradient normal to that constriction. The effect of this gradient complicates the relationship between voltage drop across the contact, electric current, and dimension of the *a*-spot. This section examines the nature of this relationship.

On passing an electric current, the temperature of an *a*-spot rises to a near-equilibrium temperature very quickly. Consider, for example, the heating of a circular constriction of radius a located between two semi-infinite conducting bodies of electric resistivity ρ and thermal conductivity λ. If ρ and λ are independent of temperature, the differential equation describing the temperature rise in the bodies in spherical coordinates is given as

$$\lambda \left\{ \frac{\partial^2 T}{\partial r^2} + \frac{1}{r}\frac{\partial T}{\partial r} + \frac{1}{r^2 \sin^2 \theta}\frac{\partial^2 T}{\partial \theta^2} \right\} = c\frac{\partial^2 T}{\partial t^2} + \rho j^2$$

where j is the current density at the location (r, θ) and c is the heat capacity per unit volume. The solution to this equation can only be obtained numerically. However, it may be shown [137] that the solution is a superposition of factors of the form

$$\frac{A}{c} \frac{\exp\left\{-\dfrac{cr^2}{4\lambda t}\right\}}{\left\{\dfrac{4\pi\lambda t}{c}\right\}^{3/2}}$$

Hence in the neighborhood of the constriction where $r \sim a$, the time constant with which the temperature rises is $ca^2/4\lambda$. For electrical conductors such as copper and aluminum, λ and c take values of 400 Wm^{-1}°C^{-1} (0.96 cal s^{-1} cm^{-1}°C^{-1}) and 3.44×10^6 Jm^{-3}°C^{-1} (0.82 cal cm^{-3}°C^{-1}) for copper, and 236 Wm^{-1}°C^{-1} (0.56 cal s^{-1} cm^{-1}°C^{-1}) and 2.43×10^6 Jm^{-3}°C^{-1} (0.58 cal cm^{-3}°C^{-1}) for aluminum at room temperature. The time constant is, thus, approximately 2.2×10^{-7} s

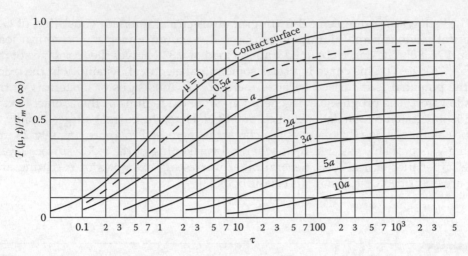

FIGURE 1.49
Supertemperature, normalized to the steady-state temperature T_m, as a function of the normalized time $\tau = (\lambda/ca^2)t$, where λ is the thermal conductivity, c is the heat capacity per unit volume, and a is the a-spot radius. The parameter μ is the distance of an isothermal surface located within the contact material, measured from the center of the constriction along the constriction cylindrical axis. The bulk temperature is taken as 0°C far from the constriction. (From R Holm, *Electric Contacts, Theory and Applications*, Berlin: Springer-Verlag, 2000 [30].)

and 2.6×10^{-7} s respectively for copper and aluminum for a constriction with a radius of 10 μm, and decreases rapidly with decreasing constriction radius. The time evolution and spatial distribution of temperature in two semi-infinite bodies separated by a circular constriction of radius a was calculated by Holm [3] and is illustrated in Figure 1.49. The small rise time associated with the heating of an a-spot forms the basis on which thermal transients are usually ignored in dealing with stationary or slowly moving contacts. In these applications, *only* the equilibrium contact temperature is of interest. Thermal transients are taken into account only in applications of rapidly moving contacts such as brush contacts, or in high-frequency, high-power electrical connections. This section focuses on electrical contacts in a thermal steady state.

1.4.1 Voltage–Temperature Relation

The conventional treatment of electrically heated contacts in thermal equilibrium assumes that the outer surfaces of the conductors are thermally insulated from the external environment. The heat produced within an a-spot can, thus, be dissipated only by conduction through the bodies in contact. Under these conditions, it turns out that the electric and thermal current flow lines follow the same paths, and hence the electric potential and isothermal surfaces within the conductors coincide [34,138,139]. This unique relationship between electric and thermal current flow led Kohlrausch (see for example [2,3]) to derive the following simple relation between the voltage drop V across the contact and the maximum temperature T_m in the contact interface as

$$V = \left\{ 2 \int_{T_1}^{T_m} \lambda_1 \rho_1 \, dT \right\}^{1/2} + \left\{ 2 \int_{T_2}^{T_m} \lambda_2 \rho_2 \, dT \right\}^{1/2} \tag{1.33}$$

where λ and ρ are respectively the thermal conductivity and electrical resistivity of the conductors, the subscripts 1 and 2 refer to the two conductors in contact, and T_1 and T_2 refer to the bulk temperatures of the contacting bodies. Thermoelectric effects are assumed to be small and are neglected. The quantities λ and ρ generally vary with temperature. Since the electric current flow lines are most constricted within the a-spot, the maximum temperature T_m occurs either within the confines of the a-spot or in its immediate vicinity. For a monometallic contact where $\lambda_1 = \lambda_2 = \lambda$ and $\rho_1 = \rho_2 = \rho$, T_m occurs precisely at the a-spot and the relation reduces to

$$V = 2 \left\{ 2 \int_{T_1}^{T_m} \lambda \rho \, dT \right\}^{1/2} \tag{1.34}$$

Over a temperature range where the quantities λ and ρ vary little with temperature, Equation 1.34 yields the well-known form of the *voltage-temperature* (*V–T*) relation for monometallic electrical contacts [3]

$$T_m - T_1 = \frac{V^2}{8\lambda\rho} \tag{1.35}$$

The quantity $(T_m - T_1)$ in the left-hand side of Equation 1.35 is defined as the *contact super-temperature*, that is, the deviation of the a-spot temperature from the bulk connector temperature. Equation 1.35 is often used in the contact literature and in designing electrical connectors to evaluate the expected supertemperature of an electrical interface during operation. Generally, connectors are designed so that the supertemperature does not exceed 1°C–3°C under extreme operating conditions. Under conditions where the supertemperature is large, that is, several tens of degrees Celsius, Equation 1.35 loses its validity since this relation was derived on the basis of constant electrical resistivity and thermal conductivity. As mentioned earlier, both ρ and λ generally vary with temperature. Under these conditions, the *V–T* relation takes on a more complicated form which will be described below.

Before deriving the more rigorous *V–T* relation for the case of elevated contact temperatures, it is important to note that the right-hand side Equations 1.33 through 1.35 involves materials properties *only* through λ and ρ, and makes no reference to contact geometry. The *V–T* relation is, thus, valid for electrical contacts of *any shape and dimensions*. An examination of the details of the mathematical derivation reveals that the *V–T* relation holds for *any thermally insulated body* capable of passing an electric current. This is illustrated in the following simple example.

Consider an electrically conductive metal bar of length $2L$ and uniform cross-section A, through which an electric current of uniform density j is passed, as shown in Figure 1.50. The temperature at both ends of the bar is held at a temperature T_1. If the conductor is thermally insulated from the surroundings, the differential equation governing the temperature distribution $T(x)$ within the solid is

$$\lambda \frac{d^2 T(x)}{dx^2} = -\rho j^2 \tag{1.36}$$

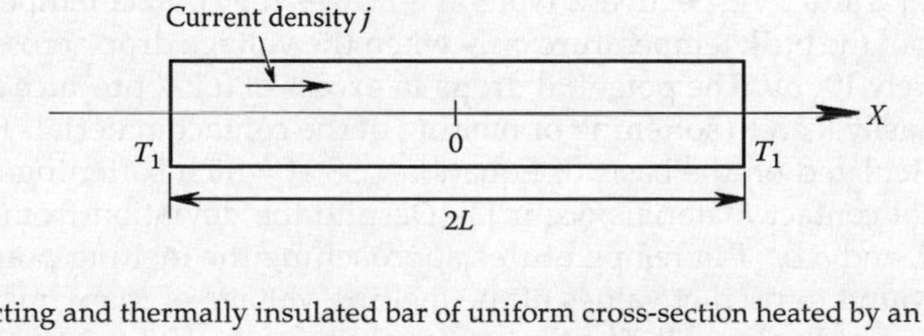

FIGURE 1.50
Electrically conducting and thermally insulated bar of uniform cross-section heated by an electrical current of density j.

where λ and ρ are the thermal conductivity and electrical resistivity of the bar material respectively, and x is the distance along the bar. If, for simplicity, λ and ρ are assumed temperature independent, the solution of Equation (1.36) is straightforward and is given as

$$T(x) = T_1 + \frac{\rho}{2\lambda} j^2 \left(L^2 - x^2\right) \tag{1.37}$$

The maximum temperature, thus, occurs at the centre $x = 0$ of the bar. Since the voltage drop across the bar is given as

$$V = 2\rho j L$$

the difference between the maximum temperature in the rod and the temperature at the ends, that is, the supertemperature, is given as

$$T(0) - T_1 = \frac{V^2}{8\lambda\rho}$$

which is identical to Equation 1.35. Thus, Equation 1.35, and more generally Equation 1.33, applies to any thermally insulated body heated by an electric current.

1.4.2 Voltage–Temperature Relation with Temperature-Dependent Electrical Resistivity and Thermal Conductivity

Usually, the dependence of the thermal conductivity and electrical resistivity on temperature can be expressed respectively as $\lambda = \lambda_0(1 - \beta T)$ and $\rho = \rho_0(1 + \alpha T)$ over a wide temperature range [3,34], where the subscript 0 refers to the value at 0°C and β and α are the temperature coefficients of thermal conductivity and electrical resistivity respectively. Note the thermal conductivity of metals generally decreases with increasing temperature whereas electrical resistivity increases. The origins of this dependence on temperature are reviewed briefly in Appendix A. The values of λ, ρ at 20°C, β and α for many metals and metallic alloys are listed in Chapter 24. Using these forms of the temperature dependence of λ and ρ produces the following V–T relation from Equation 1.34

$$V^2 = 8\lambda_0\rho_0 \left(\{T_m - T_1\} + \frac{\{\alpha - \beta\}}{2} \{T_m^2 - T_1^2\} - \frac{\alpha\beta}{3} \{T_m^3 - T_1^3\} \right) \tag{1.38}$$

Using the data of Tables 24.1 and 24.10, the calculated dependence of T_m on V for aluminum–aluminum, copper–copper and brass–brass contacts is shown in Figure 1.51 for bulk temperatures T_1 of 20°C and 100°C. The curves corresponding to aluminum–copper, copper–brass, and aluminum–brass junctions fall approximately midway between curves 1 and 2, 1 and 3, and 2 and 3 respectively. Note in all cases the contact temperature deviates significantly from the bulk temperature only when the voltage drop across the contact exceeds approximately 10 mV. The potential drops in excess of 0.1 V produce contact temperatures that can easily lead to softening or melting of the contact material. Table 1.9 lists the voltage drop calculated on the basis of Equation 1.38 at which softening and melting of common electrical contact materials occur [3]. Despite the deviation from linearity of the dependence of λ and ρ on T at temperatures approaching the melting point of metals, Equation 1.38 was found to predict values of the melting voltage of many metals in good agreement with measured values [3]. The limits of validity of the V–T relation in electrical

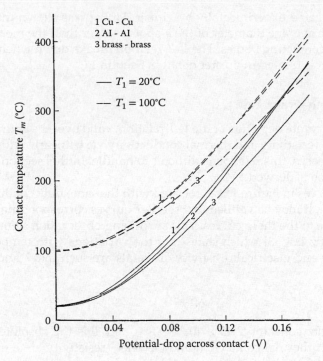

FIGURE 1.51
Voltage-temperature relationship for copper–copper, aluminum–aluminum, and brass–brass electrical contacts for bulk temperatures T_1 of 20°C and 100°C.

TABLE 1.9

Voltage for Softening (V_s) and Melting (V_m) of Common Electrical Contact Materials

Material	Heat Capacity ($Jm^{-3}°C^{-1}$) × 10^6	V_s Softening (V)	V_m Melting (V)
Al	2.4	0.1	0.3
Fe	3.6	0.19	0.19
Ni	3.9	0.16	0.16
Cu	3.4	0.12	0.43
Zn		0.1	0.17
Mo	2.6 at 20°C	0.25	0.75
	3.4 at 1500°C		
Ag	2.5	0.09	0.37
Cd	2		
Sn	1.65	0.07	0.13
Au	2.5	0.08	0.43
W	2.7 at 20°C	0.4	1.1
	3.5 at 1400°C		
	3.9 at 2100°C		
Pt	2.8	0.25	0.65
Pd	2.7		0.57
Pb		0.12	0.19
60Cu, 40Zn	3.2		0.2
60Cu, 40Sn	3		0.15
Stainless steel	3.9	0.27	0.55
WC	3	0.6	

contacts were investigated experimentally by Timsit [140]. It was shown that the relation is valid as long as the average diameter of the *a*-spot is larger than the mean free path of free electrons in the contacting bodies. The *V–T* relation breaks down when this criterion is violated. This effect is treated in greater detail in Section 1.6.

1.4.3 The Wiedemann–Franz Law

Equation 1.38 is an accurate depiction of the *V–T* relation, valid over the entire temperature range over which the resistivity and thermal conductivity vary linearly with temperature. For engineering purposes, this relation is difficult to handle. In this section, a simple version of the *V–T* relation is derived.

Note that all the curves in Figure 1.51 associated with the same bulk conductor temperature lie close together. It may be verified that the *V–T* curves corresponding to many metals also fall very close to the these curves. This surprising observation stems in part from the Wiedemann–Franz law [3] which states that the variations with temperature of the thermal conductivity and electrical resistivity of metals are such that λ and ρ are related by the expression

$$\lambda\rho = LT \tag{1.39}$$

where L is the Lorentz constant (2.45×10^{-8} V² K⁻²) and T is the absolute temperature. Equation 1.39 holds if thermal conduction and electrical resistivity arise from electronic transport in the metals. Although this relation provides a reasonable description of thermal and electrical transport properties in metals over the temperature range generally relevant to ordinary electrical contacts, it is not universally valid. The origin of the Wiedemann–Franz law is discussed in relation to the electrical and thermal conductivity of metals outlined in Appendix A. If the relation in Equation 1.39 holds, then Equation 1.34 immediately yields the rigorously valid *V–T* relation as

$$V^2 = 4L(T_m^2 - T_1^2) \tag{1.40}$$

which is indeed independent of the materials in the contact. Equation 1.40 is not as easy to use as Equation 1.35 because the supertemperature $(T_m - T_1)$ cannot be evaluated readily from the voltage drop. It is instructive to compare the values of supertemperature calculated from Equation 1.35 with those evaluated from the more accurate relation in Equation 1.40, for values of V that are generally considered excessive for electrical contacts. Consider a copper–copper contact with a bulk temperature T_1 of 20°C. Using the room-temperature values of resistivity and thermal conductivity of copper respectively as 1.75×10^{-8} Ωm and 380 Wm⁻¹°C⁻¹ in Equation 1.35, the supertemperature evaluated from Equation 1.35 and that yielded by the more accurate relation in Equation 1.40 are shown in Table 1.10 for a number of values of the voltage drop V across the contact. Note the difference is small and insignificant from the engineering viewpoint for voltage drops as large as 0.05 V. This explains the popularity of Equation 1.35 for the evaluation of supertemperature.

Equation 1.40 may be reduced to the form of Equation 1.35 by mathematical redefinition and rearrangement of terms. Defining $(T_m + T_1)/2$ as the average temperature T_{avg} in the contact, the evaluation of Equation 1.39 at T_{avg} yields

$$\lambda_{avg}\rho_{avg} = \frac{L\{T_m + T_1\}}{2}$$

TABLE 1.10

Comparison of Supertemperatures Evaluated from Equations 1.35 and 1.40

Voltage Drop (V)	Supertemperature $(T_m - T_1)$/Equation 1.36 (°C)	Supertemperature $(T_m - T_1)$/Equation 1.41 (°C)
0.005	0.47	0.44
0.01	1.9	1.7
0.02	7.5	6.9
0.03	17	15
0.04	30	27
0.05	47	41

It follows that $\{T_m^2 - T_1^2\} = \{T_m - T_1\} \times \{T_m + T_1\} = \{T_m - T_1\}2\lambda_{avg}\rho_{avg}/L$. Substituting this expression into Equation 1.40 yields

$$T_m - T_1 = \frac{V^2}{8\lambda_{avg}\rho_{avg}}$$

and the *V–T* relation in Equation 1.35 is recovered with the physical parameters λ and ρ evaluated at the average connector temperature. Because Equation 1.40 is independent of the specific electrical and thermal conductivity properties of the materials in the contact, it applies equally well to monometallic and bimetallic contacts provided the Wiedemann–Franz law applies to both contact materials. The validity of the Wiedemann–Franz law does not necessarily imply that the maximum contact temperature occurs at the location of the physical interface.

1.4.4 Temperature Distribution in the Vicinity of an *a*-Spot

The calculation of the temperature distribution within an *a*-spot is facilitated by the geometrical coincidence of the equipotential and isothermal surfaces within the volumes of the conductors. This calculation needs information additional to that for Equation 1.34. Following Holm [3], Greenwood and Williamson [34] showed that the maximum contact temperature T_m and the current I are related by the expression

$$R_C I = \rho_{1,0} \int_{T_1}^{T_m} \left\{2\int_T^{T_m} \lambda_1 \rho_1 \, dT'\right\}^{-1/2} \lambda_1 \, dT + \rho_{2,0} \int_{T_2}^{T_m} \left\{2\int_T^{T_m} \lambda_2 \rho_2 \, dT'\right\}^{-1/2} \lambda_2 \, dT \qquad (1.41)$$

where T_1 and T_2 are the bulk temperatures respectively of materials 1 and 2, R_C is the "cold" contact resistance, $\rho_{1,0}$ and $\rho_{2,0}$ are the "cold" resistivities respectively of materials 1 and 2 defined through the dependence on temperature of the electrical resistivity, that is, $\rho = \rho_0(1 + \alpha T)$, and λ_1 and λ_2 are the thermal conductivities respectively of materials 1 and 2. Equation 1.41 is remarkable as it shows the maximum temperature developed in a contact is related to the voltage drop developed *by the same current, if the identical contact had remained cold*, since $R_C I$ is the "cold" voltage drop. Using again the temperature dependence

$\lambda = \lambda_0(1 - \beta T)$, and if β is sufficiently smaller than α (e.g. $\beta \sim \alpha/20$ for copper),the integral in Equation 1.41 is given approximately as

$$\int_{T_1}^{T_m}\left\{2\int_T^{T_m}\lambda\rho\,dT\right\}^{-1/2}\lambda\,dT' = \left(\frac{\lambda_0}{\{\alpha-\beta\}\rho_0}\right)^{1/2}\left\{\frac{\alpha}{\alpha-\beta}\right\}\cos^{-1}\left(\frac{1+\{\alpha-\beta\}T_1}{1+\{\alpha-\beta\}T_m}\right) \qquad (1.42)$$

For a monometallic junction and bulk temperatures T_1 and T_2 of 0°C, R_C is given as $\rho_0/2a$, so Equation 1.41 becomes

$$I = 4a\left(\frac{\lambda_0}{\{\alpha-\beta\}\rho_0}\right)^{1/2}\left\{\frac{\alpha}{\alpha-\beta}\right\}\cos^{-1}\left(\frac{1}{1+\{\alpha-\beta\}T_m}\right) \qquad (1.43)$$

Similarly, at an equipotential surface at a temperature T and located at a finite axial distance μ from the constriction, R_C is given as $(\rho_0/\pi a)\tan^{-1}(\mu/a)$ from Equation 1.1. Equation 1.41 then becomes [34]

$$I = \frac{2\pi a}{\tan^{-1}\left\{\dfrac{\mu}{a}\right\}}\left(\frac{\lambda_0}{\{\alpha-\beta\}\rho_0}\right)^{1/2}\left\{\frac{\alpha}{\alpha-\beta}\right\}\cos^{-1}\left(\frac{1+\{\alpha-\beta\}T}{1+\{\alpha-\beta\}T_m}\right) \qquad (1.44)$$

Equating Equations 1.43 and 1.44 yields the dependence of the temperature T at the isothermal surface located at a distance μ from the constriction

$$\cos^{-1}\left(\frac{1+\{\alpha-\beta\}T}{1+\{\alpha-\beta\}T_m}\right) = \frac{2}{\pi}\tan^{-1}\left\{\frac{\mu}{a}\right\}\cos^{-1}\left(\frac{1}{1+\{\alpha-\beta\}T_m}\right) \qquad (1.45)$$

The variations of temperature both in the plane of and normal to the constriction, are illustrated in Figure 1.52 for the special case of $\beta = 0°C^{-1}$, $\alpha = 4 \times 10^{-3}°C^{-1}$ and $T_m = 1000°C$ [34]. Note the temperature decreases to that of the bulk over a distance of only several constriction diameters.

1.4.5 Deviation of the Voltage–Temperature Relation in an Assymetric Contact

1.4.5.1 Case I: Two Metals in Contact

In a bimetallic contact, the maximum temperature developed within an a-spot would be expected to occur within the material with the larger electrical resistivity, rather than at the interface between the two metals. However, simple arguments given in Appendix B show that the deviation of the interfacing temperature from the maximum temperature is not expected to be large and the maximum temperature generally occurs in the immediate vicinity of the physical interface even when the resistivities of the contacting materials differ significantly. For all practical purposes, the maximum temperature may, thus, be considered to occur at the physical interface between the two materials. This situation is descriptive of bimetallic contacts involving good electrical conductors such as aluminum,

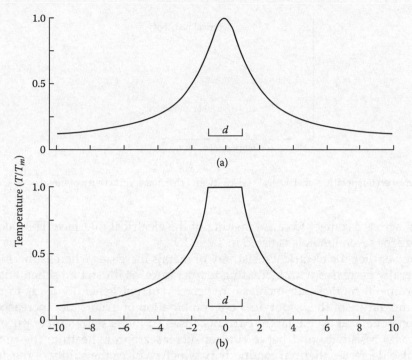

FIGURE 1.52
Steady state temperature distribution within a conductor under conditions where the thermal conductivity is constant and the resistivity varies with temperature as $\rho_0(1 + \alpha T)$, where $\alpha = 4 \times 10^{-3}°C^{-1}$, $T_m = 1,000°C$, and d is the constriction diameter $2a$ [34] (a) Temperature along the cylindrical axis of the constriction (b) Temperature in the plane of the interface. (From JA Greenwood and JBP Williamson, *Proc Roy Soc* A246: 13, 1958 [34].)

(a) Distance from constriction plane, z/a
(b) Radial distance from center of constriction, r/a, in constriction plane

copper, brasses, copper bronzes and so on, but is generally not applicable to interfaces in which highly dissimilar materials, such as copper and graphite, are used.

1.4.5.2 Case II: A Metal in Contact with a Non-Metal

In electrical contacts where the electrical conductivities of the contacting materials are vastly different, such as in copper–graphite contacts, the surface corresponding to the maximum temperature is much more deeply embedded in the material with higher resistivity than the calculation in Appendix B indicates for Case I. The location of the surface at which the maximum temperature is produced depends on the geometry of the contact. An illustrative example of this effect for the case of a copper–graphite contact is described in detail in Appendix C.

1.4.6 Special Considerations on the "Melting" Voltage in Electrical Contacts

We recall that the voltage–temperature relation was derived on the assumption of equilibrium between ohmic heating in the electrical constriction and thermal dissipation by conduction within the contacting materials. This relation was used to evaluate the potential

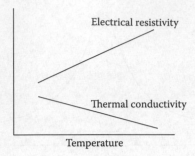

FIGURE 1.53
Generic dependence on temperature of electrical resistivity and thermal conductivity of metals.

drop required across a contact to cause melting of the electrical interface. The calculated melting voltage for several metals is listed in Table 1.9.

As mentioned earlier, the electrical resistivity of metals increases whereas thermal conductivity generally decreases with increasing temperature, as illustrated schematically in Figure 1.53. In metal contacts, an increase in electric current generally leads to a rapid increase in temperature of the a-spot due to a combination of events: the increased thermal dissipation generated by the larger current causes an increase in temperature which in turn, increases resistivity and hence further increases ohmic heating; the increased temperature also decreases thermal conductivity which reduces the ability of the junction to dissipate heat by thermal conduction. With good electrical conductors such as aluminum and copper conductor alloys, the rise in electrical resistivity and the decrease in thermal conductivity with increasing temperature are sufficiently slow so the a-spot reaches the equilibrium temperature defined by Equation 1.40. In other metals such as Ni and Fe, the increase in resistivity and decrease in thermal conductivity associated with increasing temperature is relatively rapid. In electrical contacts made from these metals, a critical voltage is found to exist at which heat can no longer be dissipated at a rate compatible with the requirements for thermal equilibrium. At this critical voltage, a small rise in temperature increases the resistivity and decreases the thermal conductivity in the a-spot sufficiently to cause significant additional heat entrapment. This, in turn, rapidly boosts the contact temperature, which leads to a further large drop in thermal conductivity, and so on. Thermal equilibrium is, thus, never attained. This condition leads to effective entrapment of ohmic heat with ensuing thermal runaway and contact melting. The critical voltage for the onset of this phenomenon is generally appreciably smaller than the conventional melting voltage [34,141].

It was stated in Equation 1.33 that the potential drop V across a thermally insulated junction is related to the electrical and thermal properties of the contact materials as

$$V = \left\{ 2 \int_{T_1}^{T_m} \lambda_1 \rho_1 \, dT \right\}^{1/2} + \left\{ 2 \int_{T_2}^{T_m} \lambda_2 \rho_2 \, dT \right\}^{1/2}$$

Similarly, it was indicated in Equation 1.41 that the total current I through the contact is related to λ, ρ and the dimensions of the contact spot as

$$IR_C = F\{\lambda_1, \rho_1, T_1, T_m\} + F\{\lambda_2, \rho_2, T_2, T_m\} \tag{1.46}$$

where

$$F\{\lambda,\rho,T_i,T_m\} = \rho_0 \int_{T_i}^{T_m} \left\{ 2\int_T^{T_m} \lambda\rho\, dT \right\}^{-1/2} \lambda\, dT$$

and R_C is the "cold resistance" of the contact, which depends on the contact geometry. The numerical evaluation of the function $F(\lambda, \rho, T_i, T_m)$ is straightforward if the dependence of λ and ρ on temperature is known. This dependence can be expressed as a power series of temperature and may include more powers of T than used in the evaluations of Equations 1.38 and 1.42. This more detailed dependence may be expressed as

$$\rho = \rho_0\{1 + \alpha T + \gamma T^2 + \eta T^3\} \text{ and } \lambda = \lambda_0\{1 - \beta T + \delta T^2 + \varepsilon T^3\}$$

so that

$$\lambda\rho = \lambda_0\rho_0\left(1 + \{\alpha - \beta\}T + \{\delta + \gamma - \alpha\beta\}T^3\right)$$

over a wide temperature range. Table 1.11 lists the values of the required constants for a number of electrical contact materials [141].

For a monometallic contact, Equation 1.46 reduces to

$$IR_C = 2F\left(\lambda_1, \rho_1, T_1, T_m\right)$$

or

$$\frac{I}{2a} = \frac{2}{\rho_0} F\left(\lambda_1, \rho_1, T_1, T_m\right) \tag{1.47}$$

Note the dependence on temperature of the right-hand side of Equation 1.47 arises solely from the parameters of the materials. Figure 1.54 shows the variation with T_m of the quantity $2F(\lambda_1, \rho_1, T_1, T_m)/\rho_0$ for the electrical couples listed in Table 1.11. The evaluations were carried out for a value of T_1 of zero, that is, where the temperature of the contacts far removed from the junction is 0°C. For the nickel–aluminum system, the curve plotted in Figure 1.54 corresponds to the quantity $2(F[\lambda_1, \rho_1 T_1, T_m] + F[\lambda_2, \rho_2, T_2, T_m])/(\rho_{1,0} + \rho_{2,0})$, where the subscripts 1 and 2 refer respectively to aluminum and nickel.

For Al, the function $F = F(\lambda, \rho, 0, T_m)$ increases up to the melting point of the metal. For Fe, Ni, and W, the function reaches a maximum at T_m^* and then decreases. The maximum is broad and its location is sensitive to the materials parameters, λ and ρ. The existence of this maximum defines a thermal instability leading to the thermal runaway effect outlined above. The nature of this instability may be understood as follows.

In the case of aluminum–aluminum contacts Equation 1.47 is always satisfied since an increase in the value of I, leading to a larger value of the left-hand side of the equation, can always be matched by an increase in the right-hand side through a larger value of T_m. Thus, an increase in electric current always leads to a higher contact temperature, up to the melting point of aluminum—660°C. In the case of iron and nickel, there is a critical temperature corresponding to the maximum of the function F at which any increase in the magnitude of I can no longer be matched by the function F, that is, Equation 1.47 breaks down. Physically, a condition is reached where $\partial T_m/\partial I \rightarrow \infty$, and thus where the slightest increase in current generates a disproportionally large increase in temperature leading to contact melting. The calculated values of the critical temperature T_m^* for the systems of interest are listed in Table 1.11.

TABLE 1.11

Electrical Resistivity and Thermal Conductivity Constants

Contact Couple	ρ_0 ($10^{-8}\,\Omega\,m$)	α ($10^{-3}\,°C^{-1}$)	γ ($10^{-6}\,°C^{-2}$)	η ($10^{-9}\,°C^{-3}$)	λ_0 ($Wm^{-1}\,°C^{-1}$)	β ($10^{-4}\,°C^{-1}$)	δ ($10^{-4}\,°C^{-2}$)	$\in$ ($10^{-9}\,°C^{-3}$)	T_m^* ($°C$)
Al/Al	2.28	6.01	0	0	240	2.24	0	0	660
Fe/Fe	8.68	6.52	4.83	5.90	79	12.2	0.789	−3.10	300
Ni/Ni	6.86	6.22	4.29	18.4	94	1.40	2.16	−2.13	348
Ni/Al									654

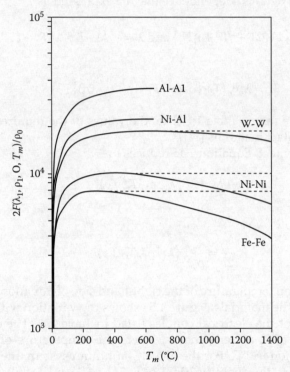

FIGURE 1.54
Dependence of the function $2F(\lambda, \rho, 0, T_m)/\rho_0$ on the contact temperature T_m. (From RS Timsit, *IEEE Trans Comp Hyb Manuf Tech* CHMT-14: 285, 1991 [141].)

This temperature is much lower than the melting point of the respective metals. The calculated melting voltages for iron–iron and nickel–nickel contacts are respectively 0.19 V and 0.16 V [141]. Figures 1.55a and 1.55b show the drop in potential across Fe/Fe and Ni/Ni contacts, and the corresponding bulk temperature T_1 in the contacting bodies, as the current is increased [141]. Note the onset of melting ("premature" melting) at 192 mV in iron and at 164 mV in nickel. These values are in excellent agreement with the calculated values [141].

Finally, note that the "melting" temperature of nickel–aluminum contacts is nearly identical with the melting point of aluminum. The effect of replacing one side of a nickel–nickel contact with a surface of large thermal conductivity, thus enhancing thermal dissipation, has a considerable effect on the "melting" temperature. "Premature" melting of electrical contacts, as described above, does not occur in common electrical contact metals such as copper, copper-based brasses and copper-based and aluminum-based bronzes.

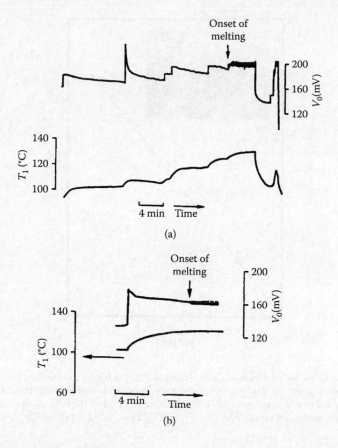

FIGURE 1.55

(a) Onset of melting in an iron–iron electrical interface; each discontinuity in the potential drop V_0 across the contact (top) is due to a steep increase in the current (b) Onset of melting in a nickel–nickel electrical interface; the sharp increase in the potential drop across the contact (top) arises from a steep increase in the current. In (a) and (b), the bottom curve represents the bulk temperature T_1. (From RS Timsit, *IEEE Trans Comp Hyb Manuf Tech* CHMT-14: 285, 1991 [141].)

1.5 Mechanics of *a*-Spot Formation

1.5.1 Smooth Interfaces

In an electrical interface, metal-to-metal *a*-spots are subjected to a variety of mechanical forces that influence the shape and dimensions of spots with time, even in apparently stationary junctions. It is well known that clean smooth surfaces pressed into intimate contact may adhere strongly under the action of intermolecular forces [15,142]. Figure 1.37b illustrated the dependence of contact resistance on mechanical load obtained with smooth oxide-free aluminum surfaces generated in ultra-high vacuum [80]. As the load was increased, the resistance dropped rapidly to a low value and remained relatively constant during the unload phase. As is evident in the figure, the oxide-free surfaces were firmly bonded together after contact was established and required the application of a relatively large tensile load to bring about separation. Figure 1.56 shows the dependence of the load required on the contact area to separate smooth and clean hemispherical AA1350 aluminum surfaces pressed into contact in ultra-high vacuum at room temperature. The aluminum

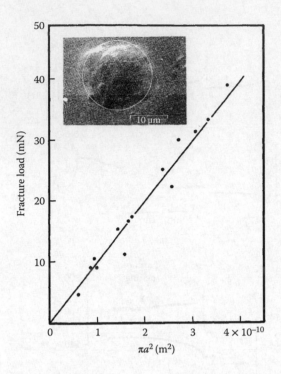

FIGURE 1.56
Load required to separate clean, oxide-free aluminum surfaces in ultra-high vacuum, as a function of the area πa^2 of metal-to-metal contact. Inset: micrograph of a contact spot after a measurement of fracture strength after applying a contact load of a few mN (see Figure 1.37); the circle was drawn on the basis of the radius a calculated from the contact resistance measured before fracture. (From RS Timsit, *Can Elect Assoc Rep 76-19, 1976* [80].)

had been annealed at 300°C. The radius of curvature R_{sur} of the surfaces was 0.254 m [80]. The contact area was varied by varying the mechanical load applied to the interface. The radius of the metal-to-metal contact area was calculated from Equation 1.4

$$R_C = \frac{\rho}{2a}\left(1 - 1.416\left\{\frac{a}{R}\right\} + \ldots\right)$$

after measuring the contact resistance at room temperature [80]. The slope of the average straight line in Figure 1.56 yields a fracture stress at the interface of 99 MPa. This is identical with the fracture stress of H12 AA1350 aluminum, suggesting cold-working of the aluminum in the a-spot. The inset of Figure 1.57 shows the micrograph of an actual contact spot after measuring the fracture strength of the hemispherical interface. The radius of the circle represents the radius calculated from a measurement of contact resistance. The agreement with the actual periphery of the contact confirms the validity of the evaluation procedure for determining the contact radius.

If two elastic surfaces of radius of curvatures, R_1 and R_2 are pressed into contact by a force P, according to Hertz [143] the surfaces make contact on a circle of radius a given by the expression

$$a = \left\{\frac{R'P}{K}\right\}^{1/3} \tag{1.48}$$

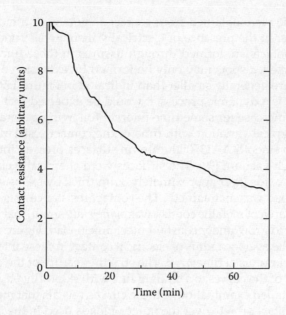

FIGURE 1.57
Typical dependence of contact resistance on time in ultra-high vacuum (pressure ~ 10^{-10} torr) immediately after rupture of interfacial oxide films. (From RS Timsit, *IEEE Trans Comp Hyb Manuf Tech* 3: 71, 1980 [99].)

with $R' = \dfrac{R_1 R_2}{R_1 + R_2}$, $K = \dfrac{4E}{3}$ and $\dfrac{1}{E} = \dfrac{\{1 - v_1^2\}}{E_1} + \dfrac{\{1 - v_2^2\}}{E_2}$

where E_1 and E_2 are Young's modulus and v_1 and v_2 are Poisson's ratio for the materials of surfaces 1 and 2 respectively. In the presence of adhesive forces between the surfaces, it was shown by Johnson et al. [144] that Equation 1.48 becomes

$$a = \left\{ \frac{R'}{K} \left[P + 3\pi\gamma_s R' + \left(6\pi\gamma_s R'P + \left\{ 3\pi\gamma_s R' \right\}^2 \right)^{1/2} \right] \right\}^{1/3}$$

where γ_s is the surface energy at the contact interface. It follows that an equilibrium contact area can be maintained even in the presence of a tensile force, up to a maximum value of

$$P = -\frac{3\pi\gamma_s R_{sur}}{4} \tag{1.49}$$

which is the force required to separate the spheres, where $R' = R_{sur}/2$. Compare now the prediction of Equation 1.49 with the data of Figure 1.56. For aluminum surfaces, the surface energy is of the order of 1 Jm^{-2} [145]. For hemispherical surfaces with $R_{sur} = 0.254$ m, the maximum tensile force given by Equation 1.49 is thus 1.2 N and is independent of the applied load. The fracture loads shown in Figure 1.56 were all considerably smaller than this value and were found to vary with applied load, since the contact radius was increased by applying larger loads. The increase in contact radius with increased load stems from plastic deformation in the contact, that is, as indicated by Equation 1.10. The observation that the fracture load was always considerably smaller than 1.2 N is due to the presence of structural defects in aluminum. This is known to lower the strength to far below the theoretical limit [146].

The data of Figure 1.56 were obtained from contact *a*-spots with a radius ranging from a few to several microns. In the presence of electrically insulative surface oxide films, the *a*-spots in aluminum contacts are formed through fissures in these films. If the films are not mechanically dispersed, *a*-spots may only be formed by extrusion of metal across the electrical interface and are generally smaller than indicated in Figure 1.56. Because the formation of metal bridges is a dynamic process, it would be expected that the area of metal-to-metal contact would increase for some time interval following generation of the *a*-spot. Figure 1.57 shows the typical variation with time of the contact resistance of an electrical interface formed by two smooth AA1350 aluminum surfaces pressed into contact at room temperature in ultra-high vacuum [99]. The surfaces were chemically clean but were covered with an aluminum oxide layer approximately 3 nm thick by exposure to oxygen , and a constant mechanical load was maintained . The contact resistance was found to decrease with time whenever the area of metallic contact was *sufficiently* small, that is, the area of electrical contact was found to grow under constant mechanical load. Figure 1.58 shows typical variations with time of the average radius of electrical contact in these interfaces, where a_0 is the initial electric contact radius at time $t = 0$. The striking feature of the data of Figure 1.58 is the eventual tendency of the curves to a straight line with slope $d\ln(a/a_0)/d\ln(t)$ of approximately 1/9 [99,147]. A detailed examination of these curves reveals that mechanical load has no apparent effect on contact growth over the range of loads used in the investigation.

The data of Figure 1.58 can be interpreted in terms of a sintering mechanism. Figure 1.59 illustrates a contact spot formed through cracks in oxide films in an aluminum–aluminum interface. For simplicity, the spot is assumed circular with a radius a. The surfaces in the immediate vicinity of the neck are assumed locally distorted approximately as spheres with an average radius of curvature r_s. Sintering describes a bonding mechanism between surfaces in which the contact area (and hence, the neck of the contact) grows as a result of

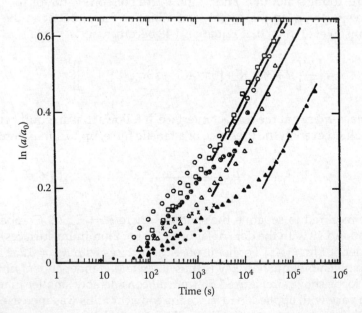

FIGURE 1.58
Typical growth of contact radius a in vacuum at low loads; the value of the initial contact radius a_0(in nanometers) and the contact load corresponding to each curve are respectively: ● 4 N and 0.19 N; ○ 100 N and 0.19 N; □ 73 N and 0.20 N; 100 N and 0.20 N; × 12 N and 0.20 N; ▲ 84 and 0.30 N; △ 96 and 0.37 N. (From RS Timsit, *IEEE Trans Comp Hyb Manuf Tech* 3: 71, 1980 [99]; RS Timsit, *Appl Phys Lett* 35: 400, 1979 [147].)

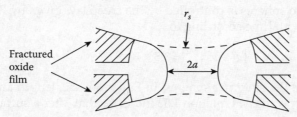

FIGURE 1.59
Schematic diagram of contact spot formed between cracks in aluminum oxide surface films.

transport of material from the immediate vicinity of the contact interface into the contact region as a result of some driving force. By far the most important driving force for mass transport in conventional sintering is that due to surface tension [148]. If S is the surface area of the contact neck of Figure 1.59, then the surface energy U is simply given as $\gamma_s S$, where γ_s is the surface energy of the solid. Since the system is in thermodynamic equilibrium only under conditions where U is a minimum, the neck grows until the surface S is minimized. The effective force F acting axially outward to induce the growth is, then, given as

$$F = -\frac{dU}{dr}$$

Early et al. [149] have shown that this force gives rise to a mean radial stress, σ on the neck surface given as

$$\sigma = \frac{3\gamma_s}{a} \tag{1.50}$$

when a is much smaller than the mean local radius of curvature r_s of the surfaces. Taking γ_s for aluminum as 1 Jm^{-2} [145] and a as 10^{-7}m yields $\sigma \sim 30$ MPa; the radial stresses induced on the contact neck by the surface tension can, thus, be large when the junction is narrow.

When the stresses induced by the surface tension are sufficiently large, the contact junction will generally grow. The transport of material required to feed the growth can occur through one or a combination of several well-established mechanisms [150] such as surface diffusion, volume diffusion, grain–boundary diffusion and creep [151]. At room temperature, sintering fed by dislocation creep represents a mechanism that can explain much of the data of Figure 1.58 [99,147]. It has been demonstrated that interfaces such as grain boundaries and free surfaces are the most common sources of dislocations [152]. A threshold stress is required to generate dislocations from such sources. In the contact situation under consideration, this threshold stress can easily be provided by the surface tension force described above.

In aluminum, at a homologous temperature above approximately 0.3 [153], that is, at temperatures above 10°C, and at relatively large stresses, the strain rate $\dot{\varepsilon}$ induced by dislocation creep is found experimentally to obey the constitutive law [154]

$$\dot{\varepsilon} = \kappa_0 \sigma^n \tag{1.51}$$

where n is nearly equal to 4.5 and where κ_0 is a constant dependent on temperature. At 300K $\kappa_0 = 1.2 \times 10^{-48}$s^{-1} Pa$^{-9/2}$. From the work of Kuczynski et al. [155], it can be shown that if sintering between two spheres is controlled by the creep law given by Equation 1.51, the neck radius increases with time according to

$$\{a^2 - a_0^2\}^{9/2} = 3.5 \times 10^{-47} r_s^{9/2} t \tag{1.52}$$

where r_s is the radius of the sintering spheres. In Equation 1.52, length and time units are meters and seconds respectively. Equation 1.52 indicates that after a sufficiently long time when $a > a_0$,

$$\ln\left\{\frac{a}{a_0}\right\} \sim \frac{1}{9}\ln t + C$$

where C is a constant. Thus, the growth curves approach a straight line with a slope of 1/9 on a log–log plot. Within the scatter of the experimental data, this is, indeed, the behavior of the a-spot data in Figure 1.58.

The effect of oxygen on the growth of a-spot in Figure 1.58 is illustrated in Figure 1.60. The properties of the contacts appeared identical at oxygen pressures of 50 torr and 100 torr. A comparison of the curves of Figures 1.58 and 1.60 indicates that contact growth is considerably hampered by the presence of oxygen: the growth rates are smaller than those obtained in vacuum and the curves generally reach a plateau in a reasonably short time. The curves bear no apparent relation to normal load. The contact growth in the presence of oxygen can be qualitatively understood in terms of the sintering model in that, in the presence of oxygen, the contact neck is covered with a surface layer of aluminum oxide with a thickness of several nanometers. The effect of the oxide film on creep is both to alter the image stresses on dislocation mobility [156] and to "strengthen" the contact neck [157], that is, the neck is effectively made from a composite material. These two factors would combine to reduce creep and could account for the relatively slow sintering rate observed with oxygen .

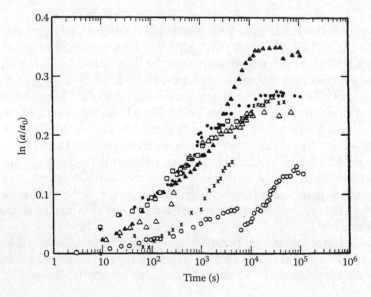

FIGURE 1.60
Typical increase in contact radius a with time in the presence of oxygen gas at a contact load of approximately 0.5 N. The value of a_0 (in nanometers) corresponding to each curve is given below. At 50: □ 49; △ 8; ● 49; ▲ 20; at 100 torr: × 5. (From RS Timsit, *IEEE Trans Comp Hyb Manuf Tech* 3: 71, 1980 [99]; RS Timsit. *Appl Phys Lett* 35: 400, 1979 [147].)

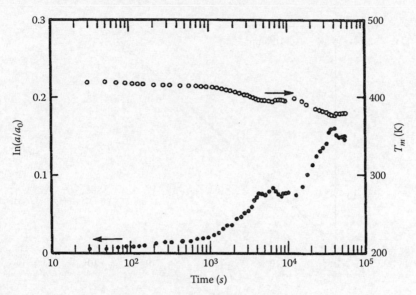

FIGURE 1.61

Typical dependence on time in oxygen gas (160 torr) of the average a-spot radius ($\bullet$) and the contact temperature T_m (O); the radius a_0 at $t = 0$ is 2.9 μm. (From RS Timsit, *Appl Phys Lett* 35: 400, 1979 [147].)

At elevated temperatures, the sintering rate increases rapidly with increasing contact temperature even with oxygen, and contact growth no longer requires a small a-spot. Figure 1.61 illustrates the growth of an a-spot in an aluminum–aluminum contact formed between two hemispheres operated in oxygen at a pressure of 160 torr and carrying a constant DC current of 13.6 A. The initial contact spot radius was determined as 2.9 μm and the initial contact temperature was approximately 232°C (405 K) [158]. Note the relatively rapid increase in a-spot radius. The radius increased to ~ 14.5 μm after approximately 14 hours, leading to a decrease in contact temperature of about 30°C. In this instance, the growth of a-spots in aluminum–aluminum contacts operated at elevated temperatures could be explained in terms of sintering driven by volume diffusion [158].

It is likely that sintering explains at least some of the contact phenomena related to the growth of a-spot observed in gold contacts [20]. Sintering is also probably responsible partially for "self-healing" of electrical contacts as observed by several workers [159,160]. On the other hand, it is unlikely that sintering plays a major role in contacts in which large-scale cold welding occurs [161], as has been reported in highly-crimped junctions at room temperature [162]. In these situations, cold-welded spots are so large that the mechanical stress on a-spot surfaces as given by Equation 1.50 would be too small to induce mass flow.

1.5.2 Rough Interfaces

It was mentioned in Section 1.2.3 that surface asperities undergo plastic deformation in most practical electrical contacts. This led to the conclusion that the true area of mechanical contact depends only on contact force, and is independent of the nominal contact area. It turns out that this relationship between contact force and contact area is more general and holds even when asperity deformation is not purely plastic. Several models of rough surfaces in contact have been proposed in which model asperities are assumed to have elastic or plastic deformation [2,163–165] or are treated in terms the random nature of the surface profile [166,167]. If it is assumed that the asperities have the same radius of curvature and *deform elastically* and

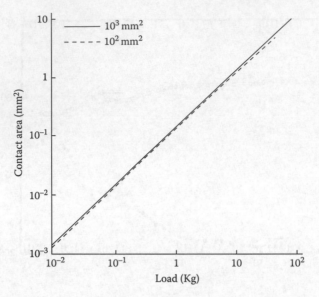

FIGURE 1.62
Relation between area of contact and mechanical load in an interface characterized by material parameters described in the text. *The contacting surface asperities are assumed to deform elastically only.* The solid curve, for a nominal area of 10 cm², and the broken curve, for a nominal area of 1 cm², show that the real area of contact is essentially independent of the nominal area. (From JA Greenwood and JBP Williamson, *Proc Roy Soc* A295: 300, 1966 [2].)

independently according to Hertz theory [143], Greenwood and Williamson [2] showed that the area A_C of true contact increases rapidly with increasing contact load and is independent of the nominal contact area. This is illustrated in Figure 1.62 where A_C curves were calculated for nominal contact areas of 1 cm² and 10 cm² using the following parameter values: density of surface asperity of 300 mm⁻², the average asperity radius of curvature β_{av} and the rms (root mean square) roughness σ_{sd} are such that $\beta_{av}\sigma_{sd}= 10^{-4}$ mm² and $E'(\sigma_{sd}/\beta_{av})^{1/2}= 245$ MPa, where E' is $E/(1 - \nu^2)$ [2]. The curves show that for a given nominal area, the area of true contact is almost exactly proportional to the load. In addition, the two curves for nominal contact areas of 1 cm² and 10 cm² are almost indistinguishable, showing that the contact area depends on the load and not on the nominal pressure. Similarly to the explanation provided in Section 1.2.3 with respect to asperities with plastic deformation, the near independence of true contact area on nominal pressure in Figure 1.62 stems from the proportional *increase* in the number of contacting asperities with nominal contact area at a given load, thus *decreasing* the pressure on each asperity. This, in turn, *decreases* the true contact area at each asperity and almost exactly offsets the area gain due to the larger nominal contact.

In order to determine the extent of plastic deformation of contacting asperities, Greenwood and Williamson [2] defined a *plasticity index* ψ as

$$\psi = \frac{E'}{H}\sqrt{\frac{\sigma_{sd}^*}{\beta_{av}}}$$

where E' is $E/(1 - \nu^2)$, σ_{sd}^* is the standard deviation of the asperity height distribution and β_{av} is the average asperity radius. According to the definition, plastic flow will be significant if $\psi > 1$, otherwise the asperities deform essentially elastically. From Table 24.1, it may be verified that the ratio E'/H is on the order of 200–400 for most electrical contact materials of interest. Since the ratio σ_{sd}^*/β_{av} is seldom smaller than about 5×10^{-3} for rough

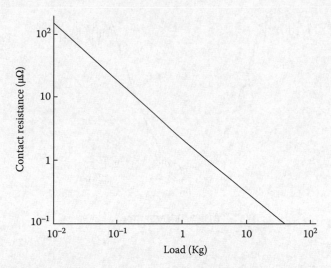

FIGURE 1.63
Dependence of electrical contact resistance on load corresponding to the data or Figure 1.62; the assumption is made that all the contact spots are electrically conducting. (From JA Greenwood and JBP Williamson, *Proc Roy Soc* A295: 300, 1966 [2].)

surfaces [2], it is clear that ψ is generally > 1 and contact asperities deform in plastic manner in electrical interfaces of practical connectors. This, again, validates the almost universal use of Equation 1.10 in relating true contact area to contact load in electrical interfaces. It is worth pointing out that contact asperities may not deform in plastic manner in devices such as Micro-Electrico-Mechanical Systems (MEMS) where the contact surfaces are very smooth and where σ^*_{sd}/β_{av} may be very small (see Chapter 10).

The decrease in electrical contact resistance associated with the curves of Figure 1.62 is shown in Figure 1.63[2]. The data of Figure 1.63 was calculated using a resistivity of 2.4×10^{-8} Ωm and on the assumption that all contacting asperities deform elastically and all contacting area are electrically conducting. The result approximates to the simple law that resistance is proportional to (load)$^{-0.9}$. In practical electrical contacts, where asperities show plastic deformation , note the area of true contact increases less rapidly than suggested in Figure 1.63 and contact resistance varies as (load)$^{-0.5}$ (see Equation 1.11).

Plastic deformation of asperities does not necessarily mean that surface asperities flatten out completely under the action of a sufficiently large mechanical compressive load. Moore [168] was the first to demonstrate the remarkable persistence of surface asperities under conditions of bulk plastic flow of a surface. In further investigations of this phenomenon, Williamson and Hunt [169] and Pullen and Williamson [170] confirmed the persistence of the asperity , as illustrated in Figure 1.64 [169], and showed the nominal and real areas of contact are proportional in local indentations. These workers also showed the real area of contact in a locally plastic indentation is affected by the hardness of surface layers relative to that of the bulk deformed material. These effects are illustrated for the case of aluminum in Figure 1.65. In these plots, the filled and unfilled data points represent respectively the nominal area of flat coupons of different dimensions and the real area of contact measured with each coupon. In specimens that were initially annealed and then bead-blasted, that is, where the hardness of the surface layers was larger than the bulk hardness of the material, the ratio of the true area to nominal area of contact was approximately 37%. In specimens that had been work-hardened and then bead-blasted, that is, where the surface roughness was approximately identical to the previous specimens but

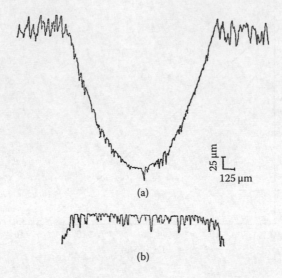

FIGURE 1.64
Two profiles of the same indentation made in aluminum by a smooth steel ball: (a) Normal profile showing the microscopic spherical indentation (b) Same profile as (a) but compensated to remove the macroscopic curvature, showing that persistence in asperity is uniform over the depression. (From JBP Williamson and RT Hunt, *Proc Roy Soc* A327: 147, 1972 [169].)

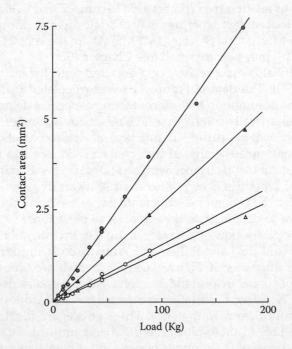

FIGURE 1.65
Dependence of true area of contact on applied load for a spherical indentation in aluminum. ● and ○ show respectively the nominal area of test coupons and true area of contact measured from the coupons: Annealed then bead-blasted specimen; the ratio of true contact area to nominal area is approximately 37%. ▲ and △ show respectively the nominal and true area of contact: Work-hardened and then bead-blasted specimen. The ratio of true area to nominal area is approximately 50%. (From JBP Williamson and RT Hunt, *Proc Roy Soc* A327: 147, 1972 [169].)

where the hardness of the surface was identical to that of the bulk metal, the contact ratio was about 50%. Similar results were obtained on gold [169]. These results were explained in terms of the movement of valley floors towards the contact interface as asperities are flattened under the compressive load [169,170]. The implications of these findings as regards electrical contact resistance are clear: the hardness of a contact surface must not be larger than that of the bulk material if the true area of contact is to be maximized.

In later years, efforts were made to generate micromechanical models to predict the evolution of microtopography of the surface in a frictional interface. Although these models have focused on metal-working interfaces, the results obviously apply to electrical contacts. For example, Wilson and Sheu [171] modeled asperity deformation produced by flattening against a smooth rigid tool surface. Upper-bound analysis was used to evaluate the resistance to flattening of the asperities as a function of the underlying bulk strain rate, and simple kinematics then related the change in real contact area between the surfaces with increasing bulk deformation. The model took account of the motion of valley floors towards the contact interface during deformation. The validity of the results of this analysis appears to be confirmed by experimental data obtained using wedge-type asperities generated on aluminum strip [171].

In more recent years, a number of different methods to model the contact of rough surfaces have been proposed, including fractal [172], and multiscale models [173–177]. The methods based on fractal mathematics were derived to account for different scales of surface features not accounted for by the statistical models. The multiscale models were developed to alleviate the assumption of self-affinity imposed by fractal mathematics and also to improve the understanding of the conditions of mechanics. Other analyses [176] have used a Fourier transform to convert two-dimensional surface roughness profiles into a series of stacked sinusoids. These methods yielded results that agree with the prediction trends of the earlier work, and differences from the earlier results were often only small [178].

The flattening of a rough surface under pressures greater than the material yield stress has been studied extensively by Wanheim [179] and Bay and Wanheim [180–182]. By flattening aluminum specimens with triangular wedge-type asperities in a coining die, these workers were able to verify a slip line field model for flattening of the asperity. Interaction of deformation zones in neighboring asperities caused the rate of flattening to decrease with increasing pressure. Makinouchi et al. [183] performed experiments in which aluminum specimens with wedge-type asperities were flattened in a rig that allowed plain strain compression of the specimen bulk. Their results indicate that no appreciable increase in flattening resistance occurs up to the maximum contact area ratio of 90%. These results are consistent with the predictions of their finite element model, which also allows for bulk plastic strain and bulk flow.

1.6 Breakdown of Classical Electrical Contact Theory in Small Contact Spots

1.6.1 Electrical Conduction in Small *a*-Spots

1.6.1.1 Contact Resistance

We recall that the classical expression for contact resistance as given by Equation 1.3 is based on a solution of the equations of classical electromagnetism where electrical flow is assumed continuous as illustrated in Figure 1.66a. On a sufficiently small scale in solids, electric current flow is not continuous, but occurs through the irregular motion of electrons as they collide with lattice imperfections, defects, and other obstacles in a conductor

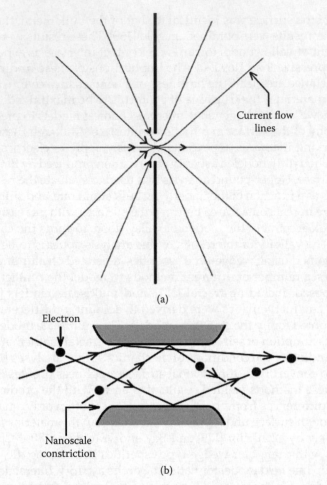

(a)

(b)

FIGURE 1.66
(a) Classical electric flow through a constriction (b) Ballistic motion of electrons through a small constriction.

or an interface. As the constriction in an electrical contact interface approaches the scale at which electronic scattering events occur, the simple concept of classical electrical flow through the constriction becomes increasingly less valid. Thus, as the constriction gets smaller, the electronic flow across may be viewed increasingly as "particulate" as illustrated schematically in Figure 1.66b.

As pointed out in earlier publications [184,185], the electrical resistance of a sufficiently small contact spot acquires a so-called Knudsen resistance component in addition to the ohmic component described by Equation 1.3. This additional component arises from the scattering of conduction electrons by the constriction boundary, and becomes prominent in junctions where the constriction radius is smaller than the mean free path l_{fp} of the electron in the conduction medium. The problem of electronic conduction through a circular constriction of radius smaller than l_{fp} resembles the well-known Knudsen problem in kinetic gas theory wherein gas molecules effuse through a small aperture. If the aperture diameter becomes comparable with the mean free path for molecular collision within the gas, the molecules no longer diffuse across the aperture but pass through the orifice ballistically. In the electrical analog, the different electronic conduction regimes are characterized by the Knudsen ratio, $K = l_{fp}/a$ where a is the constriction radius.

Consider first an electrical constriction in the limit of a large Knudsen ratio where conduction is highly ballistic. The electrons are injected into the constriction, and hence pass through the contact ballistically, as illustrated in Figure 1.66b. Sharvin [184,186,187] realized the peculiar electronic transport through a contact in the ballistic regime and calculated the resistance of a constriction of radius a as

$$R_S = \frac{4\rho l_{fp}}{3\pi a^2} \tag{1.53}$$

where ρ is the resistivity of the material. Since ρ can be written as [Appendix 1,154]

$$\rho = \frac{mv_F}{ne^2 l_{fp}}$$

where n is the free electron density in the metal, m is the electronic mass and v_F is the Fermi velocity, Equation 1.53 reduces to

$$R_S = \frac{4mv_F}{3\pi ne^2 a^2} = \frac{C}{a^2} \tag{1.54}$$

where C is a constant dependent only on the electronic properties of the conductor. Since the quantities v_F and n are relatively insensitive to temperature over the temperature range of interest here [188], C is also essentially independent of temperature. The Sharvin constriction resistance given in Equation 1.54 is thus largely independent of temperature.

Between the classical and ballistic electronic conduction regimes, the constriction resistance has been shown by Wexler [185] to be given as

$$R_B = \frac{\rho}{2a} \Gamma\{K\} + \frac{4K\rho}{3\pi a} \tag{1.55}$$

where $\Gamma\{K\}$ is a function decreasing from 1 to 0.694 as l_{fp}/a increases from 0 to ∞, as shown in Figure 1.67. It may be verified that Equation 1.55 can be written

$$R_B = \frac{\rho}{2a} \Gamma\{K\} + \frac{C}{a^2} \tag{1.56}$$

where C is defined in Equation 1.54. Thus, the second term in the right-hand side of Wexler's expression is the Sharvin resistance. Table 1.12 lists the resistivity and electronic mean free l_{fp} at room temperature of a few commonly-used unalloyed electrical contact materials of high purity [189]. Using known values of m, v_F and n, for a number of contact materials [187,188,189,190], Table 1.12 also lists the value of C in Equation 1.56 to allow an evaluation of contact resistance in the presence of the Sharvin resistance for the respective metals [191]. Figure 1.68 shows the variation of constriction resistance of a copper–copper contact consisting of a single a-spot at room temperature, as a function of the radius a [191]. The two curves correspond respectively to the classical constriction resistance evaluated from Equation 1.3 and the resistance evaluated from Equation 1.56, using the values of ρ and C listed in Table 1.12. Note, the effect of the Sharvin resistance, giving rise to a larger constriction resistance, begins to be significant for a constriction radius smaller than about 100 nm. Similar results are obtained for the other metals listed in Table 1.12.

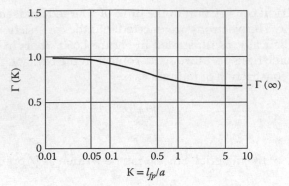

FIGURE 1.67

Dependence of Γ{K} on K. (From G Wexler, *Proc Phys Soc* 89: 927, 1966 [185].)

TABLE 1.12

Electrical Resistivity and Electronic Mean Free Path at Room Temperature

Metal	Electrical Resistivity (Ωm)	Electronic Mean Free Path l_{fp} (nm)	$C \times 10^{16}$ (Ωm^2)
Al	2.74×10^{-8}	14.5	1.70
Cu	1.70×10^{-8}	38.7	2.79
Ag	1.61×10^{-8}	52.3	3.57
In	8.75×10^{-8}	6.1	2.27
Sn	11.0×10^{-8}	4.1	1.91
Au	2.20×10^{-8}	38.3	3.58

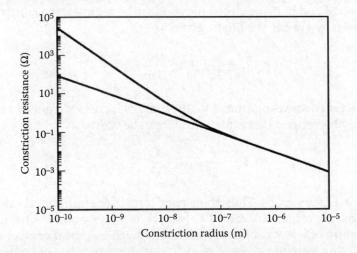

FIGURE 1.68

Constriction resistance for a copper-copper contact at room temperature, as a function of the constriction radius *a* as calculated from classical contact theory (Equation 1.3) and from Equation 1.57, using the parameter values appropriate to copper in Table 1.12. (From RS Timsit, *IEEE Trans Comp Pack Tech* 29: 727, 2006 [191].)

1.6.1.2 Joule Heat Flow Through a-Spots

An important feature of small electrical contacts where $K \gg 1$ is the so-called "non-locality" [187,192]. Non-locality means that the current density *j* within and in the immediate vicinity of the constriction cannot relate to the local electrical field *E* through the relation $j = E/\rho$ of classical electrodynamics. Hence, the local heat generated by the current density *j* in a small constriction cannot be expressed as ρj^2, within the constriction. This non-locality feature stems from the property that ballistic electrons in the constriction do not lose energy to the local lattice since they collide mostly elastically within the constriction until they clear the contact region. The electrons release their excess energy beyond the region of the potential drop across the contact [192].

In small contacts where K ≫1, some Joule-heating occurs within the constriction since the constriction resistance comprises a component of conventional ohmic resistance that is, the term $(\rho/2a)\Gamma(K)$ in Equation 1.56. In order to explore Joule heat generation in small contacts where K ≫1, it is convenient to consider Equation 1.56 for the case of a specific metal. For the benefit of this and later discussions, we shall focus on aluminum–aluminum contacts. On the basis of the data of Table 1.12, Equation 1.56 then reduces to

$$R_B = \frac{\rho}{2a}\Gamma\{K\} + \frac{1.70 \times 10^{-16}}{a^2} \tag{1.57}$$

From this expression, Joule heat will cause a resistance increase largely due to the increase in the term $(\rho/2a)$ as the temperature is increased. The heuristic picture that emerges is one where constriction resistance may be viewed as consisting of two resistances in series as illustrated in Figure 1.69 [191]. The first resistance $R_1 = (\rho/2a)\Gamma(K)$ depends on temperature. The second resistance is the Sharvin resistance (heretofore designated as R_2) $R_2 = 1.70 \times 10^{-16}/a^2$ is, thus temperature-independent. As the constriction gets smaller, the Sharvin resistance R_2 increases rapidly and eventually becomes dominant. Thus, if a voltage V is applied across the contact, only the voltage $V_1 = VR_1/(R_1 + R_2)$ is developed across the ohmic resistance R_1. In this picture, contact heating is determined largely by V_1 so melting of a contact spot would occur only if V_1 reaches the "melting voltage". It is on the basis of this argument that the breakdown of classical electrical contact behavior was detected experimentally. This is addressed in greater detail below.

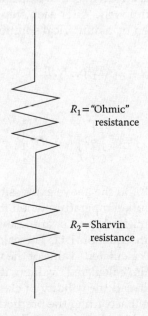

R_1 = "Ohmic" resistance

R_2 = Sharvin resistance

FIGURE 1.69
Depiction of the resistance of a small constriction consisting of a temperature-dependent "ohmic" component (R_1), and the temperature-independent Sharvin resistance (R_2). (From RS Timsit, *IEEE Trans Comp Pack Tech* 29: 727, 2006 [191].)

1.6.2 Observations of Breakdown of Classical Electrical Contact Theory in Aluminum Contacts

1.6.2.1 Experimental Data on Aluminum

Since the assumptions of classical electrical contact theory are inconsistent with the presence of the Sharvin resistance, it would be expected the V–I characteristics predicted on the basis of the Kohlrausch relation would conflict with those observed experimentally in these situations.

The validity of the voltage–temperature relation, as a function of the average a-spot dimension, was investigated in aluminum–aluminum contacts [140]. The advantage of aluminum in these investigations stems from the presence of a surface oxide layer on the contact surfaces, which allowed generating small a-spots since these spots form through cracks in the oxide films. The investigation was carried out as follows [140]. A series of aluminum–aluminum electrical couples were prepared in which the contact surfaces had been polished into optically smooth hemispheres. The use of smooth hemispheres was intended to produce a contact area that was circular and where the a-spots could be easily located within the electrical interface. In each couple, the contact load was selected to generate a well-defined area of mechanical contact. The contact surfaces of each couple were also sputter-cleaned in vacuum to remove all traces of organic contaminants, but the surfaces were re-oxidized at room temperature by exposure to clean oxygen gas. The average a-spot dimension could not be controlled by the magnitude of the mechanical contact load, and varied greatly (and seemingly randomly) from one contact couple to another.

All measurements were made under ultra-high vacuum conditions as follows [140]: the "cold" contact resistance R_c (i.e., the contact resistance at room temperature) was first determined by passing an electric current and measuring the voltage drop of a few millivolts across the contact. Following this procedure, the current was increased in stepwise fashion and the voltage drop across the electrical interface was measured at each step. The bulk temperature T_1 was monitored continuously. After each step increase of the current I, the contact temperature T_m was calculated by a numerical solution of Equation 1.47.

$$R_c I = 2F\{\lambda_1, \rho_1, T_1, T_m\} \tag{1.58}$$

with

$$F\{\lambda_1, \rho_1, T_1, T_m\} = \rho_0 \int_{T_1}^{T_m} \left\{ 2\int_{T_1}^{T} \lambda_1 \rho_1 \, dT' \right\}^{-1/2} \lambda_1 \, dT$$

using the measured values of R_c and I. In the above expressions, ρ_0 is the "cold" resistivity of aluminum, T_1 is the equilibrium bulk temperature when the current I is passed and the temperature dependence of the quantities λ_1 and ρ_1 are known. The integrals in Equation 1.58 were evaluated analytically as in Equation 1.42. The calculated value of T_m was, then, substituted into Equation 1.38 and a "calculated" value of the voltage drop V across the contact was computed. Agreement of this "calculated" value with the voltage drop measured during passage of the current I confirmed the validity of classical contact theory since the values of V, I, R_c, T_m were self-consistent according the predictions of classical theory. A discrepancy between the two values would indicate a breakdown of the theory.

Figure 1.70 shows an example of a comparison between the measured and calculated values of the potential drop V, where the initial average radius of the a-spot was approximately

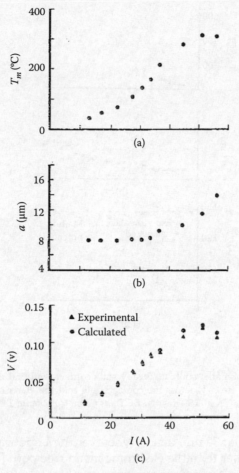

FIGURE 1.70

Dependence on electrical current (large *a*-spot): (a) Steady state contact temperature T_m (b) Mean radius *a* of electrical contact and (c) Potential drop *V* across the contact; note the electrical contact radius *a* is relatively large, varying from approximately 8 to 14 μm. (From RS Timsit, *IEEE Trans Comp Hyb Manuf Tech* CHMT-6: 115, 1983 [140].)

8 μm [140]. Note the agreement between the calculated and measured values of *V*, for values of the current *I* as large as about 35 A. It is important to point out that, as the current was increased beyond about 35 A, the temperature of the contact increased beyond about 200°C, at which juncture, the average radius *a* of the contact increased due to the sintering effect described earlier in Section 1.5.1. In these cases, the measurement of the voltage *V* across the contact was delayed until the contact resistance had stopped decreasing that is, until *a* stopped increasing. The current was, then, slowly reduced to zero to allow a measurement of contact resistance and hence an evaluation of *a* at room temperature, after which the current was increased back to next larger value.

In contrast with the data of Figure 1.70, the corresponding data of Figure 1.71 revealed a larger discrepancy between the calculated and measured *V* values, when the *a*-spot was approximately 7 nm. In Figure 1.71, note also the calculated contact temperature reached the melting point of aluminum, although there was no physical evidence of melting in the electrical interface. The divergence between experiment and theory illustrated in Figure 1.71 was always observed in aluminum contacts wherever the average *a*-spot

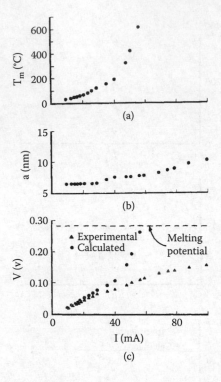

FIGURE 1.71
Dependence on electrical current (small *a*-spot): (a) Steady state contact temperature T_m (b) Mean radius *a* of electrical contact and (c) Potential drop *V* across the contact; note that the electrical contact radius *a* ranges approximately from 7 nm to 10 nm. (From RS Timsit, *IEEE Trans Comp Hyb Manuf Tech* CHMT-6: 115, 1983 [140].)

radius was smaller than about 30 nm, that is, wherever the *a*-spot radius was of the same order of magnitude or smaller than the electronic mean free path [140]. Data of the type illustrated in Figure 1.71 represent clear evidence of a breakdown of classical electrical contact theory for small values of the contact radius *a*.

1.6.2.1.1 Role of the Sharvin Resistance in Small Aluminum–Aluminum Contacts

The breakdown of classical electrical contact theory revealed in the data of Figure 1.71 stemmed from the ballistic motion of electrons in sufficiently small *a*-spots, giving rise to the Sharvin resistance, R_S. The effect of the Sharvin resistance on the voltage–temperature relation is best investigated by examining the magnitude of the two components in the contact resistance expression Equation 1.57 for aluminum–aluminum contacts. The data of Figure 1.71 indicate that the initial value of *a* in the contact was approximately 7 nm, which corresponds to a resistance R_B of nearly 2 Ω. If the *a*-spot radius was such that K= $l_{fp}/a \approx 1$, then $\Gamma(K) \approx 0.7$ and Equation 1.57 becomes

$$2 = \frac{\rho}{2a}0.7 + \frac{1.70 \times 10^{-16}}{a^2} \tag{1.59}$$

If the resistivity of aluminum at room temperature is taken as 2.74×10^{-8} Ωm [140], the *a*-spot radius is calculated from Equation 1.59 as 11.9 nm. This is larger than the value of 7 nm evaluated on the basis of classical contact theory as reported in [140]. From Equation 1.59, the Sharvin resistance R_2 is immediately evaluated as 1.20 Ω. A current of 0.15 A would be required to generate a voltage drop of 0.3 V across the 2-Ω contact. Although this would be equal to the classical "melting voltage" of aluminum [3,141], melting would not occur since the voltage drop across the "ohmic" component of resistance $R_1 = 0.80\ \Omega$ is only 0.12 V, according to the picture of Figure 1.69. In practice, the voltage drop across R_1 would be somewhat larger than 0.12 V since the resistivity would increase due to Joule heating. However, this argument shows melting would not be observed even though the conventional "melting voltage" is reached across the constriction. Note the calculated data of Figure 1.71 obtained for the 2-Ω contact *on the basis of classical contact theory*, predicted the attainment of the "melting voltage" at a current of 57 mA. This lower value of the current for generating a voltage drop of 0.3 V stems from the effect of contact temperature on the resistivity ρ in a classical contact. The above argument suggests strongly that the disagreement between the experimental and calculated data of Figure 1.71 that is, the apparent breakdown of classical contact theory, stemmed from the absence of consideration of ballistic electronic conduction in classical electrical contact theory.

1.6.3 Observations of Breakdown of Classical Electrical Contact Theory in Gold Contacts

Recently, softening of gold-gold contacts was reported under a contact force near 30 mN where the dimensions of contact asperity were comparable to the electron mean free path l_{fp} at room temperature [193]. It was found that the softening voltage under these conditions was appreciably larger than the value of 80 mV listed Table 1.9, as shown in Figure 1.72. These observations are similar to those associated with the data shown in Figure 1.71 for aluminum contacts. No melting of contact spots was observed even at a voltage drop of 450 mV across the contact, the melting voltage of gold according to classical contact theory ([3] and Table 1.9). The results were explained using a model accounting for ballistic electron transport in the contact.

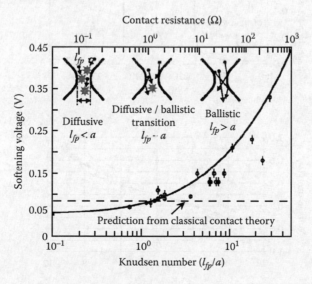

FIGURE 1.72
Measured values of the softening voltage versus the Knudsen Number l_{fp}/a in small gold–gold contacts [193]. As the Knudsen Number increases, electrons pass through the contact spots in a mode ranging from diffusive (classical) to ballistic. The measured softening voltage is compared to the predictions of both a proposed model (solid line) (see [193]) and the classical model (dashed line and Table 1.9; the classical softening voltage was identified as ~ 50 mV in [193] rather than as 80 mV). (From BD Jensen et al., *Appl Phys Lett* 86: 023507, 2005 [193].)

1.6.4 Observations of Breakdown of Classical Electrical Contact Theory in Tin Contacts

Striking evidence of the breakdown of classical theory was recorded by Maul et al. [194–196] from contact intermittencies at a contact load of 0.5 N that is, the observation of large spikes in contact resistance between tin-plated surfaces sliding in reciprocating motion. Contact intermittencies were observed after a number of cycles, thus after the probable growth of relatively thick oxide films on the surfaces due to fretting. Figure 1.73a shows an example of such intermittencies [194] as observed by passage of a current of 54 mA. Note the large voltage spikes of about 0.8 V across the sliding contact, corresponding to an electrical resistance of about 15 Ω. This voltage drop exceeds by far the melting voltage of tin (0.13 V [3] and Table 1.9). The authors reported no physical evidence of contact melting at the locations where this large voltage was detected. In this work, note that the time required to achieve thermal equilibrium in an *a*-spot (for *a* as large as 20 μm, see Section 1.4) was on the order of 1 μs. Since the sliding velocity was only 100 μm s⁻¹, the residence time over a single *a*-spot was considerably longer than 1 μs and thus more than sufficient to cause melting. Thus, sliding did not mitigate melting for the data of Figure 1.73a. Note also that the voltage often could not be sustained above 0.13 V, thus suggesting that melting had occurred over segments of the sliding path.

It is important to note in Figure 1.73a the resistance of ~ 15 Ω associated with the voltage spikes of 0.8 V could not correspond to conduction through thin tin oxide films. An estimate for the contact resistance R_{oxide} between tin oxide films may easily be made from Equation 1.11 [3]

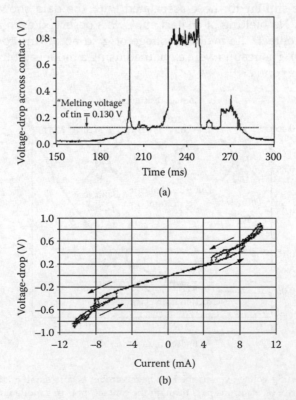

(a)

(b)

FIGURE 1.73
(a) Intermittencies in a sliding tin–tin contact (b) Voltage–current characteristic at a location of an intermittency in [194]. (From C Maul and JW McBride, *Proc 48th IEEE Holm Conf on Elect Cont*, 165, 2002 [194].)

$$R_{oxide} = \frac{\rho_{oxide}}{2} \sqrt{\frac{\pi H_{oxide}}{F}}$$

where $\rho_{oxide} \sim 1$ Ωm and $H_{oxide} \sim 9{,}800$ MPa are respectively the resistivity and microhardness of tin oxide [103]. For the contact load F of 0.5 N used by Maul et al. [195], R_{oxide} is evaluated as 1.2×10^5 Ω. Clearly, the voltage spikes could not represent conduction across a tin oxide contact.

Figure 1.73b shows the apparent reproducibility of contact resistance at one location where an intermittency had been detected and sliding had been stopped, in [195]. The direction of electric current across the contact was, then, reversed a few times. The contact resistance was about 43 Ω. Although the data of Figure 1.73b indicate that the "melting voltage" of tin was exceeded, there was no evidence of melting behavior. Melting would have caused irreversible metal flow at a-spots, and would have precluded observations of a stable and reversible voltage–current characteristic. We now interpret the data of Figure 1.73 in terms of the role of the Sharvin resistance.

First we evaluate the Sharvin and "ohmic" components of the resistance. The electronic mean free path for tin at room temperature can be calculated from the resistivity of tin and other physical parameters [188,190] and is evaluated as 4.1 nm. If it is assumed that K $\sim$ 2, then $\Gamma(K)$ $\sim$ 0.7 so that for the contact resistance of 43 Ω reported in Figure 1.73b, Equation 1.56 reduces to

$$43 = 0.7 \frac{\rho}{2u} + \frac{1.93 \times 10^{-16}}{a^2} \qquad (1.60)$$

For tin $\rho = 11.0 \times 10^{-8}$ Ωm [188,190]; from which the a-spot radius is immediately evaluated as $a = 2.61$ nm from a solution of Equation 1.60. Referring to Figure 1.69, the "ohmic" resistance and the Sharvin resistance are evaluated respectively as $R_1 = 14.7$ Ω and $R_2 = 28.3$ Ω. In this model, the voltage drop across the temperature-independent Sharvin resistance at the current of 0.008 A in Figure 1.73b is $28.3 \times 0.008 = 0.226$ V. If the effect of temperature on resistivity is neglected, the residual voltage drop across the ohmic resistance is $14.7 \times 0.008 = 0.118$ V, which is slightly below the "melting voltage" of 0.130 V for tin [3,191] and Table 1.9). Thus, no melting should have occurred at the contact spots, and none was observed [194–196]. However, the non-linear segment at the ends of the voltage–current curve in Figure 1.73b suggests that the a-spot was getting hot at a current of 0.008 A. This is consistent with the calculated value of the voltage drop across the "ohmic" component of 0.118 V. The proposed model of contact resistance, based on the increasing importance of the Sharvin resistance in small contacts, thus, again provides a rational explanation for the experimental data of Figure 1.73.

1.7 Constriction Resistance at High Frequencies

1.7.1 Skin Depth and Constriction Resistance

Because Equation 1.3 is limited to problems in DC conduction, its validity under conditions of alternating current (AC) is unclear. Holm [3] explored the effect of signal frequency on constriction resistance in a contact geometry similar to that described in [6], but his analysis was only approximate and the results can only be taken as tentative. The major difference between the DC and AC constriction resistance problems stems from the *skin effect* that occurs under AC conditions. This effect limits the penetration of the electromagnetic field to a depth of a few times the *electromagnetic penetration depth* δ within a conductor. This penetration depth δ is defined as [4]

$$\delta = \sqrt{\frac{\rho}{\pi f \mu_r \mu_0}} \tag{1.61}$$

where *f* is the signal frequency, μ_0 is the magnetic permeability of free space ($4\pi \times 10^{-7}$ H/m) and μ_r is the relative permeability of the conductor. In the DC case, constriction resistance is defined as the increase in resistance due to the replacement of a perfectly conducting interface by a constriction as described in Section 1.2. For example, the constriction resistance in a constricted cylinder is defined as the resistance of the constricted cylinder less the resistance of the identical but unconstricted cylinder [6]. Clearly, this definition of constriction resistance is not applicable to conditions of AC excitation, particularly where the frequency is large, since the electric current is concentrated near the conductor surfaces and may also be concentrated near the constriction edges. Extension of the DC definition to the AC case would lead to a situation where the resistance of a given constriction would vary with the specific geometry of the contact interface in which it is located.

With high-frequency AC, the current streamlines are tangential to insulating surfaces, except at a constriction, where the streamlines bend such that the current flow is normal to the contact area, as illustrated in Figure 1.74 [197]. For consistency with the DC case, the definition of constriction resistance at high frequencies is taken as the change in resistance due to the bending of the current streamlines at a circular contact. This resistance will be a function not only of the constriction diameter, but also of the AC excitation frequency and the electrical properties of the conductor. The total *connection resistance* of an electrical interface will then consist of the sum of (1) The constriction resistance and (2) An additional component stemming from current flow outside the constriction (i.e., from the defined boundary of the external connection). This additional component will obviously be affected by details of the contact geometry. The regions from which these two components of resistance arise are shown as A and B in Figure 1.74.

It is important to recall that the impedance of an electrical interface at high signal frequencies is determined not only by *connection resistance*, but also by interfacing capacitance and inductance. These latter effects depend on details of the surfaces in contact, such

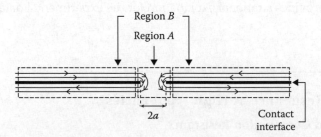

FIGURE 1.74

Schematic diagram of current flow through a circular constriction of diameter 2*a* at an electrical interface, at high frequencies; flow of electric current occurs within a surface layer determined by the electromagnetic penetration depth, δ. Only the electrical resistance of region *A* contributes to *constriction resistance*. The magnitude of the electrical resistance of region *B*, lying outside of the constriction, depends on the interface dimensions and may contribute significantly to the total *connection resistance*. (From JD Lavers and RS Timsit, *IEEE Trans Comp Packag Tech* 25: 446, 2002 [197].)

as mechanical contact load and surface roughness, and have been outlined previously [198,199]. They are not addressed in the following description.

Table 1.13 lists values of penetration within a non-magnetic material of resistivity $3 \times 10^{-8}\,\Omega$m (e.g. aluminum) for a wide range of frequencies. Typically, the net resistance of a given conductor deviates from the DC value when the penetration depth becomes smaller than the characteristic dimensions of that conductor. For example, consider a constriction with a radius of 100 µm. On the basis of the data shown in Table 1.13, the skin effect will significantly affect the resistance of such a constriction when the AC frequency is higher than 100 kHz. On the other hand, constrictions with a radius of a few to several micrometers will be affected by the penetration depth at frequencies larger than about 10 MHz. Under conditions where the penetration depth is much larger than the constriction radius, the constriction resistance will be described to a good approximation by the classical expression of Equation 1.3.

1.7.2 Evaluation of Constriction Resistance at High Frequencies

The AC constriction resistance has been evaluated numerically for the case of a constricted cylinder of material with resistivity $3 \times 10^{-8}\,\Omega$m using the Finite Element Method (FEM), for constriction radii extending to tens of micrometers and excitation frequencies extending from DC to the GHz range [197]. As anticipated, the skin effect at high frequencies causes the current to be concentrated near conductor surfaces and at insulating boundaries. This concentration is illustrated in Figure 1.75 for the case of a constriction with $a/\delta \sim 6$. Where the skin effect dominates, the entire current is confined to a layer with thickness not more than 5δ. Thus, at high frequencies, current flow to the constriction occurs along the surface defined

TABLE 1.13

Variation of Skin Depth with Frequency for a Metal of Resistivity $3 \times 10^{-8}\,\Omega$m

Frequency (Hz)	Skin Depth δ(µm)
60	11254
10^3	2757
10^4	872
10^5	276
10^6	87
10^7	28
10^8	8.7
10^9	2.8

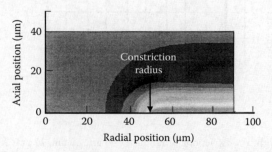

FIGURE 1.75
Illustration of current streamline distribution at 100 MHz in the vicinity of a constriction with radius 50 µm. (From JD Lavers and RS Timsit, *IEEE Trans Comp Packag Tech* 25: 446, 2002 [197].)

by the insulating layer in the electrical interface. The current streamlines remain parallel to the insulating layer except to within a distance of 2δ–3δ of the spot edge of the constriction.

Figure 1.76 shows the calculated current density within a circular constriction of radius *a* ranging from 5 μm to 50 μm for a current *I* at a frequency of 10 MHz, 100 MHz and 1 GHz. The current density was normalized to J_0 where

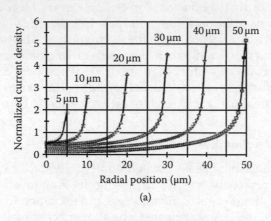

(a)

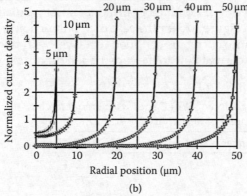

(b)

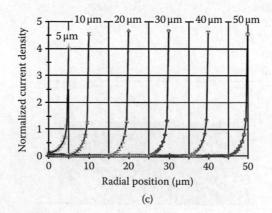

(c)

FIGURE 1.76

Current density distribution within constrictions of radius *a* of 5 μm, 10 μm, 20 μm, 30 μm, 40 μm and 50 μm at an excitation frequency of: (a) 10 MHz (b) 100 MHz (c) 1 GHz; the current density if normalized to $I/(2\pi a\delta)$ where *I* is the current and δ is the skin depth. (From JD Lavers and RS Timsit, *IEEE Trans Comp Packag Tech* 25: 446, 2002 [197].)

$$J_0 = \frac{1}{2\pi a\delta}$$

and δ is given by Equation 1.61. The current concentration at the constriction edge becomes increasingly narrow as the AC frequency increases. The constriction resistance was computed for a constriction radius of 5, 10, 20, 30, 40 and 50 μm, and for frequencies of 10, 100, 1000 MHz. The values of constriction resistance based on this evaluation are shown graphically in Figure 1.77 for the above noted set of parameters. Several features are worth noting. First, the values of constriction resistance converge as the contact radius becomes large. Second, for a given frequency, there is a critical radius a_c above which frequency ceases to be a dominating factor in determining the magnitude of the constriction resistance. Third, Figure 1.77 suggests that the critical radius can be linked to the penetration depth δ as follows

$$a_C / \delta \approx 8$$

The data of Figure 1.77 indicate clearly that for a selected value of constriction radius, the constriction resistance tends to decrease as the excitation frequency increases. At first glance, this appears to be counterintuitive. However, it must be recalled the constriction resistance describes resistance effects related to the bending of current streamlines in the vicinity of the contact spot. At very high frequencies (constriction radius $\gg$ skin depth δ), the skin effect dominates and the current flows through to a layer of width $\sim \delta$ along the perimeter of the contact spot. It is a mistake to associate this skin depth layer only with electrical resistance since, in the case of a high-frequency contact, the current streamlines must make a transition from a direction tangential to the insulating layer to one normal to the contact area. This transition, characterized by the bending of the streamlines, has an important effect on the net resistance. An analogous situation arises when a DC current flows around a bend as described earlier in Section 1.3.1. With this concept in mind, it becomes clear that at very high frequencies, the streamline pattern, and thus the constriction resistance, becomes relatively independent of frequency; that is, the pattern is set primarily by the radius of the a-spot. Thus, frequency will not have a major impact on constriction resistance at excitation frequencies much larger than 1 GHz. On the other hand, as frequency decreases, a point is reached where the current streamlines occupy more and more of the contact spot. This effect on the streamline pattern appears as an increase in

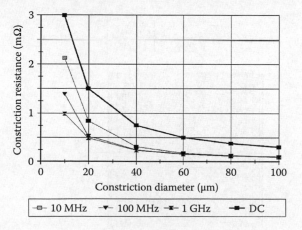

FIGURE 1.77
Dependence of constriction resistance on constriction radius for excitation frequencies ranging from DC to 1 GHz. (From JD Lavers and RS Timsit, *IEEE Trans Comp Packag Tech* 25: 446, 2002 [197].)

constriction resistance, since the bending of current streamlines occurs over a larger conductor volume. In the lower frequency limit, where the contact radius is much lesser than the skin depth δ, the contact spot tends to throttle current. This, of course, is a case where the constriction resistance is high.

The DC values of constriction resistance calculated numerically are in exact agreement with the values predicted on the basis of the classical formula $\rho/2a$ (Equation 1.3). The data in Figure 1.77 also indicate that the DC values of constriction provide an upper bound as the frequency decreases to zero. As mentioned earlier, the convergence of results from the DC and AC models takes place at frequencies where the *skin effect* at the insulating layer no longer is a dominating factor.

1.7.3 Constriction versus Connection Resistance at High Frequencies

The *connection resistance* of an electrical interface consists of the sum of the resistance of the contact region and the bulk resistance of conductors in the vicinity of the contact that is, at high frequencies, this corresponds to the sum of the resistances of region A and region B in Figure 1.74. It is the parameter usually measured in characterizing the contact properties of electrical junctions. For DC and low-frequency applications, the constriction resistance component of *connection resistance* is generally much larger than the bulk resistance so the latter component may be neglected. At low frequencies, *connection resistance* is thus taken as $\rho/2a$, and is independent of contact geometry. In contrast, *connection resistance* at high frequencies *does* depend on geometrical details of the electrical connection, as illustrated below.

Consider the *connection resistance*, R_T at high frequencies of an electrical joint generated by two cylinders of radius $R_{cylinder}$, end-butted over a constriction of radius, a, as illustrated in Figure 1.78. The *connection resistance* measured between location A1 and A2 in Figure 1.78 is

$$R_T = R_C + 2R_{bulk} \tag{1.62}$$

where R_C is the resistance of the constriction of radius a evaluated numerically and R_{bulk} is the resistance of the bulk material situated between the constriction and each resistance measurement location. R_{bulk} is the resistance of a ring of thickness given by the penetration depth δ where most of the current flows between the constriction and the cylinder edge. The ring extends from the constriction edge to the edge of the contacting cylinders and its resistance is given as

$$R_{bulk} = \frac{\rho}{2\pi\delta} \ln\left\{\frac{R_{cylinder}}{a}\right\} \tag{1.63}$$

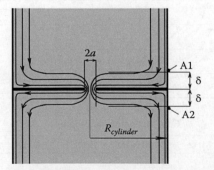

FIGURE 1.78
Two end-butted cylinders passing a high-frequency current.

TABLE 1.14

Dependence of Constriction Resistance and Connection
Resistance on Signal Frequency

Signal Frequency f (Hz)	Constriction Resistance RC (mΩ)	Connection Resistance RT (Equation 1.62) (mΩ)
10^7	2.2	3.2
10^8	1.4	4.7
10^9	1.0	11.2

Equations 1.62 and 1.63 indicate the *connection resistance* of end-butted cylinders at high frequencies increases with the dimension $R_{cylinder}$ of the contact region, for a fixed constriction radius a. Taken with Equations 1.62 and 1.63, the current streamlines illustrated in Figure 1.78 further indicate that *connection resistance* increases with increasing frequency although the constriction resistance R_C decreases as the signal frequency increases. These observations are illustrated in Table 1.14 which lists calculated values of R_C and R_T for increasing frequencies using parameter values $\rho = 3 \times 10^{-8}$ Ωm, $a = 5$ μm, and $R_{cylinder} = 100$ μm [197].

In Table 1.14, constriction resistance drops by a factor of approximately 2 but the *connection resistance* R_T increases by a factor of 3.5 as the frequency increases from 10^7 to 10^9 Hz. As indicated at the beginning of this section, the *connection resistance* data in Table 1.14 contrast strongly with *connection resistance* at DC and low frequencies where connection resistance is given by Equation 1.3, independently of the geometry and dimensions of the contact region. The difference in *connection resistance* properties at high frequencies stems from the skin effect wherein on the one hand less metal is used for electrical conduction, and on the other hand, the bending of current streamlines occurs over a relatively small volume around the constriction, as signal frequency increases.

1.8 Summary

This chapter has reviewed the origin of electrical contact resistance at solid–solid interfaces and addressed the major factors that affect the fundamental properties of electrical contacts. These factors include surface hardness, the presence of surface contaminants, contact force, and so on, because they affect the dimensions and distribution of *a*-spots in the electrical interface. Optimization of contact resistance is, thus, generally achieved by optimizing or eliminating selected physical factors through control of surface finishing, the use of relatively soft platings on the contacting surfaces, surface wipe, and so on.

To a good approximation, the true contact area is independent of the nominal contact area in metallic electrical contacts. Also to a good approximation, the true contact area is independent of details of the surface microtopography of the contacting bodies. These observations account for the measured dependence of DC contact resistance on the inverse square root of the contact force. Because electric current is severely constricted in an *a*-spot, the temperature in an *a*-spot of the operating contact may be considerably higher than the that of the bodies of the contacting components. The *universal* parameter determining *a*-spot temperature is the *voltage drop* across the electrical interface, *not* the magnitude of the electric current. Hence, quantities such as *softening voltage* or *melting voltage* can be defined for any electrical contact. However, classical electrical contact theory, and hence the *voltage–temperature* relation, breaks down when the average *a*-spot diameter is of the same order of magnitude or smaller than the mean free path of conduction of electrons in the contacting bodies. Under these conditions, electrons behave ballistically in *a*-spots and *a*-spot temperature cannot be calculated readily from the voltage drop across the contact.

Operation of a bimetallic electrical contact at consistently elevated *a*-spot temperatures may generate high-resistivity intermetallic compounds in the interface. Formation of these compounds is generally detrimental to contact stability and may eventually lead to contact failure.

Capillarity and thermally induced material flow within a small *a*-spot can have significant effects on the constriction resistance of an electrical interface. These effects are generally responsible for so-called "self-healing" in electrical contacts, that is, when the area of metal-to-metal contact increases due to Joule heating, thus decreasing contact resistance. These effects generally occur when the *a*-spot temperature reaches the softening temperature. They have been observed at room temperature in sufficiently small *a*-spots where sintering occurs.

At sufficiently high signal frequencies, constriction resistance decreases with increasing excitation frequency. This effect may be understood as follows: under DC or low-frequency conditions, the presence of a constriction at an electrical interface causes spreading of current streamlines in the immediate vicinity of the constriction. Spreading effectively increases the length of current pathways in the conductors and thus increases the resistance to electrical flow. This increase is effectively responsible for constriction resistance. At sufficiently high frequencies, current is confined near the constriction edge and near the surfaces in contact due to the skin effect. This confinement diminishes the spreading of current streamlines within the conductors and shortens the electric current pathways near the constriction. This leads to an effective decrease in constriction resistance.

Acknowledgments

The author thanks Drs Irvin H. Brockman and Robert S. Mroczkowski of AMP Incorporated (now TE Connectivity Ltd), for having supplied much of the contact resistance data associated with electroplated layers and for helpful discussions during the preparation of the first version this chapter published in this book in 1999. Much of this information has again been used in this updated version of the chapter. The author is also grateful to Dr J.B.P. Williamson for discussing details of the work described in reference 56.

Appendix 1.A Electrical Resistivity and Thermal Conductivity of Metals and the Wiedemann–Franz Law

The simple derivation of the Wiedemann–Franz law presented in this appendix requires a brief review of quantum mechanical properties of electrons in solids.

Electrons in a metal occupy a large but finite number of quantum energy states. The number of these states, ranging in energy from zero to E, is given as [188]

$$F(E) = \frac{8\pi V \{2mE\}^{3/2}}{3h^3} \tag{1.A.1}$$

where V is the metal specimen volume, m is the mass of the electron and h is Planck's constant. At a temperature of 0 K, the first electron will be located in the lowest energy state and subsequent electrons will occupy higher energy states. If there are N electrons in the metal sample, the number of states described by $F(E)$ becomes identical with N and the maximum electron energy E_{max} calculated from Equation 1.A.1 is given as

$$E_{max} = \left\{ \frac{3h^3}{8\pi} \frac{N}{V} \right\}^{2/3} \{2m\}^{-1} \tag{1.A.2}$$

For metals such as copper and aluminum, the electrons associated with each atom are arranged in closed shells with the outer shell carrying respectively one electron and three electrons. In the solid state, it is these outer shell electrons that detach themselves from the atoms and act as the "free" electrons responsible for the electrical conductivity of the metals. Since one mole of any element contains 6.02×10^{23} atoms (Avogadro's number) and the densities of copper and aluminum are respectively 8.9 g cm^{-3} and 2.7 g cm^{-3}, the density of free electrons in copper and aluminum is respectively 8.4×10^{28} and 5.0×10^{28} electrons per cubic meter. Substituting these values for the density N/V in Equation 1.A.2 yields a maximum electronic energy of 1.12×10^{-18}J (~7 eV) and 8.2×10^{-19} J (~ 5 eV) respectively for copper and aluminum. These energies are considerably larger than the energy $3kT/2$, where k is Boltzmann's constant, associated with the motion of classical particles at room temperature, since this energy is only ~ 6.2×10^{-21} J (~ 0.04 eV). This is the reason why classical physics cannot be used to treat the electrical conductivity of metals. If the electronic energy E_{max} is equated to the kinetic energy $mv^2/2$, the velocity of electrons in metals is ~ 10^6 ms^{-1} and is, thus, much larger than the velocity of ~ 10^5 ms^{-1} for classical particles at room temperature.

From Equation 1.A.1, the density of states $f(E)$ per unit energy interval obtained by differentiating with respect to E is given as

$$f(E) = \frac{dF(E)}{dE} = \frac{4\pi V \{2m\}^{3/2}}{h^3} E^{1/2} \tag{1.A.3}$$

and the distribution has a parabolic dependence on E. At 0 K, this distribution has a sharp cut-off at an energy value defined by E_{max} (~ 5–7 eV) as illustrated in Figure 1.A.1a, since no free electron can have an energy larger than this. Now, the probability that an electron has an energy E at a temperature T is given by the Fermi–Dirac distribution function $f_{FD}(E)$ defined as [188]

$$f_{FD}(E) = \frac{1}{\exp\left\{ \dfrac{E - E_F}{kT} \right\} + 1} \tag{1.A.4}$$

where E_F is defined as the Fermi energy. Inspection of expression in Equation 1.A.4 indicates that $f_{FD}(E)$ is unity at 0 K for values of E up to E_F, and is equal to zero for $E > E_F$ as shown in Figure 1.A.1b. Since the maximum electron energy at 0K is E_{max}, it follows that $E_F = E_{max}$ at 0K. Generally, the Fermi energy varies with temperature and its value is calculated by equating the total number of occupied states to the actual number of free electrons [188]. In metals, it is usually sufficient to ignore this dependence on temperature and assume that

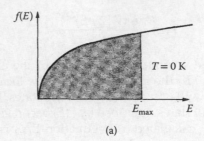

(a)

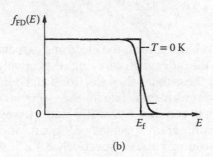

(b)

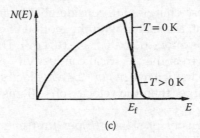

(c)

FIGURE 1.A.1
Dependence of the density of states $f(E)$ on energy for the free electron model: (a) At OK, all the states with energy up to E_{max} are occupied. (b) The Fermi–Dirac function $f_{FD}(E)$ (c) Density of occupied electronic states $N(E) = f(E)f_{FD}(E)$.

E_F is always equal to is E_{max}. The density of states accessible to electrons may now be calculated. From Equations 1.A.3 and 1.A.4, the number of states lying between energies E and $E + dE$ and occupied by electrons at any temperature is

$$N(E)dE = f(E)f_{FD}(E)\acute{a} \tag{1.A.5}$$

so that the density of occupied electronic states $N(E)$ given as $f(E)f_{FD}(E)$ varies with E as shown in Figure 1.A.1c. Note that at 0 K, the function $N(E)$ has a parabolic form with a sharp cut-off at E_F. As the temperature is raised, the energy distribution becomes slightly rounded at the cut-off and acquires a thin tail extending to higher energies. The physical significance of this change in energy distribution is that only electrons with energies near $E_F = E_{max}$ can change their state by absorption of thermal energy as this energy of order kT and is thus too small to excite electrons with energies significantly lower than E_F. The electronic specific heat may, then, be calculated by assuming that only those electrons within the thermal energy $\sim kT$ of E_F can absorb energy as the temperature is raised. The actual number of electrons involved will then be the density of states $N(E_F)$ at E_F multiplied by kT that is, $N(E_F)kT$, and the total energy absorbed will thus be $N(E_F)(kT)^2$. The electronic specific heat c_e, is then, obtained by differentiating this absorbed energy with respect to T, or

$$c_e \sim 2N(E_F)k^2T$$

The more accurate expression obtained by differentiating the integrated Expression 1.A.5 with respect to T yields [188]

$$c_e = \frac{\pi n k^2}{2E_F}T \qquad (1.A.6)$$

Now, the classical kinetic theory of gases relates the thermal conductivity λ and the specific heat per unit volume c_v as [200]

$$\lambda - \frac{c_v l_{fp} v}{3}$$

where v is the average molecular speed and l_{fp} is the mean free path between molecular collisions. In analogy, and using Equation 1.A.6, the specific heat of an electron gas in a metal may then be written as

$$\lambda = \frac{\pi^2 n k^2 v_F l_{fp}}{6E_F}T \qquad (1.A.7)$$

The electrical resistivity ρ may be related to fundamental electronic properties also by evoking the arguments in classical physics. Under the action of an electric field E_{el}, the electric current density j generated is given as

$$j = ne v_d \qquad (1.A.8)$$

where n is the electron density defined earlier as N/V, e is the electronic charge and v_d is the electron drift velocity. Now, the electric force acting on each electron is eE_{el} and this is approximately equal to mv_d/τ where τ is the mean time between collisions with atoms and m is the electronic mass, so that

$$v_d = \frac{eE_{el}\tau}{m} \qquad (1.A.9)$$

Since τ^{-1} is given as v_F/l_{fp} and v_F is defined as $[2E_F/m]^{1/2}$, Equations 1.A.8 and 1.A.9 immediately give

$$j = E_{el}\frac{ne^2 l_{fp}}{mv_F}$$

or, since j is also given as E_{el}/ρ where ρ is the electrical resistivity,

$$\rho = \frac{mv_F}{ne^2 l_{fp}} \tag{1.A.10}$$

Combining Equations 1.A.7 and 1.A. 10 and using the fact that $mv_F^2 \sim 2E_F$ yields

$$\frac{\lambda\rho}{T} = \frac{\pi^2 k^2}{3e^2} = 2.45 \times 10^{-8} \left\{ VK^{-1} \right\}^2 = L \text{ the Lorenz constant} \tag{1.A.11}$$

Equation 1.A.11 is a statement of the Wiedemann–Franz law.

The Wiedemann–Franz law was derived on the assumption that the electrical and thermal conductivities are determined by the same mean free path l_{fp} for collisions of the conduction electrons with the solid matrix. In practice, the thermal energy lost by electron collisions with the lattice is conducted by vibrating atoms in the solid. Hence, these collisions are somewhat more disruptive to thermal transfer by electrons than to electrical conduction. The effective mean free paths associated with electrical and thermal transport are, thus, not rigorously identical, with the consequence that the Wiedemann–Franz law is often not strictly valid. Deviations are most evident in the low-temperature regime (< 200 K) where scattering of conduction electrons by impurities becomes important [188,200]. In the room temperature regime and at more elevated temperatures, the Wiedemann–Franz law holds reasonably well for a wide variety of metals [188,200].

Appendix 1.B Location of the Plane of Maximum Temperature in a Bimetallic Contact

Consider the circular spot in Figure 1.B.1 involving materials 1 and 2 in which the electrical resistivities are such that $\rho_1 < \rho_2$. The maximum temperature T_m is assumed to occur in a surface located in material 2 at a distance Δz from the physical interface. The physical interface is assumed to be isothermal at a temperature T_j. If the a-spot is circular and the electric current passed through the contact is I, the electric power ΔP generated within the element of thickness Δz is given approximately as

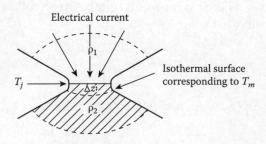

FIGURE 1.B.1
Schematic representation of isothermal surfaces in a joule-heated bimetallic interface in which the resistivities are such that $\rho_2 > \rho_1$.

$$\Delta P \approx \rho_{2,m} \frac{\Delta z}{\pi a^2} I^2$$

where $\rho_{2,m}$ is the electrical resistivity of material 2 at the maximum temperature T_m. If the thermal conductivity of material 2 at the maximum temperature is $\lambda_{,2,m}$, the rate of heat loss ΔW to material 1 from this disk is approximately

$$\Delta W \approx \lambda_{2,m} \left\{ T_m - T_j \right\} \frac{\pi a^2}{\Delta z}$$

At thermal equilibrium, $\Delta P = \Delta W$, yielding

$$T_m - T_j = \frac{\rho_{2,m}}{\lambda_{2,m}} \left\{ \frac{\Delta z}{\pi a^2} \right\}^2 I^2 \qquad (1.B.1)$$

Now from Equation 1.34, the potential drop ΔV *between the physical interface and the plane of maximum temperature* is related to the temperature difference $(T_m - T_j)$ as

$$T_m - T_j = \frac{\Delta V^2}{2\rho_{2,m}\lambda_{2,m}} \qquad (1.B.2)$$

Also, the current I is related to the total voltage drop V across the contact by the approximate relation

$$I = \frac{V}{\left\{ \dfrac{\rho_{2,m} + \rho_{1,m}}{4a} \right\}} \qquad (1.B.3)$$

Substituting Equations 1.B.2 and 1.B.3 into Equation 1.B.1 yields

$$\frac{\Delta z}{a} = \frac{\rho_{2,m} + \rho_{1,m}}{2\rho_{2,m}} \frac{\Delta V}{V} \frac{\pi}{2\sqrt{2}}$$

For the sake of illustration, if $\rho_{2,m} = 3\rho_{1,m}$ and $\Delta V/V$ takes on the large value of 0.1 ($\Delta V/V$ would be expected to be considerably smaller than this), $\Delta z/a$ would be of the order of 0.07. Thus for an a-spot of radius 10 μm, Δz would be displaced only approximately 700 nm from the physical interface, into the material of larger resistivity. If the highly conductive material is aluminum with $\rho_{l,m} = 5.5 \times 10^{-8}$ Ωm at a maximum contact temperature of approximately 400°C ($\rho_{2,m} = 1.65 \times 10^{-7}$ Ωm), using $\lambda_{2,m} = 200$ Wm^{-1}K^{-1} and $V \sim 0.2$V (or $\Delta V = 0.02$V), Equation 1.B.1 yields a difference between the temperature of the physical interface and the maximum contact temperature of only about 6°C! For all practical purposes, the maximum temperature may be considered to occur at the physical interface. However, the detailed temperature distribution within the contact region is not identical to that described by Equation 1.45 for the monometallic contact, and must be worked out on the basis of Equations 1.33 and 1.41.

Appendix 1.C Location of the Plane of Maximum Temperature in an Electrical Contact Consisting of Mating Components of Greatly Differing Electrical Resistivities

The surface at which the maximum temperature is reached in a contact consisting of mating components of vastly differing electrical resistivities would be expected to be located in the material with higher resistivity. The location of this surface may be estimated for the simple case where the contact is two-dimensional, as illustrated in Figure 1.C.1.

The contact is assumed to consist of two two-dimensional pyramids in contact over an area $2al$, where $2a$ is the width of the contact in the y-direction and l is the junction thickness. The pyramids meet at $x = 0$ and extend to the left and right of the y-axis by a distance L. The pyramid edges diverge from parallelism by an angle φ and are assumed to be thermally insulated. For the sake of simplicity, the resistivities ρ_1, ρ_2 and the thermal conductivities λ_1, λ_2 of the two materials are assumed to be independent of temperature. The equations describing the steady-state heat generation are given as

$$\lambda_1 \frac{d^2T(x)}{dx^2} = -\rho_1 j(x)^2 \tag{1.C.1}$$

and

$$\lambda_2 \frac{d^2T(x)}{dx^2} = -\rho_2 j(x)^2 \tag{1.C.2}$$

Since the cross-sectional area of the pyramid is $2al(1 + [x/a]\tan \varphi)$, where $x > 0$ is the distance from the origin, the current density function may be approximated as

$$j(x) = \frac{j_0}{1 + \left\{\dfrac{x}{a}\right\}\tan \varphi}, x > 0 \qquad j(x) = \frac{j_0}{1 - \left\{\dfrac{x}{a}\right\}\tan \varphi}, x < 0$$

where j_0 is the current density at $x = 0$. Note that this current density distribution is only approximate since there cannot be an abrupt change in the direction of current flow at $x = 0$. Also, this current distribution cannot be valid as φ approaches $\pi/2$. Substitution of

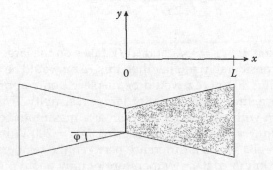

FIGURE 1.C.1
Two-dimensional electrical contact consisting of two two-dimensional pyramids touching over an area $2al$, where $2a$ is the width of the contact in the y-direction and l is the junction thickness. The pyramids meet at $x = 0$ and extend to the left and right of the y-axis by a distance L The pyramid edges diverge from parallelism by an angle φ and are assumed to be thermally insulated.

these current density distributions into Equations 1.C.1 and 1.C.2 immediately yields the following solutions for the temperature distributions:

$$T_1(x) = \frac{\rho_1}{\lambda_1}\left\{\frac{aj_0}{\tan\varphi}\right\}^2 \ln\left\{1 - \frac{x}{a}\tan\varphi\right\} + B_1 x + C_1, x < 0$$

$$T_2(x) = \frac{\rho_2}{\lambda_2}\left\{\frac{aj_0}{\tan\varphi}\right\}^2 \ln\left\{1 + \frac{x}{a}\tan\varphi\right\} + B_2 x + C_2, x > 0$$

where B and C are integration constants determined by the following boundary conditions:

$$\lambda_1 \frac{dT_1\{x\}}{dx} = \lambda_2 \frac{dT\{x\}}{dx} \text{ at } x = 0$$

$$T_1\{0\} = T_2\{0\}$$

$$T_1(-L) = T_2(L) = T_0$$

These boundary conditions immediately yield the following expressions for the constants B_1, B_2, C_1, and C_2:

$$B_1 = \frac{aj_0^2}{\{\lambda_1 + \lambda_2\}\tan\varphi}\left(\rho_1 + \rho_2 - \frac{a\lambda_2}{L\tan\varphi}\left\{\frac{\rho_2}{\lambda_2} - \frac{\rho_1}{\lambda_1}\right\}\ln\left\{1 + \tan\varphi\frac{L}{a}\right\}\right)$$

$$B_2 = -\frac{aj_0^2}{\{\lambda_1 + \lambda_2\}\tan\varphi}\left(\rho_1 + \rho_2 + \frac{a\lambda_1}{L\tan\varphi}\left\{\frac{\rho_2}{\lambda_2} - \frac{\rho_1}{\lambda_1}\right\}\ln\left\{1 + \tan\varphi\frac{L}{a}\right\}\right)$$

$$C_1 = C_2 = T_0 - \frac{\rho_2}{\lambda_2}\left\{\frac{aj_0}{\tan\varphi}\right\}^2 \ln\left\{1 + \tan\varphi\frac{L}{a}\right\} - B_2 L$$

If it is assumed that $\rho_2 \gg \rho_1$, then the isothermal surface associated with the maximum temperature is located in material 2. In this case, the condition $dT_2(x)/dx = 0$ yields the location of the temperature maximum as

$$x_{max} = -\frac{a}{\tan\varphi}\left\{\frac{aj_0^2}{B_2\tan\varphi}\frac{\rho_2}{\lambda_2} + 1\right\} \tag{1.C.3}$$

Thus, the location of the isothermal surface corresponding to the maximum temperature is sensitive to the inclination of the contacting asperities generating the *a*-spot. It may easily be shown that the voltage drop across the contact is given as

$$V = \frac{aj_0}{\tan\varphi}\{\rho_1 + \rho_2\}\ln\left\{1 + \tan\varphi\frac{L}{a}\right\}$$

For a copper–graphite contact where $\lambda_1 \sim 380$ Wm^{-1}°C^{-1} and $\rho_1 = 2.6 \times 10^{-8}$ Ω m, $\lambda_2 \sim 40$ Wm^{-1}°C^{-1} and $\rho_2 = 0.003$ Ωm, the calculated steady-state temperature distribution for the following values of contact parameters $j_0 = 2 \times 10^8$ Am^{-2}, $L = 100$ μm, $a = 10$μm

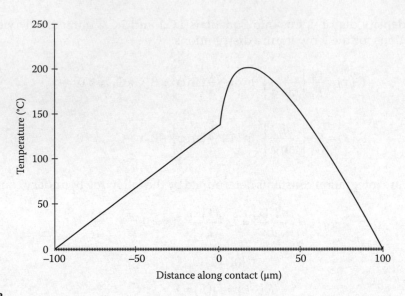

FIGURE 1.C.2

Calculated steady state temperature distribution for a two-dimensional copper–graphite contact of the geometry illustrated in Figure 1.C.1 where $\lambda_1 \sim 380\ \mathrm{Wm^{-1o}C^{-1}}$, $\rho_1 = 2.6 \times 10^{-8}\ \Omega\ \mathrm{m}$, $\lambda_2 \sim 40\ \mathrm{Wm^{-1o}C^{-1}}$ and $\rho_2 = 0.003\ \Omega\mathrm{m}$, for the following values of contact parameters: $j_0 = 2 \times 10^8\ \mathrm{Am^{-2}}$, $L = 100\ \mu\mathrm{m}$, $a = 10\ \mu\mathrm{m}$, and $\varphi = \pi/4$.

TABLE 1.C.1

Dependence of x_{max}, $T_2(0)$, $T_2(x_{max})$, and Voltage-Drop V Across the Copper–Graphite Contact on the Current Density j_0 and the Asperity Angle φ for $L = 100\ \mu\mathrm{m}$ and $a = 10\ \mu\mathrm{m}$

$j_0(\mathrm{Am^{-2}})$	φ (degrees)	V (V)	$T_2(0)$(°C)	$T_2(x_{max})$(°C)	$x_{max}(\mu\mathrm{m})$
10^8	45	0.72	54	90	22
2×10^8	45	1.44	217	360	22
3×10^8	45	2.16	489	811	22
2×10^8	30	1.99	331	623	27
2×10^8	60	1.01	137	202	18

and $\varphi = \pi/4$ is shown in Figure l.C.2. Note the maximum temperature of 360°C occurs at a distance of 22 μm from the copper–graphite interface where the temperature is 217°C. Table 1.C.1 illustrates the dependence of interface temperature, maximum temperature and the location x_{max} from the copper–graphite interface of the isothermal plane at the maximum temperature, on such factors as the current density and the asperity angle φ. Note the maximum temperature within the graphite occurs at a relatively large distance from the copper–graphite interface. Also, in all the cases discussed, the maximum temperature is significantly higher than that of the interface.

REFERENCES

1. TR Thomas, *Rough Surfaces*, New York: Longman, 1982.
2. JA Greenwood and JBP Williamson, Contact of nominally flat surfaces, *Proc Roy Soc* A295: 300, 1966.

3. R Holm, *Electric Contacts, Theory and Applications*, Berlin: Springer–Verlag, 2000.
4. WR Smythe, *Static and Dynamic Electricity*, New York: McGraw-Hill, 1968.
5. RS Timsit, The potential distribution in a constricted cylinder, *J Phys D Appl Phys* 10: 2011,1977.
6. AM Rosenfeld and RS Timsit, The potential distribution in a constricted cylinder: An exact solution, *Quart Appl Math* 39: 405, 1981.
7. RS Timsit, The potential distribution in a constricted cylinder, *Proc 14th Int Conf Elect Cont*, Paris, 21, 1988.
8. H Aichi and N Tahara, Analysis on the constriction resistance of the electric contact by the contact model using the electrolyte bath, *Proc. 20th Int Conf Elect Cont*, Nagoya, Japan, 1, 1994.
9. M Nakamura, Constriction resistance of conducting spots by the boundary element method, *IEEE Trans Comp Hyb Manuf Tech*, CHMT-16: 339, 1993.
10. TR Thomas and SD Probert, Establishment of contact parameters from surface profiles, *J Phys D: Appl Phys* 3: 277, 1970.
11. JA Greenwood, Constriction resistance and the real area of contact, *Brit J Appl Phys* 17: 1621, 1966.
12. M Nakamura and I Minowa, Computer simulation for the conductance of a contact interface, *IEEE Trans Comp Hyb Manuf Tech* CHMT-9: 150, 1986.
13. I Minowa, M Nakamura, and M Kanno, Conductance of a contact interface depending on the location and distribution of conducting spots, *Proc Elect Conf Cont, Electromech Comp Appl*, 19, 1986.
14. JF Archard, Single contacts and multiple encounters, *J Appl Phys* 32: 1420, 1961.
15. FP Bowden and D Tabor, *Friction and Lubrication of Solids*, Vol. II, Oxford: Oxford University Press, 1964.
16. A Fairweather, The closure and partial separation of an electric contact, *IEE (London)* 92: 301, 1945.
17. EL Shobert, Calculation of electrical contacts under ideal conditions, *Proc ASTM* 46: 1126, 1946.
18. P Barkan and EJ Tuohy, A contact resistance theory for rough hemispherical silver contacts in air and in vacuum, *IEEE Trans Power App Sys* 84: 1132, 1965.
19. DV Keller, Electric contact phenomena in ultra clean and specifically contaminated systems, *Proc. Holm Seminar on Electrical Contact Phenomena, (18th Holm Conf Elect Cont)*, Chicago, IL, 1, 1971.
20. RE Cuthrell and DW Tipping, Electric contacts II: Mechanics of closure for gold contacts, *J Appl Phys* 44: 4360, 1973.
21. JB Pethica and D Tabor, Characterized metal microcontacts, *J Adhes* 13: 215, 1982.
22. Y Watanabe, New instrument for measuring contact resistance developed for studying electrical contact phenomena, *Wear* 112: 1, 1986.
23. LP Solos, An approach to predicting wear and friction in contact design, *Elect Cont-1962: Eng Sem Elect Cont*, University of Maine, June 1962.
24. SW Park and SJ Na, Contact pressure and current density distribution in a circular contact surface, *IEEE Trans Comp Hyb Manuf Tech* CHMT-13: 320, 1990.
25. Y Sano, Effect of space angle on constriction resistance and contact resistance for a point contact, *J Appl Phys* 58: 2651, 1985.
26. RS Timsit, Constriction resistance of thin film contacts, *IEEE Trans Comp Pack Tech* 33: 636, 2010.
27. G Norberg, S Dejanovic and H Hesselbom, Contact resistance of thin metal film contacts, *IEEE Trans Elect Dev* 29: 371, 2006.
28. P Zhang and YY Lau, Scaling laws for electrical contact resistance with dissimilar materials, *J Appl Phys* 108: 044914, 2010.
29. P Zhang, YY Lau and RM Gilgenbach, Minimization of thin film contact resistance, *Appl Phys Lett* 97: 204103, 2010.
30. P Zhang, YY Lau and RM Gilgenbach, Thin film contact resistance with dissimilar materials, *J Appl Phys* 109: 124910, 2011.
31. P Zhang, YY Lau, and RS Timsit, On the spreading resistance of thin metal film contacts, *IEEE Trans Comp Pack Tech* 39: 1936, 2012.
32. PM Hall, Resistance calculations for thin film patterns, *Thin Solid Films* 1: 277, 1967.
33. PM Hall, Resistance calculations for thin film rectangles, *Thin Solid Films* 300: 256, 1997.
34. JA Greenwood and JBP Williamson, Electrical conduction in solids II: Theory of temperature-dependent conductors, *Proc Roy Soc* A246: 13, 1958.

35. DK Schroder, Semiconductor Material and Device Characterization, New York: Wiley, 149, 1998.
36. HH Berger, Models for contacts to planar devices, *Solid State Elect* 15: 145, 1972.
37. S Kristiansson, F Ingvarson, and KO Jeppson, Compact spreading resistance model for rectangular contacts on uniform and epitaxial substrates, *IEEE Trans Elect Devices* 54: 2531, 2007.
38. A Nussbaum, Capacitance and spreading resistance of a stripe line, *Solid State Elect* 38: 1253, 1995.
39. MS Leong, SC Choo, and LS Tan, The role of source boundary condition in spreading resistance calculations, *Solid State Elect* 21: 933, 1978.
40. RH Cox, H Strack, Ohmic contacts for Ga–As devices, *Solid State Elect* 10: 1213, 1967.
41. B Gelmont and M Shur, Spreading resistance of a round ohmic contact, *Solid State Elect* 36: 143, 1993.
42. MW Denhoff, An accurate calculation of spreading Resistance, *J Phys D Appl Phys* 39: 1761, 2006.
43. MB Read, JH Lang, AH Slocum, and R Martens, Contact resistance of flat-on-flat and sphere-on-flat thin films, *Proc 56th IEEE Holm Conf on Elect Cont jointly with 25th Int Conf Elect Cont*, Charleston, SC, 348, 2010.
44. MB Read, JH Lang, AH Slocum, and R Martens, Contact resistance in flat thin films, *Proc 55th IEEE Holm Conf on Elect Cont*, Vancouver, BC, Canada, 300, 2009.
45. S Sawada, S Tsukiji, S Shimada, T Tamai, and Y Hattori, Current density analysis of thin film effect in contact area on LED wafer, *Proc 58th IEEE Holm Conf Elect Cont*, Portland, OR, 242, 2012.
46. PF Preston, An industrial atmosphere corrosion test for electrical contacts and connections, *Trans Inst Met Fin* 50: 125, 1972.
47. RG Baker, Studies of static low voltage contacts at the Bell Telephone Laboratories, *Proc 2nd Int Symp Elect Cont Phen*, Graz, Austria, 545, 1964.
48. HW Hermance and TF Egan, Organic deposits on precious metal contacts, *Bell Syst Tech J*, 37: 739, 1958.
49. M. Antler, M. Feder, CF Hornig, and J Bohland, The corrosion behaviour of single and multiphase tin-nickel alloy electrodeposits, *Plating and Surface Finishing*, 63: 30, 1976.
50. TR Long and KF Bradford, Contact resistance behaviour of the 60Pd40Ag alloy in tarnishing environments, *Proc. Holm Sem Elect Cont*, Illinois Institute of Technology, Chicago, IL, 145, 1975.
51. M Antler, MH Drozdowicz, and CF Hornig, The corrosion resistance of worn tin-nickel and gold-coated tin-nickel alloy electrodeposits, *J Electrochem Soc* 124: 1069, 1977.
52. M Antler, Contact resistance of tin-nickel electrodeposits, *J Electrochem Soc* 125: 420, 1978.
53. M Antler, MH Drozdowicz, and CA Haque, Connector contact materials: effect of environment on clad palladium, palladium-silver alloys, and gold electrodeposits, *IEEE Trans Comp Hyb Manuf Tech* CHMT-4: 482, 1981.
54. M Braunovic, Fretting damage in tin-plated aluminum and copper connections, *Proc of Electric Contacts-1988, IEEE-Holm*, San Francisco, CA, 179, 1988.
55. M. Nakamura and I Minowa, Film resistance and constriction effect of current in a contact interface, *IEEE Trans Comp Hyb Manuf Tech* CHMT-12: 109, 1989.
56. JBP Williamson and JA Greenwood, The constriction resistance between electroplated surfaces, *Proc Int Conf Elect Cont Electromech Comp*, Appendix, Beijing, China, Oxford: Pergamon Press, 1989.
57. G Horn, Test method for cleanliness of technical contact surfaces by an automatic measurement device for contact resistance, *Proc. 8th Int ConfElect Cont Phen*, Tokyo, Japan, 499, 1976.
58. MP Asar, FG Sheeler, and HL Maddox, Measuring contact contamination automatically, *Western Electric Engineer* 23: 32, 1979.
59. ASTM B667-80, American Society for Testing Materials, Philadelphia.
60. MR Pinnel and KF Bradford, Influence of some geometric factors on contact resistance probe measurements, *Proc 28th Elect Comp Conf*, Anaheim, CA, 129, 1978.
61. J King, High-pressure electrical contacts, *Proc Elect Comp Conf*, Washington, DC, 454, 1968.
62. M Antler, WF Graddick, and HG Tompkins, Base metal contacts: An exploratory study of separable connection to tin-lead, *IEEE Trans Parts Hyb Packag* PHP-IF: 35, 1975.
63. CL Bauer and GG Lessman, Metal joining methods, *Annu Rev Mater Sci* 6: 361, 1976.

64. ER Wallach and GJ Davis, Mechanical properties of aluminum–copper solid-phase welds, *Metals Tech, April,* 183, 1977.
65. JA Rayne and CL Bauer, Effect of intermetallic phase formation on electrical and mechanical properties of flash-welded Al–Cu couples, *Proc 5th Bolton Landing Conf Weld, Phy Metall Fail Pheno,* General Electric Co., Schenectady, NY, 353, 1979.
66. JA Rayne, MP Shearer, and CL Bauer, Investigation of interfacing reactions in thin film couples of aluminum and copper by measurement of low temperature contact resistance, *Thin Solid Films* 65: 381, 1980.
67. ER Wallach and GJ Davis, Joint resistance and current paths in heat-treated aluminum–copper solid-state welds, *Metals Sci* 11: 97, 1977.
68. AW Urquhart, Interdiffusion studies on contact plating materials, *Holm Seminar on Electrical Contacts,* Illinois Institute of Technology, Chicago, IL, 185, 1976.
69. DA Unsworth and CA Mackay, A preliminary report on growth of compound layers on various metal bases plated with tin and its alloys, *Trans Inst Metal Finish* 51: 85, 1973.
70. RS Timsit. Intermetallics growth at Al–α-brass interfaces, *Acta Met* 33: 97, 1985.
71. RS Timsit. Interdiffusion at bimetallic interfaces, *IEEE Trans Comp Hyb Manuf Tech* CHMT-9: 106, 1986.
72. M. Braunovic and N. Alexandrov, Intermetallic compounds at aluminum-to-copper electrical interfaces: Effect of temperature and electric current, *IEEE Trans CompPack Manuf Tech* CHMT-A17: 78, 1994.
73. V Simic and Z Marinkovic, Review: Room temperature reactions in thin metal couples, *J Mat Sci* 33: 561, 1998.
74. M Abbasi, A Karimi Taheri, and MT Salehi, Growth rate of intermetallic compounds in Al–Cu bimetal produced by cold roll welding process, *J Alloy Comp* 319: 233, 2001.
75. KN Tu, Interdiffusion in thin films, *Ann Rev Mater Sci* 15: 147, 1985.
76. FM D'Heurle and PS Ho, *Electromigration in Thin Films: Interdiffusion and Reactions,* (eds) JM, Poate, KM Tu, and JM Mayer, New York: Wiley, 243, 1978.
77. KN Tu, Recent advances on electromigration in very–large-scale integration of interconnects, *J Appl Phys* 94: 5451, 2003.
78. U Gosele and KN Tu, Growth kinetics of planar binary diffusion couples: Thin-film case versus bulk cases, *J Appl Phys* 53: 3252, 1982.
79. D Gupta and PS Ho, Diffusion processes in thin films, *Thin Solid Films* 72: 399, 1980.
80. RS Timsit, The connectibility properties of aluminum: Contact fundamentals, *Can Elect Assoc Rep 76-19,* 1976.
81. DM Rabkin, VR Ryabov, AV Lozovskaya, and VA Dovzhenko, Preparation and properties of copper—aluminum intermetallic compounds, *Sov Powd Met Metal Ceram* 8: 695, 1970.
82. M Braunovic and N Alexandrov, Intermetallic compounds at aluminum-to-copper and copper-to tin electrical interfaces, *Proc 38th Holm Conf Elect Cont,* Philadelphia, PA, 25, 1992.
83. JT Wilson, Report of the Commission of Inquiry on Aluminum Wiring, Queen's Printer for Ontario, Part 2, Ontario, Canada, March 22, 1979.
84. PE Tegehall, Review of the impact of intermetallic layers on the brittleness of tin-lead and lead-free solder joints, IVF Project Report 06/07, IVF Industrial Research and Development Corporation, SE-431, 53 Moldal, Sweden, 2006 (available on the internet at http://extra.ivf.se/eqs/dokument/7%20pet6005.pdf).
85. CD Breach and FW Wulff, A brief review of selected aspects of the materials science of ball bonding, *Microelectron Reliab* 50: 1, 2010.
86. G Zeng, S McDonald, and K Nogita, Development of high temperature solders: review, *Microelectron Reliab* 52: 1306, 2012.
87. RJK Wassink, *Soldering in Electronics,* Scotland: Electrochemical Publications, p. 99, 1984.
88. GA DiBari, Nickel Electroplating for engineering electroforming and decorative purposes, *Plating and Surface Finishing* 70: 24, 1983.
89. D Olsen, R Wright, and H Berg, Effects of intermetallics on the reliability of tin-coated Cu, Ag and Ni parts, *Proc 13th Ann Reliab Phy Symp,* Las Vegas, NV, 80, 1975.

90. PJ Kay and CA Mackay, Barrier layers against diffusion, *Trans Inst Metal Finishing* 57: 169, 1979.

91. TK Do, A Lund, A reliability study of new nanocrystalline nickel alloy barrier layer for electrical contacts, *Proc 56th IEEE Holm Conf. Electrical Contacts and 25th Int Conf Elect Cont*, Charleston, SC, 73, 2010.

92. MA Nicolet, Diffusion barriers in thin films, *Thin Solid Films* 52: 415, 1978.

93. B Chudnovsky, V Pavageau, and M Rapeaux, The quality of tin plating on aluminum exposed to elevated temperatures: The role of underplating, *Proc 24th Int Conf Elect Cont*, Saint-Malo, France, 495, 2008.

94. M Runde, E Hodne, and B Tøtdal, Experimental study of the conducting spots in aluminum contact interfaces, *IEEE Trans Comp Hyb Manuf Tech* 13: 1068, 1990.

95. C Ruppert and M Runde, Thermally induced mechanical degradation of contact spots in aluminum interfaces, *IEEE Trans Comp Pack Tech* 29: 833, 2006.

96. HB Huntington, Electro- and thermomigration in metals, In: *Diffusion, Proc 1972 Seminar of the ASM*, American Society of Metals, 1973, p. 155.

97. H Gan and KN Tu, Polarity effect of electromigration on kinetics of intermetallic compound formation in Pb-free solder V-groove samples, *J Appl Phys* 97: 063514, 2005.

98. RJ Osias and JH Tripp, Mechanical disruption of surface films on metals, *Wear* 9: 388, 1966.

99. RS Timsit, Some fundamental properties of aluminum-aluminum electrical contacts, *IEEE Trans Comp Hyb Manuf Tech* 3: 71, 1980.

100. A Ronnquist and H Fischmeister, The oxidation of copper, *J Inst Metals* 89: 65 1960.

101. MG Hapase, MK Gharpurey, and AB Biswas, The oxidation of vacuum deposited films of copper, *Surf Sci* 9: 87, 1968.

102. W Campbell and UB Thomas, The oxidation of metals. *Trans Electrochem Soc* 91: 623, 1947.

103. GV Samsonov, *The Oxide Handbook*, New York: Plenum Press, 1973.

104. JS Anderson and NN Greenwood, The semiconducting properties of cuprous oxide, *Proc Roy Soc* A215: 353, 1952.

105. G. Honjo, Electron diffraction studies on oxide films formed on metals and alloys, Part 1: Oxidation of pure copper, *J Phy Soc Japan* 4: 330, 1949.

106. G Honjo, Electron diffraction studies on oxide films formed on metals and alloys, Part 2: Selective oxidation of alloys, *J Phy Soc Japan* 8: 113, 1953.

107. KR Lawless and AT Gwathmey, The structure of oxide films on different faces of a single crystal of copper, *Acta Met* 4: 153, 1956.

108. EG Clarke and AW Czanderna, Optical transmittance and microgravimetric studies of the oxidation of < 100 > single crystal of copper, *Surf Sci* 49: 529, 1975.

109. M Batzill and U Diebold, The surface and materials science of tin oxide, *Prog Surf Sci* 79: 47, 2005.

110. U Kuxmann and R Dobner, Untersuchungen im system zinn-zinn(IV)-oxid im temperaturbereich der mischungslucke, *Metallwissenschaft und Technik* 34: 821, 1980.

111. JH Lee and SJ Park, Temperature dependence of electrical conductivity in polycrystalline tin oxide, *J Am Ceram Soc* 73: 2771, 1990.

112. T Tamai, S Sawada, and Y Hattori, Deformation of crystal structure and distribution of mechanical stress in tin-plated layer under contact loading, *Proc 55th Holm Conf Elect Cont*, Vancouver, 160, 2009.

113. Y Nabeta, Y Saitoh, S Sawada, Y Hattori, and T Tamai, Growth law of the oxide film formed on the tin plated contact surface and its contact resistance characteristic, *Proc 55th Holm Conf Elect Cont*, Vancouver, 174, 2009.

114. T Tamai, Y Nabeta, S Sawada, and Y Hattori, Property of tin oxide film formed on tin-plated connector contacts, *Proc 56th Holm Conf Elect Cont* and *25th Int Conf Elect Cont*, Charleston, SC, 295, 2010.

115. EI Alessandrini, Effect of temperature and environment on the structure of thin single-crystal nickel oxide films, *J Appl Phys* 35: 1606, 1964.

116. SH Kulpa and RP Frankenthal, Tarnishing of nickel in air at temperatures from 23°C to 200°C and relative humidities from ambient to 95%. %, *J Electrochem Soc* 124: 1588, 1977.

117. MJ Graham and M Cohen, On the mechanism of low-temperature oxidation (23°C-450°C) of polycrystalline nickel, *J Electrochem Soc* 119: 879, 1972.

118. I Fasaki, A Koutoulaki, M Kompitsas, and C Charitidis, Structural, electrical and mechanical properties of NiO thin films grown by pulsed laser deposition, *Appl Surf Sci* 257: 429, 2010.

119. H van Oosterhout, Evaluation of electroplated nickel phosphorous with high phosphorous content, *AMP J Techn* 2: 63, 1992.

120. WE Campbell, Lectures on tarnishing, friction and wear in electrical contacts, given at the *Sym Elect Cont*, Pennsylvania State University, 1956.

121. P Vassiliou and CT Dervos, Corrosion effects on the electrical performance of silver metal contacts, *Anticorros Meth* M46: 85, 1999.

122. RV Chiarenzelli, Tarnishing studies on contact materials, *IEEE Trans PartsMat Packag* 3: 89, 1967.

123. WH Abbott, Effects of industrial air pollutants on electrical contact materials, *IEEE Trans Parts Hyb Packag* 10: 24, 1974.

124. J Guinement and C Fiaud, Laboratory study of the reaction of silver and copper with some atmospheric pollutants: Comparison with indoor exposition of these materials, *Proc 13th Int Conf Elect Cont*, Lausanne, 383, 1986.

125. H Kim, Corrosion process of silver in environments containing 0.1 ppm H_2S and 1.2 ppm NO_2, *Mater Corros* 54: 243, 2003.

126. MH Hebb, Electrical conductivity of silver sulfide, *J Chem Phys* 20: 185, 1952.

127. M Myers, Overview of the use of silver in connector applications, Technical Paper 503–1016, Tyco Electronics, 2009.

128. P Johannet, Le problème des contacts électriques à courant fort, Report from Electricité de France 1995.

129. N Cabrera and NF Mott, Theory of the oxidation of metals, *Rep Prog Phys*: 12, 163, 1948–49.

130. N Cabrera, J Terrien, and J Hamon, Sur l'oxydation de 1'aluminium en atmosphère sèche, CR Acad Sci, Paris 224: 1558, 1947.

131. WD Treadwell, and Obrist, Uber die bestimmung und bildung von oxydischen deckschichten auf Aluminium, *Helv Chim Acta* 26: 1816, 1943.

132. SJ Bushby, BW Callen, K Griffiths, RS Timsit, and PR Norton, Associative versus dissociative adsorption of water on Al[100], *Surf Sci* 298: L181, 1993.

133. S Janz, HM van Driel, K Pedersen, and RS Timsit, Structural transformations in adsorbed oxygen layers on Al surfaces observed using optical second harmonic generation, *J Vac Sci Tech* A9: 1506, 1991.

134. K Thomas and MW Roberts, Direct observation in the electron microscope of oxide layers on aluminum. *J Appl Phys* 32: 70, 1961.

135. JG Simmons, Generalized formula for the electric tunnel effect between similar electrodes separated by a thin insulating film, *J Appl Phys* 34: 1793, 1963.

136. D Meyerhofer and SA Ochs, Current flow in very thin films of Al_2O_3 and BeO, *J Appl Phys* 34: 2535, 1963.

137. HS Carslaw and JC Jaeger, *Conduction of Heat in Solids*, Oxford: Clarendon Press, 1973.

138. F Kohlrausch, Uber den Stationaren Temperaturzustand eines elektrisch Geheizten Leiters, *Ann Phys*, Leipzig 1: 132, 1900.

139. H Diesselhorst, Uber das Problem eines elektrisch erwarmten Leiters, *Ann Phys* Leipzig 1: 312, 1900.

140. RS Timsit, On the evaluation of contact temperature from potential-drop measurements, *IEEE Trans Comp Hyb Manuf Tech* CHMT-6: 115, 1983.

141. RS Timsit, The "melting" voltage in electrical contacts, *IEEE Trans Comp Hyb Manuf Tech* CHMT-14: 285, 1991.

142. J Israelachvili, *Intermolecular and Surfaces Forces*, London: Academic Press, 1992.

143. S Timoshenko and JN Goodier, *Theory of Elasticity*, New York: McGraw-Hill, 1951.

144. KL Johnson, K Kendall, and AD Roberts, Surface energy and the contact of elastic solids, *Proc Roy Soc* A324: 301 1971.

145. H Jones, The surface energy of solid metals, *Met Sci J* 5: 15, 1971.

146. MT Sprackling, *Liquids and Solids*, London: Routledge & Kegan Paul, 1985.

147. RS Timsit, Evidence of room-temperature sintering in aluminum contacts in vacuum and in oxygen, *Appl Phys Lett* 35: 400, 1979.

148. C Herring, Surface tension as a motivation for sintering, In: WE Kingston, (ed.) *The Physics of Powder Metallurgy*, New York: McGraw-Hill, 143, 1951.

149. JG Early, FV Lenel, and GS Ansel, The material transport mechanism during sintering of copper powder compacts at high temperatures, *Trans Met Soc AIME* 230: 1641, 1964.

150. JGR Rockland, The determination of the mechanism of sintering, *Acta Met* 15: 277, 1967.

151. FV Lenel and GS Ansel, Creep mechanisms and their role in the sintering of metal powders. In HH Hausner, (ed.) *Modern Developments in Powder Metallurgy*, Vol. I. New York: Plenum, 281, 1966.

152. JP Hirth, The influence of surface structure on dislocation nucleation. *Proc NPL Conf Relation bet Struct Mech Prop Mat*, London, 218, 1963.

153. MF Ashby, A first report on deformation mechanism maps, *Acta Met* 20: 887, 1972.

154. J Weertman, Creep of polycrystalline aluminum as determined from strain rate tests. *J Mech Phys Solids* 4: 230, 1956.

155. GC Kuczynski, B Neuville, and HP Toner, Study of sintering of polymethyl methacrylate, *J Appl Polym Sci* 14: 2069, 1970.

156. JP Hirth, The influence of grain boundaries on mechanical properties, *Met Trans* 3: 3047, 1972.

157. RK Hart, The oxidation of aluminum in dry and humid atmospheres. *Proc Roy Soc* A236: 68, 1956.

158. RS Timsit, Evidence of sintering in aluminum electrical contacts at high temperatures, *Proc 10th Int Conf Elect ContPhen*, Budapest, 679, 1980.

159. JBP Williamson, The self-healing effect: Its implications in the accelerated testing of connectors, *Proc 10th Int Conf Elect Cont Phen*, Budapest, 1089, 1980.

160. RS Timsit, Electrical instabilities in stationary contacts: Al-plated brass junctions, *IEEE Trans Comp Hyb Manuf Tech* CHMT-11: 93, 1988.

161. JA Schey, *Introduction to Manufacturing Processes*, New York, McGraw-Hill, 1987.

162. RS Mroczkowski and RJ Geckle, Concerning cold welding in crimped connections, *Proc 41st Holm Conf Elect Cont*, Montreal, Canada, 154, 1995.

163. VA Zhuravlev, The theoretical justification of Amonton's law for the friction of unlubricated surface, *Zh Tekh Fiz* 10: 1447, 1940.

164. JF Archard, Elastic deformation and the laws of friction, *Proc Roy Soc* A243: 190, 1957.

165. FF Ling, On asperity distributions of metallic surfaces, *J App Phys* 29: 1168, 1958.

166. DJ Whitehouse and JF Archard, The properties of random surfaces of significance in their contact, *Proc Roy Soc* A316: 97, 1970.

167. PR Nayak, Random process model of rough surfaces, *J Lub Tech Trans* ASME 93 (F): 398, 1971.

168. AJW Moore, Deformation of metals in static and in sliding contact, *Proc Roy Soc* A 195: 231, 1948.

169. JBP Williamson and RT Hunt, Asperity persistence and the real area of contact between rough surfaces, *Proc Roy Soc* A327: 147, 1972.

170. J Pullen and JBP Williamson, On the plastic contact of rough surfaces, *Proc Roy Soc* A327: 159, 1972.

171. WRD Wilson and S Sheu, Real area of contact and boundary friction in metal forming, *Int J Mech Eng Sci* 30, 475, 1988.

172. A Majumdar and B. Bhushan, Fractal model of elasticplastic contact between rough surfaces, *ASME J of Tribol* 113: 1, 1991.

173. M Ciavarella, G Demelio, JR Barber, and YH Jang, Linear elastic contact of the Weierstrass Profile. profile, *Proc Roy Soc* 456: 387, 2000.

174. RL Jackson, SH Bhavnani, and TP Ferguson, A multiscale model of thermal contact resistance between rough surfaces, *ASME J Heat Transfer* 130:081301, 2008.

175. RL Jackson and JL Streator, A multiscale model for contact between rough surfaces, *Wear* 261:1337, 2006.

176. BNJ Persson, Elastoplastic contact between randomly rough surfaces, *Phys Rev Lett* 87:116101, 2001.

177. RD Malucci, Multispot model showing the effects of nano-spot sizes, *Proc 51st IEEE Holm Conf on Elect Cont*, Chicago, IL, 291, 2005.
178. RL Jackson, RD Malucci, S Angadi, and JR Polchow, A simplified model of multiscale model of electrical contact resistance and comparison to existing closed form models, *Proc 55th IEEE Holm Conf on Elect Cont*, Vancouver, BC, Canada, 27, 2009.
179. T Wanheim, Friction at high normal pressures, *Wear*, 25: 225, 1973.
180. T Wanheim, N Bay, and AS Peterson, A theoretically determined model for friction in metal working processes, *Wear* 28: 251, 1976.
181. T Wanheim and N Bay, Real area of contact between a rough tool and a smooth workpiece at high normal pressures, *Wear* 38: 225, 1976.
182. N Bay and T Wanheim, Real area of contact and friction stress at high pressure sliding contact, *Wear* 38:201, 1976.
183. A Makinouchi, H Ike, M Murakawa, and N Koga, A finite element analysis of flattening of surface asperities by perfectly lubricated rigid dies in metal working processes, *Wear* 128: 109, 1988.
184. AGM Jansen, FM Mueller, and P Wyder, Normal metallic point contacts, *Science* 199: 1037, 1978.
185. G Wexler, The size effect and the non-local Boltzmann transport equation in orifice and disk geometry, *Proc Phys Soc* 89: 927, 1966.
186. YV Sharvin, A possible method for studying Fermi surfaces, *Sov Phys* 21: 655, 1965.
187. IK Yanson and OI Shklyarevskii, Point-contact spectroscopy of metallic alloys and compounds (review), *Sov J Low Temp Phys* 12: 509, 1987.
188. NW Ashcroft and ND Mermin, *Solid State Physics*, New York: Holt, Rinehart, and Winston, 1976.
189. H Kanter, Slow-electron mean free paths in aluminum, silver, and gold, *Phys Rev* B 1: 522, 1970.
190. GT Meaden, *Electrical Resistance of Metals*, New York: Plenum Press, 1965.
191. RS Timsit, Electrical conduction through small contact spots, *IEEE Trans Comp Pack Tech* 29: 727, 2006.
192. M Rokni, and Y Levinson, Joule heat in point contacts, *Phys Rev* B 52: 1882, 1995.
193. BD Jensen, K Huang, LW Chow, and K Kurabayashi. Low-force contact heating and softening using micromechanical switches in diffusive-ballistic electron-transport transition. *Appl Phys Lett* 86: 023507, 2005.
194. C Maul, and JW McBride, A model to describe intermittency phenomena in electrical connectors, *Proc 48th IEEE Holm Conf on Elect Cont*, Orlando, FL, 165, 2002.
195. C Maul, JW McBride, and J Swingler, Intermittency phenomena in electrical connectors, *IEEE Trans Comp Packag and Manuf Tech* 24: 331, 2001.
196. C Maul, JW McBride, and J Swingler, Infuences on the length and severity of intermittences in electrical contacts, *Proc 46th IEEE Holm Conf on Elect Cont*, Chicago, IL, 240, 2000.
197. JD Lavers and RS Timsit Constriction resistance at high signal frequencies, *IEEE Trans Comp Packag Tech* 25: 446, 2002.
198. RD Malucci, High-frequency considerations for multipoint contact interfaces, *Proc 47th IEEE Holm Conf Elect Cont*, Montréal, Quebec, Canada, 175, 2001.
199. RS Timsit, High speed electronic connectors: A review of electrical contact properties, *IEICE Trans Electron* E88-C: 1532, 2005.
200. FW Sears, *Thermodynamics*, U.K.: Addison-Wesley, 1959.

2

Introduction to Contact Tarnishing and Corrosion

Paul G. Slade

How dull is it to pause, to make an end,
  To rust unburnished not to shine in use!

Ulysses, **Alfred Tennyson**

CONTENTS

2.1 Introduction 114
2.2 Corrosion Rates 114
2.3 Corrosive Gases 117
2.4 Types of Corrosion 117
    2.4.1 Dry Corrosion 117
    2.4.2 Galvanic Corrosion 118
    2.4.3 Pore Corrosion 120
    2.4.4 Creep Corrosion 122
    2.4.5 Metallic Electromigration 123
    2.4.6 Stress Corrosion Cracking 124
    2.4.7 Contacts under Mineral Oil 125
2.5 Gas Concentrations in the Atmosphere 126
2.6 Measurements 129
    2.6.1 Weight Gain Measurement 129
    2.6.2 Visual Inspection 129
    2.6.3 Cathodic Reduction 129
    2.6.4 Scanning Electron Microscopy with Energy-Dispersive X-Ray
            Spectroscopy (SEM/EDAX) 129
    2.6.5 X-Ray Photoelectron Spectroscopy (XPS) 129
    2.6.6 Other Techniques 130
    2.6.7 Contact Resistance Measurements 130
2.7 Mixed Flow Gas Laboratory Testing 131
2.8 Electronic Connectors 131
    2.8.1 Background 131
    2.8.2 MFG Test Results 132
2.9 Power Connectors 133
2.10 Other Considerations 134
Acknowledgments 135
References 135

2.1 Introduction

The closed electrical contact has to perform its function of passing electric current reliably and with no change in its contact resistance (R_c) for the entire duration of its life. The contacts have to perform this function even though the ambient atmosphere in which they reside contains pollutants which can cause the contacts to tarnish or corrode. While most ambients no longer show the obvious pollution levels of Pittsburgh on 5 November 1945, where at 11:00 AM the city was in complete darkness [1], electric contacts are used in many environments that can create a severe reduction in their useful life. Again the polluted cities such as Los Angeles, New Delhi, Tokyo, and so on, where temperature inversions prevent the dispersion of the exhaust gases from the internal combustion engine, and where you can smell and taste the air, present an obvious problem. Contacts are used, however, in ambients where the levels of pollution are not so obvious. Many of these ambients still do give rise to corrosion of contacts. While power connectors have been used for many years in harsh outdoor environments such as steel mills and paper mills, it is only in the past 25 years that research has intensified on connectors used in indoor environments. This is especially true for electronic connectors that have found increasing use in equipment for controlling the processes of industrial production that create these harsh outdoor environments. This chapter serves as a brief introduction to the subject of contact corrosion. It initiates a discussion on business and industrial environments and how these are reproduced in laboratory testing. It discusses briefly the types of observed corrosion, and how they are measured. The discussion of environmental effects on electronic connectors and the laboratory experiments on analyzing the effects of mixed gas atmospheres and the effects of atmospheric dust are presented in more detail in Chapters 3 and 4. The effects of corrosion on power connections are discussed in Chapters 5 and 6. The effects of plating are presented in Chapter 8. Finally, Chapters 5–7 give a detailed discussion of fretting corrosion. The effects of the ambient atmosphere on opening contacts will be discussed in Chapter 10 and 19 and the effect on sliding contacts in Chapter 22.

2.2 Corrosion Rates

Clean surfaces of many metals exposed to air oxidize quickly. The initial rapid reaction rate decreases as protective films are formed. The usual tarnish films formed in air at ambient temperatures are mostly composed of mixtures of oxides, sulphides, and chlorides, as well as carbonates and sulphates. The tarnish film reaction can be illustrated by

$$P(\text{solid}) + Q(\text{gas}) \rightarrow PQ(\text{solid}) \tag{2.1}$$

The rate of reaction of metals with formation of corrosion product coatings depends in most instances on the relative permeability of the coating to the reactants. A porous corrosion product is less protective than a non-porous one. Whether or not the corrosion films are porous depends upon the relative volume of the corrosion product compared to the volume of the metal consumed to form it, i.e.,

$$\alpha = v_c / v_m = Md / mD \tag{2.2}$$

where v_c is the volume of the corrosion product, v_m is the initial volume of the metal in v_c, M is the mass of the corrosion product (kg mol^{-1}), d the density (kg m^{-3}), m the molecular mass of metal (kg mol^{-1}), and D its density (kg m^{-3}). If $\alpha > 1$, the coating will be protective, and if $\alpha < 1$, it is non-protective. If more than one reacting gas is present, it is possible to have gas corrosion reactions that can cause the reduction of the corrosion product. Thus, even though α may be greater than one for a particular metal corrosion product, the presence of a reducing agent in the gas mixture can cause the film to continue growing.

There are three equations by which common metals are known to tarnish under ordinary conditions: (1) linear, (2) parabolic, and (3) logarithmic.

The linear equation is

$$y_1 = K_1 t + A_1 \tag{2.3}$$

The parabolic equation is

$$y^2 = K_2 t + A_2 \tag{2.4}$$

and the logarithmic equation is

$$y_3 = K_3 \log(A_3 t + C) \tag{2.5}$$

where y is film thickness, t is the exposure time and K, A, and C are constants. Figure 2.1 shows how tarnish film can obey all the three equations in turn. Figure 2.2 shows experimental data illustrating change from parabolic equation $t^{1/2}$ to one in which the exponent is $t^{2/3}$ [2].

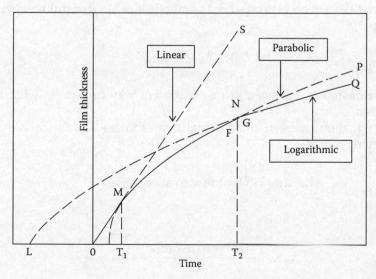

FIGURE 2.1
A hypothetical diagram showing how a metal may tarnish or corrode so as to obey the three kinetic equations in turn: O to M, linear; M to N, parabolic; and G to Q, logarithmic.

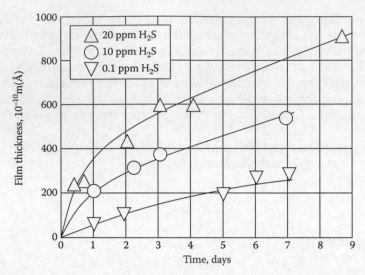

FIGURE 2.2
Growth of Ag_2S film on Ag at 25°C in humid air (PH 75%) shows an initial $t^{1/2}$ kinetics which changes to $t^{2/3}$ at low H_2S concentrations [2].

Abbott [3] showed that the kinetic growth of a corrosion film on Cu is a harsh indoor environment (Class III, see Section 2.5) in a controlled flowing gas environment is

$$y^2 = 6.72 \times 10^3 t - 3.2 \times 10^4 \tag{2.6}$$

where y is the film thickness in nanometers and t is the time in hours. In heavy industrial environment (Class IV) Abbott also observed parabolic kinetics. For milder environments (Class II) he observed

$$y^3 = 7.0 \times 10^5 t + 2.2 \times 10^7 \tag{2.7}$$

This cubic relationship is, however, just a convenient way to express what is really a logarithmic process.

The Arrhenius equation has been found in many cases to describe the change of the rate constant K with temperature,

$$K = a \exp(-\{\text{heat of activation}\}/RT) \tag{2.8}$$

or

$$\ln K = -\frac{\{\text{heat of activation}\}}{RT} + a \tag{2.9}$$

where K is the specific reaction rate, R the gas constant, T the absolute temperature, and a is a constant. Abbott showed that the Arrhenius equation operated up to about 50°C in Class III and Class IV environments [3].

2.3 Corrosive Gases

Sulfur dioxide is produced in large quantities during combustion of fuels containing sulfur.

Hydrogen sulfide occurs in natural putrefaction processes and in the production of artificial fibers such as rayon, wood pulp, refineries, and steel mills.

Nitrogen dioxide is produced when N_2 and O_2 react with each other. The main source of NO_2 is as a combustion product of the internal combustion engine.

Chlorine is a gas emitted during the manufacture of plastics, insecticides, incineration of garbage, plating, and galvanizing operations.

Considerable quantities of peroxides and *ozone* are produced by the interaction of oxygen and nitrogen oxides and, again, a major source is the internal combustion engine.

Ammonia is a product of the decomposition of nitrogenous and organic compounds.

The presence of water vapor can greatly complicate corrosion [4]. The thickness of water films adsorbed on a metal surface is a function of the relative humidity and temperature, and varies greatly with the condition of the metal surface. However, a temperature drop of 5°C from 25°C to 20°C in a container of volume 200 l in air with a relative humidity of 75% causes an invisible film of water of thickness ≈ 0.03 μm to be condensed on the walls of the chamber. This condition is commonly achieved in rooms without air conditioning and in electrical equipment cabinets. The reaction with the water film and the ambient gas can be very critical with the formation of reacting acids [4], the growth or the reduction or even the passivation of existing tarnish films and the effects of galvanic corrosion. Some researchers have discussed a critical humidity below which corrosion does not occur. It is now recognized, however, that in electrical contacts, corrosion can occur over a very wide range of humidities. In fact, the corrosion is such a complex function of (1) Humidity, (2) Materials that make up the contacts, and (3) The environmental gases and other pollutants, that the concept of a unique critical humidity is not valid; in general, however, the higher the relative humidity, the faster the growth of corrosion products [3].

2.4 Types of Corrosion

Table 2.1 gives examples of corrosion products identified during the analysis of malfunctioning equipment in different industrial environments [4]. This table illustrates how complex the reactions can be and how the environments and materials can result in quite different corrosion products. In the literature of electrical contact corrosion research classifications usually describe the type of corrosion product or film that is observed. The four main types that occur are: dry corrosion or tarnish films, pore corrosion, creep corrosion and electromigration. A further corrosion process that can occur in some contact systems is stress corrosion cracking.

2.4.1 Dry Corrosion

Oxidation or dry corrosion occurs when a metal is exposed to an "oxidizing gas" such as oxygen or sulfur dioxide; no water film is necessary for this to take place. The tarnish film will continue to grow until it is limited by the rate of diffusion of the oxidizing gas

TABLE 2.1

Examples of Corrosion Products

Type of Equipment	Place of Exposure	Material Attacked	Corrosion Product	Pollutant Type and Origin
Connectors	Computer room (refinery)	Ag undercoating (gilded)	Ag_2S	H_2S mercaptans (industrial process)
	Computer room (steel plant)	Ag undercoating (gilded)	AgCl	Cl_2, (or HCl, HClO) (industrial process)
	Indicating cabinet (motor way)	Ni undercoating (gilded)	$NiSO_4.6H_2O$	SO_2, exhaust gas
	Indicating cabinet (airfield)	Ni undercoating (gilded)	$NiSO_4.6H_2O$	SO_2, exhaust gas
Sealed relays	Electric power plant (control room)	Ag contact (gilded)	AgCl	Cl_2, (or HCl, HClO) plastics (PVC)
Unsealed relays	Seaside (pumping station)	Ag contact (gilded)	Ag_2S	H_2S mercaptans (organic fermentations)
Fusible contacts	Process equipment	Ag coating	AgCl	Cl_2, (or HCl, HClO)
Printed circuit boards (contact file)	N_2O_4 storage station (leakage)	Ni undercoating (gilded) copper substrate	$Cu(NO_3)_2.3H_2O$	Red vapor cloud + HNO_3 condensation
High-voltage bushing	Process equipment	Cu	$Cu(NO_3)_2.3H_2O$	Air ionization → nitrogen oxides
Static power contact	Ammonia plant	Cu–Sn–P alloy base for an Ag contact	Cu_2O Surface migration on Ag	Ammonia vapors + air oxygen
Commutator segment	Process equipment	Cu	CuCl	Cl_2 (or HCl, HClO) (industrials process)

through the film if $\alpha > 1$ (see Section 2.2). For some metals like aluminum and nickel even a very thin film forms an extremely protective layer. A majority of these films are insulating and so care must be used when mating contacts with such a film. The film must be ruptured before good metal-to-metal contact is made; see for example Chapters 1 and 5.

2.4.2 Galvanic Corrosion

Most corrosion processes observed in electrical contacts involve a thin layer of water on the contact surfaces. This layer is frequently quite invisible and is usually only a few monolayers thick. This film can give rise to a phenomenon known as galvanic corrosion. Figure 2.3a illustrates a water film around a contact pair of the same metal. This type of galvanic cell is called a differential electrochemical cell. The cell operates by differences in the electrolyte concentration caused by differences in dissolved electrolytes from the environment or by differences resulting from a temperature gradient. The water can become positively charged with respect to the microregion of the contact resulting in corrosion and deposition for the insulating products (see Figure 2.3a). If there are dissimilar metals involved, a common galvanic cell can be formed (see Figure 2.3b). Here, the potential available to promote the electrochemical corrosion reaction between the metals is suggested by the galvanic series shown in Table 2.2. Here a list of common metals and alloys is arranged according to their tendency to corrode when in galvanic contact. Metals close to

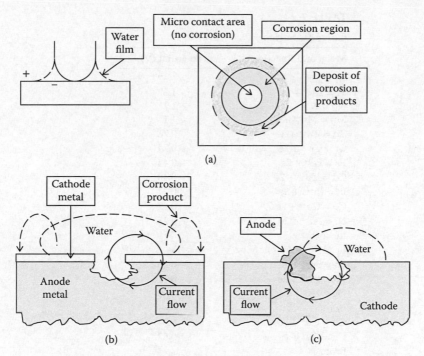

FIGURE 2.3
Examples of galvanic corrosion: (a) Galvanic corrosion, the same metal, (b) Galvanic corrosion different metals, the more anodic metal as substrate, (c) Galvanic corrosion from an embedded particle, dust.

one another on the table generally do not have a strong effect on each other, but the farther apart any two metals are, the stronger the corroding effect on the one higher in the list. The order given in Table 2.2 is maintained in natural water and the atmosphere. Although the galvanic series shown in Table 2.2 represents the potential available to promote a corrosive reaction, the actual rate of corrosion is difficult to predict. The conductivity of the water can vary depending on the solute from the atmosphere. The metal reactions also continuously dissipate energy and this can change the potentials. Finally, other foreign materials such as ambient dust deposits may alter the chemistry considerably. Another common galvanic cell can be formed if a particle from an external source becomes embedded in the contact surface (see Figure 2.3c). This type of cell can lead to a pitting of the metal surface.

In each of these cells, there is a flow of current from one metal to the other, or from one part of the metal surface to another part of the same surface, through the electrolyte. For most electric contact applications, the electrolyte is a few monolayers of water into which a polluting gas is dissolved. A small amount of oxygen dissolved in the water, for example, can greatly enhance the corrosion activity. Other atmosphere gases such as SO_2, H_2S, NO_2, and Cl_2 can also react with water to form the electrolyte. It is also possible that surface deposits such as fingerprints or dust can deliquesce to form the electrolyte even at quite low relative humidities. For example, nitrate in the Los Angeles dust requires only a 30% RH while NaCl from fingerprints only requires 60% RH. In these galvanic cells, a chemical reduction occurs at the cathode region of the metal surface. In this reaction, anions (electrons/negative ions) are released. At the same time, a chemical "oxidation" reaction occurs at the anode region that releases cations (positive metal ions) to balance the reduction reaction. The electrolyte allows the cations to flow, which completes the electrical circuit.

TABLE 2.2

Galvanic Series

Metal and Reaction[a]	Potential (V)
$Al = Al^{+++} + 3e$	− 1.67
$Ti = Ti^{+++} + 3e$	− 1.37
$Nb = Nb^{+++} + 3e$	− 1.1
$Mn = Mn^{++} + 2e$	− 1.1
$Zn = Zn^{++} + 2e$	− 0.76
$Cr = Cr^{++} + 2e$	− 0.74
$Fe = Fe^{++} + 2e$	− 0.44
$Cd = Cd^{++} + 2e$	− 0.40
$In = In^{+++} + 3e$	− 0.34
$Co = Co^{++} + 2e$	− 0.27
$Ni = Ni^{++} + 2e$	− 0.25
Inconel	—
Lead–tin solders	—
$Sn = Sn^{++} + 2e$	− 0.13
$Pb = Pb^{++} + 2e$	− 0.12
$H_2 = 2H^+ + e$	0.00
Brass	—
$Cu = Cu^{++} + 2e$	0.34
$Cu = Cu^+ + e$	0.52
Bronzes	—
Ag braze alloy	—
$Ag = Ag^+ + e$	0.80
$Pd = Pd^{++} + 2e$	0.92
Graphite	—
$Pt = Pt^{++} + 2e$	1.20
$Au = Au^{+++} + 3e$	1.52
$Au = Au^+ + e$	1.83

[a] The metals are arranged according to their tendency to corrode when in galvanic contact.

2.4.3 Pore Corrosion

This type of corrosion is discussed extensively by designers of electronic connectors, who use a thin layer of gold which is nearly always porous (see Sections 3.2.4.6 and 8.3) over a metal base or substrate. If a thin layer of water in which ionizable dissolved gas is present, it results in the galvanic cell illustrated in Figure 2.4a. The corrosion products are transported from the reactive base or substrate metal through the hole in the plating to the contact surface. This type of corrosion is called *pore corrosion* [5] and is usually associated with a synergistic effect between chlorides, oxygen, and/or sulfates. The resulting surface appears pockmarked, as illustrated in Figure 2.4b and Figure 3.11. If the corrosion products are more voluminous than the mass of metal being corroded, a mound of product will appear at the pore site. Also it is possible for the corrosion product between the metal substrate and the plated layer to exert enough force to separate the two layers. While the appearance of pore corrosion can result in a failure of a connection, predicting its occurrence and development depends upon many parameters:

a. The thickness of the final plating metal
b. The substrate metal or substrate metal layers

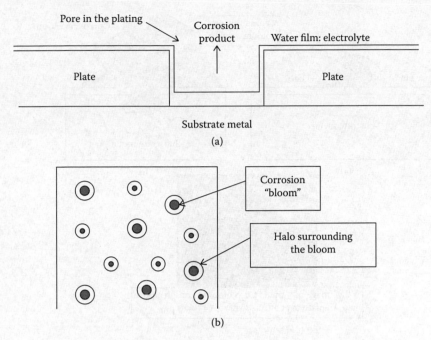

FIGURE 2.4
Pore corrosion: (a) Cross-section of a pore in a metal plating showing the corrosion path, (b) Image of a plated coupon showing the appearance of the pore corrosion.

c. The roughness of the substrate metal below the final plating metal

d. The ambient atmosphere and temperature

e. The connection's exposure to the ambient, that is, how sheltered is the actual connection

These are discussed fully in Section 8.3. One study by Sun et al. [6] using Au (thickness 1.18 µm) on Cu coupons provides an interesting insight into the development of pore corrosion. They exposed the Cu coupons with the Au layer to a flowing gas stream (air plus 10 ppb H_2S and a relative humidity of 70%). In this experiment, the pore corrosion resulted in a bloom surrounded with a thin halo of corrosion, see Figures 2.5a and 3.11. The growth of the bloom above the pore depended upon the pore's diameter and the plating thickness. They showed that the growth height of the corrosion is proportional to the square root of time for the column growth through the plate and cube root of time for the bloom above the plate. As the Cu required to grow the bloom has to diffuse through the cross-section of the pores, the height of the bloom is a function of the pore's area. They also showed an interesting result (Figure 2.6) that the bloom's density is a linear function of exposure time. Also, even though the bloom density increased with time, its size distribution remained relatively constant, as shown in Figure 2.7. They attributed this result to the required transfer of the Cu substrate to the CuS_2 of the bloom. As the Cu from the substrate moved to the corrosion product it left voids in the substrate. As the corrosion proceeded, these voids coalesced until the corrosion product is no longer in contact with the substrate and the growth of the bloom ceased (see Figure 2.5b). They were able to model the effect on the expected contact resistance, but showed that their model required that they bias it by using the larger blooms and to include the effect of the halo films surrounding these blooms.

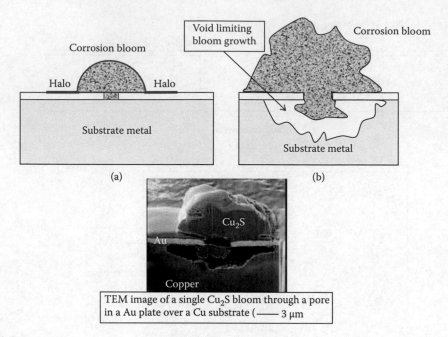

FIGURE 2.5

(a) Diagram showing the bloom growth of the corrosion above the pore with the surrounding halo (b) Diagram showing the end of bloom growth as the substrate separates from the corrosion bloom.

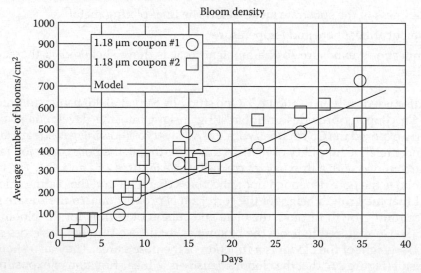

FIGURE 2.6

Pore corrosion site density as a function of time [6].

2.4.4 Creep Corrosion

It occurs (if the conditions are favorable, see Section 3.2.4.7) when a reactive substrate metal like silver [7] or copper [8] is located next to and in physical contact with a noble metal (e.g., gold) or a noble alloy (e.g., gold, silver, platinum [9] inlay or plating). The corrosion

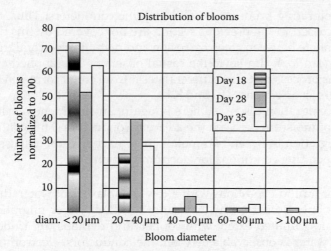

FIGURE 2.7
Normalized size distribution of bloom diameters on a gold-plated (1.18 μm thickness) film on a copper substrate [6].

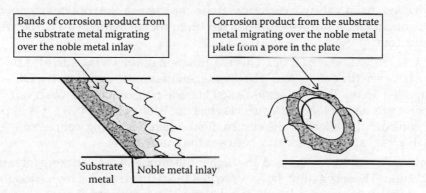

FIGURE 2.8
Examples of creep corrosion.

products of the substrate metal creep over the surface of the noble metal, as shown in Figures 2.8 and 3.14. The gold, which on its own does not form "oxides", offers little resistance to the creep of the corrosion products. Creep can also be initiated from the pores in the thin gold plating. It is usually associated with corrosion films of copper sulfide and silver sulfide.

It is important to mention that the word "creep" is also used by material scientists to describe the flow of bulk metal as a result of relaxation of stress. This type of creep occurs when a metal subjected to a constant external force over a period of time shows dimensional changes [10], as shown in Section 5.4.14, and should not be confused with "creep corrosion".

2.4.5 Metallic Electromigration

Krumbein [11] defined metallic electromigration as all such phenomena that involve the transport of a *metal* across a non-metallic medium under the influence of an electric field. That is, the migrating material is in its metallic state both at its source (e.g., base-metal substrate, a plating, an inlay or even a metal-loaded ink), and in its final form, that is after the migration, the migrating products are still metallic conductors. Thus, the creep corrosion products discussed in the previous section are not covered by this definition. It is important to differentiate between creep corrosion and metallic electromigration, because the two phenomena can look similar to the casual observer. The applied electric field is also important to this process as it also differentiates it from creep corrosion and also from formation of whiskers (see Sections 8.4.3 and 8.4.4).

The type of electromigration discussed here is, again, an electrochemical phenomenon that require a layer of moisture between the adjacent conductors. This contrasts with the solid state electromigration between two metals in contact already discussed in Section 1.3.3. The primary operating conditions for electromigration are:

1. *Moisture.* Under humid conditions with a few monolayers of water, the phenomenon normally only occurs with silver (and is called "silver migration"), but it can also occur to a limited extent with copper, and perhaps tin. Under wet conditions, where water is observable between the conductors, electromigration has been observed for a number of metals—copper, gold, tin, nickel, lead, palladium, and solder—but not for metals that form protective oxide films such as chromium, aluminum, and tungsten.

2. *Contamination.* As in galvanic corrosion, these can change the conductivity of the water. Contamination can also attract water on to the insulating surface between the conductors.

3. *Electric field between the conductors.* This is a function of the voltage divided by the distance between the conductors. The phenomenon is most prevalent if a DC voltage is applied across the conductors. At 60 Hz, the phenomenon is observed, but is not as severe, and at higher frequencies has not been observed [11]. It is important to consider the effect of the electric field when designing connectors with extremely close, insulating spacing between the conductors.

4. *Temperature.* As would be expected, the higher the temperature, the more intensive the migration. There is a limit, however, when the temperature is high enough to evaporate the moisture covering the insulation between the conductors.

The growth of metal dendrites is illustrated in Figure 2.9. The ions tend to deposit at localized sites on the cathode and produce needle-like projections. Once these projections form, the high fields at their tips greatly increase the probability of further deposition there. The silver filaments from the cathode grow back to the anode. When the filament bridges the gap, a sudden drop in resistance occurs. Continued build-ups of these filaments eventually cause a circuit failure.

2.4.6 Stress Corrosion Cracking

This type of corrosion does not usually affect the electrical contact itself, but it is an important phenomenon that the contact designer should consider, because it can affect the parts to which the contacts are attached. Stress corrosion cracking (SCC) results from a combination of (1) A tensile stress, (2) A specific environment, and (3) A susceptible metal, usually an alloy. All the three conditions should be present for failure and if only two of them were met, failure would most probably not happen. The general process initially requires an alloy which is susceptible to selective corrosion along a more or less continuous path, for example, grain boundaries. The metal has to be the anode in a galvanic cell with a

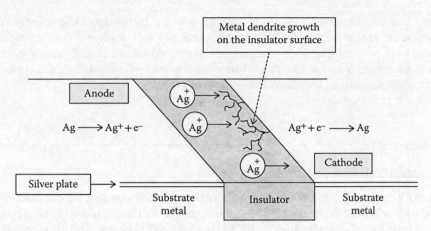

FIGURE 2.9
Metallic electromigration.

specific corrosive electrolyte. There must also be a high stress in a direction tending to pull the metal apart. The corrosion forms a notch which will, in turn, concentrate the stress at the bottom. This process will continue and the crack will propagate perpendicular to the stress until the metal fails. The stresses that can produce SCC need to be sustained, but need not be constant. They may be residual stress left in the alloy after a manufacturing process, for example, a rivet joint, or they may result from service duty, for example, as a leaf spring. The mechanisms of growth of cracks on the atomic level are complex and not well understood. As a result, the environmental effects on a given stressed alloy are difficult to predict. In fact, failures through SCC tend to be unexpected.

One good example of an unexpected failure that resulted from SCC developing in a contact system was experienced by telephone relays operating in Los Angeles [12]. Here, nitrates in the atmospheric dust deposited between the nickel–brass spring wires where they emerged from the clamp plate. The dust layer deliquesced to give a nitrate electrolyte between the wires, and SCC occurred at the high-stress region in the anode wire. When a cupro-nickel alloy was substituted, no SCC was observed.

2.4.7 Contacts under Mineral Oil

It might be expected that contacts vulnerable to tarnishing in air would be protected by immersing them in mineral oil. In fact, quite the opposite takes place. Slade [13] studied the effects of surface changes on Cu and Cu–W contacts in transformer tap changers [14] in a mineral oil ambience. Even highly refined mineral oil contains sulfur compounds, some of which are highly corrosive. Oxygen compounds can also be present especially as the oil ages [15–19]. In fact, there is evidence that while copper oxide can form from the residual oxygen available, the presence of sulfur can facilitate its formation. Resistive films begin to form on the contact surfaces as soon as they are exposed to the various types of mineral oil used in transformers. For Cu-based contacts not subjected to arcing or to arc products, the resistive films mainly consist of oxides and sulfides, but it is possible that polymeric films can also form. For contacts exposed to arc products, the resistive films will also contain carbonaceous products. In a transformer tap changer, these films can build up and give unacceptable contact resistances after a few years. However, servicing to clean the contact surfaces will return the contacts to their original contact resistance values. Plating of non-arcing Cu contacts with 15–20 μm of Ag can slow the build-up of the resistive films. Also as silver oxide and silver sulfide films are relatively soft, a strong contact force can easily rupture them resulting in a low contact resistance. Arnell et al. [20] have shown that it is possible to maintain a low contact resistance for Ag- plated, non-arcing Cu contacts in mineral oil by applying silver iodide to the contact surfaces.

2.5 Gas Concentrations in the Atmosphere

Before laboratory experiments on the corrosion of static contacts can be conducted, it is essential to know what concentrations of corrosive gases will be found in actual business and industrial environments. Graedel in 1973 [21] analyzed outdoor and indoor environments; the summary of his work is shown in Table 2.3. It is interesting to note that chlorine is not shown as an important gas to consider. Research since that time, however, has shown the crucial effect of even very low concentrations of chlorine on the mechanisms that result in the corrosion of electrical contacts [3]. Cosack produced Table 2.4 in 1986 [22]. This table shows that no single gas is involved, but usually the mixture of a number of gases in the environments is responsible. Also, some of the gases exist at very low concentrations. One of the most comprehensive studies of outdoor and indoor industrial environments is the Nordic Project on Corrosion [23–28]. Table 2.5 [24] gives an example for the average values of relative humidity (RH), temperature and pollutants measured during a

TABLE 2.3

Summary of Expected Maximum Average Value for Air Contaminants, 1973

Contaminant	Expected Maximum Average Value		Primary Ambient Air Quality Standard (1973)	Expected Long-Term Trend
	Outdoor	Indoor		
Particulate matter	185 μg m^{-3}	150 μg m^{-3}	75 μg m^{-3}	↓
Nitrate in particulate matter	4.5 μg m^{-3}	3.6 μg m^{-3}	None	→
Sulfur dioxide	0.20 p.p.m.	0.16 p.p.m.	0.03 p.p.m.	↓
Oxides of nitrogen	0.20 p.p.m.	0.16 p.p.m.	0.05 p.p.m.	↓
Total oxidants	0.06 p.p.m.	0.05 p.p.m.	0.08 p.p.m.	→
Hydrogen sulfide	0.03 p.p.m.	0.02 p.p.m.	None	→
Total hydrocarbons	5 p.p.m.	4 p.p.m.	0.24 p.p.m.	↓

Source: HW Hermance, CA Russell, EJ Bauer, TF Egan, and HV Wadlow, Regulation of air-borne nitrate to telephone equipment damage, *Env Sci Tech*, 5(2): 781–5, 1971.

TABLE 2.4

Significant Concentration Values (p.p.m.)

Industrial Air	SO$_2$	H$_2$S	NO$_2$	Cl$_2$	O$_3$	NH$_3$
Average value	0.04	0.01	0.1	0.005	0.02	0.02
Extreme value	0.22	0.4	1.0	—	0.2–0.6	0.2
Olfactory threshold	0.18	0.02	0.1	0.05	0.02	5
Danger to life	400	700	200	3	—	5000

TABLE 2.5

Characterization of Test Sites: Relative Humidity, Temperature, Concentration of Pollutants, Corrosion Product Thickness on Copper after Two Years of Exposure, and Classification of Reactive Environment According to ISA

Name	Environment	Exposure Place	Average of Climatic Parameters		Average of Concentration of Pollutants ($\mu g\ m^{-3}$)								Thickness (Å)	Classification
			RH (%)	T(°C)	Soot	SO_4^{2-}	Cl^-	NH_3 / NH_4^+	NO_2	SO_2	H_2S	Cl_2		
Sarpsborg	Pulp and paper	Bleaching, sulfite	75	22	1.59±3.3	1.0±0.1	0.1±0.1	3.1±0.4	25.0±0.5	108±37	1.1±0.5	>30	1×10^6	GX
Moss	Pulp and paper	Boiler, sulfate	35	27	9.8±2.8	11.1±6.6	0.7±0.2	25.6±4.6	25.5±4.9	44±13	>200	0.2±0.1	1×10^4	G3-GX
Sarpsborg	Pulp and paper	Boiler, sulfite	55	21	14.7±9.1	9.3±4.3	0.3±0.3	1.5±1.1	17.0±5.8	2980±720	1.5±0.1	0.5±0.1	5×10^3	G2
Iggesund	Pulp and paper	Bleaching	42	31	7.5±0	28.5±2.64	27.9±12.2	45.0±8.1	12.8±4.3	13.3±8.3	>>200	>>30	1×10^6	GX
Timra	Pulp and paper	Brown fiber production	37	23	7.1±2.0	3.1±0.9	2.1±0.4	19.8±6.0	26.0±5.5	4.0±1.0	>>200	14.7±4.0	6×10^3	G2
Stockholm	Traffic	Luggage boot in car	45	5	12.7±4.4	13.5±7.7	0.3±0.1	7.2±1.4	15.7±4.5	0.3±0.1	0.7±0.3	0.3±0.3	3×10^2	G1
Helsinki	Traffic	Outdoor, sheltered	81	4	9.3±1.1	2.2±1.2	0.3±0.1	0.7±0.3		7.0±6.7	0.5±0.2	1.1±0.2	2×10^3	G2
Holbaek	Agricultural	Piggery	83	20	25.2±6.1	19.7±7.2	6.6±2.2	670±60	19.8±11.1	0	»30	0.8±0.6	6×10^3	G2
Holbaek	Agricultural	Chicken barn			5	1	2	€10	35	0	18	1	5×10^3	G2

Source: With permission from S Arnell and G Andersson, Silver iodide as a solid lubricant for power contacts, *Proc IEEE Holm Conf Elect Cont*, 239–44, 2001.

six-week study. Again, the complex mixture of gases should be noted. The thickness of the corrosion film on copper after a two-year exposure at those sites is also shown. Table 2.5 also gives a qualitative assessment of each test site's severity based upon a standardized procedure for classifying the reactivity of atmospheric environments [28]: this ranges over GX "severe industrial," G3 "harsh industrial," G2 "moderate industrial," and Gl "weak" as you might find in a business office. The IBM field measurements of the Gl and G2 gaseous environment limits are given in Table 2.6 [30].

A similar classification for indoor environments was developed by Abbott at Battelle [3,31,32], for example Table 2.7 and also Section 3.2.3. Abbott concluded that real indoor environments for electronics can differ by at least four or five orders of magnitude in chemical severity. The degree of exposure to severe versus benign applications can differ widely among manufacturers, even within the same industry. At least four fundamentally different classes of field environment can be distinguished. He calls these: Class I (least severe environment), Class II, Class III, and Class IV (most severe environment). The comparison with this classification and the Gl to GX classification is further discussed in Chapter 3.

TABLE 2.6

G1 and G2 Gaseous Environmental Limits

Concentration Limits	Percentile	Subclass G1, Business Severity		Subclass G2, Industrial Severity	
		ppb	$\mu G/m^3$	ppb	$\mu G/m^3$
Hydrogen	50	2.2	3.1	15	21
sulfide (H_2S)	95	3.3	4.6	45	63
Sulfur dioxide	50	150	390	200	520
(SO_2)	95	500	1300	650	1700
Chlorine	50	ND	ND	5	15
(Cl_2)	95			10	29
Nitrogen dioxide	50	54	100	TBD	TBD
(NO_2)	95	170	320	TBD	TBD
Ammonia	50	35	24	TBD	TBD
(NH_3)	95	80	56	TBD	TBD
Oxidant	50	37	73	TBD	TBD
(as O_3)	95	130	260	TBD	TBD

ND = non-detectable; TBD = to be defined.

TABLE 2.7

Dominant Chemistries and Mechanisms for Degradation of Gold, Nickel, and Copper Field Samples

Reactivity Level	Chemistry[a]		Mechanism(s)	
	Metallic	Non-Metallic	Major	Minor
Class I (e.g., telephone central offices)	—	—	None	None
Class II (e.g., business offices)	Ni, Cu	Cl, 0, (S)[b]	Pore	—
Class III (e.g., controlled industrial)	Cu, Ni	S, Cl, 0	Pore	Creep
Class IV (e.g., uncontrolled industrial)	Cu, (Ni)	S, (Cl), (0)	Creep	Pore

[a] Listed in order: largest to smallest
[b] Indicates small to trace amounts

2.6 Measurements

2.6.1 Weight Gain Measurement

One of the oldest analytical tools is the use of a microbalance. The specimen is weighed after cleaning and before it goes into the test chamber. It is weighed once again after the experiment. The resulting difference in weight corresponds to the corrosion product produced during the experiment. If done carefully, this is a reliably reproducible method [33].

2.6.2 Visual Inspection

This is another very old, but useful technique when used with an optical microscope to see the extent of severe corrosion and tarnish films. It is commonly used, for example, to assess the extent of tarnish creep over the surface of a specimen. It can also be used to give a first-order assessment of how well the contact will perform electrically. One study [27] used the following naked eye correlation:

Visual Inspection	Electrical Measurements
No corrosion	Excellent
Very little corrosion	Very good
Little corrosion	Good
A little more corrosion	Rather good
Much corrosion	Poor
Very much more corrosion	Very poor

2.6.3 Cathodic Reduction

It is possible to determine film thickness to about ±15% for solid coupons of copper and silver. In some cases, it can also give an analysis of the film's composition. The accuracy, however, is not very good for gold flash over nickel over a substrate. The method applies Faraday's law for transforming charge to the thickness of the corrosion products. The method rests on several numerical assumptions; see [34,35] for more details.

2.6.4 Scanning Electron Microscopy with Energy-Dispersive X-Ray Spectroscopy (SEM/EDAX)

This technique provides information on elements with an atomic number greater than six and the morphology of the corrosion products on the surface having a thickness down to a few micrometers.

2.6.5 X-Ray Photoelectron Spectroscopy (XPS)

This technique provides information on relative abundance and chemical status of elements with atomic number greater than three from a narrow surface region with a thickness of a few nanometers.

2.6.6 Other Techniques

Electron spectroscopy for chemical analysis (ESCA) can be used to analyze the elemental species on the surface. Auger analysis can also be used for this and also to determine composition and the thickness of the tarnish films. Electron microprobe is useful to assess the elemental species that exist in a film.

2.6.7 Contact Resistance Measurements

The observation and analysis of the tarnish films gives the contact designer knowledge of how the surface of the contact can change under different classes of ambient atmospheres. It is the effect of these changes on the ability of the contacts to pass current, however, which is ultimately the most important consideration. The ability to pass current is usually assessed by change in contact resistance (R_c). Care must be used while making these measurements. The best way of measuring R_c is illustrated in Figure 2.10a. This method was developed by Holm [36] and is called "crossed rod method." Figure 2.10a shows the wiring diagram and Figure 2.10b illustrates the equipotential surfaces in one of the cylinders. The voltage measured in this example is the voltage at the constriction (V_c) and is not affected by any voltage drop resulting from the passage of current through the bulk of the conductor. If the current is I then

$$R_c = V_c / I \tag{2.10}$$

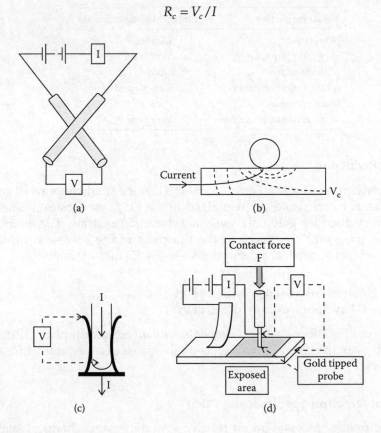

FIGURE 2.10
The crossed rod method of measuring contact resistance (R_c): (a) The wiring diagram for the measurement of contact resistance between crossed rods, (b) The equipotential surface close to the contact spot for the crossed rods, (c) R_c measurement for a stab connector, (d) R_c measurement on a coupon after exposure to a FMG environment.

If possible, R_c measurements should be made with the crossed rod concept in mind. Figures 2.10c and d show how this can be realized in a practical connectors and when measuring the R_c of a film on an experimental coupon specimen. The ASTM B667 standard gives a practical reproducible way of measuring R_c [37].

2.7 Mixed Flow Gas Laboratory Testing

The major advance in laboratory testing of environmental effects on contact materials in the past 20 years has been the development of the mixed flowing gas (MFG) test chamber, as shown in [38]. This work has been championed by Abbott [3], who will discuss this work in detail in Chapter 3. He concluded that no single test environment will satisfy the requirements for Class I to Class IV operating environments seen in practice. He also concluded from his own work and reviewing the work of others before him that:

1. Single gas mixtures such as air with hydrogen sulfide or sulfur dioxide do not simulate indoor field reactions on relevant metals such as silver and copper.
2. Arbitrarily high concentrations of gases such as hydrogen sulfide or sulfur dioxide may produce
   a. Unrealistic kinetics
   b. Unrealistic chemistries or
   c. A complete reversal in the order of material performance
3. SO_2 may not be a significant indoor pollutant.

Most researchers now agree, in general, with the first two conclusions. The third conclusion, however, still gives rise to considerable discussion and argument, so much so that sulfur dioxide is now usually included in MFG gas mixtures; see Section 3.3.5.

2.8 Electronic Connectors

2.8.1 Background

One of the main reasons that so much effort has been expended on contact corrosion research in the past 25 years has been the increased use of electronic equipment, especially computers and other electronic control equipment, in all types of environments. Computers once confined to carefully regulated, environmentally controlled rooms, for example, are now found in all industrial environments from the relatively benign environment of the business office to the Class IV environment of a steel company's rolling mill. This increased use of electronic equipment has, in turn, resulted in an increased use of connectors to transfer the electronic signals. Initially, connectors were made from a base material onto which a thick layer of gold was plated. This gold plate had a thickness of 2.5–7.7 µm. Gold plates of this thickness were, for the most part, free of porosity and were also very resistant to corrosion. However, the use of gold bonded to aluminum or thin gold plate over aluminum led to unexpected failures of electronic circuits in the 1960s [39].

The intermetallics, $AuAl_2$ (known as the purple plague or purple death) or Au_5Al_2 (known as the white plague) could form, both of which caused problems with conduction and brittleness in gold to aluminum bonds. As the price of gold increased during the 1970s, the plating thickness was decreased to approximately 0.75 µm or less. Figure 2.11 illustrates a typical cross-section of a modern connector contact. The substrate material may be copper, a copper alloy or steel. There is usually a barrier layer of about 1.25 µm of nickel to retard the diffusion of substrate metals such as copper or zinc (from brass) through the noble metal layer. Recently, the use of a nickel–phosphorus barrier layer has been proposed. This mixture has shown some promise of being somewhat more resistance to corrosion than pure nickel [40]. Further, when the nickel or nickel–phosphorus substrate is treated with a chromate and some fluorinated compounds, even greater protection is obtained [41]. A flash layer of gold is put over the nickel. This thin layer of gold typically contains 150–200 pores/cm^2, (see Section 8.3), where contact corrosion can originate. Again there has been recent work on putting a thin layer of palladium–nickel or palladium between the gold and nickel barrier layer. This also seems to slow down the corrosion processes.

2.8.2 MFG Test Results

Since 1984, there has been an increasing body of research that has demonstrated the usefulness of MFG laboratory environments to analyze corrosion and tarnishing of electronic connector contact materials and the connectors themselves. This is discussed in detail in Section 3.3.6.

Geckle and Mroczkowski presented an excellent description, based on detailed SEM and electron microprobe analysis, of corrosion films on coupons placed in a Class III, MFG environment [42]. The authors clearly show the complexity of the chemical reactions between the MFG and substrate materials. Possible paths for the corrosive atmospheres to reach the substrate even occur at pores sites a few microns in diameter, or at places where the gold plate has become thinner as a result of a defect in the nickel barrier layer. Svedlung et al. [43] have shown that the chlorine in the test environment may also accelerate the defects in the thin plating areas and open the substrate to the environment. Once the noble metal plating is penetrated, the nickel barrier is susceptible to corrosion via sulfur dioxide and chlorine attack, exposing the copper alloy substrate to the environment. A complex chemical reaction then takes place, resulting in differential attack on the copper and other alloy constituents, producing etch pits and separation of layers. The chemical reaction products migrate back through the imperfections (enlarging the defect in the process) producing a

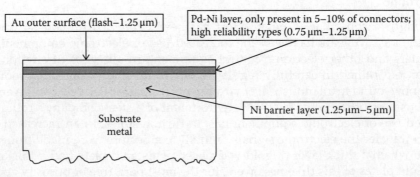

FIGURE 2.11
Schematic of the surface of an electronic connector.

corrosion mound. Oxygen is found in, as well as distributed around, the corrosion mound. The copper, sulfur, and chlorine, on the other hand, are found distributed around the corrosion mounds in a complex film structure. Similar films, generated under different exposure conditions, have been described by Abbott [3] as $Cu_2OCu_2S + Cu_xO_2Cl_2$, and by Gore et al. [34] as principally a Cu_2O with a mixture of Cu_x (OH) Cl_2, and Cu_2S. This paper again confirmed the earlier work by Abbott [23] that the chlorine reduces potentially protective oxide films which, in turn, enhance the rate of formation of copper sulfides.

2.9 Power Connectors

The use of power connectors to pass current in harsh environments has been practiced since electricity was first used in the industry. Chapters 5 and 6 discuss the use of copper and aluminum plated by silver, tin, and nickel. The analysis of actual field environments, their classification into Class I to Class IV groupings, and the development of MFG laboratory testing for electronic connectors, has certainly led to a greater appreciation of corrosion mechanisms in all electrical connectors including power connectors. This is especially true for the generally very reliable copper–copper power connections where it is common to use a silver plate over the copper. Silver itself, however, is easily attacked by reactive sulfide pollutants. The corrosion mechanism for a silver plate (< 5 μm) over copper is complex [43,44], since two chemistries (copper and silver) can be active at the same time. If silver films of thickness greater than 20 μm are used, then the copper has little or no influence on the surface films observed because of the nearly total absence of porosity in the thick silver layers [45]. Also, it is possible to have migration of silver through the silver sulfide films, which gives rise to a decrease in observed contact resistance in corroded contacts, which are then mated so that current passes through the joint [46]. This phenomenon results in a gradual decrease in R_c to a value close to that observed with clean silver contacts.

Another plating material that is widely used in power connectors is tin. Tin, of itself, can form a good bolted joint and, if the joint is mechanically stable, it can be used in a spring-loaded socket type of connector. It is especially useful in an aggressive atmosphere such as a paper mill where a silver plate would corrode very rapidly. Whitley proposed the "Tin Commandments" when using tin-plated contacts [47]. Further research since 1989 has led to the following modified version:

1. The contact must be mechanically stable. A motion of ≈ 5 μm is sufficient to break the gastight seal at the tin–tin interface. If this seal is broken, an insulating oxide film will form.

2. A contact force greater than or equal to 1N is required.

3. Tin-plated contacts require a protective lubricant to slow or even limit the ingress of the ambient atmosphere.

4. Do not use above 100°C. At this temperature, tin loses its mechanical strength and the rate of copper diffusion in tin increases rapidly.

5. Both bright tin and dull tin give similar R_c. Bright tin is aesthetically more pleasing and is easier to solder. Dull tin is less subject to formation of whiskers, see Section 8.4.3.

6. The tin plate should have thickness of at least 5 μm.

7. Tin should not be mated to gold. If this combination is unavoidable, then a lubricant inhibiting corrosion should be used on the contacting surfaces(see Section 3.4).

8. When tin contacts mate initially, there should be a wiping action.

9. Tin contacts should *never* be used to break an electric current.

10. Tin–tin contacts can used in microvolt/milliampere circuits as well as in circuits of hundreds of volts and hundreds of amperes.

2.10 Other Considerations

A discussion of the major causes of failure of static contacts is made in Chapters 5–7. It was alluded to in the first of the "Tin Commandments" in Section 2.9, and is commonly known as *contact fretting*. Here a change in temperature, mechanical vibration, or vibration that results from the interaction of current-carrying conductors with electric or magnetic fields will cause one contact to move slightly with respect to its mate. This motion results in contact wear and permits the ambient atmosphere to corrode the contact material closer to the contact spot. The circuit fails when the R_c of the contact increases beyond an acceptable level. A strong argument can be made that most connector failures occur because fretting has taken place.

One way of preventing this type of failure is to lubricate the contacts (see Chapters 3, 5, and 7). In fact, lubrication with a gold flash over nickel in a Class II FMG environment has been shown to protect electronic connectors [48]. Further research on field-tested and laboratory samples [49] has shown the beneficial effect of using specific lubricants. This work also shows the use of a lubricant must be approached with care. For example, at low temperatures, many lubricants appear to solidify and can develop films with high shear strength, which can lead to high R_c if movement occurs. On the other hand, the solidified lubricants can severely limit micromotion between the contacts, and thus limit fretting corrosion. At high temperatures, lubricants may be lost. They may also oxidize. This work, however, shows much promise and I would expect effective lubricant materials to become common in the future for low-cost electronic connectors. Chapters 3, and 7 discuss the use of lubricants for electronic connections in more detail. Chapter 5 presents data on the use of lubricants for power connections.

The last subject I would like to touch upon is "dust", which will be discussed in Chapter 4. While most contacts and connectors have some degree of shielding and the ambient gas has some degree of filtering, there are enough situations where dust can be deposited in the region of the electrical connection. This subject is extremely complex and effective research in this area is in its infancy. In a very wide-ranging study in the Chicago area, Bayer et al. [50] analyzed the dust in a larger number of industrial sites: steel mills, textile factories, paper mills, power plants, railroad yards, and metal process manufacturers. They placed dusts into four categories: minerals, metals, organics, and fibers. They also discussed the shapes: spheres, flakes, chunks, scrapings, chips, grains, and charred pieces. They showed that the upper limits for these environments were as follows:

	Suspended	Settleable
Light industry	150 μg m^{-3}	0.5 mg cm^{-2} month
Heavy industry	1000 μg m^{-3}	No upper limit

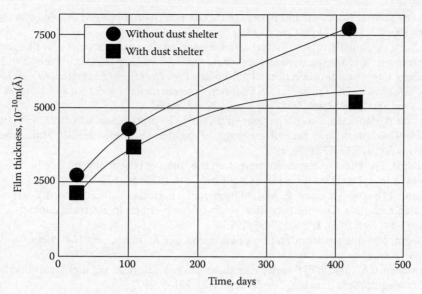

FIGURE 2.12
The effect of a dust shield on corrosion products for copper exposed to a paper factory ambient atmosphere [28].

Values for light industry are similar to the indoor value shown in Table 2.3. In a study of corrosion products on copper coupons, some with a dust shelter and some without [28] in a pulp and paper factory, seems to indicate that the dust chemistry also has an effect on the development of the tarnish film; see Figure 2.12. As I discussed in Section 2.4, dusts can have a marked effect on the conductivity of the electrolyte required for the majority of the corrosion processes. Studies on the dust in the environment has continued [51–53] and this has continued to show how complex the dust environment can be. There have been some attempts at developing laboratory dust environments (e.g., [53,54]); however, at present, there is no consensus on what constitutes a valid laboratory test. It is interesting to note in an earlier study the effects of cigarette and cigar smoke on dust concentrations was discussed. This particulate source of pollution must now be severely diminished with the advent of the smoke-free workplace.

Acknowledgments

I thank the late Mort Antler for reviewing this introduction to corrosion, for offering constructive criticism, for his suggestions for improvement based upon his lifetime of research experience, and for his interesting discussions on this subject.

References

1. S Lorant, *Pittsburgh: The Story of an American City*, Lenox, MA: Authors Edition Inc., 376–77, 1975.
2. D Simon, C Ferris, D Mollimard, MT Bajard, and J Bardolle, Study of contamination of contact materials, silver and silver palladium alloys in air containing hydrogen sulphide and nitrogen dioxide, *Proc 13th Int Conf Elect Cont*, Lausanne, 333–47, 1986.

3. WH Abbott, The development and performance characteristics of mixed flowing gas test environments, *Proc 33rd IEEE Holm Conf on Elect Cont*, IEEE, 63–78, 1987.

4. M Carballeira, A Carballeira, and JY Gal, Contribution to the study of corrosion phenomena in industrial atmosphere: Thermodynamic approach, *Proc 14th Int Conf Elect Cont*, Paris, 239–45, 1988.

5. SJ Krumbein, Corrosion through porous gold plate, *Proc Holm Conf Elect Cont*, 67–83, 1968.

6. AC Sun, HK Moffat, DG Enos, and CS Glauner, Pore corrosion model for gold-plated copper contacts, *Proc 51st IEEE Holm Conf Elect Cont*, 232–37, 2005.

7. TF Egan and A Mendizza, Creeping silver sulphide. *J Electrochem Soc*, 107(4): 353–54, 1960.

8. V Tierney, The nature and rate of creepage of copper sulphide tarnish films over gold, *J Electrochem Soc*, 128(6): 1321–26, 1981.

9. DW Williams, The effect of test environment on the creep of base metal surface films over precious metal inlays, *Proc 33rd IEEE Holm Conf Elect Cont*, IEEE, 81–5, 1987.

10. Ch Poulain, L Boyer, Ph Sainsot, MH Maitournam, F Houze, M Leclercq, J Guery, and JP Charpentier, Experimental and theoretical study of creep effects in electrical contacts. *Proc 41st IEEE Holm Conf Elect Cont*, IEEE, 147–53, 1995.

11. SJ Krumbein, Metallic electromigration phenomena, *IEEE Trans Comp Hyb Manuf Tech*, 11(1): 5–15, 1988.

12. HW Hermance, CA Russell, EJ Bauer, TF Egan, and HV Wadlow, Regulation of air-borne nitrate to telephone equipment damage, *Env Sci Tech*, 5(2): 781–5, 1971.

13. PG Slade, Interpretation of low DC current measurements of a transformer's winding resistance and the effect of the tap-changer's contacts, *Proc 56th IEEE Holm Conf Elect Cont*, 1–10, 2010.

14. EPRI Publication 1012350. New equipment and performance design review LTC management course material, Dec. 2006: J Harlow, Development of load tap changing technology, A11–A25.

15. H Kawarai, Y Fujita, J Tanimura, S Toyama, N Yamada, and E Nagao, Role of dissolved copper and oxygen on copper sulfide generation in insulating oil, *IEEE Trans Dielect Elect Ins*, 16(5), 1430–5, Oct. 2009.

16. EPRI Publication, op. cit., R Crutcher, D Hanson, and L Savio, Development of a load tapchanger monitoring technique to identify potential coking. 57–67.

17. T Amimoto, N Hosokawa, E Nagao, J Tamimura, and S Toyama, Concentration dependence of corrosive sulfur on copper sulfide deposition on insulating paper used for power transformer insulation, *IEEE Trans Dielect Elect Ins*, 16(5), 1489–95, Oct. 2009.

18. L Lewand, Passivators: What they are and how they work, *NETA World*, Spring 2006.

19. R Kurz and J Leedy, Stabilization of transformer oil against oxidation: Old practice with renewed interest for preventive maintenance, *Proc Elect Manuf Wind Conf*, 529–36, Sept. 1997.

20. S Arnell and G Andersson, Silver iodide as a solid lubricant for power contacts, *Proc IEEE Holm Conf Elect Cont*, 239–44, 2001.

21. TE Graedel, The atmospheric environments encountered by electrical components, *Proc Holm Conf Elect Cont*, Illinois Institute of Technology, 62–79, 1973.

22. U Cosack, Survey of corrosion tests with pollutant gases and their relevance for contact material, *Proc13th Int Conf Elect Cont*, Lausanne, Switzerland, 316–25, 1986.

23. JF Henriksen, C Leygraf, S Zakipour, P Villien, and M Wagner, Nordic project on corrosion of electronics, I. Analysis of environment and corrosivity of 38 field test sites, *Proc 13th Int Conf Elect Cont*, Lausanne, 326–32, 1986.

24. Z Saeid, L Christofel, H Jan, H Risto, and Z Kim, Nordic project and corrosion of electronics, II. Analysis of corrosion film chemistry on field and laboratory exposed Cu coupons and Au-plated samples, *Proc 14th Int Conf Elect Cont*, Paris, 255–60, 1988.

25. K Zachariassen, P Villien, C Leygraf, S Zakipour, JF Henriksen, and R Heinonen, Nordic project on corrosion of electronics, III. Evaluation of contact reliability of field exposed eurocard connectors, *Proc 14th Int Conf Elect Cont*, Paris, 261–64, 1988.

26. R Hienonen, T Saarinen, JF Henriksen, C Leygraf, and K Zachariassen, Nordic project on corrosion in electronics, IV. Atmospheric corrosion in gold plated connector contacts, *Proc 14th Int Conf Elect Cont*, Paris, 265–70, 1988.

27. R Hienonen, J Rakkolainen, T Saarinen, and M Aberg, Nordic project on corrosion in electronics, V. Atmospheric corrosion in electronics: A comparative study on field and laboratory test results of various electronic contacts, *Proc 14th Int Conf Elect Cont*, Paris, 271–75, 1988.

28. S Zakipour and C Leygraf, Nordic project on corrosion of electronics, VII. Improvement of method for environmental classification, *Proc 16th Int Conf Elect Cont*, Loughborough, 47–51, 1992.

29. ISA-S71.04–1985 Standard: Environmental conditions for process measurement and control systems: Air-borne contaminants, *Instru Soc Am, Res Triangle*, NC, 27709, 1985 and IBM Engineering Specification C-H 1-9700-0100.

30. JL Chao and RR Gore, Evaluation of a mixed flowing gas test, *Proc 37th IEEE Holm Conf Elect Cont*, IEEE, 216–28, 1991.

31. WH Abbott, The effect of test environment on the creep of surface films over gold, *Proc 30th Holm Conf Elect Cont*, Illinois Institute of Technology, 47–52, 1984.

32. WH Abbott, The corrosion of porous gold platings in field and laboratory environments, *Proc 13th Int Conf Elect Cont*, Lausanne, 343–7, 1986.

33. WE Campbell and UB Thomas, Tarnishing and contamination if metals, *Proc 14th Holm Conf Elect Cont*, Illinois Institute of Technology, 233–65, 1968.

34. R Gore, R Witska, JR Kirby, and J Chao, Corrosive gas environmental testing for electrical contacts, *Proc 35th IEEE Holm Conf Elect Cont*, IEEE, 123–31, 1989.

35. AT Kuhn, MS Chana, and GH Kelsall, Methods for the analysis and assessment of tarnish, *Br Corrosion J*, 18: 174, 1982.

36. R Holm. *Electric Contacts, Theory and Application*, New York: Springer-Verlag, 42–3, 1967.

37. M Antler, Contact resistance probing: Development of a standard practice by the American Society of Testing Materials, *Proc 10th Int Conf Elect Cont Phen*, Budapest, 13–21, 1980.

38. M Carballeira, G Duboy, and A Carballeira, Some parameters influencing the reproducibility of low concentration atmospheric test, *Proc 39th IEEE Holm Conf Elect Cont*, IEEE, 69–74, 1984.

39. B Seelikson and TA Longo, A study of purple plague and its role in integrated circuits, *Proc IEEE*, 52(12), 1638–41, 1964.

40. HH Law, CA Holden, J. Sapjeta, GR Crane, and S Nakamura, Electric contact phenomena of nickel electrodeposits with sharp asperites, *IEEE Trans Comp Hyb Manuf Tech*, 14(3): 585–91, 1991.

41. HH Law, J Sapjeta, and ES Sproles, Protective treatments for gold flashed contact finishes with a nickel substrate, *Proc 37th Holm Conf Elect Cont*, IEEE, 199–202, 1991.

42. RJ Geckle and RS Mroczkowski, Corrosion of precious metal plated copper alloys due to mixed flowing gas exposure, *Proc 36th IEEE Holm Conf Elect Cont*, IEEE, 193–202, 1990.

43. OA Svedlung, L Johansson, and N Vannerberg, The influence of NO_2 and Cl_2 at low concentrations in humid atmospheres on the corrosion of gold-coated contact material, *IEEE Tran* CHMT, 9: 286–92, 1986.

44. T Imrell, The importance of the thickness of silver coatings in the corrosion behaviour of copper contacts, *Proc 37th IEEE Holm Conf Elect Cont*, 237–243, 1991.

45. T Imrell, R Sjovall, and A Kassman-Rudolphi, The composition of the silver coating strongly influences the stability of stationary electrical contacts, *Proc 17th Int Conf Elect Cont*, 447–54, 1994.

46. A Kassman-Rudolphi, C Bjorkman, T Imrell, and S Jacobson, Conduction through corrosion films on silver-plated copper in power contacts, *Proc 41st IEEE Holm Conf Elect Cont*, IEEE, 124–33, 1995.

47. J Whitley, The tin commandments, *Conn Tech*, April: 27–28, 1989.

48. WH Abbott and M Antler, Connector contacts: Corrosion inhibiting surface treatment for gold plated finishes, *Proc 41st IEEE Holm Conf Elect Cont*, IEEE, 97–123, 1995.

49. WH Abbott, Field and laboratory studies of corrosion inhibiting lubricants for gold-plated connectors, *Proc 42nd IEEE Holm Conf Elect Cont*, IEEE, 414–28, 1996.

50. RG Bayer, R Ginsburg, and RC Laskey, Settleable and air-borne particles in industrial environments, *Proc 35th IEEE Holm Conf Elect Cont*, IEEE, 155–66, 1989.

51. JG Zhang and P Yu, Electric contact performance: Effects of contact surface morphology and the size of dust particles, *Proc 36th IEEE Holm Conf Elect Cont*, IEEE, 402–9, 1990.
52. JG Zhang, Effect of dust contamination on electrical contact failure, *Proc 53rd IEEE Holm Conf Elect Cont*, IEEE, 1–8, 2007.
53. Y Zhou, Y Lu, and L Lui, The role of organic compounds in simulation of dust environments, *Proc 26th Int Conf Elect Cont*, 45–9, 2012.
54. C Azeez Haque, MD Richardson, and ET Ratliff, Effects of dust on contact resistance of lubricated connector contact materials, *Proc 31st IEEE Holm Conf Elect Cont*, IEEE, 141–6, 1985.

3

Gas Corrosion

William H. Abbott and Paul G. Slade

If gold rust, what shall iron do

Canterbury Tales, **Geoffrey Chaucer**

CONTENTS

3.1 Introduction...140
    3.1.1 Scope...140
    3.1.2 Background...140
3.2 The Field Environments for Electrical Contacts...143
    3.2.1 Environmental Variables 144
    3.2.2 Corrosion Rates...145
        3.2.2.1 Copper and Silver...145
        3.2.2.2 Other Metals...145
        3.2.2.3 Film Effects...146
        3.2.2.4 Shielding Effects..147
    3.2.3 Reactivity Distributions..148
        3.2.3.1 Severity versus Performance...149
        3.2.3.2 Environmental Classes..150
        3.2.3.3 Specifications..150
    3.2.4 Corrosion Mechanisms 151
        3.2.4.1 Silver 151
        3.2.4.2 Copper...151
        3.2.4.3 Nickel..152
        3.2.4.4 Tin..152
        3.2.4.5 Porous Gold Coatings...153
        3.2.4.6 Pore Corrosion...153
        3.2.4.7 Corrosion Product Creep 154
3.3 Laboratory Accelerated Testing...157
    3.3.1 Objectives...157
    3.3.2 Definition of Acceleration Factor 158
    3.3.3 Historical Background 159
    3.3.4 Single-Gas Corrosion Effects..159
        3.3.4.1 Hydrogen Sulfide..159
        3.3.4.2 Sulfur Dioxide (SO_2) 163

        3.3.4.3   Nitrogen Dioxide (NO₂)... 163
        3.3.4.4   Chlorine... 163
        3.3.4.5   Mixed-Gas Sulfur Environments 164
        3.3.4.6   Humidity... 164
        3.3.4.7   Temperature... 164
        3.3.4.8   Gas Flow Effects... 165
    3.3.5   Mixed-Gas Environments... 167
        3.3.5.1   Test Systems 169
        3.3.5.2   Monitoring Reactivity 169
    3.3.6   Test Applications... 171
        3.3.6.1   Electronic Connectors 171
        3.3.6.2   Mated versus Unmated Exposures 172
        3.3.6.3   Other Considerations 173
3.4   Lubrication and Inhibition of Corrosion 174
Acknowledgment.. 181
References.. 181

3.1 Introduction

3.1.1 Scope

This chapter deals with the broad subject of the corrosion of electrical contacts in the field and laboratory environments in which they operate. Emphasis will be on natural, long-term, ageing reactions. More specifically, it will highlight the effects of mostly inorganic chemical species as found in natural operating environments. This will exclude various forms of manufacturing- and/or handling-induced contamination and will also exclude the almost endless possibilities of organic species in real environments. There are various reasons for this approach. One is to focus on the effects of those few, critical inorganic pollutants which modern research has identified as having the first-order effects on contact corrosion. Secondly, it is to emphasize the important effects which even extremely low levels of these pollutants may have on the corrosion process. And finally, it is to demonstrate the dramatic changes in corrosion mechanisms which can occur on contact materials as pollutant levels change.

An understanding of the field environments for the electrical contacts is essential to the second objective of this chapter. This is to examine modern trends in the development of accelerated, environmental or corrosion tests. It will be shown a single test can never address the wide range of environmental conditions which exist in practice. However, an understanding of probable field conditions for specific products together with the significance of the variety of environmental tests which are presently available may provide the basis for a more informed selection of the most appropriate tests.

3.1.2 Background

It may be questioned whether the term "corrosion" is most appropriate for the subject under discussion. This is due to the fact this term often evokes images of strong visual effects; e.g., the rusting of iron. This is dramatically different from the corrosion in the contact field which often involves the formation of thin, surface films. It is well established that thin films, even at thicknesses of the order of lesser than 100 Å may be virtually insulating [1]. Films in this range are well below the limits of visual detection which may be in the range of 300–400 Å. For this reason, it is appropriate to set forth several of the key concepts involving the corrosion of electrical contacts. One is that the user may often not perceive a corrosion problem; that is, it is subtle, and second (as will be shown) it may be produced by environments which are often perceived to be "benign."

The effects of surface films on several contact materials are shown in Figure 3.1 [2]. There is a considerable amount of relevant information in this data which will be discussed in later sections of this chapter. However, for the moment Figure 3.1 should serve to illustrate the important principles of (1) Magnitude of effects and (2) Strong differences among materials.

The latter effect is particularly relevant to the subject of accelerated environmental testing. Part of the reason for the differences shown in Figure 3.1 is believed to be the variations in the mechanical properties of surface films, which in turn are known to be related to the film chemistry. There does not appear to be much published information on this subject but one example is shown in Figure 3.2 for copper [3]. The differences shown, which approach several orders of magnitude for the same film thickness, could be attributed to the environments in which the films were formed. The specific environment is known to affect film chemistry and, in turn, fits properties. Figure 3.2 demonstrates the historical problem associated with the accelerated environmental testing of electrical contacts. These data illustrate the following principle, which is one of the major themes of this chapter:

Arbitrary tests often produce arbitrary results.

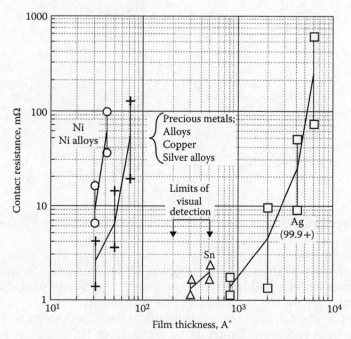

FIGURE 3.1
Effects of film thickness on contact resistance 100 g: 95th percentile.

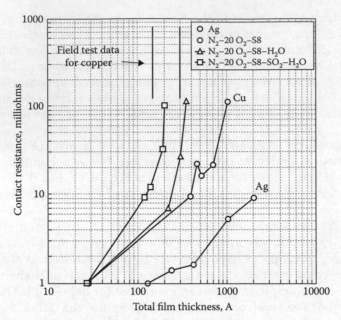

FIGURE 3.2
Contact resistance of copper and silver versus total film thickness in three environments at 30°C, 70% RH.

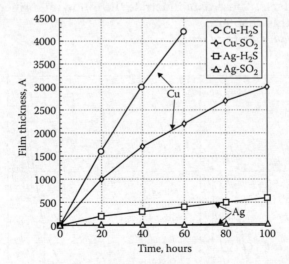

FIGURE 3.3
Tarnish of silver and copper in high concentration, single pollutant mixtures of H_2S and SO_2 in air.

It is clear from data such as those shown in Figure 3.2 that widely different conclusions or reliability estimates could result from tests conducted in different environments. This fact alone argues strongly for the need for test procedures which more accurately reflect the chemistries and mechanisms of corrosion found in real environments. The problem just discussed has been recognized since at least the 1930s [4,50] and is illustrated in Figures 3.3 and 3.4. The data in Figure 3.3 show the typical kinetic responses of two relevant materials exposed to typical, high-concentration, single-gas tests which have long

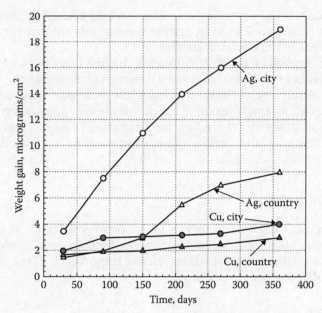

FIGURE 3.4
Relative tarnish behavior of silver and copper in the two indoor atmospheres.

been, used for the accelerated ageing of electrical contacts. In fact, such tests still exist in national and international standards [5]. These tests have historically emphasized the effects of H_2S and/or SO_2 which have been "known" to be the major pollutants causing the corrosion of electrical contacts.

Figure 3.4 shows data from some of the earliest field studies on these materials. While the data are from telephone central offices, they are similar to effects reproduced in later studies for a broad range of environments. Specifically, there is a total reversal in the order of performance of materials compared to the laboratory data of Figure 3.3. This brief discussion should serve to illustrate the problems which have been associated with the accelerated environmental testing of electrical contacts as well as components and systems employing electrical contacts. The use of unrealistic tests may produce any or all of the following: (1) A complete reversal in the order of materials performance, (2) The acceptance of materials which may perform poorly in the field (and the reverse), and (3) Overestimation of reliability.

3.2 The Field Environments for Electrical Contacts

Part of the reasons for the state of affairs just discussed has been a lack of comprehensive and relevant field data. The term "comprehensive" refers to data from a wide variety of environments according to industries, types of applications, etc. The term "relevant" refers mainly to the environments indoors, in enclosures, and so on. It will be shown that such environments, which are largely the domain of the electrical contact, differ in many important ways from the outdoor environment. For example, the early work of Vernon [6], Thompson [7], and Blake [8] described many of the effects of various environmental variables on the corrosion of copper and other metals, and the corrosion literature is replete with data on outdoor corrosion. However, almost none of this information is relevant to the indoor environments of the electrical contact.

The early work of Campbell [4] developed possibly the first body of good indoor data. It was not until many years later that studies were undertaken to define the characteristics of the latter [9–16]. Typically, each of these studies, while making important contributions to the field, examined only a small portion of the distribution of application environments around the world. Later studies by the author have examined several thousand field sites representing many industries and generic types of application environments. Many of these findings form the basis for the following sections.

3.2.1 Environmental Variables

The corrosion of electrical contacts is driven by a complex interaction of a multitude of interacting variables, all of which may be time-variant. In recent years, the critical variables have been defined together with their relative importance. These are listed below in *approximate* order of decreasing importance:

- Relative humidity
- Reactive chlorides
- Reduced sulfides (H_2S)
- Air exchange rate
- Humidity cycling
- Linear velocity
- Oxidizers (NO_2, O_3)
- Sulfur dioxide
- Temperature

Modern findings have also helped to modify certain beliefs about indoor corrosion and to establish certain important principles about the interactions among these variables. These will be listed without detailed discussion, since it is beyond the scope of this chapter.

1. Humidity is important to the corrosion process, but on its own is not a corrosive agent; that is, humidity tests alone are not ideal accelerated environmental ageing tests for electrical contacts.
2. Humidity does accelerate most corrosion processes, but high humidity (>60%–70% RH) is not required to cause relevant corrosion.
3. Corrosion may be "driven" by gas concentrations measured at levels equal to or lesser than 1 ppb of critical pollutants.
4. Corrosion is the result of the synergistic effects among a host of environmental variables.
5. Corrosion control is a trade-off between humidity and critical pollutants.
6. Corrosion usually occurs in environments which may be perceived by human senses to be "benign."

3.2.2 Corrosion Rates

Measurement of the rate of corrosion of specific metals provides important information for both technical and practical purposes. Such data are an important means of quantitatively, accurately, and inexpensively describing an environment, whether in the field or a laboratory test. The reason for this approach of measuring net corrosion rates is, it has proved impossible to develop algorithms to relate the multitude of relevant environmental variables to corrosion. Also, the latter approach would be expensive in most cases.

3.2.2.1 Copper and Silver

For the purposes of monitoring environments, copper has evolved as the primary testing material. Following the work of the author and his colleagues, the use of copper for field monitoring has been developed into a US National Standard [17]. Copper is also required for test documentation in at least two recent Mixed Flowing Gas (MFG) laboratory test specifications [18,19].

In recent years, silver has also been widely used in studies at field sites, since this material can give important information about the chemical composition of those environments. Figures 3.5 and 3.6 show typical data for both the materials. The data serve to illustrate typical kinetic responses and the magnitude of film thicknesses. These and other data will be used in a following section to introduce the concepts of reactivity distributions and environmental classes.

3.2.2.2 Other Metals

Figure 3.7 shows data for a number of other materials of relevance to the contact field. It is important to note that these data were obtained for only one of the several classes of environments which will, for the moment, be described as "worst-case office" conditions. These data are useful only to illustrate the principle that all materials of interest to the contact field can form potentially harmful surface films, even in environments which may be considered benign. The data also provide useful calibration points about the magnitude of films which could be expected under certain conditions.

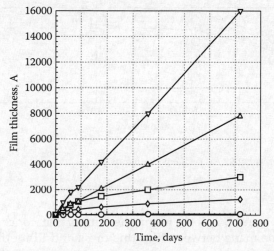

FIGURE 3.5
Reaction kinetics of copper sensors in five examples of indoor electronics operating environments.

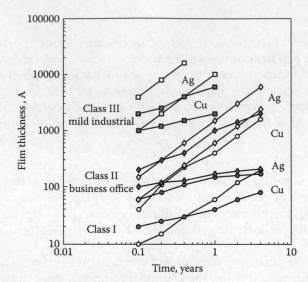

FIGURE 3.6
Tarnish of silver and copper in three classes of field environments.

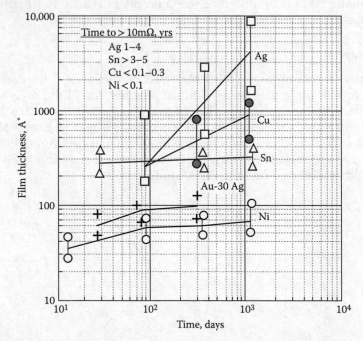

FIGURE 3.7
Kinetics of film formation in typical office/Class II environments.

3.2.2.3 Film Effects

The curves in Figures 3.1 and 3.7 point to another important concept. It refers to the distinction which must be made between film thickness and film effects. This effect is best illustrated by the data for the extremes of silver and nickel. On the basis of kinetic data alone, reference to Figure 3.7 might lead to the conclusion that for the same exposure period, silver would give the worst electrical performance, as shown by contact resistance, and nickel the best. In reality, just the reverse may often be found. The reasons for this may be found in Figure 3.1 which indicates the *effects* of surface films on nickel are far more deleterious than films on silver. In simple terms, the films on nickel are tenacious or very hard and more difficult to mechanically rupture to establish contact with the metallic asperity. In contrast, films on *pure* silver appear to be relatively soft. In summary, there is always a trade-off between kinetics of film formation and the effects of the resultant films.

3.2.2.4 Shielding Effects

It is important to note data such as those shown in Figure 3.7 are for materials directly exposed to an environment. They are useful for describing the environment, but fortunately they do not necessarily indicate the magnitude of films which will form at a contact interface *within* a device. Such values will usually be far lower—possibly by one or two orders of magnitude. This is due to the very important effect often referred to as "shielding" by structures, housing, and so on. The shielding effect is dramatically illustrated in Figure 3.8. In this case, the test sample was a relatively open edge card connector into which a simulated PC board of solid silver was inserted during a field exposure. It is clear from the visual effects that the corrosion film formed below the plane of the connector housing is far lower than that on the silver which was freely exposed to the field environment. This effect is often not fully appreciated as a major factor in the protection of electrical contacts in service. The mechanism of shielding is not fully understood but is believed to involve environmental attenuation by adsorption of pollutant onto surrounding structures as well as a limitation on air exchange rates.

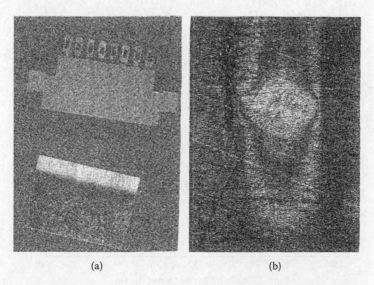

(a) (b)

FIGURE 3.8
Example of shielding by a connector housing: Silver one-year field exposure. (a) Board unmated from connector: after mated exposure (b) Shielding at connector interface by spring.

3.2.3 Reactivity Distributions

The results for copper in Figures 3.5 and 3.6 begin to demonstrate a fact which should be obvious. It is the severity levels of operating environments differ greatly between locations, see Section 2.5. An understanding of how broad this distribution may be together with the effects of these environments on relevant materials is of great commercial and technical significance, including its importance for design of realistic accelerated tests. Data of the type shown in Figure 3.5 have been obtained from various sources over the last decade at thousands of sites worldwide. When such data are normalized to a consistent exposure time; for example one year, one month, and so on, the distribution of such values can be plotted to give a statistical "picture" of the severity levels of the world's operating environments for electrical and/or electronic application environments including the electrical contact. Typical distribution plots are shown in Figures 3.9 and 3.10, taken from [20–22]., This graph reveals, in one concise set of data, many important principles related to field applications and the design of accelerated laboratory tests for electrical contacts. First, Figure 3.9 shows that actual severity levels of environments differ over a broad range

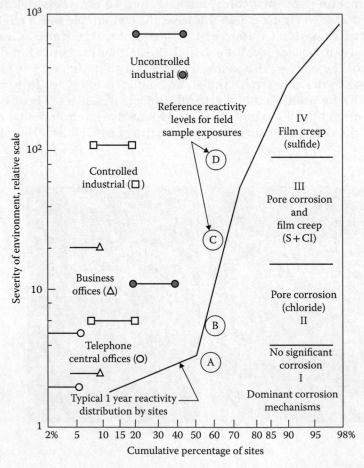

FIGURE 3.9
Reactivity of distribution and metallic corrosion mechanisms for indoor electrical and electronics environments.

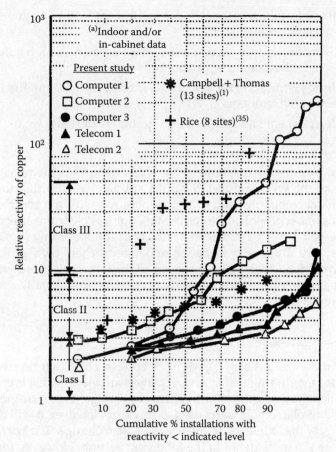

FIGURE 3.10
Typical reactivity distributions for individual manufacturers and/or product lines.

of at least four or five orders of magnitude. At first glance, this is the potential range into which manufacturers must design and qualify products, since in many instances there is little knowledge of where the products will be used in practice.

3.2.3.1 Severity versus Performance

The vertical axis of Figures 3.9 and 3.10, which would normally be expressed by some measure of corrosion (film thickness, weight gain, and so on), is a direct measure of environmental severity. These data alone are relatively meaningless to the equipment manufacturer or end user until such data are related to failure thresholds or failure rates. A discussion of this subject is beyond the scope of this chapter. In fact, such information is often regarded as proprietary. It is appropriate, however, to state some general principles relating to Figure 3.9 and the concept of environmental classes (to be defined):

1. There are many application environments in which corrosion does not occur, that is, corrosion does not affect reliability: Class I environments.
2. Beyond Class I levels, corrosion does often begin to affect reliability with effects possibly beginning in two-to-three years' time: Class II.

3. Beyond Class II levels, failure rates on some equipment may be high with failures beginning in an year's time: Class III.

4. Beyond Class III, most commercial equipment may experience high failure rates in the absence of special hardening or protection.

5. There is often little justification for designing or attempting to qualify equipment for the most severe environments.

6. Corrosion of electrical contact and electromigration (external) remain major reliability concerns.

The basis for item 5 is found in Figure 3.10. This attempts to demonstrate that the overall distribution as shown in Figure 3.9 can be subdivided into distributions which are unique to individual companies and even product lines within a company. There are many reasons for this (most of which are accidental), but it is clear the effective corrosion risk differs greatly as shown. This has been a subject of intensive investigation in recent years as companies have attempted to better understand their own product distributions. Economies have driven such questioning, since the costs associated with attempted equipment hardening for severe environments may escalate dramatically as one proceeds up the scale in Figures 3.9 and 3.10.

3.2.3.2 Environmental Classes

In Figures 3.9 and 3.10, the words Class I, II, III, or IV appear. This is a term advanced by the author to acknowledge the finding that as severity levels are increased, it is typically found the dominant corrosion mechanisms (and chemistries) change. This terminology is based on results obtained with plated gold parts used as sensors. However, it will also be shown that corrosion chemistry and kinetics on other metals also change. The corrosion mechanisms will be discussed in more detail in the following section. However, for the moment it is important to consider the implications of this class structure. First, it is clear since these dramatic changes occur there can never be a single environmental or corrosion test that will address all of these concerns. This is one of the reasons there is a shift towards multiple types of MFG tests to simulate the various reactions. Second, and because of the results in Figure 3.9, not all of the corrosion mechanisms are of equal relevance to all companies. This conclusion has enormous implications for both materials selection and qualification testing, see Section 2.5.

3.2.3.3 Specifications

Reference has already been made to one recent specification for field environmental monitoring—ISA SP71.04. This specification relies on copper as the primary means for environmental monitoring. More specifically, it relies on film thickness values derived from about one month of exposure to determine one of the four environmental classes designated G1, G2, G3, or GX. While these designations were derived from the Battelle studies, these class designations do not correspond exactly to the original Battelle Class I, II, III, and IV limits. Since some confusion has existed on this matter, an approximate translation has been given in Table 3.1 (see also Section 2.5). It should be noted that the ISA specification refers only to a method for defining the environment. No information is given or implied as to acceptable limits for operating the equipment. Furthermore, this specification relates in no way to laboratory qualification tests or requirements.

TABLE 3.1

Comparison of Battelle and ISA Environmental Classes

Battelle	ISA Classification			
	G1	G2	G3	GX
I	X			
II	X			
III		X		
IV			X	X

3.2.4 Corrosion Mechanisms

This section will not deal with a detailed discussion of true corrosion mechanisms. Instead, it will discuss some of the practical aspects of the corrosion processes.

3.2.4.1 Silver

Silver is somewhat unique amongst the materials of contact technology. One virtue of silver is, for practical purposes, it does not form oxide films as do other materials such as copper and nickel. In fact, its oxide tends to decompose above about 150°C in air. The long recognized deficiency of silver is its tendency to react with reduced sulfide species (H_2S, S_8) and form dark tarnish films of Ag_2S. In fact, silver may react with another important type of contaminant in the environment—the reactive halides (Cl, Br, F). In reality, it is mainly chlorides that arise from a wide variety of organic and inorganic sources. In many cases, these reactions may be largely "cosmetic" and have no adverse effect on reliability due to the unique properties—softness of films—on silver. Unfortunately, a unique offsetting characteristic of silver is shown in Figure 3.6. It reacts according to nearly linear kinetics; that is, the rate of film formation never slows down. For all these factors combined, it may be concluded that the performance of silver will be determined by the balance between a high rate of film formation in polluted environments and the normal force on the contact to mechanically "manage" the relatively soft films. The latter conclusion may be placed in proper perspective by noting a common observation from industrial experience. It is common to observe silver contacts in power equipment (higher force) which have turned black (thick Ag_2S/AgCl films) but continue to perform reliably. It is unlikely a similar statement could be made for any other material used in the contact field. Silver in a sulfide ambient also can form silver whiskers (see Section 8.4.4).

3.2.4.2 Copper

The reaction kinetics and film chemistries on copper are quite varied as shown in Figure 3.6. The film properties on copper are quite unlike those for silver. In fact, as shown in Figure 3.1, its characteristics represent the norm of most contact materials. It is evident from the combined effect and kinetic data why copper is rarely used directly as a contact surface except in some high-force devices and in mild environments. Some of the characteristics of copper are summarized in Table 3.2. These results indicate that in the benign or Class I environments copper forms a self-limiting of Cu_2O of about 300 Å; that is, the reaction stops. At the other extreme of

TABLE 3.2

Definition of Experimental Classes by Dominant
Mechanism(s) and Chemistries on Control Coupons

	Au/Ni/Cu (Porous)		
Class	Mechanism	Highest to Lowest	Copper Chemistry
1	None	—	Cu_2O
II	Pore	Ni, Cu, Cl, O, (S)[b]	Cu_2O, $Cu_xO_yCl_z$
III	Pore and creep	Cu, S, Ni, Cl, O	Cu_2S, Cu_2O (?)[a]
IV	Creep	Cu, S, Ni, O, Cl	Cu_2S, (Cu_2O), (?)

[a] ? indicates unknowns.
[b] () indicates minor amounts (<10%).

severe industrial environments, the kinetics approach linear behavior and the films
are often dominated by Cu_2S. The reactions on copper are strongly dependent on the
interactions of humidity and levels of pollutants.

3.2.4.3 Nickel

Nickel must be used with care as a material for a contact surface. This is due to its unique position in Figure 3.1. In the milder environments, the film on nickel is largely, if not exclusively,
NiO of perhaps a limiting thickness of 30–50 Å. It has generally been assumed this remains
a passivating film which grows no further. In fact, this is often used as the explanation of
why a nickel underplate should serve as a "corrosion barrier" under gold platings. This
apparent benefit can actually be "demonstrated" in older laboratory tests which rely only
on gases such as H_2S and/or SO_2 as the major pollutants. Modern studies have shown, however, even in Class II environments where chlorides are present, the nickel oxide film may
be attacked to render nickel susceptible to corrosion and resumed film growth. In summary, nickel may sometimes be used successfully as a contact surface in mild environments
and where sufficient force is available to mechanically manage its thin oxide film. However,
in more aggressive environments, the use of nickel is somewhat limited.

3.2.4.4 Tin

It is appropriate to consider tin separately, since it is also a very unique material. It
is both a base metal and a metal which has been used very successfully as a contact
finish. As shown by the data in Figure 3.7, tin develops a limiting film (oxide) in the
range of about 300 Å in nearly all classes of environments. This fact has been demonstrated in many field trials which have generally shown that tin may remain virtually
unaffected in nearly all classes of indoor environments. In spite of the fact this oxide
film is relatively hard, the data of Figure 3.1 show that it is quite easy to establish low-
resistance metallic asperity contact to tin. The reasons for this are well known. The
preceding discussions might appear to be close to describing the ideal contact material.
Unfortunately, the features which may make tin a very corrosion-resistant material also
make it susceptible to fretting failure as discussed in Chapters 5 through 7. In summary,
tin (and solder) can be used very successfully as a contact finish even in aggressive
environments. However, its successful use tends to be in higher force (>200–250 g) and
low-durability applications.

3.2.4.5 *Porous Gold Coatings*

This type of coating will be discussed since it has been mentioned in connection with Figures 3.9 and 3.10 to describe different corrosion mechanisms by environmental class. The mechanisms to be discussed are not restricted to gold coatings and may be found on a wide range of precious metal coatings and even tin coatings to various degrees.

3.2.4.6 *Pore Corrosion*

The mechanism of pore corrosion is discussed in Chapters 2 and 8. Field studies have confirmed that this corrosion mechanism is real, as reflected in the legends in the right-hand column of Figure 3.9. It is important to first note, while pore corrosion is a viable failure mechanism, it begins to occur only above some threshold level of severity (Class II). Below this (Class I), environments are so well controlled (by accident or design) that corrosion does not affect reliability. This is an important concept, since there are many application environments where the issues of numbers of pores, materials selection, environmental testing, and so on may be relatively unimportant. Within the last decade, field studies have revealed the importance of reactive chlorides in the corrosion processes on contact materials [13–15,23–25]. One example of this is shown in Figure 3.11. This shows one typical surface view of pore corrosion development on the surface of a typical gold plating (Au/Ni/Cu) exposed in a Class II field environment. The accompanying X-ray data show corrosion products rich in chloride (and oxygen, not shown) that have reacted with underlying metals (Ni and Cu). A point which should also be noted is the general absence of sulfur (sulfides or sulfate) in this corrosion process. Pore corrosion in Class II environments can develop rather quickly (six months to two years) depending on details or severity of the environment.

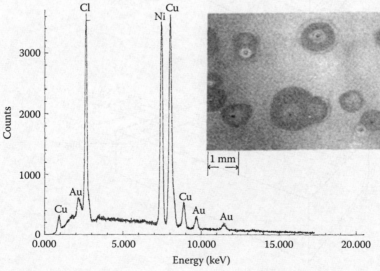

FIGURE 3.11
Class II, pore corrosion.

3.2.4.7 Corrosion Product Creep

Creep corrosion has been introduced in Chapter 2. Above Class II levels, several important changes occur. One is the appearance of a new corrosion mechanism; the second is a change in film chemistry. The new corrosion mechanism is one which has long been known [26] and which was originally associated with the movement of silver sulfide films over gold surfaces. More recent studies have demonstrated the movement of corrosion films from various base metals over precious metal surfaces [27–31].

The creep process is an interesting one and involves the movement by surface diffusion of base metal corrosion products across a more noble metal surface. The process is shown schematically in Figure 3.12 [31] in which a test sample was designed to simulate situations such as the sheared edge of a composite metal strip, a plated edge card finger, and so on. In these cases, a corrosion product may form on the copper and quickly migrate up onto and across the gold surface. This process may be quite rapid as the author has observed migration distances on actual components measured in a few millimeters within one to two years in some Class III conditions.

Examples of these features taken from actual field samples are shown in Figures 3.13 and 3.14. These indicate while creep can occur from exposed base metal edges, it is more pronounced due to the mechanism by which corrosion products spread away from pore corrosion sites (Figure 3.13). Figures 3.13 and 3.14 also show the significant change in corrosion chemistry to sulfide-rich products. In fact, it is the sulfide products (Cu_2S, Ag_2S) which

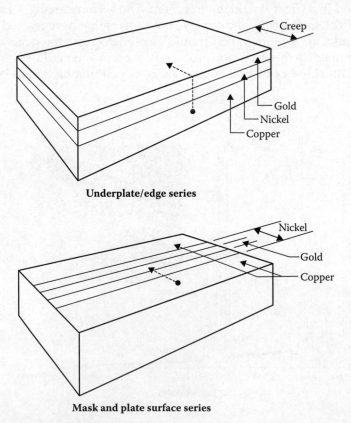

FIGURE 3.12
Test samples for creep studies.

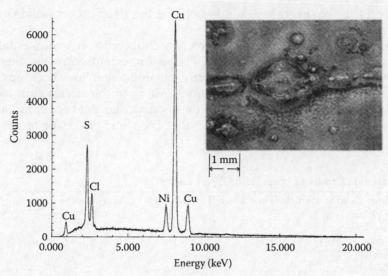

FIGURE 3.13
Class III, pore corrosion and creep.

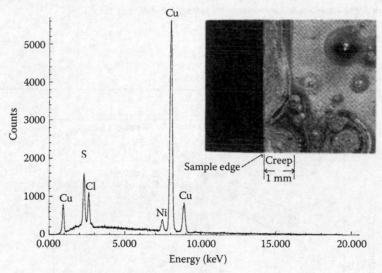

FIGURE 3.14
Class IV, edge creep.

are largely responsible for observed creep. This is one of the reasons why creep tends to be observed more frequently in the industrial environments where sulfide (not sulfate) concentrations are higher. From the preceding discussions it should not be assumed that only sulfides are important in the creep process. In fact, recent studies have shown the strong synergistic effects in which chlorides greatly accelerate the creep process but do not significantly appear in the corrosion chemistry of creep [31].

Figure 3.9 describes the Class III conditions as being characterized by the dual mechanisms of pore corrosion and creep. Class IV is characterized by creep and almost exclusively sulfide products. The latter is relatively descriptive, but it should be noted these differences are perhaps less dramatic than those between II and III. In the case of Class IV, environments tend to be so severe and sulfide-rich that the creep process tends to obscure the fact that pore corrosion still occurs.

Table 3.3 summarizes some of the known materials and environmental factors associated with the creep process. While this is an important corrosion mechanism, it is again important to recognize it may only be important for those applications which will experience Class III and IV conditions. Many of the effects summarized in Table 3.3 are shown in Figure 3.15 [31]. While these particular data are from laboratory exposures, it has been the author's experience that the same effects or order occur in

TABLE 3.3

Features of Corrosion Product Creep: Field Observations

1. Sulfide products creep at the highest rate. The relative order of common products may be $Ag_2S = 100$; $Cu_2S = 20–30$; $NiS = 1$
   Silver sulfide = 100
   Copper sulfide = 20–30
   Nickel (sulfide?) = 1–2
   Copper oxide = <0.1

2. Other reaction products (chlorides/oxychlorides) do creep but at rates several orders of magnitude lower than that of sulfides.

3. Creep of all products is the highest across pure gold.

4. The relative order of creep rates across common contact finishes may be
   Pure gold = 100
   Gold alloys = 5–50
   Palladium = 30–40
   Tin = 1–2

5. Base metal alloys may play a major role in altering creep kinetics.

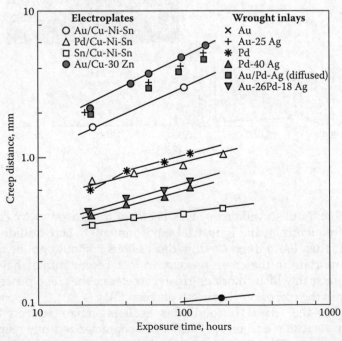

FIGURE 3.15
Effects of materials on corrosion product creep in $H_2S–NO_2–Cl_2$.

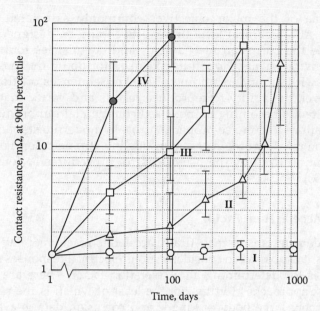

FIGURE 3.16
Corrosion of porous gold coupons in four classes (I–IV) of indoor electronics sites.

field environments. Finally, many of these findings may be placed in perspective by the data in Figure 3.16 [23]. These present good estimates of the rates at which conventional gold-plated samples may degrade in four classes of field environments and as measured by contact resistance probing. It should again be emphasized these data are for materials freely exposed to the environment and without the important attenuating effects of device shielding. For this qualification, it is clear that probable degradation times may be shorter compared to the life of the equipment design.

3.3 Laboratory Accelerated Testing

This section will deal with the subject of gaseous corrosion testing and will emphasize modern developments of the mixed gas environments. It will not discuss common tests such as temperature–humidity or even single-gas tests in any detail, since there is sufficient information in the literature that such procedures have no great relevance to the field effects already reviewed in this chapter.

3.3.1 Objectives

It is important to establish what the objectives and expectations may be as associated with the modern environmental tests of the type discussed in this chapter. There are several reasons for this. One is most of these tests are complex, require specialized equipment, are possibly of long duration, and as a result somewhat costly to perform in comparison to tests such as temperature–humidity. Second, there is an increasing need for realistic tests with known acceleration factors as a means for evaluating new materials and designs. Therefore, the outcome from such tests and the decisions reached can have significant economic consequences.

Table 3.4 lists some of the major objectives commonly associated with modern environmental tests. At a technical level these tests should be designed to reproduce as closely as possible the corrosion mechanisms and chemistries found in the various classes of field environments. Hopefully, the need for doing this has already been established together with an understanding of the consequences of "doing it wrong." At an operational level, the ability to reproduce results and verify that the reactions produced are as desired is becoming of increased commercial importance. The reason is very simple. The failure to exercise proper controls in modern tests can easily lead to differences in reaction rates by orders of magnitude in extreme cases. Fortunately, this possibility is being minimized by the use of control procedures as found in modern specifications [19,32,33].

3.3.2 Definition of Acceleration Factor

One objective of modern tests is to realistically accelerate reactions relative to the field. For electrical contacts, it is possible to consider various electrical responses such as contact resistance as the basis for determining an acceleration factor. There are problems and limitations with most approaches to this, not the least of which is the absence of data. Fortunately, a reasonably large body of kinetic and corrosion data is available and for a number of materials. For this reason, acceleration factors are commonly defined as illustrated in Figure 3.17.

TABLE 3.4

Objectives of Laboratory Accelerated Corrosion Tests

Technical
- Reproduce corrosion kinetics
- Reproduce corrosion mechanisms
- Reproduce corrosion chemistries

Commercial
- Reproduce results
- Verify test run correctly
- Stay within capabilities/cost of laboratories

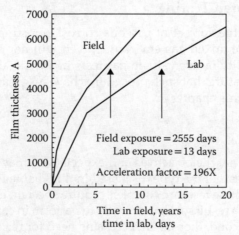

FIGURE 3.17
Example of acceleration factor determination.

The calculation process might proceed as follows. First, a determination must be made of the appropriate environmental class for the material, component, or device to be studied. This does nothing more than to define an appropriate test. Next, a target field life is defined. On the basis of this information and assuming a field database is available, a value of film thickness (or some other measure of corrosion) on a sensor material exposed for the specified life could be determined. Finally, the objective of the laboratory test would be to expose the test article to the designated environment for a timeperiod sufficient to grow at least the same amount of film.

In practice, the acceleration factor of the designated test would already be known. Since the field equivalent life would already be designated, a calculation could then be made of the exposure time. It is, of course, assumed the test selected would meet the requirements of Table 3.4. Almost any type of test can produce corrosion, and as a result an acceleration factor could be "calculated." However, reference to the quotation at the beginning of this chapter indicates why this calculation may have no meaning.

3.3.3 Historical Background

The environmental tests which have been historically used in the contact field have had two features. They have used single gases. The gases have been at high concentrations by modern standards. The gases have been mainly H_2S or SO_2. Humidity levels have been very high; and the tests have largely been static, that is, there was no flow or replenishment of gases. Discussions earlier in this chapter, and in particular those involving Figures 3.2 and 3.3, have reviewed the problems with such tests, and their unrealistic nature. This situation has been "known" for many years. In spite of this such tests are included in formal specifications [5,34] and their use continues. One reason for this may be they can be performed at a relatively low cost. We will, therefore, review some of the effects of single-gas exposures with the intent of leading to the conclusion unless the synergistic effects of multiple pollutants are incorporated into tests, there is little hope of realistically simulating field corrosion.

3.3.4 Single-Gas Corrosion Effects

Table 3.5 presents a concise summary of the effects of single pollutants on the important materials which have already been discussed when used for sensor purposes. It also begins to introduce the very important synergistic effects among pollutants and in particular those which incorporate the reduced sulfide–chloride interactions. In one respect, Table 3.5 traces from top to bottom the historical developments in environmental testing of electrical contacts.

3.3.4.1 Hydrogen Sulfide

Hydrogen sulfide is an important source of the sulfur which causes the familiar tarnishing reactions on silver, and the sulfur-containing compounds on electrical contacts. It may arise from a number of natural sources and industrial emissions. While the median levels of H_2S in indoor environments appear to be on the order of 1 part per billion (p.p.b.) or less [16,24], such concentrations are sufficient to cause rather serious corrosion. The last statement is an important one in order to place in proper perspective what constitutes "severe" environments for the electrical contact or electrical and electronic equipment in general. Pollutant are found in indoor environments in concentrations sufficient to cause

TABLE 3.5
Summary of Laboratory Gas Effects on Corrosion

| Primary Pollutant(s) | | Reactions On | | | Dominant Chemistry | |
| | | | Porous Gold | | | |
Type	Concentration[a]	Copper	Pore	Creep	Copper	Porous Gold
H_2S	<100	Low rate; «parabolic	No	No	Cu_2O	—
	>1000	Linear—exponential	Low	Low	Cu_2S	Cu_2S
SO_2	<100	Low rate; «parabolic	No	No	Cu_2O	Cu_2O
	>1000	Linear—exponential	Severe	No	$Cu_xS_yO_z$	$Cu_xS_yO_z$
NO_2	<100	Unreactive	No	No	—	—
Cl_2	<10	Low	Yes	No	$Cu_xCl_yO_z$	$Cu_xCl_yO_z$
	>30	Moderate	Yes	No	$Cu_xCl_yO_z$	$Cu_xCl_yO_z$
$H_2S + SO_2$	<100	Low	Low	No	Cu_2O	Cu_2S
	>100 <1000	Linear—exponential	Low	Low	Cu_2S	Cu_2S
$H_2S + NO_2$	<100	Low	Low	Low	Cu_2O	Cu_2S
	>100 <1000	Linear—exponential	Low	Low	Cu_2S	Cu_2S
$H_2S + SO_2 + NO_2$	<100	Moderate—parabolic	Low	Low	Cu_2O	Cu_2S
	>100 <1000	Linear—exponential	Low	Low	Cu_2S	Cu_2S
$H_2S + Cl_2$	<10	Parabolic	High	Low	$Cu_xCl_yO_z$	$Cu_xCl_yO_z$
	<100 + <10	Parabolic	Severe	High	$Cu_2S + Cu_xCl_yO_z$	$Cu_2S + Cu_xCl_yO_z$
$H_2S + NO_2 + Cl_2$	<10	Parabolic	High	Low	$Cu_xCl_yO_z$	$Cu_xCl_yO_z$
$H_2S + SO_2 + NO_2 + Cl_2$	<100 + <10	Parabolic	Severe	Severe	$Cu_2S + Cu_xCl_yO_z$	$Cu_2S + Cu_xCl_yO_z$

[a] parts per billion

serious corrosion, yet are usually well below the limits of sensory perception. This is often the case except possibly at the extremes of Class IV conditions. These findings from modern research are often difficult for many people to accept, yet in recent years our notion of what truly constitutes severe conditions have changed dramatically. It is not surprising then to find modern developments of highly accelerated environmental tests to incorporate gas concentrations measured in a few parts per billion!

Table 3.5 shows that H_2S as a single-gas pollutant even at high concentrations is not only realistic, but also relatively mild in its reactions on precious metals such as porous gold platings! H_2S at high concentrations will corrode copper, but it is not effective in promoting either pore corrosion or creep at useful rates. These conclusions are supported by the data in Figures 3.18 through 3.20, to which we will often refer for considerations of other effects of the pollutants. In summary, H_2S is an important pollutant. However, by itself it is totally unrealistic of effects found in real-world indoor environments. Its use will result in surprising mild test conditions.

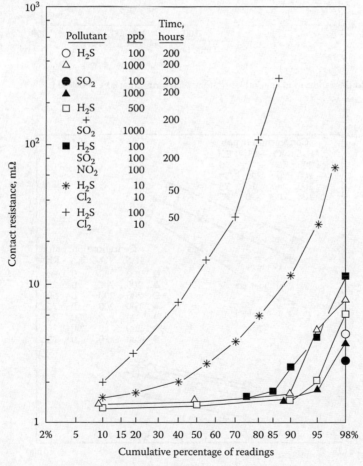

FIGURE 3.18
Contact resistance distribution of porous gold after laboratory exposures: 30°C; 70% RH.

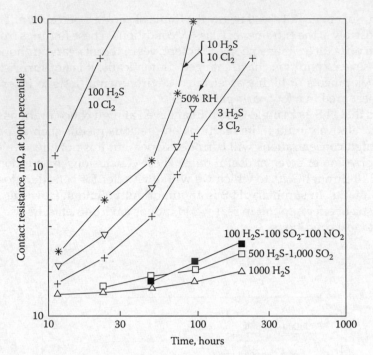

FIGURE 3.19
Kinetics of porous gold degradation in several laboratory environments: 30°C; 70% RH.

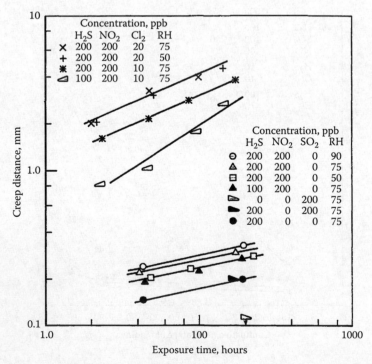

FIGURE 3.20
Effects of environmental test variables on film creep over gold from Cu–Ni–Sn alloy.

3.3.4.2 Sulfur Dioxide (SO₂)

Sulfide dioxide is a somewhat controversial pollutant. There is a perception since SO_2 is in the environment, is a pollutant that is constantly measured, and has historically been used for test purposes [34,35] then it "must", in some manner be important for use in accelerated tests. In this respect, SO_2 may be regarded as at least a "psychologically important" pollutant (authors' emphasis). We will, in fact, find SO_2 being incorporated into a number of modern mixed gas standards as the "fourth gas" [19,36–38]. This issue will be discussed in a later section of this chapter. For the present, however, Table 3.5 indicates the following: SO_2 at concentrations even significantly above field levels is essentially unreactive with materials of interest to the contact field. It is relatively unreactive towards copper (Table 3.5); it does not cause corrosion product creep (Figure 3.20); and it is nearly unreactive towards porous gold (Figure 3.18). There is increasing evidence SO_2 does not significantly participate in or influence reactions in the indoor environments. However, it is widely recognized as being of importance in true outdoor environmental exposures. In summary, SO_2 is not a particularly corrosive agent. Single-gas tests based on the use of SO_2 may be surprisingly mild and are unrealistic of any effects found in practice.

3.3.4.3 Nitrogen Dioxide (NO₂)

Nitrogen dioxide which is present in all environments is essentially unreactive towards contact materials as shown in Table 3.5. The single exception to this is from the work of Tompkins [39] who demonstrated that NO_2 may react directly with solder coatings. However, to our knowledge there are no tests which have been used or proposed based on single-gas NO_2 exposure. The importance of NO_2 derives from its almost universal use as a third gas in modern MFG tests. Its function appears to be to act as an oxidizing agent to accelerate sulfidation reactions [16,20,31,40–43].

3.3.4.4 Chlorine

Perhaps the single most important development in the field of environmental testing of electrical contacts and equipment has involved findings about the role of chlorides in the corrosion process and its incorporation into modern tests [13,14,20,21,24,44–47]. Without the proper incorporation of some form of reactive chloride into a test, it is probably impossible to even begin to simulate field reactions at useful kinetics. It is reasonable to conclude the synergistic effects of H_2S and Cl_2 are the most important of virtually all degradation mechanisms found in the field. It is important to note there is no claim that chlorine is the active species in natural environments which causes the multitude of chloride-containing corrosion products found on field components. In fact, chlorine is so reactive its life in real environments is likely to be very short. It has been argued by some that hydrogen chloride is a more likely pollutant; however, at least one study has shown that hydrogen chloride does *not* produce the chloride-based reaction found on palladium which is found as a major reaction product in the field [46]. That same study indicated that Cl_2 does produce the "correct" results.

These findings are important for purposes of laboratory simulation for reasons related to implementation. Most potential Cl sources including HCl, are difficult to introduce, control, and analyze. Fortunately, Cl_2 has proved to be the least difficult and in this respect is a "convenient" source which has been shown to correctly accelerate reactions. The effects and potent nature of Cl_2 are shown in Table 3.5 and Figures 3.18 through 3.20. This species clearly accelerates all reactions to useful (accelerated) levels. It is capable of doing so at extremely low concentrations which, in fact, push the limits of analytical capabilities. In summary, Cl_2 is probably the single most critical agent in modern environmental tests. However, it must be used in combination with other relevant pollutants. Tests without adding Cl_2 should be regarded as unrealistic and probably very mild.

3.3.4.5 Mixed-Gas Sulfur Environments

The beginnings of more modern developments involved the use of environments containing H_2S, SO_2, and NO_2 [41–47]. Much of this work brought with it the use of somewhat more realistic concentrations into the range of hundreds of parts per billion as opposed to p.p.m. (or higher) levels of the past. It was found that two-gas and three-gas tests incorporating at least H_2S and NO_2 could begin to produce more realistic results on materials such as copper and silver. However, as shown in Figures 3.18 through 3.20, such tests were eventually shown to be both very mild and often unrealistic.

3.3.4.6 Humidity

At this point it would be appropriate to consider the role of humidity in the type of corrosion reactions being considered. There is no question humidity accelerates most corrosion reactions. Earlier in this chapter, this was discussed to hopefully dispel the common belief that humidity alone is a corrosive agent. There also appears to be the common belief there is a "critical humidity" for corrosion, whereby if humidity is maintained below some critical value, corrosion will not occur. Many of these issues have been reviewed relative to indoor environments [12,15,16,20]. While relative humidity clearly accelerates corrosion in the mixed-gas environments and in an exponential manner, there is little evidence for a critical value. This is consistent with field experience by the author which has shown numerous instances of corrosion-related failures on equipment (and test samples) at relative humidity levels as low as 35%–40% annual median! Again, it is the combined effects of humidity and critical pollutants which dominate the corrosion processes.

A good summary of some of the effects discussed in the last few sections can be shown in Figure 3.21 taken from unpublished work by the author. This involves the corrosion of plain carbon steel by low-level pollutants in combination with humidity. While Figure 3.21 may not relate directly to electrical contact materials, it does dramatically illustrate the effects being discussed and on a material to which most people can relate; that is, steel is known to rust in most environments. Three important conclusions can be reached from Figure 3.21. First, humidity alone does little to steel—either visually or quantitatively. Second, most common pollutants, including SO_2 and in combination with moderate humidity, have little effect. Third, even trace levels of Cl_2 with moderate humidity accelerate corrosion by orders of magnitude. These effects directly correlate with many of the conclusions already reached. These results demonstrate that relevant corrosion can be produced at humidity levels well below saturation levels. Modern test developments have, therefore, been able to work at humidity levels in the range of 70%–75% RH and above. This is a fortunate result, since there are practical limitations which prohibit the use of some required analytical equipment at humidity levels much above this range.

3.3.4.7 Temperature

The effects of temperature have been reviewed elsewhere [20]. The reactions of the three-component and four-component MFG tests appear to follow a classical Arrhenius

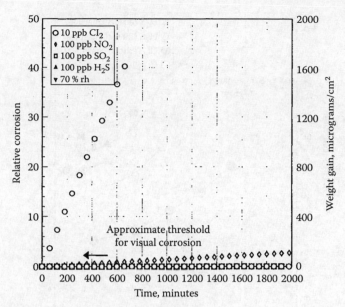

FIGURE 3.21
Effects of single pollutants on the corrosion of 1010 steel at 70% RH, 25°C.

relationship but with a relatively low activation energy of 16–17 kcal. Data for Class III and IV conditions indicate that these tests can be accelerated by temperature in theory. For example, compared to a normal test temperature of 30°C, the calculated acceleration values are ~50% at 40°C and 100% at 50°C. In practice, however, an upper temperature limit exists, and it is questionable whether the added acceleration by this means is worth the experimental difficulty of working at elevated temperatures. The problems arise not by any unrealistic change in corrosion mechanisms, but by the problems associated with the transport of moisture-laden gases from the test environment to external analytical equipment without condensation and the associated gas losses. For this reason, few, if any, modern tests operate at elevated temperature.

3.3.4.8 Gas Flow Effects

All modern developments involve the use of continuous gas flow systems in which pollutants are continually replenished. The goal is to establish conditions of dynamic equilibrium in which constant gas concentrations are maintained. The reason for this approach is shown in Figures 3.22 and 3.23. These data show that within any test chamber there will be multiple gas loss mechanisms by which pollutants will be irreversibly lost. These loss rates are dependent on both humidity and rates of air exchange. Provisions must be made to make up these losses in order to maintain a *constant* and *specified* concentration *in the gas phase*. Figure 3.22 shows that high air exchange rates are favored; however, there are practical limitations which may limit the high end to 10–20 times.

A clear distinction should be made between air exchange rate and linear velocity across samples. For the air exchange variable, we may conclude there is no fundamental effect of this variable on reaction rates provided all other variables remain constant. The linear

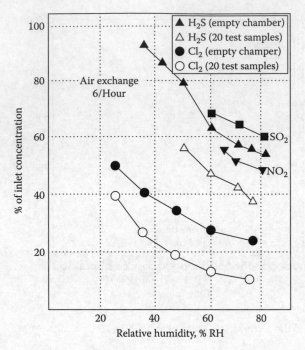

FIGURE 3.22
Analyzed gas concentration in test chamber versus humidity.

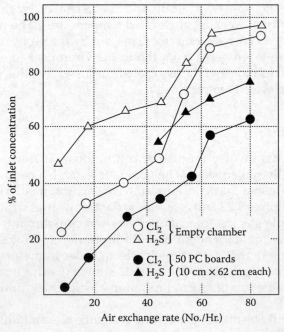

FIGURE 3.23
Analyzed gas concentration in test chamber versus air exchange.

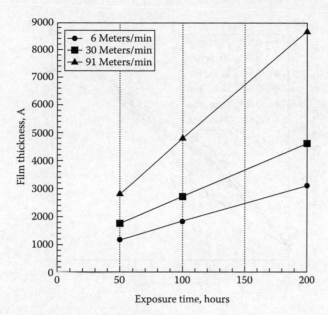

FIGURE 3.24
Effects of air flow velocity on copper Class II FMG.

velocity is not generally addressed or specified with one notable exception [38]. However, there is a clear effect of velocity which may be found on certain types of samples as shown in Figure 3.24 [48]. These data were obtained on copper, which might be used for control or verification samples in a test chamber. It is useful to note the range between about 6 m and 30 m/min probably represents the actual operating ranges of most test systems. In reality, the vast majority operate within an even narrower range towards the lower end of Figure 3.24.

In consideration of these data, it is possible that variation of as much as 50% in reported control rates could exist among laboratories due to velocity effects. This number could be even greater in those cases where samples were artificially shielded by other test hardware. There are very few published data to indicate whether any significant velocity effects exist beyond those which might occur on control materials freely exposed to the test environment. One published study by the author [48] has concluded that on actual hardware there is little, if any, effect as shown in Figure 3.25. We may conclude while linear velocity may have little effect on test hardware, it is still a variable which should be controlled and reported. The immediate reason involves the commercial implications due to possible variations on copper control samples which are being required in most modern specifications.

3.3.5 Mixed-Gas Environments

From information already presented, it should be clear that the mixed-gas environments represent the state of the art for realistic environmental simulation. These environments are often referenced as flowing mixed gas (FMG) or mixed flowing gas (MFG), with the latter terminology being the most common. All recent tests and which are in most common use have employed at a minimum the three-gas interactions involving $H_2S + NO_2 + Cl_2$ at various levels. Some require a fourth gas, which is always SO_2. All operate at humidity

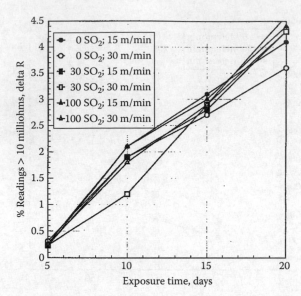

FIGURE 3.25
Effects of SO_2 and velocity on connector reliability: Au/Ni; Class II FMG.

levels in the range of 70%–80% RH and temperatures in the range of 25°C–30°C. These developments represent a significant advance in the state of the art of accelerated testing. Unfortunately, there has been a proliferation of three-gas and four-gas tests developed by different organizations and even individual companies. Right now, there are probably three major documents which are relevant to formal MFG specifications. These are:

Reference	Number of Tests
IEC 68-2-60; Test Ke	4
ASTM B845	12
Bellcore GR-63-CORE, Issue 1	2

While there is some duplication of environments among these options, the user may be left with a bewildering array of possibilities. In addition to the issue of the specific test to select, there is also the issue of duration(s) of the tests, which in most cases, is not specific. To this list must be added unique specifications from several major companies and the two or three "unusual" four-gas environments recently coming into use from the automotive sector. It would be an understatement to say there is at present no consensus regarding the "best" test(s). In fact, Table 3.6 lists those tests which are presently among the most popular and are grouped by categories for which they are ostensibly appropriate. While some of these may appear quite similar, they can unfortunately produce reaction rates which differ by a wide margin.

Typical exposure times required by individual users vary greatly. These are typically in the range of 10–20 days with occasional requirements to 40 days. As mentioned earlier, there is no consensus view regarding the "best" test(s). However, it appears that there is some convergence of views in this direction toward the four-gas tests. Specifically, and with reference to Table 3.6, item 2 in the first group and item 2 in the second group come the closest to common usage. These tests are being widely used for qualifying both components and equipment.

TABLE 3.6

Examples of Current Mixed-Gas Environments

Mild business office conditions

- 10 H_2S-200 NO_2-10 Cl_2 – 70% RH.
- 10 H_2S-200 NO_2-200 SO_2-10 Cl_2 – 70% RH
- 10 H_2S-200 NO_2-200 SO_2-10 Cl_2 – 75% RH
- 40 H_2S-500 NO_2-300 SO_2-3 Cl_2 – 70%RH
- 36 H_2S-200 NO_2-113 SO_2-10 Cl_2 – 70% RH

Mild industrial or uncontrolled conditions

- 100 H_2S-200 NO_2-20 Cl_2 – 75% RH
- 100 H_2S-200 NO_2-200 SO_2-20 Cl_2 – 70% RH

3.3.5.1 Test Systems

Modern mixed-gas tests must be conducted in continuous flow systems as discussed earlier. At the present time, there are no known commercial sources for such systems, and all are probably "home-grown." There are, however, general guidelines for these systems as described in ASTM B827, IEC 68-2-60 and [49]. Schematics of a typical system are shown in Figures 3.26 and 3.27. It is important to note there is no single correct way for designing such a system. However, there are several key requirements which are mandatory for the successful operation of these systems. These requirements represent some of the most significant advances in specifications for this field. One of these requirements involves the use of analytical equipment for monitoring gas concentrations. It is appropriate to forcefully state that without such equipment, it is totally impossible to conduct such tests in the intended and reproducible manner. The basic reason for this is found in the data of Figures 3.22 and 3.23. On this subject, there is an important definition which may not be fully articulated in the existing specifications. Gas concentrations in such tests are defined as:

> Concentrations existing in the gas phase *in the chamber volume* and with all test samples and hardware in place and with the chamber at the intended or specified humidity.

Any deviation from the intent of this definition usually results in gas concentrations far different from the target values.

3.3.5.2 Monitoring Reactivity

A second requirement found in all referenced (recent) specifications involves test documentation and verification by some form of monitoring of the corrosion or reactivity. The significance of this requirement cannot be overstated. In theory, it should be possible to control tests and obtain reproducible results with the use of gas analysis equipment. In practice, there are so many test variables to control that small variations in critical variables can cause significant differences in reaction rates. This coupled with the possibility of instrument failure mandate the use of some method to document the real intent of such tests—to produce a certain level of corrosion. The common requirement among specifications is for the use of copper coupons at a minimum. These are typically reported as weight gain values although other methods such as cathodic reduction [50] can be used as an inexpensive method of determining the approximate corrosion chemistry.

At present, there are no precise requirements for the required reaction from the prominent tests. IEC 68-2-60, Test Ke specifies a range of expected values although this range is, unfortunately, very broad. Bellcore GR-63-CORE specifies a *minimum* level of

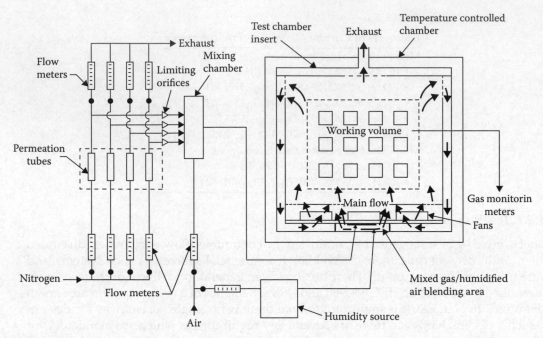

FIGURE 3.26
Schematic vertical re-circulating mixed flowing gas (MFG) tests system (with permission of ASTM).

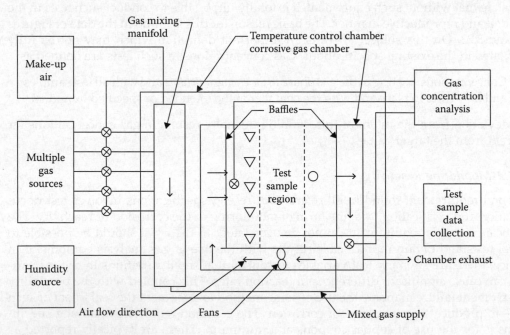

⊗ Gas concentration measurement points
◯ Temperature and humidity measurement
▽ Gas velocity measurement

FIGURE 3.27
Schematic horizontal recirculating mixed flowing gas (MFG) test system (with permission of ASTM).

reactivity which, from an operational viewpoint, is quite useful. ASTM B845 has no such requirements. In summary, results from reactivity monitoring are probably the single most important reportable data from MFG tests.

3.3.6 Test Applications

Much of the development of modern MFG test methods has been done by individuals with interests in the electrical contact field. Reference to the literature will show that major efforts have come from the electronic connector, computer, and telecommunications industries. Many of the early applications of the MFG tests have indeed involved studies of electrical contact materials in small test chambers with internal volume of a few liters. Such work is continuing particularly in the electronic connector field. However, in recent years there have been applications at all levels up to complete operating systems in chambers with volumes exceeding a few cubic meters. Specifications do not restrict the use of these tests to any areas of applications or materials. In fact, few if any, guidelines are provided for how the tests should be used. This is largely a matter of agreement between the parties involved.

3.3.6.1 Electronic Connectors

Probably the largest databases involving the MFG tests are for electrical connectors with precious metal platings, even though there is no suggestion that the tests should be restricted to such metals. Figure 3.28 shows several examples for connector reliability

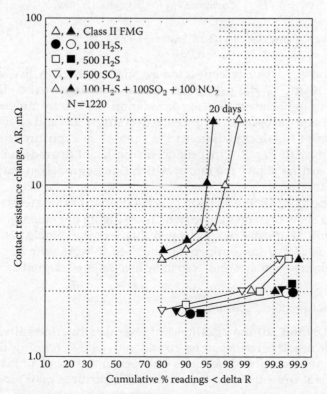

FIGURE 3.28
Edge card connector performance in dilute, sulfide-only environments: 20-day exposures, 30°C, 70% r.h.

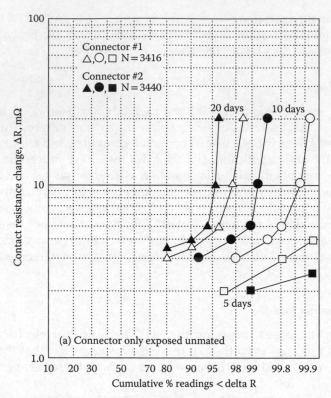

FIGURE 3.29
Performance of mated edge connector in Class II FMG environment.

studies conducted on a particular connector system exposed in five environments. The data illustrates many of the principles already discussed, since this particular connector was known to be susceptible to corrosion in field environments. These data show even very long exposures in high-concentration single-gas and sulfide-only mixed gas tests produce very little degradation. At the same time exposures in a very dilute Class II environment (with 10 ppb Cl_2), produced significant degradation (by pore corrosion). Figure 3.29 further shows the rate at which this degradation may develop in a more realistic MFG test.

These data show one more feature related to the application of these tests which involves sample size. The data show even on samples which are highly degraded as reflected in the "tails" of the distributions, the instantaneous percentage of the sample population in that tail is quite small. This fact argues strongly for the need to work with large sample sizes both to support reliability estimates and even to be able to "see" degradation.

3.3.6.2 Mated versus Unmated Exposures

The data in Figures 3.30 and 3.31 show examples of several of the effects already discussed. In addition they address the factors to be considered while choosing the application of these tests. This involves the differences between mated and unmated connector exposures. It is evident from the data that unmated exposures may produce degradation of orders of magnitude greater or faster than for mated exposures. There are even

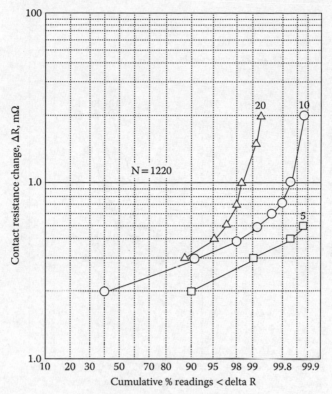

FIGURE 3.30
Performance of mated pin and socket (25 DB) connectors in Class II FMG environment.

differences according to whether the exposures are done on the pin or socket halves of the connector. These effects are due to the shielding effects discussed earlier. These examples are shown to stress the fact that application of the tests should be carefully mapped to the requirements. For example, in this case, there could be a question whether unmated exposures are really necessary and, if so, what equivalent exposure time is necessary. All of these considerations are a matter of individual choice but can have large commercial implications.

3.3.6.3 Other Considerations

Environmental tests of the MFG type are being widely used in studies far beyond the evaluation of electrical contact materials and electronic connectors. There appears to be a growing trend among companies to incorporate MFG requirements into component qualification programs. These requirements have recently moved to large-scale equipment exposures and in situ performance evaluations particularly for telecommunications equipment. In the field of industrial controls, an increasing amount of work on operating equipment appears to be done in an attempt to meet demands for reliable operation in severe environments (ISA G3 and GX). In summary, the MFG environments are being increasingly recognized and used, since they appear to be the only means for exercising the multitude of corrosion-related failure mechanisms found in practice.

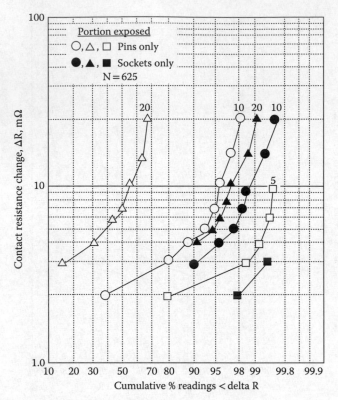

FIGURE 3.31
Performance of unmated pin and socket (25 DB) connectors in Class II FMG environment.

3.4 Lubrication and Inhibition of Corrosion

There is a continuing interest in using specialized lubricants for inhibition of corrosion and for the maintenance of a satisfactory contact resistance particularly for low-current and electronic connectors. The use of lubricants in power connectors is discussed in Section 5.5.10 and as a palliative for contact wear in Section 7.4. The trend of using thinner gold plating (i.e., <1 μm) over a non-noble substrate has made the use of protective lubricants increasingly important especially for electronic connectors. These thinner platings result in more porous coatings which can lead to pore corrosion discussed in Sections 2.4.3 and 8.3. Indeed there is an increasing suggestion to use a "flash" gold coating (i.e., <0.5 μm). This should in general be avoided unless it is coated over another noble metal such as Pd or Pt. Even then a satisfactory long-term connection cannot be guaranteed [51,52]. That being said, there is a continuing use of flash gold coatings over non-noble metals in many electronic connector systems. It is only specialized, high-reliability connectors that use a gold plate of thickness greater than or equal to 3 μm [53]. At higher currents using Sn, Sn–Pb or Ni platings, lubrication is required especially if there is a possibility of a small motion (a few μm) between the contacting surface, which can give rise to fretting corrosion (See Sections, 5.4.11, 6.4.2, and 7.3). The addition of a lubricant to a connection may, therefore, offer a hope of long-term performance of the connector. This is especially true as both the design and development of contact materials and connector is being pushed to the limits.

Lubricants have had a long history of use in electrical contacts where reduction of friction and wear are important design parameters. Their use in preventing corrosion is only now becoming prevalent. Zhang [54] has shown it is important for the lubricant to cover the contact surface well for good protection. He shows that the effective coverage depends upon: (a) The contact angle between the lubricant and the contact metal, (b) The thickness of the lubricant after its application, (c) The adhesion of the lubricant to the metal surface, and (d) The roughness of the metal surface. For electronic connectors, most lubricants are dissolved in a solvent. This solution is then coated on to the metal surface. When the solvent evaporates, the lubricant remains, coating the metal with a thin lubricant film, see, for example, Zhou et al. [55] and Abbott et al. [56]. Once the solvent has evaporated, the lubricant layer may have a thickness ranging from a fraction of a micrometer to a few micrometers [55]. If a lubricant layer of this thickness could be maintained between two metals, then a highly resistive contact would result. Fortunately as the contacts are brought together even with a low contact force, the thickness of the lubricant is reduced and good contact is established. Thus, potential users who know that in bulk the lubricant is an insulator but are unfamiliar with contact theory should not be concerned it will interfere with conduction when it is spread thinly on a contact surface. A comprehensive discussion of making contact between two metals coated with a lubricant is given by Braunovic et al. [57]. Figure 3.32 illustrates the possible stages of contacts coming together in the presence of a lubricating film. In Figure 3.32a, a thick film exists between the two contact surfaces. This can occur at very low contact forces with a thick grease film. If the film is thick enough (greater than a few nanometers), there will be little or no conduction: that is, the film will insulate one contact from the other. As the contact force increases, as in

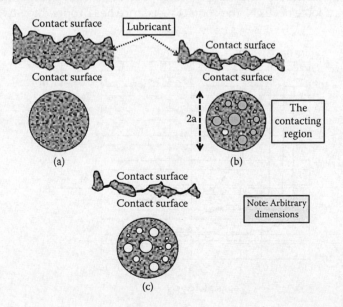

(a)

(b)

(c)

FIGURE 3.32
Stages of making contact between two metals in the presence of a lubricant: (a) A thick layer of lubricant and no conduction. (b) At higher contact forces, the lubricant is displaced and conduction between the metal contacts is established. (c) As the contact force increases further, a stage will be reached when only a very thin layer of lubricant exists between the contact spots and satisfactory conduction is established.

Figure 3.32b, the lubricant layer becomes thinner as it is forced away from the contact region. A stage will be reached when the micro-peaks in the contact region are separated by a thin layer of lubricant. The contact resistance will then be

$$R_C = \frac{\rho_m}{2a} + R_f \qquad (3.1)$$

Where ρ_m is the resistivity of the contact material and R_f is the total resistance of the film between the contact spots and a is the contact spot radius. The exact determination of R_f is not obvious. As can be seen from Figure 3.32b, there will be varying spacing between the contact micro-peaks resulting in differing thickness of the lubricant. Braunovic et al. [57] present an analysis that determines the effective R_f resulting from field emission of electrons from one contact to the other through a thin lubricating film of less than 1 nm. A discussion of the tunnel effect for conduction through thin contact films can also be found in Holm [58] and Tamai [59]; see also Figure 20.6. Other mechanisms for conduction through thin films such as A-fritting and B-fritting may also come into play; see Holm [60] and Section 1.3.6. As the contact force increases, the metallic high spots will engage; also see Figure 3.32c. However, as the lubricant is non-compressible it is expected that a very thin layer of lubrication (maybe only a layer one molecule thick) will still remain between the contacts. So the question to be answered is, "How will this affect the contact resistance?" Figure 3.33 illustrates the effect of lubricating films on two contacts (a gold probe onto platinum) closing with very low contact forces (0.1 cN to 60 cN where 1 cN ≈ 1gF):

i. *No lubrication*: Conduction begins once contact is made at ~0.17 cN with a contact resistance of ~6 × 10^{-2} Ω. As the contact force increases, the contact resistance decreases as would be expected from analyzing the formation of the contact region, see Section 1.2. For a contact force of 60 cN, the contact resistance has dropped to ~6 × 10^{-3} Ω.

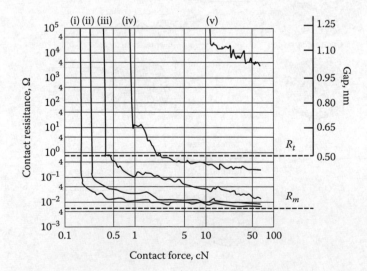

FIGURE 3.33
Dependence of the contact resistance as the normal force on a gold probe against a platinum contact is increased: (i) No lubrication (ii) Platinum surface lubricated with oleic acid (iii) Platinum surface lubricated with polyglycol diproxamine D157 (iv) Platinum surface with a lubricant thickened with lithium stearate (v) Platinum surface with a lubricant thickened with lithium 12 hydroxysterate. The calculated constriction resistance for the two metals with no lubrication is given by R_m and R_t is the calculated tunnel resistance for a 0.5-nm lubricant film [57].

ii. *Lubrication with oleic acid (a thin monounsaturated omega-9 fatty acid, biodegradable oil: see for example, Kano et al. [61]):* Conduction begins for a contact force of ~0.23 cN, but now the contact resistance is ~10^{-1} Ω. Thus, the thin layer of lubricant has increased the initial contact resistance by about 66%. As the contact force is increased to 60 cN, the contact resistance drops to ~8 × 10^{-3} Ω; about 33% greater than for the unlubricated contacts. Thus, even though a thin lubricating layer exists between the contacting spots, the conduction through the contact is satisfactory even at these very low contact forces.

iii. *Lubrication with polyglycol diproxamine D157 (a thin oil, see for example, Konchits et al. [62]):* Conduction with this somewhat denser oil now begins at a contact force of ~0.5 cN, with a contact resistance of ~8 × 10^{-1} Ω. This is the value calculated for tunneling of electrons though a 0.5-nm film [59]. It is 13 times the initial value measured for the unlubricated contact pair. As the contact force is increased and the oil is displaced, the contact resistance drops until at 60 cN, its value is ~4 × 10^{-2} Ω; about seven times greater than for the unlubricated contacts.

iv & v. *Lubrication with thicker greases:* The contact resistance values at the contact forces in this data are high, indicating that the lubricant is not displaced effectively at these low contact forces. However, if the contact force is increased to values you would expect from power contacts then the grease film would again be displaced until only a very thin layer is left and the contact resistance will again drop to the acceptable levels seen with the thin oils.

If contact is made with some sliding motion, as is common with some electronic connectors, some of the lubricant may well be displaced from some of the contact high spots and metal-to-metal contact may well be established. Even though a lubricant properly applied does not interfere with conduction through the contact spot, the potential user has to be aware of a number of technical parameters before using it. For example:

1. What is the lubricant's long-term stability?
2. Will the lubricant migrate or be lost to the contact region?
3. How is the lubricant affected by the connector's operating environment?
4. Will the addition of a lubricant attract dust?

A number of these questions have been addressed by Abbott and Antler [56] for a wide range of "conventional" lubricants and in field and laboratory studies of some unique lubricants by Abbott [63]. Figures 3.34 and 3.35 show examples of data from these two references. The conclusions that can be drawn from these studies can be summarized as follows:

1. Lubricants cannot be used in an arbitrary way, that is, they must be thoroughly tested before they can be qualified for a particular application.
2. Most lubricants do, in fact, provide some short-term benefits. These do not necessarily imply a long-term satisfactory connection.
3. Very few lubricants which are known to the contact community and which have been designed for reduction of friction and wear will survive comprehensive environmental testing.
4. Some lubricants appear to even introduce long-term risk, that is, add to the corrosion of the contact region.

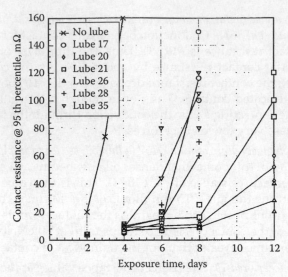

FIGURE 3.34
Contact resistance of best lubricated gold-gold connectors exposed unmated in Class II FMG.

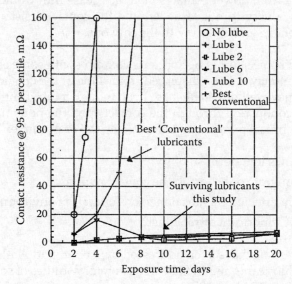

FIGURE 3.35
Contact resistance of unmated gold-gold connectors versus time; Class II after thermal ageing, 80°C, 1000 h.

5. There are a few lubricants that have been identified as providing exceptional protection from corrosion and have no known engineering risk on the contacts' surfaces.

6. Some lubricants appear capable of enhancing reliability to a far greater degree than is likely possible by changes in contact material or by changes in connector design.

As shown in Figures 3.34 and 3.35 for unmated coupons with a thin Au plate (~0.7 μm) exposed to a Class II FMG, samples with no lubrication may degrade very rapidly. The time to failure (one to two days) is considerably less than the one industry standard for unmated exposures that requires at least five days. In fact, it is Abbott's experience that the results shown in Figure 3.34 are somewhat typical of many commercial products, that is, very few will survive exposure for more than a few days even at Class II levels.

The results with lubrication are somewhat unprecedented. Figure 3.34 shows survival times exceeding five days and approaching ten days for the best of the lubricants in this study. Unfortunately, many of these same lubricants suffer degradation when subjected to other environmental testing such as thermal aging, low temperatures, and exposure after durability cycling. In these data, the best lubricant has been defined as one containing a mixture of:

2% 6 ring polyphenylether (PPE), 0.5% microcrystalline wax (MP 70°C–75°C) in a volatile carrier of mineral spirits.

This type of lubricant is somewhat familiar to the contact community. Experiments have been conducted with this type of lubricant applied to edge card connectors of polished (CLA 0.18 ± 0.03 μm), solid Cu plated with Ni (1.3 μm) then Au (0.8 μm) making contact with double sided phosphor-bronze springs plated again with Ni (1.3 μm) then Au (0.8 μm). These connections have a carefully controlled contact force of 200–300 gF/connector [56]. The work shows that the lubricant offers substantial protection to the porous Au. One lubricant with the smallest amount of PPE and microcrystalline wax mentioned above has good performance down to –55°C and one of the best overall protections both before and after thermal aging. However, from this work it appears that the wax content must be closely controlled for the best performance. It has also been reported by Garte et al. [64] that this type of lubricant offers good tribological properties. The application of PPE or polyalkyline glycol (PAG) can also alleviate the affects of fretting corrosion in connections [65]. Fretting experiments using two pairs of brass-plated crossed rods connections (see Section 2.6.7) show encouraging results. The first pair uses two brass rods plated with 3.7 μm of Cu and then with 3.7 μm of tin–lead. The second pair uses one brass rod with the same tin–lead over Cu against a brass rod plated with 1.5 μm of Ni then with 0.6 μm of Au. The contact pairs are then subjected to 20,000 fretting cycles. A substantial increase in contact resistance is observed for both contact pairs. When PPE or PAG is carefully applied to the wear tracks after this initial fretting operation, both contact pairs return to their original low contact resistance values and remain low for another 80,000 fretting cycles. This experiment shows that some lubricants can even be used to recover severely degraded connections. An excellent listing of lubricant studies up to 2005 for various substrates and platings is given by Chudnovsky [66] and Braunovic et al. [67].

The use of quite a different class of lubricant for use in connectors has been introduced to alleviate the failure of Sn-plated steel pins mated to Au-plated sockets that control a fuel shut-off valve in the F-16 Fighting Falcon aircraft. The failure of this connection has been implicated in a number of crashes of this airplane. The lubricants studied are a class of Military-Specification (M-S) materials that had previously been specified for corrosion protection of components such as jackscrews, cadmium-coated steel pins, and ejector seat mechanisms [68]. These lubricants were quite unknown to the contact community at the time of the initial study. The materials studied have the M-S codes: (a) MIL-L-87177A and (b) MIL-C-81309C. Twelve of these M-S materials (one from M-S (a) and 11 from M-S (b)) have been evaluated for connector protection under a number of environmental and flowing gas conditions. In general, nine of them provide some corrosion protection and three show excellent protection. These three are D5026NS from Zip Chemo Products (M-S (b)), Super Corr from Lektro Tech (M-S (a)) [69] and S-Sure from LHB Industries (M-S (b)). However, three of the other (M-S (b)) lubricants actually seem to promote corrosion; some in more severe environments than others. These results clearly demonstrate the dangers of using a lubricant that has not been thoroughly evaluated for a particular use. Details of this work are given by Hanlon et al. [70] and a brief description of the data for the successful (M-S (a)) lubricant is given by Abbott [71]. He also shows the failure of this connector is due to fretting corrosion caused by vibrations induced in the connector whilst the aircraft is flying and not from the environmental ambient when the aircraft stays in its hangar. Even though the (M-S (a)) lubrication successfully reduces the failure rate, he concludes the connector should be inspected and lubrication applied at an interval of 100 hours of flight time. Hanlon et al. conclude the M-S (a) lubricant should be applied to all new connectors that go into KC stores. They also recommend it be applied to all newly built cables and equipment that recycles through manufacturing locations from the field. Research continues on new lubricant materials to inhibit contact corrosion and maintain a low contact resistance. Vapor phase corrosion inhibitors (VCIs) have shown some promise for inhibiting corrosion of contact surfaces [66,72–74]. Noël et al. have experimented with various compositions of perfluorinated polyether (PFPE) with additives and they have shown some promise [75–77]. They show a PFPE lubricant can form a thin fluorinated film on metals such as Ni and thus postpone their degradation and oxidation [76]. Also nanotechnology has begun to be investigated for contact maintenance [77,78].

The experimental data on the use of lubricants in connector contacts presented in this chapter should be treated with care before using them for application to a particular connector or contact system. The question that must be asked is," Will the experimental data apply to a quite different connector system?" It is, therefore, important to compare the proposed application, connector design and environment to that described in the experiment. For example, the crossed rod fretting experiment and data on the recovery of low contact resistance [65] is obtained in an open air laboratory environment. Also extreme care is taken to place the lubricant in the wear tracks. The experiment continues for only a short time afterward (80,000 more fretting cycles). So, the question arises are the data relevant for enclosed connections with the need for a longer life and in a different ambient environment? The fuel shut-off valve for the F-16 aircraft is another good example. Many of the lubricants tested show excellent protection for the Sn to Au contact match under many different adverse environments. However, the adverse effects only show when the connector is subjected to the vibration expected from normal flight experience. Then, only three of the lubricants are shown to have some protective properties [70,71]. This should be a major consideration for connectors that will be used in all forms of transportation, but the reader can think of many other applications where vibration should be taken into account [79–83]. It is also important to choose a lubricant that will not be affected by the environment the connector will be used in. One obvious example is temperature [51,52,70]. Lubricants can be subjected to four temperature aging effects: evaporation, surface migration, polymerization, and degradation. Lubricants may degrade in particular ambients. Some lubricants may polymerize in the presence of Cu or a Cu-based alloy. Others may induce galvanic corrosion between contacts that have different electrolytic potential (see Table 2.2). Dust is also an important consideration: the effects of dust will be discussed in Chapter 4. Unfortunately, research has shown that there is no perfect lubricant to protect contacts from corrosion and from fretting. Fortunately most contact systems are in an enclosed casing and experience very low levels of corrosive gas ambients. Lubricants, however, have been shown to be effective and if chosen carefully will provide a connector a longer life than would have been the case without their use. One important result from this research is that a Au-plated or Au flash contact against a Sn or Sn–Pb contact should be avoided if possible. If, however, the use of this combination is unavoidable a suitable lubricant applied to the contact surfaces is recommended.

Acknowledgment

We appreciate the permission given by ASTM to use Figures 3.26 and 3.27.

References

1. R Holm, *Electric Contacts*, 4th ed. (reprinted), New York: Springer-Verlag, 2000, 135.
2. WH Abbott, Materials, environment, motion, and electrical contact failure mechanisms, *Proc 35th IEEE Holm Conf on Elect Cont*, 1989, 3.
3. WH Abbott and HR Ogden, The influence of environment on tarnishing reactions, *Proc 4th Int Symp Elect Cont Phen*, 1968, 35.
4. WE Campbell and UB Thomas, Tarnishing studies on contact materials, *3rd Int Res Sympo Elect Cont Phen*, Orono, 1966, 83.
5. IEC 68-2-42, IEC 50B 272 (1986), DIN 40 046 Part 36, Jeida 30.
6. WHJ Vernon et al., Laboratory study of the atmospheric corrosion of metals, *Trans Far Soc* 27: 255, 1931.
7. DH Thompson et al., The atmospheric corrosion of copper results of 20-year tests, ASTM Spec Tech Publ No. 175, 1955, p 77.
8. BE Blake, Summary Report of the ASTM Section 6 Field Tests, *Proc. 2nd Int Sym Elect Cont Phen*, Graz, 1964, 531.
9. JC Mollen and M Trzeciak, Analysis of copper films after field exposure, *Proc 16th Holm Conf on Elect Cont*, 1970, 37.
10. RV Chiarenzelli, Tarnishing studies on contact materials, *Proc 2nd Holm Sem Elect Cont*, 1965, 63.
11. B Wiltshire, A survey of contact atmospheres in UK telephone exchanges, *Brit Telecom Tech J*, 21(1): 74, 1984.
12. DW Rice et al., Atmospheric corrosion of nickel, *J Elect Soc* 127: 563, 1980.
13. SP Sharma, Atmospheric corrosion of Cu, Ni, and Ag, *J Elect Soc* 125: 2005, 1979.
14. CA Haque and M Antler, Atmospheric corrosion of clad Pd and Pd–Ag alloys, Part I: Film growth and contamination effects, *Proc 27th IEEE Holm Conf Elect Cont*, 1981, 183.
15. DW Rice et al., Atmospheric Corrosion of Copper and Silver, *J Electrochem Soc* 128: 275, 1981.
16. DW Rice et al., Indoor Corrosion of Metals, *J Electrochem Soc* 127: 891, 1980.
17. ISA Specification, ISA SP71.04.
18. ASTM B827-93.
19. IEC 68-2-60; Test Ke.
20. WH Abbott, The development and performance characteristics of mixed flowing gas test environments, *Proc 33rd IEEE Holm Conf Elect Cont*, 1987, 63–39.
21. WH Abbott, The measurement of equipment operating environments to evaluate corrosion related failure mechanisms, IEC Document 65, Montreal/USA, May 7, 1965.
22. WH Abbott, A review of flowing mixed gas test developments, *Brit Corr J* 24(2): 153, 1989.
23. WH Abbott, The corrosion of porous gold platings in field and laboratory environments, *Proc 13th Int Conf Elect Cont*, 1986, 343.
24. M Antler, Field studies of contact materials in telephone central offices, *Proc 11th Int Conf Elect Cont*, 1982, 297.
25. SP Sharma et al., Development of Pd and Pd–Ag alloy, Part II: Film chemistry, *Proc 27th IEEE Holm Conf Elect Cont*, 1981, 203.
26. TF Egan and A Mendizza, *J Electrochem Soc* 107: 353, 1960.
27. RV Chiarenzelli, Air pollution effects on contact materials, *Proc Holm Sem Elect Cont Phen*, 1965, 63.
28. M Antler, Plating, 53: 1431, 1966.

29. WH Abbott, Effects of industrial air pollutants on electrical contact materials, *Proc Holm Sem Elect Cont Phen*, 1973, 96.
30. WH Abbott, Recent studies of tarnish film creep, *Proc 9th Int Conf Elect Cont Phen*, 1978, 117.
31. WH Abbott, The effects of test environment on the creep of surface films over gold, *Proc 12th Int Conf Elect Cont Phen*, Chicago, 1984, 47.
32. ASTM B845-93.
33. ASTM B827-93.
34. IEC 68-2-43, IEC 50B 272 (1986), DIN 40 046 Part 37, Jeida 38.
35. British Standards Institution, Sulphur dioxide test for contacts and connections, B52011, 1977.
36. Bellcore GR-63-Gore, Issue 1, October 1995, 5–40.
37. R Gore et al., Corrosion gas environmental testing for electrical contacts, *Proc IEEE Holm Conf Elect Cont*, 1989, 123.
38. IBM spec. C-Hl-9900-010.
39. HG Thompkins, The interaction of some atmospheric gases with a tin–lead alloy, *J Electrochem Soc* 120(5): 651, 1973.
40. U Cosack, Survey of corrosion tests with pollutant gases and their relevance for contact materials, *13th Int Conf Elect Cont*, Lausanne, 316, 1986.
41. J Potinecke, Behavior of contact surfaces consisting of gold alloys in H_2S–NO SO_2 atmospheres, *8th Int Conf Elect Cont Phen*, Tokyo, 1976, 275.
42. KL Schiff, Technical realization and experimental performance of corrosive gas test equipment for accelerated testing of electrical contacts, *Proc Holm Conf Elect Cont*, Chicago, 1977, 43.
43. WH Abbott, Effects of industrial air pollutants on electrical contact materials. *Proc Holm Conf Elect Cont*, Chicago, 1973, 94.
44. WH Abbott, Studies of natural and laboratory environmental reactions on materials and components, 9th Progress Report, unpublished data, August 1986.
45. ES Sproles et al., Results of exposure of palladium-based contact materials to Cl and H_2S ambients, *11th Int Conf Elect Cont Phen*, Berlin, 1982, 143.
46. WA Crossland et al., The accelerated tarnish testing of contacts and connectors employing silver alloy contacts, *Proc Holm Conf Elect Cont*, Chicago, 1973, 265.
47. J Potinecke, Behavior of contact surfaces consisting of Ag and Pd–Ag alloys in tarnishing environments, *Proc Holm Conf Elect Cont*, Chicago, 1975, 139.
48. WH Abbott, Effects of velocity and SO_2 on the reactions of a flowing mixed-gas environment, *Proc16th Int Conf Elect Cont*, Loughborough, 1992, 35.
49. WJ Curren et al., Design and operating characteristics of a mixed gas environmental chamber, *Corrosion/85*, Boston, 1985, Paper 326.
50. WE Campbell and UB Thomas, Tarnish studies: Electrolytic reduction method for the accurate analysis of thin films on metal surfaces, *Trans Elect Soc* 76: 303, 1939.
51. N Aukland et al., A comparison of a clad material (65Au21Pd1Ag) and an electroplated palladium material system under fretting conditions at elevated temperatures, *Proc 45th IEEE Holm Conf. on Elect Cont*, 1999, 213.
52. N Aukland et al., The effect of fretting at elevated temperatures on a clad material (65Au21Pd1Ag) and an electroplated soft gold over palladium-nickel material system, *IEEE Trans Comp Packag*, 23(2): 252, 2000.
53. M El-Hadachi et al., Contact resistance of gold coating contaminated by organic pollutant, *Proc 24th Int Conf on Elect Cont*, 2008, 384–89.
54. JG Zhang, The application and mechanism of lubricant on electrical contacts, *Proc 41st IEEE Holm Conf on Elect Cont*, 1994, 145–54.
55. YL Zhou and JG Zhang, Investigation of the thickness of lubricant film on gold-plated surfaces, *Proc 44th IEEE Holm Conf on Elect Cont*, 1998, 159–65.
56. WH Abbott and M Antler, Connector contacts: Corrosion-inhibiting treatments for gold-plated finishes, *Proc of the 41st IEEE Holm Conf Elect Cont*, 1995, 97.
57. M Braunovic et al., *Electrical Contacts: Fundamentals, Applications and Technology*, Boca Raton, FL: CRC Press, 2007, 419–38.

58. R Holm, op cit, 118–32.
59. T Tamai, Recovery of low contact resistance due to electrical and mechanical breakdown of contact films, *Proc of the 56th IEEE Holm Conf Elect Cont*, 2010, xxiii–xl.
60. Holm, op cit, 135–152.
61. M Kano and K Yoshioda, Ultra-low friction of DLC coating with lubricant, *Intl Conf Sci Frict, J Phy*, Conference Series 258012009, 2010, 1–8.
62. VV Konchits et al., Micromechanics and electrical conductivity of point contacts in boundary lubrication, *Trib Int*, 29(5): 365–71, 1996.
63. WH Abbott, Field and laboratory studies of corrosion inhibiting lubricants for gold- plated connectors, *Proc 42nd IEEE Holm Conf Elect Cont*, 1996, 414.
64. SM Garte and H Steadly, Protection of gold-plated connector contacts from wear and corrosion, *Proc Conn Symp II CIT*, 1973, 230–40.
65. M Antler et al., Recovery of severely degrade tin–lead plated connector contacts due to fretting corrosion, *Proc 43rd IEEE Holm Conf Elect Cont*, 1997, 20–32.
66. B Chudnovsky, Lubrication of electrical contacts, *Proc 51st IEEE Holm Conf Elect Cont*, 2005, 107–14.
67. Braunovic et al., op cit, 449–55.
68. KM Martin https://www.corrdefense.org/Academia%20Government%20and%20Industry/XVIII%20-%20MARTIN%20-%20MIL-L-87177A%20Tri-%20Service%20Paper.pdf
69. http://www.lektro-tech.com/supercorraorb/infoapplications.html
70. JT Hanlon et al., MIL-L-87177 lubricant bullet-proof connectors against chemical and fretting corrosion, Sandia Report SAND 2002-1454, 2002. (This report has 13 appendices from a number of authors going into great detail the experimental work performed before the recommended lubrication was chosen. In particular, it gives the details of Abbott's work at Battelle that forms the basis of his study referenced below)
71. WH Abbott, Performance of the gold–tin connector interface in a flight environment, *Proc 44th IEEE Holm Conf Elect Cont*, 1998, 141–50.
72. BA Miksic et al., Efficacy of vapor phase corrosion inhibitor technology in manufacturing, *J Sci Eng: Corr Sci* 60(6): 515–22, 2004.
73. S Noël et al., Lubrication mechanisms of hot-dipped tin separable electrical contacts, *Proc 47th IEEE Holm Conf Elect Cont*, 2001, 197–202.
74. S Noël et al., A new mixed organic layer for enhanced corrosion protection of electric contacts, *Proc 50th IEEE Holm Conf Elect Cont*, 2004, 274–80.
75. S Noël et al., Effect of fluorinated lubricants on the friction modes of tin electrical contacts submitted to fretting, *Proc 24th Int Conf Elect Con*, 2008, 278.
76. S Noël et al., Influence of contact interface composition on the electrical and tribological properties of Ni electrodeposits during fretting tests, *Proc 26th Int Conf Elect Cont*, 2012, 2221.
77. S Noël et al., Nanocomposite thin films for surface protection in electrical contact applications, *Proc 53rd IEEE Holm Conf Elect Cont*, 2007, 160–66.
78. S Noel et al., Electrical conduction properties of molecular ultrathin layers in a nanocontact, *Proc 56th IEEE Holm Conf Elect Cont*, 2010, 380–86.
79. GT Flowers et al., Vibration thresholds for fretting corrosion in electrical connectors, *Proc 48th IEEE Holm Conf Elect Cont*, 2002, 133–39.
80. GT Flowers et al., Modeling early stage fretting of electrical connectors subjected to random vibration, *Proc 49th IEEE Holm Conf Elect Cont*, 2003, 45–50.
81. C Chen et al., Modeling and analysis of a connector system for the prediction of vibration-induced fretting degradation, *Proc 55th IEEE Holm Conf Elect Cont*, 2009, 129–35.
82. J Gao et al., The influence of particulate contaminants on vibration- induced fretting degradation in electrical connectors, *Proc 56th IEEE Holm Conf Elect Cont*, 2010, 108–12.
83. R Fu et al., Experimental study of the vibration-induced fretting of silver-plated high-power automotive connectors, *Proc 56th IEEE Holm Conf Elect Cont*, 2010, 113–20.

4

Effect of Dust Contamination on Electrical Contacts

Ji Gao Zhang

The scepter, learning, physic must
   All follow this and come to dust.

Fear no more, **William Shakespeare**

CONTENTS

4.1 Introduction 186
    4.1.1 Background 186
    4.1.2 The Importance of the Dust Problem 187
    4.1.3 The Complexity of the Problem 189
    4.1.4 The Purpose of the Studies 189
4.2 Dusty Environment and Dust Composition 190
    4.2.1 The Source of Dust 190
    4.2.2 The Collection of the Dust Particles for Testing 191
    4.2.3 The Shape of the Dust Particles 191
    4.2.4 The Identification of the Inorganic Materials 192
    4.2.5 The Organic Materials in Dust 192
    4.2.6 The Water Soluble Salts in Dust 192
4.3 The Characteristics of Dust Particles 193
    4.3.1 The Electrical Behavior 193
        4.3.1.1 Measurement of the Electric Charge 193
        4.3.1.2 The Electrostatic Attracting Force on the Particle 195
    4.3.2 Mechanical Behavior 198
        4.3.2.1 Load Effect 198
        4.3.2.2 For Stationary Contacts 199
        4.3.2.3 For Sliding Contacts 201
        4.3.2.4 The Effect of Lubricants Coated on Contact Surface 201
        4.3.2.5 Sliding Contacts on Lubricated and Dusty Contacts 203
        4.3.2.6 Fretting (Micro Motion) on Lubricated and Dusty Contacts 204
    4.3.3 Chemical Behavior 205
        4.3.3.1 Dust Particles Create Pores 205
        4.3.3.2 Corrosion Appears as a Result of Dusty Water Solutions 205
        4.3.3.3 Indoor Exposure Results 206
        4.3.3.4 Construction of the Corrosion Stain 208
        4.3.3.5 Fretting Experiments on Dust Corroded Coupon Surfaces 209

4.4   Application Conditions in Dusty Environment...211
       4.4.1   Explanation of the Special Features..211
                4.4.1.1   Covered by Accumulated Small Particles.........................211
                4.4.1.2   Accumulative Particles Caused by Micro Motion..............211
                4.4.1.3   High and Erratic Contact Resistance.................................212
                4.4.1.4   The Element of Si Causes High Contact Resistance.........213
                4.4.1.5   Organics Act as Adhesives...214
                4.4.1.6   Corrosion Products Trap the Dust Particles.....................216
                4.4.1.7   Difference Between Short Life and Longer Life Contacts.................218
                4.4.1.8   Large Pieces of "Stepping Stones".....................................218
                4.4.1.9   The Performance of Failed Mobile Phones.......................218
       4.4.2   Other Examples..218
4.5   Theoretical Analysis of Connector Contact Failure Due to the Dust....................219
       4.5.1   Two Micro Worlds in Contact..219
                4.5.1.1   Particles Get into the Contact Interface............................219
       4.5.2   "Preliminary Attachment"...220
                4.5.2.1   Adhesive Effect..220
                4.5.2.2   Trapping Effect of Corrosion Products.............................221
       4.5.3   Contact Failure Mechanism..222
                4.5.3.1   Single Particle and Ideal Model...222
                4.5.3.2   Complicated Model – Number of Particles and Morphology of
                           Contact Pairs...223
       4.5.4   Micro Movement...223
                4.5.4.1   Contact Failure..223
4.6   Future Work...224
       4.6.1   Dust Test for Connectors...224
       4.6.2   Suggestion of the Dust Test...224
       4.6.3   Minimizing the Dust Problem...225
                4.6.3.1   Cleaning the Samples...225
References...226

4.1 Introduction

4.1.1 Background

The fact that dust can cause contact failure was well known to practicing electricians, engineers and scientists even in the middle of the 20th century. It was formally reported in 1956 by Williamson et al. [1] and in 1981 by Mano [2]. From 1970 to 1985, more papers on the dust problem were published. A review in 1985 given by Reagor and Russell [3] showed that dust contamination of contacts was a problem that could cause contact failure in telecommunication systems. They also described erratic contact resistance found on lubricated, dusty surfaces. Mano and Honma et al. [4,5] published research papers on testing and simulation methods. Papers were also published that showed contact failure may have led to unreliable electronic and electrical systems [6,7]. From the 1970s to the mid-1990s, advanced industrial countries recognized the serious problems resulting from air pollution. They then made large investments and much research to reduce the air pollution and purify the air [8]. Thus, electrical contact problems due to corrosive air and dust contamination were substantially reduced in these countries and the research effort was considerably reduced. During this time, simulation of the corrosive gases on the connector contact surfaces was developed by using the flowing mixed gas FMG system [9] (see Sections 2.7 and 3.3.5), but the problem of dust particles was quite different. Although several theoretical and experimental studies have been made [10,11], this subject is so complex [12] that research for a clearer understanding was still necessary.

4.1.2 The Importance of the Dust Problem

Research on the effects of dust is now becoming more important and more urgent due to the huge market that is growing rapidly in the developing countries, where contact failure caused by dust contamination is one of the major problems. Furthermore, developing countries such as China are already becoming global manufacturing bases and their components are being incorporated into products in many other countries including the advanced industrial nations. Although contamination due to dusty air is very serious in China, it also unfortunately affects neighboring countries such as Korea and Japan. It is, however, difficult for the developing countries to strictly control and reduce the air contamination quickly.

To illustrate the specific situation existing in China, the following failure examples have been selected to demonstrate the scope of the problem.

Example 1: High porosity is found on gold plating surfaces caused by the deposition of dust particles before or during the electroplating process. Serious corrosion (the dark spots) occurs through the pores and spreads on the gold-plated surface as seen in Figure 4.1 [13].

Example 2: Connectors mounted in telecommunication systems suffer from serious contact failures. Figure 4.2a illustrates the contamination on a connector contact surface taken by a scanning electron microscope (SEM). Elements of a spot at

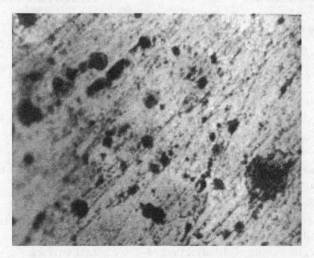

FIGURE 4.1
Corrosive products have grown up through the pores as a result of dust particle deposition.

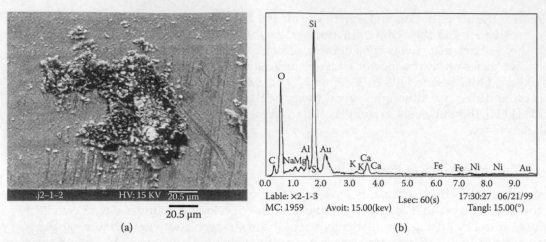

FIGURE 4.2
(a) Accumulated particles (b) Elements in the particles.

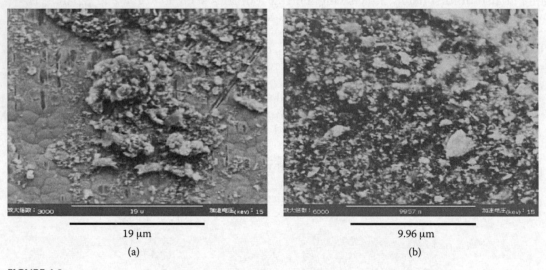

FIGURE 4.3
(a) Numerous dust particles accumulated on the failed contact surface (b) Photo taken from part of (a), at higher magnification.

the bright area were tested by X-ray energy spectroscopy (XES) and illustrated in Figure 4.2b. The contact surface is covered by tiny particles. Very high atomic percentage (a/0) of silicon and oxygen are found at the spot. Other elements such as Na, Ca, Al, K, and Mg are also included. This shows a mixture of elements in a typical dust composition.

Example 3: A mobile phone had failed after only 10 months and 22 days. The failed contact surface is covered by numerous particles (see Figure 4.3). Figure 4.3b which is a higher magnification section of Figure 4.3a shows these particles are extremely small but they seem to be tightly attached on the contact surface.

4.1.3 The Complexity of the Problem

The complexity of the effect of dust particle contamination can be summarized by three major influencing factors: (1) The environment which includes the size and composition of particles. It can change from place to place and can vary in different seasons. (2) Different materials contained in the dust have their own inherent characteristics with respect to the performance of electrical contacts such as electrical and mechanical behavior, chemical corrosion and so on. (3) The various application conditions, for example, the micromovement at the contact interface; the special materials involved; the conditions that cause the dust particles to adhere together; the corrosion products that trap the particles and attach them on the contact surfaces. These kinds of "preliminary attachment conditions" can cause a number of particles to embed into the surface. Local humidity and temperature cycles can also enhance the effect of the attachment. When the above factors combine together, harmful particles can enter and contaminate the contact interface and cause high and unstable contact resistance leading to failure. From this point, the very simple procedure that had been used in dust testing for connectors did not fully simulate the dust effect.

Contact failure of electromechanical components involved in the dust contamination is seen in both relays and connectors. Relay contacts have open and close movement. Connector contacts are static, but can have minute up and down as well as laterals movement. Since late 1990's, most relays have been covered or are hermetically sealed so the failure resulting from dust ingress is significantly reduced. Most connector contacts, on the other hand, are exposed to air. Thus, they suffer from the ingress of dust particles and serious corrosion. Since micro movements resulting from external vibration and impact at the connector contact can often occur, the mechanism of connector contact failure is quite complicated.

Questions arise from this research are: What kind of particles can cause contact failure? What is the mechanism of failure when lateral movement occurs? How can the particles get into the connector contact interface, but not be pushed away when lateral movement occurs? How can they persistently remain at the interface and create high resistance? Why are the past testing methods not workable for connectors? Why does a certain failure level is still remained in sealed relays?

To answer these questions, research studies have been carried out from two directions: (1) The investigation of the failed contact surfaces collected from practical field applications, and to the statistical summarization of the testing result. Based on the analysis of the special features of the failed contact surfaces resulting from particle contamination, reasonable deductions have been carried out for explaining the failure. (2) The study of dusty environments, the source of dust, its composition, the size of the particles, the materials contained in the dust and their characteristics.

4.1.4 The Purpose of the Studies

The purpose of the research is to thoroughly understand the mechanism of failure caused by particle contamination. A workable and effective dust test system for connector application had to be established in order to determine the optimum structure and packaging design for connectors and other equipment working in dusty environments. This research is expected to set standards in the future for manufacture, storage, transportation processes and to improve electrical contact reliability.

4.2 Dusty Environment and Dust Composition

4.2.1 The Source of Dust

According to the China's official data, the area of desert and sandy regions area is about 1.74 million km²: that is 18.2% of the area of China. Serious winds start from Mongolia; they pass the Gobi Desert and other large desert areas including various dried salt lakes all the way to the north, north-west and north-east of China. Not only sand and salt, but also dust particles from construction materials used to build new cities and highways, smoke particles from exhaust pipes of motor vehicles, and particles from coal burning are all blown up from the ground into the air. Thus dusty air contains large number of particles made up from a variety of materials. Each of them may contribute different effects on contact surfaces. Unfortunately, there are few obstacles such as forests and grass-lands in the path of the blowing wind. Consequently the wind and the ground materials from a large region causes frequent dust storms. The dusty air can affect almost all the northern part of China. Even coastal cities, such as Shanghai, experience the precipitation of large amounts of fine particles. Figure 4.4a illustrates a dust storm in Beijing, China [14]. Figure 4.4b shows that after rain, the dusty air is washed away. A mountain from far west of Beijing can be seen. Figure 4.4c in normal days, dusty air is covering the whole city.

(a)   (b)

(c)

FIGURE 4.4
(a) Dust storm in Beijing (b) After raining at the same site (c) Same site at the normal days.

4.2.2 The Collection of the Dust Particles for Testing

Beijing has been chosen as a typical city in the north regions for dust collecting. Not only is it the capital of China with a high population, but it is also on the wind-path from the desert and other desert areas. It has a great deal of construction using concrete and has a large number of automobiles. Within China, it is considered to only be a moderately dusty city, and can thus represent the situation of many cities in the region. The western part of Beijing does not have high industrial contamination, and so the dust particles are collected indoors in a building on the university campus. Shanghai has also been chosen for the long time indoor air exposure and testing since it is a large commercial and industrial city located on a coastal region of China.

4.2.3 The Shape of the Dust Particles

Dust particles have been collected indoors from the west of Beijing. The photos of shapes and size of particles were taken by scanning electron microscope (SEM) are shown in Figure 4.5a through f [14]. It is seen from these photos that, most particles from Beijing are not simply minerals with polyhedral shapes like sand particles, but they are constructed from various tiny particles. Organics play an important role in causing them to adhere together. The particle shown in Figure 4.5e contains biotite as its core material, which is covered by gypsum and has organic materials coating its outside. Spherical particles come from the vapor of metals such as iron and copper. When the vapor cools down in air; spherical droplets are formed and are soon covered with oxide. After the steel and copper companies had been moved away from the suburb of Beijing, spherical particles have not been present.

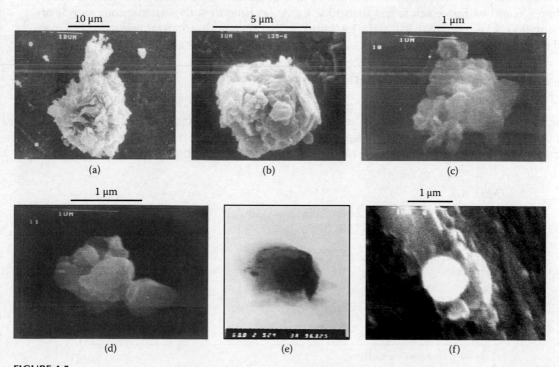

FIGURE 4.5

(a) Particle with size of 30-35 μm (b) Particle with size of 5-6 μm (c) Particle with size of 2-3 μm (d) Particle with size of 1-2 μm (e) Multilayer of tiny particle (f) Spherical particle formed by iron, covered by oxide.

4.2.4 The Identification of the Inorganic Materials

Dust particles have been collected and separated into three groups from fine to large particles by suspension in distilled water. By using automatic x-ray powder diffraction (XRPD), electron probe microanalysis (EPMA), and supplementary test by transmission electron microscopy (TEM), it has been possible to obtain a comprehensive record of the inorganic compounds in fine dust particles [15]. The (XRPD) result is estimated and calculated by the "relative intensity method" with quartz as the internal standard. The result is listed in Table 4.1; the results given by (EPMA) are similar with only a few exceptions. For fine particles, the inorganic material is more than 70% by weight of the dust particles, and clearly comes from the sandy areas, construction materials discussed before. The large sized particles are mainly quartz and feldspar.

4.2.5 The Organic Materials in Dust

Dichlormethane has been used as a solvent for the extraction of the organic compounds. The function groups of the organics are examined by infrared spectrometry. Organic compounds are then identified usually by gas chromatography/mass spectrometry (GC/MS), and finally the weight percentage of the organic compounds as well as carbon black in the dust is determined by thermo gravimetric analysis [16]. The results show that the main constituents of the organic compounds are a series of alkanes ($C_7 - C_{40+}$) and some ortho-benzendicarboxylic acid esters ($C_{16}H_{22}O_4$ and $C_{24}H_{38}O_4$). At room temperature, aliphatic hydrocarbons from ($C_9 - C_{16}$) are liquid and those greater than C_{16} are solid. Ortho-benzendicarboxylic acid esters are oily liquids. These organic materials are soft and often sticky, and generally are not conductive. By means of the thermo gravimetric analysis, the weight percentage of the total evaporable and combustible materials in the dust under 700°C is about 28.6%, of which water content is less than 5% and carbon black is less than 15%. In the original dust the organic compounds are estimated to be more than 9% of the total dust. Fibers are seen at room temperature, but most of them are burnt away below 350°C, and the few remaining ones are burnt below 550°C.

4.2.6 The Water Soluble Salts in Dust

Dust particles have been collected indoors from both Beijing and Shanghai. They are dispersed in distilled water and gently heated to obtain samples of dust solution. By means of an inductive-coupled plasma spectrometer (ICP), elements of positive ions are tested. Negative ions in the dust solution are measured by an ion chromatographic spectrometer (IC). Both

TABLE 4.1

Composition of Inorganic Materials in Fine Dust Particles (Size < 10 micron), Compared With that of All Dust (By Weight)

Materials	Quartz	Feldspar	Calcite	Gypsum	Mica
Fine particles	1	0.8	1.91	0.72	2.24
All dust	1	0.77	0.48	0.05	0.41
Materials	dolomite	kaolinite	lime	illite	hornblende
Fine particles	0.10	+	+	+	+
All dust	+	+	+	+	+

(+) means traces are detected.
Feldspar includes mainly orthoclase ($KAlSi_3O_8$), and albite ($NaAlSi_3O_8$)
Mica includes mainly muscovite $KAl_2Si_3AlO_{10}$ (OH,F), and biotite $KMg_3AlSi_3AlO_{10}$ (OH,F)

TABLE 4.2

Fractional Weight Percentage of Elements for Ions in Dust Solution (mg/g)

	Dust Solution (Beijing)	Dust Solution (Shanghai)
i) Positive Ions		
Na	2.11	3.75
K	1.50	1.77
Ca	20.55	27.54
Mg	1.85	2.09
Zn	0.05	0.01
Fe	0.09	0.04
Cu	0.04	0.02
Mn	0.05	0.13
Si	-	0.18
ii) Negative ions		
Cl	6.34	17.35
SO_4^2	18.90	32.25
F	0.58	2.30
NO_3	3.86	5.22

positive and negative ions are listed in Table 4.2 [17]. Clearly, water soluble salts are involved in the dust. The salts are mainly sulfates and chlorides of elements such as Ca, Na, Mg, NH_4 and K, Surprisingly the salt is hardly be visible using an optical microscope or even an SEM. It is believed that most of them have an micro crystal shape attached onto other particles. In a humid environment, the water film may dissolve the salt and make the solution an electrolyte. Thus corrosion products mainly are chlorides and sulfates of nickel and copper. They grow and even surround a particle. Table 4.2 shows the positive ions and negative ions detected from Shanghai and Beijing. From the list, it is reasonable to deduce that the corrosion products of sulfates and chlorides are more severe in Shanghai than in Beijing as a result of the dust particle corrosion. Since the materials in the dust collected from Beijing and Shanghai are now almost known, our studies on failed contact surfaces are able to use the combination of scanning electron microscope and x-ray energy spectroscope (SEM/XES) from the detected elements to deduce the possible materials involved in the dust corrosion.

4.3 The Characteristics of Dust Particles

4.3.1 The Electrical Behavior

4.3.1.1 Measurement of the Electric Charge

Dust particles carry electric charges that can be measured by the Millikan method. A brief introduction of the measurement is shown in paper [18]. Test results show that particles carry similar amounts of positive charge and negative charge as illustrated in Figure 4.6. In a DC electric field, the amount of fine particles deposited on a negative polarity surface and on a positive polarity surface is closely similar. Thus, when the two surfaces are in contact, the number of particles at the interface is almost doubled. Since dust particles comprise of various materials, each of them may carry different levels of electric charge. Measurements are also made of the electric charge carried by particles of pure materials that are contained in the dust.

A sequence of materials is listed with typical amounts of electric charge from high to low as follows: Organic materials, TiO_2, gypsum, mica, quartz, feldspar, calcite, aluminum oxide, solid particle with salt coverage. For organic materials such as polystyrene particles with a radius of 1 μm to 2 μm, the average electrical charge carried by these particles is about two times higher than that of the same sized dust particles. Experiment results further show that the amount of charge is closely related to the effective surface areas and to the resistivity of the particles as well as to the humidity of the environment. High humidity will greatly reduce the electric charge see Figure 4.7. On average, the electric charge carried by dust particles with 1.5 μm of radius in wet condition of 75% RH, is 50% less than that in dry condition of 25% RH.

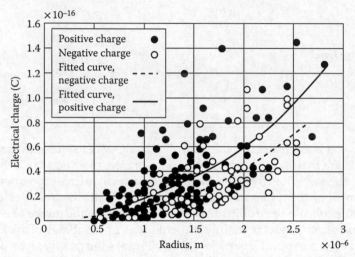

FIGURE 4.6
Electric charge carried by dust particles.

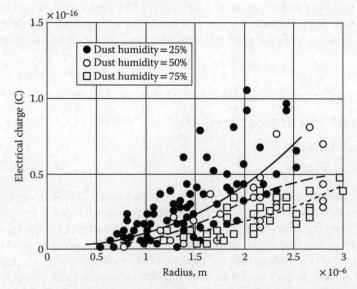

FIGURE 4.7
Electric charge carried by the particles at different humidities.

4.3.1.2 The Electrostatic Attracting Force on the Particle

4.3.1.2.1 No External Electric Field

A dust particle carrying an electric charge may build up a tiny static electric field at the metal surface. The attracting force F_e between particle and the metal surface is expressed as the following equation (4.1):

$$F_e = -Q^2/16\pi\varepsilon_0(d_0)^2 \tag{4.1}$$

Where, Q is the electric charge carried by the particle. ε_0, is the permittivity of air, and d_0 is the distance from the center of particle to the metal surface. When the distance is short, the attracting force can be very strong.

4.3.1.2.2 The Effect of Surface Orientation and Surface Roughness

Experimental results show that when a contact surface is set vertically to the direction of gravity force on the dust particle, both the gravity force and the tiny field attractive force are in same direction and are cumulative; i.e., they are added together. Thus, dust particle deposition density reaches a maximum value and appears little effected by the surface roughness. But when contact surface is parallel to the gravity force on the particle, deposition of particle is a function of the contact surface's roughness: the rougher the surface, the lower the dust deposition density [20] as illustrated in Table 4.3a. The percentage of different sized particles in the total dust density is quite different; the percentage of large particles increase with the contact surface roughness, but the percentage of small size particle in the total dust density is reduced, see Table 4.3b.

The explanation of the above observation is that the density of the larger particles increases with the surface roughness since rough surfaces may accommodate larger particles, but at the same time they may cause the small particles to deposit around them. Because the number of small particles is the majority, the total dust density is reduced with the increase of contact roughness. By using a least square method, dust density is fitted to a log-normal distribution which is in agreement with the testing result made by L. White of IBM [10], As the surface roughness increases (from #3 polished by Grit 1000 to #2 by Grit 600 and #1 by Grit 400), the particle size distribution curves become broader and the peaks of the curves move to larger size of particles as shown in Figure 4.8. More recent work raises, however, raises questions about the validity of the log-normal distribution. The high magnification achievable with modern SEMs makes it possible to observe the presence of the large number of very small particles.

TABLE 4.3a

Contact Surface Roughness versus Particle Deposition Density

Surface Polished by Grit Number	Surface Roughness in Arbitrary Unit	Dust Particle Density in Arbitrary Unit
Grit 1000 smooth	0.1	5.7
Grit 600 medium	0.8	1.2
Grit 400 rough	1.8	1.0

TABLE 4.3b

The Percentage of Different Particle Size in the Total Dust Density vs. Surface Roughness

Surface Polished by Grit Number	Percentage% Total Dust Density	Percentage% Total Dust Density	Percentage% Total Dust Density
Polished by grit no.	Size D < 0.4 μm	0.4 < D < 1.1 μm	D > 1.1 μm
Grit 1000 smooth	98	1.8	0.2
Grit 600 medium	82	14	4
Grit 400 rough	52	38	10

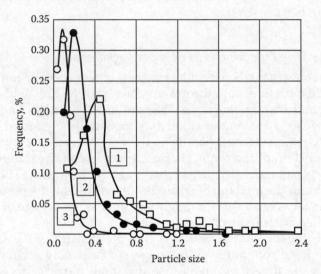

FIGURE 4.8
Particle size (μm) distribution fits a log-normal curve.

4.3.1.2.3 The Effect of an External Electric Static Field

An external static electric field may attract more particles as shown in the well known equation:

$$F = QE \qquad (4.2)$$

Where F is the attracting force, E is the external field intensity, Q, the electric charge carried by the particle. An experiment with two vertically positioned coupons, with parallel surfaces has been used to investigate the effect of an external static field on dust collection; see Figure 4.9a. Dust density on the positive surface increases almost linearly with the applied voltage. The linear increase of the total dust density with increasing dc voltage is shown in Figure 4.9b. Table 4.4 indicates that application of a static voltage enhances the deposition of smaller particles. The highest dust density, however, is found at the edges of the coupon, where the field strength is highest. This is shown in Figure 4.10.

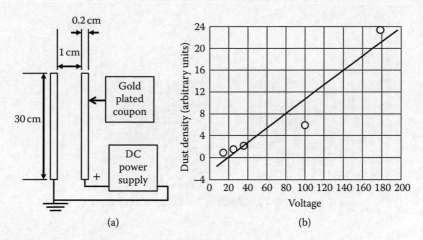

FIGURE 4.9

(a) The experimental set up (b) The dust density vs. voltage.

TABLE 4.4

Voltage vs. Particle Density (Number of Particles/mm²)

Voltage	<1 μm	1 – 2.5 μm	2.5 – 5 μm	5 – 10 μm	>10 μm	Sum
0 v	58	22	14	2	1	96
30 v	222	50	26	14	9	321
90 v	193	83	46	16	10	348

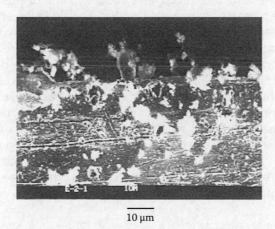

10 μm

FIGURE 4.10

An example of very dense particle precipitation at the high field region on the edge of coupon.

4.3.1.2.4 *The Attraction of Particles to an Open Contact Gap*

Figure 4.11 shows a gap between a probe and a flat plate such as can be seen for an open contact. FEM indicates that the field intensity in the gap is greatly enhanced. Experimental results verify that a high particle density occurs on the flat plate in the region of the gap [23]. The particle density for a 1 mm gap is higher than that for a 5 mm gap and the particle density for a 5 mm gap is higher than that for a 10 mm gap. The particles are moved from

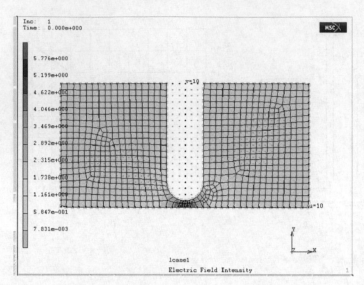

FIGURE 4.11
The electric field intensity at the open gap between the probe and the sheet.

the area away from the probe to the higher field region between the probe and the plate. This shows that it is possible for particles to move to a region of higher field intensity. Thus in a relay where vibration takes place during contact opening and closing, it is possible to dislodge particles that have adhered to the relay's walls. Some of these may settle close to the contacts. When the contacts are opened the electric field between them may attract these particles to the contact surfaces. Failures have been seen even with relays that have been carefully cleaned and checked before sealed. One example has found particles such as quartz and feldspar located at the contacts [21]. Another one found that plastic fibers caused the failure [22]. This shows that particles that carry a high level of electric charge may attach and hide on the relay's walls or on other parts. Once the relay experiences an impact, the internal electric field between the contacts may attract particles to move to the contact surfaces, which may eventually result in device failure. Dust deposition is highly selective [23]. Under the same electric field intensity the deposition depends greatly on the amount of electric charge carried by the particles. Some materials, such as organic materials, and titanium oxide are rare in dust, but if they do occur they have a high probability of being found on the contact surfaces.

4.3.2 Mechanical Behavior

4.3.2.1 Load Effect

When particles are located at the contact interface, two interacting conditions between dust and contact commonly occur: hard particles on a soft surface, and soft particles on a hard surface. The hardness of various materials is listed in Table 4.5 [24], using the Moh's scale. This scale is based on the ability of a harder material to scratch a softer material. A Moh's hardness above 6, such as feldspar, quartz, or hornblende, can be considered as hard relative to a gold plated surface, while hardness below 2-3, such as illite, kaolinite, calcite, gypsum, mica, and organics such as fiber are soft.

TABLE 4.5

Hardness of Material in Mohs

Hardness Mohs	Minerals & Metals	Formula
1	Talc	$Mg_3Si_4O_{10}(OH)_2$
1-2	Illite	$K_{0.6}(H_3O)_{0.4}Al_{1.3}Mg_{0.3}Fe^{2+}_{0.1}Si_{3.5}O_{10}(OH)_2 \cdot (H_2O)$
1.5-2	Kaolinite	$Al_2Si_2O_5(OH)_4$
2	Gypsum	$CaSO_4\ (2H_2O)$
2-2.5	Muscovite	$KAl_2Si_3AlO_{10}(OH,F)_2$
2-3	Biotite	$KMg_3AlSi_3O_{10}(OH)_2$
2.5-3	Gold, silver	Au, Ag
3	Calcite	$CaCO_3$
3-4	Lime	CaO
3.5-4	Dolomite	$CaMg(CO_3)_2$
4	Fluorite	CaF_2
4-5	Iron	Fe
5	Apatite	$Ca_5(PO_4)_3(OH-,Cl-,F-)$
5-6	Nickel	Ni
5-6	Hematite	Fe_2O_3
5-6	Hornblende	$(OH)_2Ca_2(Mg,Fe)_4(Si_6Al_2)O_{22}$
6	Orthoclase	$KAlSi_3O_8$ Albite $NaAlSi_3O_8$
6.5	Steel	
6-7	Epidote	$4CaO \cdot 3(Fe,Al)_2 \cdot O_36SiO_2H_2O$
7	Quartz	SiO_2

4.3.2.2 For Stationary Contacts

1. A hard particle pressed the soft surface of the stationary contacts

   Figure 4.12 illustrates a model with a hard particle (e.g., quartz or feldspar) which is much harder than that of the contact interface (e.g., a gold plated surface). The particle is embedded into the gold plated surface under normal force [25]. For simplicity, assume that the particle shape is a sphere. The embedded depth r' is expressed in Equation 4.3 [26]

   $$r' \approx P/(\pi D H_{Au}) \tag{4.3}$$

   Where, P is the normal force. D is diameter of the spherical particle, r is an average micro peak height and H_{Au} is the gold plating hardness. Assume that the upper surface and lower surface are similar, then the distance H' between the micro peaks of two surfaces can be expressed as Equation 4.4.

   $$H' = D - 2(r + r') \tag{4.4}$$

   When the normal force increased to an extent that $H' = 0$, the two contact surfaces touch each other, electric current can then pass through the connection of the micro peaks as illustrated in Figure 4.13. For this one micro-spot the contact resistance R_1 is:

   $$R_1 = \rho/(2a_1) \tag{4.5}$$

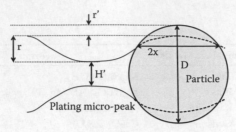

FIGURE 4.12
Model with a particle embedded at the contact interface.

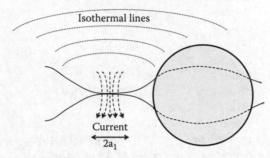

FIGURE 4.13
The electric current passes through the peak to peak contact.

Where a_1 is the radius of the micro-spot. As normal force is increased, more micro-peaks are in contact and the total contact resistance R can be expressed as (see Chapter 1, Section 1.2):

$$R = \rho / \left(\left(1 / \left\{ 2 \sum_i a_i \right\} \right) + 1 / \{ 2\alpha \} \right) \tag{4.6}$$

Where, a_i is the radius of a contact i^{th} spot and α is the Holm radius of the spot cluster.

When the particle is fully buried into the surface, contact resistance is quite low. Only when the particle is partially embedded and little metallic contact is made will the contact resistance be high. If sufficient current passes through the contact spots that the isothermal representing the metal's softening temperature (see Chapter 1, Section 1.4) reaches to vicinity of the particle, the embedding depth may well increase as shown in Figure 4.13. More contact spots may then be created and the contact resistance may be reduced even further. Experimental data support the above deduction. Figure 4.14 shows a number of pits visible on a gold plated sheet after an electric current has passed through a dusty contact surface with the resultant plastic deformation around the particles' location. The particles have been removed to reveal the pits.

2. A soft particle pressed onto the comparatively hard surface of the stationary contacts

Materials frequently found in dust are inorganic particles such as calcite, mica, gypsum [27], and carbon black. Organic particles such as fabric fiber or some kinds of organic particles are much softer than silver and most other metals. This type of particle can deform under normal contact force and remain in place on the contact

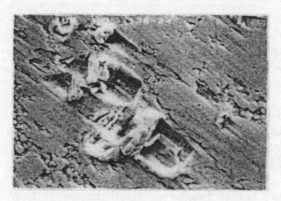

FIGURE 4.14
Pits on gold plated contact surface.

surface. Large particles can be very harmful in a relay application, but in connectors they are usually pushed away from the contacting region and are less harmful to the contact. Calcite particles are easily to be cracked under a force. For static contacts it appears to do little harm, but for sliding or fretting the particles tend to adhere together, form a pancake, which can cause failure. Mica is a laminate material, which can be broken into small pieces and spread over the contact. They are difficult to dislodge, because they are very thin and flat. If these tiny flat pieces cover all the initial metallic contact spots, the contact could fail. The testing results for gypsum show that it is less harmful than calcite, because even though it can crack into pieces, they do not adhere together. They also can slightly dissolve in water, but there is a lack of evidence to show that gypsum can cause serious corrosion. Carbon black is soft. It is likely that carbon particles may fill in the micro valleys or holes and make the surface smoother. It is also a low conductivity conductor.

4.3.2.3 For Sliding Contacts

When a strong lateral force is applied, the pits caused by embedding of the hard particle could be prolonged to form scratches or wear tracks. The metal from these wear tracks usually becomes wear debris and can be pushed away from the tracks. The debris volume V is expressed as,

$$V = kPX/H_m \tag{4.7}$$

Where P is the contact force, X is the sliding distance, H_m, is the metal's hardness, k is wear coefficient, see Horn [28]. When contact force is below 0.04 N, k approaches to zero. From 0.04 to 0.5 N, k is very changeable. Beyond 0.5 N, k is stable and equal to 3×10^{-4}. Thus when contact force is in between 0.04 to 0.2 N, the tracks can be very light.

4.3.2.4 The Effect of Lubricants Coated on Contact Surface

The presence of lubricants affects the results of sliding wear in dusty environments. Studies by Mottine and Reagor [29], Hague et al. [11] give specific data. It must be noted that the environments and applications are very complicated. There is no "universal lubricant" which can fit all kinds of requirements [30] (see Chapter 3, Section 3.4). Many lubricants can attach to the contact surface thus are remain in place and will not be pushed away. The permittivity of the lubricant can be an important consideration. Figure 4.15 shows a schematic of a model of a particle above a lubricated layer. Assume a dust particle acts as a point-source carrying a charge Q and is located at a distance d_0 above the surface of a dielectric layer of thickness d_1. The dielectric layer is supported on a grounded conductor. With the permittivity of vacuum, ε_0, and the permittivity of the dielectric layer, ε_1, the force F acting on the particle is,

$$F = -Q^2 / 4\pi\varepsilon_0 \left[2d_0 + 2(\varepsilon_0 + \varepsilon_1)d_1 \right]^2 \qquad (4.8)$$

The force ratio F/F_0 as function of the ratio $\varepsilon_1/\varepsilon_0$ is plotted in Figure 4.16. F_0 represents the force acting on the particle for a dielectric layer with $\varepsilon_1 = \varepsilon_0$. Properties of several wax and liquid lubricants are listed in Table 4.6.

A wax lubricant coated on contact is a layer that covers the whole surface. Wax with a high permittivity attracts more dust particles than the one with a lower permittivity. Liquid lubricants coated on a contact surface generally appear as droplets spread over that surface. Liquids with high permittivity also attract more dust particles, but the particles only concentrate and float on the droplets. They are less prevalent on a surface with no lubricant. Thus, dust density for the surface coated with a liquid is much less than that of the surface coated with wax. The experimental data is shown in Table 4.7.

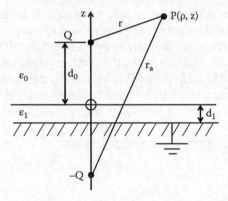

FIGURE 4.15
A model of a particle carrying a charge Q above a lubricated layer.

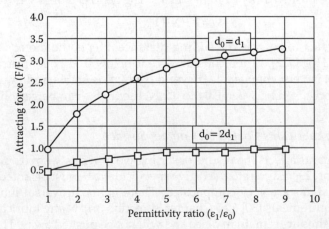

FIGURE 4.16
Permittivity vs. attracting force.

TABLE 4.6

Selected Properties of the Lubricants

| Identification | | | | Selected Properties | | |
Lubricant	Origin	State	Melting Point in (°C)	Permittivity at 20 (°C)	Value in (cSt)	Temp. at (°C)
					Viscosity	
Paraffin	China	Wax	50-55	1.98(2.06)	/	/
By-2	China	Wax	70-80	4.5(5.6)	/	/
OS-124	USA	Liquid	/	4.7(4.5, 5.6)	366	38
					13	93
9200	Japan	Liquid	/	1.8(2.25)	420	40
					35	100
9300	Japan	Liquid	/	2.6(2.18)	400	40
2000	Japan	Liquid	/	3.2(2.83)	60	40
					8	100

Number in the bracket () is another test result.

TABLE 4.7

Permittivity versus Dust Deposition Density

Lubricant	State	Permittivity at 20C	Dust Density for Liquids Relative to Lubricant 9200	Dust Density for Waxes Relative to Paraffin	Dust Density Relative to Lubricant 9200
9200	Liquid	1.8	1	-	1
OS-124	Liquid	4.7	1.6	-	1.6
Paraffin	Wax	1.98	-	1	2.16
By-2	Wax	4.5	-	1.8	4.29

4.3.2.5 Sliding Contacts on Lubricated and Dusty Contacts

During sliding on a wax lubricated surface, particles mix with the wax. Particles cannot leave this mixture. Thus the lubricated surface can result in a high contact resistance. BY-2 possesses a high permittivity and collects a high density of dust particles. Thus the probability of getting contact trouble with this lubricant would be very high. OS-124 is a liquid lubricant also with a high permittivity and thus can attract dust particles. However, dust particles float on the liquid which has a high contact angle. During sliding dust particles can be pushed away with less harmful effect [31]. Experimental results are summarized and illustrated in Figure 4.17a for wax lubricants and Figure 4.17b for liquid lubricants [32]. Contact resistance above 10 mΩ is considered failure for gold plated connector contacts. The sum resistance of failed contacts is expressed as "resistance" with respect to the permittivity for various lubricants. The number of markers of the same kind expresses the number of experiments. In these experiments the contacts are made to slide a distance of 2 cm for 50 operations. Contact resistance is measured 8 times for each sliding operation. Conclusions from the sliding experiments: (a) The sum of failure resistance is higher for wax than for liquid lubricants. (b) High permittivity of the lubricants results in high sum of failure resistance. (c) A high contact force or high current reduces the sum of failure resistance for lubricated contacts.

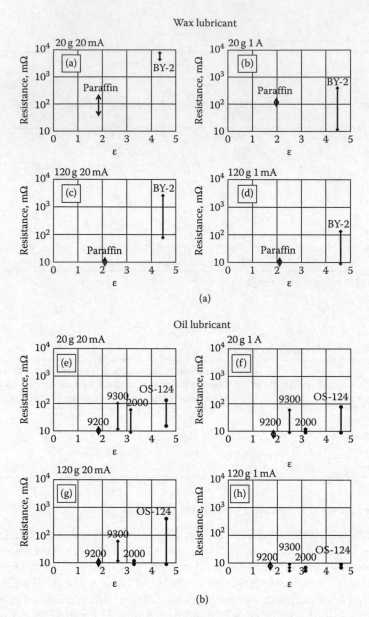

FIGURE 4.17
Contact resistance for dust on lubricated surfaces. (a) Wax lubricants. (b) Liquid lubricants.

4.3.2.6 Fretting (Micro Motion) on Lubricated and Dusty Contacts

The effect of the five lubricants, BY-2, Paraffin, OS-124, C-2000, and C-9200, on the contact failure under micro-motion has been studied [32]. As found in the experiments with sliding contacts, permittivity and the state of the lubricants are also important with micro motion. The results are presented in Figure 4.18a, where the total number of failures as a function of the permittivity of the lubricants is plotted. They are compared with results of the sliding experiment shown in Figure 4.18b. The trends in the two figures are similar. Notice that in Figure 4.18a the vertical axis is the total number of failures, while in Figure 4.18b

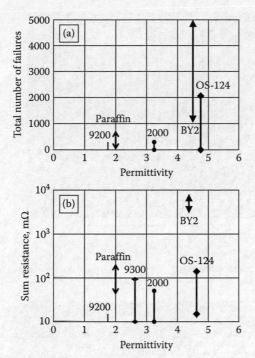

FIGURE 4.18
(a) The fretting experiment. (b) The sliding experiment.

it is the sum of resistance failures. Evidently, the resistance of the total number of failures is much higher than the sum of resistance failures. Therefore, as a result of the experiments, failures caused by micro motion are more serious than that of sliding contacts. In micro movement experiments on a BY-2 lubricated wax surface show that it performs badly compared to the liquid. Dust particles attach strongly to the wax in some places in the moving track and they remain there.

4.3.3 Chemical Behavior

4.3.3.1 Dust Particles Create Pores

When the electro plate is exposed in air, dust particles can easily deposit on the substrate metal before and during plating process. Gold ions can only reduce on a conducting metal surface but not on insulating particles. As the gold plating surrounding the particle gets thicker tiny pores are created in the gold plated layer. The number and size of pores resulting from dust deposition are more serious than those caused by surface roughness. Figure 4.19 shows very poor gold plating with numerous holes (dark spots) on the surface. XES detection at the holes can find elements of dust composition in them. Such gold plating is very susceptible to sliding or fretting wear. In fact, the gold plate can be easily peeled off and cause further corrosion, see Figure 4.20.

4.3.3.2 Corrosion Appears as a Result of Dusty Water Solutions

Experiments show that an aqueous solution of dust collected from Shanghai and Beijing can cause serious corrosion on thin gold plated surfaces. For accelerating the process, scratches are made to expose the nickel underplate. Dusty water solution is dripped onto

20 µm

FIGURE 4.19
High porosity as a result of dust contamination.

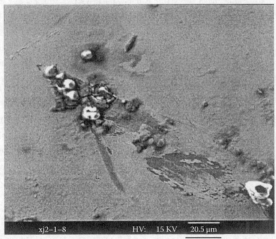

20.5 µm

FIGURE 4.20
Corrosion at the end of a wear track after the gold layer was peeled off.

the thin gold plated surface several times after the solution evaporates. Corrosion is clearly seen as shown in Figure 4.21. XES results show that not only elements of sulfates and chlorides of nickel and copper are present, but also elements of water soluble salts such as Ca, K, Na can be detected [33]. The result is consistent with previous experiments that show the elements appear as both negative and positive ions in the dust solution.

4.3.3.3 Indoor Exposure Results

Several long term exposure experiments have been carried out in Shanghai and Beijing. Total indoor exposure duration on gold plated coupons in Shanghai is 22 months and in Beijing is 10 months. Figure 4.22 shows the corrosion stain formed on thin gold plated coupon surface after 15 months indoor exposure in Shanghai [34]. As it is described in

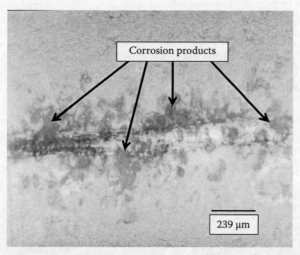

FIGURE 4.21
Corrosion on the scratches.

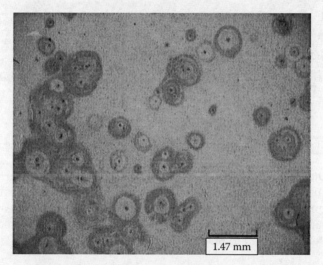

FIGURE 4.22
The corrosion stains on a thin gold-plated surface.

Chapters 2 and 8, thin gold layers are always porous and can be corrosion will be initiated at the pore. The corrosion product appears as stains on the plating's surface. Evidently at least one pore is underneath each stain. Thus the porosity of thin gold layer can be counted approximately by counting the number of stains. The major components of the electrolytes that cause the corrosion are the water soluble salts carried by dust particles. The salts dissolve in the water film formed on the plated surface when there is a humid ambient. Dust particles are therefore also associated with pore corrosion, and they may be found located on the stain area or in a region close by the stain. The number of corrosion stains and number of particles can be counted using optical microscope. Note that using an optical microscope it is only possible to observe particle sizes greater than ~ 4 µm. Table 4.8 lists number of stains/per cm² and number of particles/per cm² as a function of exposure time over a gold plated area of 64 cm² and an average surface roughness R_a of 0.2 µm. It can be seen

TABLE 4.8

Number of Stain and Particles with Exposure Time

Exposure Time Number of Months	Number of Stains/cm² Thin Au Plating		Particle Density Number of Particles/cm²
	0.48 µm	1.55 µm	
3	132	41	920
8	139	51	1220
15	139	75	/
22	180	80	/

51.4 µm   14.5 µm

(a)   (b)

FIGURE 4.23
(a) The structure of corrosion stain (b) A higher magnification of part of the rings in (a).

from the Table that the number of corrosion stains and the number of particles increases with the exposure time. This does not take into account the number of small particles (i.e., < 4 µm), which can greatly add to this count. As the number of stains increases with exposure time, each stain area also grows larger. The growth rate of the stain area (mm²) as well as average diameter (µm) is found to fit a log distribution curve [35].

4.3.3.4 Construction of the Corrosion Stain

Figure 4.23a illustrates an SEM photo of a typical construction of a stain after 22 months indoor exposure in Shanghai. There is a central core located at the pore site surrounded by several rings. Figure 4.23b shows a part of the rings, they were constructed mainly by many tiny islands. The elements found at the core and in the rings are mainly chlorides, sulfates, hydrates and oxides of nickel and copper. In some places elements of dust are found to be mixed with this corrosion. The special construction of the corrosion stain of a central core surrounded by corrosion rings can be imagined as seaweed pushed radially by a tide. When the tide ebbs the seaweed is left on the surface as in the form of rings. This will be repeated from tide to tide leaving a succession of rings. Corrosion products formed the pore and starting from the pore float away from the core on electrolytes or on dusty water during a cycle of high humidity. When this liquid layer dries in a low humidity cycle the corrosion rings are left behind [36]. After 22 months of indoor exposure, the core size can be as large as 20 µm when tested by 3-D profilometry. Rings can be as thick as 0.3 µm

TABLE 4.9

The Difference Between Corrosion Products and Dust

	Dust Particles	Island-Like Corrosion Products
Elements in the composition	Si Al, C, O, Na K, Mg, S	Mainly Ni, Cu, S, O, Cl
Shapes	Polygon, comparative sharp edges	No special shape, no sharp edge
Distribution	Separated with each other, usually with different elements	Connected with each other, almost the same elements
Surface	Very rare cracks	Cracks on the surface

when checked by Auger electron spectrometry AES (sputtered by Argon ions for 10 min with speed of 300A/min). The growth mode of these corrosion products is not a layer type of growth but a tiny-island ring growth instead. The construction of the corrosion stains on gold plated coupons exposed in Shanghai and Beijing appear to be similar. The corrosion starts at the first month to form at the pore and gradually forms a closed ring which then extends outward to form rings surrounding a central core. Ring-like corrosion products look similar to the dust particles. The difference between them can be observed by SEM/XES and is summarized in Table 4.9.

It is not easy to find the contour of the particles that are so small that might be partly buried or surrounded by corrosion products. In other words, it is difficult to verify their existence using an SEM. It is possible, however, using XES to identify the elements of dust composition, which, in turn, will show the existence of the very small dust particles. In order to distinguish whether the corrosion product is mainly caused by corrosive gases in the air or by dust particles, XES analysis can again be helpful. If the corrosion products are created by SO_2 and Cl_2 gases in air, they contain only elements of Ni, Cu, S, O, and Cl. If, however, they are caused by dust particles, elements of the positive ions of water soluble salts such as Ca, Na, and K will be present in the products. Silicates as well as calcium compounds may also appear within the corrosion products.

4.3.3.5 Fretting Experiments on Dust Corroded Coupon Surfaces

Fretting experiments have been carried out using a fretting machine. The moving length is in a range of 30-50 µm with fixed direction and length. The normal force is set at 0.5 Newton. Contact resistance is measured during the experiment. The probe is initially placed in contact close to the core and the fretting cycle passes through the core area. The initial resistance is few hundred mΩ. As the number of fretting cycles increases, the resistance drops to a value lower than 10 mΩ as illustrated in Figure 4.24a [37]. Such a reduction is caused by the cracking and scattering of the large area of corrosion at the core, and also its weak attachment to the contact surface. It can be seen in Figure 4.24b after the fretting experiment, the core has not remained in the track area. The smooth region B is the fretting area, A shows a remnant of the core, C is the wear debris. The debris has been pushed away from the fretting area. The elements left in the core and in the wear debris are listed in Table 4.10. They are mixture of silicates and chloride of nickel and copper.

Fretting on various ring areas gives an initial low contact resistance. However, in some places, the resistance can increase to a very high value and remain there as the fretting experiment continues. This is illustrated in Figure 4.25a. The corrosion product surrounding the particles creates the attachment for the particle to the contact surface. When the fretting movement occurs, the particle becomes embedded into the surface and becomes

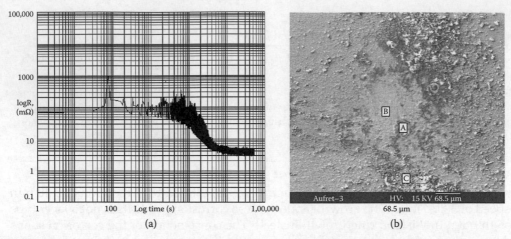

(a)   (b)

FIGURE 4.24

(a) An example of the resistance dropping during fretting on core area (b) The worn area and remnant of the core.

TABLE 4.10

The Atomic Percentage of the Elements in the Remnant of the Core and in the Wear Debris in Figure 4.24b

	Ni	Cu	Au	Si	Cl	O	S
Core area, spot A	22.68	3.44	2.88	1.61	9.39	60.01	/
Wear debris, spot C	29.60	6.91	1.68	0.48	5.17	54.55	1.49

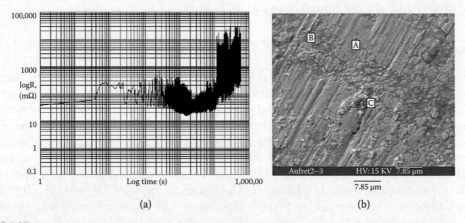

(a)   (b)

FIGURE 4.25

(a) Fretting on ring area (b) An example of a particle embedded in the fretting region.

tightly attached as the fretting cycling continues [38]. Figure 4.25b shows an enlarged figure of embedded particle. Table 4.11 shows the XES testing result at four spots. At spot A, XES shows only the gold plated surface. At spot B, chlorides of nickel and copper are seen. Spot C, shows a mixture of silicate particles, carbon and chlorides of nickel and copper. C spot also shows a particle with a high silicon concentration. At spot D, chlorides

TABLE 4.11

The Atomic Percentage of the Elements at Different Spots in Figure 4.25b

	Cu	Ni	Au	C	O	Al	Si	Cl
A	17.08	16.14	66.78	/	/	/	/	/
B	21.48	5.00	26.36	22.13	16.43	/	/	8.59
C	13.23	4.29	17.33	31.00	22.53	4.23	5.48	3.63
D	24.96	3.45	12.45	33.56	16.98	/	/	8.60

and carbon cover the surface. The wear track on Figure 4.25b shows a dust particle has been strongly pressed and tightly attached to the surface.

4.4 Application Conditions in Dusty Environment

4.4.1 Explanation of the Special Features

Different applications of connectors may show different features at the failed contact surfaces. Since large numbers of failed mobile phone connectors have been analyzed and reviewed statistically [38,39], detailed features of failed mobile phone connectors can be presented as a typical examples of dust particle contamination. Connectors in mobile phones consist of a spherical probe in contact with a printed circuit board (i.e., a ball-on flat design). Normal force applied on the contacts can vary from 0.5 N to 1 N. In most cases, the contacts, substrate metal is phosphor bronze, coated with nickel as barrier layer to prevent diffusion of copper and to harden the surface. A layer of gold plating is coated on top of the contacts. Field failed contact surfaces are tested by optical microscope and SEM and by XES. For organic materials, they are tested by Infrared Fourier Spectrometry. Special features of failed contact surfaces are identified and described as follows.

4.4.1.1 Covered by Accumulated Small Particles

Figure 4.26a shows contaminants covering the failed contact's surface. Figure 4.26b displays the enlarged portion of the accumulated particles marked in the rectangular square in Figure 4.26a. The contaminants are actually composed of numerous tiny particles.

4.4.1.2 Accumulative Particles Caused by Micro Motion

The examples below show that micro movement plays an important role in the failure of movable contacts. It is found that about 42% of the contact surfaces on the PWB appeared to have sliding wear tracks; however, 58% are without clear wear tracks or have no tracks at all. Figure 4.27a illustrates the micro wear tracks on a failed contact surface. Clearly, the motion is not in a regular straight line with a fixed length as had expected and as had been used in the laboratory experiments. On the contrary, it is erratic both in direction and in distance [40]. Figure 4.27b illustrates the accumulated particles found at the end of the wear track. The effect of micro motion at the contacts can be summarized as producing wear debris both on the corrosion products and on the plated metals. Hard particles such as quartz are embedded into the plating's surface. The laminate particles such as mica and

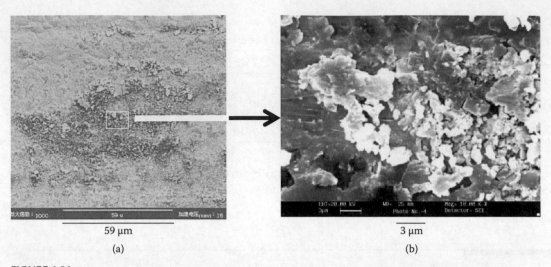

59 µm

(a)

3 µm

(b)

FIGURE 4.26
(a) Contaminants on the contact surface (b) Enlarged portion of particles from (a).

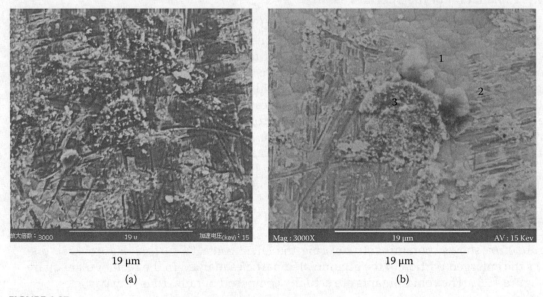

19 µm

(a)

19 µm

(b)

FIGURE 4.27
(a) Wear tracks as result of micro motion (b) Particles piled at the end of the wear track.

soft particles such as calcite and gypsum are pressed onto the contact surface. An important consequence of this micro motion is the collection and accumulation of contaminant particles and the formation of tiny high resistance regions. Once the contact probe climbs onto the surface of such a region, contact failure can occur.

4.4.1.3 High and Erratic Contact Resistance

Figure 4.28a shows contaminants covering the area of failed contact. Figure 4.28b shows the enlargement of the lower region in Figure 4.28a. Figure 4.28c presents the enlarged portion of the right side region. Figure 4.28d shows the upper region. From these pictures,

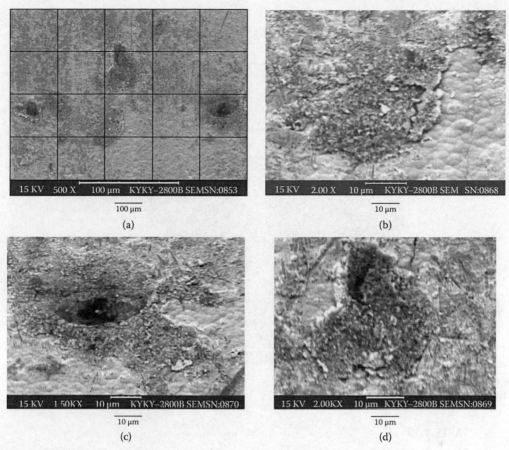

FIGURE 4.28

(a) Contaminants on the failed contact: (b) Lower left hand side region in (a): (c) Right side region of (a): (d) Upper region of (a).

the surface of the particles looks to be flattened and tightly attached to the contact's surface. This indicates that particles have undergone not only with the high normal force but also lateral shearing force. Table 4.12 shows the contact resistance of the contaminated area in Figure 4.28a. Figure 4.28a is divided into 20 square regions named as X (x axis) and Y (y axis). Resistance testing spots are located at the center of the square regions. Distance between each spot is about 50 μm. The resistance at different spots can vary a lot. Testing many other failed contacts, resistance as high as that expected from insulation has been found. It is easy to predict when micro motion has occurred if the probe stops on a high resistance region; it is equivalent to a high resistance connecting in series with the electronic circuit which may well result to the circuit failure. When the probe moves to another region with a lower resistance the contact resistance becomes acceptable. Thus, during micro movement, the performance of circuit can be very unstable.

4.4.1.4 The Element of Si Causes High Contact Resistance

Testing results show that accumulated particles are composed of dust, corrosion products, and wear debris of surface materials. Testing of high resistance regions has found that 56 out of 74 testing spots contain silicon. This is much higher than has been observed for

TABLE 4.12

The Contact Resistance (mΩ) in the Contaminated Regions in Figure 4.28a

Test Spot	X1	X2	X3	X4	X5
Y1	13	165	125	245	526
Y2	74	81	55	216	3188
Y3	60	66	133	53	113
Y4	53	372	150	47	61

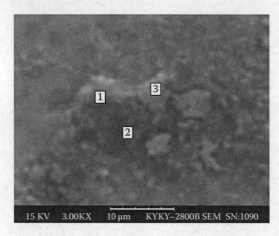

FIGURE 4.29
Illustrates the contaminants with 3 testing spots.

TABLE 4.13

The Atomic Percentage of Elements at Different Spots

Fig	Spot No	C	O	Al	Si	S	Cl	Ni	Au
	1	36.9	52.4		2.8			4.0	3.8
	2	82.4	13.3	0.1	0.8	0.2	0.1	2.4	0.6
	3	39.5	46.2		1.4			2.3	10.7

other elements found in dust. Silicates in the dust such as quartz, several kinds of feldspar and mica may therefore contribute more than other materials in developing high resistance contact surfaces. Figure 4.29 shows contamination on a failed contact surface. The life time for this mobile phone is 11 months and 20 days. Three testing spots are randomly selected at the contaminant with high resistance see Figure 4.29. The elements are listed in Table 4.13. Apparently carbon and organics heavily cover the surface. Although corrosion products seem to be very light, silicates are involved in all three testing spots.

4.4.1.5 Organics Act as Adhesives

XES inspection of failed contacts has found that 252 out of 363 total testing places (i.e., about 70%) have very a high carbon concentration plus a certain amount of oxygen. Contaminants in Figure 4.30 show an example, from a mobile phone whose life is only 3 month and 11 days in Beijing. Once the phone is used it cannot shut down, and

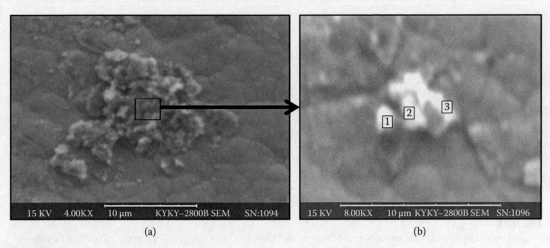

(a)   (b)

FIGURE 4.30
(a) The frame scanned by XES, see Table 4.14: (b) Detailed spots analysis by XES, see Table 4.14.

TABLE 4.14

Frame Scanning and Spot Test Result by XES Atomic Percentage

Fig	Test spot	C	O	Na	Mg	Al	Si	Ti	Fe	Ni	Cu	Au
a	frame	56.5	32.9			0.4	0.2			4.4		5.6
b	No.1	84.2	13.7	0.1		0.5	0.1	0.1	0.5	0.5		0.3
b	No.2	72.3	26.7			0.3	0.1	0.3	0.2			0.1
b	No.3	78.4	20.8			0.3	0.1	0.3	0.2			

sometimes it will not open either. Contamination has been found on battery contacts. In Table 4.14 of XES data taken on Figure 4.30a and region 1, 2 and 3 on Figure 4.30b show a lot of carbon and other organics mixed with silicate particles. Nickel oxides may also possibly exist, but are not observed in this analysis. Figure 4.31a shows the contaminant photographed using an SEM. Further investigation by precise infrared spectrometry on Figure 4.31a reveals a certain level of lactates of sodium and calcium. Figure 4.31b shows specific spectra. The upper spectrum is the standard for sodium lactate. The lower one is the test sample. The sample fits the standard well. Lactate is part of the human sweat. It is also found in hand lotion. In a certain temperature range and with certain water content, it can act as an adhesive to adhere separate particles together. It can also attach the particles to the contact surface. This can be an extremely destructive condition when sliding or micro motion is present. The consumer product of sodium lactate is a 60% aqueous solution miscible in water. Tests shows that this solution can prevent particles from scattering when a pushing force is applied. Lactates can also perform as a lubricant on the metallic interface between a probe and a PWB, since they possess a polar head, which can attach to a metal's surface, (see construction of sodium lactate where COONa is the polar head)

$$H_3C-CH-COONa$$
$$|$$
$$OH$$

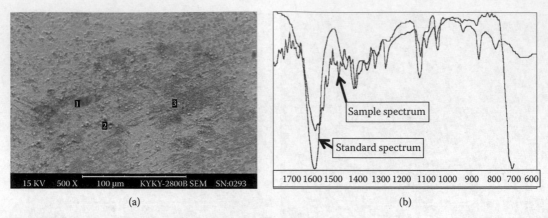

(a)   (b)

FIGURE 4.31
(a) Contaminated area with high carbon concentration. (b) Upper curve is standard spectrum of sodium lactate.

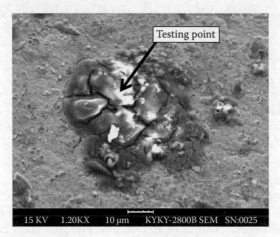

FIGURE 4.32
The contaminant on the contact surface.

4.4.1.6 *Corrosion Products Trap the Dust Particles*

Using the SEM/XES it is seen that elements of nickel and copper can combine with the dust contaminants to form corrosion products that can strongly attach to the nickel under plate. Thus the dust particles can be trapped by and mixed with the corrosion product. The following example shows this effect. When the accelerating voltage of the XES is increased, the x-ray penetration depth deepens according to the equation derived by Andersen and Hasler [41]:

$$R(x) = \frac{0.064}{\rho}(E_0^{1.68} - E_c^{1.68}) \tag{4.9}$$

where, *R(x)*, the penetrating distance of characteristics of x-ray in μm, E_0 the accelerating voltage in kV, E_c the critical exciting energy of the characteristic x-ray for the element, necessary to eject an electron from the inner K, L, M shell in kV, ρ the atomic density in g/cm³. Figure 4.32 shows the tested contaminated surface. Table 4.15 give the atomic percentage of materials found at the testing point.

TABLE 4.15

Elements Atomic Percentage versus Accelerating Voltage E_0 ((kV))

E_0(kv)	C-(Ka)	O-(Ka)	Na-(Ka)	Al-(Ka)	Si-(Ka)	P-(Ka)	S-(Ka)	Cl-(Ka)	K-(Ka)	Ca-(Ka)	Ni-(Ka)	Cu-(Ka)	Au-(Ma)
7	80.97	13.56	0.39		0.08			1.27	0.52			3.16	0.05
10	80.77	11.84	1.64	0.18	0.28		0.24	0.81	0.68			3.56	0.02
12	72.26	15.25	1.92	0.11	0.12		0.38	1.10	0.77	0.42	4.27	3.36	0.04
15	58.03	25.08	2.45	0.08	0.18		0.45	1.71	0.99	0.33	6.13	4.48	0.09
17	40.03	18.77	11.84	0.68	0.38		1.01	3.90	1.33	0.50	10.47	10.60	0.51
20	29.26	22.67	11.72	0.71		0.34	1.34	4.58	1.46	0.31	14.49	11.62	1.50
22	10.28	15.75	7.81	1.53	1.64		1.02	10.05	1.96	0.54	25.66	22.02	1.76
25	17.49	23.90	13.04	0.92		3.02	1.06	4.64	1.40	0.76	22.79	9.42	1.55

As the accelerating voltage is increased to $E_0 = 12$ kV, the Ni concentration jumps from zero atomic percentage to 4.27 atomic percentage. As the accelerating voltage continues to increase the concentration of Ni also increases, but there is no step jump in the Ni concentration. Thus it can be seen that at $E_0 = 12$ kV can be considered the turning point at which the penetrating x-ray has reached the top edge of Ni plating. It can be seen that dust elements such as Si, Al, K, Na, Ca are accompanied by S, Cl, and Ni and are gathered at this surface. Using the above equation, thickness of the contaminants can also be roughly calculated. The result is very close to the measurement of 3D profilometry. Thus the corrosion products can trap the particles and attach them to the Ni surface. If a normal force is applied to particles, hard particles in dust can be embedded into the metal surface. The trapping effect following the embedding effect make the particles tightly fixed to the metal. This can result in a strong friction effect when a force attempts to push the particles away. After cleaning off the contaminated particles at a failed contact surface, pits and tracks can be seen that shows the hard particles have been embedded into the surface of the Ni under plate, see Figure 4.33a and Figure 4.33b. The grey stains display the Ni under plate where the gold plated layer has been almost completely worn away.

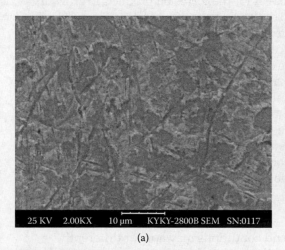

25 KV 2.00KX 10 µm KYKY-2800B SEM SN:0117

(a)

25 KV 8.00KX 10 µm KYKY-2800B SEM SN:0124

(b)

FIGURE 4.33
(a) Irregular tracks and loss of the gold areas (b) Pits with irregular sharp edges.

4.4.1.7 Difference Between Short Life and Longer Life Contacts

SEM/XES testing has been made on contaminated contact surfaces for 4 failed mobile phones with a short life time (~3 months) and for 4 failed mobile phones with a longer life time (~11 to 12 months). 404 testing spots are examined for the short life phones contacts and 122 testing spots are examined for the longer life phones. Statistical results show that for the short life phones, more carbon, organics and silicates are present, but fewer corrosion products. The longer life mobile phones show the opposite result: more corrosion and less carbon, organics, and silicates are seen. This shows that although corrosion trapping may occur at the same time as the adhesion of the organics on the contact, the corrosion trapping probably needs a longer time to attach the particles onto the contact surface. The fact also shows that those mobile phones with very short life duration must have been working in conditions with a high density of particles and a high concentration of organics such as sweat or lactates. The attached particles either by trapping or adhesive effect may have helped the particles to be embedded into the contact surface. Even if they are only partially buried the two sides of contact could be isolated by the exposed particles. Thus, contact resistance greatly depends on the volume resistivity and surface resistivity of the exposed particles.

4.4.1.8 Large Pieces of "Stepping Stones"

Particles embedded into the contact surface not only are difficult to be pushed away, but they also provide a blockage preventing other particles to be moved forward. As more particles are embedded, they become a fence to stop and accumalate even more particles. Thus a large growth of embedded particles appears. During micro motion these large particles can serve as "stepping stones" to decrease the inclined angle for the probe to climb over a surface lump.

4.4.1.9 The Performance of Failed Mobile Phones

The above special features of failed connector contact surfaces due to particle contamination can cause serious result to the performance of the mobile phone: the electrical contacts are different from that of the other electrical components. Failure of others can be tested before they start to operate. But for the connectors, there is a time delay for the contacts to fail because they need time to accumulate particles and organics or for the corrosion products to grow up. Therefore, when connectors are new, even the gold plating has serious defect, the problems can not be discovered. Contact failure can be occurred suddenly without any alert. And some times they can be self healed to operate again since micro motion at the interface can change the location of the contact. To investigate the micro contamination and find out the high resistance region is a very difficult task, because the particle contamination is not visible by bare eyes or even by optical microscope. High magnification and precision equipment are required.

The special features due to particle contamination for different kinds of connectors and with various application conditions may not be the same. However, there are some factors in common: Irregular micro motion, tiny particles accumulation especially including silicates, adhesives and attachments such as corrosion products and organic materials, humidity and temperature cycling, defects of the surface plating etc.

4.4.2 Other Examples

Various connectors with different applications and conditions may appear differently.

An example of failed bus bar and clips caused by serious dust particles contamination were found. Strong corrosion products mixed with heavy dust particles were covered on contact surfaces of both bus bar and the clips. The thick insulated layer caused arcing and stress relaxation of the substrate metal.

Another example was the electric bolt type connectors used in the outdoor power line system in China. Serious stress corrosion was found on the washer which was used to tighten the bolt screw with the nuts. However since the washer was totally cracked and broken, so the screws were loosened, dust particles could squeeze into the contact area. The contact areas were filled with thick layer of contaminants. They were mixture of corrosion products and dust particles. Interesting thing is that the wear tracks were not only in x and y directions, but also something like twisting and rocking. It is probably due to the swinging of wire in windy days and different seasons [42]. Arcing is also found at the contact surface which shows that the contaminant covered on the contact is almost insulated.

4.5 Theoretical Analysis of Connector Contact Failure Due to the Dust

4.5.1 Two Micro Worlds in Contact

As it is shown by J.B.P. Williamson [43], a pair of metallic surfaces in contact actually is two micro worlds in contact. The real contact area depends on the deformation of the micro peaks of two surfaces in connection; Influencing factors are the normal force, micro peaks height and their distribution, depth of micro valleys as well as the surface hardness of the metals. Therefore, from a micro point of view, there is a large gap or space between the two contact surfaces which can accommodate small particles. Figure 4.34, illustrates a new contact pair with average roughness Ra = 0.26 µm, 29 micro peaks are spread on a length of 149 µm (for simplicity, only x profile is considered). The average distance between peaks is 5.15 µm. Maximum peak height is 0.65 µm; maximum depth of micro valley is 0.82 µm. Assume the particle is a sphere, thus the largest particle size accepted by the contact pair can not exceed 1.6 µm. Practically the size of particles should be much smaller. When the contact surface has been worn out to make the surface rougher, accepted particle size can be larger.

Some analysis for the harmful particle size in the previous literature are based on the macro point of view, the calculation result is around several tens of microns which is much larger than the practical particle size tested on the failed contact surface.

4.5.1.1 Particles Get into the Contact Interface

As illustrated above, a spherical shape of probe is in contact with a plain coupon surface. When the contact pair is exposed in dusty environment, particles with various sizes are deposited on the coupon surface. During sliding or micro movement in various directions, the large particles will be blocked by the micro peaks of the surface and stay outside of the contact, only the small particles are able to move into the gap of the interface. In other words, the size of the particles that can get into the contacts and accommodate at the contact interface depends on the morphology of both contact surfaces. New connector contact with very low surface roughness say 0.2 µm can reject the particles larger than 2-3 µm. But a frequently used contact with worn out and rough surface may accept larger particles to get into the interface.

4.5.2 "Preliminary Attachment"

4.5.2.1 Adhesive Effect

As micro movement occurs at the contact interface, particles can be pushed by the micro peaks of one side of the contact surface as shown in Figure 4.34. When particles meet micro peaks of the other side of the surface which can block the particle to move forward, particles could turn to other directions since particles are free from moving or even rolling around. If particles are adhered by some kind of organics, then separated particles are connected together as a group by the adhesives. Thus the probability of meeting the opposite micro peaks is greatly increased, and there may not provide enough space for the group of particles to move away. Some of the particles may be attached by the other side of micro peaks at the interface; this is considered as "preliminary attachment." The micro peaks of one side of the contact may climb on the particles, and part of the normal force can be transferred to the particles. If the normal pressure applied on the particles is higher than the yield strength of the surface metal, particles can be embedded into the surface, thus "final attachment" may happen. The only way to move such kind of particle away is to dig the particles out of the surface to create wear debris and remain pits or wear tracks on the contact surface.

A simulation experiment has been done to verify the above description: small SiO_2 particles were spread on clean coupon surface without any organics, fretting was applied on the coupon. Figure 4.35a illustrates that the particles were scattered away. Figure 4.35b shows that contact resistance was kept at very low and stable rate. Results verified that no particle was inserted at the interface.

Another clean coupon was dipped in sodium lactate at room temperature. After dry up process, same amount of small SiO_2 particles were spread on the coupon. During fretting, particles were adhered together as shown in Figure 4.36a. Contact resistance during fretting could be as high as it was insulated, illustrated in Figure 4.36b.

Experimental result showed that the evaporation of lactate kept at 60°C for 40 days was equivalent to dipping the contact samples into the alcohol solution with 0.5% of lactate and then dried up at room temperature [44]. Further test verified that 0.5% of lactate could act as both adhesive for dust particles and lubricant for metal surfaces. Concentration of lactate higher than 3%, the film was too thick so that the particles could be floated and moved away during fretting and kept very low contact resistance [45].

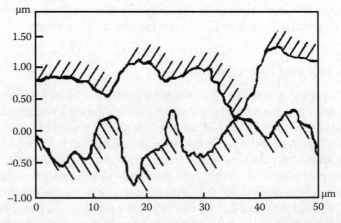

FIGURE 4.34
Illustration showing two sides of a contact pair just touching.

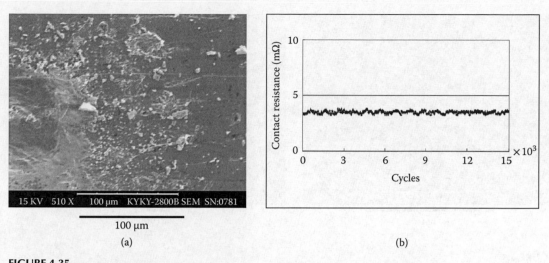

FIGURE 4.35
(a) Particles are scattered away with no lactate on the contact surface (b) Contact resistance versus fretting cycles no lactate.

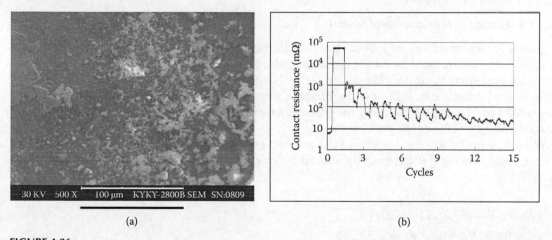

FIGURE 4.36
(a) Particles are adhering in groups with lactate on the contact surface (b) Contact resistance versus fretting cycles with lactate.

4.5.2.2 Trapping Effect of Corrosion Products

Trapping effect was also simulated in the lab for verification of its preliminary function of attachment on the contact surface. Clean coupons were dipped in 1% of NaCl water solution. After dried up, small SiO_2 particles were spread on the coupons. Six cycles of humidity and dry up procedure were applied on the coupons and then the coupons were mounted on the fretting machine for experiment and contact resistance measurement. During fretting for 5,000 cycles, contact appears insulated. Figure 4.37a illustrates the corroded dusty contact after fretting. Figure 4.37b shows the enlarged area of rectangular square. The dark small areas were the particles embedded in the contact. Another experiment with same kind of dusty coupons but without dipping NaCl solution were undergone the same humid cycling and measurement of contact resistance. No apparent corrosion was found, and contact resistance was kept at 1-2 mΩ.

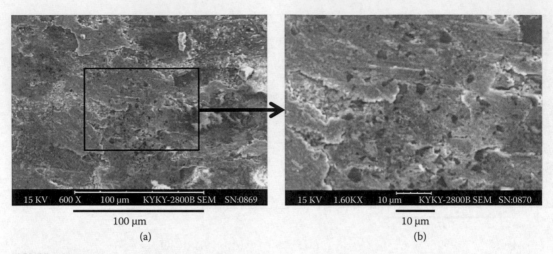

100 μm   10 μm

(a)   (b)

FIGURE 4.37
(a) A corroded dusty contact after fretting. (b) The enlarged area of rectangular square in (a).

4.5.3 Contact Failure Mechanism

4.5.3.1 Single Particle and Ideal Model

Figure 4.38, illustrates a single and spherical particle in contact with an ideal probe and coupon [46]. Failure condition is, if probe is pushed and climbed on the particle, lifted up from coupon surface, contact resistance depends on the particle resistance that may cause failure. In the figure, F_1, F_2, F_3 are the pushing forces. F_1', F_2', F_3' are frictional forces. N is normal force. N = Ns + Np. Ignore the surface deformation, F is the force to push the probe to move and climb over the particle.

When the probe moves away from the original position and climbs on the particle surface, the particle itself should not be pushed away, i.e., the following conditions have to be satisfied.

$$F_3 > F_3', F_1 > F_1' \text{ and } F_2' > F2, (1)$$

$$F_3' = \mu_3 Ns, F_1' = \mu_1 N', F_2' = \mu_2 Np, (2)$$

$$F > F_3 + F_1 \cos \alpha, F_2' > F, (3)$$

Where, μ_3 is friction coefficient between probe and coupon. If a thin layer of lubricant is coated on the metal surface, the friction coefficient μ_3 could be lower than 0.1.

μ_1 is friction coefficient between probe and particle. This friction coefficient depends on particle surface roughness and lubrication effect. When micro valleys of particle surface are filled with tiny carbon particles and particle surface is also coated with a thin layer of lubricant, friction coefficient between probe and particle can be very low either.

μ_2 is friction coefficient between particle and the coupon. If particle is trapped by the corrosion product, frictional coefficient depends on shearing strength between corrosion product and the metal. When the particle is in "preliminary attachment" to the coupon, the probe may partly climb on the particle, part of the normal force could be transferred to the particle thus the particle may be embedded into the coupon surface, the friction coefficient could be very high.

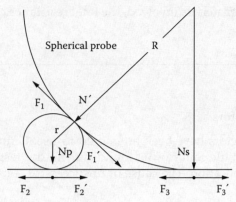

FIGURE 4.38
The condition for a probe to climb on a particle (exaggerated).

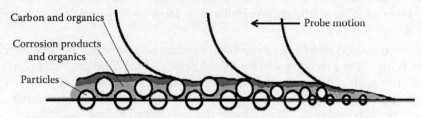

FIGURE 4.39
Layers of particles piled on top of each other may act as stepping stones.

4.5.3.2 Complicated Model – Number of Particles and Morphology of Contact Pairs

Based on the description of "preliminary attachment" and "final attachment," a more complicated model is attempted for simulation as shown in Figure 4.39. The first layer of particles are half buried in the contact surface. The second and third layer of particles are piled upon them. All the particles are surrounded by corrosion products, also mixed with very small amount of organic adhesives. Carbon particles are covered on top. Figure 4.39, illustrates that the accumulated particles are acted as "stepping stones" for the probe to climb.

The height of accumulated particles is not as high as it was expected. Testing result shows that it is around 1.5 to 3 μm. That is because the contact force is repeatedly pressed upon the particles, the indentation depth is getting deeper.

4.5.4 Micro Movement

The importance of micro movement for the connector is that during the movement, particles of high resistance can be collected from vicinity areas which depend on the tolerance of the moving direction and changeable moving distances. The fretting machine should be thoughtfully designed. From micro point of view, moving distance can be ranged from several tens of microns to few hundreds of microns.

4.5.4.1 Contact Failure

Since lactates can dissolve in water, the resistivity of lactates will not be very high as that of the other organic materials. However, only very small amount of lactates are acting in the accumulated particles, and when the environment is very dry, with certain amount of silicates and other insulated materials involved, the total resistance can be high enough to cause contact trouble.

4.6 Future Work

4.6.1 Dust Test for Connectors

Several dust tests for connectors have been attempted to use. Unfortunately, the results were not satisfactory. Very little failure was found during the test which could hardly reflect the practical field application. The cause of unsuccessful tests can be summarized as follows:

1. Particle size used in the test was too large which can hardly be accepted by the connector contact interface. For instance, Arizona dust which is mainly sand particle has been used as artificial dust in the test. Most of the particle size is larger than 5 μm.

2. Fretting machine and sliding machine used for dust test provide only one moving direction with a fixed moving distance. Wear debris and dust particles could be collected only within a straight moving line. It is not possible to collect particles with high resistivity from the vicinity areas. Even though particles are pushed to the end of the path and piled up as a lump but the probe or rider is restricted to go further to climb on the particles. In fact, the more accurate the moving, the less the failure could possibly appear.

3. Particles are separated with each other, during sliding or micro motion, they are scattered away. There is lack of materials to hold them together.

4. Water soluble salts are not contained in the material list of artificial dust. Therefore, the effect of trapping the particles by corrosion product will not appear. Even the salt is consisted in the list, but they are separated large crystals. In humid environment, corrosion products can be grown up separately with dust particles, no trapping effect is occurred.

5. The test procedure is too simple, which is not the combination and interaction effect of many parameters that should be considered.

4.6.2 Suggestion of the Dust Test

It has to prepare and study the influencing factors prior to set up the test system. Still take China as an example, due to the complexity of dusty environment; the environments can be sorted as: north and northwest of China, coastal areas, industrial cities, acid rain areas, crowded cities consuming large quantities of petroleum and coal etc.

The application conditions are even more complicated, roughly they can be sorted as indoor and outdoor, low signal control or telecommunication, high current applications, kinds of organic materials involved etc.

As for mobile phone used in north area of China, the special influencing factors are: salty dust particles, lactates as adhesives and lubricant, micro movement etc. A common artificial particle list can be given comprising particles with quartz, feldspar, mica, calcite, gypsum, etc. Particle size ranges from 0.5 to 3 μm. All the particles should be dipped in the salted distill water before spreading on the test coupons or PWB. The coupons are coated with diluted lactates preferable 0.5% in alcohol solution before located in the dust chamber. A dust chamber with humidity and temperature cycling control system is provided. Fully dry up is a very important stage in the cycling procedure. Fretting machine should be redesigned and manufactured. The machine can be controlled with x and y directions and changeable distances according to the tolerances of the tested connector contacts and possible effect of external impact and vibration. A new software system controls the moving action which may collect the least required particles and testing the contact resistance as well.

4.6.3 Minimizing the Dust Problem

Some approaches include protecting the contacts and making them less sensitive to the presence of environmental dust. Some specific methods, addressing improvement of local environmental conditions during manufacture and operating conditions are briefly discussed.

4.6.3.1 Cleaning the Samples

4.6.3.1.1 Air Blown

Air blown cleaning the relay contact as part of the final assembling stage before sealing seems to be insufficient, because particles may attach on the wall and other parts inside the relay. After external impact or vibration as well as internal electric field attraction, particles can move to the contact surface. Therefore, it needs longer time for the air blown from various directions to clean away all the particles from every corner inside the relay.

4.6.3.1.2 Ultrasonic Cleaning

Experimental result shows that the substrate of surface exposed to ultrasonic cleaning in acetone and alcohol for more than 15 minutes, the dust particles were removed noticeably. It was also found that surface was almost dust-free after 30 minutes of ultrasonic cleaning in water, in alcohol, and in acetone. There seem to exist opportunities to optimize ultrasonic cleaning. The common frequency range of ultrasonic cleaners is 17.2 to 21.8 kHz, which may be too narrow for efficient removal of dust particles of various sizes. Optimization of ultrasonic cleaning must also consider the amplitude of the ultrasonic field, the orientation of the surface to be cleaned relative to the field, the duration of ultrasonic exposure, and the fluid used for carrying energy to the surface. Acetone and alcohol are used to get rid of the organic materials such as wax and liquid lubricants remained on the substrate surface after punching process, Water is superior for removing particles out of the surface.

4.6.3.1.3 Local Environment

In China, for a factory which manufactures electronic equipment, the following principles are proposed: Double layer of glass windows, rubber or plastic seals for the door frames are necessary; the wall including ceiling and floor must not create particles; forests and grasslands should be planted surrounding the factory and in neighboring community; Air condition, filters are needed, room humidity should be controlled. It is suggested to build a fountain at the outdoor to keep the air wet and reduce the floating dust particles in air. The fly ash emission from all smoke stacks should be strictly controlled.

4.6.3.1.4 *Sealed the Whole Plating Process*

Exposure of the electro plating process in dusty environment is extremely harmful. It is necessary to seal the plating baths and filter the plating solutions. The atmosphere close to the plating lines must be kept dust-free.

4.6.3.1.5 *Improvement of Packaging Design for Electronic Equipment*

An effective damping system to mount on both sides of the connector contacts is necessary. It is expected to reduce the micro movement caused by external impact and vibrations.

Avoid or shield the electric field closed to the contact area.

Filters close to the inlet should be easy to change and clean.

The path of air flow from outside through the slots to inside and to the connector or near the contact area should be designed as a zigzag way or a maze with number of blocks. Since this subject involves various fields of science: electronics, tribology, surface material science, electro chemistry, organic chemistry, etc. To analyze and solve the problems need scientists and engineers of different fields to work and cooperate together.

References

1. J.B.P. Williamson, J.A. Greenwood and J. Harris, "The Influence of Dust Particles on the Contact of Solids," Proc. Royal Society of London, UK, 237, 1956, p.560.
2. K. Mano, "Contact Failure by Dust Contamination," Reliability of Contact Components, General Electronics Publisher, Japan, 3rd edition, June 15, 1981, p.92.
3. B.T. Reagor and C.A. Russell, "A Survey of Problems in Telecommunication Equipment Resulting from Chemical Contamination," Proc. Holm Conf., 1985, p.157.
4. K. Mano, "Testing method of Environmental Powder for Electro-mechanical Components" Proc. Relay Research Meeting, Japan 1989, p.393.
5. N. Honma, H. Okamoto and M. Yamaguchi, "SOx and Dust Concentration in Contact Atmosphere," Proc. Holm Conf., 1980, p.287.
6. J.T. Ma, "Determination of Equipment Reliability Rating by Calculating Statistically the Failure in the Field Operation," Proc. IC-CEMCA, Nagoya, Japan 1986, p.703.
7. J.L. Mezec,"Electrical Contacts for use in Tele-Communication Equipment," Proc. 14th ICEC, 1988, p.81.
8. R.C. Robbins, "Economic Effects of Air Pollution on Electrical Contacts," Proc. Holm Seminar, 1973, p.80.
9. W.H. Abbott, "The corrosion of Copper and Porous Gold in Flowing Mixed Gas Environments," Proc. Holm Conf., 1989, p.141.
10. L. White, "Development of a New Connector Dust Test," Proc. Holm Conf., 1987, p.87.
11. C.A. Hague, M.D. Richardson and E.T. Ratliff, "Effects of Dust on Contact Resistance of Lubricated Connector Materials," Proc. Holm Conf., 1985, p.141.
12. P. Slade. *Electrical Contacts, Principles and Applications*. Chapter 2. Introduction to contact tarnishing and corrosion. 1st edition, ISBN 0-8247-1934-4, Marcel Dekker, Inc., 1999, p.109.
13. J.G. Zhang, K.D. Zhou, C.X. Du, "The Porosity of Gold Plating by Dust Contamination," Proc. 34th IEEE Holm Conf., Sept 1988, p.301.
14. Ji Gao Zhang, "Particle Contamination, the Disruption of Electronic Connectors in the Signal Transmission System," J. Zhejiang University Science A ISSN 1862-1775, 2007, 8(3): 361–369.
15. Y.N. Liang, J.G. Zhang, J.J. Liu, "Identification of Inorganic Compounds in Dust Collected in Beijing and Their Effects on Electric Contacts," 43rd IEEE Holm Conf. 1997, p.20.

16. J.G. Zhang, Y.N. Liang, J.W. Wan, B.S. Sun, "Analysis of Organic Compounds in Airborne Dust Collected in Beijing," Proc. 44th IEEE Holm Conf., 1998, p.166.
17. J.W. Wan, J.C. Gao, X.Y. Lin and J.G. Zhang, "Water Soluble Salts in Dust and their Effects on Electric Contact Surfaces," Proc. ICECT. Nagoya, Japan, July 1999, p.37.
18. J.C. Gao and J.G. Zhang, "Measurement of Electric Charges Carried by Dust Particles," Proc. 48th IEEE Holm Conf., 2002, p.191.
19. J.G. Zhang, C.H. Mei and X.M. Wen, "Dust Effects on Various Lubricated Sliding Contacts," *IEEE Trans. CHMT*, Vol.13, No.1, March 1990, p.46.
20. J.G. Zhang and P. Yu, "Electric Contact Performance: Effect of Contact Surface Morphology and Size of Dust Particles," Proc. 15th ICEC/36th Holm Conf., Montreal, Canada, 1990, p.402.
21. Z.K. Chen and G. Witter, "Contact Resistance Failure in Relay Assembly Process," Proc. 1st International Conf. on Reliability of Electrical Products and Electrical Contact ICREPEC, Suzhou, China, Aug. 2004, p.68.
22. C. Leung, V. Behrens and G. Horn, " Review of Automotive Electric Contact Reliability and Research," Proc. 1st International Conf. on Reliability of Electrical Products and Electrical Contact ICREPEC, Suzhou, China, Aug. 2004, p.17.
23. J.G. Zhang, J.C. Gao and C.F. Feng, "The Selective Deposition of Particles on Electric Contact and their Effect on Contact Failure," Proc. 51th IEEE Holm Conf., USA, 2005, p.127.
24. Appendix: "Hardness of Materials in Mohs" in "Effect of dust contamination on electrical contact failure," Proc. IEEE Holm Conf., 2007, p.xxix.
25. J.G. Zhang and X.M. Wen, "The Effect of Dust Contamination on Electric Contacts," *IEEE Trans. CHMT*, Vol.9, No.1, March 1986, p.53–58.
26. R. Holm Appendix I, F, "The Ball and Pyramid Indentation Test for Hardness," *Electric Contacts, Theory and Applications*, 4th edition Sprinber-Verlag Berlin/Heidelberg/New York, 1967, p.373.
27. Y.L. Zhou, Y. Lu, H. Wang, A.L. Ma and H.D. Liu, "Electrical Contact Behavior of Several Typical Materials in Dust," Proc. 56th IEEE Holm Conf./25th ICEC, USA, 2010, p.520.
28. R.H. Horn Chapter 10, "Semipermanent and Permanent Pressure Connectors" *Physical Design of Electronic Systems Vol III*, ISBN 0-13-666370-2 Prentice-Hall Inc. p.595.
29. J.J. Mottine and B. Reagor, "The Effect of Lubrication on Fretting Corrosion at Dissimilar Metal Interfaces in Socket Device Applications," Proc. Holm Conf., 1984, p.171.
30. J.G. Zhang, "The Application and Mechanism of Lubricant on Electrical Contacts," Proc. 41st Holm Conf., Chicago, IL.USA. Oct. 17–19, 1994, p.145.
31. J.G. Zhang, L. Tang, C.B. Shao and J.T. Ma, "Floating and Moving Abilities of Particles on Lubricated Sliding Contact Surfaces," Proc. 17th ICEC, Nagoya, Japan, July 1994, p.381.
32. J.G. Zhang and W. Chen, "Wipe on Various Lubricated and Non-Lubricated Electric Contacts in Dusty Environments," *IEEE Trans.CHMT*, Vol.14, No.2, June 1991, p.309.
33. X.Y. Lin and J.G. Zhang. "Dust Corrosion," Proc. 50th IEEE Holm Conf./ICEC, 2004, p.255.
34. X.Y. Lin and J.G. Zhang, "Properties of Corrosion Stains on Thin Gold Platings," Proc. 48th Holm Conf., Orlano, Fl, USA, Oct 2002, p.156.
35. X.Y. Lin, "Dissertation "Atmospheric Corrosion on Connector Contact Surface" PhD Dissertation, BUPT, 2009, p.59.
36. J.G. Zhang, X.Y. Lin, Y.L. Zhou and J.B.P. Williamson, "Tidal Corrosion and Concentric Rings on Gold Plated Contacts," Proc. 20th ICEC, Stockholm, Sweden, June 2000, p.303.
37. Yi-Lin Zhou, Xue-Yan Lin and Ji-Gao Zhang, "The Electrical and Mechanical Performance of the Corroded products on the Gold Plating after Long Term Indoor Air Exposure," Proc. 46th Holm Conf. Chicago, IL, USA, Sept. 2000, p.18.
38. J.G. Zhang, "Effect of Dust Contamination on Electrical Contact Failure," 53rd IEEE Holm Conf. on Electric Contacts, Pittsburg, PA, USA, Sept. 2007, p.xiv 43.
39. J.G. Zhang, "A Summary Report on the Mechanism of Electric Contact Failure due to Particle Contamination," 57th Proc. IEEE Holm Conf., USA, 2011, p.197.
40. C. Feng, J.G. Zhang, G. Luo and V. Halkola, "Inspection of the Contaminants at Failed Connector Contacts," 51th IEEE Holm Conf. on Electric Contacts, Chicago, IL, USA, Sept 2005, p.115.

41. J.I. Goldstein and H. Yakowitz, "Practical Scanning Electron Microscopy," ISBN 0-306-30820-7, Plenum Press: NY, 1975, p.83.

42. G.P. Luo, J.G. Lu and J.G. Zhang, "Failure Analysis on Bolt-Type Power Connector's Applications," 45th IEEE Holm Conf. USA, 1999, p.77.

43. J.B.P. Williamson, "The Microworld of the Contact Spot," 27th Proc. Holm Conf., USA, 1981, p.1.

44. Y.L. Zhou, Y. Lv and H. Wang, "The Influence of the Organic Compounds on the Contaminated Electrical Contacts," Proc. IEEE 57th Holm Conf., USA, 2011, p.317.

45. Y.L. Zhou, Y. Lv, L.B. Lin, "The Role of Organic Compounds in Simulation of Dust Environments for Electric Contacts," Proc. 26th ICEC/4th ICREPEC, Beijing, China, 2012, p.45.

46. J.G. Zhang, L.J. Xu, J.C. Gao and C.F. Feng, "Further Studies of the Mechanical Behavior of Particles on Electric Contact Failure," Proc. 24th ICEC, France, 2008, p.284.

Part II

Nonarcing Contacts

5

Power Connectors

Milenko Braunovic

Power like a desolating pestilence,
Pollutes what e're it touches ....

Queen Mab, **Percy Bysshe**

CONTENTS

5.1   Introduction 233
5.2   Types of Power Connectors 234
    5.2.1   Plug-and-Socket Connectors 236
    5.2.2   Wire Connectors 236
    5.2.3   Bolted Connectors 236
    5.2.4   Insulation Piercing Connectors 236
5.3   Properties of Conductor and Connector Materials 237
    5.3.1   Definition of Conductor and Connector Systems 237
    5.3.2   Factors Affecting Conductivity 237
        5.3.2.1   Effect of Temperature 238
        5.3.2.2   Effect of Lattice Imperfections 238
        5.3.2.3   Magnetoresistance 239
        5.3.2.4   Skin Effect 240
    5.3.3   Conductor Materials 240
        5.3.3.1   Copper and Copper Alloys 240
        5.3.3.2   Aluminum and Its Alloys 243
    5.3.4   Materials for Connector Systems 244
        5.3.4.1   Pure Metals and Alloys 244
    5.3.5   Electroplating and Cladding 246
5.4   Parameters Affecting Performance of Power Connections 247
    5.4.1   Factors Affecting Reliability of Power Connections 247
    5.4.2   Contact Area 249
    5.4.3   Plastic Deformation 252
    5.4.4   Elastic Deformation 253
    5.4.5   Plated Contacts 253
    5.4.6   Oxidation 256
    5.4.7   Corrosion 260
        5.4.7.1   Atmospheric Corrosion 260
        5.4.7.2   Localized Corrosion 261
        5.4.7.3   Crevice Corrosion 261

5.4.7.4   Pitting Corrosion 262
5.4.7.5   Pore Corrosion 262
5.4.7.6   Creep Corrosion 262
5.4.8   Dust Corrosion 263
5.4.9   Galvanic Corrosion 263
5.4.10 Thermal Expansion 264
5.4.11 Fretting 267
   5.4.11.1   Factors Affecting Fretting 268
   5.4.11.2   Mechanisms of Fretting 270
   5.4.11.3   Examples of Fretting Damage in Power Connections 274
   5.4.11.4   Compression Connectors 274
   5.4.11.5   Bus-Stab Contacts 276
   5.4.11.6   Plug-In Connectors 278
   5.4.11.7   Bolted Connections 278
   5.4.11.8   Fretting in Aluminum Connections 283
   5.4.11.9   Effect of Electrical Current 285
   5.4.11.10 Fretting in Coatings (Platings) 286
   5.4.11.11 Fretting in Circuit Breaker Terminal
                 Contact Materials 287
5.4.12 Intermetallic Compounds 288
5.4.13 Intermetallics in Copper–Tin Systems 296
   5.4.13.1   Example of Intermetallics Formation in Power Connections 298
5.4.14 Stress Relaxation and Creep 299
5.4.15 Nature of the Effect of Electric Current 301
5.4.16 Effect of Electric Current on Stress Relaxation 302
5.4.17 Creep 305
5.5 Palliative Measures 306
5.5.1   Contact Area 307
5.5.2   Contact Pressure 313
5.5.3   Mechanical Contact Device 320
5.5.4   Disc-Spring (Belleville) Washers 322
5.5.5   Wedge Connectors 325
5.5.6   Automatic Splices 326
5.5.7   Dead-end Connectors 327
5.5.8   Shape-Memory Alloy Connector Devices 327
5.5.9   Coating (Plating) 331
5.5.10 Lubrication—Contact Aid Compounds 334
5.5.11 Bimetallic Inserts 336
5.5.12 Transition Washers 337
5.5.13 Multilam Contact Elements 338
5.5.14 Welded Connections 339
   5.5.14.1 Thermite (Exothermic) Welding 339
   5.5.14.2 Friction Welding 339
   5.5.14.3 Explosion Welding 340
   5.5.14.4 Resistance Welding 341
   5.5.14.5 Resistance Brazing 341
5.5.15 Connector Design 342
   5.5.15.1 Fired Wedge-Connectors 343
   5.5.15.2 Stepped Deep Indentation Connectors 344

5.6   Connector Degradation...345
      5.6.1   Economical Consequences of Contact Deterioration.......................346
      5.6.2   Power Quality...346
5.7   Prognostic Models 349
      5.7.1   Prognostic Model 1 for Contact Remaining Life............................349
      5.7.2   Prognostic Model 2 for Contact Remaining Life............................350
      5.7.3   Physical Model 351
5.8   Shape-Memory Alloys (SMA) 351
      5.8.1   Origin of Shape-Memory Effect..351
              5.8.1.1   One-Way Memory Effect...352
              5.8.1.2   Two-Way Memory Effect...353
      5.8.2   Applications of SMA in Power Connections 353
      5.8.3   Electrical Connections...353
      5.8.4   Temperature Indicators..355
5.9   Metal Foam Materials 355
      5.9.1   Aluminum Foam Materials...355
              5.9.1.1   Electrical and Thermal Properties of Foam Materials......357
              5.9.1.2   Power Connection Applications 357
      5.9.2   Copper Foam Materials 358
              5.9.2.1   Applications of Copper Foam Materials..............................358
      5.9.3   Silver Foam Materials..360
5.10 Installation of Power Connections 361
      5.10.1 Examples of Improper Installations...361
5.11 Accelerated Current-Cycling Tests (Standards)......................................363
      5.11.1 Present Current-Cycling Tests 363
References...366

5.1 Introduction

Steadily increasing energy consumption in densely populated areas imposes severe operating conditions on transmission and distribution systems, which have to carry greater loads than in the past and operate at higher temperatures. Also, new economical designs of power connectors have pushed the connector closer to the limits permitted by the standards. The connector is generally a weak link in the power grid, which raises doubts about the ability of some of the connector designs to provide effective long-term connections. This, and poor installation practices, can be the most frequent sources of detrimental performance in a power system. The degradation rate of power connectors in service cannot be determined precisely, which makes maintenance scheduling difficult. There are two main reasons for this: first, there is a general lack of awareness of the problem, since connection deterioration is a time-related process; and second, the specific features of connection deterioration are not readily recognizable, because the failure of a power connection is usually associated with thermal runaway, thus making identification of the degradation mechanism difficult. The adverse consequences of this situation are reflected in the materials specifications for electrical joints, their use and care, and the general reliability of the entire power network. In addition, mounting evidence demonstrates that the measures currently used by manufacturers to qualify connectors do not satisfactorily reflect their ability to perform under varied and adverse field conditions. These standards have a number of shortcomings: the tests are long and costly; the deterioration mechanisms cannot be clearly detected under the prescribed test conditions; and, lastly, most connectors can pass the tests without difficulty and yet can fail under normal service conditions.

The primary purpose of an electrical connection is to permit the uninterrupted passage of electrical current across the contact interface. It is clear that this can only be achieved if a good metal-to-metal contact is made. The processes occurring in the contact zone are complex. Although the nature of these procedures may be different, they are entirely governed by the same underlying phenomena, the most significant being the degradation of the contacting interface and the associated changes in contact resistance, load, temperature, and other parameters of a multipoint contact. Although the outcome of different parameters on the contact behavior has been investigated in the past, a unified model describing the complex processes occurring in the contact zone is still lacking. In recent years, the use of new contact materials and ever-increasing tendency towards higher current capacities has emphasized the effects of oxide layers, surface roughness, diffusion physico-chemical and structural transformations, fatigue, creep, electroplasticity, etc. The complexity of these processes cannot be answered by experimental approach. Rather, theoretical modelling is required to elucidate their interdependence. It should be borne in mind, however, that development of a model which will adequately describe the processes in electrical contacts, has to include specific operating conditions imposed by a particular type of contact. Thus, it is of considerable interest to:

- Identify the major parameters determining the character and lifespan of power connections in terms of their mechanical and metallurgical metal-to-metal characteristics.

- Quantify the limits of these parameters and establish reliability criteria under different operations and environmental conditions (high current, high temperature, accelerated aging).

- Provide practical palliative measures for power connectors susceptible to degradation under different operations and environmental conditions and determine their effectiveness to assure satisfactory connector performance.

- Review the existing testing methods and introduce readers to new tendencies in the development of accelerated testing procedures which allow a better connector life specifications.

- To review recent advances in the materials technologies applied to power connections.

5.2 Types of Power Connectors

In order to meet the mechanical and electrical requirements and also assure reliable performance of a connector during its expected service life, various designs have been developed and used in the field with varying degree of success. The generic connector designs commonly used on distribution network are illustrated in Figure 5.1. Field experience has shown that a wide variety of connector designs have given good service over several years, but the factors contributing to their success and to failure has not be determined with any degree of certainty. However, the ever-increasing demand for electricity in recent years has increased

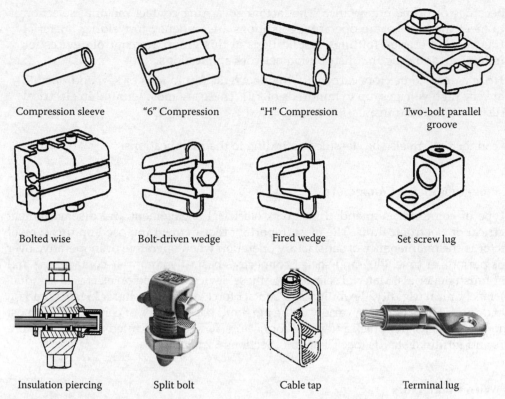

| Compression sleeve | "6" Compression | "H" Compression | Two-bolt parallel groove |

| Bolted wise | Bolt-driven wedge | Fired wedge | Set screw lug |

| Insulation piercing | Split bolt | Cable tap | Terminal lug |

FIGURE 5.1
The generic connector designs commonly used on power network.

electrical loading on power transmission and distribution lines, which, in turn, has raised the average operating temperature of conductor lines, up to 130°C during times of peak force transmission. These operating conditions may expose connectors to temperatures exceeding the operating range for which many connecting devices were initially designed and tested to the existing standards. Thus new performance measures are being demanded.

It is generally accepted that a good mechanical joint is also a good electric joint. In many cases, this may not be so owing to the intrinsic nature of current transfer across the contacting members, and also the design features of a particular connection and deterioration mechanisms that may impact on the connection performance. Hence, it is of importance to review the basic design characteristics of the most common types of power connections and the factors that influence their performance since only with proper attention to these factors can a good electrical connection be made consistently. Connectors can be broadly classified into three groups: light-, medium-, and heavy-duty connectors according to their current-carrying capacity and their functional operation.

1. *Light-duty connectors are devices carrying very low flows (below 5 A), operating at voltages up to 250 V.* The successful operation of these devices depends mainly on maintaining relatively low and stable contact resistance, and also on the selection of the contact materials. This type of connector is discussed in Chapter 6.

2. *Medium-duty* connectors are those carrying appreciably higher currents (above 5 A), and operating at voltages up to 1,000 V. For this group electrical wear

becomes of prime importance. The factors governing contact material selection to meet the very severe operating conditions are tendency to welding, material transfer and erosion (pitting). Applications of this group are control devices for industrial, domestic and distribution network applications.

3. *Heavy-duty* connectors carry very high currents (up to tens of kA) and operate at very high voltages (up to hundreds of kV). These are mostly found in electrical distribution and transmission systems.

The connectors can also be classified according to their applications:

5.2.1 Plug-and-Socket Connectors

This type of connector is intended for quick electrical engagement and disengagement of electrical or electronic units. The most important requirement imposed on this type of connector is the maintenance of satisfactory operation when operative or inoperative over various periods of time. Plug-and-socket connectors basically comprise contact base and contact-finish materials. Material selection for these devices is based on electrical (conductivity), mechanical (rigidity, flexibility) and contact force deflection characteristics, and on contact design. There is a wide variety of the plug-and-socket types of connector, the most common being jacks, pins, rack and panel connectors, IC sockets, printed-circuit edge connectors and terminal boards (See Chapter 6, Sections 6.2 and 6.3).

5.2.2 Wire Connectors

These are devices intended for connecting wires and cables to connection points of electrical equipment. In general, wire connectors can be permanent such as welded, crimped or thermo compression bonds; or semi-permanent such as soldered, wrapped or screwed joints. The most commonly used are lugs, crimps, splices, compression-screw lugs, set screws, binding-head screw terminals, eyelets, clamp-on types, split-bolt types, and straight-coupling types.

5.2.3 Bolted Connectors

Since bolted and compression connections are widely used on the power network, this chapter will focus on some basic feature of these connections, degradation mechanisms affecting their performance, palliative measures used to suppress the adverse effects of the contact deterioration, installation practices, testing methods and also recent developments in connector materials and design.

5.2.4 Insulation Piercing Connectors

Insulation piercing connectors are designed to operate with hermetically-sealed electrical contacts to prevent moisture ingress. After installation, the perforated insulator on the conductor presses on the sides of connector teeth with sufficient force to prevent ingress of harmful environmental contaminants through the perforations. The shape and number of teeth on the jaw are designed to optimize the grip on the conductors.

5.3 Properties of Conductor and Connector Materials

5.3.1 Definition of Conductor and Connector Systems

A conductor system is defined as an assembly comprising the current-carrying member (conductor), insulation, protection, shielding and termination. The current-carrying member is normally a solid wire or a combination of wires (strands) not insulated from one another. A cable on the other hand is a stranded conductor or a combination of conductors insulated from one another. The main function of the current-carrying member is to yield an uninterrupted passage of electrical current through the system, and the insulation serves to restrain the current flow in the system to the conductors. The protective function of the conductor system isolates the current-carrying member from external influences. The basic function of shielding is to reduce the effect of electric and magnetic fields on the cable.

A connector serves to provide connection between electrical circuits, to carry the current for the required period without overheating. The execution of a connector is defined by the following factors: tendency to oxidation (tarnishing), welding, and corrosion; hardness; melting point; resistance to wear and friction; electrical and thermal conductivity; and also by the operating conditions such as interrupting voltage, current to be carried, contact pressure, contact size, frequency of operation, and rapidity of interruption. The most significant feature of a connector is the contact resistance. There are a number of factors affecting the contact resistance; for example, mechanical, physical, and electrical properties of the connector materials, the tendency of the contact material to oxidation, contact pressure and contact area, see Chapter 1.

In ordinary engineering usage, a solid conductor is a material of high conductivity. The electrical conductivity of metallic conductors is of the order of 10^6–10^8 S/m at ambient temperature (see, for example, Table 24.1C). Solid metallic conductors can be generally classified into two groups according to their applications:

- Pure metals, the most common and widely used of which are Cu and Al, sometimes alloyed with other metals to improve their mechanical properties; and
- Alloys, used as conductors with particular properties such as wear resistance, magnetic properties and friction, the most common materials being bronzes, brasses and some aluminum alloys.

Practical application of solid metallic conductors requires a detailed knowledge of various properties of conductor materials, such as electrical, thermal physical, chemical, mechanical and tribological characteristics, because in service, solid conductors are subjected to various mechanical and thermal stresses and also environmental effects.

5.3.2 Factors Affecting Conductivity

Modern electron theory of metals asserts that conduction is to be understood in terms of the effective number of free electrons and the mean free path of those electrons. It is the large number of free electrons which makes elements such as silver, copper and aluminum good conductors. On the other hand, for a given conductor, the deleterious effects on conductivity imparted by alloying, plastic deformation, structural defects and heating result from a reduction of the mean free path of the electrons. Contributions to the resistivity coming from the sources of electron scattering are additive. This is known as Matthiessen's rule, which can be summarized as

$$\rho = \rho_i + \rho_o \tag{5.1}$$

where ρ_i is a temperature-dependent term reflecting the thermal vibration of the lattice ions and is known variously as the "ideal," intrinsic, lattice or phonon resistivity; ρ_o usually called the "residual" resistivity, arises from electron scattering by the lattice imperfections, and is generally independent of temperature.

Although the rule gives good agreement with experimental data, especially at higher temperatures, there is increasing evidence that it is not strictly valid and that deviations are to be expected at lower temperatures. The nature of these deviations depends on the type of impurity atoms or other defects present as well as on their quantity; this is particularly evident when transition metal impurity atoms are present. Deviations from Matthiessen's rule found experimentally may be generally expressed as a temperature-dependent measure.

5.3.2.1 Effect of Temperature

Over the moderate temperature range of most common service operations, such as 0°C–150°C, the properties of conductor materials vary linearly with temperature. The changes in conductivity or resistivity and also physical dimensions with temperature are appreciable and should be taken into account in many engineering calculations. In the case of linear conductors these changes can be expressed as:

$$R_T = R_o[1 + \alpha_R(T - T_o)] \tag{5.2}$$

$$l_T = l_o[1 + \alpha_L(T - T_o)] \tag{5.3}$$

where R_T and l_T are respectively the resistance and length of a conductor material at some temperature T; R_o and l_o are the resistance and length at 20°C and α_R and α_L are respectively the coefficients of electrical resistance and linear expansion (See Tables 24.1B and 24.1C).

5.3.2.2 Effect of Lattice Imperfections

The effect of lattice imperfections on the resistivity of a pure metal manifests itself through an increase in the residual resistivity ρ_o which is very sensitive to the presence of imperfections in the lattice. In studying the effects of imperfections, to minimize the effects of thermal vibrations (phonon scattering), electrical resistivity measurements are usually carried out at very low temperatures, where the thermal part of the resistivity may be considered negligible. In this instance, at low defect concentrations, the residual resistivity increases with the concentration of defects since the interferences among the defects themselves can be overlooked. The change in the residual resistivity is then an appropriate measure of the defect concentration in the metal.

(a) *Impurities and solutes.* The presence of impurities or solutes (alloying additions) in the lattice decreases of the conductivity of a conductor much more than any other lattice imperfection. The extent of the reduction depends on the type, concentration and the metallurgical state in which the impurities are present. The impurities are more effective in reducing the conductivity when present in solid solution than as, or incorporated in, a second phase of the microstructure. This is understandable, since disturbances of the lattice periodicity on an atomic scale, as produced by impurities in solid solution, more effectively increase the electrical resistance than perturbations on a macro scale caused by the presence of a second phase. Within the solubility limit of a particular impurity there is a linear relationship between the concentration of this impurity and the increase in electrical resistivity. It is clear that purity of solid conductors is of prime importance for electrical purposes. However, producing high-purity conductor materials involves higher processing costs, perhaps not be economical for practical use. Increasing the purity of conductor materials weakens them mechanically. Therefore deliberate addition of a particular solute or solutes in limited amounts may considerably improve a conductor's the mechanical response without degrading its conductivity.

(b) *Deformation dislocations.* Plastic deformation of a metal tends in general to harden it, reduce its ductility and increase its tensile strength and electrical resistivity. The increase in tensile strength is useful and thus many types of conductors are finished by cold working. All or at least an appreciable fraction of the increase in electrical resistivity is caused by the scattering of conduction electrons by dislocations introduced into the lattice by plastic deformation. Broadly speaking, the resistance increase $\Delta\rho$ owing to plastic strain γ is given as

$$\Delta\rho = a\gamma^n \qquad (5.4)$$

where a and n are constants characteristic of a particular conductor material. Annealing of a plastically deformed conductor reduces electrical resistance and tensile strength but increases ductility.

(c) *Vacancies.* An appreciable concentration of vacancies can be made in a solid conductor by rapid quenching from an elevated temperature and also by irradiation with high-energy particles. The effect of vacancies produced in this way on the electrical resistivity is more pronounced in very pure metals, since in less pure metals both the number of vacancies created and their resistance contributions are affected by the solute.

(d) *Grain boundaries.* In an ideally pure metal, the contribution to the resistivity owing to the grain boundaries arises from the electron scattering at these boundaries. An electron crossing the boundary enters a region where it cannot continue in the same direction and with the same velocity. This is owing to the anisotropy of elastic and electronic properties of the solid in the grain boundary region. With advances in microelectronics, considerable attention has recently been given to the gist of the grain boundaries on electrical conductivity of thin solid films, not only at lower but also at higher temperatures. This is because it has been found that the grain boundaries and also the segregated alloy or impurity species at the grain boundaries can significantly affect the performance and transport properties of thin, solid film conductors.

5.3.2.3 Magnetoresistance

When a magnetic field applied in the same direction as the electric field causes a current to flow through a specimen, the conduction electrons are constrained to follow helical instead of linear paths between collisions. Because of this effect, the impedance is usually higher than that obtained in the absence of a magnetic field. The fractional change in resistance that occurs is called longitudinal magneto resistance. On the other hand, application of the external magnetic field in a transverse direction to the electric field results in a different current density parallel to the applied electric field. The fractional change in resistance that occurs is called transverse magneto resistance. In addition, a potential gradient is made perpendicular to both the applied magnetic and electric fields, resulting in the phenomenon called the Hall Effect.

Except in certain ferromagnetic metals and alloys, longitudinal and transverse magneto resistances are positive, that is, the electrical resistance increases with increasing magnetic field strength. In the case of ferromagnetic materials, iron for instance, resistivity, particularly at low temperatures, initially decreases, reaches a minimum value and then gradually increases as the field strength increases. Such behavior of the magneto resistance at lower fields is associated with changes in magnetic domain configurations, but the detailed electron scattering mechanism is not fully understood. The relative magnitude of the resistivity change $\Delta\rho$ owing to the magnetic field usually increases with purity. This rule holds well for most metals, but there are some exceptions, such as high-purity aluminum, for which the magneto resistivity tends towards a saturation value.

5.3.2.4 Skin Effect

The conductivity of a solid conductor when an alternating current field is applied is controlled at higher frequencies by the phenomenon called the skin effect. This is an electro-dynamic effect arising from the way in which the time-varying electric and magnetic fields and electric current are interrelated. The distribution of the current in a specimen is nonuniform; for example, a 50 or 60 Hz alternating current flows mainly in a 10 mm layer at the surface of the conductor. It is important for nonmagnetic materials such as copper or aluminum, the magnetic permeability of which is 1, so that the increase in AC resistance at higher frequency is entirely owing to the increase in the frequency of the passing AC current. In the ferromagnetic materials, the conduction is characterized by numerous unusual features arising from the presence of magnetic ions in a conductor influencing the resistivity, regardless of the external field.

5.3.3 Conductor Materials

Generally speaking, a conductor consists of a solid wire or an assembly of wires stranded together and used either bare or insulated. In this article, only bare conductors will be considered. The most extensively used materials for electrical conductors are copper and aluminum. The basic properties and applications of these materials are summarized in Table 5.1.

5.3.3.1 Copper and Copper Alloys

Copper is a soft, malleable and ductile metal with high conductivity and excellent weldability and solderability. By rolling and drawing, a variety of electrical products such as wires, sheets, tubes, shaped bars, and flat busbars can be manufactured. The high-conductivity copper useful for electrical applications must be brought about by careful refining treatments such as electrolytic refining which removes Ag, Au, As, Sb and other impurities. The most common copper used in the power industry is electrolytically tough pitch copper ETP or C11000 made by electrolytic refining of copper. The main shortcoming of ETP copper is the embrittlement to which it is subject when heated in hydrogen to temperatures of 370°C or more. This results from the presence of oxygen in the metal which reacts with the hydrogen, making steam and leading to internal cracking. The solution to

TABLE 5.1

Selected Properties of Copper and Aluminum Conductors

	Aluminum			Copper		
	EC-0 (A)	Al-Mg (5005) (B)	Al-MgSi (6201) (C)	OFHC (D)	Phosphor Bronze95/5 (E)	Brass (70/30) (F)
Density (g cm^{-3})	2.7	2.7	2.69	8.94	8.86	8.53
Melting Point (°C)	660.0	652.0	654.0	1083	1060	955.0
Coeff. Linear thermal expansion (10^{-6} K^{-1})	23.6	23.70	23.40	17.0	17.8	20.30
Thermal Conductivity (W cm^{-1} K^{-1})	2.34	2.05	2.05	3.91	0.84	1.2
Electrical Resistivity (μΩcm)	2.80	3.32	3.20	1.70	8.7	6.4
Thermal Coeff. of Electrical Resistivity (10^{-3} K^{-1})	4.46	4.03	4.03	3.93	4.0	1.0
Elastic Modulus (GPa)	69.0	69.6	69.6	115.0	110.0	110.0
Yield Strength (MPa)	28.0	193.0	310.0	69.0	140.0	110.0
Tensile Strength (MPa)	83.0	200.0	330.0	220.0	345.0	330.0
Specific Heat Capacity (J/g/K)	0.9	0.9	0.9	0.38	0.38	0.38
Current Carrying Capacity (%)	80.0			100.0		
Hardness (× 10^2 N mm^{-2})	2.3	5.1	9.5	4.2	5.0	6.0

A—Annealed;
B—0.8% Mg Fully cold worked (H19);
C—0.7% Si—0.8% Mg, Solution treated Cold worked, Aged (T81);
D—Annealed, grains size 0.05 mm;
E—94.8% Cu-5% Sn-0.2% P, Annealed, grain size 0.035 mm;
F—70% Cu-30% Zn, Annealed.

this problem is to use copper of substantially lower oxygen content. While phosphorus is an effective deoxidizer for copper, it degrades the conductivity too much to give a product suited for electrical applications. Instead, electrolytic slabs are melted and refined in a special process using oxygen-free inert gas and no metallic oxidizers. The result, a 99.98% pure copper with essentially no oxygen and <0.005% of any one impurity, is known as oxygen-free high-conductivity copper (OFHC). The conductivity of copper is frequently referred to in terms of the international annealed copper standard (IACS), thus % IACS equals 100 (resistivity IACS/resistivity of sample). In absolute terms, the IACS has resistivity of 1.7241 μΩ cm. A common criterion for defining the purity of metals is the ratio of their resistivity at 273 and 5.2 K. This ratio varies between 150 and 400 for OFHC copper but can reach 1000–5000 and higher for zone-refined materials.

Under normal atmospheric conditions, copper is comparatively resistant to corrosion. At room temperature, an oxide layer, Cu_2O, forms which protects the surface from further oxidation and is a semiconductor. At higher temperatures as a result of exposure to air, a CuO oxide layer is formed. Considerable corrosion of copper may be produced by air containing ammonia or chlorine compounds. The use of copper near the sea coast is undesirable, since the salts present in the air can cause severe corrosion. Moist atmospheres containing sulfur dioxide attack copper, resulting in the formation of a mixed oxide sulfide scale. For electrical applications, the mechanical properties of copper have to be improved, but in doing so the electrical conductivity is often reduced. Strengthening can be achieved by cold working and/or alloying copper with various elements. Cold-drawn pure copper can be softened by annealing at 200°C–325°C, but previous cold deformation and the presence of impurities can alter this annealing range. The higher the level of prior cold deformation, the lower is the range of annealing temperature, whereas the presence of impurities or addition of various elements raises the annealing temperature.

5.3.3.1.1 Cu–Ag

The amount of silver added is in the range 0.030–0.1% and results in improved creep strength and resistance to softening at elevated temperatures without appreciable sacrifice of electrical conductivity. This alloy is commonly used for current collectors of electrical machines.

5.3.3.1.2 Cu–Be

This contains 0.5–2% Be as the principal alloying element, but Ni and Co are also often added so as to achieve desirable properties. It is nonmagnetic and has excellent mechanical (elastic) properties. Its primary application is for springs, diaphragms, switch parts and electrical connectors. The precipitation hardening alloy is heat-treated by annealing at 900°C followed by water quenching and subsequent aging at 425°C.

5.3.3.1.3 Cu–Cd

This alloy contains 0.0~1.0% Cd and has excellent capacity for cold working and hot forming and also for soldering, brazing and gas-shielded arc welding. It is widely applied in fine wire applications for airplane electric circuitry, as well as in commutators segments and other applications.

5.3.3.1.4 Cu–Cd–Sn

The total amount of Cd and Sn may reach 2%. The primary applications are for telephone lines, electric motor brushes, and parts for switching devices.

5.3.3.1.5 Cu–Cr

The Cr concentration is in the range 0.150.9%. This precipitation-hardened alloy has a large part of the solute contained in the second phase, which imparts excellent mechanical resistance at higher temperatures. Its main applications include electrode materials for welding machines, heavy duty electric motors and circuit breaker components, switch contacts, current-carrying arms and shafts and electrical and thermal conductors requiring more force than is provided by unalloyed copper.

5.3.3.1.6 Cu–Te

The amount of tellurium added is 0.30.7%, to improve machinability while retaining ~90% IACS. This alloy also has excellent solderability and corrosion resistance. It can also be applied at relatively high temperatures. Typical uses include electrical connectors and motor and switch parts.

5.3.3.1.7 Cu–Zr

This contains 0.1–0.2% Zr. Because of its low tendency to embrittlement and improved creep behavior at higher temperatures and mechanical stresses, it is used for switches and circuit breakers for high-temperature and high-vibration service, commutators and studs and bases for power transmitters and rectifiers.

5.3.3.1.8 Bronzes

This class of copper encompasses high Cu-Sn alloys. The tin concentration ranges from 5% to 15%. All bronzes have superior mechanical properties but inferior electrical properties relative to copper. The electrical resistivity of bronze can be 2–20 times that of electrolytic copper. Bronzes are frequently ternary or quaternary alloys containing third elements such asp. Si, Mn, Zn, Al, Cd or Ni; the third element is normally expressed in the name of the alloy. Bronzes for electrical applications contain less tin and other metals than bronzes for structural applications, for which mechanical properties and corrosion resistance are determining factors. Typical applications of the bronzes are springs, diaphragms, bushings, face plates, connectors, and electrical machinery parts.

5.3.3.1.9 Brasses

These are alloys containing nominally 15–40% Zn. Addition of other metals such as Mn Ni and Al improves their mechanical strength. Brasses are seldom used for electrical conductors, owing to their low conductivity. Typical electrical uses are conduits, screw shells, sockets and receptacle contact plates where formability is an important consideration. When using some types of brass intended for mechanical or structural applications, care should be taken to avoid dezincification and stress-corrosion cracking (See Section 2.4.3.4), which occur under certain conditions.

5.3.3.2 Aluminum and Its Alloys

In recent years, for a number of economic and engineering reasons, aluminum has gained an ever-increasing application, Because of its light weight, relatively good electrical and thermal properties, availability and moderate cost, aluminum is being seen as a viable alternative to copper for many conductor applications in electrical systems. In substituting aluminum for copper, however, due account should be taken of their differences in resistivity, mechanical strength and density. For the same resistance and length, an aluminum conductor should have a cross-sectional area 60% larger than that of an equivalent copper conductor, whereas the weight of the aluminum conductor is 48% of that of the copper conductor. The current-carrying capacity of aluminum is 80% of that of copper. Aluminum is a ductile metal with relatively high thermal and electrical conductivity. It is softer than copper and can be rolled into thin foils several μm thick. However, because of its low mechanical strength, aluminum cannot be drawn into very fine wires. The resistivity and mechanical durability of aluminum depend on its purity and grade of cold work. By selecting the proper fabrication process, aluminum containing 10 ppm of impurities can be obtained with a resistivity ratio surpassing 1,000. Higher resistivity ratios, >30,000, can be obtained by zone melting.

The resistivity of high-purity aluminum (99.999%) is 2.635 $\mu\Omega$ cm at 20°C, whereas that of the commercial grade is in the region of 2.78 $\mu\Omega$ cm. The commercial grade aluminum contains nominally <0.1% Si and <0.015% (Mn, Ti, Cr. V). To minimize further, the effect of the impurities (Ti and V in particular) on the conductivity of the aluminum, 0.02% boron is often added, leading to the transformation of these impurities (except Mn) into borides that have very little effect on electrical conductivity, as they are not in the dissolved phase. Pure aluminum, even if hard drawn, possesses inadequate mechanical properties. This shortcoming can be somewhat overcome by alloying with a variety of other metals, resulting in improved tensile and creep strengths. The alloys most frequently used for electrical applications are Al–Mg or Al–Mg–Si also containing Fe or Co.

Broadly speaking, there are three main categories of application of aluminum and its alloys. These are overhead transmission lines and underground cables, coil winding (magnet wire) and busbar conductors. For overhead transmission lines, the aluminum alloy used generally contains 0.8% Mg or 0.5% MgSi and has high strength combined with relatively good electrical conductivity. However, the mechanical strength of this alloy may not always be sufficient, for example, for long spans in overhead lines, in that case, conductors of composite construction are used, in which the core of the line is composed of steel wires. Alloys for coiled winding wire (magnet wire) have a relatively high concentration of Fe and a low Si content. This ensures rather high elongation values for the wire in the annealed condition, a higher recrystallization temperature and a higher tensile strength at elevated temperatures. Further improvements in the mechanical strength of these alloys can be achieved by adding small amounts of Mg or Cu. The use of aluminum alloys for coil winding wires requires some design alterations: motors and transformers have to have larger slots to accommodate the larger gauge size of the aluminum wire if this is to have the same conductance as copper magnet wire.

For busbar, Al–Mg–Si alloys are mainly used, because of their excellent corrosion resistance and their good workability and electrical and mechanical properties. When jointing the busbars, care must be taken in order to minimize the effect of stress relaxation (See Section 5.4.14). One of the most important drawbacks that prevent even wider use of aluminium as a conductive material is the lack of a truly reliable and economic method of termination. To overcome this problem many methods such as welding, plating, ultrasonic bonding, plasma spraying, bolting, clamping, and brazing have been adopted, but most of them are relatively expensive and require greater operator care; in some cases they are marginal in electrical or mechanical operation.

5.3.4. Materials for Connector Systems

5.3.4.1 Pure Metals and Alloys

Pure metals and alloys in this group are used as plating materials for copper and aluminium substrates.

5.3.4.1.1 Silver

It is widely used as plating or coating material for the contact parts of the connectors (See also Chapter 8). The main drawbacks of silver are low melting and boiling points, low mechanical strength, possible contact welding, and a tendency to form sulfide films (tarnishing). Other problems with silver are whisker growth and the diffusion of silver atoms through certain electrical insulation materials, such as phenolic fiber, under the influence of applied electrical fields, which may cause failure of the insulation to occur (See Chapters 2 and 8). Improved mechanical properties and a higher resistance to tarnishing may be achieved by alloying with copper, cadmium, gold, palladium or platinum. The addition of copper decreases the electrical conductivity, resistance to oxidation and corrosion, melting point and cost, but increases hardness. Small amounts of nickel (0.2–3%) improve the wear rate and decrease the chance of welding and tarnishing. The addition of cadmium (~5%) decreases the electrical conductivity, melting point and oxidation resistance, but improves the resistance to tarnishing. The additions of platinum, palladium, or gold all harden silver, decrease the electrical conductivity and improve the resistance to wear, tarnishing, and metal transfer.

5.3.4.1.2 Copper

Copper and its alloys have been described in the previous section. It is sometimes used as a plating material.

5.3.4.1.3 Platinum

Platinum has exceptional resistance to tarnishing, oxidation and corrosion. It is suitable for light-duty applications where operating currents are below 2 A, contact-pressure is low and where reliability is the most important parameter. However, under fretting conditions, platinum contacts are susceptible to frictional polymerization. Additions of iridium, ruthenium, and osmium to form binary or ternary alloys increase hardness, mechanical strength, melting point, resistivity, and wear resistance.

5.3.4.1.4 Palladium

Palladium has a lower resistance to corrosion, oxidation, and tarnishing than platinum. It begins to tarnish at 350°C but the tarnish film formed decomposes at 900°C. The additions of copper, ruthenium, silver, or combinations of other metals improve the mechanical properties of palladium with some decrease in corrosion resistance, and lower cost. In atmospheres containing traces of organic compounds, palladium contacts subjected to motion relative to each other (fretting) tend to form an insulating frictional polymer.

5.3.4.1.5 Gold

Gold has an excellent tarnish and oxidation resistance, but is very soft and susceptible to mechanical wear, metal transfer and welding. It is widely applied in computers and telecommunication and data transmission devices where operating currents are not more than 0.5 A. The addition of copper, silver, palladium, or platinum, forming binary and ternary alloys, improves the hardness without loss of tarnish resistance, but usage is restricted to low-current applications. Hard gold contains a low percentage of nickel or cobalt (See Chapter 8).

5.3.4.1.6 Rhodium

Rhodium is also very resistant to tarnishing, but is very strong and extremely useful as a contact material. Nevertheless, owing to difficult fabrication it is used solely as a plating material in light-duty electrical contacts where reliability is of the utmost importance.

5.3.4.1.7 Tungsten

Tungsten is a very hard metal with excellent resistance to wear, welding, and material transfer, with high melting and boiling points. Its main disadvantages are low corrosion and oxidation resistance, high electrical resistivity and poor formability. Tungsten for contact applications is generally combined with silver made by powder metallurgical processes (See Chapter 16).

5.3.4.1.8 Nickel

Even though nickel forms a protective oxide film, it and its alloys are suitable for a wide variety of applications, the majority of which involve corrosion and/or heat resistance. It is widely used in as a plated substrate for other metal platings such as tin, silver and gold. Other applications require low-expansion, electrical resistance, soft magnetic and shape-memory nickel-base alloys. When used in a clad composite with copper, it can provide composite with a very well controlled-expansion characteristics. The low expansion of Invar with other alloys of different expansion can provide a series of thermomechanical control and switchgear devices. The electrical resistance nickel alloys are commonly used in instrumentation and control equipment to measure and regulate electrical characteristics or to generate heat in furnaces and appliances. The most common alloys in this form of alloys are Cu–Ni (2–45% Ni), Ni–Cr–Al (35–95% Ni) and Ni–Cr–Si (70–80% Ni). The permeability properties of soft magnetic nickel-iron alloys are used in switchgear and direct current motor and generator designs. The lower-nickel alloys (<50% Ni) with a fairly constant permeability over a narrow range of flux densities are primarily used in the rotors, armatures and low level transformers. High-nickel alloys (~77% Ni) are used for applications in which power requirements must be minimized such as transformers, inductors, magnetic amplifiers and shields, memory storage devices and tape recorder heads.

5.3.5 Electroplating and Cladding

Electroplating and cladding are common methods of covering large contact surfaces (See also Chapter 8). The most common and widely used plating materials are silver, gold, tin, and nickel. Base materials commonly used as the current-carrying member of a connector, such as copper and its alloys, can readily be plated with any of these plating materials. Tin has also come to be considered as a suitable substitute for gold plating on electrical connectors used in low-voltage low-current operations. Nevertheless, the primary applications of tin in the electronic and electrical industries are either as solders or coatings to aid soldering or improve the connectability of wires and cables with the electrical equipment. Tin coatings are soft and ductile, and with a thick coating soldering can easily be done. However, owing to its low hardness and tendency to oxidize readily to yield a self-healing film, tin is less satisfactory as a contact finish material.

According to the available data, nickel appears to be the most practical coating material from the point of view both of its cost and the significant improvements to the metallurgical and contact properties of electrical connectors. The resistance of nickel to form intermetallic phases with copper, aluminum, and other metals, makes it as a very effective diffusion barrier in a variety of electrical and electronic devices where diffusion between the coating and substrate base represents a significant problem. In recent years, nickel was successfully employed for coating aluminum conductors and power connectors. However, nickel does not protect aluminum galvanically and presents subsurface corrosion problems when plated over aluminum. Furthermore, fretting produces considerable degradation of the contact zones in nickel-coated aluminum contacts. Lubrication and higher loads are found to mitigate these adverse effects. Despite these disadvantages, nickel-plating is still becoming more attractive for contact applications in electric power applications.

Cladding of common base metals (such as copper and its alloys [brass, bronze], steel, and aluminum) with precious metals such as gold and silver in order to obtain the optimum combination of functional properties is now a well-established technique. Cladding permits properties such as thermal and/or electrical conductivity, high strength, corrosion wear and high temperature resistance, weldability, light weight and springiness to be merged. Cladding can be done in the form of inlays, toplays, overlays and edge lays. Typical applications of clad metals include contacts, thermostats, blades, springs, contact brackets, bonding pads, lead frames and connectors.

5.4 Parameters Affecting Performance of Power Connections

The widespread use of aluminium in a variety of electrical applications has prompted numerous studies into the processes occurring in aluminium connections. Published experimental evidence suggests that reliable aluminium connections cannot be obtained by the routine application of the practices and methods established for joints with copper conductors. The problems with interfacing aluminium derive from the fact that whenever two dissimilar metals are brought together, the differences in their physical, mechanical and metallurgical properties as well as the manner in which they react under specific conditions determine their level of compatibility. The primary purpose of an electrical connection is to allow uninterrupted passage of electrical current across the contact interface, which can only be achieved if a good metal-to-metal contact is made and maintained. However, in the case of aluminum connectors, this requirement cannot always be met owing to the ever-present insulating oxide layer at the surface, the propensity to undergo creep and stress relaxation, the susceptibility to galvanic corrosion, and the large thermal expansion coefficient, which may lead to fretting at the contacting interfaces. This situation can result in failure of a connection. The complexity of failure mechanisms in aluminum power connections is best depicted in the course of a cycle, as shown in Figure 5.2. These concerns will be discussed later in this chapter.

5.4.1 Factors Affecting Reliability of Power Connections

Reliability is most commonly defined as the probability of equipment or a process to function without failure, when operated correctly, for a given period of time, under stated conditions. Anything out of that is a failure. One of the most significant problems in providing reliability to electrical contact is the discrete nature of the interface. An electrical contact between slides is formed in discrete regions within the contacting interface and these areas (*a*-spots) are the only current conducting paths (See Chapter 1). The formation of the real and conductive contact areas controls the reliability and efficiency of the electrical contact. These processes depend on a great number of independent or interrelated

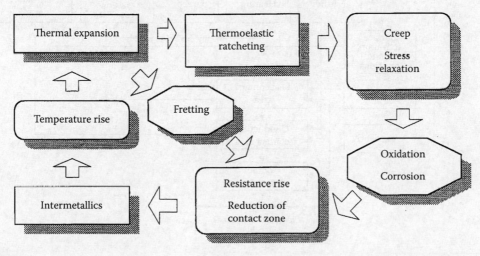

FIGURE 5.2
Schematic of degradation mechanisms in aluminum connections.

factors. The variety of the factors can be conventionally divided into the *performance* factors governed by the operating conditions and the *design-technological* factors determined by the fabrication characteristics of a contact unit. The performance factors (parameters) are divided basically into two groups: internal and external (Figure 5.3). Figure 5.4 shows schematically the influence of the design-technological factors on the reliability and quality of electrical contacts. The selected kind of contact materials, the contact geometry, the intermediate layers separating the contacting surfaces, the quality of the deposited coatings and the contact surface microrelief determine the apparent contact area, the size, number, and distribution of contact spots. This, in its turn, influences the real and electrical contact areas, the constriction and surface film resistances, and, lastly, the electrical contact reliability. The internal factors are the mechanical (the contact load, the type and characteristics of motion such as case, the sliding velocity, and reciprocation) and electric (type and strength of current, operating voltages) factors.

The external factors may be temperature-time variation, humidity, atmospheric pressure, effect of aerosols etc. and these are often uncontrollable. The performance factors affect the properties of contact materials and surface films, the occurrence of physical and chemical processes in the contact zone, wear particle formation thus influencing the state of the interface and, finally, the contact resistance and reliability of electrical contacts

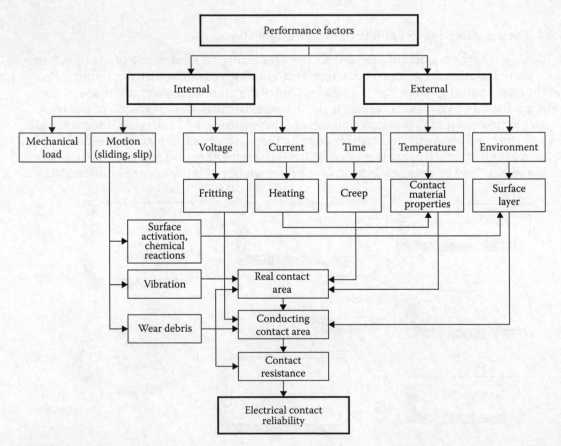

FIGURE 5.3
Effect of performance factors on the reliability of electrical contacts.

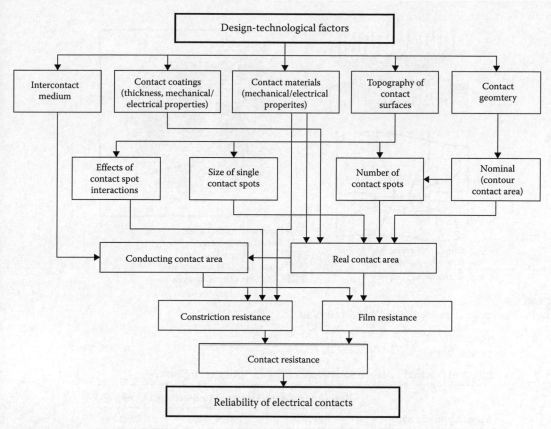

FIGURE 5.4
Effect of design-technological factors on the performance of electrical contacts.

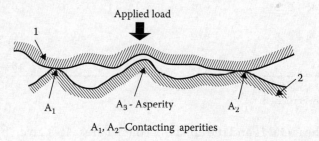

FIGURE 5.5
Contacting asperities.

5.4.2 Contact Area

It has been established [1,2] and presented in Chapter 1 that real surface is not flat but comprises many asperities (Figure 5.5). Hence, when contact is made between two metals, surface asperities of the contacting members will penetrate the natural oxide and other surface contaminant films, establishing localized metallic contacts and, thus, conducting paths. As the force increases, the number and the area of these small metal-metal contact spots will increase as a result of the rupturing of the oxide film and extrusion of metal through the ruptures [3–5]. The real contact area A_r is only a fraction of the apparent contact area A_a, as illustrated in Figure 5.6. The relationship between the applied normal

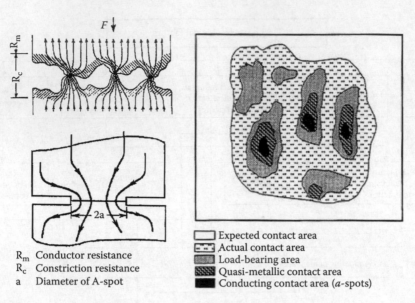

R_m Conductor resistance
R_c Constriction resistance
a   Diameter of A-spot

Expected contact area
Actual contact area
Load-bearing area
Quasi-metallic contact area
Conducting contact area (*a*-spots)

FIGURE 5.6
Schematic of current constriction and real contact area.

TABLE 5.2

Effect of Normal Load on Real Area of Contact for Clean Surfaces

Alloy/Applied Load	Real Contact Area/Apparent Contact Area (A_r/A_a) (%)		
	10 N	100 N	1000 N
Al (H-19)	0.01	0.1	1.0
Al (H-0)	0.05	0.5	5.0
Al + 0.75% Mg + 0.15% Fe (H-19)	0.01	0.1	1.0
Al + 0.75% Mg + 0.15% Fe (H-0)	0.02	0.2	2.0
Cu (H-0)	0.008	0.08	0.8

(H-0)—Fully annealed;
(H-19)—Fully hardened.

load F_c, hardness of the metal H and the apparent contact area A_a is given by the following expression

$$F_c = \varepsilon H \, A_a \tag{5.5}$$

on the amount of deformation of the asperities and is equal to 1 in most practical contact systems. On the other hand, Holm [1] has shown that the hardness (H) is related to the yield stress (σ_y) by the following expression:

$$H = 3\sigma_y \tag{5.6}$$

The results, shown in Table 5.2 show the real contact area as a percentage of the apparent contact area A_a at various normal loads. It should be noted, however, that the real contact area calculated in this manner includes the load-bearing area which can be covered with the oxide film and is not, therefore, a dependable path for transfer of electrical current. Hence, the conducting contact area will be a fraction of the calculated real contact area. Current passing across a contact interface is therefore constructed to flow through these a-spots. The electrical resistance of the contact owing to this constricted flow of current is called "constriction resistance." Since the metals are not clean, the passage of electric current may be affected by thin oxide, sulfide and other inorganic films usually present on metal surfaces, the total resistance of a joint (R_{ab}) is then a sum of the constriction resistance (R_c), the resistance of the film (R_f) and the bulk resistance (R_b)

$$R_{ab} = R_c + R_f + R_b \tag{5.7}$$

R_c = Constriction resistance = $\rho/2a = \rho(\pi H/F)^{1/2}$,

R_f = Film resistance = $\sigma_f\, d_f/n\pi a^2$, d_f film thickness

R_b = Bulk resistance

ρ = resistivity of the contacting asperity material

n = number of contacting asperities

σ_f = Film conductivity

In most practical applications, the contribution of these films to the total contact resistance is of minor importance, since the contact spots are usually created by the mechanical rupture of surface films. Both tunnelling and fritting are also considered as operative mechanisms for the current transfer across the film. Figure 5.7a illustrates schematically the current transfer across the contact interface whereas in Figure 5.7b is shown variation of the contact resistance with applied contact load. It should be pointed out that the electrical interface of an a-spot is far different from the single circular contact spot. In fact, the true metal-to-metal contact is limited to a cluster of microspots, within the nominal contact spot, where the contacting materials extrude to touch each other through cracks in their oxide film as demonstrated by Williamson in the case of aluminum connections [3].

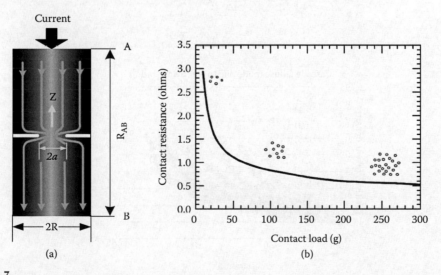

(a)   (b)

FIGURE 5.7
(a) Schematic of current transfer across the contact interface and (b) Contact resistance variation with applied contact load.

The effect of current density variations in power contacts has been investigated by Malucci [4]. The Greenwood approach to estimating the interaction between current carrying contact spots was used to study the degradation of individual spots. The degradation is assumed to occur from electro-migration which causes a non-uniform increase in the effective resistivity across each contact spot. The latter results were used to assess the degradation of a simulated multi-point interface to demonstrate the cascade failure mode believed to come in power contacts. In addition, factors such as spot size, position and interaction with nearby spots were assessed in their impact on current density variation across the contact region. Figure 5.8 shows the random array cascading degradation of the power connection interface.

Joint resistance of copper–copper busbar joints as a function of contact load at 105°C was investigated by Schlegel et al. [6]. It was demonstrated that in pure copper joints the ageing mechanism of joint force reduction is determined by different physical mechanisms: setting process, dynamic recovery, dynamic primary recrystallisation, and grain coarsening (secondary recrystallisation). The acting mechanisms depend on time, temperature, grain size at initial conditions, the degree of cold work and foreign particles. The result of the long-term tests showed that the force reduction is critical at temperatures ≥140°C in joint systems with washers and material compositions allowing large deforming degrees of the Cu-ETP. For Cu-ETP the material condition having higher mechanical properties at initial state showed a worse long-term behavior. For practical applications at high temperatures it is recommended to use the less cold-deformed material.

5.4.3 Plastic Deformation

For power connectors where the contact force is much higher than a few Newtons plastic deformation of asperities that form the *a*-spots occurs. As indicated in Chapter 1 if the film resistance is ignored, the constriction resistance will be:

$$R = \frac{\rho}{2}\sqrt{\left(\frac{\pi \xi H}{F}\right)} = \frac{C_1}{F^{1/2}} \quad C_1 = \frac{\rho}{2}\sqrt{\pi \xi H} \tag{5.8}$$

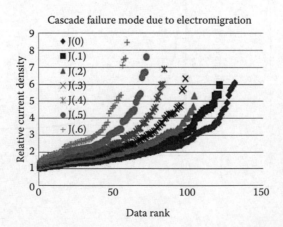

FIGURE 5.8
Random array cascading degradation.

5.4.4 Elastic Deformation

For very light contact forces that are presented in Chapter 12, the deformation of the asperities is an elastic deformation. This gives a constriction resistance proportional to the (1/3) the power of the applied load (F) and sphere radius (d) if all elastic deformation is assumed.

$$R = \frac{C_2}{d^{1/3}F^{1/3}} \tag{5.9}$$

There can be regions of contact force where both regimes combine [7,8].

5.4.5 Plated Contacts

In the case of plated contacts, when the plating thickness is comparable to the a-spot diameter, plating introduces an additional component to the contact resistance. This is because the constriction lies partly in the plating and partly in the bulk conductor. As a result, the current flow is diffracted as it crosses the boundary, as schematically illustrated in Figure 5.9 [5]. *Electrically conductive coatings* produced by *electroplating* reduce the contact resistance owing to, among other factors, decrease in hardness, higher conductivity of plating compared to the substrate, prevention of insulating film formation, corrosion, reduction in mechanical wear, etc. In case of conductive plating, the electrical resistance depends on the relation between the coating thickness and the a-spot diameter as well as in the ratio of conductivities of plating and substrate. Figure 5.10 depicts the current distribution in the metal surface film when the film conductivity is smaller than that of the substrate and a-spot radius is more or less the same size as the film thickness. If the resistivity of the plating is larger than that of the substrate and the a-spot radius is close to the film thickness, the electrical current from the a-spot spreads out easier into the substrate than into the plating. The potential drop in the vicinity of the a-spot in the substrate is negligible in comparison with the potential drop across the film [1]. Williamson [3] has shown that the constriction resistance of a single copper conductor with current flowing through a single contact spot is increased by a factor of five if the conductor is plated with a layer of tin as thick as the contact spot radius.

$$R_t = R_s + \frac{\rho_f t}{\pi a^2} \qquad R_t = \frac{\rho}{4a}\left[1 + \frac{4}{\pi}\frac{\rho_f}{\rho}\left(\frac{t}{a}\right)\right] \tag{5.10}$$

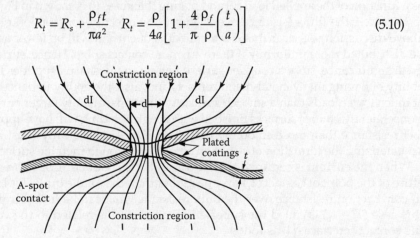

FIGURE 5.9
Schematic of formation of a-spot between plated surfaces.

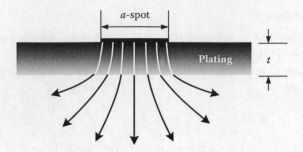

FIGURE 5.10
Current distributions in metal surface film when the film conductivity is smaller than that of the substrate and a-spot radius is of the same size as the film thickness.

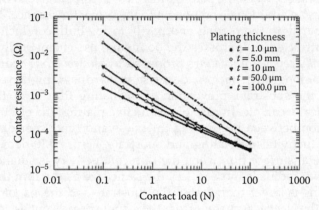

FIGURE 5.11
Contact resistance variation of tin-plated contact with applied load and plating thickness and ξ = 1.0.

If a-spot radius and the plating thickness do not differ greatly, the spreading resistance increases approximately linearly with plating thickness. For relatively thick film, the spreading resistance deviates from the above expression and approaches $\rho_f/4a$. The contact resistance as a function of the applied load (F) and plating thickness (t) is shown in Figure 5.11. If the ratio $(\rho_f/\rho)(t/a) \gg 1$ the film resistance overshadows the effect of constriction resistance. From the above discussion it is clear that electrical paths of current will be fewer and the current will be distributed more uniformly if there are more contact spots. Hence, surface roughness is of great significance, since a rougher surface having many sharp asperities has a greater probability of having many metal-metal contacts and, also, the ability to penetrate its counterpart at much lower loads than a smooth surface with a consequently larger current-carrying area. Consequently, contact surfaces finished with rough abrading will have appreciably lower contact resistance than those smoothly machined [6,9]. This is clearly illustrated in Figure 5.12 schematizing the formation of a-spot on smooth and rough metallic surfaces [9,10].

When the current is confined to flow through the conducting spots (a-spots), the temperature of the point of contact (T_c) may be higher than that of the bulk (T_b). Hence, the increase in constriction resistance over the bulk resistance that can be found from Equation 5.2. In this case ($T_c - T_b$) is called the supertemperature and is related to the voltage drop across the contact interface (U) as follows

$$T_c^2 - T_b^2 = U^2/4L \tag{5.11}$$

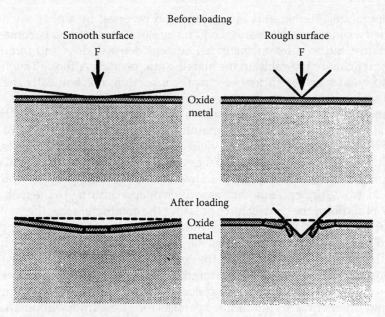

FIGURE 5.12
Schematic of a-spot formation on smooth and rough metallic surfaces.

where L is the Wiedemann-Franz Lorenz number with the value of 2.45×10^{-8} $(V/K)^2$ [11]. It is clear that even a relatively modest increase in the contact voltage drop (U) can raise the supertemperature considerably enough to produce basic metallurgical changes such as softening or even melting of the conducting spheres. Equation 5.11 is valid under the following conditions [11]:

- Constriction is purely metallic without any interference from the oxide or contamination layers.
- Power losses owing to constriction are removed by thermal conduction.
- Current equipotential and heat equithermal flows are identical.
- Contact voltage U and temperature T are measured at points with small voltage and temperature gradients.

In a good connection, the temperature of the interface is just slightly higher than the bulk temperature but in a poor connection the supertemperature increases the bulk temperature and accelerates deterioration of the contact areas. The deterioration is cumulative resulting higher resistance, in increasingly higher temperatures and ultimate failure of the connection. In certain circumstances, such as short-circuit conditions, melting of the contact zones can occur even in well designed joints. As further deterioration occurs, these molten zones coalesce into larger areas in as the whole joint assembly becomes overheated. It might seem that the creation of welded contacts would improve the connection stability. However, on subsequent cooling and hardening, the metal contracts and cracks owing to the internal stresses set up in the process. Oxidation of the contact zones further reduces the number of available electrical conducting paths, with the result that overheating and ultimate mechanical failure of the joint occur.

The above-described scenario for contact deterioration has been questioned by Williamson [12,13] who argues that the degradation of a connector does not progress inexorably to ever-higher temperatures. Rather, it is interrupted and reversed by a phenomenon called self-healing. When a connector deteriorates and its metallic contact area becomes smaller, its supertemperature, induced by an increased current density, rises and intensifies the stresses in the joint or, more precisely, in the metallic microwelds. Although such build-up of stress tends to build a more efficient electrical connection, mechanically the accumulated stress will seek relief in the paths of least resistance, that is, in the axial direction of the joint. If the magnitude of this tension exceeds the elastic limit of the contacting members, the material yields. The contact force, maintained by the spring element of the joint, will be opposed by material with a reduced yield strength thus causing the contact area to grow. As a result, both the constriction resistance and the supertemperature will decrease. If, during softening, the metal flow fails to follow the self-healing process which causes the contact area to grow, higher temperatures will develop, resulting in the melting of the contact region. At this point the contact force pressing the conductors together will extrude the molten metal through the fissures in the contact interface and, therefore, increase the contact area and reduce the constriction resistance. The contact region cools to a lower temperature as the liquid metal freezes and the contact self-heals.

It is also important to note that the deterioration of a connector proceeds slowly at a pace determined by the nature of different processes operating in the contact zone and, likewise, in the environment. This initial stage persists for a long time without making any noticeable changes because it is an intrinsic property of clusters of *a*-spots that their overall constriction resistance is not sensitive to small changes in their size. However, when the contact resistance increases sufficiently to cause the local temperature to increase, a self-accelerating deterioration resulting from the interaction of thermal, chemical, mechanical and electrical processes will be triggered and the contact resistance will rise abruptly. Hence, no deterioration will be noticeable until the final stages of the connector life. This is clearly illustrated in Figure 5.13 depicting the variation of contact resistance with time for different values of current [13].

5.4.6 Oxidation

The oxidation of the metal-metal contacts within the contact interface is widely considered as the most serious degradation mechanism occurring in mechanical connectors. Oxidation is a chemical process that increases the oxygen content of a base metal and as a result, base metal or radical looses electrons. Oxidation of the metal-metal contacts

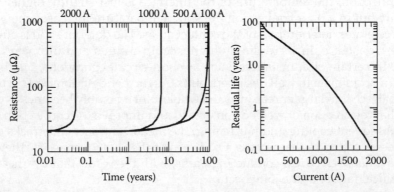

FIGURE 5.13
Variation of connector resistance with time for different values of the applied current.

within the contact interface is widely accepted as the most serious degradation mechanism occurring in mechanical connectors [10]. In the case of aluminum contacts, it is generally considered a less likely mechanism of degradation, since oxide growth is self-limiting and reaches a limiting thickness of about 10 nm within a very short period of time. This is very much less than the diameter of the contact spots, generally considered being much more than 10 nm for rough surfaces. Aluminum oxide forms as a duplex film when the bare aluminum surface is exposed to the oxygen-containing atmosphere. This duplex film consists of a very thin non-porous, inner barrier layer next to the metal with a thicker, more or less porous outer bulk layer on top. The barrier layer, which is temperature dependent, reaches a maximum thickness within microseconds, whereas the bulk film, which develops more slowly, is dependent on the relative humidity and the temperature. Up to and beyond the melting point of aluminum, the oxide formed on the metal surface remains intact and protective thus providing aluminum with a greater corrosion resistance. Oxidation kinetics of some common contact materials is presented in Table 5.3. Aluminum oxide is hard, tenacious and brittle, with a high resistivity of 10^{24} $\mu\Omega$ cm. It is also transparent so that even the bright and clean appearance of an aluminum conductor is no assurance that a low contact resistance can be achieved without appropriate surface prep. The oxide film may be broken electrically as well as mechanically. If the oxide film is thick it can be broken down by A-fritting at relatively high voltages (see Chapter 1). In electrical contacts having one or both contact members of aluminum, the current flow is restricted to flow through the areas where the oxide film is ruptured.

In the case of copper it was shown that in the presence of oxygen-bearing atmospheres the continuous oxidation of the metal–metal contacts by oxidation can cause rapid increase in the contact resistance to a high value after remaining relatively low for a considerable length of time. The oxides of copper grow, flake, and spall off from the base metal. From about 40°C to about 200°C in air, there is a continual temperature dependent thickness growth of the Cu_2O oxide. At about 200°C and above other copper oxides form while continually consuming metal. Copper oxides are softer as compared to aluminum oxides and more easily disrupted by the applied contact force. They are also semiconducting and copper contacts

TABLE 5.3

Oxidation Kinetics of Some Common Electrical Contact Materials

Metal	Ambient	Product	Characteristic Features	Thickness (nm) At	10^3 hr	10^5 hr
Cu	Air	Cu_2O	Oxide forms immediately	20°C	2.2	4.0
			Temperature-dependent	100°C	15.0	130.0
			Initially slow growth rate	20°C	4.2	6.1
Sn	Air	SnO	Weak temperature-dependence	100°	25.0	36.0
			Self-limiting	20°C	1.6	15.0
Ni	Air	NiO	Weak temperature-dependence	100°C	3.4	34.0
Al	Air	Al_2O_3	Oxide forms immediately (2 nm in seconds)	Self-limiting growth		
			Humidity and temperature-dependent	Very hard and insulating		
Ag	Sulfur	Ag_2S	Depends of sulfur-vapor concentration Remains thin and decomposes at 200°C	Humidity-dependent No effect on contact		
	Ozone	Ag_2O				

with an initially high resistance, due for instance to poor surface preparation, can show a steady decrease in contact resistance with time as a result of the growth of semiconducting layer over a large area. The electrical resistivity of Cu_2O is 10^{10} $\mu\Omega$ cm. In the presence of a sulfur-bearing atmosphere tarnishing of the copper surface is normally observed owing to sulfide formation from hydrogen sulfide in the air. The development of tarnish film is strongly dependent on the humidity which can reduce it if a low sulfide concentration prevails or increase it if the sulfide concentration is high. The effect of humidity on the maturation of the oxide layer in aluminum is shown in Figure 5.14 [14].

Campbell [15] estimated that, for an oxide film on copper in contact with gold, semiconduction would start at ~0.4 V and increase as the contact voltage rose to 1.5 V, when molten metallic junctions could be created (Figure 5.15). However, Holm [1] felt that melting was not necessary for A-fritting. The fritting voltage for a 100 nm layer of Cu_2O is less than 0.0001 V, while the fritting voltage for the same thickness of Al_2O_3 is 40 V. New contacts can be formed by A-fritting, while B-fritting (see Chapter 1), resulting from plastic flow at a-spots, enlarges the contact area and reduces the constriction resistance. B-fritting can occur when the contact voltage required to cause metal softening (~0.1 V) is reached [1].

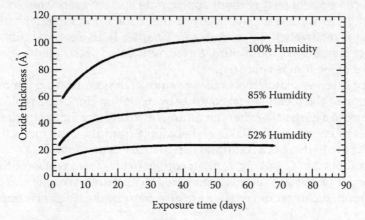

FIGURE 5.14
Effect of humidity on the oxidation kinetics of aluminum ($\mathring{A} = 10^{-10}$ m).

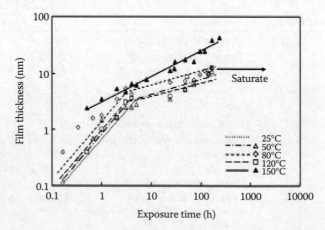

FIGURE 5.15
Growth of oxide film for exposure to elevated temperature (data of 25°C is indicated only a line).

There have been many references in the literature to the hardness and tenacity of the Al_2O_3 film and to the difficulty of making a contact through this film as compared to the relative ease of forming metal-to-metal contacts in copper through a Cu_2O film. Tylecote [16], however, provided many examples of cold welding of aluminum as well as of copper showing that a lower percent deformation is required to initiate welding in aluminum than in copper, and that welding is initiated at a lower deformation in cold-forged metals.

It is commonly stated that, because annealed metals have greater ductility, plastic flow occurs more easily through fractures in the oxides of such metals, thus forming larger *a*-spots. It was suggested that, under plane-strain conditions, deformation was more concentrated, and cracks in the oxides were, therefore, larger, in cold-worked aluminum. Braunovic [17] showed that in high-purity aluminum (99.999% pure) and an aluminum −0.5 at. % magnesium alloy, thermal cycling caused impurities to segregate to free surfaces and affected significantly the hardness, contact resistance, and resistivity. It was suggested that vacancies, formed near the coherent metal-oxide interface, diffused into the metal and resulted in a flow of solute from the metal towards the loose surface. Auger electron spectroscopy confirmed higher magnesium content both in the oxide and at the surface in the aluminum–magnesium alloy. The solute segregation to the oxide lowered the contact resistance, but the mechanism is unknown since many other complex reactions which could affect mechanical and electrical properties (e.g., clustering of vacancies, dislocations, polygonization) could occur simultaneously.

Property of the tin oxide film formed on tin-plated connector contacts was investigated by Tamai et al. [18]. To clarify the properties of the oxide film formed on a tin plated surface, oxide film thickness, chemical composition, crystal structures and contact resistance characteristics for temperatures over 25°C to 150°C were studied. These attributes of a tin-plated surface are very important for applications of small size connectors. At temperatures lower than 150°C, the growth law of the oxide film showed linear law until 5 mm thick at the initial level of exposure. After this point, the growth of the oxide film indicated a 1/4 law until 15 mm in thickness, and then, the film thickness was saturated. However, at an exposure temperature of 150°C, the growth law showed a parabolic law for the entire exposure to 150°C Morphology of the tin oxide film formed at a temperature lower than 120°C was mostly an amorphous SnO flat layer, whereas at temperatures higher than this, island like crystallized SnO_2 formed on the SnO layer was found by high magnification TEM. The molecular composition of the film was obtained by XPS. For low temperature, SnO is dominant in an amorphous tin oxide layer, but SnO_2 existed slightly in the layer. For high temperature such as 150°C, SnO_2 is dominant, but SnO exists slightly. Furthermore, for the 300 nm constant grain size of tin plated surface increased, if the temperature exceeds 80°C. The grain size increased up to 1000 nm at 120°C. After this, the grain disappeared, indicating melting of the tin plated layer. Under these conditions an intermediate compound growth occurred just below the surface layer. The relationship between contact resistance and exposure time which is directly related to film thickness, was clarified for an exposure temperature up to 25°C. Contact resistance characteristic indicated low constant level until approximately 10 nm in thickness of the film. As the film thickness increased thicker than this, contact resistance increased to a high value such as 5 Ω and the high contact resistance depends on the contact load. The same characteristics were recorded for an exposure temperature lower than 120°C. However, at the temperature of 150°C, a very different contact resistance characteristic was obtained, see Figure 5.16. The contact resistance increased until 17 nm in film thickness almost the same as the tendency of low temperature mentioned above. However, after this thickness, the contact resistance decreased. This fact is owing to growth of an intermediate compound of copper and tin alloy.

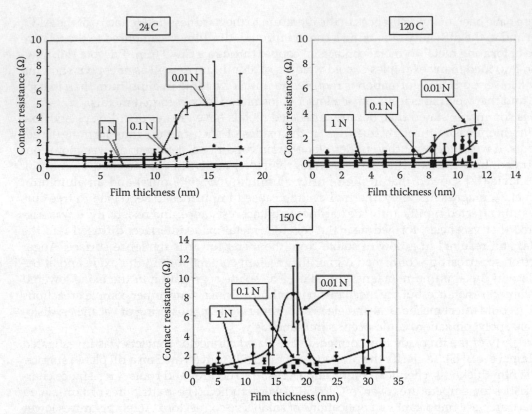

FIGURE 5.16
Relationship between contact resistance and oxide film thickness for tin exposure at 25°C, 120°C, and 150°C.

5.4.7 Corrosion

The subject of corrosion was introduced in Chapters 2, 3, and 4. Corrosion is a chemical or electrochemical reaction between a metallic component and the surrounding environment causing detectable changes that lead to a deterioration of the component material, its properties and function. It begins at an exposed metal surface altering progressively the geometry of the affected component without changing the chemical composition of the material or its microstructure. Degradation initiates with the formation of a corrosion product layer and continues as long as at least one of the reactants is able to spread through the layer and sustain the reaction. The composition and characteristics of the corrosion product layer can have a significant influence on the corrosion rate. Among the many forms of general corrosion that could potentially affect the power equipment metallic components atmospheric, localized, crevice, pitting and galvanic are probably the most common.

5.4.7.1 Atmospheric Corrosion

Atmospheric corrosion is the gradual degradation or alteration of a material by contact with substances such as oxygen, carbon dioxide, water vapor, and sulfur and chloride compounds that are present in the air. Uniform thinning of component material is probably the most common form of general corrosion. Due to the electrolytic nature of corrosion, only a very thin film of water is needed to accelerate degradation. Although the rate of atmospheric corrosion is dependent on the humidity, temperature, and levels of sulfate, chloride, and other atmospheric pollutants, it is usually not constant with time and tends to decrease as the duration of exposure increases.

5.4.7.2 Localized Corrosion

Localized corrosion is similar to general corrosion except the rate of attack is usually much faster and the size of the affected area is significantly smaller. Damage caused by localized corrosion is often difficult to detect and quantify because visible surface flaws tend to be small and often do not provide a good indication of the extent of damage that has occurred under the surface. Specific forms of localized corrosion include crevice, pitting, and localized biological.

5.4.7.3 Crevice Corrosion

Crevice corrosion is a form of localized attack of a metal surface adjacent to an area that is shielded from full exposure to the environment because of the close proximity between the metal and the surface of another material. Narrow openings or spaces between metal-to-metal or non-metal-to-metal components, cracks, lines, or other surface flaws can serve as sites for corrosion initiation. Humidity and pollution can penetrate into a crevices and cavities inside mechanical and compression connectors not filled with contact lubricant. In the bolted connectors, the bolts made of stainless steel are more prone to crevice corrosion than those of carbon steels, especially in the presence of chlorides. The initiation of crevice corrosion is illustrated schematically in Figure 5.17 [19]. The simplest method for preventing crevice corrosion is reduced crevices in the design of the structure, improving drainage and sealing of edges or keeping crevices as open as possible thus preventing entry of moisture. A protection method called "hot wax dip," commonly used in the automotive industry, involves the painting of surfaces before assembly. Cathodic protection is also found to be an effective method against crevice corrosion, but anodic protection is often wrong. The use of alloys which are less vulnerable to crevice corrosion is another protection method. Also, addition of inhibiting substances to bulk solution is found to be a very effective protection method. Application of passivating compounds such as chromate and nitrate is another effective and practiced method. Another protective measure is overlaying susceptible areas with an alloy which is more resistant to crevice corrosion. The use welds rather than bolted or riveted joints is also a way of limiting crevice corrosion.

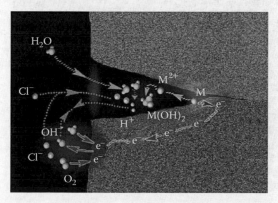

FIGURE 5.17
Schematic of crevice corrosion initiation. (Courtesy of Laboratoire de Physicochimie Industrielle.)

5.4.7.4 Pitting Corrosion

Pitting corrosion is localized degradation of a metal surface confined to a point or small area that takes the form of cavities. The pits are generally irregularly shaped and may or may not get filled with corrosion products. Pitting usually affects metals that are covered with a very thin coating with the pits forming at weak spots in the coating and at sites where the coating is damaged mechanically under conditions where self-repair will not take place. The stainless steels are particularly sensitive to pitting corrosion, but other metals, such as passive iron, chromium, cobalt, aluminum, copper, and their alloys are also prone to this pattern of damage [19]. Typical case of pitting corrosion is shown in Figure 5.18. Pitting corrosion is frequently observed in CO_2 and H_2S environments. Pits are generally initiated as a result of local breakdown of corrosion product films on the surface and corrosion will continue at an accelerated pace. Pits may become connected as the corrosion damage increases. Corrosion products are dark brown to grayish black and loosely adhering. In H_2S systems, the pits are usually shallow round depressions with etched bottoms and sloping sides. Broadly speaking, the pits are not connected, and corrosion products are black and tightly adhering to the metal surface.

5.4.7.5 Pore Corrosion

Pore corrosion occurs in thin porous plating in the as a result of galvanic cell formed in the presence of a thin water layer containing ionizable gas. The corrosion products are transported from the reactive base or substrate metal though the hole in the plating to the contact surface. This type of corrosion is usually associated with a synergistic effect between chlorides, oxygen and/or sulfates and is evidenced by the appearance of the pores which are defects in a coating which expose the underlying metal, underplate or underplate and substrate (see Chapters 2, 3, 4, and 8).

5.4.7.6 Creep Corrosion

Creep corrosion can occur when a reactive substrate metal like silver on copper is located next to and in physical contact with a noble metal or a noble alloy inlay of plating. The substrate metal corrosion products creep over the noble metal surface. Creep can also

FIGURE 5.18
Example of pitting corrosion in a copper tube. (Courtesy of Laboratoire de Physicochimie Industrielle.)

be initiated from the pores in the thin gold plating. This corrosion process is usually associated with copper sulfide and silver sulfide corrosion films. The appearances of pore and creep corrosion are shown in Figure 5.19.

5.4.8 Dust Corrosion

This type of corrosion occurs owing to presence of water-soluble salts in the dust. Such solutions form electrolyte and causes metal to rust. This problem has been extensively studied in China by Zhang and his associates [20] who have indicated that the relative humidity and in particular the pH factor was one of the most important parameters affecting dust corrosion. It appears the corrosion of dust particle increases almost linearly with relative humidity as seen in Figure 5.20 whereas the typical appearance of corrosion product around the dust particle is indicated in Figure 5.21 (See Chapter 4).

5.4.9 Galvanic Corrosion

In a bimetallic system, galvanic corrosion is one of the most serious degradation mechanisms (See Chapters 2 and 8). Whenever dissimilar metals are coupled with the presence of solutions containing ionized salts, galvanic corrosion will occur. The driving force behind the flow of electrons is the difference in voltage between the two metals with the direction of flow depending on which metal is more active. The more active (less noble) metal becomes anodic and corrosion occurs while the less active metal becomes cathodic (see Table 2.2). Figure 5.22a shows schematic of galvanic corrosion in aluminum-to-copper joints whereas Figure 5.22b shows an example of typical corrosion damage in aluminum-to-copper compression connector. In the case of aluminum-to-copper connections, aluminum (the anodic component) dissolves and is deposited at the copper cathode in the form of a complex hydrated aluminum oxide, with a simultaneous evolution of hydrogen at the cathode (copper). The process will proceed as long as the electrolyte is present or until all the aluminum has been consumed, even though the build-up of corrosion products may limit the rate of erosion at the surface. The aluminum-to-copper connection is affected by corrosion in two ways: either the contact area is drastically reduced, causing an electrical failure, or the connector is severely corroded, causing a mechanical failure. In most cases, failure is owing to a combination of both effects. The factors that determine the level or severity of galvanic corrosion are numerous and complex but probably the most important

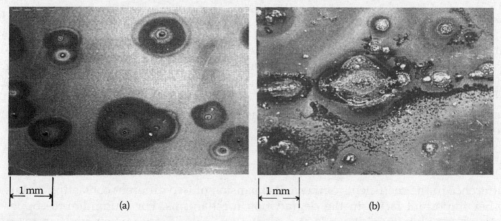

1 mm   1 mm

(a)   (b)

FIGURE 5.19
(a) Pore corrosion and (b) Pore and creep corrosion.

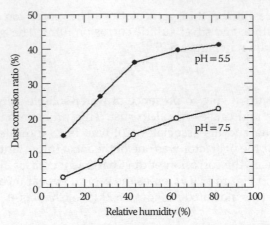

FIGURE 5.20
Corrosion ratios of dust particles with different pH as a function relative humidity.

FIGURE 5.21
SEM images of the corrosion product around dust particle.

is humidity. In order to limit the damaging result of galvanic action in corrosive environments and maintain a low contact resistance, various palliative measures such as plating with a metal of intermediate galvanic potential, contact aid-compounds and transition washers have been used. Corrosion behavior of different material combinations in power connectors and recommended mitigating measures taken to suppress the deleterious effect of corrosion are summarized in Table 5.4.

5.4.10 Thermal Expansion

The difference in the coefficients of thermal expansion of two different contacting metals is another important factor in the degradation mechanisms. For aluminum-to-copper connections, the aluminum expands at a greater rate than copper when exposed to an increase in temperature. As a result, either large lateral movements occur in the contact

FIGURE 5.22
Schematic of galvanic corrosion in aluminum-to-copper joints (a) and (b) example of corrosion damage in aluminum-to-copper compression connector.

TABLE 5.4

Corrosion Behaviors of Different Contact Materials Combinations and Mitigating Measures Required to Suppress the Effect of Corrosion

Material Combination	Corrosion Behavior	Mitigating Measures
Aluminum/Copper (alloy)	Severe corrosion of aluminum surfaces in saline environment	Lubrication and use of Al transition washers
Aluminum/Tin-plated Copper (alloy)	Plating thickness < 10 μm severe corrosion	Lubrication
	Plating thickness > 10 μm no severe corrosion	Use lubricated Al transition washers.
Aluminum/Silver-Plated Copper (alloy)	Plating thickness > 5 μm	Lubrication not required
	No severe corrosion of the contact zone	Current-carrying contact-pairing should be avoided due to formation of intermetallics.
Tin-Plated Aluminum/Copper (alloy)	In industrial environment corrosion of tin and its peeling intensifies corrosion of aluminum	Remove tin and lubricate Use lubricated Al transition washers.
Tin-Plated Aluminum/ Tin-Plated Copper (alloy)	Plating thickness < 10 μm no severe corrosion of aluminum surface	Remove tin and lubricate Use lubricated Al transition washers.
Nickel Plated Aluminum/ Copper (alloy)	If not protected corrosion at nickel-aluminum interface in saline environment can occur	Lubrication
Nickel-Plated Aluminum/ Nickel-Plated Copper	If not protected corrosion at nickel-aluminum interface in saline environment can occur	Lubrication

zone shearing the metal-contact bridges, thus reducing the contact area, or plastic deformation takes place in a region adjacent to the contact interface. This loss increases the contact resistance which, in turn, causes the connection temperature to rise. At higher temperatures, the stresses may be relieved by a recovery in the matrix. On cooling, however, the thermal stresses build up again and further interracial shearing and/or plastic deformation occur since at lower temperatures there will be very little recovery in the matrix and thus little stress reliever. When the process is repeated many times, considerable plastic deformation in the contact zone will occur if the thermal stresses generated are greater than the yield stress of the aluminum. The end result is a cascading effect which accelerates the degradation of the connection until failure. A simple estimate of the magnitude of the maximum elastic stresses generated at the peak temperature in a thermal cycle for aluminum-to-copper connections can be calculated by assuming that the thermal stresses generated in the aluminum are negligible, since the sample is effectively annealed. On cooling, both aluminum and copper will contract by the amount Δd given by

$$\Delta d = \varepsilon_t \Delta T \tag{5.12}$$

where ε_t is the coefficient of thermal expansion and ΔT is the temperature change. The differential strain owing to the constraint is then

$$\Delta d\varepsilon = \Delta T[\varepsilon_t(Al) - \varepsilon_t(Cu)] \tag{5.13}$$

Since for aluminum $\alpha_t = 24.0 \times 10^{-6}$ (1/°C) and for copper $\alpha_t = 17.2 \times 10^{-6}$ (1/°C) the corresponding differential strains for aluminum-copper contacts at 100°C, 150°C and 200°C are respectively 6.8, 10.2, and 13.6×10^{-4} (1/°C). The tensile yield strength of aluminum is 55 MN/m^2 and the elastic modulus 70 GN/m^2, hence the yield strain is 7.8×10^{-4} (1/°C). Comparison with the calculated differential strain values above indicates that the aluminum should yield during cooling of the contact.

Another consequence of the greater thermal expansion of aluminum is thermo elastic ratcheting [21]. In a bolted aluminum-to-copper joint where a steel bolt is used, excessive tightening of the bolt can plastically deform the aluminum and copper conductors during the heating cycle which cannot regain their original dimensions during the cooling cycle. Repeated heating and cooling cycles can thus cause loosening of the joint, which in turn will increase the joint temperature and contact resistance. The effect of thermoelastic ratcheting on the mechanical integrity of a bolted joint with different types of mechanical contact devices under current cycling conditions was investigated by Braunovic [21]. It was shown that the detrimental effect of thermoelastic resulting in the form of loosening of joints can be considerably reduced by disc-spring (Belleville) washers in combination with thick flat washers. The least satisfactory performance was observed in joints comprising lock-spring (Grower) and thin flat washers. This effect, illustrated in Figure 5.23, shows a

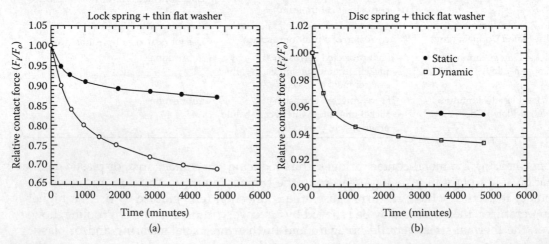

FIGURE 5.23
Effect of thermo-elastic ratcheting on the mechanical integrity of bolted aluminum joints.

comparison of the loss in contact force during the static relaxation test and the residual contact force remaining in the joint during the cooling periods (OFF) of current cycling.

The difference between the static and dynamic (current cycling) stress relaxation is considered as a measure of the thermoelastic ratcheting effect. The results, presented as changes in the relative contact force with time, show that the thermoelastic effect is considerably reduced when disc-spring and thick flat washers were used. Joints with lock spring and thin flat washers are strongly affected by the thermoelastic ratcheting as manifested by a substantial loss in the contact load during current cycling and loosening of a joint, which, in turn, increases the contact resistance and temperature possibly to the point of failure. It is important to note that the use of lock-spring and thin flat washers is a common practice in the electrical industry; In view of these results it this practice is not recommended.

Recently Schlegel et al. [6] showed that different ageing mechanisms are responsible for the long-term behavior of bolted joints used in high current systems. One of these mechanisms is the debasement of the joint force depending on temperature and time. If the joint force falls below a critical value, the joint resistance, the thermal dissipation and thereby the temperature of the joint can increase to a critical level. According to IEC 61439-1 the highest accepted temperature of joints is 140°C. The influence of the force reduction to the ageing of joints is tested on current-carrying, bolted Cu-ETP (CW004A) busbar joints using washers and spring washers. The temperatures used were up to 160°C. The joint force and the joint resistance are measured as a function of time. The possible physical mechanisms for the long-term tests and the results of microscopic investigations of the busbar material are discussed. On the basis of this experimental data an extrapolation of the lifetime of such electrical joints is discussed in relation to the practical operating lifetime of more than 50 years.

5.4.11 Fretting

Fretting (see also Chapters 6 and 7) is a common problem of significant practical importance that can affect a wide range of electrical equipment and can incur costly component replacement and even more expensive equipment downtime. The process is defined as accelerated surface damage occurring at the interface of contacting materials subjected to small oscillatory movements. The required oscillatory movement of the contacting members can be produced by mechanical vibrations, differential thermal expansion of contacting metals, load relaxation, and by junction heating as the power is turned on and off. It is generally accepted that fretting is concerned with slip amplitudes not greater than 125 μm. The sequence of events leading to the development of the fretting damage at the contact interface is described in Figure 5.24. Although the adverse effects of fretting were observed as early as 1911 [22] at the contact surfaces of closely fitting machine elements subjected to vibration and correctly diagnosed as mechanical in origin. The phenomenon was given little attention until 1927 when Tomlison [23] coined the term "fretting corrosion" to handle this kind of surface damage. This definition includes fretting wear, fretting fatigue, and fretting corrosion. The evolution of fretting damage at the contact interface of different contact materials is illustrated in Figure 5.25 As it can be understood irrespective of the contact metals used, the occurrence of fretting damage is quite obvious. Systematic studies of electronic connectors having tin- and solder-plated contacts and numerous reports of failures in service indicate that fretting is one of the prime failure mechanisms. In the event of power electric connections, however, very little published information or reports of failure owing to fretting are available, for two main reasons. First, there is a general lack of awareness of the problem, since fretting is a time-related processes. Second, the effects of fretting are not readily recognizable, since the failure of a power connection

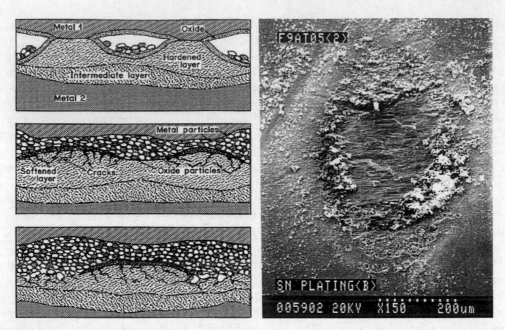

FIGURE 5.24
Schematic of evolution of fretting damage in electrical contacts and a SEM image of a typical fretting wear damage of the contact zone.

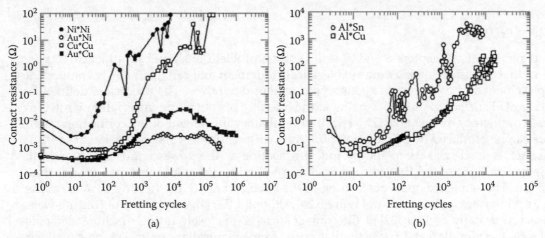

FIGURE 5.25
Evolution of the fretting damage at the contact interface of different contact materials.

is commonly associated with the destruction of the contact zone by arcing, thus making identification of the fretting products rather difficult.

5.4.11.1 Factors Affecting Fretting

The nature of fretting depends on a multitude of variables. Many hypotheses have been proposed to account for the effects observed but no unified model for the process has yet emerged and no single theory has yet been established as correct to the exclusion of any other. Detailed explanations of the present state of the art have been given by Campbell [24], Waterhouse [25], Golego [26], and Hurrics [27].

Factors known to affect fretting may be divided into three broad categories, namely:

- Contact conditions,
- Environmental conditions
- Material properties and behavior.

Factors affecting fretting are schematically illustrated in Figure 5.26. As shown, these ingredients may interact with one another and influence both the nature and the extent of fretting damage. Under certain conditions, the effects of environment may be excluded from the contact area and, therefore, will have no strong influence on fretting. But then, under different contact conditions the same environment may have ready access to the contact zone and have a strong influence. It is clear that adequate simulation of any practical fretting wear problem, contact conditions, fretting load levels and amplitudes, materials, and environmental conditions must be studied. The fretting process is too complex to enable extrapolation with confidence from one set of conditions to another very different set of conditions. Two basic conditions for fretting to occur are relative movement or slip and amplitude of motion sufficient to cause the damage. Experimental evidence shows that amplitudes of the order of 10^{-8} cm (<100 nm) are sufficient to produce fretting [25]. Thus, from a pragmatic standpoint, there seems to be no minimum surface slip amplitude below which fretting will not occur. Although there may be a debate as to the upper limit which may still modify the process as fretting, there is no doubt that in situations where microslip prevails, that is, where slip occurs over only part of the contacting surface, the

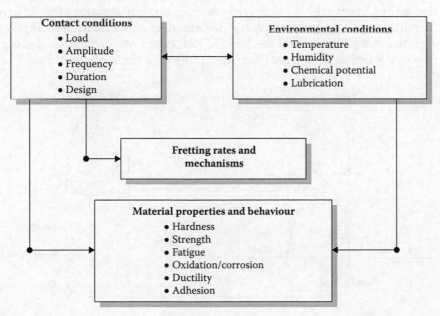

FIGURE 5.26
Schematic representations of the factors affecting fretting.

cause is only characteristic of fretting. Mindlin [28] has demonstrated that the minimum slip amplitude for fretting to occur is given by

$$\delta = \frac{[3(2-\upsilon)(1+\upsilon)]}{8Ea}\mu P\left[1-\left(1-\frac{T}{\mu P}\right)^{2/3}\right]$$

(5.14)

where a is the diameter of the contact outer radius, E is the Young modulus, ν is Poisson's ratio, P is the normal force, μ is the static coefficient of friction between the contact surfaces and T is the tangential force ($T < \mu P$). Figure 5.27 illustrates a classic example of microslip occurring between a steel ball and flat, where the ball has an oscillating tangential force. It is generally agreed that fretting damage increases with increasing amplitude and that the mechanical properties of the contacting materials significantly affect the threshold value for fretting to occur [28].

5.4.11.2 Mechanisms of Fretting

One of the first theories for fretting corrosion was advanced by Tomlison [23] who suggested that fretting corrosion is caused by molecular attrition. The cohesion between atoms and molecules which arises as they approach causes them to detach from the surfaces and subsequently become oxidized. Tomlinson argued that fretting corrosion is not influenced by the normal load because, according to him, molecular attrition is independent of external forces. Godfrey [29] proposed that fretting damage occurs as a result of adhesion between the surfaces. The wear debris is extruded from the contact area and reacts with the environment. Nonetheless, good adhesion between two oxide-covered surfaces is questionable. Feng and Rightmire [30] stated that fretting begins with adhesive wear followed by a transition period in which accumulation of the trapped wear particles gradually contributes to abrasive action. Eventually the damage is solely caused by abrasion. The occurrence of loose wear particles is attributed to plastic deformation at the contacting high spots. Uhlig [31] proposed that chemical and mechanical factors are responsible for fretting corrosion. An asperity rubbing on a metal surface produces a track of clean metal, which

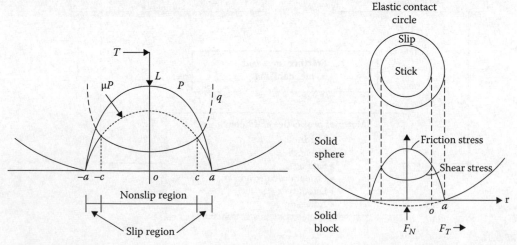

FIGURE 5.27
Microslip occurring between a flat and steel ball under oscillating tangential force.

immediately oxidizes or upon which gas molecules rapidly adsorb. The following asperity wipes off the oxide or initiates the reaction of metal with adsorbed gas to form oxides. This is the chemical factor of fretting. In addition, asperities penetrate below the surface to cause wear by a welding or shearing action by which metal particles are dislodged. This is the mechanical factor of fretting. Stowers et al. [32] identified two different regimes of fretting as a function of amplitude. In the low-amplitude regime, the volumetric wear per cycle is a function of the amplitude squared, whereas in the high-amplitude regime it is directly proportional to the amplitude. They suggested that material loss resembles that produced by unidirectional adhesive wear much more closely than that produced by other modes of wear. Consequently, the amount of wear can be computed using Archard's adhesion model for wear [33] which requires that when asperities come into contact and adhere strongly to each other, the subsequent separation occurs in the bulk of the weaker asperity in a single action. This procedure is assumed to produce an atom from the softer surface which adheres to the harder surface. When these transferred particles become free, loose wear particles are formed and wear, observed as weight loss, is assumed to occur. Oding and Ivanova [34] stated that fretting is associated with a thermoelectric effect that causes electro-erosive action at the contacting surfaces. Material from one surface, anodic with respect to the other, is being removed from atoms torn from the surface by the direction of the electric field. Atoms moving in to fill the vacant sites at the surface resulting in vacancy diffusion in the region below the surface. The concentration of vacancy increases until a critical concentration is achieved when they coalesce as micropores and microcracks. This means that only one of the surfaces undergoes damage. By passing the countercurrent or by choosing suitable pairs of metals for the contact, this could be reversed. Although attractive, this theory cannot explain why surface damage arises between two similar metals, unless the thermoelectric effect of an oxide film is invoked. Suh [35] proposed a delamination mechanism of wear which involves the initiation of subsurface cracks which propagate parallel to the surface and lead to detachment of flakes 0.1–20 µm thick. These fractures are thought to initiate at voids and vacancies developing from dislocation pileups below the surface layer, and at a critical length, the cracks shear to the surface. It should be emphasized that the delamination theory of wear as it stands is applicable only for the case of low-speed sliding, where the temperature rise at the contacting surface is so low that diffusion and phase transformation are not involved in the wear process.

The delamination mechanism of wear was used by Waterhouse and Taylor [36] to explain the formation of loose wear particles and the propagation of subsurface cracks in the fretting surfaces. The end result of this process is the detachment of oxide-coated plates of metal about 1.3–3.5 µm thick. The continuing fretting action grinds the initial wear particles down to particles of smaller size with higher oxide content. They also divided the fretting corrosion process into two stages. Adhesion occurs in the early stage of fretting and it is more significant with noble metals or in an inert environment. When this stage is passed, the surfaces become smooth and removal of material from the surfaces occurs by delamination. The transition from adhesion to delamination is a function of the material and the nature of the surroundings. Sproles et al. [37] concluded from the observed mode of metallic material removal that fretting wear by a delamination mechanism is predominant, rather than an abrasive wear mechanism or a welding and material transfer mechanism. Oxide debris is formed by the oxidation of metallic debris or by the constitution and subsequent scraping away of a thin oxide film from the metal surfaces. Godet [38] proposed a third-body concept and a velocity accommodation mechanism. The debris formed in the contact remains in the interface for several cycles, and fretting has to be viewed as a three-body contact. This approach focuses on the role of the third-body material which separates the two first bodies rubbing through its load-bearing capability. Directly transposed from the lubrication theory, this concept emphasizes notions of third-body flow and velocity accommodation. Noël et al. [39] examined the fretting behavior of nickel coatings with two types of deposits for electrical contact applications Sulfate nickel layers are electrodeposited in different conditions and show very different behaviors during fretting tests. The characteristics of the layers are analyzed and show different compositions and microstructures. The compositions are measured by X-Ray Photoelectron Spectroscopy (XPS) which allows determining the chemical nature of the compounds formed during exposure to air. Topography is measured by AFM and the roughness and grain characteristics are measured. Electrical properties at the micro/nanoscale are measured with the CP-AFM technique. Various loads are applied to the cantilever beam; the electrical characterization is performed versus the load.

The results of fretting experiments are analyzed in terms of fretting regimes. The fretting regimes occurring during the test of nickel layers involve partial slip which delays the occurrence of contact resistance (Rc) increase (Figure 5.28). Gross slip in the interface is shown to create very poorly conducting wear debris leading to drastic increase of Rc. Bright nickel is compared to matte nickel in a fretting test simulating the micro-displacements caused by vibrations. A drastic increase of the resistance is observed with bright nickel while with matte nickel a partial slip regime sets up which postpones the failure of the contact. This behavior is related to the microstructure of the coating which depends on the mechanisms of evolution of texture in nickel electro-deposits. Bouzera et al. [40] determined the minimum fretting amplitude in medium force for connector coated material and pure metals. For automotive applications, the mechanical behavior of the contact area under vibrations is one of the key components for connector reliability. Such vibrations are typically in the range of 10–2000 Hz and result in displacements of only a few microns, at the contact interface. In the present study, a bench test has been developed to control more representative motions down to 1 μm. The objective is to determine the minimum amplitude for fretting-corrosion degradation on the basis of the evolution of contact resistance and to study the effects of the material, the contact force, the coating for these low displacement amplitudes.

To obtain the limit of the appearance of fretting, a sub-micrometer incrementing displacement amplitude methodology was used on high stiffness bench test including a double PZT actuator. It was found that the fretting degradation starts to occur from 2 to 6 μm when the contact force is from 0.5 to 2.5 N with a tin coated terminal (Figure 5.29).

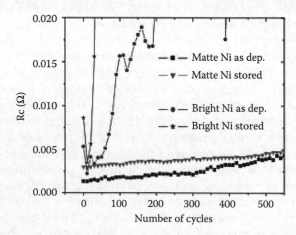

FIGURE 5.28
Contact resistance R_c during the first cycles of fretting experiments performed on matte and bright coatings as received and after storage.

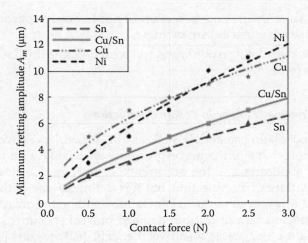

FIGURE 5.29
Minimum fretting amplitude A_m versus normal force for copper tin-coated and pure metals Sn, Ni, and Cu-(symbol) experimental data and (line) theoretical data.

Moreover pure copper, tin, and nickel have similar amplitudes fretting limits while noble metals confirm the absence of fretting up to 10 µm amplitude and for a large number of operations (10^6 cycles). Best fitting of the obtained minimum fretting amplitude data to Mindlin equation is discussed and amended by correcting factor. It was found that hard materials like Ni need higher amplitude than softer material like Sn to enable fretting whereas coating material in connector needs higher amplitude than pure metal of coating For non-noble materials as (Cu, Sn, Ni), the lifetime of the electrical contact is limited, the electrical contact resistance is sharply increased at amplitude as low as few µm, and its lifetime is dependent on the kinetics of forming an insulating layer of oxidized debris. However a noble metal as (Ag), their lifetime of the electrical contact is unlimited and no minimum fretting amplitude up to 12 µm is observed, its electrical contact resistance is stable and low because its debris does not oxidize. The theoretical data of minimum amplitude calculated from Mindlin model seems to be different with experimental data for the used materials. Since the experimental procedure used does not completely respect the Mindlin assumptions of the theoretical model, a correcting factor C is then introduced into the Mindlin equation to improve the convergence. Future work could discuss and determines the way this discrepancy depends on other contact materials and the test conditions. It is necessary to quantify the amplitude of transition at which fretting starts when an electrical contact is subjected to micro-displacement tests. A method of incrementing amplitude has been applied to different types of materials. This technique easily allows the detection of the transition between P.S. regime and G.S. regime for each material that corresponds to fretting appearance.

Although there is still no perfect unanimity on the mechanisms of fretting, particularly with respect to the relative importance of the processes involved, it can be safely assumed that the following processes are present:

1. Disruption of oxide film on the surface by the mechanical action exposes clean and strained metal which will react with the environment and rapidly oxidize.

2. The removals of material from the surfaces by adhesion wear delamination or by shearing the microwelds formed between the asperities of the contacting surfaces when the contact was made.

3. Oxidation of the wear debris and formation of hard abrasive particles that will continue to damage the surfaces by plowing.

4. Formation of a thick insulating oxide layers and wear debris (a third body) between the contacting surfaces.

5.4.11.3 Examples of Fretting Damage in Power Connections

Aluminum, copper, and plating materials such as tin, nickel, and silver are the most common contact materials. One of the major problems associated with the use of these materials in electrical applications is the occurrence of fretting. Fretting may not contribute directly to the failure of a connection, but it is definitely one of the prime factors causing electrical instability and subsequent joint failure. Although contact resistance was used to monitor the development of fretting damage in steel specimens as long ago as 1956 [41], it was only eight years later that Fairweather et al. [42] revealed how fretting can cause considerable instability and serious degradation of telephone relays and switches. The deleterious effect of fretting, however, was not widely recognized as a serious factor in the degradation of electrical connections until 1974 when Bock and Whitley [43] clearly showed its importance. Since then, systematic studies of electronic [44–50] and automotive [51] connector systems and reports on in-service [52,53] failures have established that fretting is one of the major contact deterioration mechanisms in dry connections. The phenomenon appears to be inherent in all contact and conductor materials such as gold, palladium, tin, nickel, silver, copper, and aluminum [47–61]. A comprehensive review of fretting in electrical connections has been given by Antler [62,63], while Mallucci [64] and Bryant [65] provided comprehensive models to predict the contact resistance behavior under fretting conditions. The result of fretting is of considerable importance to power connections involving aluminum. This is because dry aluminum contacts are most susceptible to fretting damage since hard aluminum-oxide particles easily abrade the contacting metals. This, in turn, will initiate the oxidation of exposed materials and accumulation of highly insulating fretting debris at the contact interface. As a result, the contact resistance will increase at a very high rate thus raising the joint temperature which, in turn, will accelerate the deterioration of the contact members.

5.4.11.4 Compression Connectors

In the case of compression connectors, such as shown in Figure 5.30, when two dissimilar metals are pressed against each other, surface asperities will penetrate the natural oxide films providing a good metallic contact. Also, because of the highly localized contact stresses, the asperities will be severely deformed and there will be the mechanical seizure of the contacting metallic surfaces. In the case of unprotected (non lubricated) aluminum over unprotected copper (Figure 5.30a), an increase in temperature causes the outer aluminum portion of the connection to expand at a rate greater than that of the copper thus generate fretting motion at the contact interface and shear the metallic bridges. This, in turn, will cause a loss of contact surface area. Exposed to the atmosphere, aluminum oxidizes resulting in further reduction of the contact area when the connection returns to the initial temperature.

With unprotected copper over unprotected aluminum (Figure 5.30b), the latter expands in the axial direction since the residual stress in the copper and its lower rate of thermal expansion restrict the wire to expand circumferentially. If the axial stress exceeds

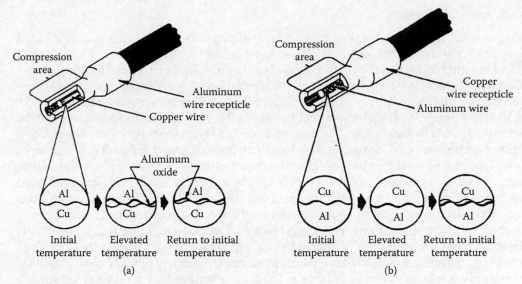

FIGURE 5.30
Compression connection between (a) aluminum terminal and copper conductor; (b) copper terminal and aluminum conductor.

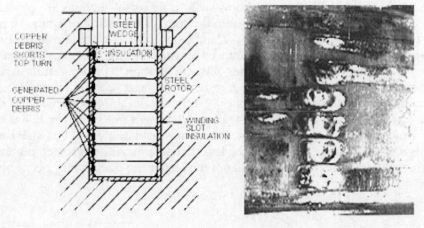

FIGURE 5.31
Fretting damage in power generators.

the elastic limit of aluminum, a permanent deformation result, and the wire will not revert to its original compressed position contact. Dang and Braunovic [66] used wire-to-sleeve resistance mapping together with SEM (Scanning Electron Microscopy) and X-Ray Analysis (EDX) techniques to study the contact resistance and metallurgical changes of compression sleeve connectors in aluminum cable splices following accelerated aging tests. It was shown that fretting and thermomechanical movements resulting from current cycling are the most probable cause for the degradation and overheating of these connectors. Another example of the effect of fretting is shown in Figure 5.31. It depicts the fretting damage originated at the copper windings that resulted in the loss of insulation and accumulation of the copper debris between the transformer windings [67].

5.4.11.5 Bus-Stab Contacts

Field experience with separable power contacts involving aluminum busbars has indicated failures in certain industrial locations. These were evidenced by mechanical erosion and electrical burning of metal at the contact areas resulting ultimately in open-circuit condition. Johnson and Moberly [68] have investigated the effect of fretting on the performance of bus-stab contacts with tin- and silver-plated aluminum busbars under normal operating conditions. It is now recognized that bus-stab contacts are subjected to three modes of mechanical motions, illustrated in Figure 5.32. The first mode occurs when the busbars are subjected to variations in electrical load, changing their length owing to thermal expansion. This results in slow slide motions with respect to the stab contact (direction A-A, Figure 5.32a) which, in turn, causes elongation of the stab contact area along the busbar. This was confirmed by the burnished track on an actual aluminum bar. These relatively slow induced motions are a function of bus-run length and temperature changes and, consequently, will be relatively short in many applications. The second mode of motion (direction B-B, Figure 5.32a) is attributed to electromagnetically induced vibrations. The driving force for these motions is created by currents flowing in adjacent bus members. In common busway configurations, the forces between busbars can be as large as 100 N/m of busbar length under rated current conditions. Such forces can cause busbars displacement of the order 20 μm perpendicular to those induced by thermal expansion. The third way of bus-stab contact motion is transverse displacement perpendicular to both thermal and electromagnetically induced motions (direction C–C, Figure 5.32a and b). It is owing to rigid stab mounting and the vibratory motions discussed above. An illustration of how such "transverse" motions are generated is shown in Figure 5.32b. As the busbar vibrates to the right of its resting or neutral position, owing to the motion shown as B-B in Figure 5.32a, the stab's two contacts are constrained to move in a transverse direction. One contact moves upward and the other downward along the bar, as they follow their respective arc lines. Busbar vibration to the left results in reverse transverse motion of the stab contacts. The frequency of these rubbing transverse movements is identical to that of the electromagnetically induced vibratory motions, 120 Hz, from which they are generated. On the basis of the physical dimensions of a typical 400 A stab, and under normal electrical load conditions, the relative displacement of bus-stab contacts owing to transverse motions is required to be less than 25 μm. Although such motion displacements are small, significant fretting wear of the bus or the plating may occur because

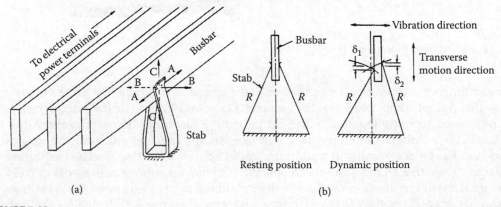

(a)   (b)

FIGURE 5.32
(a) Direction of mechanical motion acting on bus-stab: (b) Transverse motion of slab contact along busbar width contacts.

of the increasing number of slides. The resulting degradation of protective platings would lead to the establishment of high-resistant films, surface damage and arcing, which, in turn, favors more chemical attack, burning and failure of the contact metals.

Contact area distress of 95% and 15% was observed following tests of tin-plated and silver-plated stabs, respectively. Distress percentages were assigned by the authors and they represent judgments on the basis of the preliminary contact failures being 100%. An indication of contact distress related to the bus-stab interface contact materials may be obtained from the micrographs in Figure 5.33. Contact distress becomes very small when transverse sliding motions of the stab are reduced to 5 μm. It is most likely that, at this level of motion, inherent resiliency of the stab actually prevents relative slide thus reducing abrasive rolling or rocking motion. Direct correlation with the above observations of distress is found in the contact voltage and bulk temperature values shown in Table 5.5. When a tin-plated stab was combined with the silver-plated aluminum bar, contact voltages greater than 270 mV were measured. However, in the case of silver-plated bus-stab contacts subjected to pulsed

FIGURE 5.33
Contact zone after distress caused by transverse sliding motion of bus-stab interface.

TABLE 5.5

Bus -Stab[a] Electrical Contact Performance Characteristics

Part	Stab Plate	Bar Motion[b] (μm)	Contact Voltage (mv)		Contact Bulk Temp (°C)		Contact Distress (%)
			Pulse	Steady	Pulse	Steady	
1	Sn	12	>270	>290	52	>156	95
	Ag	12	31	12	31	41	15
2	None[c]	8	134[d]	–	30	–	75
	Sn	8	42[d]	38	26	32	20
	Ag	8	13[d]	7	24	25	2
	Sn	5	7[d]	3	24	25	8
3	Sn	5	42	15	31	36	11
	Ag	5	33	13	29	41	
4	None[e]	"static"	174	180	70	168	90
	None	"static"	35	15	33	44	1
	Sn	"static"	43	17	29	36	2
	Ag	"static"	29	11	30	41	1
5	Ag[f]	8	112	162	52	108	90

[a] BusBar: Ag plated Al: Electrical load 1,000 A(rms)—46 hr., 100 A(rms)—4 hr. Stab: Cu-Cd alloy: Contact load 70 N
[b] 120 Hz double motion frequency
[c] 3 hour run
[d] Electrical load 250 A(rms)–46 hr., 100 A (rms)–4 hr.
[e] Unplated Al busbar
[f] Sn plated Al busbar

and steady electrical loads of 1,000 and 400 A, relatively low values of contact voltage of 31 and 12 mV respectively were measured. The best performance is obtained when the contact members are plated with silver, rather than tin, and mechanically held together with forces of at least 70 N. Nevertheless, when the contact members are subjected to 120 Hz, relative slide motions with amplitudes greater than 5 μm the time to failure is greatly shortened. Hence, unless transverse motions can be minimized below the critical amplitude level, bus-stab protective metal platings will wear at appreciable rates and eventually be worn out, resulting in higher contact resistance and temperature and final "burn" failure.

5.4.11.6 Plug-In Connectors

Plug-in connectors are in common use for connecting equipment to a busway. Devices connected to busways cover a wide range of industrial control equipment ranging from individual circuit breakers to feeders for panel boards. Under certain conditions, fretting can occur in the connector-busbar interface owing to the same type of relative motions as in the bus-stub contacts as discussed above. The connector assemblies used in these applications must reliably conduct currents ranging from 100 A to several thousand amperes throughout the lifespan of the bus duct system. Thiesen and Forsell [69] designed a special fretting testing system to investigate the performance of a typical 200 A plug-in connector under fretting conditions. Tests were carried out on the silver-, tin-, and cadmium-plated contact surfaces showed that fretting produced significant contact damage as manifested by the large accumulation of the fretting debris in the contact zones which resulted in high contact resistance. The authors concluded that this fretting testing system can be an excellent tool in comparative analysis of the design requirements for connector systems.

5.4.11.7 Bolted Connections

It is generally accepted that the reliability of bolted joints is attributed to high contact forces and large apparent contact areas with virtually no relative displacement between the contacting members. Although this may hold for copper-copper joints, it is certainly not necessarily the case for aluminum-copper connections. This is because the coefficient of thermal expansion of aluminum is 1.36 times that of copper and when a bolted aluminum-copper joint is over heated by the passage of current, aluminum will tend to expand relative to copper causing displacement of the contact interface. The shearing forces generated by differential thermal expansion will rupture the metallic bonds at the contact interface and cause significant degradation of the joint. Published experimental evidence and reports of trouble in service show that bolted joints may not be as impervious to degradation and failure. The effect of relative motion in aluminum-copper bolted joints has been investigated in the laboratory by Bond [70], Naybour [71], Jackson [14], and Roullier [72] who have shown that the relative displacement induced by differential thermal expansion forces is one of the major degradation mechanisms of aluminum-copper joints. Yet, despite the seriousness of this problem, there is very little published information linking field failures with the degradation effects of relative motion (fretting) in bolted joints.

Recently, fretting damage in real life tin-plated aluminum and copper connectors, commonly used for distribution transformers, has been reported by Braunovic [73,74]. The connectors examined had been removed from the service after 7–10 years of service owing to either overheating, as discovered by routine thermography inspection, or unstable performance on the network under normal operating conditions. The connectors were bolt-type tin-plated aluminum or copper busbars jointed with either tin-plated or bare aluminum cable terminals. A typical example of the connectors with signs of severe fretting damage

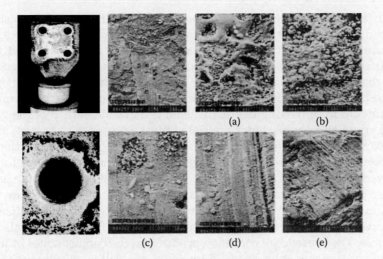

FIGURE 5.34
Typical example of a tin-plated connector removed from service with signs of severe fretting damage as evidenced by a characteristic band of accumulated fretting debris and oxides and SEM surface analysis of typical fretting damage in aluminum-to-tin-plated copper connection: (a) Electrical erosion (region 1); (b) Accumulation of fretting debris and oxides (region 2); (c) Delamination wear (region 3); (d) Abrasion (region 4); and (e) Combined delamination-abrasion wear.

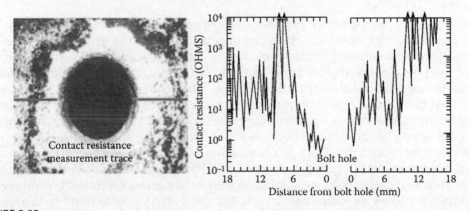

FIGURE 5.35
Variation of contact resistance with the distance from the bolt hole in the contact zone of a tin-plated copper connector.

as manifested by a characteristic set of accumulated fretting debris and oxides is shown in Figure 5.34. The significant feature of these events is that fretting causes serious damage to the contact areas. Examination of these areas reveals the presence of four distinct forms of fretting damage: electrical erosion by melting/arcing (region 1, Figure 5.34a), accumulation fretting debris, delamination and abrasion. The results of contact resistance measurements across the contact zone of a tin-plated copper connector are shown in Figure 5.35. The most important feature is that the contact resistance of the zones severely damaged by fretting increases rapidly to very high values, and in some locations, open-circuit conditions developed. These results correlate well with those obtained under controlled laboratory conditions showing that fretting causes the contact resistance of tin-plated contact to increase at very rapid rates. The cyclic nature of the contact resistance, manifested by sharp fluctuations from one site to another, indicates that the fretting debris formed on the connector surface is not a continuous layer but rather porous, since the transition from high- to low-conducting current paths occurs over relatively short distances. Another interesting feature of these events is the presence of a region around the bolt with very low contact resistance values. This can be associated with the formation of good metallic contact with the load-bearing area under the flat washer where the contact pressure is highest. Detailed SEM examination of these zones shows no evidence of severe fretting damage or accumulation of fretting debris. Instead, the grainy texture of the tin-plating was heavily deformed and compacted, as revealed by the presence of areas with smooth surfaces with clearly visible scars of mild abrasion. The results of x-ray diffraction analysis of the fretting debris removed from the damaged contact zones of tin-plated aluminum connectors showed that the fretting debris was composed mainly of tin but the presence of $Al(OH)_3$ was also detected, indicating that the corrosion products of aluminum were also prominent constituents. This suggests that fretting was not the sole mechanism responsible for the observed degradation of tin-plated connectors and that corrosion also played a very important role. The presence of corrosion products in the contact zones indicates that the introduction of an aqueous solution to a fitting environment may influence the fretting process in one or both of two distinct ways. Firstly, the liquid can serve as a lubricant, so separating the metal surfaces and reducing adhesion, friction and wear rate. Secondly, a liquid may induce an anodic corrosion reaction within the fretting scar, trapping the corrosion products inside the fretting scars and therefore increasing the rate of wear.

5.4.11.7.1 Tin and Tin Alloys

Tin and tin alloys are used extensively in a variety of applications in the electronic and electrical industries either as solders or as coatings to aid soldering or to mitigate galvanic corrosion when dissimilar metals like copper and aluminum are in contact, Such widespread use prompted numerous investigations into determining the parameters for ensuring satisfactory and reliable operation of these joints. As a consequence, it is now well-established that the formation of intermetallics, corrosion and fretting are the most serious degradation mechanisms impairing the reliable operation of tin and tin alloy connections. Although the nature of these degradation mechanisms is different, they all have an adverse effect the contact resistance of a connection.

Systematic studies of dry tin-plated connection failures have established that fretting is probably the foremost detrimental mechanism to the performance of a connection involving tin and tin alloys. Dry tin-plated contacts are the most susceptible to fretting damage, since hard tin-oxide particles easily abrade the contacting, metals thus alleviating the oxidation of exposed materials and accumulation of highly insulating fretting debris at the contact interface. As a result, the contact resistance will increase at very high rates thus raising the joint temperature which, will accelerate the deterioration of the contact members. A typical example of the deleterious effects of fretting in tin-plated contacts is shown in Figure 5.36. It is seen that contact resistance increases rapidly after only a few hundred cycles, eventually leading to an open circuit condition. This takes place over a wide range of contact loads and becomes worse with increasing slip amplitude. The consequences of such a dramatic increase in the contact resistance are, initially, Joule heating of the contact spots followed by melting, sublimation and decomposition of the oxides and vaporization of the tin. This is illustrated in Figure 5.37 showing the contact voltage as a function of the fretting cycles. Two sustained plateaus in the contact resistance characteristics are present: one coinciding with the melting voltage of tin, corresponding to the melting temperature of tin, that is, 232°C, the other in the range where tin oxides melt and tin vaporizes. Note that the voltage range for the second plateau corresponds to the temperature range of 1000–3000°C.

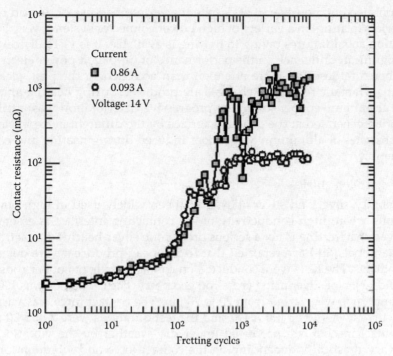

FIGURE 5.36
Typical examples of deleterious effect of fretting in tin- and solder-plated contacts.

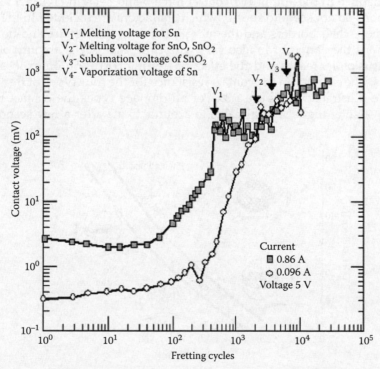

FIGURE 5.37
Contact voltage as a function of fretting cycles.

It should be pointed out, however, that despite the widespread use of tin and tin-alloys as contact materials (plating) in a variety of high power connector systems, very little published information about failures owing to fretting is available. It is virtually certain that fretting has a significant influence on the performance of tin-plated power electrical connections and therefore deserves more attention than accorded in the past. Jackson [67] who found that a dramatic resistance changes are produced during heating and cooling of a tin-plated aluminum–copper joint. The progressive deterioration was attributed to the relative motion generated at the contact interface by the differential expansion forces between the substrates of aluminum and copper induced during heating and cooling of the contact assembly.

5.4.11.7.2 *Silver and Silver Alloys*

Owing to its nobility, silver and silver alloys have been widely used in high power electrical applications where high conductivity of the contacting interface is essential. It is generally believed that fretting is not a serious problem in silver-bearing contact, although recently Kassman et al. [61] have reported that fretting can produce severe deterioration of the power contacts. The tests were conducted on contacts made of copper rods electroplated with 5 and 17 μm of silver using cross-rod geometry. The frequency was 100 Hz and the vibration amplitude was varied from 20 to 70 μm. The contact force was varied from 5 to 50 N. The current was 10 or 100 A DC and the test durations were 2 and 20 minutes. The type of contact mechanism produced under dry weather by the vibration motion was found to vary drastically depending on the contact force and vibration amplitude. In the case of unlubricated contacts after 2 minutes of vibration four contact deformation mechanisms were distinguished, in the case of unlubricated contacts. Their occurrence could be mapped in the amplitude/contact force plane, seen in Figure 5.38. Coating thickness variations do not seem to specify which characters are included in the map, but serve rather to move their borders and the margin between the seizure and non-seizure zones. Furthermore, the damage can also vary with time, that is, the amount of material transfer and extrusion increases and the effect of fretting corrosion on exposed copper becomes obvious. The changes in the contact resistance for the four types of fretting damage observed are illustrated in Figure 5.39. For all damage types the contact resistance dropped from a slightly higher initial value to around 15 μΩ after a few seconds. After

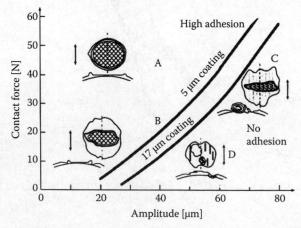

FIGURE 5.38
Mapping of the damage mechanisms as a function of contact force and vibration amplitude 52.

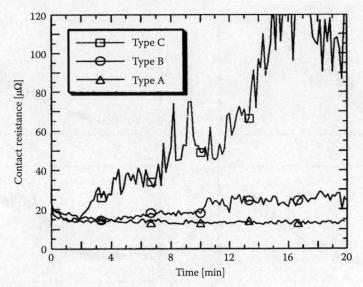

FIGURE 5.39
Contact resistance variation with fretting time in silver-plated copper.

some 2 minutes the resistance starts to escalate for Types C and D. This is because both types quickly expose copper in the contact area, which is twisted into an insulating copper oxide layer by fretting corrosion. After 20 minutes, the contact resistance had increased approximately 1000% for Type C, which was the worst type. Type B showed a moderate 70% increase while Type A showed no increase even after trying for a long time and remained still consistently low after 2000 minutes.

5.4.11.8 Fretting in Aluminum Connections

The deleterious effects of fretting are of particular interest in aluminum connections since aluminum exposed to the atmosphere rapidly oxidizes, forming an oxide film within a very short time. Hence, under fretting conditions, accumulated wear debris and oxides cannot be effectively removed from the contact zone and a highly localized, thick insulating layer is formed leading to a rapid increase in contact resistance and, subsequently, to virtually open circuits. The performance of aluminum conductors in contact with different plating and base metals under fretting conditions has been extensively studied by Braunovic [54–58,73,74,77–79] who has shown that fretting adversely affects the contact behavior of aluminum in contact with practically all common contact materials. Some typical examples of such behavior are shown in Figure 5.40. The effect appears to be characterized by three points:

1. The contact resistance remains virtually unaffected by the fretting action and can be explained as follows. When the contact is made, surface asperities will penetrate the natural oxide films, thereby establishing localized metallic contacts and, conducting paths. The number of conducting paths thus established will not be greatly affected by the fretting-induced displacement of the contact interface, since the wear products formed will be predominantly composed of metallic particles, thus a relatively good electrical contact will be maintained.

2. Further fretting will force the wear debris to escape and allow the establishment of a larger number of metallic contacts. Consequently, a good metallic contact

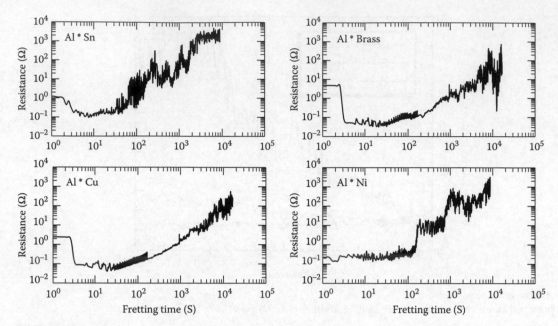

FIGURE 5.40

Typical contact resistance behaviour of aluminum in contact with common contact materials under fretting conditions: slip amplitude 25 μm; contact load 2 N; frequency 0.005 Hz

between contacting surfaces will be established and maintained, resulting in a very low, practically fluctuation-free contact resistance. The cyclic nature of the contact resistance, manifested by short-term fluctuations, is most likely owing to temporary rupture of the oxide layers and the appearance of local metallic contacts, which provide good conducting paths. The persistence of these fluctuations indicates that the generation of conducting spots by mechanical rupture of the intermediate layers of oxides and fretting debris is counterbalanced by the closure of the paths newly formed by the fretting action.

3. After prolonged exposure to the fatigue-oxidation process, metallic layers in the contact zones soften and progressively separate. The contact zone now consists of a thick insulating layer containing oxides and wear debris, and any remaining metallic contact is lost, causing a sharp increase in the contact resistance. The cyclic nature of the latter is owing to the temporary rupture of the insulating layer by fritting and the appearance of localized metallic conducting paths formed as a result of contact self-healing [12]. Subsequent wiping will fracture these conducting bridges and the contact resistance will rise, which in turn will further increase the contact spot temperature up to melting, sublimation and decomposition of the oxides and even vaporization of the contact materials.

Fretting of aluminum causes an extensive exchange of materials in the contact zones. Some representative examples of the Scanning Electron Microscope (SEM) and Energy Dispersive X-Ray Analysis (x-ray mapping, EDX) of material transfer in the aluminum-to-tin plated brass connections are shown in Figure 5.41. The deleterious effects of fretting can be greatly reduced by applying higher contact loads and/or lubricants to the contact zones.

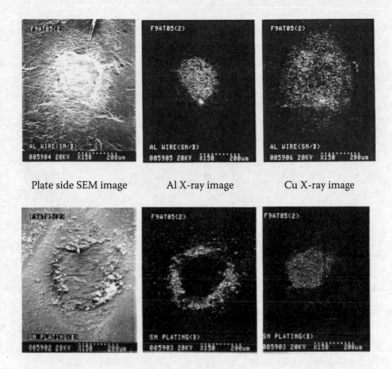

Plate side SEM image   Al X-ray image   Cu X-ray image

FIGURE 5.41
SEM micrographs of fretting damage in contact zones of an aluminum conductor in contact with tin-plated connections. The X-ray images of Al, Cu, and Sn exemplify the transfer of materials in both directions. Light areas depict the fretting debris.

When higher load is applied, the onset of the thermal runaway condition can be delayed since much higher contact resistance is needed to reach the contact voltage or spot temperature required for this to happen. The effect of contact load is of considerable practical importance because of aluminum has an intrinsic tendency to creep and stress relaxation, which can result in the loss of a contact and consequently accelerate the degradation of a connection with the ever-present fretting action. It is therefore essential to maintain high contact pressures in all electrical connections in order to assure sufficiently large current-carrying areas and the mechanical integrity of a joint.

5.4.11.9 Effect of Electrical Current

The result of fretting in copper-to-copper contacts under AC (60 Hz) and DC current conditions was investigated by Gagnon and Braunovic [78]. It was demonstrated that the overall contact resistance behavior of copper-to-copper wire-plate couples under AC and DC current was practically the same as pictured in Figure 5.42. Fretting debris in the samples fretted under DC current are compacted without flake-like plates, Figure 5.43. Furthermore, the fretting damage in samples operating under AC current is more pronounced and characterized by the presence of flake-like debris that was not observed in the samples operating under DC current as clearly depicted in Figure 5.43. The characteristic feature of the samples under AC current conditions is a pronounced distortion of the contact voltage (see Figure 5.44) and the presence of large amounts of flake-like fretting debris widely scattered around the contact zone.

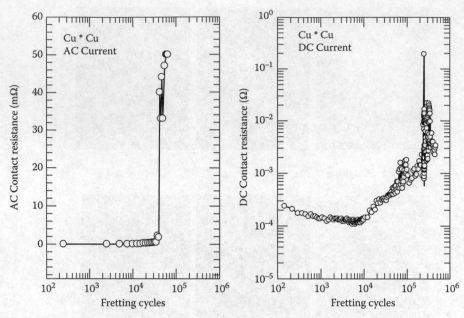

FIGURE 5.42
Effect of fretting on the AC and DC contact resistance of copper-to-copper wire-plate contacts. Fretting conditions: contact load 400 g, fretting frequency 1 Hz, slip amplitude 100 μm current 50 mA.

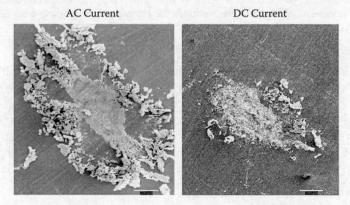

FIGURE 5.43
SEM images of the contact zones after 40,000 fretting cycles under AC and DC current conditions. (From M. Braunovic, *Electric Contacts-1989*, San Francisco, 1989, p. 179.)

5.4.11.10 Fretting in Coatings (Platings)

Presently, electronic and electrical industries are experiencing a relentless increase in the use of protective and wear-resistant coatings for electrical contact applications. Such movement is driven not only by the cost benefit demands but also useful functions offered by the coatings such as: corrosion and wear protection, diffusion barriers, conductive circuit elements, fabrication of passive devices on dielectric surfaces and others (See Chapter 8). The performance of an electronic/electrical connector is essentially controlled by the surface phenomena such as pollution, oxidation, re-oxidation, sulfide-formation, corrosion, etc. The presence of these contaminants on the surface increases the contact resistance and

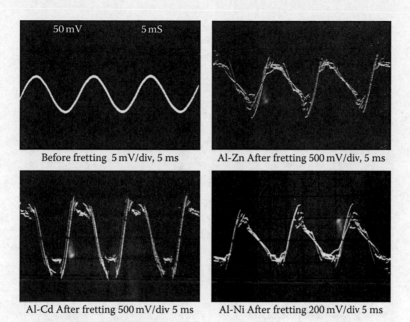

FIGURE 5.44
Effect of fretting on the AC waveform signals.

is detrimental to the connection reliability. Examples of the effect of coating materials on the contact resistance and coefficient of friction on the fretting behavior of different plating materials are indicated in Figure 5.45 [79]. It should be noted that despite the obvious advantages of the coatings, there are still a number of factors such as surface imperfections (porosity), hardness (soft), adherence to the substrate and resistance to oxidation and the effects of a corrosive environment are probably the most important affecting the characteristics of practically all types of coated electrical connections.

5.4.11.11 Fretting in Circuit Breaker Terminal Contact Materials

Although there is only limited data described in the open literature on fretting problems with circuit-breaker contact materials, the above discussion on the deleterious effects of fretting in power contacts may be equally applied to circuit-breakers. Ambier and Perdigon [81] have designed a special test system to simulate the real-life operation of a circuit breaker and used it to investigate the degradation of the contact materials by fretting. The contact materials examined were copper (OFHC and copper-tellurium alloy) and tin-, nickel-, and silver-coated copper and aluminum alloy (6061 grade). Tests were conducted at contact loads of 10–30 N, oscillation frequency 20–300 Hz, slip amplitude 25–200 μm and with current in the range 10–250 A. It was shown that as a result of fretting severe damage occurred in the contact zones of almost all the materials tested. The most stable performance was observed in the aluminum contacts coated with 150 μm thick silver layer. However, the stability of silver-coated contacts was greatly influenced by the contact load, coating thickness and processing methods and slip amplitude. Commercially available silver-coated aluminum with 7 μm layer of silver and with 15 μm nickel underlayer showed an erratic behavior under different fretting conditions. At low contact loads, a very unstable contact resistance behavior was observed whereas at higher loads higher wear rates dominated.

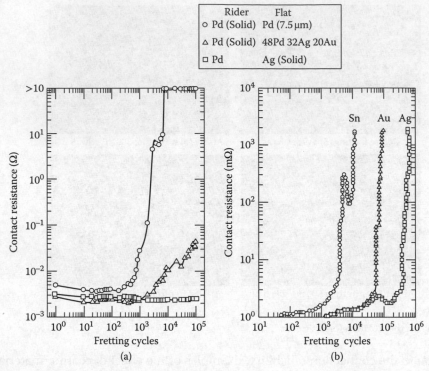

Rider	Flat
○ Pd (Solid)	Pd (7.5 μm)
△ Pd (Solid)	48Pd 32Ag 20Au
□ Pd	Ag (Solid)

FIGURE 5.45
Contact resistance of (a) solid palladium fretting against palladium- and PdAgAu-plating and solid silver and (b) tin-plated spherical pin (rider) fretting against tin-, gold-, and silver-plated flats.

5.4.12 Intermetallic Compounds

Bimetallic welds, particularly, aluminum-to-copper are increasingly being utilized in a variety of electrical applications. Such joints, caused by friction welding, pressure welding, diffusion and roll bonding, flash welding and explosion welding, are characterized by a relatively stable joint interface and negligible intermetallic formation. In service, however, frequent current surges on the network may generate favorable conditions for inter-diffusion to occur and, thus, nucleation and growth of intermetallics at or near the initial interface. This, in turn, can seriously impair the overall electrical stability and mechanical integrity of bimetallic joints since intermetallic phases have much higher electrical resistance and lower mechanical strength (See Chapter 1, Section 1.3.3). Some authors found that tensile strength, ductility, impact resistance and electrical resistance of flash-welded aluminum-to-copper joints are practically unaffected by thermal treatment for two years at 149°C to 5 minutes at 371°C [82]. Investigations of roll-bonded [82,84], hot-pressed [75] and flash-welded [75,76] aluminum-to-copper joints show that the mechanical and electrical properties are significantly affected by the formation and growth of intermetallics at the joint interface. It was shown that, when the total width of intermetallic phases exceeds 2–5 μm the aluminum-to-copper joint rapidly loses its mechanical integrity [85,86].

Figure 5.46a depicts the intermetallic phases formed at the interface of an aluminum-copper contact while in Figure 5.46b is shown the nanohardness traverse of the intermetallic phases formed at the aluminum-copper interface in samples diffusion annealed.

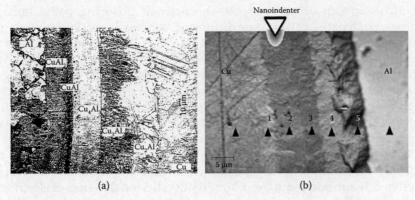

FIGURE 5.46
(a) Microstructure of the intermetallic phases formed at the interface of Al–Cu contact under the influence of thermal gradient; (b) Nanohardness traverse of the intermetallic phase formed at the copper-aluminum joint interface.

TABLE 5.6

Important Characteristics of Intermetallic Phases Formed in Bimetallic Al–Cu

Phase	Symbol	Composition	Cu (wt%)	Al (wt%)	Hardness ×102 (N/mm²)	Resistivity (µΩ cm)	D_o^2 (cm/s)	Q (kcal/mol)
Phase 1	γ_2	Cu_2Al	80	20	3.5	14.2	3.2×10^{-2}	31.6
Phase 2	δ	Cu_3Al_2	78	22	18.0	13.4	2.6×10^{-1}	33.5
Phase 3	ζ_2	Cu_4Al_3	75	25	62.4	12.2	2.7×10^6	61.2
Phase 4	η_2	$CuAl$	70	30	64.8	11.4	1.7×10^{-6}	19.6
Phase 5	θ	$CuAl_2$	55	45	41.3	8	9.1×10^{-3}	29.3

The composition and characteristics of intermetallic phases formed in the aluminum-to-copper joints are shown in Table 5.6. To illustrate the importance of interdiffusion and the formation of intermetallic compounds, let us use the following example. The process of diffusion is controlled by the following expression:

$$D_1 = D_o \exp(-Q/RT_1) D_2 = D = D_o \exp(-Q/RT) \tag{5.15}$$

where D_o is a constant, Q is the activation energy for diffusion and R is the universal gas constant. In the case of aluminum-copper, the activation energy for diffusion of aluminum is $Q = 40$ kcal/mole. Hence the diffusion rates at temperatures T_1 and T_2 ($T_2 = T_1 + \Delta T$) will be

$$D_o \exp(-Q/RT_2) D_2/D_1 = \exp Q/R(1/T_1 - 1/T_2) \tag{5.16}$$

Using Equation 5.16 the temperature rise required to double the diffusion rate $D_2/D_1 = 2$ at $T_1 = 60°C$ is calculated to be $\Delta T = 4°C$. Bearing in mind the fact that, owing to the current constriction, the *a*-spot may develop a higher temperature, the diffusion rates at the *a*-spot can be considerably higher than in the bulk. To prove this point, let us assume that the *a*-spot is at $T_2 = 300°C$ and the bulk at $T_1 = 60°C$. Then the calculated diffusion rate at the *a*-spot will be 10^{12} faster than that in the bulk. Under these conditions, the formation of intermetallics is very likely to happen.

Accelerated diffusion can occur via some short-circuit diffusing paths rather than through the lattice. Such paths, called "pipe diffusion," can be dislocations, grain and sub-grain boundaries. Diffusion along these paths is found to be considerably faster (a few orders of magnitude) than diffusion of the same species through the lattice [87]. Another mechanism that can enhance mass transfer across the contact sport is electromigration. Runde [88] has shown that high-current density can provoke mass transport by electromigration through the contact spots. As a result, accumulation of vacancies at some distance inside the contact will occur, leading to the formation of vacancy clusters and voids which, in turn, will reduce the mechanical strength of the contact area.

Gusak and Gurov [89] and Pimenov et al. [90] who have shown that the current has a considerable effect not only on diffusion but also on the kinetics of nucleation and formation of phases in bimetallic systems. Braunovic and Alexandrov [91] investigated the effect of electric current on the morphology and the dynamics of formation of intermetallic compounds of bimetallic friction-welded aluminum-copper joints, see Table 5.6. The establishment and growth of intermetallic compounds were analyzed in the temperature range 200°C–500°C realized by heating Al–Cu joints with an AC current of different intensities (400–1000 A). It was demonstrated that electrical current accelerates the kinetics of formation of intermetallic phases and significantly alters their morphology in bimetallic aluminum-to-copper friction-welded joints.

The growth kinetics of intermetallic phases under the influence of electrical current is much higher than under diffusion annealing in a temperature gradient. The growth rate of the intermetallic phases was determined by measuring the thicknesses of the interdiffusion layers (x) after selected time intervals (t) at each of the diffusion-annealing temperatures using the following expression (see Section 1.3.3):

$$x^2 = Dt \tag{5.17}$$

where D is the interdiffusion rate constant at the selected temperature. The thicknesses of intermetallic phases formed by diffusion annealing by an electrical current and in a temperature gradient are given in Tables 5.7 and 5.8. The activation energy characterizing the rate of formation of the intermetallic phases is given as

$$D = D_o \exp\left(-Q/RT\right) \tag{5.18}$$

where D_o is a constant, Q the activation energy, T the temperature and R the universal gas constant. The reaction rate constant (D) calculated from the slope of $(D)^{1/2}$ of each curve is plotted versus $1/T$ in Figure 5.47. In both cases, the rate of formation of intermetallic

TABLE 5.7

Thickness (μm) of Intermetallic Phases Formed in Aluminium–Copper Bimetallic Joints After Diffusion Annealing by an Electric Current

Time (Hours)	Diffusion Annealed by Electric Current						
	200°C	250°C	300°C	350°C	400°C	450°C	500°C
1	0.3	0.5	1	3.5	5.5	10	20
2	0.45	0.7	1.5	5	10	25	30
5	0.7	1.2	3	8	15	30	50
24	1.6	2.5	5	17	40	65	100
D (cm²/s)	3.0×10^{-13}	7.3×10^{-13}	3.0×10^{-12}	3.3×10^{-11}	2.1×10^{-10}	4.8×10^{-10}	1.2×10^{-9}
Q/D_o	$Q = 13.2$ kcal/mole		$D_o = 5.9 \times 10^{-7}$		$Q = 26.5$ kcal/mole		$D_o = 3.3 \times 10^{-2}$

TABLE 5.8

Thickness (μm) of Intermetallic Phases Formed in Aluminum–Copper Bimetallic Joints After Diffusion Annealing in a Furnace

| Time (Hours) | Diffusion Annealed in Furnace | | | | | |
	250°C	325°C	380°C	425°C	470°C	520°C
1	0.5	1	3.5	9	10	20
5	1	2	8.6	18	20	55
24	1.5	4	22	30	60	102
120	3	10	30	60	120	250
D (cm²/s)	1.6×10^{-13}	1.3×10^{-12}	1.9×10^{-11}	6.9×10^{-11}	3.5×10^{-10}	1.4×10^{-9}
Q/D_o	$Q = 17.2$ kcal/mole		$D_o = 2.2 \times 10^{-6}$	$Q = 32.7$ kcal/mole		$D_o = 0.8$ (cm²/s)

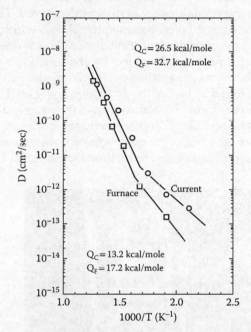

FIGURE 5.47

Arrhenius plot of the rate constant for growth of the diffusion layers in aluminum–copper bimetallic contacts diffusion annealed in temperature and electrical field gradients.

phases in samples diffusion-annealed in an electrical field gradient could not be identified by the single activation energy over the entire temperature range studied but rather by considering the two separate areas, that is, 200°C–300°C and 350°C–500°C.

The activation energy corresponding to temperatures below approximately 350°C is considerably lower ($Q_C = 13.3$ kcal/mol) than that obtained for higher temperatures ($Q_C = 20.5$ kcal/mol). It is also important to note that for the same temperature range, these values are significantly lower than those obtained for the samples diffusion annealed in a furnace, that is, $Q = 17.2$ kcal/mol and $Q = 32.2$ kcal/mol. The same type of behavior was reported by Timsit [92] in aluminum-brass contacts diffusion annealed in a temperature gradient. The activation energies in the temperature range 150°C–300°C and 350°C–450°C were almost identical to those measured in the Al–Cu bulk diffusion couples. It was demonstrated that both the pace and the total diffusion bandwidth of the intermetallic phases formed under the influence of electric current are higher than in samples diffusion-annealed in furnaces. It is interesting to note, that these results are in contrast with those obtained in the case of roll-bonded aluminum-brass couples since no effect of electrical current on the intermetallic growth was found in this system [93].

The most characteristic feature of the microstructure of the Al–Cu interface after diffusion annealing by an electric current is that, over the entire temperature range investigated, the interdiffusion layer consisted basically of four major bands (Cu_2Al, Cu_4Al_3 CuAl and $CuAl_2$), as seen in Figure 5.48. The compositions formed are given in Table 5.6. The impaired mechanical integrity of the Al–Cu bimetallic joints treated by an electric current is clearly evidenced by extensive cracking not only across the whole intermetallic bandwidth but also within different phases and at a neighboring interface as seen in Figure 5.49. It is clear that complex structural processes occur at the contact interface during the action of the electric current, following which the mode, the kinetics of formation, and the development of the intermetallics are significantly altered. The question immediately arises as to whether the formation of intermetallic phases within the allowable operating temperature range in an electrical connection, generally reckoned to be 100°C–150°C, is sufficiently fast to produce an appreciable effect on the functioning of a connector that otherwise follows the requirements generally determined by the existing standards.

Using the experimentally found values of the activation energies (Figure 5.47), the thicknesses of the intermetallic layers formed in aluminum-copper were calculated for different time intervals at 100°C, 150°C, and 200°C and are depicted in Figure 5.50. Since electrical connections operating under network overload conditions often experience temperatures of 200°C and above, these temperatures were included in the calculation. It is clear that

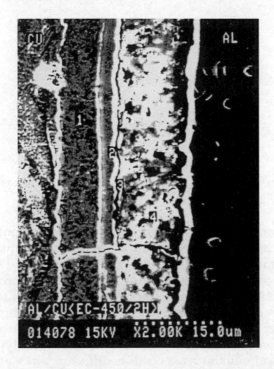

FIGURE 5.48
SEM micrograph of the intermetallic phases formed in the bimetallic Al–Cu contacts after 2 hours of diffusion annealing at 450°C by an electrical current.

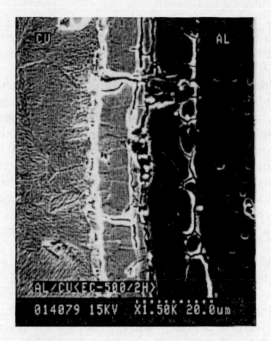

FIGURE 5.49
Cracks formed in bimetallic layers illustrating the fragility of intermetallic phases formed at bimetallic Al–Cu contact. Microhardness of copper and aluminum are respectively 42 and 38 kg/mm².

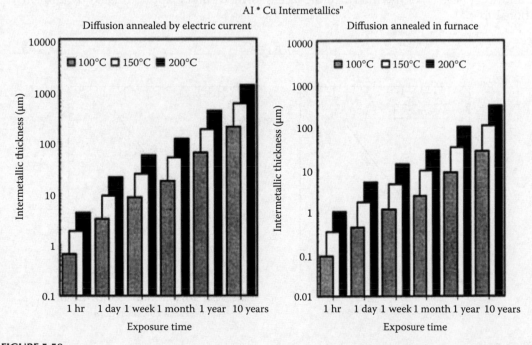

FIGURE 5.50
Calculated thicknesses of the intermetallic phases formed in Al–Cu bimetallic contacts following diffusion-annealing in thermal and electrical field gradients for different times at 100°C, 150°C, and 200°C.

at 100°C, the growth kinetics is relatively slow. However, at 150°C and 200°C, the rate of intermetallic growth is quite appreciable. It is interesting to note that the rate of intermetallic growth in samples treated by an electric current is roughly twice as high as that of samples diffusion-annealed in a furnace.

The mechanical properties of aluminium–copper are strongly affected by the presence of the intermetallics when their thickness at the interface reaches the critical value of 2 μm. At this thickness, the interface between the two metals in contact becomes brittle, therefore creating the interface highly porous, more susceptible to adverse environmental effects and generation of numerous fissures in the interdiffusion layer [94]. From Figure 5.50 it appears that the critical thickness of 2 μm can be reached after a few days of exposure to 100°C produced by the electric current. If such is the case, this finding can have serious implications for the mechanical integrity of Al–Cu electrical connections. The effect of electric current on the formation and growth of intermetallic phases therefore deserves more attention because the electrical connections are constantly subjected to the action of electric current.

The formation of intermetallic phases has an extremely detrimental effect on the electrical resistance, as demonstrated by dramatic increases in the contact resistance. A direct consequence of higher resistance is heating of the contact, which results in an increased rate of formation of intermetallic phases and as well as other degradation processes such as creep, stress relaxation, fretting, oxidation, etc. The effect of intermetallic phases formed at the Al–Cu interface on the contact resistance is shown in Figure 5.51.

$$R = (1/A)\rho_i x_i \tag{5.19}$$

Within the scatter of experimental data, the rate of resistance rise across the Al–Cu interface and thickness (x) of the intermetallic layers formed can be approximated by a linear relationship

$$(R_F - R_o)/R_o = Cx \tag{5.20}$$

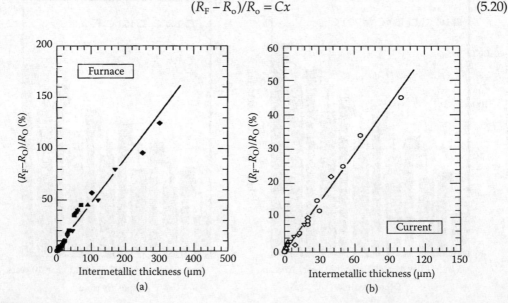

FIGURE 5.51
Growth of contact resistance with thickness of the Intermetallic phases formed in Al–Cu bimetallic contacts after diffusion annealing by a temperature and electrical field gradients.

Where R_o and R_F are respectively the resistances measured across the Al–Cu interface before and after diffusion treatment of an electric current; C is the proportionality constant having the value 0.4 for the samples diffusion annealed in a temperature gradient and 0.48 for the samples annealed by an electric current. The resistivity of the intermetallic layers in the aluminum-to-copper bimetallic joints can be expressed as

$$\rho = (1/x_t)\, [RA - (d/2)(\rho_{Al} + \rho_{Cu})] \tag{5.21}$$

where x_t is the thickness of the total intermetallic layer (microns); $\rho_{Al} = 2.4\ \mu\Omega$ cm and $\rho_{Cu} = 2.0\ \mu\Omega$ cm—are the resistivity of aluminum and copper respectively; $d = (D - x_t)$; $D = 0.3$ cm is the spacing between the potential probes. The calculated value of the resistivity of the total intermetallic layer in Al–Cu heat-treated solid-phase joints is $\rho_t = 18.5\ \mu\Omega$ cm. The resistivity values for the individual phases range between 8 and 14.2 $\mu\Omega$ cm (Table 5.6). The observed difference exceeds the experimental error indicating that there are additional contributions to the total resistivity.

It is believed that these contributions can be attributed to porosity, cracking, grain size changes and increased dislocation density in the diffusion zone. These defects constrict the current flow across the interface and thus considerably reduce the effective contact area. As a result, the resistivity of the total intermetallic layer increases. One plausible explanation is that the observed effect might be associated with the material transport by electromigration [94]. The transport phenomenon, first reported in 1861, and was termed as electromigration. Conceptually, electromigration may be considered as current-induced diffusion. In any region in the material where conditions exist that make an atomic jump in one direction more probable than another, a net mass flow may occur. This process of current-induced mass flow takes place only in the immediate vicinity of the contact spots. In electromigration material transport occurs via interaction between the atoms of a conductor and a high density current of the order of 10^3–10^5 A/cm^2. The drift velocity varies strongly with temperature and is particularly high along grain boundaries, dislocations and other lattice defects. The migration velocity v of the atom is given by

$$v = (j\, \Delta Z^*\ eD)/kT \tag{5.22}$$

where j is the current density, Δ the electrical resistivity, Z^* the dimensionless term regarded as the effective valence or charge number of the ion, D the diffusion coefficient, k the Boltzmann constant and T the absolute temperature.

Modern studies [94,95] of aluminum electromigration have centered on the reliability of thin-film interconnections in microcircuit applications. Reduction of the conductor line width to enable the circuit density to be increased has brought the current density in aluminum thin-film "wiring" up to a range where electromigration failure (open circuit) has become a reliability concern. Investigations in this regard have generally focussed on DC applications. Although electromigration has not previously been associated with the deterioration of contact spots, the current density in the a-spots of a practical contact can be substantially higher than in the thin-film interconnections of an integrated circuit. It is therefore suggested that electromigration is of considerable importance for the reliability and degradation of bulk electric contacts.

Electromigration deterioration of a-spots in aluminum contacts has been studied by Runde [96] in Al–Al and Al–AlZn alloys using both AC and DC currents. It was shown that mass transport by electromigration appears to influence both the mechanical and the electrical properties of a contact spot. In aluminum contacts, the atoms tend to migrate toward the anode, and hence an accumulation of vacancies occurs a short distance inside the cathode contact member. The high vacancy concentration causes considerably lower mechanical strength and higher electrical resistivity in these areas. Degradation with DC was demonstrated, but AC results were inconclusive owing to the unstable electrical behavior of the specimens.

Aronstein [97,98] has investigated AC electromigration and compared the results with DC under otherwise identical conditions. The specimens tested were two-dimensional models of asperity contact junctions, made of solid metal configured to avoid experimental scatter owing to many of the variables inevitably encountered in experiments utilizing actual contact interfaces. In order to obtain the most reproducible and meaningful results, the actual metallic current-carrying area of the "*a*-spot" constriction was carefully controlled. The experimental results demonstrated electromigration failure with AC as well as DC. No significance was attached to the observed difference in median time-to-failure, since the sample size for each condition tested was relatively small. Electromigration in AC applications is expected in aluminum connections when current density through the *a*-spots is high. With AC, the aluminum atoms in *a*-spots at the contact interface may be considered to have double the number of available paths (relative to DC) by which to move out of the high-current-density neighborhood, but only half the current-application time in each direction. Conceptually, all other factors being equal, AC electromigration deterioration of *a*-spot should therefore be about the same as DC.

Electromigration is not expected under low-current density conditions. Therefore, a comparatively low density AC current ($<10^3$ A/cm^2) alone are unlikely to cause electromigration, thus, to be the operating mechanism responsible for the observed accelerated formation of the intermetallic phases in electrically heat-treated Al–Cu bimetallic joints. An alternative is that the accelerated formation of the intermetallic phases is a result of an enhanced diffusion of some species, most probably copper. The enhancement may result from the concentration of vacancies and interstitials exceeding their respective equilibrium values, thus increasing proportionally the contributions from the vacancies and interstitials to the diffusion rates.

Another possible mechanism is that accelerated diffusion occurs via some short-circuit diffusing paths rather than through the lattice. Such short diffusion paths, called "pipe diffusion" can be dislocations, grain and subgrain boundaries [87]. If "pipe diffusion" is the operating mechanism, then the kinetics of intermetallic phase formation and growth leads to the conclusion that the mobility of the diffusing species, probably copper, is much greater in the presence of an electric field than in the temperature gradient. Hence, it is believed that interaction between the applied electric field and lattice defects notably dislocations and grain boundaries, enhance the migration of the diffusing species along these short-circuit diffusion paths and thus accelerate the formation of intermetallics. The absence of certain phases in samples heat-treated by an electrical current can be attributed to the increased incubation time of some phases owing to the presence of an electrical field. In other words, the growth of these phases from their critical nuclei is suppressed by diffusion interaction with the critical nuclei of neighbouring phases having greater diffusion permeability [87].

5.4.13 Intermetallics in Copper–Tin Systems

In view of the fact that the power connections are operating at relatively high temperature, possibly 100°C, the formation of intermetallics at the copper-tin plating interface becomes an important issue. The formation and development of these phases will affect both electrical and mechanical properties of the contact interface. Figure 5.52a illustrates some typical morphological characteristics of the intermetallic phases formed during diffusion annealing by thermal gradient. The EDX elemental composition analysis of the intermetallic phases formed indicated that the phase 1 corresponds to Cu_3Sn while phase 2 to Cu_6Sn_5. Figure 5.52a shows the SEM image of these phases formed at the interfaces of tin-coated copper while Figure 5.52b shows the elemental X-ray line scan of Sn cross the tin-coating. Flexible connectors often referred to as expansion joints, jumpers, braids, braided shunts are widely used in power generating stations and substations. Their primary purpose is to establish continuous current transfer, for example, between rigid busbars, tubular conductors or current-carrying parts of other electrical equipment such a power meter, disconnect switches etc. Flexible connectors are created by compressing the copper or tinned copper ferules with inserted ends of stacks of copper or tinned copper laminates or bundles of braided wires. In order to maintain mechanical and electrical uniformity the sides of ferrule are often soldered. The characteristic feature of the intermetallic phases formed in tin-plated copper is that these phases are significantly harder than the base material (copper) thus making these susceptible to brittle facture. The composition, physical and mechanical characteristics of intermetallic phases formed in the tin-copper systems are shown in Table 5.9.

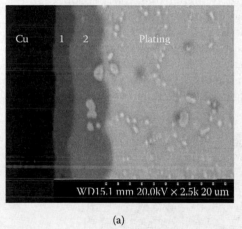

(a)

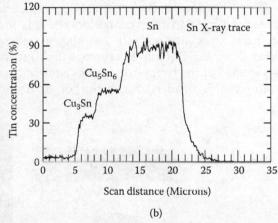

(b)

FIGURE 5.52
(a) SEM micrograph of the intermetallic phases formed at the copper-tin interface in tin-plated copper wires. Note the presence of two distinct phases (1) Cu_3Sn and (2) Cu_6Sn_5 (b) Tin x-ray scan across the copper-tin interface of tin-plated copper wires showing the phases Cu_3Sn and Cu_6Sn_5.

TABLE 5.9

Room Temperature Properties of Intermetallic Compounds, Copper and Tin

	Cu_6Sn_5	Cu_3Sn	Cu	Sn
Hardness VHN (MPa)	378.0	343.0	369	51
Elastic modulus (GPa)	85.56	108.3	115	47.0
Thermal expansion (PPM/°K)	16.3	19.0	16.5	23.0
Heat capacity (J/g/°K)	0.286	0.326	0.385	0.226
Resistivity (µΩ cm)	17.5	8.93	1.65	11.6
Density (g/cm³)	8.28	8.90	7.96	6.97
Thermal conductivity (W/m K)	34.1	70.4	401	66.8

5.4.13.1 *Example of Intermetallics Formation in Power Connections*

In flexible connectors made of bare copper wires or laminates, temperature raise increases the rate of oxidation of wires or laminates while in the tinned copper flexible connectors it accelerates the rate of formation of intermetallic phases at the tin-copper interfaces. In both cases the end results are the same—reduction of the current-carrying cross section area, overheating and connector failure. Typical examples of flexible tinned copper connectors, intact and failed connectors, are shown in Figure 5.53. A scrutiny of the copper–tin interface with new (intact) flexible connector reveals the presence of intermetallic phases. The resistance measured across the copper–tin interfaces of failed connector shows that their resistance values are almost half that of the intact flexible connector (Table 5.10). Lower bulk resistance of damaged connectors is a result of annealing of cold drawn copper laminates as the connector overheats. The difference in resistance measured across Cu–Sn–Cu interface and inside Cu–Sn intermetallic bands of intact and damaged connectors can be attributed to the presence of different amounts of fractured intermetallics and porosity in these zones. Higher concentration of fractured intermetallics and porosity reduce the effective conducting area and increase the bulk resistance.

The results of microhardness measurements, shown in Table 5.11 indicate clearly that the microhardness of the intermetallic phases is considerably higher than that of copper laminates (70–$100\,kg/mm^2$) and tin plating ($6\,kg/mm^2$). In addition, the microhardness of intermetallic phases formed in intact connectors is higher than that in a damaged one. The observed difference is associated with the morphology of the phases present in these samples. The results of Scanning Electron Microscope (SEM) surface analysis show that the intermetallic phases formed at the copper-tin interface not only in the damaged, but also the intact flexible connector (Figure 5.54) during the hot-dip tinning process. Detailed examination of the

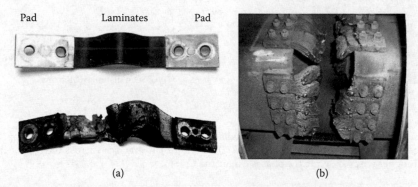

(a)   (b)

FIGURE 5.53
(a) Typical examples of flexible tinned copper connectors: intact and failed connector: (b) Example of failed flexible power connector.

TABLE 5.10

Resistance Measured in the Bulk of Connector Laminates, Across Cu–Sn Interface and Inside of Cu–Sn Intermetallic Phases

Sample	Laminate Bulk (µΩ)	Cu–Sn–Cu Interface µΩ	Inside Cu–Sn Band (µΩ)
Intact	3	9.35	16.78
Damaged	2.2	15.25	25.65

TABLE 5.11

Microhardness Measured in Copper Connector Laminates, within Tin Plating and Inside Cu–Sn Intermetallic Phases

Sample	Laminate Bulk (kg/mm²)	Tin Plating (kg/mm²)	Inside Intermetallics (kg/mm²)
Intact	100	6	345
Damaged	70	–	266

Intact connect

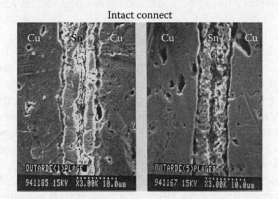

FIGURE 5.54
SEM micrographs and EDX spectra of intermetallic phases formed in intact and damaged flexible tinned copper connectors.

Amaged connector

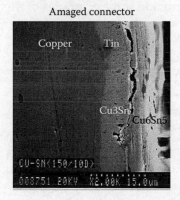

FIGURE 5.55
SEM micrograph of cracks at interface of intermetallic phases Cu_3Sn and Cu_6Sn_5.

intermetallic phases reveals considerable cracking at the copper–tin and tin–tin interfaces which is a clear indication of their low mechanical strength, hence their brittle nature. The brittleness of these phases is illustrated by the presence of fractured debris at these interfaces (Figure 5.55) and also by their substantially higher microhardness values.

5.4.14 Stress Relaxation and Creep

Creep, or cold flow, occurs when a metal is subjected to a constant external force over a period of time. The rate of creep depends on stress and temperature and is higher for aluminum than for copper. Stress relaxation also depends on time, temperature and stress but, unlike creep, is not accompanied by dimensional changes. It occurs at high stress levels and is evidenced by a reduction in the contact pressure owing to changes in metallurgical structure. The change from elastic to plastic strain has the effect of significantly reducing the residual contact pressure in the joints, resulting in increased contact resistance, possibly to the point of failure. The loss of initial contact pressure can be accelerated at elevated temperatures, making a loss of contact area in a relatively short time. Hence, excessive conductor deformation and high stresses produced by certain connector systems with no way of providing residual mechanical loading to the contact interface cause accelerated stress relaxation and eventual failure of a joint.

The gist of the metallurgical state and temperature on the strain relaxation of aluminum and copper was investigated by Naybour and Farrell [71,99] and Atermo [100]. It was shown that, for electrical grade aluminum (EC-1350 grade), increasing the temperature and the amount of hardening augmented the rate of stress relaxation. It was also revealed that stress relaxation in hard-drawn EC-grade aluminum conductors was anisotropic, being much faster the transverse than in the longitudinal direction, and a great deal faster in hard-drawn than annealed wire. This is shown in Figure 5.56. Mounting evidence indicates that mechanical properties such as creep [101–103], stress relaxation [104–106], flow stress [107–109], metal forming [109–117], etc. can be significantly affected by the action of electric current. This so-called electroplasticity manifests itself in a spectacular increase in the ductility of a material. It is generally believed that the outcome is a result of the interaction between electrons and dislocations, although some controversy still exists regarding the precise nature of this electron-dislocation interaction. Even though there is a scarcity of data, I shall briefly review the nature of the effect of electric current on and provide some examples of this effect on stress relaxation and creep of some metals.

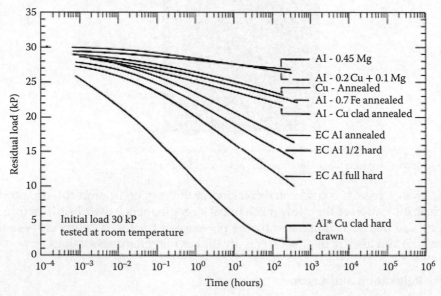

FIGURE 5.56
Effect of metallurgical state on the stress relaxation of aluminum.

5.4.15 Nature of the Effect of Electric Current

It is now well established that the electrons in a metal can exert a drag on dislocations, particularly those moving at high velocity at very low temperatures [94–96]. The electron drag coefficient B_{ew} is given by

$$(f/l) = \vartheta\ b = B_{ew}V_d \tag{5.23}$$

where (f/l) is the force per unit length acting on the dislocation, ϑ is the resolved shear stress, b the Burgers vector and V_d the dislocation velocity. The electron drag coefficient Bew is of the order of 10^{-5} dyne/cm^2 [103,104]. The interaction of moving electrons with the dislocations led Troitskii and Likhtman [115] to conclude that drift electrons can exert a force ("electron wind") on dislocations and, therefore, that such a force should occur during the passage of an electric current through a metal when plastically deformed. A comprehensive reassessment of the extensive study of Russian scientists on the electroplastic effect is given by Trotsky and Spitsyn [104]. Theoretical considerations of the power exerted by drift electrons on dislocations are given in the papers by Kravchenko [116], Klimov et al. [117] and, more recently Roschupkin et al. [118]. The force is considered to be proportional to the difference between the drift electron velocity v_e, and the dislocation velocity v_d thus indicating that the electron wind force (f_{ew}/l) is proportional to the current density,

$$\vartheta_{ew}b = K_{ew}J \tag{5.24}$$

where ϑ_{ew} is the stress acting on the dislocation owing to an electron wind, b the Burgers vector, K_{ew} the electron force coefficient and J the current density. These theories give a different magnitude of the constant K_{ew} which, in turn, is reflected in the value of the electron drag coefficient B_{ew} given by:

$$B_{ew} = \vartheta_{ew}b/v_e = K_{ew}e\ n \tag{5.25}$$

where e is the electron charge and n the electron density. The values of B_{ew} predicted by Kravchenko for metals are of the order of 10^{-6}dyn-s/cm^2, those by Klimov et al. 10^{-6} to 10^{-4} dyn-s/cm^2 and those by Roschupkin et al., 10^{-5} to 10^{-3} dyn-s/cm^2. These values are all within the range of those measured experimentally for B_{ew} although the theory of Klimov et al. gives the values in best accord with those most generally accepted [117,118].

In recent years, the electroplastic effect has been studied in the United States by Conrad and coworkers [119–125] who have carried out a series of investigations into the effects of single high-density DC pulses (~10^3 A/mm^2 for ~60 μs duration) on the flow stress of a number of polycrystalline metals (Cu, Pb, Sn, Fe and Ti), representing a range of crystal structures and valences. The primary objectives of the work of this group were to determine:

- The magnitude of the drift electron-dislocation interaction which occurs with the application of a current pulse during plastic flow,
- The physical basis of this interaction.

In order to ascertain the magnitude of the electron-dislocation interaction, attention was given to separating the side effects of the current, such as Joule heating, pinch, skin and magnetostrictive effects. This was an indispensable first step since there was considerable skepticism [126] as to whether the electroplastic effect is a consequence of the electron-dislocation interaction. It has been argued that the observed changes in the mechanical properties such as flow stress, creep etc. are caused by a side effect of current such as Joule heating and pinch effect. However, Cao et al. [127] has clearly demonstrated that the observed changes in the mechanical properties cannot be explained in terms of the current side effects but rather are the result of the electron-dislocation interaction.

5.4.16 Effect of Electric Current on Stress Relaxation

Troitskii and co-workers [104] have used stress relaxation to study the electroplastic effect in a number of metals such as zinc, lead and cadmium under the influence of high-density (10^4 A/cm^2) current pulses. The frequency of the pulsing current was 100 Hz and the pulse duration 65 μs. They found that the application of current increases the rate of stress relaxation and significantly affects its onset; see Figure 5.57. Figure 5.58a depicts stress relaxation $\Delta\tau$ with time with and without current. Figure 5.58b illustrates the rate of stress relaxation $\Delta\tau$ with and without current as a function of applied stress. Note that the onset of stress relaxation with and without electric current is different: it is significantly lower in the samples under the influence of electric current.

The mechanical properties of metals may also be tempted by low-density electric current as reported by Silveira and co-workers [128,129]. They have demonstrated that the rate of stress relaxation of polycrystalline Al and Cu near 0.5 T_m (where T_m is the melting temperature) can be increased with the application of a small (1.6 A/mm^2 continuous AC or DC) current, the effect being greater for the DC current. The effect decreased with the number of current cycling times. Also, they found that the DC current altered the dislocation arrangement in the Cu specimens; partial destruction of the cell structure occurred even in the first relaxation cycle. No change in the dislocation structure was observed for Al. The effect of DC electric current on stress relaxation is shown in Figure 5.59. Braunovic [130] compared the stress relaxation of aluminum conductor wire at room temperature and low initial stress (20 N) under the influence of a low-density electric current (3 A/mm^2) with that of a wire

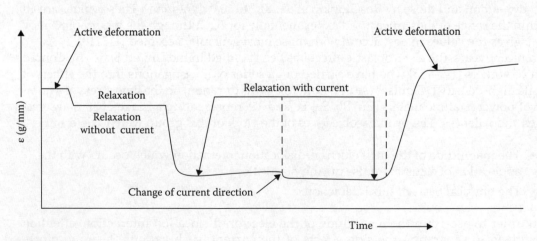

FIGURE 5.57
Stress relaxation of zinc crystal without and the intercept when the current was applied.

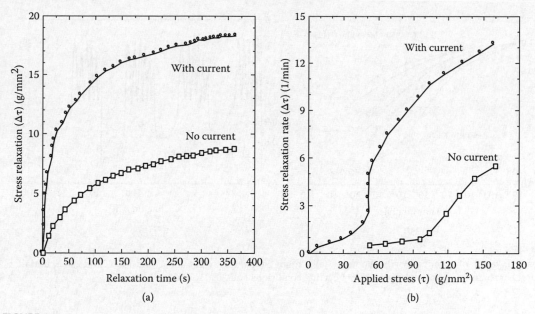

FIGURE 5.58
(a) Variation of stress relaxation with time with and without current; (b) variation of the rate of stress relaxation with applied stress for samples with and without current.

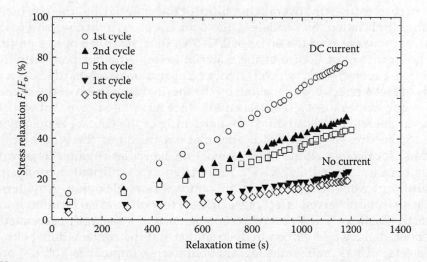

FIGURE 5.59
Effect of DC electric current on stress relaxation of copper.

relaxing at higher temperature (150°C) and higher initial contact stress (260 N); see Figure 5.60. Here the relative force changes (F_i/F_o) versus time, where F_i is the contact force at the instant i and Fo. is the initial contact force. It is clear that the stress relaxation of an aluminum wire conductor at elevated temperature (150°C) and under current cycling conditions is similar. In other words, the initial contact force ($F_i = 20$ N) of a wire subjected to current cycling decreases at virtually the same rate as a wire subjected to stress relaxation at 150°C

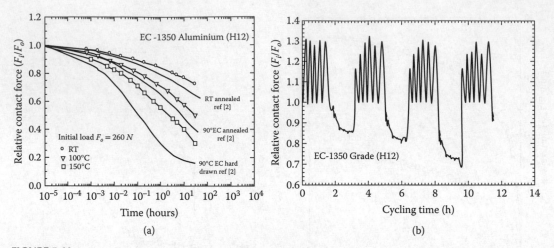

FIGURE 5.60
Stress relaxation of aluminum wire conductor (a) at room temperature (RT), 100°C and 150°C and 260 N initial load; (b) at room temperature under current cycling conditions and 20 N initial load.

and higher initial stress $F_o = 260$ N. The observed effect of current cycling on stress relaxation cannot be justified in terms of the temperature effect alone, since the temperature rise produced by the passage of a low-density current through the wire conductor is merely 2°C–3°C.

One possible explanation is that current cycling causes thermal fatigue in aluminum wires which in turn reduces the overall durability of the material and thereby increases the rate of stress relaxation. Nevertheless, the temperature rise in the wire resulting from the current passage is of the order of 2°C–3°C, thus, insufficient to significantly change the properties or structure of the material (wire). Another possibility is that during cycling the contact zone, which is under compression, work-hardens as a result of increased contact stresses brought about by the thermal expansion of the wire. This, in turn, can augment the rate of stress relaxation, since it is well known that work hardening increases the rate of relaxation. Work-hardening of the contact zone may occur in response to the stresses generated by thermal expansion of the wire being high enough to cause localized yielding of the contacting asperities. Again this is unlikely because a temperature exceeding 2°C–3°C cannot generate sufficient thermal stresses to create significant work-hardening and, therefore, stress relaxation. An alternative explanation is that the observed effect of electric current on stress relaxation is associated with electroplasticity. The stress relaxation of a metal is generally associated with the arrangement, density and motion of dislocations and their interactions with other structural defects such as grain and subgrain boundaries, impurities, solutes, precipitates, etc. Hence, when the electric current passes across the wire, it will not only cause heating and, hence, the elaboration of the wire but also weaken the binding forces between dislocations and obstacles impeding dislocation motion, which otherwise cannot be overcome by thermal activation alone. As a result, nucleation and multiplication of dislocations will be alleviated this, in turn, will increase the density of mobile dislocations and alter their arrangement. Repeated passage (current cycling) of current through the wire will free more and more dislocations from pinning defects, enhance mobility and reduce their density in the material, thus, resulting in an increased rate of stress relaxation. It should be pointed out, however, that the effect of temperature on the processes involved was not excluded. However, because of the relatively low heating generated by the passing current, thermal activation alone is insufficient to significantly affect the dynamics of dislocation motion and their arrangements in the material.

5.4.17 Creep

The elementary processes of plastic deformation of metal during creep can be described by the phenomena related to the dislocation dynamics and viscous diffusion flow. The latter takes place at temperatures close to melting point. At medium temperatures and low applied stresses the creep rate gradually diminishes and vanishes, as reported by the logarithmic law

$$\varepsilon = \varepsilon_o + \alpha \ln(\beta t + 1) \tag{5.26}$$

where α and β are constants and ε_o is the initial deformation. The logarithmic law of creep is valid for metals and alloys with different types of crystal lattice at temperatures below (0.2–0.3) T_m, (T_m is the melting temperature) and can be used to study the mechanisms responsible for the rate of plastic deformation of a metal. At low temperatures, creep is dominated by the dislocation mechanism which reflects the combined action of thermal fluctuations and applied stress, resulting in an unimpeded motion of dislocations throughout the lattice. Figure 5.61 depicts the evolution of creep in power connections and accompanying processes occurring in the contact. The effect of electric current on creep in different metals has been studied extensively by Klypin [102] at a low-density DC current, (0.25 A/mm^2) and different temperatures. It was demonstrated that the creep rate is significantly affected by the action of the electric current, the more so at lower creep rates and higher current densities (see Table 5.12). Increasing the number of current applications (cycles) was found to reduce the effect of electric current. A typical creep curve for copper tested at 400°C and a stress $\sigma = 6$ kgf/mm^2 is shown in Figure 5.62. It is clear that when the current is applied the creep rate sharply increases.

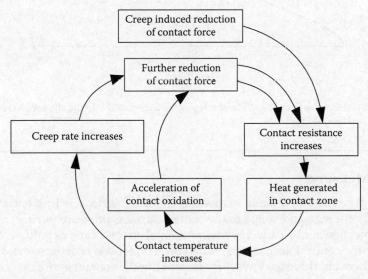

FIGURE 5.61
Schematic of creep evolution.

TABLE 5.12

Effect of DC Current (0.15 A/mm²) on the Creep of Different Materials

Material	Temperature (°C)	Stress (kgf/mm²)	Creep Rate (%/h)		Creep Strain Polarity (+)	Creep Strain Polarity (−)
			Current OFF	Current ON		
Copper	400	6.0	0.55	1.80	0.50	0.20
	400	4.0	0.20	1.20	0.20	0.05
	600	2.0	0.30	1.00	0.20	0.0
	800	0.2	0.10	1.00	0.20	0.15
Nickel	800	5.0	1.00	−0.30	−0.30	−0.25
	800	3.0	0.06			
	500	20.0	0.06	2.20	0.90	0.12
Cobalt	700	10.0	0.50	0.80	0.90	0.20
	700	6.0	0.30	1.00	0.25	0.16
	700	0.12	0.05	0.36	0.08	0.08
	800	5.00	0.60	1.20	0.25	0.12
Titanium	600	6.00	0.15	1.20	0.45	0.0
Aluminum	300	2.00	0.03	1.50	0.30	0.16
	500	20.00	0.50	3.60	0.50	-
Steel	500	8.00	0.20	1.70	0.30	0.20
	600	5.00	0.30	1.80	0.45	0.40

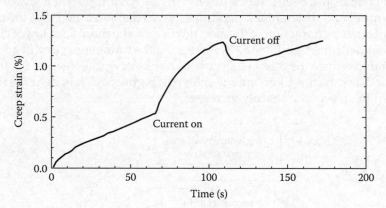

FIGURE 5.62

Effect of current on the creep of copper at 400°C and stress $\sigma = 6$ kgf/mm². Note the change in creep rate when the current was switched ON and OFF.

5.5 Palliative Measures

The uninterrupted passage of current across the contact interface is influenced by many parameters such as the size and shape of the contact surface, pressure (torque) and contact resistance and protection against adverse environmental effects. Various palliative measures are used to ensure the continuous passage of current, A brief review of those most frequently used as follows. There are three basic ways for improving the contact performance: (a) development of new contact materials, coatings, and lubricants; (b) special techniques involving the structure and state of the interface; (c) improvement in contact design (Figure 5.63).

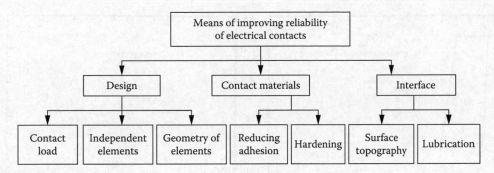

FIGURE 5.63
Main means of improving reliability of sliding contacts.

5.5.1 Contact Area

The size of the contact area is generally defined by the stiffness of the contacting members, applied force, and the rated current density. It should be sufficiently large to prevent the rise of the contact interface temperature under normal and emergency conditions. However, the contact temperature itself is not a critical parameter since it is an effect rather than a cause of the processes occurring in a joint, being a function of the current density, the geometrical dimensions of the contact, and the voltage drop across the contact. The contact temperature can exceed that in the bulk of a conductor or connector without causing electrical instability at the contact interface. On the other hand, the contact voltage cannot increase to even moderate values without causing the contact temperature to exceed that of the conductor or connector bulk temperature. It is therefore imperative that changes in the contact temperature and voltage with the connector operating time remain very small; ideally <10 mV. This requirement can be met providing the area of real contact is sufficiently large so that, despite initial and long-term deterioration, there is still a reserve of contact spots (a-spots) to ensure that the overheating conditions in the joints are not reached.

Practice and published experimental evidence has shown that mechanical abrasion (brushing) and lubrication of aluminum is the simplest and most efficient method of obtaining a large number of contact points [2,9]. In order to prevent oxidation of the exposed metal, brushing should be done with a contact aid compound (grease) applied to the contact surfaces. Serration of the connector contact surfaces is a very efficient method and most desirable when splicing conductors because it assures a large contact surface area between connector and conductor. Figure 5.64 shows the result of surface preparation and lubrication on the contact resistance of aluminum bolted joints [14,131].

In the event of a bolted joint, it has been indicated that current lines are distorted at the joints as a result of which, the resistance of even a perfectly made overlapping joint (no interface resistance) is higher than that of a bar of the same duration as the joint. This is known as "streamline effect" and is determined by the ratio between the overlap and busbar thickness. Figure 5.65 illustrates the action of current flowing through an overlapping busbars of uniform width (w), and thickness (t). Thus, the strip is shaped like a lapped joint but is void of contact surfaces along AB, except through part of its length AB where the thickness is $2t$. A current (I), is passed through the busbar from end to end. The voltage drop is measured between A and B, and between B and C whereby distance AB to BC. Hence,

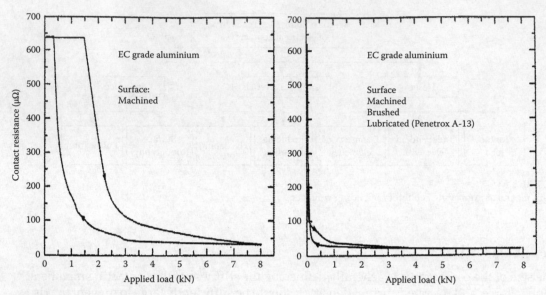

FIGURE 5.64
Effect of contact load and surface finish on the contact resistance of aluminum.

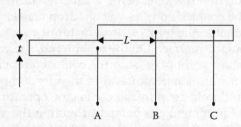

FIGURE 5.65
Schematic of an overlap joint without contact interface.

$$V_{AB} = RI \text{ and } V_{BC} = R_B I \tag{5.27}$$

$$e = V_{AB}/V_{BC} = R/R_B$$

where e is the streamline effect, R is the total resistance of a joint and R_B is resistance of equal length of a busbar and is given as:

$$R_B = \rho(L/wt) \tag{5.28}$$

Hence, the streamline effect e becomes a function of the ratio L/t, i.e

$$e = R(w/\rho)\,(1/L/t) \tag{5.29}$$

Melsom and Booth et al. [132] have tested a number of different busbar combinations and obtained the results shown in Figure 5.66 which allows calculating the resistance caused by the streamline effect of a perfectly made overlap in relation to the resistance of an equal length of a single busbar. It appears that the streamline effect rapidly decreases until the overlap/thickness ratio reaches a value of 2 when its decrease is slowed down and practically

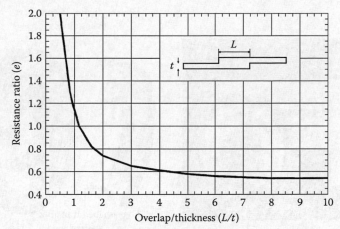

FIGURE 5.66
Effect of current distortion ("streamline effect") on the contact resistance of an overlapping joint.

stabilized upon passing a value of 6. In other words, upon reaching this value, the current distortion is minimized and the current lines become parallels. Consequently, the optimal overlap length in a bolted joint should be 5–7 times the busbar thickness. However, to meet all requirements, and for the sake of convenience, it is common practice to make overlaps equal to the width of the bus bar. Although this procedure is usually satisfactory for busbars of ordinary dimensions, it may not provide sufficient joint contact area on busbars that have a high thickness-to-width ratio. Hence, as a rule of thumb, the minimum overlap should be from 8 to 10 times the bar thickness. Furthermore, since the actual area of contact is much less than the total area of overlap, the determination of overlap on the basis of the width alone is reasonably safe practice only if the current density in the contact surface does not exceed 1/3 to 1/4 of the current density in the busbar cross section.

Donatti [133] has studied extensively bolted overlapping joints and found that the current tends to traverse the contact surface at the end points of the joint while the passage of current in the intermediate region is minimal. It was also demonstrated that when the two conductors of substantially different resistance values are in contact, the current will pass across the contact interface almost exclusively at the end of the better conductor, the remaining contact interface being almost inactive. Therefore, to reduce current density in the joints, it is useless to increase the contact surface by expanding its length, since this does not alter the current distribution to an appreciable extent. Rather, increasing the width of the contact surface was found to be more beneficial. Jackson has shown [14] that in a bolted joint, unless it is grown in a perfect manner, it is not possible to hold a uniform distribution of the current on the contact surface. Furthermore, it was also indicated that the current tends to traverse the contact surface at the end points of the joint while the intermediate area is used very little for the passage of current. This feature was exploited in the edge-shaped transition washer as seen in Figure 5.67 [134]. In the case of bolted connections, it was shown that the contact area can be increased by changing the busbar design. In other words, cutting the slots in the busbar in a manner as shown in Figure 5.68, the actual surface area of a joint can be increased by 1.5–1.7 times that of without slots. The contact resistance of a joint configuration without slots (A) is 30–40% inferior to that of (B), and is mechanically and electrically more stable under current cycling conditions [135]. The beneficial effect of sectioning the busbar is attributed to a uniform contact pressure distribution under the bolt and a larger contact area.

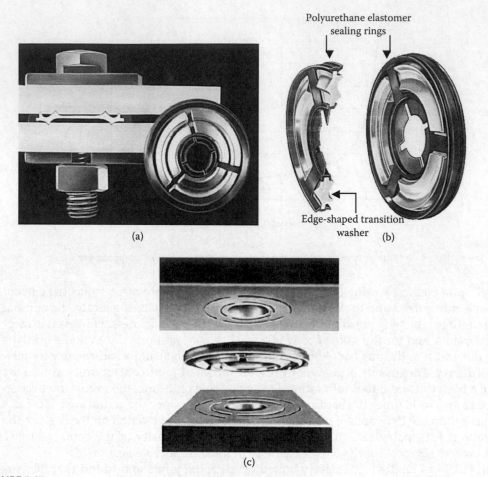

Polyurethane elastomer
sealing rings

Edge-shaped transition
washer

(a)   (b)

(c)

FIGURE 5.67
Pfisterer Transition washer (Elast contact disk) (a) Cross section of the busbars with the transitions washer showing the contact points, (b) Cross section of a transition washer, (c) Busbar contact surface showing the impressions left by the sharp edges of transition washer.

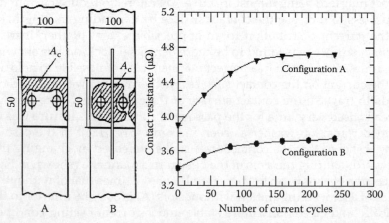

FIGURE 5.68
Contact area of a joint with slots (B) is 1.5 to 1.7 times larger than without slots.

This idea has been explored by Braunovic [136] in the case of high voltage (>700 kV) power connectors used for connecting stranded 4000 MCM conductors to a variety of power equipment at the sub-station site. The connectors used to be with intact current-carrying pads (no slots) and with pads modified by cutting slots seen in Figure 5.69.

The contact resistance was derived from the voltage drop measured between the potential leads positioned on the connector pads and busbars as shown in Figure 5.69. Although this figure depicts only the location of the voltage drop leads for the connectors with slots, the same system was used for the connectors without slots. The results in Table 5.13 show the mean values of joint contact resistance of connectors with four- and six-bolt configurations whereas Table 5.14 shows calculated composite radius of a-spot (a_c) and composite contact area (A_c). To estimate the size of contact zone where conduction takes place, this zone can be envisaged as a large circular composite area comprising several discrete small areas (a-spots) with composite electrical resistance R_c (see Equation 5.7). Where ρ is the bulk resistivity of the connector aluminum alloy and a_c is the composite radius. Hence, using the resistivity value $\rho = 50$ nΩ m typical for the aluminum commonly used for this type of connectors, the composite radius a_c and real conducting area A_c; see Section 5.2.2. From the results presented in Tables 5.13 and 5.14 it is clear that sectioning the overlapping

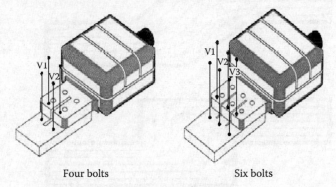

Four bolts   Six bolts

FIGURE 5.69
Schematic of the bolted joint configurations and positioning of the voltage probe leads for the contact resistance measurements.

TABLE 5.13

Mean Values of Joint Contact Resistance of Connectors with Four- and Six-Bolt Configuration

Condition	No Slots	With Slots	R_s/R_{ns}
Four-Bolt Configuration			
As received	9.19	6.70	0.73
Brushed	5.48	3.97	0.72
Brushed + Lubricated	0.39	0.27	0.69
Six-Bolt Configuration			
As received	22.52	15.20	0.68
Machined	2.74	1.55	0.56
Machined + Lubricated	0.32	0.18	0.56

R_s— Contact resistance of contacting members with slots.
R_{ns}—Contact resistance of contacting members without slots.

TABLE 5.14

Calculated Values of Joint Contact Resistance for
Composite Radius (a_c) of Composite Contact Area (A_c)

Surface Finish	No Slots		With Slots	
	a_c	A_c	a_c	A_c
Four-Bolt Configuration				
As received	0.27	0.23	0.37	0.43
Brushed	0.46	0.66	0.63	1.25
Brushed + Lubricated	6.29	1124.21	8.85	245.92
Six-Bolt Confuguration				
As received	0.11	0.04	0.16	0.08
Machined	0.91	2.60	1.61	8.14
Machined + Lubricated	7.72	187.14	13.89	605.81

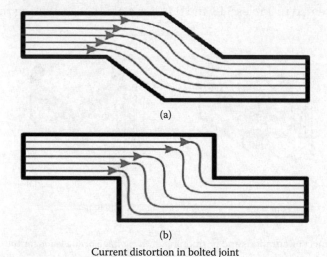

(a)

(b)

Current distortion in bolted joint

FIGURE 5.70
Current lines of an idealized joint without (a) and with (b) slanted busbar ends.

bolted joints combined with machining and lubricating the contact surface assured very pronounced enlargement of the actual conducting area and hence the lowest contact resistance. The slotted approach may prevent bolts from loosening with the improved contact area. On the other hand, connectors whose current-carrying parts (pads) were not given any surface treatment nor were sectioned, had extremely small conducting area and, thus, highest contact resistance.

Boychenko and Dzektster [135] have shown that the current distortion in the joint can be reduced by slanting the busbar ends at 45°; see Figure 5.70. Slanting the busbar ends will make the current flow across the joint more uniform. The initial contact resistance of bolted joints is 15% lower; see Figure 5.71. Also, the rate of contact resistance increase in a joint with slanted busbars (a) during current-cycling is 1.3–1.5 times slower (Figure 5.71) than that without slanted busbars (b). Lower contact resistance in a joint with slanted busbar is owing to uniform current distribution at the contact interface which eliminates localized overheating, thus improving the contact thermal stability. The contact resistance

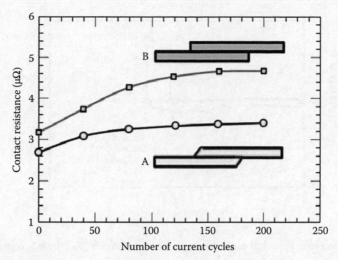

FIGURE 5.71
Effect of current-cycling on contact resistance of bolted joints with and without slanted busbar ends. Contact load 100 kN.

was derived from the voltage drop measured between the potential leads positioned on the connector pads and busbars as shown in Figure 5.69.

5.5.2 Contact Pressure

A well-designed connector should have adequate mechanical strength to maintain its mechanical integrity under normal and overload conductor operating conditions. It should also have sufficient contact pressure to maintain an ample contact area, thus allowing uninterrupted passage of current across the contacting interfaces. However, this pressure should never exceed elastic limits of the joint components since plastic deformation can increase stress relaxation and creep of the contacting members leading to eventual loss of a contact.

From the results presented in Tables 5.13 and 5.14 it is clear that connectors whose current carrying parts (pads) were not given any surface treatment nor were sectioned, had extremely small conducting area and, thus, highest contact resistance. The effect of contact load on the contact resistance of closed current-carrying copper contacts such as in the power disconnect switches, was investigated by Bron et al. [137]. The evolution of contact resistance of spherical copper contacts under 50 and 100 N contact force and a current of 1000 A was measured as a function of time. Note that the contact voltage of 100 mV is much greater than the 10 mV required to give a reliable contact joint. The results, shown in Figure 5.72 demonstrate clearly that increasing the contact load from 50 to 100 N extended the onset of the contact resistance increase, that is, the contact lifetime by 5 times. Furthermore, it was also shown that by increasing the applied contact force, the permissible temperature of electrical contacts can also be increased.

In the case of compression-type connectors, such as those used for splicing aluminum or copper conductors or terminating these at a particular electrical equipment such as transformer, disconnect switch etc. the force required to bring two or more conductors into contact and maintain this contact is provided through the deformation of a portion of the connector. This portion, usually referred to as the barrel or sleeve, is twisted in a predetermined shape and to a predetermined extent by a special tool, which may be a

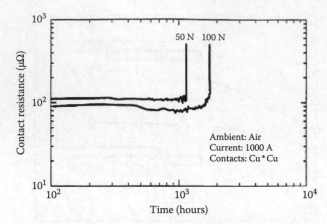

FIGURE 5.72
Evolution of contact resistance of spherical copper contacts with time under 50 and 100 N contact force.

simple hand-actuated tool or a fully automatic power-operated machine. The deformation of the connector barrel is permanent so that the clamping force is maintained. The degree of compression is determined by looking at several factors. Mechanically, the compression force has to be sufficient to satisfy the mechanical requirement for the association. Electrically, the impedance of the connection has to be small enough so that the joint complies with the electrical demands. Simultaneously, the deformation of the connector sleeve has to be sufficient to sustain the required compression force for good. The force needed to execute this procedure has to be within the capabilities of the installation tool, allowing a sufficient margin of safety.

There are definite relationships between the force necessary to achieve a given deformation or compression of any connector barrel, the conductor, and the installation die configuration. The same is true of the relationship of the degree of compression to the resulting electrical and mechanical characteristics of the connection. In establishing contact between a connector sleeve or cylinder, and a cable, deformation of the cylinder walls exerts pressure on the wire. The level of electrical conductivity of the connection is related to the amount of deformation. Put differently, in the initial stages of this deformation, the contact resistance is practically proportional to the deformation, that is, it decreases as the distortion increases. However, once the conductor is fixed firmly in the sleeve, the decrease in contact resistance levels off until no decrease is observed with further deformation. At this point a satisfactory electrical connection has been established.

The electrical and mechanical characteristics of a connector are equally important aspects that determine its reliability. Hence, to obtain a satisfactory electrical and mechanical connection, both characteristics have to be considered. Unlike electrical resistance which reaches a practically steady state with progressive deformation, the mechanical strength of a connection gradually decreases with increasing deformation which, in turn, will increase the rate of stress relaxation and creep of a connector. Closely related to this is the effect when dissimilar metals are used for the conductor and in the connector. Heating the joint by passing current through it creates a difference in the relative coefficient of expansion which may reduce the contact pressure, particularly if the conductor has a higher coefficient of expansion than the connector. There are other factors that can cause loss of mechanical load on the contacts. In a compression joint, a stranded conductor may not align itself perfectly during assembly and any subsequent realignment will make severe reductions in mechanical load. Furthermore, since aluminum is susceptible to creep, some functions will be under much greater strain than others and some stressed far beyond the yield point thus creating the conditions for extrusion to occur. Temperature will affect the rate of creep and, if increased far enough, will reduce the mechanical strength of a joint by annealing.

Experience has shown that a properly designed compression connector can exhibit a very good performance under high temperature cycling conditions. This is owing to the mechanical restraint provided by the structural configuration of the connector and the nearly complete enclosure of the conductor, which partially seals the contact area from adverse effects of the environment. The effect of conductor hardness on the quality of different compression-type connector joints has been investigated by Jesvold et al. [138]. The tests, conducted according to standard International Electrotechnical Commission IEC 1238-1 on both soft and hard conductors, have demonstrated that the conductor hardness and the extent of the mechanical deformations created during assembly strongly influence the quality of a compression joint. It was demonstrated that reliable connections between soft (annealed) aluminum conductors can be obtained simply by using connectors that mechanically deform the strands so heavily that the strand hardness becomes comparable with that of hard-drawn conductors. Braunovic and Dang [139] made a comparative evaluation of different compression-type underground connections. The connectors were evaluated according to their performance under current-cycling conditions. The connectors evaluated were compression sleeve and stepped-deep indentation connectors as shown in Figure 5.73. The results of the current-cycling tests in Figure 5.74 show changes in the maximum and minimum temperature and normalized resistance with a number of current cycles. The data were obtained by averaging the readings from the four connectors of the same kind under tests. The normalized resistance with respect to 20°C was obtained using the expression

$$R(20°C) = R(T)/[1 + 0.0036\ (T - 20)] \tag{5.30}$$

where $R(T)$ is the contact resistance measured at temperature T. The most significant feature of these events is that the performance under current-cycling conditions is significantly affected by the type of connection. This is mainly reflected in the changes in contact resistance with current-cycling time whereas the connector temperature was affected to a lesser degree. The stepped-deep indentation connector had the most stable performance: both the temperature and contact resistance remained practically untouched by the

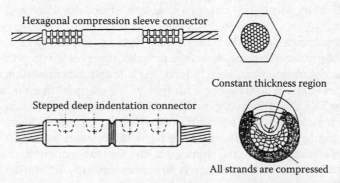

FIGURE 5.73
Schematic of compression sleeve and stepped deep indentation connectors.

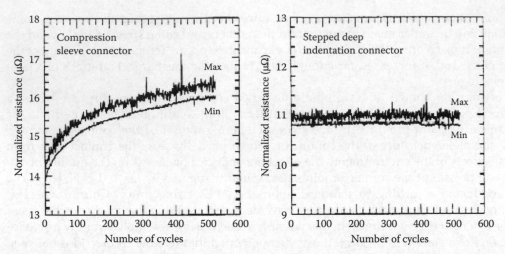

FIGURE 5.74
Effect of current-cycling on the normalized contact resistance of compression sleeve and stepped deep indentation connectors.

current-cycling. However, the compression sleeve connector showed a different tendency: although its temperature remained fairly stable, its contact resistance was greatly affected by current-cycling. The increase in the contact resistance was 13%, thus exceeding the acceptance resistance criterion limit [140–142].

There are several plausible explanations for the observed inferior performance of compression sleeve connectors. One possibility is that the mechanical properties (hardness in particular) of the connector and the conductor are different, which, during mechanical deformation, will cause both connector and conductor to work-harden at different rates. As a consequence, during current-cycling, both connector and conductor will relax at different paces. The net result is a progressive loss of contact load and a gradual increase in the joint resistance, although the bulk temperature may not have been moved to the point of bankruptcy. This is because the contact resistance is more sensitive to the changes occurring at the *a*-spots than to the bulk connector temperature. Another possibility is that current-cycling produces thermal fatigue of the *a*-spots. The magnitude of the thermal strains induced will depend on the contact spot temperature rise and the coefficient of thermal expansion of the joint materials. The magnitude of the corresponding stresses generated by the thermal strains is influenced by the reduction in the yield strength with temperature as well as by structural changes that can affect ductility and resistance to thermal shock.

It is also possible that the design of the compression sleeve connector does not provide sufficient mechanical deformation to bring down the initial contact resistance to a level where it is independent of the deformation applied. Indeed, the initial contact resistance of compression sleeve connectors is about 30% higher than that of the stepped-deep indentation ones as pictured in Figure 5.73. Put differently, it is the ability of the connector contact surface to envelope the conductor and provides solid support which is important. It seems that in the compression sleeve connection this may not have been achieved, resulting in a greater tendency for substantial relaxation from the slippage of strands subsequent to installation. The advantages and disadvantages of compression type connectors are summarized in Table 5.15.

For bolted joints the contact resistance in clean contacts is determined by the resistivity of the contact members, the contact area and distribution of the conducting spots in the

TABLE 5.15

Summary of Advantages and Disadvantages of Compression-Type Connectors

Advantages	Disadvantages
Low cost, relatively reliable performance, use of recommended tools and/or dies removes the human element during installation.	Proper installation tooling for a compression system program involves potentially high capital investments due to a large variety of different types of compression tooling to select from.
Connector construction provides better conductor encirclement while retained oxide inhibiting compound protects the contact area from the atmosphere, thus assuring a maintenance free connection.	Accurate die and tool selection is essential for proper installation of a compression connection.
High localized and consistent forces imparted by the installation tool break down the oxides and establish contact points (*a*-spots) for reduced contact resistance, thus providing electrically and mechanically sound connection.	Due to the need for specific tools and dies to install a compression connection, installers must be trained how to use the proper techniques and maintain these tools.
The softness of compression connector material relative to the conductor prevents spring back and contact separation.	In some compression connections, manually operated tools require greater physical exertion to install, thus when installing numerous connections, installers can become fatigued and possibly not complete the specified number of crimps.
Due to their geometry, compression connectors are considerably easier to insulate or tape than mechanical connectors.	Not as robust as soldering fittings.
These connectors are most suitable in areas of wind, vibration, ice build-up and other stress-associated tension applications.	Should be used in applications where fitting will not be disturbed and subjected to flexing or bending.
Require no soldering.	Compression fittings are much more sensitive to dynamic stresses.
Require no special tools or skill to operator.	They are also bulkier and less aesthetically pleasing than a neatly soldered joint.
Work at higher pressures and with toxic gases.	The use of a special tool is disadvantage in any situation, likewise, not having to use a special tool is an advantage.
Useful when occasional disassembly is required for maintenance.	
They can be assembled and disassembled without affecting the joint integrity.	
Can be used where heat source (soldering) is prohibited.	

interface between the contact members. The area and distribution of the conducting spots are generally determined by the surface finish and the magnitude of the applied force and the manner in which it is applied. Therefore, the quantitative application of laboratory data to an actual bolted joint is possible only in so far as the laboratory conditions duplicate the activity of the fastener. Every fastener (bolt with washers or clamp assembly) produces conducting spots that are allotted according to the characteristics of the fastener and the busbar.

The relationship between the contact resistance (R_c) and the applied force (F) can be described by;

$$R_c = BF^{-n} \tag{5.31}$$

where B is the proportionality constant and n is the exponent. The values of the constants a and n depend on the joint configuration, type of lubricant and coating used. The values of n for bare, coated and lubricated aluminum-to-aluminum and aluminum-to-copper connections were found to vary between 0.1 and 1.0. The effect of contact load on the contact resistance of aluminum bus bar connections is shown in Figure 5.75 [143]. For practical purposes, the problem of adequate contact force for a joint of satisfactorily low initial resistance can

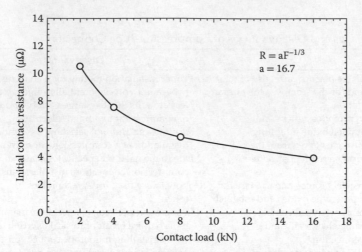

FIGURE 5.75
Effect of contact load on the contact resistance of aluminum-to-aluminum bolted joints.

be simplified by considering the applied force as uniformly distributed over the apparent contact area. Naybour and Farrell [71] have shown that, provided the initial stress at 20°C is less than 60 N/mm², stress relaxation is not likely to cause significant contact deterioration on aluminum conductors if the initial load is 2000 N. This value of contact load is considered as a practical minimum although added efficiency can be realized with higher contact loads which can maintain points of intimate contact even with some stress relaxation.

It should be emphasized for bolted joints that the use of contact force as an indicator of a satisfactory joint assembly is impractical. Instead, tightening torque is almost exclusively used in practical situations. An empirical relationship commonly used to relate the tightening torque to the applied force generated by the bolt in a joint is given:

$$T = K D F \tag{5.32}$$

T—Tightening torque (Nm)
D—Bolt diameter (m)
K—Constant ("nut factor" – dimensionless)
F—Contact force (N)

The "nut factor" K is greatly dependent on the coefficient of friction, finish and lubrication of the threads and other bearing surfaces. For specified conditions of finish and lubrication there is considerable variation in this relationship. Also both the tightening torque must be specified and the rate of tightening must be controlled. The rate of relaxation from cold flow of aluminum conductor is quite large under the pressures encountered in most bolted connectors. Hence a slow rate of application of torque will result in more relaxation being taken up before the final installation torque is reached. Although continuous torque application produces more rapid failure, a stepped application of torque was selected as the standard of tests. Continuous torque application was considered impractical for a standard procedure since multi-bolt connectors must be installed by tightening each bolt in turn by steps.

Tightening torques depend upon the contact pressure required for the particular busbar application. The amount of pressure for a given torque value varies over a wide range;

depending upon whether threads are dry, lubricated, hot galvanized, or otherwise treated. From the standpoint of the intensity of the bolts, a maximum tightening torque equal to 95% of the yield strength of bolt is recommended. Hence, it is essential to maintain the tightening torque within this limit since upon exceeding it, the busbar will undergo plastic deformation which, in turn, will increase the stress relaxation and creep and thus cause loosening of a joint. The end result is an increased contact resistance and temperature causing the eventual joint to fail. A recap of the tightening torques specified by different manufacturers for bolting different power equipment reveals a great diversity in the recommended torque values. For instance, the range of tightening torque for a 12.7 mm (½ in.) diameter bolt is 50–80 Nm (35–60 lb-f). Hence, the forces generated by these torques may exceed the yield strength of certain bolt material.

To illustrate the importance of tightening torque and also the "nut factor" K, let us calculate the stresses generated by the torque range 50–80 Nm (35–60 lb-f) in 12.7 mm diameter (½-13 in.) bolts made of silicon bronze and stainless steel 304 and 316. The results are shown in Table 5.16. From this table it can be inferred that for lower values of "nut factor" K ($K = 0.15$) certain bolt materials cannot sustain the forces generated by the applied tightening torque and would deform plastically. The lower values of the "nut factor" K ($K = 0.15$) are generally attributed to the effect of lubrication on the coefficient of friction. Hence, it appears that same tightening torque may induce much higher force in the lubricated than in non lubricated bolts. This is very important since, it may be contended, that the use of contact aid compounds may adversely affect the mechanical integrity of a bolted joint by exceeding the permissible force limits in the bolts. The habit of common contact aid compounds (see also Section 5.6.5) did not produce any significant increase in the contact forces neither in the bolts nor in the joint. However, when boundary lubricant, such as Mo_2S ($K = 0.15$), was used, much higher force was generated in the bolt. Table 5.17 shows the range of the "freak factor" when jointing was made with dry bolts and also when

TABLE 5.16

Characteristics of Bolt Materials and Calculated Forces in Bolts Generated by Tightening Torques for Different Values of "Nut Factor" K

Bolt	σ_y MPa (KSI)	A_b, mm² (in.²)	F_{max} (95%)σ_y kN (lb)	Torque Nm (lb·ft)	Calculated Force in Bolt kN (lb)		
					$K = 0.3$	$K = 0.2$	$K = 0.15$
Silicon Bronze	365 (53)	91.5 (0.142)	31.7 (7 144)	50 (35)	12.4 (2 800)	18.7 (4 200)	24.8 (5 600)
C651000	365 (53)	91.5 (0.142)	31.7 (7 144)	60 (45)	16.0 (3 600)	23.9 (5 400)	31.8 (7 200)
ASTM 468-93	365 (53)	91.5 (0.142)	31.7 (7 144)	70 (50)	17.8 (4 000)	26.5 (6 000)	35.5 (8 000)
	365 (53)	91.5 (0.142)	31.7 (7 144)	80 (60)	21.3 (4 800)	31.8 (7 200)	42.5 (9 600)
Stainless steel 304	207 (30)	91.5 (0.142)	17.9 (4 044)	50 (35)	12.4 (2 800)	18.7 (4 200)	24.8 (5 600)
ASTM 593	207 (30)	91.5 (0.142)	17.9 (4 044)	60 (45)	16.0 (3 600)	23.9 (5 400)	31.8 (7 200)
	207 (30)	91.5 (0.142)	17.9 (4 044)	70 (50)	17.8 (4 000)	26.5 (6 000)	35.5 (8 000)
	207 (30)	91.5 (0.142)	17.9 (4 044)	80 (60)	21.3 (4 800)	31.8 (7 200)	42.5 (9 600)
Stainless steel	414 (60)	91.5 (0.142)	35.8 (8 088)	50 (35)	12.4 (2 800)	18.7 (4 200)	24.8 (5 600)
316 cold worked	414 (60)	91.5 (0.142)	35.8 (8 088)	60 (45)	16.0 (3 600)	23.9 (5 400)	31.8 (7 200)
ASTM 593	414 (60)	91.5 (0.142)	35.8 (8 088)	70 (50)	17.8 (4 000)	26.5 (6 000)	35.5 (8 000)
	414 (60)	91.5 (0.142)	35.8 (8 088)	80 (60)	21.3 (4 800)	31.8 (7 200)	42.5 (9 600)

σ_y—Yield stress,
A_b—Bolt cross section,
F_{max}—Maximum force supported by the bolt.

TABLE 5.17

Values of "Nut Factor" When Dry and Lubricated Bolts were Used
for Assembling the Bolted Busbar Joints

Nut Factor	Dry Bolts	Contact Aid Compounds	Boundary Lubricants
K	0.20–0.22	0.19–0.21	0.15–0.16

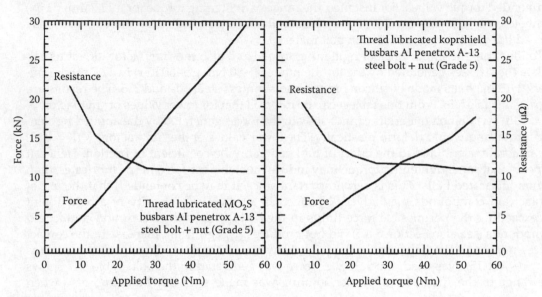

FIGURE 5.76
Effect of the "nut factor" on the contact force generated in bolted joints.

lubricated with the contact aid compounds and boundary type lubricant. It appears that
there is a very small difference between the values of the "nut factor" K for dry bolts and
those lubricated with the contact aid compounds. The effect of lubrication that is the "nut
factor" K on the contact force generated in bolted joints is shown in Figure 5.76.

5.5.3 Mechanical Contact Device

Although a bolted type connection is widely used method of joining aluminum and copper conductors, there are doubts concerning its reliability under the operating conditions,
owing mainly to the fact that when two dissimilar metals are used, the difference in their
physical, mechanical and metallurgical properties results in difficulties to arrive at a satisfactory joint. In order to shorten this problem, various palliative measures in the design and
joint configuration were considered. One of these is the use of suitable mechanical contact
devices combined with appropriate tightening of a joint, thus ensuring that all members of
a joint remain within their elastic limits under all anticipated operating conditions.

To illustrate the importance of these mechanical contact devices, let us examine the tensions generated in a bolted joint with no current passing through the joint. The joint configuration consists of two aluminum or copper busbars measuring $38 \times 12.7 \times 250$ mm, two
38 mm diameter flat washers and a 12.7 diameter steel or bronze bolt. During assembling,
the bolt is subjected to a simple tension while the busbars to compression. The "nut factor"

TABLE 5.18

Stresses Generated in the Bolt and Busbar Conductors

Torque Nm (lb.ft)	A_b, mm² (in.²)	A_w mm² (in.²)	Contact Force kN (lb)	Bolt Stress MPa (PSI)	Busbar Stress MPa (PSI)
50 (35)	91.5 (0.142)	824 (1.276)	18.7 (4 200)	204 (29 577)	23 (3 290)
60 (45)	91.5 (0.142)	824 (1.276)	23.9 (5 400)	260 (38 028)	29 (4 230)
70 (50)	91.5 (0.142)	824 (1.276)	26.5 (6 000)	290 (42 253)	32 (4 700)
80 (60)	91.5 (0.142)	824 (1.276)	31.8 (7 200)	348 (50 704)	39 (5 650)

A_b—Bolt cross section,
A_w—Apparent contact area under washer.

used for calculating the forces generated in the bolt by applying torque was $K = 0.2$. The stress distribution in the busbars under the washer was assumed to be uniform. The calculation results are shown in Table 5.18. If the busbars were made of aluminum EC-0, the stresses generated by tightening torque of 60 Nm (45 lb-f) would exceed their yield strength which is 28 MPa (4000 PSI). It should be pointed out, that the above calculations were made assuming a uniform stress distribution in the contact zone under the washers. However, since the real contact area is much smaller than the apparent contact area, the stresses generated at the *a*-spots, where the actual contact is made will much higher.

The heating and cooling of a bus joint as the result of normal changes in electrical load cause corresponding increases and diminutions in the thickness of the bolted joint, varying in magnitude with the thermal expansion coefficient of the material, Moreover, the bolts are usually not heated by the current in the conductors to the same extent as the conductor itself. The result may be a temperature differential of 10°C or more between the bolt and the conductor, which tends to narrow over a period of time. During the hot portion of the load cycle, the differential expansion of the bus conductor and the bolts add additional stresses to the original stresses in the conductor and the bolt. The increase in the stresses in the busbar conductors (F_{sup}) and the bolt are is proportional to the temperature rise (ΔT) and is given as

$$F_{sup} = k_1 \Delta T \quad k_1 = \frac{(\alpha_a - \alpha_b) A_b E_b}{1 + \dfrac{t}{a}\left(1 + \dfrac{A_b}{A_w}\right) + \dfrac{A_b E_b}{A_a E_a}} \tag{5.33}$$

α_a—Coefficient of thermal expansion of the busbar conductor

E_a—Elastic modulus of the busbar conductor

α_b—Coefficient of thermal expansion of the bolt

t—Thickness of the washer

A_b—Bolt cross section

a—Thickness of the busbar conductor

E_b—Elastic modulus of the bolt

A_w—Apparent area under the washer

A_a—Apparent contact area of overlapping joint

For the joint assembled initially at 55 Nm (40 lb-f), that is, contact force of 22 kN (4800 lb) the calculated additional force, generated in the busbars and the bolt when the joint temperature increased to 120°C is given in Table 5.19. The coefficients of thermal expansion used to

TABLE 5.19

Increase in Stresses in the Bolt and the Busbar Conductors Due to Temperature Rise of 120°C

Bolt	Busbars	F_{sup} kN (lb)	F_{tot} kN (lb)	Bolt Stress MPa (PSI)	Busbar Stress MPa (PSI)
Steel	Aluminum	19.0 (4 327)	41.0 (9 127)	443.0 (64 319)	40.0 (5 810)
Steel	Copper	6.7 (1 510)	28.0 (6 310)	306.0 (44 487)	28.0 (4 017)
Bronze	Aluminum	8.0 (1 794)	30.0 (6 594)	320.0 (46 469)	29.0 (4 197)

TABLE 5.20

Recommended Tightening Torques for Bolted Joints Assembled with Different Bolts Sizes

Bolt	Tightening Torque		
M	Nm	Kgf.m	lbf.ft
10 (3/8)	35–50	3.5–5	26–37
12 (1/2)	60–85	6–8.5	45–63
16 (5/8)	150–200	15–20	110–150

calculate the additional force in the joints were: steel—11, aluminum—23.6 and copper—16.5 ($K^{-1} \times 10^{-6}$). The results shown in Table 5.19 clearly show that the stresses and the resulting deformation are much greater for aluminum than for copper busbar conductor for the same temperature rise. This illustrates the importance of maintaining reasonably constant pressure in bolted joints during the heating and cooling cycles of operation. Hence, the higher yield strength and better creep characteristics of busbar conductors and the use of suitable mechanical contact device are very important in order to obtain stable electric joints. Recommended tightening torques for bolted joints assembled with different bolt sizes are given in Table 5.20.

5.5.4 Disc-Spring (Belleville) Washers

The mechanical integrity of the bolted connector is strongly affected by the joint configuration, which can differ according to the mechanical device used. The combination of a disc-spring (Belleville) and thick flat washers was found to assure the most satisfactory mechanical and electrical integrity of bolted joints under current-cycling and stress relaxation conditions [58]. One of the most important characteristics of disc-spring washers is their ability to elastically absorb deformation caused by an outside load. In order to comply with the space limitations imposed by the geometry of a bolted joint, disc-spring washers have to be as modest as possible and made from materials having a high tensile strength and high elastic limit. These materials should also have high dynamic fatigue resistance, sufficient plastic deformation ability exceeding the elastic limit to permit manufacture of cold worked springs as well as to minimize failure of springs under sudden and sharp load changes. The effect of different types of mechanical contact devices on the performance of bolted aluminum-to-aluminum joints under current cycling and stress relaxation conditions was investigated by Braunovic [58,130]. It was demonstrated that the mechanical integrity of the connector is strongly affected by the joint configuration, which differs according to the mechanical device used. The combination of a disc-spring (Belleville) and thick flat washers assured the most satisfactory mechanical and electrical integrity of bolted joints under current-cycling and stress relaxation conditions; see Table 5.21.

TABLE 5.21

Forces Generated in the Busbar Conductors without and with the Disc-Spring (Bellevile) Washers

Bolt	Busbars	Force in Busbars without Disc-Spring Washer kN (lb)	Force in Busbars with Disc-Spring Washer kN (lb)
Steel	Aluminum	19.0 (4 327)	1.2 (276)
Steel	Copper	6.7 (1 510)	0.4 (95)
Bronze	Aluminum	8.0 (1 794)	0.8 (173)

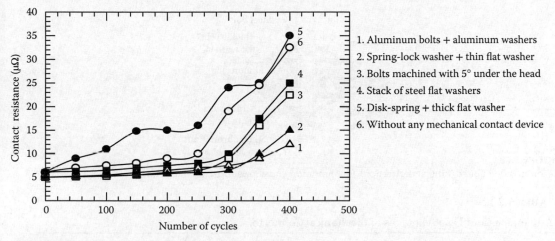

1. Aluminum bolts + aluminum washers
2. Spring-lock washer + thin flat washer
3. Bolts machined with 5° under the head
4. Stack of steel flat washers
5. Disk-spring + thick flat washer
6. Without any mechanical contact device

FIGURE 5.77
Effect of different mechanical contact devices on the contact resistance of a bolted joint under current-cycling conditions.

Disc-spring washers are generally made of spring steel such as high-carbon, chrome-vanadium, chrome-vanadium-molybdenum, tungsten-chrome-vanadium and stainless steels. However, other resilient materials, such as silicon and phosphor bronze, beryllium copper, inconel and nimonic can also be employed. Effect of different mechanical contact devices on the contact resistance of a bolted joint under current-cycling conditions is illustrated in Figure 5.77. Figure 5.78 shows joint configurations for jointing aluminum or copper busbar conductors. In all cases, it is recommended to use flat washers at least 3–4 mm thick. With steel bolts, disc spring washers with a high spring constant and flat washers with thickness at least twice that of the disc-spring washer should be applied.

One method of avoiding failures owing to differential thermal expansion consists of using aluminum alloy bolts. Since the coefficients of expansion of the aluminum conductor and the bolts are essentially the same, high contact pressure is held during both heating and cooling cycles. The modulus of elasticity of aluminum alloy bolts is about one-third that of steel bolts. Thus a large amount of "spring-follow" is provided in the bolts. Since the compressive stress in a bolted joint is concentrated under the head and nut of the bolt, aluminum bolts made of alloys with yield strength of 400 MPa and tensile strength of 470 MPa, such as 7075 T-72 alloy, are recommended. Equally, for jointing the copper busbars, bolts made of a high-strength bronze (everdur) are preferable over that of steel. Aluminum alloy and bronze bolts are nonmagnetic and, therefore, not subject to heating from hysteresis losses in ac fields, as is the case with steel bolts. Advantages and disadvantages of mechanical connectors are given in Table 5.22. Another important factor

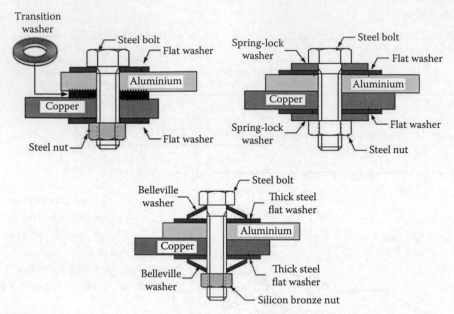

FIGURE 5.78
Recommended bolted joint configurations for aluminum and copper busbar conductors.

TABLE 5.22

Advantages and Disadvantages of Mechanical Connectors

Advantages	Disadvantages
Inherent resilience of the connector components permits follow-up of creep and reduces the stresses due to thermal expansion that tend to cause excessive creep	Specific torque requirements must be followed to provide the proper clamping force needed for a sound electrical connection
Ease of installation (sockets, wrenches, screwdrivers, etc) and removal, simple to use, require minimal training to install properly	Inconsistency of forces applied over identical mechanical installations is not generally repeatable due to use of uncalibrated torque wrenches
Can be disassembled without damage to the connection components and may be reusable if in good condition	Because of relatively low mechanical holding strength, these connectors cannot be used as full tension connections and in areas of high vibration; more maintenance and periodic inspection may be required
Electrical performance of mechanical connectors meets or exceeds the industry requirements for which they are designed, thus, performance is not compromised when using mechanical connectors in tested applications	Owing to their geometry, installing mechanical connectors on insulated conductors is usually difficult and awkward

affecting the performance of bolted joints is the selection of the bolts. The use of steel bolts in aluminum and copper bolted joints without mechanical contact devices such as disc-spring (Belleville) to compensate for the difference in the thermal expansion between aluminum, copper and steel, may result in thermoelastic ratcheting which, in turn, will adversely affect the joint integrity and performance. The beneficial effect of spring washer is shown in Figure 5.79 as a function of time and under static and dynamic operating conditions. In the case of static operating conditions the loss of relative contact force was simply measured as a function of time. However, in the case of dynamic operating conditions, the joint were subject to current cycling and the remaining contact force was measured at the preselected time intervals.

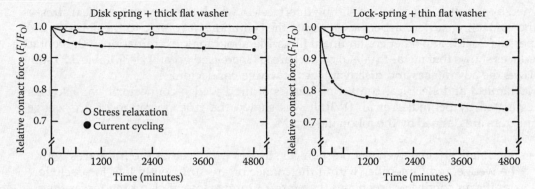

FIGURE 5.79
Effect of mechanical contact devices on the relative contact force in bolted joints.

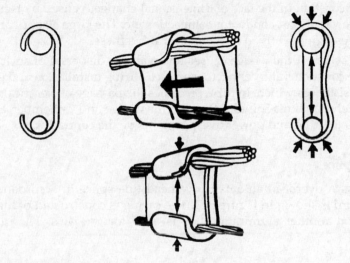

FIGURE 5.80
Schematic of the wedge connector assembling.

5.5.5 Wedge Connectors

The wedge connector incorporates a wedge component and a tapered, C-shaped spring body (or C-body). Installation is made by driving the wedge between two conductors into the C-member deforming it plastically and generating the force that secure the installed wedge and conductors in place. The mechanical properties of the C-member are the most important to the reliable mechanical function of wedge connector. The C-body is characterized by an elastic compliance capable of paying for cable compaction and is therefore responsible for maintaining nearly constant mechanical load on conductors during the lifetime of the connectors. In addition, the relatively small mechanical stresses generated in the electrical contact interfaces prevent significant conductor creep assure the dependable operation of a wedge connector under severe operating and environmental conditions. Figure 5.80 depicts schematically assembling of the wedge connector and the forces generated in the joint. Driving action is produced either by firing a cartridge to propel the wedge at high velocity (powder actuated) or by tightening mechanically driven bolt which drives the wedge between the conductors. As a result, the electrical interfaces are formed by the shearing of rough sliding interfaces in which the large mechanical stresses at the contacting surface asperities generate the conditions favorable to the abrasion and dispersal of oxide and other contaminant films. Metallographic examination of the contact interfaces shows that metal-to-metal contact area is increased greatly [143]. Table 5.23 summarizes the advantages and disadvantages of wedge-connectors.

Mechanical and electrical contact properties of fired wedge-connector have also been investigated by Schindler et al. [143]. It was shown the true operation of fire wedge-connectors are defined by the following factors:

- The relatively low averages mechanical stresses produced at the interfaces with the wedge and C-member within the connector are sufficiently low to preclude significant conductor creep and therefore, minimize loss of clamping force along the conductors.

- The relatively large elastic compliance of the C-body allows the contact forces to remain nearly constant in the face of dimensional changes caused by factors such as temperature variations, conductor compaction etc. The large elastic compliance thus contributes significantly to the connector robustness.

- The abrading action of the wedge causes removal and dispersal of surface oxide films from the conductor and connector surfaces during installation and produces low contact resistances which, in turn, promotes the passage of electrical currents of approximately equal magnitudes through the wedge and C-member, and thus, minimize joule heating and power dissipation inside the connector.

5.5.6 Automatic Splices

A full tension Class A overhead automatic overhead line splice in accordance with the ANSI C119.4 standard is shown in Figure 5.81. The casing is constructed of high strength aluminum alloy and another aluminum alloy is used for the jaws. The inside of the

TABLE 5.23

Advantages and Disadvantages of Wedge-Connectors

Advantages	Disadvantages
Powder actuation provides consistent, uniform performance and requires low physical exertion from an operator to complete a connection	Dedicated nature of powder actuation requires full support from the user in terms of training, maintenance and service
Rapid mechanical wiping action as the wedge is driven between the conductors breaks down surface oxides and generates superior contact points thus reducing overall contact resistance	To ensure a safe and proper installation, precautions are and specially trained and qualified installers are required for installing wedge connections
Installation is accelerated with the use of lightweight, portable tooling with simplified loading and engaging mechanisms	Mechanical wedge connectors are installed with wrenches, require more physical exertion for installation, and show more inconsistent performance due to discrepancies caused by contaminants on the hardware and wide tolerances of shear-off bolts
The spring effect of the "C" body maintains constant pressure for reliable performance under severe load and climatic conditions whereas a large connector mass provides better heat dissipation	Mechanical wedge spring bodies are typically manufactured by casting which produces much less spring action to maintain the connection
Electrical performance of fired-on wedge connectors are excellent due to the low contact resistance developed during installation	Wedge connectors are restricted to non-tension, outdoor applications and suited only for a limited range of conductors; due to their geometry, full insulation of wedge connector is difficult

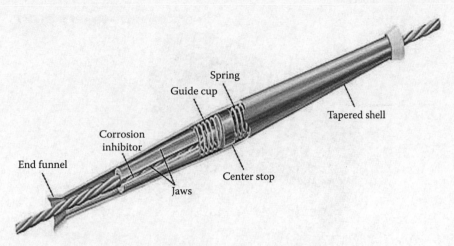

FIGURE 5.81
Automatic overhead line splice. (Courtesy of Hubbell Power Systems.)

connector is filled with a factory inhibitor to protect the contact zones from the adverse environmental effects. The intent of this connector assures a very fast installation of aluminum, aluminum alloy and ACSR conductor materials. For easy identification, the end funnel guides are color coded.

5.5.7 Dead-end Connectors

This connector has no bolts to slow installation. Installation is simple requiring only locking the jaws of the dead-end open, inserting the conductor between the jaws and tapping the back of the jaws to lock the conductor in place. Once contact between the clamp and conductor is made, the applied spring tension holds the conductor securely thus tightening errors are eliminated; see Figures 5.82 and 5.83. The automatic clamping creates a permanent installation owing to the wedge action which produces the full tension of the conductor. The quality of the installation is not contingent on the workmanship. There's no feeding of the conductor through the dead-end. Installation of the wedge dead-end is easy because the connector faces the installer. There's no need for extra tools or special skills and adjustability is easy. The connector configuration resists wind induced motion, galloping, Aeolian vibration and sub-conductor oscillations. The connector construction comprises two standard round conductors twisted around each other at nine foot intervals to change the air foil on the suspension bridge. Full tension dead-end assemblies for ACSR conductors consist of an aluminum deadened body, steel dead-end eye, 15° jumper terminal and terminal mounting hardware. Terminal and tongue have NEMA hole spacing. Figure 5.84 shows this type of connector. The body is made of seamless extruded aluminum alloy tube. The eye is made of galvanized forged steel whereas terminal is made of seamless extruded aluminium alloy tube. Connector terminal hardware is made of ½-13 aluminum alloy.

5.5.8 Shape-Memory Alloy Connector Devices

The shape-memory effect (SME) refers to the ability of certain materials to "remember" a shape, even after severe deformation. When a material with the material body-memory ability is cooled below its transformation temperature (martensite phase), it has a very low

FIGURE 5.82
Fargo GDW 2000 wedge Deadend connector. (Courtesy of Anderson Fargo-Hubbell Power Systems Inc.)

FIGURE 5.83
Anderson SD Deadend Strain Clamp. (Courtesy of Anderson Fargo-Hubbell Power Systems Inc.)

yield strength and can be deformed quite easily into a new physical body. When heated above its transformation temperature, it undergoes a change in crystal structure which causes it to return spontaneously to its original shape (austenite phase). During this isotropic transformation process, as the atoms shift back to their original positions, a substantial amount of energy is released. The SMA has high sensitivity to deformation over a narrow temperature range which makes them ideal as disc-spring washers. Furthermore, these alloys have the ability to alter the configuration of the load-deflection curve which can be used to advantage in many applications. Even in improperly installed bolted joints (low contact force), the heat developed by the joint looseness can be used to lower and stabilize the contact resistance in a bolted joint [144–147]. Table 5.24 shows an overview of identifying SMA products developed and used in different spheres. A detailed review of shape-memory materials is given in CEA Report No. SD-294A entitled "Use of shape-memory materials in distribution power systems" [144]. Comprehensive reviews of general and electrical applications of SMA materials are given in [144,147–150].

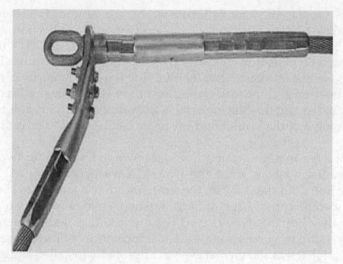

FIGURE 5.84
Full tension deadend assemblies.

TABLE 5.24

Shape-Memory Alloy Applications in Power and Electronics Industries

Electrical Power Industry	Electronics Industry
Connectors, circuit breakers, fuses	Integrated circuit connectors (IC)
Switches/switching devices	Zero insertion force connectors (ZIF)
De-icing/sagging of transmission lines	Dual in-line pin connectors (DIP)
High-power heat engines, robotic devices	Pin grid array package (PGAP)
Electrical/thermal actuators, thermo-markers	Micro-strip connectors
Overcurrent/overheating protection	Locking rings for braided terminals
Optical fibre splices	Disk drive lift/lower recording heads
Contact bounce dampers	
Nuclear power plant applications	

Shape-memory alloys (SMAs) can be used as disc-spring material. SMAs have a high sensitivity to deformation over a narrow temperature range which makes them ideal as disc-spring washers. The beneficial effect of SMA disc-spring washers is emphasized by the fact that, even with a low initial contact force (faulty installation), the heating caused by the reduced contact force generates an extra strain on the joint, thereby preventing the joint from going bad. Furthermore, these alloys have the ability to alter the configuration of the load-deflection curve which can be used to advantage in many applications. This feature of SMA washer was used by Oberg and Nilsson [147] to demonstrate that even in improperly installed bolted joints (low contact force), the heat developed by the joint looseness can be used to lower and stabilize the contact resistance in a bolted joint. One of the most telling features of shape-memory alloys is that they can be used in many different ways to a broad range of merchandise. These alloys are basically functional devices in that they are more important for what they can do (action) than for what they are (property). Their practical applications are numerous: they have been used successfully as thermal and electrical actuators, thermo-mechanical energy converters, electrical connectors, circuit breakers, and mechanical couplers, as well as in robotic applications, medicine, and other fields. The high sensitivity of shape-memory alloys over a narrow temperature range makes them ideal as a replacement for Belleville washers. Example of the beneficial effect of shape memory alloy mechanical device (washer) in power connection application is depicted in Figure 5.85 depicting the load-deflection dependence of NiTi shape-memory flat washer at 20°C and 65°C [144]. It is clear that as the contact temperature increases the contact force in the joint also increases thus maintaining the joint integrity. Since the material at 20°C is fully martensitic, the load-deflection curve is nonlinear whereas at 65°C it is essentially linear indicating that the material is fully austenitic. This ability of an SMA washer to vary the shape of the load-deflection curve can be used to advantage in many power applications.

The use of SMA washer in power connections is shown in Figure 5.86. Two busbars are joined by a bolt, two flat washers, a nut and two SMA washers interposed between the flat washers (Figure 5.86 Cold state). When the temperature of a joint rises above the SMA transformation temperature, the shape of SMA washers change into a more heat-stable state (austenitic) as they become more curved (Figure 5.86 Elevated temperatures). As a result the contact pressure in a joint increases. The importance of this feature has recently been demonstrated [144]. It was shown that even in improperly installed joints (low contact

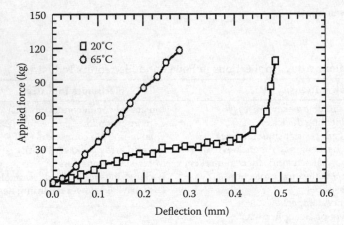

FIGURE 5.85
Load-deflection characteristics of a NiTi shape-memory alloy washer at 20°C and 65°C.

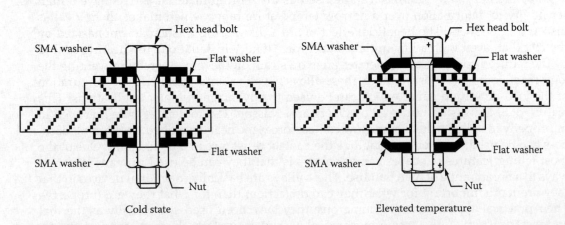

Cold state   Elevated temperature

FIGURE 5.86
Effect of SMA flat washer on the joint configuration at low and high temperatures. Note the change in the SMA washer shape at high temperature.

force), the heat, induced by a loosened joint causes the SMA washer to change its shape and increases the contact pressure. The net result is a stable joint with a low contact resistance.

The effect of shape-memory (SMA) and disc-spring (Belleville) washers on the performance of aluminum-to-aluminum bolted joints was under current-cycling conditions has been investigated by Braunovic and Labrecque [145] and Labrecque et al. [146]. It was shown that the use of SMA disc-spring washers produced significant improvement in electrical and mechanical stabilities of bolted joints during current-cycling conditions. The great flexibility and adaptability of SMA technology in a wide variety of electrical applications will eventually make it a part of the electrical industry.

5.5.9 Coating (Plating)

This subject was introduced in Section 5.4.5. The coating of aluminum or copper with different metals is one of the most common commercial practices used to improve the stability and suppress the galvanic corrosion of aluminum-to-copper connectors. The most widely used coating materials are tin, silver, copper, cadmium, and nickel. One of the first comparative studies on the efficiency of coating materials was by Bonwitt who in 1948 [151] investigated the electrical performance of plated and bare bolted aluminum-to-copper connections exposed to elevated temperature and a saline environment. The principal measurement parameter, contact resistance, was measured after the bolted joints were exposed to high temperatures and salt-spray. The coating materials investigated were cadmium, tin and zinc applied to copper busbars. Tin-plated copper followed by bare copper showed the best performance whereas zinc-plated copper was the worst. However, the contact resistance of both tinned and bare copper increased after exposure to high temperature. It was also shown that galvanic corrosion was not eliminated by the coating and that lubrication of bolted busbars was essential in a saline environment. Hubbard et al. [152] evaluated the effect of cadmium, tin, and zinc platings on the performance of aluminum-to-copper joints of different design under current-cycling conditions and in a saline environment. Cadmium on either aluminum or copper and hot-flowed electro-tin on copper were ranked as the most efficient plating materials. Again, lubrication was cited as essential in prolonging the useful life of plated aluminum-to-copper connections in a saline environment. Bond and McGeary [153] evaluated the effectiveness of cadmium, nickel, tin, silver and tin as coating materials in maintaining the stability of aluminum-to-copper bolted-type connections subjected to heat/current cycling conditions and also in a saline environment. It was demonstrated that the nickel-coated connections were superior in performance to another plating material, as evidenced by their stable contact resistance behavior under simulated service conditions. Although widely used in electrical industry, there is mounting evidence indicating that tin neither effectively prevents galvanic corrosion nor ensures the stability of aluminum-to-copper connectors. Tin-plating, traditionally used to curb the adverse effects of galvanic corrosion, requires no special surface preparation prior to assembly and improves the functioning of joints at higher temperatures. There are two main reasons for this:

1. Tin-plated contacts are very susceptible to fretting, which causes severe degradation of contacting interfaces and leads to unacceptably higher contact resistance, instability and, finally, an open circuit.

2. Tin easily forms intermetallic phases with copper even at room temperature, rendering the contact interface very brittle, highly resistant and susceptible to the influence of the surroundings. Lubrication of bolted busbars was essential for reducing the corrosion damage in a saline environment as illustrated in Figure 5.87.

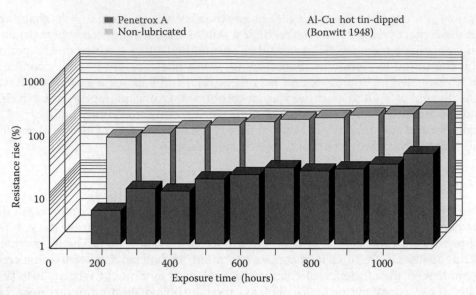

FIGURE 5.87
Effect of saline environment on the contact resistance of lubricated (Penetrox) and non-lubricated aluminum-to-hot-tin dipped copper bolted joints.

Silver plating thickness of 5–15 μm for coating switchgear and enclosed bus is generally considered adequate. Because of the porosity of silver plating, however, a thicker deposit is required where abrasion and weather resistance are factors. Where the joint will be connected and disconnected many times during the life, for instance, of the disconnect switch, the silver plating must be thick enough to prevent exposure of the aluminum or copper through wear. Although silver plating of electric contacts is beneficial in maintaining low electric resistance, it has a potential disadvantage: silver, like copper, is cathodic to aluminum (see Section 2.4.2) and may, therefore, cause galvanic corrosion of aluminum. Furthermore, owing to sensitivity of silver to tarnishing and whisker formation the use of silver-plating in environments where sulfurized contaminants are present has to be avoided; see Chapter 8. The purpose of protective contact aid compounds for optimum performance is essential for silver-coated joints that are exposed to high humidity or moisture.

Most switchgear is installed in relatively clean air with normal humidity. In such cases, lubrication of the silver-plated joints will seal the joints adequately and protect them against the entrance of an electrolyte and the possibility of galvanic corrosion. Some manufacturers use lubrication as a regular practice while others use high temperature wax, such as ceresin wax, dissolved in naphtha, which is easily applied to coat the silver-coated contact surfaces. In outdoor seacoast exposures and in other strongly corrosive atmospheres where galvanic corrosion prevails, inhibitor paints should be used in addition to the contact aid compounds or wax coating. The problem of porosity and excessive wear of silver-plated contacts during make and break operations, as in the type of disconnect switches, can be limited and practically eliminated by welding thin solid silver plates or strips to the copper or aluminum contact blades or jaws [154]. The welded solid silver contact interfaces are impervious to the creep or pore corrosion, less susceptible to fretting, have superior wear resistance and are significantly less affected by the sulfur-bearing environment since, as shown by Kassman-Rudolphi [155] the presence of tarnish films can have a beneficial effect on the contact behavior. Put differently, an appropriate combination of silver and sulfurized film properties can have positive effects on the wear and deformation of the contact and thus outweigh the negative effects of tarnishing on the electrical behavior of contact systems where sliding, fretting and make-and-brake actions are taken.

Nickel-coated connections were superior in performance to another plating material, as evidenced by their stable contact resistance behavior under simulated service conditions. From the available data, nickel appears to be the most practical coating material from the standpoint both of its economy and the significant improvements to the metallurgical and contact properties of aluminum-to-copper connectors. The superiority of nickel to other coating materials was recently confirmed by Jackson [156] who performed current cycling tests on tin-, silver-, and nickel-plated copper busbars bolted to 1350 grade aluminum. The nickel coatings on copper connections showed excellent stability and low initial contact resistance. The poor performance of tin- and silver-plated connections was attributed to the effect of differential thermal expansion between the substrates of aluminum and copper that promotes progressive loss of the contact spots and, thus, deterioration of the contact. Lefebre et al. [157] showed that the instability of aluminum connections can be significantly reduced by a newly developed technology of direct nickel plating. The results of severe current-cycling tests demonstrated excellent stability of the contact resistance of nickel-plated aluminum. However, Aronstein [158] and Hare [159] have demonstrated that although nickel-coated aluminum wire conductors under different operating and environmental conditions had a better connectability than uncoated aluminum wire conductors, the coating/conductor interface was very susceptible to erosion. Braunovic [57] and Bruel and Carballeira [160] on the effect of fretting on the contact properties of nickel-coated aluminum conductors revealed that, despite marked improvement brought about by the nickel coating, fretting still produces a considerable degradation of the contact zones. Lubrication, higher loads and shorter slip amplitudes are found to mitigate these adverse effects.

Braunovic [161] made a comparative evaluation of tin-, silver- and nickel-plated copper-to-aluminum connections on the basis of the performance of coated joints under current-cycling and fretting conditions and ability to protect the contact against corrosion in saline and industrially polluted environments. The results, summarized in Table 5.25, provide an overall assessment of the various coatings. The INDEX was obtained by averaging the absolute values assigned to the performance of joints under different laboratory (current-cycling and fretting) and environmental (saline and industrial pollution) conditions. The lower the INDEX numbers the more efficient the coating material. From this table, it appears that nickel plating significantly enhances the stability of aluminum-to-copper connections (the lowest INDEX), while tin and silver coating show the poorest performance (the highest INDEX) under different operating and environmental conditions.

TABLE 5.25

Summary of Comparative Evaluation of Different Coating Materials for Aluminum-to-Copper Connections

Contact Pairs		Index
Aluminum (Nickel-plated)	Copper (Nickel-plated)	0.7
Aluminum (Copper-plated)	Copper (Bare)	1.0
Aluminum (Bare)	Copper (Nickel-plated)	1.3
Aluminum (Bare)	Copper (Silver-plated)	2.0
Aluminum (Bare)	Copper (Bare)	2.4
Aluminum (Bare)	Copper (Tin-plated)	2.7

5.5.10 Lubrication—Contact Aid Compounds

It has been known for some time that the use of suitable lubricant (contact aid compound) improves the performance of an electric contact. When the contact is made, the lubricant is forced out from the peaks of high pressure and hence the metallic conduction through the contact is not disturbed (see Section 3.4). As a result, the oxidation of clean metal surfaces is virtually prevented and a high area of metallic contact, hence, low contact resistance, and protection of the contact zone from adverse environmental effects are maintained. Factors affecting lubrication of power connections are shown in Figure 5.88. It is a common practice that for both aluminum and copper busbar connections, the contact surfaces should be abraded through the suitable contact aid compound with a wire brush or abrasive cloth. Due to a more rapid formation of an initial oxide film on aluminum, this procedure is more important for aluminum conductors than for copper. When electrical equipment is furnished with either silver- or tin-plated terminals, the plated contact surfaces should not be rubbed or brushed. Only the contact surfaces of the aluminum or copper bus that are to be bolted to these surfaces should be fixed. It is recommended that such plated surfaces, before being bolted to aluminum or copper bus, be cleaned with cotton waste and then coated with a suitable compound to serve as a sealer only.

Various contact aid compounds have been used for ensuring reliable operation of aluminum and copper connectors. Although many such compounds are commercially available, very little published information relevant to their functioning and efficiency under accepted utility service conditions can be found in the literature. The advantages, properties, functions, and performances of a variety of lubricants for separable and sliding electronic contacts is well documented [162–166]. The contact aid compounds commonly used for aluminum-to-aluminum and aluminum-to-copper connections were evaluated by Braunovic [167] on the basis of their effect on the performance and stability of a bolted joint under current cycling and fretting conditions, contact resistance force relationships, stability to thermal degradation, spreading tendency and ability to protect the contact against corrosion in saline and industrial pollution environment. The results, summarized in Table 5.26

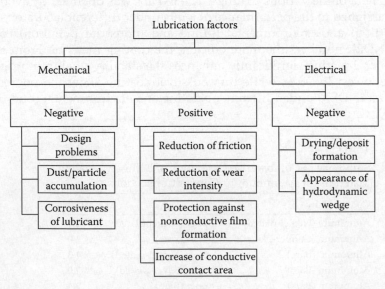

FIGURE 5.88
Factors affecting lubrication of power connections.

TABLE 5.26

Summary of Comparative Evaluation of Different Contact Aid Compounds for Aluminum-to-Aluminum and Aluminum-to-Copper Connections

Aluminum-to-Aluminum		Aluminum-to-Copper	
Contact-Aid Compound	Index	Contact-Aid Compound	Index
Penetrox A-13	0.7	Nikkei S-200	0.7
Silicon Vacuum Grease	0.7	Koprshield CP-8	1
Penetrox A	0.8	Penetrox A-13	1
Aluma Shield	1.2	Pefco	1
Fargolene GF-138	1.2	No-Oxid-A	1.3
Fargolene GF-158	1.2	Fargolene GF-158	1.4
Contactal HPG	1.3	Silicone Vacuum Grease	1.5
Petroleum Jelly	1.3	Non-lubricated	1.6
ZLN 100	1.5	Contactal HPG	1.8
Alcan Jointing Compound	1.8	Petroleum Jelly	1.9
AMP Inhibitor	2.5		
Kearnalex	2.5		
Non-lubricated	2.6		

provide an overall assessment of the various compounds and a more quantitative method of comparison. The INDEX is defined as an average of the numerical values assigned to the performance of joints under different laboratory and environmental conditions. The lower the INDEX number, the more efficient is the compound. It was shown that the compounds rendering a low initial contact resistance and having the performance INDEX < 1.0 enhance the stable performance of these connections.

One possible explanation is that the observed difference is connected with the shear strength of the compound film, that is, compounds with higher shear strength will require higher contact forces to bust the compound film on the contacting surfaces and establish metal-to-metal contacts. Hence, it appears that lower shear strength of contact aid compounds is very important in establishing and maintaining a very low joint contact resistance. Another possible explanation is the presence of metallic and/or non-metallic particles added to the compounds to improve their ability to shear the oxide films on the contacting surfaces. Although the results of the analysis of these additives were not sufficient to determine the attributes of such particles, it is most likely that their composition, size, shape, concentration, etc., is most likely responsible for the observed difference. The results are sufficiently consistent to conclude that certain contact aid compounds commonly used for aluminum-to-aluminum connections can also be used for aluminum-to-copper joints. From these results it can be concluded that contact aid compounds with the performance INDEX < 1.0, such as Penetrox A-13 and Nikkei, can be recommended for all connections involving aluminum. It is also important to note that most of the lubricants recommended contain additives in the form of oxide of metallic particles such as illustrated in Figure 5.89.

It was found [131] that the presence of additives in the lubricating oil has significant effects of the connector performance. The connections lubricated with grease An example of the beneficial effect of lubrication (Koprshield) is depicted in Figure 5.90 depicting the variation of the contact resistance of Al–Cu and Cu–Cu wire-plate combinations under fretting conditions. Note the substantial improvement in the contact resistance performance of lubricated wire-plate joints under fretting conditions. In the case of all-copper connections, it is

Failed current cycling test   Passed current cycling test
(a)   (b)

FIGURE 5.89
Scanning Electron Microscopy of particles added to lubricants. The lubricant A with sharp-edged particles failed the current cycling test; whereas the lubricant B with uniformly dispersed spherical particles passed the current cycling test.

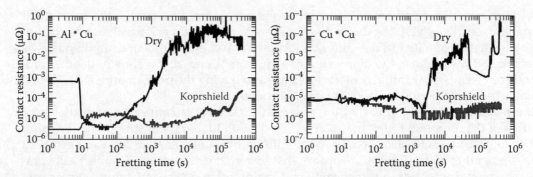

FIGURE 5.90
Effect of lubrication on the contact resistance performance of dry Al–Cu and Cu–Cu wire-plate combinations under fretting conditions.

generally accepted practice not to use any contact aid compound. Nevertheless, copper connections are also susceptible to degradation during their service life although their deterioration can proceed for a long time without any appreciable changes in their operation. This is a false sense of security since the experience has shown that the deterioration of copper connections occurs rather abruptly triggered by the accelerated interaction of chemical, thermal, mechanical and electrical processes at the contact interface. Hence, in order to extend the useful service life of copper connections, the use the contact aid compounds such as Nikkei or Koprshield or equivalent having the performance INDEX # 1.0 is strongly recommended.

5.5.11 Bimetallic Inserts

Another method of circumventing the incompatibility between aluminum and copper is the use of bimetallic aluminum-copper transition contact plates. These bimetallic inserts are copper and aluminium plates joined together by roll-bonding or other jointing

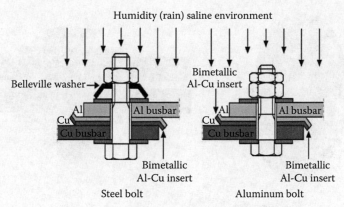

FIGURE 5.91
Typical configuration for aluminum-to-copper bolted joints with bimetallic insert.

methods which insure a continuous interface and good transition bond between those two metals. The bimetallic plates are primarily intended for the busbar type connectors and require the use of contact-aid compound (grease) to prevent the galvanic corrosion at the copper-aluminum interface. Figure 5.91 shows typical configurations for aluminum-to-copper bolted joints when bimetallic inserts are used. To prevent galvanic interaction between aluminum and copper, it is essential to place aluminum busbar and also aluminum side of the bimetallic insert above the copper conductor and also apply the suitable contact aid compound. It should be pointed out, however, that although the bimetallic contact plates reduce the risk of gross corrosion, the problem of jointing aluminum to copper still remains since the two new interfaces aluminum-aluminum and copper-copper, created in the contact zone, may introduce additional impediment to the enactment of the current across the junction. Also, these new contact surfaces have to be processed in the same manner as in the case of aluminum-to-aluminum and copper-to-copper connections that are to be brushed and greased.

5.5.12 Transition Washers

An alternative method of joining aluminum-to-copper is the use of a transition washer inserted between the contacting aluminum and copper surfaces. This method of joining is recommended for situations where plating is inconvenient, or where improvements are required for existing installations. The material used for transition washers is either 60/40 brass or high-strength Al–Mg–Si alloy. The sharp surface profile of these washers ruptures the oxide films without the need for further cleaning and establishes a substantially larger contact area that is more resistant to aging than direct surface contact. Examples of these transition washers commonly used in Germany and Great Britain, France, and Canada are shown in Figure 5.92. It is interesting to note that a transition washer made of high strength bronze and having the same geometry is currently used in Germany for jointing copper busbar conductors. It should be pointed out, however, that although these washers do not require surface preparation of the busbar conductors, the use of contact aid compounds is essential. Furthermore, in the case of brass transition washers, there is always a danger of the formation of intermetallics at the contact interface for, if not properly installed, the washer/conductor interface temperature may rise high enough as to elicit the formation of intermetallics.

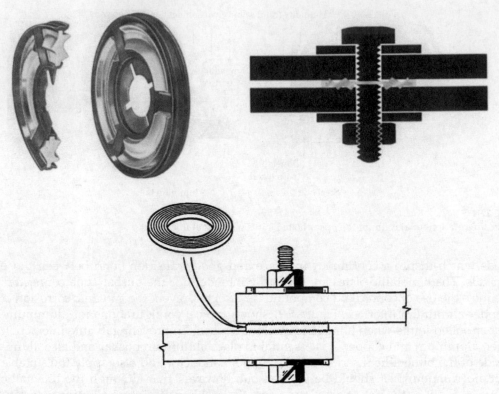

FIGURE 5.92
Transition washers made of Al–Mg–Si alloy (used in Germany) and brass (used in the UK, France and Canada).

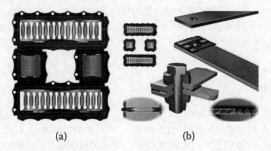

(a)   (b)

FIGURE 5.93
(a) Multilam contact element, (b) Applications of multilam contacts in busbar connection.

5.5.13 Multilam Contact Elements

The multilam contact elements are another type of transition washers made of copper-beryllium alloys. The basic characteristic of this connector type is a large number of defined lover-type contact points. Each louver form an independent current bridge thus creating a multitude of the current ways that substantially reduce the overall contact resistance. Figure 5.93 shows the application of the multilam contacts in the busbar joints [168]. The advantages of multilam contacts are high-current transmission in hermetically sealed chambers thus eliminating the need for plating, cleaning or lubricating of bus bars, ease of installation, modular design, substantially prolonged service life, cost effectiveness and high operating reliability.

5.5.14 Welded Connections

Welding is a highly acceptable method for making connections with all types of aluminum and copper conductors. It creates highly efficient electrical connections which are permanent, economical, have good appearance, and particularly suitable for connecting two members of different cross-section. A properly welded joint is most reliable joint from the electrical standpoint since there is an essentially homogeneous union with no contact resistance to generate heat from high currents. There are a multitude of welding processes that can be used for joining aluminum and copper but only those most commonly used for welding power connections and conductors will be drawn. A detailed description of different welding processes can be found in *ASM Handbook Volume 6* "Welding, Brazing and Soldering" [169].

5.5.14.1 Thermite (Exothermic) Welding

Thermite welding is a fusion welding process in which two metals become bonded over being heated by superheated metal that undergoes an aluminotermic reaction. The liquid metal resulting from the reaction between a metal oxide and aluminum acts as the filler metal and hangs around the conductors thus making a molecular weld. Thermite welding is extensively used for making grounding connections between copper conductors. The advantages of this process are excellent current-carrying capacity equal to or greater than that of the conductor, high stability during repeated short-circuit current pulses and excellent corrosion resistance and mechanical strength. Advantages and disadvantages of exothermic (termite) connections are presented in Table. 5.27. Figure 5.94 show typical crucible-mold setup for welding copper conductors.

5.5.14.2 Friction Welding

Friction welding is a solid-state welding process in which the passion for welding is produced by direct conversion of mechanical energy to thermal energy at the contact interface without the application of external electrical energy or heat from other sources. Friction

TABLE 5.27

Advantages and Disadvantages of Exothermic (Thermite) Connections

Advantages	Disadvantages
Excellent current-carrying capacity equal to or greater than that of the conductors, high stability during repeated short-circuit current pulses, excellent corrosion resistance and mechanical strength	Cost, lack of repeatability, numerous mold requirements, potential down-time caused by inclement weather or wet conditions, safety risks to personnel and equipment
	The intense heat damages both the conductor and its insulation, anneals the conductor so that exothermic connections cannot be used in tension applications
	The resultant weld material exhibits lower conductivity and physical properties than the conductor, being similar to cast copper

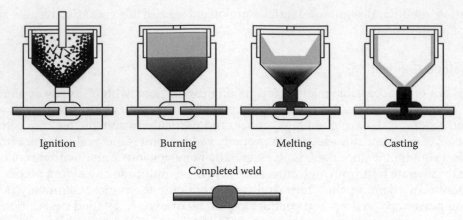

Ignition   Burning   Melting   Casting

Completed weld

FIGURE 5.94
Typical crucible-mould set-up for thermite welding of copper conductors.

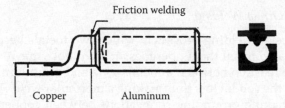

Friction welding

Copper   Aluminum

FIGURE 5.95
Typical friction-welded terminals.

welds are created by taking a non-rotating work piece in a contact with a rotating work piece under constant or gradually increasing pressure until the interface reaches welding temperature and then the rotation is interrupted to complete the weld. The frictional heat produced at the interface rapidly raises the temperature of the work pieces, over a very short axial distance, values approaching but bellow the melting range. During the final stage of welding process, atomic diffusion occurs while the interfaces are in contact, allowing a metallurgical bond to form between the two fabrics. This welding process is used to make bimetallic aluminum-to-copper terminals and thus avoid direct contact between aluminum and copper. These terminals are usually copper flat plates or tubes friction welded to compression-type aluminum connectors. When installed, the friction welded terminals convert the aluminum-to-copper joint into an aluminum-aluminum and copper–copper joint. This type of a connector is widely used in Europe. Figure 5.95 illustrates a typical friction welded terminal.

5.5.14.3 Explosion Welding

Explosive welding is a cold weld pressure welding in which the controlled energy of a detonating explosive is used to create a metallurgical bond between two or more similar or dissimilar metals. No intermediate filler metal, that is, brazing compound or soldering alloy is needed to promote bonding, and no external heat need be given. Diffusion does not occur during bonding. As a consequence of the high-velocity collision of the two metals, the contaminant surface films are jetting off the base metals. The best attribute of the explosive welding is that dissimilar metal systems can be bonded even when conventional fusion welding techniques are metallurgically unacceptable because of the formation of intermetallic compounds. The most common use of explosive welding is the production of corrosion resistance clad metals and transition joints used to aid dissimilar metal welding. This technique has been used for splicing the overhead transmission lines [170]. The explosively welded connectors are completely metallic, void free and offer a totally compressed uniformly smooth and straight fitting without bird caging the conductor. The explosively welded connectors are completely metallic, void free, and offer a totally compressed uniformly smooth and straight fitting without bird caging the conductor. However, despite the apparent advantages of this technique for splicing the transmission lines, there is a number serious disadvantage such as the safety of the personnel during installation and considerable noise which limits the use of this technique in highly remote geographic locations. During the explosive welding of high residual stresses are brought forth in the bond zone that can significantly increase the creep and stress relaxation of a joint, hence, annealing is required to remove these stresses. Furthermore, because of the high pressures generated in the bond zone will exceed the dynamic yield stress of the metals and they will flow plastically during the procedure.

5.5.14.4 Resistance Welding

Resistance welding is a procedure in which varying surface are joined in one or more spots by the heat generated by resistance to the flow of electrical current through busbars or conductors that are held together under force by electrodes. The contacting surfaces in the area of current concentration are heated by a short-time pulse of low-voltage, high-amperage current to form a fused nugget of weld metal. When the flow of current ceases, the electrode force is preserved while the weld metal rapidly cools and solidifies. This method of welding is widely used for joining copper and copper alloys. The resistance weldability of any copper alloy is inversely proportional to its electrical and thermal conductivities. Most copper alloys require a short weld time, low electrode force, and high current and different electrode materials that are compatible with the alloy being welded.

5.5.14.5 Resistance Brazing

Resistance brazing is essentially a resistance welding process in which the work pieces are heated locally and filler metal, inserted between the work pieces, melts by the heat obtained from the resistance to the flow of electric current through the joint. Parts of any shapes can be resistance brazed, provided the surfaces to be joined are either flat or conform over a sufficient contact area in order to maintain the pressure permitting the heating current to flow through the joint, thus, enabling the filler metal to be distributed throughout the joint by capillary action. This welding process is very suitable for joining copper and its alloys. The use of high-resistitivity electrodes provides an efficient method of localized heating at the joint but the avoiding fusion of the copper base metal. A flux is used in almost all resistance brazing since it provides a coating that prevents or minimizes oxidation of the joint during heating, dissolves oxides present or formed during heating and assists molten filler metal in wetting the contact zone to promote capillary flow. Somewhat a similar resistance brazing method has been recently developed by Nguyen-Duy et al. [171] to join solid silver plates to copper for the contacting members (jaws and blades) of high-voltage disconnect switches. Instead of traditional filler metals, the amorphous silver was used. The effects of severe wear, current-cycling and short-circuit tests, conducted on the switches having solid silver contacts, showed significant superior performance when compared with those having silver coated contact surfaces.

5.5.15 Connector Design

In order to match the mechanical and electrical requirements and also assure reliable performance of a connector during its expected service life, various plans have been developed and used in the field with varying degree of success. The generic connector designs commonly used on distribution network are illustrated in Figure 5.1. The field experience has shown that a wide variety of connector designs have given good service over several years, but the factors contributing to their success and to failure elsewhere could not be ascertained with any degree of certainty. However, the ever-increasing demand for electricity in recent years increased electrical loading on power transmission and distribution lines by many utilities which, in turn, raised the average operating temperature of conductor lines, up to 130°C during times of peak force transmission. Connectors are now required to run reliably at relativistic elevated temperatures.

The field and laboratory tests showed that most of the compression-type connectors and non-lubricated aluminum-to-copper or aluminum-to-tin plated copper busbar joints showed the greatest instability, whereas the fire-driven wedge and bolted type with adequate surface preparation and lubrication showed the most stable contact performance under different operating (environmental) and current-cycling conditions. Typical effects of current-cycling and field exposure tests are presented in Figures 5.96 and 5.97. The result of 10 months exposure in saline and industrial environments on the performance of lubricated and non lubricated bolts-type connectors are shown in Figure 5.98. The contact configurations Al–Ni, Al–Ag, and Al–Sn are Al-base conductors in contact with Ni-, Ag-, and Sn-plated copper-conductors. The superior performance of nickel-plated connections is evident. Detailed surface analysis of the connectors exposed to a saline environment indicated that severe corrosion had occurred and that corrosion products accumulated between the connector aluminum body and copper conductor. Accumulation of these

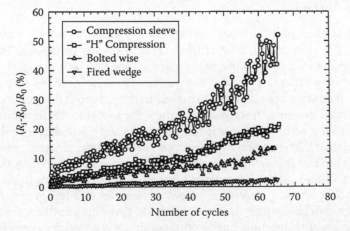

FIGURE 5.96

Effect of current-cycling on the contact resistance of different types of connectors. (R_i is the maximum contact resistance after the i-th cycle; R_0 is the initial contact resistance at zero cycle).

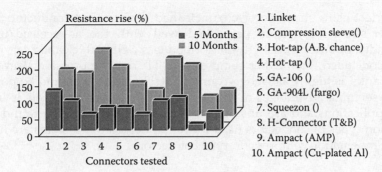

FIGURE 5.97
Effect of field exposure in a saline environment on the contact resistance of different connectors.

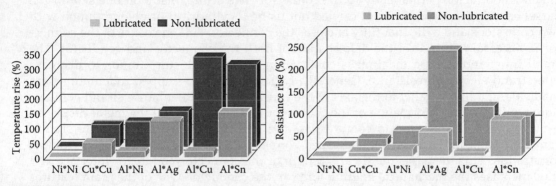

FIGURE 5.98
Effect of 10 months of exposure to saline and industrial environments on the performance of lubricated and non-lubricated bolted-type connections. Note: Al–Ni, Al–Ag, and Al–Sn are Al-base conductors in contact with Ni-, Ag-, and Sn-plated copper-conductors [172].

products was most pronounced in compression-type connectors. The severity and tenacity of the saline environment could not prevented by using the water-sealing products (mastic, tapes) since the severe corrosion and accumulation of corrosion products were clearly evident in the connectors completely covered by this supposedly water-sealing mastics or tapes. Covering the connectors with the products that cannot insure a complete hermetic seal would result merely in an accelerated corrosion since the water-sealing agents will prevent the escape of corrosion products from the contact zone.

5.5.15.1 Fired Wedge-Connectors

Jondhal et al. [173] and Sprecher et al. [174] evaluated a number of different tap-connectors commonly used on the web to provide an electrical connector between a main power cable and a tap conductor. The evaluation was prompted by the results of infrared surveys of the overhead facilities indicating that a substantial number of the tap-connectors were running unacceptably hot. The evaluation was implemented in accordance with the accepted testing procedure as specified by the ANSI C119.4 standard. The results showed that fired wedge-connectors perform best overall in all the tests carried out. Fired wedge-connector technology appears to be superior to all other joining technologies. Although compression "H" connectors performed well with service-aged/cleaned conductor, they did not pass the more severe test using the as-received/uncleaned service-aged conductor. Similarly, the compression sleeve connectors performed well with the as-received/uncleaned service-aged conductor but, surprisingly, performed less well with cleaned serviced-aged conductor. It is concluded that neither compression "H" nor compression sleeve connector technology ranks as highly as wedge-connector technology. It should be pointed out that despite the seeming advantages of firing wedge-connectors, their role is limited to overhead line applications and are unsuitable for the role of the underground systems. The most common types of connectors used for the underground systems are hexagonal or stepped deep indentation compression-type connectors.

5.5.15.2 Stepped Deep Indentation Connectors

Stepped deep indentation connectors use a specially-designed crimping technique to attach a conductor permanently to the connector. The crimp, made on one side of the barrel carrying the conductor, is carried out using a cylindrical or oblong punch with two cone steps and a die that fully encloses the connector. Full enclosure of the connector is necessary to control the deformation of both the connector and conductor. The small inner indentation maximizes compaction of the conductor strands to decrease interstrand electrical resistance. Generally, two separate two-stepped indentations are made along a line parallel to the axis of the connector bore to enhance strand compaction. Owing to the deformation of the connector and conductor, all the metal displaced by the punch is subjected to compressive and shearing forces that generate significant cold welding between the deformed connector walls and the displaced conductor strands. This increases the shear strength of the connector/conductor interface and thus increases the mechanical holding force on the conductor. One of the main features of stepped deep indentation connectors is that all the conductor strands participate in current conduction, thus yielding exceptionally low contact resistance. Since the walls of the small deep indentation can only be deformed in compression, the two cone steps generate a deformation geometry that resists the metal flow in the connector after crimping. Thus the electrical interfaces do not degrade owing to creep flow or stress relaxation in the connector.

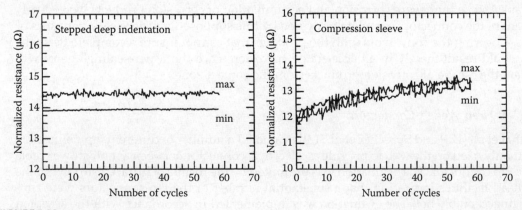

FIGURE 5.99
Effect of current-cycling on the normalized contact resistance of 500 kcmil stepped deep indentation and compression sleeve connectors.

The reliability of different types of connectors commonly used for the underground network has been evaluated by Braunovic [175]. The connectors evaluated included compression sleeve, bolted, elbow (load break) and steeped deep indentation connectors. The effects of accelerated current-cycling tests showed that steeped deep indentation and bolted (with proper surface preparation and lubrication) connectors showed superior performance over that of the compression sleeve and elbow connectors. The results of current-cycling tests for stepped deep indentation and compression sleeve connectors are shown in Figure 5.99. It is clear that the contact resistance of stepped deep indentation connector was not affected by the current-cycling while the compression sleeve connector showed unstable behavior as manifested by a gradual increase of its contact resistance with the number of current cycles. The inferior performance of the compression sleeve connector is attributed to its pattern which does not provide sufficient mechanical deformation to bring down the initial contact resistance to a level where it is independent of the deformation applied. Actually, the initial contact resistance of compression sleeve connector was about 30% higher than that of the stepped deep indentation connector. It seems that in the compression connector the contact surface does not fully envelope the conductor which leads to a greater tendency to substantial relaxation from the slippage of strands subsequent to installation.

5.6 Connector Degradation

Electrical connections are designed to operate over a long period of time. However, their age and their life span and optimum life cycle managements are affected by a number of factors, such as original design criteria, manufacturing process, operating conditions, maintenance procedure and safety consideration. Hence, the accumulated damage can be measured, their history analyzed and an expected remaining life calculated. The remaining life is the period of time after which the probability of failure becomes unacceptably high. A common feature in usefulness and life expectancy of any component is expresses by the well-known "bathtub" reliability curve, shown in Figure 5.100. As can be seen, there is a high probability of failure during the first few hours or weeks of operation, break-in

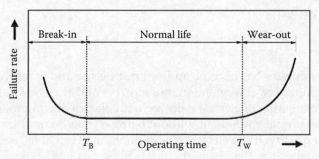

FIGURE 5.100
Failure probability ("bathtub") curve.

(start-up) period, usually caused by manufacturing or installation problems. Following this initial period, the probability of failure is relatively low for an extended period (normal life) until it increases sharply with elapsed time or hours of operation (wear-out period). The sum of operating time and the most probable expected remaining life may or may not match the original design life. Hence, the expected remaining life, while still a range or a probability distribution, will probably be more precise. Every component of an electrical system can reach the failure situation, which will likely occur as the operating time approaches the end of its life span. The useful lifetime of an electrical component or a device is determined by the design and manufacture and is affected by a series of service conditions considered as normal. Clearly, if during the lifespan of a component the operating conditions are or become more severe than expected, the useful life will be shortened at a higher rate.

5.6.1 Economical Consequences of Contact Deterioration

Electrical connections are designed conservatively. The properties of the materials used in the connections may deteriorate owing to in-service aging or under the influence of deterioration mechanisms that can reduce their useful lifespan and/or reduce their operating safety margin. Stretching out the use of older connections beyond their originally anticipated life can offer major economic benefits. In order to draw out their operating life, however, their remaining useful lifetime must be evaluated to ensure that safety and structural integrity are held during the extended operating period. Managing and extending the life of critical power equipment require reliable and continuous monitoring as the validity of any action taken regarding the lifetime of a component. Once an error has been discovered and its evolution is monitored, the severity of the defect can be assessed and decisions made what actions to be considered [176]. Potential damages will be restricted when incipient faults are detected and timely actions implemented. Early detection limits the amount of adjacent damages and confines the area calling for repair and maintenance. The early detection of incipient faults will greatly reduce unplanned power outages and improve the reliability of the power and service supplied. Furthermore, since fault conditions often lead to catastrophic failures, their early detection will limit these consequences and thus insure the safety of the substation personnel. Monitoring a fast developing fault and evaluating its progress provides the necessary information to apply all the substantive resources to react on time and reduce the overall damages. Itemized power equipment/components susceptible to ageing and different modes of deterioration, identify the cause and indicates potential impact and cost are shown in Table 5.28.

5.6.2 Power Quality

In recent years, power quality has become an important issue and is receiving increasing attention by utility companies, equipment manufacturers and the end-users. Power quality is defined as the interaction of electrical power with electrical equipment. Correct and reliable operation of electrical equipment without being damaged or stressed qualifies as good electrical power quality. If the electrical equipment malfunctions, is unreliable, or is damaged during normal usage, the power quality is poor. Generally speaking, any deviation from the normal circuit voltage source (either DC or AC) can be classified as a power quality. Most power quality problems are chain reactions. An incipient event causes an

TABLE 5.28

Itemized Power Equipment/Components Susceptible to Ageing

Application	Cause	Impact
Power distribution	Poor breaker connections	Overheating, arcing, burning, fire
Circuit breakers	Overheating, overloading	Conductor strands broken—
Conductors	Conductor strands broken	overhead line could come down
Splices	Loose/corroded/improper	Expensive repair and replacement
Disconnect switches	connections and splices	Safety considerations
Miscellaneous power	Loose/corroded connections	Arcing, short-circuiting, burning, fire
components	Poor contacts	25% of all power equipment failures
Switches breakers	Overloading	are caused by loose electrical
	Overheating	connections
		Cost of repair and replacement very
		expensive
		Safety considerations
Transformers	Loose/deteriorated connections	Arcing, short-circuiting, burning,
	Overheated bushings	fires
	Poor contacts (tap changer)	Expensive rewinding and
	Overloading	replacement

Note: Different modes of deterioration identifies the cause, potential impact, and cost.

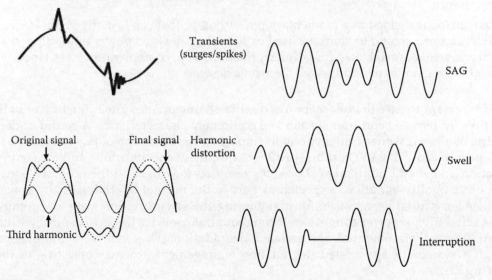

FIGURE 5.101
Most common electrical disturbances found in the electrical system.

electrical disturbance, which is conducted by electrical system and eventually reaching all electrical equipment from power connectors to sensitive electronics. The most common power quality issues shown in Figure 5.101 are as follows:

Disturbances refer to power quality variations that occur at random intervals but are not associated with the steady-state characteristics of the voltage.

Transients, also called surges and spikes, are distortions of electricity caused by lightning, large motors starting, routine utility activities and other conveniences.

Harmonic distortion is deviation from a perfect sine wave represented by sinusoidal components having a frequency that is an integral multiple of the fundamental frequency caused by the power supplies of certain electronic appliances, including televisions, fax machines, and especially personal computers. Harmonic distortion can overheat building transformers, building wiring, wiring in modular office panels, motors, and components in some appliances. This results in an increase in power consumption. The immediate effects of harmonic distortion are an increases in peak voltage, malfunctioning of control and/or regulation circuits, false switching of electro-technical and electronic equipment and interfere with the neighboring telecommunication lines.

Sag is a short (less than a second) decrease in the normal voltage level caused by faults on distribution and transmission circuits. Sags do not damage equipment, but can cause computers to restart or lock up and other appliances to lose memory. Even though the effects of these disturbances can be the same as long-duration outages, voltage sags can be more important because they occur much more frequently.

Swell is a short (less than a second) increase in the normal voltage level. Mostly caused by motors stopping, swells generally do not upset or damage appliances but can initiate the failure of a stressed component in an electronic appliance.

Interruption is defined as a momentary power outage that can last anywhere from fractions of a second to hours. Caused by lightning, downed power lines, tripped circuit breakers and blown fuses, interruptions disrupt computer processes, clocks and the memories of unprotected electronic devices.

It is important to note that many power quality characteristics are a function of both the supply system and end-user system and equipment characteristics. A sound understanding of electrical power quality and its impact on the operation of power systems is a vital part of acquiring the best blueprint for equipment specifications and for facility protection. In general, equipment should be designed to withstand the normal steady-state power quality variations expected as part of the normal operation of the power system. An essential prerequisite for maintaining the structural integrity and retaining the reliability of power connectors throughout their service life is to control within defined limits their potential aged-related degradation. This can be accomplished through a systematic age-related degradation management process consisting of the following tasks:

- Understanding deterioration mechanisms—Failure modes
- Minimizing degradation of power connectors
- Inspection, monitoring and assessment
- Maintenance, repair and replacement
- Development of utility centered maintenance program

Although much progress has already been made in diagnostic and monitoring of functioning and efficiency of power

5.7 Prognostic Models

The most important prognostic model requirement is the estimation of the remaining life of individual components such as connectors and disconnects switches, rather than that of a population of components. Accurate prognostication enables different palliative measures to be developed and trained towards power equipment most susceptible to deterioration owing to ageing, environmental and other adverse influences affecting its performance and properties. One of the most important prerequisites for developing prognostic model is a collection of information on the performance of a powerful component of its initiation through final state. Development of a prognostic model comprises two important but separate parts: model derivation and validation. Deriving and validating a model along the same dataset leads, by definition, to over optimistic estimates of the model's accuracy. A statistically valid model provides unbiased predictions when applied to new datasets, but this quality does not necessarily mean that the model has applicable validity. To be applicable the model must be precise enough to serve the purpose for which it was acquired. Models based only on genes which are highly related to the component performance are more likely to be similarly predictive in some other context. From a practical viewpoint simple models are more likely to be readily integrated into utility maintenance and predictive practices with minimal disruption.

5.7.1 Prognostic Model 1 for Contact Remaining Life

For the prognostic model of remaining life of power connection it was assumed that the contact interface is homogeneous with a ring shaped a-spot as shown in Figure 5.102 [177]. For the purpose of this example, it was assumed that ingress of oxygen and oxide film growth is one of the key factors affecting the conductive area in contact between two metallic surfaces. A full description of this model can be found in Reference [2]. Figure 5.103 shows the calculated contact lifetime of a bolted connection as compared with the observed contact resistance variation with time.

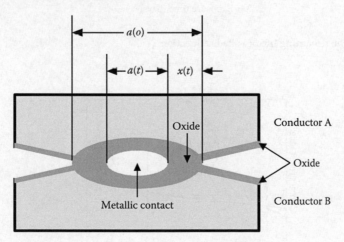

FIGURE 5.102
Schematic view of the contact spot.

5.7.2 Prognostic Model 2 for Contact Remaining Life

The contact ageing process is schematically illustrated in Figure 5.104. Stage 1 is a running-in period, stage 2 is a normal operation period and stage 3 is the intensive ageing period. One of the essential tasks of reliability theory is the prediction of the residual lifetime of an article at the melting stage. According to theoretical approach to reliability, the proper methods for forecasting residual life are based on the physical statistical evaluations. These methods are based both on the statistical mathematical methods (time series analysis) and the physical modeling of the object ageing. It should be mentioned that a time dependence of a contact resistance $R(t)$ is not a smooth function. It is a sum of a regular component (trend) and stochastic fluctuations that are shown in insert in Figure 5.105. From the statistical point of view the contact resistance time dependence $R(t)$ is a typical example of a non-stationary stochastic process [178].

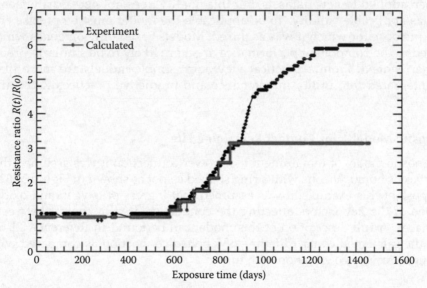

FIGURE 5.103
Example of predicting the remaining life of bolted connection.

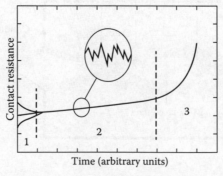

FIGURE 5.104
Schematic of the ageing process of a permanent electrical contact.

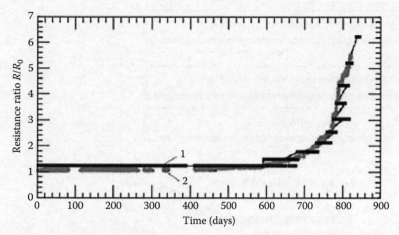

FIGURE 5.105
Estimated and experimentally observed contact resistance of a bolted joint (1) Estimated; (2) Experimentally observed.

5.7.3 Physical Model

Takano and Mano [179] were the first to propose a diffusion model of ageing of an electrical connection; see Figure 5.106. According to this model, the insulation surface film grows into the contact interface as a result of diffusion of an oxidizing agent within the contact area. In the central zone of a contact spot there is a quasimetallic area, where the surface film is negligibly thin. The conductivity of this quasimetallic area is rather high owing to tunnel and thermo-ionic mechanisms of current transfer. As the film thickness increases the radius of quasimetallic spot decreases.

5.8 Shape-Memory Alloys (SMA)

Shape-memory alloys were introduced in Section 5.5.8. Although they were observed as long ago as 1938, they were not widely recognized until 1962 when the Ni-Ti alloy was discovered by Buchler et al. [180] of the U.S. Naval Ordnances Laboratory. The alloy was named "Nitinol" for *(NI)*ckel-*(TI)*tanium Naval Ordnance Laboratory *(NOL)*. The SME is unique to two groups of alloys: nickel-titanium (Ni-Ti) and copper-based alloys (Cu–Zn–Al and Cu–Ni–Al) whose transition temperature is highly sensitive to the alloy composition and thermo-mechanical treatment. In the case of Ni-Ti, the temperature can be varied from –200°C to +100°C, whereas with copper-based alloys it can range from –105°C to +200°C. Their practical applications are numerous. Comprehensive reviews of the general applications of shape-memory materials and in the electrical industry are given by Schetky [149] and Braunovic [150].

5.8.1 Origin of Shape-Memory Effect

The shape-memory effect is a consequence of the continuous appearance and disappearance of martensite with rising and falling temperature [180]. The martensitic transformation is spontaneous and reversible, and occurs during cooling from the parent phase or austenite to the martensitic state or martensite. This phase transformation is a first-order displacive

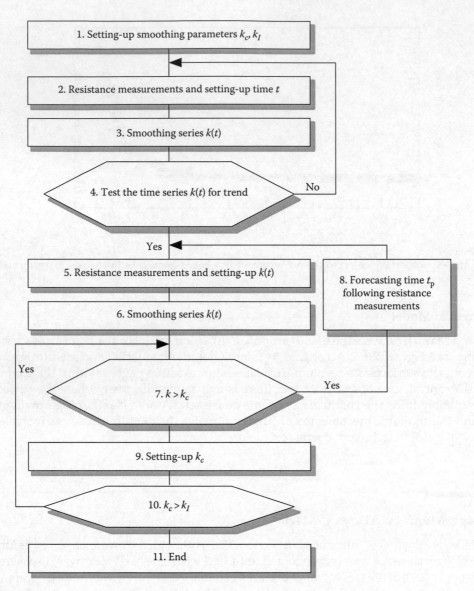

FIGURE 5.106
Residual life flow chart for electrical contacts.

transformation in which a body-centre cubic phase (austenite), on cooling, transforms by a shear-like mechanism to martensite, which is both ordered and twinned. The martensitic transformation is diffusionless, that is, it involves a cooperative rearrangement of atoms over a short distance into a new stable crystal structure without changing the chemical nature of the matrix; see Figure 5.107.

5.8.1.1 One-Way Memory Effect

The one-way memory effect can be explained simply, using two examples: a straight SMA wire fixed at one end and a bent wire (Figure 5.108). Stretching or bending the wire beyond its yield point at room temperature results in deformation after unloading

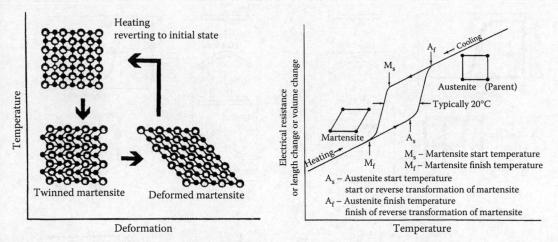

FIGURE 5.107
Schematic of one-way and two-way shape-memory effect.

and the wire will remain in the stretched or bent condition until it is heated to above the transformation temperature of the alloy (A_f) no load is applied, it will return to its original shape. Subsequent cooling below the transformation temperature (M_f) causes no macroscopic shape changes. The one-way effect can be repeated many times. Reactivation of the shape-memory effect can only be done by deforming the wire in the martensitic state. Since no special treatment is necessary, cyclic use of the one-way memory effect with external reset force in many instances is the more economical solution.

5.8.1.2 Two-Way Memory Effect

The two-way memory effect refers to the ability of SMA to assume one shape in the martensitic state (low temperature) and then spontaneously change to a second shape when heated to above A_f (high temperature) (Figure 5.107). This process can be cycled fairly reproducibly between different shapes by simply changing neither the temperature, providing the strain level involved is not excessive nor the exposure temperature too high. The two-way effect can only be produced by special thermomechanical treatment ("training") which comprises multiple heating/cooling cycles under the same applied external force, thus allowing the material to "remember" the training process.

5.8.2 Applications of SMA in Power Connections

The SMA technology has not yet penetrated the electrical industry significantly, in part owing to a general lack of awareness as well as to the rather limited engineering data available about shape-memory materials. However, a growing use of such materials indicates a wider recognition of the advantages and potential of this technology.

5.8.3 Electrical Connections

The first SMA connector, called *Cryocon* [181] was a pin-and-socket type developed by Raychem Corporation in the mid 1970s for the Trident Missile Program. The connector forms a high-compression fit yet can be quickly released and recoupled. The design is in the form

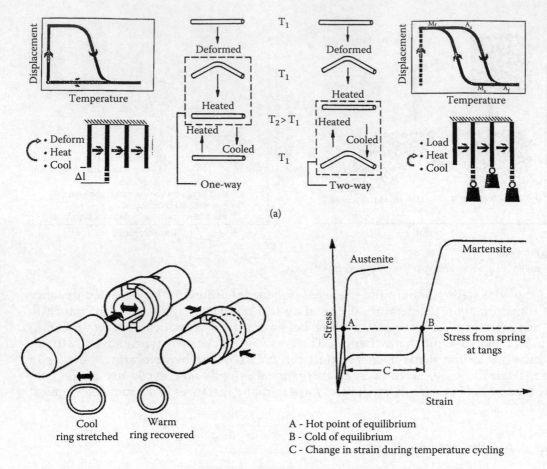

(a)

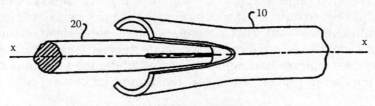

A - Hot point of equilibrium
B - Cold of equilibrium
C - Change in strain during temperature cycling

The cryocon SMA connector

The Souriau SMA connector

FIGURE 5.108
Crycon pin-and-socket connector with its stress-strain characteristics and the Souriau pin-and-socket connector.

of a heavy walled tube referred to as a biasing spring (typically of made of Be-Cu alloy or other conductive spring material) and a ring made of Ni-Ti, called the driver; see Figure 5.108.

The *Souriau connector* [182] is a pin-and-socket type and operates using the two-way memory effect. After training, the SMA (Cu-Zn-Al) socket can function in two states: open when cold and closed when hot (ambient temperature). The make/break operations can be performed many times without destroying the electrical continuity or mechanical strength of the connector. The gripping forces developed in the Souriau connector are five times higher than those exerted in a conventional connector thus providing good resistance to shock and vibration. A simplified schematic of the Souriau connector is also seen in Figure 5.108.

5.8.4 Temperature Indicators

In recent years a number of very simple SMA-based temperature indicators have been developed and used on the network. The advantage of this type of temperature indicator over infrared (thermography) detection is the possibility of continuous supervision of the condition of the joints, electrical contacts of different units of power system equipment and devices, etc. Furthermore, owing to intrinsic characteristics of the shape-memory materials, these indicators can be used as activators which can collect and transmit the collected temperature data to a remote computer-based data acquisition system, located usually, but not necessarily, at substation. These activators can also provide on-site monitoring of the equipment performance and interpretation of this data in terms of Pass/Fail warning and event marking such as date, duration and number of temperature excursions and, thus, a more detailed assessment of the state of various devices and contact junctions on the power system [183,184]; see also Reference [2].

5.9 Metal Foam Materials

In recent years there has been considerable increase in interest for metal foams Such an unabated interest in these materials is a result of not of only what these materials are but more so what they can offer in terms of a variety of properties [185–189]. Metals foams are a new class of materials with low densities and novel physical, mechanical, electrical, thermal and acoustic properties. Owing to their intrinsic structure and attributes, the metallic foams offer significant performance gains as the efficient energy absorbers for light, stiff structures, efficient absorption of energy, sound and vibration, and other unique and specialized applications.

5.9.1 Aluminum Foam Materials

The most widely used and investigated metallic foams are aluminum- and aluminum alloy-based foams. Aluminums foams have very interesting properties, in that they preserve many of the general characteristics of the parent metal, while offering the very high strength to weight ratios of honeycomb type structures and enhanced abilities for energy and sound absorption, as well as enhanced impact, insulating and shielding characteristics. Figure 5.109 illustrates some typical aluminum foam-based materials. The foam material is enclosed between solid aluminum plates. It can be closed cell or open cell: see Figure 5.109a and b. Mechanical properties of aluminum foams are best illustrated by their stress-strain characteristics as shown schematically in Figure 5.110. The enhanced properties of aluminum foams generate new substitution opportunities for aluminum,

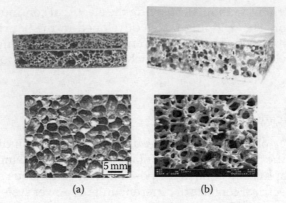

(a)   (b)

FIGURE 5.109
(a) Typical foam materials (b) Foam material is enclosed between solid aluminum plates. (Courtesy of A.G.S. Taron Technologies.) and examples of (a) Closed cell and (b) Open cell aluminum foams.

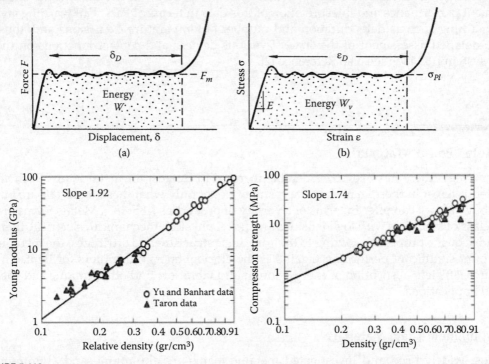

FIGURE 5.110
Schematic of compression stress-strain characteristics of aluminum foam [188] and Young modulus and compression strength as a function of foam relative density and density, respectively.

steel, wood sheathing, other composites and materials in market applications where their characteristics ensure new added values:

- *Ultra light weight* of aluminum foams results in a factor of 5 in terms of reduction in weight compared to steel of equal stiffness.
- *Energy and vibration absorption* properties of aluminum foams allow absorption of a large quantity of mechanical energy at almost constant pressures and

TABLE 5.29

Mechanical Properties of Foam Materials Produced by Different Manufacturers of Aluminum Foam Materials

Source	Density (gr/cm³)	Elastic Modulus (GPa)	Compression Strength (MPa)	Tensile Strength (MPa)	Densifcation Strain (%)	Absorption Capacity (Nm)
Alporas	0.2–0.25	0.4–1.0	1.3–1.7	1.6–1.8	0.7–0.82	
Alullight	0.3–1.0	1.7–12	1.9–14.0	2.2–30.0	0.4–0.8	
Cymat (SiC reinforced)	0.07–0.56	0.02–2.0	0.04–7.0	0.05–8.5	60–90	
Fraunhofer	0.38–0.54	0.61–1.33	1.88–2.82			
Taron Bare/Natural Skin	0.59–0.85	4.1–7.5	6.94–11.7		6.2–50.0	20.6–120
Taron Hard Wire Reinforced	0.71–0.87	7.1–9.2	17.9–23.9		20.5–35.0	65–240
Taron SiC Reinforced	0.39–0.4	7.0–12.0	3.4–6.7		45.8–52.9	131–135

thus open up significant potential in armouring applications for commercial, industrial and military uses.

- *Mechanical damping* capacity of metallic foams is omni-directional and larger than that of solid metals by up to a factor of 10; see Table 5.29.
- *Thermal management* characteristics of metallic foams provide exceptional heat transfer ability and resistance to direct flame and combustion reduce system costs when used as wall structure in thermally controlled containers or rooms.
- *Acoustic absorption* capacity of metallic foams is owing to their reticulated structure.

5.9.1.1 Electrical and Thermal Properties of Foam Materials

Figure 5.111 depicts the variation of electrical conductivity of a number of aluminum foam materials as a function of relative density [188]. The resistance data were obtained by placing the micrometer probes inside the foam and measuring the resistance sampled by the potential probes. From the data shown, it appears that the conductivity data can be approximated by the following expression:

$$\sigma/\sigma_{Al} = \alpha \, (\rho/\rho_{Al}) + (1-\alpha) \, (\rho/\rho_{Al})^{3/2} \tag{5.34}$$

Where α is a coefficient $\alpha = 0.05$. This expression is shown as a continuous line on Figure 5.111a. Figure 5.111b shows the thermal conductivities of aluminum foam and dense aluminum foam precursor. As it can be seen thermal conductivity of aluminum foams is several orders of magnitude lower than that of precursor. Therefore, owing to such low thermal conductivity, aluminum foams are generally not suited for simple thermal insulation but can be used for fire protection.

5.9.1.2 Power Connection Applications

The mechanical and physical properties of metal foams are determined by their cellular structure and relative density that are strongly dependent on the production method and the production parameters; see Table 5.30. It should be borne in mind, that although

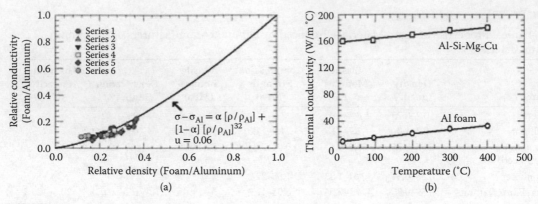

FIGURE 5.111

(a) Relative electrical conductivity of aluminum foams as a function of relative density, (b) Thermal conductivity of aluminum foam and aluminum precursor.

TABLE 5.30

Properties of Metallic Foam Materials

	Ni Foam	Cu Foam	NiCr Foam	Al Foam	FeNi Foam	Al Oxide Foam
Composition	Ni 99.9%	Cu 99.9%	Ni 75% Cr 25%	Al 96% Si 4%	Fe 65% Ni 35%	Al_2O_3 98% Ca, K, Cr
Density (g/cm³)	0.35–0.7	0.45–0.8	0.4–0.6	0.16–0.4	0.4–0.6	0.2–0.8
Porosity (%)	85–97	85–95	90–95	88–97	85–95	80–95
Melting point (°C)	1445	1080	1400	660		
Application point (°C)	650 (450)	250 (100)	800	250 (140)	400 (250)	1350

Notes: Density depends on production technology, composition and porosity. Temperature applications depend on production technology, composition and aggressive environment.

the specific mechanical and physical properties of cellular metals are inferior to their bulk properties, the metallic foams can, nevertheless, be efficient when their structural properties are explicitly used. The current transfer across the contact with aluminum foam washer is shown schematically in Figure 5.112 [190].

5.9.2 Copper Foam Materials

The pore size of copper foam materials ranges from 0.1 to 10 mm. The porosity is in the range of 50–98%. The volume density is in the range 0.1–0.8 g/cm³. Metallic foams typically retain some physical properties of their base material. Foam made from non-flammable metal will remain non-flammable and the foam is generally recyclable back to its base material. Coefficient of thermal expansion will also remain similar while thermal conductivity will likely be reduced. Figure 5.113 illustrates some shapes of copper foam materials.

5.9.2.1 Applications of Copper Foam Materials

The performance of copper-based connections can be greatly improved with the use of Ecocontact [191] inserts made of copper-based foam materials; see Figure 5.114. The foam enhances contact between contacting members and eliminates electric overload. It

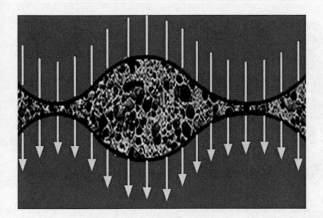

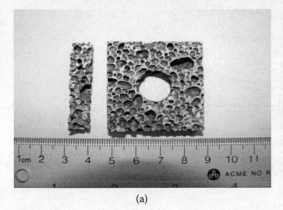

(a)

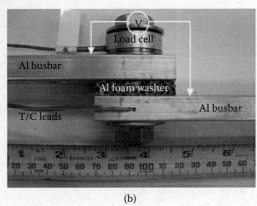

(b)

FIGURE 5.112
Schematic of current transfer across the contact with aluminum foam washer and (a) Aluminium foam, (b) Bolted joint configuration with Al foam washer and load cell.

FIGURE 5.113
Shapes of copper foam materials.

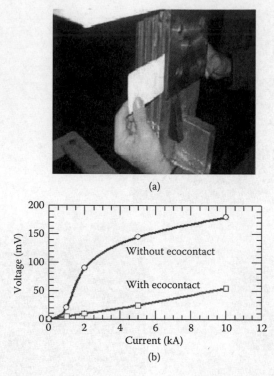

(a)

(b)

FIGURE 5.114
(a) Installation of Ecocontact copper-based foam insert; (b) Voltage-current dependence of connections with and without the use of Ecocontact.

TABLE 5.31

Effect of Ecocontact on the Performance of Copper-Based Bolted Joints

	Temperature	ΔU	Power Loss
Without ecocontact	>250°C	350 mV	3500 W
With ecocontact	40°C	3 mV	30 W
After 1 month of operation	40°C	2 mV	20 W

ameliorates the current transfer across the contact interface of a used or damaged connection with no surface renovation beforehand. Installing Ecocontact foam does not require any specific operation. Properties of different types of metallic foam materials are shown in Table 5.30. The performance of bolted joints with the use of Ecocontact copper-based joints under normal loading conditions with the use of foam insert is shown in Table 5.31. As it can be seen the use of Ecocontact inserts considerably improves the performance of bolted joints as manifested by a substantially lower operating temperatures and power losses as compared with those without these inserts.

5.9.3 Silver Foam Materials

The contact foams are produced by evaporation of pure silver in a vacuum. These foams, with their honeycomb structure, offer good porosity and malleability characteristics. The honeycomb structure enables production of surface comprising thousands of micron-sized contact points (up to 130, 000 points/in.² for a foam thickness of 0.08 in. [20 cm]). These micro-points perforate the oxide layers, thus creating micro-welds and eliminating any current concentration in the contact zones. The Ecocontact silver foams allow to increase the service life of high-amperage connections by slowing down contact surface degradation and to restore good conductivity to worn connections. The reduction in connection maintenance costs comes from eliminating resurfacing operations. Installation of Ecocontact foam inserts requires no special procedures. Users simply slightly separate the connection, slide in the interface foam and retighten the connection. Ecocontact foam is supplied ready to install in the form of flat plates or sheets cut to the size of the contact surface. These plates are designed for electrical connections operating from 100 to 100,000 A.

5.10 Installation of Power Connections

To ensure proper function of a busbar joint, certain measures must be taken during installation. The most important items for proper bus connections include the correct selection of connecting bolts; washers, deburring, tightening of bolts and nuts to the correct torque and surface preparation. For permanent low-resistance bus connections, the surface preparation of the connection is as important as, if not more important than, the selection of the proper joint compound. The measures to be taken to prevent the adverse effect of environment on the functioning of the joint depend upon the busbar material and the environment in which the busbars are installed. In the case of bolted joints, the connectors should comply with the requirements determined with respect to the rated conductor current-carrying capacity and be able to carry the continuous currents without exceeding the temperature rise of the weakest conductor or equipment terminal with which it is intended to be used. The hardware used in connectors must be compatible with the connection material, have high mechanical strength, and be corrosion resistance and correspond to NEMA recommendations. Practice has shown that the rigidity of power connector current-carrying pads and misalignment when bolted to another rigid pad is often a cause of impaired current transfer and thus higher contact resistance. This, however, can be circumvented by sectioning the current-carrying pads in a manner as shown in Figure 5.115 which brings in significant improvement to the mechanical integrity and electrical efficiency of the bolted joint. The beneficial effect of sectioning is owing to a reduced rigidity of the sectioned pads, thus lower tendency to misalignment and a more uniform stress distribution under the bolts that significantly enlarges the contact area. Establishing and maintaining required contact load is essential for reliable performance of an electrical connection. Hence, on-line monitoring contact load would provide a possibility to survey the behavior of power connections under normal and overload operating conditions. Unfortunately, monitoring of contact loads can only be applied to bolted-type connections and not to the other types such as compression or insulation piercing connectors.

5.10.1 Examples of Improper Installations

Improper installation of bolted joint is most likely the main cause for the faulty operation a joint. Typical cases are illustrated in the following figures. The use of oversized flat washers and bolts is shown in Figure 5.116a. Clearly this joint configuration cannot

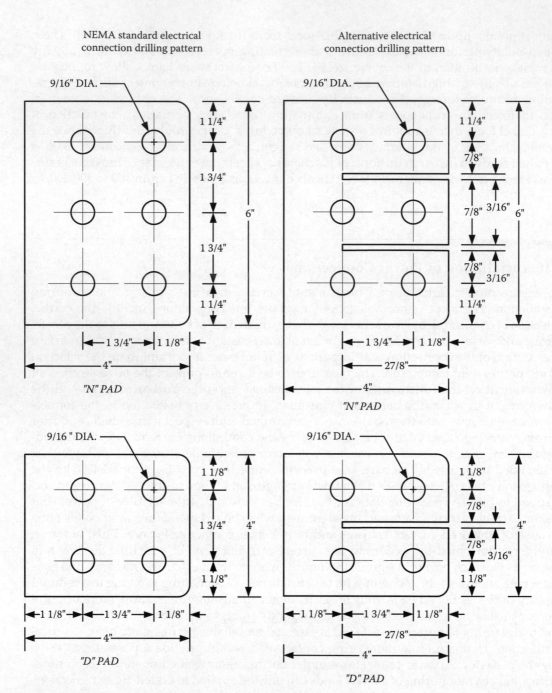

FIGURE 5.115
NEMA and alternative connection drilling patterns.

assure any control of the contact force yet the least to assure the reliable operation of the bolted joint. As a result, overheating and melting of a joint occurred as manifested by traces of molten metal as seen in Figure 5.116b. Another example of improper installation is depicted in Figure 5.116c. As seen not only the use of wrong installation but more so the selection of the joint hardware and inadequate installation of the joints were

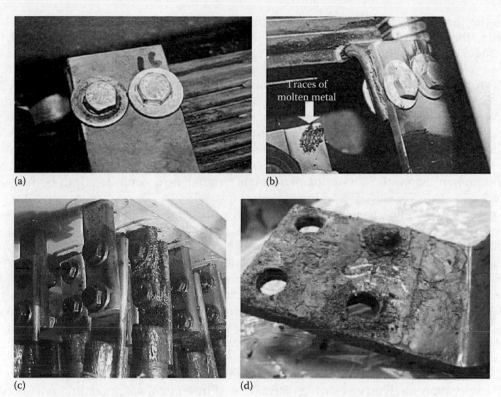

FIGURE 5.116
(a) Improper installation with large thin flat washers and no Belleville washers. (b) Joint overheating and products of melting. (c) Unacceptable selection of joint hardware and improperly assembled joints. (d) Tin-plated busbar showing the results of an excessive heating and formation of a thick insulating layer.

responsible for the overheating of the joints. Clear visible signs of connector overheating resulting in melting of the busbar contact zone are obvious. Figure 5.116d shows a tin-plated busbar with clear signs of excessive overheating and formation of thick insulating layer on the bus terminal. Clearly this type of tin-plated connector assembly and its use on the network is not recommended. It is, indeed, unexplainable and worrying that despite the installation procedures set by different international and domestic standards, the joint configuration as seen in this figure are often found on the power network.

5.11 Accelerated Current-Cycling Tests (Standards)

5.11.1 Present Current-Cycling Tests

Designing accelerated tests is a very complex task requiring a complete understanding of the operation and the failure mechanisms of the electrical connection. The complexity of accelerated testing results from the fact that there are many unanswered questions on contact phenomena still not fully explained within the limits of present knowledge which can result in misleading conclusions. A review of the existing current-cycling tests used throughout the world to characterize the connector performance and contact materials reveals that practically every industrial country has its own testing procedure. In some countries, the United States for example, there are several organizations such ANSI, IEEE, Underwriters Laboratory and military, that are involved in standardization of the current-cycling testing procedure and contact material characterization. The majority of these testing procedures are concerned with the determination of connector performance under "artificial conditions." In other words, the performance data is obtained under accelerated/simulated conditions of one sort or another that was then used to extrapolate to the expected field conditions. Unfortunately, these extrapolations are often questionable since the presence of more than one deterioration mechanism implies that a very detailed knowledge of each deterioration process, with respect to the acceleration factors, must be known before the connector performance under field conditions can be fully assessed.

The present accelerated heat cycle tests derive the accelerated deterioration from current overload and severe ambient conditions, that is, still air. These two factors combine to produce higher body temperatures and greater contact point super-temperatures. These higher temperatures produce conditions under which deterioration mechanisms such as stress relaxation, oxidation, fretting or formation of intermetallics are enhanced. Furthermore, there are some mechanisms which may have a definite inception temperature. Connector compounds will not flow away from the contact points at lower temperatures. Chemical breakdown of the compounds at the contact may commence only at elevated body temperatures and contact super-temperatures. If this occurs there will be not only a marked reduction in the protection against oxidation but there may be corrosive attack as well. Also, if the degree of load cycling is limited by a limitation of temperature some deterioration mechanisms such as formation of intermetallics or fretting may not commence.

Existing current-cycling testing procedures are shown in Table 5.32. These procedures are based on the temperature cycles and some include application of short-circuit currents. It is clear the performance criteria differ from one standard to another. From the point of view of the connector performance these tests provide quite reliable and repeatable results on the ability of connectors to carry load-current. However, these tests suffer from several shortcomings. They are long, costly, and may not be satisfactorily related to field conditions since the mechanisms responsible for the deterioration of the connector in the laboratory under accelerated conditions may well differ from those causing the field failures. Practice has shown that most of the connectors tested according to the conditions set by these standards can easily satisfy their performance criteria, yet fail when exposed to real operating conditions. This is because these performance criteria are not based on the theories of electrical contacts but rather on arbitrarily selected testing conditions designed to simulate field conditions. Consequently, these criteria are misleading since they fail to adequately describe the state of deterioration of the conductor/connector interface, which ultimately determines the overall performance of a connection.

Measurement of the contact resistance of a connector in a cold state—as required by the standards—is marginally affected by current-cycling and, hence, misleading, since it fails to adequately describe connector performance. Continuous measurement of the contact resistance during the heating cycle provides a much better indication of the state of deterioration of the *conductor/connector interface,* which ultimately determines the overall performance of a connection. It is therefore very important to measure contact resistance continuously at the end of the ON (hot-resistance) and OFF cycles (cold-resistance). In order to overcome these shortcomings and examine the performance of aluminum

TABLE 5.32

Most Common Current-Cycling Testing Procedure Used by Utilities and Manufacturers for Evaluating the Connector Performance

	ANSI C119.4	CSA C-57	IEC	UK	France	Germany	Sweden
Temperature	100°C Rise	100°C Rise	120°C	80°C Rise	120°C	100°C	125°C
Time Hot (minute)	60	60	10+	5	15	0	36087
Short-circuit (no of cycles)	0	0	6 (1 second)	3 (1 second)	8 (1 second)	10 (1 second)	3 (1 second)
Short-Circuit (Max T°C)	–	–	250°C	160°C	170°C	160°C–210°C	250°C
Short-circuit (after x cycles)	–	–	200	0	200	750	750
Criteria (T°C)	<10°C	<10°C	$T_{max} = T_{ref}$				
Criteria (R%)	<5%	<5%	$k/k_o < 1.5$				

k—Connector resistance factor for each connector at any stage of measurement series,
k_o—Connector resistance factor of same connector measured at cycle zero,
T_{ref}—Temperature of reference conductor,
T_{max}—Max connector temperature.

connections under different operating conditions, a new accelerated test, radically different from the present ones, was developed and incorporated into the revised CSA C-57 standard. The test is based on the cyclic heating and cooling of a connector by a current several times higher than rated and continuous measurement of the contact resistance and temperature.

To avoid overheating of a conductor and at the same time allow the connector to attain the same temperature regime, short conductor lengths welded to large equalizers are used. This allows the temperatures of both the connectors and conductor to be kept at approximately the same level. The use of currents substantially higher than those under which connectors normally operate, is warranted since the current surges on the network are not infrequent which, in turn, can increase the current loads to very high levels thus creating necessary conditions for the accelerated deterioration of a connector. It is worth noting that Williamson [12] questioned the validity of using a higher current to accelerate the connector degradation. He argues that higher current may exceed the failure limit of the bulk temperature before self-healing process can occur. Accordingly, the connector would operate in two different modes, that is, at normal and high currents, and its inability to pass a high current gives no information at all about its reliability under normal conditions. Williamson concluded that this is an insidious pitfall inherent in the practice of accelerated testing and that in certain circumstances the use of high currents can inhibit the very phenomena on which the reliability of the connector normally depends. This argument may be applicable if the self-healing mechanism is operative. However, the evidence so far indicate that self-healing was observed in just a few instances [12] and thus cannot be universally applied. Hence, the use of higher currents is justified because their main function is to produce large differential thermal expansion at the contacting interface which, in turn, shears the conducting bridges and initiate fretting.

References

(In this listing Electrical Contacts = Proc. IEEE Holm Conf. on Electrical Contacts and ICEC = Proc. Int'l Conf. on Electrical Contacts)

1. R. Holm, *Electric Contacts*, Reprinted 4th Edition, Springer Verlag, New York (2000).
2. M. Braunovic, V.K. Konchits and N.K. Myshkin, *Electrical Contacts: Fundamentals, Applications and Technology*, CRC Press, Baco Raton (2007).
3. J.B.P. Williamson, "The Microworld of the Contact Spot", Electrical Contacts-981, IIT (1981), p. 1.
4. R. Malucci, "Comparison of Model Prediction of Current Density Variations across Simulated Contact Regions", Electrical Contacts-2010 (2010), p. 230.
5. R. Mroczkowski, "Connector Contact Surfaces—Where the Action Is", AMP Technical Paper, pp. 268–83.
6. S. Schlegel, S Grossman, H. Lobl, M. Hoidis, U. Kaltenborn, T. Magier, "Joint Resistance of Bolted Copper—Copper Busbar Joints Depending on Joint Force at Temperatures beyond 105°C", Electrical Contacts-2010, p. 444.
7. R. Jackson and I. Green, "A Finite Element Study of Elasto-Plastic Hemispherical Contact Against a Rigid Flat", *J. Tribol.*, vol. 5, (2005), p. 343.
8. S. Wadwalkar, R. Jackson and L. Kogut, "A Study of the Elasto-Plastic Deformation of Heavily Deformed Sphaerical Contacts", *Proc. Inst. Mech. Eng.Part J. Eng. Tribol.*, vol. 224, (2010), p. 3.
9. A. Oberg, A. Bohlin and K.E. Olson, "The Influence of Contact Surface Preparation on the Performance of Copper and Aluminium Connectors", *Proceedings of the 16th ICEC*, Loughborough, UK (1992), p. 476.
10. R.D. Naybour and T. Farrell, "Mechanical Connectors for Aluminum Cables: An Investigation of Degradation Mechanisms in Connector Clamps and Bolted Terminations", ECRC/R426, Capenhurst, Electricity Council Research Centre (1971).
11. F. Kohlrausch, "Uber den Stationaren Temperaturzustand eines Elektrish Geheizten Leiters", *Ann. Phys.*, vol. 1, (1900), p. 132.
12. J.B.P. Williamson, "The Self-Healing Effect: Its Implications in the Accelerated Testing of Connectors", Proceedings of the 10th ICEC, Budapest, Hungary (1980), p. 1089.
13. J.B.P. Williamson, "Deterioration Processes in Electrical Connections", *Proceedings of the 4th ICEC*, The University of Swansea, July (1968).
14. R.L. Jackson, "Electrical Performance of Aluminum/Copper Bolted Joints", *IEEE Proc.*, vol. 129, Pt. C, No. 4, (1982), p. 177.
15. W.E. Campbell, "Reduction of the Rate of Film Formation on Silver and Brass by Purification of the Atmosphere", Electrical Contacts-1977, Chicago, IIT (1977), p. 185.
16. R.F. Tylecote, "The Solid Phase Welding of Metals", Edward Arnold, London (1968).
17. M. Braunovic, L. Pomathiod and J-P. Bailon, "The Effect of Surface Segregation of Mg on the Contact Resistance of Al-0.5%Mg Alloy", *IEEE Trans.*, PHP-13, No. 3 (1977), p. 31.
18. T. Tamai, Y. Nabeta, S. Sawada, Y. Hattori, "Property of Tin Oxide Film Formed on Tin-Plated Connector Contacts", Electrical Contacts-2010 (2010), p. 294.
19. "Crevice corrosion", INSA, Lyon, Laboratoire de Physicochimie Industrielle, http://www.cdcorrosion.com/index-gb.htm.
20. J.-G. Zhang and X.M. Wen, "The Effect of Dust Contamination on Electric Contacts", *IEEE. Trans. CHMT*, vol. 9, No. 1, March (1986), pp. 53–8.
21. M. Braunovic and M. Marjanov, "Thermoelastic Ratcheting Effect in Bolted Aluminum-to-Aluminum Connections", *IEEE Trans. CHMT*, vol. 11, No. 1, (1988), p. 54.
22. E.M. Eden, W.N. Rose and F.L. Cunningham, "The Endurance of Metals", *Proc. Inst. Mech. Eng.*, vol. 4, (1911), p. 875.
23. G.A. Tomlison, "Rusting of Steel Surfaces in Contact", *Proc. R. Soc.*, vol. A115, (1927), p. 472.
24. W.E. Campbell, Symp. "Fretting Corrosion", ASTM Spec. Tech. Publ. No. 144, (1953), p. 3.
25. R.B. Waterhouse, *Fretting Corrosion*, Pergamon Press, Oxford (1972).

26. N.L. Golego, A.Ya. Alyabev and V.V. Shevelya, *Fretting Corrosion of Metals*, Tekhnika, Kiev (1974) (In Russian).
27. P.L. Hurricks, "The Mechanism of Fretting—A Review", *Wear*, vol. 15, (1970), p. 389.
28. R.D. Mindlin, "Compliance of Elastic Bodies in Contact", *J. Appl. Mech.*, vol 16, (1949), p. 259.
29. D. Godfrey, "Investigation of Fretting Corrosion by Microscopic Observation", NACA Tech. Note No. 2039 (1950).
30. I.M. Feng and B.G. Rightmire, "The Mechanism of Fretting", *Lubr. Eng.*, vol. 9 (1953), p. 134.
31. H.H. Uhlig "Mechanism of Fretting Corrosion", *J. Appl. Mech.*, vol. 21, (1954), p. 401.
32. L.F. Stowers and E. Rabinowicz, "The Mechanism of Fretting Wear", *J. Lubr. Tech.*, vol. 95, (1973), p. 65.
33. J.F. Archard, "Contact of Rubbing Flat Surfaces", *J. Appl. Phys.*, vol. 24, (1953), p. 981.
34. I.A. Oding and V.S. Ivanova, "Thermoelectric Effects in Fretting Wear", *Proceedings of the International Conference on Fatigue of Metals*, (1956), p. 408.
35. N.P. Suh, "The Delamination Theory of Wear", *Wear*, vol. 25, (1973), p. 111.
36. R.B. Waterhouse and D.E. Taylor, "Fretting Debris and Delamination Theory of Wear", *Wear*, vol. 29, (1977), p. 337.
37. E.S. Sproles Jr., D.J. Gaul and D.J. Duquette, "A New Interpretation of the Mechanism of Fretting and Fretting Corrosion Damage", in *Fundamentals of Tribology*, Ed. N.P. Suh and N. Saka, MIT Press, Cambridge (1978), p. 585.
38. M. Godet, "The Third Body Approach: A Mechanical View of Wear", *Wear*, vol. 100, (1994), p. 437.
39. S. Noël, D. Alamarguy, S. Correia and P. Laurat, "Fretting Behavior of Nickel Coatings for Electrical Contact Applications", Electric Contacts-2011 (2011), p. 75.
40. A. Bouzera, E. Carvou, L. Tristani and E.M. Zindibe, "Minimum Fretting Amplitude in Medium Force for Connector Coated Material and Pure Metals", Electrical Contacts-2010 (2010), p. 101.
41. A.J. Fenner, K.H. Wright and J.Y. Mann "Fretting Corrosion and Its Influence on Fatigue Failure", *Int. Conf. Fatigue of Metals*, Inst. Mech. Eng., London (1956), p. 11.
42. A. Fairweather, F. Lazenby and A. Parker, "Development of Resistance and Microphone Noise at a Distrurbed Contact" Proceedings of the 2nd International Symposium on Electrical Contact Phenomena, Graz, May 4–6, Technische Hochschule, Graz (1964), p. 316.
43. E.M. Bock and J.H. Whitley, "Fretting Corrosion in Electric Contacts", Electric Contacts-1974, IIT, Chicago (1974), p. 128.
44. S.J. Krumbein, "Contact Properties of Tin Plates", Electric Contacts-1974, IIT, Chicago (1974), p. 38.
45. M. Antler, W.F. Graddick and H.G. Tompkins, "Base Metal Contacts: An Exploratory Study of Separable Connection to Tin-Lead", Electric Contacts-1975, Chicago (1975), p. 25.
46. N. Tian, N. Saka and E. Rabinovicz, "Friction and Failure of Electroplated Sliding Contacts", *Wear*, vol. 142, (1991), p. 57.
47. M. Antler and E.S. Drozdowicz, "Fretting Corrosion of Gold-Plated Connector Contacts", *Wear*, vol. 74, (1981–1982), p. 27.
48. J.W. Souter and W. Staunton, "The Fretting Wear of Electrodeposited Contact Coatings", *Trans. Inst. Met. Finish*, vol. 66, No. 1, (1988), p. 8.
49. M. Antler, "Fretting Corrosion of Solder-Coated Electrical Contacts", *IEEE Trans. CHMT*, vol. 7, (1984), p. 129.
50. J.M. Hooyer and K. Peekstok, "The Influence of Practical Contact Parameters on Fretting Corrosion of Tin-Base Low-Level Connector Contacts", Electric Contacts-1987, *IEEE-Holm*, Chicago (1987), p. 43.
51. J.H.M. Neijzen and J.H.A. Glashorster, "Fretting Corrosion of Tin-Coated Electrical Contacts", *IEEE Trans. CHMT*, vol. 10, (1987), p. 68.
52. A. Lee and M. Mamrick, "Fretting Corrosion of Tin-Plated Copper Alloy", *IEEE Trans. CHMT*, vol. 10, (1987), p. 63.
53. J.J. Motine and B.T. Reagor, "Investigation of Fretting Corrosion at Dissimilar Metal Interfaces on Socketed IC Device Applications", Electric Contacts-1983, IIT, Chicago (1983), p. 61.
54. M. Braunovic and P. Gervais, "Fretting Problems in Electrical Industry", *Proceedings of the 17th Inter-Ram*, Hershey (1990), p. 356.

55. M. Braunovic, "Fretting Corrosion Between Aluminum and Different Contact Materials", Electric Contacts-1977, IIT, Chicago (1976), p. 223.

56. M. Braunovic, "Degradation of Al–Al and Al–Cu Connections due to Fretting", *Wear*, vol. 125, (1988), p. 53.

57. M. Braunovic, "Fretting in Nickel-Coated Aluminum Conductors", *Proceedings of the 15th ICEC, Montreal* (1990), p. 461.

58. M. Braunovic, "Effect of Different Types of Mechanical Contact Devices on the Performance of Bolted Aluminum-to-aluminum Joints Under Current-cycling and Stress Relaxation Conditions", Electrical Contacts-1986, IIT, Boston (1986), p. 133.

59. H. Kongsjorden, J. Kulsetas and J. Sletbak, "Degradation of Electrical Contacts Caused by Oscillatory Micromotion Between the Contact Members", Electrical Contacts-1978, Chicago (1978), p. 87.

60. H. Tian, N. Saka and E. Rabinovicz, "Fretting Failure of Electroplated Gold Contacts", *Wear*, vol. 42, (1991), p. 265.

61. A. Kassman, T. Imrell and S. Jacobson, "Fretting of Powered Silver Plated Connectors in a Corrosive Environment", *Proceedings of the 17th ICEC*, Nagoya (1994), p. 327.

62. M. Antler and E.S. Sproles, "Effect of Fretting on the Contact Resistance of Palladium", *IEEE Trans. CHMT*, vol. 5, No. 1, (1982), p. 158.

63. M. Antler, "Electrical Effects of Fretting Connector Contact Materials: A Review", *Wear*, vol. 106, (1985), p. 5.

64. R.D. Mallucci, "Dynamic Model of Stationary Contacts Based on Random Variations of Surface Features", Electrical Contacts-1991 (1991), p. 90.

65. M.D. Bryant, "Resistance Build-Up in Electrical Connectors due to Fretting Corrosion of Rough Surfaces", Electrical Contacts-1993, Pittsburgh (1993), p. 178.

66. C. Dang and M. Braunovic, "Metallurgical and Contact Resistance Studies of Sleeve Connectors in Aluminum Cable Splices", *IEEE Trans CHMT*, vol. 13, (1990), p. 74.

67. R.L. Jackson, Private communication.

68. J.L. Johnson and L.E. Moberly, "Separable Electric Power Contacts Involving Aluminum Bus-Bars", Electric Contacts-1975, IIT, Chicago (1975), p. 53.

69. P.J. Thiesen and K.A. Forsell, "Connector Dependent Fretting Corrosion Test System", Electric Contacts-1979, IIT, Chicago (1979), p. 109.

70. N. Bond, "Alumiunm Contact Surfaces in Electrical Transition Interfaces", *IEEE Trans. PMP*, vol. 5, No. 2, (1969), p. 104.

71. R.D. Naybour and T. Farrell, "Connectors for Aluminum Cables: A Study of the Degradation Mechanisms and Design Criteria for Reliable Connectors", *IEEE Trans. PHP*, vol. 9, No. 1, (1973), p. 30.

72. L. Roullier, "Contact Interfaces in Aluminum Mechanical Joints", *Proceedings of the 2nd International Symposium on Electric Contact Phenomena*, Graz, Austria, Technische Hochschule (1967).

73. M. Braunovic, "Fretting Damage in Tin-Plated Aluminum and Copper Connections", Electric Contacts-1989, San Francisco (1989), p. 179.

74. M. Braunovic, "Effect of Fretting in Aluminum-to-Tin Connections", *IEEE Trans. CHMT*, vol. 13, No. 3, (1990), p. 579.

75. M. Antler, "Tribology of Electronic Connectors: Contact Sliding Wear, Fretting and Lubrication", in *Electrical Contacts: Principles and Applications*, Ed. P.G. Slade, Marcel Dekker, New York (1999), p. 309.

76. A. Kassman and S. Jacobson, "Surface Damage, Adhesion and Contact Resistance of Silver Plated Copper Contacts Subjected to Fretting Motion", *Wear*, vol. 165, (1993), p. 227.

77. M. Braunovic, "Fretting in Electrical/Electronic Connections: A Review", *IEICE Trans.*, vol. E85, No. 1, January (2002), p. 1.

78. D. Gagnon and M. Braunovic, "Fretting in Copper-to-Copper Contacts Under AC and DC Current Conditions", *IEEE Trans. CPT.*, vol. 24, (2001), p. 378.

79. M. Braunovic, "Aluminum Connections: Legacies of the Past", Electrical Contacts-1994, Chicago (1994), p. 1.

80. J.L. Queffelec, N. Ben Jemaa, D. Travers and G. Pethieu, "Materials and contact shape studies for automobile connector development", *Proc. 15th ICEC*, Montreal (1990), p. 225.

81. J. Ambier and P. Perdigon, "Fretting Corrosion of Separable Electrical Contacts", Electric Contacts-1984, IIT, Chicago (1984), p. 105.

82. C.R. Dixon and F.G. Nelson, "The Effect of Elevated Temperature on Flash-Welded Aluminum-Copper Joints", *Trans. AIEE II*, vol. 78, (1960), p. 491.

83. E.R. Wallach and G.J. Davis, "Joint Resistance and Current Paths in Heat-Treated Aluminum/Copper Solid-State Welds", *Met. Sci.*, March (1977), p. 97.

84. E.R. Wallach and G.J. Davis, "Mechanical Properties of Aluminum-Copper Solid-Phase Welds", *Met. Tech.*, April (1977), p. 183.

85. D.M. Rabkin, V.R. Ryabov, A.V. Lozovskaya and V.A. Dovzhenko, "Preparation and Properties of Copper-Aluminum Intermetallic Compounds", *Sov. Powder Met. Ceram.*, vol. 8, No. 92, (1970), p. 695.

86. M.M. Nakamura, Y. Yonezawa, T. Nakanishi and K. Kondo, "Durability of Al–Cu Hot-Pressure Welding Joints", *Wire J.*, April (1977), p. 71.

87. N.A. Gjosten, *Diffusion*, ASM, Metals Park, OH (1973), p. 41.

88. M. Runde, "Material Transport and Related Interfacial Phenomena in Stationary Aluminum Contacts", PhD Thesis, The Norwegian Institute of Technology (1987).

89. A.M. Gusak and K.P. Gurov, "Kinetics of Phase Transformations in a Diffusion Zone During Interdiffusion Phase Formation", *Phys. Met. Metallogr*, vol. 53, No. 5, (1982), p. 12.

90. V.N. Pimenov, K.P. Gurov, K.I. Khudyakov, S.S. Dol'nikov, R.A. Milievskii, V.S. Khlomov and Yu.G. Miller, "Effect of Electrical Current on the Phase Formation in the Diffusion Layer", *Fiz Khim Obrabotki Materialov*, vol. 1, (1978), p. 107 (in Russian).

91. M. Braunovic and N. Alexandrov, "Intermetallic Compounds in Aluminum-to-Copper Electrical Interfaces: Effect of Temperature and Electric Current", *IEEE Trans. CPMT-17, Part A*, vol. 17, (1994), p. 78.

92. R. Timsit, "Electrical Instabilities in Stationary Electrical Contacts: Al-Plated-Brass Junctions", *IEEE Trans. CHMT*, vol. 11, No. 1, (1988), pp. 43–5.

93. J.A. Rayne and C.L. Bauer, "Effect of Intermetallic Phase Formation on Electrical and Mechanical Properties of Flash-Welded Al–Cu Couples", Proceedings of the 5th Bolton Landing Conference on Weldments, General Electric (1979), p. 353.

94. H.B. Huntington, "Electromigration in Metals" in *Diffusion in Solids: Recent Developments*, Ed. A.S. Nowick, and J.J. Burton, Academic Press, New York (1974), Chapter 6.

95. A. Sorzoni, B. Neri, C. Caprile and F. Fantini, "Electromigration in Thin-Film Interconnection Lines: Models, Methods and Results", *Mater. Sci. Rep.*, vol. 7, No. 4–5, (1991), p. 143.

96. M. Runde, E. Hodne and B. Totdal, "Experimental Study of the Conducting Spots in Aluminum Contact Interface", Electrical Contacts-1986, Chicago (1986), p. 213.

97. J. Aronstein, "Electromigration Failure of Aluminum Contact Junctions", Electrical Contacts-1995, Montreal (1995), p. 10.

98. J. Aronstein, "AC and DC Electromigration in Aluminum Contact Junctions", Electrical Contacts-1996, Chicago (1996), p. 311.

99. R.D. Naybour and T. Farrell, "Degradation Mechanisms of Mechanical Connectors on Aluminum Conductors", *Proc. IEE*, vol. 120, No. 2, (1973), p. 273.

100. R. Atermo, "A Method for Testing Compressive Relaxation in Aluminum Wire Conductors", *Wire J.*, September (1973), p. 127.

101. O.A. Troitsky and A.G. Rozno, "Electroplastic Effect in Metals", *Sov. Phys. Solid State*, vol. 12, (1970), p. 161.

102. A.A. Klypin, "Creep of Metals Under the Influence of Electric Current", *Probl. Prochnosti*, vol. 9, (1972), p. 35.

103. V.I. Stashenko, O.A. Troitskii and V.I. Spitsyn, "Action of Current Pulses on Zinc Single Crystals During Creep", *Phys. Status Solidi A*, vol. 79, (1983), p. 549.

104. O.A. Troitskii and V.I. Spitsyn, "Investigation of the Electroplastic Deformation of a Metal by the Method of Stress Relaxation and Creep", *Sov. Phys. Dokl.*, vol. 21, No. 2, (1976), p. 111.

105. O.A. Troitskii and V.I. Stashenko, "Stress Relaxation Investigation of the Electroplastic Deformation of a Metal", *Phys. Met. Metallogr*, vol. 47, (1979), p. 149.

106. O.A. Troitskii, V.I. Spitsyn and V.I. Stashenko, "The Effect of an Electric Current on the Relaxation of Stresses in Crystals of Zinc, Cadmium and Lead", *Sov. Phys. Dokl.*, vol. 23, No. 7, (1978), p. 509.

107. O.A. Troitskii and A.G. Rozno, "Effect of Electric Current on the Plastic Deformation of a Metal", *Phys. Met. Metallogr*, vol. 30, No. 4, (1970), p. 153.

108. V.P. Lebedev and V.I. Hotkrvich, "Effect of Electric Current pulses on the Low-Temperature Deformation of Aluminum", *Phys. Met. Metallogr.*, vol. 54, No. 2, (1982), p. 23.

109. S.K. Varma and L.R. Cornwell, "The Electroplastic Effect in Aluminum", *Scripta Met.*, vol. 13, (1979), p. 733.

110. O.A. Troitskii, "Effect of the Electron State of a Metal on Its Mechanical Properties and the Phenomenon of Electroplasticity", *Strength Met.*, vol. 9, No. 1, (1977), p. 35.

111. V.I. Spitsyn and O.A. Troitskii, *Electroplastic Deformation of Metals*, Nauka, Moscow, (1985), (In Russian).

112. O.A. Troitskii, V.I. Spitsyn and V.G. Ryzhkov, "Electroplastic Drawing of Steel, Copper and Tungsten", *Sov. Phys. Dokl.*, vol. 23, No. 11, (1978), p. 861.

113. O.A. Troitskii, "Pressure Shaping by the Application of High Energy", *Mater. Sci. Eng.*, vol. 75, (1985), p. 37.

114. V.I. Bazaykin, V.E. Gromov, V.A. Kuznetsov and V.N. Peretyatko, "Mechanics of Electrostimulated Wire Drawing", *Int. J. Solids Struct.*, vol. 27, No. 11, (1991), p. 1639.

115. O.A. Troitskii and V.I. Likhtman, "Anisotropy of Electron and -Radiation on the Deformation of Brittle Zinc Single Crystals", *Sov. Phys. Dokl.*, vol. 148, (1963), p. 332.

116. V.Ya. Kravchenko, "Effect of Directed Electron Beam on Moving Dislocations", *Sov. Phys. JETP.*, vol. 51, (1966), p. 1135.

117. K.M. Klimov, V.D. Shnyrev and N.I. Novikov, "The Electroplasticity of Metals", *Sov. Phys. Dokl.*, vol. 19, (1975), p. 787.

118. A.M. Roschupkin, O.A. Troitskii and V.I. Spitsyn, "Development of the Concept of the Effect of Current with High Density on the Plastic Deformation of Metals", *Sov. Phys. Dokl.*, vol. 286, No. 3, (1986), p. 633; (1979), p. 532.

119. K. Okazaki, H. Kagawa and H. Conrad, "A Study of the Electroplastic Effect in Metals", *Scripta Met.*, vol. 12, (1978), p. 1063.

120. K. Okazaki, M. Kagawa and H. Conrad, "Additional Results on the Electroplastic Effect in Metals", *Scripta Met.*, vol. 13, (1979), p. 277.

121. K. Okazaki, M. Kagawa and H. Conrad, "An Evaluation of the Contribution of Skin, Pinch and Heating Effects to the Electroplastic Effect in Titanium", *Mater. Sci. Eng.*, vol. 45, (1980), p. 109.

122. A.F. Sprecher, S.L. Mannan and H. Conrad, "On the Mechanisms for the Electroplastic Effect in Metals", *Acta Metall.*, vol. 34, No. 7, (1986), p. 1145.

123. H. Conrad and A.F. Sprecher, "The Electroplastic Effect in Metals", in *Dislocations in Solids*, vol. 8, Ed. F.R.N. Nabarro, Elsevier (1989), p. 498, Chapter 43.

124. H. Conrad, J. White, W.D. Cao, X.P. Lu and A.F. Sprecher, "Effect of Electric Current Pulses on Fatigue Characteristics of Polycrystalline Copper", *Mater. Sci. Eng.*, vol. A145, (1991), p. 1.

125. H. Conrad, A.F. Sprecher, W.D. Cao and X.P. Lu, "Electroplasticity—The Effect of Electricity on the Mechanical Properties of Metals", *J. Mater.*, September (1990), p. 28.

126. R. Timsit, "Remarks on Recent Experimental Observations of the Electroplastic Effect", *Scripta Met.*, vol. 15, (1981), p. 461.

127. W.D. Cao, A.F. Sprecher and H. Conrad, "Measurement of the Electroplastic Effect in Nb", *J. Phys. E Sci. Instrum.*, vol. 22, (1989), p. 1026.

128. V.L.A. Silveira, M.F.S. Porto and W.A. Mannheimer, "Electroplastic Effect in Copper Subjected to Low Density Electric Current", *Scripta Met.*, vol. 15, (1981), p. 945.

129. V.L. Silveira, R.A.F.O. Fortes, and W.A. Mannheimer, "Modification of the Dislocation Substructure in Copper During Electroplastic Treatment", *Beitr. Elektronenmikroskop. Direktabb. Oberfl.*, vol. 15, (1982), p. 217.

130. M. Braunovic, "The Effect of an Electric Current on the Stress Relaxation of Aluminum Wire Conductors", *Proceedings of the 7th International Conference on Strength of Metals*, ICSMA-7, Pergamon Press (1985), p. 619.

131. M. Braunovic, "Evaluation of Different Contact Aid Compounds for Aluminum-to-Copper Connections", *IEEE Trans. CHMT*, vol. 15, No. 2, (1992), p. 216.

132. S.W. Melsom and H.C. Booth, "The Efficiency of Overlapping Joints in Copper and Aluminum Busbar Conductors", *J. IEE.*, vol. 60, No. 312, (1922).

133. E. Donati, "Overlapping Joints in Electric Furnace Circuits", *L'Energia Elettrica*, vol. 12, No. 6, (1935), p. 433.

134. "Al Elastic Contact Disks", Karl Pfisterer Techn. Bulletin, No. 2 (1991).

135. V.I. Boychenko and N.N. Dzektser, "Busbar Connections", *Energiya*, 1978, (In Russian).

136. M. Braunovic, "Effect of Connection Design on the Contact Resistance of High Power Ovelapping Bolted Joints", *IEEE Trans, CPT*, vol. 25, No. 4, (2002), p. 1.

137. O.B. Bron, M.G. Myasnikova, I.P. Miroschnikov and V.N. Fyodorov, "Permissible Temperatures for Electric Contacts", *Proceedings of the 10th ICEC*, Budapest (1980), p. 71.

138. H. Jesvold, M. Runde and G.I. Tveite, "The Effect of Conductor Hardness on Aluminum Compression Joints", *Proceedings of the 17th ICEC*, Nagoya (1994), p. 497.

139. M. Braunovic and C. Dang, "Comparative Assessment of Different Current-Cycling Procedures for Testing Power Connections", Electrical Contacts-1997, Philadelphia (1997), p. 87.

140. "Connectors for Use between Aluminum-to-Aluminum or Aluminum-to-Copper Bare Overhead Conductors", July 1991, ANSI C119.4-1991.

141. "Electric Power Connectors for Use in Overhead Line Conductors", CSA C57-1966.

142. M. Braunovic, Unpublished work.

143. J.J. Schindler, R.T. Axon and R. Timsit, "Mechanical and Electrical Contact Properties of Wedge Connectors", *IEEE Trans. CPMT-A*, vol. 19, (1996), p. 287.

144. M. Braunovic, "Use of Shape-Memory Materials in Distribution Power Systems", CEA Report No. SD-294A.

145. M. Braunovic and C. Labrecque, "Shape-Memory Alloy Mechanical Contact Devices", Electrical Contacts-1994, Chicago (1995), p. 21.

146. C. Labrecque, M. Braunovic, F. Trochu, V. Brailovski, Y.Y. Quian, T. Terriault and L.M. Schetky, "Experimental and Theoretical Evaluation of the Behaviour of a Shape-Memory Alloy Belleville Washer Under Different Operating Conditions", Electrical Contacts-1995, Montreal (1996), p. 195.

147. A. Oberg and S. Nilsson, "Shape-Memory Alloys for Power Connector Applications", Electrical Contacts-1993, Pittsburgh (1993), p. 225.

148. T. Waram, "Design Consideration for Shape-Memory Belleville Washers", *Raychem Tech. Doc.*

149. McD. Schetky, "The Present Status of Industrial Applications for Shape-Memory Alloys", *EPRI TR*, vol. 105072, (1995), p. 41.

150. M. Braunovic, "Use of Shape-Memory Materials in Distribution Power Systems", Canadian Electricity Association, CEA Report SD-294A (1994).

151. W.F. Bonwitt, "An Experimental Investigation of the Electrical Performance of Bolted Aluminum-to-Copper Connections", *AIEE Trans.*, vol. 67, (1948), p. 1208.

152. D.C. Hubbard, R.W. Kunkle and A.B. Chance, "Evaluation of Test Data in Determining Minimum Design Requirements for Aluminum-Copper Connectors", *AIEE Trans.*, vol. 73, (1954), p. 616.

153. N. Bond and F.L. McGeary, "Nickel Plating for Improved Electrical Contact to Aluminum", *IEEE Trans.*, IA-9, (1973), p. 326.

154. C.L. Bauer and G.G. Lessmann, "Metal-Joining Methods", *Annu. Rev. Mater. Sci.*, vol. 6, (1976), p. 361.

155. A. Kassman-Rudolphi, "Tribology of Electrical Contacts—Deterioration of Silver Coated Copper", Doctorate Dissertation, Upsala University, Sweden (1996).

156. R.L. Jackson, "Significance of Surface Preparation for Bolted Aluminum Joints", *Proc. IEE*, vol. 126, Pt. C, No. 2, (1981), p. 45.

157. J. Lefebre, J. Galand and R.M. Marsolais, "Electrical Contacts on Nickel Plated Aluminum: The State of the Art", Electrical Contacts-1990, Montreal (1990), p. 454.

158. J. Aronstein, "Evaluation of a plated Aluminum Wire for Branch Circuit Applications", Electrical Contacts-1987, Chicago (1987), p. 107.

159. T.K. Hare, "Investigation of Nickel-Plated Aluminum Wire Using Analytical Electron Microscope", Electrical Contacts-1987, Chicago (1987), p. 113.

160. J.F. Bruel and A. Carballeira, "Durabilité des Contacts Électriques en Aluminum", *Proc. SEE*, December (1986), p. 139.

161. M. Braunovic, "Evaluation of Different Platings for Aluminum-to-Copper Connections", *IEEE Trans. CHMT*, vol. 15, No. 2, (1992), p. 204.

162. M. Antler, "The Lubrication of Gold", *Wear*, vol. 6, (1963), p. 44.

163. W.O. Freitag, "Wear, Fretting and the Role of Lubricants in Edge Connectors", Electrical Contacts-1975, IIT (1975), p. 17.

164. W.E. Campbell and M.J. Trcka, "Development of Lubricant Composition for Sliding Switch Contacts", Electrical Contacts-1974, IIT, Chicago (1974), p. 258.

165. W.H. Abbott and J.H. Whitley, "Lubrication and Environmental Protection of Alternatives to Gold for Electronic Connectors", Electrical Contacts-1975, IIT, Chicago (1975), p. 9.

166. A. Huber, "The Lubrication of Electrical Contacts", *Proceedings of the 13th ICEC-1986*, Lausanne, Switzerland (1986), p. 447.

167. M. Braunovic, "Evaluation of Different Types of Contact Aid Compounds for Aluminum-to-Aluminum Connectors and Conductors", *IEEE Trans. CHMT*, vol. 8, No. 3, (1985), p. 313.

168. Multilam Contact Elements, http://www.multi-contact-usa.com/products/multilamline/1.

169. ASM Handbook Volume 6 "Welding, Brazing and Soldering" p. 164.

170. "Xeconex-Implosive Connectors", Implo Technologies Inc. Tech, Bulletin, (1997).

171. P. Nguyên-Duy, S. Boisvert, C. Lanouette and Y. Blanchette, "Optimisation des plages de contact de l'ensemble marteau-mâchoire des sectionneurs GEC Alsthom type BCVB 735 kV 4000A2", Technical Report IREQ 96-013 (1996).

172. M. Braunovic, "Effect of Different Types of Mechanical Contact Devices on the Performance of Bolted Aluminum Joints Under Current Cycling and Stress Relaxation Conditions", *Proceedings of the 32nd IEEE Holm Conference on Electrical Contacts* (1986), p. 133.

173. D.W. Jondhal, M.L. Rockfield and G.M. Cupp, "Connector Performance of New vs Service-Aged Connector. Part I"; J.D. Sprecher, M.L. Rockfield and T. Nobelfer, "Connector Performance of New vs Service-Aged Connector. Part II", *Proceedings of the 1994 Power Engineering Society IEEE Transactions and Distribution Conference*, 94CH3426-0 (1994).

174. J. Sprecher, B. Royce, "Higher Load Demands New Tap Connector Practices," *Electrical World*, vol. 208, 48–55, 1994.

175. M. Braunovic, "Reliability of Underground Connections", Canadian Electricity Association Report, CEA 227 D 916 (1997).

176. B.W. Callen, B. Johnson, P. King, R.S. Timsit and W.H Abbott, "Environmental Degradation of Utility Power Connectors in a Harsh Environment", Electrical Contacts-1999 (1999), p. 63.

177. M. Braunovic, V.V. Izmailov and M.V. Novoselova, "A Model for Life Time Evaluation of Closed Electrical Contacts", Electrical Contacts-2005 (2005), p. 217.

178. V.V Izmailov, "Conductivity of Long-Term Closed Electrical Contacts", Proceedings of the 10th International Conference on Electrical Contact Phenomena, Budapest (1980), p. 93.

179. E. Takano and K. Mano. "The Failure Mode and Lifetime of Static Contacts". *IEEE Trans. Pt. Mater. Pack.*, vol. 1, (1968), p. 51.

180. W.J. Buehler, J.V. Gilfrich and K.C. Weilley, "Shape-Memory Effect in Ni-Ti Alloys", *J. Appl. Phys.*, vol. 34, (1963), p. 1457.

181. "Use of Shape Memory Alloys in High Reliability Fastening Applications", Intrinsic Devices Inc., http://www.intrinsicdevices.com.

182. http://www.souriau.com/.

183. "Shape-Memory Temperature Indicator", Furukawa Electric Co., Jap. Patent No. 2-70218 (1990).

184. "Shape-Memory Temperature Indicator to Monitor Overheating of Electric Power Cables", Tohoku Denryoku KK-Fujikura Ltd., Jap. Patent No. 3-201313 (1991).
185. L. Lischina, Private communication.
186. M. Ashby, A. Evans, N.A. Fleck, L.J. Gibson, J.W. Hutchinson and H.N.G. Wadley, *Metal Foams: A Design Guide*, Butterworth (2000).
187. M. Braunovic and S. Vatchiantz, Unpublished work.
188. S. Vatchiantz and N. Manukyan, "Structure Dévelopment in Aluminum Foam Manufactured by Hot Rolling", SMA ICAA11, Conference, Japan (2010).
189. J. Jerz, P. Simancik, M. Bortel, S. Kubo and J. Kovacik, "The Design of Lightweight Armour Sheets", in *Proceedings of the International Conference on Cellular Metals and Metal Foaming Technology*, Ed. J. Banhart and N.A. Fleck, MIT Publishing, Bremen (1999), p. 235.
190. M. Braunovic and D. Gagnon, "Aluminum Foam Transition Washer for Power Applications", *IEEE Holm Conference Proceedings* (2008), p. 258.
191. Ecocontact http://www.amcetec.com/ecocontact.

6

Low-Power Commercial, Automotive, and Appliance Connections

Anthony Lee and George Drew

They weighed me, dust by dust—
They balanced film by film

Emily Dickinson

CONTENTS

6.1 Introduction 376
6.2 Connectors 377
    6.2.1 Functional Requirements 377
    6.2.2 Types of Connectors 378
    6.2.3 Mechanical Considerations 381
6.3 Contact Terminals 384
    6.3.1 Contact Physics 384
    6.3.2 Terminal Types 384
    6.3.3 Other Electrical Contact Parameters 392
6.4 Degradation of Connector Contact 393
    6.4.1 Surface Films 393
    6.4.2 Fretting Corrosion of Tin-plated Contacts 396
    6.4.3 Examples of Contact Failures 402
        6.4.3.1 Automotive Position Sensor Connector 402
        6.4.3.2 Fuel Injector Connector 403
        6.4.3.3 Glowing Contacts 403
        6.4.3.4 Electrolytic Corrosion 404
        6.4.3.5 Incompatible Plating and Low Contact Force 404
6.5 Automotive Connector Contacts 405
    6.5.1 Vehicle Conditions 405
    6.5.2 High Power Connectors for Electric and Hybrid Vehicles 405
    6.5.3 Aluminum Wiring Connections 406
    6.5.4 Connections for High-Vibration Environment 408
6.6 Summary 408
References 409

6.1 Introduction

Perhaps the most useful function of electrical connection systems is to serve as a convenience for assembly, installation, and servicing of an electrical device or system. For example, the internal connections in a timer control module of a dishwasher enable the assembly of such a module. In addition, the electrical connections of the harness to the motor, valves, and power sources complete the entire dishwasher assembly. Likewise, the physical installation of a component such as a motor into a clothes washer drive assembly or a radio into the dashboard of an automobile is done first, and then the electrical circuits are completed by mating the connectors or installing other required electrical connections. As each subassembly is linked, each harness is connected with other circuits and to the power source. Depending on the complexity of the entire electrical system, a variety of connectors such as in-lines, junction splices, device connectors, and headers are used. The overall system architecture of the entire "wiring harness" determines the number of points and the location where the circuits will be connected. Also, depending on the type of components in the circuits, connection systems may serve to connect low-current or low-voltage, high-frequency electronics circuits, high-current power circuits, timer and control circuits, sensor circuits, and communication signal circuits.

In this chapter, we deal mainly with electrical contacts of the separable type such as the male and female terminals in connection systems. We will also briefly examine the semi-permanent type of crimp and insulation displacement connections because they are intimately related to connectors and are subject to similar physical principles.

As with all electrical connection systems, the basic design of an electrical contact depends on a set of application parameters. Table 6.1 summarizes the range of circuit parameters within the scope of this chapter.

The basic design and application of the connections require the consideration of (a) The electrical devices to be connected, which define the voltage and current level, and (b) The functional, physical, electrical, mechanical, and environmental requirements of the system to determine the correct connection system to be used. For example, a connection made directly to the motor must have sufficient retention force to withstand the vibration from the motor, whereas a ribbon cable will require a different connector from a power feed line with large-gauge round cables.

TABLE 6.1

Application Parameters of Connection Systems in the Area of Low-Power Commercial, Residential, Automotive, and Appliance Circuits

Type	Circuit	Voltage (V)	Current (A)
Power	Appliance/commercial/residential	120–240 ac	10–100
Lighting	Appliance/commercial/residential	120–240 ac	< 10
Control	Appliance/commercial/residential	24–36 ac	< 2
Power	Automotive	12–36 dc	> 1
Power	Automotive, Electric Drives	~ 300 ac, dc	~ 300
Control	Automotive	5–12 dc	< 1
Low Power	Electronics	5–12 dc	< 1
Signal	All	< 1	< 0.1

The environment has a major impact on the design and material chosen for each type of connector housing and electrical contacts. For example, outdoor versus indoor applications can have a major implication on the design of the terminals and the material used (see Chapters 2, 3, 7, and 8). Other important parameters include the range of operating temperature, duty cycle, number of insertions, thermal cycling or shock, seal and protection, and the corrosion level of the environment. Section 6.5 will give examples of automotive connectors as applied to the challenging automotive environment.

Also, the initial cost of a connection system, including the cost of assembly, may determine the type of connector used. On the other hand, the cost associated with service and maintenance of such a system could be much higher than an alternative, which has a higher initial cost. If such equipment is for an application at a remote location, the total cost of the system may dictate the connection system selected.

Various industry standards and specifications have been developed, forming some common ground for both suppliers and users of connectors or connection within a given industry. For example, the American Society of Testing and Measurement (ASTM) has an extensive list of contact-related standards especially related to alloy materials, testing methodology, and recommended measurement practices. The Electronics Industries Association (EIA) also has a list of contact-related standards. Many of those standards from both EIA and ASTM have been approved as American National Standards. For appliance and consumer-related products, the connection systems are tested and listed together with the product as Underwriter's Laboratory (UL)-approved. For household and related components, the National Electrical Code (NEC) may also apply. For automotive connectors, there are a number of applicable SAE and ISO specifications. Also, the military specifications (Mil Specs) have comprehensive connector standards for devices associated with military applications in the United States.

In this chapter, Section 6.2 addresses connectors, their function, design, subcomponents, and assembly. In Section 6.3, we will focus on the electrical contact terminals of the electrical connections. The connector contact degradation mechanisms are discussed in Section 6.4. Since automotive applications have dominated the recent drastic growth in electrical and electronics systems, we shall devote a new Section 6.5 on this topic in this Second Edition.

6.2 Connectors

6.2.1 Functional Requirements

Although the basic function of connectors is to provide electrical interconnections to complete the circuit, a large number of non-electrical factors determine the type of connector or connection system used. To begin with, mechanical requirements such as size, shape, mounting, and mating and unmating force and frequency need to be considered. In addition, environmental requirements such as temperature, temperature cycle, humidity, contaminants (solid, liquid, and gaseous), shock and vibration, and sealing further complicate the design and selection process. The electrical requirement often goes beyond circuit voltage and current. Many applications also require low contact resistance, terminal impedance, polarity, and appropriate insulation level. To satisfy these often competing requirements, connection technology has evolved into a sophisticated engineering discipline. The challenge to the connection engineers is to design the lowest-cost products that meet all requirements, which in reality often work at cross purposes.

6.2.2 Types of Connectors

There are several generic types of connectors frequently found in automotive, appliance, and commercial applications: (1) Simple terminal–terminal, (2) Rack and panel, (3) Plug and receptacle, (4) Edge-on, and (5) Compliant pin.

The *terminal–terminal* type is the simplest and has been in use ever since the beginning of electricity. An example is illustrated in Figure 6.1, showing the simple blade-box connector. Clearly, manual connection is required for this type of connector, and essentially no insulation or enclosure is present to protect the terminals. This is acceptable when the connector is used within an enclosure such as the appliance's housing. This connector has both sufficient contact force to provide a low contact resistance even with non-noble coatings and sufficient mechanical force to prevent the contacts from parting.

The *rack* and *panel* type connector is typically for mounting of equipment where one side of the connector is on the removable part of the equipment and the other half on the stationary or fixed part. As the removable part is installed on the "rack", the connector is mated as well. A precision guide or a floating connector is mounted to ensure alignment of contact terminal to avoid damage due to variations in position during insertion. Figure 6.2 illustrates an example of such a connector.

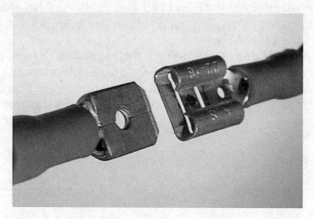

FIGURE 6.1
Example of the simple terminal–terminal type connector.

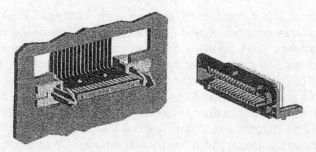

FIGURE 6.2
Example of the rack and panel type connector.

The *plug* and *receptacle* type is perhaps the most widely used. This type is very flexible, especially in the assembly of complex harness or circuits such as the installation of the automotive harness to support a large number of electrical features and functions. Figure 6.3 shows an example of this type of connector.

Depending on the specific application, the connector may be a part of a device with the mating part on the wiring harness. The component on the device side is known as a header or device connection. Also, such a connector is used as an in-line for joining one harness segment with another. Another application is a bulkhead feed-through. Here, the connector serves to permit the harness to terminate at some physical barrier and the circuit to continue on the other side such as at the wall socket, plug at the back of a control unit, or the feed-through at the firewall between the engine and the passenger compartments. When the number of lines is large, a plug-and-receptacle connector, which requires manual installation, can be difficult to mate. Some type of mechanical assistance in the form of lever or bolt-screw is often incorporated. Also, there are other assurance features to insure the quality of the installation including, lock, terminal position assurance (TPA), and connector position assurance (CPA). An example of this type of connector with the full complement of features is shown in Figure 6.4. Further, other requirements like shrouding, connector seal, and cable dressing to reduce mechanical and environmental stresses increase the complexity of the connector design.

The *edge-on* type connector is mainly for board-to-board and wire-to-board connections. In the majority of applications, a part of the terminal on the edge of the printed circuit board has its metal traces exposed. The other part of the connector has

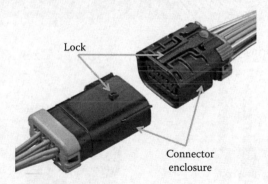

FIGURE 6.3
Example of the plug and receptacle type connector.

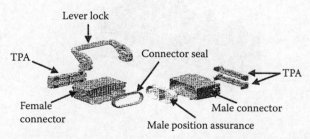

FIGURE 6.4
An automotive connector of the plug and receptacle type with TPA, CPA, lock, seal, and mechanical assist lever.

terminals and attachments to the cables. In some instances, a flexible printed circuit is used. In this case, a modified form of the edge-on connector is used as illustrated in Figure 6.5 [1,2].

Another common connection is the *compliant pin* connector [3,4,5]. A compliant pin consists of an electrical terminal incorporating a relatively stiff spring section which is designed to be press-fitted into a plated-through hole of a printed circuit board. Figure 6.6 shows examples of one of the most common compliant pin shapes—called the "eye of the needle". While it has been used in electronic components for many years, compliant pin technology has only recently (from 2005) been used in high-volume automotive electronics as an alternative to solder joints. A desirable feature in the design is to have enough contact force to promote galling (cold welding) at the contact spots during insertion of the pin into

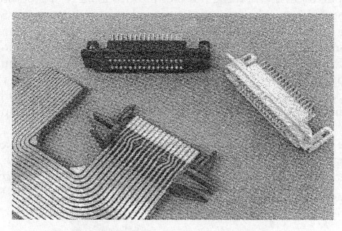

FIGURE 6.5
Edge-on connector for flexible printed circuits (FPC).

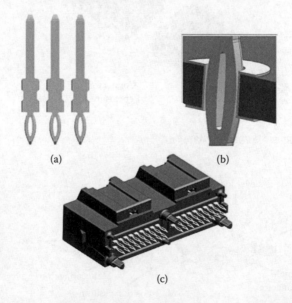

(a)

(b)

(c)

FIGURE 6.6
Compliant pin connector, (a) Typical "eye of needle" terminal design, (b) Cross-section after insertion into circuit board, and (c) Automotive compliant pin connector.

a circuit board. This provides a very stable electrical connection with a low resistance for connections that do not have to be taken apart.

On the other hand, the terminal insertion force must be low enough that the printed circuit boards, or board-mounted electronic components, are not damaged when the compliant pins are press-fitted into place [3,4]. The relatively large centerline spacing required between compliant pins can be a disadvantage compared to soldering or other termination methods. As the spacing between the plated holes decreases, the interaction of stresses around the holes can cause damage to the printed circuit board.

6.2.3 Mechanical Considerations

While the number of connector designs for different applications is large, they share several common features. Depending on the requirements and the environment the connector must service, the complexity can vary. Except for the simplest types of connectors, the essential parts of a connector are: contact terminals, line insulation, enclosure, and cable or wire termination. We will briefly discuss each of these parts.

Contact terminals: These are the conducting members and consist of two terminals, one male and another female, that pass the current when they are in the mated position. Section 6.3 covers different types of contact terminals. From the mechanical consideration, the terminals have attachment to the cable on the one end and the contact interface on the other. This interface also requires a low and stable resistance during its service life.

Line insulation: For connectors with multiple lines, electrical insulation, or isolation between adjacent lines is essential. Such insulation also serves to maintain the terminal location and alignment. As illustrated in Figure 6.3, the line insulation is a part of the plastic enclosure.

Enclosure: This is the housing for the connector where the parts of the connector are assembled. Mating is accomplished by joining the two parts of the connector enclosure together. Thus, the enclosure must provide mechanical guides for mating. Further, it also provides protection to the internal parts. For the rack and panel type connector, part of the housing is in the equipment. It can also be a header of an electronic module.

Cable termination: It forms the transition between the cable and the terminal and is often a non-separable connection. Usually, a metallic or mechanical (soldered, welded or crimped) joint serves both the electrical and mechanical connection. Although the wire–terminal interface is a permanent contact interface outside the scope of the separable contact interfaces being discussed here, it does play an important role in the assembly and performance of the terminals as a part of the connector. This interface provides the electrical and mechanical link from the terminal to the wire. Its interface resistance and contact degradation can lead to degradation of the separable contact interface at the male or female terminal. A brief discussion on the crimp termination will be presented in Section 6.3.2.

An insulation displacement connection (IDC) is also used as a method for attaching the cable to the terminal. As shown in Figure 6.7, the cable is fed into an IDC slot. During insertion, insulation to the cable is cut, the slot is forced to expand, and an electrical contact with

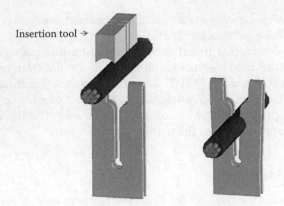

FIGURE 6.7
Insulation displacement connection (IDC) before and after cable insertion.

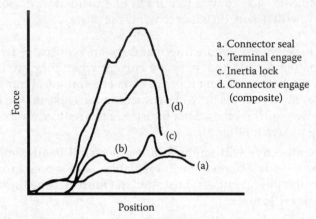

a. Connector seal
b. Terminal engage
c. Inertia lock
d. Connector engage
   (composite)

FIGURE 6.8
Connector engagement force characteristics.

the wire is made. Examples of such a termination method were reported in the literature [6–8]. The inherent advantage of IDC connection is the consistent contact force between the wire and the IDC slot. Also, such termination process is attractive for automation because the cables going to the same connector can be terminated simultaneously. On the other hand, crimping and soldering are done individually.

Other connector features illustrated in Figure 6.4 include connector seals. For connectors exposed to the elements, seals effectively prevent the intrusion of liquid, gas, and particulates into the contact terminal areas. Liquid contaminants are a major cause of contact corrosion and loss of electrical insulation between lines; dust and particulates can often lead to high-resistance contacts (see Chapters 2 and 4). Sealing against gaseous contaminants is much more difficult. However, the seals do drastically reduce the diffusion of corrosive gaseous pollutants into the terminal contact areas and greatly increase the service life of terminals. Some seal designs must also handle situations when a pressure difference exists between the inside and the outside of the connector such as during temperature cycling or submersion under water.

For manual connector engagement, one can capture the mating force as shown in Figure 6.8. This force–displacement characteristic reflects events occurring during the mating process. For manual mating, the peak force must be sufficiently low to be acceptable to the operator. Equally important is the work performed or the area under the force–displacement curve. From an operator's point of view, the connector engagement needs to relate to an "effort" of mating. The "effort" is comprised of not only the peak force and the work of mating, but also the ergonomic environment such as the reach and orientation of the mating, size and shape of the connector, visibility of the parts, and the posture of the operator.

The quantification of the "effort" of mating a connector is certainly not an exact science and often subjective. However, the design must incorporate features that result in a connector that is "easy" for the operator to put together.

When the number of contacts in the connector is in the tens, especially when seals are used, a mechanical assist engagement may be required. Such a feature may take the form of a lever as shown in Figure 6.4.

Some connector designs even go all the way to a zero-insertion-force (ZIF) concept [9,10]. The basic principle used is that during connector mating, the frictional force is nearly zero, and a subsequent clamping action presses the contacts together. Most ZIF concepts basically separate the insertion action from the application of the normal force.

Another feature is the connector lock. The holding force of a connector is often not sufficient to maintain the engagement for the service life. This is especially true for applications, where the connector experiences frequent movement or is under a vibratory environment. A case in point is the connector attached to the engine of a motor vehicle. A positive locking mechanism is essential to ensure engagement is maintained. Such a connector lock feature is shown in Figure 6.3.

As an aid to the assembly and installation of the connector, terminal position assurance (TPA), and connector position assurance (CPA) are incorporated into more sophisticated connector designs. Such features are illustrated in Figure 6.4. A common approach in TPA design is to prevent the connector from being further assembled unless all terminals are positioned correctly. Similarly, unless the connector is fully mated and locked, the CPA cannot be installed. Both TPA and CPA have greatly improved the installed quality of connection systems.

A filtered connector is a specialized connector design where a low-pass L–C circuit, for example, is incorporated into its construction. An example is shown in Figure 6.9. Here, the ferrite block provides the series inductance, and an array of chip capacitors is placed between the pins and the ground. Also, a C–L–C pi filter section can be similarly implemented by adding another capacitor array. Incorporating a filter in the connector provides high-frequency isolation between the two sides of the connector. This saves valuable board space and addresses the electromagnetic compatibility of the circuit.

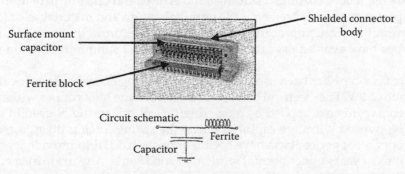

FIGURE 6.9
Filter connector with an *L-C* low-pass circuit.

6.3 Contact Terminals

6.3.1 Contact Physics

The conduction of electricity through a metallic join in physical contact occurs at small asperity points or *a*-spots. As described in detail in Chapter 1, those *a*-spots give rise to a construction resistance at the contact. Such contact resistance may be expressed as

$$R_c = [\rho^2 \eta \pi H / 4F]^{1/2} \tag{6.1}$$

where ρ is the resistivity of the contacting metal, η (≈ 1) is an empirical coefficient representing the surface cleanliness condition, H is the hardness of the metal, and F is the contact force. For connector application, the contact resistance must be low and stable for the life of the connector. The material selection will determine the resistivity, ρ and the hardness, H. The contact force, F, is a key design parameter for the terminal. The surface condition parameter, η, is determined by the level of corrosion and contamination in the environment and the degree of protection provided to the contacts. Its value is approximately one for most metal contact systems commonly used for connectors. Physically, a conductor with a low ρ will be a better contact material, and a softer material (low H) will provide larger *a*-spot size as will a greater contact force.

In practice, the desire to have connectors that are smaller, lighter, with lower cost, and operable under extreme environment offer many engineering challenges for the real world applications. For example, the contact resistance can be reduced by the use of a softer material. However, such a material can also be easily worn out after only a few insertions (see Chapter 7). Furthermore, contact terminals are often served as a spring member supplying the required contact force: a softer material tends to have lower spring constant and thus will not be a good spring member so it is often plated on to a good spring substrate (see Chapter 8).

6.3.2 Terminal Types

The basic function of an electrical terminal is to provide the electrical contacting point(s) when mated at a low and stable electrical resistance for the service life of the connector under the required operating environmental conditions. The key parameters are the electrical contact material, the contact force and wire termination. There is no unique solution for a given need among the various combinations of terminal configurations, base alloy materials, spring members, coatings, claddings, and crimps and other form of termination. In addition to performance requirements, the terminal design and material selection must consider the overall fabrication process and assembly of the connector as well. A number of terminal types have evolved over the past years. Table 6.2 summarizes the most common types.

The *wire–wire twist* type has been used extensively in joining household electrical wiring for AC of 110 and 220 V. The electrical contacts are between the bare copper wires twisted together. The contact force is applied by a tapering helical coil spring housed in an insulating shell. The twisted wires are captured inside the spring forcing them to remain in contact. The copper–copper contacts have been demonstrated [11] to provide low contact resistance for tens of years under normal residential conditions. Any oxide formed is thin and easily fritted through by the supply voltage. In addition, the number of contacting points is large, affording a reliable and durable connection. Typical circuit current is in the

TABLE 6.2

Summary of Terminal Types

Type	Contact Form	Spring Member	Base Metal	Coating	Typical Current (A)	Termination	Common Usage
Wire–wire twist	Multiple wire–wire	Coiled spring insert	Cu, Al	none	≤ 20	None	Household wiring
Wire–screw or plug–screw	Wire–flat or flat-flat	Deformed thread/ washer	Cu, Al	none, Sn	≤ 100	None or crimp	Household wiring/ appliance
IDC	Wire–blade	Cantilevered	Cu, Cu–alloys	none	≤ 20	None or crimp	Commercial, automotive
Tuning fork	Wire–blade	Cantilevered	Cu, Cu–alloys	none	≤ 20	None or crimp	Commercial, automotive
Blade-box	Row	Deformed box	Cu–alloys	none, Sn, Ni	≤ 40	Crimp or solder	Appliance, automotive
Blade-leaf	Beam	Cantilevered	Cu–alloys	Sn, Ag, Au, Ni	≤ 30	Crimp or solder	Automotive, appliance, commercial
Pin–sleeve	Multiple beam	Cantilevered	Cu–alloys	Sn, Ag, Au	≤ 10	Crimp, IDC or solder	Commercial, appliance, automotive
Pin–hyperboloid	Wire–pin	Stretched wire	Cu, Cu–alloys	Sn, Ag, Au	≤ 2	Crimp, IDC or solder	Commercial
Bump–flat	Butt	External	Cu, Cu–alloys	Sn, Pd Ag, Au	≤ 2	None or solder	Commercial, automotive
Press-fit	Pin-hole	Compliant pin	Cu–alloy	Sn, Solder	≤ 5	Crimp or solder	Commercial. automotive

range below 20 A for household type of applications. Rise in temperature at the contacts is typically low helped by low connection density even at the central distribution panel. Also, the wire gauge used is under strict requirement of the local building codes.

Aluminum alloy wires are also acceptable for similar application. However, it is more difficult to establish and maintain good electrical contact to aluminum due to aluminum oxides on the surface [12–17]. The application of contact lubricants and more stringent control of the assembly are required [18–20] to maintain a trouble-free connection.

In many respects, the *wire–screw* or the *lug–screw* type is similar to the wire–wire twist type. Both are designed for manual mating of individual circuits. In this case, the screw type of terminal is for connecting a wire to a device or a piece of equipment. In the simplest case, a wire is wrapped around a screw and a tightened nut supplies the contact force. For convenience, the wire may be crimped to a lug, which has a hole for assembling onto the screw. As such, the contacts are a wire against flat or, in the case where a lug is used, a flat against a flat contact.

The contact force is applied through the screw action compressing the contacting surfaces together. The spring member is the compliant metal parts such as the screw thread, wire wrap, or deformed washers. Because high contact force may be applied, the current-carrying capacity of this type of contact is high. It is important to recognize that the inherent spring constant of such contact system may be quite large. Any small change induced by thermal expansion and contraction, vibration, or even metal flow and relaxation under high stress, can cause the contact force to drop to an inadequately low level. It is a common practice to include a lock or star washer into such an assembly. It serves not only the purpose of preventing the screw from coming loose, but also provides a lower spring constant to the contact spring system so the contact force can be maintained as a result of small, but often unavoidable, dimensional changes. Figure 6.10 shows such a screw type contact.

The *insulation displacement connection* (IDC) terminal has been developed for ease of installation [6,21,22]. An insulated wire can be connected to a device without stripping and in a single operation. In the insertion process, the insulation is cut, and the exposed wire makes contact with the blades formed in a narrow V slot. The blades are pushed apart by the wire at the end of the insertion providing the contact force. The mating process is similar to that illustrated in Figure 6.7. The design of the IDC terminal is sensitive to gauge and insulation. The slot size and shape, the cutting edge, and the spring constant of the blade are usually tailored for small-range wire gauges. In addition, the wire-slot orientation must be maintained to avoid disconnection. This further requires proper strain relief as part of the connector design. The wire–blade contact is typically small, thus limiting the circuit current an IDC can carry, and is not suitable for high-current connector applications. Incidentally, an IDC is often used as a termination at the terminal to replace the crimp or solder joint. The electrical contacts of the *tuning fork* type terminal, shown in Figure 6.11, are similar to the IDC type. Since the tuning fork does not need to cut the insulation function, the terminal design may be adapted to engage both wire and a flat plate. The groove design can accommodate a wider range of widths and contact force to increase the number of insertions permitted.

For many years, by far the most common connector type has been the *blade-box* terminal, shown in Figure 6.12, for appliances and commercial electrical circuits. It can carry a wide range of current and has a simple, robust construction. The typical design configuration

FIGURE 6.10
Wire-screw type terminal.

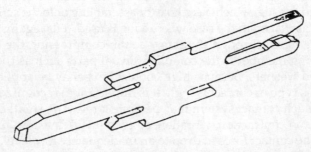

FIGURE 6.11
Tuning fork type terminal.

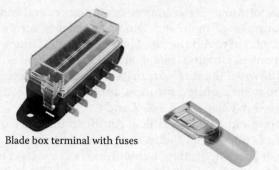

Blade box terminal with fuses

Example of the female terminal

FIGURE 6.12
A blade-box terminal.

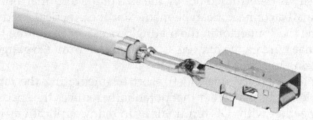

FIGURE 6.13
Example of a blade-leaf terminal.

consists of a double row of contacts. The flat blade is inserted into the rolled up sides during mating. The rolled up sides apply a strong contact force. Because of the large contacting area, this terminal may be applied for a higher current range. With high contact force and multiplicity of contacts, the blade-box terminal is resistant to vibration. With proper selection of coating, the contacts are nearly "gas-tight" and thus less susceptible to gaseous corrosion.

The insertion force is usually high, and contact wear is also high, limiting the number of insertions for this type of contacts. Together with the relatively large terminal size, applications of blade-box terminals are limited to connectors with a few circuits to keep the mating effort within reasonable limits.

The *blade-leaf* type terminal evolves from the blade-box type. As the connection density is increased, the box design cannot accommodate smaller blades. Instead of forming a rolled up box, a box with a thinner stock is formed with leaf contacts pressed towards the interior of the box. When mated, the blade makes contact with the opposing leaf contacts. Here, the leaf springs supply the contact force. A variety of leaf designs is possible. An example is illustrated in Figure 6.13.

This type of terminal has been very popular for automotive circuits where lower insertion force and higher circuit density are important. On the other hand, the automotive environment is uniquely challenging. Connector component engineers must contend with mechanical vibration, corrosion by salts, exposure to fluids from lubricating oils, high-pressure washer jets, soft drinks and coffee, and extremes of temperature outdoors and under the hood.

As the terminal size is further reduced, we have the *pin-sleeve* type of terminal. Here, contacts are even smaller and contact forces are much lower, but circuit density is certainly higher. Pin contacts have been widely used in the electronics and aerospace industries.

It is natural to adapt these terminals for automotive, appliance, and commercial electronics. Figure 6.14 shows examples of male and female pin-sleeve terminals. Cantilevered beams form multiple contacts surround the pin. Since most applications deal with some or all electronic signal and control circuits, voltage and current can be quite low where fritting or electric field breakdown of the surface film (see Chapter 1) is less likely. Noble metal coating for the terminal may be required for most applications. The effects of thin noble metal coatings are discussed in Chapters 2, 3, 4, 7, and 8.

The *pin-hyperboloid* type is rather unusual [23]. As illustrated in Figure 6.15, it uses a cross-mesh of fine gauge of copper–alloy wire as the female terminal against a pin. The wire mesh is stretched over the pin, forming a multitude of contacts. The flexibility of the wire mesh also allows the contacts to float along with the pin. This type of contact is least susceptible to vibration and allows perfect alignment with the pin. Because of its higher cost and low current capacity, this type of terminal is used only in exclusive circumstances depending on the requirements of the environment and reliability.

The connection density is determined not by the size of the terminal alone, but the cable size also limits how small a connector can be made. For the practical needs of assembling the connector, handling the connector in the mating process, or installing the harness into the equipment, a circular wire gauge may be as small as 0.13 mm² (26 gauge). This sets the lower limit of line spacing to about 1.5 mm.

To further increase line density, one must resort to integrating the terminal into the wire. IDC is an example discussed earlier that potentially reduces the space between lines. The use of flex cable together with IDC terminals as in many applications in the computer industry is another example. In appliance and automotive applications, the flexible printed circuit (FPC) wiring offers similar advantages of integrating terminals with the wiring. The FPC has a well-defined pattern of copper conductor on a flexible dielectric material

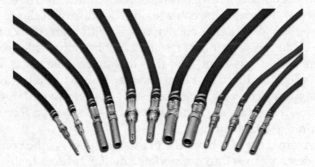

FIGURE 6.14
Examples of pin and sleeve terminals.

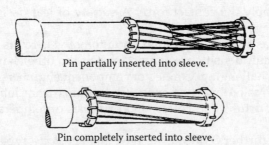

Pin partially inserted into sleeve.

Pin completely inserted into sleeve.

FIGURE 6.15
Example of a pin-hyperboloid terminal.

such as a polyimide or polyester. The copper conductors are typically sandwiched between two layers of flexible dielectric material. FPC can also come in single-sided, double-sided, or even multilayered construction. The conductors in the FPC have accesses for attachment of circuit components and for making direct connections.

Conventional connections to an FPC are by soldering the connector pins directly to the termination points on the FPC often in a form of a plated through-hole. To facilitate assembly, an example is shown in Figure 6.5 of a recent development of direct connection onto FPC. Here, the bare traces on the FPC are backed by a part of the connector structure, transforming them into rigid terminals which mate with a sleeve type female terminal.

Another type of terminal that is ideal for high density and FPC is the *bump-flat* type shown in Figure 6.16. The bump is part of the trace on the FPC. Pressing the bumps against the flat traces of another FPC makes the connections. Here, an external force must be applied, holding the FPCs together. Because of the high stresses at the bump–flat interface, the contact is very stable. However, the high stress also requires careful selection of contact material and bump geometry. The current-carrying capacity of such a small bump is limited to a couple of amperes. However, multiple bumps may be used for higher current requirements. The current requirement and the line density set the limit of the connection density. In principle, very fine pitch bump connection is possible. In addition, the flat side may be a hard circuit board also. The use of bumps on circuit board opens up the possibility of making non-permanent type connections anywhere on the board, and not limited only along the edge as in the case of edge board connectors.

Cable termination. This transition between the cable and the terminal is a non-separable connection. Usually, a metallic or mechanical (soldered or crimped) joint serves both the electrical and mechanical connection. Although the wire–terminal interface is a permanent contact interface it does play an important role in the assembly and performance of the terminals as part of the connector. This interface provides the electrical and mechanical link from the terminal to the wire. Degradation of the terminal or cable interface can lead to degradation of the separable contact interface at the male or female terminal.

The most common type of wire–terminal interfaces for automotive and appliance wiring is F-crimps and barrel crimps, as shown in Figure 6.17. F-crimps are named for their resemblance to the letter F when viewed from the side.

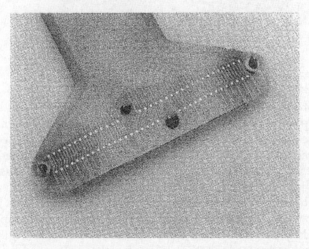

FIGURE 6.16
An FPC with bump contacts.

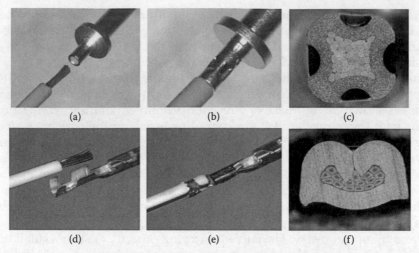

FIGURE 6.17
Photos of barrel and F-crimps (a) Cable insertion into barrel crimp, (b) Finished barrel crimp, (c) Barrel crimp cross-section, (d) Cable insertion into F-crimp, (e) Finished F-crimp, and (f) F-crimp cross-section.

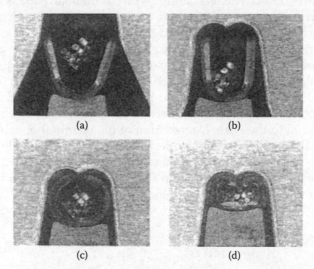

FIGURE 6.18
Stages of the crimping process.

In the termination process, the wire insulation is stripped, the bare section of the wire is inserted into the crimp wing or barrel region, and a crimp tool presses the barrel or rolls the wings that collapse around the wire creating both mechanical and electrical bonding. Crimps have been used for decades joining wires and terminals of various sizes with extremely good results. To understand the physics of crimping, one must look at how a crimp is formed (See also Chapter 5).

Figure 6.18 shows the stages of the crimping process. First, the wire is positioned at the inside of the crimp wing (Figure 6.18a, and then the barrel or the crimp wings forming the barrel is collapsed by the crimp tool trapping the wire inside (Figure 6.18b and c). Then, the wires are further pressed together, loosening the surface oxides, inducing some metal flow, and forming a tight bundle (Figure 6.18d). The last stage is the formation of mechanically and electrically good metal-to-metal bond as clean metals are forced together. This is often referred to as cold welding. Finally, as the tool is released, the assembly that was under high compression "rebounds." The crimp wings or the barrel actually relax and the metal bounces back from its compressed state. Note the contact force between the wire bundle and the crimp wings actually reduces a little at the bounce back. The design of the crimp must allow for this bounce back such that sufficient residual force still remains to maintain a good mechanical joint and low contact resistance [24,25].

Unfortunately, the optimum electrical performance and the highest mechanical strength do not occur at the same level of crimp compaction. Producing a crimped connection with the maximum pull strength often results in poor long-term electrical stability, while a crimp with the best long-term electrical performance can have poor mechanical properties. Therefore, a trade-off must be found between the best combination of electrical and mechanical performance. Figure 6.19 shows this relationship for a typical crimped connection to a stranded cable. The dashed line in Figure 6.19 shows how the pull strength of the crimp varies with different levels of crimp compaction, while the solid line shows how the long-term electrical performance varies with crimp compaction. A freshly made crimp has relatively low resistance over a wide range of compaction levels, so discrimination between compaction levels is performed after environmental conditioning. One good reference for environmental exposure tests is the SAE USCAR-21 specification for automotive crimped connections [26]. Based on Figure 6.19, the best combination of electrical and mechanical performance for a crimped connection occurs at compaction levels which are tighter than the point where the highest pull strength occurs, yet loose enough that strands are not damaged. A well-controlled crimping process is very critical for

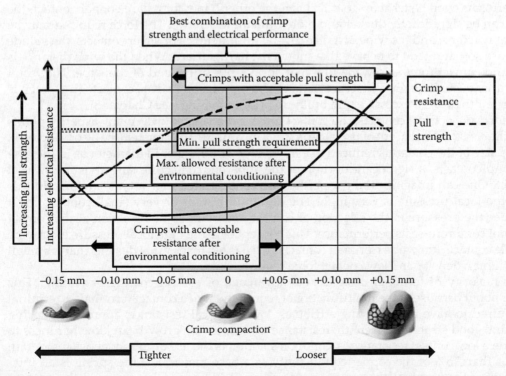

FIGURE 6.19
Crimp mechanical and electrical performances versus crimp compaction.

making reliable wire terminations. For the case shown in Figure 6.19, it is ± 50 microns. To maximize the operating window where a crimped connection can be used, terminal manufacturers use a judicious selection of material, gauge, coating, internal surface serrations of the barrel, tool geometry, profile of the pressing force, and final crimp deformation [24]. In many applications where small gauge wire is used or the crimp experiences high operating temperatures, the crimping may no longer be reliable, so alternative termination processes, such as welding and soldering, may be used. They are more expensive, but the resulting joints provide electrically stable interfaces with a good mechanical strength.

6.3.3 Other Electrical Contact Parameters

In terminal designs, contact resistance is the most important consideration. However, there are other parameters, which are quite important as well. These are force, pressure, wipe, bulk resistance, and plating or coating. Although contact resistance varies inversely as the square root of the contact force (see Equation 6.1), the contact force also provides the frictional force holding the mated terminals together. Also, together with the wipe action, the contact force provides the cleaning action removing non-conducting films and particulates from the surface at mating. This cleaning action is essential for establishing a low initial contact resistance at mating. Depending on the contact geometry, contact material, and the type of surface film, the force and the amount of wipe required can vary. For example, the thin film on tin-coated contacts, the softness of the tin and the hard tin oxide, a wipe distance of few tenths of a millimeter with a force of less than 1 N is quite sufficient to establish a low contact resistance. Under a wide range of conditions, Yasuda reports that tin performs even better than gold [27]. On the other hand, for a bare copper contact, the films can be significantly thicker and more difficult to remove. The force required may be several newtons, and the wipe distance several millimeters. For silver contacts, the sulfide films are soft and tend to behave like snow on a hard surface. When the wipe distance is long and contact force is low, the film may form a pile at the end of the wipe. The result is an unexpectedly high contact resistance. Thus, protecting silver contacts in storage is important to prevent excessive film build-up prior to mating (see Chapter 8).

In normal electrical contacts, the contact force is distributed over many asperity points (see Chapter 1). The concept of pressure is still a meaningful parameter (see also Chapter 1). As the size of the contact is reduced to a fraction of a millimeter in diameter as in the case of bump contacts, a significant portion of the apparent contacting surface is in physical contact. One can imagine all the asperities are concentrated in a very small area. In this case, the local pressure is very high. The design of bumps or very small contacts must consider the pressure in the selection of contact materials, so deformation at the contact does not compromise its performance. This concept of using high local pressure to provide reliable contacts known as Hertzian contacts [1,2] has led to terminal designs that are small or with high density and low contact force requirements.

The majority of terminal designs take advantage of contact coating or plating. This comes about because of the multifunctional requirement of a contact terminal. A terminal is required to have the following attributes: low electrical resistance, appropriate spring constant, good structural strength, resistance to surface film growth, and low hardness to provide a low contact resistance. A common solution is the use of copper alloys that retain greater than 75% of the copper conductivity as a base material. If the spring is an integral part of the terminal, there are several bronzes and brasses that have excellent spring constants. At the contact area, the copper alloy is not adequate. Coating or plating the copper alloy with silver, gold, palladium, and similar noble metal or their alloys provides excellent, stable and low contact resistance with little film growth. The subject of plating is discussed in detail in Chapter 8. To lower the cost, for many applications, tin plating and sometimes nickel have been successfully used.

Tin has some very interesting properties that make it an ideal terminal coating material. It is a soft metal, which translates into a low-contact resistance contact. Although oxides and other films quickly form on its surface, these films are typically hard and moderately self-passivating. This means after initial growth, the film itself protects the metal below from additional oxidation. The hardness of the film combines with the soft metal making it easier to break the film. The contact force with a little wiping action can quickly expose the fresh metal at the contact, forming a low resistance joint. There are failure modes that relate to tin contacts. One is fretting corrosion (see Chapters 7 and 8), and the other is intermetallic formation [27–32 and Chapters 1 and 5]. As will be discussed in Section 6.4, with proper design, tin may be used with a high degree of reliability [33].

Another type of coating is cladding. This process involves mechanical bonding of a thin strip of noble alloy on copper alloy at the base. Cladding requires no wet chemistry and can be extremely thin down to the micrometer range on a selected area of the strip stock. Cladding can be an economic viable alternative coating.

Vapor deposition is another coating method that does not rely on wet chemistry. By co-depositing different materials, this process can tailor the coating composition at the atomic level. Also, such coating can be dense and pore-free and often takes on an amorphous form. Such coating is more resistant to delamination and inter-atomic diffusion, both of which can lead to contact degradation. With many regions of the world having strong environmental restrictions for plating, both cladding and physical vapor deposition are viable coating methods without the environmental concerns of wet chemistry of the plating process. These coating methods are further discussed in Chapter 8.

Bulk resistance of a terminal, especially those for higher current ratings, is an important consideration. In relative terms, the bulk resistance is in the micro-ohm to tens of micro-ohm range while the contact resistance is 10–100 times larger. Keeping the bulk resistance low reduces the overall resistance of the circuit. More importantly, the wiring and the terminal often form the major heat sink for that part of the circuit. This is particularly true when the harness is attached to a heat-generating device such as a motor or a light source. The terminal should not be a heat source or cause impedance to heat flow into the wiring. In some circuits, a slight imbalance in the heat flow can raise the contact temperature significantly, so that degradation from stress relaxation of the contact spring or the crimp can lead to failure or shorten the life of the electrical contact.

6.4 Degradation of Connector Contact

6.4.1 Surface Films

Within the scope of this chapter, those contacts which are applied in commercial, appliance, and automotive wiring harnesses, typically have a copper-alloy base material and a non-noble coating. For certain connections at low voltage or in electronic applications, a thin layer of noble coating is used. Depending on the environmental condition, it is unavoidable that insulating surface films form during the fabrication process, while the terminal is in storage, during installation of the connection system, and while the connection is in service. Surface films represent a major contact degradation mechanism. Chapters 2, 3, 4, and 8 discuss surface films and corrosion effects that lead to the growth of such films. Here, we focus on how to avoid excessive film growth and how to permit current flow in the presence of films under the operating conditions of the type of circuits.

First, we need to understand the transformation of the contact surface with exposure. Clearly, chemical reactions with gas molecules and water vapor in the atmosphere are very difficult to avoid. Thus, galvanic corrosion is the major mechanism for growth of films. In recent years, extensive investigation done at Battelle [34] and others [35,36, and Chapter 3] on corrosion have provided not only an understanding of the mechanisms of film growth, but also the characterization of the environment and methods of simulating the corrosion in the laboratory. The most challenging environment to characterize is the automotive environment. The large fluctuations in temperature and humidity during the course of each day, the seasonal changes, and the reactive gaseous compositions can change drastically. The other uncontrollable factor is the location of the vehicle. Exposures to the polluted air of Los Angeles are quite different from exposures to the hot humid air with salty mist at Miami Beach. Even within the vehicle itself, the region inside the engine compartment has an entirely different environment from that inside the trunk, for example. From a Delphi study conducted by the authors, as shown in Figure 6.20, these are examples of metal coupons exposed to the atmosphere for nine months in different parts of a vehicle. The photograph clearly shows a large range of film types and film thicknesses as reflected by the color and microstructure of the films. These films were analyzed by cathodic reduction method (see Chapter 2, Section 2.6.3) to determine the types and to estimate the film thicknesses. Figure 6.20 gives representative data showing the films on the metal coupons from different location within the vehicle. We are able to establish the range of film thickness and composition at different locations within the vehicle. By varying the exposure time, the growth dynamics of these films can also be characterized (see Figure 6.21).

The main goal of this sequence of study at Delphi was to develop a laboratory simulation of contact corrosion. Under a controlled flowing mixed gas environment at higher

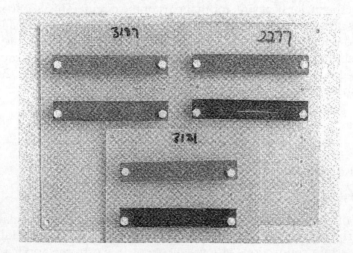

FIGURE 6.20
Photographs of silver-plated and copper coupons exposed to air for nine months in different locations of a vehicle in actual field operation. #3187 is located in the passenger compartment, #2277 is in the trunk, and #3181 in the engine compartment.

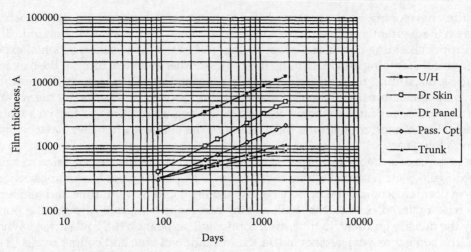

FIGURE 6.21
Projected film thickness on copper coupons based on field data.

TABLE 6.3

Environmental Classification ([34] and see Chapters 2 and 3) for Different Locations Based on Copper Reactivity

Location	Classification	One-Year Film Growth (Å)
Under hood	IV	4,000
Door skin	IV–	1270
Door panel	III–	500
Passenger compartment	III+	730
Trunk	III–	560

concentration, an accelerated test condition was successfully developed to represent exposures in the actual environment. Table 6.3 shows the environmental classification that covers the range of exposures in one year for various locations within the vehicle.

Of course, the contacts are normally located inside a connector. The environment is quite different from an open exposure. The shrouding and shielding of the connector housing does influence the film growth by restricting the movement of the gas. The test condition only provides the outside environment for which a connector is exposed. Also, the degradation of the contacts under test is monitored through the contact resistance of the interface, but not by measuring the film growth.

Dust and particulates are another source of contaminant of the insulating surface [37,38, see Chapter 4]. Proper protection of the contacts during storage and assembly can eliminate a part of the problem. It is more difficult to guard against dust and particulates during operation. The two common approaches to reduce the effects of dust are connector seals and a high contact force. Seals, as discussed in Section 6.2, isolate the contacts from fluid and dust. High contact force provides a more stable interface. Even if small interfacial movement occurs, the contact can effectively push the dust particles away. This is one reason the robust blade-box type of contact can survive years out in the open even without the benefit of a connector housing.

Fluid exposures are another source of insulating films. Contacts are often successfully applied in fluids such as inside the fuel tank and the transmission of a car in spite of organic films developing on the contacts [39]. However, certain household fluids such as coffee, when inadvertently spilled on the contact, can cause a non-conducting organic film to form. Proper shielding of contacts in appliances and in the car to such accidental exposure is required. Stamping and machining fluids are another source of films. If the contacts retain excess machining fluid or are not properly cleaned after processing, the residual fluid may not cause immediate problems because it is easily pushed aside by the wiping action of the contacts. However, some of these fluids degrade in time to form a varnish coating which can interfere with the functioning of the contacts after they have been in service for a few years.

Another source of surface films is pore corrosion derived from the base metal. An example is a thin gold plate terminal with a copper alloy. The plating does have imperfections in the form of microscopic holes or pores (see Chapters 2 and 8). With time and temperature, the base material components such as zinc can readily migrate through the pores driven by the density gradient. In time, even copper will appear on the gold surface. When oxidized, the contact resistance rises due to the formation of zinc and copper oxides. The application of a thin barrier layer such as nickel between the base metal and the gold surface is an effective preventive measure (see Chapter 8). Of course, a thicker or pore-free coating can also reduce pore diffusion.

6.4.2 Fretting Corrosion of Tin-Plated Contacts

All base metal contacts are susceptible to what is known as fretting corrosion. Briefly, fretting refers to a minute relative movement between mated contacts. This relative motion may be due to mechanical vibration, shock, or differential thermal expansion and contraction. In the comprehensive review articles [40,41] Antler and McBride provided a clear understanding of the fretting corrosion mechanism. Also, both Chapters 5 and 7 discuss various aspects of fretting corrosion. In the commercial, appliance, and automotive connections, tin coating is widely used. Here, we would like to focus on fretting corrosion of tin-plated contacts related to specific applications such as those described in Table 6.1.

In fretting of base metals such as tin contacts, the repetitive exposure of fresh metal and its subsequent oxidation with each movement and the accumulation of oxides at the contact interface are believed to be the main mechanism responsible for the corrosion and failure of these contacts. The result is a steady rise in the contact resistance leading eventually to high resistance levels and even open circuits.

Extensive experimental work has led to an understanding of fretting corrosion [42–45] of tin, and in particular for simulating conditions often encountered in the automotive application [46,47]. We would like to summarize the major findings of the studies done by the authors here. Table 6.4 gives the ranges and the parameters studied. The fretting motion can be simulated by controlling the frequency and the length of the fretting track. The entire experiment is controlled by a PC, which also measures the contact resistance periodically as schematically illustrated in Figure 6.22. A stepping motor driving a linear precision stage provides the cyclic fretting motion. After a preset number of fretting cycles, the motion is temporarily halted to allow for contact resistance measurements made at ten points along the track length in both directions and at both polarities.

The contact pair consists of a rider dimple or a rod against a flat coupon. The contact force is applied by a dead weight placed directly over the rider. To vary the contact temperature, the flat contact support structure is maintained at the desired temperature.

TABLE 6.4

Fretting Corrosion Experiment Parameter Range

Parameter	Range Studied
Track length	12–80 μm
Frequency	10 Hz
Contact resistance	4 wire, dry circuit (20 mV, 100 mA maximum) or 5 V and 14 V at 0.1 A and 1.0 A
Contact material	3 μm thick matte tin plated CA65400
Contact configuration	Flat coupon against 3.2 mm diameter dimple or flat coupon against plated rods with various shapes
Contact force	0.5–4.0 N
Contact temperature	25–110°C

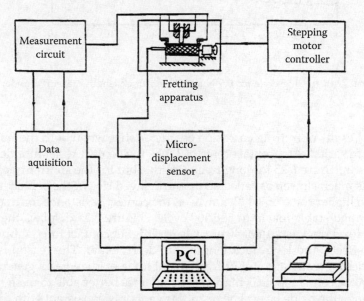

FIGURE 6.22
Schematic of the fretting corrosion experimental set-up.

Figure 6.23 captures the fretting corrosion of tin-plated contacts for the case of a flat coupon against the dimple at a track length of 20 μm and a contact force of 0.5 N. Under these conditions, the dry circuit contact resistance is initially in the milliohm range. At around 100 cycles, the resistance starts to rise reaching the 100 mΩ range after 1,000 cycles. For all practical purposes, the contact's electrical performance is no longer reliable after a few hundred fretting cycles. It has been well established [40] this corrosion characteristic of tin is a result of enhanced oxide formation and accumulation in the contact region as fresh metal is exposed to oxygen at each fretting cycle.

An understanding of the corrosion mechanism is crucial in successful application of tin with its many extremely attractive properties including low cost as a contact material. The challenge is to design connection systems such that fretting corrosion is minimized and, if it occurs, the contact performance is not compromised.

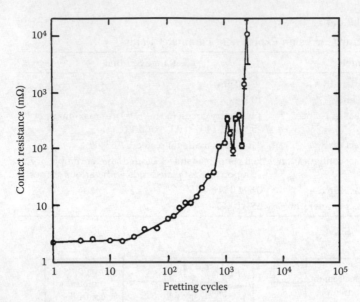

FIGURE 6.23
Rise in contact resistance with fretting cycle for 0.5 N and 20 μm track length measured under dry circuit condition.

Figure 6.24 gives results of fretting corrosion characteristics when both the track length and contact force are varied. At a contact force of 1.0 N, the contact resistance degrades in a similar fashion as in Figure 6.25 for long track lengths, but for the shorter track lengths, fretting corrosion is much slower, as reflected by the delayed rise in the contact resistance characteristics. At a higher force of 2.0 N, however, the contact resistance for tracks below 32 μm remains low and stable for up to 500,000 cycles. Figure 6.25 displays the data in a different way for a fixed track length of 40 μm. Clearly, higher contact force suppresses the fretting corrosion effects, and long track lengths are detrimental. The reason is a longer track length provides a greater amount of fresh metal to be oxidized and generates more oxide debris. When the contact force is higher, the contact is better able to push the oxides aside or break through the oxide layer. The results are not only consistent with the fretting corrosion mechanism, but also allow design guidelines to be established.

The mechanics of pushing the oxide aside as well as breaking the layer clearly depends on the contact geometry. To determine the best contact geometry, a number of contact configurations including hemispherical, conical, and wedge contacts have been studied. At a contact force of 1 N, hemispheres of different sizes do not show any major effect on the fretting corrosion characteristic as shown in Figure 6.26. Similarly, cones with included angles from 60° to 120° have shown only minor impact. We can conclude conical contacts are better than hemispherical ones. For wedge contacts, however, the 60° wedge contact provides the most stable contact resistance even at a low contact force of 0.5 N, as shown in Figure 6.27.

With fretting, the oxide debris builds up. The wedge shape, especially at sharp 60°, provides the best configuration to push away the oxides. This is believed to be the reason for the good performance of wedge contacts.

At elevated temperatures, which occur in many applications such as near an operating electrical motor or near the engine under the hood of a car, the conventional wisdom is to assume fretting corrosion will be the worst because the oxidation process is faster at higher

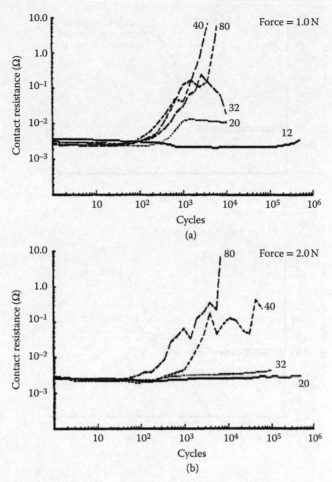

FIGURE 6.24
Fretting corrosion characteristics for contact force of (a) 1.0 N and (b) 2.0 N for track lengths up to 80 μm.

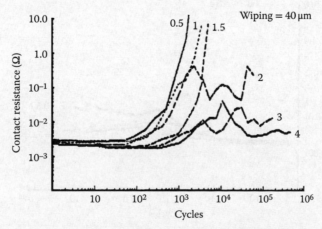

FIGURE 6.25
Fretting corrosion characteristics for track length of 40 μm at different contact forces.

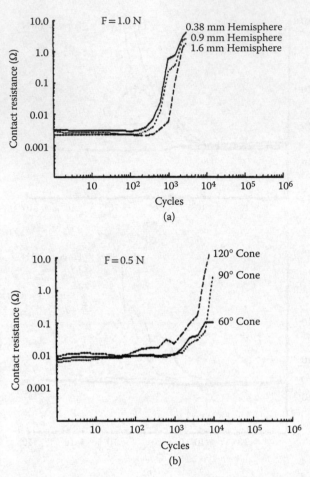

FIGURE 6.26
Fretting corrosion characteristics for the hemispherical and conical contacts: (a) F = 1.0 N, (b) F = 0.5 N.

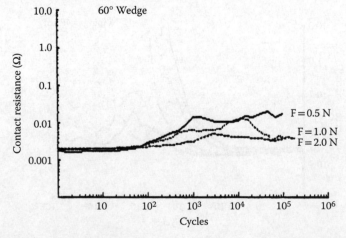

FIGURE 6.27
Fretting corrosion characteristics for wedge contacts.

temperatures. To quantify the effects of high temperature environment on fretting corrosion, the entire contact sample is maintained at a desired temperature by controlled heating of the entire contact support structure. Figure 6.28 shows duplicate contact resistance characteristics for the case of 0.5 N at 60°C in a dimple against a flat contact. The results confirm the rise in resistance rise is similar to but occurs at fewer cycles than at room temperature.

If 20 mΩ is selected as a failure point on the resistance characteristics, one can determine the number of fretting cycles for the contact to reach that failure point or "cycles to failure" (CTF). A higher corrosion rate corresponds to a low CTF. In terms of CTF, the contact effect of temperature is captured in Figure 6.29. The error bars represent the scatter in the data and the + is the average at that temperature. At temperatures just above room temperature, the corrosion rate increases with temperature, as expected. However, for temperatures above 60°C, this trend reverses. At 110°C, the contact resistance stays low up to 50,000 cycles for one of the runs. This is unexpected. One possible reason is the softening of the tin at higher temperatures makes it easier to break through the oxide layers.

The series of detailed studies described above provided design guidelines for effective use of tin for electrical contacts in many applications. Clearly, relative movement must be kept below 10–20 μm. This result is particular to this experiment. Usually researchers have limited movement to about 1–2 μm. Contact force must be above the 1–2 N level, if relative contact motion is not avoidable; and for temperatures in the 30°C –70°C range, a higher corrosion rate is expected. In certain instances, the application of contact lubricant greatly

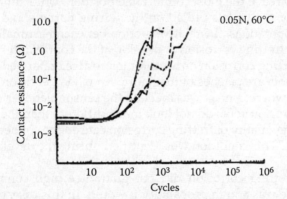

FIGURE 6.28
Fretting corrosion for 0.5 N at 60°C, four consecutive runs.

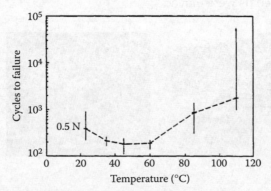

FIGURE 6.29
CTF as a function of contact temperature for a contact force of 0.5 N.

protects the contacts from fretting corrosion [47,48 and Chapter 3]. The lubricant isolates the fresh tin from being oxidized at each fretting cycle.

6.4.3 Examples of Contact Failures

The failure mechanisms for connectors are well known; however, it is sometimes difficult to identify which mechanism, or their combination, is responsible for causing the actual contact failure. The following examples describe real-world connector failures, along with explanation of the degradation mechanisms and potential solutions.

6.4.3.1 Automotive Position Sensor Connector

Years ago, cars used a mechanical linkage between the accelerator pedal and the engine to open or close the throttle and control the speed of the vehicle. Today pedal position sensors, electrical actuators, and power train computers are used to control speed without a mechanical linkage between the accelerator pedal and engine. Pedal position sensors typically send two different voltage signals to the computer module. As the accelerator is pressed down, one voltage signal increases and the other decreases. The computer, then, cross-checks the signals to verify the actual pedal position. In this case, the vehicle had adjustable pedals, as shown in Figure 6.30a. The driver controls the forward and backward position of the pedals to reach the most comfortable position. Unfortunately, the forward and backward motion of the pedals pulled on the wiring harness and connector to the sensor. With repetitive operations, the relative motion between terminals at the tin-plated contact interface caused fretting corrosion to develop at the contact interface. Figure 6.30b shows an example of fretting corrosion on an automotive accelerator pedal position sensor connection. The dark spots are patches of insulating tin oxide debris on the contact spots of a receptacle terminal where contact was made to the sensor blade terminals.

While the fretting corrosion spots do not look much different than the rest of the terminal surface, the insulating quality of fretting corrosion caused the connection resistance to increase to several ohms. This condition was identified during the development stage and was quickly remedied.

The appearance of fretting corrosion did not guarantee high connection resistance would develop, but there was a high probability it would. In this case, gold-plated terminals and improved wire routing were used to solve the problem. The gold-plated terminals were resistant to fretting corrosion and the improved wire routing reduced the motion of the terminals when the pedals were moved.

(a)   (b)

FIGURE 6.30
Automotive pedal position sensor (a) Sensor located at the top of the accelerator pedal in the circled area, and (b) Close-up view of electrical terminal contact showing fretting corrosion inside dashed circle.

6.4.3.2 Fuel Injector Connector

An automotive fuel injector connector developed an open circuit after operating under high vibration conditions during the product development stage (see Figure 6.31a). The terminal contacts and the surrounding plastic connector melted, so there was no way to check for fretting corrosion on the terminal contact interfaces. However, the fuel injector connector terminals nearby that had not meled did show evidence of fretting corrosion (see Figure 6.31b). After more investigation and laboratory experiments, the cause for the melted open circuit connection was identified. High vibration caused the terminals in the harness connector to move relative to the male blade terminals of the fuel injector. Over time, fretting corrosion developed from the relative motion, and the connection resistance increased.

A combination of the injector current (a few amps) and the high connection resistance (a few ohms) caused the connection to generate excessive heat. In this case, the local temperature at the terminals was high enough to cause the terminal contact spring to relax. As the terminal contact force decreased, the local terminal temperature increased even further. The local temperature eventually exceeded the melting temperature of the tin plating and the copper alloy used for the terminals and the terminals melted. The surrounding plastic also melted. With no remaining contact force and a high resistance, the oxidized copper contact interface developed an open circuit. The solution was to clamp the harness better to reduce connector motion during vibration. The tin-plated terminals were replaced with silver-plated terminals for better resistance to fretting corrosion. See also Section 3.4 where the failure of gold-to-tin connections resulted in the crash of the F-16 aircraft.

6.4.3.3 Glowing Contacts

The National Fire Prevention Association reported home fires caused by refrigerators, freezers, dishwashers, and other kitchen appliances not used for cooking resulted in property damage estimated at US$ 73 million during 2005–2009 [49]. While the fires often destroyed the evidence needed to determine the exact cause, faulty or damaged electrical wiring was suspected in many cases. The "glowing contact" degradation mechanism is a possible cause for some appliance and house wiring fires. Loose electrical connections or broken wires can form a molten oxide bridge across a small gap after repeated make and break cycles. The molten oxide bridge can become hot enough to glow bright orange and reach temperatures over 1000°C. The high temperature can then ignite combustible

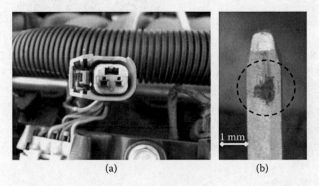

(a)   (b)

FIGURE 6.31
Fuel injector connector, (a) Melted harness connector, and (b) dark area on injector terminal blade due to fretting corrosion.

materials in the area. This mechanism has been studied and reported by Shea, and others [50–52]. Figure 6.32 shows the start of melting on a household 110 V outlet caused by loose connections to the solid core, copper wire (see also Chapter 15).

6.4.3.4 Electrolytic Corrosion

A large dump truck was unable to unload when an open circuit occurred in the environmentally sealed connection to the actuator for the dumping mechanism. Inspection of the connector in Figure 6.33 indicated one of the terminals in the connector was corroded away, while other terminals in the connector were only discolored and covered with dark corrosion products. Further inspection showed the connector seal was damaged—allowing water to enter the connector. In the presence of an electrolyte like salt water or road splash, the 12-V positive terminals in a connector will lose material, which is then deposited on nearby ground connections or circuits operating at lower voltage. The plating corrodes away first. Corroded plating can lead to high contact resistance or open circuits, but a loss of contact force due to corrosion of the terminals is a more common cause of corrosion-related connector problems.

6.4.3.5 Incompatible Plating and Low Contact Force

The horn on a prototype vehicle stopped working after a few months of ground testing. The same horn had been used successfully on many different types of vehicles for many years.

FIGURE 6.32
A 120-V household outlet showing overheated screw terminal due to loose wire connections and signs of arcing.

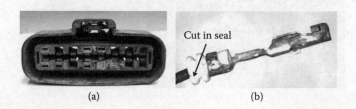

(a)   (b)

FIGURE 6.33
Severe connector damage due to water leakage through the cut seal (a) Corroded connector housing and (b) Cut in ribbed seal and corroded terminal.

The electrical terminals on the horn used unplated brass blades. The mating connector on other vehicles had always used unplated brass terminals with a relatively high contact force of 25 N, but this particular vehicle used a new connector design. The harness terminals were tin-plated, and they were designed with a contact force of 3 N. Lower contact force terminals were required to meet the connector plugging force requirements of an automotive assembly plant. Evaluation of the un-plated brass horn contacts showed the contact resistance of the un-plated brass horn terminals was relatively low when the terminals were new. However, high resistance oxide films grew on the terminals after environmental exposure. It was found that terminals with 25 N contact force were able to displace the high resistance oxide films, but the films could not be penetrated with terminals having only 3 N of contact force. In this case, the solution was to add tin plating to the horn terminals and use mating tin-plated terminals with a slightly higher contact force (7 N). Unlike bare brass terminals, the tin plating on the horn blades is less likely to develop high resistance films over time. The combination of higher contact force and improved terminal contact materials resulted in a more reliable horn connection.

6.5 Automotive Connector Contacts

6.5.1 Vehicle Conditions

Modern vehicles have seen tremendous growth in the use of electrical and electronic devices over the past several decades. In the 1940s, vehicles used a 6-volt DC electrical system which provided power for a starter motor, wiper, ignition, and lighting circuits. Vehicles now require multiple voltage levels, and currents range from nano-amps to hundreds of amperes. Some typical automotive voltage levels are described in Table 6.1. All of these electrical systems require electrical connections to aid in assembling and servicing the vehicles.

Automotive connections need to perform in extreme environments. Temperatures can be as low as −40°C in arctic conditions or over 150°C at some engine-mounted sensors. The following sections discuss the requirements of some more specialized automotive connections.

6.5.2 High Power Connectors for Electric and Hybrid Vehicles

On a typical passenger car with a 12-volt electrical system and an internal combustion engine, the peak electrical power for a high-power electrical circuit is typically under 2 kW. Starter connections are required to carry hundreds of amperes for a few seconds at a time during cranking, and the charging circuits can carry about 100 A under some conditions. Simple, bolted ring terminals are used for most of these connections.

Hybrid or electric vehicles can operate at hundreds of volts and some circuits may carry hundreds of amps for extended periods. Connections must maintain low, stable resistance over time, but there are several additional requirements. Many of the connections require electromagnetic shielding to prevent high-power circuits from influencing the performance of nearby electronic components. In addition, the isolation resistance between neighboring circuits must be high enough to prevent dielectric breakdown between circuits, and the connectors may need to be environmentally sealed. Battery packs, electric motors, DC/DC converters, and DC/AC inverters all require connections designed for reliable performance in the demanding environment in the vehicle. A high-power electric vehicle connection is shown in Figure 6.34. Some of the special features are highlighted, including electromagnetic shielding, HVIL (high voltage interlock circuit), and isolated terminal cavities. The HVIL is required to prevent hot disengagement of contacts because high voltage arcing can severely damage the contact terminals and perhaps the operator.

Another type of connection is used for charging the batteries on plug-in hybrids and electric vehicles. The charger connectors need to survive thousands of mating cycles without degrading the connection. Figure 6.35 shows a charge plug made to the SAE J1772 or IEC 62196-2 standards [53,54]. The terminal contacts are typically plated with silver or gold, and they are designed to meet requirements for low connection resistance after thousands of mating cycles in a relatively dirty environment.

6.5.3 Aluminum Wiring Connections

Un-plated copper has been used for most automotive wiring from the beginning. Copper has high conductivity along with good strength and resistance to corrosion. It is also relatively easy to make electrical connections to copper wiring with mechanical means, crimping, soldering, welding or brazing. Volatile copper prices, along with a need to reduce vehicle mass, make aluminum an attractive alternative to copper wiring. Aluminum wiring has been used in specialized applications on vehicles since the 1970s, but is now being used more often. There is great potential for savings in mass and cost of material by replacing copper with aluminum wiring, but there are also a number of

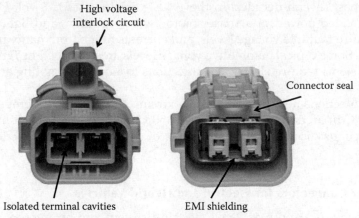

FIGURE 6.34
Example of a high-voltage, in-line connector.

FIGURE 6.35
Charge port coupler and vehicle connector for electric and plug-in hybrid vehicles.

significant challenges related to the strength, conductivity, resistance to corrosion, and electrical contact properties of aluminum.

Aluminum wiring has roughly one-half the conductivity of copper, so a copper cable must be replaced by an aluminum cable of a larger diameter for equivalent current-carrying capacity. At the same time, aluminum has about one-third the density of copper, so the mass of an aluminum cable is less than the mass of a copper cable with the same current-carrying capacity as illustrated in Figure 6.36.

It is already difficult to package the wiring in the limited space of a modern vehicle, so increasing the size of aluminum cables to match the current-carrying capacity of copper will require more space. Connector housings may also need to be bigger to make room for cables with larger diameters. Corrosion resistance is another factor to consider when converting from copper to aluminum cable. The latter is susceptible to galvanic corrosion when it is attached to copper alloy terminals and in the presence of an electrolyte. Figure 6.37 shows a mechanically crimped connection between a copper alloy terminal and an aluminum alloy cable before and after a severe corrosion conditioning. Most of the aluminum corroded away after conditioning. This type of corrosion can be controlled by sealing the copper alloy–aluminum junction from electrolytes or locating the terminations in a relatively dry environment. Aluminum cable has approximately one-half the tensile

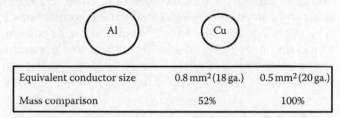

	Al	Cu
Equivalent conductor size	0.8 mm² (18 ga.)	0.5 mm² (20 ga.)
Mass comparison	52%	100%

FIGURE 6.36
Comparison of copper and aluminum cable core size of the same ampacity.

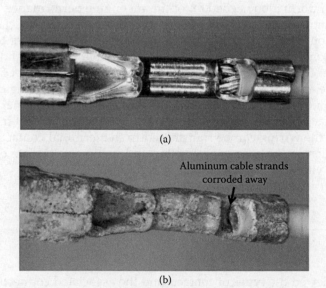

FIGURE 6.37
Galvanic corrosion of a tin-plated copper alloy terminal crimped to aluminum cable; (a) New sample and (b) No aluminum remains after severe corrosion conditioning.

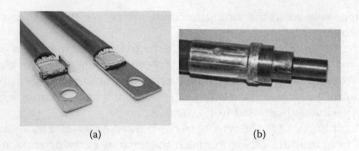

(a)   (b)

FIGURE 6.38
Welded connections between copper alloy and aluminum (a) Ultrasonic welds of stranded aluminum cable to brass terminals and (b) Friction welded copper to stranded aluminum cable.

strength of copper wiring. If aluminum cable is sized to carry the same current as a copper cable, the break strength of the aluminum cable will be lower than strength of the equivalent copper cable. Aluminum alloys are available with higher strength, but the electrical conductivity drops as strength increases.

Electrical connections to aluminum cable are challenging, since aluminum forms a hard, insulating oxide film on its surface. This insulating film needs to be ruptured or displaced in the contact areas in order to make an electrical connection with good long-term stability. Mechanical connections can be used to make stable electrical connections to aluminum, but welding may also be used. Aluminum cannot be soldered with conventional soldering fluxes, but several welding processes can join aluminum to aluminum or copper to aluminum. Figure 6.38 shows examples of ultrasonically welded and friction welded connections. Both of these welding methods produce a low resistance, metallurgical bond.

6.5.4 Connections for High-Vibration Environment

Modern internal combustion engines use a variety of sensors mounted directly on the engine to supply the information needed for optimum engine performance. Throttle position sensors, temperature sensors, knock sensors, and many other specialized sensors and actuators are mounted on all kinds of engines—from lawnmowers to giant diesel generators. Connections to automotive sensors must be able to withstand vibration for the life of the engine. The key to good performance of a connector subject to high vibration is to eliminate relative movement at the contact interfaces so fretting corrosion does not occur (see Section 6.4.2). Effective relief of the strain in the wiring harness is one of the most important factors to minimize relative movement at the contact interfaces. Even small movements of the wiring can be transmitted to the terminal contacts. In addition, minimizing the motion in the connector housing and using appropriate contact plating and lubricant for the vibration environment is very important.

6.6 Summary

This chapter has reviewed the types of contacts and the associated connection systems in low-power commercial, automotive, and appliance circuits. The complexities of the connection system and the performance requirement have challenged the industry to design smaller and environmentally robust components. The engineering of these products requires understanding of the physics of the contacts and of the chemistry of degradation processes. The selection of material, contact configuration, coating, sealing, lubricant, and many other parameters to satisfy a given requirement is not straightforward. There is still an element of artistry in the design and application of connection systems. Validation testing under realistic conditions is still required. However, testing often takes too long and is cost prohibitive. Developments in the area of better materials and testing methodologies are needed.

References

1. R Campel and H Feigenbaum, Shaped contacts expand horizon for flex circuits, *Proc Interconn Tech*, October 1993, 26.
2. E Jensen, G Drew, and C Schreiber, Apparent contact stress in high-density flex interconnects, *Proc of the 4th Int Conf Flex Circ*, FlexCon 97, September 1997, 37.
3. IEC 60352-5, Press-in connections:General requirements, test methods and practical guidance, International Electrotechnical Commission, 2003.
4. K Ring and T Schreier-Alt, Reliability of press-fit contacts and adjoining SMD components on printed circuit boards, *Proc 55th IEEE Holm ConfElect Cont*, 2009, 223.
5. Y Nomura, et al., Press-fit connector for automotive ECU's, *Proc 52nd IEEE Holm Conf Elect Cont*, 2006, 211.
6. M Hatanaka and J Eilers, High-density insulation displacement connector system, *Proc 22nd Ann Conn Interconn Tech Sym*, Philadelphia, October 1978, 277.
7. Stefan Jorgen, Insulation displacement technology as technically equivalent and more cost-effective alternative compared to crimp technology, *Proc 50th Ann IEEE Holm Conf Elect Cont*, 2004, 421.
8. Stefen Jorgens and Hennig Taschke, Modeling the insulation displacement process – the results of a FE model study, *Proc 52nd IEEE Holm Conf Elect Cont*, 2006, 217.
9. Y Mizusawa, Miniature ZIF contact system using leverage, *Proc 17th Int Conf Elect Cont*, Nagoya, Japan, July 1994, 303.
10. A Ohno, Development of zero insertion force connector for automobile, *Proc 17th Int Conf Elect Cont*, Nagoya, Japan, July 1994, 313.
11. PMA Sollars and IH Corby, Stress relaxation processes in wire wrapped joints, *Proc 20th Ann Holm Sem Elect Cont*, October 1974, 156.
12. J Aronstein, Behavior of overheating aluminum wire branch circuit connections under normal loads, *Proc 26th Ann Holm Conf Elect Cont*, October 1980, 117.
13. J Aronstein, Failure and overheating of aluminum wire twist-on connection, *Proc 27th Ann Holm Conf Elect Cont*, October 1981, 259.
14. J Aronstein, Al conductor with Mg alloy on surface:Evaluating conductor-to-conductor splices, *Proc 40th Ann Holm Conf Elect Cont*, 1994, 119.
15. J Aronstein, Conduction through surface films on aluminum wire, *Proc 34th Ann Holm Conf Elect Cont*, 1988, 173.
16. J Aronstein, An updated view of the aluminum contact interface, *Proc 50th IEEE Holm Conf Elect Cont*, 2004, 98.
17. RS Timsit, Electrical instabilities in Al–Al and Al–brass stationary point junctions, *Proc 34th Ann Holm Conf Elect Cont*, 1988, 151.
18. M Braunovic, Evaluation of contact-aid compounds for Al-to-Al connectors and conductors, *Proc 30th Ann Holm Conf Elect Cont*, 1984, 97.

19. KE Olsson, Long-term stability of Al–Al contacts and Ni-plated Cu–Al contacts, *Proc 34th Ann Holm Conf Elect Cont*, 1988, 167.
20. J Aronstein, An updated view of the aluminum contact interface, *Proc 50th Ann Holm Conf Elect Cont*, 2004, 98.
21. J Legrady, A new type of very high reliability torsion IDC which can accept a large range of wire gauges, *Proc 22nd Ann Conn Interconn Tech Conf*, 1989, 337.
22. NK Mitra, An evaluation of the insulation displacement electrical contact, *Proc 25th Ann Holm Conf Elect Cont*, 1979, 99.
23. E Dalrymple and R Downey, 0.5″ × 0.5″ connector with hypertac LIF contacts, *Proc 21st Ann Conn Interconn Tech Conf*, 1988, 235.
24. R Mroczkowski, Mechanical permanent connections to wire and cable, In *Electrical Connector Handbook*, McGraw-Hill, 1998, 9.1–9.23.
25. R S Timsit, Contact properties of of tubular crimp connections: elementary considerations. *Proc 24th Int Conf Elect Cont*, 2008, 151–56.
26. SAE/USCAR-21, Performance specification for cable-to-terminal electrical crimps, Revision 2, Society of Automotive Engineers, Warrendale, PA, 2008.
27. KT Yasuda, S Umemura, and T Aoki, Degradation mechanisms in tin- and gold-plated connector contacts, *Proc 32nd IEEE Holm Conf Elect Cont*, 1986, 27.
28. T Tamai, Recovery of low contact resistance due to electrical and mechanical breakdown of contact films, *Proc 56th IEEE Holm Conf Elect Cont*, 2010, xxiii–xl.
29. S. Noel, et al., Study of thin underlayers to hinder contact resistance increase due to intermetallic compound formation, *Proc 55th IEEE Holm Conf Elect Cont*, 2009, 153.
30. M Braunovic and D Gagnon, Formation if IM in lead-free system, *Proc 52nd IEEE Holm Conf Elect Cont*, 2006, 267.
31. A Ober and K Olsson, Computer modeling of contact degradation by IM growth, *Proc 43rd IEEE Holm Conf Elect Cont*, 1997, 41.
32. A Nakamura, et al., Contact characteristics of tin-plated copper alloys in high temperature environments, *Proc ICEC*, 1994, 151.
33. F Ostendorf, et al., There is tin, and there is tin:Characterization of tribological and electrical properties of electroplated tin surfaces, *Proc 57th IEEE Holm Conf Elect Cont*, 2011, 261.
34. WH Abbott, The development and performance characteristics of mixed flowing gas test environments, *Proc 33rd IEEE Holm Conf Elect Cont*, 1987, 63.
35. C Perrin and D Simon, Atmospheric corrosion on some materials used in electrical contacts, *Proc ICEC*, 1998, 147.
36. IO Wallinder and P Eriksson, Characterization of the corrosiveness of an automotive environment, *Proc ICEC*, 1998, 157.
37. C Haque, Effect of dust on contact resistance of lubricated connectors, *Proc* of the 31st IEEE Holm *Conf Elect Cont*, 1985, 141.
38. Dong Wang, et al, Investigation of dust test and modeling of dust ingression, *Proc 55th IEEE Holm Conf Elect Cont*, 2009, 218.
39. Ed Smith and Hugh Ireland, Contact system for use in sulfur-containing reformulated gasoline, *Proc 50th IEEE Holm Conf Elect Cont*, 2004, 289.
40. M Antler, Survey of contact fretting in electrical connectors, *Proc 30th IEEE Holm Conf Elect Cont*, 1984, 3.
41. JW McBride, Development in fretting studies applied to electrical contacts (review paper), *Proc 52nd IEEE Holm Conf Elect Cont*, 2006, 170.
42. S. Masui, et al., Measurement of contact resistance distribution in fretting corrosion track for the tin-plated contacts, *Proc 57th IEEE Holm Conf Elect Cont*, 2011, 92.
43. Chen Chen, et al., Modeling and analysis of a connector system for the prediction of vibration-induced fretting degradation, *Proc 55th IEEE Holm Conf Elect Cont*, 2009, 129.
44. Ito Tetsuya, et al., Three-dimensional structural study of tin-plated fretting contacts, *Proc 55th IEEE Holm Conf Elect Cont*, 2009, 316.

45. Robert Malucci, Characteristics of film development in fretting experiments on tin-plated contacts, *Proc 45th IEEE Holm Conf Elect Cont*, 1999, 175.
46. A Lee and MS Mamrick, Fretting corrosion of tin-plated electrical contacts, In *Material Issues for Advanced Electronic and Opto-electronic Connectors*, TMS Publication, 1991, 71.
47. A Lee and MS Mamrick, Fretting corrosion of tin-plated copper alloys, *IEEE Trans*, CHMT 10: 63, 1987.
48. C.H. Leung and A. Lee, Thermal cycling-induced wiping wear of connector contacts, *Proc 43rd IEEE Holm Conf Elect Cont*, 1997, 132.
49. National Fire Protection Association, Home structure fires involving kitchen equipment other than cooking equipment, Fire Analysis and Research Division, Quincy, 2009.
50. Dongwei Li et al., A method for residential series arc fault detection and ID, *Proc 55th IEEE Holm Conf Elect Cont*, 2009, 7.
51. JJ Shea, RF currents produced from AC arcs with asymmetrical electrodes, *Proc 56th IEEE Holm Conf Elect Cont*, 2010, 188.
52. JM Martel et al., Study of arcing faults in the low-voltage electrical installations, *Proc 56th IEEE Holm Conf Elect Cont*, 2010, 199.
53. SAE J1772, SAE electric vehicle and plug-in hybrid electric vehicle conductive charge coupler, Society of Automotive Engineers, Warrendale, 2012.
54. IEC 62196-2, Plugs, socket-outlets, vehicle connectors and vehicle inlets:Conductive charging of electric vehicles, International Electrotechnical Commission, 2011.

7

Tribology of Electronic Connectors: Contact Sliding Wear, Fretting, and Lubrication

Roland S. Timsit and Morton Antler

Adde parvum parvo magnus acervus erit
(Add little to little and there will be a great heap)

– Ovid

CONTENTS

7.1 Introduction 414
7.2 Sliding Wear 415
    7.2.1 Early Studies 415
    7.2.2 Adhesion 416
        7.2.2.1 "Wiping" Contaminant from Contact Surfaces 417
        7.2.2.2 Mild and Severe 418
        7.2.2.3 Prow Formation 419
        7.2.2.4 Rider Wear 421
        7.2.2.5 Gold Platings: Intrinsic Polymers and Junction Growth 422
        7.2.2.6 Electroless Gold Plating 423
    7.2.3 Abrasion 424
    7.2.4 Brittle Fracture 426
    7.2.5 Delamination and Subsurface Wear 427
    7.2.6 Effect of Underplate and Substrate 428
        7.2.6.1 Hardness 428
        7.2.6.2 Roughness 431
    7.2.7 Electrodeposited Gold: Relationship of Wear to Underplate Hardness 431
        7.2.7.1 Hardener Metal Content 432
    7.2.8 Clad Metals 435
    7.2.9 Tin and Tin–Lead Alloys 437
    7.2.10 Silver 439
7.3 Fretting 441
    7.3.1 Background 441
    7.3.2 Fretting Regimes 442
    7.3.3 Static versus Dynamic Contact Resistance 445
    7.3.4 Field and Laboratory Testing Methodologies 447
        7.3.4.1 Generation of Fretting Displacement 447
        7.3.4.2 Determination of Contact Resistance 449

    7.3.5   Materials Studies...449
         7.3.5.1   Apparatus...449
         7.3.5.2   Metals Having Little or No Film-Forming Tendency.......................450
         7.3.5.3   Non-Noble Metals/Fretting Corrosion...............................452
         7.3.5.4   Frictional Polymer-Forming Metals.....................................457
         7.3.5.5   Dissimilar Metals on Mating Contacts...............................460
    7.3.6   Wear-Out Phenomena...464
         7.3.6.1   Gold-Based Systems...464
         7.3.6.2   Palladium-Based Systems.....................................465
         7.3.6.3   Tin and Tin–Lead Alloy Systems.......................466
         7.3.6.4   Role of Underplate and Substrate.......................467
    7.3.7   Parametric Studies..468
         7.3.7.1   Cycle Rate...468
         7.3.7.2   Wipe Distance...470
         7.3.7.3   Force...472
    7.3.8   Environmental Effects...474
    7.3.9   Thermal..475
    7.3.10 Effect of Current...476
    7.3.11 Surface Finish and Contact Geometry...479
    7.3.12 Material Transfer, Wear, Film Formation, and Contact Resistance 479
         7.3.12.1 Summary of Physical Processes.......................479
7.4   Lubrication..481
    7.4.1   Introduction..481
    7.4.2   Metallic Films...481
         7.4.2.1   Principles of Metallic Film Lubrication...............................481
         7.4.2.2   Sliding and Wiping Contacts...482
         7.4.2.3   Fretting Contacts...487
    7.4.3   Fluid Lubricants...488
         7.4.3.1   Background..488
         7.4.3.2   Some Fundamental Properties of Lubricants 488
         7.4.3.3   Requirements...491
         7.4.3.4   Types of Fluid Lubricants: A Sliding Contact Investigation.............492
         7.4.3.5   Control of Fretting Degradation...496
    7.4.4   Grafted and Self-Assembled Lubricant Layers.................................500
    7.4.5   Greases and Solid Lubricants...503
         7.4.5.1   Greases...503
         7.4.5.2   Solids...504
    7.4.6   Lubricant Durability..504
    7.4.7   Other Considerations..505
References...506

7.1 Introduction

Sliding wear and fretting are key determinants of the performance of connectors, printed circuit boards with edge contacts, switches, instrument slip rings, and other electrical components. Unlike the technology of relays and circuit breakers, which are designed for current make and break and where the physics of the arc and erosion are the focus of contact research, the technology of connectors is concerned with metal transfer, material removal, friction, and electrical noise. Connector life, cost, and especially reliability are determined in large part by sliding wear and fretting phenomena.

In this chapter, we discuss the tribology of electrical contacts under three headings: sliding, fretting, and lubrication. In electrical connections, wear is generally associated with separable rather than permanent interfaces because surface damage is often visible to the naked eye after separation. Although interfacial defects introduced by extensive mechanical wear are usually detrimental to the function of an electrical contact, there are many instances where a limited amount of wear actually benefits an electrical connection. For example, initial sliding in a separable connection often disrupts electrically insulating surface films that interfere with electrical current flow and is thus beneficial to contact performance. Over the years, a number of wear mechanisms have been identified and are described in detail in the literature [1,2].

As this chapter reviews tribological data recorded over many years, some of which were obtained well before adoption in the scientific literature of SI units, we have opted to reproduce the units used in the original publications. Conversion of earlier data to SI units would have necessitated a great deal of reworking of graphs and tables, which we decided was not necessary to fulfill the major objective of the present review. Thus, the units of contact force will include gf, kgf, N, and so on. Similarly, units of length will include mm, cm, m, and so on.

7.2 Sliding Wear

7.2.1 Early Studies

When two surfaces are brought together, they touch at the tip of surface asperities (i.e., at small contact spots or *a*-spots) and the area of true contact is small. This area is determined by the contact force and the microhardness of the contacting materials [3–7]. When the surfaces slide over each other at a relatively low speed, mechanical damage to the surfaces stems from the action of one of several major wear mechanisms. Whatever the wear mechanism, mechanical wear leads to the removal of material from surface asperities in contact with each other. The inescapable consequence of this asperity interaction is that the amount of material removed from a sliding surface depends on the area of true contact A_C. Since $A_C = F/H$ where F is the mechanical contact load and H is the microhardness of the metal, as described in Equation 1.10 in Chapter 1, the amount W of material worn off a surface over a sliding distance s is

$$W = K \frac{F}{H} s, \tag{7.1}$$

where K is the wear coefficient associated with the materials in sliding contact. Theoretical support for Equation 7.1 shows that the wear coefficient is associated with the probability that an interaction between asperities leads to the production of a wear particle [8–10]. It also should be noted that Equation 7.1 applies to the steady state and does not address run-in wear, which has been found usually to dominate the sliding of many contact devices. Separable connectors as they are usually used do not attain equilibrium wear because the total sliding distance in their lifetime is relatively small.

Wear phenomena that have been observed in practical devices are complex and resist simple categorization. For example, mechanical debris produced as a result of adhesive transfer causes secondary wear by abrasion. Wear mechanisms that were considered to be significant in the early days of electrical contact research, such as Holm's "molecular wear" and "interlocking" [11], have fallen out of favor. Nevertheless, it is convenient to after relevant describe the wear of metallic contacts by processes that today are recognized to be relevant to electrical connectors: adhesion, abrasion, brittle fracture, fretting, and, in special cases, subsurface wear with layered materials and delamination. There are still other categories of sliding wear phenomena, especially outside of the electrical contact field such as chemically induced wear.

7.2.2 Adhesion

Adhesive wear occurs when sliding surfaces experience metal transfer, and is generally characterized by the generation of fine particles from one or both of the surfaces. These particles may come off the surface from which they are formed and may either adhere to the mating surface as a transfer layer or accumulate as loose debris on the sliding surfaces. Adhesive wear stems initially from adhesive bonding between the touching surface asperities, under conditions where this bonding is stronger than the cohesive strength of the metal.

The adhesive wear mechanism is illustrated schematically in Figure 7.1 [2]. It is initiated when localized adhesion occurs at contact spots between sliding surfaces as illustrated in Figure 7.1a. This is followed by repeated transfer of material from one asperity to the other and by growth of the transferred material (Figure 7.1b through d). Transferred particles eventually detach themselves and are subsequently flattened and dispersed as loose wear debris (Figure 7.1e). The rate of debris generation increases where sliding motion consists of reciprocating displacements over the same wear track. Under conditions where sliding motion is unidirectional and the contact load is sufficiently large, adhesive wear in chemically clean sliding interfaces can lead to so-called prow formation, as described later.

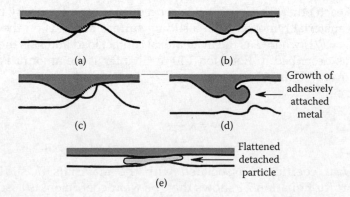

FIGURE 7.1
Schematic representation of adhesive wear in a sliding interface: (a) localized adhesion between contacting surface asperities during sliding, (b) transfer of material from one asperity to the other, (c) repetitive adhesion process, (d) growth of transferred particles, and (e) particle detachment and subsequent debris flattening.

For non-noble materials, metal debris generated by adhesive wear undergoes oxidation to form highly electrically insulating material.

7.2.2.1 *"Wiping" Contaminant from Contact Surfaces*

Although adhesive wear is generally deleterious to the long-term contact performance, it can also act to reduce rather than increase contact resistance for cases of weakly adhering material in a sliding interface, such as organic deposits. In such cases, adhesive wear reduces contact resistance by displacing electrically insulating layers such as oxides, sulfides, and other materials formed from chemical reactions with air or environmental pollutants, without appreciable removal of the underlying metal. This action is often identified as "wipe" cleaning motion [12,13]. "Wiping" is usually recommended as a first step in generating low contact resistance in separable electrical connections such as pin–socket contacts. This precaution is recommended even for gold-plated contacts to enhance contact reliability, since gold has a tendency to absorb thin carbonaceous layers [14,15]. An example of organic contamination on a gold surface exposed to laboratory air for a few days, after having been cleaned chemically, is illustrated in Figure 7.2. The typical effect of wipe on contact resistance between contaminated gold contacts is illustrated in Figure 7.3 [13]. The graphs show the decrease in contact resistance with increasing contact load when the gold surfaces are stationary and are subsequently set into sliding motion. The data of Figure 7.3a relate to the case where the surfaces were not severely contaminated, thus showing a small resistance decrease during sliding. In contrast, the data of Figure 7.3b show that the contamination of the flat gold surface was so severe that the contact was essentially open at a contact load as large as ~1.2 N. However, the contact resistance decreased rapidly after sliding motion was initiated. The sliding motion was clearly effective in squeezing contaminant material out of *a*-spots. Continuing wipe-travel led to a further decrease in contact resistance as additional contaminant was pushed out to the side of the wear track.

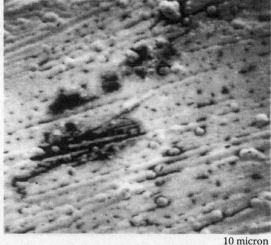

10 micron

FIGURE 7.2
Dark areas are organic contamination on an initially clean gold surface after exposure to laboratory air for several days. Scanning electron microscope (SEM) at low (2 kV) beam energy.

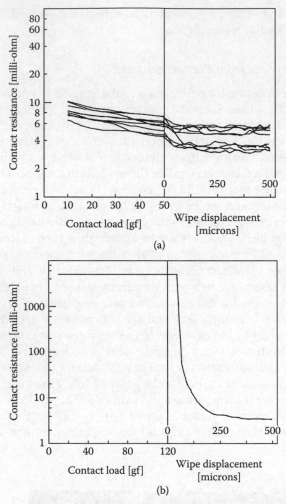

FIGURE 7.3
Contact resistance versus normal contact load and versus wipe distance for a gold probe sliding on (a) a contaminated gold-plated surface and (b) a highly contaminated gold-plated surface. (After, IH Brockman et al., *IEEE Trans. Components, Hybrids and Manuf. Technology*, CHMT-11: 393–400, 1988 [13].)

7.2.2.2 Mild and Severe

Several researchers in the 1950s [16–18] observed that many systems operate at widely different rates of adhesive wear, called *mild* and *severe*, and that the scale of wear may change in a narrow range of load, as in Figure 7.4. Below the transition load with base metals, wear debris is usually oxidized and finely divided (10^{-4}–10^{-2} mm); above it, debris is largely metallic and coarse (10^{-1}–1 mm). The transition occurs due to the increasingly rapid rate of formation of wear particles with increasing load or from lubrication failure. With oxide-free noble metals, sliding is in the severe regime at all loads when the surfaces are very clean; in practice transition loads do occur, but they seem to be related to the removal of adventitious contaminants during sliding. Such materials, as in Figure 7.2, can be strikingly protective; but when repeat-pass transversals occur on too short a timescale

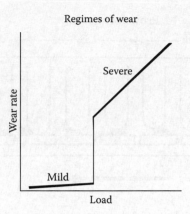

FIGURE 7.4
Regimes of wear.

for the contaminant to be renewed, the transition to severe wear occurs. For this reason, the durability of unlubricated separable connector contacts is related to the rate of mating and unmating [19].

The transition load in air for clean electrodeposited cobalt–gold is about 25 gf and decreases to about 5 gf for clad Au99Co1, as measured using a gold hemispherical dimple sliding on a flat surface in laboratory sliding experiments.

7.2.2.3 Prow Formation

At relatively large contact loads (say > 100 gf) and if sliding occurs in one direction or is reciprocating over a sufficiently long distance, adhesive wear of most clean contact metals proceeds by a severe adhesive mechanism called prow formation [20,21]. It may be described phenomenologically as follows for contacts of the same material: when the mating bodies are of different size, there is net metal transfer from the part with the larger surface to the part with the smaller surface, for example, from flat to rider in a rider–flat apparatus. The process is similar to that illustrated in Figure 7.1, but the debris particles are larger due to the long sliding distance in the same direction. A lump of severely work-hardened metal (the prow) builds up and in turn wears the flat by continuous plastic shearing or cutting, while the rider does not wear. Prows then detach from the rider either by back-transfer to the flat or as loose debris. If the rider always traverses virgin metal, prow formation continues indefinitely. This is shown schematically in Figure 7.5. When contact members of the same material are identical in size and shape, the initial prow can come from either member. However, once transfer occurs, the receptor then becomes the rider and the opposing surface wears. With dissimilar metals, prows form even when the flat is harder than the rider, provided this difference is not greater than a factor of about 3. Friction and contact resistance characteristics of the system are determined by the prow metal; thus, a palladium rider sliding on a gold flat behaves as does gold on gold, while a gold rider on palladium slides like palladium on itself [22].

Prows from soft ductile metals can be observed with the unaided eye (the size of prows appears to be roughly inversely related to hardness), examples of which are given in Figures 7.6 and 7.7. Prow formation occurs with gold, palladium, palladium–nickel and

Prow-formation

◄———— Growth ————►|◄————— Loss ————►|◄— Growth

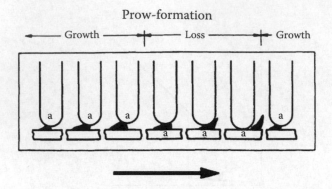

FIGURE 7.5
Prow formation mechanism (schematic). Letter "*a*" designates the surface to which prow is attached. Arrow indicates direction of movement of flat.

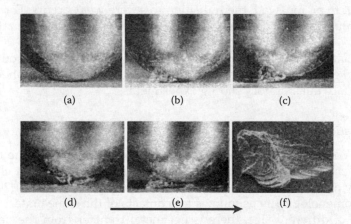

FIGURE 7.6
Prow formation mechanism for a rider, 3.2 mm in diameter, sliding on a flat specimen. Solid gold contacts, 500 gf load: (a) start of run; (b) well-developed prow; (c and d) loss of portion of prow by back-transfer to flat; (e) newly formed prow; and (f) prow consisting of overlapping thin layers of metal back-transferred to flat. The arrow indicates the direction of movement of the flat. (After M Antler, *Wear,* 7: 181–204, 1964 [20].)

palladium–silver alloys, platinum, silver, copper, copper–beryllium alloys, phosphor bronzes, lead, aluminum, bismuth, iron, titanium, and many other metals. Metals that do not work-harden, such as indium at room temperature, fail to form prows [23]. This is because prows must become harder than the flat by metalworking in order to wear the latter, particularly by cutting. Tin–lead (50/50) alloy appears to be a special case; prows from it have been found [20] to become work-softened (the Knoop Hardness at 25 gf (KHN_{25}) was found to vary from 0.6×10^2 to 0.9×10^2 N mm^{-2}, compared with 1.9×10^2 N mm^{-2} for the initial material). Perhaps there is an initial hardening followed by softening as the prow continues to be worked while it is between the contact members. Prows have been found to be layered [24], as shown in Figures 7.6f and 7.7. A number of mechanisms have been proposed for the formation of prows [20,21,25], but none has yet been confirmed experimentally.

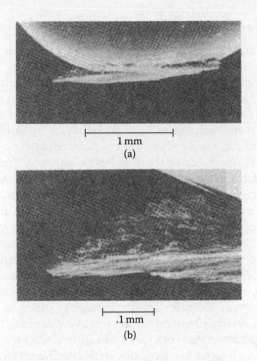

1 mm
(a)

.1 mm
(b)

FIGURE 7.7
SEM micrograph of gold rider with adherent prow after unlubricated sliding against solid gold flat. Note the layered structure of the prow. Conditions: 300 gf, 500 passes, reciprocation. (After H Fry, HG Feller, *Prakt Metallographie* 9: 182–197, 1972 [24].)

7.2.2.4 Rider Wear

When the rider traverses the same track repetitively in the same direction, or in reciprocating motion over a large displacement, prow formation eventually ceases and is replaced by "rider wear", a mechanism in which the smaller contact surface wears. This occurs primarily by transfer to the flat with some loose particle formation; in rider wear, the flat may gain mass relatively rapidly after an interval of mass loss. As explained earlier, the initial mass loss stems from occasional attachment (initially) of prows or particle debris to the rider. The transition to a mass gain mode is due to the accumulation of back-transfer prows to the flat. This back-transfer also causes an increase in the effective surface hardness of the flat to a level of extreme work-hardening along the entire wear track. When the surface of the wear track on the flat reaches the hardness of prows, wear of the flat effectively ceases and the rate of rider wear increases rapidly.

Experimental observations indicate that there is an initial roughness of the flat above which prow formation does not occur. For pure gold, the critical roughness is in the range of 0.6–1.3-µm center line average (CLA) at 100 gf [26]. The harder and less ductile the metal, the smaller is the critical roughness, which may explain why prow formation is unlikely to occur with rhodium, ruthenium, equiatomic tin–nickel electrodeposit, and other hard brittle materials. It has been argued [27,28] that adhesive wear is affected by the mutual solubility of the materials in the sliding interface, whereby higher mutual solubility leads to higher adhesion and hence to increased wear rates. However, investigations of sliding gold surfaces on platinum group metals (low mutual solubility with gold) did not support this premise [29].

7.2.2.5 Gold Platings: Intrinsic Polymers and Junction Growth

Adhesive mechanisms dominate the wear rate of gold electrodeposits in high-reliability electrical connectors. Much of the work on adhesive wear in gold electrodeposits has been conducted with practical connector components. A recurring theme in this work has been the identification of deposition conditions and gold composition that optimize wear resistance of the electrodeposit [26,30–33].

The fundamental reasons for differences in wear behavior in different gold electrodeposits were not understood for a long time. When electrodeposited hard golds became available, it seemed that the resistance of these plates to adhesive wear relative to that of the soft pure gold deposits could be attributed to hardness. It later became apparent that this explanation was incomplete, because tests using wrought (hardened) gold alloys of hardness comparable with that of electroplated golds (e.g., from cyanide-base bath) showed the wrought alloys to be inferior [34]. Observations that a relatively pure hard gold such as metal-worked 24-karat hard gold wears poorly further emphasized the inadequacy of attributing wear resistance solely to hardness. At this historical juncture, one potential explanation for the superior durability of the cyanide gold plates was a return to an early suggestion [34] that codeposited polymeric materials [35] in the cobalt (and nickel) golds act as lubricants to mitigate wear. Despite some experimental evidence that conflicted with this explanation [36], the cause of differences in the relative wear resistance of gold electrodeposits was eventually elucidated via work that focused initially on cobalt-hardened gold.

It was confirmed that cobalt–gold electroplates deposited from cyanide baths include polymer particles consisting in part of a complex cobalt cyanide compound [35]. The particles range in average dimensions from 2 nm to 7 nm and are uniformly distributed in the gold deposit [37]. Some polymer particles may be larger than this [38]. These observations are in general agreement with a later investigation that concluded [39] that a significant fraction (perhaps a dominant fraction) of the cobalt in cobalt-hardened gold characterized by superior resistance to adhesive wear is present in the form of intrinsic polymers rather than in solid solution. The experimental evidence indicates that the metallurgical structure, the hardness, and the ductility of gold electrodeposits are strongly coupled to the composition, dimensions, and distribution of the codeposited material. Observations indicate that the most adhesive wear-resistant deposits are those characterized by low ductility, for example, with elongations less than 1%. In addition to acid bright cobalt-hardened gold, nickel-hardened gold plates containing 0.28 wt% nickel and cobalt-hardened gold containing 1.3 wt% cobalt plus 0.55 wt% indium have similar durability [26] and comparable brittleness. The role of polymers in cyanide nickel-hardened gold plates is presumed to be similar to that in the cobalt-hardened version.

It is possible to understand the correlation of low ductility with increased resistance to adhesive wear in terms of fundamental properties of contact interfaces and, specifically, in relation with the phenomenon of junction growth. This phenomenon has been found to occur ([9], p 56) where adherent surface asperities undergo tangential stress in a sliding interface. In these situations, large increases in the area of asperity contact can occur before the junction shears, particularly with ductile materials, as shown schematically in Figure 7.8. The formation of prows with attendant transfer and their eventual loss to form loose debris is closely related to junction growth, since a prow in the initiation and growth stage can be considered to be a single large junction. A material that is resistant to adhesive wear must then be characterized by a capability to resist junction growth. Greenwood and Tabor [40] confirmed the validity of this premise by modeling junctions that simulate perfect adhesion. Sheets of metal were V-notched at opposite sides, then sheared. Brittle

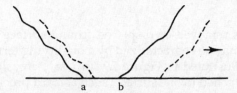

FIGURE 7.8
Schematic representation of junction growth. The normal load deforms the asperity so that contact occurs along *ab*. When a tangential stress is imposed in the direction of the arrow, the junction grows, as shown by the dashed lines, and the two surfaces move together slightly.

materials, like work-hardened copper, gave little transfer, while a ductile metal, lead, produced a large fragment akin to a prow.

In summary, the experimental data reviewed in the above paragraph suggest that an interplay of three factors controls adhesive wear. In order to minimize adhesive wear, the contact surfaces should have the following properties:

- Be hard so as to limit the area of initial contact.
- Be characterized by low ductility to limit junction growth.
- Be lubricated to inhibit asperity adhesion.

Thus, high hardness alone, with or without codeposited polymers, is insufficient to provide good adhesive wear resistance, while hardness coupled with low ductility enhances wear resistance. The efficacy of contact lubricants in reducing wear in ductile load plate may be understood as due to compensating for the low wear resistance of ductile gold plates, both hard and soft, by mitigating adhesion and hence minimizing junction growth [41]. The action of lubricants is discussed in greater detail in Section 7.4.

7.2.2.6 Electroless Gold Plating

Since the mid-1990s, Electroless Nickel and Immersion Gold (ENIG) plating has been used in many electronic components as a metal finish alternative to gold electroplate. An ENIG layer is formed by the deposition of electroless nickel–phosphorus on a catalyzed copper surface followed by a thin layer of immersion gold. The specification covering ENIG requires 3–6 μm of nickel–phosphorus and 0.05–0.1 μm of immersion gold. ENIG is a versatile surface finish and was introduced initially as a viable surface finish on printed circuit boards (PCBs) for surface-mounting microchip packages. The gold layer thickness is about 0.1 μm since the main original objective of the gold was to protect the underlying nickel from oxidizing and becoming difficult to solder. The use of ENIG is being extended to provide a gold coating on contact surfaces of an increasing number of electrical connectors because the deposition process is relatively rapid and inexpensive. This application of ENIG has often proven unsatisfactory because the maximum gold layer thickness achievable is 0.1–0.2 μm, and gold layers of this thickness are not necessarily impervious to air and environmental contaminants and generally do not pass the requirements of a "classic" porosity test. ENIG layers on connectors are thus prone to pore corrosion. In addition, the ENIG deposition process often leads to the formation of black deposits stemming from corrosion of the underlying nickel during deposition and leading to poor adhesion of the ENIG layer to the substrate [42–44]. Properties of ENIG coatings will be addressed in greater detail in Chapter 8.

7.2.3 Abrasion

In abrasive wear, material is removed or displaced from a surface by hard particles. In electrical contacts, abrasion may be characterized by a qualifying terms: *two-body* abrasion and *three-body* abrasion as illustrated in Figure 7.9. *Two-body* abrasion is caused by hard particles that are strongly attached to one surface, such as hard precipitates on the surface of an alloy, whereas *three-body* abrasion stems from hard particles such as sand trapped in a sliding contact and free to move around in the interface.

A simple model for abrasive wear involves the removal of material by plastic deformation as illustrated in Figure 7.10. Consider the action of a large number of abrasive particles with each represented as a cone of semi-angle θ_i dragged across the surface of a ductile material to form a groove. In accordance with Figure 7.10b, the width of the groove generated by the ith particle is $2a_i$ where a_i is the cone radius at the top of the groove of depth D_i.

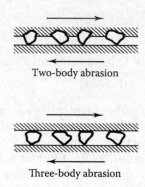

FIGURE 7.9
Schematic illustrations of two types of abrasion: two-body and three-body abrasions.

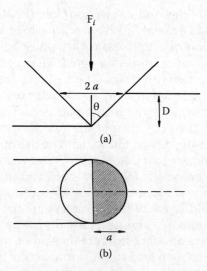

FIGURE 7.10
Simple abrasive wear model in which a cone removes material from a surface: (a) elevation view of a conical abrasive particle forming a groove in a substrate and (b) plan view of the abrasive particle at a distance D from its apex.

If F_i is the average normal applied force on the ith particle and σ is the maximum mechanical stress on the area $\pi a_i{}^2$, then F_i is estimated as

$$F_i = \sigma \frac{\pi a_i^2}{2} = \frac{1}{2}\sigma \pi D_i^2 \tan^2 \theta_i \tag{7.2}$$

Since each hard particle is pushed into the surface and deforms it plastically, we estimate $\sigma \approx H$ where H is the hardness of the worn material. In the simple picture that the volume of material removed from a path of length L is $\varepsilon a_i D_i L = \varepsilon L D_i^2 \tan \theta_i$, where ε is a probability of removal, the wear rate q_i defined as the material volume displaced per unit sliding distance by each particle is

$$q_i = \frac{\varepsilon L D_i^2 \tan \theta_i}{L} = \frac{2\varepsilon F_i}{\pi \tan \theta_i H} = \left(\frac{2\varepsilon k_i}{\pi \tan \theta_i}\right) \frac{F}{H}$$

since F_i is proportional to the total applied force F as $F_i = k_i F$. Summing over all abrasive particles, the total wear rate is $Q = \sum_i q_i$, which reduces to

$$Q = K\frac{F}{H} \tag{7.3}$$

where $K = \sum_i \dfrac{2\varepsilon k_i}{\pi \tan \theta_i}$.

Thus, K is a constant that depends on material properties (via the parameter ε) and the geometry of the abrasive particles. It is noteworthy that Equation 7.3 is identical with Equation 7.1 that was derived on the basis of a simpler model than that of Figure 7.10 [5–7]. In electrical junctions, abrasion occurs either through the intentional introduction of hard particles into interfaces to abrade insulating surface layers such as oxide films or through the ingress of free-moving foreign hard particles such as sand. For these reasons, abrasive wear in electrical contacts arises largely from *three-body* abrasion. The mechanics of abrasive wear and the considerable volume of experimental data related to this subject are provided by Rabinowicz [1], Hutchings [2], and Suh [45].

Abrasive particles are intentionally introduced in aluminum connectors, and particularly aluminum power connectors, to abrade aluminum oxide surface films and promote the formation of metal–metal contacts in electrical connections involving aluminum. In these applications, the abrasive medium often consists of particles of a hard metal such as steel or ceramic powder suspended in a lubricant or mixed into a grease. The lubricant or grease is applied to the contact surfaces of the connector. In connector applications, the choice of hard abrasive grit must be made judiciously since the abrasive medium is generally a poor electrical conductor and must not interfere with electrical contact formation. Since hard abrasive particles are not easily deformable, the particle size must be smaller than the roughness of the contacting surfaces to avoid interfering with electrical contact formation. Abrasion of insulating surface oxides is generally achieved during installation when conductor and connector are mated under the action of normal force and shear forces.

In electrical connections where contact surfaces are coated with a soft metal such as tin, for example, in tin-coated copper busbars, the use of abrasive grit to break up tin oxide films requires special precautions. In these applications, the grit must consist of relatively soft material such as zinc to minimize damage to the coating and preclude significant tin loss. Although abrasion caused by zinc is limited, the relatively good electrical conduction properties of zinc enhance the electrical contact quality since grit particles act as conducting bridges through the contaminant surface layers. One practical advantage of zinc is a less stringent requirement on particle size control in the lubricant than with hard grit. Relatively large zinc particles can be tolerated since they are easily deformed during connection assembly and do not interfere seriously with the mechanics of electrical contact formation.

The ingress of hard foreign particles, such as sand, into electrical interfaces involving connectors coated with a noble metal, such as gold, generally leads to severe abrasion of the coating. Since the thickness of noble metal electroplates is small, generally in the range of ~ 0.2–1.5 μm, abrasion leads to rapid loss of electroplate with subsequent devastating consequences for the integrity of the electrical connection. It is the present authors' experience that electrical failures occur too frequently in electronic connectors deployed in a marine, desert, or highly polluted industrial environment due to inadequate protection measures against the action of hard particulate matter in these environments.

7.2.4 Brittle Fracture

Brittle fracture wear occurs with materials having small tensile strength compared to their compressive strengths. The surface develops cracks across the wear track during sliding, as shown in Figure 7.11 [46]. The hardness of many contact materials increases with increasing hardener concentration as shown in Figure 7.12 for the case of Ni-hardened gold [47], but an additive content of cobalt or nickel larger than 0.5–1.0 wt% leads to high brittleness. This increases susceptibility to brittle fracture wear. These observations apply to rhodium and gold electrodeposits from cyanide baths. The most detrimental effect of brittle fracture to contact performance is the exposure of substrate metal to air at the cracks. This exposure may permit corrosion to occur in aggressive environments. Neither abrasive nor brittle fracture wear can be significantly reduced by lubrication.

It is difficult to develop crack porosity in ductile contact materials, even with repeated wipes. This is evident in the examples of Figure 7.13 that compares the changes in wear patterns undergone by brittle and ductile gold electrodeposits in two-body abrasive wear tests under identical conditions.

As a class, wrought noble contact metals such as inlays are more ductile than their electroplated counterpart. Therefore, inlay abrasion resistance should be superior.

In the majority of situations where electronic connectors with noble metal contacts are used, adhesive wear processes are more important than abrasive wear and brittle fracture.

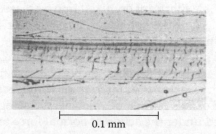

0.1 mm

FIGURE 7.11
Wear track on 12–μm-thick tin–nickel electrodeposit having an overplate of 0.43 μm dc-plated cobalt–gold. Diamond rider; 90° conical tip with a radius of 0.1 mm. Single pass at 200 gf. Note the brittle fracture in the deposit. (After M Antler, *IEEE Trans Parts, Hybrids, Packaging* PHP-9(1): 4–14, 1973 [46].)

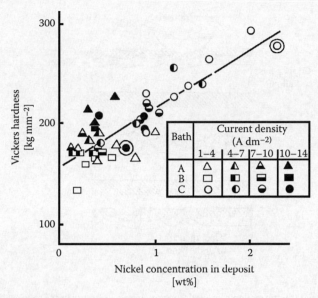

FIGURE 7.12
Variation of hardness of nickel-hardened gold with nickel content, for a wide range of values of electrical current densities during electroplating. (After LG Liljestrand et al., *IEEE Trans. Components, Hybrids and Manufacturing Technology* CHMT 8: 123 128, 1985 [47].).

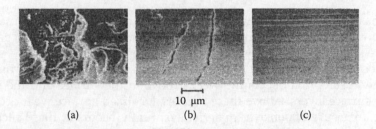

FIGURE 7.13
Two-body abrasion. Worn gold-plated flats from single pass runs at 200 gf with diamond stylus under the same conditions as in Figure 7.11: (a) 3.3 μm hard cobalt–gold on copper; (b) same as (a) with 2.5-μm nickel under-plate; (c) 3.3 μm pure soft ductile gold with 2.5-μm nickel underplate on copper. (After M Antler, *IEEE Trans Components, Hybrids, and Manufacturing Technol.* 4(1): 15–29, 1981 [65].)

The selection of a contact material should generally be made by first considering its adhesive wear properties. If it is practical to employ contact lubricants to mitigate adhesive wear, then materials with superior resistance to abrasion and brittle-fracture wear can sometimes be used.

7.2.5 Delamination and Subsurface Wear

Delamination is a proposed wear mechanism based on the concept of accumulation of near surface plastic strain. In the model proposed originally [48], delamination wear occurs in repeat-pass sliding when cracks become nucleated below the surface, and finally lead to the loosening of thin sheets of metal that turn into wear debris. Crack nucleation originates in dislocation pile-ups, a short distance below the surface, particularly at a hard second phase in the matrix [48–50]. However, unambiguous experimental evidence in support of

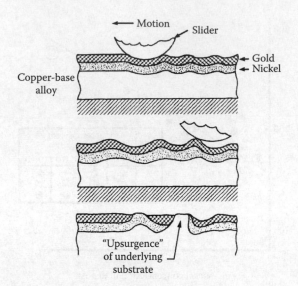

FIGURE 7.14
Steps of the upsurgence mechanism on a gold/nickel/copper surface: *upper sketch:* generation of pressure on the nickel layer underneath the gold, which buckles the nickel layer and forms surface protuberances above the buckle; *middle sketch:* preferential wear of the gold occurs as the contact passes repeatedly over the region, leading to local thinning of the gold; *bottom sketch:* with repeated reciprocating motion, the amount of buckling and preferential wear can increase, leading to localized exposure of nickel and eventual localized exposure of the copper. (After Ref. 58).

the basic model has been less than convincing [51–55]. Conclusions that delamination wear was responsible for damage after extended unlubricated repeat-pass sliding of gold-plated contacts [56] were never unequivocally supported by other workers.

In certain systems, sliding can lead to progressive subsurface plastic deformation [57]. This material flow results in a deepening wear groove in the surface and the formation of buckles in the surface layers. Above these buckles, localized surface wear occurs. With repeated cycling, exposure of sublayer material eventually occurs at these sites, with or without a general thinning of the surface layers [58]. The appearance is suggestive of local "upsurgence" of the sublayers, through the surface, as shown schematically in Figure 7.14. This mechanism is affected by load, number of cycles, and the thickness and stiffness of the surface layers. It was also found that with sufficiently thick and stiff layers, the mechanism could be eliminated.

7.2.6 Effect of Underplate and Substrate

7.2.6.1 Hardness

In general, hard materials are more wear resistant than soft materials. This property originates in the reduction of real area of contact associated with increased hardness and hence in the smaller probability of material removal from the surface during sliding. Area of true contact is thus a basic determinant of all wear processes. It is obvious, therefore, that when considering layered materials, like noble metal contacts made by electroplating or cladding, increasing underplate and substrate hardness may mitigate wear, particularly where the coatings are thin.

Examples of the beneficial role of a hard nickel underplate in reducing wear in gold electrodeposits are given in Figure 7.15 for adhesive wear and Figure 7.16 for abrasive wear [59].

In Figure 7.15, hemispherically ended solid gold riders with a diameter of 3.2 mm were mated in reciprocal motion to pure and to cobalt–gold-plated flats at a series of loads. The tests were carried out for various numbers of reciprocating passes on 1-cm-long tracks. In Figure 7.16, the rider was a 90° conical diamond having a rounded tip with a radius of 0.1 mm, and individual runs were made for a single pass at a series of loads. Wear was determined by paper electrography. The Wear Index is the ratio of the sum of the lengths of the decorated features in the electrograph (i.e., the lengths of separate wear deposits along the wear track) to the total track length multiplied by 100. It is evident that the nickel underplate, which was the hardest material in all cases, was highly effective in reducing wear rates in the gold.

Figure 7.16 illustrates the advantage of combining a hard substrate, copper–beryllium, with a nickel underplate. The superiority of a cobalt–gold electrodeposit compared to pure gold in adhesive wear is also shown in Figure 7.15, as discussed earlier in connection with junction growth. On the other hand, pure ductile gold on a nickel underplate is superior to cobalt–gold in abrasive wear (Figure 7.16b and d) because it does not degrade by brittle fracture; junction growth plays no role in abrasive wear.

Additional examples of the beneficial value of multilayer coatings were reported in studies with electroplated gold on palladium and electroplated gold on clad Pd70Ag30

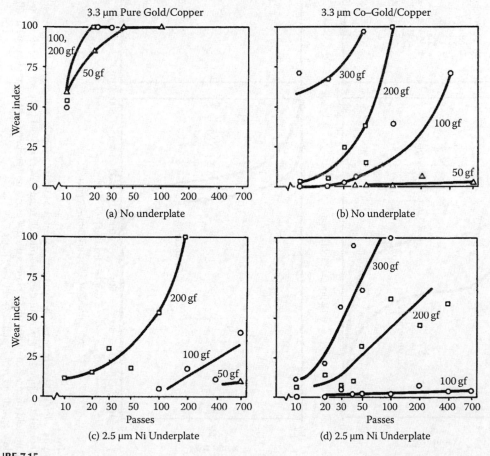

FIGURE 7.15
Electrographic wear indexes from unlubricated adhesive wear runs with 3.3-μm-thick gold electrodeposits: (a), (c) ductile pure gold, with and without 2.5-μm nickel underplate; (b), (d) hard cobalt–gold, with and without 2.5-μm nickel underplate. (After M Antler, MH Drozdowicz, *Bell Sys Tech J* 58(2):323–349, 1979 [59].)

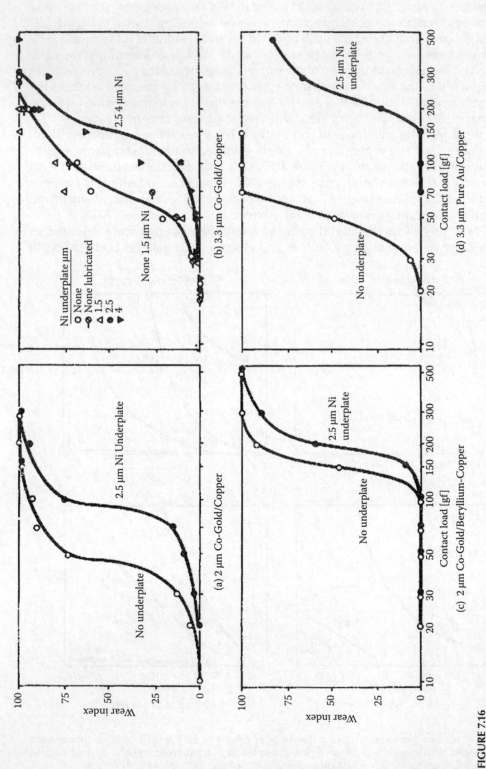

FIGURE 7.16

Electrographic wear indexes from two-body abrasive wear runs with various gold electrodeposits, substrates, and underplates. Conical diamond rider, (a), (c) a hard substrate (beryllium copper) or a hard underplate (nickel) reduce wear of gold plate, and the effects are additive; (b), (d) ductile soft gold plate is superior to hard brittle cobalt–gold plate when nickel underplate is used. Results on specimen without underplate in (d) are poor because gold was non-adherent. (After M Antler, MH Drozdowicz, *Bell Sys Tech J* 58(2):323–349, 1979 [59].)

alloy [60]. In these cases, the second layers were less noble but harder than gold. The advantage of composites, aside from superior durability, is that they may prove to be lower cost replacements for thick gold electroplates. This development is discussed in Section 7.4.

Another advantage of hard ductile underplates is that they can reduce the tendency of a brittle overlayer to crack at high contact force. This has been demonstrated with rhodium plate where nickel was found [61] to be a superior underplate to the one that was soft, such as silver. This concept has been used with various three-layer finishes, such as Au–Rh–Ni [62] and Au–Pd–Ni [63]. The system Au–SnNi–Ni would be an interesting finish where the brittleness of SnNi might be circumvented by the hard nickel; SnNi is a desirable underplate because of its inertness [64], which effectively eliminates corrosion in aggressive atmospheres due to pores in the gold.

Yet another advantage of a hard underplate is in sliding friction. Friction is reduced, as well as durability improved, with an increase in sublayer and substrate hardness [65].

7.2.6.2 Roughness

Surface roughness has little effect on the wear of unlubricated, identical unplated solid metals [1], except that very smooth clean surfaces may experience excessive junction growth and increased adhesive wear. Actually, in repeat-pass sliding, clean surfaces tend to become roughened as a consequence of transfer and wear to an extent that depends on the material and conditions of the experiment, such as load. In the mild wear regime, particularly with ductile metals, surfaces become burnished during sliding. In lubricated interfaces, surfaces also become increasingly burnished and the lubricant trapped in the valleys between the high spots emerges and continues to protect. When the entire surface has become smooth, lubricant failure and high wear ensue [46].

Gold-plated smooth contacts have been found to wear less than rough surfaces in both adhesive and abrasive wear [66]. This is illustrated in Figures 7.17 and 7.18 where substrates were prepared to three values of roughness and plated with the same mass coverage of gold. One important cause for the increase in wear rate of platings with increasing roughness is that wear tends to occur preferentially on surface high spots [67]. Thus, the gold is lost by wear first from the asperities. With smooth surfaces, wear is less localized than with rough surfaces, and more passes are required to achieve a given level of wear. The wear rate of a clad surface may be affected by the hardness of the mating surface [68].

7.2.7 Electrodeposited Gold: Relationship of Wear to Underplate Hardness

Some studies of the sliding wear of various golds are summarized in reference [57] where it is shown that both high hardness and low ductility are required to minimize the rate of adhesive wear. On the other hand, in Section 7.2.4 it was also indicated that abrasive wear resistance is improved where both hardness and ductility are increased. Subsurface wear resistance (Section 7.2.6) appears also to be increased with increased hardness. Deposit composition, the plating process, topography, underplate-substrate thickness and mechanical properties, and surface cleanliness (lubrication) likewise play a significant role in determining the sliding wear rate of a contact materials system. Our understanding of the interplay of these various factors is incomplete, and much more work is needed to develop a comprehensive quantitative view of contact wear. In this regard, significant investigations of the correlation of wear properties of electroplated gold with deposit composition and plating process are described in [69,70].

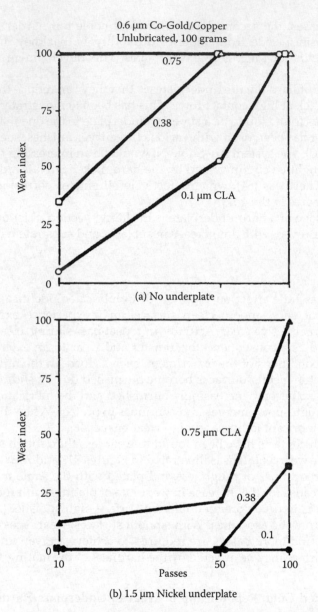

(a) No underplate

(b) 1.5 µm Nickel underplate

FIGURE 7.17
Unlubricated adhesive wear. Wear indexes from runs on 0.6-µm cobalt–gold-plated copper samples having three degrees of roughness. Solid gold riders. Smooth surfaces gave less wear: (a) no underplate; (b) 1.5 jam nickel underplate. (After M Antler, *Insulation/Circuits*, 26(1): 15–19, 1980 [66].)

7.2.7.1 Hardener Metal Content

A limited number of investigations have been made to relate the adhesive wear resistance of acid hard gold deposits to cobalt or nickel content [46]. Gold deposits with as little as 0.06% cobalt [71] or as much as 4% cobalt or nickel [72] were found to have acceptable wear behavior. The dependence of hardness on nickel content in nickel-hardened gold has already been shown in Figure 7.12 from the comprehensive investigation by

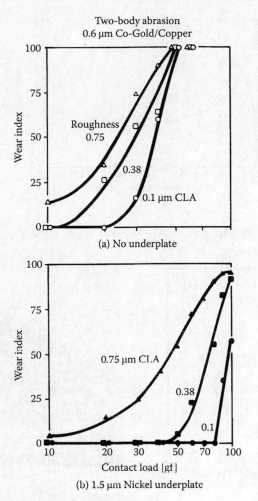

FIGURE 7.18
Two-body abrasion. Wear indexes from single-pass runs with diamond stylus on same samples as in Figure 7.17. Wear is inversely related to surface roughness: (a) no underplate; (b) 1.5-μm nickel underplate. (After M Antler, *Insulation/Circuits*, 26(1): 15–19, 1980 [66].)

Liljestrand et al. [47]. For the sake of completeness, the corresponding data for cobalt–gold is shown in Figure 7.19 [36]. We note that hardness increases somewhat more rapidly with increasing cobalt than with nickel concentration. Figure 7.19 also shows the decrease in friction coefficient associated with increasing hardness. In recent years, the use of cobalt–gold has fallen into some disfavor due to cost factors and less favorable contact properties than nickel–gold, including the tendency of cobalt in gold to segregate to the surface beyond about 160°C [73]. We shall focus on the wear properties of nickel–gold, since the wear properties of cobalt–gold follow similar trends [65].

The data of Figure 7.12 indicate that hardness increases linearly with nickel content and does not level-off at nickel contents larger than about 0.5 wt% as is the case for cobalt-hardened gold. Figure 7.20a is reproduced and annotated from Antler's investigations [20,46] and illustrates the durability of 0.1% nickel–gold deposits ($H = 150$–200 kg mm^2) in reciprocating sliding as a function of the thickness of electroplates deposited on oxygen-free,

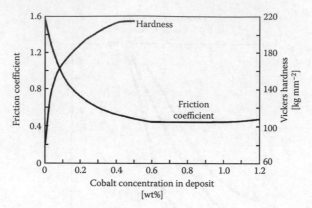

FIGURE 7.19
Variation of hardness of cobalt-hardened gold, with cobalt content. (After M Antler, *IEEE Trans Parts Hybrids, Packaging,* 10: 177–17, 1974 [36].).

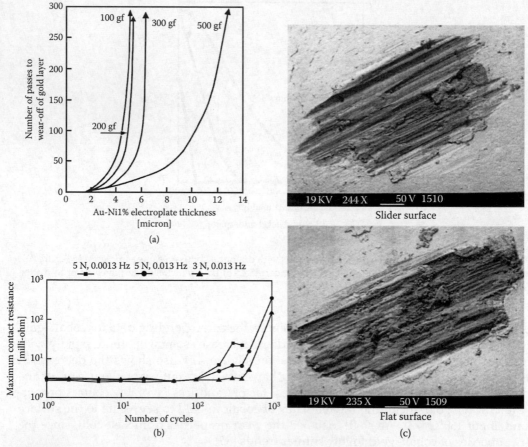

FIGURE 7.20
(a) Dependence of durability of hard gold on electrodeposit thickness (After M Antler, *IEEE Trans Parts, Hybrids, Packaging* PHP-9(1): 4–14, 1973 [46].), (b) contact resistance versus number of low-frequency cycles at 150°C and contact loads of 3 N and 5 N; 0.2% cobalt–gold on copper alloy (Cu95.5, Si3.0, Sn1.5, Cr0.1) (After CH Leung, A Lee, *Proceedings of the IEEE Holm Conference on Electrical Contacts,* Philadelphia, 1997, pp 132–137 [74].), (c) SEM micrographs of wear tracks on the slider and the flat surface associated with (b). The light and dark areas consist, respectively, of gold and nickel (or copper) due to larger electron reflectivity of gold.

high-conductivity (OFHC) copper flats. The mating rider was 3.18 mm in diameter and was similarly plated. Reciprocation sliding was carried out for 700 passes over a 10-mm track length. The durability of the deposits was determined by the number of passes required for wear off from the flat, for contact loads ranging from 50 gf to 500 gf. The data of Figure 7.20a indicate that durability increased with increasing gold thickness but the rate of increase was not constant. In particular, durability increased sharply as the electroplate thickness increased beyond a critical value. This critical value was found to vary with applied contact load. For example, at a contact load of 100 gf, durability increased sharply for deposit thicknesses larger than about 4 μm. Similarly at contact loads of 200 gf, 300 gf, and 500 gf, there was a rapid increase in durability for gold thicknesses larger than 5 μm, 6 μm, and 12 μm respectively. The sharp durability increase was due to a significant decrease in wear rate after initial running-in. The decrease in wear rate stemmed, in part, from hardening of the gold through deformation and cold-working during running-in. The data of Figure 7.20a also indicate that, for a given electroplate thickness, durability decreased with increasing contact load. This is consistent with the trend predicted by Equation 7.1.

Over the last decade, the requirement on connectors located under automobile engine hoods to operate reliably at elevated temperatures has led to investigations of wear induced by slow reciprocating displacements due to cyclic thermal expansions and contractions. Such slow reciprocating displacements would occur where the contact load is sufficiently large to mitigate effects due to small high-frequency displacements caused by mechanical vibrations in the engine compartment. Although non-noble metal contacts would be expected to degrade relatively rapidly at elevated temperatures due to increased oxidation rates, it was unclear how gold-plated contacts would perform. All contact degradation mechanisms are thermally activated and become more severe at higher ambient temperature. For gold-plated contacts, these mechanisms include metal softening, possible rapid interdiffusion with substrate materials and diffusion to the surface of hardeners in gold (see Section 7.4), and other possible effects. Figure 7.20b shows the variation in contact resistance of a dimpled surface sliding over a flat, with a number of reciprocating cycles. The mating surfaces consisted of 0.8 μm cobalt–gold on 2-μm nickel on a C65400 copper alloy substrate, and the sliding distance was 300 μm [75]. The work was carried out at 150°C with contact loads of 3 N and 5 N, and frequencies of 0.0013 Hz and 0.013 Hz. Contact resistance increased more rapidly at 5 N than at 3 N. Degradation stemmed from wear through of the electroplate to the underlying underplate and substrate and accumulation of oxidized wear debris. The trend with increased load is consistent with the data of Figure 7.20a. The increased degradation with smaller frequency was attributed to increased time for oxide formation during the reciprocating motion once the protective gold-plating layer had worn through. The SEM micrographs of the contact surfaces in Figure 7.20c show the wear through of the gold electroplates. It was verified that wear led to exposure of the nickel barrier as well as the copper substrate, with subsequent oxidation of the exposed surfaces. It is worth emphasizing that none of the investigations of hardened-gold layers have included a correlation of wear properties with structural or other physical property of the deposits. As indicated earlier, hardener metal content and hardness are by themselves unreliable predictors of the performance of a gold coating in a sliding contact application.

7.2.8 Clad Metals

As a class, clad and electrodeposited metals with the same composition are tribologically similar. However, there is a major exception with cobalt and nickel-hardened cyanide gold platings compared to virtually all clad noble metals that have been considered for electronic connector separable contacts. Some plated hard golds are considerably less prone to adhesive wear than corresponding clad materials [76]. This is due to the combination of high hardness and low ductility of electrodeposits, which discourages junction growth and prow formation.

Comparative examples of unlubricated sliding are shown in Figure 7.21. The micrographs compare the wear damage generated between a rider electroplated with cobalt–gold sliding over a flat coated with the same electrodeposited material, with wear produced by sliding the same rider over a clad Au69Ag25Pt6 flat [77]. In the case of the worn electrodeposits shown in Figure 7.21a, sliding remained in the mild regime with relatively small mechanical deformation and small wear debris particles and flakes. In contrast, wear on the clad surface was severe in Figure 7.21b, with a high coefficient of friction. In addition, prows of the clad material were evident on the electroplated-gold rider surface, while the flat clad contact surface was covered with prominent back-transfer prows after repeat traversals. Clearly, the results of these investigations point to the superior wear resistance of the electrodeposited gold coating over the mechanically clad material. The susceptibility of clad material to wear stems largely from weak bonding of the clad to the substrate. Since bond weakness generally occurs locally over a substrate, the wear resistance of a clad surface should vary inversely with the clad area since the wear path over a small clad is short and has a smaller probability of crossing weak bonding regions. This in turn leads to a smaller probability of cladding detachment during sliding. Thus, the use of clad metals should be restricted to the smaller contact component in a connector, in order to minimize the probability of wear due to clad detachment as was verified in practical hardware studies [77].

Guidelines for the use of clad metal contacts in connectors can be summarized as follows:

- Clad metal contacts should not be mated to themselves.
- Clad metals may be mated to a hard gold electroplate or a hard gold-flashed palladium or palladium alloy finish, provided that the clad metal has the smaller swept area.

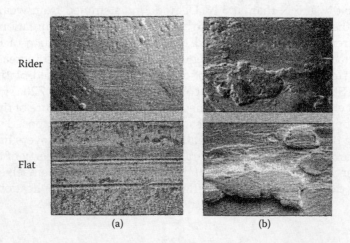

(a)   (b)

FIGURE 7.21
Worn specimens from unlubricated sliding with electrodeposited cobalt–gold riders and flats of (a) cobalt–gold electrodeposit and (b) clad Au69Ag25Pt6. 100 gf load, 20 cycles. Coefficients of friction at ends of runs: (a) 0.3, (b) 1.2. (After M Antler, *Connection Technol* 6(3): 29–33, 1990 [77].)

- The plated surface should be smooth, that is, free of surface protuberances and other features which may abrade the cladding.
- It is desirable to use a good boundary contact lubricant.

7.2.9 Tin and Tin–Lead Alloys

Tin and tin–lead alloy finishes are widely used for connector contacts. However, these finishes are limited to applications where few, usually only 1 or 2, but generally not over 10 matings, are required in their lifetime. The adhesive wear rate of these materials is estimated to be 100 times greater than that of hard gold plate. These tin-based platings are also easily abraded.

Although pure tin coatings are sometimes specified, their tendency to form whiskers, unless fused, makes tin–lead alloys the preferred finish. Also, tin–lead coatings are more easily soldered at the terminal end of the contact due to their lower melting points than pure tin. Thicknesses of 2.5–5 μm are normally specified, usually on a nickel or copper underplate. Normal loads should be larger than about 1 N for gently curved contacts, because the oxide on these metals must be broken or displaced to achieve a low contact resistance.

The inferior durability of tin and tin–lead alloys is due to low hardness and therefore easy displacement of these materials in sliding applications. Adhesive wear occurs by a pseudo prow formation process, involving metal transfer from the larger to the smaller contact surface involved in sliding; but transferred metal is work softened (observed for 50–50 tin–lead) rather than work hardened. This adhesive wear mechanism has been termed "wedge flow" because the transferred metal on the rider flows *in the direction of sliding*, as shown in Figure 7.22 [20]. Rider wear soon follows, as in prow formation.

An important aspect of the sliding wear of tin and tin–lead alloys is the effect of intermetallics due to interdiffusion of the tin and the substrate material on the coefficient of

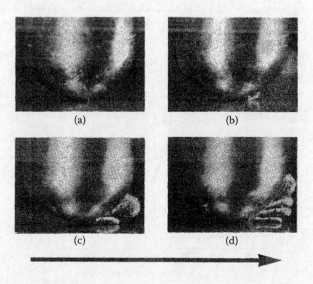

(a)   (b)

(c)   (d)

FIGURE 7.22
Wedge-flow mechanism. Solid 50/50 tin–lead alloy sliding on itself, 3.2-mm diameter rider, 2.5-cm diameter circular track, 180 gf: growth and loss of wedge by back-transfer to flat shown. Note metal emerging at rear of rider. A, after 1/4 rev.; B, at 10 revs.; C, 40 revs.; D, 60 revs. Back-transfer solid appears at right in D. Note separation of rider and flat during sliding. Wedge-flow metal originates in flat. Arrow indicates direction of movement of flat. (After M Antler, *Wear*, 7: 181–204, 1964 [20].).

friction and contact resistance. Intermetallic materials, such as Cu_6Sn_5 and Cu_3Sn [78,79] for a copper-based underplate or substrate, are hard and characterized by a much larger electrical resistivity than tin. It has been found [80] that there are three stages in wear, as shown in Figures 7.23 and 7.24, described as (i) plowing of the tin—actually wedge flow adhesive wear (Stage 1), (ii) sliding on the intermetallics (Stage II), and (iii) penetration of the intermetallic layers with sliding on the substrate (Stage III). The termination of Stage 1 marks the end of the most useful life of a tin-plated contact. Since the rate of formation of intermetallics is high [78,79], the lifetime of contacts with a tin finish is quite short. In the

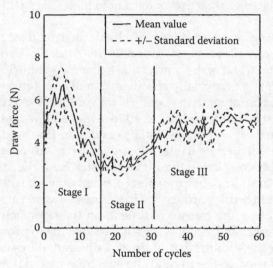

FIGURE 7.23
Variation in draw force during a test sequence and identification of the three stages of wear. Electrodeposited dull tin (10 μm) with a normal force of 1 kgf. (After T Hammam, *Proceedings of the 18th International Conference on Electrical Contacts*, Chicago, 1996, pp 321–330 [80].)

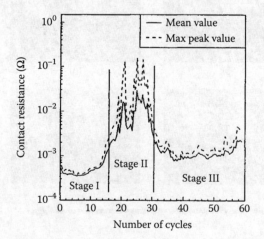

FIGURE 7.24
Variation in contact resistance during the test sequence shown in Figure 7.23, for electrodeposited dull tin (10 μm) with a normal force of 1 kgf. (After T Hammam, *Proceedings of the 18th International Conference on Electrical Contacts*, Chicago, 1996, pp 321–330 [80].)

case of contact surfaces coated with a tin–lead alloy layer, residual lead remains once the tin is consumed in forming intermetallics. Perhaps the sliding of tin and tin–lead should be considered to involve a succession of interfacial materials, which changes the characteristics of the sliding interface as these materials appear in the contact due to wear.

The degradation of tin-based contacts by fretting corrosion has been the focus of much attention, to be discussed in Section 7.3. In recent years, the RoHS legislation (Restrictions of the Use of Certain Hazardous Substances) [81] for the use of lead-free materials has required the removal of lead, including lead from tin alloys used in electronic devices [82]. This has necessitated the development of techniques to minimize whisker formation in tin finishes, such as annealing at temperatures slightly below melting, the use of appropriate underplate structures, and so on [83,84]. Because tin and tin-alloy electroplates are soft and prone to mechanical wear and abrasion, they are limited to applications where relatively few matings (e.g., on the order of 10) are expected in their lifetime. For this reason, connectors designed for frequent matings in service are plated with hard gold which is considerably more resistant to wear than tin. Other wear characteristics of tin and tin-alloy electroplates, and in particular fretting wear, will be described in a detailed description of wear mechanisms.

7.2.10 Silver

With the price of gold continuing to climb, silver is increasingly regarded as a relatively low-cost alternative to gold. As will be described later in relation to fretting wear, clean silver contact surfaces have a tendency to *stick* or *weld* under a sufficiently large contact load. This *stick* tendency leads to a high coefficient of friction which in turn leads to poor durability performance and poor fretting properties where sliding displacements are large. In contrast with gold, silver readily forms surface tarnish films when exposed to sulfur-containing and chlorine-containing atmospheric contaminants. Field exposure studies of silver-plated connectors have shown that tarnish films on silver consist primarily of α-silver sulfide (Ag_2S) and to a lesser extent silver chloride ($AgCl$) which is more insulating and potentially harder to displace than the sulfide [85–94]. These tarnish films are generally semi-conductive at ambient temperatures, mechanically soft, and relatively easily displaced with contact interface wipe at sufficient normal loads. The presence of tarnish films has a significant effect on the wear properties of silver.

Figure 7.25a shows an example of the wear sustained in reciprocating motion by clean/untarnished silver surfaces in a sliding couple consisting of crossed copper rods plated with a 4-μm thick silver coating. The reciprocating displacement frequency was 0.2 Hz at a large contact force of 60 N [95]. As expected for clean silver, the dominant deterioration mechanism in untarnished contacts is adhesive wear and mechanical deformation increases with increasing contact force. Although tarnished silver coatings with a thickness of 2 or 4 μm show no clear evidence of adhesive wear, with the wear track always smooth as illustrated in Figure 7.25b, the amount of mechanical wear depends on the silver coating thickness. For example, the degree of mechanical wear of tarnished silver coatings with a thickness of 19 μm was found to be large and similar to the wear observed for untarnished silver as shown in Figure 7.25a [95], that is, adhesive wear was dominant. The adhesive wear observed with thick tarnished silver stems from breakup and dispersal of the tarnish layers due to the relatively large deformation allowed by the thick metal coating, and ensuing metal flow and exposure of fresh silver leading to sticking and subsequent severe wear. Some measurements have revealed adhesion at contact

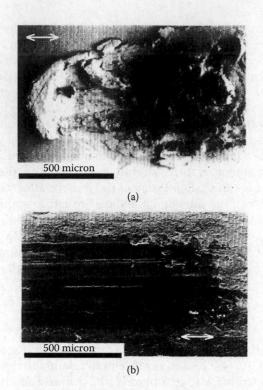

(a)

(b)

FIGURE 7.25
Surface appearance after 20 reciprocating cycles on a silver coating 4 μm thick at a contact load of 60 N: (a) untarnished surface and (b) tarnished surface. The sliding direction is parallel to the bottom edge of the micrograph. (After A Kassman Rudolphi, S Jacobson, *Tribology Inter.* 30: 165–175, 1997 [95].)

forces as small as 10 cN on tarnished silver [96]. This is illustrated in Figure 7.26 where the adhesion coefficient α (= adhesion force/contact force) is shown as a function of the sliding distance of hard gold on tarnished solid silver. The larger adhesion with increased sliding distance stems from increased breakup of the tarnish layers, also evidenced by the notable drop in contact resistance. The surface layers of the worn tarnished silver surfaces on copper are often found to consist of a mix of copper oxides, copper sulfide, and silver sulfide [95].

The data of Figures 7.25 and 7.26 illustrate the following important contact properties of silver coatings:

  i. Tarnish films on silver are relatively easily displaceable.
 ii. The presence of tarnish does not necessarily imply high contact resistance, if sufficient relative interfacial displacement is allowed to disrupt the tarnish layers [97,98]; we will address the issue of tarnish film disruption in greater detail later in description of fretting wear.
iii. Relatively thick silver coatings generally lead to lower contact resistance than thin coatings; the superior electrical contact properties of thicker coatings are attributed to easier metal deformation in thick metal coatings and, hence, easier disruption and dispersal of tarnish films on the surfaces of thick silver.

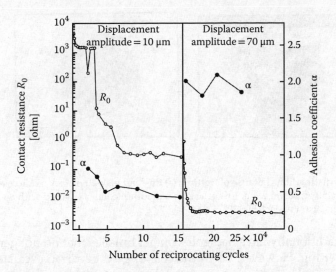

FIGURE 7.26
Solid gold (hardness 200 kg mm^{-2}) sliding on tarnished silver (hardness 80 kg mm^{-2}): contact resistance R_0 and adhesion coefficient α as a function of the number of reciprocating cycles. The displacement amplitude was initially 10 μm and was then increased to 70 μm after about 15.4 × 10^4 cycles. (After H Bresgen, *Wear* 69: 157–165, 1981 [96]).

7.3 Fretting

7.3.1 Background

This subject is also discussed in Chapters 5 and 6. An understanding of the mechanisms that contribute to connector degradation is necessary to facilitate the correct selection of contact materials and component designs, and to establish the conditions in which the materials must be used in order to assure reliable performance. One of these contact failure processes is fretting or small amplitude contact movement. The small displacements may range from a few micrometers to as much as 100 μm in electronic connectors [99] and are caused by external vibrations [100] or changing temperature [101], for example, due to differences in the coefficients of thermal expansion of the elements to which mating contacts are mounted. Electromagnetically induced vibrations are significant in some types of high-power bus connections [102,103].

Fretting causes metal transfer and wear. Base metal contacts produce insulating oxide debris such as the particles shown in Figure 7.27a from tin–lead solder plate. Palladium and other platinum group metals catalyze the formation of insulating frictional polymers on the surfaces in the absence of significant metallic wear [104], like that illustrated in Figure 7.27b. The precursors to these polymers are adsorbed organic air pollutants from the atmosphere. The end result of these surface-contaminating processes is the onset of variable contact resistance (electrical noise) during fretting and the attainment of an unacceptably high contact resistance even when such motions cease.

Field studies of electronic connectors, relays, switches, and power connections have identified fretting as a certain [103–107] or a probable [100,108,109] cause of contact failure. Although degradation due to fretting had long been recognized in relays and switches in the telephone industry [104,110], it was not considered to be significant in electronic

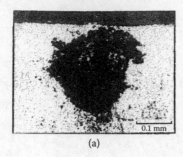 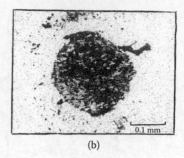

(a)   (b)

FIGURE 7.27
Contacts that acquire insulating films because of fretting: (a) tin–lead plate (the black oxide covered surface and debris should be noted) (150 gf; 20 μm wipe; 10^4 cycles); (b) palladium contact covered with and surrounded by frictional polymers (50 gf; 20 μm wipe; 8 Hz, 10^5 cycles).

connectors which traditionally have had gold contact finishes that do not generate significant films due to fretting. However, when the cost of gold began to escalate first in 1972 and later after about 2008, attention began to be directed to thinner gold coatings and to alternative materials [98,111,112]. Some of these alternatives, such as tin and tin–lead alloys, are particularly prone to fretting failures. This is the reason for the increasing tempo of investigations of fretting.

The objective of Section 7.3 is to consider mechanisms of fretting degradations, to survey contact material behavior and the effect of operational parameters on contact resistance, and to provide design guidelines for fretting control in electronic connectors. The role of contact lubricants in preventing fretting problems and in recovering failed connections in favorable cases is considered in Section 7.4.

7.3.2 Fretting Regimes

Fretting motion in a contact interface occurs when a force applied in a direction parallel to the interface exceeds the friction between the mating surfaces. Coefficients of friction typically are the same as those observed for metals under adhesive sliding conditions, about 0.15–1.0 or lower [113]. A high normal force can minimize the possibility of motion. Contact springs can be designed to be very flexible and minimize fretting displacements [114].

Assuming that an interfacial displacement of amplitude δ occurs, the resulting contact surface damage will depend on the magnitude of δ. This has been categorized by various authors [115–117] into four regimes (the first three for fretting and the fourth for gross movement):

i. If δ is small and directly proportional to the tangential force (T) and there are no indications of microscopic interfacial sliding, the sliding regime is defined as *sticking*. The linear dependence of T on δ illustrated schematically in Figure 7.28 corresponds to an elastic shear stress–strain curve for the two mating surfaces. This indicates that the macroscopic displacement between the contacting surfaces is mainly accommodated by elastic deformation in the near-surface regions of the two components. The interface is maintained under stick contact conditions by the adhesively joined asperities; larger displacements will cause plastic deformation and shearing in the fretting direction. The stick regime may occur for movements of about 1 μm depending on the material, contact geometry, and other factors. Figure 7.29 illustrates stick for contacts in a crossed cylinder device [115]. Although

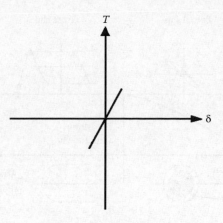

FIGURE 7.28

Plot of tangential force T versus the interfacial displacement δ in the *sticking* regime. The linear relationship indicates that macroscopic displacement between the contacting surfaces is accommodated by elastic deformation. (After O Vingsbo, S Soderberg, *Wear* 126: 131–147, 1988 [115].)

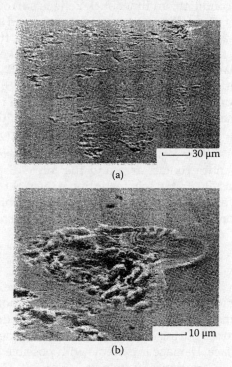

FIGURE 7.29

(a) Fretting wear scar, characteristic of stick contact conditions and (b) detail of the fretting wear scar in (a): material, niobium, 2 µm wipe; 1.09 kgf, 100 Hz; 10^6 cycles. (After O Vingsbo, S Soderberg, *Wear* 126: 131–147, 1988 [115].)

essentially no detectable surface damage may be generated initially, the *sticking* regime cannot be dismissed as non-fretting since reciprocating motion may lead to the nucleation and propagation of surface fatigue cracks, particularly along the rim of the contact area, which would lead to wear debris formation.

ii. If δ is smaller than a threshold transition amplitude δ_t as illustrated in the left-hand side of Figure 7.30, the sliding regime is defined as *partial slip* [117]. Under

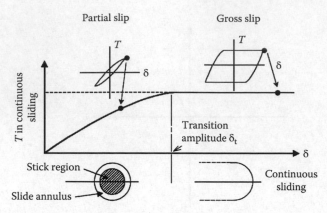

FIGURE 7.30
Sliding regimes for a contact interface consisting of a sphere or the hemispherical surface of an asperity tip in contact with a flat with a normal force. (After S Hannel et al.,*Wear* 249: 761–770, 2001 [117])

this interfacial motion condition, an inner stick zone is surrounded by an annulus where displacement is small but where slip occurs and the force–displacement curve resembles a hysteresis curve centered around the origin. At small partial slip amplitudes, the contact stick zone remains relatively undamaged, but the surrounding annular slip region may host areas of crack formation, fretting fatigue, and wear debris. This is shown in Figure 7.31 for A1S1 304 stainless steel. Movements are of the order of ~1–3 μm. Clearly, there is an inverse relationship between the width of the slip annulus and the dimensions of the inner stick zone, that is, a large stick zone is associated with a narrow slip region.

iii. If δ is larger than $δ_t$, the sliding regime corresponds to *gross slip*. In this case, sliding occurs across the entire contact area and the force–displacement curve is essentially trapezoidal as illustrated schematically in the right-hand side of Figure 7.30. The tangential force T is independent of the displacement amplitude δ and is related to the normal contact force P in accordance with the conventional relation

$$T = \mu P,$$

where μ is the friction coefficient. Initial gross slip favors the elimination of surface native oxides and promotes strong metal–metal interaction. The friction coefficient increases until the partial slip condition is reached, that is, *gross slip*. All asperity contacts are broken during each cycle. Asperities slide across several others of the opposing surface. Damage is extreme with possible delamination wear. Movements of 10–100 μm are typically involved. Figure 7.32 for AISI 1018 steel illustrates this regime [115].

iv. Clearly, there will be an intermediate sliding region where displacement occurs via both partial slip and gross slip. This sliding regime is appropriately defined as *mixed slip*. Fretting associated with the *mixed slip* regime is generally characterized by an initial *gross slip* condition followed by a *partial slip* situation. Initial *gross slip* favors the elimination of surface native oxides and promotes the formation of metal–metal metallurgical bonds. The friction coefficient increases until the partial slip condition is reached. For intermediate displacement amplitudes and with increasing number of fretting cycles, the variation of tangential force

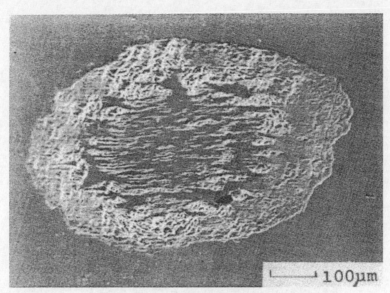

FIGURE 7.31
Fretting wear scar characteristic of mixed stick and slip contact conditions: material, AISI 304; 4 µm wipe; 1.15kgf, 100Hz; 10^6 cycles. (After O Vingsbo, S Soderberg, *Wear* 126: 131–147, 1988 [115].)

with reciprocating displacement evolves from that typical of *gross slip* to one characteristic of *partial slip*. This evolution of the tangential force–displacement curve with increasing number of fretting cycles is illustrated in Figure 7.33b. In contrast, the corresponding curves for *partial slip* and *gross slip* remain stationary as represented schematically in Figure 7.33a and c.

v. *Reciprocating sliding.* With large displacements, say more than 100–200 µm, there is sliding wear with mechanisms and rates that have been described in Section 7.2.

Degradation mechanisms of materials for relatively long traversals of contact insertion and withdrawal in a connector are not fundamentally different from those in the gross slip regime. Although wear debris, corrosion products, and frictional polymers tend to accumulate more readily on surfaces with the smaller fretting movements, wear rates and the utility of lubricants in mitigating failures are similar for fretting and for gross sliding. Thus, many investigators have conducted studies with large movements (e.g., 0.3 mm in [75], 0.4 mm in [118], and 10 mm in [119]) and obtained results that are directly applicable to the micro-movements of fretting. The so-called fretting maps describe the relationship between tangential force and displacement [115,117] for categories (i)–(iv) listed above.

7.3.3 Static versus Dynamic Contact Resistance

Fretting leads to the formation of metallic wear and oxidized debris by adhesion, abrasion, and delamination processes. When the contact metal is non-noble or catalytically active (examples are discussed later), insulating materials appear in the contact interface. If these substances are inorganic solids, like oxides, the process is called "fretting corrosion." Catalytic materials such as the platinum group metals yield organic contaminations, or

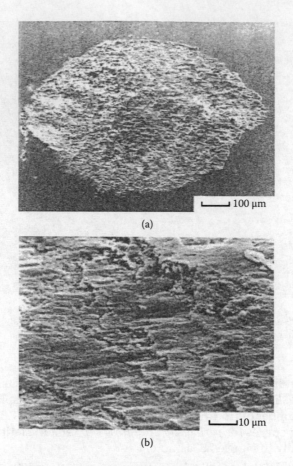

(a)

(b)

FIGURE 7.32
(a) Fretting wear scar, characteristic of gross slip contact conditions and (b) detail of scale-like topography of wear scar: material, AISI 1018; 15 μm wipe; 570 gf; 20 kHz; 10^6 cycles. (After O Vingsbo, S Soderberg, *Wear* 126: 131–147, 1988 [115].)

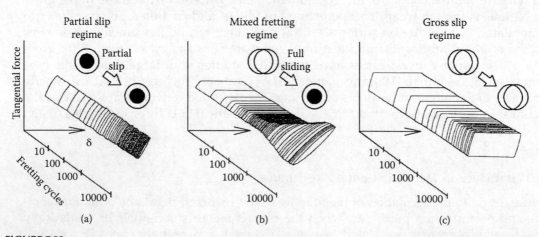

FIGURE 7.33
Schematic representation of the three slip regimes in interfacial sliding. (From S Hannel et al. *Wear* 249: 761 – 770, 2001, Figure 7.3 [117]. With permission)

"frictional polymers." The accumulation of insulating solids in the contact zone causes the electrical resistance to increase. Figure 7.34 is a representation of this process.

There are two aspects of electrical resistance degradation: (1) the "static" contact resistance of the system at rest, and (2) the variability of contact resistance that may reach high values during fretting, even open circuit, for nanoseconds up to much longer times depending on the velocity of the reciprocating displacements, cycle rate, contact materials, and the physical properties and thickness of the insulating layer. The practical consequences of fluctuating contact resistance in digital circuitry may be errors in signed transmission. An elevation of static contact resistance in power circuits can lead to electrical failure of the connection by Joule heating.

There has been considerable speculation about the mechanisms responsible for contact resistance changes in fretting corrosion, and several [120–123] models have emerged. Two of these models are summarized below. The first model is the so–called *asperity model*. This model proposes that the area of metal–metal contact decreases continually during reciprocating motion as metallic asperities and wear debris are transformed into oxides, whether layered and adherent or loose and particulate. There is also a volume growth, from metal to oxide, which further tends to separate the contact surfaces. The reduction of contact area causes contact resistance to increase, and motion eventually leads to momentary loss of asperity contact followed by the re-establishment of contact with another array of asperities. However, an analysis [124] of the relation between the duration of contact resistance excursions and fretting velocity for copper contacts in a particular application showed that excursions are generally too brief to be explained by the asperity model. The second model proposed to account for contact resistance changes in fretting corrosion is the *granular interface model*. In this concept [123], fretting debris is composed of metal particles, oxide-covered debris, and fully oxidized material. The contact interface is represented schematically in Figure 7.34c. The resistance of such a debris-filled interface would be very high, but small displacements may alter the resistance if an electrical conduction path is developed via contacting metal debris particles. Changes in contact resistance due to interfacial displacements would be momentary.

It is likely that both the asperity and granular interface models have elements of validity for actual electrical contacts, and even coexist with a relative importance that depends on the material system involved. It has been shown [118] that with the increase in static contact resistance from fretting, the frequency of short-term resistance excursions increases as well.

7.3.4 Field and Laboratory Testing Methodologies

In fretting test methodologies, the extent of fretting depends on the method used to induce movement and on the apparatus employed. Ensuing contact resistance changes may depend on the electrical current level and on the procedure for measuring contact resistance. This was not always recognized, and an early example of variable results involved a laboratory determination of contact resistance changes of palladium contacts due to fretting [108,125,126].

7.3.4.1 Generation of Fretting Displacement

All separable connectors can be made to fret, provided that the stresses are applied which exceed their contact withdrawal (retention) force [100]. Realistic hardware evaluation depends, however, on imposing only as much external force as is involved in the

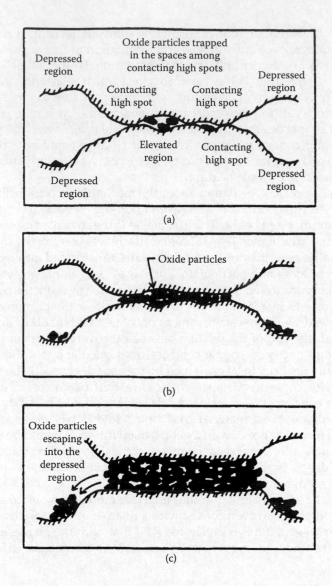

FIGURE 7.34
Schematic representation of formation and accumulation of fretting corrosion solids: (a) accumulation of oxide particles in the spaces among the high spots; (b) integration of a company of high spots into one single united area, after the space among the high spots is filled by oxides; (c) spilling of oxide particles into the adjoining depressed regions. (After IM Feng, BG Rightmire, *Lubrication Engrg* 9: 134–136, 1953 [120].)

application [102]. The origin of the force may be thermal, mechanical, or electromagnetic, as described above. Nevertheless, hardware studies are sometimes conducted by forcing displacements, as when a printed circuit board inserted in a connector is rocked [111,127]. Laboratory hardware testing has also been conducted by using severe external vibrations [100,105], by thermal cycling [111], and by imposing a reversing force of constant magnitude to the samples [128]. Materials studies are usually made using equipment that utilizes idealized contact geometries and where fretting is caused by either thermal changes [101,129–133] or mechanical means [101,102,104,110,125,129,133–141], or is electromagnetically induced [102,117,139].

7.3.4.2 Determination of Contact Resistance

In laboratory testing, contact resistance may be measured by dc or ac methods at low voltage and current in order to preclude heating or electrical breakdown of insulating films that may be present [142,143], and with no current flowing except during the periodic resistance measurement. Alternatively, the circuit may be continuously powered during fretting at levels consistent with the intended application. Static contact resistance changes due to fretting were found [118,127] to be much less than the changes observed during transient operation. Another method [128] for evaluating contact resistance variability during fretting obtains the spectral content of the voltage drop using a high-speed recorder when passing a dc current through the contacts.

7.3.5 Materials Studies

7.3.5.1 Apparatus

The testing apparatus used for investigations of fretting corrosion of electrical interfaces is generally designed to generate small reciprocity displacements reproducibly and allow simultaneously for measurements of contact resistance along the wear track. A stationery rider (usually a hemispherically ended rod, rivet, short segment of a cylinder, or connector contact) is usually dead-weight loaded to produce a force ranging from 5 to 500 gf against a flat or other suitable specimen located on a precision slide table. The table is often driven by a dc stepping motor through a micrometer screw. The motor is interfaced to a computer that controls track length and other test parameters. Investigations of the effects of fretting corrosion are generally made in an uncontrolled environment.

The accurate measurement of electrical contact resistance requires the use of a dc four-wire dry circuit technique [143] (see Section 2.7.7) with an open circuit voltage of 10–20 mV and current limited to 100 mA. Current and voltage leads are clamped to the rider and flat. The computer monitors and samples contact resistance at preprogrammed numbers of cycles. A variety of testing devices have been described over the years [117,137,139–141,150].

A useful categorization of contact resistance stability in comparing materials in the work carried out by Antler divided fretting behaviors into three classes: *stable* or *acceptable* (Type III), *unstable* (Type I), and *intermediate* (Type II). These behaviors are defined in Table 7.1 according to both the numbers of cycles required to attain 10 mΩ and the value of the contact resistance after 10^5 reciprocating cycles. Figure 7.35 illustrates representative behaviors for various materials from a study [144] of frictional polymerization.

TABLE 7.1

Classification of Contact Resistance Behaviors

	Type I (Unstable)	Type II (Intermediate)	Type III (Stable)
Cycles to attain 10 mΩ	<5000	>5000	>10^5
Contact resistance by 10^5 cycles	>1 Ω	<1 Ω	<10 mΩ

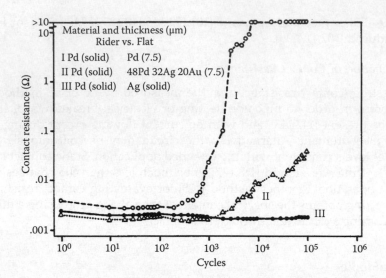

FIGURE 7.35
Contact resistance behaviors (typical) from fretting for 10^5 cycles at 50 gf and 8 Hz with a 20 μm wipe: curve I, unstable behavior, rise to high value after little fretting; curve II (intermediate), delayed rise and maximum value below 1 Ω; curve III, stable behavior; details of rider-flat combinations are given in the graph, and Table 7.1 defines the contact resistance behaviors. (After M Antler, *Wear* 81, 159–173, 1982 [144].)

7.3.5.2 Metals Having Little or No Film-Forming Tendency

7.3.5.2.1 Silver versus Silver

The earlier data of Figure 7.25 illustrated the mechanical wear properties of silver–silver contacts and showed that tarnish layers are easily displaceable. Similar effects are observed for large displacements in the fretting regime. Figure 7.36 illustrates the evolution with fretting time of the contact resistance between silver-plated copper surfaces in air, at contact loads of 20 N and 60 N for relatively large fretting displacements of 20 μm and 60 μm (i.e., *gross slip* regime) at a reciprocation frequency of 100 Hz [95]. The measurements were made for both tarnished and untarnished contacts. For untarnished contacts, Figure 7.36a, contact resistance increased gradually with fretting time. This is as expected since, as explained earlier in relation to adhesive wear, clean silver contacts tend to weld. In fretting motion characterized by relatively large displacements, mechanical damage in fretting was similar to that observed under full sliding conditions (Figure 7.25a), thus increasingly exposing the underlying oxidation-prone copper [95]. The rate of contact resistance increase was found to increase with fretting displacement but was relatively independent of contact load. The observations for tarnished silver were different as illustrated in Figure 7.36b. Initially, contact resistance decreased rapidly due to breakup and displacement of the tarnish layers and subsequent exposure of fresh metallic silver. Exposure of metallic silver led to contact sticking and thus to a sharp drop in contact resistance. Eventually, fretting displacements caused silver removal and subsequent exposure of the underlying copper, from which oxidized particle debris was generated. This eventually led to an increase in contact resistance.

The effects of fretting displacement amplitude on contact resistance in silver–silver contacts under relatively light contact loads (50 gf), compared with such effects in other materials at this contact load, are shown in Figure 7.37. These data show that the stability of silver is superior to that of most other electrical contact metals. Silver thus displays stable

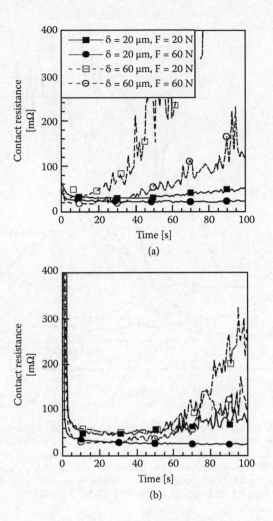

FIGURE 7.36
Typical trends of contact resistance during fretting (100 Hz) between copper surfaces plated with 4-μm-thick silver layers. Fretting was conducted at a contact load of 20 N or 60 N and with a displacement of 20 μm or 60 μm: (a) untarnished silver and (b) tarnished silver. (After A Kassman Rudolphi, S Jacobson, *Tribology Inter.* 30: 165–175, 1997 [95].)

(Type III) behavior when mated to itself. This stability stems from the attributes that silver is low wearing, oxide free, and does not form frictional polymers [104]. Because silver is prone to tarnish in atmospheres containing certain sulfur or chlorine compounds, it finds only limited application in electronic connectors. It is used successfully as a finish on aluminum busbar contacts [103] and is widely used as a finish on copper. As described in Chapter 8, the thickness of silver coatings on copper must generally be larger than about 5 μm in electronic connectors and 15 μm in high power connectors in order to mitigate the deleterious effects tarnish (e.g., silver sulfide) growth in a contaminated environment.

7.3.5.2.2 Gold versus Gold
Although solid gold has been found to form a trace of polymer when fretted in benzene vapor [104] or immersed in an oil [99], this contaminant has no detectable effect on connector contact resistance due to its small amount. Polymers have also been detected on gold

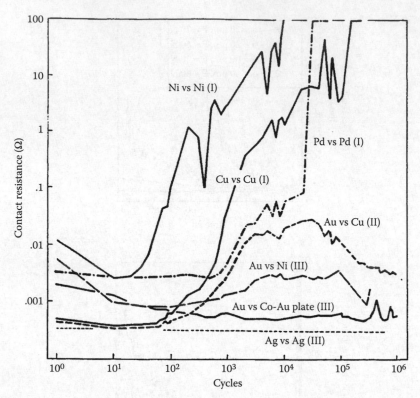

FIGURE 7.37

Contact resistance behaviors for various combinations of materials fretted at 50 gf with a 20 μm wipe at 4–8 Hz
(contact resistance behaviors according to Table 7.1). Type I (Unstable): solid nickel versus nickel plate 2.5 μm
thick on copper; solid copper vs. solid copper; solid palladium versus clad palladium 5 μm thick on nickel.
Type II (Intermediate): solid gold vs. solid copper. Type III (Stable): solid gold vs. nickel plate 2.5 μm thick on
copper; solid gold vs. 0.2% cobalt–gold plate 0.6 μm thick on nickel underplate and 2.5 μm thick on copper; solid
silver vs. solid silver. (After M Antler, MH Drozdowicz, *Wear* 74: 27–50, 1981–82 [99].)

surfaces in continuous sliding at 5–15 gf, as in instrument slip rings [119,145,146]. Gold surfaces when scratched have been found to catalyze the decomposition of adsorbed organic
compounds [147]. Figure 7.37 presents data for the fretting of solid gold against a 99.0%
pure hard gold plate. Fretting displacements in contacts consisting of gold sliding over
gold lead to stable (Type III) behavior [99].

7.3.5.3 Non-Noble Metals/Fretting Corrosion

7.3.5.3.1 Nickel versus Nickel and Copper versus Copper

Non-noble metals form fretting corrosion products that may have a significant effect on
contact resistance after few cycles of operation. This is illustrated in Figure 7.37 for nickel
and copper contacts mated to themselves [99]. Mechanical disruption of superficial initial
oxides, if present, occurs within 10 cycles with falling contact resistance, and then contact
resistance rises rapidly, attaining levels of 1–10 Ω in only 10^3 cycles.

7.3.5.3.2 *Tin-Base versus Tin-Base Surfaces and Electrodeposited*
Thick Sn60–Pb40 Solder versus Solder

Contact resistance problems due to fretting between surfaces coated with tin or a tin-base coating are often encountered. Tin-base coatings and solder are used in low-cost connectors for consumer applications [101] on the contacts of pluggable integrated circuits and their sockets [105,109,148] and in some computer products [149]. Examples of contact resistance traces during fretting motion in tin-plated copper contacts at room temperature and at a relative humidity (RH) of 55% are illustrated in Figure 7.38, where the tin layer thickness was 3 μm [150]. In the two contact resistance traces, the initial large resistance stemmed from the presence of a thin film of insulating tin oxide on the surface of the tin-plated copper alloy contact, which was removed after the first few reciprocating displacements. Following the initially large resistance decline, the gradual increase in contact resistance was attributed to the slow buildup of tin oxide between the contact surfaces.

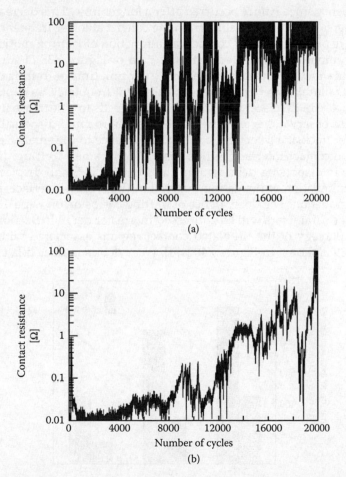

FIGURE 7.38
Change in contact resistance of tin-plated copper contacts as a function of fretting cycles obtained at different frequencies (a) 3 Hz and (b) 20 Hz. (tin layer thickness: 3 μm, amplitude: 25 μm, temperature: 22°C, humidity: 55% RH, normal load: 0.5 N (~50 gf), current: 0.1 A). (After YW Park et al., *Tribology International* 41: 616 - 628, 2008 [150].)

The subsequent rapid increase in contact resistance was due to the large accumulation of wear debris as illustrated schematically in Figure 7.34, consisting of tin metal and oxidized tin. The buildup of oxidized tin reduces the electrical conducting area.

The data of Figure 7.38a and b were recorded at a frequency of reciprocating motion of 3 and 20 Hz, respectively, but the wear path length was maintained at 25 μm. In these examples, the contact was deemed to have failed when the contact resistance reached a value of 100 mΩ. A comparison between Figure 7.38a and b reveals that failure was reached after about 20,000 cycles at 20 Hz, but only after about 7,000 cycles at a frequency of 3 Hz. Thus, the effect of increasing the frequency of reciprocating motion over a wear track of fixed length was to decrease the degradation rate of contact resistance per fretting cycle. However, there is another way of interpreting the fretting corrosion data. At 20 Hz, the total time of wear generation and oxidation of the tin surfaces was about 20,000/20 = 1,000 s, whereas this time interval increased to about 2,300 s at 3 Hz. Thus, the rate per unit time of contact degradation due to fretting corrosion actually decreased at the smaller displacement frequency since failure occurred after a longer time. This decrease may stem from effective dispersal of oxidized debris from the contact at lower frequencies.

The time to failure of the tin-plated contacts as a function of fretting motion frequency and displacement amplitude associated with the data of Figure 7.38 is summarized in Figure 7.39. The data show the clear decrease in rate per unit time of fretting corrosion (an increase in time to failure) with decreasing displacement frequency, as explained above. In addition, an increase in displacement amplitude also leads to a decrease in the fretting corrosion rate. These observations are attributed to the more rapid dispersal rate of wear debris from a longer track at a selected frequency, and hence to the smaller rate of debris accumulation in the contact interface in this case. Clearly, small fretting displacements cannot be effective in dispersing debris out of the wear track. Small displacements thus lead to quicker accumulation of insulating material in the contact interface, and hence to quicker contact degradation. The dependence of fretting corrosion on wipe distance, wipe frequency, and other parameters will be addressed in greater detail in Section 7.3.7.

The surface morphology of the tin-plated contact regions associated with the data of Figures 7.38 and 7.39 is shown in Figure 7.40 [151]. Wear debris is ejected laterally outside

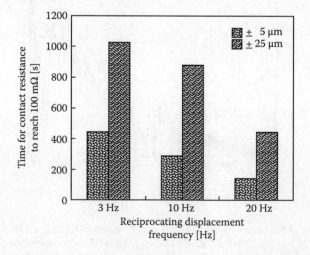

FIGURE 7.39

Time required for the contact resistance to reach a threshold value of 100 mΩ for track lengths of 5 μm and 25 μm at fretting frequencies of 3, 10, and 20 Hz (temperature: 22°C, humidity: 55% RH, normal load: 0.5 N (~ 50 gf), current: 0.1 A). (After YW Park et al., *Tribology International* 41: 616 - 628, 2008 [150].)

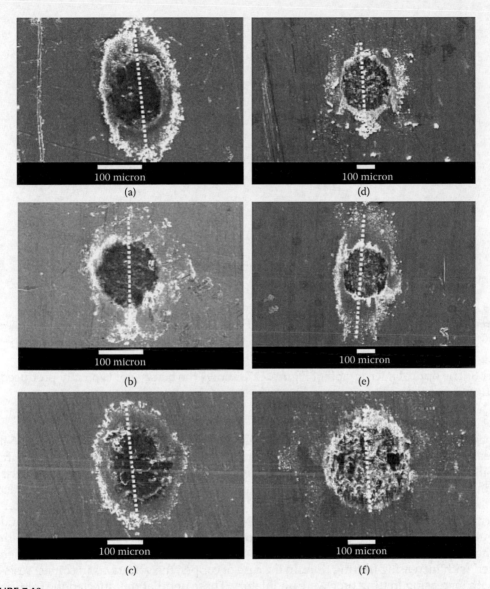

FIGURE 7.40
Surface morphology of the worn areas on the tin-plated contacts associated with the data in Figures 7.38 and 7.39, after 20,000 fretting cycles. The micrographs show the fretted region for the following test conditions: (a) amplitude: 5 μm and 3 Hz; (b) amplitude: 5 μm and 10 Hz; (c) amplitude: 5 μm and 20 Hz; (d) amplitude: 25 μm and 3 Hz; (e) amplitude; 25 μm and 10 Hz; and (f) amplitude: 25 μm and 20 Hz. The dotted line identifies the fretting direction. (After YW Park et al., *Surface and Coatings Technology* 201: 2181 - 2192, 2006 [151].)

of the fretted zone. Although the shape of the fretted zone was not affected significantly by the fretting conditions, the dimensions of the damaged contact zone increased in accordance with an increase in track length. The debris particles were found to consist of a mixture of tin metal, tin oxide, copper, and copper oxide, with the copper originating from the plated substrate. Electrical contacts coated with tin–lead are characterized by poor contact resistance performance in fretting (Type I, Unstable) and the performance of tin–lead versus gold is even worse (Figure 7.41). Reasons for this behavior are discussed later in

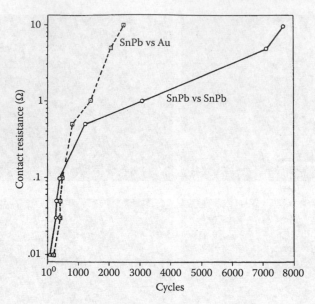

FIGURE 7.41
Contact resistance vs. number of fretting cycles for tin–lead vs. tin–lead compared with gold vs. tin–lead-plated contacts (50 gf; 8 Hz, 10 μm wipe). (After M Antler, *IEEE Trans Comp, Hybrids, Manuf Technol* 7:129–138, 1984 [180].)

Chapter 8. The optical micrograph of Figure 7.27a shows the black spot typically produced by fretting corrosion.

7.3.5.3.3 Other Base Metals Mated to Themselves

Many base metals such as aluminum [130,142,152] and various copper alloys [131,133,153,154] have been found to have unstable contact resistance due to fretting. Although the failure trends illustrated for tin-plated copper in Figures 7.38 and 7.39 are typical of the effects of fretting corrosion for many materials, the trends do not necessarily apply to all contact materials or to all fretting conditions. For example, Figure 7.42a illustrates the effects of fretting corrosion for monometallic contact interfaces comprising bare copper, brass, nickel, and aluminum where reciprocating motion was imposed at 16 Hz, at a contact load of 0.3 N with a fretting amplitude of 30 μm [153]. In all cases, contact resistance first decreased and then increased with increasing fretting time. In the case of copper, the variations in contact resistance with increasing fretting time were undulatory. These undulations affected the fretting time necessary to reach a selected high contact resistance. This is illustrated in Figure 7.42b where we note that the time interval to generate a contact resistance of 100 mΩ in the copper contacts associated with the data of Figure 7.42a first decreased, that is, the fretting corrosion rate increased, and then the time to 100 mΩ decreased. This decrease occurred as the fretting amplitude was extended beyond 40 or 60 μm, depending on the fretting frequency.

The data of Figures 7.38, 7.39, and 7.42 illustrate the observations that the effects of fretting corrosion on contact resistance are generally strongly influenced by the frequency and the amplitude of the reciprocating motion. Similar data associated with copper–copper and gold-plated copper contacts sliding in reciprocating motion will be described later. These observations are particularly relevant to the performance of electronic connectors in the transportation industry where fretting is an established connector degradation mechanism [155]. However, the data of Figures 7.38 through 7.42 also illustrate frequent observations that different non-noble contact materials do not necessarily behave in the same way

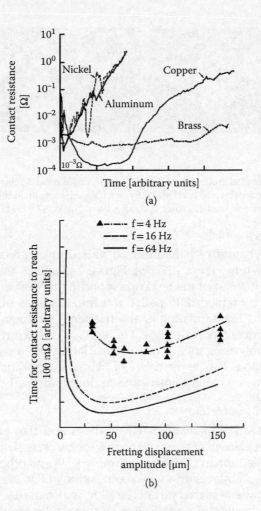

FIGURE 7.42
(a) Effect of fretting time on contact resistance in monometallic contacts of bare copper, brass, nickel, and aluminum, at a fretting frequency of 16 Hz, a contact load of 0.3 N and a displacement amplitude of 30 µm and (b) dependence on displacement amplitude of the time required for contact resistance to reach a value of 100 mΩ in copper–copper contacts. (After PH Castell et al., *Proceedings of the 12th International Conference on Electric Contact Phenomena*, Chicago, pp 75–82, 1984 [153].)

in fretting interfaces. Conflicting trends regarding the effects of fretting can only stem from differences in the mechanisms responsible for the generation of metal and oxidized metal detritus for different materials in sliding. On this premise, the effects of fretting corrosion on contact resistance may be expected to depend not only on chemical properties of the surfaces in contact, such as susceptibility to oxidation, but also on the physical properties such as metallurgical structure, surface hardness, and so on of the mating materials.

7.3.5.4 Frictional Polymer-Forming Metals

Polymer-forming materials include [104] the four platinum group metals having technological importance in contacts, platinum, palladium, rhodium, and ruthenium, and many of their alloys. In addition, several other metals yield polymers when fretted in benzene vapor including tantalum, molybdenum, and chromium [104].

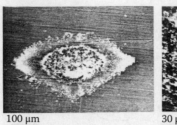

FIGURE 7.43

Wear track on the palladium flat after fretting for 10^5 cycles against a solid palladium rider. The debris consists of mixture of palladium wear particles and frictional polymer. The micrograph at the right provides a magnified view of the wear track shown on the left. (After M Antler, *Wear* 81, 159–173, 1982 [144].)

7.3.5.4.1 Palladium versus Palladium

The contact resistance behavior of palladium on palladium in reciprocating motion is unstable (Type I), as shown in Figure 7.37 [99]. Figure 7.43 shows the wear region on the flat palladium surface at the end of the test corresponding to Curve I in Figure 7.35 [144]. The visible loose debris contained only palladium and residues of polymeric material, that is, frictional polymer. The formation of the frictional polymer during fretting was responsible for the rise in contact resistance. The surface was worn only slightly, which indicates that frictional polymers are electrically insulating but provide both lubricating function and wear inhibition. The performance of platinum, rhodium, and ruthenium on themselves under the same test conditions would undoubtedly be Type I.

7.3.5.4.2 Palladium Alloys Mated to Themselves

Clad [156] and electroplated [157] palladium–silver alloys and clad palladium–gold–silver alloys [158] are potential replacements for gold connector contacts. These materials are also less reactive than pure palladium to chlorine-containing [159] and other [160] compounds in the atmosphere. DG R-156, an inlay with a wt% composition of Pd60Ag40 having a gold-rich surface with a diffusion gradient in the body of the alloy, is also used in connectors [161]. The low polymer-forming tendency of materials from this group, especially DG R-156 and the palladium–silver alloys from Pd70Ag30 to Pd30Ag70, compared with that of palladium, is significant [136,144]. Electroplated palladium–nickel alloys containing 70–85 wt% palladium are also widely used, but perform poorly in fretting [136,144,157,162–167]. Flash gold over-platings can improve the performance of these metals until the gold is lost by wear.

Promising substitute materials for hard gold as a contact finish also include palladium–cobalt with cobalt concentrations ranging from about 10 to 20 wt%, and palladium-nickel [167–169]. Figure 7.44 is adapted from investigations of these materials [169] and illustrates the evolution of the friction coefficient and contact resistance of sliding contacts with gold-flashed (GF) palladium–nickel, palladium–cobalt, and nickel–gold, all on a nickel under-plate and a gold flash. The data suggest superior wear properties for GF palladium–cobalt.

7.3.5.4.3 Mechanisms of Frictional Polymerization

The mechanisms of frictional polymerization are not well understood. Despite considerable speculation [147,170–173], the hypothesis advanced by Hermance and Egan [104] is plausible since it is consistent with most observations of the composition and physical properties of the polymeric materials [119,146]. This hypothesis states that polymer precursors are strongly adsorbed on the sliding surfaces of the catalytic metals and react among themselves to form high-molecular-weight, cross-linked solids. This simple model of polymerization

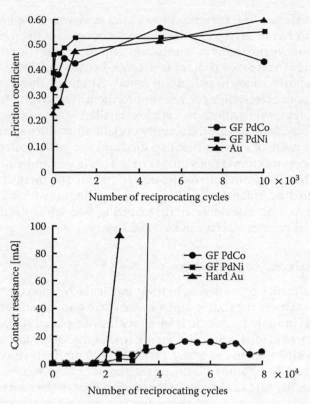

FIGURE 7.44
Evolution of the friction coefficient and contact resistance of sliding contacts with gold-flashed (GF) palladium–nickel, palladium–cobalt, and nickel gold, all on a nickel underplate and a gold flash. The data suggest superior wear properties for GF palladium–cobalt. (After G Holmbom et al., *Proc. 31st Annual IICIT Connector and Interconnection Symposium*, Danvers, MA, pp. 313–320, 1998 [168].)

[172–174] of adsorbed organic species is consistent with observations of the locations where polymeric material accumulates at the periphery of a contact undergoing reciprocating sliding motion. "Wiping" may also activate the surface to cause polymerization.

Poisons for conventional catalysis reactions are ineffective in stopping frictional polymerization, although the rate of reaction can be reduced when non-polymerizable low-molecular weight compounds are present that compete for surface sites with polymer precursors. The alloying of catalytic metals with non-catalytic elements reduces the population density of active catalytic sites and thus reduces susceptibility to frictional polymer formation. The particularly low susceptibility of Pd60Ag40 to frictional polymer formation [174] suggests a special role of silver in mitigating polymer growth.

Although most laboratory work has been conducted with artificial atmospheres containing high levels of pollutants, any open-air environment contains enough organic material for frictional polymers to form in sufficient quantity to cause electrical contact problems at some juncture in separable palladium–palladium contacts.

More recently [175,176], studies of relays indicate that palladium oxides, or related palladium compounds, can be formed on palladium–palladium contact surfaces operating in air in inactive (zero discharge) circuits when the level of organic pollutants is very low. Relay contacts experience both impact and slight wipe, but it appears that the wipe action is largely responsible for the oxidized palladium debris. The accumulated debris leads to a significant increase in contact resistance. The presence of organics and water vapor suppresses the reaction [177]. Oxidation has been hypothesized to occur because of the mechanical energy involved in contact closure, hence the term "mechanochemical reaction" for the phenomenon.

The passage of current stabilizes contact resistance because of Joule heating, which decomposes the palladium oxides to palladium metal. Although these studies were not directed to electronic connectors, they are relevant to them. The chemistry of this oxidation and of organic polymer formations is complex. Further studies are required to provide a full mechanistic explanation for the activity of palladium during fretting.

From an engineering point of view, the use of palladium and palladium alloys as contact materials for electronic connectors is not without risk. This is shown by palladium–nickel alloy electrodeposits that also form both oxides and frictional polymers of unknown composition [165]. Flash gold overplates on palladium and palladium–nickel are marginally effective in stabilizing contact resistance during fretting (see later). Thorough testing is required before catalytic contact metals can be considered for service.

7.3.5.5 Dissimilar Metals on Mating Contacts

Cold welding and transfer may occur during fretting, particularly when the metals are noble, ductile, or soft, but only where the sliding surfaces are clean and the surface layers are sufficiently thick. This was indicated earlier in relation with silver-plated surfaces (Figures 7.25 and 7.26). Otherwise, gross cold welding or even micro-welding at contact asperities is generally negligible in electrical contacts. Many practical systems involve contacts made of materials having different composition, form, or mechanical characteristics. In these situations, fretting will generally lead to changes in the composition of the contact interface from the composition that existed initially, and thus affect the ensuing contact behavior. Several important systems have been studied with contact finishes of sufficient thickness to preclude wear-out during testing. Important such systems are discussed below.

7.3.5.5.1 *Palladium versus Gold or Gold Alloys*

Gold and gold–silver surfaces sliding on palladium were found to be satisfactory [178,179] in early investigations of electrical contact applications, in contrast to all-palladium interfaces that produce frictional polymers. This is because pure gold [137], high-karat gold alloys [137], and gold–silver alloys [144] are softer than most forms of palladium. Consequently, transfer takes place primarily to the palladium surface, so that the system becomes an all-gold or all-gold alloy. Type III (Stable) contact behavior then occurs, identical to that in Figure 7.35 for the palladium versus silver couple.

Table 7.2 lists various materials that were mated to palladium in an extensive study [144] of fretting. Type III alloys include thick gold electrodeposits containing about 0.2 wt% cobalt, Au99Co1, Au98Ni2, Au70Ag30, Au69Ag25Pt6, and several silver–palladium and silver–palladium–tin alloys.

Figure 7.45 illustrates one of these systems, solid palladium versus gold, where the palladium surface became covered by transferred gold. There was little debris and no polymer. On the other hand, if a metal that is significantly harder is mated to palladium, transfer will be from the palladium surface, thereby making the system all-palladium with resultant Type I (Unstable) contact resistance behavior. An example of this type of material combination is Au75Cu25 (KHN_{25} = 24.3 × 10^2 N mm^{-2}) mated to palladium (KHN_{25} = 19.2 × 10^2 N mm^{-2}). Figure 7.46 shows the Au75Cu25 surface to which solid palladium has transferred and produced frictional polymers. Other examples from Table 7.2 of hard metals to which solid palladium transfers include Au55Ag39Cd3In3 and Pd60Ag40.

TABLE 7.2

Materials Mated to Solid Palladium (50 gf, 20 μm Wipe, 8 Hz, 10^5 Cycles)

Metal	Form[a]	Thickness (μm) and Behavior Type		
		I (Unstable)	II (Intermediate)	III (Stable)
Gold and gold alloys				
Au	Solid			>1000
Au	Clad		1.25	3.8, 7.5
Au	Clad		0.75/Pd	
Au	Clad		0.75/Pd60Ag40	
Au	Electrodeposit, 0.2%Co	0.6[b]		3.3
Au	Electrodeposit soft/Ru	0.5/0.5		
Au99 Co1	Clad			2.5
Au98 Ni2	Clad			2.5
Au90 Ni10	Clad		1.8	
Au75 Cu25	Clad	2.5		
Au70 Ag30	Clad			1.25, 7.5
Au69 Ag25 Pt6	Clad			3.8
Au55 Ag39 Cd3 In3	Clad	1.2[c]		
Au40 Pd36 Ag24	Clad	1.25	7.5	
Palladium and palladium alloys				
Pd	Solid	>1000		
Pd	Clad	1.25, 3.8, 5.0, 7.5		
Pd	Electrodeposit	2.5		
Pd80 Au20	Clad	1.25, 7.5		
Pd80 Ni20	Electrodeposit	2.5		
Pd60 Au40	Clad	1.25, 7.5		
Pd60 Ag40	Clad	1.25, 3.8, 7.5		
Pd50 Ni50	Electrodeposit	2.5		
Pd48 Ag32 Au20	Clad	1.25	3.8, 7.5	
Silver and silver alloys				
Ag	Solid		1.5	>1000
Ag	Clad		1.5	7.5
Ag92 Sn8	Clad			1.5, 7.5
Ag75 Pd25	Clad		1.5	7.5
Ag75 Pd23 Sn2	Clad		1.5	7.5
Ag75 Pd17 Sn8	Clad		1.5	7.5
Ag60 Pd40	Clad	4.0		
Other metals				
Rh	Solid	>1000		
Ru	Solid	>1000		
Ru	Electrodeposit	1.0		

[a] Electrodeposits on a Ni underplate, except where indicated by slash (/). Clad meals on Ni inter-liner, except where indicated by slash (/).

[b] 15,000 cycles to attain 10 mΩ.

[c] 20,000 cycles to attain 10 mΩ.

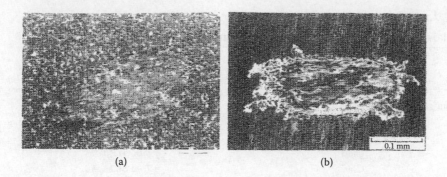

(a)   (b)

FIGURE 7.45
Specimens from fretting (a) solid palladium rider vs. (b) gold flat (3.8 μm cladding). Gold had transferred to the palladium, making the system all-gold; there was little wear debris. Type III (Stable) contact resistance behavior: 50 gf, 8 Hz, 20 μm wipe, 10^5 cycles. (After M Antler,*Wear* 81, 159–173, 1982 [144].)

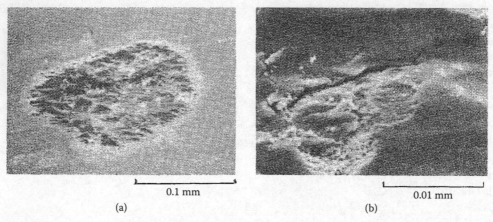

0.1 mm   0.01 mm
(a)   (b)

FIGURE 7.46
(a) Au75Cu25 flat (2.5 μm cladding) after fretting against solid palladium rider (palladium had transferred to the flat, making the system all-palladium); debris consisted largely of palladium particles and frictional polymers; Type I (Unstable) contact resistance behavior; (b) as (a), but at a higher magnification (wear of palladium occurs initially by adhesive transfer followed by delamination): 50 gf, 8 Hz, 20 μm wipe, 10^5 cycles. (After M Antler, *Wear* 81, 159–173, 1982 [144].)

Metal transfer does not always occur in one direction. If materials are not too unlike in hardness, say, when the hardness differs by about 10% or less, there may be significant metal transport both ways. In these instances, low contact resistance can be maintained if the composition of the rubbing surfaces from initiation of fretting remains predominantly that of non-polymer forming materials.

7.3.5.5.2 *Gold versus Nickel and Gold versus Copper*

The nickel–nickel and copper–copper systems are characterized by unstable contact resistance behavior (Figure 7.37) because of fretting corrosion [99]. However, the contact resistance behavior improves dramatically when these metals are mated to pure gold [99]. In the case of electrodeposited nickel ($KHN_{25} = 44 \times 10^2$ Nmm^{-2}), which is considerably harder than gold, transfer occurs nearly entirely from gold to the nickel with Type III (Stable) behavior, as shown in Figure 7.47. When solid gold is coupled to solid copper ($KHN_{25} = 11.7 \times 10^2$ Nmm^{-2}), there is some contact resistance instability (Figure 7.48) with Type II (Intermediate) behavior.

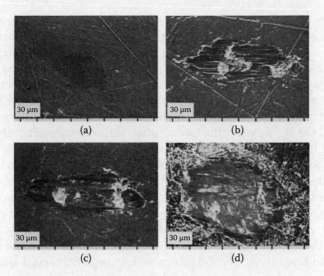

FIGURE 7.47

Scanning electron micrographs of worn flats from runs of increasing duration with solid gold riders on 0.05 μm Co–Au-plated copper flats: (a) 10^3 cycles; (b) 10^4 cycles; (c) 10^5 cycles; (d) 10^6 cycles. (After M Antler, MH Drozdowicz, *Wear* 74: 27–50, 1981–82 [99].)

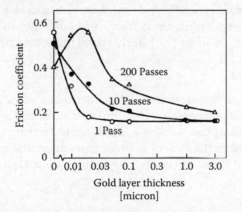

FIGURE 7.48

Variations of friction coefficient measured from cross-rod assemblies between two identical gold-plated palladium surfaces after 1 pass, 10 passes, and 200 passes, as a function of the thickness of pure gold coatings. The palladium consisted of a layer 0.5 μm thick deposited on 1.5 μm of nickel coated on phosphor-bronze rods. (After T Sato et al., *Proceedings of the Holm Conference on Electrical Contacts*, Chicago, pp 41–47, 1980 [63].)

7.3.5.5.3 Gold versus SnPb Solder

Gold versus electroplated Sn60Pb40 system is characterized by Type I (Unstable) contact resistance behavior [180] (Figure 7.41), because of the transfer of tin–lead to the gold, with resultant fretting corrosion of the base material on both surfaces. Transfer of hard gold does not occur because of the large difference in the bulk hardnesses of the contact metals (electroplated gold containing 0.2% cobalt, $KHN_{25} = 18 \times 10^2$ Nmm^{-2}; Sn60Pb40, $KHN_{25} = 1.2 \times 10^2$ Nmm^{-2}). Surprisingly, however, contact resistance begins to rise after fewer fretting cycles than in the all-tin–lead contact pair. This result is consistent with hardware experience [105,109,148] and may stem from the formation of hard intermetallic particles or patches in the contact regions due to gold–tin interdiffusion (see Chapter 1). This would affect the mechanics of disruption of surface films on the metals in contact [180,181], as tentatively explained below.

Oxide films fracture when subjected to mechanical deformation. These films thus fracture more easily when grown on a substrate that deforms relatively easily (e.g., a soft substrate) than on a hard surface [182]. During fretting between gold and tin–lead, the harder gold member becomes thinly coated with tin–lead that interdiffuses with the gold, thus forming a thin layer of tin–gold intermetallic compound [the deleterious effects of intermetallic layer growth on contact resistance are addressed in Sections 1.3.3 and 8.4.2]. Because intermetallic compounds are hard, their formation further hardens the gold surface. During fretting, the tin–lead on the two sliding surfaces subsequently oxidizes, but the crack network on the surfaces is less extensive than in the case of tin–lead on tin–lead because the oxide on the tin transferred to the gold cracks less readily. This leads to a more rapid increase in contact resistance than in the tin–lead on tin–lead case, as illustrated in Figure 7.41. This hypothesis is supported by important results of other work on gold-tin contacts [180]. In this work, gold surfaces thinly coated with tin–lead and mated to heavily oxidized tin–lead without wipe were characterized by a significantly higher contact resistance than a contact consisting of a thick tin–lead plating joined to the same oxidized tin–lead specimen. Additional data of the effects of fretting at tin–gold sliding interfaces will be shown later.

7.3.5.5.4 Other Systems

The performance of a number of other dissimilar metal contacts has been described, including aluminum versus brass plated with thick silver, cadmium, nickel, tin, or zinc [130], copper versus aluminum plated with silver or tin [102], silver-plated aluminum versus a copper–cadmium alloy plated with tin or silver [103], and several miscellaneous systems [101,104].

7.3.6 Wear-Out Phenomena

Fretting wear may lead to the penetration of a contact material of small thickness, as is often the case with electrodeposited and clad finishes. Contact resistance may change when this occurs, particularly if the substrate is subject to fretting corrosion or frictional polymerization. Examples of variations of contact resistance due to wear-out in sliding interfaces involving different types of coatings and substrate materials are given below.

7.3.6.1 Gold-Based Systems

The following examples illustrate typical observations of wear-out in a gold–copper sliding system. The investigations of interest focused on wear of solid gold riders sliding in reciprocating motion on copper flats and on gold plated copper flats [99]. The results of sliding on bare copper flats have already been shown in Figure 7.37 and indicate that contact resistance increased with increasing reciprocating cycles, reaching the beginning of a maximum plateau after about 2000 cycles and subsequently decreasing after 10^4–10^5 cycles from 50–100 mΩ to about 4 mΩ after 10^6 cycles. In contrast to these observations, the reciprocating sliding of solid gold riders on copper flats plated with a 0.6 μm thick hard gold layer containing 0.2 wt% cobalt, led to a contact resistance that remained low near 1 mΩ until completion of 10^6 reciprocating cycles [99]. In these two sliding systems, analyses using scanning electron microscopy and energy dispersive X-ray (EDX) spectroscopy [193] revealed that the sliding surfaces consisted of a mixture of gold, copper, and copper oxide.

The contact resistance behavior just described was explained from examinations of sliding surfaces after various numbers of reciprocating cycles involving a solid gold rider sliding on copper plated with only 0.05-μm Co–Au [99]. The thin gold layer was intended to provide good imaging contrast between worn and unworn regions of the copper surface after sliding, in examinations using scanning electron microscopy. Electron micrographs of worn areas on the copper flat, after various number of reciprocating cycles, are shown in Figure 7.47 [99]. With such thin gold layers on the copper, the wear data were essentially identical with those of fretting on unplated copper but the contact resistance did not rise to 50–100 mΩ level. Consistent with the data of Figure 7.37, the worn surface after 10^3 cycles shown in Figure 7.47a was only lightly burnished, with little gold transfer to the copper and a slight increase in contact resistance probably due to copper oxide formation. Figure 7.47b shows a surface after 10^4 cycles (contact resistance had increased to 6 mΩ). Adhesive transfer had clearly occurred with the original thin gold coating having been worn significantly from the copper, and the surface scratches of the unworn contact having been obliterated. The rider and the flat had identical appearances. The few loose particles evident on the flat were large and found to contain both copper and gold. Figure 7.47c shows the flat surface after 10^5 cycles (contact resistance had increased to 3.3 mΩ). The flat was smoother than that shown in Figure 7.47b, which indicates that adhesive transfer between the sliding surfaces had diminished significantly. Figure 7.47d shows a worn copper surface after very long sliding, 10^6 cycles (contact resistance, 3.0 mΩ). At equilibrium, the worn areas of the two sliding surfaces consisted of a mixture of gold, copper, and copper oxides and copious loose debris had formed. Relative to the concentration of other materials, the gold content on the worn copper surface was found to increase with the thickness of the original gold deposit on the copper. The worn area on copper plated with 0.6 μm of hard gold had an appearance similar to that of Figure 7.47d after 10^6 cycles, although the resistance had remained small throughout the entire fretting run.

The micrographs of Figure 7.47 show that gold remained in the contact area after 10^6 cycles, thus contributing to maintaining a low contact resistance (3 mΩ). The data of Figure 7.37 corresponding to solid gold sliding over copper also correspond to a system where the gold supply was copious and, thus, capable of maintaining a low contact resistance for considerably longer than 10^6 cycles.

7.3.6.2 Palladium-Based Systems

As will be described in detail in Section 7.4.2, thin gold layers or small gold particle dispersions can be effective lubricants. The lubrication of palladium layers by pure gold as measured by Sato et al. [63] is reproduced in Figure 7.48. These data were obtained at a contact load of 300 gf from crossed phosphor-bronze cylinders electroplated successively with a 1.5-μm-thick nickel underplate, palladium to a thickness of 0.5 μm, and gold to the thickness shown in the graph. The data showed that lubrication effectiveness is optimal for a gold thickness of ~0.03–0.2 μm if the number of passes is relatively small. This is consistent with other data for the lubrication by gold on palladium and palladium alloys [63,183–185] which, by themselves, are characterized by low resistance to adhesive wear. These observations have stimulated the use of palladium-based finishes that are less expensive than the equivalent thickness of plated hard golds.

If the usual connector durability requirement of up to 200 insertions was the only requirement for resistance to fretting, then flash thicknesses of gold of about 0.2 μm would be adequate for lubrication on palladium. Under practical situations, the number of fretting cycles is generally much greater than 200. As illustrated in Figure 7.48, it is generally found that palladium mated to palladium plated with pure gold (0.05 μm thick) is characterized by Type I (Unstable) contact resistance behavior in fretting at a load as small as 50 gf with a 20 μm wipe. The thin gold layer is readily penetrated for a number of fretting cycles exceeding 200, as suggested by the data of Figure 7.48. Figure 7.49a illustrates the

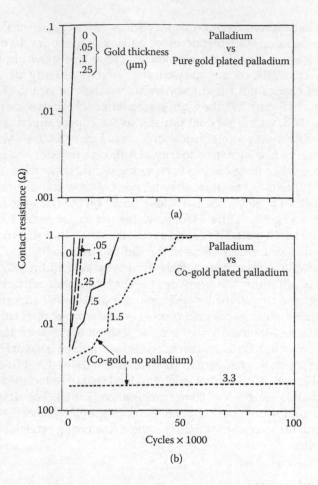

FIGURE 7.49
Contact resistance vs. number of cycles for solid palladium mated to palladium plate 1.5 μm thick on 1.25 μm of nickel underplate on a hard copper alloy substrate: (a) palladium plate, uncoated or coated with pure gold electrodeposit that is 0.05–0.25 μm thick; (b) palladium plate, uncoated or coated with 0.2 wt% cobalt–gold electrodeposit that is 0.05–0.5 μm thick (results from fretting solid palladium against electrodeposited cobalt–gold 1.5 and 3.3 μm thick on a nickel underplate are included for comparison) (50 gf; 8 Hz, 20 μm wipe). (After M Antler, *IEEE Trans. Components, Hybrids, and Manufacturing Technology* CHMT-8: 87–104, 1985 [195].)

relationship of contact resistance to thickness of electroplated pure gold on a palladium contact mated to solid palladium [186,195]. Even gold layers that were as thick as 0.25 μm altered the contact resistance behavior little from that of palladium versus palladium. Hard golds on palladium (Figure 7.49b) and palladium alloy platings are slightly more durable during fretting than pure gold. The data of Figure 7.49 show clearly the devastating effects on contact resistance of wear-out of the gold layer. In using gold as a protective/lubricant layer on palladium, it is thus important to be aware of durability limit of gold electodeposits.

7.3.6.3 *Tin and Tin–Lead Alloy Systems*

Because of the importance of tin as a contact material, particularly in the automotive industry, fretting in tin-plated contacts has been a topic of intense investigation over many years [80,106,118,127,137,150,151,154,155,184,187–192]. The surface morphology of tin-plated copper contacts after fretting has already been illustrated in Figure 7.40 [151]. The probable stages of fretting wear were as follows [150,151]: (i) cracking and partial removal of tin oxide to allow metal-to-metal contact and an initially low contact resistance, (ii) generation of tin oxide debris particles in the contact zone as illustrated in Figure 7.34 simultaneously with slow re-oxidation of the underlying tin, leading to an increase in contact resistance, (iii) steady removal of tin by a combination of adhesive wear and abrasion wear by hard oxide debris articles simultaneously with re-oxidation of tin, thus leading to a further increase in contact resistance, (iv) complete removal of the tin by adhesive and abrasive wear, thus leading to the presence of large amount of oxide debris and an ensuing large contact resistance increase, and (v) exposure of the underlying copper with ensuing fretting corrosion of the copper and the generation of copper oxide debris leading to catastrophic contact failure.

Investigations of fretting in sliding contacts involving tin–lead alloys have been made but have been limited in scope. A limited scanning electron microscope-EDX analysis [193,194] was made of contact surfaces for a system involving plated Sn60Pb40 contacts mated to themselves or to hard gold on nickel underplate. The substrates were brass. During extended fretting runs, contact resistance rose, then fell sharply, and finally increased again. Changing compositions of the surfaces due to wear may have been responsible for these results. An interesting finding was limited transfer of gold to the original tin–lead opposing contact. This was attributed to the much harder copper–tin intermetallics [78] that formed spontaneously on that member on aging, and that became a contact surface after the tin–lead had worn through.

7.3.6.4 Role of Underplate and Substrate

The beneficial effects of an underplate, particularly a nickel underplate, in increasing resistance to wear have been described in Section 7.2.7 (Figures 7.15 through 7.18). The durability of a thin plated layer in sliding contact over a long track is determined in large part by the composite hardness of the system, as indicated in Equation 7.1. For example, the incorporation of a nickel plate characterized by a bulk hardness of about 44.1×10^2 N mm^{-2}, between a gold deposit and a substrate that is considerably softer such as copper (about 11.8×10^2 N mm^{-2}) is advantageous as was illustrated in Figure 7.15 [59]. The same consideration applies to fretting wear, and Figure 7.50 is an example [99] of the improvement

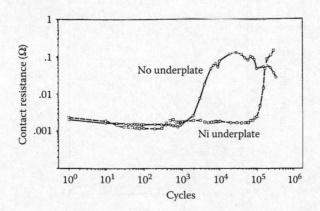

FIGURE 7.50

Contact resistance vs. the number of fretting cycles for solid Au70Ag30 alloy against cobalt–gold-plated copper 0.05 μm thick with and without nickel underplate (2.5 μm). The contact resistance behavior is Type I (Unstable) when cobalt–gold-plated copper is used and Type III (Stable) with the sample having a nickel underplate (50 gf, 8 Hz; 20 μm wipe). (After M Antler, MH Drozdowicz, *Wear* 74: 27–50, 1981–82 [99].)

in contact resistance for Au70Ag30 versus 0.05 μm cobalt–gold-plated copper with and without nickel underplate. When an underplate is not used, wear-out of the gold occurs quickly and fretting corrosion ensues (Type I, Unstable contact resistance behavior). With a nickel underplate, contact resistance remains stable for about 100 times more fretting cycles (Type III, Stable contact resistance behavior). It has been shown [99] that Au70Ag30 does not transfer to nickel as readily as does solid gold, and so following wear-out of the 0.05-μm cobalt–gold layer beyond 10^5 cycles, fretting corrosion of nickel was probably responsible for the insulating materials at the interface which caused the contact resistance to rise. The improvement due to nickel underplate has also been found [99] for larger thicknesses of cobalt–gold-plated layers than those cited here.

7.3.7 Parametric Studies

The rate of electrical contact degradation due to fretting corrosion is influenced by the magnitude of contact parameters such as the amplitude and frequency of fretting displacements, the contact force, the magnitude of electrical current through the contact, environmental factors such as humidity, and other factors. The effects on contact resistance of the amplitude and frequency of the fretting motion in tin–tin contacts have already been illustrated in Figures 7.38 through 7.40. Similar effects on other contact materials, and the influence of other contact parameters in sliding interfaces involving a variety of materials, are reviewed in this section.

7.3.7.1 Cycle Rate

Contact resistance changes due to fretting corrosion and frictional polymerization can be expected to be influenced by displacement frequency because both contact degradation mechanisms involve rate-dependent surface chemical reactions. Figure 7.51 shows variations in contact resistance for copper versus copper and for gold versus gold-flashed copper systems in fretting through a range of frequencies. The test conditions used in this work [99]

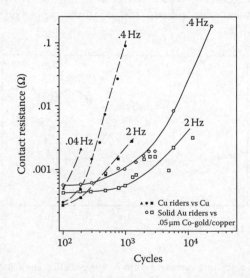

FIGURE 7.51
Effect of the cycle rate on contact resistance: ▲, ●, ■, copper vs. copper; ○, □, solid gold vs. 0.05 μm cobalt–gold on copper; 50 gf, 8 Hz, 20 μm wipe. (After M Antler, MH Drozdowicz, *Wear* 74: 27–50, 1981–82 [99].)

consisted of 50 gf (~0.5 N) load, 20 μm wipe, and a cycle rate ranging from 0.04 to 2 Hz. The contact resistance trends in Figure 7.51 are similar to those noted in the case of tin–tin contacts in Figures 7.38 through 7.40 where the rate of increase in contact resistance (per unit time) was found to be smaller at the smaller cycling frequencies, and thus the total time interval to reach a selected value of resistance (say to failure) actually increased at smaller frequencies. In the copper–copper contacts in Figure 7.51, it may be verified that the time required to reach a selected contact resistance increased with a decrease in cycling frequency.

As stated earlier with respect to tin–tin, the contact resistance behavior is explained as follows: the major cause for the behavior illustrated in Figure 7.51 is in the kinetics of oxidation whereby at lower frequency, oxide thicknesses can grow larger between wipes, but the newly formed oxide is more effectively displaced during successive wipes at low frequency than at high frequency. The net result is a slower increase in contact resistance (per unit time) at low cycle frequencies. In the case of the gold–copper pair, contact resistance degradation is delayed because of the requirement that the gold coating be worn through to underlying copper before oxides can form, but the trends in the cycle rate curves are the same as in the all-copper case,

Although the effects of fretting corrosion on contact resistance are influenced by the frequency of the reciprocating motion, the effect of cycle rate varies with the combination of materials in the sliding couple. Conflicting trends regarding the effects of fretting stem from the differences in oxidation rates and in the mechanisms responsible for the generation of metal and oxidized metal detritus for different materials in sliding.

The degradation of palladium–palladium contacts due to fretting (Figure 7.37) stems largely from frictional polymer formation. In these cases, degradation rate is expected to be affected not only by displacement frequency and amplitude in the way illustrated in Figures 7.39 and 7.51, but also by the concentration of polymer-forming contaminants in the service or test environment [167]. Figure 7.52 shows the effect of benzene vapor concentration on frictional polymer formation [104]. Frictional polymer yield was determined with palladium–palladium contacts as a function of benzene vapor concentration at reciprocating frequencies of 6 Hz and at 120 Hz. Clearly, the rate dependency for generation of polymer is related to the rate at which benzene molecules reach the surface when benzene concentration ranges from tens to hundreds of parts per million. The data indicate that the reaction rate of benzene vapor with palladium on the wear track decreased with

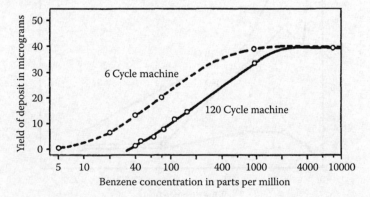

FIGURE 7.52
Yield of frictional polymer as a function of concentration of benzene vapor and wipe frequency. Palladium vs. palladium: 30 gf, 6 Hz and 120 Hz, 170 μm wipe, 4×10^6 cycles. (After HW Hermance, TF Egan, *Bell Syst Tech J* 37: 739–777, 1958 [104].)

increasing cycling frequency. At the high-level benzene vapor concentration of 10^4 ppm, the polymer formation rate reached a plateau. Also, these data show that there is a minimum concentration of vapor below which polymer could not be detected.

7.3.7.2 Wipe Distance

The data of Figures 7.39 and 7.42b illustrated observations that the amplitude of fretting displacements influences contact resistance but that these effects may lead to conflicting contact resistance trends, that is, the time required to reach a selected contact resistance value generally increases but it may also decrease with increasing fretting amplitude. Conflicting trends can only stem from differences in the mechanisms responsible for the generation of metal debris and oxidized metal detritus for different materials in sliding. On this premise, the effects of fretting displacement amplitude on contact resistance may be expected to depend on chemical properties of the surfaces in contact, such as susceptibility to oxidation, and on physical properties such as metallurgical structure and surface hardness, as well as on the thickness of the mating materials. Contact resistance trends also depend on the contact resistance value that defines the "time or number of cycles to reach this value."

Figure 7.53 shows the number of cycles to reach a contact resistance of 0.1 Ω, 1 Ω and 10 Ω from fretting of tin–lead on the same alloy, plotted against wipe length from 10 to 240 μm [180]. On the 0.1 Ω curve, the number of wiping cycles to reach 0.1 Ω was found to increase with increasing wipe amplitude. This initial trend is thus similar with the data of Figure 7.39 for tin–tin contacts. For displacements that far exceeded those indicated in

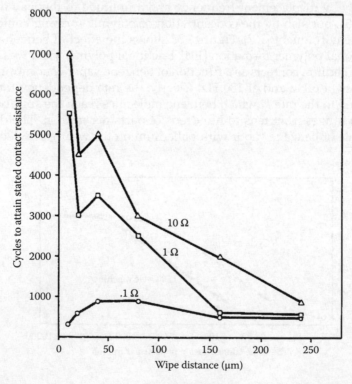

FIGURE 7.53
Number of fretting cycles vs. distance of wipe for Sn60Pb40 vs. Sn60Pb40 required to attain 0.1 Ω, 1 Ω, and 10 Ω: 50 gf, 8 Hz. (After M Antler, *IEEE Trans Components, Hybrids, Manuf Technol*, 7: 129–138, 1984 [180].)

Figure 7.39, the time required to reach 0.1 Ω (and 1 Ω, 10 Ω) increased with increasing wipe distance in conflict with the data of Figure 7.39. It is likely that other factors such as metallurgical structure and hardness were responsible for the difference in the trends of contact degradation rate for large wipe displacements.

Results also showing a decrease in the time required for contact resistance to increase to 0.1 Ω, as the wipe amplitude was increased, were obtained in frictional polymerization studies with palladium–palladium contacts [125,174]. The data are shown in Figure 7.54 and were collected under dry and lubricated conditions under test conditions similar to those associated with the data of Figure 7.53. The data of Figure 7.54 are easy to understand since the formation of frictional polymers aligned along the sliding direction is presumably enhanced with increased reciprocating displacement. The typical appearance of the frictional polymers formed under these conditions is shown in Figure 7.55a. Frictional polymer formation rates may be greatly enhanced by depositing a thin layer of an organic material on one of the palladium surfaces before reciprocating sliding, as shown in Figure 7.55b.

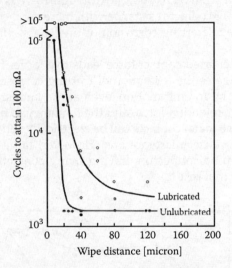

FIGURE 7.54
Number of fretting cycles vs. distance of wipe for palladium–palladium contacts required to attain 100 mΩ: 50 gf, 8 Hz. Data for unlubricated surfaces and for flats lubricated with polyphenyl ether obtained by immersion in a 5 wt% solution of 1,1,1-trichloroethane. (After M Antler, *ASLE Trans* 26(3): 376–380, 1983 [174].)

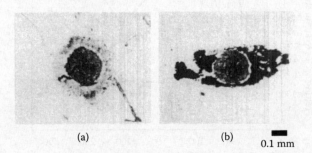

FIGURE 7.55
Optical micrograph of frictional polymers associated with the data in Figure 7.54: (a) unlubricated flat and (b) flat lubricated with a thin film of synthetic hydrocarbon obtained by immersion and withdrawal from 0.5 wt% solution in 1,1,1-trichloroethane. (After M Antler, *ASLE Trans* 26(3): 376–380, 1983 [174].)

The fretting data presented above makes it clear that contact resistance degradation rates due to the amplitude of fretting displacements cannot be predicted unambiguously. However, the data also suggest that if the contact resistance increase is to be maintained smaller than about 100 mΩ, fretting displacement amplitudes must be maintained smaller than a few to several micrometers.

7.3.7.3 Force

The rate of contact degradation due to fretting corrosion generally decreases with increasing contact load [150,180]. The influence of an increased force on contact resistance in fretting interfaces can be understood if the larger force enhances penetration of fretting corrosion products or of frictional polymers. In addition, a larger area of mechanical contact subjected to interfacial displacement is more effective in dispersing surface insulating layers, and thus is less prone to entrapping electrically insulating particle debris than a relatively small contact area at smaller loads. Figure 7.56 presents the results [180] of tests with tin–lead fretted on itself from 30 to 300 gf. Although there is data scatter, the trend indicates increasing stability of contact resistance with increasing contact force. However, even at 300 gf, the resistance to fretting corrosion of tin–lead sliding on tin–lead is relatively low.

In a sliding interface, an increased contact force leads to a larger friction force, which in turn resists motion and mitigates the formation of fretting corrosion products. The data of Figure 7.57 were obtained by forcing motion, even at the largest loads. In a connector, raising normal load increases retention force, and the tendency to fret is thereby reduced. This is why highly loaded base metal contacts can be electrically satisfactory when stresses occur that might otherwise cause micromotions. However, large normal loads cannot usually be designed into wiping connectors having many contacts because of the large engagement forces and wear that result.

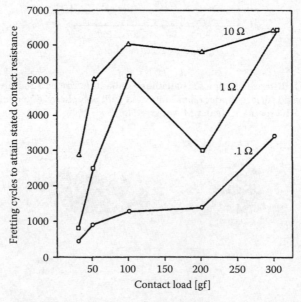

FIGURE 7.56
Number of fretting cycles for Sn60Pb40 vs. Sn60Pb40 required to attain 0.1, 1, and 10 Ω as a function of load: 20 µm wipe, 8 Hz. (After M Antler, *IEEE Trans Components, Hybrids, Manuf Technol*, 7: 129–138, 1984 [180].)

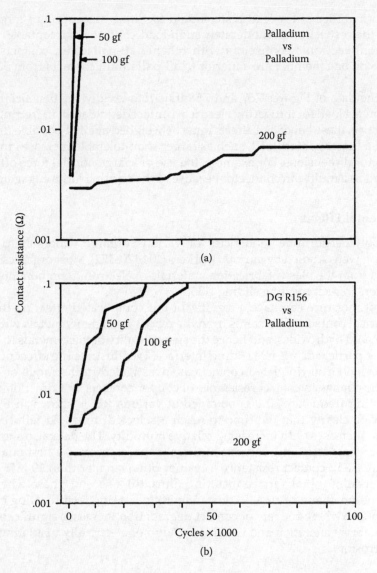

FIGURE 7.57
Contact resistance from fretting of (a) solid palladium and (b) DG R-156 mated to palladium electroplate as a function of load. The palladium electroplate was 1.5 μm thick on 1.25 μm nickel. Data are worst-case results from four runs at each condition. Contact resistance was stable at 200 gf, especially with DG R-156, compared to runs at lower load: 20 μm wipe, 8 Hz. (After M Antler, *IEEE Trans. Components, Hybrids, and Manufacturing Technology* CHMT-8: 87–104, 1985 [195].)

The effect of load on the contact resistance of palladium-based contacts is presented in Figure 7.57 for palladium–palladium and DG R-156-palladium [195]. The data describe the results of four runs made at a contact force, respectively, of 50, 100, and 200 gf with the two pairs of materials. Figure 7.57a shows Unstable Type I contact resistance behavior at 50 and 100 gf, and Stable (Type III) behavior at 200 gf. Thus, the palladium versus palladium system is similar to the tin–lead versus tin–lead couple at low load, yet it is superior at high (200 gf) load. This suggests that the resistance to penetration of frictional polymers is appreciably smaller than that of surface oxides on tin–lead, although the mechanical deformability of tin–lead plate should facilitate disruption of foreign films. Thus, provided the restriction of moderately high-load contacts is acceptable, it may be possible to design electronic connectors with reliable all-palladium contacts. Systems having DG R-156 on one member are superior to all-palladium contacts from 50 to 200 gf (Figure 7.57b).

In summary, the data of Figures 7.57 and 7.58 illustrate vividly the beneficial effects of increasing the contact load for mitigating electrical contact degradation in fretting contacts. However, in practice, these beneficial effects must be weighed against possible undesirable consequences of a larger contact load, such as larger-than-tolerable increases in retention force and connector dimensions. For example, the use of a large normal force often cannot be accommodated in a multi-pin connector because of the ensuing large engagement force.

7.3.8 Environmental Effects

The frictional polymer-forming tendencies of various organic vapors in palladium–palladium fretting were studied by various workers [104,170–172]. Most organics including those that outgas from the plastic fabrication materials of electronic components can form frictional polymers in reciprocating sliding [136].

The presence of moisture can have a significant but complicated effect on the fretting behavior of electrical contacts. This stems from the reactivity of many metals with water to form oxides form and hydroxides, and hence the susceptibility of these metals to form electrically insulating particle debris in a sliding interface. We illustrate the effect of humidity on the fretting behavior of tin-plated contacts as investigated in the range of dry air to 80% RH [196]. The change in contact resistance of copper contacts as a function of fretting cycles (reciprocating frequency 50 Hz) obtained at various RH is shown in Figure 7.58. These data indicate clearly that the time to reach a selected threshold failure value for contact resistance increases with increasing relative humidity. The data also show that the evolution of the contact resistance consists of three main features: (i) a first phase characterized by a decrease in contact resistance to values often smaller than 10 mΩ and stemming from dispersal of initial surface insulting films, (ii) a second phase where contact resistance remains small and relatively constant with increasing reciprocating cycles, and (iii) a last third phase where the rate of contact degradation increases significantly due to increased wear debris generation and ensuing buildup of electrically insulating material, and galvanic corrosion.

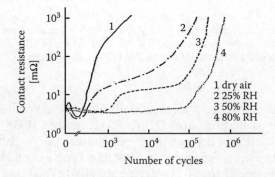

FIGURE 7.58
Contact resistance versus number of fretting cycles in copper–copper contacts under various conditions of relative humidity; 0.5 N, 50 Hz, 10µm fretting amplitude. (After JF Bruel et al., *Proc. 14th Int. Conf. on Electrical Contacts*, Paris, France, 1988, pp. 219–223 [196].)

The increasing delay in the onset of the last phase with increasing relative humidity is attributed to the increasing lubrication effect of moisture, which reduces adhesion and thus mitigates mechanical wear. As already mentioned, the rapid electrical degradation in the last phase under each of the conditions 1, 2 and 3 in Figure 7.58 stems from the access of debris from galvanic corrosion products into the electrical contact regions. This increases surface separation and introduces material of higher electrical resistivity into the sliding interface.

It was earlier shown (Figure 7.37) that solid silver does not degrade in air due to fretting. However, if the environment contains a high concentration of gaseous pollutants which can tarnish silver, such as hydrogen sulfide and chlorine at elevated relative humidity, then films may form so rapidly that fretting corrosion occurs. This was demonstrated in a laboratory test at a low cycle rate and relatively small contact normal load [92] using the Battelle Class III flowing mixed gas atmospheric test (see Section 3.3.5). At higher loads and higher fretting velocities, there was little or no effect from these pollutants [116].

7.3.9 Thermal

The extent of fretting corrosion can be expected to vary with temperature if there is a change in the physical properties of the contact metals or protective lubricant coatings that may be used. A significant change in oxidation rate or other film-forming reactions will also affect the contact behavior. This has been of interest in the automotive sector for under-hood connectors. Figure 7.59 shows the number of cycles to failure [197] at particular test conditions for tin-plated contacts. It was concluded that with rising temperature to about 60°C, an increase in the oxidation rate of tin was responsible for the reduction in cycles to failure. At a still higher temperature of 110°C, the tin softened. This resulted in an enlargement of the contact area, confirmed from an examination of samples after test, with subsequent improvement in fretting corrosion behavior. The illustrative data in Figure 7.59 confirm observations of the effects of increasing temperature on fretting corrosion in tin contacts by other workers [122,197–200] and in more recent investigations [150].

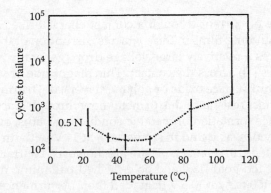

FIGURE 7.59
Cycles to failure as function of temperature as derived from fretting corrosion characteristics at 50 gf. Failure is defined by the contact resistance attaining 20 mΩ for the first time; 20 μm track, 10 Hz; 3-μm tin plate on C65400 copper alloy. (After A Lee et al., *Proceedings of the IEEE Holm Conference on Electrical Contacts*, pp 87–91, 1988 [197].)

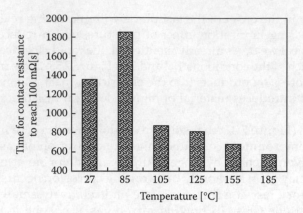

FIGURE 7.60
Time required for the contact resistance to reach a threshold value of 100 mΩ at various temperatures in tin-plated copper contacts; 0.5 N, ±90 μm track; 10 Hz; humidity: 55% RH; 0.5 A.). (After YW Park et al., *Tribology International* 41: 616–628, 2008 [150].)

The formation of intermetallic compounds (IMCs) between tin and the substrate material, such as copper or a copper alloy, may delay the deleterious effects of fretting corrosion at elevated temperatures if the intermetallics grow as a reasonably uniform layer. This effects stems from the high hardness of Cu–Sn IMCs formed at elevated temperatures. A thick intermetallic layer increases the number of cycles required for the rider to penetrate the intermetallic compound and reach the underlying substrate [190]. However, IMCs are notorious for their inferior electrical conductivity properties. The growth of too thick IMCs thus leads to an increase in contact resistance, which may offset the benefits of high hardness. Such effects are shown in Figure 7.60 for tin-plated copper contacts, where the number of reciprocating cycles reach to 100 mΩ, as a function of the temperature at which fretting was carried out between two tin-plated copper surfaces [150]. These data indicate that tin-plated contacts are increasingly less reliable for applications at temperature exceeding about 100°C [189,190,192].

7.3.10 Effect of Current

One effect of the passage of current through a contact characterized by high resistance due to the presence of insulating films in the contact is disruption of the films. Disruption of insulating films requires a relatively large voltage drop (generally a few to about 20 V for many insulating layers [3]) across the contact. This magnitude of voltage drop causes electrical breakdown of thin surface oxide or polymer layers, or "fritting." "Fritting" leads to a rapid decrease in contact resistance due to metal flow through cracks in the disrupted insulating films, and hence formation of metallic conducting paths, even without appreciable current. Such behavior was found in studies at 5 at 14 V with tin [199] plate and at 1 and 2 V with tin–lead-plated [201] contacts. There are, however, contrary results in limited studies with electrodeposited gold [138] and gold-flashed palladium nickel [165] contacts in which a small current, 100 mA, at 10.5 V degraded their performance. Still other experiments [99] showed no effect at dry circuit currents and voltages (100 mA, 20 mV maximum) for some base and noble metal contacts. The range of observed effects of current on contact behavior in fretting are explained in terms of the dependence of the *a*-spot temperature on the voltage drop across the contact.

The temperature T_M of a Joule-heated a-spot is determined by the voltage drop V across the contact in accordance with the relation (see Chapter 1):

$$T_M = \sqrt{\frac{V^2}{4L} + T_0^2} \tag{7.4}$$

where L is the Lorenz constant (2.45×10^{-8} V^2 K^{-2}) and T_0 is the bulk temperature of the bodies in contact (i.e., far from a-spots). Concomitantly, T_M is determined by the current. Equation 7.4 applies to all conductors due to the Wiedemann–Franz law which relates the electrical resistivity and the thermal conductivity of metals. The variation of T_M with V is shown in Figure 7.61 for values of T_0 of 27°C and 100°C. For purposes of illustration, we now consider the sequence of events in contact spots as current is increased across an electrical interface.

For a contact initially at room temperature, an increase in current leading to a voltage drop of less than about 40 mV leads to an a-spot temperature of about 50°C according to Figure 7.61. Such a voltage drop across the contact will have a negligible effect on contact resistance for many contact materials but may lead to a slight decrease in contact resistance in interfaces involving metals such as tin and indium since a-spots in such interfaces may deform relatively rapidly near 50°C, thus increasing the area of true contact. However, because the oxidation rate of non-noble metals increases rapidly with increasing temperature, the imposition of a steady voltage drop of 40 mV involving metals such as tin will activate degradation due to oxidation. In this illustrative example, the dominance of the creep mechanism over oxidation will lead to acceptable contact performance whereas dominance by oxidation will eventually cause catastrophic contact failure. In practice, the dominance of one mechanism over the other will be determined by external factors such as contact load, fretting displacements, and so on.

As current is increased and the voltage drop across a contact exceeds 40 mV, the temperature of a-spots quickly approaches the softening point of the materials in contact. According to the data in Chapter 1, contact asperities of most materials (other than tin) reach their softening temperature at a voltage drop of between 70 mV and 110 mV, corresponding to contact-spot temperature ranging from about 90°C to 150°C. In the vicinity of the softening temperature, the contact asperities of metals flow relatively rapidly and lead to an increase in area of true contact. However, the nefarious effects of thermally activated

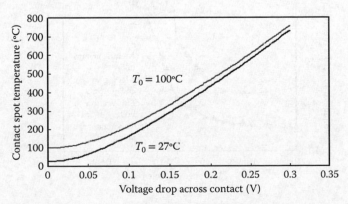

FIGURE 7.61
Dependence of a-spot temperature on voltage drop across a contact (Chapter 1, Equation (1.40)).

mechanisms, such as oxidation, interdiffusion, galvanic corrosion, and so on, are also enhanced. In contacts involving noble or near-noble metals such as gold and platinum, the imposition of a voltage drop corresponding to or exceeding the softening temperature will cause rapid interdiffusion of the noble metal either with the underplate material or with alloying elements [202,203] (see Timsit in Chapter 1). For example, the generation of high temperature in hard gold causes migration to the gold surface of hardeners such as nickel and cobalt [203]. This is deleterious to contact stability since nickel and cobalt oxidize readily and may render the gold surface highly susceptible to fretting corrosion.

As current is further increased to generate a voltage drop exceeding 110 mV, the temperature of a-spots approaches the melting point of the materials in contact. According to the data in Chapter 1 (Timsit), contact asperities of most materials reach their melting temperature at a voltage drop of between 130 mV (tin) and 430 mV (copper and gold), corresponding to contact-spot temperature, respectively, of 232°C (tin) 1060°C (gold) and 1084°C (copper). The generation of such large voltage drops in stationary or separable contacts eventually leads to catastrophic failure. Increasing the current to reach a voltage drop beyond the melting voltage first leads to boiling of metal in a-spots, and then to melting and possible boiling of surface layers such as oxides, sulfides, and so on located at spots of mechanical contact.

Figure 7.62 illustrates the degradation of tin–tin contacts subjected to fretting motion but with the contact resistance data recast as voltage drop across the contact, as a function of fretting cycles [204]. The two sets of data were obtained at respective currents of 0.86 A and 0.096 A. The two data sets show an increase in contact resistance, and hence an increase in voltage drop across the contact, with increasing fretting cycles due to fretting corrosion. For the data corresponding to 0.86 A, we note a plateau identified as V_1 at about 130 mV. This plateau corresponds to the melting voltage for tin. Other voltage plateaus that were identified as V_2, V_3, and V_4 were assigned, respectively, to melting of tin oxide, sublimation of SnO_2, and vaporization of tin. We note that the voltage range for the second plateau at about 600–700 mV corresponds to the temperature range of 1660–1990°C according to the temperature–voltage relation of Equation 7.4. At the smaller current of 0.093 A, thermal runaway did not happen until later during fretting since the generation of a voltage drop of 130 mV required reaching a higher contact resistance than at 0.86 A. Nevertheless, the

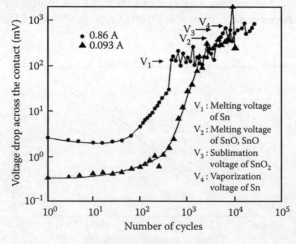

FIGURE 7.62
Voltage drop across the contact versus fretting cycles for two tin-plated copper surfaces, for two different values of electrical current. (After A Lee, MS Mamrick, *Hybrids and Manufacturing Technology* 10: 63–67, 1987 [204].)

voltage drop eventually reached the four plateaus V_1–V_4, as it did at the higher current. Auger depth profiles carried out after recording the data in Figure 7.62 confirmed the predominant presence of tin oxides in contact spots. Data similar to those of Figure 7.62 were reported by other investigators [191,205,206].

7.3.11 Surface Finish and Contact Geometry

Contact geometry was found [132] to affect the stability of contact resistance of fretting tin-alloy plated surfaces. In this early work, contacts adapted with a 90° or 120° wedge-shaped surface were compared with domed contacts when mated to tin-alloy-plated flats. Motion was mechanically or thermally forced at right angles to the line of contact of the wedge. The resistance of the wedge contacts was more stable than that of the domed contacts. These results were later confirmed in an independent study [198].

In other work, a tin–lead-coated checkerboard surface consisting of rectangular mesa-like high spots, which were about 100 μm on one side alternating with cavities 50 μm deep of similar shape and size, was tested as a model contact surface [207]. This surface geometry led to a Stable (Type III) contact with a consistently low contact resistance. It was suggested that fretting corrosion debris was pushed into the depressions during movement, thereby removing it from the contact surface and allowing the initially low contact resistance to be maintained.

A description of surface finish covers quantification of surface roughness and the characterization of platings and underplates on the surface. Measurements of a direct correlation between initial surface roughness and the evolution of contact resistance under fretting conditions have not been reported. One of the difficulties for such a correlation is the rapid change in surface roughness of a fretted zone even after a few reciprocating cycles [208]. Nevertheless, on the basis of the data shown in Figures 7.15 and 7.16 illustrating the increase in adhesive wear rate with increasing surface roughness, it would be expected that fretting corrosion rates are also accentuated with increased surface roughness.

7.3.12 Material Transfer, Wear, Film Formation, and Contact Resistance

7.3.12.1 Summary of Physical Processes

Contact resistance varies during fretting by a variety of mechanisms involving material transfer, metal removal, and chemical transformations at the sliding surfaces. Many of these mechanisms have been cited in preceding sections, but it is useful to summarize them in order to unify this treatment of contact behavior. Important examples of fretting include the following:

1. *Identical contact metals; non-film-forming (e.g., gold versus gold).* Gold and high-karat gold contacts form little surface contamination and are generally prone to wear-out (material transfer to a mating surface) and back-transfer from softer mating metals. Material transfer and back-transfer rates are smaller for electrodeposits containing transition metals that originate from gold cyanide baths [65], mostly due the lubricating effect of small polymer particles present in these deposits. Transfer metal becomes work-hardened and few loose wear particles are produced, especially with the ductile wrought forms of this metal. Contact resistance remains relatively stable during fretting and may even fall as the area of contact grows.

2. *Identical base contact metals; fretting corrosion (e.g., tin and tin-base finish versus itself and tin–lead solder versus itself).* Two series of events can be recognized, depending on whether oxidation happens before or after wear debris forms; first, metal

transfer and back-transfer between the surfaces can occur with the generation of occasional particles which then oxidize; and, second, preliminary oxidation of the metal surface, followed by removal of oxide to expose the original surface, which also oxidizes in a repeating process. In both situations, loose oxide accumulates and contact resistance rises.

3. *Identical frictional polymer-former metals (e.g., palladium versus palladium).* Organic materials are transformed on the surface to frictional polymers under the action of reciprocating sliding. These polymers may be scraped loose. There is little metal transfer and wear. Polymers accumulate and contact resistance increases.

4. *Dissimilar contacts; one of the metals forms electrically insulating films.* Two examples may be distinguished. Firstly, if the metal that does not form an electrically insulating film is significantly softer than the mating contact material (e.g., gold versus palladium), the base surface becomes covered with the noble metal, thus converting the sliding interface to that described in the above-listed item 1. Secondly, if the metal that forms an electrically insulating film is significantly softer (e.g., tin or tin–lead solder versus gold), an all-base metal system as described in the above-listed item 2 is formed due to metal transfer. The mechanics of oxide film fracture may be less favorable with dissimilar contact metals than when both surfaces are initially identical.

5. *Contact material wear-out (e.g., gold-flashed palladium versus itself).* Wear and transfer occur with contact resistance determined by the metal originally at the surface. Continued wear-out leading to the appearance of underlying material also leads to a changing contact resistance. Gold-flashed palladium is characterized initially by a low and stable contact resistance, as described in the above-listed item 1, but this is followed by rising contact resistance when palladium is exposed, as in above-listed item 3.

6. *Delamination wear.* Delamination wear is a mechanism of metal removal that occurs after prolonged fretting in which thin layers are lost by the worn surface to become loose debris.

The capability to predict the effects of changes in operational parameters on fretting corrosion rates from these examples is limited. For example, increasing wipe amplitude may increase metal transfer and wear in some contact materials, and hence increase the quantity of insulating oxide debris that is produced. This leads to contact resistance increases due to oxidized particle pile-up. In contrast, increasing the wipe amplitude in other contact materials may increase the efficacy of oxidized debris dispersal and hence mitigate the rapid buildup of oxidized material in the wear track. This mitigates rapid electrical contact degradation. The effects of frictional polymers on metals can be predicted more reliably than effects of oxidation because transfer and wear are not required for frictional polymer films to form. Decreasing cycle rate increases the time for the deposition of organic matter and thus increases susceptibility to frictional polymer formation on catalytic-type surfaces from lower concentrations of atmospheric organics. Decreasing cycle rate therefore leads to a greater increase in contact resistance for a given number of fretting cycles. Whether dealing with frictional polymer formation or oxidation, increasing the contact force increases material transfer and wear rate and enlarges the contact area. Consequently, more polymers or oxide debris form depending on the contact metal. On the other hand, increasing force enhances the penetration of insulating layers, and on balance, contact resistance diminishes.

In Section 7.4, it will be shown that use of fluid and grease lubricants in electrical contacts reduces base metal transfer, wear rate, wear debris generation, and oxide thickness due to shielding of the mating surfaces from the atmosphere. Lubricants therefore tend to stabilize contact resistance. With catalytic metals, the frictional polymers are readily dispersed in excess lubricant.

7.4 Lubrication

7.4.1 Introduction

Section 7.4 considers the lubrication of separable electronic connectors. Lubrication is broadly defined as the practice of coating contact surfaces to reduce mechanical wear and/or friction, and degradation due to fretting. In electrical contacts, the objective is to preserve the physical integrity of the contact surfaces and to maintain a stable contact resistance.

Lubricants may be described as (1) thin metallic films and organic (2) fluids, (3) greases, and (4) solids. There are still other purposes for some lubricant coatings such as a reduction of the oxidation and corrosion of contact materials, and these will be cited later.

7.4.2 Metallic Films

7.4.2.1 Principles of Metallic Film Lubrication

When two touching surfaces are in a relative motion, shearing forces develop leading to surface deformation, friction, mechanical wear, and so on, all of which are affected by surface topography, physical and chemical properties of the materials, the environmental conditions, and other factors. Friction originates from the deformation and shearing of surface asperities [2,5]. Adhesive wear occurs when both the surface and subsurface of mating bodies are subjected to mechanical stresses that lead to material loss. According to the adhesion theory of friction, the frictional force, F, is determined by the shear strength, τ, and the real area of contact, A_C, namely $F = A_C\tau$, as illustrated schematically in Figure 7.63. For friction force to be low, A_C and/or τ must be small. This means that materials associated with low friction should be characterized by high hardness and/or low shear strength. However, this generally is not achievable with monolithic materials; hard metals are generally characterized by high shear strength. However, by using thin layers of soft metallic films on a hard, smooth surface, friction and wear can usually be reduced.

In thin metallic film lubrication, the normal load is supported by the area of true contact and friction is determined by the yield point, p, of the support material and shear strength of the film. The coefficient of friction, μ, is a function of τ/p. In practice, a thin, easily

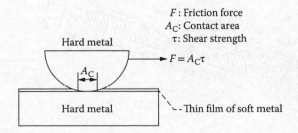

FIGURE 7.63
The friction between metal surfaces can be lowered by depositing a thin film of a soft or easily sheared metal on a hard metal substrate.

sheared metallic film is applied to a hard substrate, and if the surfaces are brought into contact the load will deform the surface asperities of the film but the film will remain supported by the substrate. The area of true contact A_C with the film is given as [2,5]

$$A_C = \frac{F_N}{H},$$

where F_N is the normal contact force and H is the hardness of the softer of the two materials in contact (H is usually the hardness of the thin metallic film). In electronic contacts where F_N is on the order of 0.10 kgf and H of a thin film is on the order of 100 kg mm^{-2}, A_C is on the order of 10^{-3} mm^2 and the contact area is thus small. When a tangential force is imposed in a sliding contact, shear takes place easily in the film. Gold is nearly always the thin film metallic lubricant used on electronic connector contacts because it is chemically inert. As discussed later, transition-metal gold-alloy electrodeposits are an exception; they are both relatively hard and easily sheared.

Another requirement of metallic film lubricants is that mating surfaces be relatively smooth and conform well so that the load is supported without major plastic deformation of surface asperities, and little permanent deformation of the substrate. For effective lubrication, there can be no slip between the metallic film and its substrate. Thus, the lubricating film must be strongly bonded so as not to be detached by the shear stresses that develop in the lubricant layer. The extent to which the coefficient of friction can be lowered depends primarily on film thickness, surface roughness, the degree of localized deformation, and the mechanical properties of the film relative to those of its substrate.

7.4.2.2 Sliding and Wiping Contacts

A widely used contact system in telecommunications connectors is illustrated in Figure 7.64. The socket contacts are inlaid with 2.5 µm of DG R-156 and the backplane pin contacts were plated with 1.5 µm of cobalt–gold and lubricated with a polyphenyl ether oil. For cost reduction, it was desired to replace the gold plating with palladium. However, this was not satisfactory, because the palladium plate, although harder than gold, is a poor adhesive wearing material. It displays excessive junction growth, described earlier (Figure 7.8). Since thin gold layers on harder palladium plate were already recognized as capable of providing lubrication function in connectors [63] and on steel in aerospace-bearing applications [209], it was clear that they should be explored further in this application. A bench modeling study [210] was conducted with DG R-156 connector contacts as the rider, and

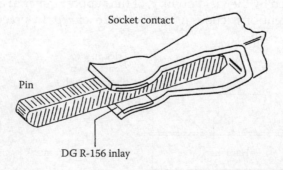

FIGURE 7.64
An individual pin–socket contact from an electrical connector consisting of a pin plated with gold or with palladium having a thin gold overplate and a socket contact with DG R-156 inlaid on the springs. (After M Antler, M Feder, *IEEE Trans Components, Hybrids, and Manufacturing Technology* 9(4): 485–491, 1986 [210].)

palladium-plated flats were obtained with pure (soft) gold and cobalt (hard) gold platings in a range of thicknesses. A parallel study was conducted with cobalt–gold-plated riders (2 μm thick on 1.25-μm nickel underplate). Both unlubricated and lubricated flats were examined. The lubricant was a polyphenyl ether fluid. Runs were conducted at a load of 200 gf for 200 cycles and a track length of 1 cm. These conditions simulate those of the actual connector.

Figures 7.65 and 7.66 are the friction data from this study. Figure 7.67 shows surfaces from some of the worn flats after 200 reciprocation cycles. Figures 7.68 and 7.69 are plots

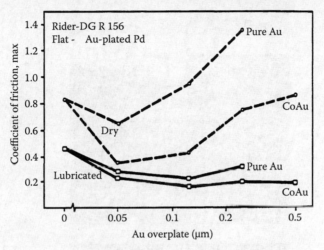

FIGURE 7.65
Relationship between the maximum values coefficient of friction in runs with DG R-156 coated riders versus gold on palladium-plated coupons (using a nickel subplate). The variables were the type of gold plate (either pure or cobalt-hardened) and its thickness, and whether the flats were dry or were lubricated with a thin layer of a polyphenyl ether fluid. (After M Antler, M Feder, *IEEE Trans Components, Hybrids, and Manufacturing Technology* 9(4): 485–491, 1986 [210].)

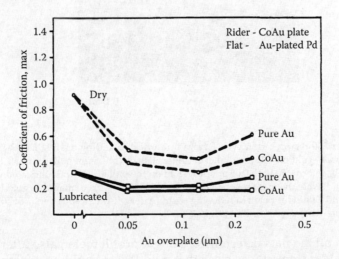

FIGURE 7.66
Dependence of the coefficient of friction on the type of gold overplate on the palladium-plated flat and on whether the surfaces were dry or were lubricated with a thin layer of polyphenylether fluid. Cobalt–gold-plated riders were used for comparison with the data in Figure 7.67 with DG R-156 coated riders. (After M Antler, M Feder, *IEEE Trans Components, Hybrids, and Manufacturing Technology* 9(4): 485–491, 1986 [210].)

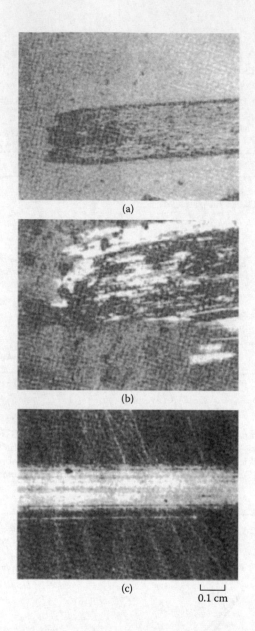

(a)

(b)

(c)   ⊢——⊣
                                       0.1 cm

FIGURE 7.67
Wear of plated flats after 200 reciprocation cycles, from unlubricated sliding against DG R-156-coated riders plated with (a) 1.5-μm cobalt–gold on 1.25 μm of nickel underplate or (b) 1.5 μm palladium on 1.25 μm of nickel underplate. (c) Same as (b), but coated with 0.05 μm pure gold on the palladium and lubricated with a thin layer of a polyphenyl ether fluid. Note the more severe wear of the palladium (b) compared to gold (a) and the effectiveness of a gold flash and a fluid lubricant in reducing wear. (After M Antler, M Feder, *IEEE Trans Components, Hybrids, and Manufacturing Technology* 9(4): 485–491, 1986 [210].)

of wear track depth on the flats determined using a profilometer, also after 200 reciproaction cycles. For this system, it was clear that gold layers could be effective lubricants, and that there was an optimal ("critical") layer thickness of about 0.05 μm for minimal friction and wear. Thicker coatings, especially of pure gold, were characterized by higher friction, stemming from sliding on the bulk gold rather than from shear of a thin layer

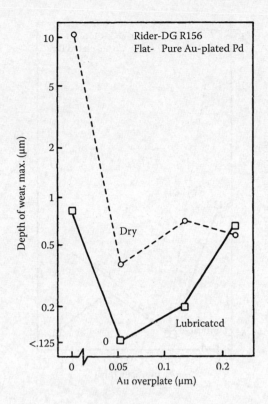

FIGURE 7.68

Wear of flats plated with a 1.5-μm-thick palladium plate on 1.25 μm of nickel from dry and lubricated sliding against DG R-156 riders, after 200 reciprocation cycles. The thickness of the pure gold overplatings on the flats was varied from 0 to 0.25 μm. Wear is expressed as the depth of the deepest scratch from three parallel traversals across the track using a profilometer. (After M Antler, M Feder, *IEEE Trans Components, Hybrids, and Manufacturing Technology* 9(4): 485–491, 1986 [210].)

supported by the hard substrate. Furthermore, the cobalt–gold coating was found superior to the pure gold coating. This is illustrated in Figure 7.67. Although the micrographs in Figure 7.67a and b relate to relatively thick metal coatings, thus not optimized for metal-film lubrication, the difference in the wear marks from unlubricated sliding after 200 passes illustrate the inferior lubrication properties of palladium plate compared to cobalt–gold plate.

The data of Figures 7.65 and 7.66 are explained briefly as follows: at small film thicknesses, lubrication by the metal film is negligible because it is too thin and too patchy to affect the shear strength of the interface with the substrate material, and friction remains high; with increasing film thickness, the low film shear strength begins to influence the shear strength of the sliding interface and the friction coefficient decreases. Minimum friction is reached where the film is sufficiently thick that ploughing friction begins to be significant. Beyond this point, friction continues to increase with increasing film thickness since ploughing friction becomes dominant.

The value of a supplementary organic lubricant to a metallic "lubricant" film was also demonstrated, as shown in Figures 7.65, 7.66 and 7.67c [210]. Commercial exploitation of the cobalt–gold-flashed, palladium-plated connection was made with a 2% coating of a six-ring polyphenyl ether fluid.

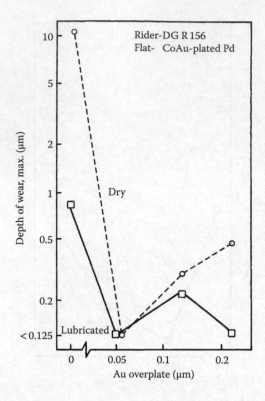

FIGURE 7.69
Wear of palladium-plated flats from dry and lubricated sliding against DG R-156 riders, after 200 reciprocation cycles. The palladium platings were overplated with various thickness of cobalt–gold. These experiments were similar to those relating to Figure 7.68 in which pure gold overplatings were used. Note that the maximum depth of the wear tracks was equal to or less with cobalt–gold than where pure gold was used. (After M Antler, M Feder, *IEEE Trans Components, Hybrids, and Manufacturing Technology* 9(4): 485–491, 1986 [210].)

Additional examples of the use of thin-film gold lubricant coatings for wiping contacts are for electrodeposited palladium–nickel [165,167], palladium–silver [157], tin–nickel [64], ruthenium [211], rhodium [62], nickel and nickel–phosphorus [212], and for clad noble metals [213]. Developments with thin gold plate on various metals are summarized in [214].

Guidelines for the use of flash gold platings on common contact materials in typical electronic connector are as follows:

1. Substrates and underplatings should be smooth and hard. Nickel underplate at a thickness of 2–3 μm is effective in improving durability.

2. The thickness of the contact finish when combined with the gold overplate should be sufficient to (a) resist corrosion in the environments of interest and (b) mitigate instability of contact resistance due to substrate or underplate diffusion in elevated temperature applications. It may be necessary to avoid porous deposits when the contacts are to be used in aggressive atmospheres.

3. Although pure gold may be employed as a metallic film lubricant, cobalt–gold is preferred to pure gold. A 0.05 μm thickness is adequate.

4. All deposited metal layers should be adherent, consistent with good plating practices.

5. A supplementary fluid lubricant is desirable in order to achieve the lowest friction and to maximize durability.

7.4.2.3 Fretting Contacts

Because of concerns that frictional polymerization can seriously limit the application of palladium, palladium–nickel, and other catalytic metals in connectors, there has been interest in exploring means for inhibiting polymer formation. If gold is mated to palladium (which is harder), as discussed earlier, the connection behaves as though it were all gold because of the transfer of gold to palladium during fretting. This effectively arrests the formation of deleterious polymers, and was found to be a method for utilizing palladium contacts in wire spring telephone relays in the 1970s and 1980s. In such contact assemblies, one of the palladium contacts required a thick coating of gold or high-gold alloy so that gold transfer to the opposite initially bare surface did not deplete the available gold. Coatings of Au75Ag25 and Au69Ag25Pt6 with a thickness as large as 20 μm have been used for this purpose [215]. Unfortunately, the commercial use of such thick gold-alloy layer can no longer be contemplated in light of the high price of gold. *Thin* gold films, say, less than 0.25 μm, are relatively ineffective in controlling frictional polymerization on these catalytic substrates since the large number of fretting cycles in many applications leads to the relatively early wear-out of gold. This has already been discussed on earlier and Figure 7.49. However, thin gold films may be used on contacts made from palladium or palladium alloys if the number of fretting cycles expected in service is not expected to be large.

Another example is illustrated in Figure 7.70 [195], from work that was part of the development of a telephone connector (Figure 7.64). In this case, the effects of fretting modeled in the laboratory using gold-flashed palladium mated to itself were compared with those in unlubricated palladium–palladium contacts. The data indicated that the gold flash on palladium only marginally improved its fretting behavior. It was necessary to employ a fluid lubricant together with the gold flash in order to achieve acceptable performance. Fluid lubricants are discussed more fully later. Subsequent studies with gold-flashed palladium–nickel electrodeposits have shown it to have behavior similar to that of gold-flashed palladium [165,216].

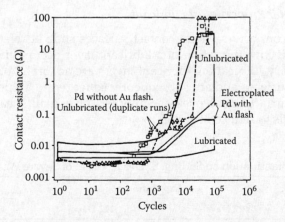

FIGURE 7.70
Contact resistance from fretting of samples with pure gold flash on 1.5-μm palladium having a 1.25-μm nickel subplate mated to itself, clean and lubricated with a polyphenyl ether obtained by immersion and withdrawal of both contacts from a 0.5% solution in a volatile solvent. All data from four replicate runs at each condition fall within their respective bands. Data for duplicate runs with unlubricated palladium without gold flash is indicated by dashed lines. The gold flash had limited durability under the test conditions. Contact resistance was more stable when fluid lubrication was used: 50 gf, 20 μm wipe. (After M Antler, *IEEE Trans. Components, Hybrids, and Manufacturing Technology* CHMT-8: 87–104, 1985 [195].)

7.4.3 Fluid Lubricants

7.4.3.1 Background

The use of fluid lubricants on connector contacts is by no means universal. When few connector insertions are required and wear is of little concern, there generally is no need to lubricate. This is because exposure to the atmosphere and ordinary handling may deposit sufficient adventitious contaminations to provide adequate protection. This is illustrated in Table 7.3, which shows results of an earlier laboratory study [217] that examined the onset of severe wear in sliding contacts that had been exposed to air in several locations of the same building. It was found that severe wear was initiated after as few as four reciprocation cycles in some connectors to more than 200 cycles in other contacts. In Figure 7.2, the scanning electron micrograph of a typical contaminated gold surface was shown. In the work reported in Table 7.3 [217], contamination occurred relatively quickly due to the absence of shielding of the exposed connector surface. For the connectors investigated, the variability of natural contaminations, the need with multicontact connectors for significant reduction of insertion force, and the large number of required engagements with associated risks of wear-out, eventually led to the intentional use of fluid and grease lubricants to achieve acceptable performance [217]. It is worth recalling that the practice of lubricating electrical contacts is as old as the earliest developments of electrical connectors [218].

Some commonly used lubricants that were originally intended for applications other than in connectors have often led to highly unsatisfactory results. Examples of such lubricants are silicone fluids and water-soluble polyalkylene glycols. The former are incompatible with many metals, and furthermore can contaminate nearby exposed relay contacts with disastrous consequences [219] (see Chapters 10 and 19); the latter "contact enhancers," not only are poor lubricants [75] but also can corrode metals, although they may have a temporary beneficial effect in reducing contact resistance [220,221]. These comments apply as well to greases and solid lubricants that have been used for connectors.

7.4.3.2 Some Fundamental Properties of Lubricants

In separable electronic connectors where contact surfaces slide at relatively low speeds and lubricant viscosity is low, the action of a lubricant layer occurs via boundary lubrication [2,222]. In boundary lubrication, lubricant layers are adsorbed or chemisorbed on one or the two mating surfaces and are subsequently squeezed to a thickness of molecular dimensions as the surfaces slide over one another. The small shear strength of the squeezed fluid films leads to reduced friction.

TABLE 7.3

Effect of Atmospheric Contamination on Sliding: Cycles to Onset of Severe Wear

Location	Exposure		222 Day Samples After 350 h at 50°C
	22 Days	**222 Days**	
Humidity-controlled room	4	4	4
Chemical laboratory	8	95	37
Telephone switching equipment room	14	>200	42

Conditions: DG R-156 inlay flats; cleaned; then exposed before wear test in several locations. Sliding versus cobalt–gold-plated riders at 100 gf on a 1-cm-long track until a coefficient of friction of 0.4 was reached.

The efficacy of lubrication has been explored as a function of the lubricant thickness down to sub-monolayer levels. Many of these investigations have relied on ultra-high vacuum techniques and low temperatures to allow both controlled lubricant-layer deposition and the subsequent characterization of the interaction of the layers with the substrate using surface analytical techniques [223 and references within, 224]. Figure 7.71 shows results that typify observations in such investigations. In this example, the static friction coefficient was measured between two sliding Cu(111) surfaces modified by adsorbed layers of 2,2,2-trifluoroethanol (CF_3CH_2OH) [223]. The coverage of CF_3CH_2OH ranged from 0 to 14 monolayers (ML) on each Cu(111) surface at a temperature of 120 K. At lubricant coverages <1 ML, the friction coefficient between the surfaces was observed to be very high (> 4). It was only when the lubricant coverage on the two surfaces reached or exceeded 1 ML that the chemisorbed layers acted as a lubricant and friction began to decrease. Ultimately, the friction coefficient fell to a limiting value of ~0.3–0.4 for an accumulation of approximately eight molecular layers of the lubricant.

The formation of *a*-spots between two lubricated surfaces in an electrical contact must begin with a reversal of the mechanism associated with the data of Figure 7.71, whereby bulk lubricant is first displaced from contacting regions so that only a few molecular layers of residual lubricant remain in the contact, as the contact load is increased. Furthermore, experimental observations that the use of lubricant in a clean or otherwise stable electrical contact has a negligible effect on contact resistance [225] (e.g., see also Figure 7.70) suggest that an appropriate electrical contact lubricant is one where molecules can be totally squeezed out of *a*-spots to allow metal–metal contact. In this way, the lubricant layers do not interfere significantly with electrical current flow in the contact. It is worth pointing out

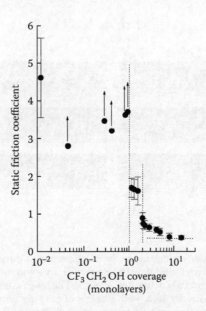

FIGURE 7.71

Static friction coefficient between two Cu(111) surfaces lubricated by thin layers of 2,2,2-trifluoroethanol (CF_3CH_2OH) versus the CF_3CH_2OH coverage. The data points with vertical arrows at a coverage of less than 1 ML are also averages of 10 measurements but represent lower limits of the static friction coefficients. Load ≈ 40 mN, sliding velocity = 1, 2, or 3 μm/s, temperature = 120 K. (After AJ Gellman, JS Ko, *Tibology Letters* 10: 39–44, 2001 [223].)

that displacement of large lubricant accumulations may not occur in an electrical contact if the sliding speed is sufficiently large because the lubricant causes hydrodynamic lift in this case [41].

Starting in the 1990s, computer simulations of the behavior of molecular films in a squeezed confined interface, and attendant experimental work, have provided information on mechanisms responsible for the displacement of thin lubricant films from a-spots in an electrical contact. Typical results of these simulations are illustrated in Figure 7.72 [226]. The data show views of the calculated atomic and molecular configurations at different times during molecular dynamics (MD) simulations of the sliding process of two rough Au(111) crystal surfaces with an entrapped hexadecane layer. Each of the two gold surfaces carries a single asperity, with a height of only six atom planes. The displacement sequence (a) to (f) shows the calculated asperity deformation and the attendant expulsion of the hexadecane molecules as the asperities collide. The behavior of the lubricant thin film in the narrow interface in Figure 7.72 stems from the loss of isotropy of the liquid film in the vicinity of the gold surfaces due to forces originating in the solid surfaces [227]. This leads to oscillatory solvation forces when the surfaces are pressed together, accompanied by a series of layering transitions, each corresponding to a stepwise decrease in film thickness by the expulsion of a discrete amount of liquid lubricant.

Experimental work has confirmed some major results of MD simulations. For example, Figure 7.73 shows snapshots of the contact area between two crossed mica cylinders entrapping a thin layer of 1-undecanol ($C_{11}OH$) lubricant [227]. The diameter of the contact area is about 100 μm. In images (a) to (f), the contact load is steadily increased from 0 to about 20 mN and the darker color corresponds to smaller film thickness. The micrographs show clearly that as the contact load is increased the dark area (from which the lubricant has been expelled) increases, while the remaining light-colored area (i.e., the remaining

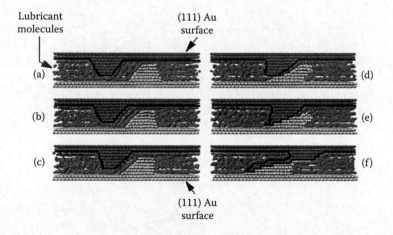

FIGURE 7.72
Side views of the atomic and molecular configurations at different times during molecular dynamics (MD) simulations of the sliding process of two rough Au(111) crystal surfaces with an entrapped hexadecane layer. The profile of the upper gold contact surface is delineated in black. The lower gold surface is shown in light gray. The two gold surfaces carry a single asperity each, with a height of only six atom planes: (a) the two asperities on the sliding gold surfaces are approaching each other, trapping, and starting to squeeze the hexadecane layer to a smaller thickness between them; (b) the two asperities are beginning to deform due to the shear force transmitted by the thin hexadecane film trapped between them; (c) same as (b) but the asperities undergo additional deformation; (d) the asperities are in contact and are severely deformed, and begin to squeeze out hexadecane molecules from the contact; (e), (f) continued severe deformation of the asperities with total expulsion of hexadecane from the contact. (After J Gao et al., *Science* 270: 605–608, 1995 [226].)

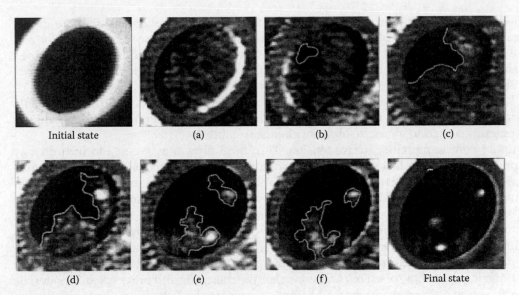

FIGURE 7.73
Snapshots of the contact area between two crossed mica cylinders entrapping a thin layer of 1-undecanol ($C_{11}OH$). The diameter of the contact area is about 100 µm. In images (a) to (f), the contact load is steadily increased from 0 to about 20 mN. Darker color corresponds to smaller film thickness. As the contact load is increased, the dark area located within the contact region and delineated by the white-dashed curve, increases while the remaining light area (i.e., the remaining trapped lubricant) decreases. The decreasing light area reveals steady expulsion of lubricant molecules from the contact region. (After F Mugele, M Salmeron, *Phys. Rev. Letters* 84: 5796–5799, 2000 [227].)

trapped lubricant) decreases. The dark area is delineated by the white-dashed curve. The decreasing light-colored area reveals steady expulsion of lubricant molecules from the contact region and relates to a specific layering transition undergone by the trapped fluid.

The data in Figures 7.71 through 7.73 indicate that effective lubrication in a squeezed interface can be achieved with as little as one lubricant molecular layer, and that lubricant films can be expelled from the interface under a sufficiently large contact pressure. Clearly, these observations explain why it is possible to generate metal–metal *a*-spots in a lubricated electrical contact.

7.4.3.3 Requirements

This section briefly summarizes the requirements of a lubricant, methods of application, and provides examples of widely used fluids in current practice.

Fluid lubricants for connector contacts are used nearly always as thin layers. Coverages range from 10–1,000 µg cm^{-2}, which correspond to thicknesses of the order of 100–10,000 µm (assuming that the layer is uniform), depending on the lubricant density and contact surface roughness. At a small thickness, lubricants that are transparent and colorless in bulk—as is usual—are barely visible.

As indicated earlier, the sliding of lubricated electrical contacts is largely in the boundary lubrication region [2,222], where boundary (interactive surface) effects determine lubricating behavior. Other lubricant properties such as viscosity, vapor pressure, and so on are relevant to issues such as the effectiveness of lubricant deposition and retention on a surface, lubricant spreading, lubricant evaporation, and other factors. This is discussed later.

There are properties that are unassociated with lubricant function, but that are highly relevant to the suitability of a candidate fluid as a lubricant in a connector. One major requisite for an acceptable lubricant is low volatility, especially if the connector is designed for use at elevated temperatures. It is usually impractical to relubricate the connector once it is in service.

A lubricant should also be thermally stable and resistant to oxidation, an especially important consideration for elevated temperature service. Oxidation-inhibiting additives are sometimes used to increase lubricant chemical stability.

Another consideration is surface tension. Thin fluid films may creep and in time may disappear from the contact surface. Contamination of nearby electronic components may occur. Creep rate is also determined by the lubricant viscosity. Low surface energy barrier coatings, such as poly-IH,IH-pentadecafluoroctylmethacrylate, are sometimes applied to relay contacts to protect them from silicone contamination.

A lubricant should not react with any connector component to form intractable solid residues. Some contact metals, especially copper and aluminum, may promote both lubricant degradation and oxidation of the base metal [221,228,229,230]. A related requirement is that the lubricant not soften or otherwise alter the dielectric and other organic materials of construction of the connector and adjacent components.

The lubricant should not interfere with the method for making a permanent connection of a conductor to the contact by soldering, crimping, and so on.

The lubricant should be non-hygroscopic, because such fluids can become conductive, allowing electrical leakage between adjacent contacts and conductors, and hence promoting corrosion.

The lubricant should be able to disperse undesirable solid contaminants on the surfaces, such as wear debris and dusts.

Non-toxicity and dermatologic acceptability are also an obvious requirement.

7.4.3.4 Types of Fluid Lubricants: A Sliding Contact Investigation

The lubrication of base metals, especially ferrous metals, has been much studied. An important mechanism of anti-wear effectiveness with these metals stems from the reaction of additives in inert fluids with the metal to form thin layers of easily sheared solids on the surface. For example, fatty acids produce protective soap films. Other compounds that contain chlorine, sulfur, or phosphorus, decompose at hot spots and react with the sliding surface to form metal chlorides, sulfides, or phosphates that reduce friction and increase the transition load from mild to severe wear by the so-called EP (extreme pressure) mechanism. Both soap formation and EP processes are, however, unavailable for gold and other inert noble metals. Examples of this non-effectiveness was demonstrated in early studies [231] of gold sliding using dilute mineral oil solutions of lauric acid, tricresyl phosphate, and carbon tetrachloride.

Various fluids have been tested, at least preliminarily, as lubricants for gold and other connector contact materials. These range from primary (uncompounded) materials to complex mixtures incorporating antioxidants, viscosity index improvers, detergents, and other additives that are used for mechanical devices. An early study [41] of the lubrication of gold determined sliding characteristics with a number of fluids that were available at the time (1963) including dimethylpolysiloxanes, phenylmethylpolysiloxanes, silicate esters, polychlorotrifluoroethylenes, diesters, fluorinated esters, polyalkylene glycols, chlorinated hydrocarbons, phosphate esters, a polyphenyl ether, petroleum oils, and several metalorganics dissolved in mineral oil. The study was continued in 1987 [217] under slightly different test conditions with additional materials, including another polyphenyl ether, perfluoroalkyl polyethers, and poly-alpha-olefins. Products of unknown composition were excluded from these studies.

The investigation involved initially a determination of the coefficient of sliding friction, wear rate, and contact resistance using solid gold specimens in a rider-flat bench apparatus. Contacts were submerged in the fluids. It was found that the effectiveness of the lubricant depended on its chemical structure and viscosity.

The friction data are summarized in Figure 7.74, which also shows whether contact resistance varied during sliding and the cause of the resistance variations. Noise was found to originate from (1) erratic sliding due to stick-slip, surfaces that became rough due to severe wear, and the presence of wear debris; (2) thick film lubrication where the high lubricant viscosity prevented good metal–metal contact (identified as hydrodynamic lift in Figure 7.74); and (3) the formation of insulating layers from lubricant decomposition products ("non-corrosive film formation" [231]). Figure 7.75 shows photomicrographs of typical worn specimens and profilometer traces across the wear tracks of the flats.

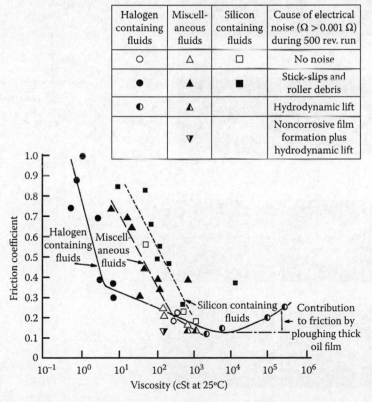

FIGURE 7.74
Coefficient of friction versus viscosity for fluid lubricants. Friction determined near end of 500 rev runs at 100 gf and 1 cm s^{-1} on 2.5-cm circular track using solid gold contacts. Friction is related both to fluid type and to its viscosity. Electrical noise with poor lubricants is due to hydrodynamic lift or sliding-generated insulating solids which adhere to surface. (After M Antler, *IEEE Trans Components, Hybrids and Manuf Technol.* 10(1): 24–31, 1987 [217].)

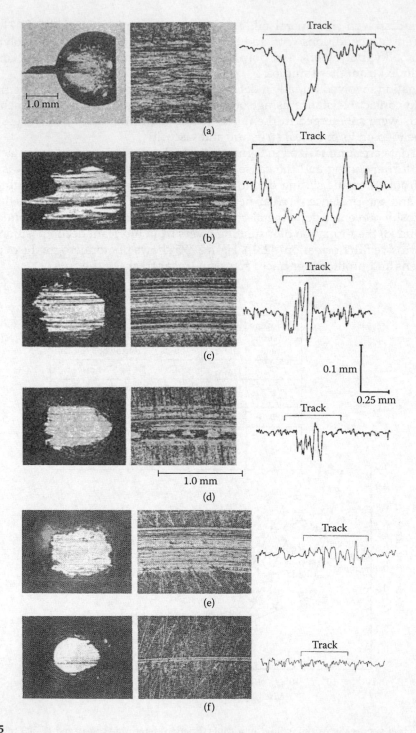

FIGURE 7.75
Photomicrographs of representative worn riders and flats from 500 rev. runs at 100 gf, 1 cm s⁻¹. Stylus profiles made across track: a = unlubricated; b = OS-59; c = Dow–Corning 200 (500 cSt grade); d = Ucon LB-135; e = Versilube 81717; f = Fluoroester P.

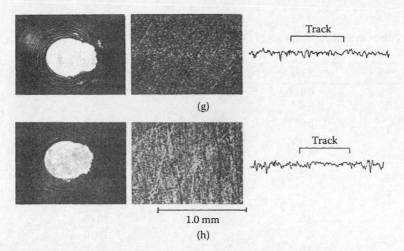

FIGURE 7.75 (*Continued*)
Photomicrographs of representative worn riders and flats from 500 rev. runs at 100 gf, 1 cm s^{-1}. Stylus profiles made across track: g = Fluorolube LG-160; h = OS-124. Direction of motion or rotating flat: left to right. Poor lubricants (b)–(e) allow metal transfer, prow buildup on rider against the direction of sliding and roughening of flat.

The salient findings are as follows:

1. Halogen-containing fluids are better lubricants than those which incorporate silicon. Oxygen-containing lubricants and hydrocarbons have intermediate behavior.

2. Electrical noise can occur during sliding with fluids characterized by either low or high viscosity; low-viscosity fluids are poor lubricants. High-viscosity fluids promote hydrodynamic separation of the contacts. The viscosities where these effects occur depend on contact normal load, sliding velocity, and other factors. However, the absence of electrical noise during engagement is not normally a requirement of connectors.

3. Lubricants characterized by a favorable combination of thermal, chemical, and lubrication properties for gold-plated electrical contacts include the five- and six-ring polyphenyl ethers and viscous perfluoroalkyl polyethers [232].

Reasons for differences in the effectiveness of lubricants according to chemical class at lower viscosities are not yet established. It is possible that the more effective lubricants are adsorbed more strongly on gold surfaces than those that are less effective.

A key part of the work described in reference [232] was a determination of the time intervals to reach sliding failure of freely exposed contacts that were aged through a range of temperatures in a moving air stream. The results are summarized in Table 7.4 for two polyphenyl ethers. The times to sliding failure in installed connectors would be expected to be longer than those listed in Table 7.4 since a liquid lubricant evaporates less quickly from a mated interface in an installed connector than from contact surfaces that are exposed to open air. High-molecular weight perfluoroalkyl polyether fluids are less volatile than polyphenyl ethers with the same viscosities, but tend to migrate from the contact surface because of low surface tensions. On the other hand, perfluoroalkyl polyether fluids are characterized by viscosity indices that are superior to those of the polyphenyl ethers, that is, their viscosities change less with changing temperature. Both classes of lubricants are characterized by high thermal stabilities to about 290°C for the perfluoroalkyl polyethers

TABLE 7.4

Calculated Times to Sliding Failures of Contacts Coated with Polyphenyl Ether Lubricants (Years)

Temperature (°C)	2% Five-Ring Polyphenyl Ether		2% Six-Ring Polyphenyl Ether	
125	0.023	(8 days)	0.6	(220 days)
120	0.03	(111 days)	0.8	(290 days)
110	0.05	(18 days)	2.0	
100	0.09	(34 days)	3.8	
90	0.18	(66 days)	7.6	
80	0.4	(148 days)	17.0	
70	0.84	(306 days)	36.0	
60	1.8		66.0	
55	3.0		105.0	
50	4.4		170.0	
40	12.0		450.0	
30	32.0		1,200.0	
25	62.0		2,000.0	

Note: Unmated surfaces in moving air. Sliding at 200 gf for 200 cycles. Cobalt–gold-plated contacts vs. DG R-156 coated with the PPE. Criterion of failure: coefficient of friction of 0.4 which occurs at the transition from mild to severe wear. The times to failure for mated connectors are 2–100 times greater than the values in this table, depending on their design.

and 250°C for the polyphenyl ethers. At temperatures higher than 250°C, five-ring polyphenyl ethers undergo chemical breakdown, which may lead to corrosion of an underlying non-noble metal surface [233].

An interesting characteristic of polyphenyl ethers is that they do not wet surfaces due to their high surface tension (about 50–54 dynes cm^{-1} at 25°C). The beading tendency of polyphenyl ethers facilitates visual detection of the lubricant in ultraviolet light if a fluorescent dye [234] in small amount is incorporated in the lubricant–solvent mixture. Polyphenyl ethers represent a lubricant class of considerable interest for applications in many subsequent studies of fretting control, corrosion inhibition, and so on. Other lubricants have come into use as the connector market has grown, but there is relatively less engineering literature on them because of the secrecy that surrounds lubrication practice for electronic connectors.

7.4.3.5 Control of Fretting Degradation

Lubricants can significantly improve the contact resistance behavior of fretting electrical contacts. The amount of lubricant required depends on the contact metals and the mechanism(s) by which they degrade. Also, only fluids and greases are useful; solid lubricants, such as microcrystalline wax, are of little value in fretting because they are quickly displaced from the contact.

7.4.3.5.1 Reduction in Wear Rate

Oxidized metal that forms during the fretting of base metals originates, in part, in loose wear debris. A lowering in wear rate therefore tends to stabilize contact resistance. When the contact finish is gold, a decrease in the wear rate is beneficial since the time until the base underlying metal is exposed is extended. An example of the life extension of a contact under reciprocating displacement through the use of a lubricant is shown in Figure 7.76 [99],

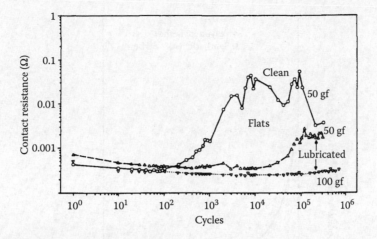

FIGURE 7.76
Contact resistance vs. fretting cycles for solid gold riders against 0.23-µm cobalt–gold-plated copper flats, with and without polyphenyl ether obtained by immersion and withdrawal from a 0.5% solution of the contact lubricant in a volatile solvent: 20 µm wipe. (After M Antler, MH Drozdowicz, *Wear* 74: 27–50, 1981–82 [99].)

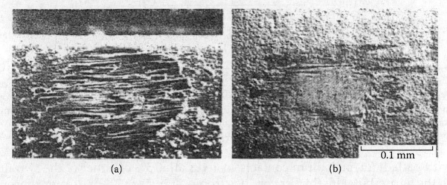

(a)   (b)

FIGURE 7.77
Worn specimens from fretting with 2.5-µm cobalt–gold electroplate mated to itself: (a) unlubricated; note rough surface and dark region in center which is exposed nickel underplate; (b) coated with polyphenyl ether obtained by immersion and withdrawal from a 0.5% solution in a volatile solvent surface is burnished: 50 gf, 20 µm wipe, 10^5 cycles. (After M Antler, *IEEE Trans Components, Hybrids, Manuf Technol,* 7(4): 363–369, 1984 [186].)

in which the contact resistance behaviors of gold-plated copper with and without a lubricant are compared. The lubricant consisted of a five-ring polyphenyl ether.

Visual confirmation that introduction of the lubricant led to the decrease in gold wear rate indicated in Figure 7.76 is shown in Figure 7.77. In this figure, the optical micrographs compare the surface damage of identical cobalt–gold-plated surfaces that had been subjected to the same fretting wear exposure but that had been, respectively, unlubricated and lubricated [186]. In dry fretting (Figure 7.77a), the gold surface was worn to the nickel underplate whereas the gold surface was only burnished and wear was negligible (Figure 7.77b) with introduction of the lubricant.

7.4.3.5.2 Effect on Frictional Polymerization

Selected lubricants are capable of stabilizing the contact resistance between metals that catalyze frictional polymer formation. Lubricant effectiveness is usually strongly dependent on the thickness of the lubricant, as shown in Figure 7.78 [235]. Some benefit was

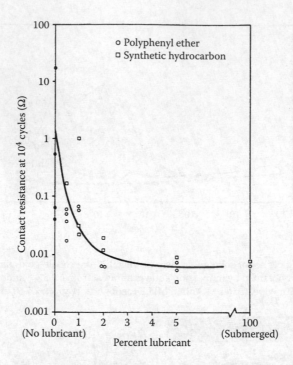

FIGURE 7.78
Contact resistance after 10^4 cycles of fretting in palladium–palladium contacts as a function of amount of lubricant on the surface. Coatings obtained by immersion and withdrawal of the flat contact in solutions of the lubricants having concentrations of 0.5, 1, 2, and 5% in a volatile solvent, or by placing a drop of lubricant on the mated contacts. Each data point is the result from a single run. Solid circles are data from unlubricated samples. Contact resistance stability is improved with increasing amount of lubricant: 50 gf, 20 µm wipe. (After M Antler, *IEEE Trans Components, Hybrids, Manuf Technol*, 7(4): 363–369, 1984 [186].)

obtained with palladium–palladium contacts coated with 0.5% of either of the two fluids indicated in the figure; but with thicker coatings from solutions of 2% or greater concentration, the beneficial effect was particularly striking.

Another example was given earlier in Figure 7.70 with fretting palladium contacts covered with a gold flash where a five-ring polyphenyl ether coating from a 0.5% solution was used. The maximum contact resistance was three orders of magnitude smaller with lubricant than without it after about 10^5 fretting cycles, and the initiation of contact resistance rise was also significantly delayed. Probably, the effectiveness of the lubricant was due to both reduction in the rate of wear of the gold flash and contact resistance stabilization once the gold had been worn away to the underlying palladium. It is recommended that lubricants be employed in sliding systems such as gold-flashed palladium, palladium–nickel and other palladium group metals, and more generally in contact interfaces susceptible to rapid degradation by fretting corrosion due to a special or a harsh environment in service.

7.4.3.5.2.1 Mechanisms Figure 7.79 is a striking example of polymer dispersal from a long run with palladium contacts immersed in USP mineral oil [174]. In this case, the polymer debris consisted of a voluminous filmy semi-transparent black solid sheet located alongside the wear scar. The polymer debris was convoluted, which suggests that the frictional polymers were produced continuously during fretting, and that successive slides pushed away the freshly formed material. The polymer film was semi-transparent, which indicates that its thickness ranged from a few tens to hundreds of nanometers thick. It is

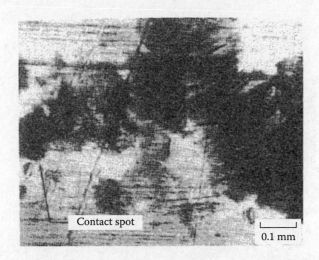

FIGURE 7.79
Frictional polymer from fretting palladium–palladium contacts. Flat lubricated by immersion in USP mineral oil. Lubricant not removed. Note thin-convoluted dark polymer film dispersed in excess lubricant. 50 gf, 20 μm wipe, 5.5×10^5 cycles. (Optical micrograph, after M Antler, *ASLE Trans* 26(3): 376–380, 1983 [174].)

surmised that the film grew to this thickness on the surface before being pushed aside. However, the film must have been readily penetrated by the mating asperities, since contact resistance was little different during the entire run from the value measured before fretting was initiated [174].

7.4.3.5.3 Effect on Fretting Corrosion

Lubricants have long been recognized to be effective in inhibiting the increase in contact resistance between base metal surfaces that degrade by fretting corrosion [101,134,135] and an example [180] is given in Figure 7.80 for tin–lead to tin–lead contacts. In this case, a 2-μm coating from a 20% solution of mineral oil in a volatile solvent stabilized contact resistance for at least 10^5 cycles, whereas a coating extracted from a 5% solution stabilized contact resistance for only ~10^3 cycles. These data are typical of many findings relating to the dependence of lubricant effectiveness on the quantity of fluid used than on its composition.

Fluids studied for the fretting corrosion problems of tin and tin–lead contacts include mineral oils, poly-alpha-olefins, polyphenyl ethers, polyalkylene glycols, and poly-chlorofluoroethylenes in [180]; and diesters, silicones, and phosphate esters in [134]. In order to achieve uniformity of lubricant distribution in a mated system, it is generally advisable to apply the lubricant to the two contacting sliding surfaces rather than to only one.

7.4.3.5.3.1 Mechanisms
The mechanisms by which lubricants act to mitigate an increase in contact resistance during fretting displacements include wear reduction, dispersion of insulating solids, and enhanced shielding the surface from air, thereby retarding the rate of oxide formation. The enhanced resistance to fretting corrosion due to the presence of an appropriate lubricant is illustrated in Figure 7.81 where worn tin–lead surfaces coated with different thicknesses of lubricant are compared. The surface treated with the 20% polyphenyl ether solution is bright, whereas the comparable contact coated with 5% polyphenyl ether is oxide-covered, much like surfaces from unlubricated fretting.

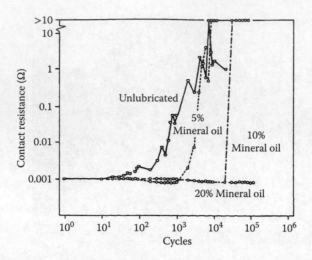

FIGURE 7.80
Contact resistance vs. fretting cycles for Sn60Pb40 vs. Sn60Pb40-plated contacts. Flats lubricated by immersion and withdrawal from 5, 10, and 20% solutions of USP mineral oil in a volatile solvent. The thicker the lubricant layer, the greater the stability of contact resistances: 50 gf, 20 μm wipe. (After M Antler, *IEEE Trans Components, Hybrids, Manuf Technol,* 7: 129–138, 1984 [180].)

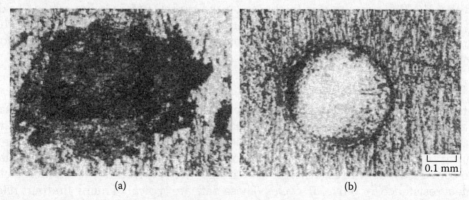

FIGURE 7.81
Lubricated Sn60Pb40-plated flats after fretting against Sn60Pb40-plated riders: (a) coated with 5% polyphenyl ether; (b) coated with 20% polyphenyl ether: 50 gf, 20 μm wipe, 10^5 cycles. The flat in (a) is covered with an insulating black oxide (contact resistance, 37 Ω) and in (b) the surface is bright (contact resistance, 0.9 mΩ). (After M Antler, *IEEE Trans Components, Hybrids, Manuf Technol,* 7: 129–138, 1984 [180].)

7.4.4 Grafted and Self-Assembled Lubricant Layers

In extreme environmental conditions such as high vacuum, high temperature, or under conditions of frequent interfacial displacements during extended service where high-lubricant durability is required (e.g., in microelectromechanical systems, aerospace applications, and so on), the use of a liquid lubricant in an electrical contact is often unsatisfactory. This stems from a number of factors such as eventual displacement of the lubricant from the contact region at an early juncture in service, lubricant evaporation, or even crystallization. One possible approach to resolving the problem of lubricant depletion is the use of chemical grafting of lubricant layers to the underlying metal surface. Although there are limitations to the use of grafted lubricant layers for many contact applications, as will be addressed later, the technology is relatively novel and promising and deserves description.

The term "grafting" simply means attaching molecular components to the lubricant molecules (i.e., polymerize the lubricant) to achieve a strong chemical attachment to the substrate, as illustrated schematically in Figure 7.82 [236]. An excellent review of chemical grafting technology is given in reference [237]. Chemical attachment to the solid (also described as "grafting" to the substrate) must not be so strong as to prevent displacement of lubricant molecules during contact to allow a-spot formation. One side benefit of molecular grafting is mitigation of oxidation, sulfidation, and other corrosion reactions of the metal substrate with the atmospheric environment, since the grafted layers minimize exposure of the underlying surface to air.

Although there are several grafting techniques [236,237], electropolymerization is becoming the preferred approach since it also allows deposition of the structurally modified lubricant on the desired surface [238–243]. The process of electropolymerization under cathodic polarization consists in starting an acid–base reaction between one end of a monomer in solution and the metallic surface acting as electrode ("grafting" the monomer to the metal). This process is then followed by initiating a reaction between the monomer and the graft (i.e., the lubricant molecule) to start the growth of the molecularly modified lubricant film. An organic layer strongly linked to the substrate is thus obtained, as illustrated schematically in Figure 7.82. The lubricant film thickness may vary from a few nanometers to a few micrometers. Figure 7.83 shows an example of contact resistance properties of polyacrylonitrile layers grafted onto nickel-plated brass. The cumulative values of contact resistance were obtained for a number of samples before and after heat treatment in air under six different conditions with a load of 50 gf [243]. The decrease in contact resistance with increasing treatment temperature was associated with changes in the molecular structure and molecular length of the grafted polymeric

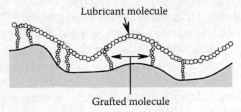

FIGURE 7.82
Schematic diagram of "grafting" molecular components to a lubricant molecule (i.e., polymerization of the lubricant) to achieve a strong chemical attachment to the substrate.

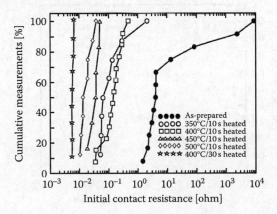

FIGURE 7.83
Cumulative distribution of contact resistance for polyacrilonitrile layers in interfaces consisting of surfaces electroplated with nickel and subjected to six different heat treatments, a contact load of 50 gf. (After F Houzé et al., *IEEE Trans Components, Hybrids, and Manuf. Technol.-Part A* 18: 364–368, 1995 [243].)

structure on the surface [243]. The friction coefficient following some of the thermal treatment conditions are shown in Figure 7.84. These data indicate clearly that optimal results combining low contact resistance and low friction required longer heating at 400°C in this case. Aside from the practical challenge of subjecting connectors to such an elevated temperature to optimize lubricant function, these data also point to one of the limitations of grafted lubricants that their durability may not be adequate where a large number of reciprocating cycles are expected.

Attachment of a lubricant to a surface may also be achieved by modifying one or more end groups of a lubricant molecule before deposition to increase bond strength to the surface. This technique is also often identified as grafting, but this description is not strictly applicable since the lubricant structure is modified chemically in the absence of the surface for which it is intended. Bare metal surfaces tend to adsorb organic materials readily because these adsorbates lower the free energy of the interface between the metal and the ambient environment [244]. Self-assembled monolayers (SAMs), or self-assembled multilayers, provide a convenient, flexible, and simple system with which to tailor the interfacial properties of metals (as well as metal oxides and semiconductors). SAMs are organic assemblies formed by the adsorption of molecular constituents from solution or the gas phase onto the surface of solids and the subsequent spontaneous organization of the adsorbates into crystalline or semicrystalline structures [245]. The molecules or ligands that form SAMs have a chemical functionality, or "headgroup," with a specific affinity for a substrate. There are a number of headgroups that bind to specific metals but one of the most extensively studied SAM class is derived from the adsorption of alkanethiols (i.e., sulfur-containing alkanes) on gold, silver, copper, palladium, and platinum [246]. The high affinity of thiols for the surfaces of these metals makes it possible to generate well-defined organic surfaces with useful and highly alterable chemical functionalities displayed at the exposed interface. This affinity is often so high that SAMs of alkanethiols and dialkanethiols on gold are becoming key elements for building many systems and devices with applications in the wide field of nanotechnology, thus extending SAMs to applications beyond those of lubricants on gold [247].

As already mentioned, one of the major current obstacles to the widespread use of grafted or functionalized lubricant layers in practical electrical contacts, at least as regards lubricants for which there exists test data [248,249], is the lack of durability over several millions (or more) reciprocating cycles. Durability of this magnitude is required for most anti-wear or

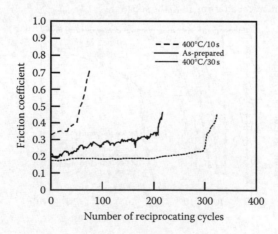

FIGURE 7.84
Evolution of the mean friction coefficient versus the number of reciprocating cycles, corresponding to three of the experimental conditions of polyacrilonitrile layers on nickel shown in Figure 7.83. (After F Houzé et al., *IEEE Trans Components, Hybrids, and Manuf. Technol.-Part A* 18: 364–368, 1995 [243].)

anti-fretting applications in practical separable electronic connectors. The maximum durability of grafted or functionalized lubricants reported so far does not exceed approximately 2×10^5 cycles [250]. For functionalized SMAs, another potential obstacle is the challenge of designing a chemical headgroup that does not promote corrosion under any circumstance. For example, the introduction of a thiol (sulfur) group in selected lubricants raises concerns about possible detachment and subsequent migration of sulfur to underlying copper alloy or Ni materials via pores in the gold, to promote pore corrosion (see Chapters 2 and 8). At present, the most promising applications for grafted and functionalized SMA lubricant layers is in selected MEMS structures where mechanical wear must be minimized and the number of reciprocating cycles during the expected device life may be relatively small.

7.4.5 Greases and Solid Lubricants

7.4.5.1 Greases

Greases are semi-solids consisting of a thickening agent and a fluid lubricant. Many fluids have been incorporated into greases and used in connectors for reducing contact wear and for controlling fretting corrosion. The thickeners include finely divided silica, soaps (sodium, lithium, barium, and other fatty acid salts), clays, Teflon powder, and various gelling agents, usually at 5–10% concentration. Mixtures of fluids and microcrystalline petroleum or synthetic waxes have also had extensive use. Petrolatum is a mixture of microcrystalline wax and mineral oil. Other additives, such as antioxidants, may be incorporated in the grease. Greases that contain graphite, or metallic particles such as zinc or silver, are not used in electronic connectors.

Grease structures are complex, consisting of interlocked fibrous crystallites (Figure 7.85a) or finely divided particles having a large surface area (Figure 7.85b) [251]. The solid component of greases forms a mixture in which the fluid is entrapped.

An advantage of greases is that they retain a fluid on the contact surface, thus minimizing loss by creep. Some "bleeding" may still occur, and, at worst, leave behind the solid phase that can cause the contact resistance of a connector to increase. The solid component must not agglomerate, or else contact resistance may be degraded. Clay- and Teflon-thickened

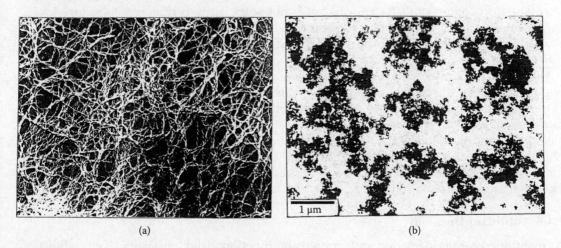

(a) (b)

FIGURE 7.85
Grease structures: (a) lithium hydroxystearate; (b) silica base. (After JH Harris, In ER Braithwaite, ed. *Lubrication and Lubricants*, Amsterdam: Elsevier, 1967, pp 197–268 [251].)

greases have been troublesome, especially for lightly loaded devices; but this was often attributable to deficiencies in the manufacture of the greases rather than in anything fundamental. Greases must not be too hard, since this may interfere with surface asperity mating. "Self healing," or the ability of flow back to recoat the wear track after a pass, is also less efficient with grease than with the base fluid of the grease [46]. One of the potential hazards of using grease in a separable electronic connector is the inadvertent introduction of copious amounts of corrosion-promoting materials, such as sulfur, since greases contain materials derived from petroleum. The presence of such contaminants defeats one of the main objectives of using grease, which is to protect the contact from corrosive atmospheric pollutants.

Greases are applied to surfaces by brushing, injection through nozzles, or, for thin coatings, by dilution in a volatile liquid and immersing the contact in the stirred mix. Alternately, the diluted grease may be sprayed with a propellant on the contacts. Agitation is necessary to ensure that the solid, fluid lubricant, and carrier components are uniformly mixed.

The technical literature on greases, their compositions, manufacture, and application, is extensive; but very little has been published that is specific to connectors. A few examples of work with greases from the literature are as follows:

1. Reference [194] used a polyalkylene glycol—10% fumed silica grease to restore tin–lead contacts to service that had failed by fretting corrosion.

2. Reference [74,75] used a halogenated silicone oil gelled with a fluorocarbon polymer to stabilize the contact of fretting tin-plated contacts.

3. Reference [252] evaluated a lithium soap and clay-thickened grease in a switch. Although this was not a connector application, an analysis of contact problems is given.

4. Reference [180] tested greases made from microcrystalline wax with these fluids: polyphenyl ether, polyalkylene glycol, poly-alpha-olefin, mineral oil, and a grease containing a diester, lithium soap, and a synthetic wax in a tin–lead fretting study.

7.4.5.2 Solids

Solid lubricants have generally been unsatisfactory for connector contacts. Graphite burnished on contacts lubricates well but is not sufficiently conductive to assure low contact resistance. Connectors with graphite-lubricated contacts were commercially available at one time [253]. Tough low-friction solids, such as Teflon powder, dusted into contact surfaces are unsuccessful [180].

Microcrystalline wax, which may be viewed as a highly plastic solid, can be used on wiping contacts if the contact surfaces are rough. The wax is entrained into the low spots of the surface roughness, and as the surface becomes smoother through use, the wax lubricates effectively [46]. Once the surface is burnished, any residual wax is displaced from the wear track, and sliding may convert from the mild to the severe regime. Since the porosity of platings increases with initial roughness, the trend to thinner precious metal deposits for economic reasons has increased awareness of the advantage of preparing contact surfaces that are smooth. With this, interest in wax lubricants has essentially disappeared.

7.4.6 Lubricant Durability

As mentioned earlier in relation with grafted and functionalized lubricants, the durability, or persistence of lubricant coatings on contact surfaces, is of concern when many mating cycles are required during the life of a separable connector. Unlike most fluid-lubricated bearings that are adapted with sumps that provide a continuous copious supply of lubricant to the surfaces, electric contact interfaces carry lubricant in thin coatings and thus in highly limited supply. If only a few matings are involved, very thin lubricant layers are generally satisfactory, for example, grafted layers or lubricants layers obtained by dipping from an appropriate solution for both noble and non-noble metal platings [232,248–250]. Adventitious contamination is often beneficial if the contaminant is neither corrosive nor chemically reactive with the contact surfaces, and acts as a lubricant. In such cases, and if mating and re-mating does not involve sliding along precisely the same track [254] to disperse or remove the contaminant, the effective supply of adventitious lubricant available to the sliding interface is large and effective lubricant durability is high.

In the case of fretting, however, the numbers of wear cycles may be large, and thin lubricant layers (whether flash gold coatings,fluids, greases or adventitious contaminants) are relatively ineffective. There have been few quantitative studies of lubricant wear-out processes. Operating conditions, contact shape and surface roughness, the contact material, and the physical properties of the lubricant coating determine its persistence. Figures 7.78 and 7.80 illustrate examples from fretting where the effectiveness of a lubricant was related to the amount deposited on the surface.

For materials that are thermally stable and relatively resistant to oxidation, lubricant wear-out stems from evaporation, film creep, and mechanical loss [22,41,232]. Mechanical loss is due to removal by sliding wherein wear debris carries the lubricant away. In the case of solid lubricants such as thin gold layers, the lubricant is lost more directly via layer wear-out and subsequent displacement from the wear track. When chemical effects occur, such as oxidation of the lubricant at excessively high temperatures, the viscosity changes. This leads to sludge formation and even contact corrosion; these mechanisms then conspire to deplete the lubricant supply from the sliding interface.

On the basis of the literature on lubricant durability reviewed above, and the authors' experience, the following recommendations can be made for the required concentration of lubricant in a carrier solvent for forming an acceptably thick layer of residual lubricant on a contact surface by dipping, for a variety of practical electrical contact applications:

1. for wiping and sliding contacts (friction and wear reduction), 1–2%;
2. for fretting (control of corrosion and frictional polymerization), 2–20%;
3. for tarnish and galvanic corrosion inhibition, discussed later, 2–5%;
4. for dispersing particulate matter on the contact (see below), 1–10%.

Thicker lubricant coatings are not deleterious, and neat fluids as droplets or smears of greases, have been used with good results.

7.4.7 Other Considerations

7.4.7.1 Dispersal of Particulates

Contamination of contact surfaces by dust can interfere with the formation of a reliable electrical contact. The severity of the deleterious effect of dust relates to the sliding/wiping distance in the connection. Long wipes are more effective in displacing contaminants than short movements, as in fretting. Lubricated surfaces are more prone to entrapping particulates than clean surfaces, and the severity of the resulting problem depends on the physical state of the lubricant. Waxes and greases are more likely to retain particles than fluids, which tend to disperse these contaminations. Dielectric constant also plays a role;

lubricants with low constants attract dusts less [255–257]. Particle size, normal load, and thickness of the lubricant layer also are variables.

An interesting investigation of the effectiveness of selected contact surface shapes in displacing dust during sliding was reported in [258]. It was found that hemispherical dimples are most effective in dispersing particulate matter such as sand from a sliding connector interface due to the ease with which particles can be shoved aside by a moving hemisphere.

7.4.7.2 Corrosion Inhibition

The benefits of organic lubricants in reducing friction and wear and the control of fretting degradations in connectors are well recognized, as discussed earlier. Concurrently, there has been a reduction in the average thickness of noble metals used in connectors because of cost. An increase in electrical contact problems associated with corrosion, and particularly pore corrosion (see Chapters 3, 4, and 8) in thin finishes has resulted in a renewed interest in tailoring lubricants so that they also can inhibit corrosion and tarnishing. Most attention has been directed to the development of lubricants of increased effectiveness on electrodeposited gold finishes.

Observations of the property of petrolatum to reduce the sulfidation of silver appeared in the early contact literature. An expansion of this concept led to the development of a grease consisting of microcrystalline wax and a five-ring polyphenyl ether fluid [259] that turned out to be an effective lubricant and tarnish inhibitor for silver. Sporadic work followed, for example [111,260,261]. A review of several thin coatings and other surface treatments, and their evaluation in a flowing mixed gas corrosion test coupled with other physical studies, are provided in [220]; however, tribological studies were not part of that investigation. Coatings formed from microcrystalline wax and polyphenyl ethers appear to have broad utility. However, these layers must be carefully chosen based on the requirements of the application, especially service temperature requirements. Even more effective coatings undoubtedly can be developed [262] (see the discussion in Section 3.4).

References

1. E Rabinowicz. *Friction and Wear of Materials*, ASM, 1981.
2. IM Hutchings. *Tribology: Friction and Wear of Engineering Materials*, Boca Raton: CRC Press, 1992.
3. R Holm. The friction force over the real area of contact (in German). Wiss Veroff *Siemens-Werk* 17(4): 38–42, 1938.
4. RF Tylecote. *The Solid Phase Welding of Metals*, London: Edward Arnold (Publishers) Ltd., 1968.
5. FP Bowden, D Tabor. *The Friction and Lubrication of Solids, Part 1*, London: Oxford University Press, 1950.
6. FP Bowden, D Tabor. *The Friction and Lubrication of Solids, Part II*, London: Oxford University Press, 1964.
7. R Holm. *Electric Contacts*, Berlin: Springer-Verlag, 1946.
8. E Rabinowicz, D Tabor. Metallic transfer between sliding metals: An autoradiographic study. *Proc Roy Soc A* 208: 455–475, 1951.
9. JF Archard. Contact and rubbing of flat surfaces. *J Appl Phys* 24(8): 981–988, 1953.
10. JF Archard. Single contacts and multiple encounters. *J Appl Phys* 32(8): 1420–1425, 1961.
11. IM Feng. Metal transfer and wear. *J Appl Phys* 23: 1011–1029, 1952.
12. EW Glossbrenner. Sliding contacts for instrumentation and control. In PG Slade, ed, *Electric Contacts: Theory and Applications*, New York: Marcel Dekker, Inc., 1999, p 885.

13. IH Brockman, CS Sieber, RS Mroczkowski. A limited study of the effects of contact normal force, contact geometry and wipe distance on the contact resistance of gold-plated contacts. *IEEE Trans Comp, Hybrids, Manuf Technol* CHMT-11: 393–400, 1988.

14. T Smith. The hydrophilic nature of a clean gold surface. *J Colloid Interf Sci* 75: 51–55, 1980.

15. RA Coutu Jr, PE Kladitis, KD Leedy, RL Crane. Selecting metal alloy electric contact materials for MEMS switches. *J Micromech Microeng* 14: 1157–1164, 2004.

16. JT Burwell, CD Strang. On the empirical law of adhesive wear. *J Appl Phys* 23(1): 18–28, 1952.

17. JT Burwell, CD Strang. Metallic wear. *Proc Roy Soc A* 212: 470–477, 1952.

18. JF Archard, W Hirst. The wear of metals under unlubricated conditions. *Proc Roy Soc A* 236: 397–410, 1956.

19. R Bruckner, T Lapham. The effect of cycle rate and lubrication on the wear of electrical contacts. *Proceedings of Annual Electronic Connector Symposium*, Cherry Hill, NJ: Electronic Connector Study Group, Inc., 1969, pp 351–368.

20. M Antler. Processes of metal transfer and wear. *Wear* 7: 181–204, 1964.

21. M Cocks. Interaction of sliding metal surface. *J Appl Phys* 33: 2152–2101, 1962.

22. M Antler. Wear, friction, and electrical noise phenomena in severe sliding systems. *ASLE Trans* 5: 297–307, 1962.

23. M Antler. Metal transfer and the wedge forming mechanism. *J Appl Phys* 34(2): 438–439, 1963.

24. H Fry, HG Feller. Investigation of wear with the scanning electron microscope. *Prakt Metallographie* 9: 182–197, 1972.

25. KI Molanin. In VD Kuznetsov, ed, *Metal Transfer and Build Up in Friction and Cutting*, New York: Pergamon, 1966, p 283 (transl. from Russian).

26. M Antler. Wear of electrodeposited gold. *ASLE Trans* 11(3): 148–260, 1968.

27. N Ohmae, E Rabinowicz. The wear of the noble metals. ASLE Preprint No. 78-LC-2C-1.

28. SA Barber, E Rabinowicz. Material selection for noble metal slip rings. *Proceedings of Holm Conference on Electrical Contacts*, Chicago, 1980, pp 33–40.

29. DH Buckley. *Friction Behavior of Members of the Platinum Metals Group with Gold*. NASA Tech Note D-7896, NASA, Langley Research Center, 1975.

30. AJ Solomon, M Antler. Mechanisms of wear of gold plate. *Plating* 57: 812–816, 1970.

31. FI Nobel, DW Thomson, JM Leibel. An evaluation of 18 karat and 24 karat hard gold deposits for contact applications. *Proceedings of the 4th Plating in Electronic, Industry Symposium*, American Electroplaters' Society, Indianapolis, 1973, pp 106–130.

32. FB Koch, Y Okinaka, W Wolowodiuk, DR Blessington. Additive-free hard gold plating for electronics applications, Part 1. *Plat Surf Finish* 67(6): 50–54, 1980.

33. FB Koch, Y Okinaka, W Wolowodiuk, DR Blessington. Additive-free hard gold plating for electronics applications, Part II. *Plat Surf Finish* 67(7): 43–45, 1980.

34. WH Abbott. Precious metal electrodeposits for electrical contact applications. *Proceedings of the Symposium on Electrodeposited Metals as Materials for Selected Applications*, NTIS, US Dept. Commerce, Springfield, VA, 3–4 November, 1971, pp 32–37.

35. Y Okinaka, FB Koch, C. Wolowodiuk, DR Blessington. "Polymer inclusions" in cobalt-hardened electroplated gold. *J Electrochem Soc* 125(11): 1745–1750, 1978.

36. M Antler. Wear of gold plate; effect of surface films and polymer codeposits. *IEEE Trans Parts Hybrids, Packag* 10: 11–17, 1974.

37. Y Okinaka, S. Nakahara. Structure of electroplated hard gold observed by transmission electron microscopy. *J Electrochem Soc* 123(9): 1284–1289, 1976.

38. M Antler. Structure of polymer codeposited in gold electrodeposits. *Plating* 60(5): 468–173, 1973.

39. JP Celis, J Roos, J Van Humbeek. Cobalt hardened gold layers for electrical connectors: optimization of wear properties. *Trans Inst Met Fin* 67: 70–72, 1989.

40. JA Greenwood, D Tabor. The properties of model friction junctions. *Conference on Lubrication and Wear*, London, 1957, Inst. Mech. Eng., paper 92.

41. M Antler. The lubrication of gold. *Wear* 6: 44–65, 1963.

42. N Biunno, M Barbetta. A root cause failure mechanism for solder joint integrity of electroless nickel/immersion gold surface finishes, *IPC EXPO Conference Proceeding*, Long Beach, CA, 1999, S18-5-1–S18-5-8.

43. F Cordes, R Huemoeller. Electroless nickel-gold: is there a future? Electroless Ni/Au plating capability study of BGA packages, *IPC EXPO Conference Proceeding*, Long Beach, CA, 1999, S13-1-1–S13-1-6.

44. M Walsh. Electroless nickel immersion gold and black pad. *Galvanotechnik* 93(9): 2281–2286, 2002.

45. NP Suh. *Tribophysics*, Englewood cliffs, NJ: Prentice-Hall, Inc., 1986.

46. M Antler. Tribological properties of gold for electric contacts. *IEEE Trans Parts, Hybrids, Packag* PHP-9(1): 4–14, 1973.

47. LG Liljestrand, L Sjogren, LB Revay, B Asthner. Wear resistance of electroplated nickel-hardened gold. *IEEE Trans Comp, Hybrids, Manuf Technol* CHMT-8: 123–128, 1985.

48. NP Suh. The delamination theory of wear. *Wear* 25: 111–124, 1973.

49. N Saka, JJ Pamies-Teixeira, NP Suh. Wear of two-phase metals. *Wear* 445: 77–86, 1977.

50. NP Suh et al. Series of papers published with a number of co-authors in *Wear* 44: 1–162, 1977. (Also see references in these papers to earlier works on delamination).

51. P Heilmann, J Don, TC Sun, DA Rigney. Sliding wear and transfer. *Wear* 91: 171–190, 1983.

52. DA Rigney. Sliding wear of metals. *Ann Rev Mater Sci* 18: 141–163, 1988.

53. J Don, TC Sun, DA Rigney. Friction and wear of Cu-Be and dispersion-hardened copper systems. *Wear* 91: 191–199, 1983.

54. AR Rosenfield. Wear and fracture mechanics. In DA Rigney, ed, *Fundamentals of Friction and Wear of Materials*, Metals Park, OH: American Society for Metals, 1981, pp 221–234.

55. JW Ho, C Noyan, JB Cohen, VD Khanna, Z Eliezer. Residual stresses and sliding wear. *Wear* 84: 183–202, 1983.

56. H Tian, N Saka, E Rabinowicz. Friction and failure of electroplated sliding contacts. *Wear* 142: 57–85, 1991.

57. RG Bayer. A general model for sliding wear in electrical contacts. *Wear* 162–164: 913–918, 1993.

58. RG Bayer, E Hsue, J Turner. A motion induced sub-surface deformation wear mechanism. *Proceedings of the International Wear of Materials Conference*, Orlando, FL, 1991, pp 489–496.

59. M Antler, MH Drozdowicz. Wear of gold electrodeposits: Effect of substrate and of nickel underplate. *Bell Syst Tech J* 58(2): 323–349, 1979.

60. KL Schiff. Precious metals cost reduction without loss of reliability. *Proceedings of the 9th Annual Connector Symposium*, Electronic Connector Study Group, Inc., Cherry Hill, NJ, 20–21 October, 1976, pp 131–149.

61. A Keil, E Mahle. Wear behaviour of rhodium coatings on variously hard underlays. *Metalloberflache* 2: 44–48, 1968.

62. CA Holden. Wear study of electroplated coatings for contacts. *Proceedings of the Engineering Seminar on Electrical Contact Phenomena*, Chicago, 1967, pp 1–20.

63. T Sato, Y Matsui, M Okada, K Murakawa, Z Henmi. Palladium with a thin gold layer as a sliding contact material. *Proceedings of the Holm Conference on Electrical Contacts*, Chicago, 1980, pp 41–47.

64. M Antler. Corrosion control with tin nickel and other intermetallics. In H Leidheiser Jr, ed, *Corrosion Control by Coatings*, Princeton, NJ: Science Press, 1979, pp 115–133.

65. M Antler. Sliding wear of metallic contacts. *IEEE Trans Comp, Hybrids, Manuf Technol* 4(1): 15–29, 1981.

66. M Antler. Wear of gold contact finishes: The importance of topography, underplate, and lubricants. *Insulation/Circuits* 26(1): 15–19, 1980.

67. SM Garte. Effect of substrate roughness on the porosity of gold electrodeposits. *Plating* 53: 1335–1339, 1966.

68. M Antler. Sliding wear and friction of electroplated and clad connector contact materials: Effect of surface roughness. *Wear* 214: 1–9, 1998.

69. WF Fluehmann, FH Reid, PA Mausli, SG Steinemann. Effect of pulsed current plating on structure and properties of gold cobalt electrodeposits. *Plat Surf Finish* 67(6): 62–65, 1980.

70. KJ Whitlaw, JW Souter, IS Wright, MC Nottingham. Wear properties of high speed gold electro-deposits. *IEEE Trans Comp, Hybrids, Manuf Technol* 8(1): 46–51, 1985.

71. CA Mattoe. Effect of low levels of codeposited cobalt on electroplated gold deposits. *Proceedings of the Annual American Electroplaters and Surface Finishers Conference*, Orlando, FL, 1982.

72. G Horn, W Merl. Friction and wear of electroplated hard gold deposits for connectors. *Proceedings of the 6th International Conference on Electrical Contact Phenomena*, Chicago, 1972, pp 65–72.

73. M Antler. Gold plated contacts: effect of thermal aging on contact resistance, *Proceedings of the IEEE Holm Conference on Electrical Contacts*, Philadelphia, PA, 1997, pp 121–131.

74. CH Leung, A Lee. Thermal cycling induced wiping wear of connector contacts at 150°C. *Proceedings of the IEEE Holm Conference on Electrical Contacts*, Philadelphia, 1997, pp 132–137.

75. CH Leung, A Lee. Thermal cycling induced wiping wear of connector contacts at 150°C. *IEEE Trans Comp Packag Technol* 22: 72–78, 1999.

76. M Antler, ET Ratliff. Sliding wear of inlay clad metals and electrodeposited cobalt-gold. *IEEE Trans Comp, Hybrids, Manuf Technol* 6(1): 3–7, 1983.

77. M Antler. Guidelines in the selection of noble metal connector contact finishes: Sliding friction and wear. *Connection Technol* 6(3): 29–33, 1990.

78. PJ Kay, CA Mackay. The growth of intermetallic compounds on common basis materials coated with tin and tin-lead alloys. *Trans Inst Met Fin* 54: 68–74, 1976.

79. RS Timsit. Interdiffusion at bi-metallic electrical interfaces. *IEEE Trans Comp, Hybrids Manuf Technol* CHMT-9: 106–116, 1985.

80. T Hammam. Friction, wear, and electric properties of tin-coated tin bronze for separable electric connectors. *Proceedings of the 18th International Conference on Electrical Contacts*, Chicago, 1996, pp 321–330.

81. RoHS, Directive of the European Parliament on the restriction of the use of certain hazardous substances. http://www.rohs.eu/english/index.html, 2009.

82. M Warwick. Implementing lead free soldering-European Consortium Research. *J. Surface-Mount Technol* 12: 1–12, 1999.

83. JW Osenbach, RL Shook, BT Vaccaro, BD Potteiger, AN Amin, KN Hooghan, P Suratkar, P Ruengsinsub. Sn whiskers: material, design, processing, and post-plate reflow effects and development of an overall phenomenological theory. *IEEE Trans Electron Packag Manuf* 28: 36–62, 2005.

84. M Dittes, P Oberndorff, L Petit, CFT Philips. Tin whisker formation – results, test methods and countermeasures. *Proceedings of the 53rd Electronic Components & Technology Conference*, New Orleans, 2003, pp 822–826.

85. RV Chiarenzelli. Tarnishing studies on contact materials. *IEEE Trans Parts, Mater Packag* PMP-3: 89–96, 1967.

86. WH Abbott, HR Ogden. The influence of environment on tarnishing reactions. *Proceedings of the 4th International Research Symposium on Electrical Contact Phenomena*, Swansea, UK, 1968, pp 35–39.

87. WH Abbott. The effects of test atmosphere conditions on the contact resistance of surface films on silver. *Proceedings of the 11th International Conference on Electrical Contact Phenomena*, Berlin, Germany, 1982, pp 294–296.

88. DW Rice, P Peterson, EB Rigby, PBP Phipps, RJ Cappell, R Tremoureaux. Atmospheric corrosion of copper and silver. *J Electrochem Soc* 128: 275–284, 1981.

89. M Antler. Field studies of contact materials: contact resistance behavior of some base and noble metals. *IEEE Trans Comp, Hybrids, Manuf Technol* CHMT-5: 301–307, 1982.

90. J Guinement, C Fiaud. Laboratory study of the reaction of silver and copper with some Atmospheric Pollutants – Comparison with indoor exposition of these materials. *Proceedings of the 13th International Conference on Electrical Contacts*, Lausanne, Switzerland, 1986, pp 383–390.

91. WH Abbott. The development and performance characteristics of mixed flowing gas test environments. *Proceedings of the 33rd IEEE Holm Conference on Electrical Contacts*, Chicago, IL, 1987, pp 63–78.

92. WH Abbott. Materials, environment, motion, and electrical contact failure mechanisms, *Proceedings of the 35th IEEE Holm Conference on Electrical Contacts*, Chicago, IL, 1989, pp 3–11.

93. T Imrell, R Sjovall, A Kassman. The composition of the silver coating strongly influences the stability of stationary electrical contacts. *Proceedings of the 17th International Conference on Electrical Contacts*, Nagoya, Japan, 1994, pp 447–454.

94. TE Graedel. Corrosion mechanisms for silver exposed to the atmosphere. *J Electrochem Soc* 139: 1963–1970, 1992.

95. AK Rudolphi, S Jacobson. Stationary loading, fretting and sliding of silver coated copper contacts – Influence of corrosion films and corrosive atmosphere. *Tribology Inter* 30: 165–175, 1997.

96. H Bresgen. Contact resistance, adhesion and wear of tarnished and untarnished silver subjected to an oscillating friction load. *Wear* 69: 157–165, 1981.

97. M Myers. Overview of the use of silver in connector applications. http://www.tycoelectronics.com/documentation/whitepapers/pdf/Ag_use_connectors_503-1016.pdf, Technical Paper 503-1016, 5 February 2009.

98. M Myers. The performance implications of silver as a contact finish in traditionally gold finished contact applications. *Proceedings 55th IEEE Holm Conference on Electrical Contacts*, Vancouver, BC, Canada, 2009, pp 310–318.

99. M Antler, MH Drozdowicz. Fretting corrosion of gold-plated connector contacts. *Wear* 74: 27–50, 1981–1982.

100. WB Fagerstrom, ET Nicotera. Fretting corrosion in connectors. *Proceedings of the Connectors and Interconnections Symposium*, Electronic Connector Study Group, Inc., Philadelphia, 1981, pp 303–312.

101. EM Bock, JH Whitley. Fretting corrosion in electric contacts. *Proceedings of the Holm Seminar on Electrical Contacts*, Chicago, 1974, pp 128–138.

102. PJ Thiesen, JA Forsell. Connector dependent fretting corrosion test system. *Proceedings of the Holm Conference on Electrical Contacts*, Chicago, 1979, pp 109–115.

103. JL Johnson, LE Moberly. Separable electrical power contacts involving aluminum bus bars. *Proceedings of the Holm Seminar on Electrical Contacts*, Chicago, 1975, pp 53–59.

104. HW Hermance, TF Egan. Organic deposits on precious metal contacts. *Bell Syst Tech J* 37: 739–777, 1958.

105. JJ Mottine, BT Reagor. Investigation of fretting corrosion at dissimilar metal interfaces in socketed IC device applications. *Proceedings of the Holm Conference on Electrical Contacts*, Chicago, 1983, pp 61–71.

106. JH Whitley. Investigation of fretting corrosion phenomena in electric contacts. *Proceedings of the 8th International Conference on Electrical Contact Phenomena*, The Institution of Electrical Engineers of Japan, Tokyo, 1976, pp 659–665.

107. JH Keefer, RH Gumley. Relay contact behavior under noneroding conditions. *Bell Syst Tech J* 37: 778–814, 1958.

108. ES Sproles. A comparison of Pd–Pd and Pd–Au connector contacts in air and toluene saturated air. *Proceedings of the Connectors and Interconnections Symposium*, Electronic Connector Study Group, Philadelphia, 1981, pp 267–280.

109. GW Mills. Preliminary evaluation of DIP/socket connection reliability. *IEEE Trans Comp, Hybrids, Manuf Technol* 2(4): 476–489, 1979.

110. A Fairweather, F Lazenby, AE Parker. Development of resistance and microphonic noise at a disturbed contact. *Proceedings of the 2nd International Symposium on Electrical Contact Phenomena*, Technische Hochschule, Graz, Austria, 1964, pp 316–338.

111. WH Abbott, JH Whitley. The lubrication and environmental protection of alternatives to gold for electronic connectors. *Proceedings of the Holm Seminar on Electrical Contacts*, Chicago, 1975, pp 9–16.

112. M Antler. Gold connector contacts: developments in the search for alternate materials. *IEEE Trans Parts, Hybrids, Packag* 11: 216–220, 1975.

113. N Aukland, H Hardee, A Wehr, P Lees. An examination of the metallic bonding of a clad material and two gold plating systems under constant force fretting conditions. *Proceedings of the IEEE Holm Conference on Electrical Contacts*, Philadelphia, 1997, pp 7–19.

114. P van Dijk, F van Meijl. A design solution for fretting corrosion. *Proceedings of the 18th International Conference on Electrical Contacts*, Philadelphia, 1997, pp 375–382.
115. O Vingsbo, S Soderberg. On fretting maps. *Wear* 126: 131–147, 1988.
116. AK Rudolphi. Tribology of electrical contacts, deterioration of silver coated copper. PhD Thesis, Uppsala University, Sweden, 1996.
117. S Hannel, S Fouvry, Ph Kapsa, L Vincent. The fretting sliding transition as a criterion for electrical contact performance. *Wear* 249: 761–770, 2001.
118. WH Abbott. Time distribution of intermittents versus contact resistance for tin–tin connector interfaces during low amplitude motion. *Proceedings of the Holm Conference on Electrical Contacts*, Chicago, 1983, pp 55–59.
119. WH Abbott, WE Campbell. Frictional polymer formation on precious metals - experimental observations. *Proceedings of the 9th International Conference on Electric Contact Phenomena*, Chicago, 1978, pp 359–362.
120. IM Feng, BG Rightmire. The mechanism of fretting. *Lubrication Engrg* 9: 134–136, 1953.
121. MD Bryant. Assessment of fretting failure models of electrical connectors. *Proceedings of the IEEE Holm Conference on Electrical Contacts*, Chicago, 1994, pp 167–175.
122. RD Malucci. Impact of fretting parameters on contact degradation. *Proceedings of the 18th International Conference on Electrical Contacts*, Chicago, 1996, pp 395–403.
123. R Schubert. Degradation and regeneration of copper junctions. *Phys Rev B* 43: 1433–1440, 1991.
124. SR Murrell, SL McCarthy. Intermittance detection in fretting corrosion studies of electrical contacts. *Proceedings of the IEEE Holm Conference on Electrical Contacts*, Philadelphia, 1997, pp 1–6.
125. M Antler, ES Sproles. Effect of fretting on the contact resistance of palladium. *IEEE Trans Comp, Hybrids, Manuf Technol* 5(1): 158–166, 1982.
126. ES Sproles. The effect of frictional polymer on the performance of palladium connector contacts. *Proceedings of the IEEE Electronic Components Conference*, San Francisco, CA, 1980, pp 317–325.
127. WH Abbott, RL Schreiber. Dynamic contact resistance of gold, tin, and palladium connector interfaces during low amplitude motion. *Proceedings of the Holm Conference on Electrical Contacts*, Chicago, 1981, pp 211–219.
128. HS Blanks. Detection and accelerated testing of vibration-induced connector wear. *Proceedings of the Holm Conference on Electrical Contacts*, Chicago, 1983, pp 45–53.
129. M Braunovic, NS McIntyre, WJ Chauvin, I Aitchison. Surface analysis of fretting damage in electrical contacts of aluminum with different contact materials. *Proceedings of the Holm Conference on Electrical Contacts*, Chicago, 1983, pp 231–242.
130. M Braunovic. Effect of fretting on the contact resistance of aluminum with different contact materials. *IEEE Trans Comp, Hybrids, Manuf Technol* 2(1), 25–31, 1979.
131. H Kongsjorden, J Kulsetas, J Sletbak. Degradation of electrical contacts caused by oscillatory micromotion between the contact members. *Proceedings of the 9th International Conference on Electrical Contact Phenomena*, Chicago, 1978, pp 87–92.
132. SM Garte. The effect of design on contact fretting. *Proceedings of the Holm Seminar on Electrical Contacts*, Chicago, 1976, pp 65–70.
133. D Gyurina, EF Smith. A laboratory study of the electrical properties of copper alloys in electric contact applications. *Proceedings of the Holm Conference on Electrical Contacts*, Chicago, 1980, pp 85–93.
134. WO Freitag. Lubricants for separable connectors. *Proceedings of the Holm Seminar on Electrical Contacts*, Chicago, 1976, pp 57–63.
135. WO Freitag. Wear, fretting and the role of lubricants in edge card connectors. *Proceedings of the Holm Seminar on Electrical Contacts*, Chicago, 1975, pp 17–23.
136. WA Crossland, PMK Murphy. The formation of insulating organic films on palladium-silver contact alloys. *IEEE Trans Parts, Hybrids, Packag* 10(1): 64–73, 1974.
137. M Antler. Fretting of electrical contacts. In SR Brown, ed, *Materials Evaluation Under Fretting Conditions*, Special Technical Publication, STP780, West Conshohocken, PA: ASTM, 1982, pp 68–85.

138. NR Aukland, HC Hardee. A statistical comparison of gold plating systems under specific fretting parameters. *Proceedings of the IEEE Holm Conference on Electrical Contacts*, Chicago, 1994, pp 177–188.

139. P Jedrzejczyk, S Fouvry, P Chalandon. A fast methodology to quantify electrical-contact behaviour under fretting loading conditions. *Wear* 267: 1731–1740, 2009.

140. C Maul, JW McBride, J. Swingler. Inermittency phenomena in electrical connectors. *IEEE Trans Comp Packag Technol* 24: 370–377, 2001.

141. EM Chow, K Klein, DK Fork, T Hantschel, CL Chua, L Wong, K Van Schuylenbergh. Intermittency study of a stressed-metal micro-spring sliding electrical contact. *Proceedings of the 53rd Electronic Components and Technology Conference*, New Orleans, LA, 2003, pp 1714–1717.

142. M Braunovic. Aluminum connections: legacies of the past. *Proceedings of the IEEE Holm Conference on Electrical Contacts*, Chicago, 1994, pp 1–31.

143. *Standard Methods for Measuring Contact Resistance of Electrical Connections (Static Contacts)*. B539-80, West Conshohocken, PA: ASTM.

144. M Antler. Fretting of electrical contacts: An investigation of palladium mated to other materials. *Wear* 81: 159–173, 1982.

145. E Rabinowicz, SW Webber. The formation of frictional polymers on noble metal surfaces. *Proceedings of the 11th International Conference on Electrical Contact Phenomena*, Berlin, 1982, pp 98–107.

146. WH Abbott. Frictional polymer formation on precious metal alloys. *Proceedings of the Holm Conference on Electrical Contacts*, Chicago, 1979, pp 11–16.

147. S Mori, Y Shitara. Mechanochemical activity of nascent surfaces of gold, silver, and copper. *Proceedings of the IEEE Holm Conference on Electrical Contacts*, Philadelphia, 1997, pp 183–189.

148. JJ Mottine, BT Reagor. The effect of lubrication on fretting corrosion at dissimilar metal interfaces in socketed IC device applications. *Proceedings of the 12th International Conference on Electric Contact Phenomena*, Chicago, 1984, pp 171–181.

149. WO Freitag. Wear, fretting and the role of lubricants in edge card connectors. *IEEE Trans Parts, Hybrids Packag* PHP-12(1): 40–44, 1976.

150. YW Park, TSN Sankara Narayanan, KY Lee. Fretting corrosion of tin-plated contacts. *Tribol Int* 41: 616–628, 2008.

151. YW Park, TSN Sankara Narayanan, KY Lee. Effect of fretting amplitude and frequency on the fretting corrosion behaviour of tin plated contacts. *Surf Coat Tech* 201: 2181–2192, 2006.

152. M Braunovic. Evaluation of different types of contact aid compounds for aluminum to aluminum connectors and conductors. *Proceedings of the 12th International Conference on Electric Contact Phenomena*, Chicago, 1984, pp 97–104.

153. Ph Castell, A Menet, A Carballeira. Fretting corrosion in low-load electrical contacts: a quantitative analysis of the significant variables. *Proceedings of the 12th International Conference on Electric Contact Phenomena*, Chicago, 1984, pp 75–82.

154. J Ambier, P Perdigon. Fretting corrosion of separable electrical contacts. *Proceedings of the 12th International Conference on Electric Contact Phenomena*, Chicago, 1984, pp 105–112.

155. GT Flowers, X Fei, MJ Bozack, RD Malucci. Vibration thresholds for fretting corrosion in electrical connectors. *IEEE Trans Comp Packag Technol* 27: 65–71, 2004.

156. P Capp, A Epstein. An evaluation of palladium silver inlay as an alternative to gold plated BeCu spring clips used in screw machine DIP sockets. *Proceedings of the National Electronic Packaging Conference*, Anaheim, CA, 1982, pp 238–252.

157. FI Nobel. Electroplated palladium silver alloys. *Proceedings of the 12th International Conference on Electrical Contact Phenomena*, Chicago, 1984, pp 137–154.

158. H Harmsen, H Thiede. Precious metal alloys as wear resistant contact materials for connectors. *Proceedings of the Connector Symposium*, Electronic Connector Study Group, Inc., Cherry Hill, NJ, 1977, pp 392–402.

159. TR Long, KF Bradford. Contact resistance behavior of the 60Pd40Ag alloy in tarnishing environments. *Proceedings of the Holm Seminar on Electrical Contacts*, Chicago, 1975, pp 145–154.

160. CA Haque, M Antler. Atmospheric corrosion of clad palladium and palladium silver alloys: Film growth and contamination effects. *Corros Sci* 22(10): 939–949, 1982.

161. FE Bader. Diffused gold R156 – A new inlay contact material for Bell System connectors. *Proceedings of the 11th International Conference on Electric Contact Phenomena*, Verband Deutscher Elektrotechniker, Berlin, 1982, pp 133–137.

162. KJ Whitlaw. How effective is palladium nickel as a replacement for gold? *Trans Inst Met Finish* 60: 141–146, 1982.

163. AH Graham, MJ Pike-Biegunski, SW Updergraff. Evaluation of palladium substitutes for gold. *Plat Surf Finish* 70: 52–57, 1983.

164. MP Toben, JL Martin, RA Russo. Palladium-nickel electrodeposits—properties and selection. SUR/FIN 97. *Proceedings of the American Electroplaters and Surface Finishers Conference*, Orlando, FL, 1997.

165. CA Morse, NR Aukland, HC Hardee. A statistical comparison of gold and palladium-nickel plating systems for various fretting parameters. *Proceedings of the IEEE Holm Conference on Electrical Contacts*, Montreal, 1995, pp 33–49.

166. N Aukland, I Leslie, H Hardee, P Lees. A statistical comparison of a clad material and gold flashed palladium-nickel under various fretting conditions. *Proceedings of the IEEE Holm Conference on Electrical Contacts*, Montreal, 1995, pp 52–63.

167. JP Bare, AH Graham. Connector resistance to failure by fretting and frictional polymer formation. *Proceedings of the IEEE Holm Conference on Electrical Contacts*, Chicago, IL 1985, pp 147–155.

168. G Holmbom, F Humiec, JA Abys, EJ Kudrak, GF Breck, I Boguslavsky. Materials properties of palladium alloy electrodeposits. *Proceedings of the 31st Annual IICIT Connector and Interconnection Symposium*, Danvers, MA, 1998, pp 313–320.

169. AH Graham. Wear resistance characterization for plated connectors. *30th Holm Conference on Electrical Contacts*, Chicago, IL, 1984, pp 61–67.

170. SW Chaikin. On frictional polymer. *Wear* 10: 49–60, 1967.

171. WE Campbell, RE Lee. Polymer formation on sliding metals in air saturated with organic vapors. *ASLE Trans* 5: 91–104, 1962.

172. BT Reagor, L Seibles. Structural analysis of deuterated and nondeuterated frictional polymers using Fourier transform infrared spectroscopy and pyrolysis gas chromatography/mass spectroscopy. *IEEE Trans Components, Hybrids, Manuf Technol* 4(1): 102–108, 1981.

173. A Shinchi, Y Imada, F Honda, K Nakajima. A tribochemical investigation of the reaction products on Pd plated contacts. *Proceedings of the IEEE Holm Conference on Electrical Contacts*, Montreal, 1995, pp 64–70.

174. M Antler. Effects of lubricants on frictional polymerization of palladium electrical contacts. *ASLE Trans* 26(3): 376–380, 1983.

175. M Hasegawa, K Sawa., K Miyachi. Remarkable increase of contact resistance by mechanochemical reaction on Pd contacts under no discharge condition. *Proceedings of the 14th International Conference on Electric Contacts*, Paris, 1988, pp 233–237.

176. K Karasawa, Z-K Chen, K Sawa. Effect of impact and wipe action on contact resistance increase in Pd material. *Proceedings of the IEEE Holm Conference on Electrical Contacts*, Montreal, 1995, pp 282–288.

177. A Shinchi, Y Imada, F Honda, K Nakajima. Electric contact surface of Pd-plated metal in organic gas/air atmospheres. *Wear* 230: 78–85, 1999.

178. FH Reid. Palladium, plating—processes and applications in the United Kingdom. *Plating* 52: 531–539, 1965.

179. DT Napp. Substitution of palladium for gold in IBM products. *Proceedings of the 5th Plating in the Electronic Industries Symposium*, American Electroplaters' Society, New York, 1975, pp 28–33.

180. M Antler. Fretting corrosion of solder-coated electrical contacts. *IEEE Trans Comp, Hybrids, Manuf Technol* 7: 129–138, 1984.

181. RJ Osias, JH Tripp. Mechanical disruption of surface films on metals. *Wear* 9: 388–397, 1966.

182. JBP Williamson. The microworld of the contact spot. *Proceedings of the Holm Conference on Electrical Contacts*, Chicago, 1981, pp 1–20.

183. H Grossmann, M Huck. Tribological studies of electroplated sandwich layers on connectors. *Proceedings of the 10th International Conference on Electrical Contact Phenomena*, Vol 1, Hungarian Electrotechnical Association, Budapest, 1980, pp 401–411.

184. JH Whitley, I-Y Wei, S Krumbein. A cost-effective high performance alternative to conventional gold plating on connector contacts. *Proceedings of the IEEE Electronic Components Conference*, Orlando, FL, 1983, pp 404–417.

185. GJ Russ. A comparative study of electroplated palladium as a contact finish. *Proceedings of the IEEE Electronic Components Conference*, Orlando, FL, 1983, pp 394–399.

186. M Antler. Effect of fretting on the contact resistance of palladium electroplate having a gold flash, cobalt gold electroplate, and DG R-156. *IEEE Trans Comp, Hybrids, Manuf Technol* 7(4): 363–369, 1984.

187. JHM Neijzen, JHA Glashorster. Fretting corrosion of tin-coated electrical contacts. *IEEE Trans Comp, Hybrids, Manuf Technol* 10(1): 68–74, 1987.

188. O Schneegans, F Houzé, P Chrétien, S Noël, C Bodin, L Boyer, Jl Tristani, EM Zindine. Fretting degradation of tin plated contacts studied by means of a new electrical cartography technique based on atomic force microscopy, *Proceedings of the 19th International Conference on Electrical Contacts*, Nuremberg, Germany, 1998, pp 187–193.

189. T Hammam, R Sundberg. Heat-treatment of tin-coated copper base alloy and the subsequent effect on friction, wear and electric properties. *Proceedings of the 20th International Conference on Electrical Contact Phenomena*, Stockholm, Sweden, 2000, pp 291–296.

190. T Hammam. The impact of sliding motion and current load on the deterioration of tin-coated contact terminals. *IEEE Trans Comp Packag Technol* 23: 278–285, 2000.

191. D Alamarguy, N Lécaudé, P Chretien, S Noel, Ph Teste. Current effect on fretting degradation of hot dipped tin contacts. *Proceedings of the 21st International Conference on Electrical Contacts*, Zurich, Switzerland, 2002, pp 179–184.

192. T Hammam, A Kassman-Rudolphi, P Lundstrom. Vibration-induced deterioration of tin-coated connectors studied by using a force controlled fretting bench-test. *Proceedings of the 51st IEEE Holm Conference on Electrical Contacts*, Chicago, IL, 2005, pp 97–106.

193. J Goldstein, DE Newbury, DC Joy, CE Lyman, P Echlin, E Lifshin, L Sawyer, JR Michael. *Scanning Electron Microscopy and X-Ray Microanalysis*. New York: Kluwer Academic/Plenum Publishers, 2003.

194. M Antler, N Aukland, H Hardee, A Wehr. Recovery of severely degraded tin-lead plated connector contacts due to fretting corrosion. *Proceedings of the IEEE Holm Conference on Electrical Contacts*, Philadelphia, 1997, pp 20–32.

195. M Antler. Survey of contact fretting in electrical connectors. *IEEE Trans Comp, Hybrids, Manuf Technol* CHMT-8: 87–104, 1985.

196. JF Bruel, P Smirou, A Carballeira. Gas environment effect on the fretting corrosion behaviour of contact materials. *Proceedings of the 14th International Conference on Electrical Contacts*, Paris, France, 1988, pp 219–223.

197. A Lee, A Mao, MS Mamrick. Fretting corrosion of tin at elevated temperatures. *Proceedings of the IEEE Holm Conference on Electrical Contacts*, San Francisco, CA, 1988, pp 87–91.

198. A Lee, A Mao, MS Mamrick. Effect of contact geometry on fretting corrosion of tin-plated copper-alloy. *Proceedings of the 14th International Conference on Electric Contacts*, Paris, 1988, pp 225–231.

199. A Lee, M Mamrick. Fretting corrosion of tin with electrical load. *Proceedings of the 13th International Conference on Electric Contacts*, Lausanne, Switzerland, 1986, pp 1176–1480.

200. RD Malucci. Characteristics of films developed in fretting experiments on tin-plated contacts. *IEEE Trans Comp Packag Technol* 24: 399–407, 2001.

201. NA Stennett, J Swingler. The effect of power on low frequency fretting corrosion. *Proceedings of the IEEE Holm Conference on Electrical Contacts*, Pittsburgh, 1993, pp 205–210.

202. HG Tompkins, MR Pinnel. Low temperature diffusion of copper through gold. *J Appl Phys* 47: 3804–3812, 1976.
203. HG Tompkins, MR Pinnel. Relative rates of nickel diffusion and copper diffusion through gold. *J Appl Phys* 48: 3144–3146, 1977.
204. A Lee, MS Mamrick. Fretting corrosion of tin-plated copper alloy. *IEEE Trans Comp, Hybrids Manuf Technol* 10: 63–67, 1987.
205. CE Heaton, SL McCarthy. High cycle fretting corrosion studies on tin-coated contact materials. *Proceedings of the 47th IEEE Holm Conference on Electrical Contacts*, 2001, pp 209–214.
206. M Braunovic. Fretting damage in tin-plated aluminium and copper connectors. *IEEE Trans Comp, Hybrids Manuf Technol* 12: 215–223, 1989.
207. N Saka, MJ Liou, NP Suh. The role of tribology in electrical contact phenomena. *Wear* 100: 77–105, 1984.
208. YW Park, TSN Sankara Narayanan, KY Lee. Fretting corrosion of tin-plated contacts: evaluation of surface characteristics. *Tribol Int* 40: 548–559, 2007.
209. M Antler, T Spalvins. Lubrication with thin gold films. *Gold Bull* 21(2), 59–68, 1988.
210. M Antler, M Feder. Friction and wear of electrodeposited palladium contacts: Thin film lubrication with fluids and with gold. *IEEE Trans Comp, Hybrids, Manuf Technol* 9(4): 485–491, 1986.
211. RG Baker, TA Palumbo. The potential role of ruthenium for contact applications. *Plat Surf Finish* 69(11): 66–68, 1982.
212. CA Holden, RL Opila, HH Law, GR Crane. Wear resistance of nickel and nickel phosphorus alloy electrodeposits. *Proceedings of the IEEE Holm Conference on Electrical Contacts*, San Francisco, 1988, pp 101–107.
213. PW Lees, DWM Williams. Characterization of composite clad-electroplated contact materials. *Proceedings of the 23rd Annual Connector and Interconnection Technology Symposium*, Electronic Connector Study Group, Inc., Toronto, 1990, pp 133–148.
214. M Antler. Thin gold contact finishes for electronic connectors: A perspective. *Proceedings of the SUR/FIN 84*, American Electroplaters and Surface Finishers Society, Indianapolis, 1994.
215. M Antler. Gold in electrical contacts. *Gold Bull* 4: 42–46, 1971.
216. M Antler, unpublished results.
217. M Antler. Sliding studies of new connector contact lubricants. *IEEE Trans Comp, Hybrids Manuf Technol* 10(1): 24–31, 1987.
218. R Holm. *Electric Contacts Handbook*, 3rd ed., Berlin: Springer Verlag, 1958, see esp. p 400.
219. GJ Witter, RA Leiper. A comparison for the effects of various forms of silicon contamination on contact performance. *Proceedings of the 9th International Conference on Electric Contact Phenomena*, Chicago, 1978, pp 371–376.
220. WH Abbott, M Antler. Connector contacts: Corrosion inhibiting surface treatments for gold-plated finishes. *Proceedings of the IEEE Holm Conference in Electrical Contacts*, Montreal, 1995, pp 97–123.
221. RS Timsit, EM Bock, NE Corman. Effect of surface reactivity of lubricants on the properties of aluminum electrical contacts. *IEEE Trans Comp, Packag Manuf Technol Part A* 21(3): 500–505, 1998.
222. WE Campbell. The lubrication of electrical contacts. *Proceedings of the Holm Conference on Electrical Contacts*, Chicago, 1977, pp 1–17.
223. AJ Gellman, JS Ko. The current status of tribology surface science. *Tribol Lett* 10: 39–44, 2001.
224. JB Adams, LG Hector, DJ Siegel, H Yu, J Zhong. Adhesion, lubrication and wear on the atomic scale. *Surf Interface Anal* 31: 619–626, 2001.
225. J Aronstein, WE Campbell. Contact resistance and material transfer of soft-metal microcontacts. *Proceedings of the Holm Seminar on Electric Contact Phenomena*, Chicago, 1969, pp 7–27.
226. J Gao, WR Luedtke, U Landman. Nano-Elastohydrodynamics: structure, dynamics, and flow in non-uniform lubricated junctions. *Science* 270: 605–608, 1995.
227. F Mugele, M Salmeron. Dynamics of layering transition in confined liquids. *Phys Rev Lett* 84: 5796–5799, 2000.

228. RV Steenstrup, VN Fiacco, LK Schultz. A comparative study of inhibited lubricants for dry circuits, sliding contacts. *Proceedings of the Holm Conference on Electrical Contacts*, Chicago, 1982, pp 59–68.

229. SL McCarthy, RO Carter, WH Weber. Lubricant-induced corrosion in copper electrical contacts. *Proceedings of the Holm Conference on Electrical Contacts*, Philadelphia, 1997, pp 115–120.

230. JG Zhang, YL Zhou, K Sugimura, A Nagae. A comprehensive experiment of water soluble lubricant covered on gold plated surface. *Proceedings of the 18th International Conference on Electrical Contacts*, Chicago, 1996, pp 444–454.

231. M Antler. Organometallics in lubrication. *Ind Engrg Chem* 51: 753–758, 1959.

232. M Antler. Electronic connector contact lubricants: The polyether fluids. *IEEE Trans Comp, Hybrids, Manuf Technol* 10(1): 32–41, 1987.

233. HK Trivedi, CJ Klenke, CS Saba. Effect of formulation and temperature on boundary lubrication performance of polyphenylethers (5P4E). *Tribol Lett* 17 (1): 1–10, 2004.

234. FD Messina. Fluorescence detection of thin film electrical contact lubricants. *IEEE Trans Comp, Hybrids, Manuf Technol* 7: 47–55, 1984.

235. M Antler. Electrical effects of fretting connector contact materials: a review. *Wear* 106: 5–33, 1985.

236. A Bhattacharya, BN Misra. Grafting: a versatile means to modify polymers, techniques, factors and applications. *Prog Polym Sci* 29: 767–814, 2004.

237. S Palacin, C Bureau, J Charlier, G Deniau, B Mouanda, P Viel. Molecule-to-metal bonds: electrografting polymers on conducting surfaces. *ChemPhysChem* 5: 1468–1481, 2004.

238. C Boiziau, G Lécayon. Le greffage des polymères sur les métaux. *La Recherche* 19: 888–897, 1988.

239. G Lécayon, P Viel, C le Gressus, C Boiziau, S Leroy, J Perreau, C Reynaud. A metal-polymer interface study using electropolymerized acrylonitrile on nickel surfaces. *Scanning Microsc* 1: 85–93, 1987.

240. G Lécayon, P Viel, J Boissel, J Delhalle. Contact pour connecteur électrique revêtu d'un film de polymère et son procédé de fabrication. French Patent 9 101 604, 2 December, 1991.

241. HH Law, J Sapjeta, CED Chidsey, TM Putvinski. Protective treatments for nickel-based contact materials. *J Electrochem Soc* 141: 1977–1982, 1994.

242. HH Law, J Sapjeta, ES Sproles. Protective treatments for gold-flashed contact finishes with a nickel substrate. *IEEE Trans Comp, Hybrids, Manuf Technol-Part A* 18: 405–408, 1995.

243. F Houzé, S Noël, P Newton, L Boyer, P Viel, G Lécayon. Elaboration of heat-treated thin films of polyacrylonitrile for connector application. *IEEE Trans Comp, Hybrids, Manuf Technol-Part A* 18: 364–368, 1995.

244. AW Adamson, AP Gast. *Physical Chemistry of Surfaces*, 6th ed., New York: Wiley-Interscience, 1997.

245. F Schreiber. Structure and growth of self-assembling monolayers. *Prog Surf Sci* 65: 151–256, 2000.

246. JC Love, LA Estroff, JK Kriebel, RG Nuzzo, GM Whitesides. Self-assembled monolayers of thiolates on metals as a form of nanotechnology. *Chem Rev* 105: 1103–1169, 2005.

247. C Vericat, ME Vela, G Benitez, P Carrob, RC Salvarezza. Self-assembled monolayers of thiols and dithiols on gold: new challenges for a well-known system. *Chem Soc Rev* 39: 1805–1834, 2010.

248. S Noël, D Alamarguy, A Benedetto, P Viel, M Balog. Influence of grafting properties of organic thin films for low level electrical contacts protection. *Proceedings of the 54th IEEE Holm Conference on Electrical Contacts*, Orlando, 2008, pp 249–257.

249. D Alamarguy, S Noël, O Schneegans, R Meyer, L Tristani, A Di Meo. Surface investigations of bonded perfluoropolyether monolayers on gold surfaces. *Surf Interface Anal* 36: 1210–1213, 2004.

250. S Noël, D Alamarguy, N Lécaudé, O Schneegans, L Tristani. Multi-scale study of the electrical properties of organic layers grafted on gold surfaces. *Proceedings of the 51st IEEE Holm Conference on Electrical Contacts*, Chicago, 2005, pp 245–254.

251. JH Harris. Lubricating greases. In ER Braithwaite, ed. *Lubrication and Lubricants*. Amsterdam: Elsevier, 1967, pp 197–268.
252. K Klungtvedt. A study of the effects of silicate thickened lubricants on the performance of electrical contacts. *Proceedings of the 18th International Conference on Electrical Contacts*, Chicago, 1996, pp 262–268.
253. D Tarquin. Graphite coating extends contact life. *Proceedings of the Annual Connector Symposium*, Electronic Connector Study Group, Inc., Cherry Hill, NJ, 1994, pp 40–44.
254. PW Lees. The influence of physical separation between cycles on the wear behavior of connector contact materials. *Proceedings of the IEEE Holm Conference on Electrical Contacts*, Pittsburgh, 1993, pp 211–217.
255. JG Zhang, CH Mei, XM Wen. Dust effects on various lubricated sliding contacts. *Proceedings of the IEEE Holm Conference on Electrical Contacts*, Chicago, 1989, pp 35–42.
256. JG Zhang, W Chen. Wipe on various lubrication and non-lubricated electric contacts in dusty environments. *Proceedings of the 15th International Conference on Electric Contacts*, Montreal, 1990, pp 401–416.
257. JG Zhang. Influence of contaminents in connector performance. *Proceedings of the 16th International Conference on Electrical Contacts*, Loughborough, England, 1992, pp 111–121.
258. IH Brockman, CS Sieber, RS Mroczkowski. The effects of the interaction of normal force and wipe distance on contact resistance in precious metal plated contacts. *Proceedings of the IEEE Holm Conference on Electrical Contacts*, San Francisco, 1988, pp 73–83.
259. RV Chiarenzelli, BC Henry. Lubricating separable electric contacts and tarnish prevention. *Lubrication Engrg* 22: 174–180, 1966.
260. SJ Krumbein, M Antler. Corrosion inhibition and wear protection of gold plated connector contacts. *IEEE Trans Parts, Mater, Packag* 4(1): 3–11, 1968.
261. M Antler. Corrosion control and lubrication of plated noble metal connector contacts. *IEEE Trans Comp, Packag, Manuf Technol Part A* 19(3): 304–312, 1996.
262. WH Abbott. Field and laboratory studies of corrosion-inhibiting lubricants for gold-plated connectors. *Proceedings of the 18th International Conference on Electrical Contacts*, Chicago, 1996, pp 414–428.

8

Materials, Coatings, and Platings

Morton Antler and Paul G. Slade

Gold is the fool's curtain, which hides all his defects from the world

Feltem

CONTENTS

8.1 Introduction 520
    8.1.1 Scope 520
    8.1.2 Requirements of Contact Finishes and Coatings 520
    8.1.3 Terminology 520
8.2 Metallic Finishes 520
    8.2.1 Wrought Metals 521
    8.2.2 Electrodeposits and Electroless Deposits 522
        8.2.2.1 Thickness of Platings 522
        8.2.2.2 Plating Hardness 523
        8.2.2.3 Classification of Platings 523
    8.2.3 Contact Finishes Produced by Non-Chemical Methods 525
    8.2.4 Metal-in-Elastomer Materials 525
    8.2.5 Overview 526
8.3 Properties Related to Porosity 526
    8.3.1 Origins of Porosity 527
    8.3.2 Tests of Porosity 527
    8.3.3 Relationships Between Porosity, Thickness of Finish, and
        Substrate Roughness 529
    8.3.4 Effect of Underplatings, Flash Coatings, and Strikes on the Porosity
        of Electrodeposits 530
    8.3.5 Reduction in the Chemical Reactivity of Finishes by the Use of
        Underplates 531
8.4 Metallurgical and Structural Properties 532
    8.4.1 Thermal Diffusion 532
    8.4.2 Intermetallics 539
    8.4.3 Tin Whiskers 541
    8.4.4 Silver Whiskers 542
8.5 Physical and Mechanical Properties 543

8.5.1   Characteristics of Layered Systems...543
     8.5.1.1   Hardness 543
     8.5.1.2   Contact Resistance 543
8.5.2   Topography...545
Acknowledgement...546
References...546

8.1 Introduction

8.1.1 Scope

This chapter concerns finishes and coatings that are used on the contacts of separable electronic connectors. Because of the relatively small size of these connectors it is difficult to design their contact springs in such a way as to supply high normal loads that could facilitate the fracture and displacement of insulating oxide and corrosion films on uncoated base metal substrates. It is, therefore, necessary to modify the contact metals so as to be able consistently provide low contact resistance during the lifetime of the connector. This is achieved with suitable contact finishes and coatings. The terms "finishes" and "coatings" mean any material that is applied to a substrate. Thus, it includes electrolytic and electroless deposits, clad metals, fused metal layers, weldments, and materials from physical processes such as ion implantation and vacuum evaporation. These terms also embrace lubricants and chemical treatments for inhibiting the formation of corrosion and tarnish films on separable contact surfaces.

8.1.2 Requirements of Contact Finishes and Coatings

The purpose of a contact finish is to upgrade the contact substrate's properties such as its oxidation, tarnish, and corrosion resistance, electrical conductivity, durability, frictional properties, and solderability. In order to be useful, finishes and coatings should not have concomitant deleterious properties. For example, while silver-plated contacts resist degradation by fretting corrosion, they can be tarnished in mildly polluted environments [1]. The proper selection of contact finishes and coatings requires an understanding of the engineering requirements of the application. Additional considerations include material cost and availability, and ease of manufacture .

8.1.3 Terminology

Common terms that describe coatings and substrates are given in Table 8.1.

8.2 Metallic Finishes

Although several finishes and surface treatments have been evaluated for connector contacts, only a few are widely used. These include some multilayered electrodeposits, electroless deposits, claddings, and fluids besides greases on the contact metal or as secondary coatings on metal finishes.

TABLE 8.1

Terminology

Contact or Basis Metal	Substrate
Base metal	Non-noble metal
Finish	Layer(s) on substrate
Plating	Electro- or electroless deposit
Underplate	A layer under the top plating
Subplate(s)	When there are more than two layers of plating
Strike	Thin electrodeposit between two thicker electrodeposits or an electrodeposit and the substrate
Flash	Thin electrodeposit at outer surface
Cladding	Rolled metal, metallurgically bonded to the substrate
Liner	Clad metal underlayer (usually nickel) between surface layer of metal and the substrate; also called interliner

8.2.1 Wrought Metals

Claddings are made by compressing two or more metals at a very high force which creates metallurgical bonds at their interfaces and elongates the composite. The top layers are usually a noble metal on a nickel liner, and most substrates are copper alloy spring metals such as 98.1 wt% Cu, 1.9 wt% Be (Unified Numbering System [UNS] designation 17200); 94.8 wt% Cu, 5.0 wt% Sn, 0.19 wt% P (phosphor bronze, C51000) and 88.2 wt% Cu, 9.5 wt% Ni, 2.3 wt% Sn (C72500); see Chapter 24, Table 24.9. With a flat stock, the cladding is obtained by squeezing between rolls (see Chapter 17, Section 17.5). Clad inlays are produced by first skiving a groove in the substrate into which a strip of the precious contact metal on the nickel liner is placed and then rolled. For clad wires of circular or other shapes, the substrate in the form of a rod is pressed into the hollow tubes of nickel and the noble metal, which then is drawn through a die. Heat is released during reduction which facilitates interdiffusion at the layer interfaces. Work-hardening occurs, and so annealing in an inert atmosphere, followed by further reduction, may be needed to obtain the desired thickness. Sheet clad metal is, then, slit to width and is the raw material for stamping and forming the contact. The amount of required precious metal is minimized by selectively locating it on the substrate so to achieve the ideal "contact engineering surface." Excess material, cut away in the stamping, is reclaimed. Thicknesses of the noble metal in the inlay for contacts typically range at 0.75–5 µm. Multilayer inlays can also be obtained. The nickel liner is usually several micrometers thick.

Common claddings include pure gold, 75Au25Ag, palladium, 80Pd20Ni, 60Pd40Ag, 69Au25Ag6Pt, and DG R-156 (diffused gold 60Pd40Ag). DG R-156 is made by partially diffusing a golden top layer on palladium–silver to give a resulting surface composition of about 80% gold [2]. The ratio of thicknesses of the gold and 60Pd40Ag in the starting material is about 1/10, although different ratios of the metals are sometimes used [3]. Pure palladium and palladium–nickel alloys with a diffused gold surface have also been made.

Round "dots" of solid noble contact metals are produced by welding the end of a wire to the substrate, and then clipping, and coining. The weldment is usually applied to a contact spring. Pure gold, gold alloys hardened with 25%–30% silver, and palladium have been used. Typical weldment thicknesses are 25–75 µm.

Fusion coatings are made by contacting the sheet substrate with molten tin after prior treatment to remove superficial oxides. Thickness is controlled by wiping the molten surface with blades or with blowing air. Typical coating thicknesses are 2.5–10 μm. The sheet metal is then stamped and formed to the finished part. Fusion-coated contacts are attractive because of their low cost.

8.2.2 Electrodeposits and Electroless Deposits

Electrodeposits are produced by the electrolysis of a solution containing ionic species with electrons supplied by an external power source. Electroless plating involves the use of a chemical reducing agent in solution which provides the electrons for converting metal ions to the elemental form. This process requires a catalytic surface, and once plating is initiated, the metal being deposited catalyzes the process further. Electroless nickel contains 5%–10% phosphorus.

Important contact finishes are listed in Table 8.2 together with other deposits that are no longer used in significant amounts. The most common finishes are silver, tin, nickel and hard golds (golds that contain about 0.1%–0.5% cobalt, nickel, or iron and plated from acid or neutral cyanide chemistries). Gold flashes deposited on to nickel, palladium and 80Pd20Ni alloy are cheaper replacements for hard golds in many applications.

8.2.2.1 Thickness of Platings

Table 8.3 describes the thicknesses of common contact platings in electronic connectors.

TABLE 8.2

Some Electronic Connector Contact Finishes[a]: Electrolytic and Electroless Deposits

In Current Commercial Use
Sn
Hard golds
Gold flash/80Pd20Ni[b]
Gold flash/Pd[b]
Ag
Au, pure
Rh
Ni & electroless Ni (Contains 5-10% P) with no noble top layer
Obsolete
Pd
65Sn35Ni
Gold flash/65Sn35Ni[b]
AuAg (max. 5% (Ag) alkaline cyanide process
AuCuCd alloy, 18 karat cyanide process
Hard gold on an Ag underplate[b]
AuAg on an Ag underplate
Sn–Pb alloys (usually containing about 40–60% or 5–10% Pb)[c]

[a] Usually on a Ni (sometimes electroless Ni or electrolytic NiP) underplate. Supplementary lubricant coatings are common.
[b] The flash is usually hard gold, although pure gold is sometimes used.
[c] Pb is now considered a hazardous material and is banned for general use

TABLE 8.3

Nominal Plating Thicknesses

Hard golds	1.25 µm; military, aggressive environments, high-reliability, life-threatening applications 0.75 µm; common industrial applications < 0.75 µm; short-lived products, primarily for consumer applications in benign or protected environments, and where durability requirements are modest.
Palladium and 80Pd20Ni alloy	Generally used with a gold flash. 0.75–1.25 µm; normal industrial applications where high levels of durability are required.
Tin	2.5–5 µm; non-critical applications; some usage below 2.5 µm, but not recommended.
Nickel	5 µm; where relatively high forces can be used; protected environments; battery contacts 1.25–2.5 µm; common as an underplate.
Silver	5–15 µm; when plated on copper, 5 µm results in a mixed Cu/Ag chemistry. At 15 µm Ag chemistry dominates

TABLE 8.4

Electrodeposit Hardness ($\times 10^2$ Nmm^{-2})

Tin	0.8–3.5
Gold, pure	4–9
Silver	4–18.5
Hard golds	16–22
Palladium	20–30
80Pd20Ni	35–55
Nickel	25–55
Ruthenium	80–100

Typical values for these electrodeposits used in electric contacts are at the mid-range.

8.2.2.2 Plating Hardness

The hardness of platings is generally slightly greater than that of wrought metals having the same composition. When the layer is thinner than about 10–20 µm, its intrinsic hardness is determined on polished cross-sections by metallography using microhardness indenters, such as a Knoop diamond at 0.25 N, or optical microscopy. Smaller indentation loads and measurement of the indentation at high magnification with a scanning electron microscope is sometimes employed for very thin deposits. *The hardness of a multilayered contact normal to its surface is different from the bulk hardness of the finish* (see Sections 1.3, 8.5.2 and Table 24.1C). Table 8.4 gives the intrinsic hardness of common electrodeposits.

8.2.2.3 Classification of Platings

A brief overview of the properties of the most common electrodeposited coatings is given here.

8.2.2.3.1 Hard Golds

They are the best choice for the finish in most applications. As they have noble metal properties, they are less prone to formation of films than any other metal. They are the most "forgiving" finish available. However, they are subject to pore corrosion and surface film creep in aggressive environments when in contact with certain base metals. They are also difficult to form without cracking because of their low ductility.

8.2.2.3.2 *Palladium and Palladium–Nickel Alloy*

These metals form tarnish films and frictional polymers in polluted environments, which can be minimized with a gold flash. Although previously these metals cost about one-third to one-half the cost of gold of the same thickness, it is no longer the case. Thus, the tendency to use them in place of gold in some applications has lessened. The gold flash reduces the wear due to adhesives, and instability of the contact resistance from fretting of the palladium and palladium–nickel, but its durability may be limited unless the connection is appropriately lubricated [4]. Now, there are several improved techniques available for the process of plating Pd–Ni to a substrate [5,6]. Older, now obsolete, palladium plating processes tend to give microcracked finishes. Now that the price of palladium has risen, it is no longer being considered as a plating material for connectors.

8.2.2.3.3 *Silver*

With the continued rise in the price of gold, efforts are on in exploring the suitability of silver as a substitute plating material [7]. Although it is only a "near noble" metal, its much lower cost and excellent electrical and thermal characteristics make it a potentially attractive alternative to gold. Unlike gold, silver is susceptible to forming tarnish films of Ag_2S and $AgCl$ if exposed to ambients where sulfur or chlorine compounds are present (e.g., H_2S, PVC packaging and even some cardboard packaging). Thus, a gold plate cannot be replaced straightway with a silver one. There are several points of care required. Myers [7] has considered the performance of a silver plate as replacement for gold using a 2.4-μm silver layer over a 1.7-μm nickel underplate on a rolled phosphor bronze strip substrate. She has found that a coating of a palliative or inhibitor compound such as Evabrite [8] provides some long-term protection from tarnishing under normal temperatures and humidities. The study results in a number of conclusions:

1. For storage and shipping, silver-plated connectors should be placed in a sulfur absorbent packaging. PVC packaging must be avoided as it can result in severe chloride corrosion of the silver. Also some cardboard packagings can contribute to formation of sulfides.

2. Silver requires a higher contact force than gold in order to break through any light tarnish film that may be present. It is also useful to have a wiping action when making contact. Thus, this could well place a limit on the number of mating cycles for silver-plated connectors.

3. The higher contact force and the greater coefficient of friction increase the insertion force required for silver-plated connectors.

4. There are two other concerns regarding the use of silver in plating. The first is electromigration ([9] and also see Chapter 2, Section 2.4.3) and the second is the formation of silver whiskers from tarnish layers, that is Ag_2S (see Section 8.4.4). Both of these effects can result in failure especially where separated conductors are closely spaced.

8.2.2.3.4 *Nickel*

Nickel is inexpensive, but is covered with a tough oxide that must be broken to make a good electrical contact. It requires a relatively high force for forming a reliable connection. It is resistant to corrosion except at high humidity conditions; further, when in contact with more noble metals (gold, silver, and so on), it may corrode severely. It is susceptible to fretting corrosion, but there is some evidence that this can be lessened for up to 10,000 cycles [10]. It can also be used with care in highly polluting industrial environments [11].

8.2.2.3.5 *Tin*

Tin coatings are relatively inexpensive, but not durable. They are prone to degradation due to fretting corrosion, especially when mated to other metals. Pure tin can form "whiskers" (see Section 8.4.3). Degradation is also possible due to formation of intermetallic compounds. While lead is added to prevent the growth of whiskers, it cannot be used as a coating due to its hazardous nature; that is, the use of tin–lead alloys has mostly disappeared from electrical contacts and solders, because of the RoHS (Restrictions of Hazardous Substances) regulations.

8.2.2.3.6 *Intrinsic Polymers in Hard Gold Plates*

The unique properties of hard golds are attributable to the metalloorganic solids that are termed "intrinsic polymers" [12]. These polymers are fine particulates, of the order of tens of angstroms, and are also deposited (co-deposited) primarily at grain boundaries [13]. Some polymeric material may be present as larger islands in the deposit [14]. The solids have been hypothesized to be potassium–cobalt (or nickel)–gold–cyanide complexes [15]. Part of the co-deposited metal is present in the gold. The density of hard gold is 5%–10% less than that of pure gold. Unusual properties of hard gold compared to wrought gold that contains a small (< 1%) amount of alloy metals, are its high hardness, low ductility, and superior resistance to wear due to adhesion.

8.2.2.3.7 *Other Deposits*

Many other electrodeposited connector contact finishes have been described and some have been used commercially to a limited degree. A few examples are "pure" hard golds (24-carat) hard gold [16] and additive-free hard gold [17], golds containing arsenic, cadmium, or other metals from sulfite-plating solution chemistries [18], ruthenium with and without a gold flash [19], gold-flashed PdCo [20] and PdNiCo [21] alloys, heat-treated pure gold plate on solid nickel or a thick nickel underplate [22], and heat-treated polyacrylonitrile on nickel [23]. The absence of compelling technical advantages, process limitations, inadequate deposit characterization, the highly proprietary aspects of the finishes, the lack of industry standards, and the absence of suppliers who are able to provide them have been impediments to their widespread acceptance.

8.2.3 Contact Finishes Produced by Non-chemical Methods

Some connector contact finishes obtained by physical processes, besides cladding, have been evaluated to a limited extent. These include ion-plated and vapor-deposited silver, sputtered gold, silver, and silver–tin alloy [24], flame-sprayed tungsten carbide and other "conductive hard metals" [25], titanium nitride [26], and gold or palladium [27] or silver [28] implanted with nitrogen, boron [29], or other elements [30]. The high cost of the equipment required to prepare these finishes on contact substrates has been the most serious impediment to their commercial use. However, there seems to be a growing interest for using a suitable deposition technology for making silver coatings to replace gold [31].

8.2.4 Metal-in-Elastomer Materials

Metal-in-elastomer materials consist of metallic conductors, embedded in sheet elastomers, that are oriented in the Z (thickness) direction [32,33]. The conductive elements are gold-plated or silver-plated tiny metal spheres or fine wires that protrude from the surface. The elastomer provides insulation in the X–Y plane and is usually a silicone rubber. These materials are used to interconnect components such as fine line printed circuit boards. They have the advantage of not requiring precise orientation, and are made as part of a connector structure involving clamps and fixtures for holding the elements of the system together.

8.2.5 Overview

Hard golds, tin, gold-flashed palladium and palladium–nickel alloys, electrolytic and electroless nickel, and some clad noble metals constitute most of the contact finishes for commercial electronic connectors. There is an increasing use of silver (for its high conductivity and lower cost than gold) in some connector applications. The use of these lower cost materials is far more common in higher voltage/higher current circuits (see Chapter 5).

In many cases, the designer has the choice of specifying either clad or plated contact materials for his product. The comparative properties and fabrication characteristics of these metals can be summarized as follows:

1. Spring materials with a clad noble metal or tin layer can be fabricated into the finished product without further processing of the plating.
2. Certain materials can be obtained only by cladding, that is, DG R-156, 69Au25Ag6Pt, while others, for example, hard golds and electroless nickel, are unique platings.
3. Both cladding (inlay, top-lay, edge-lay) and electrodeposition processes can selectively locate a material on the contact.
4. Clad metals are ductile and lend themselves to forming. A hard gold, when thicker than a flash, is prone to cracking during fabrication. Some electrodeposits, such as tin, are sufficiently ductile to permit forming after plating. Hollow tubes can more readily be made with a finish of the clad stock on the inner surface than by plating the formed part.
5. Plated layers can be made thinner than claddings. It is usually more economical to clad when a large thickness is required.
6. Hard golds have an unique ability to resist adhesive wear compared to clad metals. A gold flash can be put on a clad metal to improve its resistance to wear due to adhesion.
7. There are relatively few fabricators of clad contact metals, despite the fact that processing technology, although primitive by today's standards, has been known since the early 1960s. Plating is much more deeply entrenched, having been practiced for some metals since long before the birth of the electronic connector. There are numerous captive and independent plating organizations that are supplied by many chemical and process equipment firms. The scientific and trade literature on electrodeposition is large. Organic lubricants have been used to some extent in conjunction with metallic finishes for separable contracts. They are discussed in Chapters 3, 5, and 7.

8.3 Properties Related to Porosity

The selection of contact finishes for electronic connectors is based on a number of factors. The main technical considerations are the chemical, metallurgical, physical, and mechanical properties, and the commercial aspect relates to the feasibility of production. When cost is a major consideration, precious metals are used in small thicknesses and selectively placed on the substrate. Porosity in the finish and the migration (creep) of surface films over the precious metal from adjacent base metal in aggressive environments may become a problem (see also Sections 2.4.4 and 3.2.4).

A few base metal finishes are suitable for connector contacts in applications that have low current. Thick, relatively pore-free, tin coatings are an exception because of the fairly low reactivity, proneness to deformation which facilitates the disruption of surface films, and solderability. However, poor resistance to sliding wear, susceptibility to fretting corrosion, formation of intermetallic compounds , and whiskering of pure tin can degrade contact reliability. Even some precious and semi-noble finishes, like silver and the platinum group metals, can acquire insulating films by catalytic effects or by sulfiding, and are therefore severely limited. Considerations involved in the selection of finishes are the subject of this section. Creep and electromigration are discussed in Chapters 2 and 3. Catalytic effects, or frictional polymerization, are described in Chapter 7 as part of the problems of fretting.

8.3.1 Origins of Porosity

Pores are defects in a coating which expose the underlying metal, the underplate, or the underplate and substrate ([34,35]; see Sections 2.4.3 and 3.2.4.6). They may be intrinsic in the layer, that is, present in the coating when it is made. Subsequent forming of the coated metal can also produce cracks in the finish. Wear from service of the component may result in penetration of the layer. Intrinsic porosity in electrodeposited coatings originates in occluded foreign materials such as oxide particles in the substrate surface, even after rigorous cleaning. It has been found that these substances may be layered in "zones" in the substrate and appear during etching. Particulates from inadequately filtered plating solutions may co-deposit and cause porosity. Scratches on the substrate are often the sites of pores in the deposit. Co-deposited particulate material in the coating's surface may cause localized corrosion and degradations due to pore formation. Clad metal finishes likewise may become porous if dust is embedded in the surface during fabrication. High spots on a rough substrate may penetrate the clad layer during rolling or drawing and create pores.

8.3.2 Tests of Porosity

Except for base metal coatings that are so thick that porosity is virtually absent, porosity testing is quite common. Its main objectives are to:

1. Determine the quality of the finish—how well it protects the contact from exposure to a potentially aggressive atmosphere.
2. Assess the effects of mechanical processing: stamping, forming, chamfering of the edge contacts of a printed circuit board.
3. Determine the effects of wear due to engagement.
4. Assist in the development of new coating processes.
5. Facilitate control of the coating process because:
   a. Porosity is easy to determine.
   b. Porosity is a sensitive indicator of the stability of the process.
6. Satisfy customer requirements of excluding porosity in a contact finish.

Many porosity tests have been devised. Nearly all involve exposure of the contact to aggressive chemical reagents which develop corrosion "decorations" at the defect sites. The requirements of porosity tests, therefore, are that they:

- Should be fast—minutes to hours in duration
- Should be easy, and of low cost
- Should be reproducible
- Should be standardized (ASTM and other organizations are active in standardization)
- Should be non-reactive with the finish
- Should enable pores to be counted or sized
- Should have good pore delineation; minimal spreading of decoration (little tendency to "halo" formation)
- Should have the tendency for minimal formation of hygroscopic corrosion solids (sample can be heated after gas exposure to eliminate moisture).

Among the most useful porosity tests for precious metal coatings are the following:

Test Method	ASTM
Nitric acid vapor	B735
Sulfur dioxide (sulfurous acid method)	B799
Flowers-of-sulfur	B809
Paper electrography	B741

The gas exposure tests are suitable for contacts of any size and shape. Paper electrography is limited to flat or gently curved substrates, but provides a permanent record of defects in the surface. Method B735 should not be employed for palladium-based finishes because nitric acid vapor attacks palladium. Method B799 is suitable for all common precious metal coatings except alloys that are high in base metal content. Immersion in ammonium polysulfide solution after a gas exposure can darken corrosion solids to make them more visible. The decorations, whether corrosion solids or colored spots on paper in electrography, are not the actual sizes of the defects, but are considerably larger. Since porosity is usually of microscopic dimensions, magnification—produced by the spreading or growth of the corrosion material—is necessary in order to facilitate observation of the pores. The magnification factor diminishes with increasing size of the defect. Porosity tests are destructive since the contact cannot be used after testing. A model of pore corrosion development for a 1.18-μm gold plate over copper with a roughness of 262 nm shows a general agreement with experimental data [36]. However, the measured electrical contact resistance values show a bias towards the larger bloom not taken into account by the model.

The maximum acceptable number of pores in a separable connector contact finish of a precious metal depends on the severity of the service environment, shielding inherent in the connector structure, the ability of the contact to displace insulating corrosion products during engagement, the criterion for circuit failure, and the reliability that is required in the application. A typical specification for a high-reliability, 20-year-life, 40-contact connector limits the numbers of pores in a sample lot of contacts to no more than two to four on the engineering surface, that is, 0.05–0.1 defects per contact. This requirement can usually be satisfied with a 1-μm thick plated or clad precious metal coating that is made by state-of-the-art manufacturing processes.

8.3.3 Relationships Between Porosity, Thickness of Finish, and Substrate Roughness

The porosity of a finish diminishes with increasing thickness. Porosity rises with increasing substrate roughness. Typical data are given in Figure 8.1 for electrodeposited gold on copper substrates that were abraded with metallographic papers before plating. The sensitivity of porosity to roughness is often overlooked by connector designers and manufacturers. This is unfortunate because, besides porosity, a contact's durability is degraded as the roughness is increased.

The reason for the dependence of porosity on substrate topography is that thickness (as given in Figure 8.1) is generally determined by X-ray fluorescence because it is nondestructive, rapid, and inexpensive. However, this method actually determines the mass of metal on the surface being studied, not thickness. Since a rough surface has a larger area, the real thickness of the coating will be less than if the surface were smooth. The sections in Figure 8.2 [37] show contacts having the same mass of deposit (identical "apparent" thickness). In addition, a non-uniform thickness distribution of plating on the surface (compare Figures 8.2a and b) exaggerates the discrepancy between the actual and apparent thicknesses. A larger surface will have a larger total number of defects, such as occluded impurities, which also contributes to porosity in the plating.

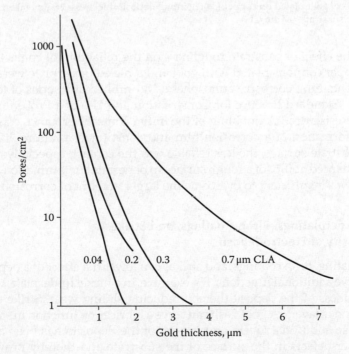

FIGURE 8.1
Dependence of porosity on deposit thickness and substrate roughness [38] (CLA–center line average); pure gold with 1.5 μm nickel underplate deposited on copper coupons that were abraded with metallographic papers prior to plating; porosity determined by electrography.

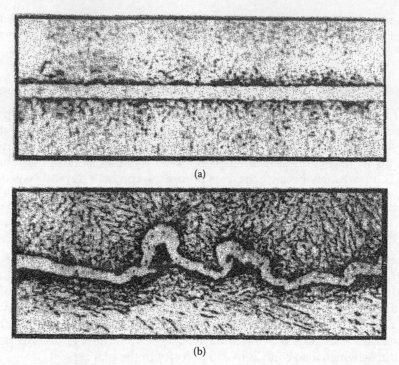

(a)

(b)

FIGURE 8.2
Metallographic sections of gold plated on copper [38]; nominal deposit thickness, 2.9 μm: (a) Smooth substrate, 0.04 μm CLA (b) Rough substrate, 0.7 μm CLA.

An example of the effect of substrate roughness on the reliability of contacts is given in Figure 8.3 [38]. Copper coupons plated with gold and a nickel underplate were exposed to a sulfur dioxide-containing corrosive atmosphere, and probed at a series of loads according to ASTM B667, "Standard Practice for Construction and Use of a Probe for Measuring Electrical Contact Resistance." A doubling of the initial contact resistance was considered to be evidence of formation of a corrosion film and constituted a failure. It is clear that the rougher the substrate surface, the less reliable was the contact, especially at low loads. Apparently, the enhanced ability of a rough surface to penetrate contamination and to lower contact resistance was insufficient to overcome the larger amount of corrosion products.

8.3.4 Effect of Underplatings, Flash Coatings, and Strikes on the Porosity of Electrodeposits

In general, underplating, flash coatings, and strikes can lower the effect of a deposit's porosity. These layers serve additional functions. For example, in a nickel underplate, they increase the composite hardness of the deposit thereby reducing sliding wear. Strike coatings can improve the adhesion of layers. A gold flash can serve a lubricating function in some systems.

The effect of these additional layers on porosity of the deposits has been attributed to their ability to cover defects in the surface of the substrate and thereby provide a better surface for plating. As these layers are better for closing the pores as compared to the original deposit as they thicken during plating, a given total thickness of a multilayered material will be less porous than a single layer of the same thickness. The multilayers consist of one or more of the strike, underplate, and flash, along with the deposit. A pore at the

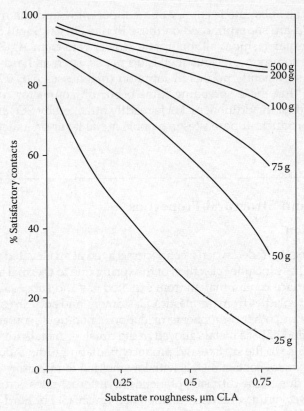

FIGURE 8.3
Effect of substrate roughness on contact reliability of plated contacts after exposure to a corrosive atmosphere: 1% SO_2, 95% RH, 45°C, 120 hours [39]; pure gold plate (1 µm) on nickel underplate (2.5 µm); copper substrate. Probed according to ASTM B667. Criterion of satisfactory contact: contact resistance, not in excess of 1 mΩ.

surface of a layer, that is, the underplate, is considered to be a defect. This analysis assumes that a pore in the deposit is smaller than an original defect of the underlying metal. Even so, a nickel-flashed underplate, a 0.46-µm gold top layer exposed to air for 15 months still shows considerable pore corrosion [39].

8.3.5 Reduction in the Chemical Reactivity of Finishes by the Use of Underplates

Some underplates can passivate a contact with a noble metal finish that is prone to corrode due to porosity. Electrodeposited 65Sn35Ni becomes passive after exposure in air, and as an underplate does not corrode at pores in a gold coating [40,41]. This alloy is not a satisfactory contact material, because its protective oxide resists penetration, and therefore, its contact resistance is unacceptably high. However, a flash of gold converts the system to one that is conductive, yet inert. Nevertheless, gold-flashed 65Sn35Ni has not been successful as a connector contact finish on substrates which flex during use, such as springs, due to its brittleness. It has been used as an edge contact finish for printed circuit boards.

An analogous situation is the use of nickel in place of silver as an underplate for gold, the latter a common practice until the 1960s. Silver sulfide, which forms at pores in gold plated on silver in atmospheres that contain traces of H_2S or elemental sulfur, spreads over the surface and degrades contact resistance. Nickel does not produce a spreading film on the gold. However, if the substrate is copper or a copper alloy and the nickel underplate is porous (pores in gold are superimposed on those in nickel), Cu_2S will form and spread on the surface in a similar fashion, although at a slower rate than Ag_2S. Unfortunately, nickel underplate and common precious metal top plates, such as hard golds and gold-flashed 80Pd20Ni, are sufficiently porous in common thicknesses (i.e., 0.75 μm hard gold on 1.25 μm nickel), so that these spreading films occur on connector contacts at severe environmental conditions. In addition, nickel is readily attacked by SO_2 and may produce mounds of corrosion products at pore sites in a noble metal layer on a nickel underplate.

8.4 Metallurgical and Structural Properties

8.4.1 Thermal Diffusion

Contact resistance can be degraded when a connector is aged at an elevated temperature due to a variety of causes. These include relaxation of the spring due to thermal stress which lowers the contact force; surface contamination from expelled gas and subsequent condensation on the contact surface of volatiles from the plastics, elastomers, and adhesives of the connector, surrounding structures and other components; oxidation of non-noble contact materials; and, for noble metals, oxidation of base metals alloyed in the finish or metals from the underplate or substrate, which diffuse to the surface and produce insulating films. Additional degradation factors are: (1) Sticking of contact metals with low melting temperature such as tin plate; (2) Kirkendall porosity due to the diffusion of metals at different rates across their interface which can lead to the weakening and delamination of a finish; (3) For hard golds, the movement of potassium and carbon compounds to the surface which originates in the thermal degradation of its intrinsic polymers (in addition to oxidation of the diffused base metal hardeners); and (4) Formation of intermetallic compounds that can change the mechanical and electrical properties of the contact. The latter process is described in Sections 1.3.3 and 8.4.2.

Thermal diffusion has been studied in some detail for gold platings and, more recently, palladium-based finishes. The earliest concerns with gold plate were in the telephone industry during the 1960s with the need for connectors and printed circuit boards having gold-plated contacts. The changes in packaging came with the move to circuitry with solid state components. Although contact temperatures rarely exceeded 65°C, the traditional 40-year-life requirement raised questions about long-term effects. A study was conducted with gold-plated copper that was aged above 300°C [42]. The thickness of the copper oxide films that formed on the surface was determined by cathodic reduction. Copper diffusion rates were plotted and extrapolated to room temperature by the Arrhenius relationship (log diffusion rate vs. reciprocal of absolute temperature), and lifetimes of a finish were estimated—making assumptions about the maximum amount of copper oxide that could be tolerated. This approach indicated that contact resistance instabilities due to thermal aging at 65°C would not occur for thousands of years. However, what was overlooked was there are several diffusion paths, besides the mass diffusion which predominates above 300°C. Diffusion via these other paths is of the orders of magnitude faster than mass diffusion at lower temperatures, and includes diffusion along grain boundaries, dislocations, and other defects. Gold-plated connectors ordinarily are not used at very high temperatures. The linear extrapolation of diffusion rates from high temperatures to relevant use temperatures is incorrect.

Contact resistance is dependent on many things besides the finish including contact design, normal load, topography, wipe, and the composition and mechanical properties of the underplate and substrate. It is difficult to equate contact resistance determined under a given set of conditions to values obtained with different test parameters. Nevertheless, trends in contact resistance have been obtained as a function of a number of variables [43]. Table 8.5 indicates these trends, and Figures 8.4 through 8.11 are some data obtained by the contact resistance probing of coupons with a gold probe according to ASTM B667. Samples were aged in ovens in air. The rate of rise of contact resistance increases with increasing temperature as shown in Figure 8.4 [44,45].

TABLE 8.5

Effect of Heating: Contact Resistance Trends and Electrodeposits

For Contact Resistance Rise

Increase time of heating (constant temperature)

Increase temperature (constant time)

Gold-plated copper

1. Co–Au[a] > Ni-Au[a] > pure Au and Au with noble metal hardener
2. Pure Au: thin > thick
3. Co-Au[a]: thick > thin
4. %X (X − Co or other base metal in hard gold): high%X > low%X (depends on atmosphere sinking effect kinetics)
5. Au Porosity: high > low (flash Au thicknesses)
6. Pure Au: no underplate > Ni underplate
7. Co-Au[a]: Ni underplate > no underplate (depends on Au thickness)

Gold-plated brass

8. No underplate > Cu underplate > Ni underplate
9. Grain size of Au or Ni underplate: small > large

Palladium platings on copper

10. 80Pd20Ni > Au and Au flashed/80Pd20Ni > Pd and Au flashed/Pd

[a] Hard golds.

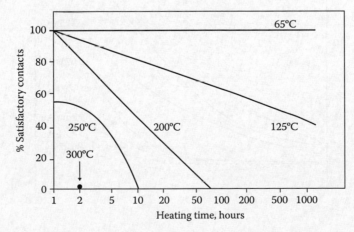

FIGURE 8.4

Effect of heating on contact reliability of 0.5 μm thick pure gold plating on a polished copper substrate [45]; probed at 1 N; criterion of satisfactory contact: Contact resistance not in excess of 1 mΩ.

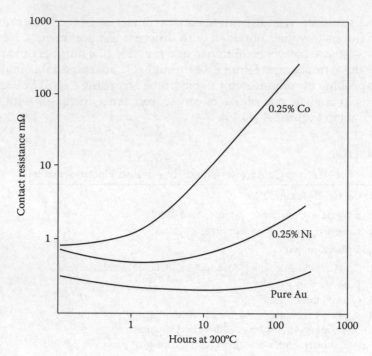

FIGURE 8.5
Thermal stability of contact resistance at 1 N of 50 μm thick gold plate; aged at 200°C; effect of cobalt and nickel additions (0.25%).

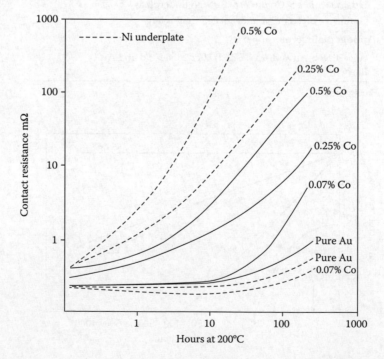

FIGURE 8.6
Thermal stability of contact resistance at 1 N of 2.5 μm hard gold plates containing 0.07, 0.25, or 0.5% cobalt without (solid, lines) and with (dashed lines) 2.5 μm nickel underplate; copper substrate, aged at 200°C.

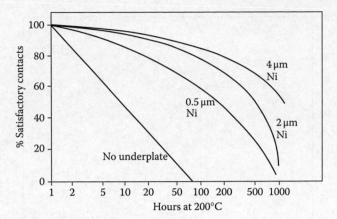

FIGURE 8.7
Effect of heating at 200°C on the contact reliability of 0.5 μm thick pure gold plate on a polished copper substrate without and with various thicknesses (0.5, 2, 4 μm) of nickel underplate. Probed at 1 N. Criterion of satisfactory contact: contact resistance not in excess of 1 mΩ.

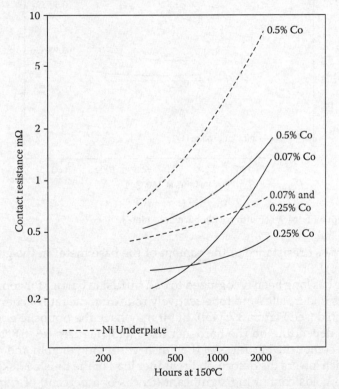

FIGURE 8.8
Thermal stability of contact resistance at 1 N of thin (0.5 μm) hard gold plates on copper containing 0.07, 0.25, or 0.5% cobalt without (solid lines) and with (dashed lines) 2.5 μm nickel underplate. Aged at 150°C.

In the gold-plated copper system, hard golds degrade more quickly than pure gold and gold alloys with noble metals such as palladium (Figure 8.5). This is due to the diffusion of the base metal hardener and polymer constituents to the surface of the hard gold [43]. Golds hardened with nickel show more stable contact resistance on thermal aging those hardened with cobalt as shown in Figure 8.5 [43,46]. There is an expected dependence of

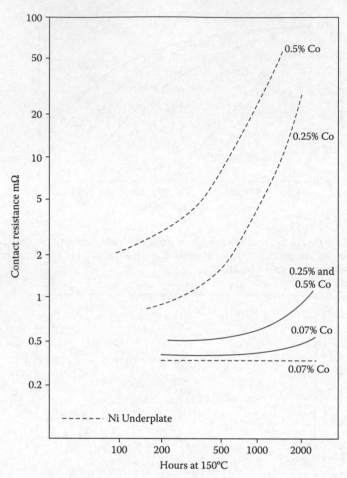

FIGURE 8.9
Same as Figure 8.8, except with thick (2.5 μm) cobalt–hard gold plates.

rate of change of contact resistance to the content of the base metal in the gold at a given thickness (Figure 8.6) [43].

Nickel underplate has long been recognized to be a diffusion barrier for copper, zinc, and other metals used as underplates and for effectively retarding the rate of change of contact resistance of pure gold (see Figure 8.7; [45]). But it may have the opposite effect when the finish is hard gold, depending on the hardener used ([43]; Figures 8.6, 8.8, and 8.9). This may be because the composite hardness of the finish, consisting of gold and nickel underplate, is greater, which lowers the deformability under load of the finish. Nickel underplate is harder than hard golds containing significant (tenths of a percent) of cobalt or nickel. The films originate in base metals in the hard golds rather than as diffused species from the substrate as with pure gold without an underplate. With reduced deformability, insulating films are less readily displaced. An alternative hypothesis is without the barrier layer, the hardener species can diffuse both ways, to the external surface and into the substrate. The latter would reduce the concentration of the oxidizable species on the external surface, and thus, retard the growth kinetics of the resistive film. In the presence of the nickel barrier, the diffusion is only towards the external surface. Therefore, the amount of diffused metal at the gold surface (in the form of an insulating film) would be greater with

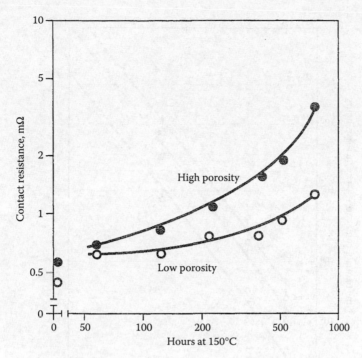

FIGURE 8.10

Contact resistance of pure gold platings, 0.15 μm thick, on copper: ○ low porosity gold; ● high porosity gold. Aged at 150°C. Probed at 0.25 N. The low porosity deposit had about 25–80 pores cm^{-2}, and the high porosity plating had > 1000 pores cm^{-2}. (See Ref. [48] for details.)

the nickel underplate, consequently with a more pronounced degrading effect on contact resistance. This apparent adverse effect of nickel underplate for hard golds is surprising, for its role as a diffusion barrier has been considered to be a reason for its use.

Thin, pure gold degrades more rapidly than thick deposits [43] because the diffusion path for the underplate and substrate metals is shorter. The reverse is true for hard golds, perhaps because thick deposits have more total base metal. This effect is especially pronounced with hard golds on a nickel underplate, as can be seen in Figures 8.8 and 8.9.

It has also been found that the environment plays a role in changes in contact resistance. If oxidation of diffused base metal at the surface is the rate-limiting step rather than diffusion, then reactions other than oxidation may increase the rate of change of contact resistance. It has been found that chlorine in the atmosphere can cause more rapid increases of contact resistance than clean air [47]. Printed circuit boards with fire retardants are a source of chlorine. The dependence of rate of change of contact resistance on grain size is predictable, as is porosity of the deposit (Figure 8.10), for grain boundaries and pores are the diffusion paths for the defects [48]. Figure 8.11 (not probed according to ASTM B667) plots the times to obtain given increases in contact resistance for a hard gold on phosphor bronze as a function of temperature with an Arrhenius extrapolation to longer times [49].

Table 8.6 is a short list of maximum temperatures to which finishes in their usual thicknesses on nickel underplate can be heated, based on the author's experience. Silver does not oxidize, and therefore is a preferred material, notwithstanding its tendency to tarnish if unprotected.

Pure palladium does not oxidize below 350°C and, being free of intrinsic polymers, has a considerably higher limit of thermal stability than hard golds. Palladium is also a diffusion

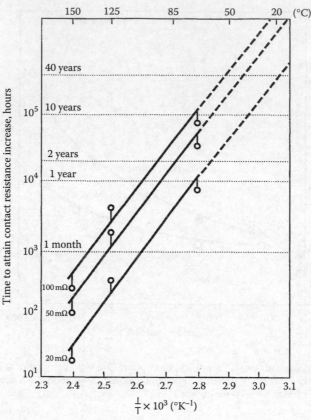

FIGURE 8.11
Time to attain contact resistance increases of 20, 50, and 100 mΩ for 0.3 μm cobalt–hard gold plated on phosphor bronze [49]. Probed at 0.1 N.

TABLE 8.6

For High-Reliability Applications, Maximum Temperatures at Which Coatings Can Be Heated for Extended Periods[a]

Material	Approximate Temperature (°C)
0.38 μm plated pure Au on plated or solid Ni or Ni alloy spring materials; heat treated [12]	>200
Ag	>200
Pd, 60Pd40Ag, Rh (with and without Au flash)	200
DG R-156	150
Hard golds	125[b]
80Pd20Ni	<100[c]
Au–Cu–Cd alloy electrodeposits	<100
Sn, Sn–Pb alloys	<100[d]

[a] Recommended maximum temperatures are highly dependent on contact design, normal load, environment, coating, underplate and liner thicknesses, underplate–substrate composition and mechanical properties, and contact surface roughness.

[b] Temperature limit may be higher depending on hardener metal, its amount, gold thickness, and other factors (see Table 8.5).

[c] Increase to 125°C when gold flash is used.

[d] Higher temperatures may be satisfactory for non-critical applications.

barrier for copper and other metals. DG R-156 and 60Pd40Ag similarly have a higher temperature limit than hard golds. On the other hand, 80Pd20Ni is relatively poor due to its large nickel content, although a flash of pure or a hard gold can retard this degradation.

AuCuCd platings (cyanide chemistry) are typical of many alloy golds that contain considerable base metal, and have relatively poor thermal stability [45]. Bright tin platings contain co-deposited organics which degrade their thermal stability, and matte (pure) tin has been found to be better. After extended aging, the tin can be entirely consumed in forming intermetallics with the copper or nickel on which it is coated. This is discussed in Section 8.4.2.

The development of coatings having stable contact resistance at high temperatures has been stimulated by the proliferation of connectors for automotive underhood applications. Recent reports on some of these coatings discuss: clad thin layers of 69Au25Ag6Pt on 60Pd40Ag with a nickel liner [50], vapor-deposited thin gold or 80Pd20Ni on vapor-deposited nickel [51] and gold flash on a multilayered palladium-based finish [52].

8.4.2 Intermetallics

The effect of growth of intermetallic compounds on the contact properties of electronic connectors with tin finishes has been determined. Extensive studies have also been made with tin and other metals for utility connectors (Chapters 1 and 5). These studies were prompted by the realization that intermetallic compounds can form rapidly, even at low temperatures and especially at elevated temperatures [53]. Contact finishes are thin and the electrical and mechanical properties of intermetallics are very different from those of the parent metals. Intermetallics are less conductive (by a factor ~ 10) and harder (by a factor ~ 5), and these changes can degrade contact resistance [54].

Figure 8.12 shows the growth of intermetallics on copper for some tin and tin–lead alloy coatings at 24°C and 50°C [55,56]. The rate of growth on nickel is slightly lesser [57,56]. The intermetallics from tin and copper are Cu_6Sn_5 (for temperatures <150C) [53] and Cu_3Sn

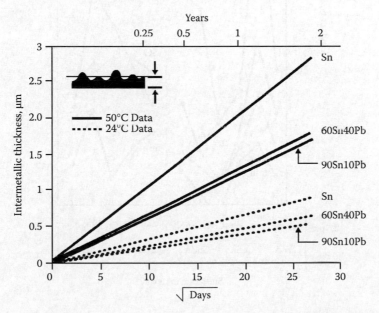

FIGURE 8.12
Growth of intermetallic compound layers at interface for tin and tin–lead alloys at 24 and 50°C [55]. Inset shows typical convoluted thickness of the layers.

and Cu_6Sn_5 (for temperatures >150°C) [53] and for tin and nickel Ni_3Sn_4 (with possibly other compounds) [57,58]. The demarcations between the intermetallic layers are highly irregular.

The contact resistance of tin on various substrates with copper and nickel underplatings after seven years of exposure at 50°C, followed by one year at 20°C, is plotted in Figure 8.13 [59]. Contact resistance was determined by probing. In general, it increased significantly, although the data are scattered. A sulfur dioxide exposure after aging attenuated the increase in contact resistance. Fine-grained, bright tin was poorer than matte tin. The contact resistance of tin–lead alloy after similar aging was much better behaved.

The conclusion from these experiments, coupled with other considerations, is that bright tin plating is an inferior metal for contact applications. Although matte tin is, it is not prudent to use it for high-reliability, long-duration applications. The susceptibility of these finishes to film formation in atmospheres polluted with chlorine and nitrogen dioxide is well known [60], and they are also prone to degradation by fretting corrosion. The solderability

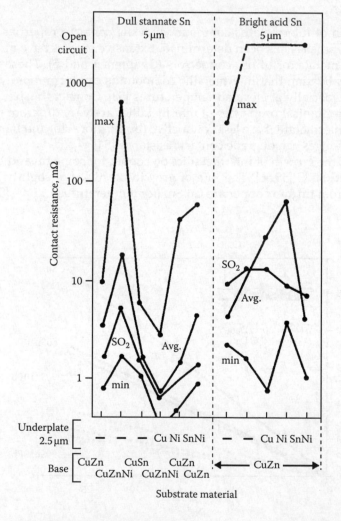

FIGURE 8.13
Contact resistance of dull and bright tin coatings aged at 50°C on various substrates and underplates [59]. Average, minimum, and maximum values are shown after 7 years in air and the average after an additional exposure to 0.15 p.p.m. SO_2 at 70% RH, 25°C, for 35 days. Probed at 1 N.

of tin coated contacts is degraded on aging, which limits the shelf life of contacts with these finishes.

8.4.3 Tin Whiskers

Whiskers are filamentary growths that often develop spontaneously on the surface of pure tin, zinc, and cadmium [1,61–65]. They are single crystals which grow by diffusion at their base from the surface of the finish. Whiskers may be straight or convoluted, are of varied length, often to 1 cm, and typically are 1–5 μm in diameter. Their population density may be as high as 10^4 cm^{-2} of surface. Figure 8.14 [64] illustrates typical whiskers from a tin plating on copper. Although a complete understanding of growth of tin whiskers is, at present, not available, stress in the deposit certainly promotes the process [66]. Whiskers grow more rapidly on substrates in compression [65]. Their formation may depend on the composition of the substrate and the underplate, and on environmental factors. Whisker density and length is greater on a smooth tin surface than on a rough one [67]. Their growth is enhanced in an ambient with very high humidity: 65%–90%RH.

Whiskers may be detrimental if they bridge adjacent contacts (Figure 8.15) [68]. They can carry currents to about 25 mA; higher currents cause them to fuse. Reflowed deposits do not develop whiskers. In earlier times, growth of tin whiskers could be avoided with finishes that contained 5% or lesser of lead. Thus, in the past it was considered to be poor practice to use pure tin platings on contacts, since tin–lead alloy deposits were a practical solution to the whisker problem. Now, however, lead is banned from almost all contact applications to eliminate problems of lead poisoning. So, there has been a change from tin–lead solders and the use of lead in tin plate. This has led to increased research into prevention and control of formation of tin whiskers [66].

The answer to the problem of uncontrolled automobile acceleration , that was seen a few years ago, was eventually traced to shorting of the Accelerator Pedal Position (APP) sensor due to tin whiskers [69]. This sensor controlled the vehicle's acceleration, but when shorted by whiskers 11 μm diameter, it could no longer function to prevent the uncontrolled acceleration. Leidecker et al. [69] showed that to test for these very fine whiskers, one had to be careful to not apply too high a voltage. If such a high voltage is applied across these

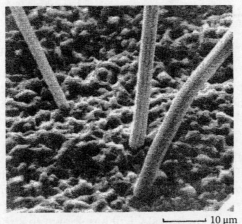

└─────┘ 10 μm

FIGURE 8.14
An array of closely spaced tin whiskers. Finish is 5 μm of tin on a copper substrate [64].

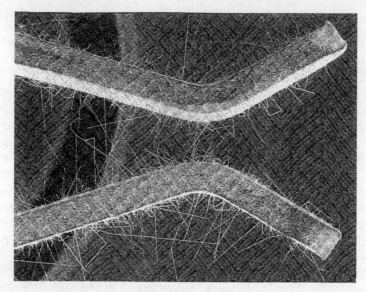

FIGURE 8.15
Whisker growth on tin-plated contacts. (From RP Diehl and NA Cifaldi, *Proc 8th Ann Conn Sym*, Cherry Hill, 327–36, 1975.)

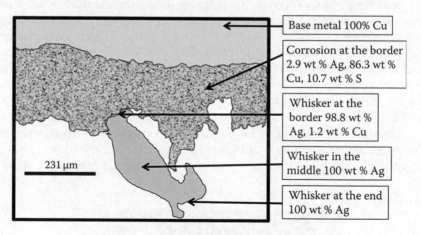

FIGURE 8.16
Diagram of a cross-section through a silver whisker from a silver plated copper substrate showing its composition [70].

whiskers, they would be ruptured. The researchers said this was one reason that the whisker problem was not identified sooner.

8.4.4 Silver Whiskers

Silver Sulfide (Ag_2S) is a major corrosion product on silver-plated surfaces in an ambient with concentrations of a sulfuric gas such as H_2S [1]. Once the Ag_2S layer is in place, whiskers of silver can begin to form [1,70]. Unlike the tin whiskers that form from a surface of pure tin, the silver whiskers require a layer of Ag_2S before they begin to form. Although the corrosion layer is necessary, Chudnovsky [70] has shown that the whisker is made up almost entirely of silver; see Figure 8.16. The metallic nature of this whisker means that it is highly conductive and can easily cause adjacent conductors to short and fail.

8.5 Physical and Mechanical Properties

8.5.1 Characteristics of Layered Systems

8.5.1.1 Hardness

The hardness of a contact finish depends on the deformability of its individual layers and the substrate, their thicknesses, and the load. Hardness is an important characteristic of contacts because it is one of the controlling parameters of the size of an *a*-spot and constriction resistance. When base metal contact finishes are covered by insulating films such as oxides, the extent of the disruption of these films in connections also depends on hardness; the success of soft base coatings such as tin lies in the ease with which its superficial oxides can be penetrated. In addition, the durability of finishes that are subjected to sliding and fretting is in part controlled by hardness. Examples of these effects are illustrated in Chapter 6. A method for calculating the hardness of layered systems has been proposed [71]. Examples of composite hardnesses are given in Tables 8.7 and 8.8.

8.5.1.2 Contact Resistance

The contact resistance of finishes depends on the hardness, thickness, and conductivity of the constituent layers, as well as on the load. This is discussed in Chapter 1. In the case of base metal contacts, whether plated or solid, that are covered with insulating films, the geometry and dimensions of the opposing contact are also the determining factors. So-called, "sharp" contacts are better able to penetrate these films. Probe contacts made of hard metals resist deformation better than those that are soft, and therefore exert higher pressure. This is illustrated in Figure 8.17, a determination of the contact resistance of oxide-covered 65Sn35Ni electrodeposit, copper, and nickel [72]. Hard, sharp contacts,

TABLE 8.7

Hardness of Plated Contact Materials[a]

Cobalt–gold (μm)	Nickel Underplate (μm)	Substrate (thickness, mm)	Depth of Penetration of Identer (μm)[b]	×10² Nmm⁻² Knoop, 0.25 N
2	none	Cu (3.2)	2.6	5.6
2	2.5	Cu (3.2)	2.5	6.3
3.3	2.5	Cu (3.2)	2.0	9.3
3.3	4	Cu (3.2)	1.7	13.1
2	2.5	Be–Cu (4.9)	1.1	28.8

[a] Hardness determined normal to surface.

[b] Depth of penetration $= \dfrac{\text{length of diagonal, Knoop}}{30.53}$

Hardness of bulk materials from metallographic sections (×10² Nmm⁻²)		
Co-gold	18	(Knoop, 0.25 N)
Ni underplate	55	(Knoop, 0.1 N)
Substrate, Cu	4–6	(Knoop, 0.25 N)
Substrate, Be–Cu	26.6	(Knoop, 0.25 N)

TABLE 8.8

Hardness of Plated Contact Materials

Soft Gold (μm)	70Pd30Ni (μm)	Ni Underplate (μm)	Substrate (μm)	×10^2 Nmm^{-2} Vickers, 1N
0.15–0.40	2.5	–	22.5	12.9
0.15–0.40	2.5	1.2	22.5	15.2

Hardness of bulk materials from metallographic sections		
Soft gold	85	kg mm^{-2} (Vickers, 1 N)
70Pd30Ni	400	kg mm^{-2} (Vickers, 1 N)
Ni underplate	230	kg mm^{-2} (Vickers, 1 N)
Substrate, Cu	125	kg mm^{-2} (Vickers, 1 N)

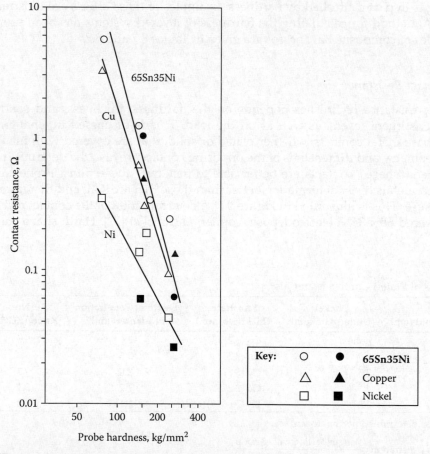

FIGURE 8.17
Median values of contact resistance at 0.5 N for aged 65Sn35Ni electrodeposit, solid copper, and solid nickel determined with various probe metals (hemispherically ended, 3.2 mm diameter [72]. Open symbols are for noble metal probes and solid symbols for copper and nickel probes.

generally plated with hard golds, are used in commercial probe testers for determining the functioning of integrated circuits in production control. The slopes of the curves showing contact resistance versus force characteristics for plated contacts illustrate the relative contributions of the finish and the substrate (Figure 8.18) [73].

8.5.2 Topography

The topography of most electrodeposited contact finishes is generally similar to that of the surfaces on which they are plated. However, some processes have a leveling effect in that deposition occurs preferentially in the valleys of the substrate surface, while other processes plate preferentially on the high spots. This is illustrated in Figure 8.19. Copper leveling

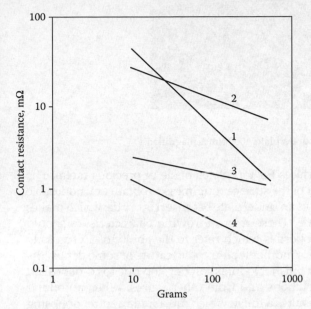

Curve	Conductivity Finish	Substrate
1	Low	High
2	Low	Low
3	High	Low
4	High	High

FIGURE 8.18
Contact resistance-force characteristics for some plated contact materials. (Schematic adapted from HN Wagar, In *Physical Design of Electronic Systems*, Vol. III, [eds. D Baker, D Koehler, W Fleckenstein, C Roden, R Sabia] Upper Saddle River, NJ: Prentice-Hall, 439–99, 1971 [73].)

Curve	Conductivity finish	Substrate
1	Low	High
2	Low	Low
3	High	Low
4	High	High

(a)

(b)

(c)

■ Deposit
▨ Substrate

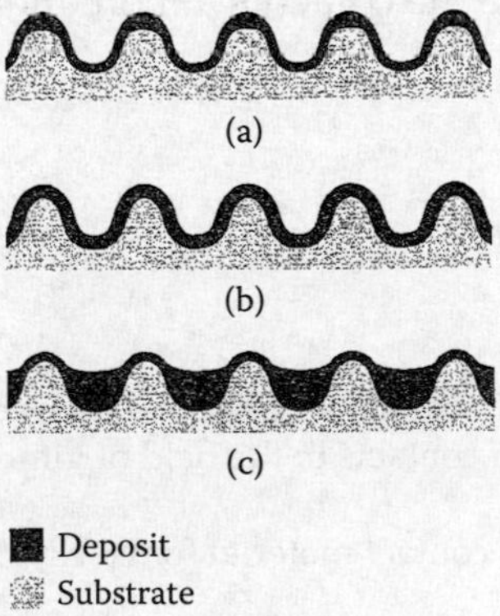

FIGURE 8.19
Schematic illustrations of various microthrowing characteristics of electrodeposits: (a) negative throw; (b) geometric throw; (c) leveling throw.

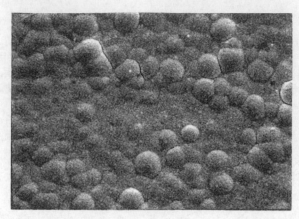

FIGURE 8.20
Nodular finish. Gold-flashed palladium on a nickel underplate. 1000 times magnified.

underplates are used to obtain smooth surfaces for top plates made of precious metal. The thickness of the deposit must be of the order of the surface roughness to obtain any noticeable improvement. Leveling nickel is unsuitable as an underplate because it is brittle, which makes it impractical to form a contact after plating. These micro-throwing characteristics should be distinguished from macro-throwing properties which refer to the ability of a process to plate into recesses such as connector sockets. Another topographic feature of some deposits is nodularity. Figure 8.20 illustrates a nodular finish. Nodules can develop due to co-deposited particulates from poorly filtered plating solutions and from other causes. When a nodular deposit is due to its nickel underplate, the result is a finish which can abrade a softer opposing surface during engagement. The composite hardness of finishes with a nickel underplate is greater than that of common top plates, such as hard golds, as shown in Table 8.7.

Acknowledgment

I (PGS) am very grateful to Marjorie Myers who critically reviewed this chapter and who suggested additions that have helped make the contents a more thorough review of coating and plating technology.

References

1. W Schmitt, S Franz, J Heber, O Lutz, and V Behrens, Formation of silver sulfide layers and their influence on the electrical characteristics of contacts in the field of information technology, *Proc 24th Int Conf Elect Cont*, 489–94, 1982.
2. FE Bader, Diffused gold R156: A new inlay contact material for bell system connectors, *Proc 11th Int Conf Elect Cont*, 133–37, 1982.
3. N Aukland, H Hardee, A Wehr-Aukland, and P Lees, A comparison of a clad material (65Au21Pd14Ag) and an electroplated gold over palladium material system under fretting conditions at elevated temperatures, *Proc 45th IEEE Holm Conf Elect Cont*, 213–25, 1999.

4. M Myers, Comparison of hard Au versus hard Au-flashed PdNi as a Contact finish, *Proc 56th IEEE Holm Conf Elect Cont*, 49–57, 2010.

5. JA Abys, HK Straschil, I Kadija, EJ Kudrak, and J Blee, The electrodeposition and material properties of palladium–nickel alloys, *Metal Finish* 89 (7), 43–52, 1991.

6. W Zhang, M Clauss, J Guebey, and F Schwager, Low ammonia, high-speed palladium–nickel electroplating process for connector applications, *Metal Finish* 65 (1), 33–37, March 2009.

7. M Myers, The performance implications of silver as a contact finish in traditionally gold finished contact applications, *Proc 55th IEEE Holm Conf Elect Cont*, 307–15, 2009.

8. There are a number of inhibitor materials that claim to protect Ag from tarnishing, Here is a very non-complete list of examples: Evabrite, Tarnaban, Silverbrite W TPS, *Antitarnish 616*.

9. S Yang, J Wu, and M Pecht, Electrochemical migration of land grid array sockets under highly accelerated stress conditions, *Proc 51st IEEE Holm Conf Elect Cont*, 238–45, 2005.

10. C Dervos, J Novakovic, and P Vassiliou, Electroless Ni–B and Ni–P coatings with high-fretting resistance for electrical contact applications, *Proc 51st IEEE Holm Conf on Elect Cont*, 281–88, 2005.

11. B Chudovsky, Degradation of power contacts in industrial atmosphere: Plating alternative to Ag and Sn, *Proc 49th IEEE Holm Conf Elect Cont*, 98–104, 2003.

12. Y Okinaka, et al., Polymer inclusions in cobalt-hardened electroplated gold, *J Electrochem Soc* 125 (11): 1745–50, 1978.

13. Okinaka and S Nakahara, Structure of electroplated hard gold observed by transmission electron microscopy, *J Electrochem Soc* 123 (9): 1284–89, 1976.

14. M Antler, Structure of polymer co-deposited in gold electrodeposits, *Plating* 60 (5): 468–74, 1973.

15. Electrodeposition of hard golds: A hypothesis to explain its special features, *Gold Bulletin* 11 (4): 132–33, 1978 (according to research by ET Eisenmann).

16. FI Nobel, et al., An evaluation of 18 karat and 24 karat hard gold deposits for contact applications, *Proc 4th Plat Electro Ind Sym*, American Electroplaters' Society, Indianapolis, 106–30, 1973.

17. FB Koch, et al., Additive-free hard gold plating for electronics applications (in two parts), *Plat Surf Fin* 67 (6): 50–54, 1980; 67 (7): 43–45, 1980.

18. BDT plating processes, Sel-Rex Corporation, Nutley, NJ.

19. RG Baker, and TA Palumbo, The potential role of ruthenium for electronic applications, *Plat Surf Fin* 69 (11): 61–68, 1982.

20. I Boguslavasky, et al., Electroplating of PdCo alloy for connector applications, *Proc Int Inst Conn Interconn Tech Conf*, 217–36, 1996.

21. M Naillat, et al., Tribological behavior of Pd–Ni–Co electrodeposited alloys, *Proc 13th Int Conf Elect Cont*, 468–75, 1986.

22. JAE Bell and D Hope, Corrosion resistant high temperature contacts, *Proc Conn Interconn Tech Sym* International Institute of Connector and Interconnection Technology Inc., San Jose, 345–52, 1992.

23. F Houze, et al., Electrical characterization of thin polyacrylonitrile electropolymerized on nickel for connector applications, *Proc 16th Int Conf Elect Cont*, 267–71, 1992.

24. G Herklotz, et al., Application of sputtered layers in contact technology, *Proc 15th Int Conf Elect Cont*, 182–92, 1990.

25. M Antler and SJ Krumbein, Contact properties of conductive hardmetals and tin nickel plate, *Proc 11th Eng Sem Elect Cont*, 103–38, 1965.

26. S Benhenda, TiN coating in electrical contacts applications, *Proc 16th Int Conf Elect Cont*, 181–84, 1992.

27. PW Leech, The effect of nitrogen implantation in the tribological properties of gold-based alloys and electroplated palladium, *Proc 33rd IEEE Holm Conf Elect Cont*, 53–62, 1987.

28. M Sun, et al., Modification of Ag-plated contacts by nitrogen ion implantation, *Proc 18th Int Conf Elect Cont*, Chicago, 467–71, 1996.

29. M Antler, et al., The effect of boron implantation on the sliding wear and contact resistance of palladium, 60Pd40Ag, and a CuNiSn Alloy, *IEEE Trans Comp Hyb Manuf Tech* CHMT–5 (1): 81–84, 1982.

30. H Anli, Investigation on improving wear resistance of electric contact materials for aviation by ion implantation, *Proc 16th Int Conf Elect Cont*, 235–39, 1992.

31. An example can be found at www.impactcoatings.se

32. CA Haque, Characterization of metal-in-elastomer connector contact material, *Proc 35th IEEE Holm Conf Elect Cont*, Chicago, 117–22, 1989.

33. G Rosen, et al., A comparison of metal-in-elastomer connectors: The influence of structure on mechanical and electrical performance, *Proc 15th Int Conf Elect Cont*, 151–65, 1990.

34. M Clarke, Porosity and porosity tests, In *Properties of Electrodeposits, Their Measurement and Significance,* (eds) R Sard, H Leidheiser Jr, and F Ogburn, Princeton: The Electrochemical Society, 122–41, 1975.

35. SM Garte, Porosity, In *Gold Plating Technology,* (eds) FH Reid and W Goldie, Ayr: Electrochemical Publications, Ltd, 295–315 and 345–59, 1974.

36. A Sun, H Moffat, D Enos, and C Glauner, Pore corrosion model for gold-plated copper contacts, *Proc 51st IEEE Holm Conf Elect Cont*, 232–37, 2005.

37. SM Garte, Effect of substrate roughness on the porosity of gold electrodeposits, *Plating* 53: 1335–40, 1966.

38. M Antler, Gold-plated contacts: Effect of substrate roughness on reliability, *Plating* 56: 1139–44, 1969.

39. YL Zhou, XY Lin, and JG Zhang, The electrical and mechanical performance of corroded products on gold plating after long-term indoor air exposure, *Proc 46th IEEE Holm Conf Elect Cont*, 18–26, 2000.

40. M Antler, et al., The corrosion resistance of worn tin–nickel and gold coated tin–nickel alloy electrodeposits, *J Electrochem Soc* 124 (17): 1069–75, 1997.

41. M Antler, Corrosion control with tin nickel and other intermetallics. In *Corrosion Control by Coatings,* (ed.) H Leidheiser Jr, Princeton: Science Press, 115–33, 1979.

42. UB Thomas (Bell Labs). Unpublished work.

43. M Antler, Gold-plated contacts: Effect of thermal aging on contact resistance, *Proc 43rd IEEE Holm ConfElect Cont*, Philadelphia, 121–31, 1997.

44. M Pinnel, et al., Oxidation of copper from gold alloy solution at 50°C–150°C, *J Electrochem Soc* 126 (10): 1798–805, 1979.

45. M Antler, Gold-plated contacts: Effect of heating on reliability, *Plating* 57: 625–28, 1970.

46. M Huck, Kontakteigenschaften galvanischer hartgoldschichten bei erhöhten temperaturen, *Metallwissenschaft und Technik* 46(1): 32–35, 1992.

47. HG Tompkins and MR Pinnel, Low-temperature diffusion of copper through gold, *J Appl Phy* 47(9): 380–3812, 1976.

48. M Antler and MH Drozdowicz, Gold-plated contacts: The relationship between porosity and contact resistance on elevated temperature aging, *Plat Surf Fin* 63 (9): 19–21, 1976.

49. L Revay, Surface accumulation rate of diffusing species on thin gold electroplates, *IEEE Trans Comp, Hyb Manuf Tech* CHMT–15 (5): 870–75, 1992.

50. PW Lees, Connector contact finishes for high temperature automotive applications, *Proc Int Inst Conn Interconn Tech Conf*, Boston, 241–50, 1996.

51. YT Cheng, et al., Vapor-deposited thin gold coatings for high-temperature electrical contacts, *Proc 42nd IEEE Holm Conf Elect Cont*, 404–13, 1996.

52. E Kudrak, et al., A high-temperature electrical contact finish for automotive app lications, *Proc SUR/FIN 1996*, American Electroplaters and Surface Finishers Society, 1996.

53. WM Wang, AQ He, Q Liu, and DG Ivey, Solid state interfacial reactions in electrodeposited Cu–Sn couples, *Trans Non-ferr Met Soc* China, 20: 90–96, 2010.

54. M. Braunovic, Effect of intermetallic phases on the performance of tin-plated copper connections and conductors, *Proc 49th IEEE Holm Conf Elect Cont*, 124–31, 2003.

55. DA Unsworth and CA Mackay, A preliminary report on growth of compound layers on various metal bases plated with tin and its alloys, *Trans Inst Met Fin* 51: 85–90, 1973.

56. PA Kay and CA Mackay, The growth of intermetallic compounds on common basis materials with tin and tin-lead alloys, *Trans Inst Met Fin* 54: 68–74, 1976.

57. S Noël, N Lécaudé, S Correia, P Gendre, and A Grosjean, Electrical and tribological properties of tin-plated copper alloy for electrical contacts in relation to intermetallic growth, *Proc 51st IEEE Holm Conf Elect Cont*, 1–10, 2006.

58. S Noel, D Alamarguy, S Correia, and P Gendre, Study of thin underlayers to hinder contact resistance increase due to intermetallic compound formation, *Proc 55th IEEE Holm Conf Elect Cont*, 153–59, 2009.

59. U Lindborg, et al., Intermetallic growth and contact resistance on tin contacts after Aging, *Proc 21st Holm Sem Elect Cont*, 25–31, 1975.

60. M Antler, et al., Base metal contacts: an exploratory study of separable connection to tin–lead, *IEEE Trans Parts, Hyb Packag* PHP–11: 35–44, 1975.

61. KJ Compton, et al., Filamentary growth on metal surfaces: Whiskers, *Corrosion* 7: 327–34, 1951.

62. RM Fisher, et al., Accelerated growth of tin whiskers, *Acta Met* 2: 368–73, 1954.

63. GT Galyon, Annotated tin whisker bibliography and anthology, *IEEE Trans Electron Packag Manuf* 28: 94, 2005.

64. PL Key, Surface morphology of whisker crystals of tin, zinc and cadmium, *Proc IEEE Electron Comp Conf*, 155–60, 1970.

65. E Crandall, G Flowers, R Jackson, P Lall and M Bozack, Growth of Sn whiskers under net compressive and tensile stress states, *Proc 57th IEEE Holm Conf Elect Cont*, 304–08, 2011.

66. G Gaylon, Whisker formation concepts: The end game, *IEEE Trans Comp, Packag Tech* 1 (3): 1098–109, 2011.

67. C Rodekohr, G Flowers, J Suhling, and M Bozack, Influence of substrate surface roughness on tin whisker growth, *Proc 54th IEEE Holm Conf Elect Cont*, 245–48, 2008.

68. RP Diehl and NA Cifaldi, Elimination of tin whisker growth on interconnections, *Proc 8th Ann Conn Sym*, Cherry Hill, 327–36, 1975.

69. H Leidecker, L Panashchenko, and J Brusse, Electrical failure of an accelerator pedal position sensor caused by tin whisker and discussion of investigative techniques used for whisker detection, *5th Int Tin Whisk Sym*, 2011. http://ne nasa.gov/whisker

70. B Chudnovsky, Degradation of power contacts in industrial atmosphere: silver corrosion and whiskers, *Proc 48th IEEE Holm Conf Elect Cont*, 140–50, 2002.

71. PA Engel, Vickers microhardness evaluation of multilayer platings for electrical contacts, *Proc IEEE Holm Conf Elect Cont*, Chicago, 66–72, 1991.

72. M Antler, Contact resistance of oxidized metals: dependence on the mating material, *IEEE Trans Comp, Hyb Manuf Tech* CHMT–10 (3): 420–24, 1987.

73. HN Wagar, Principles of conduction through electrical contacts, In *Physical Design of Electronic Systems*, (eds) D Baker, D Koehler, W Fleckenstein, C Roden, and R Sabia, Vol. III, Upper Saddle River, NJ: Prentice-Hall, 439–99, 1971.

Part III

The Electric Arc and Switching Device Technology

9

The Arc and Interruption

Paul G. Slade

All the elements of earth except Hydrogen and some Helium … were made in the interiors
of collapsing stars. We are made of star stuff.

Cosmos, **Carl Sagan**

CONTENTS

9.1 Introduction 554
9.2 The Fourth State of Matter 554
9.3 Establishing an Arc 558
    9.3.1 Long-Gap Gas Breakdown 558
    9.3.2 Vacuum Breakdown and Short-Gap Breakdown 566
    9.3.3 The Volt–Current Characteristics of Separated Contacts 569
9.4 The Formation of the Electric Arc 570
    9.4.1 The Formation of the Electric Arc During Contact Closing 570
    9.4.2 The Formation of the Electric Arc During Contact Opening 571
9.5 The Arc in Air at Atmospheric Pressure 578
    9.5.1 The Arc Column 578
    9.5.2 The Cathode Region 581
    9.5.3 The Anode Region 584
    9.5.4 The Minimum Arc Current and the Minimum Arc Voltage 585
    9.5.5 Arc Volt–Ampere Characteristics 588
9.6 The Arc in Vacuum 592
    9.6.1 The Diffuse Vacuum Arc 592
    9.6.2 The Columnar Vacuum Arc 595
    9.6.3 The Vacuum Arc in the Presence of a Transverse Magnetic Field 596
    9.6.4 The Vacuum Arc in the Presence of an Axial Magnetic Field 596
9.7 Arc Interruption 597
    9.7.1 Arc Interruption in Alternating Current Circuits 597
        9.7.1.1 Stage 1 – Instantaneous Dielectric Recovery 600
        9.7.1.2 Stage 2 – Decay of the Arc Plasma and
                    Dielectric Reignition 602
        9.7.1.3 Thermal Reignition 603
    9.7.2 Arc Interruption in Direct Current Circuits 604
    9.7.3 Vacuum Arc Interruption in Alternating Circuits 607
    9.7.4 Arc interruption of Alternating Circuits: Current Limiting 608

9.7.5   Interruption of Low Frequency and High Frequency Power Circuits...........609
9.7.6   Interruption of Megahertz and Gigahertz Electronic Circuits 609
Acknowledgments.. 612
References.. 612

9.1 Introduction

The first eight chapters of this book have discussed fixed contacts. You have explored the physical effects of two metals in contact and the effect of chemical reactions at the contact interface. By using these fundamental principles, you have been shown how it is possible to design fixed contacts for a very wide range of currents: e.g., from connectors for power circuits that pass many thousands of amperes and must survive short circuit currents of over 10^5 A to connectors for electronic equipment that can see currents as low as 10^{-9} A. With this chapter, you will now begin to study the equally interesting effects that occur as you separate two contacts in order to interrupt the flow of current in a circuit. For the most part, I will be concerned with currents greater than 10 mA and circuit voltages greater than a few volts. The range of currents and voltages will be discussed in detail later in this chapter. This range of course covers a vast array of switching devices from small ac and dc relays and switches to large circuit breakers that interrupt high-power circuits.

This chapter concentrates upon the electric arc, its structure, how is it formed, and how it is used to interrupt electric current. Chapter 10 will discuss the general effect of arcing upon the electric contacts and Chapters 11–15 will show how these effects can be used to design a wide range of switching devices. Chapters 16–19 then will discuss the range of contact materials used for switching, their manufacture, their construction and attachment, how they are tested, and the chemical effects of contaminants.

9.2 The Fourth State of Matter

The ancient Greeks thought the world was composed form four states of matter: "earth," "water," "vapor," and "fire." Three thousand years later, we now believe the universe is still composed of four states of matter, but now they are: "solids," "liquids," "gases," and "plasmas." The first three states of matter are to be found only on cold planets such as the Earth and as meteoric matter. The fourth, the plasma state, is the predominant state of matter in the universe. The stars, the nebulae, the intergalactic gas, the solar winds and the earth's upper atmosphere are all in the plasma state. The first three states of matter, however, are common to our daily experience and will serve as a starting point to develop some basic concepts related to the ideas of "states of matter."

A solid object such as a copper rod, for example, has a shape. It can be squeezed, hammered, bent, etc., into a shape and will retain its shape if left undisturbed. A liquid has no shape. It is characterized by volume, i.e., a liquid flows and assumes the shape of the container that holds it. A gas has no shape or volume. A gas expands to fill the container that holds it and exerts a pressure on the walls of the container. As a solid is heated it melts to form a liquid, and when the liquid is heated it eventually forms a molecular gas. We know that temperature is associated with the average kinetic energy of the molecules, and that the kinetic energy of the molecules is associated with their average velocity. The shape of the solid is defined by long-range intermolecular forces which overcome the kinetic energy of the molecules. As the kinetic energy of the molecules is increased by adding heat the intermolecular forces still hold the molecules together, but when the thermal energy is greater than the energy needed to hold collective groups of molecules in a shape, a liquid is formed. As more heat is added the average kinetic energy exceeds any intermolecular binding energy, and all molecules act independently, and the liquid becomes a molecular gas. As still more heat is added the gas molecules bounce off the walls of the container in which they are contained. They also collide with each other. It is now important to have some knowledge of the kinetic theory of gases. This is a very well-developed subject and if you are unfamiliar with this subject, I would recommend that you consult one of the many excellent books on the subject [1,2]. Here, only a very brief summary is given.

The molecules or atoms of a gas are in random motion and are colliding frequently with each other and the walls of the vessel in which they are contained. These collisions are elastic, i.e., energy and momentum are conserved and they have no preferred direction, all directions being equally probable. The collisions cause a continual interchange of energy. This energy has a steady-state distribution which can be derived using statistical techniques [2]. It can be shown that the overall kinetic energy of the gas and the average kinetic energy of each atom depend upon the absolute temperature T. The kinetic energy is divided equally between the degrees of freedom, each degree of freedom having $(1/2)kT$, where k is Boltzmann's constant $(1.38 \times 10^{-23} \text{ J K}^{-1})$. Thus for a monatomic gas with three degrees of freedom:

$$\frac{1}{2}m\overline{c^2} = \frac{3}{2}kT \tag{9.1}$$

where m is the mass of the gas atom, $\overline{c^2}$ is the mean square speed. Thus the total kinetic energy per cubic meter of a monatomic gas is

$$\text{Total kinetic energy} = \frac{3}{2}nkT \tag{9.2}$$

where n is the number of gas atoms per cubic meter. For a diatomic gas such as N_2, which has two more degrees of freedom, rotational and vibrational, the total kinetic energy is $(5/2)nkT$. It follows from Equation 9.1 that the mean square speed for a monatomic gas is given by $3kT/m$. A useful parameter is the mean speed $\overline{c}$, which has a similar value to the square root of $\overline{c^2}$, but is *not* equal to it (since "mean" and "root mean square; RMS" are defined differently)

$$\overline{c} = \left(\frac{8kT}{\pi m}\right)^{1/2} \tag{9.3}$$

By using the term mean speed $\overline{c}$ we are assuming that some atoms travel slower than $\overline{c}$ and others faster. By considering the random nature of these collision processes, Maxwell and Boltzmann showed that the number dn of atoms out of a total n that have speed between c and $c + dc$ is given by (2):

$$\frac{dn}{dc} = \frac{4n}{\pi^{1/2}}\left(\frac{m}{2kT}\right)^{3/2} c^2 \exp\left(-\frac{mc^2}{2kT}\right) \tag{9.4}$$

The Maxwell–Boltzmann distribution function is shown in Figure 9.1. As can be seen, while some atoms travel very much faster and others very much slower than mean velocity $\bar{c}$, nearly 90% of them have speeds between $(1/2)\bar{c}$ and $2\bar{c}$ at any given time. The most probable speed is $c_0 = (2kT/m)^{1/2}$. The pressure p of the gas is given by

$$p = \frac{nm\overline{c^2}}{3} \text{ or } nkT \tag{9.5}$$

If the gas is molecular, and still more heat is added, the average kinetic energy of the molecules will exceed the binding energy of the molecules and the molecules will dissociate into individual atoms. Upon further heating the atoms will be ionized, i.e., they will start to lose an electron. This will leave the normally neutral atoms with a positive charge; we call an atom in this state a positive ion (or, in most cases in this text, an ion for short). The mixture of ions and electrons is called the fourth state of matter or "plasma state." A plasma can be defined as:

> Any system in which there are free electrons and ions where the spatial and temporal average density of charges of each sign are approximately equal.

In the context of electric contacts, the plasma is ionized gas that 'burns' between parted contacts and is called an electric arc, or arc for short. The equality of the charge densities (i.e. quasineutrality) is an extremely important property and it will be used later in this chapter when discussing the current interruption process using the electric arc.

The term "arc" was first introduced by Sir Humphrey Davy [3] in the early 19th century, who, observing a discharge between two horizontal conductors, saw the hot luminous portion in the middle rise by convection while the ends remained fixed. The term plasma was introduced over 100 years later by I. Langmuir to give the matter inside the arc a name. A qualitative understanding of a plasma and of an arc in particular had to wait until the discovery of the electron by Thompson in 1897 [4] and the Bohr model of the atom in 1901 [5]. The subject finally built a solid foundation with the publication of Townsend's seminal work, "Electricity in Gases" in 1915 [6]. Since that time, many books have been written describing arcs, e.g., References [7–11], and on the more practical aspects of the use of arcs

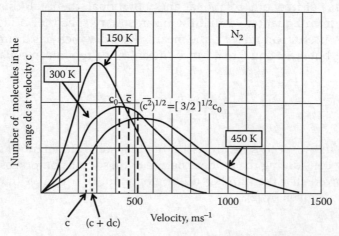

FIGURE 9.1
Example of Maxwell–Boltzmann speed distribution for nitrogen at three temperatures [17].

to interrupt electric circuits, e.g., References [12–17]. Unfortunately, only Holm's book [14] and Browne's book [17] are still in print. It is, therefore, my purpose in this chapter to give you a general understanding of the arc as it applies to electric contacts without exploring the detailed physics, which can be found elsewhere. My discussion will, of course, be based upon the vast body of research that has already been done and I will provide detailed references for those of you who wish to study this subject further.

Without asking for the moment how we would do it, suppose we heat dry, atmospheric air to 25,000 K. With the help of Figure 9.2, we will follow what happens to the nitrogen and oxygen molecules as the temperature increases. At about 1,500 K the oxygen molecules begin to form monatomic oxygen atoms and for nitrogen molecule nitrogen molecules begin to dissociate and forms monatomic nitrogen atoms at about 2,500 K. At about 6,500 K both the oxygen and nitrogen atoms start to ionize and a plasma begins to form. By about 20,000 K there is almost no neutral gas left and the plasma is said to be fully ionized. As the temperature increases further, the oxygen and nitrogen ions gradually lose another electron and becomes doubly ionized. To strip all the electrons from these atoms, however, would require a plasma temperature of millions of degrees. For arcs that occur between electric contacts in switching devices, the temperature range is between approximately 6,000 K and approximately 20,000 K depending upon the arc current and other factors resulting from the design of the switch. At first sight, these temperatures seem very forbidding. Let us, for the moment, explore this aspect of the arc.

We know from Equation 9.1 that the kinetic energy of a gas is related to the absolute temperature T. Therefore, for a nitrogen ion ($m = 2.3 \times 10^{-26}$ kg) at 20,000 K its root mean square velocity is

$$(\overline{c^2})^{1/2} = 4.9 \times 10^3 \text{ ms}^{-1} \tag{9.6}$$

This seems to be a formidable velocity but it is easily achievable in the laboratory. If, for example, we take the glass tube shown in Figure 9.3, evacuate the gas from it, place a potential U across the electrodes and introduce a nitrogen ion at the positive electrode (the anode) it is possible to calculate the value of the potential drop U required to accelerate the ion to a velocity of 4.9×10^3 ms^{-1} by the time it reaches the negative electrode (the cathode). Kinetic energy of the ion at the negative electrode = potential energy of the ion at the positive electrode, that is,

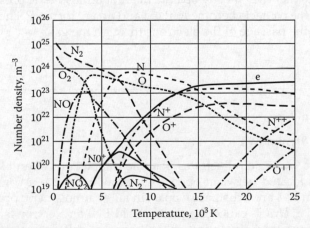

FIGURE 9.2
The number of particles per m^3 for dry air at atmospheric pressure as a function of temperature.

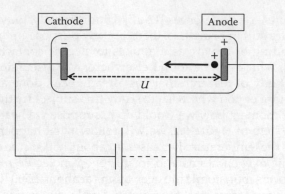

FIGURE 9.3
The acceleration of a nitrogen ion in vacuum.

$$\frac{1}{2}m\overline{c^2} = eU \qquad\qquad (9.7)$$

where e is the charge which is 1.6×10^{-19} coulombs for an electron or for a singly ionized atom. Then for the nitrogen ion

$$\frac{1}{2}\times 2.3\times 10^{-26}\times 24.0\times 10^6 = U\times 1.6\times 10^{-19}$$

that is,

$$U = 1.7\,\text{V} \qquad\qquad (9.8)$$

So, using electrodes with a very low voltage across them, it is possible to accelerate an ion to a very high velocity, indeed a velocity that gives a very high equivalent temperature. This can only occur, however, if the tube is under high vacuum, where the nitrogen ion has a very low probability of colliding with another gas molecule. As soon as we introduce gas to the tube the elastic collisions between the ion and the gas quickly reduce the velocity of the ion to be in equilibrium with the background gas. We know from our experience with switching devices that they mostly operate in air at atmospheric pressure. How then does the gas between the contacts become ionized and an arc formed? The answer is to be found by considering the passage of the electrons through the gas.

9.3 Establishing an Arc

9.3.1 Long-Gap Gas Breakdown

Let us consider Figure 9.4, which illustrates possible interactions of an electron with gas molecules and atoms located between open contacts which have a constant voltage U impressed across them and are a distance d apart in air at atmospheric pressure. The voltage per unit distance i.e., U/d, is called the electric field E. If an electron is introduced into the gas between the contacts, being negatively charged, it drifts towards the anode and gains energy before it collides with a gas molecule or a gas atom. The energy an electron

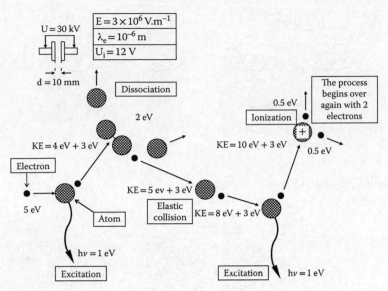

FIGURE 9.4
Possible interactions with a gas of an electron accelerated by an electric field.

gains by travelling through a one volt drop in potential is 1 eV (1.6×10^{-19} J). The maximum energy the electron gains between collisions is:

$$E\lambda_e = \frac{U\lambda_e}{d} \tag{9.9}$$

where λ_e is the average distance travelled by the electrons between collisions, and is called the *electron mean free path*. In the simplified example shown in Figure 9.4, the average energy gained by the electron between collisions with the gas is 3 eV (i.e. $3 \times 10^6 \times 10^{-6}$ eV). The electron collides with the gas in two ways. The first is called an *elastic collision* [1,18]. Here the electron bounces off the gas molecule in the same way a small glass marble would off a large bowling ball. The collision preserves momentum and, because the electron mass m_e is much less than the gas's mass, the electron energy loss is close to zero. After the collision, the electron continues to pick up energy from the electric field.

The second type of collision is called an *inelastic collision*. Here the electron transfers some of it energy to the gas atom or molecule. It is these collision processes that eventually lead to the formation of an electric arc. There are many types of inelastic collisions [18]. I shall only briefly describe three of them.

1. *Dissociation* is the process of splitting a molecule. For example, a nitrogen molecule can be dissociated into two nitrogen atoms. Dissociation is possible if the electron energy is greater than the bond strength of the molecule.

$$e + N_2 \rightarrow e + N + N \tag{9.10}$$

2. *Excitation and relaxation* is the process by which light is emitted from the gas. Figure 9.5 illustrates the process. The impacting electron forces an electron in the atom to a higher energy level, leaving the atom in an excited state. The excited state

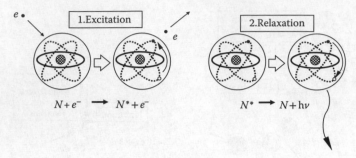

FIGURE 9.5
Excitation and relaxation produces radiation.

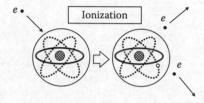

FIGURE 9.6
Ionization releases an electron from the atom leaving a positive ion.

is usually very unstable and the excited electron returns to its original energy state in less than a nanosecond. The energy difference between the states is given off as light.

$$e + N \rightarrow N^* + e \tag{9.11}$$

then

$$N^* \rightarrow N + h\nu \tag{9.12}$$

where h is Planck's constant and v is the frequency of the light.

3. *Ionization* is the process that directly results in an electric arc. If the impacting electron has an energy equal to or greater than the energy to remove the most weakly bound electron from the atom, then there is a chance that the impacting electron will remove this electron from the atom, see Figure 9.6. This process results in a positive ion and in two electrons; the original impacting electron and the electron freed from the atom.

$$e + N \rightarrow 2e + N^+ \tag{9.13}$$

The minimum energy to remove the weakest bound electron is called the *ionization energy*. Values of ionization potentials, V_i, for atoms useful for electric contacts, are given in Table 9.1. The energy an electron receives depends upon the strength of the electric field E and upon the distance it is accelerated between collisions. In Figure 9.4 there are three collisions before an atom is ionized (there may well be many more). When this occurs, two electrons are now free to interact with the gas. It only takes about 30 such doublings to

TABLE 9.1

Values of Ionization Potential

Gas	Ionization Potential (V)
Air	14
A	15.7
CO_2	14.4
H	13.5
N	14.5
O	13.5
Ag	7.6
Al	6.0
Cu	7.7
Ni	7.6
Mo	7.2
Sn	7.3
Pd	8.3
Pt	9.0
W	8.0

produce 10^9 electrons: this is termed a Townsend avalanche. The number of new ions produced per meter of path by the accelerated electron is inversely proportional to its mean free path λ_e. If α is the number of ionizing collisions per meter in the direction of the field, then:

$$\alpha = \frac{f(E\lambda_e)}{\lambda_e} \tag{9.14}$$

now $f(E\lambda)$ can be represented by

$$f(E\lambda_e) = \exp\left(-\frac{V_i'}{E\lambda_e}\right) \tag{9.15}$$

where V_i' is an effective ionization potential

$$\frac{1}{\lambda_e} = Ap \tag{9.16}$$

where p is the gas pressure and A is a constant. Thus

$$\alpha = Ap\exp\left(-\frac{AV_i'}{E/p}\right) \tag{9.17}$$

α is called the first Townsend ionization coefficient [1,18]. Values for the mean free path of molecules (λ_g), ions (λ_i) and electrons (λ_e) are given in Table 9.2.

If two contacts are separated by a distance d and if n_0 electrons per cubic meter are liberated from the cathode initially then the number n_1 of electrons per cubic meter arriving at the anode is

$$n_1 = n_0 \exp(\alpha d) \tag{9.18}$$

TABLE 9.2

Mean Free Path of Molecules, Electrons, and Ions at Atmospheric Pressure and Room Temperature

Gas	Molecular Mean Free Path, λ_g (nm)	Approximate Ion Free Path, $\lambda_i \approx \sqrt{2}\lambda_g$ (nm)	Approximate Electron Mean Free Path, $\lambda_e \approx 4\sqrt{2}\lambda_g$ (nm)
Air	96	135	180–545
A	99	140	295–560
CO_2	61	86	185–345
H_2	184	260	550–1,045
He	296	418	890–1,675
H_2O	72.2	102	215–410
N_2	93.2	132	280–530
Ne	193	273	580–1,095
N_2O	70	99	210–395
O_2	99.5	141	300–565
SO_2	45.7	65	135–260

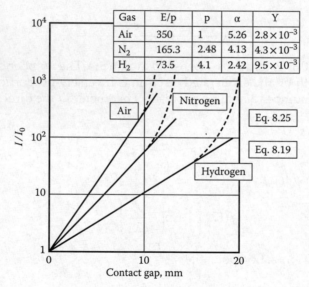

FIGURE 9.7

Typical $\log(I/I_0)$ versus contact gap curves for Townsend breakdown in air, nitrogen, and hydrogen [18].

or, converting this to current [18],

$$I = I_0 \exp(\alpha d) \tag{9.19}$$

However as Figure 9.7 shows, if the ratio I/I_0 exceeds a given value, the increase in I exceeds the value calculated from Equation 9.19. As the current increases by the ionization process, the ions, ever increasing in number, drift toward the cathode. When these ions reach the cathode, they help liberate more electrons and thus increase I_0. The extra electrons liberated give rise to a secondary ionization process that causes the current to increase faster than in Equation 9.19. There have been other secondary processes formulated for the further liberation of electrons from the cathode such as photoelectric emission as well as enhanced ionization in the gas itself [18].

If we assume that the extra electrons are liberated by ion bombardment, then let us for a moment consider a number γ (called the second Townsend coefficient) of new electrons emitted from the cathode for each of the incoming positive ions. Again, let us also assume n_1 electrons per square meter reach the anode per second, that initially only n_0 electrons leave the cathode per square meter per second and that n_c is the total number of electrons per square meter per second that are liberated from the cathode from all the effects combined. The number of positive ions in the gas is equal to $n_1 - n_c$. The number of electrons leaving the cathode is

$$n_c = n_0 + \gamma(n_1 - n_c) \tag{9.20}$$

$$n_c = \frac{n_0 + \gamma n_1}{1 + \gamma} \tag{9.21}$$

Using Equation 9.18, the number of electrons reaching the anode is

$$n_1 = \frac{n_0 + \gamma n_1}{1 + \gamma} \exp(\alpha d) \tag{9.22}$$

$$n_1 = \frac{n_0 \exp(\alpha d)}{1 - \gamma[\exp(\alpha d) - 1]} \tag{9.23}$$

or converting to current,

$$I = \frac{I_0 \exp(\alpha d)}{1 - \gamma[\exp(\alpha d) - 1]} \tag{9.24}$$

Now
$$\exp(\alpha d) \gg 1$$
so:

$$I = \frac{I_0 \exp(\alpha d)}{1 - \gamma \exp(\alpha d)} \tag{9.25}$$

Thus as the electric field in the electrode gap increases, the current increases and there is a sudden transition from a "dark discharge" to one of a number of forms of sustained discharge. The initiation of a discharge based directly upon the mechanisms using the two Townsend coefficients is called *Townsend breakdown*. This transition, sometimes called a spark, consists of a sudden increase of current in the gap and is accompanied by a sudden increase in light visible between the contacts. It is this spark that initiates the arc. From Equation 9.25,

$$I \to \infty \ as \ \gamma \exp(\alpha d) \to 1$$

Let us assume the contact gap breaks down or sparks when

$$In\frac{1}{\gamma} = \alpha d \tag{9.26}$$

When the gap d breaks down:

$$E = \frac{V_B}{d} \tag{9.27}$$

where V_B is the breakdown voltage or sparking potential. From Equation 9.17, we know that

$$\alpha = Ap \exp\left(-\frac{AV_i'}{E/p}\right) \tag{9.28}$$

Thus, at breakdown using, Equations 9.27 and 9.28,

$$\alpha = Ap \exp\left(-\frac{AV_i'pd}{V_B}\right) \tag{9.29}$$

Now, using Equation 9.26, and assuming breakdown occurs when I suddenly increases

$$ln\frac{1}{\gamma} = Apd \exp\left(-\frac{AV_i'pd}{V_b}\right) \tag{9.30}$$

$$V_B = \frac{-AV_i'pd}{ln\left(\dfrac{Apd}{ln[1/\gamma]}\right)} \tag{9.31}$$

Thus the breakdown voltage V_B for a given gas with an effective ionization potential V_i' is a function of the gas pressure multiplied by the contact gap (pd) alone [11]. This is known as Paschen's law, and was discovered in 1889 [19]. From Equation 9.5, $p = nkT$, so

$$V_B = f(nd) \tag{9.32}$$

Figure 9.8 shows a typical Paschen curves for air and SF_6 [20]. A qualitative explanation an easily be applied to it. For a given contact gap, at higher the pressures (to the right of the minimum value) the electron's mean free path λ_e is smaller. The electrons, therefore, lose energy through more frequent collisions. In order to ensure a breakdown, the electric field must be high enough for the electrons to gain sufficient energy between collisions. As $E = V_B/d$ then V_B has to increase. At low values of pressure (to the left of the minimum value) the electron can now travel further before hitting an atom, but the probability of impact has decreased enough that each collision requires a higher probability for ionization. For this to occur, the electrons must gain more energy from the electric field and thus V_B has to increase again. Table 9.3 gives the minimum breakdown voltage $(V_B)_{min}$ and minimum $(pd)_{min}$ values for various gases. Note, that for a limited range of V_B there are two values of pd. Thus for a given V_B and gas pressure there are two possible electrode gaps at which breakdown can occur. Under some circumstances, therefore, the intuitive answer to an unwanted breakdown of increasing the contact gap and thus increasing the breakdown distance will not always apply.

In most careful experiments for measuring α and γ the initiating current I_0 is by photo emission or thermionic emission from the cathode. In most electric contact applications the initial electron current results from electrons that are liberated from the cathode by field emission (see Section 9.3.2) or by other random physical processes. There may well be a period of time when V_B across an open contact is exceeded (see Section 9.4.1), before breakdown begins, but once it is initiated, it can occur extremely quickly, see for example Figure 9.9.

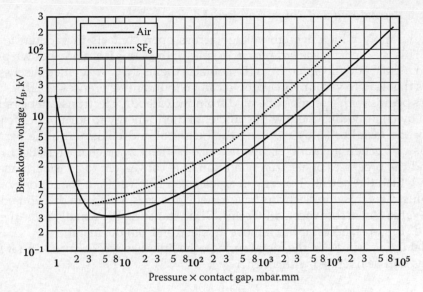

FIGURE 9.8
Paschen curves for air and SF$_6$.

TABLE 9.3

Minimum Breakdown Voltage $(V_B)_{min}$ and $(pd)_{min}$ Values for Various Gases

Gas	$(V_B)_{min}$ (V)	$(pd)_{min}$ (10^{-3} torr, m)	d_{min}, cm (p = 1 atmosphere i.e., 760 torr)
Air	327	5.7	0.75×10^{-3}
A	137	3.9	1.18×10^{-3}
N$_2$	251	6.7	0.83×10^{-3}
H$_2$	273	1.5	1.51×10^{-3}
O$_2$	450	7.0	0.92×10^{-3}

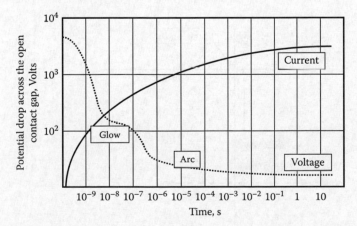

FIGURE 9.9
Time for contact gap breakdown and arc formation for a 1 mm contact gap.

9.3.2 Vacuum Breakdown and Short-Gap Breakdown

It is possible to apply a high enough voltage across open contacts in vacuum to cause breakdown. This is called *vacuum breakdown*. In spite of the name it is a breakdown process that occurs in metal vapor liberated from the contacts themselves. A detailed discussion of the processes that lead to vacuum breakdown can be found in References [20,21].

The process begins with electrons liberated from the cathode. Electrons can be removed from a metal by (1) providing them with sufficient kinetic energy to surmount the potential barrier of the metal or (2) by reducing the height of and/or thinning the barrier so that electrons can penetrate it and escape because of their wave characteristic. In order to understand the processes of electron liberation, it is convenient to consider a metal as a box within which the potential energy of an electron is lower than that of one outside. Figure 9.10 illustrates this, the electrons have kinetic energies which are distributed up to a maximum value ξ in accordance with Fermi–Dirac statistics. We usually refer the work function φ, i.e., the work necessary to remove an electron from the metal.

If ultraviolet light is shone on the metal surface, the light has an energy hv and if $hv > \varphi$ then electrons will be liberated with an energy E'

$$E' = hv - \phi \tag{9.33}$$

This is called photoelectric emission. If the metal is heated to a temperature T, electrons will be emitted. Richardson and Dushman gave the equation

$$J = BT^2 \exp\frac{\phi}{kT} \tag{9.34}$$

where J is the current density and B is a constant whose value is between 30 and 100 A cm^{-2} K^{-2} for most metals, see Table 9.4. Under a strong electric field the work function is effectively reduced to

$$\phi' = \phi - e^{3/2}E^{1/2} \tag{9.35}$$

See Figure 9.11. Experimental measurements in vacuum at room temperature, however, show that currents considerably higher than are calculated from Equations 9.34 and 9.35 could be emitted from cold cathodes with high electric fields applied. Fowler and

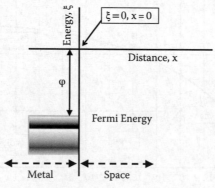

FIGURE 9.10
Potential energy versus distance for an electron near a metal surface.

TABLE 9.4

Thermionic Emission B Values for Some Metals

Contact Material	B (A cm^{-2} K^{-2})
Ag	60
Au	60
C	6
Cr	50
Cu	65
Mo	85
Ni	30
Pt	32
W	65

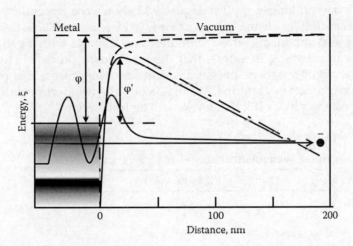

FIGURE 9.11

An illustration of the quantum-mechanical tunneling mechanism by which a field-emitted electron may escape, or tunnel, from the surface of a metal. It shows the modified high-field potential barrier outside the surface (solid line), where the contributions of the image potential and the applied field ($E = 3 \times 10^9$ V m^{-1}) are shown by the broken and the broken–solid lines, respectively. In addition, the wave function of a tunneling electron is depicted [21].

Nordheim [22] used a wave theory to show that electrons would tunnel across the potential barrier and not have to surmount it at all, see Figure 9.11. They showed that

$$J = \left(\frac{B_1 E^2}{\phi} \right) \exp\left(-\frac{B_2 \phi^{3/2}}{E} \right) \tag{9.36}$$

where B_1 and B_2 are known fundamental physical constants [20,21]. It is, therefore, possible to liberate electrons from microscopic projections on the cathode. If the voltage imposed across the electrode is high enough for two contacts separated by a distance d in vacuum. It is also possible for these electrons, in turn, to heat the metal below the anode's surface [20]. Metal vapor is liberated into the contact gap when the current density in a cathode projection causes it to evaporate and/or when the anode's subsurface reaches it boiling point [20]. The ionization of this metal vapor then results in a breakdown of the contact gap. It is now recognized that the breakdown voltage is very dependent upon the microscopic surface conditions of the cathode. It is thought that for a given metal there is given critical breakdown field given by E_c but the breakdown voltage V_B is given by

$$V_B = \frac{E_c d}{\beta} \tag{9.37}$$

where β is an enhancement factor which depends upon the geometry of the contacts and the microscopic variations of the cathode surface [20]. Table 9.5 gives values of E_c for a number of metals. Equation 9.37 is only valid for contact gaps less than about 0.3 mm [20]. At longer gaps $V_B \approx K d^{1/2}$, where K is a constant [20]. Vacuum breakdown has also been observed between contacts in atmospheric air at room temperature if the distance between the contacts is less than 5–10 electron mean free paths, i.e. $d \leq 10^{-5}$ m. At these small contact gaps Paschen's law no longer applies. Figure 9.12 shows the dependence V_B of as a function of contact spacing for clean contacts in air [23]. Here $V_B \approx 100d$ (when d is in µm). This relationship is also valid for V_B in vacuum between contacts with a gap of less than about 0.3 mm. The inference is, therefore, that the breakdown process between closely spaced contacts in air at atmospheric pressure is similar to the process that results in the breakdown between contacts in vacuum [23]. It has been shown experimentally for closely spaced, palladium contacts in air [13] that an arc will be initiated if

TABLE 9.5

Critical Vacuum Breakdown Field for Various Metals

Metal	E_c (× 10^8 Vm^{-1})	φ (eV)
Cr	53	4.6
Mo	54	4.4
Stainless steel	59	4.4
Au	64	4.8
W	65	4.5
Cu	69	4.5
Ni	104	4.6

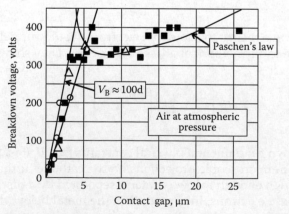

FIGURE 9.12
Breakdown voltage as a function of contact gap in the range 0.2–40 µm [23].

$E > 2 \times 10^8$ V m^{-1} clean contacts

$E > 30 \times 10^8$ V m^{-1} stringently clean contacts

$E > 0.5 \times 10^8$ V m^{-1} organically contaminated or activated contacts

Thus, from Figure 9.12 if the circuit voltage is 40 V for clean contacts the arc will be initiated in air at atmospheric pressure with a contact spacing of approximately 4×10^{-7} m. Once the arc is initiated the pressure from the arc on the contacts may prolong the closing time (e.g., see Figure 13.40). Figure 9.13 puts together the gas breakdown and the small gap breakdown data.

9.3.3 The Volt–Current Characteristics of Separated Contacts

Take the circuit shown in Figure 9.14, where it is hypothetically possible to gradually increase the circuit voltage from zero. When the various states of the breakdown process occur the series resistor permits fine control of the current. The progression of the curve is summarized below [13].

A. In this region, electrons emitted at the cathode are collected at the anode.

B. At a certain voltage level all the electrons emitted at the cathode are collected and current saturation is reached. Beyond this point further current increase requires either interaction of the electrons and the gas, or much stronger fields; e.g., fields at the cathode surface that are found for vacuum breakdown.

C. Electron collisions with the gas between the contacts create additional electron–ion pairs, which increase the current, and this marks the beginning of the breakdown.

D. When the electron multiplication reaches a critical stage a Townsend avalanche begins.

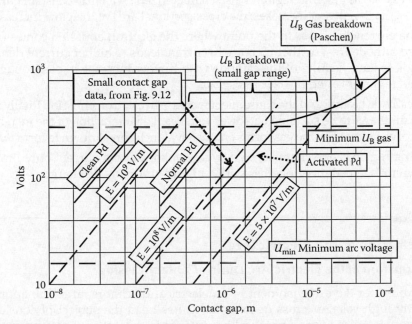

FIGURE 9.13
Combined voltage and contact gap values for long contact gap and short contact gap breakdown [13].

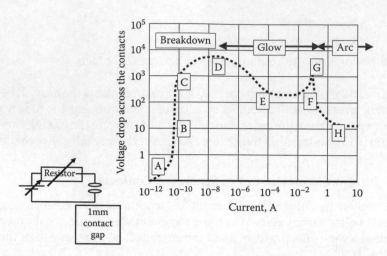

FIGURE 9.14

Voltage–current characteristics for gas breakdown and arc formation between open contacts [13].

E. After breakdown, the current is sustained by the same avalanche mechanism, except that a smaller voltage is required to maintain the necessary ionization. This steady discharge is characterized by a relatively constant voltage drop over about two decades of low current ($<10^{-1}$ A), a cold cathode and a number of luminous zones. Hence it is called a *glow discharge*. For any particular gas there is a characteristic sustaining voltage. The voltage also depends upon the cathode material. The principal voltage drop occurs in the space between the cathode and start of the luminous region, called the *cathode fall*. This voltage is required to produce the discharge sustaining electrons from the cathode by field emission enhanced by secondary emission processes. The glowing region is also characterized by a fairly constant current density, i.e., as the current increases, the cross-sectional area of the glow increases.

F. When the current increases to the point where the electron emission comes from the entire cathode area, further increase in current leads to higher current density. This leads to joule-heating of the cathode with further increase in electron emission. This region is called the *abnormal glow*.

G. Intense cathode heating and the consequent increase in electron emission finally permits the discharge to be sustained at lower voltages. The interaction of the increased number of electrons with the gas forms another avalanche breakdown to form *the arc*.

H. The arcing region is one characterized by low voltage and a current only limited by the circuit's impedance, together with a highly luminous discharge.

9.4 The Formation of the Electric Arc

9.4.1 The Formation of the Electric Arc During Contact Closing

The time required for the establishment of an electrical discharge, after the application of a sufficiently high voltage across open contacts, has been the subject of considerable scientific interest. It is also of great practical importance in contact applications: especially when they are closing.

Once a voltage is impressed across an open contact gap, there are two important time periods. The first is the statistical time lag, t_{st}. Here the first electron becomes available to begin the breakdown process. The second is the formative time lag t_f. This is the time required for the discharge to become established. The t_f depends upon the over-voltage, the gas, the contact geometry, and the number of initiating electrons [24].

If the threshold voltage for breakdown is V_B and a voltage $U > V_B$ is impressed across the contact gap, then it has shown that

$$t_f \propto \theta^{-1} \tag{9.38}$$

where $\theta = (U - V_B)/V_B$. Figure 9.15 shows the compilation of experimental data [24]. From this figure, it can be seen that t_f can range from about 40 ns when $U = 2\ V_B$, 400 ns when $U = 1.1\ V_B$, 4 μs when $U = 1.01\ V_B$ and 40 μs when $U = 1.001\ V_B$. Therefore, once the voltage is impressed across the contacts and the first electron is initiated, all the processes discussed in Figure 9.14 can occur very quickly: see for example Figure 9.9. For example, if a contact is closing at 1 ms^{-1} it only travels 1 μm in 1 μs, so even if the contacts are 10^{-4} m apart an arc established between them will burn for 100 μs before they actually touch.

9.4.2 The Formation of the Electric Arc During Contact Opening

If the circuit current is greater than a minimum value and the voltage that appears across the opening contacts is also greater than a minimum value (see Section 9.5.4), an arc will always form between them. The arc formation depends entirely upon the properties of the contact material and the arc always initiates in metal vapor from the contacts themselves. From Chapter 1 (Equation 1.11) the contact resistance R_c is given by

$$R_c = \frac{\rho}{2a} = \frac{\rho}{2}\sqrt{\frac{\pi H}{F}} \tag{9.39}$$

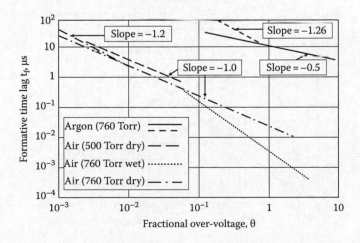

FIGURE 9.15
Plots of formative time lag, t_f, versus fractional over voltage using data supplied by a number of researchers [24].

where 'a' is the radius of the real area of contact, H is material hardness, ρ is resistivity and F the holding force. As the contact begin to open $F \to 0$ and so $a \to 0$ and so R_c will increase. As R_c increases, U_c, the voltage drop across the contacts, given by

$$U_c = IR \tag{9.40}$$

also increases. Chapter 1 (Equation 1.40) also showed that the temperature of the contact spot T_c will increase because

$$T_c^2 = T_0^2 + U_c^2 \times 10^7\, K \tag{9.41}$$

where T_0 is the ambient temperature. A stage will be reached when the temperature of the contact spot equals the melting point, T_m, of the metal [25]. Figure 9.16 compares calculated values of U_c at T_m with measured values for a wide range of metals used as electric contacts. Table 9.6 gives values taken from Holm [14]. Once the contact spot has melted and the contacts continue to part, they draw a molten metal bridge between them. This molten metal bridge always forms between the contacts even at low currents [26–29], even when the contacts open with a very slowly [28] or with high acceleration [26,30,31], and even when they open in vacuum [32,33]. For example, Figure 9.17 shows three voltage traces of molten metal bridges forming between Au contacts in a 10 mA, 4 V circuit [29] and Figure 9.18 [27] shows the length of the molten metal bridge over the current range 1 to 79 mA. The molten metal continues to bridge the opening contacts until it becomes unstable and ruptures. After the rupture of the molten metal bridge an arc forms in its vicinity. A typical change in the initial voltage drop across the contacts is shown in Figure 9.19. Figure 9.20a–c shows examples of the measured voltage as the contacts open and form the molten metal bridge, as the bridge ruptures and the arc is initiated between three different contact materials and for a wide range of currents: 20–1,000 A and three different contact metals [30,34,35].

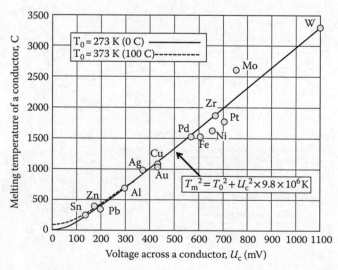

FIGURE 9.16
Relationship between the measured voltage drop across a conductor and its calculated and measured melting temperature.

TABLE 9.6

Melting Voltages for Various Metals

Metal	Melting Temperature (K)	Measured Melting, U_m (V)
W	3,683	1.1
Mo	2,883	0.75
Ni	1,726	0.54
Cu	1,356	0.43
Au	1,336	0.43
Ag	1,234	0.37
Al	933	0.3
Pt	2,042	0.71

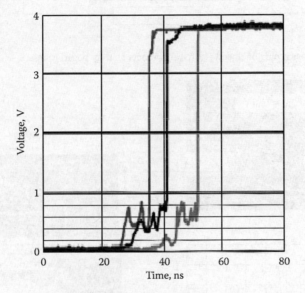

FIGURE 9.17
The voltage characteristics of molten metal bridges from three openings of Au contacts in a 10 mA, 4 V circuit [29].

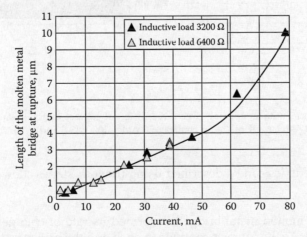

FIGURE 9.18
The length if the molten metal bridge for currents in the range 1–79 mA [27].

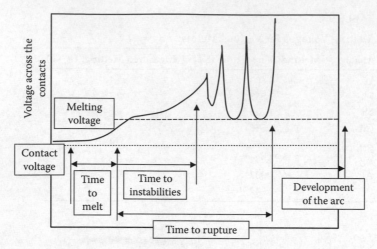

FIGURE 9.19
The initial voltage drop across a pair of closed electrical contacts as they begin to open.

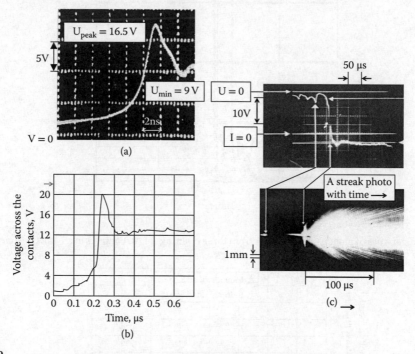

FIGURE 9.20
Actual initial voltages across contacts opening over three different contact materials and a wide range of currents; (a) Au contacts, 20 A[dc], 6 V [35], (b) Cu contacts, 50 A[dc], 100 V [34] and (c) Ni contacts, 1,000 A[ac], 440 V [30].

These voltage characteristics can be described using the four stages shown in Figure 9.21 [36,37]:

Stage (a): Once the molten metal bridge has formed its rate of change of voltage is about 2×10^3 V s^{-1}. As the contacts continue to open and the bridge is drawn further it becomes unstable. There are a number of physical reasons for this instability,

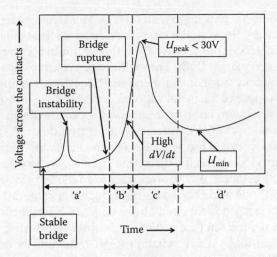

FIGURE 9.21
The four stages of molten metal bridge rupture and metal phase arc formation.

TABLE 9.7
Comparison of the Boiling Voltage (U_{bl}) with the Break Voltage (U_b) for Various Metals

Metal	Breaking Voltage (U_b)	Calculated Boiling Voltage (U_{bl})	Boiling Temperature T_{bl} (K)
Ag	0.75	0.77	2,485
Cu	0.8	0.89	2,870
W	1.7	1.82	5,800
Au	0.9	0.97	3,090
Ni	1.2	0.97	3,140
Sn	0.7	0.87	2,780

including surface tension effects, boiling of the highest temperature region, convective flows of molten metal resulting from the temperature variation between the bridge roots and the high-temperature region. Figure 9.20c shows this instability for Ni contacts opening a current of 1,000 A. The streak photograph of this opening shows that the unstable bridge is incandescent (i.e. >2,900°C). In this case the high temperature results in more metal entering the bridge and it cools. The bridge will eventually rupture, releasing metal vapor into the contact gap when the voltage across it, U_b, is close to the calculated boiling voltage U_{bl} of the contact materials, i.e.:

$$U_b = \sqrt{L(T_{bl}^2 + T_0^2)} \qquad (9.42)$$

where L is the Lorenz constant and T_{bl} K is the boiling temperature see Table 9.7.

Stage (b): Once the bridge ruptures the voltage across the contacts rises very rapidly without a discontinuity from about 10^3 V s^{-1} to about 10^9 V s^{-1}: at 10 mA for Au contacts $dV/dt \approx 0.5 \times 10^9$ V s^{-1} [29], at 150 A for Cu contacts $dV/dt \approx 5 \times 10^9$ V s^{-1} [34], at 1 kA for Cu contacts $dV/dt \approx 10^9$ V s^{-1} [31] and for Ni contacts $dV/dt \approx 2 \times 10^9$ V s^{-1} [31]. This rate of rise of the voltage will depend upon the dimensions of the molten metal bridge just before its rupture. After the bridge rupture a very high pressure, perhaps as high as 100 atmospheres [36,37], very low electrical conductivity, metal vapor exists between the contacts. This region can then be considered to be a capacitor with a very small capacitance. Because the circuit's inductance prevents a rapid change in current (see Figure 9.20c), charge flows from the circuit inductance into this small capacitor causing the very high dV/dt. The metal vapor volume expands rapidly into the surrounding lower pressure ambient and as it does its pressure also decreases rapidly. When the pressure of the metal vapor decreases to 3–6 atmospheres conduction is initiated with a voltage across the contacts of a few 10's of volts, see Figure 9.20. At these pressures the discharge that forms is the "pseudo arc" [38,39] where the current is conducted by ions. During this stage the electrons required for charge neutrality will be introduced into the discharge from secondary emission resulting from ion impact at the cathode. As the original molten metal bridge will have material from both the cathode and the anode, net transfer of material from the anode to the cathode is expected and is indeed observed [36,37]: see Section 10.3.5. At this stage radiation from neutral Cu (CuI radiation) is observed during the initial rapid rise of the voltage, but ionized Cu radiation (CuII radiation) is not observed, see Figure 9.22a and b [34]. Once the pseudo arc forms there is a large increase of the CuI radiation and a slowly rising increase of the CuII, see again Figure 9.22 a and b.

Stage (c): As the pressure of the metal vapor continues to decrease to about 1–2 atmospheres, the pseudo arc transitions into the usual arc discharge with an arc voltage impressed across the contacts whose value is about that of the minimum arc voltage expected for an arc operating in the contacts' metal vapor (i.e., $U_{min} \approx 10–20$ V, see Section 9.5.4). Here again net material transfer will be from anode to cathode (see Section 10.3.5). Thus all arcs formed between opening contacts in all ambient will operate in metal vapor and only transition to one operating mainly in the

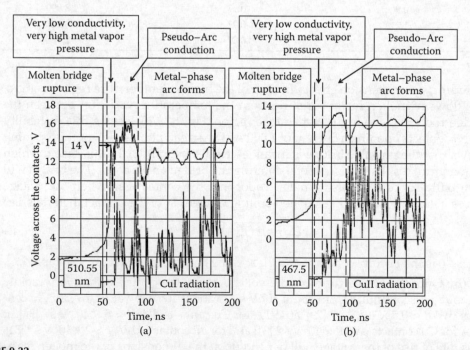

FIGURE 9.22
The transition from molten metal bridge to the metal phase arc [34]: (a) Voltage, neutral copper radiation (Cu I), Cu contacts, 50 A, 100 V and (b) voltage, ionized copper radiation (Cu II), Cu contacts, 50 A, 100 V.

ambient gas as the contacts continue to open (see Section 10.3.5). It is only in a vacuum ambient that the arc continues to operate in metal vapor evaporated from the contacts themselves [20,32,33]. In order to sustain this arc a minimum arc current is also required (see Section 9.5.4).

Stage (d): At this stage as the contacts continue to open the arc between them gradually transitions from the metallic phase arc to the ambient, gaseous phase arc with most of the current now carried by electrons. Once the metallic phase arc forms there is a large increase in the CuII radiation, see Figure 9.22b. The whole sequence is illustrated in Figure 9.23. As the contacts continue to open this metallic phase arc transitions into an arc operating in the ambient atmosphere (see Section 10.3.5). The parameters required to form and sustain an arc are summarized in Table 9.8.

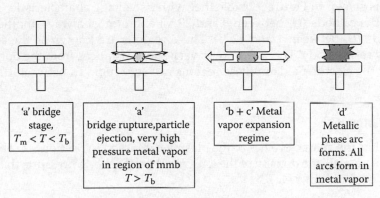

| 'a' bridge stage, $T_m < T < T_b$ | 'a' bridge rupture, particle ejection, very high pressure metal vapor in region of mmb $T > T_b$ | 'b + c' Metal vapor expansion regime | 'd' Metallic phase arc forms. All arcs form in metal vapor |

FIGURE 9.23
The opening sequence of an electric contact; formation of the molten metal bridge, its rupture and arc formation.

TABLE 9.8

Parameters Required to Form and to Sustain an Arc

Condition	Voltage Across Contacts	Circuit Current
To Initiate Breakdown		
Gas breakdown:	V_B (gas) $\approx > 329 + 3 \times 10^6 \times d$ for air	$I \approx > 10$ and < 600 mA
No conduction to glow	(d in m.)	
No conduction to stable or unstable arc	For air at atmospheric pressure	$I \approx > 20$ mA
Small gap or vacuum breakdown for	V_B (vacuum) $> E_c d$: i.e., $\approx > 10^8 d$	A critical pre-breakdown
$U \leq U_B$ (gas)	$E_c \approx 0.5 \times 10^8$ V m^{-1} "activated" contacts	emission current:
	$E_c \approx 2.0 \times 10^8$ V m^{-1} "normal" contacts	$I \approx > 50$ mA to 1.0 A
	$E_c \approx 30 \times 10^8$ V m^{-1} "clean" contacts	$I \approx > 1$ A in vacuum
Conduction to arc stable or unstable		$I \approx > 50$ mA in Air
To Sustain Breakdown		
Glow (gas dependant)	V_G (gas) $= 280 + 10^6 \times d$ for Air (d in m.)	$I_{G(min)} \approx > 10$ mA
Arc (contact material dependant)	$V_A = V_m + P \times 10^3 d$, where P ranges from 1 to 10 and is $f(I)$ V_m ranges from 10 to 20 V is f (metal and I)	$I_{min} = 0.5$ A to 1.0 A (metal surface condition)
To Initiate and Sustain Arc During Contact Opening		
Contact opening	$V_A \geq V_m$	$I_{min} = 0.5$ A to 1.0 A (metal surface condition)

9.5 The Arc in Air at Atmospheric Pressure

This section will concentrate upon the arc in air at atmospheric pressure, because this is the arc that the vast majority of switching contacts experience. It occurs during the closing and opening of most relays, contactors, switches, and circuit breakers that operate in air. There are some high-voltage circuit breakers and contactors that use the gas such as H_2 and SF_6 [17] or operate under oil (i.e., high-pressure H_2). The general description of the arc in air also applies to the arc in any high-pressure gas including SF_6 and H_2. The other form of arc is the so-called vacuum arc, which occurs in a special class of interrupters where contacts operate in a high-vacuum environment [20]. The vacuum arc does exhibit special characteristics different from the arc in air and will be described briefly in Section 9.6.

A generic arc is shown in Figure 9.24 together with its voltage characteristic. There is a voltage drop at the cathode (U_c between 8 and 20 V) and almost always another voltage drop at the anode (U_a between 1 and 12 V). These regions are known as the *cathode fall* and the *anode fall* respectively. They occur over very short distances from the open contact surfaces, so that the electric fields in these regions are very high. In between them is the *arc column*.

9.5.1 The Arc Column

The arc column has the characteristics of the plasma discussed in Section 9.2. The density of the electrons, n_e, equals the density of the ions, n_i. At atmospheric pressure, there is local thermodynamic equilibrium, i.e.,

$$T_e(\text{electron temperature}) = T_i(\text{ion temperature}) = T_g(\text{gas temperature})$$

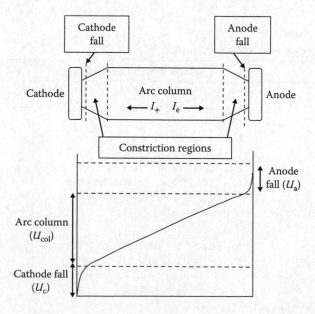

FIGURE 9.24

Schematic of an arc with the column constricted at the contacts with the corresponding voltage distribution (note: the length of the cathode and anode sheath regions are not to scale).

For the low-current, free-burning arcs (approximately 5 A), T_g is between 6,000 and 7,000 K and the temperature for high-current, free-burning arcs (approximately 1,000 A) is about 20,000 K. It is unusual in free-burning arcs to obtain a temperature much higher than this value, thus the average energy of the ions and gas atoms in arcs has a range from about 0.85 eV for low-current arcs to about 2.5 eV for high-current ones. The arc temperature can be determined spectroscopically [40]. Figure 9.25 shows the temperature profile of a 1,000 A, free-burning arc between Cu contacts [41]. The current density in the arc core is

$$J = (n_e e M_e + n_i e M_i)E \tag{9.43}$$

where M_e and M_i are the electron and ion mobilities and E is the column's electric field strength. Since, $M_e \gg M_i$, greater than 99% of the current in the arc column is carried by electrons. The ion concentration in the ac column at atmospheric pressure can be calculated using a simplified form of Saha's equation [9,42,43]:

$$\log_{10} \frac{n_i^2}{n} = -5400\frac{V_i}{T} + \frac{3}{2}\log_{10} T + 15.385 \tag{9.44}$$

where V_i is ionization potential of the gas, $n = n_n + n_i$ (n_n density of the neutral atoms). In most arcs between metal contacts that are far enough apart, there is a mixture of the atmospheric gases and metal vapor from the contacts. The gas having the lowest ionization potential in the mixture is the most readily ionized. As most metal vapors have much lower ionization potentials than nitrogen or oxygen, any metal vapor in the air will be preferentially ionized. The metal vapor concentration can also be calculated using the Saha equation together with a spectroscopic analysis of the arc [41]. In recent years the arc column has been modeled very successfully using the calculating power of computers [44]. The use of suitable software to design and analyze switching devices will be discussed in

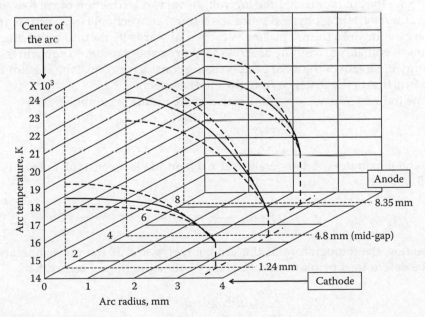

FIGURE 9.25
Temperature profile for a 1,000 A free-burning arc in air between copper contacts [41].

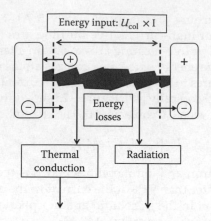

FIGURE 9.26
The Elenbaas–Heller arc model.

Chapter 14. A simple, steady state, arc model given by Elenbaas [45] and Heller [46] has found wide use to explain arcing phenomena. This model is illustrated in Figure 9.26. Here the energy input is given by the current flowing in the column times the voltage drop (i.e., Joule heating), is only balanced by the radial thermal losses and radiation, i.e.:

$$\sigma E_{col}^2 = -\frac{1}{r}\frac{d}{dr}\left(r\kappa\frac{dT}{dr}\right) + P_r \tag{9.45}$$

where r is the arc radius, κ thermal conductivity, P_r the radiation loss, σ the electrical conductivity and E_{col} is the column's electric field. Using this equation it is possible to determine at least qualitatively how an equilibrium arc will behave.

Example 1: The low current arc (1–30 A): here, using Lowke's analysis [47], it is possible to determine how the arc radius and the arc voltage vary as a function of current. In this current range, the controlling physical process is natural convection. For a vertical arc, the electrical energy going into the arc plasma is carried upwards by natural convection. The integrated flow of enthalpy across any arc cross section can be considered equal to the total electrical energy upstream of the axial position being considered. Thus, the arc radius will increase as a function of the distance from the lower contact.

If we assume the arc temperature as a function of arc radius is parabolic, i.e.

$$T(r) = T_{mx} - Dr^2 \tag{9.46a}$$

where T_{mx} is the maximum arc temperature at $r = 0$ and D is a constant. Substituting in Equation 9.45

$$T(r) = T_{mx} - \frac{\sigma(E_{col})^2 r^2}{4\kappa} \tag{9.46b}$$

Then neglecting the temperature at the arc boundary and if the arc temperature is assumed to be determined primarily by the energy balance at the arc center then:

$$\frac{I^2}{\sigma A^2} = \frac{4\pi\kappa T_{mx}}{A} + P_r \tag{9.46c}$$

where $\sigma(E_{col})^2$ is expressed in terms of the current using Ohm's law $I = \sigma E_{col}A$ and A is the arc area. Lowke [46] showed that if the radiation losses were negligible, then the arc radius r and the arc voltage U at a height z above the lower contact are given by:

$$r = \frac{0.56}{(\delta \hbar \sigma)^{1/4}} \left(\frac{\sigma z}{\delta_b g} \right) I^{1/2} \tag{9.46d}$$

$$U = \int E \, dz = 1.37 \left(\frac{\delta \hbar}{\sigma} \right)^{1/2} \left(\frac{\delta_b \, g}{\delta} \right) z^{3/4} \tag{9.46e}$$

where δ is the average arc plasma density, δ_b is the density of the gas at the arc boundary, and $\hbar$ is the enthalpy. Calculating the value of U as a function of I over this current range gives the familiar negative volt–ampere characteristic that will be discussed in Section 9.5.5. It is interesting to note that, as the arc radius (and hence the arc area) increases as the current increases, then at least qualitatively, one would have expected from Ohm's law that the arc voltage would decrease.

Example 2: Assume the cooling rate at the arc boundary is increased while the current is kept constant. This is commonly done in practical circuit breakers and dc contactors (see Chapter 14). The arc will respond by reducing its diameter, because the conducting radius of the gas will have decreased. Because the arc current remains constant, the current density will increase, i.e., $J = \sigma E$ will increase and the power density σE^2 will also increase. The increase in the power density will increase the temperature of the arc. Thus, in order to satisfy the Elenbaas–Heller equation, the temperature gradient at the arc boundary will increase at a faster rate than the decrease in arc diameter. As soon as an equilibrium state is reached, the consequence of cooling the outer edges of the arc will be at a higher temperature than at the arc's center.

Example 3: A higher thermal conductivity will also tend to increase the radial heat loss from the arc, and the arc will again respond with a smaller diameter and higher peak temperatures. This effect of gas with different thermal conductivities on arc diameter is well illustrated in Figure 9.27 [48].

Example 4: The high current arc (>100 A): here the arc properties are also largely determined by convection, but at these currents the convection is driven by the self magnetic field generated by the current flowing in the arc. Other physical processes also become more dominant as the arc current increases: radial pressure gradients become significant; radiation losses dominate at the arc center; viscous and turbulent forces can result in further energy losses. Lowke [48] analyzed of these effects and developed the positive volt–ampere characteristic observed for high current arcs, see Section 9.5.5. The self-magnetic also causes a drastic reduction in arc radius as the current increases from ~50 A to ~1,000 A, see Figure 9.28.

9.5.2 The Cathode Region

The cathode contact provides the electrons, the fuel that allows the arc to continue burning between the contacts. If the source of electrons from the cathode is terminated, then the arc will extinguish. The cathode region illustrated in Figure 9.29 is usually characterized by having high electric fields of 10^8–10^9 V m^{-1}, high thermal gradients (e.g., see Figure 9.25),

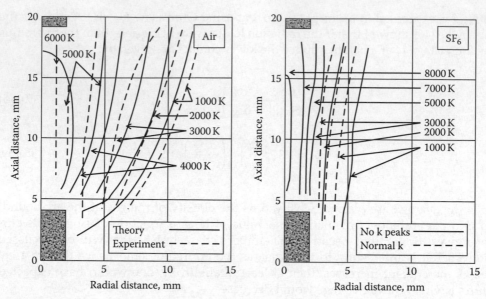

FIGURE 9.27
Temperature profiles for free burning 10 A arcs in air and SF_6 [47].

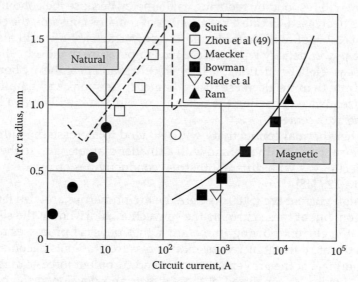

FIGURE 9.28
The radius of a free-burning arc column as a function of current [48].

contraction, i.e., the current density is higher than that of the arc column (maybe as much as 100 times greater) and at high currents, plasma jets [9] can form in this region. Two separate descriptions of electron emission have been observed, the first on refractory metals with high melting points such as tungsten. These readily emit electrons when heated to a temperature less than the melting point of the metal. The current density is given by the Richardson–Dushman equation, see Section 9.3.2, Equation 9.34. Most practical contact systems, however, use metals that boil at temperatures well below the temperature required to produce enough electrons to sustain the arc. Many researchers have studied the possible mechanisms to product the electrons from so called "cold cathodes." An

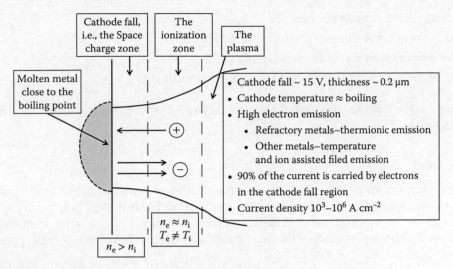

FIGURE 9.29
The cathode region of the arc.

excellent review is presented by Guile [50]. A interesting description and model of the cathode fall for tungsten contacts (i.e. thermally emitting contacts) is given by Zhou et al. [51]. However, as Kesaev [52] remarked ...

> this branch of physics is a collection of a large number of isolated pieces of work conducted under conditions difficult to compare and, as a rule, giving equivocal results.

In general, there is a consensus that the electron emission from "cold cathodes" involves a combination of thermally enhanced field emission (called T–F emission) and the effects of ion bombardment [12,53,54]. In the cathode fall region only about 90% of the current is carried by electrons, so 10% is carried by the ions.

There must be an energy balance at the cathode. Among the components that supply energy are:

1. Thermal energy from ions and neutral atom bombardment
2. Kinetic energy from ions dropping through the cathode fall
3. Condensation of ions and neutral atoms on the cathode
4. Radiation from the arc plasma
5. Chemical reactions in the cathode surface
6. Joule-heating of the cathode

Among the components that cause energy loss are:

1. Energy for electron emission which is approximately $eU\varphi$ for each thermionically emitted electron but is zero for each electron emitted by pure field emission. For cold cathodes the energy loss is somewhere in between.
2. Evaporation of cathode material.
3. Ejection of metal globules and particles.

4. Radiation from cathode surface.

5. Dissociation of molecules at the cathode surface.

6. Heat conduction by the bulk metal of the cathode contact.

7. Heat conducted or convected away by the surrounding gas.

In Figure 9.30, Cobine illustrated typical energies involved in some of these mechanisms [55].

9.5.3 The Anode Region

The anode region shown in Figure 9.31 can be either active or passive. In the passive mode, it serves only to collect the electrons carrying the current from the arc column. A space charge in front of the anode accelerates electrons from the column (i.e. $n_e \neq n_i$). If, however, the thermal boundary layer between the arc column and the anode surface is small enough

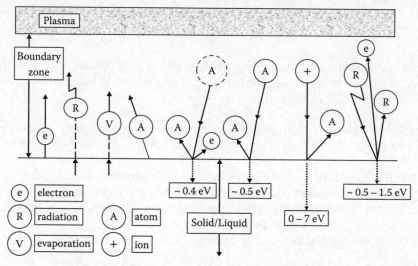

FIGURE 9.30
Possible mechanism of energy transfer at the cathode and/or anode [55].

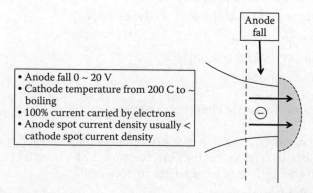

FIGURE 9.31
The anode region of the arc.

and the electron density gradients are high enough so that a substantial electron diffusion flow exists, then a space charge region is unnecessary. The anode fall voltages can thus range from close to zero to as high as 15–20 V. In most of the arcs which exist for times >1 ms between opening contacts with currents higher than a few amperes, an anode fall will exist with fields between 10^8 and 10^7 V m^{-1} and the anode fall thickness will be 10^{-4} to 10^{-6} m. An excellent review of the anode fall region is presented by Heberlein et al. [56]. Although much of their analysis is for arcs in an argon ambient, the general principles they present are relevant for arcs in air and other gases. The components that supply energy at the anode are:

1. Thermal energy from neutral atom bombardment
2. Kinetic energy from the electrons dropping through the anode fall
3. Condensation of neutral atoms on the surface
4. Radiation from the arc plasma
5. Chemical reactions on the anode surface
6. Joule-heating of the anode
7. The energy gained by entering electrons from the work function, which is approximately $eU\varphi$

These are balanced by:

1. Evaporation of anode material
2. Ejection of globules and particles
3. Radiation from the anode surface
4. Dissociation of molecules at the anode surface
5. Heat conduction into the bulk of the contact
6. Heat conducted or converted away by the surrounding gas

9.5.4 The Minimum Arc Current and the Minimum Arc Voltage

Once an arc has been established between opening contacts, it requires a continuous supply of electrons from the cathode to sustain it. These electrons will only be liberated in sufficient quantity if the cathode temperature remains high enough to liberate thermionic electrons for refractory contacts, or to liberate electrons by T–F emission for low melting point contacts. These electrons will ionize the metal vapor that the arc initially forms in. Intuitively it is reasonable to assume that if the arc current drops below a minimum value, i.e.,

$$I_A \leq I_{min} \tag{9.47}$$

the energy losses discussed in Section 9.5.2 will exceed the energy input to the cathode and electron production will cease, causing the arc to be extinguished or if the current is low enough the arc may not form at all after the rupture of the molten metal bridge. Detailed experiments on opening contacts for currents less than 1 A using oscilloscopes with a high frequency resolution by Ben Jemma et al. [57–60] and extended by Hasagawa et al. [61–63] have shown that the definition of I_{min} needs to be revised. Figure 9.32 shows that an arc discharge can occur between Ag contacts at currents well below the former I_{min} of 0.4 A given in Table 9.9. This figure shows that the actual arc duration at any given current has considerable variation. While the average variation with current is generally

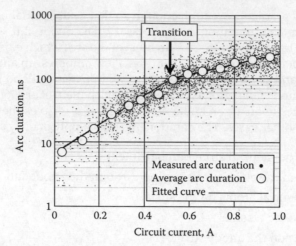

FIGURE 9.32

The arc duration for Ag contacts opening a 14 V dc circuit as a function of current, showing the wide statistical distribution of the measured values [57–59].

TABLE 9.9

The Minimum arc Voltage (U_{min}) and Minimum Arc Current (I_{min}) and Limiting Arc Current (I_{lim}) for Various Contacts and Comparison of U_{min} with Sum of the Ionization Potential V_i and the Work Function Potential $U\varphi$ (i.e. $V_i + U\varphi$)

Contact Material	V_i (V)	$U\varphi$ (V)	$V_i + U\varphi$ (V)	U_{min} (V)	I_{min} (former data, clean) A	I_{min} (new defn. 58, 59, 62) A	I_{min} (no arc current 58, 59, 62) mA
Al	5.98	4.10	10.08	11.2	0.4		
Ag	7.57	4.74	12.31	12	0.4	0.37	60
Au	9.22	4.90	14.12	12.5	0.35	0.4	80
Cu	7.72	4.47	12.19	13	0.4	0.29	45
Fe	7.90	4.63	12.53	12.5	0.45		
Ni	6.63	5.05	12.68	13.5	0.5		
Pd	8.33	4.97	13.30	14	0.8	0.8	100
Pt	8.96	4.60	13.56	14	0.9		
Rh	7.70	4.57	12.27	13	0.35		
W	7.98	4.49	12.47	13.5	1.0		
Sn	7.3	4.64	11.94	13.5		0.2	15
C	11.27	4.6	15.87	20	0.02		<20
Ag (In/SnO$_2$)				11.0		0.5	60

linear, there is a transition close to 0.5 A where the slope increases. Figure 9.33 gives the average arc times for five different contact materials. Examples of the measured voltage wave forms across Ag contacts show three separate regimes, Figure 9.34. (A) Where no arc occurs and a voltage transient determined by the inductance and capacitance in the circuit. (B and C) Here a short duration arc is seen with a voltage close to the expected minimum value, U_{min}. This discharge is unstable and 'chops' out without reaching a value close to the circuit's emf (see the requirements for interrupting a dc circuit in Section 9.7.2). Because the current chops to zero in this way, there is still current trapped in the circuit's inductance. Thus the voltage wave form after the arc has extinguished is similar

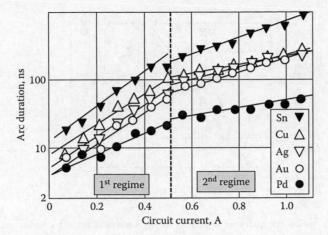

FIGURE 9.33
The average arc duration as a function of current for five different contact materials opening a 14 V dc circuit [57–59].

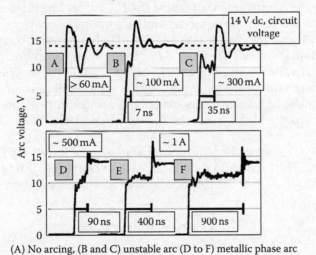

(A) No arcing, (B and C) unstable arc (D to F) metallic phase arc

FIGURE 9.34
Measured voltages across opening Ag contacts as a function of current a 14 V dc circuit: (A) no arcing, (B) and (C) a low voltage unstable arc and D–F, the formation of the metallic phase arc [57–59].

to that shown in (A) and depends upon the residual current and the circuit's inductance and capacitance. This unstable condition is seen more clearly at higher currents for an arc between contacts opening in vacuum and is there termed "The chop current" [20]. (D to F) At these currents a stable arc forms in the metal vapor from the ruptured molten metal bridge. It operates until the arc voltage has driven the circuit current down to a value where the arc can no longer exist. The resulting voltage wave form after the arc has extinguished still shows the oscillations that you see in (A) and (B and C). As the circuit current is reduced arcing only occurs 100% of the time for currents greater than a given value which is strongly dependent upon the contact material, see Figure 9.35. Ben Jemma used this result to define as I_{min} the current above which an arc occurs 100% of the time. The values of this I_{min} are compared to the traditional values in Table 9.9. The lowest current that arcing can occur I_{lim} will be much lower than this I_{min}. For switch designers it is

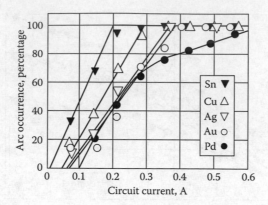

FIGURE 9.35
The probability of arc formation for currents less than 0.6 A for five different contact materials a 14 V dc circuit [57–59].

important to note that contact erosion and a change in the contact surface resulting from the electric arc will occur at currents well below the I_{min} value. The duration and stability of these short duration arcs show a strong dependence upon the contacts' opening speed, the circuit's inductance and the circuit's dc emf [58–63]. Note that this discussion of I_{min} is only valid for the initial contact opening. After the molten metal bridge has ruptured the arc forms in metal vapor between very closely spaced contacts. The energy loss in this arc is mostly through thermal conduction, with a very small percentage lost through radiation. When opening higher currents, when the arc column is longer, the arc can become unstable at currents higher than the I_{min} discussed here. This will be discussed in Section 9.7.2.

Also it is reasonable to assume that a minimum voltage is required across the open contacts to sustain the arc. The arc would require at least a voltage, U_{min}, that corresponds to the work function voltage, $U\varphi$, of the cathode contact and the ionization potential of the gas, V_i. Thus,

$$U_{min} \approx V_i + U_\varphi \tag{9.48}$$

Table 9.9 compares measured U_{min} with $V_i + U\varphi$ for low current arcs between contacts made from a number of different metals. This table also presents values for I_{min}. If a contact becomes contaminated with carbon deposits (this is called "activation", Section 9.2.3) then the I_{min} could decrease to a value you would associate with carbon contacts. Figure 9.36 [64] and Table 9.8 [13] summarize the parameters required to form and sustain an arc.

9.5.5 Arc Volt–Ampere Characteristics

A great deal of data has been developed to show how the arc voltage varies with arc current for a given contact gap. For low dc currents the arc voltage has the negative characteristic shown in Figure 9.37. It tends to go asymptotic at I_{min} and U_{min} [14,65–67]. Actual data for Cu and AgCdO contacts are shown in Figures 9.38 and 9.39. These curves show us negative volt–ampere characteristic, i.e., as the current in the arc increases, so the voltage drop across the arc decreases. The total arc voltage U_A is given by (see Figure 9.24)

$$U_A = U_c + U_{col} + U_a \tag{9.49}$$

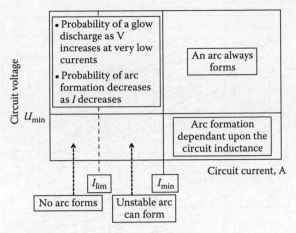

FIGURE 9.36
Voltage and current ranges required to sustain an arc.

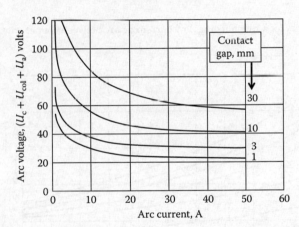

FIGURE 9.37
Voltage–current characteristics for a free-burning, low current dc arc between copper contacts, showing the effect of contact gap.

The form of U_{col} is complicated by the fact that the arc column usually tends to constrict as it gets closer to the cathode and anode. Figure 9.40 shows the measured total arc voltage U_A as a function of contact gap [68].

Here

$$U_A = \Delta + (\tau + d)E(I) \tag{9.50}$$

where d is the arc length and I the arc current.

$$E(I) = b\left(\frac{\ln I}{q}\right)^{-3} \tag{9.51}$$

and $\Delta = 26$ V, $\tau = 1.1$ cm (for silver contacts), 1.3 cm (for copper contacts) and 1.6 cm (for tungsten contacts), $b = 5,400$ V cm^{-1} $q = 7.4 \times 10^{-3}$ A.

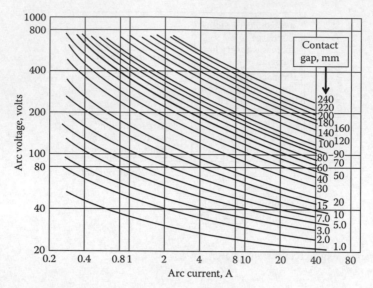

FIGURE 9.38

Voltage–current characteristics for a free-burning, low current dc arc between copper contacts as a function of the contact gap [65].

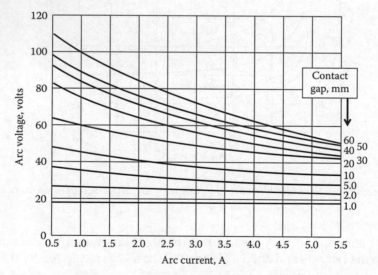

FIGURE 9.39

Voltage–current characteristics for a free-burning, low-current dc arc between AgCdO contacts, as a function of contact gap [66].

At higher currents the arc voltage levels out and then increases as a function of current. Figure 9.41 shows the how arc's gross electric field (i.e. V/d) varies over 8 decades of current [69]. Lowke modeled the arc voltage 1 cm above the cathode and compared it to experimental data; this is shown in Figure 9.42 [48] (see also Examples 1 and 4 in Section 9.5.1). The important aspect of these data is to note that the arc voltage is *not* a function of the system or circuit voltage, but instead is determined by the power input required to sustain the arc.

Because the energy stored in the arc is associated with its conductance and with finite rates of energy flow, the arc conductance cannot respond instantaneously to current changes.

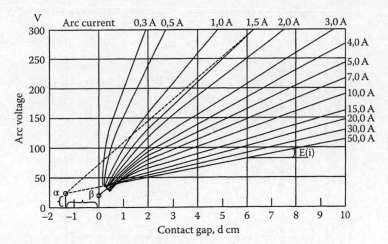

FIGURE 9.40
Arc voltage as a function of contact gap (or arc length) for a free-burning arc in air [68].

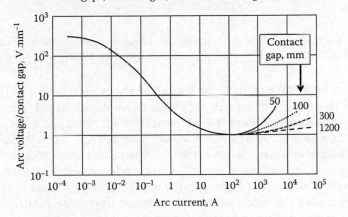

FIGURE 9.41
Voltage–current characteristics as a function of contact gap in air for currents in the range 10^{-4}–10^4 A [69]).

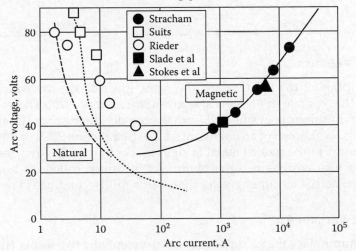

FIGURE 9.42
Voltage–current characteristic for a free-burning arc in air [48].

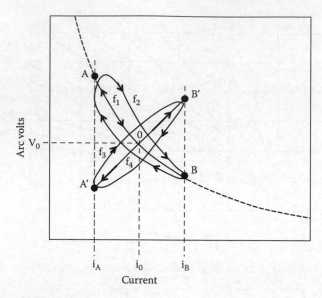

FIGURE 9.43

f_1 is the static dc arc characteristic and f_4 is the characteristic for a resistance obeying Ohm's law; the dynamic arc characteristics change from f_1 to f_4 as the frequency of the current increases. Here $f_1 < f_2 < f_3 < f_4$ [71].

The conductance and arc voltage tends to lag behind changes in current [70,71]. The degree of lag depends upon the rate of change in arc current and the electrical inertia of the arc, called the arc's *time constant*. In Figure 9.43, the effect of different ac frequencies is illustrated. The frequency f_1 is low enough that the static arc voltage–current characteristic is followed; f_4 on the other hand is so fast that the arc conductance cannot follow the current at all and it shows a resistive characteristic. Frequencies f_2 and f_3 have intermediate values [72]. In the case of alternating power frequency arcs (50 or 60 Hz), in the range of ~10 A to a few hundred amps, the arc voltage tends to remain constant during each current half cycle, because the arc in a regime where the arc voltage does not appreciably vary with current (see Figure 9.41). It is only near current zero where an increase in voltage may be observed.

9.6 The Arc in Vacuum

A complete description of the vacuum arc, the name given to the arc which is formed between contacts that operate in a vacuum, is given in Reference [20]. The vacuum arc is an arc that forms in metal vapor evaporated from the contacts themselves. It can be formed either from vacuum breakdown of an open contact gap (see Section 9.3.2) or during contact parting by the rupture of the molten metal bridge (see Section 9.4.2). After the rupture of the molten metal bridge, a high-pressure column arc is formed called the bridge column [20,72,73] which can persist for times greater than half a millisecond, see Figure 9.44.

9.6.1 The Diffuse Vacuum Arc

After the bridge column stage the vacuum arc exists in essentially two forms: (1) diffuse and (2) columnar. For currents up to approximately 5 kA, the vacuum arc is diffuse [73,74] and can be characterized by a multiplicity of rapidly moving cathode spots (the number of spots

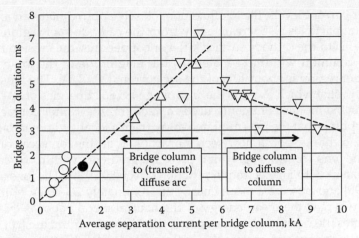

FIGURE 9.44
Duration of the bridge column formed after the rupture of the molten metal bridge for Cu–Cr contacts opening in vacuum [20, 73].

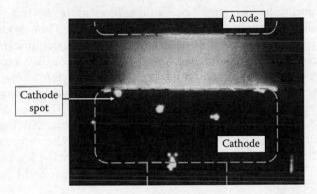

FIGURE 9.45
The diffuse vacuum arc.

being approximately proportional to the current and dependent on the contact material), a diffuse inter-contact plasma, and a diffuse collection of current at the anode (see Figure 9.45). These cathode spots are essential to the maintenance of the discharge. Most of the arc voltage drop occurs across a space-charge sheath at the cathode. The maximum current conducted by each of the multiple cathode spots varies with the contact material, and is related [20,75,76] to the thermal parameter $T_b\kappa^{1/2}$ of the cathode, where T_b is the normal boiling-point temperature of the contact material, and κ is its low-temperature thermal conductivity. The current density at the cathode surface of an individual cathode spot is extremely high, and this region is therefore one of high power density. For arcs on copper contacts, the cathode fall is approximately 18 V, the maximum current per cathode spot is approximately 100 A, and the cathode spot current density at the surface, based on crater sizes, is 10^8 A cm^{-2} [77,78].

All the material required to maintain the arc comes from the cathode spots. Of particular importance is the fact that the cathode spots are regions of intense ionization [20,52,75,79]. It is in this region that both ions and electrons are produced, both of which move away from the cathode spot with high energy. For current levels of several hundred amperes, the ion flux leaves the cathode spots with a spatial distribution approximating a cosine law

[80–83]. For higher current levels, the ion flux more closely approximates an isotropic distribution [84]. Metal particles also stream away from the cathode spot regions, primarily in a direction parallel to the cathode surface [81]. For copper arcs, the erosion rate is about 10^{-4} g C^{-1} and, in common with most materials, the magnitude of the ion current leaving the cathode spot region is about 10% of the arc current [20,78,84]. The ions possess an energy (in electron volts) which exceeds the arc voltage [85,86]. Energies of 50–100 eV have been measured for the metals commonly used in vacuum interrupter contact materials. Initially it was proposed that the ions migrated away from a localized potential maximum [85–88] within the cathode spot. It now seems probable that this ion acceleration results from a force on the ions from the very high pressure gradients within the cathode spot coupled to some extent with an electron-ion friction effect [20,79,89]. The ions are multiply charged [86,87,89,90] and possess a mean energy approximately 20–40 eV (with velocities about 10^6 cm s^{-1}), whereas the mean energy of the neutral metal vapor is less than 1 eV. Figure 9.46 illustrates the complex cathode spot of the diffuse vacuum arc [20].

The primary cathode spot parameters that influence successful interruption at current zero are (1) the high velocity of the ionized metal vapor away from the cathode surface, which leads to a rapid decrease in the inter-contact plasma density, and (2) the absence of a cathode spot at the new cathode (see Section 9.7.3).

For certain arcing conditions, there may be significant evaporation from the anode due to the formation of a single, grossly evaporating anode spot [91]. This anode spot operates near the normal boiling temperature of the anode material, and can significantly increase the inter-contact plasma density. Furthermore, since the erosion is not distributed, anode spot formation can cause gross melting of the contact. Finally, a plasma jet from the anode spot can cause the cathode spots to bunch together [92], with resulting gross erosion on the cathode.

Anode spots can form from an initially diffuse vacuum arc. The probability of anode spot formation [92,93] increases with increasing contact separation, with increasing circuit current and arc duration, and with decreasing anode area. Furthermore, the probability of spot formation increases with a decrease in the anode thermal parameter $T_m(\kappa \delta C_0)^{1/2}$, where T_m is the melting temperature of the anode material and κ, δ, and C_0 are the thermal

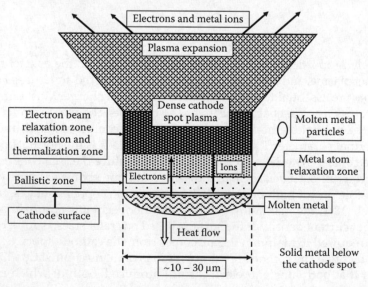

FIGURE 9.46
A model of the cathode spot showing its complexity (not to scale) [20].

conductivity, density, and specific heat respectively. For small anode areas and long contact spacing, when most of the ions in the plasma stream from the cathode spots are no longer incident on the anode, the arc voltage increases as a result of the formation of an anode sheath and the heating of the anode surface.

9.6.2 The Columnar Vacuum Arc

If the current at the instant of contact separation approaches approximately 15 kA, the bridge rupture leads to formation of a single, high-vapor-pressure arc column. This arc is a high pressure arc and is similar to the arc described in Section 9.5 (see Figure 9.47). The appearance of the high-current vacuum arc has been extensively studied [72,93,95,96]. It has been shown that, depending on the current level and contact spacing, the diffuse arc can also form an anode spot and a columnar arc before going diffuse again close to current zero. It is possible, however, to form a columnar arc (Figure 9.48) from the initial bridge arc which will stay columnar until just before current zero. At the highest currents, the electrode regions of this columnar column exhibit intense activity, with jets of material being ejected from the contact faces [72,95]. In spite of this severe contact activity, even this arc mode can sometimes return to the diffuse mode just before current zero. Figure 9.49 shows an example of an arc appearance diagram developed by Heberlein and Gorman for a vacuum arc between butt contacts [72].

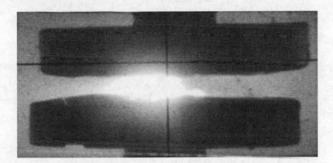

FIGURE 9.47
Photograph of a high-current vacuum arc.

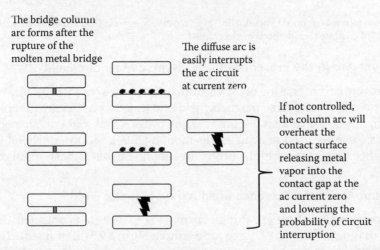

The bridge column arc forms after the rupture of the molten metal bridge

The diffuse arc is easily interrupts the ac circuit at current zero

If not controlled, the column arc will overheat the contact surface releasing metal vapor into the contact gap at the ac current zero and lowering the probability of circuit interruption

FIGURE 9.48
Possible modes of the vacuum arc appearance for high current vacuum arcs formed between opening contacts.

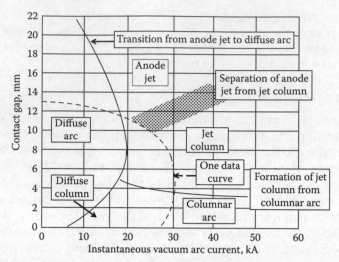

FIGURE 9.49
Vacuum arc appearance diagram for high-current vacuum arcs between butt contacts [72].

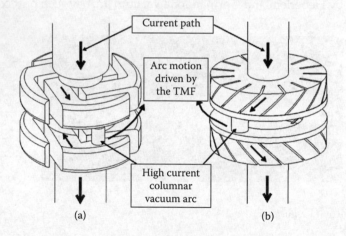

FIGURE 9.50
The motion of the high current columnar vacuum arc on transverse magnetic field contact structures, (a) spiral shaped contacts and (b) contrate cup-shaped contacts [20,94].

9.6.3 The Vacuum Arc in the Presence of a Transverse Magnetic Field

It is possible to prevent severe heating of the vacuum interrupter's contact surfaces by subjecting the columnar arc to a transverse magnetic field and, thus, force the arc to rapidly travel around the perimeter of the contact. Figure 9.50 illustrates spiral and cup shaped contacts that have been successfully used to achieve this motion [20,94]. Schulman [95,96] has developed the appearance diagrams for high current arc motion between these spiral contacts.

9.6.4 The Vacuum Arc in the Presence of an Axial Magnetic Field

One method of creating a diffuse arc at high currents is to design a contact structure that applies an axial magnetic field [20,99], see for example Figure 9.51. For a sufficiently high axial field, the vacuum arc can be maintained in the diffuse mode to very high currents [97–99]. After the rupture of the molten metal bridge, a bridge column forms, and this arc

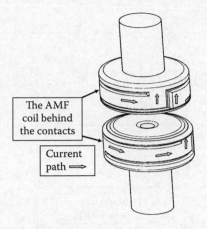

FIGURE 9.51
One contact design to force a high current vacuum arc into the diffuse mode as a result of an axial magnetic field structure behind the contact faces [20,94].

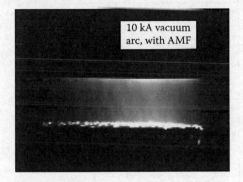

FIGURE 9.52
Photograph of a high-current (10 kA), diffuse vacuum arc formed with an axial magnetic field across the open contacts.

slowly expands into a diffuse arc [99]. Once the arc has gone diffuse, the axial magnetic field forces the arc to remain diffuse. The electrons are confined by the magnetic field lines in the inter-contact region and, because of the associated creation of radial electric fields, the ions are also confined to the inter-contact region [20]. An example of a 10 kA diffuse arc is shown in Figure 9.52. During this high-current arcing the diffuse arc distributes the arc energy over the whole contact surface and thus prevents gross erosion of the contacts.

9.7 Arc Interruption

9.7.1 Arc Interruption in Alternating Current Circuits

A single phase ac circuit is shown in Figure 9.53. Here the ac power supply is connected to an inductive load by a cable and current is permitted to flow by closing the switch, *S*. Also shown is the circuit's resistance and the inductor's stray capacitance to ground. The analysis of the currents and voltages that occur when such a circuit is switched is beyond the scope of this book: it is the subject that requires a volume of its own [17,100]. It is enough for this section to use the example shown in Figure 9.54. Here the circuit has a large inductive

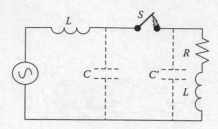

FIGURE 9.53
The alternating current ac circuit.

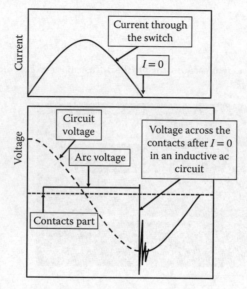

FIGURE 9.54
Interruption of an inductive ac circuit at current zero, showing the change in voltage across the open contacts.

component, i.e., the circuit current lags the system voltage. When the contacts in the switch open an arc is formed and burns to the first current zero. After the arc extinguishes at the natural ac current zero, the voltage across the contacts rises to about 1.7 times the circuit's peak voltage in a few tens of microseconds to the line voltage and then exceeds it [17,20]. The voltage that appears across the contacts is called *transient recovery voltage* (TRV). Both the TRV and its rate of rise depend upon the circuit inductance and the stray capacitance in the circuit [20,100]. The process of arc interruption at current zero is of critical importance. The ac circuit reverses $2f/s$ where f is the power frequency. So in a 60 Hz circuit the current passes through a current zero 120 times each second (at 50 Hz it is 100 times each second). When the contacts part, the arc which forms allows the circuit current to flow to a natural current zero. As the current goes to zero, the energy input to the arc also goes to zero. For a brief time at current zero ($I = 0$) the energy lost by the arc and by the cathode and anode regions exceeds the energy input. This gives the gap between the contacts the opportunity to change from a reasonable electrical conductor to an insulator and thus prevent the continued flow of the current. If this occurs, the circuit is interrupted. All ac switches depend upon the arc to extinguish for the interruption of the current flowing in the circuit. It is important for the designers of these switches to understand the phenomena that occur at the contacts that cause the arc to extinguish.

Slepian [101] first described the arc extinction process in his "'race theory," where the recovery of the gap between the contacts depends upon two competing voltage values: (1) the TRV of the circuit, which is impressed across the contacts at current zero and *is only dependent* upon the circuit inductance, resistance, and capacitance, and (2) the increase in the dielectric strength of the residual arc channel. This is characterized by the voltage required to reignite the arc, called the *reignition voltage*. The reignition voltage increases with time. It is *only dependent* upon the contact gap parameters, such as the arc current before current zero, the gap length, contact material, arc chamber design, and the ambient gas, but it is *independent* of the circuit parameters. No dielectric breakdown and reignition of the arc can occur if the TRV across the contacts is below the reignition curve of the contact gap. Two empirical reignition curves are shown in Figure 9.55 [102,103]. The full reignition curve has the four distinct stages illustrated in Figure 9.56 [104].

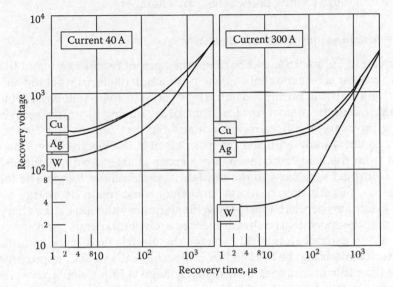

FIGURE 9.55
Recovery of Cu, Ag, and W contacts, with a 2 mm contact gap, after one ac half-cycle of arcing at 40 A and 300 A [102,103].

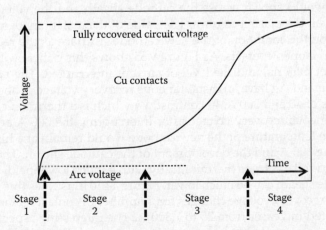

FIGURE 9.56
The four stages of recovery, an illustration of the free recovery/reignition characteristic.

1. instantaneous recovery that occurs in <1 µs;
2. a slight increase in the dielectric strength followed by a plateau;
3. a steady increase in recovery that lasts between a few tens or a few hundred microseconds;
4. the full dielectric strength of the "cold" contact gap.

This recovery process takes place in three stages.

- *Stage 1*. The rapid adjustment of the plasma at $I = 0$ and the cessation of electron emission from the cathode.
- *Stage 2*. The cooling of the arc column and the recombination of the ionized gas.
- *Stage 3*. The long-term cooling of the whole arc chamber.

9.7.1.1 Stage 1 – Instantaneous Dielectric Recovery

Stage 1 is illustrated in Figure 9.57a and b. When the current reaches zero and the reverse voltage begins to appear across the contacts, the production of electrons at the old cathode stops. If no electrons emission is initiated at the new cathode, there will be a redistribution of the arc plasma close to the contact surfaces. First of all, during the initial recovery stage, the slow moving ions can be considered stationary, Figure 9.57a. As the TRV is applied across the open contacts (now with the reverse polarity) the electrons will be repelled instantaneously from the new cathode leaving a space charge sheath, Figure 9.57b. The cathode sheath is formed very close to the cathode approximately 10^{-7} m. The former arc column is a plasma, the 4th state of matter and thus must retain its charge neutrality, i.e., $n_e = n_i$. The electrons therefore must stay with stationary the ions. It has been shown experimentally that the space charge sheath reaches an instantaneous recovery voltage of about 300 V for a cold cathode. It is at this voltage the electric field at the cathode is high enough for electron emission to be initiated from the cold cathode: i.e., approximately 3×10^9 V m^{-1}. If less than this instantaneous recovery voltage is impressed across the contact gap, all the voltage appears across the cathode sheath and none across the plasma column. If the voltage U_R appearing across the contact gap is greater than this instantaneous recovery voltage, then the voltage distribution shown in Figure 9.58 would occur. That is for a cold cathode, 300 V would appear across the cathode sheath and the voltage $U_R - 300$ V would be impressed on the plasma column. The value of the instantaneous recovery voltage does depend upon the condition of the new cathode's surface as the recovery voltage is impressed across the open contacts. As Figure 9.55 shows this critical voltage is dependent upon the contact material and the level of current interrupted. The non thermionic emitting materials Cu and Ag have an instantaneous recovery voltage of about 300 V after interrupting currents of 40 and 300 A. In contrast, W, which is a thermionic emitter, has a much lower instantaneous recovery voltage after interrupting the 300 A arc. This can be expected, because the temperature of the new cathode would remain at a higher temperature after interrupting 300 A and the development of the cathode sheath. The new cathode would then be able to emit electrons thermionically with a much lower field. Thus, the voltage across the cathode sheath can be much lower. Figure 14.10 illustrates this how the value of the instantaneous recovery voltage changes for a number of contact materials as the circuit current interrupted increases from 25 to 1,300 A. The interesting aspect of these data is that instantaneous recovery voltage for the contact materials Ag-CdO (10 and 15 wt%), Ag-SnO$_2$(12 wt%) and Ag-Ni(10 wt%) is not greatly reduced for currents up to 1,300 A.

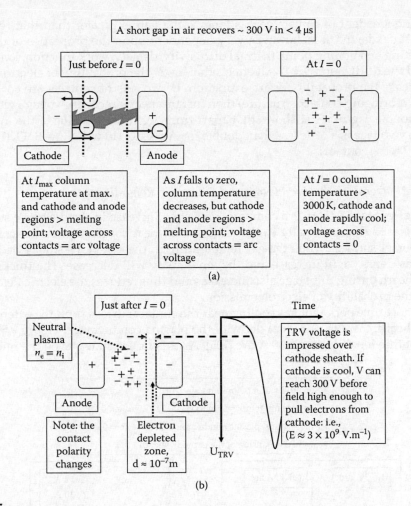

A short gap in air recovers ~ 300 V in < 4 μs

Just before $I = 0$

At $I = 0$

Cathode

Anode

At I_{max} column temperature at max. and cathode and anode regions > melting point; voltage across contacts = arc voltage

As I falls to zero, column temperature decreases, but cathode and anode regions > melting point; voltage across contacts = arc voltage

At $I = 0$ column temperature > 3000 K, cathode and anode rapidly cool; voltage across contacts = 0

(a)

Just after $I = 0$

Time

Neutral plasma $n_e = n_i$

Anode

Cathode

U_{TRV}

Note: the contact polarity changes

Electron depleted zone, $d \approx 10^{-7}$m

TRV voltage is impressed over cathode sheath. If cathode is cool, V can reach 300 V before field high enough to pull electrons from cathode: i.e., $(E \approx 3 \times 10^9 \text{ V.m}^{-1})$

(b)

FIGURE 9.57
Stage 1: The initial rapid adjustment of the inter-contact plasma immediately after current zero in an ac circuit showing the development of the cathode sheath (not to scale) in front of the new cathode.

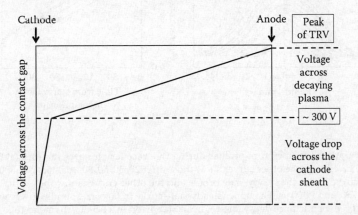

Cathode

Anode

Peak of TRV

Voltage across decaying plasma

~ 300 V

Voltage across the contact gap

Voltage drop across the cathode sheath

FIGURE 9.58
The voltage distribution across the contact gap (note the width of the cathode sheath is exaggerated).

Stage 1 is independent of contact gap as long as the gap is greater than that required for "vacuum breakdown." It is, however, very dependent upon the properties of the contacts, e.g., melting and boiling point, thermal diffusivity, density, work function, ionization potential, and the dI/dt before $I = 0$. All effects that lower the probability of electron emission will increase the probability of arc extinction. If two sets of contacts are connected in series (e.g., a bridging contact structure) then instantaneous recovery voltage would be expected to double. Figure 14.11 shows that apart from Ag contacts value of the instantaneous recovery voltage is a little less than double for Ag–CdO (10 wt%), Ag–SnO$_2$ (12 wt%) and Ag–Ni (10 wt%) contacts.

9.7.1.2 Stage 2 – Decay of the Arc Plasma and Dielectric Reignition

Immediately after Stage 1 there is a comparatively slow increase in the dielectric strength of the contact gap. This is usually explained by axial heat conduction to the comparatively cool contact surfaces. Thus the gas temperature in these regions slowly decreases, the neutral gas density will increase and the ion density will decrease. The thickness of the cathode sheath during this stage also increases and thus reduces the electric field at the cathode and the probability of electron emission.

After this short time period of gas cooling near the contact surfaces and the extension of the cathode sheath, there is a gradual decay of the plasma conductance. Figure 9.59 gives an example of this for a 5 A and a 30 A arc [105]. As this figure shows even the middle of

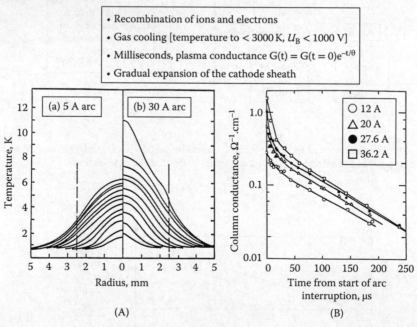

FIGURE 9.59

Stage 2: (A) Measured transient temperature profiles during free recovery for arcs in air of (a) 5 A: top curve shows the steady-state profile and the other curves show those after 20, 30, 50, 80, 105, 155, 200, 300, 500, 750, and 1,000 μs; (b) 30 A: top curve shows the steady-state profile and the other curves show those after 20, 50, 75, 100, 150, 200, 300, 500, 700, 750, and 1,000 μs. The upper steady-state curve in (a) corresponds closely with the profile at 100 us in (b) [74]. (B) Free conductance decay for arcs in air stabilized in a 0.5-cm-diameter tube for initial currents ranging from 12 to 36 A. For conductances of less than 0.2 Ω^{-1} cm, the rate of decay for a given conductance is insensitive to the initial conditions [105,106].

the plasma column is cooled down to a critical temperature where the thermal ionization according to Saha's equation becomes negligible. In air at atmospheric pressure this critical temperature is on the range approximately 2,000–3,000 K [106]. At gas temperatures below this value the TRV will stress the whole gap between the contacts, i.e., the voltage is homogeneously distributed along the total contact gap and the breakdown voltage depends upon the gap length and the gas pressure or gas number density, n, which is proportional to (T^{-1}) as discussed in Section 9.3.

The time it takes for the plasma column to cool depends strongly upon the design of the arc chamber. For example, it depends upon the contact gap; the shorter the contact gap the more effectively is the gas in the gap cooled by axial heat flow towards the "cool" contact surfaces which act as heat sinks. Experimentally it has been shown that an optimum gap for the low current arcs (~<30 A) appears to be approximately 0.5 mm. It also depends upon the ionization potential of the ambient gas and is affected by impurities such as residual metal vapor. Some arc chambers have a magnetic drive to lengthen the column and perhaps to drive the arc into metal plates such as deion plates [107], which enhance the recovery effect (see Chapter 14 Sections 14.1.1 and 14.1.2). In high-voltage circuit breakers the cooling of the arc column is of paramount importance because in these devices the TRV exceeds the 300 V in a few microseconds after $I = 0$ [17]. A breakdown of the contact gap after a current pause at current zero and the appearance of the TRV across the contact gap is called a *dielectric reignition*.

9.7.1.3 Thermal Reignition

The voltage drop across the space charge sheath of approximately 300 V that forms in front of the cathode produces a field in front of the cathode that permits the field emission of electrons. If the TRV has a voltage greater than 300 V then the excess voltage appears across the decaying plasma, see Figure 9.58. Under the correct conditions, the electrons liberated from the new cathode by field emission can be accelerated by the voltage drop across the plasma and may cause the arc column to heat up again. This phenomenon is called *thermionic reignition*. If this happens a post-arc current can be sustained which might be able to heat up the arc column again. Thermal reignition is not decided by a race between two independent voltage curves as it is for dielectric reignition, but by a race between the electrical power input into the cooling arc column and the rate of energy loss by the arc column. The electrical power input is controlled by the supply voltage, the circuit impedance and the continuous change in the arc conductance (see Figure 9.59). The rate of energy loss by the arc column depends upon the ambient gas, the contact gap and the whole design of the arc chamber.

A more extreme case of reignition occurs if the new cathode is able to emit electrons right after $I = 0$ and the instantaneous formation of the insulating positive space charge layer in front of the cathode is limited or is prevented. This is called *thermal reignition*. Electron emission will occur most easily if the old anode (or new cathode) is very hot just before $I = 0$ and electrons are thermionically emitted, e.g., a recovering arc between refractory contacts (C, W, Mo) or the contacts are covered by a refractory surface layer, e.g., oxides of Mg, Al, Mn, Mo. Compare the recovery curves at 40 and 300 A for tungsten contacts in Figure 9.55. An excellent example of an arc chamber specifically designed for reignition of the electric arc at current zero is a high-pressure sodium street lamp [108]. Here the electrodes are made from tungsten coils in which is deposited a Thoria paste to ensure the emission of copious electrons at each $I = 0$. In this case the thermionic electron emission certainly prevents the initial recovery process (Stage 1). Thermionic reignition will occur, however, even with non-refractory contact materials if the arc root has caused gross melting at the anode and the voltage across the cathode sheath is enough to initiate T–F electron production (see Section 9.5.2). Thermal reignition and thermionic reignition can occur between non-refractory contacts that try to interrupt currents greater than a few hundred amperes in medium voltage circuits unless measures are taken to rapidly cool the arc column.

Simplified power balance equations were formulated by Cassie [109] and Mayr [110] and later improved by Frost and Browne [111]. Much of the design of high current switching devices used in circuits where the voltage is greater than 300 V is dependent upon optimizing the rate of decay of the residual arc plasma after $I = 0$ [17] (see also Chapter 14).

9.7.2 Arc Interruption in Direct Current Circuits

The arc interruption process follows the same physical principles as has already been discussed for ac circuits (see Section 9.7.1). The major difference is that a dc switch has to develop its own current zero before electron emission at the cathode ceases and the space charge sheath can develop. If we take a typical dc circuit (Figure 9.60),

$$U_C = L\frac{dI}{dt} + RI + U_A \tag{9.52}$$

where U_C is the voltage of the dc power supply, L is the circuit inductance, R is the circuit resistance, I is the current and U_A is the arc voltage across the opening contacts. Rearranging Equation 9.52,

$$\frac{dI}{dt} = \frac{1}{L}([U_C - RI] - U_A) \tag{9.53}$$

Before the contacts begin to open $U_C = RI$ and U_A equals zero, so $dI/dt = 0$. As soon as the arc is initiated at $t = 0$, dI/dt must have a negative value equal to $-U_A/L$, i.e., the current initially decreases. If the term $(U_C - RI)$ at some point equals U_A, then di/dt again will equal zero and the current will cease decreasing. This is a stable point and in the circuit shown in Figure 9.61, a stable arc will exist. The basic criterion for arc interruption, therefore, is that U_A must at least equal $(U_C - RI)$. Figure 9.61 shows two examples of voltage–current arc characteristics: (see Section 9.5.5) (1) U_{ARC1} intersects $U_C - RI$ and the arc in this condition is stable at the lower value where the curves intersect; (2) A characteristic like U_{ARC2} would always ensure arc interruption.

For dc interruption not only must you establish the arc once the contacts part, but you must also ensure that the arc voltage is forced to reach the system voltage. This will ensure that

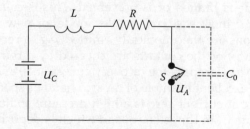

FIGURE 9.60
The direct current dc circuit.

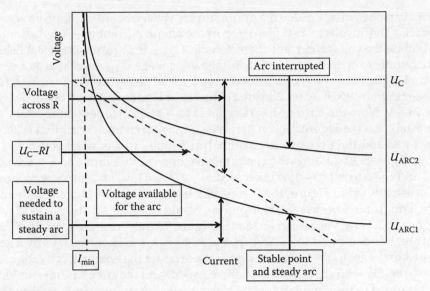

FIGURE 9.61
The conditions for a stable arc and for interruption of a dc circuit (arbitrary scales for voltage and current).

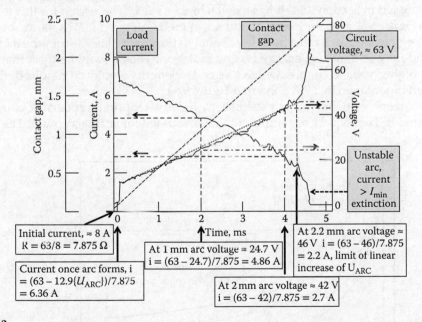

FIGURE 9.62
The voltage and current for AgCdO contacts opening at 0.5 m.s⁻¹ and interrupting a 8 A, 60 V dc mostly resistive circuit [112].

the current flowing in the circuit is forced to zero and the contact gap recovery processes can be established. The arc voltage can be increased by lengthening the arc, or by cooling it is a number of ways, see Chapters 13 and 14. The recovery process follows the same stages as have been discussed for ac current interruption. Figure 9.63 is an example of AgCdO contacts interrupting an 8 A, 63 V dc circuit [112]. Using this figure and Figure 9.62 the interruption of this circuit can easily be explained. It can be seen in Figure 9.62 that once the arc forms after the rupture of the molten metal bridge the minimum arc voltage of ~12.9 V appears across the opening contacts. This results in a sudden drop in the circuit current to ~6.36 A. A momentary stable point on the voltage/current characteristic, U_{ARC1}, in Figure 9.63 is established As the contacts continue to open the arc lengthens, the arc voltage U_{ARC} increases as a result of the arc column lengthening and the transition from a metal phase arc to an arc in the ambient gas, air (see Section 10.3.5). After 2 ms the contacts in Figure 9.62 have opened to 1 mm and $U_{ARC2} \approx 24.7$ V. Now the current has decreased to ~ 4.86 A. A new momentary stable point is now established in Figure 9.63 on the voltage/current characteristic that reflects the higher contact gap and the decrease in current. Incidentally this is what you might have expected from Figure 9.39. At 4 ms the contacts have opened to 2 mm and the arc voltage $U_{ARC3} \approx 42$ V and the current has now decreased to ~ 2.7 A. Thus, another new momentary stable point is established in Figure 9.63 that results from the increased arc length and the lower current. I might be expected that this process would continue until $U_{ARC4} + RI \approx U_C$ as shown in Figure 9.63. However, as can be seen in Figure 9.62, at a contact spacing of 2.2 mm when the current has decreased to ~ 2.2 A the current begins to decrease rapidly. Here the energy supplied to the arc ($U_A \times I_C$) can no longer make up for the energy lost by the arc. The $-dI/dt$ increase rapidly with rapid increase with dU_{ARC}/dt and the circuit is interrupted. The small current trapped in the circuit inductance results in a small spike in the voltage across the contacts before the open circuit voltage of ~63 V appears across them. It is important to note that when observing this opening process in dc circuits that the I_{min} is quite different from that discussed in Section 9.5.4. It has a much higher value. The reason for this is (1) the arc shown in Figure 9.62 is operating in air and not in metal vapor and (2) as the arc becomes longer the energy to sustain it must be at least equal to the energy losses from the increased arc length and may also include energy loss by radiation. Thus, I would expect that when interrupting higher voltage circuits where longer arc lengths would be required, the I_{min} would be even higher than in 2.7 A shown in Figure 9.62.

Once the current zero has been established, the interruption process is shown in Figure 9.64a and b. The sequence is similar to that discussed for ac interruption. The major

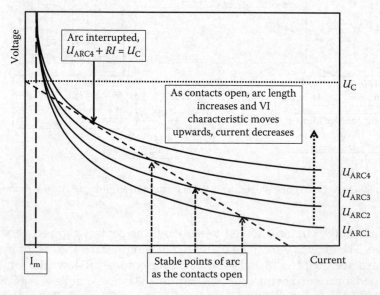

FIGURE 9.63
Illustration of dc circuit interruption for use with Figure 9.62 (arbitrary scales for voltage and current).

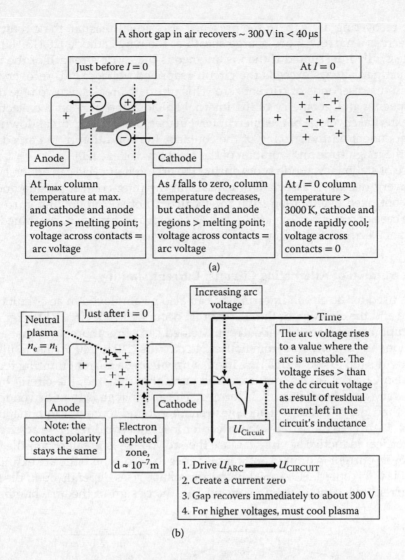

FIGURE 9.64
Stage 1: The initial rapid adjustment of the inter-contact plasma immediately after the forced current zero in a dc circuit showing the development of the cathode sheath (not to scale) in front of the cathode.

difference being the cathode remains the cathode and the anode remains the anode. Stage 1 is illustrated in Figure 9.64a and b. Stages 2–4 are similar to those discussed for ac current interruption.

9.7.3 Vacuum Arc Interruption in Alternating Circuits

All successful vacuum interrupter designs have contacts that force the vacuum arc to go diffuse before the current reaches zero [20,94]. At current zero, the vapor-producing cathode spots extinguish. The residual vapor and plasma within the inter-contact region rapidly condense and recombine on both the shield and contact surfaces, and the original vacuum condition is rapidly approached. Analysis of the recovery processes directly following current zero is complicated by the non-uniform distribution of the reapplied voltage in the recovering arc gap. In the presence of residual plasma, the circuit voltage is impressed across a narrow space-charge sheath at the new cathode [20,113,114]. The ac circuit is successfully interrupted if the instantaneous dielectric strength of the recovering inter-contact gap always exceeds the circuit reapplied voltage. Full recovery can be attained within microseconds of current zero. This ultimate breakdown voltage depends on both the spacing and geometry of the internal shields, and also on the electric field stress on the external envelope of the interrupter. Further, the ultimate breakdown voltage is critically dependent on the spacing of the contacts, the condition of the arced contact surfaces, and the magnitude and duration of the recovery voltage [20].

The presence of stationary anode spots during the arcing half cycle can adversely affect dielectric recovery due to (1) associated increases in the inter-contact plasma and vapor densities, (2) continued evaporation from the localized hot spot following current zero, and (3) for the case of refractory materials such as carbon and tungsten, continued thermionic emission of electrons following contact polarity reversal [20].

9.7.4 Arc interruption of Alternating Circuits: Current Limiting

The technique used for dc circuit interruption can also be employed in ac circuits to limit the magnitude and the duration of the current that occurs when a circuit breaker is operating to interrupt a short circuit fault current. Molded case circuit breakers that are used to protect a wide range of circuits, from household circuits (110 to 240 V rms) to industrial distribution circuits (up to 600 V rms), are increasingly using a current limiting technique [115,116]: see also Chapter 14. As soon as a fault current is initiated, the circuit breaker's contacts open very quickly and an arc is formed. The arc voltage is then driven to a high value, greater than the ac circuit voltage, by stretching it and forcing it into metal plates: see Chapter 14. Figure 9.65 illustrates the effect for a symmetrical resistive ac circuit and for an asymmetrical inductive ac circuit. Once the arc voltage is greater than the circuit's system voltage the current is limited and rapidly goes to zero, where the arc extinguished and the circuit is interrupted. How quickly the arc voltage rises depends upon the sensing of the fault current, the contact opening speed and the design of the arc chamber. Even

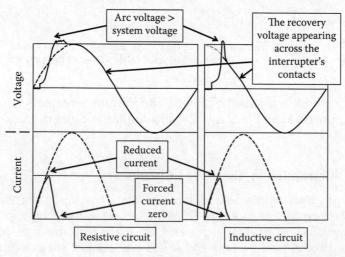

FIGURE 9.65
Illustration of current limiting in resistive and inductive ac circuits.

gassing from the arc chamber walls can greatly influence the voltage rise [117]. Any gap in the arc runners can also affect the speed of the arc's motion [118]. Many circuit breaker designs now use the principle of dc interruption (i.e., of driving the arc voltage above the circuit voltage) in ac circuits to create a new current zero. In doing so the fault current is greatly reduced both in magnitude and duration.

9.7.5 Interruption of Low Frequency and High Frequency Power Circuits

Most electrical ac power generation systems have a frequency of 50 or 60 Hz. This depends upon a country's historical development of its electrical grid. There are some systems, however, that use lower frequencies. Some electric locomotive systems use 16 2/3 Hz and others 25 Hz. Higher frequencies up to 400 Hz are used for electric ship development. When such circuits install a circuit breaker designed to protect 50/60 Hz circuits, the question arises, "How well will they perform?" There is a dearth of published literature on this subject. One study [119] is shown in Figure 9.66. Here the fault current interrupting ability of a vacuum interrupter has been evaluated for frequencies from 16 2/3 to 800 Hz. It shows that the maximum performance is close to its design value at 80 Hz and that its performance drops as the frequency drops or increases. Thus a circuit breaker designed to protect a circuit at 50/60 Hz should be de-rated for use in circuits with lower or higher ac frequencies.

9.7.6 Interruption of Megahertz and Gigahertz Electronic Circuits

The use of mechanical switches to control very high frequency circuits at first sight seems unusual. Solid state devices such as MOSFETS [120] have been successfully employed and have performed well for many decades. The main task of any switching device is to conduct the high frequency current when closed with a low insertion loss and, when open, to provide complete isolation. Table 9.10 [121] compares the MOSFET with three mechanical switches: (a) MEMS [122] (see Chapter 12), (b) a reed relay (see Chapter 11) and (c) an electro-mechanical relay. As the comparison shows the electro-magnetic relay has the

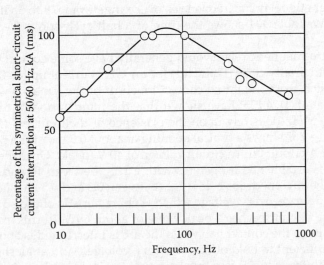

FIGURE 9.66
The interruption ability of a vacuum interrupter as a function of the electrical system's frequency [117].

TABLE 9.10

Comparison of the Characteristics of Solid State and Mechanical Switches for Controlling High Frequency Circuits

Characteristic	MOSFET	RF MEMS	Reed Switch	Electro-Mechanical RF Relay
Size	Very small	Very small	Small	Larger
Contact resistance	1–50 Ω	0.5 Ω	0.1 Ω	0.05 Ω
Breakdown, V (rms)	<100	300–500	300–500	700–1,000
Mechanical life	∞	10^7–10^8	10^7–10^8	10^7
Electrical life	∞	10^5–10^6	10^6	10^6
Power, W	0.5	3	10	10
Switching speed	<5 ns	1–20 µs	<2 ms	<5 ms
Frequency, GHz	<8	dc – 5	dc – 5	dc – 5
Power consumption				
Off state, mW	1–20	0	0	0
On state, mW	–	<10	0–70	0–140
Drive voltage, V	5–8	5–35	1–24	1–24
Assembly	Very good	Very good	Good	Good
Cost	Low	High	Low	Low

TABLE 9.11

The Durations of One Half Cycle of Current for Frequencies from 50 Hz to 10 GHz

Frequency	Time 1/2-Cycle	Frequency	Time 1/2-Cycle
50/60 Hz	10/8.3 ms	500 kHz	1 µs
500 Hz	1 ms or 1,000 µs	1 GHz	0.5 µs or 500 ns
100 kHz	5 µs	10 GHz	50 ns

lowest insertion loss (i.e. contact resistance) and the highest potential for circuit isolation (i.e. breakdown voltage). Johler [123] has shown indirect evidence of the switching ability of the electro-magnetic relay for frequencies over range from 0.8 to 3 GHz and power ranges from 1 to 37 W. Table 9.11 shows the time of a half cycle for ac frequencies from 50 Hz to 10 GHz.

On closing an electro-magnetic relay would generally have some contact bouncing and may take up to 40 µs to finally close [123]. Thus for a 1 GHz current up to 40 cycles would pass through the switch from the initial contact touch until the contacts would be completely closed. Miki et al. [124,125] have shown that the interruption of high frequency currents in up to 800 MHz does rely upon the existence of their current zero occurring every half cycle. Figure 9.67 shows a typical opening at about 0.5 ms^{-1} for a 500 Hz current in a mostly resistive 2 A circuit with a circuit voltage of 40 V (peak). Once the contacts part and the metallic arc is established it operates with a U_{min} between 10 and 15 V. The arc remains stable until the current drops below I_{min} (between 0.3 and 0.5 A) when in becomes unstable and chops out. A recovery voltage which is characteristic of the stray inductance and capacitance of the circuit reaches a peak value of about 60 V. In this example the arc is stable for about 0.5 µs so the contact gap when the arc is interrupted is about 0.25 µm. In this case the gap is sufficient to hold off the recovery voltage. Miki et al. show that interruption of circuits with frequencies less than or equal to 500 kHz always interrupt at the first current zero if the arcing time is greater than 1/4 cycle. Above 500 kHz the arc is not

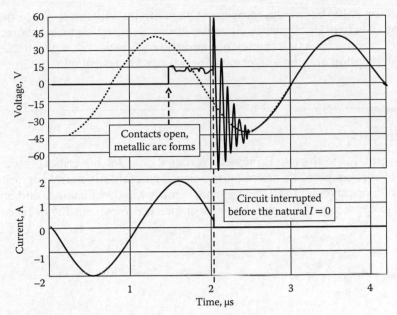

FIGURE 9.67
Arc behavior and interruption of a 2 A, 40 V (peak), 500 kHz circuit [124,125].

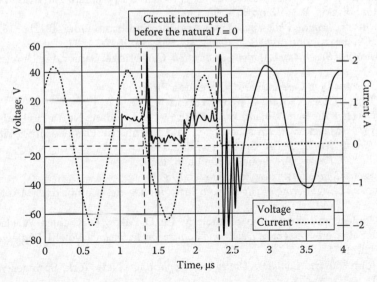

FIGURE 9.68
Arc behavior, reignition and interruption of a 2 A, 40 V (peak), 1 MHz circuit [124,125].

interrupted at the first current zero. Figure 9.68 illustrates this for a 1 MHz circuit. The contact gap at the first current zero after 0.28 μs of arcing is about 0.14 μm, which is insufficient to hold off the recovery voltage. The researchers show that for frequencies above 500 kHz in their circuit about 2 μs of arcing is needed or a contact gap of about 1 μm is required before the circuits are completely interrupted. Johler [121] also finds that making or breaking higher frequency RF loads (up to 3 GHz) requires the arc to operate over several periods of the current before interruption is achieved. Again the contact gap must achieve a minimum value before the recovery voltage can be withstood. These data show that interruption of very high frequency currents is possible using electrical contacts. Thus the advantage of the low insertion impedance and the high isolation is attainable as long as the longer interruption and the longer switch-on times are acceptable.

Acknowledgments

I am extremely grateful to the late Professor Werner Rieder for his critical review of the initial version of this chapter, his attention to scientific detail, his many suggestions for changes, and his encouragement. All of which have improved the substance and breadth of the chapter. I also appreciate the use of the photographs taken by Erik Taylor (Figures 9.45, 9.47, and 9.52).

References

1. JM Meek, JD Craggs. *Electrical Breakdown of Gases*. Oxford: Clarendon Press, 1953.
2. For the reader who wishes to pursue further the kinetic theory of gases, a good place to begin is a good University level Physics text such as: G Joos and I Freeman. *Theoretical Physics*. 3rd edition. London: Blackie, Part VI, pp 563–598, 1960.
3. H Davy. *Elements of Chemical Philosophy*, Vol. 1. London: Smith and Elder, 1812, p. 152.
4. JJ Thompson. On the cathode rays. *Proc. Cambridge Phil. Soc* 10: 74, 1899.
5. N Bohr. *The Theory of Spectra and Atomic Constitution*. Cambridge: Cambridge University Press, 1922.
6. JS Townsend. *Electricity in Gases*. Oxford: Clarendon Press, 1915.
7. A von Engel. *Ionized Gases*. Oxford: Clarendon Press, 1955.
8. W Elenbaas. *The High Pressure Mercury Vapor Discharge*. Amsterdam: North-Holland, 1951.
9. W Finkelnburg, H Maecker. Elektrische Bogen und thermische Plasmen. In: S Fluegge, ed. *Handbuch der Physik*, Vol. 22. Berlin: Springer-Verlag, 1956.
10. JM Sommerville. *The Electric Arc*. London: Methuen, 1959.
11. JD Cobine. *Gaseous Conductors*. New York: Dover, 1958.
12. TH Lee. *Physics and Engineering of High Power Switching Devices*. Cambridge, MA: MIT Press, 1975.
13. HN Wagar. Contact and connection technology. In: D Baker, D Koehler, W Fleckenstein, C Roden and R Sabia, eds. *Integrated Device and Connection Technology*, Vol. III. Englewood Cliffs, NJ: Prentice Hall, 1971.
14. R Holm, E Holm. *Electric Contacts: Theory and Application*. New York: Springer-Verlag. 3rd Printing, 2000.
15. JM Lafferty, ed. *Vacuum Arcs, Theory and Applications*. New York: Wiley-Interscience, 1980.
16. MF Hoyaux. *Arc Physics*. New York: Springer-Verlag, 1968.
17. TE Browne. *Circuit Interruption, Theory and Techniques*. New York and Basel: Marcel Dekker, 1984.
18. F Llewellyn-Jones. *Ionization and Breakdown in Gases*, London: Methuen/New York: Wiley, 1957.
19. F Paschen. Über die zum Funkenübergang in Luft, Wasserstoff, und Kohlensäuer bei verschiedenen Drucken erforderliche Potentialdifferenz. *Wied Ann* 37(3): 69–96, 1889.
20. PG Slade. *The Vacuum Interrupter, Theory, Design and Application*, Boca Raton, FL: CRC Press, 2008.
21. RV Latham. *High Voltage Vacuum Insulation*. London: Academic Press, 1995.

22. RR Fowler, LW Nordheim. Electron emission in intense electric fields. *Proc. R. Soc.* A119: 173–181, 1928.
23. PG Slade, ED Taylor. Electrical breakdown in atmospheric air between closely spaced (0.2 µm–40 µm) electrical contacts, *IEEE Trans. Compon. Pack. Tech.* 25: 390–396, 2002.
24. WB Maier, A Kadish, CJ Buchenauer, RT Robiscoe. Electrical discharge initiation and a microscopic model for formative time lags. *IEEE Trans. Plasma Sci.* 21(6): 676–683, 1993.
25. N Wakatsuki, H Homma. Melting phenomena and arc ignition of breaking relay contacts, *Proceedings of the 54th IEEE Holm Conference on Electrical Contacts:* 15–20, 2008.
26. T Utsumi. Theoretical and experimental investigations of the dynamic molten bridge. *IEEE Trans. Parts Mater. Pack.* PMP 5(1): 62–68, 1969.
27. K Miyajima, S Nitta, A Mutoh. A proposal on contact surface model of electro-magnetic relays—based on the change of showering arc wave formations with the number of contact operations. *IEICE Trans. Electron.* E81-C(3): 399–407, 1998.
28. H Ishida, Y Watanabe, M Taniguchi, H Inoue, T Takago. Observation of contact bridge phenomena at transient and steady state. *Proceedings of the 50th IEEE Holm Conference on Electrical Contacts:* 519–522, 2004.
29. JW McBride, L Jiang, C Chianrabutra. Fine transfer in electrical switching contacts using gold coated carbon nano-tubes. *Proceedings of the 26th International Conference on Electrical Contact Phenomena:* 353–358, 2012.
30. PG Slade, MD Nahemow. Initial separation of electrical contacts carrying high currents. *J. Appl. Phys.* 42(9): 3290–3297, 1971.
31. PP Koren, MD Nahemow, PG Slade. The molten metal bridge state of opening electric contacts. *IEEE Trans. Parts Hyb. Pack.* PHP–11(1): 4–10, 1975.
32. PG Slade, MF Hoyaux. The effect of electrode material on the initial expansion of an arc in vacuum. *IEEE Trans. Parts Hyb. Pack.* PHP–8: 35–47, 1972.
33. PG Slade. The transition from the molten metal bridge to the metallic phase bridge column arc between electrical contacts opening in vacuum, *Proceedings of the 23rd International Symposium on Discharges and Electrical Insulation in Vacuum,* 198–201, 2008.
34. R Haug, T Kouakou, J Dorémieux. Phenomena proceeding arc ignition between opening contacts: experimental study and theoretical approach. *Proceedings of the 36th IEEE Holm Conference on Electrical Contacts:* 543–549, 1990.
35. M. Hopkins, R. Jones, J. Evans. Short time transients in electrical contacts. *Proceedings of the 8th International Conference on Electrical Contacts:* 270–272, 1976.
36. PG Slade. Opening contacts: the transition from the molten metal bridge to the electric arc, *IEICE Trans. Electron.* E93-C(6): 1380–1386, 2010.
37. PG Slade, The transition to the metal phase arc after the rupture of the molten metal bridge for contacts opening in air and vacuum. *Proceedings of the 54th IEEE Holm Conference on Electrical Contacts:* 1–8, 2008.
38. V Puchkarev, M. Bochkarev. High current density spotless vacuum arc as a glow discharge. *IEEE Trans. Plasma Sci.* 25(4): 593–597, 1997.
39. W. Ebling, A. Foster, R. Radtke, *Physics of Non-Idealized Plasmas.* Stuttgart: Teubner Verlagsgesellschaft, 1991.
40. L Holtgreven. *Plasma Diagnostics.* New York: Wiley Interscience, 1968.
41. PG Slade, E Schulz-Gulde. Spectroscopic analysis of high-current, free-burning ac arcs between copper contacts in argon and air. *J. Appl. Phys.* 44(1): 157–162, 1973.
42. MN Saha. Ionization in solar chromosphere. *Philos. Mag.* 40: 472, 1920.
43. CG Suits. The temperature of the copper arc. *Proc. NAS, Phys.* 21(1): 48–50, 1935.
44. JVR Heberlein, CW Kimblin, A Lee. Nature of the electric arc. In: TE Browne, ed. *Circuit Interruption, Theory and Technique,* New York: Marcel Dekker, 1984, pp. 135–157.
45. W. Elenbaas. *Ähnlichkeitsgesetze der Hochdruckentladung.*: Physica 2, p. 169, 1935.
46. G Heller. Dynamic similarity laws of the mercury high pressure discharge. *Physics.* 6: 389, 1935.
47. JJ Lowke. Calculated properties of vertical arcs stabilized by natural convection. *J. Appl. Phys.* 50(1): 147–157, 1979.

48. JJ Lowke. Simple theory of free burning arcs. *J. Phys. D Appl. Phys.* 12: 1873–1885, 1979.
49. X Zhou, L Qi, B Liu, G Zhai. Study on static arc behavior in a relay considering influence of the silver alloy. *Proceedings of the 26th International Conference on Electrical Contact Phenomena:* 353–358, 2012.
50. AE Guile. Arc-Electrode phenomena. *Proc. IEEE (Lond.) IEEE Rev.* 118(9R): 1131–1154, 1971.
51. X Zhou, J Heberlein, E Pfender. Theoretical study of factors influencing arc erosion of cathode. *Proceedings of the 38th IEEE Holm Conference on Electrical Contacts:* 71–78, 1992.
52. IG Kesaev. *Cathode Processes in the Mercury Arc.* Moscow: State Power Engineering Press, 1961 (translated and published in English by Consultants Bureau, New York, 1964).
53. TH Lee, A Greenwood. Theory for the cathode mechanism in metal vapor arcs. *J. Appl. Phys.* 32: 916–923, 1961.
54. C Spatami, D Teillet-Billy, JP Gauyacq, Ph Teste, JP Chabrier. Ion assist emission from a cathode in an electric arc, *J. Phys. D Appl. Phys.* 30: 1135–1145, 1997.
55. JD Cobine. Introduction to vacuum arcs. In: JM Lafferty, ed. *Vacuum Arcs Theory and Application.* New York: Wiley-Interscience, 1980, pp. 8–12.
56. J Heberlein, J Mentel, E Pfender. The anode region of electric arcs: a survey. *J. Phys. D Appl. Phys.* 43: 1–31, 2010.
57. N Ben Jemaa, JL Queffelec, D Travers. Theoretical multisite arc model applied to statistical arc duration measurement on low break at low electrical level. *Proceedings of the 30th Holm Conference on Electrical Contacts:* 573–579, 1984.
58. N Ben Jemaa, L Lehfaoui, L Nedelec. Short arc duration at low current (<1 A) and voltage (14–42 Vdc), *Proceedings of the International. Conference on Electrical Contacts:* 379–384, 2000.
59. N Ben Jemaa. Short arc duration laws and distributions at low current (<1 A) and voltage (14–42 Vdc). *IEEE Trans. Compon. Packaging Technol.* 24(3): 358–363, 2001.
60. N Ben Jemaa. Contacts conduction and switching in dc levels, *Proceedings of the 48th IEEE Holm Conference on Electrical Contacts:* 1–15, 2002.
61. M Hasegawa, Y Kamada. An experimental study on re-interpretation of minimum arc current of electrical contacts. *IEICE Trans. Electron.* E88–C(8): 1616–1619, 2005.
62. M Hasegawa, Y Tamaki, Y Kamada. An experimental study of minimum arc current of relay contacts and possible re-interpretation of the meaning thereof. *Proceedings of the 52nd IEEE Holm Conference on Electrical Contacts:* 153–158, 2006.
63. M Hasegawa. An experimental study of minimum arc current of Ag contacts with different opening speeds. *Proceedings of the. 53rd IEEE Holm Conference on Electrical Contacts:* 294–299, 2007.
64. K Suhara. Arc extinction and reignition at a fixed short gap—a research on arc V-I characteristics. *IEICE Trans. Electron.* E87–C(8): 1348–1355, 2004.
65. W Reider, H Schneider. Ein Beitrag zur Physik des Gleichstromlichtbogens. *Elin-Z* 5: 174–187, 1953.
66. JW McBride, SM Sharkh. The influence of contact opening velocity on arc characteristics. *Proceedings of the 16th International Conference on Electrical Contacts:* 395–400, September 1992.
67. J Sekikawa, T Kubono. An experimental equation of V-I characteristics of breaking arc for Ag, Au, Cu and Ni electrical contacts. *IEICE Trans. Electron.* E86-C(6): 926–931, 2003.
68. W Reider. Leistungsbilanz der Electroden und Charakteristiken frei brennender Bogen. *Z Physik.* 146: 629–643, 1956.
69. LA King. The voltage gradient of the free burning arc in air or nitrogen. *Proceedings of the 5th International Conference on Ion Phenomena in Gases:* 871, 1961.
70. GR Jones, GH Freeman, H Edels. Transient temperature distributions in cylindrical arc columns following abrupt current changes. *J. Phys. D Appl. Phys.* 4: 236–245, 1971.
71. RG Colclaser. Electrical and system aspects. In: TE Browne, ed. *Circuit Interruption: Theory and Techniques.* New York: Marcel Dekker, 1984, pp. 11–38.
72. JVR Heberlein, JG Gorman. The high current metal vapor arc column between separating electrodes. *IEEE Trans. Plasma Sci.* PS–8, 283–289, 1980.

73. MB Schulman, PG Slade. Sequential modes of drawn vacuum arcs between butt contacts for currents in the range 1 kA to 16 kA. *IEEE Trans. Compon. Packaging Manuf. Tech.* 18(1): 417–422, 1995.

74. BE Djakov, R Holmes. Cathode spot division in vacuum arcs with solid metal cathodes. *J. Phys. D Appl. Phys.* 4: 504–509, 1971.

75. LP Harris. Arc cathode phenomena. In: JM Lafferty, ed. *Vacuum Arcs: Theory and Application.* New York: Wiley, 1980, pp. 120–168.

76. B Juttner. Formation time and heating mechanism of arc cathode craters in vacuum. *J. Phys. D Appl. Phys.* 14: 1265–1275, 1981.

77. JE Daalder. *Cathode-Erosion of Metal Vapor Arcs in Vacuum.* Thesis, University of Eindhoven, The Netherlands, 1978.

78. CW Kimblin. Erosion and ionization in the cathode spot regions of vacuum arcs. *J. Appl. Phys.* 44: 3074–3081, 1973.

79. E Hantzsche. Theories of cathode spots. In: RL Boxman, PJ Martin and DM Sanders, eds. *Handbook of Vacuum Arc Science and Technology.* New Jersey: Noyes, 1995, pp. 151–208.

80. I. Beilis. Theoretical Modeling of Cathode Spot Phenomena. In: RL Boxman, PJ Martin and DM Sanders, eds. *Handbook of Vacuum Arc Science and Technology.* New Jersey: Noyes, 1995, pp. 151–256.

81. JE Daalder. Erosion and the origin of charged and neutral species in vacuum arcs. *J. Phys. D Appl. Phys.* 8: 1647–1659, 1975.

82. DT Tuma, CL Chen, DK Davies. Erosion products from the cathode spot region of a copper vacuum arc. *J. Appl. Phys.* 49: 3821–3821, 1978.

83. Z Zalucki, J Kutzner. Ion currents in the Vacuum arc. *Proceedings of the 8th International Symposium Discharges Electrical Insulation Vacuum,* August 1976, pp. 297–302.

84. JVR Heberlein, DR Porto. The interaction of vacuum arc ion currents with axial magnetic fields. *IEEE Trans Plasma Sci.* PS–11: 152–159, 1983.

85. CW Kimblin. Vacuum arc ion currents and electrode phenomena. *Proc. IEEE.* 59: 546–555, 1971.

86. WD Davies, HC Miller. Analysis of the electrode products emitted by dc arcs in a vacuum ambient. *J. Appl. Phys.* 40: 2212–2221, 1969.

87. AA Plyutto, VN Ryzhkov, AT Kapin. High speed plasma streams in vacuum arcs. *Sov. Phys. JETP* (New York) 20: 328–337, 1965.

88. LP Harris. A mathematical model for cathode spot operation. *Proceedings of the 8th International Symposium Discharges Electrical Insulation in Vacuum,* 1978, pp. F1–F18.

89. G Yushkov, A Bugaev, I Krinberg, E Oks, On a mechanism of ion acceleration in vacuum arc-discharge plasma. *Doklady Phys.* 46(5): 307–309, 2001.

90. E Byon, A Anders, Ion energy distribution functions of vacuum arc plasmas. *J. Appl. Phys.* 93(4): 1899–1906, 2003.

91. HC Miller. A review of anode phenomena in vacuum arcs. *Contrib. Plasma Phys.* 29(3): 223–249, 1989.

92. CW Kimblin. Anode voltage drop and anode spot formation in vacuum arcs. *J. Appl. Phys.* 40: 1744–1752, 1969.

93. GR Mitchell. High current vacuum arcs: Part I and II. *Proc. IEEE* (London). 177: 2315–2332, 1970.

94. PG Slade. The vacuum interrupter contact. *IEEE Trans. Compon. Packaging Manuf. Tech.* 7(1): 25–32, 1984.

95. MB Schulman. Separation of spiral contacts and the motion of vacuum arcs at high AC currents. *IEEE Trans. Plasma Sci.* 21(5): 484–488, 1993.

96. MB Schulman. The behaviour of vacuum arcs between spiral contacts with small gaps. *IEEE Trans. Plasma Sci.* 23(6): 915–918, 1995.

97. CW Kimblin, RE Voshall. Interruption ability of vacuum interrupters subjected to axial magnetic fields. *Proc. IEEE* (London). 119: 1754–1758, 1972.

98. S Yanabu, S Souma, T Tamagawa, S Tamashita, T Tsutsumi. Vacuum arc under an axial magnetic field and its interrupting ability. *Proc. IEEE* (London). 126(4): 313–320, 1979.

99. MB Schulman, PG Slade, JVR Heberlein. Effect of an axial magnetic field upon the development of the vacuum arc between opening electric contacts. *IEEE Trans. Compon. Hyb. Manuf. Tech.* 16(1): 180–189, 1993.

100. A Greenwood. *Electrical Transients in Power Systems*, 2nd edition. New York: Wiley, 1991.

101. J Slepian. Extinction of an a-c Arc. *AIEE Trans.* 47: 1398–1408, 1928.

102. HA Bentounes. Influence de la nature des electrodes sur la vitesse de retablissement de la rigidite dielectrique d'un milieu post-arc. Thesis Universite de Paris-Sud Centre d'Orsay, No 651, 1984.

103. HA Bentounes, J Muniesa. Electrode materials and recovery of the post-arc dielectric strength. *Proceedings of the 13th International Conference on Electrical Contacts:* 176–180, September 1986.

104. H Edels and FW Crawford. Arc interruption. *J. IEE.* 2: 712–716, 1956; 3: 88–93, 1957.

105. CW Kimblin, JJ Lowke. Decay and thermal reignition of low-current cylindrical arcs. *J. Appl. Phys.* 44(10): 4545–4547, 1973.

106. FW Crawford, H Edels. The reignition voltage characteristics of freely recovering arc. *Proc. IEEE.* 107A: 202–212, 1960.

107. A Lee, PG Slade. Molded-case, low-voltage circuit breakers. In TE Browne, ed. *Circuit Interruption, Theory and Techniques,* New York: Marcel Dekker, 1984, pp. 527–565.

108. JF Waymouth. *Electric Discharge Lamps.* Cambridge, MA: MIT Press, 1971.

109. AM Cassie. *Arc Rupture and Circuit Severity: A New Theory. CIGRE.* Report 102, Paris, 1939.

110. O Mayr. Beitrage zur Theorie des statischen und des dynamischen Lichtbogens, *Arc Elektrotech.* 37: 588–608, 1943.

111. LS Frost, TE Browne. Calculation of arc-circuit interaction. In TE Browne, ed. *Circuit Interruption, Theory and Techniques,* New York: Marcel Dekker, 1984, pp 187–240.

112. SMS Sharkh, J McBride. Voltage steps in atmospheric low current arcs between opening silver metal oxide contacts. *Proceedings of the 43rd IEEE Holm Conference on Electrical Contacts:* pp. 233–237, 1997.

113. M Glinkowski and A Greenwood. A computer simulation of post-arc plasma behaviour at short contact separation in vacuum. *IEEE Trans. Plasma Sci.* 17: 45–50, 1989.

114. B Fenski, M Lindmayer. Post arc currents of vacuum interrupters with radial and axial magnetic field contacts – measurements and simulations. *Proceedings of the 19th International Conference on Electrical Contact Phenomena,* pub. VDE: 259–267, 1998.

115. A. Lee, PG Slade. Molded-case, low-voltage circuit breakers. In TE Browne, ed. *Circuit Interruption, Theory and Techniques.* New York and Basel: Marcel Dekker, Chapter 14, 1984, pp. 155–561.

116. D Chen, H Liu, H Sun, Q Liu, J Zhang. Effect of magnetic field of arc chamber and operating mechanism on current limiting. *IEICE Trans. Electron.* E86-C, 6: 915–920, 2003.

117. C Rümpler, H Stammberger, A Zacharias. Low-voltage arc simulation with out-gassing polymers. *Proceedings of the 57th IEEE Holm Conference on Electrical Contacts:* 1–8, 2011.

118. R Guan, H Liu, Y He, D Chen. Influence on the current limiting resistance on the arc commutation process across the gap of a separated runner. *IEICE Trans. Electron.* E94-C, 9: 1416–1421, 2011.

119. PG Slade. *The Vacuum Interrupter, Theory, Design and Application,* Boca Raton, FL: CRC Press, 466–469, 2008.

120. http://en.wikipedia.org/wiki/power_MOSFET.

121. W Johler. RF performance of ultra-miniature high frequency relays. *Proceedings of the 49th IEEE Holm Conference on Electrical Contacts:* 179–189, 2003.

122. http://en.wikipedia.org/wiki/microelectromechanical_systems.

123. W Johler. Basic investigations for switching of RF signals. *Proceedings of the 53rd IEEE Holm Conference on Electrical Contacts:* 229–238, 2007.

124. N Miki, K Sawa. Breaking arc characteristics in various power frequencies. *Proceedings of the. 49th IEEE Holm Conference on Electrical Contacts:* 284–288, 2007.

125. N Miki, K Sawa. Arc extinction characteristics in power supply frequencies from 50 Hz to 1 MHz. *Proceedings of the 24th International Conference on Electrical Contact Phenomena:* 13–18, 2008.

10

The Consequences of Arcing

Paul G. Slade

Thunder is good, thunder is impressive; but it is the lightning that does the work.

Mark Twain

CONTENTS

10.1 Introduction.. 618
10.2 Arcing Time... 618
    10.2.1 Arcing Time in an AC Circuit.. 618
    10.2.2 Arcing Time in a DC Circuit.. 618
    10.2.3 Activation of the Contact.. 622
    10.2.4 Arcing Time in Very Low-Current DC Circuits:
            Showering Arcs... 624
10.3 Arc Erosion of Electrical Contacts.. 628
    10.3.1 Erosion on Make and Erosion on Break... 631
    10.3.2 The Effect of Arc Current.. 631
    10.3.3 The Effect of Contact Size.. 633
    10.3.4 Determination of Contact Size in AC Operation... 635
    10.3.5 Erosion of Contacts in Low-Current DC Circuits... 636
    10.3.6 Erosion of Contacts in Low-Current AC Circuits... 644
10.4 Blow-Off Force.. 646
    10.4.1 Butt Contacts.. 647
10.5 Contact Welding.. 651
    10.5.1 Welding of Closed Contacts... 651
    10.5.2 Welding During Contact Closure.. 654
    10.5.3 Welding as Contacts Open.. 657
10.6 Changes in the Contact Surface as a Result of Arcing... 657
    10.6.1 Silver-Based Contacts... 659
    10.6.2 Silver-Refractory Metal Contacts... 659
    10.6.3 Other Ambient Effects on the Arcing Contact Surface: Formation
            of Silica and Carbon and Contact Activation....................................... 665
Acknowledgments... 667
References... 667

10.1 Introduction

Slepian once said if Nature had not given us an arc, then we would have had to invent one. The electric arc allows a smooth transition of the circuit current to current zero. If the current were to suddenly chop to zero once the contacts had parted, then the energy stored in the circuit inductance, L, would result in large overvoltages given by

$$V = -L\frac{dI}{dt} \tag{10.1}$$

Switch designers would have to design for these voltages. Fortunately, for the most part, the presence of the arc usually limits the values of voltage to a maximum of up to two times the circuit voltage.

The consequences of this arc, however, affect all of our choices for the application of contact material. This chapter describes the effects of arcing on electrical contacts; Chapters 11 through 15 will discuss design considerations for switching devices that produce arcs as they close and open, and Chapters 16 through 19 will further discuss the consequences of arcing on electrical contacts.

10.2 Arcing Time

10.2.1 Arcing Time in an AC Circuit

When switching normal load current, most switching devices have a maximum arcing time, t_h, of one half-cycle of current. Thus, for a 60-Hz circuit, the maximum arcing time is 8.3 ms and for a 50-Hz circuit, it is 10 ms. There could be the occasional operation when the contacts open just before a current zero and the arc will persist until the following current zero. If the contacts open at random then the average arcing time t_b over the life of the switch will be

$$t_b = t_h/2 \tag{10.2}$$

For contacts used in devices that switch large overload currents or short-circuit currents where asymmetrical values will occur [1–3], the arcing time can be longer than one cycle. If, however, the arcing time is more than 23 ms for a 50-Hz circuit or 19 ms for a 60-Hz circuit (i.e., a little more than one whole cycle of current), there is usually something wrong with the design and operation of the switch.

10.2.2 Arcing Time in a DC Circuit

The duration of the arc in a DC circuit is much more complex. As the arc voltage, U_A, increases to the circuit voltage, U_C, the circuit current decreases to a value at which the arc can no longer be sustained. As we have seen in Section 9.7.2, at this value, the arc extinguishes and the circuit current is interrupted. However, even if $U_A > U_C$ for an infinitesimally small instant of time, a finite time period is required to dissipate the ½LI^2 energy stored in the circuit inductance. If we consider the circuit shown in Figure 10.1, then

$$U_C = L\frac{dI}{dt} + RI + U_A(t) \tag{10.3}$$

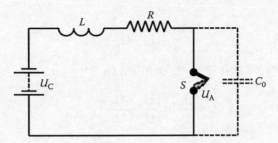

FIGURE 10.1
The direct current (DC) circuit.

where R is the circuit resistance. In principle, the exact function of time for U_A can be substituted in Equation 10.3 and I as a function of time can be calculated. If the voltage of the arc formed after the rupture of the molten metal bridge is constant, Equation 10.3 can easily be solved. This case always occurs in a low-voltage DC circuit where U_C is less than the minimum arcing voltage U_{min}, that is, where U_C is less than about 12 V. A single operation of the contacts follows this sequence [4]:

1. The contacts close allowing current to flow and discharging the local contact capacitance, C_0. If an arc discharge occurs at all, it will be of extremely short duration assuming there is no contact bounce.

2. The current builds up to its final value $I_0 = U_C/R$ with a time constant L/R.

3. When the contacts begin to separate, the molten metal bridge is formed. As they continue to separate, the voltage across the contacts reaches the rupture voltage of the bridge, U_b. The current is now $I_b = I_0\{1 - U_b/U_c\}$. The time to reach the rupture voltage depends upon I_0 and the opening velocity of the contacts (see Section 9.4.2).

4. The bridge ruptures and creates a region of high pressure and low conductance metal vapor between the contacts. The voltage across the contacts begins to rise at a rate which is $dU/dt \sim I_b/C_0$. The value of dU/dt depends upon C_0 which, in turn, depends not only upon the contact material, but also the length and diameter of the molten metal bridge, that is, upon I_0. Figure 9.20 shows that for Au contacts with $I_0 = 20$ A, $dU/dt \approx 2.8 \times 10^7$ Vs^{-1}, for Cu contacts with $I_0 = 150$ A, $dU/dt \approx 2.5 \times 10^7$ Vs^{-1}, and for Ni contacts with $I_0 = 1,000$ A, $dU/dt \approx 10^7$ Vs^{-1},

5. The voltage rises to a value of a few tens of volts and then drops to a value close to the minimum arc voltage associated with the contact material, U_{min}. (see Section 9.5.4).

6. When the gap voltage reaches U_{min}, an arc is initiated in the metal vapor left after the rupture of the bridge.

7. For a short arc, the arc voltage U_A remains constant, and the supply voltage U_C is effectively opposed by a back voltage U_A. The circuit current I, thus decreases, and it is given by [4]

$$I(t) = \left\{\frac{U_C - U_A}{R}\right\} + \left\{\frac{U_A - U_b}{R}\right\}\exp\left\{-\frac{Rt}{L}\right\} \qquad (10.4)$$

8. The arc will continue until the current has fallen to the minimum arcing current I_{min}; if we substitute $U_0 = I_{min}R$, then the arcing time t_a is given by

$$t_a = \{L/R\}\ln\left\{\frac{U_A - U_b}{U_A - U_C + U_0}\right\}$$ (10.5)

In order to simplify this, we can usually assume $I_{min} \approx 0$.

$$t_a = \frac{LI_0}{U_C}\ln\left\{\frac{U_A - U_b}{U_A - U_C}\right\}$$ (10.6)

Figure 10.2 shows the arc duration, t_a for gold contacts in a 6-V circuit for various values of L [5]. In Figure 10.3, t_a is plotted against LI for currents ranging from 1 A to 15 A. It can

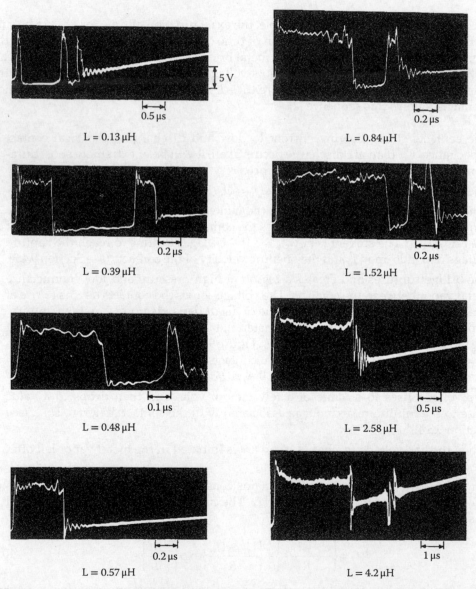

FIGURE 10.2
Arc duration for gold contacts interrupting a 6 V, 10 A, DC circuit [5].

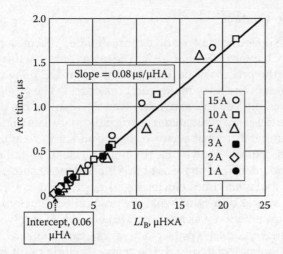

FIGURE 10.3
Arc duration as a function of circuit inductance × current for gold contacts a 6-VDC circuit [5].

TABLE 10.1

Arc Duration and the Effect of the Circuit Voltage for Low-Voltage DC Circuits

Contact material	Ag		Cu		Pd		Au	
U_A	12		13		17.5		15	
U_b	0.75		0.8		1.5		0.9	
$U_A - U_b$	11.25		12.2		16.0		14.1	
U_C	6	12	6	12	6	12	6	12
$U_A - U_C$	6	0	6	1	11.5	5.5	9	3
$\log_e \dfrac{U_A - U_b}{U_A - U_C}$	0.6	100	0.6	2.5	0.3	1.1	0.4	1.5
t_a μs ($R = 1\,\Omega, L = 50\,\mu H$)	30	5000	30	125	15	55	20	75

be seen from Equation 10.6 that, even though the $\log_e$ term does not change rapidly as $(U_A - U_C) \to 0$, t_a will increase considerably; this is illustrated in Table 10.1. Thus, in DC circuits where the voltage can vary such as in the nominally 14-V automotive electrical circuits, it is important to design the switching system so that the U_A is greater than the maximum possible value of U_C. This will necessitate choosing a material with the correct U_{min}, as shown in Table 9.9, and/or stretching the arc to drive $U_A > U_{min}$, because U_A will increase both from an increase in length as well as from the substitution of O_2 and N_2 for metal vapor as the arcing medium (see Section 10.3.5).

For higher voltages (say U_C in the range 100–1,000 V) the design of the switch again must ensure the development of a high $U_A \to (U_C - RI)$. This usually cannot be achieved by just opening the contacts further. In order to achieve a higher U_A, the arc must be cooled or split into a degree of activation it into deion plates [6] or into narrow insulating slots (see Section 14.1.1). The designer has to be careful, however, not to increase U_A too rapidly, because it is still possible for the arc to re-strike in the residual hot gas remaining between the contacts [7]. This will be discussed again in Section 10.3.5. For circuit voltages greater than ~1,000 V, other means have to be used to drive the current in the switch to zero. These are beyond the scope of this book, but examples of successful very high-voltage DC switches are given in [8,9].

10.2.3 Activation of the Contact

Contact activation will be introduced here and a more detailed discussion of this phenomenon will be presented in Chapter 19 (Section 19.2). The term contact activation was first used by Germer et al. [10] as "the process of particle formation on DC relay contacts that gives an increase in the duration of the relay make and break arcs." The pioneering research by Germer and his colleagues showed this activation was the result of the deposit of carbonaceous particles. These particles are generated from organic materials that decompose under the conditions of high temperature that occur in the arc and at the contact surfaces. Visual examination of activated contacts shows that the normal, clean metallic contacts become covered with a blackish powder. SEM analysis of this black deposit shows it to be small submicron carbon particles that sometimes cluster together giving a cauliflower appearance. I suspect that Germer et al. was observing what are now called "Buckyballs" or Fullerenes [11], but in 1957, his instrumentation was not powerful enough to resolve them.

The organic vapors that produce contact activation can come from many sources. External sources—those not associated with the relay—include paint fumes, floor polish fumes, exhaust fumes, and some so-called contact cleaners. After-shave lotions and hair sprays have also been able to cause activation in the laboratory. Nature can contribute to the problem with, for example, pine vapor from pine trees. Internal sources—those associated with the relay—include plasticizers in the cellulose acetate of the coil winding, phenol from the relay structure and printed wiring boards, and organic lubricants used in conjunction with connector contacts. For relays operating in an automobile environment, the source of organic vapors is enhanced by the lubricating oils and fuel that are required to allow the automobile to function.

For relays operating in an organic ambient, it has been shown the total exposure of the contacts to organics is critical, that is the arc duration as the contacts open multiplied by the organic concentration in the ambient atmosphere is the important factor in determining the degree of activation [12]. In Figure 10.4, arc duration in microseconds is plotted as a function of the contact exposure in pascal seconds (1 pascal (Pa) $= 10^{-2}$ mbar $= 10^{-5}$ atmosphere at STP). Three regions are defined from the activation curve. The first is the region of limited ion drift at low exposure values, where the arc durations are independent of exposure and give values of arc duration similar to the clean activated contact. In

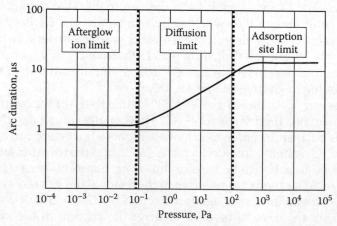

FIGURE 10.4
Note: Arc duration for activated contacts as a function of contact exposure to organic vapor [12].

this example, for a DC circuit of 0.5 A and 50 V, it is about 1.5 µs. The second is the region of limited diffusion at intermediate values where the durations are proportional to the square root of the exposure. Finally, there is the region of adsorption sites at high exposure values where, once again, arc durations are independent of exposure. The actual critical exposure values between the regions depend upon the organic activant and the metal being examined. A model of contact activation [12] predicts the form of the curve in Figure 10.4. From a practical point of view, the diffusion-site-limited and adsorption-site-limited regions are the most damaging, because the activation results in increased arcing time: in the example shown here, this is by a multiple of 10.

A unique activation curve occurs for each metal–organic pair. Predicting how any particular combination will behave, therefore, is very difficult. However, by examining the performance of a number of metals in a number of activants, Gray et al. [12] showed that comparisons can be made. Metals examined are Au, Ag, Pd, Re, Pd–Ag and Pd–Ni, and organics used are benzene, fluorinated hydrocarbons, *n*-hexane, isopropanol, phenol, cyclohexane, and diethylphthalate. Rankings in terms of arc duration are shown in Table 10.2 [13]. Of the metals studied for telephone switching applications, Pd–Ni and Au appear to be the most and least resistant to activation respectively. Differentiating between organics is much more difficult; however, Gray showed that the lower critical exposure point depended upon the number of carbon atoms per molecule of the activating organic. The larger the organic molecule, in terms of the number of carbon atoms, the lower the

TABLE 10.2

Statistical Ranking of Metals and Organic Activants on Relay Contacts

	Metals	Activants
Ion-drift-limited region	PdAg	Benzene
	Pd	n-Hexane
	Au	Isopropanol
	Ag	Phenol
	Re	Cyclohexane
	PdNi	Freon Diethylphthalate
Diffusion-limited region	Au	Phenol
	Ag	n-Hexane
	Pd	Diethylphthalate
	Re	Benzene
	PdAg	Cyclohexane
	PdNi	Isopropanol
		Freon
Adsorption-site-limited region	PdAg	Benzene
	Au	Phenol
	Pd	Freon
	Ag	Isopropanol
	Re	n-Hexane
	PdNi	Diethylphthalate
		Cyclohexane

Note: The bracket indicates that statistically no difference could be detected between members of a group. Metals and activants are listed in order of decreasing arc duration.

exposure required to activate the contacts. The smaller number of organic molecules is compensated by the larger number of carbon atoms in the molecule. This is as would be expected, because the amount of carbon on the contact surfaces determines the arcing duration. The sticking coefficient of the organics on the metal surface also plays a role, as does the effect of oxygen content of the molecule [14]. Some organic lubricants, which contain a large number of carbon atoms, have been observed to cause activation at concentrations of only a few parts per trillion! Thus, while helping one system to operate, lubricant vapors can be detrimental to other nearby systems with open contacts. It has been shown, however, the electric arc is required to activate contact surfaces. Exposure to organic vapors without arcing will not result in activation [15].

10.2.4 Arcing Time in Very Low-Current DC Circuits: Showering Arcs

There are DC circuits where the currents are closer to or below the minimum arc current I_{min} (see Section 9.5.4), for example, in the range ~1 mA to ~100 mA. These circuits can still form a series of transient arcs and at currents <60 mA, transient glow discharges (see Section 9.3.3) that can affect the performance and life of the electrical contact. This discharge is called a *showering arc* and is described in detail by Wagar [16,17] and Miyajoma et al. [18]. A typical circuit is shown in Figure 10.5. Once the contacts open, the DC current between the contacts chops to zero if its value is low enough. The circuit's inductance, however, will not allow the circuit's energy to instantaneously drop to zero. The result is a transient charging and discharging of the circuit capacitance, both lumped and that distributed in the wiring. The initial rate of rise of voltage is again given by I_0/C; examples are presented in Table 10.3 [16,17].

The peak voltage U_p is a complex function of the circuit load, inductive energy and capacitive energy stored in the circuit [16,17]. A typical transient is shown in Figure 10.6. In Section 9.3, it has been shown if the voltage exceeds the breakdown voltage of the contact

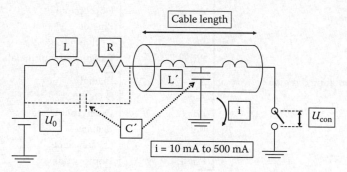

FIGURE 10.5
Low-current DC circuit [16,17].

TABLE 10.3

Typical I_0/C_0 Values for Low-Current Switches

I_0/C_0 (Vs^{-1})	Value of C(F)	Possible value of I(A)
10^{11} (high)	10^{-10} (2 m of cable)	10
2×10^8 (typical)	10^{-9} (20 m of cable)	0.2
5×10^4 (low)	10^{-6} (large protection network)	0.05

Activation makes contact surfaces more cathodic.

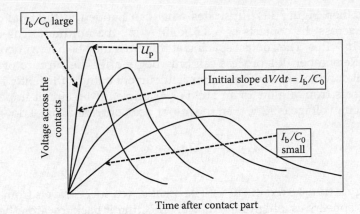

FIGURE 10.6
Voltage transient generated across the open contacts on interrupting current in a low-current DC circuit [16,17].

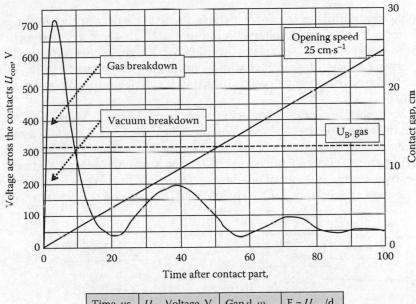

Time, μs	U_{con} Voltage, V	Gap d, m	$E = U_{con}/d$
1	225	2.5×10^{-7}	9×10^{-8}
10	290	2.25×10^{-7}	1.3×10^{-8}

FIGURE 10.7
Breakdown as a result of the voltage transient developed across the open contacts in a low-current DC circuit [16,17].

gap, then a discharge can be initiated in fractions of a microsecond. Whether or not breakdown occurs can be determined by matching the contact's breakdown characteristic with the available circuit voltage. If we take the short gap and gas breakdown characteristics illustrated in Figures 9.12 and 9.13 and superimpose typical circuit voltage characteristics, it can be seen in Figure 10.7 that it is possible for the circuit voltage curves to exceed the voltage required to initiate a discharge. If the circuit current exceeds I_{min}, a sustained arc will occur between the contacts as has been discussed in Section 9.3. If the circuit current is lesser than I_{min}, then Figure 10.7 illustrates what can happen. If the total capacitance produces a dV/dt across the contacts of $\sim 3.5 \times 10^8$ V s^{-1}, the minimum gas breakdown (350 V) is reached in ~ 1 μs. The contact spacing at this time may be a few micrometers for a typical contact. Remember this process can only begin after the rupture of the molten metal bridge. Breakdown occurs with the transient current being sufficiently high for the discharge to progress from a glow to an arc. For example, if the circuit impedance Z is ~ 100 Ω and the circuit voltage is 12 V, then the initial peak current at breakdown, I_p is

$$I_p = \frac{350-12}{100} = 3.4\,\text{A} \tag{10.7}$$

The stored energy is transferred in $\sim 10^{-9}$ s. The breakdown surge travels from the contact to the load, which appears as a high impedance, reflecting it back towards the contact in reverse polarity. Upon arrival at the contact after a time

$$t = 2l\sqrt{LC} \tag{10.8}$$

this reverse voltage extinguishes the arc and starts a new cycle of charging in the positive direction, as before. If the reverse voltage is high enough, an arc may even form in the reverse direction. A succession of very short duration discharges occurs and it is clear why this has been termed a *showering arc*. It continues until the energy stored in the circuit is dissipated or until the opening contact gap causes the transient circuit voltage to fall below U_B (see Figure 10.8).

This qualitative description of the showering arc was modeled by Mills [19]. One of his results is shown in Figure 10.9 and it shows a remarkable correspondence to the measured voltage transients. Miyajima et al., who used a high inductance 3.3 mA, 10 V circuit, show that the showering arc produces so much contact erosion as to change the roughness of the contact surfaces over a large number of operations. From one to 10^5 operations, the roughness does not change appreciably, but from 10^5 to 10^7 operations the surface roughness

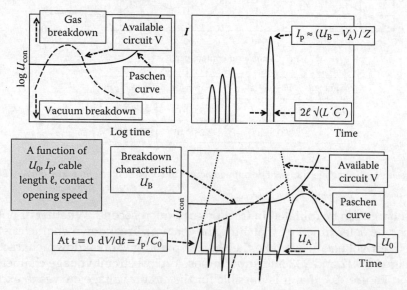

FIGURE 10.8
The development and cessation of the showering arc [16,17].

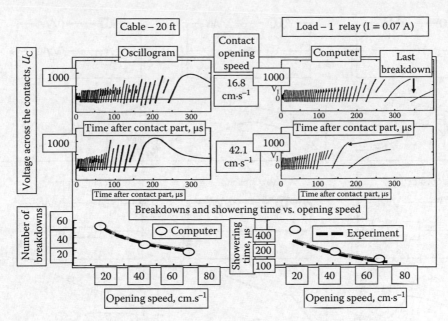

FIGURE 10.9
Comparison of the measured showering arc voltage transients with a computer model of the showering arc process [19].

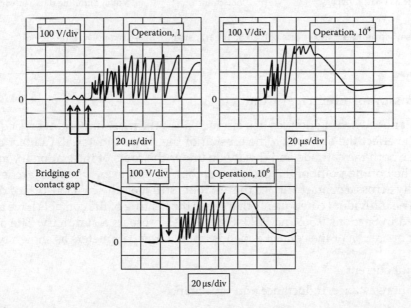

FIGURE 10.10
Changes in showering arc voltage with number of operations showing the initial bridging of the contact before the showering arc develops [18].

doubles [18]. For the very low current in their experiments the molten metal bridge that forms as the contacts open is about 0.3 μm when it ruptures; see Figure 9.18. It is, thus, possible that metal from this rupture will again bridge the contact gap and the showering arc will be delayed; see Figure 10.10. It is possible to modify the voltage transients across the contacts by using the different protective circuits shown in Figure 10.11 [16].

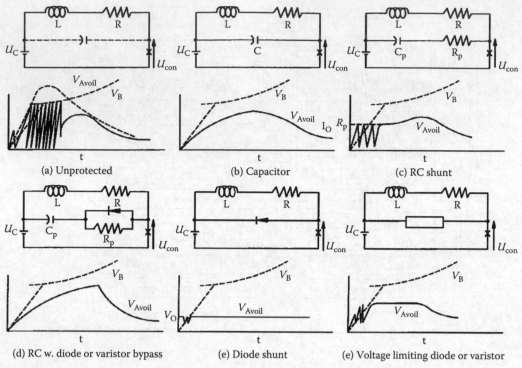

FIGURE 10.11
Possible methods of minimizing the effects of the voltage transient in low-current DC circuits [16].

10.3 Arc Erosion of Electrical Contacts

For the designer of switching devices in a circuit, one of the most important consequences of arcing is the effect the arc has on the erosion of the contact material. Contact erosion occurs because both the cathode and the anode under the roots of the stationary arcs can be heated to the boiling point of the contact material. In fact, even when the arc is forced to move rapidly across a contact surface the arc roots still melt the contact surface directly under them. The individual components that cause the heating of the contacts have already been discussed in Sections 9.5.2 and 9.5.3. Figure 10.12 presents a summary. The amount of erosion per operation of the contact depends upon many parameters as shown below:

1. The circuit current
2. The circuit resistance, inductance and capacitance
3. The arcing time
4. The contact material
5. The structure of the contact material
6. The shape and size of the contact
7. The contact's attachment
8. The opening velocity of the contact
9. The bounce on making contact

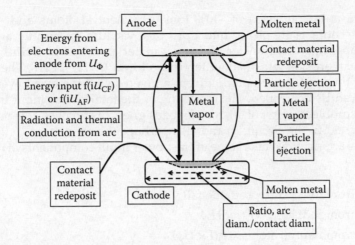

FIGURE 10.12
Energy balance at the contacts from the arc and the effect on contact erosion.

10. The open contact gap
11. Motion of the arc on the contacts
12. The design of the arc chamber
    (a) Gas flow
    (b) Insulating materials used

The phenomenon of contact erosion is further complicated by mechanical stresses on the contact as a result of the impact on closing. These stresses caused by the switch mechanism affect the contact materials in widely different ways which depend upon the material and the manufacturing process (see Chapter 16).

The consequences of this is while it is possible to present some general principles to let you appreciate the qualitative aspects of arc erosion, most of the design information given can only be used as a guideline. Only electrical testing of the final design of the switch will give the actual erosion the contacts will experience in that device.

To illustrate this, we can show in principle the mass lost per operation of the contact should be given by

$$\text{mass loss} = f\left(\text{total power input into the contacts}\right) \tag{10.9}$$

Both sides of this seemingly simple equation present complexities that prevent it from ever being established, let alone solved. For example, what is meant by the mass loss? The total mass loss from a contact is a mixture of the following components:

- Metal vapor evaporated from the arc roots +
- Metal droplets ejected from the arc roots −
- Metal re-deposited back onto the contact faces −
- Metal deposited from the opposite contact

Even in the case of a diffuse vacuum arc with Cu contacts where the erosion of the cathode [20] is given by:

$$w = 120 \, \mu gC^{-1} \tag{10.10}$$

it has been shown w is composed of ~10% ions, ~1% neutral atoms and ~89% metal particles [21]. If the contacts are placed into a practical vacuum interrupter and operated at close contact gaps, then the transfer of contact material back and forth between the contacts results in an erosion rate of about ten times lower (× 0.1) [22,23]. The actual erosion products are, thus, difficult to predict from a given switching event even when the parameters are carefully controlled. The modeling of the total power input into the contacts also presents problems. First of all, for most designers of switches, it is only possible to measure the current in the circuit $I(t)$ and also perhaps the arc voltage $U_A(t)$ across the contacts during the arcing. If we take some of the power input components at the cathode, (Section 9.5.2):

- Power input from ions $= f_1(I \times \text{cathode fall voltage})$
- Power input from radiation $= f_2(I \times U_A)$
- Power input from neutral atoms $= f_3(I \times U_A)$
- Joule-heating of contact $= f_4 (I^2)$

Again, this is very difficult to model. Most researchers try to perform comparative studies on one contact apparatus and most designers of switches ultimately test their device to show the current interruption and number of operations that are satisfactory for a particular application. There is, however, an enormous literature on contact erosion [24], but most of it applies to specific situations and should only be used as a guide by the designer. Examples of erosion models that take into account vaporization of metal and splashing of material can be found in [25,26]. The reader is referred to Chapter 16 for a detailed discussion of contact erosion for various material compositions and for various testing conditions. The measurement of contact erosion is also briefly discussed in Sections 13.5 and 18.4.4. An example of comparative erosion data for a large number of contact types is given in Figure 10.13 [27]. Here some seemingly similar contact materials have considerably different erosion rates. The following sections give you a few parameters you need to consider and the difficulty in generalizing the measurement of arc erosion from any one set of experimental data.

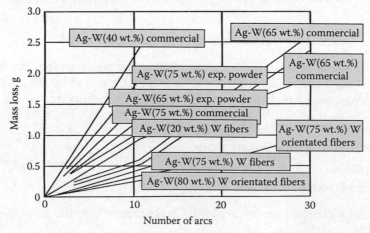

FIGURE 10.13
Comparative erosion of electric contracts [27].

10.3.1 Erosion on Make and Erosion on Break

Arcing occurs as the contacts come together by the breakdown of the contact gap before they touch each other. This arcing period depends upon the circuit voltage and closing speed of the contact, but it is usually of very short duration. Once the contacts touch, they may bounce apart again and an arc will form between them. Figure 10.14 illustrates the contact closing event: (a) An expanded view of the closing movement of the contact just before they touch, (b) The possible variations in voltage across the closing contacts with an expanded view as the contacts come together and (c) The initiation of the circuit current. As the contacts close, a pre-strike arc can occur just before the contacts touch (See Section 9.3). Once the contacts touch, bouncing can occur. In some switching devices, the total bounce time may last for milliseconds (see Section 13.7). The characteristics of the bounce will also influence the erosion of the contacts [28]. It is, therefore, possible to have as much contact erosion on make as it is on break. For the special case of contactors, the type of device that switches motors (see Sections 14.3.4 and 18.3) where the current at make is normally six times (6×) the break current, it is even possible for the erosion at make to be higher than that at break. An example of material loss or arcing time for break-only and make-only contact pairs is shown in Figure 10.15 [29]. To further complicate this discussion it is entirely possible for make and break erosion taken together to be lower than the sum of the make-only erosion and break-only erosion (see Section 18.4.4).

10.3.2 The Effect of Arc Current

Figure 10.16 shows a discontinuous erosion at ~1,000 A [30], that is, at about this current the erosion rate can increase by as much as 30 times (30 ×), so

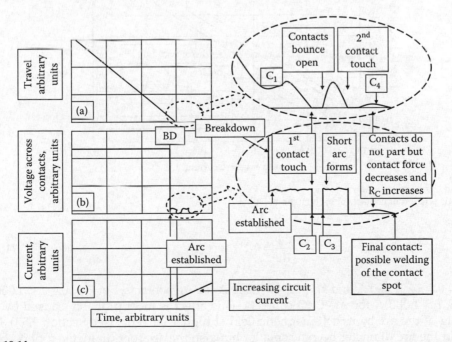

FIGURE 10.14
Schematic diagrams of contact closure showing (a) the travel and contact bounce, (b) the gap as flowing through the contacts during contact closing [91].

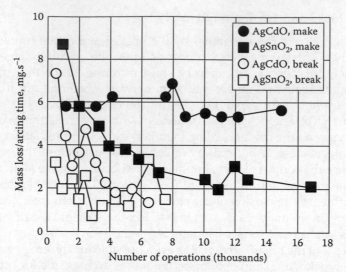

FIGURE 10.15
Erosion of silver-metal oxide contacts, break only and make only [29].

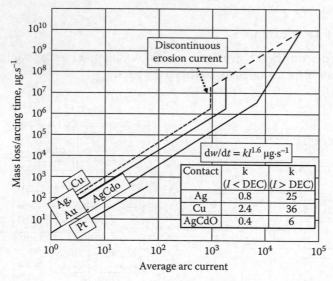

FIGURE 10.16
Discontinuous erosion of contacts at high currents [30].

$$\frac{dw}{dt} = k_I I^{1.6} \ \mu gs^{-1} \tag{10.11}$$

where $k_I = 0.8$, $I < 1,000$ A and $k_I = 25$, $I > 1,000$ A for Ag contacts, but $k_I = 0.4$, $I < 1,000$ A and $k_I = 6$, $I > 1,000$ A for Ag–CdO contacts. This effect is most probably caused by the constricting of the arc by its self-magnetic field at higher currents (see Section 9.5.1). Not only would the arc diameter become smaller but also the arc root diameter spot on both the cathode and the anode will also become smaller. The increase in the power density in the arc spots will give rise to excessive heating of the spot with the consequence that jets of metal will be ejected from the spot regions. The exponent in Equation 10.11 seems to be very dependent upon test conditions and various authors have published values in the range 1.0–3.7.

10.3.3 The Effect of Contact Size

The conduction of heat away from the contact surface is an important parameter that affects erosion. The contact's dimensions (i.e., area, length and breadth, and thickness) are all important in assessing the probable rate of erosion. Figures 10.17 through 10.19 [31] show effects of shape and size on the erosion of Ag–W contacts one half cycle of unidirectional AC current with the polarity of the contacts fixed. It has also been shown that the discontinuous erosion can also occur at currents less than 1,000 A if the contact diameter is too small [31,32]. Borkowski et al. [33,34] again using Ag–W contacts opening a unidirectional, half cycle of AC current, show that the discontinuous erosion depends upon the contact diameter for a given current, but they plotted their data as a $f(\int Idt)$. They show for volume of erosion ΔV of 10-mm diameter Ag contacts:

(a) For cathode erosion, $\Delta V = 0.076(\int Idt)^{1.8}$ before the discontinuity and $\Delta V = 0.36(\int Idt)^{1.28}$ after.

(b) For anode erosion, $\Delta V = 1.95(\int Idt)^{1.4}$ before the discontinuity and $\Delta V = 0.68(\int Idt)^{1.03}$ after.

They also show the temperature behind the contact face shows a steady increase as the $(\int Idt)$ increases until the discontinuous value of $(\int Idt)$ when it increases at a slower rate.

For currents less than 50 A, it is possible to make a conservative assessment of the contact volume required for a given application. Figure 10.20 shows an arc root on a contact surface. A rule of thumb says that the contact diameter D_c should be three times (3×) the

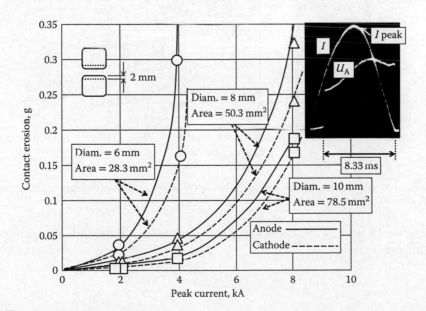

FIGURE 10.17
Contact erosion as a function of peak current, the effect of contact area, for Ag–W (50 wt.%) contacts opening at 0.5ms⁻¹, after 5 operations and showing a typical current pulse and arc voltage characteristic [31].

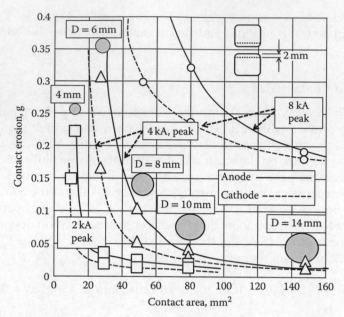

FIGURE 10.18
Contact erosion as a function of contact area for Ag–W (50wt.%) contacts for three separate peak current pulses, opening at 0.5ms⁻¹, after 5 operations [31].

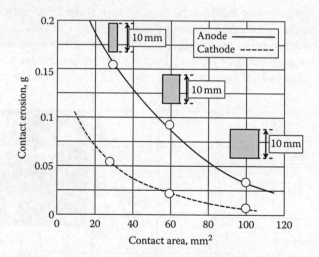

FIGURE 10.19
Contact erosion as a function of contact area and shape for Ag–W (65 wt.%) contacts, opening a current of 8kA peak, at 0.5ms⁻¹, after 20 operations [31].

arc root's diameter D_A and its thickness should be greater than the depth at melting of the arc which is approximately

$$T_m = \frac{D_A}{4}$$

(10.12)

Experimentally determined arc root current densities are given in Figure 10.20, from which an arc root diameter for a given current can be calculated.

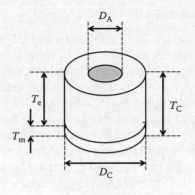

Total contact thickness

$T_C = T_e + T_m$

$T_m = D_A/4$ and $D_C = 3D_A$

$$T_C = \frac{\text{Erodable} - \text{contact} - \text{volume}}{\pi D^2/4} + \frac{D_A}{4}$$

Contact material	Arc spot current density A·cm^{-2}
Fine Ag	4500
Cu (36)	5500
Ag-CdO	6000
W-Ag	5500 [35]

FIGURE 10.20
Erosion at the arc root and the necessary contact dimensions.

10.3.4 Determination of Contact Size in AC Operation

Let us accept the erosion data given in Figure 10.15. Also assume the arcing time at break t_b on the average is one-half of the period of one half-cycle of current (i.e., 4 ms for a 60-Hz circuit and 5 ms for a 50-Hz circuit) and the bounce time on closing is t_m. If the erosion rate is

$$\frac{dw}{dt} = k_I I^{1.6} \ \mu g s^{-1} \tag{10.13}$$

Then the average erosion per operation is

$$w = \{k_m I_m^{1.6} t_m + k_b I_b^{1.6} t_b\} \ \mu g \tag{10.14}$$

If we assume that the contact diameter is D_c and its height is T_c, and assume the erosion is uniform, then the number of operations that can be sustained by the contact is:

$$n = \frac{V\delta}{w} \tag{10.15}$$

where δ is the density

$$V = \pi \frac{D_c^2}{4} \left\{ T_c - \frac{D_A}{4} \right\}$$

giving

$$n = \pi \frac{D_c^2}{4} \left(\frac{\left\{ T_c - \dfrac{D_A}{4} \right\} \delta}{\{k_m I_m^{1.6} t_m + k_b I_b^{1.6} t_b\}} \right) \tag{10.16}$$

10.3.5 Erosion of Contacts in Low-Current DC Circuits

A great deal of work was performed in the 1950s and 1960s on erosion in low DC currents [36]. A lot of this work was driven by the telephone industry's need to have relays with very long lives. In recent years, even though the logical functions of telephone relays have been taken over by electronic and optical–electronic systems, the relays still retain their safety functions (i.e., disconnecting circuits, protection from electromagnetic interference, and protection from stresses in voltage, current and temperature). In fact, the number of telecommunications relays may well be higher today than ever before. A major difference is the signal relays these days has a volume of only approximately 10 cm³ and a weight of l g compared to the 1950s relays of approximately 230 cm³ and 250 g. During this change in function and size, the relay bodies became manufactured from organic materials and the contact forces have decreased. The research activity has focused, therefore, more and more upon reliability rather than contact erosion.

In recent years, there has also been a considerable research effort to produce very reliable DC relays for automobile applications (see Sections 13.6 and 16.2.3). This has been driven by the ever increasing number and complexity of automobile electrical systems, which require reliable, long-life, low-cost relays that operate in nominally 14-V circuits and must switch currents in the range 0.1–100 A. The increase in electrical systems in cars has led to a corresponding increase in the weight of the wiring. In an attempt to reduce this added weight, the automobile companies have considered increasing the battery voltage from 14 V to 42 V. This has resulted in considerable R and D effort to develop reliable switches, relays and connectors for this DC circuit with higher voltage [37]. The 42-VDC circuit presents many challenges for the control and connecting systems. In a 14-V DC system, designing a switch to interrupt the circuit is relatively straightforward. Once the contacts have parted and the molten metal bridge has ruptured, an arc with a voltage $U_A \approx 13$ V appears between the contacts in less than a microsecond. This forces the circuit current to drop to the unstable I_{min} in a few tens of microseconds and the circuit is easily interrupted. In a 42-V DC circuit, the contacts have to open to a gap larger than 1 mm, as shown in Figure 10.21, to achieve a $U_A \approx 42$ V which will result in the interruption of the circuit, see Section 9.7.2 [37–40]. Figure 9.28 shows for a given contact gap, the voltage needed to sustain the arc increases as the current decreases. This figure shows for copper contacts and a 1mm contact gap, that if the current falls below

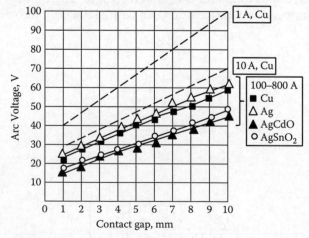

FIGURE 10.21
Expected arc voltage as a function of contact gap for currents of 1 A, 10 A (Figure 8.38) and 100-800 A [40].

about 0.5A in a 42V circuit the arc will become unstable and the circuit will be interrupted [41]. Sugiuri et al. [41] also show that a high transverse magnetic field (~200 mT) can reduce the interruption time for a slowly opening contact (10 mms^{-1}) from 78 ms to 10 ms where the contact gap is only 0.1 mm. The disconnecting and reconnecting of connectors under load also presents a considerable challenge [42–44]. Not only will the resulting arc cause damage to the connector, but also to the automotive technician. Disconnecting and reconnecting loaded connectors in a 14-V circuit is relatively safe and only results in minor arcing damage.

In spite of the above challenges, there continues to be a need to interrupt DC currents in the range of a few tens of amperes in circuits up to and beyond 500 V [45–48]. This need is driven by the increasing application of photovoltaic solar power systems and by the expected demand for electric and hybrid vehicles. Thus, to produce switches that can generate arc voltages high enough to interrupt these circuits, renewed research and development efforts are taking place. One study by Yoshida et al. [45] uses a double-break contact system with a permanent magnet (20–30 mT) to provide a transverse magnetic field across the contacts in order to drive the series arcs away from their respective contact gaps. Figure 10.22 shows a diagram of structure of the series contact with the magnetic field and the direction of motion of the arc. It also shows the combined arc voltages and their effect on the current leading to the interruption of a 400-V, 30.2-A DC circuit with a resistive load of 13.2 Ω. At 1 ms, there is an immediate drop in current to about 28 A as soon as the two series arcs are formed and inject a back voltage of 32 V into the circuit. As the contacts continue to open, the combined arc voltage slowly increases with a corresponding slow decrease in the current. From 1 ms to 3 ms (i.e., for the first 2 ms of arcing and where each contact gap <0.72 mm), the voltage increases at a rate that is expected for a stationary arc between the two pairs of contacts opening at 0.36 ms^{-1}. Thus, at 3 ms the current has only decreased to about 26 A. It is only when each contact gap is greater than 0.72 mm that the 20–30 mT magnetic field begins to drive the arcs off their respective contacts. This causes the arcs to lengthen, resulting in a

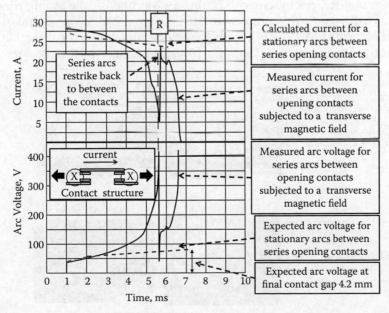

FIGURE 10.22
Interruption of a 400 V/30.2 A DC circuit using a bridging contact opening at 0.36 ms^{-1} to produce two series arcs [45].

rapid increase in the arc voltage and a corresponding rapid decrease in the circuit current. At about 5.6 ms, the arc voltage reaches 400 V and the current is being driven to zero. However, at this time (R in Figure 10.22), the field between each contact pair ($\sim 1 \times 10^5 \text{Vm}^{-1}$) is high enough for the arc to re-strike and be re-established back between those contacts with a combined arc voltage of ~75 V as would be expected from the open contact gaps at that time. The circuit current now returns to the expected value of ~24.5 A. This shows an optimum arc voltage, the contact gap and the arc movement for interruption of these higher voltage DC circuits arise to prevent this re-strike phenomenon. Yoshida et al. [45] indeed show for their contact system, this re-strike phenomenon does not occur for circuit voltages up to 300 V. However, even with re-striking, they do successfully interrupt DC circuits up to 500 V with currents from 5 A to 30 A. As would be expected with a two-break system, they find that for a 50-VDC circuit, the transverse magnetic field does not have an effect on the arcing time.

In Section 10.2.2 the consequences of arc voltage U_A, circuit voltage U_C, arc current I and circuit inductance on the duration of the arcing time t_a on the contact erosion has been explored. First of all, erosion in DC circuits is complicated by the fact, for a given contact material, switching a given set of circuit conditions, will always cause a net transfer of contact material from one contact to the other. The net transfer will eventually result in a build-up of a pip on one contact and a crater on the other. Figure 10.23 [49] shows a good example of this phenomenon for an Ag contact pair. Sometimes, the net metal transfer is from the anode to the cathode (i.e., a net cathode gain). In this case, the arc is sometimes called an *anode arc* [36]. Sometimes, the net metal transfer is in the reverse direction (i.e., a net anode gain). In this case, the arc is sometimes called a *cathode arc* [36]. The direction of net metal transfer is also dependent on the contact material; so under a given set of conditions, one material will have a net transfer to the cathode (sometimes called *anodic material*) while another material will have a net transfer to the anode (sometimes called *cathodic material*).

Let us explore a qualitative explanation of these erosion characteristics. In Figures 10.2 and 10.3, we saw that the arcing time in a 6-V circuit for the gold contacts depends on the circuit inductance and the circuit current. In those experiments, the arcing time ranged from about 100 ns to 3.5 μs. As the contacts part, the molten metal bridge is drawn between them and eventually ruptures. The effects of the bridge rupture on the direction of metal transfer were hotly debated in the 1950s and 1960s. A number of thermoelectric effects

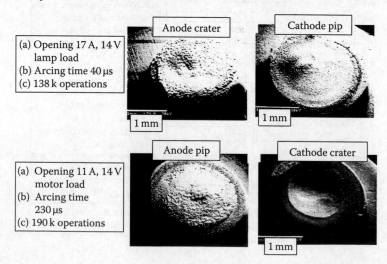

(a) Opening 17 A, 14 V lamp load
(b) Arcing time 40 μs
(c) 138 k operations

Anode crater

Cathode pip

1 mm

1 mm

(a) Opening 11 A, 14 V motor load
(b) Arcing time 230 μs
(c) 190 k operations

Anode pip

Cathode crater

1 mm

FIGURE 10.23
Pip and crater formation on a silver contact pair [49].

such as the Thompson effect and the Kohler effect were postulated to cause a net metal transfer of eroded metal from the anode to the cathode [50,51]. Another effect postulated by Bühl discussed the possibility that the fine metal droplets that formed after the rupture of the molten metal bridge would become positively charged [51]. Electro-migration has also been postulated as a possible explanation of net metal transfer of contact material to the cathode [52]. It is now generally agreed for currents greater than I_{lim} (see Section 9.5.4) the effect of bridge erosion is overwhelmed by short arc erosion. It is only when the circuit current is below I_{lim}, that contact erosion resulting from the molten metal bridge is observed [53,54]. After the rupture of the molten metal bridge, there is a very small gap between the contacts with a metal vapor with high pressure between them. As I have discussed in Section 9.4.2 (Figures 9.17 through 9.21), initially after the rupture of the bridge, there is no conduction across the metal vapor and the voltage rises rapidly across the contact gap. It is only when the metal vapor pressure reduces to a few atmospheres that a pseudo arc forms with the current carried mainly by metal ions. Thus, during this stage the net transfer of metal would be expected to be from anode to cathode. Figure 10.24 shows that while metal

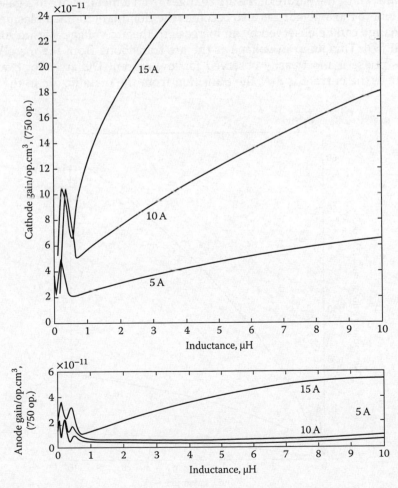

FIGURE 10.24
Absolute cathode gain and anode gain for gold contacts in a 6VDC circuit using a radioactive tracer method showing more metal transferring from the anode to the cathode than transferring from the cathode to the anode [5].

is transferred both from the cathode and the anode immediately after the rupture, the net transfer is from the anode to the cathode; see Figure 10.25. This data is made possible using a radioactive tracer technique [5,55]. In Figure 10.26, the effect of the metal transfer from the pseudo arc is seen in the first 400 µs. As the contacts continue to open and the cloud of metal vapor expands, the pressure of the metal vapor continues to decrease and a normal arc will form with 99% of the current now carried by electrons [56–58], see Section 9.5. In this arc, the electrons liberated by the cathode will ionize the metal vapor, creating positive metal ions and will also evaporate anode material by electron bombardment. Figure 10.27 illustrates the effects at the cathode and anode [59]. The metallic ions move toward the cathode and impact on the cathode. This may erode some cathode material, but because these ions are the same material as the cathode, they have a high probability of sticking, so you would expect an accumulation of material on the cathode, that is, a net cathode gain. Figure 10.26 illustrates this metallic arc erosion during the metallic arc phase. Although both cathode and anode do gain some material from the other contact, the net effect is for the cathode to gain material.

If the arc lasts for a long enough time, the contact gap increases and the ambient gas will enter the arc. Thus, the nature of the arc changes from a metallic arc to a gaseous arc, that is, an arc where the ambient gas also becomes the dominant ionized plasma column [56,57]. This change can be observed by an increase in the arc voltage; Figure 10.28 illustrates this well [60]. This jump in voltage as the arc transitions from the metallic phase to the gaseous phase is most easily observed for low-current DC arcs. For example, in Figure 9.62 where the current is 8 A, the transition from the metallic arc to the gaseous

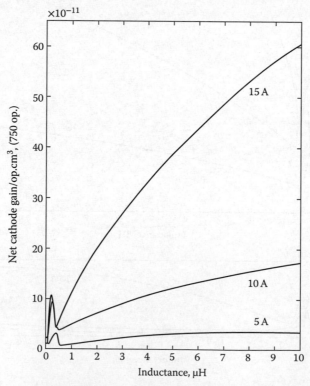

FIGURE 10.25
Net cathode gain for gold contacts in a 6-VDC circuit, from Figure 10.24.

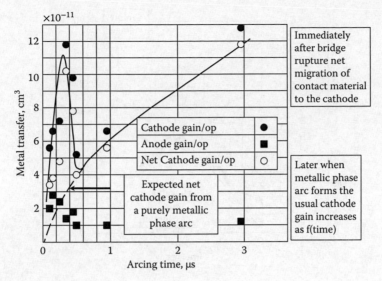

FIGURE 10.26
Cathode gain, anode gain and net cathode gain for gold contacts switching a 10 A, 6 V circuit as a function of arcing time from Figures 10.24 and 10.25.

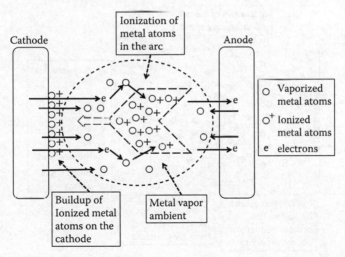

FIGURE 10.27
Contact erosion in a low-current metallic phase arc.

arc is not as evident from the measurement of arc voltage. Some researchers relate further steps in this low-current arc voltage to other electron–gas interactions as the contacts continue to part [60–62]. In this state, the ambient gas increasingly becomes ionized. When the gas ions reach the cathode, they gain energy as they drop through the cathode fall, but now they erode cathode material and do not add material to it (see Figure 10.29). As a result as the gaseous arc continues, more material is now eroded from the cathode than from the anode and a net anode gain begins to be observed, as for example in Figure 10.30 [63]. The generalized net transfer to the cathode for arcs in DC circuits of short duration and low current is shown in Figure 10.31.

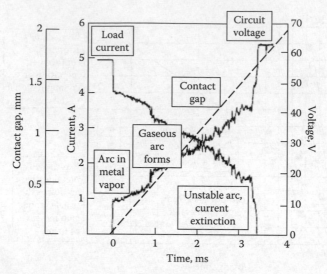

FIGURE 10.28
Voltage drop across opening silver contacts (0.5 ms^{-1}) in a 4.9 A, 62 VDC circuit showing the transition from a metal vapor arc to a gaseous arc [60].

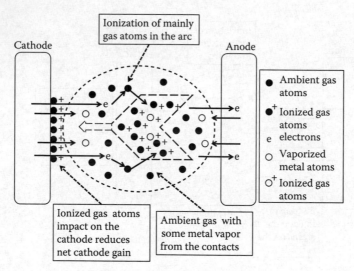

FIGURE 10.29
Contact erosion in a low-current gaseous phase arc.

The effect of this material transfer when switching DC circuits is it produces cathode pips and anode craters or anode pips and cathode craters on contacts. The exact nature of the erosion depends upon the arc duration (which, in turn, depends also upon the circuit current, inductance, and voltage) and on the contact material. Figure 10.23 [49] gives an example for Ag contacts. For an arcing time of 40 μs, a cathode pip and a slight anode crater develops, but for a longer arcing time of 230 μs (~6× longer), a very visible anode pip develops with a well-defined cathode crater. Figure 10.32 illustrates one experimental study of contact erosion for relatively long duration DC arcs switching a 60-VDC circuit [64]. At low currents (short arcing times), the metallic phase arc usually shows a net cathode gain. As the arcing time

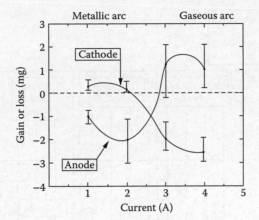

FIGURE 10.30
Net erosion characteristics of a metal vapor arc and the effect of the transition into a gaseous arc [63].

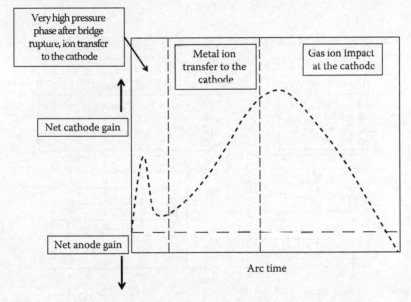

FIGURE 10.31
Generalized net cathode gain for contacts switching a short duration, low-current DC arc [5].

increases and the arc forms the gaseous phase, the cathode loses material and the anode gains. This continues until at the higher currents there is intense removal of metal from both the anode and the cathode. The metal transfer is also dependent upon the contact material, the contact opening speed and the finished cleanliness of the contacts. Figure 10.33 shows the effect of these parameters for one set of experimental conditions for Ag–SnO$_2$ powder metal contacts and for Ag–SnO$_2$–InO$_2$ internally oxidized contacts [65].The transfer process is made more complicated if organic vapor is present, because surface activation tends to make the contacts more cathodic. Activation also tends to cause the arc roots on the contacts to jump to the carbon sites, allowing more uniform erosion over the contact surfaces and thus minimizing the effects of formation of pips and craters. Holmes and Slade [66] achieved a similar effect with silver contacts by adding a small percentage of tungsten.

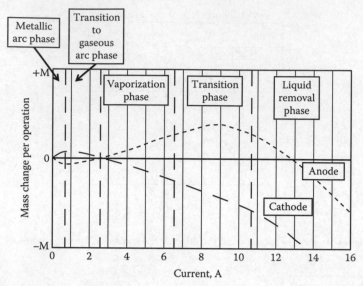

FIGURE 10.32
Mass transfer between contacts opening a 60 VDC circuit as a function of current i.e. arcing time [64].

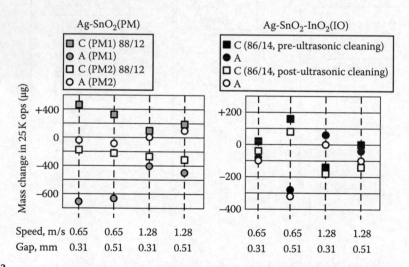

FIGURE 10.33
Comparison of material transfer of silver metal oxide contacts made by different processes and compositions opening a 20 A, 13.5 V, 0.44 mH circuit as a function of two contact opening speeds and two final contact gaps [65].

10.3.6 Erosion of Contacts in Low-Current AC Circuits

In Section 10.2.1, I stated that the average arcing time for contacts opening in AC circuits is $t_b = t_h/2$, where t_h is one half-cycle of current. The arcing time in AC circuits is not usually dependent upon the arc voltage, because most AC switches operate at voltages greater than 100 V (RMS). For some switches, the arcing time may depend upon the point on the AC wave at which the contacts open and the time from there to the first current zero when the circuit is interrupted. As shown in Section 10.3.3, at high currents there is considerable erosion from both contacts and neither has a net material transfer from the opposing contact. At lower currents, the discussion in Section 10.3.5 may have some relevance. Sharkh et al. [67] have studied the erosion of Ag–CdO contacts for determining the wave point at switching for one half-cycle of a 50-Hz unidirectional current. Their data, however, is taken with very slow contact opening of 0.1ms^{-1}. The change in mass of the cathode and the anode contact is measured after a few thousand opening operations for contacts opening 1 ms, 3 ms, 5 ms and 7 ms after the initiation of the AC current. This results in a contact spacing at the first current zero of 0.9 mm, 0.7 mm, 0.5 mm, and 0.3 mm, respectively. Care should be taken in applying this data for general switching in AC circuits, because most AC switches open with faster opening speed (up to 1ms^{-1}) and have larger contact gaps at current zero. The slow opening speed and the small final contact gap is most probably required to obtain consistent erosion data as Slade et al. found when measuring contact erosion opening contacts at 880 A (RMS) [68]. Figure 10.34 does show a marked dependence of the change in mass of the anode and cathode as a function of the point on the wave at which the contacts open. In an AC half-cycle, the current goes from zero to zero in a sinusoidal wave, so the point at which the contacts open not only determines the arcing time, but also determines the current level. The data in Figure 10.34 shows a general loss of metal from the cathode and a general anode gain up to 5 ms of arcing. For longer arcing times, the anode gain and the cathode loss are both reduced. In general, most AC switches have a random opening. One exception would be a switch that is opened by an AC magnetic coil, such as a motor contactor (see Chapter 14). Here, the magnetic force on the actuating armature would begin to open or release the contacts at a set position on the AC current passing through the coil. Shea [69] shows that contact erosion in a three phase vacuum contactor will be completely different from one phase to the next for a point on wave opening, but will be uniform across all three phases for a random opening (see also Section 13.6.4).

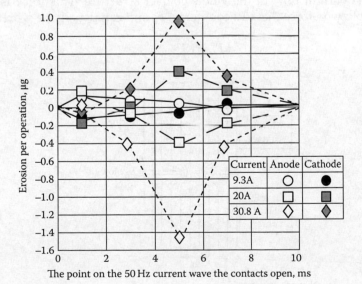

FIGURE 10.34
Effect on point-on-wave on erosion for AgCdO contacts, 240 V (RMS), opening at 0,1 ms^{-1}, for three currents, 9.3 A, 20 A, and 30 A [67].

10.4 Blow-Off Force

From Chapter 1 we know that the current flow through a pair of contacts is constricted to flow through a very small area. This is illustrated in Figure 10.35. The flow of the current along the contact surfaces gives rise to a force F_B that tends to blow the contacts apart. This force F_B is given by [70,71].

$$F_B = \frac{\mu_0 I^2}{4\pi} \ln \frac{R}{a} \text{ or}$$

$$F_B = 10^{-7} I^2 \ln \frac{R}{a} \tag{10.17}$$

where μ_0 is the magnetic permeability in vacuum ($\mu_0 = 4\pi \times 10^{-7}$ VsA^{-1}m^{-1} ≈ 1.25663706 × 10^{-6} Hm^{-1} or NA^{-2} or TmA^{-1} or WbA^{-1}m^{-1}), I is the instantaneous current, R is the measured contact radius, and a is the average radius of the constriction. Using Equation 10.17 and noting that $\ln[R/a] = \frac{1}{2}\ln[R/a]^2$

$$F_B = \frac{\mu_0 I^2}{8\pi} \ln \frac{HA}{F} \tag{10.18}$$

where A is the total area of the contact face. The flaw in using Equation 10.18 is that as the blow-off force on the contacts increases, the contact force F decreases. Barkan [72] has considered this problem. Given that the total force on the contacts F is

$$F = F_S + F_A - F_B \tag{10.19}$$

where F_S represents the spring force holding the contacts together and F_A represents a force resulting from the current flow in the whole contact structure. This force is very often arranged to be a blow-on force that can counteract F_B and can usually be represented by

$$F_A = KI^2 \tag{10.20}$$

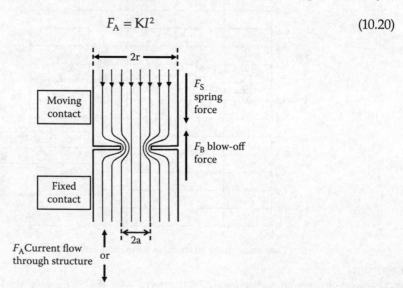

FIGURE 10.35
Contact blow-off force from the passage of current through the contact spot.

where K is a constant. Substituting for F in Equation 10.18 from 10.19 yields

$$F_B = \frac{\mu_0 I^2}{8\pi} \ln \frac{HA}{F_S + F_A - F_B} \tag{10.21}$$

F_B appears in both sides of this equation and the equation cannot be solved explicitly for F_B. Barken put Equation 10.21 into the form

$$\exp\left\{-\frac{8\pi F_B}{\mu_0 I^2}\right\} = \frac{F_S + F_A - F_B}{HA} \tag{10.22}$$

and showed that if the right-hand side $= \alpha_1$ and the left-hand side $= \beta_1$, a solution of (10.22) can be found when

$$\alpha_1 = \beta_1$$

and

$$\frac{d\alpha_1}{dF_B} = \frac{d\beta_1}{dF_B}$$

The second expression gives

$$F_B = \frac{\mu_0 I^2}{8\pi} \left\{ \ln \frac{8\pi HA}{\mu_0 I^2} \right\} \tag{10.23}$$

Substituting for F_B in Equation 10.23 into 10.22 gives the minimum spring force required:

$$F_{Smin} = \frac{\mu_0 I^2}{8\pi} \left\{ 1 + \ln \frac{8\pi HA}{\mu_0 I^2} - \frac{8\pi}{\mu_0 I^2} F_A \right\} \tag{10.24}$$

10.4.1 Butt Contacts

Let us take the contact shown in Figure 10.35. Here $F_A = 0$, so Equation 10.24 reduces to

$$F_{Smin} = \frac{\mu_0 I^2}{8\pi} \left\{ 1 + \ln \frac{8\pi HA}{\mu_0 I^2} \right\} \tag{10.25}$$

Figure 10.36 shows data for the contact force F_{Smin} required to balance the blow-off force F_B using butt contacts in high vacuum. It can be seen that Equation 10.25 correlates well with the experimental data. The curve for Equation 10.25 shown in Figure 10.36 can be represented (within the experimental error) by

$$F_S = 4.45 \times 10^{-7} I^2 \, \text{N} \tag{10.26}$$

High currents such as these are encountered by vacuum interrupters in vacuum circuit breakers, because the testing standards require them to withstand (i.e., their single contact pair stay closed) while passing the full fault current for up to 3 s. It is, therefore, critical to have a contact force in excess of F_{Smin} not only to keep the contacts closed, but also to provide a high enough force to maintain a low contact resistance. If the contact resistance increases too much, the contact region may melt and the contacts will weld [73]; see Section 10.5.1. The increased application of photovoltaic solar panels and the development of hybrid and

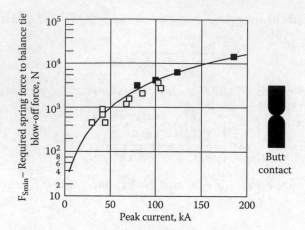

FIGURE 10.36
Minimum spring force that counters the blow-off force from the passage of high current through the contact spot [72].

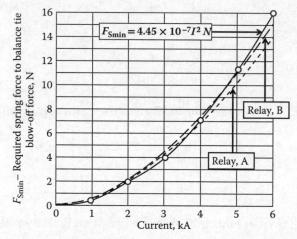

FIGURE 10.37
Minimum spring force that counters the blow-off force from the passage of currents in the range 1 kA–6 kA through the contact spot [74].

electric automobiles has not only raised the interest in switching DC currents of a few hundred amperes and circuit voltages of a few hundred volts, but these switches and relays and their connections now have to withstand fault currents of a few thousand amperes without welding [74]. Figure 10.37 shows the F_{Smin} for two relays with different contact materials: one with Cu contacts and the second with a harder Ag–Cu–Cr material. The small difference in the values would be expected from the difference in the hardness of the two contact materials (see Equation 10.25). It is interesting to note Equation 10.26 is still a good approximation for these lower currents to the values calculated from Equation 10.25.

If the contacts are permitted to blow open, the blow-off force will rapidly be reduced as the contacts begin to open, and other forces resulting from the formation of the arc will add to the opening force. Figure 10.38 illustrates the forces involved. Once the contacts begin to open, the contact area decreases, the region of contact melts and a molten metal bridge forms; see Section 9.4.2. If the bridge ruptures a very high pressure metal vapor region is formed between the contacts [75], which will provide sufficient force to continue the opening of the contacts. Even if the bridge does not form, a dense high pressure region can still exist between the contacts [76]. This is sometimes termed a floating arc [77]. As the contacts continue to open, an arc can form between them. The pressure from this arc can continue the contact parting [76, 78–81]. Figure 10.39 illustrates these effects [76]. In this example a high current is passed through the bridge contact shown. The spring force on the closed contacts is about 1.4 N and when they are open, it is about 1.55 N. The effect of the passage of the high current is shown in six stages: (a) The contacts are closed and the current begins to flow across them. (a)–(b) The blow-off force initially parts the contacts, and begins to diminish when the molten metal bridge forms. (b) The bridge ruptures and the high pressure metal vapor between the contacts forces one side of the bridge contact

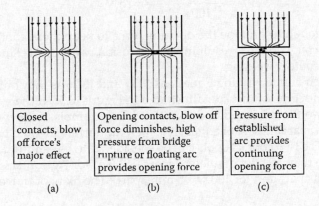

Closed contacts, blow off force's major effect	Opening contacts, blow off force diminishes, high pressure from bridge rupture or floating arc provides opening force	Pressure from established arc provides continuing opening force
(a)	(b)	(c)

FIGURE 10.38
The forces resulting from the passage of high current through a closed contact: (a) the blow off force resulting from the passage of current through one contact region, (b) the continued opening force resulting from the formation of a high pressure metal vapor volume between the contacts in the region of the contact, (c) extension of the opening force provided by the arc which forms between the contacts.

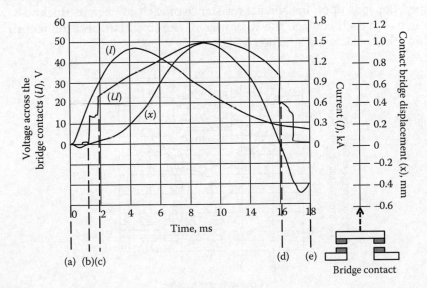

FIGURE 10.39
Contact blow open showing the voltage U across the contacts as the contacts part 'x' under the influence of the high current I, [76].

to open initiating a metallic phase arc. (c) The second contact opens forming a second series arc. (c)–(d) The pressure from these arcs continues to force the contact apart. (d)–(e) The spring force on the contacts overcomes the effect of the lower current arc and the contacts close. The repulsion force from the arc for closely spaced contacts can be roughly estimated using the analysis presented by Holmes et al. [66]. At the initial formation of the arc in Figure 10.39, (b)–(c), the arc voltage $\approx$15V, the average current $\approx$500A and its duration $\approx$1ms, which gives the average arc energy as $15 \times 500 \times 10^{-3} = 7.5$ J. If we conservatively suppose that 10% of this energy goes into heating and expanding the ambient gas and that 1% of this gas expansion is confined by the inertia of the cold gas, then as the contacts move about 100 µm in this time the average force on the contacts is:

$$F = \frac{7.5 \times 0.1 \times 0.01}{100 \times 10^{-6}} = 75\,\text{N}$$

This is certainly enough to provide the opening force to keep these contacts open. This contact blow opening may not only result in severe erosion of the contact material, but also may result in severe contact welding.

Care should be taken in interpreting Equations 10.17 and 10.26 as they only strictly apply to the butt contacts with one contact spot illustrated in Figure 10.35. If there are two contact spots, the initial blow-off force would be reduced [82]. For example, if the two contact spots are placed far enough apart and are of equal area, then the current I would be split and each spot would pass $\frac{1}{2}I$. The two contact spots can be considered to be semi-independent when they are spaced more than 10 spot diameters apart [83] (i.e., for Ag contacts with a contact force of 10 N, ~1.6 mm apart and for 100 N, ~5 mm apart). The blow-off force would be:

$$F_B = 4.45 \times 10^{-7}\{I/2\}^2 + 4.45 \times 10^{-7}\{I/2\}^2 = 4.45 \times 10^{-7}\{I^2/2\}$$

Some high-current switches use multiple current paths in parallel to lower the effect of the blow-off forces at each contact. A discussion of these multiple paths can be found in references [72,82,84]. Figure 10.40 shows the current distribution for ten-finger contacts. The distribution depends upon the contact resistance of each contact, its thickness, width, the distance between each contact, the thickness of the fixed bar and the length of the fingers to which the contacts are attached [85].

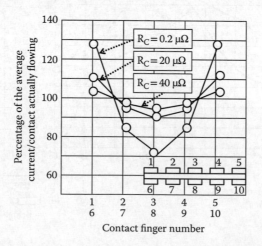

FIGURE 10.40
The current distribution of the average current in 10 contacts arranged in a knife-edge structure [85].

10.5 Contact Welding

All contacts weld or stick to some extent. It is only when the strength of the weld prevents the proper operation of the contacts that it presents a serious problem. The formation of a weld is a complex function of the circuit current, depending on factors like whether or not arcing occurs, the contact material, the structure of the contact surface, the structure and nature of tarnish films, and the design of the structure in which the contacts operate. The weld force F_w is given by

$$F_W = \Gamma A_W \qquad (10.27)$$

where Γ is the tensile strength of the material and A_w is the area of the weld. Now as $\Gamma \approx H/3$, where H is the hardness, then

$$F_W = H A_W / 3 \qquad (10.28)$$

The tensile strengths or breaking strength of a material is not an exact number, and is usually given as a range (see, for example, Table 10.4).

The elastic limit is always exceeded before the breaking stress is reached. The process of drawing a material into a wire increases its strength: in fact, the finer the wire, the greater its breaking stress. Cold-working generally tends to increase the breaking stress of the material. So even with simple equations like Equation 10.27 and Equation 10.28 you can see for any given contact system there will be a rather large variation in the measured weld force F_w.

Contact welding can occur if a high enough current passes through closed contacts and causes the contact spot to melt. Welding can also occur after an arc is initiated between contacts as they close. This arc can result from the electrical breakdown of the closing contact gap (Section 9.4.1) and it also can be continued by the contacts bouncing open once they have initially touched (see Figure 10.14).

10.5.1 Welding of Closed Contacts

If we assume that welding occurs when the contact area reaches the melting temperature T_m, then using Equation 9.41 the voltage across the closed contacts when they weld will be:

$$V_m = \left(10^{-7} \{ T_m^2 - T_0^2 \} \right)^{1/2} \qquad (10.29)$$

TABLE 10.4

Hardness and Tensile Strength

Metal	Hardness (10^8N m^{-2})	Tensile Strength (10^8 Nm^{-2})
Cu (cast)	4–7	1.2–1.7
Cu (rolled)	4–7	2.0–4.0
Cu wire (hard drawn)	4–7	4.0–4.6
Cu wire (annealed)	4–7	2.8–3.1
Ag wire	3–7	2.3–3.5
W wire	12–40	15–35

$$V_m = I_{weld} \times R_c \tag{10.30}$$

$$I_{weld} = \frac{1}{R_c}\left\{10^{-7}\sqrt{T_m^2 - T_0^2}\right\} \tag{10.31}$$

$$I_{weld} = \frac{1}{9\rho}\sqrt{\frac{F}{H}}\left\{10^{-7}\sqrt{T_m^2 - T_0^2}\right\} \tag{10.32}$$

if ρ is in Ωcm, H in N mm^{-2}, F in N. Equation 10.32 can be further refined by taking into account how the physical constants change as a function of the temperature.

For example, by combining straightforward contact equations with experimental data for the contact blow-off force (Equation 10.26) and contact material hardness and resistivity data as a function if temperature, it is possible to develop an easily usable equation for the threshold welding current, I_{weld}, for a single region of contact as a function of the applied contact force, F_s. (where the total contact force is $F_t = F_s - F_B$) for a single region of contact and for a current pulse of a few milliseconds: e.g. an ac current half cycle [86], see Eq. 10.33

$$I_{weld} = \frac{2U_m\sqrt{F_s}}{\left[\left\{\rho_0\left[1+\frac{2}{3}\alpha(T_1-T_0)\right]\right\}^2 \pi(0.1H_0) + 4.45\times10^{-7}\times4U_m^2\right]^{\frac{1}{2}}} \tag{10.33}$$

Where I_{weld}, is in amperes, F_s in Newtons, U_m is the contact material's melting voltage (See Table 24.1A), ρ_0 the resistivity in Ω.mm and H_0 is the contact material's hardness in N.mm^{-2} at temperature T_0. Also T_1 is a temperature close to, but less than the contact material's melting temperature T_m and α is the materials temperature coefficient of resistivity. Using Equation 10.33 together with values for $\rho_0 H_0 U_m$ and α found in contact material properties (See Tables 24.1A&C) values of I_{weld} compare well with experimental values given by in Figure 10.41.

I have found the following expression gives a quick way of making a conservative estimate for the required extra force needed to keep butt contacts together and still prevent them from welding.

$$F_S = 8\times10^{-5}I^{1.54}\,\text{N} \tag{10.34}$$

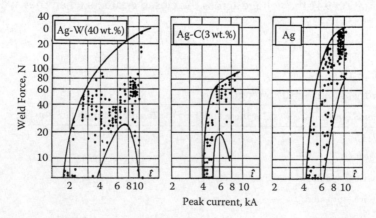

FIGURE 10.41
Welding of closed contacts showing the effect of the threshold current and the variability of the weld force for a given current [87].

Figure 10.41 illustrates the threshold current effect for a number of different contact materials [87]. Figure 10.41 also illustrates the extreme variability of the weld strength at a given current. For this reason, researchers frequently plot their weld data as a cumulative probability for a given set of circuit and contact conditions (see, for example, Figure 10.42 [88,89].

It is possible to obtain an estimate of the maximum weld force by making some simple assumptions [90]. Let us consider Figure 10.43; the volume of metal in the contact region that is melted is assumed to be a sphere whose radius a is the radius of the weld in area A_w. It is also assumed that the energy from the welding current is used for adiabatic heating of this spherical melted region. The energy W_c to melt the contact region is

$$W_C = m\left(C_V\{T_m - T_b\} + C_L\right) \tag{10.35}$$

where m is the mass of material melted, C_V the specific heat, T_m the melting temperature, T_b the initial temperature, and C_L the latent heat of fusion. Now if δ is the material density, then:

$$m = \tfrac{4}{3}\{\pi a^3 \delta\}$$
$$W_C = \tfrac{4}{3}\{\pi a^3 \delta\}\left(C_V\{T_m - T_b\} + C_L\right) \tag{10.36}$$

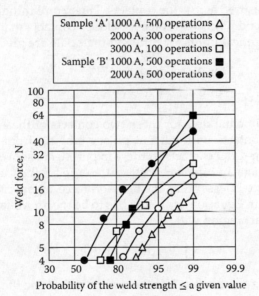

FIGURE 10.42
Cumulative weld strength of an Ag–W contact pair [88].

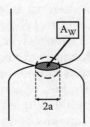

FIGURE 10.43
The weld area and the assumed spherical molten region.

Now if the weld area is assumed to be the same as the cross-section of the molten sphere, then,

$$A_W = \pi a^2 \tag{10.37}$$

The weld force F_w is

$$F_W = \Gamma \pi a^2 \tag{10.38}$$

Using Equations 10.36 and 10.38, eliminating a,

$$F_W = K W_C^{2/3} \tag{10.39}$$

where

$$K = \Gamma \pi \left[\frac{3}{4\pi\delta[C_V(T_m - T_b) + C_L]} \right]^{2/3} \tag{10.40}$$

If all the energy is used to melt the contact spot, then:

$$W_C = \int V_C I dt \tag{10.41}$$

where V_C is the voltage measured across the contacts. Thus, combining Equation 10.39 with Equation 10.41 gives the maximum weld force for closed contacts that would be expected to occur. Examples of theoretical limits for different materials are given below:

$$\text{for Ag} - \text{CdO:} F_W = 67W^{2/3}$$
$$\text{for Cu:} F_W = 107W^{2/3}$$

Figure 10.44 gives experimental data for these two contacts of these two materials. The experiment used closed contacts with current pulses ranging from less than 30 kA to 180 kA. The data for the Ag–CdO contacts in these experiments shows remarkable agreement with the theoretical data, but there is a 50% difference for the Cu contact material.

If the contact holding force is lesser than the blow-off force, an arc can form between the contacts. The voltage used in Equation 10.41 now has to be the arc voltage U_A, and the total energy input into the contact region will be

$$W_C = \int_{t_c} V_C I dt + \int_{t_a} U_A I dt \tag{10.42}$$

The time t_c is the time the contacts are together and t_a is the time the contacts are arcing.

10.5.2 Welding During Contact Closure

As discussed in Section 9.4.1, it is possible to form an arc as contacts close. Once the arc has formed, it can cause melting at the contacts. When they touch, the molten spots will freeze and a weld can form. Fortunately for most switching devices, this "pre-strike" arc is of very short duration and the welds that form are usually quite weak. The problem of welding becomes more severe if the contacts bounce open once they have initially made contact (See also Section 18.4.3). An illustration of contact closing showing a pre-strike arc and contact bounce is shown in Figure 10.14 [91]. As the contacts close a pre-strike arc can occur. Once the contacts touch, they can now bounce open. The amount of bouncing depends upon the contact material, the design of the switch and is a function of the energy in the moving contact when it makes contact. At the end even though the contacts do not

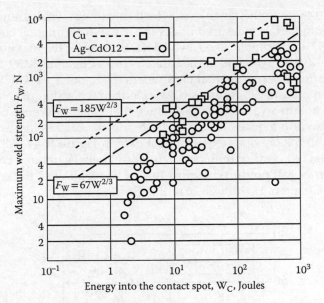

FIGURE 10.44
Weld force as a function of energy into the closed contact spot showing the maximum weld force expected for Cu and Ag–CdO(12 wt.%) contacts for high current pulses >30 kA–180 kA [92].

bounce completely open, the contact force can still be low enough for contact melting to occur before the contacts are completely closed.

In order to obtain an estimate of the maximum weld strength that can be observed it is again possible to use Equation 10.39, that is,

$$F_W = KW_C^{2/3} \tag{10.43}$$

For the case of arcing between two very closely spaced contacts, an estimate for W_c is now

$$W_C = \int_{t_a} U_A I \, dt \tag{10.44}$$

where U_A is the arc voltage and t_a is the total arcing time.

Values of F_w lesser than those calculated by Equation 10.43 are frequently observed and can be explained by a combination of a number of possible physical effects. For example, if the arc roots on the contacts move, the melted spots may not be exactly opposite to each other when the contacts close, thus reducing A_W. If the arc duration t_a is large enough, the heating of the contact region may not be adiabatic and the heat balance described in Section 10.3 comes into effect. If the arc is very long, not all the arc energy goes into heating the contact spots. The bounce time of the contacts can be complex with the contacts opening and closing a number of times during one closing operation. Thus, the exact value of t_a that affects the final melt zone is not easy to determine. Finally, the contact surface itself can be different from the bulk metal (see Section 10.6). For example, if the contact is an alloy, the composition of the surface after arc erosion may be different from the bulk contact. Thus, the value of Γ may vary considerably. Also, the surface after heating may contain inclusions of oxides or other compounds which will also change the tensile strength of the weld.

Figure 10.45 shows welding data for Ag–CdO contacts for currents ranging from 20 A to 1,200 A using a single, controlled bounce time of 0.5 ms to 5 ms [92]. Again, there is a wide range of values for weld strengths, but the maximum weld force measured gives

$F_W = KW^{2/3}$. Figure 10.46 gives the range over which maximum weld strengths vary for a wide range of contact materials measured under the same experimental conditions as the data in Figure 10.45 [92,93]. Except for the Ag–Ni data, the line for the maximum weld force here is close to $F_W \approx K(W_C)^{2/3}$ which is in the same form as Equation 10.39. Now, however, using a completely different experimental parameters the data for Ag-CdO gives the constant K to be between 5.2 and 7.6, whereas if we calculate a K value from Equation 10.40 for this material, we obtain a value of 67 [90]. Thus, there is a difference of a factor of 10 between the data shown in Figures 10.45 and 10.46 and the calculation using the model for closed contacts, and the previous experimental data in Figure 10.44 [90]. In an attempt to obtain a comparison of how different materials would behave under a constant bounce time, Turner and Turner [94] produced Figure 10.47. The welding of the contacts will be discussed further in Chapters 11 through 14 and in Section 18.4.3.

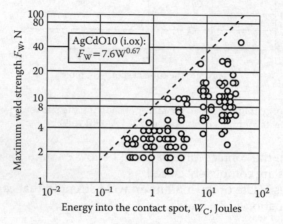

FIGURE 10.45
Weld force as a function of arc energy for a single controlled bounce, showing the maximum weld force for Ag–CdO (10 wt%) internally oxidized (i.ox) contacts [92,93].

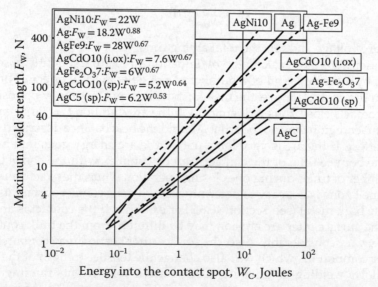

FIGURE 10.46
Maximum weld force for various metals subjected to a single controlled bounce [92,93].

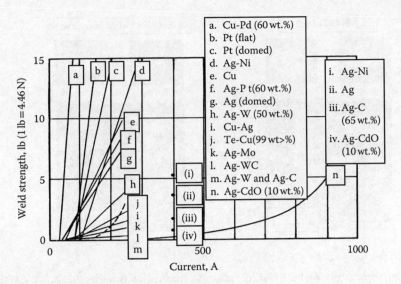

FIGURE 10.47
Comparison of weld force for various contacts of diameter 0.5 cm and with a closed contact force of 3 lb (13.4 N) subjected to a single controlled bounce of 3.8 ms [94].

10.5.3 Welding as Contacts Open

It is possible to observe contact welding as contacts open. This is an unusual event, but has been known to occur. It can result from a poor mechanical design for the switch that allows contact "chattering" as they open: that is, there can be contact sliding or repeated open and close operations before the contacts finally part. It can also occur from bridging of the contact gap by molten particles from the rupture of the molten metal bridge, if the contacts open too slowly (i.e., less than $0.1\,\text{ms}^{-1}$). Zhao et al. [95] have demonstrated the effect of welding as contacts open for load currents of 50 A and 60 A in a 28-V DC circuit. They show the occurrence of such welds increases after a few thousand open and close operations.

10.6 Changes in the Contact Surface as a Result of Arcing

Each arcing event will change the structure of the contact surfaces. We have already discussed contact erosion which is usually the most visible and disruptive effect of arcing. However, other effects can also be identified. The arc melts the contact surface and when the surface solidifies, its make-up can change. This is well illustrated for the arc in atmospheric air for $Ag–SnO_2$ and $Ag–SnO_2–InO_2$ contacts; see Figure 10.48 [96]. This figure shows the surface and the internal structure of the two contacts after they have interrupted the current five times. The surface of the $Ag–SnO_2$ contacts shows gas voids and a pronounced segregation of the Ag from the SnO_2. The surface of the $Ag–SnO_2–InO_2$ contacts again exhibits a porous surface, but now the oxides are still embedded in the Ag matrix. Similar changes in the surface structure of other contacts after arcing in air have also been seen: Ag–Cd–O is a good example [97].

For contacts opening in air, the possible changes in the surface structure of the contacts can become quite complex. There are many possible interactions between the arc, the contact surface and the constituents of the air. Figure 10.49 illustrates some of the possible interactions. In "clean" air, the hot contact surface and the air can form oxides, nitrides and

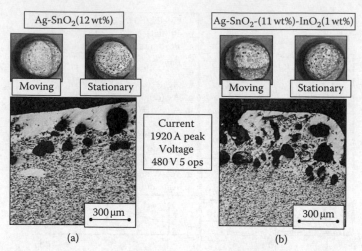

FIGURE 10.48

Change of the surface structure of Ag-SnO$_2$ and Ag-SnO$_2$-InO$_2$ contacts interrupting 1920 A, 400 V AC in air [96].

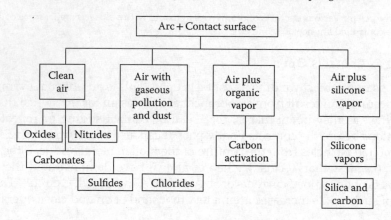

FIGURE 10.49

Possible interactions of electric contacts and the ambient air during arcing.

carbonates (from CO$_2$). If the air has industrial pollution such as SO$_2$, H$_2$O, chlorine compounds and dust, it is possible to form sulfides and chlorides. Organic vapor, as we have already seen (Section 10.2.3) can result in activation of the surface. Finally, if vapor from high-temperature silicone oils, grease and sealing compounds interacts with the arc, it is possible to form silica (SiO$_2$) and C on the contact surfaces.

These surface interactions generally result in a change in the contact resistance. Possible changes in contact resistance R_c are illustrated in Figure 10.50. This figure shows how complex the changes can be. Depending upon the actual mechanical operating system, the ambient atmosphere, the contact material and the arc characteristics, it is possible for R_c to increase, decrease, or remain unchanged. For example, oxide films formed by arcing during interruption of the current can be ruptured mechanically when the contacts close when sufficient force is applied and if there is a sliding action after the contacts touch. This is what high-power molded case circuit breakers rely upon when they use Ag–W as a contact material (see Chapter 14). Contact erosion also can drastically change the R_c. For example, it is possible to erode so much of the contact surface during each arcing operation that a new contact surface of virgin metal is exposed each time, and will thus maintain

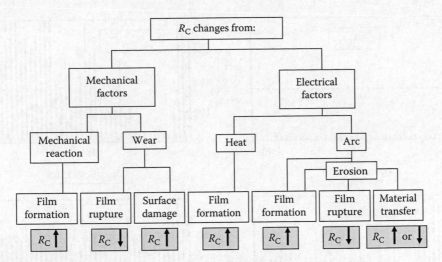

FIGURE 10.50
Possible changes in constant resistance, R_c.

a low R_c. Once a surface film has formed on the contact and the R_c increases, even a low current flowing through the closed contact can result in an increased temperature at the contact spots and this, in turn, can result in additional film formation.

10.6.1 Silver-Based Contacts

For pure silver contacts and silver metal oxide contacts (Ag–MeO), such as, Ag–CdO and Ag–SnO$_2$, the effects shown in Figure 10.50 depend greatly upon the power level (i.e., current and voltage) of the electric circuit and the forces exerted by the switch mechanism. For most devices operating at 50 A and greater, the closed contact force and the contact wipe on closing is usually enough to rupture any surface oxides and sulfides that may form. The major concern for switch designers for devices that operate in this power range are contact erosion (i.e., operating life) and contact welding.

At lower currents, where the contact forces are lower, it is possible to form surface films that result in high R_c. Figure 10.51 shows an example of changes in R_c for silver contacts switching a low current in N$_2$ and N$_2$ plus O$_2$ [98]. These authors attributed these variations in R_c to the formation of an oxide layer. In an interesting set of experiments comparing filtered laboratory air and artificial air (N$_2$ and O$_2$ mixture), Witter and Polevoy [99] have shown that it is possible to form silver carbonate on the contact surface from the CO$_2$ present in air. By varying the arc duration by alternatively switching an inductive and resistive load, these authors showed that at long arc times where a gaseous arc predominates, the probability that carbon and oxygen in the arc will react with either the small amount of silver vapor in the arc or with the hot silver in the arc root region to form AgCO$_2$ is high. If the arc is of short duration, however, and the metallic arc dominates, the presence of greater density of silver vapor in the arc limits the formation of AgCO$_2$ and, indeed, the arc roots also evaporate any carbonate that has already formed on the contact surface (see Figure 10.52) (see also Section 16.2).

10.6.2 Silver-Refractory Metal Contacts

The advantages and disadvantages of silver–refractory contacts are illustrated in Figure 10.53 [100]. It is generally agreed this class of material is more resistant to arc erosion than all other contact materials and the brittle nature of the refractory metal and

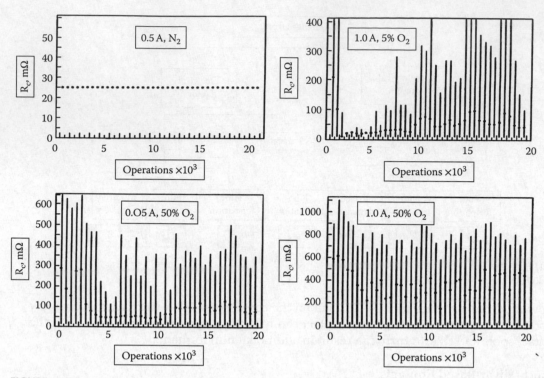

FIGURE 10.51

Example of contact resistance variations for silver contacts operating in nitrogen and nitrogen containing oxygen as a function of the number of switching operations [98].

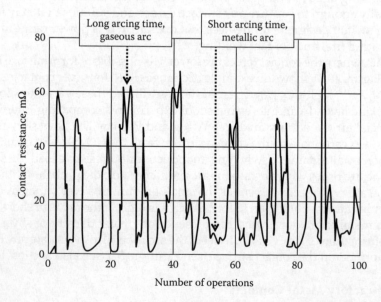

FIGURE 10.52

Contact resistance changes of closed silver contacts resulting from the formation and evaporation of $AgCO_2$ films after long arc times and short arc times [99].

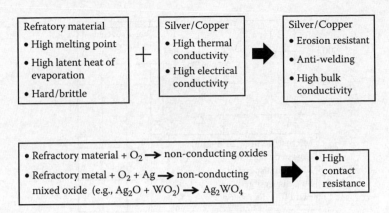

FIGURE 10.53
The silver-refractory metal contact system.

the metal matrix structure makes it weld-resistant. This mixture is, therefore, the contact of choice for those devices that have to interrupt very high short-circuit currents. The most common contact mixture is a mixture of silver and tungsten. The advantages of this class of contact material, the erosion and weld resistance, however, have to be balanced against their tendency for formation of oxides and the resulting high R_c. When W–Ag materials are switched in air, tungsten oxides and silver tungstates are formed on the contact surfaces [101–103]. Their presence is usually observed by measuring an increase in R_c. The probability of formation of oxides in the region of contact, however, can vary widely from one operation of the contact to another [103,104], and it also depends strongly upon the contact force and upon the mechanical wiping action in the switching system studied. It is not surprising then that quantitative R_c values from one researcher to another can be quite different. Witter and Abele [105] observed a monotonic decrease in R_c after 4,000 operations at 20 A as the silver content of the tungsten–silver contacts increased (see Figure 10.54). They also showed the finer the starting powders, the higher the R_c (see Figure 10.54). Leung et al. [106] also showed a decrease in R_c values for tungsten–silver contacts, but little effect of manufacturing method was evident (see also Figure 10.54). Lindmayer and Roth [104] showed how the parameter of contact diameter can markedly affect the measured R_c. When their contact erosion data (see Figure 10.55) is superimposed upon their R_c data (Figure 10.56), it can be seen immediately that there is a correlation between the current at which the onset of the excessive erosion occurs and that at which R_c begins to decrease. The increased erosion permits enough free silver onto the contact surfaces to allow a high probability of silver-to-silver contact and hence a low R_c value. These researchers confirmed the effect of erosion on R_c when they showed at 1,000 A, for 6-mm diameter contacts, W–Ag, which had a higher erosion than WC–Ag, also had a lower R_c. These data contrast with my experiments [107] using contacts with a diameter of 4 mm and a current of 20 A, where excessive erosion was not a concern. I showed no difference between R_c values for W–Ag and WC–Ag. Leung and Kim [108] found lower R_c values at 30 A for WC–Ag contacts than for W–Ag, even though the erosion rate for W–Ag is higher than that for WC–Ag contacts. When Co or Ni is added to W–Ag or WC–Ag, the R_c tends to be somewhat lower [109,110]. This may be the result of a somewhat increased erosion rate. There are other chemical effects of these additives, however, that have been identified [107,108] and which may slow down the formation of surface oxides. On the other hand, Witter [110] showed that when the additive Ni is used in high concentrations, the formation of Ni–W phases and Ni–W oxides gave rise to a rapidly deteriorating R_c.

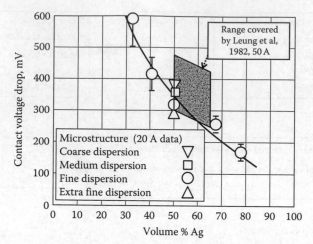

FIGURE 10.54
Contact resistance changes as the Ag content of W–Ag contacts is varied [105,106].

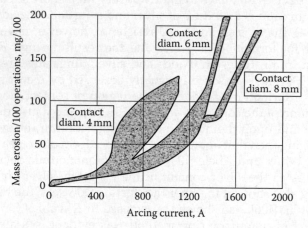

FIGURE 10.55
Erosion of W–Ag contacts and effect of contact diameter [104].

Experiments have been made with the alloy (W, Ti) C in an effort to produce a contact material which is an improvement over W–Ag [111,112]. Contacts made from (W, Ti) C–Ag showed excellent R_c values when switched up to 3,000 operations at 20 A (see Figure 10.57). Doremieux et al. [111] also showed low R_c values for this number of operations, but when they attempted to use this material for contactor applications, they obtained high R_c values after 20,000 operations at 20 A. When these contact switched high short-circuit currents [111] their surfaces were left with a thin, uniform, hard crust that contained mainly (W, Ti) with a high concentration of oxygen, either dissolved in the metal matrix or as suboxides of titanium and tungsten. This layer is practically devoid of silver and high R_c values are observed. The effects of the refractory oxides could be reduced by increasing the silver content (see Figure 10.57) and by allowing the contacts to erode more rapidly under short-circuit arcing, but these contacts had an increased probability of welding [112].

The long-term effects of passage of low current (20 A) through W–Ag contacts has also been investigated [113]. Both the contact's voltage drop V_c and the temperature rise of the stationary contact arm were continuously measured. Four characteristic voltage types were recorded (Figure 10.58): Type 1, a steady voltage less than 100 mV; Type 2, a voltage

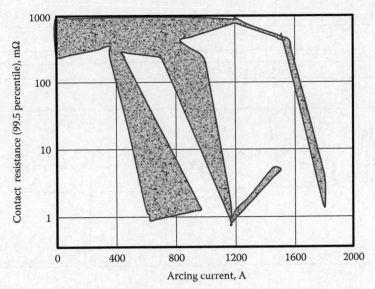

FIGURE 10.56
Effect of contact erosion from Figure 10.54 on the contact resistance of W–Ag contacts [104].

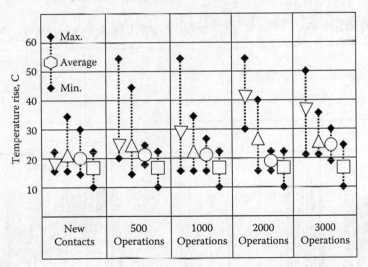

FIGURE 10.57
A comparison of the average, maximum, and minimum temperature rise for (W, Ti) C–Ag (∇, 35 wt% Ag), (Δ, 50 wt% Ag), (o, 65 wt% Ag), and ($\square$ pure Ag) contacts switching 20 A AC as the number of switching operations increases [112].

plateau (range 150–600 mV) with frequent, rapid excursions to lower values; Type 3, rapid voltage oscillations (as much as ±100 mV) superimposed upon a slower varying mean (≈400 mV); and Type 4, extreme Type 3 (peaks as high as 1.3 V) with sudden excursions to a lower plateau (≈300 mV). I analyzed the data in terms of the expected temperature of the contact spot, the formation and decomposition of silver tungstates, and the volatility of W-oxides above 1,000°C. I showed that W–Ag can exhibit a self-limiting contact resistance. In other words, no thermal runaway is observed for the limited experimental conditions used. Further experiments with high silver content in WC–Ag–C contacts showed much lesser V_c activity. The increased silver content and the reducing effect of the carbon may

The Consequences of Arcing

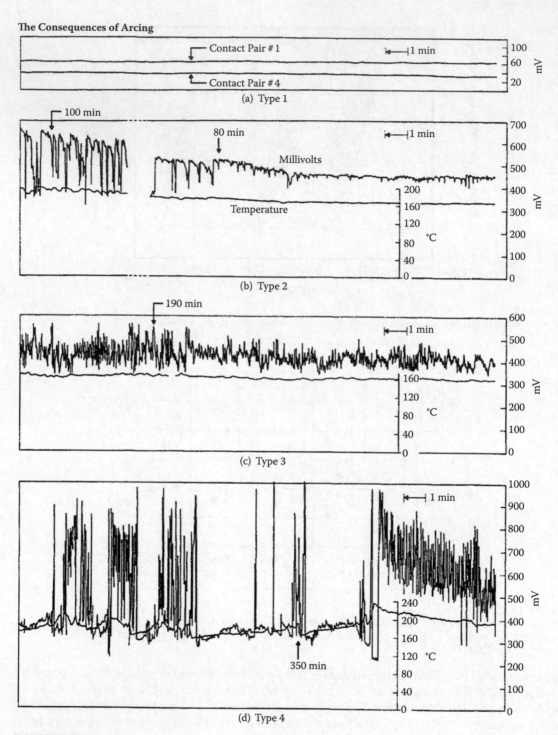

FIGURE 10.58
Possible voltage drop across closed W–Ag contacts reflecting changes in the surface chemistry of the contacts [113].

limit the formation and stability of any silver tungstate or tungsten oxide. The reducing effect of carbon has been demonstrated by Lindmayer and Shroeder [114], who investigated Ag–C (3 wt% C) operated against W–Ag (35 wt% Ag). For an Ag–C cathode the R_c remained less than 0.1 mΩ, but when Ag–C was the anode the R_c was a much higher value but still lower than that of two W–Ag contacts. They attributed this polarity effect to the fact the Ag–C as cathode had much greater erosion than the Ag–C as anode, and hence the reducing atmosphere of C and CO was much greater (see also Section 16.3). Testing of various compositions of Ag–C and Ag–WC–C contacts for anti-welding, arc erosion, mechanical impact strength and interruption, showed that no one material worked best for all categories [115]

10.6.3 Other Ambient Effects on the Arcing Contact Surface: Formation of Silica and Carbon and Contact Activation

There is a well-researched body of knowledge dating back to the 1960s [116] on the deleterious effect of silicone vapors from silicone sealants, greases and oils on switching contacts. When the adsorbed silicone molecules on the contact surfaces are subjected to the electric arc, SiO_2 and C are formed. Contacts switching in such an ambient eventually form a high-resistant surface, which eventually results in the failure of the switch [116–124]. Demethyl silicones can be represented by $((CH_3)_2SiO)_n$ where n is the degree of polymerization [123]. For D_n where $n < 3$ the silicone is highly volatile, but for D_n where $n > 5$ the evaporation rate is extremely low. Tamai [123] experimented with the D_4 compound, a liquid, which gradually evaporates at room temperature. D_4 is an unreactive residual silicone in silicone rubbers, sealants, potting compounds and greases, which gradually is released by them into the surrounding ambient. So, for sealed or even covered switches, it is possible to gradually develop a silicone vapor concentration surrounding the switch's contacts. When coupons of Au and Ag are exposed to D_4 in a sealed chamber, a silicone film gradually forms on them. Figure 10.59 shows this for two concentrations of D_4; 1,300 ppm and 7 ppm [123]. The film reaches a thickness limit of ~1.3 nm for both concentrations: in ~10 hours for

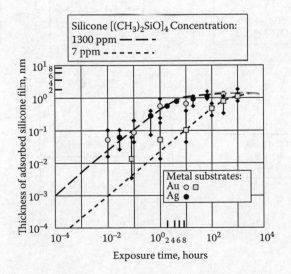

FIGURE 10.59
The growth of film thickness on the surfaces of Au and Ag coupons as a function of time in an air and two concentrations of D_4 silicone ambient [123].

the concentration of 1,300 ppm and ~800 hours for the 7ppm concentration. The process of film formation on the contact surfaces is quite complex. It has been discussed by Tamai for the D_4 compound. As shown in Figure 10.60a, when the D_4 silicone molecule lands on a metal's surface it is adsorbed as a result of its O atom attaching to the dangling bonds on the metal's surface. This new molecule on the metal's surface is most probably a volatile liquid and can easily evaporate back into the ambient environment. Thus, the process shown in Figure 10.60a can go back and forth between the metal surfaces and the ambience. However, as Figure 10.59 shows, a surface layer does eventually form on a metal's surface. Figure 10.60b shows if an O_2 molecule is absorbed by the liquid silicone on the contact

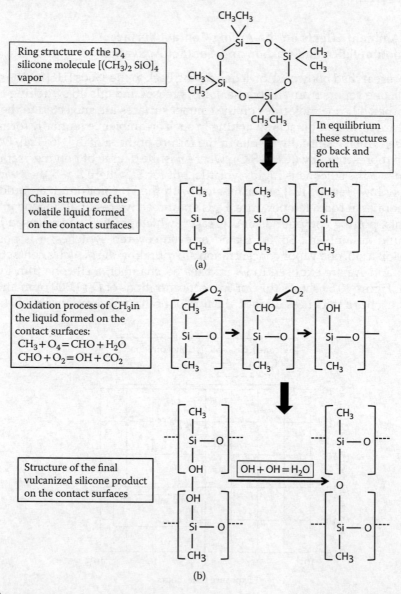

FIGURE 10.60

(a) The equilibrium reaction of D_4 on a metal's surface, (b) The oxidation and the polymerized film formation on the metal's surface. [124].

surfaces it changes the nature of the film. It eventually forms into a polymerized film that no longer evaporates. When this film is subjected to the electric arc, SiO_2 and C are formed which deposit onto the contact surfaces. Discussions on the effects of this phenomenon will also be presented in Chapter 19, Section 19.4. The effects of activation on arc duration and contact erosion have already been discussed in Section 10.2.3. Further discussion of the effects of activation will also be presented in Chapter 19, Section 19.2.

Acknowledgments

I am extremely grateful to Professor Werner Rieder who critically reviewed this chapter for the first edition. His attention to scientific detail, his many suggestions for changes, and his encouragement, improved its substance and breadth. I also wish to thank CH Leung, who gave me the photographs used in Figures 10.23 and 10.47.

References

1. A Greenwood, *Electrical Transients in Power Systems*, 2nd ed, New York: Wiley, 1991.
2. R Garzon, *High-Voltage Circuit Breakers: Design and Application*, New York: Marcel Dekker, 1997.
3. PG Slade, *The Vacuum Interrupter: Theory, Design and Application*, New York: CRC Press, 2008.
4. WB Ittner, Bridge and short arc erosion of copper, silver and palladium contacts on break, *J Appl Phys* 27 (4): 382–388, 1956.
5. PG Slade, Current interruption in low voltage circuits, *IEEE Trans Parts Hybrids Packaging* PHP-5 (1): 56, 1969.
6. A Vassa, E Carvou, S Rivoirard, L Doublet, C Bourda, D Jeannot, P Ramoni, N Ben-Jemaa, Magnetic blowing of break arcs up to 360 VDC, *Proc 56th IEEE Holm Conf Elect Cont* 96–100, 2010.
7. E Gauster, W Rieder, Back-commutation in low voltage interrupters: influence of recovery time, geometry, and materials of contacts and walls, *Proc 43rd IEEE Holm Conf Elect Cont* 91–96, 1997.
8. A Lee, PG Slade, KH Yoon, J Porter, J Vithayathil, The development of a HVDC SF_6 Breaker, *IEEE Trans Power App Syst* PAS-104 (10): 2721–2730, 1985.
9. JG Gorman, CW Kimblin, RE Voshall, RE Wien, PG Slade, The interation of vacuum arcs and magnetic fields with applications, *IEEE Trans Power App Syst* PAS-102 (2): 257–266, 1983.
10. LH Germer, JL Smith, Activation of electrical contacts by organic vapors, *Bell Syst Tech J* 36: 769–812, 1957.
11. http://en.wikipedia.org/wiki/Fullerene
12. EW Gray, TA Uhrig, GF Hohnstreiter, Arc durations as a function of contact metal and exposure to organic contaminants, *J Appl Phys* 48 (1): 1977.
13. EW Gray, How organic vapors cause contact activation, *Insulation/Circuits*, March 1977.
14. S Shimada, Y Takenaka, K Funaki, Effects of various organic vapors on contact resistance, *Proc 17th Int Conf Elect Cont*, Mano R&D Tech Center, Nagoya, Japan, 159–167, July 1994.
15. CN Neufeld, WF Rieder, Carbon contamination of contacts due to organic vapors, *IEEE Trans Components, Packaging Manuf Technol— Part A* 18 (2): 399–404, 1995.
16. HN Wagar, Contact and connection technology, In *Integrated Device and Connection Technology*, Englewood Cliffs: Prentice-Hall, 439–654, 1971.
17. HN Wagar, Predicting the erosion of switching contacts that break inductive loads, *Holm Seminar on Electric Contact Phenomena* 95–102, 1968.

18. K Miyajima, S Nitta, A Mutoh, A proposal on contact surface model of electromagnetic relays—based on the change in showering arc waveforms with the number of contact operations, *IEICE Trans on Elect* E81-C (3): 399–407, 1998.

19. CW Mills, The mechanisms of the showering arc, *IEEE Trans Parts Hybrids Packaging* PHP-5 (1): 47–55, 1969.

20. CW Kimblin, Erosion and ionization in the cathode spot regions of vacuum arcs, *J Appl Phys* 44: 3074–3081, 1973.

21. DT Tuma, CL Chen, DK Davis, Erosion products from the cathode spot region of a copper vacuum arc, *J Appl Phys* 49: 3821–3829, 1978.

22. MB Schulman, JA Bindas, PG Slade, Effective erosion rates for selected contact materials in low voltage contactors, *IEEE Trans Components Packaging Manf Techn—Part A* 18 (2): 329–333, 1995.

23. MB Schulman, PG Slade, L Loud, Influence of contact geometry on effective erosion of Cu-Cr, Ag-WC and Ag-Cr vacuum contactor materials, *IEEE Trans Components Packaging Techn* 22 (3): 405–413, 1999.

24. See for example the many papers on contact erosion published in the Proceedings of the Holm Conference on Electric Contacts 1951 to the present and the International Conference on Electrical Contacts from 1961 to the present. All of these papers now available on a DVD edited by Th, Schoepf available from VDE Verlag GMBH (ISBN 978-3-8007-3365-1). Since the late 1960s a number of these papers have been published in IEEE Trans PHP, IEEE Trans CHMT, and IEEE Trans CPMT.

25. F Pons, An electrical arc erosion model valid for high currents: vaporization and splash erosion, *Proc 54th IEEE Holm Conf Elect Cont* 9–14, 2008.

26. P Borkowski, A computer program for calculation of electrode mass loss under electrical arc conditions, *Proc 23rd Int'l Conf on Elect Cont* 354–360, 2006.

27. HW Turner, C Turner, Contact materials, their properties and uses, Electrical Times, Part 2, August, 277, 1967.

28. E Hetzmasnnseder, W Rieder, The influence of bounce parameters on the make erosion of silver metal-oxide contact materials, *IEEE Trans Components Packaging Manuf Technol—Part A* 17 (1): 8–16, 1994.

29. W Rieder, V Weichsler, Make erosion mechanism of AgCdO and AgSnO$_2$ contacts. *IEEE Trans Components, Hybrids Manuf Technol* 15 (3): 332–338, 1992.

30. HW Turner, C Turner, Discontinuous contact erosion. *Proc 3rd Int'l Research Symp on Elec Cont Phen* 309–320, 1966 and The erosion of heavy current contacts and material transfer produced by arcing. *Proc 4th Int'l Res Symp on Elec Cont Phen* 196–200, 1968.

31. E Walczuk, Arc erosion of high current contacts in the aspect of CAD of switching devices. *Proc 38th IEEE Holm Conf on Elect Cont*, IEEE Cat No 92CH31 25–2, 1–16, October 1992.

32. AM Abdul Assis, A Erk, M Schmeizle, Andert sich der Abbrand von Schaltstucken mit wachsendem Ausschaltstrom sprunghaft? ETZ-A, 94 (H.4): 239–240, 1973.

33. P Borkowsko, E Wulczuk. Temperature rise behind fixed polarity Ag-W contacts opening on a half cycle of high current and its relationship to contact erosion, *Proc 22nd Int'l Conf on Elect Cont* 334–340, 2004.

34. P Borkowski, M Hasegewa. Arc erosion of polarized contacts Ag-W by high current, *IEICE Trans on Elect* E93-C (9): 1416–1423, 2010.

35. K Ioyama, T Yanobe, Estimation for the spot size of short gap discharge in near one atmosphere pressure, *IEICE Trans on Elect* E82-C (1): 55–59, 1999.

36. LH Germer, Physical processes in contact erosion, *J Appl Phys* 29 (7): 1067–1082, 1958.

37. T Schoepf, Electrical contacts in the automotive 42VDC power net, *Proc 21st Int'l Conf on Elect Cont* 43–55, 2002.

38. L Doublet, N Ben-Jemaa, F Hauser, D Jeannot, Electrical arc phenomenon and its interaction on contact materials at 42VDC for automotive applications, *Proc 50th IEEE Holm Conf Elec Cont* 8–14, 2004.

39. T Schoepf, W Rieder, Consequences for automotive relays of a 42VDC network in vehicles, *IEEE Trans Components Packaging Technol* 23 (1): 177–182, 2000.

40. T Klonowski, R Andlauer, T Leblanc, F Faure, R Meyer, P Testé, High intensity contact opening under DC voltages, *Proc 50th IEEE Holm Conf Elec Cont* 459–466, 2004.

41. T Sugiura, J Sekikawa, T Kubono, Break arcs driven by transverse magnetic field in a DC 48V/6-24A circuit, *IEICE Trans on Elect* E94-C (9): 1381–1387, 2011.

42. T Schoepf, Disengaging connectors under 42VDC loads, *Proc 48th IEEE Holm Conf Elec Cont* 120–127, 2002.

43. T Schoepf, Unplugging of connectors under load in 42VDC power networks for vehicles, *VDE-Fachbericht* 57: 11–20, 2001.

44. J Razafiarivelo, M Porte, N Ben-Jemaa, Evaluation of arcing damage on connectors for 42VDC, *Proc 50th IEEE Holm Conf Elec Cont* 15–21, 2004.

45. K Yoshida, K Sawa, K Suzuki, M Watanabe, H Daijima, Influence of voltage and current on arc duration and energy of DC electromagnetic contactor, *Proc 57th IEEE Holm Conf Elec Cont* 17–21, 2011.

46. A Vassa, E Carvou, S. Rivoirard, L Doublet, C. Bourda, N Ben-Jemaa, DC-blowing under pulsed magnetic field, *Proc 26th Int'l Conf on Elec Cont* 23–29, 2012.

47. Y Liu, Z Li, K Wang, R Wang, Experimental study of arc duration under external transverse magnetic field in a dc 580V circuit, *Proc 26th Int'l Conf on Elec Cont* 30–34, 2012.

48. T Shrank, E-D Wilkening, M Kurrat, F Gerdinand, P Meckler, Breaking performance of a circuit breaker influenced by a permanent magnet field at DC voltages up to 450V, *Proc 26th Int'l Conf on Elec Cont* 35–40, 2012.

49. CH Leung, A Lee, Contact erosion in automotive dc relays, *IEEE Trans Components Hybrids Manuf Technol* CHMT-14 (1): 101–108, 1991.

50. R Holm, E Holm, *Electric Contacts: Theory and Applications*. New York: Springer-Verlag, 4th ed. 2000, Chap. 65.

51. F Llewellyn-Jones, *The Physics of Electrical Contacts*. Oxford: Clarendon Press, 1957, Chap. IV.

52. R Timsit, Electro-migration in a liquid metal bridge before contact break, *Proc 56th IEEE Holm Conf Elec Cont* 564–569, 2010.

53. K Hiltmann, A Schumacher, K Guttmann, E Lemp, H Sandmaier, W Lang, New micromechanical membrane switches in silicon technology, *IEEE Trans Components Packaging Technol* 25 (3): 397–401, 2002.

54. JW McBride, L Jiang, C Chianrabutra, Fine transfer in electrical switching contacts using gold coated carbon nano-tubes, *Proc 26th Int'l Conf on Electric Contact Phen* 353–358, 2012.

55. F Llewellyn-Jones, MR Hopkins, CR Jones, Measurement of metal transfer in electrical contacts by radioactive tracer method, *Brit Jour Appl Physics* 12 (9): 485–489, 1961.

56. PJ Boddy, T Utsumi, Fluctuations of arc potential caused by metal vapor diffusion in arcs in air, *J Appl Phys* 42: 3369–3373, 1971.

57. EW Gray, Some spectroscopic observations of the two regions (metallic vapor and gaseous) in break arcs, *IEEE Trans Plasma Science* 1: 30–33, 1973.

58. PG Slade, Opening contacts: the transition from the molten metal bridge to the electric arc, *IEICE Trans on Elect* E93-C (6): 1380–1386, 2010.

59. Z-K Chen, H Mizukoshi, K Sawa, Contact erosion patterns of Pd material in dc breaking arcs, *IEEE Trans Components, Packaging Manuf Technol—Part A* 17 (1): 61–67, 1994.

60. SMS Sharkh, JW McBride, Voltage steps in atmospheric low current arcs between opening silver metal contacts, *Proc 43rd IEEE Holm Conf Elec Cont* 233–237, 1997.

61. N Ben-Jemaa, JL Queflelec, D Travers, Some investigations on slow and fast arc fluctuations for contact materials proceeding in various gases and direct currents, *Proc 36th IEEE Holm Conf Elec Cont* 18–24, 1990.

62. H Hoft, N Vogel, Potential steps during initial stages of short and medium arcs, *Proc 13th Int'l Conf on Elec Cont* 59–63, 1986.

63. Z-K Chen, K Sawa, Effect of arc behavior on material transfer: a review, *Proc 42nd IEEE Holm Conf Elec Cont* 238–251, 1996.

64. J Swingler, JW McBride, A comparison of the erosion and arc characteristics of Ag-CdO and Ag-SnO$_2$ contact materials under DC break conditions, *Proc 41st IEEE Holm Conf Elec Cont* 381–392, 1995.

65. Z-K Chen, G Witter, A comparison of contact erosion for opening velocity variations for 13.5 volt circuits, *Proc 52nd IEEE Holm Conf Elec Cont* 15–20, 2006.
66. FA Holmes, PG Slade, Suppression of pip and crater formation during interruption of alternating currents, *IEEE Trans Components, Hybrids Manuf Technol* CHMT-1: 59–65, 1978.
67. SM Sharkh, JW McBride, A comparison between ac and dc erosion of Ag-CdO contacts, *17th Int'l Conf on Elec Cont* 569–576, 1994.
68. PG Slade, R Kossowsky, R Aspden, R Bratton, The use of ceramic matrices infiltrated with silver for electric contact applications, *IEEE Trans Parts, Hybrids Packaging* PHP-10 (1): 37–42, 1974.
69. J Shea, Modeling contact erosion in three phase vacuum contactors, *IEEE Trans CPMT—Part A* 21 (4): 556–564, 1998.
70. R Holm, E Holm, *Electric Contacts: Theory and Application*. New York: Springer-Verlag, 1967, Chap. 11.
71. AC Snowdon, Studies of electromagnetic forces occurring at electrical contacts, *AIEE Trans* 80: 24–28, 1961.
72. P Barkan, A new formulation of the electromagnetic repulsion phenomenon in electric contacts at very high currents, *Proc 11th Int'l Conf on Elec Cont* 185–188, June 1982.
73. PG Slade, ED Taylor, RE Haskins, Effect of short circuit duration on the welding of closed contacts in vacuum. *Proc 51st IEEE Holm Conf Elec Cont* 69–74, 2005.
74. E Carvou, N Ben-Jemaa, B Mitchell, C Gautherot, J Rivene, L Colchen, Contact behavior of electrical vehicle battery junction box under high shorting and breaking current, *Proc 26th Int'l Conf on Elec Cont* 151–155, 2012.
75. PG Slade, Opening electrical contacts: the transition from the molten metal bridge to the metallic phase, *Proc 24th Int'l Conf on Elec Cont* 1–6, 2008.
76. M Bizjak, S Karin, H Nouri, Influence of vapor pressure on the dynamics of repulsion by contact blow off, *Proc 21st Int'l Conf on Elec Cont* 268–275, 2002.
77. BF Huber, An oscilloscope study of the beginning of a floating arc, *Proc Holm Sem on Elect Cont* (IIT), 141–152, November 1967.
78. JJ Shea, Blow-open forces on double-break contacts, *IEEE Trans Components, Packaging Manuf Technol—Part A* 17 (1): 32–38, 1994.
79. X Zhou, P Theisen, Investigation of arcing effects during contact blow open process, *Proc 44th IEEE Holm Conf Elec Cont* 100–108, 1998.
80. A Krätzcschmar, R Herbst, V Kulinovich, F Nothnagel, F Berger, Gas dynamic processes as the main reason for contact levitation in switching of high operating overload currents, *Proc 24th Int'l Conf on Elec Cont* 317–323, 2008.
81. JP Chaberie, P Testé, R Andlauer, T Leblanc, J Devautour, JP Guerry, Experimental study of contact opening, *Proc 41st IEEE Holm Conf Elec Cont* 194–199, 1995.
82. Y Kawase, H Mori, 3-D element analysis of electrodynamic repulsion forces in stationary electrical contacts taking into account asymmetric shape, *IEEE Trans on Magnetics* 33 (2): 1994–1999, 1997.
83. RD Malucci, The effects of current density variations in power contact interfaces, *Proc 57th IEEE Holm Conf Elec Cont* 55–61, 2011.
84. PG Slade Chapter 15, 586-59 in TE Browne. *Circuit Interruption, Theory and Techniques*. New York and Basel: Marcel Dekker, 1984.
85. Y Yoshioka, Calculation of current distribution in heavy current contacts with many parallel fingers, *Proc 7th Int'l Conf on Elec Cont* 382–388, 1974.
86. PG Slade, The current level to weld closed contacts, *Proc. 59th IEEE Holm Conf. on Electrical Contacts*, 2013.
87. E Walczuk, Welding tendency of closed contacts, *Metall* 4: 381–384, 1977.
88. PP Koren, PG Slade, CY Lin, Welding characteristics of Ag–W contacts under high-current conditions, *IEEE Trans Components Hybrids Manuf Technol* CHMT-3 (1): 50–55, 1980.
89. E Geldner, W Haufe, W Reichel, H Schreiner, Schweisskraft von Reinsilber, Reinkupferund verschiederen Kontakwerkstoffer auf Silberbasis, ETZ-A 93: 216–306, 1972.
90. M Bet, G Souques, Behaviour of electrical contacts under current waves of great intensity. *Proc 10th Int'l Conf on Elec Cont Phen* Budapest, 23–34, 1980.

91. S Kharin, H Nouri, D Amft, Dynamics of arc phenomena at closure of electrical contacts in vacuum circuit breakers, *IEEE Trans Plasma Science* 33: 1576–1581, 2005.

92. E Walczuk, D Boczkowski, Computer controlled investigations of the dynamic welding behavior of contact materials, *Proc IEEE Holm Conf on Elec Cont* 11–16, October 1996.

93. E Walczuk, D Boczkowski, D Wojcik, Electrical properties of Ag-Fe and Ag-Fe$_2$O$_3$ composite contact materials for low voltage switchgear, *Proc 24th Int'l Conf on Elec Cont* 48–54, 2008.

94. HW Turner, C Turner, Contact materials, their properties and uses. *Electrical Times*, Part 3, 313, 1967.

95. L Zhao, Z Li, H Zhang, M Hasegawa, Random occurrence of contact welding in electrical endurance tests, *IEICE Trans on Elec* E94-C (9): 1362–1367, 2011.

96. C Leung, E Streicher, D Fitzgerald, J Cook, High current erosion of Ag-SnO$_2$ contacts and the evaluation of indium effects in oxide properties, *Proc 52nd IEEE Holm Conf Elec Cont* 143–150, 2006.

97. R Kossowsky, PG Slade, Effect of arcing on the microstructure and morphology of Ag–CdO contacts, *IEEE Trans Parts, Hybrids Packaging* PHP-9 (1): 39–44, 1973.

98. ZK Chen, K Ario, K Sawa, The influence of oxygen concentration on contact resistance behaviour of Ag and Pd materials in DC braking arcs. *Proc 17th Int'l Conf on Elec Cont* Mano R&D Tech. Center, Nagoya, Japan, 79–87, July 1994.

99. GJ Witter, I Polevoy, A study of contact resistance as a function of electrical load for silver based contacts, *Proc 17th Int'l Conf on Elec Cont* 503–568, July 1994.

100. PG Slade, Arc erosion of tungsten based contact materials: a review, *Refractory Hard Metals* 5 (4): 208–214, 1986.

101. A Keil, CL Meyer, Uber die Bildung von isolierenden Deckschickten auf Kontakten aus Verbundmetallen. *E.T.Z*, 73 (2): 31–34, 1952.

102. HD Kuhn, W Rieder, Electrotech und Machinebau, 79: 493–497, 1962.

103. PG Slade, Effect of the electric arc and the ambient air on the contact resistance of silver, tungsten and silver-tungsten contacts, *J Appl Phys* 47 (8): 3438–3443, 1976.

104. M Lindmayer, M Roth, Contact resistance and arc erosion of W-Ag and WC-Ag, *IEEE Trans Components Hybrids Manuf Technol* CHMT-2 (1): 70–75, 1979.

105. GJ Witter, VR Abele, The change in surface resistance of W–Ag contacts as a function of composition, microstructure and environment, *Proc 8th Int'l Conf on Elec Cont* 445–451, 1976.

106. CH Leung, RC Bevington, PC Wingert, HJ Kim, Effects of processing methods on the contact performance parameters for Ag–W composite materials, *IEEE Trans Components, Hybrids Manuf Technol* CHMT-5 (1): 23–31, 1982.

107. PG Slade, The switching performance of refractory carbide-silver contacts, *IEEE Trans Components, Hybrids Manuf Technol* CHMT-2 (1): 127–133, 1979.

108. CH Leung, HJ Kim, A comparison of Ag–W, Ag–WC and Ag–Mo electrical contacts, *IEEE Trans Components Hybrids Manuf Technol* CHMT-7 (1): 69–75, 1984.

109. S Kabayama, M Koysma, M Kume, Silver–tungsten alloys with improved contact resistance, *Powder Metallurgy Int'l* 5 (3): 122–126, 1973.

110. GJ Witter, The effect of nickel additions on the performance of W–Ag materials, *Proc 11th Int'l Conf on Elec Cont* 351–355, 1982.

111. JL Doremieux, G Fuzier, JP Langeron, The use of mixed titanium and tungsten carbide silver composites as contact material, *Proc 10th Int'l Conf on Elec Cont* 885–894, 1980.

112. PG Slade, CY Lin, AR Pebler, Titanium–tungsten carbide/silver, a new electric contact material, *IEEE Trans Components, Hybrids Manuf Technol* CHMT-4 (1): 176–184, 1981.

113. PG Slade, Variations in contact resistance resulting from oxide formation and decomposition in AgW and Ag–WC–C contacts passing steady currents for long time periods, *IEEE Trans Components Hybrids Manuf Technol* CHMT-9 (1): 3–16, 1986.

114. M Lindmayer, KH Schroeder, Effect of asymmetrical material combination on contact switching behavior, *IEEE Trans Components Hybrids Manuf Technol* CHMT-2 (1): 41–45, 1979.

115. C Leung, D Harman, L Doublet, C Bourda, Y Cui, L Hu, Electrical and mechanical lives of Ag-C and Ag-WC-C contacts, *Proc 26th Int'l Conf on Elec Cont Phen* 233–239, 2012.

116. LE Moberly, Performance of silver contacts in atmospheres containing silicone vapors, *Insulation* 19: April 1960.

117. A Eskes, HA Groenendijk, The formation insulating silicon compounds on switching contacts, *Proc IEEE Holm Conf on Elec Cont* 187, 1987.

118. NM Kitchen, CA Russell, Silicon oils on electric contacts—effects, sources, and countermeasures, *Proc of Holm Seminar on Elec Cont Phen* 75, 1975.

119. GJ Witter, RA Leiper, A study of the effects of thin film silicone contamination and the rate of testing on contact performance, *Proc 10th Int'l Conf on Elec Cont* 829, 1980.

120. W Trachslin, Contact behaviour and surface conditions of commercially available contact rivets, *Proc 8th Int'l Conf on Elec Cont* 131, 1976.

121. GJ Witter, RA Leiper, A comparison for the effects of various forms of silicon contamination on contact performance, *Proc 9th Int'l Conf on Elec Cont* 371, 1978.

122. GJ Witter, RA Leiper, A study of contamination levels measurement techniques, testing methods, and switching results for silicon compounds on silver arcing contacts, *Proc IEEE Holm Conf on Elec Cont* 173–180, 1992.

123. T Tamai, Adsorption of silicone vapor on the contact surface and its effect on contact failure of micro relays, *IEICE Trans on Elect* E83-C (9): 1402–1408, 2000.

124. T Tamai, S Sawada, Y Hattori, Manifold decomposition process of silicone vapor and electrical contact failure, *Proc 26th Int'l Conf on Elec Cont* 261–266, 2012.

11

Reed Switches

Kunio Hinohara

I beseech your Lordships to be merciful to a broken reed.

Francis Bacon

CONTENTS

11.1 Principles and Design of the Reed Switch 674
    11.1.1 Pull-In Characteristics of a Reed Switch 674
    11.1.2 Drop-Out Characteristics of a Reed Switch 682
    11.1.3 Magnet Drive Characteristics of a Reed Switch 683
        11.1.3.1 X–Y Characteristic H (Horizontal) 684
        11.1.3.2 X–Z Characteristic H (Horizontal) 684
        11.1.3.3 X–Y Characteristic V (Vertical) 685
11.2 Recommended Contact Plating 686
    11.2.1 Materials for Contact Plating 686
    11.2.2 Ground Plating 686
    11.2.3 Rhodium Plating 687
    11.2.4 Ruthenium Plating 687
    11.2.5 Other Platings 688
        11.2.5.1 Copper Plating 688
        11.2.5.2 Tungsten Plating 689
        11.2.5.3 Rhenium Plating 689
        11.2.5.4 Iridium Plating 689
        11.2.5.5 Nitriding the Permalloy (Ni-Fe [48 wt%]) Blade Material 689
11.3 Contact Surface Degradation and Countermeasures 689
    11.3.1 Surface Deactivation Treatment 690
        11.3.1.1 Life Test of Samples Left for 24 Hours after Sealing 691
        11.3.1.2 Life Test of Samples Left for One Week after Sealing 691
        11.3.1.3 Life Test of Samples Left for One Month after Sealing 693
        11.3.1.4 Life Test of Samples Left for Three Months, Six Months, and
                One Year after Sealing 693
    11.3.2 Prevention of Contact Adhesion 695
11.4 Applications of Reed Switches 696
    11.4.1 Reed Relays 697
    11.4.2 Applications of Magnetic-Driven Reed Switches 698
References 701

11.1 Principles and Design of the Reed Switch

A reed switch is a pair of blades in a magnetic material, such as 52-alloy (Ni 52%, Fe Bal.), sealed in a glass tube together with an inert gas (see Figure 11.1) [1–5]. As shown in Figure 11.2, the magnetic field of a coil or magnet induces a north or south pole in a blade, and consequently the contacts make as a result of the attractive force. When the magnetic field is removed, the contacts break as a result of the elasticity in the reeds.

A reed switch has the following characteristics:

1. Since the contacts, together with an inert gas, are hermetically sealed in a glass tube, the substitution operation is free from the influence of the external environment.

2. Since a reed switch undergoes no unnecessary operations because it has few mechanically moving parts, the life with no load is virtually limitless.

3. Since the moving components are lightweight with a high resonant frequency, the contacts make or break very quickly.

11.1.1 Pull-In Characteristics of a Reed Switch

The relationship between Φ_g (the magnetic flux passing through a contact gap) and Q (magnetomotive force Q when the blades are set along the axis of the coil with a certain contact gap and overlap) in Figure 11.3 [5–12] is represented by

$$\Phi_g = \frac{Q}{rL_1} \frac{\cosh \alpha L_2/2 - \cosh \alpha (L_2 - L_1)/2}{\cosh \alpha L_2/2 + (pr_g/2\alpha)\sinh \alpha L_2/2} \tag{11.1}$$

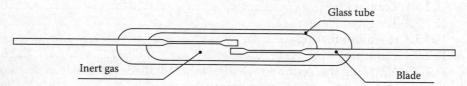

FIGURE 11.1
Basic structure of reed switch.

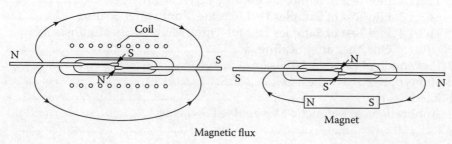

FIGURE 11.2
Principle of operating reed switch.

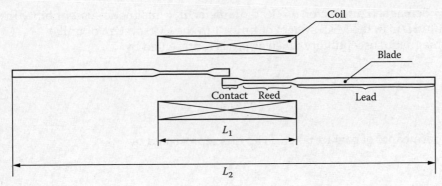

FIGURE 11.3
Magnetic analysis model of reed switch.

where, Φ_g = magnetic flux passing through the contact gap, Q = magnetomotive force, L_1 = coil length, L_2 = reed switch length, p = permeance per unit length, r = magnetic resistance per unit length, r_g = magnetic resistance per unit length of contact, $\alpha = \sqrt{pr}$

With a magnetomotive force per unit length taken as q, the number of coil turns as N, and coil current as I in Equation 11.1, the following expression holds:

$$q = \frac{Q}{L_1} = \frac{NI}{L_1} \tag{11.2}$$

Then, NI is represented by Equation 11.3 below, on rearranging Equation 11.1:

$$NI = L_1 r \frac{\cosh \alpha L_2/2 + (pr_g/2\alpha)\sinh \alpha L_2/2}{\cosh \alpha L_2/2 - \cosh \alpha(L_2 - L_1)/2} \Phi_g \tag{11.3}$$

The permeance per unit length, p, is approximated by the permeance when an elongated ellipsoid of rotation symmetry is magnetized in the direction of the length:

$$\frac{p}{\mu_0} = 2\pi \frac{1}{\varepsilon^2}\left(\frac{1}{N_d} - 1\right) \tag{11.4}$$

$$N_d = \frac{1}{\varepsilon^2 - 1}\left(\frac{\varepsilon}{\sqrt{\varepsilon^2 - 1}}\ln(\varepsilon + \sqrt{\varepsilon^2 - 1}) - 1\right) \tag{11.5}$$

$$\varepsilon = \frac{L_2}{d} \tag{11.6}$$

Where, N_d = demagnetizing factor, d = lead diameter, μ_0 = magnetic susceptibility in vacuum ($4\pi \times 10^{-7}$ H/m in the MKS system of units, 1 in the CGS system of units)

With the magnetic susceptibility taken as μ, r is represented by

$$r = \frac{1}{\mu\pi(d/2)^2} \tag{11.7}$$

With the permeance of contact taken as p_g, r_g is represented by

$$r_g = \frac{1}{p_g} \tag{11.8}$$

Next we will obtain p_g. Using the dimensions of the contact as shown in Figure 11.4. Where, g = contact gap, a = overlap, w_1 = contact width and t_1 = contact thickness. The permeance of contact, p_g, is the sum of the permeance of the side, p_{gs}, and the permeance of the facing surface, p_{gt}, and is as follows:

$$p_g = p_{gs} + p_{gt} = \mu_0\beta\frac{aw_1}{g} \tag{11.9}$$

$$\beta = 1 + k\frac{g}{a} \tag{11.10}$$

k, as a function of g/a with t_1/w_1 as a parameter, is obtained from Figure 11.5. From Equations 11.8 through 11.10, r_g is expressed by

$$r_g = \frac{g}{\mu_0\left(1 + k\dfrac{g}{a}\right)aw_1} \tag{11.11}$$

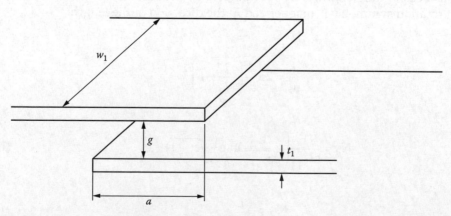

FIGURE 11.4
Dimensions of contact.

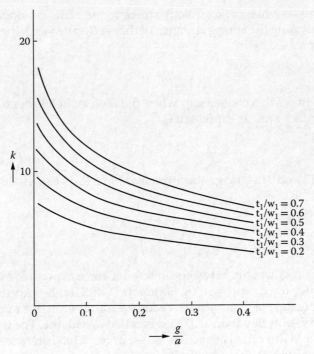

FIGURE 11.5

k vs. $\frac{g}{a}$ with $\frac{t_1}{w_1}$ as parameter.

The magnetic attractive force acting on the blades, F_m, is expressed by

$$F_m = \frac{1}{8\pi} \Phi_g^2 \frac{dr_g}{dg} \tag{11.12}$$

Applying Equation 11.11 gives

$$F_m = \frac{\Phi_g^2}{8\pi \left(1 + k\dfrac{g}{a}\right)^2 aw_1} \tag{11.13}$$

where $\mu_0 = 1$. When a reed switch is closed, $g = 0$. The magnetic attractive force in this case, F_0, is expressed by

$$F_0 = \frac{\Phi_g^2}{8\pi aw_1} \tag{11.14}$$

where the contacts are assumed to be not plated. If a reed switch, whose contacts are plated to a thickness of t_p, is closed by a saturated magnetic flux Φ_s caused by a soak current, magnetic attractive force F_s is expressed, from Equation 11.13, as follows:

$$F_s = \frac{\Phi_s^2}{8\pi \left(1 + k\dfrac{2t_p}{a}\right)^2 aw_1} \tag{11.15}$$

Equation 11.15 expresses a magnetic attractive force for an ordinary operation. With the stiffness of a reed, indicating the spring strength of the reed, taken as S, the retractive force in the blades, F_r, is expressed by

$$F_r = Sx \tag{11.16}$$

where x = displacement. With a contact gap when the reed switch is open taken as G, the retractive force for contact gap g is expressed by

$$F_r = S(G - g) \tag{11.17}$$

From Equations 11.13 and 11.14, magnetic attractive force F_m is expressed by

$$F_m = \frac{F_0}{\left(1 + k\dfrac{g}{a}\right)^2} \tag{11.18}$$

Figure 11.6 shows the relationship between a retractive force expressed by Equation 11.17 and a magnetic attractive force expressed by Equation 11.18. The abscissa represents a contact gap g, and the ordinate, magnetic attractive force F_m and retractive force F_r.

The straight line represents Equation 11.17 and is called a load line. The three curves represent Equation 11.18 and are called attractive force curves. The difference between them is caused by F_0. Each of them is described below.

F_{m_1}: when $g = g_1$, $F_{m_1} = F_r$.
   However, when $g < g_1$, $F_{m_1} < F_r$.
   Consequently, the reed switch cannot close, and the apparent gap becomes g_1 only.

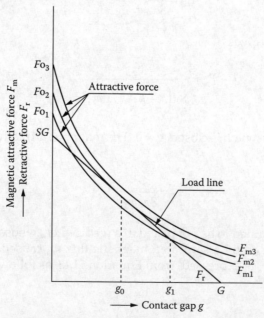

FIGURE 11.6
Attractive and retractive characteristics.

F_{m_2}: The load line touches the attractive force curve when $g = g_0$

  When $g < g_0$, $F_{m_2} > F_r$.

  Consequently, the reed switch closes.

F_{0_2}: The point at which the attractive force curve touches the load line is the attractive force required to pull the reed switch. The magnetic flux at this time is pull-in magnetic flux Φ_{PI}. In general, contact force F_c is expressed by

$$F_c = F_m - F_r \tag{11.19}$$

And contact force F_{c_2} is expressed by

$$F_{c_2} = F_{0_2} - SG \tag{11.20}$$

This represents the contact force when the contacts pull in.

F_{m_3}: when $0 \leq g \leq G$, $F_{m_3} > F_r$.

  Consequently, the reed switch is always closed.

  Contact force F_{c_3} is expressed by

$$F_{c_3} = F_{0_3} - SG \tag{11.21}$$

Applying Equation 11.15 to F_{0_3} results in

$$F_{c_3} = F_s - SG \tag{11.22}$$

This represents a contact force during ordinary operation.

We will now obtain the pull-in conditions for a reed switch. A reed switch pulls in at the point where the attractive force curve touches the load line. At this point, with a contact gap of g_0, the retractive force expressed by Equation 11.17 becomes equal to the magnetic attractive force expressed by Equation 11.18. Therefore,

$$F_m = F_r(g = g_0) \tag{11.23}$$

Then

$$\frac{F_0}{\left(1 + k\dfrac{g_0}{a}\right)^2} = S(G - g_0) \tag{11.24}$$

Since the attractive force touches the load line, the slopes become identical as expressed below:

$$\frac{dF_m}{dg} = \frac{dF_r}{dg}(g = g_0) \tag{11.25}$$

Then

$$-\frac{2\frac{k}{a}F_0}{\left(1+k\frac{g_0}{a}\right)^3}=-S \tag{11.26}$$

From Equations 11.24 and 11.26,

$$g_0=\frac{2k\frac{1}{a}G-1}{3k\frac{1}{a}} \tag{11.27}$$

Substituting pull-in ampere-turns NI_{PI} for NI and pull-in magnetic flux Φ_{PI} for Φ_g in Equation 11.3, and using

$$1 \text{ ampere turn (AT)} = \frac{4\pi}{10} \text{gilbert} \tag{11.28}$$

gives

$$NI_{PI}=\frac{10}{4\pi}L_1 r\frac{\cosh\alpha L_2/2+(pr_g/2\alpha)\sinh\alpha L_2/2}{\cosh\alpha L_2/2-\cosh\alpha(L_2-L_1)/2}\Phi_{PI}(AT) \tag{11.29}$$

A reed switch whose contact gap is G when it breaks begins to close at a contact gap of g_0. Then, magnetic resistance r_g of the contact is obtained from Equation 11.11 as follows:

$$r_g=\frac{g_0}{\left(1+k\frac{g_0}{a}\right)aw_1} \tag{11.30}$$

where $\mu_0=1$.

Φ_{PI} is expressed from Equations 11.26 and 11.27 as follows:

$$\Phi_{PI}=\sqrt{\frac{4\left(1+k\dfrac{G}{a}\right)^3}{27k\dfrac{1}{a}}S\times8\pi aw_1\times980} \tag{11.31}$$

Now, if the dimensions of a reed switch are given, the NI-Φ characteristics of the switch can be obtained. However, it is necessary to calculate the stiffness S of the reed in advance.

To calculate the stiffness S of a reed, specify the dimensions of the reed, apply load $W(g)$, and take the deflection of the reed as x (mm) as shown in Figure 11.7.

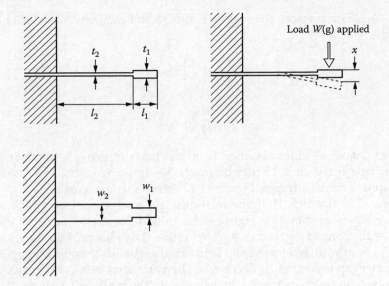

FIGURE 11.7
Dimensions of reed.

With cross-sectional secondary moments taken as I_1 and I_2, the stiffness S of a stepped cantilever as shown in Figure 11.7 is calculated by

$$
S = \frac{W}{x}
$$

$$
= \frac{E}{\dfrac{\ell_1^3}{3I_1} + \dfrac{\ell_2^3}{3I_2} + \dfrac{1}{I_2}(\ell_1^2\ell_2 + \ell_1\ell_2^2)} \tag{11.32}
$$

$$
I_1 = \frac{w_1 t_1^3}{12} \tag{11.33}
$$

$$
I_2 = \frac{w_2 t_2^3}{12} \tag{11.34}
$$

where E is Young's modulus. Young's modulus of 52-alloy annealed at 850°C in a hydrogen atmosphere for 10 minutes is as follows:

$$
E = 1.3 \times 10^7 \, (g/mm^2) \tag{11.35}
$$

Obtaining magnetic resistance r requires the magnetic susceptibility in Equation 11.7. μ is determined by magnetic flux Φ.

If the diameter, d, of wire to be used as a blade in this reed switch is 0.5 mm, it is necessary to obtain the B–H characteristics of a 0.5 mm diameter wire. Using the B and H values obtained with a B–H curve tracer, the magnetic flux Φ and magnetic susceptibility μ can be obtained from

$$\Phi = B \cdot \pi \left(\frac{d}{2} \right)^2 \tag{11.36}$$

$$\mu = \frac{B}{H} \tag{11.37}$$

Table 11.1 gives Φ and μ values obtained from the measured B–H characteristics of a 40 mm diameter ring made of a 0.5 mm diameter 52-alloy wire annealed at 850°C in a hydrogen atmosphere for 10 minutes. Figure 11.8a shows the Φ–μ curve.

From Equations 11.29 through 11.31, the relationship between contact gap G, overlap a, and pull-in value NI_{PI} is as shown in Figure 11.8b. In addition to the fact that the pull-in value increases as the contact gap increases (see Figure 11.6), Figure 11.8b shows the value of a_m at which NI_{PI} is minimized by overlap being changed with the contact gap kept constant. As the contact gap increases, a_m decreases. The variations in NI_{PI} in the manufacturing of reed switches can be made smaller by using a_m in the reed switch design stage.

11.1.2 Drop-Out Characteristics of a Reed Switch

We will now obtain drop-out ampere-turns NI_{DO} of a reed switch using blades whose contact plating thickness is t_p. Substituting NI_{DO} for NI, Φ_{DO} for Φ_g, and r_p for r_g because of the contact plating in Equation 11.3, and using Equation 11.28, gives

$$NI_{DO} = \frac{10}{4\pi} L_1 r \frac{\cosh \alpha L_2/2 + (p r_p/2\alpha) \sinh \alpha L_2/2}{\cosh \alpha L_2/2 - \cosh \alpha (L_2 - L_1)/2} \Phi_{DO} \, (AT) \tag{11.38}$$

TABLE 11.1

Magnetic Characteristics of 0.5 mm Diameter 52-Alloy Wire

H (Oersted)	B (Gauss)	Φ (Maxwell)	μ
0.10	291	0.57	2,910
0.15	523	1.03	3,487
0.20	1,168	2.29	5,840
0.25	1,915	3.76	7,660
0.30	2,555	5.02	8,517
0.35	3,830	7.52	10,943
0.40	5,910	11.60	14,775
0.45	7,025	13.79	15,611
0.50	7,820	15.35	15,640
0.55	8,460	16.61	15,382
0.60	8,630	16.94	14,383
1.1	10,540	20.70	10,540
2.0	11,650	22.87	5,825
2.5	11,970	23.50	4,788
3.0	12,440	24.43	4,147
5.0	13,250	26.02	2,650
10.0	14,200	27.88	1,420

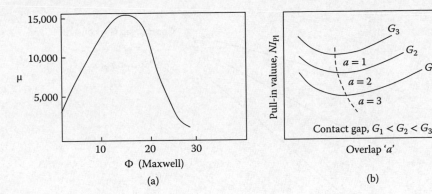

FIGURE 11.8
(a) Φ–μ curve of 0.5 mm diameter 52 alloy wire. (b) Relationship between pull-in value, contact gap, and overlap.

Magnetic attractive force F_m, as in Equation 11.15, is as follows:

$$F_m = \frac{\Phi_{DO}^2}{8\pi\left(1+k\dfrac{2t_p}{a}\right)^2 aw_1} \tag{11.39}$$

Retractive force F_r is expressed by

$$F_r = S(G-2t_p) \tag{11.40}$$

Then, drop-out magnetic flux Φ_{DO} is expressed by

$$\Phi_{DO} = \sqrt{8\pi\left(1+k\frac{2t_p}{a}\right)^2 aw_1 \times 980 \times S(G-2t_p)} \tag{11.41}$$

From Equation 11.11,

$$r_p = \frac{2t_p}{\left(1+k\dfrac{2t_p}{a}\right)aw_1} \tag{11.42}$$

In this case, $\mu_0 = 1$.

It is obvious from Figure 11.6 that the drop-out value increases as the contact gap increases. Figure 11.9 shows the change in drop-out value NI_{DO} with overlap a and contact plating thickness t_p for a constant contact gap of G. Figure 11.9 shows that the drop-out value increases as the contact plating thickness or overlap increases.

11.1.3 Magnet Drive Characteristics of a Reed Switch

As mentioned previously, one of the characteristics of the reed switch is that it can be easily driven by an external magnet as well as a coil. There are many kinds of magnets, and there are also many different magnet shapes and magnetizing methods. There are

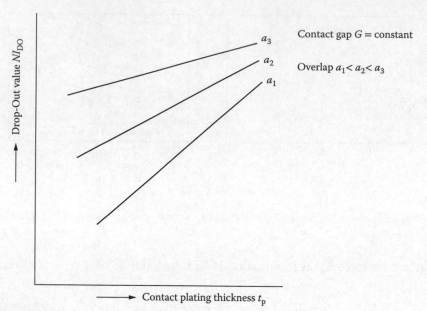

FIGURE 11.9
Relationship between drop-out value, contact plating thickness, and overlap.

also many kinds of reed switches with different pull-in values. As a result, the magnet drive characteristics becomes complicated. This section explains typical magnet drive characteristics [5].

11.1.3.1 X–Y Characteristic H (Horizontal)

Here a rectangular bar magnet is used with the magnet movement direction and the center of the reed switch to $Z = 0$ (constant) as shown in Figure 11.10a. The X–Y characteristic H is the magnetic drive characteristic of the reed switch when the magnet moves freely in the X–Y plane with the pole direction kept parallel to the reed switch. The reed switch closes and opens as the magnet moves. The "on" point is taken as the point where the reed switch closes due to the movement of the magnet, and the "off" point as the point where the closed reed switch opens due to the movement of the magnet. Plotting the "on" and "off" points results in Figure 11.10b, i.e., X–Y characteristic H. In Figure 11.10b, the area inside the solid line connecting the "on" points is called the "on" area, and the area outside the broken line connecting the "off" points is called the "off" area. The area between the "on" area and the "off" area is called the "hold" area, "hysteresis" area, or "differential" area. When the intensity of the magnet is strong, the "on" area and "hold" area marked by * are generated, resulting in three-point operation.

11.1.3.2 X–Z Characteristic H (Horizontal)

Here the magnet movement direction is shown in Figure 11.11a. The X–Z characteristic H is the magnetic drive characteristic of the reed switch when the magnet moves freely in the X–Z plane with the pole direction kept parallel to the reed switch. Connecting the on and off points, as in the case of X–Y characteristic H, results in X–Z characteristic H as shown in Figure 11.11b.

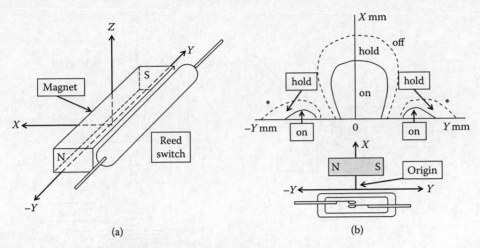

FIGURE 11.10

(a) Magnet movement direction (#1). (b) *X–Y* characteristic *H*.

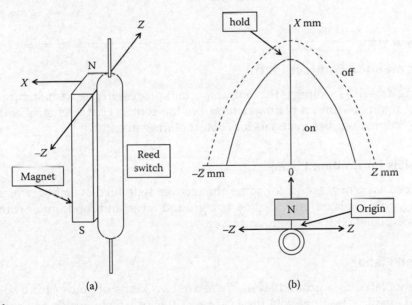

FIGURE 11.11

(a) Magnet movement direction (#2). (b) *X–Z* characteristic *H*.

11.1.3.3 X–Y Characteristic V (Vertical)

Here the magnet movement direction with the center of the reed switch to $Z = 0$ (constant) is shown in Figure 11.12a. The *X–Y* characteristic *V* is the magnet drive characteristic of the reed switch when the magnet moves freely in the *X–Y* plane with the pole direction kept perpendicular to the reed switch. Connecting the on and off points, as in the case of the *X–Y* characteristic *H* and *X–Z* characteristic *H*, results in *X–Y* characteristic *V* as shown in Figure 11.12b. In the *X–Y* characteristic *V*, the reed switch opens when the magnet comes near the contacts of the reed switch.

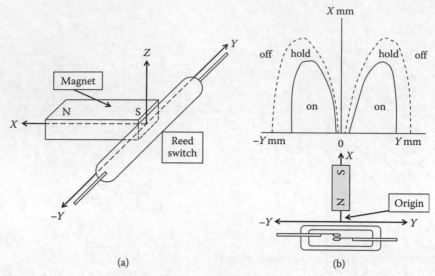

FIGURE 11.12
(a) Magnet movement direction (III). (b) X–Y characteristic V.

11.2 Recommended Contact Plating

Contact plating as well as sealing is the most important process in the manufacture of reed switches. The quality of contact plating determines the contact characteristics of the reed switch. This section describes materials for contact plating in detail [5].

11.2.1 Materials for Contact Plating

In ordinary reed switches, except for some changeover-type reed switches, which use a special plating, gold is plated onto 52-alloy as a ground layer, and rhodium or ruthenium in some cases is plated on the background layer.

11.2.2 Ground Plating

Gold is usually plated as a ground layer. There are two kinds of gold-plating solutions: pure gold and cyanic. The pure gold type is in popular use. Gold striking is performed before gold plating in many instances. Gold striking may be necessary to attain secure adhesion. If the required adhesion and other characteristics can be attained by gold plating only, gold striking can be omitted.

Palladium has been attracting attention as an alternative to gold. Palladium-plating solution of good quality is also being developed. Since the density of palladium is about 60% that of gold, the weight of palladium used is about 40% less than that of gold plating for the same plating thickness. However, the monetary value of palladium is now similar to that of gold. Consequently, the cost of using it for ground plating cannot really be shortened. Palladium is somewhat advantageous in terms of performance because of higher melting point and hardness twice that of gold. Table 11.2 lists the properties of gold and palladium.

TABLE 11.2

Properties of Gold and Palladium

	Gold	Palladium
Atomic number	79	46
Atomic weight	197.0	106.4
Density (20°C)	19.3 g/cm³	2.03 g/cm³
Crystalline structure	Face-centered cubic lattice	Face-centered cubic lattice
Melting point	1,063°C	1,555°C
Hardness	2.5~3	4.8
Specific resistance	2.2 μΩ cm (18°C)	10.4 μΩ cm

11.2.3 Rhodium Plating

Rhodium is a commonly used contact material for reed switches. Rhodium has a high melting point and hardness, and has excellent abrasion resistance, resistance against sticking, and corrosion resistance.

Initially when rhodium was first put on the market, a good plating solution was not available, and the rhodium contact surface had a problem of cracking. The cracking problem was addressed over 25 years ago and successful crack-free rhodium plating with thicknesses up to 10 μm was achieved. This success was achieved partly by the development of an excellent stress relaxing agent. This technique, together with the surface deactivation treatment described in Section 11.3.1, are core techniques for the development of reed switches' rhodium contact. Since thick plating is possible with these techniques, reed switches with wide variations, ranging from ultra-miniature reed switches employing thin plating to high-power reed switches employing thick plating, can be obtained. A proper pretreatment and post-treatment make almost maintenance-free and stable rhodium plating possible.

At present, the matter primary concerning rhodium is cost. Limited supplies of rhodium are produced and it is used as a catalyst in anti-pollution equipment. As a result, the price has risen. Good contact plating requires that the concentration of sulfuric acid in the rhodium-plating solution to be maintained below a certain point, then it is often necessary to replace the rhodium-plating solution. This also raises the cost. Research is being conducted into removing only the sulfuric acid from the rhodium-plating solution. A sulfuric acid removing system, however, has not been completed yet due to a filtration problem and other problems. In order to alleviate the cost problem, ruthenium is an attractive alternative to rhodium.

11.2.4 Ruthenium Plating

The atomic number of ruthenium is 44, adjacent to rhodium, with an atomic number of 45, in the periodic table. Ruthenium has a higher melting point and hardness and better abrasion resistance, resistance against sticking, and corrosion resistance than rhodium. Although the production of ruthenium is lower than that of rhodium, its price remains relatively low, because it is unfit for ornamentation, i.e., a poor gloss results from its low reflectance of visible light rays.

The problem of cracking in plated layers is again a grievous one. An effective stress-relaxing agent has not been developed for ruthenium in spite of the past research for a solution. Even if the plating conditions are optimized and post-treatment is executed properly, the maximum crack-free plating thickness is about 1.5 μm. Research continues on special post-plating treatments to control cracking. Some manufacturers perform crack-free plating of ruthenium by sputtering, not by electroplating. Sputtering, however, has its own problems of productivity and quality.

Thin plating of ruthenium has a life several times that of rhodium at low and intermediate loads. In this connection, the market potential associated with these loads, mainly applications for reed relays for use in ATEs (LSI testers and board tester), will probably increase.

It is said that ruthenium does not require the surface deactivation treatment (See Section 11.3.1) after plating because a thin, stable oxide film is formed on the plated surface. There are many areas to be analyzed and studied concerning ruthenium plated surfaces [13,14].

Many attempts have been made to increase the plating thickness of ruthenium. One of them is alloy plating which uses rhodium containing about 10% ruthenium. Plating solution for this alloy plating is on the market. Ruthenium-silver alloy plating has also been reported.

Table 11.3 gives the properties of rhodium and ruthenium.

11.2.5 Other Platings

In addition to the above-mentioned plating, copper plating, nickel plating, tungsten plating, titanium, and iridium plating are actually practiced or have been reported. Some of these are described below.

11.2.5.1 Copper Plating

A non-magnetic chip is welded to the normally closed blade of a changeover-type reed switch. Copper-plating (0.33 mm thick) can also be used in place of the non-magnetic chip, see Figure 11.13. The plating a thickness of copper that exceeds 100 μm is quite complex. It involves the optimum selection of copper-plating solution, diffusion treatment and frequent gold strikings, but stable contact characteristics can be obtained. In the early stages, the copper-plated portion can cause problems that result from the heat in surface treatment and sealing. However, this non-magnetization process is free of problems and provides quite stable non-magnetization.

TABLE 11.3

Properties of Rhodium and Ruthenium

	Rhodium	Ruthenium
Atomic number	45	44
Atomic weight	102.9	101.1
Density	12.41 g/cm³ (20°C)	12.30 g/cm³(19°C)
Crystalline structure	Face-centered cubic lattice	Hexagonal close-packed lattice
Melting point	1,966°C	2,350°C
Hardness (plated)	6	6.5
Specific resistance (plated)	5.1 μΩ cm (20°C)	7.46 μΩ cm
Mean reflectance of visible rays	79%	635

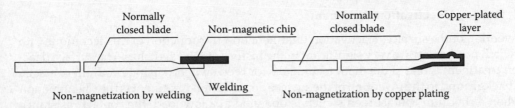

FIGURE 11.13
Non-magnetization of normally closed blade.

11.2.5.2 Tungsten Plating

Tungsten, whose melting point is much higher than that of rhodium and ruthenium, is a very useful contact material for high-power reed switches for which the user's requirements are getting more severe [15]. Tungsten is usually plated on a contact by chemical vapor deposition (CVD), which is quite costly. It has been confirmed that a tungsten layer shows excellent abrasion resistance and resistance against sticking because of the exceptionally high melting point. The tungsten layer, however, soon oxidizes in air, and consequently the initial contact resistance can be high. This problem can be solved by surface treatment after CVD and by optimizing the sealing gas.

11.2.5.3 Rhenium Plating

Rhenium, like tungsten, has a higher melting point than both rhodium and ruthenium, and is one of the quite useful contact materials for high-power reed switches. Reported is a reed switch whose contact surface is covered with a three-layer structure of Re–Au–Re to suppress metal transfer caused by a molten metal bridge when the contacts begin to open.

11.2.5.4 Iridium Plating

Iridium also has a higher melting point that rhodium and ruthenium, a resistivity close to that of rhodium and hardness close to that of ruthenium. Successful crack free contact surfaces have been obtained by using a gold substrate on the blade, a rhodium plate on the gold and a final iridium plate on the rhodium. Testing using inductive circuits of 24 V dc, 36 mA and 12 V dc, 73 mA show a superior performance over rhodium contacts [16].

11.2.5.5 Nitriding the Permalloy (Ni-Fe [48 wt%]) Blade Material

An interesting study of nitriding the Ni-Fe blade instead of electroplating it with a contact material shows that this modification of the blade can perform satisfactorily for some applications of the reed switch [17].

11.3 Contact Surface Degradation and Countermeasures

Reed switches have a very simple structure and very small size. Coupled with the developments in electronics, reed switches have greatly advanced in terms of function and performance. Two of these advances are discussed in this section.

11.3.1 Surface Deactivation Treatment

The invention of the surface deactivation treatment and introduction of this technique into reed switch production are very significant in the history of reed switches. The surface deactivation treatment is one of the main techniques for reed switches using rhodium contacts.

The background to the invention of the surface deactivation treatment is the phenomenon that when a rhodium contact reed switch is operated at a low load, the contact resistance increases during the initial stage of operations [18]. Figure 11.14 shows this phenomenon. Examination of the contact surface of a reed switch whose contact resistance had increased reveals the formation of brown insulating material. Analysis by electron probe microanalysis (EPMA) shows that the material is carbon (see Figure 11.15a and b). Rhodium is active in adsorption and catalysis. Plated rhodium on a contact surface adsorbs organic impurities from the air, and the mechanical energy generated by the collisions of contacts polymerizes them. The contact resistance seems to decrease and converge to a certain level after increasing, because the generated polymer is destroyed and polymer pieces scatter. Since the polymer pieces are present between contacts, the contact resistance is unstable (see Figure 11.16a and b).

In order to prevent this polymerization, the rhodium-plated surface must be deactivated to suppress the adsorbing and catalytic activities. The deactivation process requires the blades be treated at 450°C in an oxygen environment. Organic impurities adsorbed on the

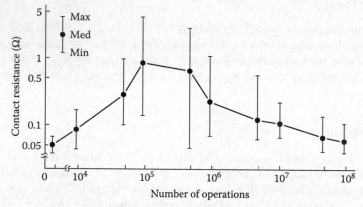

FIGURE 11.14
Increase in contact resistance of rhodium contact reed switch.

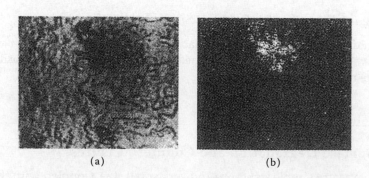

(a)   (b)

FIGURE 11.15
(a) Contact surface of reed switch after 10^5 operation ($\times$ 550); (b) X-ray image of contact surface of reed switch after 10^5 operations ($\times$ 550).

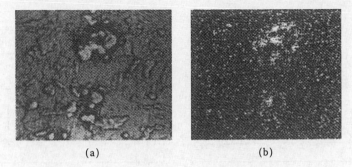

(a)   (b)

FIGURE 11.16
(a) Contact surface of reed switch after 10^7 operations ($\times$ 550); (b) X-ray image of contact surface of reed switch after 10^7 operations ($\times$ 550).

TABLE 11.4

Samples for Checking the Effect of the Surface Deactivation Treatment

		Surface Deactivation Treatment	
Left in benzene vapor		No	Yes
	No	Rh	Rh + O$_2$
	Yes	Rh + benzene	Rh + O$_2$ + benzine

blade contact surface are removed by burning, and then an oxygen film is formed on the cleaned rhodium surface to prevent organic impurities from being adsorbed again [19] (patent no. 916386 in Japan, patent no. 3857175 in USA, patent no. 2303587 in West Germany).

To evaluate the effect of the surface deactivation treatment, samples as shown in Table 11.4 were prepared. Samples which had been left in benzene vapor for 24 hours were also used. Some of these samples were left for 24 hours after sealing, some for one week, some for one month, some for three months, some for six months, and some for one year. Then, individual groups underwent a life test of 100 million operations with a resistive load of 12 V/5 mA operated by a permanent magnet.

11.3.1.1 Life Test of Samples Left for 24 Hours after Sealing

As shown in Figure 11.17a, the "Rh + O$_2$" group, i.e. the deactivated contact surface group, had a stable contact resistance throughout 100 million operations, indicating that no polymer was formed. The "Rh + O$_2$ + benzene" group also had a stable contact resistance even though the group was left in benzene vapor for 24 hours after the surface deactivation treatment. On the other hand, the "Rh" group, i.e., a group without the surface deactivation treatment, had an increase in contact resistance, indicating the formation of polymer. In particular, the "Rh + benzene" group, i.e., a group that did not receive the surface deactivation treatment and that was left for 24 hours in benzene vapor, had an extreme increase in contact resistance, indicating the formation of a lot of polymer (see Figure 11.18a and b).

For the other samples, only the "Rh" and "Rh + O$_2$" groups underwent a test.

11.3.1.2 Life Test of Samples Left for One Week after Sealing

The "Rh + O$_2$" group had a stable contact resistance throughout 100 million operations. However, the "Rh" group had a larger increase in contact resistance than the group left for 24 hours, and the variation in contact resistance was also larger (see Figure 11.17b).

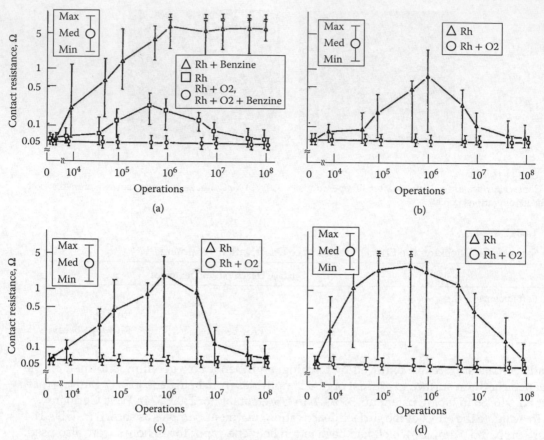

FIGURE 11.17
(a) Fluctuations in contact resistance (life test of samples left for 24 hours after sealing); (b) fluctuations in contact resistance (life test of samples left for one week after sealing); (c) fluctuations in contact resistance (life test of samples left for one month after sealing); (d) fluctuations in contact resistance (life test of samples left for one year after sealing).

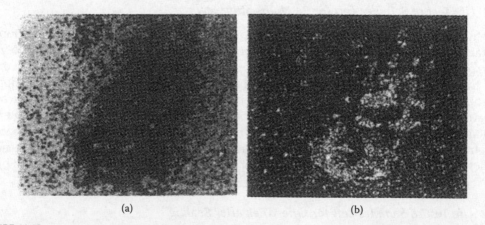

FIGURE 11.18
(a) Contact surface of reed switch after 10^5 operations (left in benzene vapor without the surface deactivation treatment); (b) X-ray image of contact surface of reed switch after 10^5 operations (left in benzene vapor without the surface deactivation treatment).

11.3.1.3 Life Test of Samples Left for One Month after Sealing

The "Rh + O_2" group had a stable contact resistance. The "Rh" group had a larger increase in contact resistance than the group left for one week (see Figure 11.17c).

11.3.1.4 Life Test of Samples Left for Three Months, Six Months, and One Year after Sealing

The data were similar to the one year data shown in Figure 11.17d, the "Rh" group experienced a further increase in contact resistance. The "Rh + O_2" group had as stable a contact resistance as the group left for 24 hours, and the formation of polymer was not mentioned in this group (see Figure 11.19).

These experiments show that a surface deactivation treatment suppresses adsorbing and catalytic activities. As rhodium generally resists oxidation, then studies of on how oxygen exists on the rhodium-plated contact surface were made [20–22]. The relationship between the contact surface treatment temperature and amount of oxygen on the contact surface was examined using Auger electron spectroscopy (AES) [23]. Figure 11.20 shows the results. Before the contacts were sealed, the amount of oxygen on the rhodium surface simply increases with the treatment temperature. However, after the contacts were sealed, the amount of oxygen increased sharply at a treatment temperature of 400°C–450°C, but the amount was almost constant at very low levels at a treatment temperature of below 400°C. This meant that virtually all oxygen on the rhodium surface treated at below 400°C dissociated. Since the temperature range of 400°C–450°C seemed to be important for the establishment of an oxygen layer on the rhodium surface, further tests were conducted using reflection high-energy electron diffraction (RHEED), molecular optical laser examiner (MOLE) [24], and X-ray photoelectron spectroscopy (XPS) [25, 26]. The RHEED electron diffraction patterns showed that rhodium oxide was not detected on the rhodium surface without deactivation treatment and the rhodium surface treated at 300°C. However, a thin broad electron diffraction pattern of rhodium oxide (Rh_2O_3) was obtained from the rhodium surface treated at 400°C, and a clear electron diffraction pattern of Rh_2O_3 was obtained from the rhodium surface treated at 450°C. In addition, a further intensive electron diffraction pattern of Rh_2O_3 was obtained from the rhodium surface treated at 500°C.

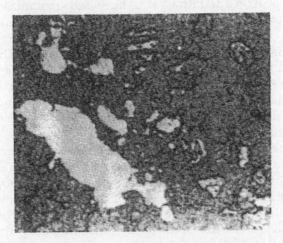

FIGURE 11.19
Deactivated contact surface of reed switch after 100 million operations.

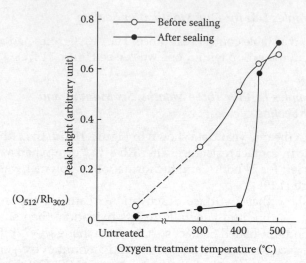

FIGURE 11.20
Amount of oxygen on rhodium surface vs. treatment temperature.

TABLE 11.5

Binding Energy of $Rh_{3d_{5/2}}$ and O_{1s}

	Binding Energy (ev)		ΔE
Sample	$Rh_{3d_{5/2}}$	O_{1s}	$O_{1s} - Rh_{3d_{5/2}}$
Untreated	307.4	531.4	224.0
Oxygen treated (400°C)	308.2	530.2	222.0
Oxygen treated (450°C)	308.6	530.2	221.6
Oxygen treated (500°C)	308.5	530.1	221.6
Standard Rh_2O_3	308.3	529.9	221.6

In other words, rhodium oxide (Rh_3O_3) with less crystallization started to emerge on the rhodium surface treated at 400°C, and Rh_2O_3 with increased crystallization existed stably on the rhodium surface treated at 450°C. A change in Rh_2O_3 on the rhodium surface in a temperature range of 400°C–450°C corresponded to the critical temperature zone in AES. More Rh_2O_3 seems to exist on the rhodium surface treated at 500°C. The Raman band of Rh_2O_3 was not observed from the untreated rhodium surface and the rhodium surface treated at 400°C, but the Raman band of Rh_2O_3 was observed from the rhodium surface treated at 450°C. In addition, a more intensive Raman band of Rh_2O_3 was observed from the rhodium surface treated at 500°C. The above analytic results showed that an Rh_2O_3 film existed on the rhodium surface treated at 450°C and a thicker Rh_2O_3 film was formed by treatment at 500°C. The existence of Rh_2O_3 film at a treatment temperature of 450°C, but not at 400°C corresponded to the critical temperature zone in the AES. Table 11.5 shows the binding energy of and difference between the $Rh_{3d_{5/2}}$ electron and the O_{1s} electron.

The results of the analysis above show that metal rhodium accounts for most of the untreated rhodium surface, but the rhodium oxide (Rh_2O_3) begins to be generated in the rhodium surface treated at 400°C and coexists with metallic rhodium. Only Rh_2O_3 is observed on the rhodium surface treated at 450°C, and the rhodium surface treated at 500°C is the same as that treated at 450°C. A change from the coexistence of Rh and Rh_2O_3 on the rhodium surface treated at 400°C to the existence of Rh_2O_3 only on the rhodium surface treated at 450°C corresponds to the critical temperature zone in AES. The RHEED, MOLE, and XPS, indeed confirm that oxygen exists on the deactivated rhodium-plated contact surface in the form of rhodium oxide (Rh_2O_3). Contact surface deactivation also improves the sealing properties through a betterment in the usability of blades. In other words, a metal surface covered with an oxide film shows better adhesion to glass than a pure metal surface. This is because metal oxide diffuses into the glass and promotes a bond between the glass and metal in sealing.

11.3.2 Prevention of Contact Adhesion

Adhesion of contacts that is called soft sticking of the contacts is caused by a friction of contact surfaces. Soft sticking has a low detection rate and is difficult to reproduce in distinction from ordinary sticking in which a load current and discharge make the contact surfaces rugged by erosion and the rugged contact surfaces stick to each other. Since it is difficult to predict when soft sticking occurs, soft sticking used to be one of the critical factors reducing the reliability of reed switches. Thus, the study of soft sticking is one of critical interest. Oxygen on the rhodium-plated contact is effective against soft sticking and has been used to suppress soft sticking [27–30].

Blades were prepared whose rhodium-plated contacts were deactivated and blades whose rhodium-plated contacts were left untreated. They were sealed in glass tubes in a nitrogen atmosphere or a highly reducing atmosphere including hydrogen (see Table 11.6). Figure 11.21a shows the sticking test circuit, and Figure 11.21b shows the sticking test mode. Adhesion was checked after every reed switch operation. In Figure 11.21b, coil excitation was attained by half-wave rectification of 50 Hz ac and the peak was 70 ampere turns (AT). A load of 50 kΩ (5 V and 100 µA) was connected to the contacts.

Figure 11.22a shows the results of the sticking test. Sticking occurred with reed switches whose contacts were deactivated and locked in a hydrogen environment and reed switches whose contacts were untreated and sealed in a nitrogen or hydrogen atmosphere. The sticking occurrence rate was roughly the same in these cases. No sticking at all was detected with reed switches whose contacts were deactivated and locked in a nitrogen atmosphere. The oxygen amount on the contact surface was analyzed of each sample using Auger electron spectroscopy (AES) to define the relationship between the outcomes of the sticking test and oxygen along the touch surface. Figure 11.22b shows the results of the analysis. Virtually no oxygen exists on the contact surface of reed switches whose contacts were deactivated and locked in a hydrogen atmosphere: the hydrogen atmosphere effectively reduced the oxygen. The reed switches whose contacts were untreated and sealed in a nitrogen or hydrogen atmosphere also showed no oxygen on the contact surfaces. There was, however, a lot of oxygen on the contact surface of reed switches whose contacts were deactivated and sealed in a nitrogen atmosphere in the

TABLE 11.6

Test Samples

O_2N_2	Contact surface deactivated, sealed in N_2 atmosphere
O_2H_2	Contact surface deactivated, sealed in H_2 atmosphere
N_2	Untreated, sealed in N_2 atmosphere
H_2	Untreated, sealed in H_2 atmosphere

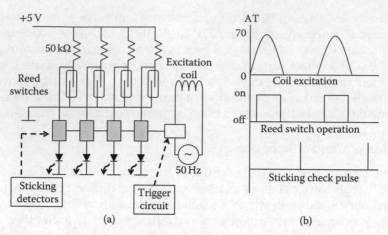

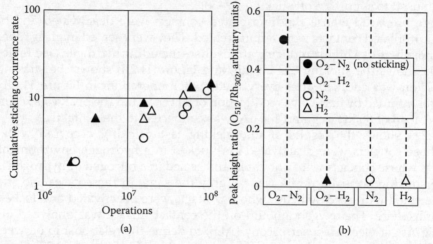

FIGURE 11.21
(a) Sticking test circuit. (b) Sticking test mode.

FIGURE 11.22
(a) Results of sticking test. (b) Amount of oxygen on contact surface.

form of Rh_2O_3 (see Section 11.3.1) Thus, only those reed switches free from sticking are those whose contacts are deactivated and sealed in a nitrogen atmosphere. Because there is a lot of oxygen on the contact surfaces of these switches, it can be deduced that oxygen on the rhodium-plated contact surface not only results in a stable contact resistance, but also is effective against contact sticking.

11.4 Applications of Reed Switches

Reed switches have been used as reed relays-reed switches with multilayer coils around their glass tubes—in many cases since their development. They have also been used, in combination with permanent magnets, with various devices and equipment. At present, many kinds of reed relays have been developed, and also diversified requirements have expanded the applications of magnet-driven reed switches [5,6]. Reed switches can also be used as thermal switches, which go on and off according to the temperature, by using thermal ferrite.

11.4.1 Reed Relays

A reed relay is an electromagnetic relay consisting of a reed switch and a driving coil around it or by using a permanent magnet as shown in Figures 11.2 and 11.3. The driving coil or magnet activates switching. There are many kinds of reed relays according to what they are to be applied for. Reed relays have the features described below.

1. They have excellent resistance to environment. Since the contacts are hermetically sealed in a glass tube, they are free from the influence of the surrounding atmosphere. An explosion-proof structure is employed.
2. They have excellent operating characteristics. The operating time, bounce time, and release time are short (see Figure 11.23). Although they vary, depending on the model, the operate time and bounce time are 0.5 ms maximum and the departure time is 0.2 ms maximum [31]. These are one tenth or less than those for mechanical relays.
3. Their input and output are completely isolated from each other electrically.
4. They are miniature and lightweight. The trend of reducing the size of devices and equipment causes reed relays to be miniaturized further. This trend is particularly noticeable with reed relays in ATEs (LSI testers and board testers), which use many reed relays. At present, miniature reed relays measure only 15 mm × 5 mm × 5 mm.
5. They have high sensitivity. Reed relays can control a large current and large voltage on the output side with a small input current of several to several tens of milliamperes.

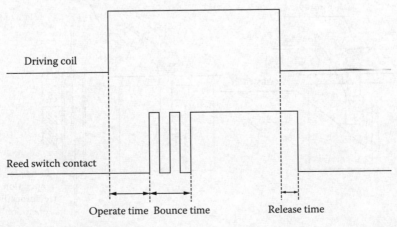

FIGURE 11.23
Operating characteristics of reed relay.

Various types of reed relays using these features have been developed along with the development of office automation (OA), factory automation (FA), and home automation (HA). Specifically speaking, they are SIP-type reed relays and DIP-type reed relays which are easily automatically mounted on printed circuit boards and can be directly driven by TTL-ICs and CMOS-ICs, surface-mounted-type reed relays, ultra-miniature reed relays, flat package reed relays, tubular-type reed relays that allow leads to be cut and bent at any point, and stand-type reed relays with a single end.

11.4.2 Applications of Magnetic-Driven Reed Switches

Figures 11.24 through 11.32 show many applications and are self explanatory.

The reed switch use is also expanding to higher power applications such as controls in hazardous atmosphere, railway control devices, railway signaling devices and electric power utility devices, elevator control, heavy industry control and machinery non-contact safety switches [32]. At the other extreme, micro-fabricated reed switches are being developed to operate with miniature electronic circuits [33].

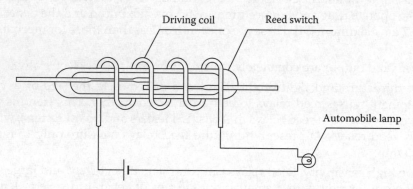

FIGURE 11.24
Lamp burnout sensor for an automobile.

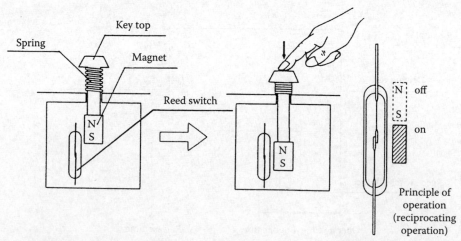

FIGURE 11.25
Key switch for use with key boards.

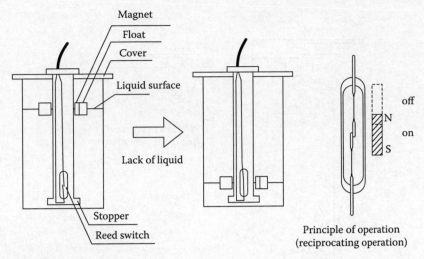

FIGURE 11.26
Float switch to detect fluid levels.

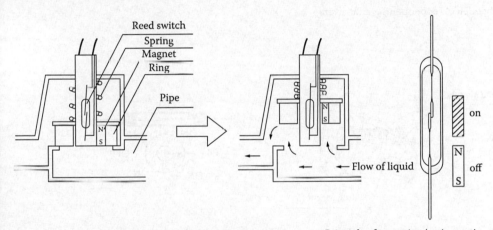

FIGURE 11.27
Flow detection switch to detect fluid motion in pipes.

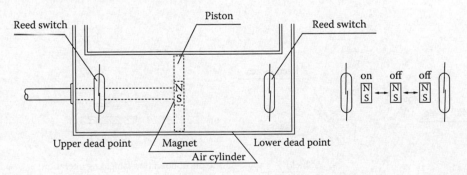

FIGURE 11.28
Air cylinder switch.

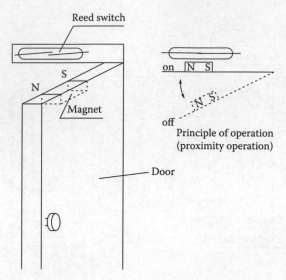

FIGURE 11.29
Door switch that detects door opening and closing.

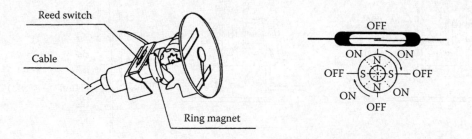

FIGURE 11.30
Speedometer switch.

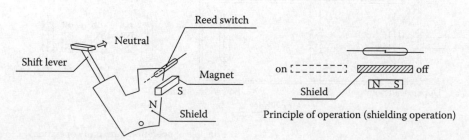

FIGURE 11.31
Neutral position detection switch.

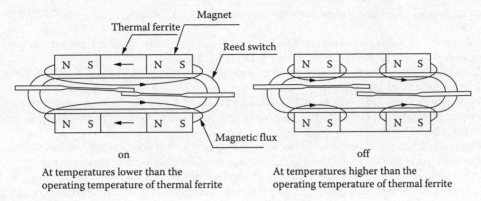

FIGURE 11.32
Principle of operating thermal switch for a wide application from electric ranges to automobile engines.

References

1. OM Hovgaard, GE Perreault. Development of reed switches and relays. *Bell Syst Tech J* 34 (2): 309, 1955.
2. LD Dumbauld. Dry reed switches. *Control Engineering*, July: 75, 1963.
3. HN Wagar. Some measurement techniques for characterizing sealed reed contact performance. *IEEE Trans Comp Parts*, December: 8, 1964.
4. R Holm. *Electric Contacts*, 4th ed. Berlin: Springer-Verlag, 3rd Printing, p 171, 2000.
5. T Yano, C Kawakita, M Yasuda, K Hinohara. *Reed Switches*. Tokyo: Oki Electric Industry, 1987.
6. C Kawakita, K Hinohara, T Kobayashi, Y Mizuguchi. Development of ultraminiature wide differential reed switches. Proceedings of the 39th Relay Conference, April 1991.
7. K Hinohara, T Kobayashi, C Kawakita. Magnetic and mechanical design of ultraminiature reed switches. *IEEE Trans Comp Hybrids, Manuf Technol* 15: 172–176, 1992.
8. RL Peek Jr. Magnetization and pull characteristics of mating magnetic reeds. *Bell Syst Tech J* 40 (2): 1961.
9. KF Bradford, FE Bader. Estimating sealed reed contact forces. *Bell Lab Rec* 44 (5): 1966.
10. K Kato, H Suzuki. Design of magnetic reed switch. *Trans IECE J51-C* (5): 211–218, 1968.
11. M Takahashi, S Yamazaki, K Kato, T Tanii, K Ono, M Nagao, M Watanabe. Development of DC-10 reed switch for DEX-2. *Elect Commun Lab Tech J* 18 (4): 79–99, 1969.
12. RJ Gashler. Magnetic force characteristics of electromechanical relays. Proceedings of the19th Relay Conference, pp 18.1–18.16, 1971.
13. T Yokokawa, T Yano, C Kawakita, K Hinohara, T Kobayashi. Contact surface conditions of ruthenium-plated contact reed switches. Proceedings of the 14th International Conference on Electric Contacts, pp 135–140, 1988.
14. T Yokokawa, T Yano, C Kawakita, K Hinohara, T Kobayashi. Thickness of ruthenium oxide film produced by the surface deactivation treatment of ruthenium-plated contact reed switches. Proceedings of the 35th IEEE Holm Conference on Electrical Contacts, pp 177–181, 1989.
15. L-J Xu, JG Zhang, B Miedzinski. Modeling of vacuum reed failure by using finite element method. Proceedings of the 46th IEEE Holm Conference on Electrical Contacts, pp 115–119, 2000.
16. Y Hashimoto, K Yokoyama, T Kobayashi. A study on the contact materials for ultra-miniature reed switches. Proceedings of the 24th International Conference on Electrical Contacts, pp 113–118, 2008.
17. K Arushanov, S Karabanov, I Zeltser, R Maizels, E Moos. New technology of modification of reed switch contact surfaces with the usage of ion-plasma nitriding. Proceedings of the 26th International Conference on Electrical Contacts, pp 284–287, 2012.

18. HW Hermance, TF Egan. Organic deposits on precious metal contacts. *Bell Syst Tech J* 37: 739–776, 1958.
19. T Yokokawa, C Kawakita. High reliability reed switches with rhodium plated contacts. Proceedings of the 21st Relay Conference, pp 13.1–13.7, 1973.
20. K Hinohara, A Nagai. Surface deactivation treatment of rhodium-plated contact reed switches. *OKI Tech Rev* 52 (123): 59–63, 1986.
21. T Yokokawa, T Yano, C Kawakita, K Hinohara, A Nagai. A study on the surface deactivation treatment of rhodium-plated contact reed switches. *IEEE Trans Comp Hybrids Manuf Technol CHMT* 9: 124–127, 1986.
22. T Yokokawa, T Yano, C Kawakita, K Hinohara, A Nagai, T Kobayashi. A study of the thickness of rhodium oxide film produced by the surface deactivation treatment of rhodium-plated contact reed switches. *IEEE Trans Comp Hybrids Manuf Technol CHMT* 10: 42–46, 1987.
23. DG Castner, GA Somorjai. LEED, AES and thermal desorption studies of the oxidation of the rhodium (111) surface. *Appl Surf Sci* June: 28–38, 1980.
24. H Ishida, A Ishitani. Raman microprobe analysis of thin films formed on the surface of silver electrical contacts utilizing the surface-enhanced Raman scattering effect. *Appl Spectrosc* 37 (5): 1983.
25. AD Hamer, DG Tisley, KA Walton. X-ray photoelectron spectra of compounds containing rhodium-halogen bonds and of rhodium (II) acetate and its derivatives: rhodium 3d and halogen np binding energies. *J Chem Soc Dalton* 116–120, 1973.
26. JS Brinen, A Melera. Electron spectroscopy for chemical analysis (ESCA) studies on catalysts. Rhodium on charcoal. *J Phys Chem* 76 (18): 1972.
27. T Yokokawa, T Yano, C Kawakita, K Hinohara, A Nagai. A study on the soft sticking of rhodium contact reed switches. Proceedings of the 34th Relay Conference, pp 12.1–12.3, 1986.
28. T Yokokawa, T Yano, C Kawakita, K Hinohara, T Kobayashi. A study on the rhodium-plated contact reed switch. Proceedings of the International Conference on Electrical Contacts and Electromechanical Components, pp 254–262, 1989.
29. T Yokokawa, T Yano, C Kawakita, K Hinohara, T Kobayashi. Contact performance degradation and surface deactivation techniques. Proceedings of the 1st International Symposium on Corrosion of Electronic Materials and Devices, Vol. 91-2, pp 423–429, 1991.
30. K Hinohara. Tribology in electrical contacts. *J Jpn Soc Tribol* 39 (5): 447–450, 1994.
31. N Wakatsuki, A Yamamoto. Bouncing of a reed switch due to coulomb's electrostatic force. Proceedings of the 52nd IEEE Holm Conference on Electrical Contacts, pp 165–169, 2006.
32. K Hamada, Y Shimizu, S Umezaki. Applications of heavy-duty reed switches. Proceedings of the 23rd International Conference on Electrical Contacts, pp 534–540, 2006.
33. S Day, T Christenson. A high aspect ratio micro-fabricated reed switch capable of hot switching. Proceedings 59th IEEE Holm Conference on Electrical Contacts, pp. 336–343, 2013.

12

Low Current and High Frequency Miniature Switches: Microelectromechanical Systems (MEMS), Metal Contact Switches

Benjamin F. Toler, Ronald A. Coutu, Jr., and John W. McBride

Transparent forms, too fine for human sight

Rape of the Lock, Alexander Pope

In small proportions we just beauties see;
And in short measure life may perfect be.

Short Measures (from an Ode), Ben Johnson

CONTENTS

12.1 Introduction 703
    12.1.1 Common MEMS Actuation Methods 704
12.2 Micro-Contact Resistance Modeling 705
12.3 Contact Materials for Performance and Reliability 713
12.4 Failure Modes and Reliability 720
12.5 Conclusion 725
References 725

12.1 Introduction

In this chapter, we will concentrate on the effect of very low contact forces on the design of microelectromechanical systems (MEMS) switch contacts and on the contact materials that have been employed in MEMS switches. We will not describe the details of MEMS switch design or their manufacture. Radio frequency microelectromechanical system (RF MEMS) switches can be used in mobile phones and other communication devices [1]. Often, micro-switches are used in phase shifters, impedance tuners and filters. Phase shifters, imped-ance tuners, and filters are control circuits found in many communication, radar and measurement systems [2]. MEMS switches offer much lower power consumption, much better isolation, and lower insertion loss compared to conventional field-effect transistor and PIN diode switches, however (see Chapter 9, Table 9.10), MEMS switch reliability is a major area for improvement for large-volume commercial applications [3]. The integrated circuit community is struggling to develop the future generations of ultralow-power digital integrated circuits and is beginning to examine micro switches [4]. Low power consumption, isolation, and reduced insertion loss are achieved by the mechanical actuation of the switch which physically opens or closes the circuit.

To enhance reliability, circuit designers need simple and accurate behavioral models of embedded switches in CAD tools to enable system-level simulations [5]. The MEMS literature indicates that changing the type of electrical load during testing reveals the physical limitation for micro-switches [6]. Rebeiz states that a good assumption for failure of the micro-switch is assumed to be when the contact resistance becomes greater than 5 Ω, which results in an insertion loss of -0.5 dB [1]. According to Rebeiz, the primary cause of micro-switch failure is due to plastic deformation in the contact interface such as "damage, pitting, and hardening of the metal contact area, which is a result of the impact forces between the top and bottom metal contacts" [1]. The description relates closely to "cold" switching mechanical failure. "Cold" switching is generally recognized to be actuating the switch repeatedly without applying RF or DC power during actuations, limiting the switch life of mechanical failures such as structural fatigue, memory effect, stiction of the actuators, etc. [6]. In "hot" switching, contributors to early micro-switch failure include "material transfer high current density in the contact region and localized high-temperature spots" [1].

12.1.1 Common MEMS Activation Methods

In terms of design, the most common form of actuation is electrostatic [7,8]. Figure 12.1 depicts an example of an electrostatic micro-switch. Electrostatic actuation offers the advantage of no power loss when the micro-switch is open. The disadvantage can be the high actuation voltages required to close the micro-switch. Other forms of actuation include electrothermal, magnetic, magnetostriction, and piezoelectric [9–11]. Electrothermal and magnetic actuation both offer the advantages of low control voltages and high contact force, but draw high current and dissipate significant levels of power when actuated [12]. Electrothermal also offers the advantage of being bi-directional with the ability to apply high force, but its disadvantages include slow actuation (millisecond range) as well as quiescent power loss. A disadvantage also shared by magnetic actuation, quiescent power loss implies the exercise of power

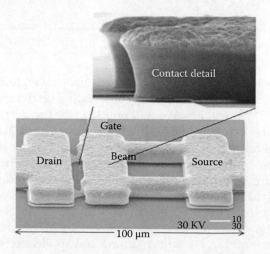

FIGURE 12.1
Electrostatic micro-switch example. (From Majumder S, McGruer N, Adams G, Zavracky P, Morrison R, Krim J, *Sensors Actuators*, vol. 93, no. 1, pp. 19–26, 2001 [7].)

at all times. Hysteresis is another disadvantage of magnetically actuated micro-switches. Comparatively, magnetic also has the advantages of high force actuation, being bi-directional but is also able to hit micro-second switching speeds. Also, due to their fabrication requirements, magnetic actuators are difficult to manufacture. Piezoelectric actuation can provide fast actuation speeds but due to the different layers of material that comprise a piezoelectric material, there is a parasitic thermal actuation caused by a differential thermal expansion of the different layers. Piezoelectric actuation has a disadvantage of "short throw" or small movement based on the number of stacked layers. Piezoelectric actuators also require high crystallization temperatures which make them difficult to integrate into a MEMS device. Given the lack of power loss when open and high isolation, electrostatic actuation is the most commonly used methods of MEMS engineers. Mechanical switch design considerations are focused on improving the functioning of the micro-switch through mechanical design innovation. From reducing the actuation voltage to decreasing the switching time, all aspects of performance focus on the beam geometry: i.e., "engineering away" shortfalls of the micro-electric contacts.

12.2 Micro-Contact Resistance Modeling

For DC micro-switches, resistance modeling requires knowledge of the surface of the two contact materials as well as their material properties. Though contaminants can have a major impact on micro-contact resistance, they are not initially considered for the description and determination of micro-contact resistance. Holm first identifies this in his example of contact resistance using two cylinders in contact at their bases [13], see also Section 2.6.7. When two surfaces meet, and because no surface is perfectly smooth, asperity peaks or "*a*-spots", from each surface meet at the interface and form contact areas, see Chapter 1. Asperities provide the only conducting paths for the conveyance of electrical current [8]. Figure 12.2 shows a graphical representation of the apparent contact area, contacting *a*-spots, and the effective radius of the actual conducting area. The effective area is utilized for making simplified contact resistance calculations. Holm also investigated contact resistance changes due to plastic and elastic deformation of *a*-spots; which greatly affects the interface of the contact areas. Resistance for the cylinders then, is simply the measured voltage between the two rods divided by the current flowing through them.

Majumder et al., modeled micro-contact resistance with three steps [7]. First, see the contact force, as a function of applied gate voltage, available for the mechanical design of the electrostatically actuated micro-switch. Second, specify the effective contact area at

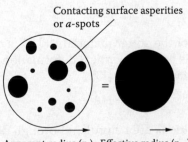

Contacting surface asperities
or *a*-spots

Apparent radius (r_a)   Effective radius (r_{eff})

FIGURE 12.2
A-spots as an effective radius. (From Coutu R, McBride J, Starman L, *Proceedings of the 55th IEEE Holm Conference on Electrical Contacts*, Vancouver, British Columbia, Canada, pp. 296–299, 2009 [14].)

the interface as a function of contact force [7]. Finally, determine the contact resistance as a function of the distribution and sizes of the contact areas. Majumder et al., like Holm, also noted that the surface profile of the contact interface is sensitive to plastic and elastic deformation. He also investigated ballistic electron transport using Sharvin's equation.

Elastic modeling is accurate for extremely low values of contact force (a few mN) where surface asperities retain their physical forms after the contact force is removed. Elastic-plastic deformation occurs at the boundary between the permanent plastic deformation and the temporary elastic deformation. Under plastic deformation, permanent surface change occurs by the displacement of atoms in asperity peaks [15] whereas neighboring atoms are retained under elastic deformation [16].

Asperity contact area under elastic deformation is given by [17]:

$$A = \pi R \alpha \tag{12.1}$$

where A is contact area, R is asperity peak radius of curvature, and α is asperity vertical deformation. Hertz's model for effective contact area for elastic deformation as:

$$r_{eff} = \sqrt[3]{\frac{3F_{cE}R}{4E^t}} \tag{12.2}$$

where E' is the effective Hertzian modulus derived from

$$\frac{1}{E'} = \frac{1-v_1^2}{E_1} + \frac{1-v_2^2}{E_2} \tag{12.3}$$

with E_1 as the elastic modulus for contact one, v_1 is Poisson's ratio for contact one, E_2 as the elastic modulus for contact two, v_2 is Poisson's ratio for contact two [7,18]. Force is related to the asperity contact area by [17]

$$F_{cE} = \frac{4}{3}E'\alpha\sqrt{R\alpha}. \tag{12.4}$$

To account for the asperity contact area and force under plastic deformation, the well known model from Abbot and Firestone that assumes sufficiently large contact pressure and no material creep is used [19]. Single asperity contact area and effective contact area are defined using Equations 12.5 and 12.6 [19]:

$$A = 2\pi R \alpha \tag{12.5}$$

$$r_{eff} = \sqrt{\frac{F_{cp}}{H\pi}} \tag{12.6}$$

where H is the Meyer hardness of the softer material [18], A is contact area, R is asperity peak radius of curvature, and α is asperity vertical deformation [19]. The effective contact area radius is then related to contact force by [13]

$$F_{cp} = HA \tag{12.7}$$

While plastic and elastic definitions are helpful, a thorough description of deformation cannot be provided without considering the elastic plastic transition between the two kinds of deformation. Elastic-plastic material deformation asperity contact area is given as

$$A = \pi R \alpha (2 - \alpha_c/\alpha) \tag{12.8}$$

where α_c is the critical vertical deformation, where elastic-plastic behavior begins [17]. Effective contact area is given by

$$r_{\text{eff}} = \sqrt{\frac{F_{cEP}}{H\pi\left[1.0.62 + 0.354\left(\frac{2}{3}K_Y - \left(\frac{\alpha_c}{\alpha}\right)\right)\right]}} \qquad (12.9)$$

where K_Y as the yield coefficient. Inclusion of Chang's force equation and rigorous mathematical manipulation provides the relationship between contact area and contact force [17]:

$$F_{cEP} = K_H A \qquad (12.10)$$

with K_H is a hardness coefficient. Equations 12.5, 12.6, and 12.10 provide the relationship between contact force and effective conducting area. It is important to note that contact force directly influences the effective bearing area and will also impact contact resistance by default.

For the standard test configuration, resistance can be modeled as shown in Figure 12.3 where R_c represents contact resistance, R_{cf} represents the resistance due to contaminate films, R_{sh} is sheet resistance, and R_{par} is the parasitic resistance from solder connections, clip leads, wires, etc. [20]. This R_{sh} and R_{par} can be eliminated if the experiment is set up using a four wire cross bar configuration [13], see also Section 2.6.7.

As mentioned in Section 1.6 of Chapter 1, based on the effective conducting area and how it compares with the mean free path of an electron, current flow is described as ballistic, quasi-ballistic, or diffusive [14]. Figure 12.4 shows a plot of Mikrajuddin's derived gamma function which describes electron flow as a function of the Knudsen number K, which is calculated by the effective radius and the electron's elastic mean free path. The significance is that it describes situations for complete diffusive electron transport or complete ballistic electron transport whereas Wexler's original derivation included all higher order effects.

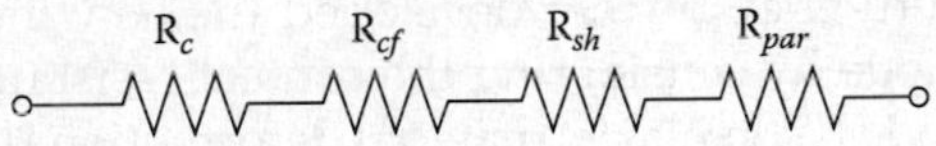

$$R_c \qquad R_{cf} \qquad R_{sh} \qquad R_{par}$$

FIGURE 12.3
Standard test configuration resistance model [20].

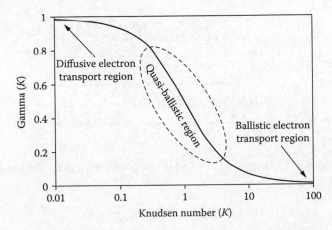

FIGURE 12.4
A plot of Mikrajuddin et al.'s derived Gamma function. (From Coutu R, McBride J, Starman L, *Proceedings of the 55th IEEE Holm Conference on Electrical Contacts*, Vancouver, British Columbia, Canada, pp. 296–299, 2009 [14].)

Previously, Wexler derived an interpolation for electron transport between ballistic and diffusive transport regions and Majumder et al. developed the following micro-contact resistance model [7,21]:

$$R_w = R_S + \Gamma(K)R_c \tag{12.11}$$

where $\Gamma(K)$ is a slowly varying Gamma function of unity order [21], R_S is the Sharvin resistance, and R_c is the constriction resistance based on diffusive electron transport. The semi-classical approximation for resistance when electrons exhibit ballistic transport behavior is the Sharvin resistance formula shown as Equation 12.12 [7].

$$R_S = \frac{4\rho K}{3\pi r_{eff}} \tag{12.12}$$

Majumder et al's model was improved when Coutu et al developed a new micro-contact resistance model for elastic deformation based on contact resistance considering elastic deformation and diffusive electron transport as well as a contact resistance model considering elastic deformation and ballistic electron transport, shown by Equation 12.13 [7].

$$R_{WE} = R_{cBE} + \Gamma(K)R_{cDE} \tag{12.13}$$

where R_{WE} is the Wexler-based resistance (using Mikrajuddin's derived Gamma function) for the elastic material deformation [18]. In this instance, $\Gamma(K)$ has been replaced by Mikrajuddin et al.'s well behaved Gamma function describing complete diffusion:

$$\Gamma(K) \approx \frac{2}{\pi} \int_0^\infty e^{-Kx} \mathrm{Sinc}(x)\,\mathrm{d}x. \tag{12.14}$$

Continuing this work, Coutu et al. further developed the new model for contact resistance in the elastic-plastic mode by considering the contact resistance equation based on ballistic electron transport and elastic-plastic material deformation [18] (shown in Equation 12.15) and contact resistance based on diffusive electron transport and elastic-plastic material deformation (shown in Equation 12.16) [18].

$$R_{cBEP} = \frac{4\rho K}{3\pi} \sqrt{\frac{H\pi\left[1.062 + 0.354\left(\frac{2}{3}K_Y - 3\left(\frac{\alpha_c}{\alpha}\right)\right)\right]}{F_c}} \tag{12.15}$$

$$R_{cDEP} = \frac{\rho}{2} \sqrt{\frac{H\pi\left[1.062 + 0.354\left(\frac{2}{3}K_Y - 3\left(\frac{\alpha_c}{\alpha}\right)\right)\right]}{F_c}} \tag{12.16}$$

With Equations 12.15 and 12.16 the new model for contact resistance for elastic-plastic deformation is then [18]:

$$R_{WEP} = R_{cBEP} + \Gamma(K)R_{cDEP} \tag{12.17}$$

Convergence of electrical current flow lines from a distance far from the constriction and the subsequent spreading out of the current from the constriction is known as constriction resistance or commonly contact resistance [13,22]. The spreading resistance inherently affects contact resistance. To model spreading resistance, Karmalkar et al. developed a simple closed-form model to predict accurate and complex calculations of circular and rectangular contact spreading resistances [23]. The method was to solve the three dimensional Laplace equation

$$\nabla^2 \psi = 0 \tag{12.18}$$

subject to the appropriate boundary conditions in several iterations to consider changing geometries. Holm, by contrast, represented spreading resistance as a 5% increase in constriction resistance [13]. By interpolating the results of the different geometric solutions, the resistance average was calculated [23]. Experimental tests revealed close (within 2%) agreement with standard numerical analysis software. Their study found that the developed model accurately predicts all the trends of resistance, to include a significant variation as a function of the smaller electrode location, dependence on the electrode separation-to-width ratio, and saturation with increase in the larger electrode area for both equipotential and uniform current density boundary conditions [23].

When considering micro-contacts, surface contamination has severe negative impacts for electrical contacts by physically separating the conductive electrode surfaces [24]. Based on thickness and composition, the adsorbed contaminants can increase contact resistance by orders of magnitude [24]. When left exposed to the ambient lab air, device surfaces can be covered with various contaminants which will affect conductivity [25]. This concept was experimentally verified by Lumbantobing et al. where during cyclical contact loading, electrical contact resistance was erratic due to the strong dependence of contact resistance on an insulating thin film at the contact interface [26].

Timsit explored the effect of constriction resistance on thin film contacts. He postulated that the spreading resistance of an asperity in a thin film will be drastically different than of an asperity in bulk material due to the different boundary conditions [22]. This convergence is visually presented in Figure 12.5.

His work revealed that the contact resistance for a contact with two identical films can be instantly calculated as twice the spreading resistance [22]. Also, the constriction resistance between two films of the same thickness L in contact over a constriction of radius a deviates greatly from the classical expression $p/2a$ for two contacting bulk solids wherever $a/L \geq 0.02$ [22]. A counter-intuitive discovery was shown revealing that spreading resistance in a radially-conducting film initially decreases with decreasing film thickness [22]. This is counter-intuitive because the resistance of a solid conductor increases with decreasing thickness [22]. Lumbantobing et al. experienced reduced electrical contact resistance on a contact with a native oxide during cyclic contact loading and attributed the reduced resistance to the local rupture of the film, resulting in asperity nanocontacts that reduced the resistance [26]. They also found that the nearly uniform thickness of the native oxide film predicted in their experiment illustrated the strength and robustness of oxide thin film under the tested loading conditions.

To examine contact resistance models based on thin films, Sawada et al. performed a current density analysis of thin film effects in the contact area on a LED wafer [27]. By utilizing a unique setup using an indium bump to match with a gallium phosphorous wafer, the team was able to examine and image current flow through the contact. Imaging was possible because the current flow was causing optical emission in the wafer. The images and results showed that the current flow in the contact was located primarily around the perimeter of the contact. The role of imaging enabled greater insight in determining actual

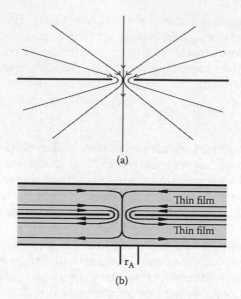

FIGURE 12.5
(a) Spreading of current streamlines in two "bulk" conductors in contact over a circular spot of radius a.
(b) Spreading of current streamlines near a constriction between two thin films. (From Timsit, R, *IEEE Transactions on Components and Packaging Technologies*, vol. 33, no. 3, pp. 636–642, 2010 [22].)

micro-contact area. The research showed that classical theory for contact resistance was sufficient provided the conducting film was sufficiently thick (200 µm) [27]. The results were in agreement with Timsit's model for the constriction resistance of thin films,. Still, if the film thickness is 50 µm or less, the value of the contact resistance is bigger than the bounds of the classical theory.

Timsit's results are very similar to those published by Norberg et al. whereby contact resistance in thin films was approximated by empirical modifications of Holm's classical relation:

$$R_s = \frac{\rho}{2a} \tag{12.19}$$

where ρ is the resistivity of the conducting material and a is the radius of the constriction [13,28]. However, Norberg et al. approximated constriction resistance for more complex geometries and the effects of bulk resistance being isolated from constriction resistance [22].

Timsit also examined the major electrical conduction mechanisms through small constrictions and concluded that the onset of the Sharvin resistance, which stems from ballistic electronic motion in a constriction, eventually invalidates the basic assumptions of classical electrical contact theory [29]. He reported through the use of a simple a-spot model and quantum mechanics that the cooling of a small a-spot due to heat loss by the surrounding electrically-insulating films is not sufficiently large enough to have an impact or account for the breakdown of classical theory. The conjecture is proposed that for metal atomic scale constrictions, a single atom corresponds to a single conductance channel which implies that the conductance would not decrease smoothly as the mechanical contact load is decreased [29]. Instead, the conductance will drop in well-defined steps since the act of contacting atoms is decreased by discrete units of one or a few at a time [29,30]. The thin film work of Timsit, Norberg, and Sawada provide necessary insight for future micro-contacts study since the thin films studied are in the same order as films routinely used to fabricate MEMS switches.

As seen in Chapter 1, assumptions about asperity size and quantity greatly impact contact resistance calculations. While most contact resistance models consider only single "small" constrictions, the typical rough surface may include many small contacts of varying sizes. However, quantum effects may be present with sufficiently small contacts. To investigate the quantum and size dependent contact mechanisms of the asperity sizes on typical surfaces, Jackson et al. examined the effect of scale dependent mechanical and electrical properties on electrical contact resistance between rough surfaces [31]. Beginning with classical contact mechanics, they used established multi-scale models for perfectly elastic and elastic-plastic contacts for the purpose of predicting electrical contact resistance between surfaces with multiple scales of roughness. They then examined scale dependent strength of the materials tin and gold and found that the yield strength varies by over two orders of magnitude as the contact diameter changes. Lastly, using an iterative multi-scale sinusoidal method to calculate the median radius of a contact at given scales, an analytical model of electrical contact resistance was broken.

Poulain et al. examined quantized conductance with micro-contacts by breaking contact in such a way that the dimensions of the conducting members of the micro-contact were much smaller than the mean free path of the electron [32]. The team found that by using a micro-switch and a nanoindentor, they were able to witness quantized conductance plateaus before separation of the two contact members. The conditions for observing the quantized conduction phenomena are a switch opening at an extremely low speed and a current limitation near 150 µA [32]. Upper and lower micro-contacts made of Au/Au as well as micro-contacts made of Ru/Ru were tested. Independent of the contact material, the quantized conductance behavior was witnessed. The plateaus were consistent with theoretical predictions for quantum ballistic transport in atomic-sized contacts. The work showed that the metallic bridge formed during contact separation, in the last stage of break (see Section 9.4.2), consists of only a few atoms and is similar to a nanowire or waveguide that reveals the wave character of the electrons [32]. The team observed that the reproducibility of the results is difficult due to the fact that the elongation of the atomic-sized bridge is difficult to control and is strongly related to atomic arrangements [32].

While dealing with the quantum theory to describe current flow through nano scale asperities is being explored, some researchers are developing methods to simulate electrical contact resistance of ohmic switches with Finite Element Modeling (FEM). Pennec et al. examined the impact of surface roughness on the electrical contact resistance under low actuation forces (from a few tens of micronewtons to 10 mN) [33]. An important aspect of their work was to clearly define the surface roughness of the contact. The common practice is to take the average radius of curvature of the asperities which is determined by a measurement of the surface profile [34]. The drawback of this method is common that the determination of the average radius is subjective to the scale of the observation, and is also limited by the measurement resolution [35–36]. In order to clearly determine the surface roughness of the contact, three methods were examined: statistical, fractal, and deterministic [33]. A statistical approach is based on a stochastic analysis which can be limited to the firmness of the measuring instrument [33]. A fractal method, on the other hand, random surface texture is characterized by scale-independent fractal parameters [33].

The deterministic approach was chosen due to its closest representation of the actual surface [33,38]. Deterministic methods capture discrete data points for real heights on the surface which avoids assumptions of the micro geometry of the *a*-spots [33]. Kogut stated that even though there are several methods to model contacting rough surfaces, the most convenient one is the probabilistic approach [38]. This approach replaces the two rough surfaces by a smooth surface in contact with an equivalent rough surface, replacing

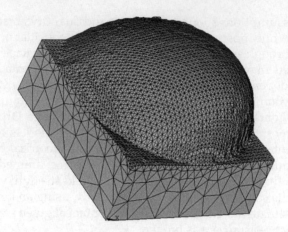

FIGURE 12.6
Geometry of finite element mesh in rough contact bump. (From Pennec F, Peyrou D, Leray D, Pons P, Plana R, Courtade F, *IEEE Transactions on Components, Packaging and Manufacturing Technology*, vol. 2, no. 1, pp. 85–94, 2012 [33].)

asperities with simple geometric shapes, and getting into a probability distribution for the asperity parameters [39]. The probabilistic model was developed by Greenwood and Williamson and was developed for elastic contacts [15].

As shown in Figure 12.6, using an Atomic Force Microscope (AFM), the team was able to capture 3-D data points of contact bumps and apply a low resolution mesh in order to quickly determine the effective contact area under 100 μN of force [33]. By stepping up the resolution for the effective contact area to the effective computation memory limits, Pennec et al. were able to model a contact resistance in agreement with literature [33]. While their method did not take into account contaminant films, the results show that including the fine-scale details of the surface roughness must be taken into account when calculating contact resistance [33]. However, while AFM's can achieve 1 nm resolution of surfaces, the number of contact elements and definition of elastic-plastic materials in the model can prevent the calculations from succeeding due to computer memory limitations [33]. Conclusive evidence is given that reducing the sampling interval from 1 to 10 nm is sufficient for the calculation of electrical contact resistance.

Proponents of fractal models, Rezvanian et al. believe that the random and the multiscale nature of the surface roughness can be better described by fractal geometry [7,37,40]. Fractal-based models have been developed by a number of researchers but lack considerations for elasticity [41,42]. Persson et al. developed a novel fractal method which is not dependent on fractal roughness and is not scale dependent like the Greenwood model [43,44]. The disadvantage is that this method is exclusive to fractal surfaces [45].

Similar in nature, Wilson et al. considered multi-scale roughness, or the description of the surface, to be sinusoids stacked into layers to represent the rough surface [46]. While quantitative discrepancies exist between the statistical methods and layered sinusoids, the team was able to show qualitative similarities for both elastic and elastic-plastic deformation [46]. In fact, until higher force loads are reached, the model is very much in agreement with standard methods [46]. At higher loads where the contact radius is large compared to the asperity tip radius, the models differ greatly [46]. This method of stacking sinusoids however is not limited to contact resistance but is also employed to model adhesion [47]. Where the classical approximation for an area is a simplified model that typically bundles asperities into a few, the stacked sinusoids allow for a more practical representation of a multiscale surface [47].

So far, it is evident that the surface of the physically connecting electrodes is a key ingredient for the determination of electrical contact resistance. Modifications to the surface via a thin film from adsorbed contaminants from either ambient air or hermetic environments will greatly decrease the conductivity of the contact. To improve electrical conduction between contacts, Jackson et al. have tried to reduce the contact resistance by applying an anisotropic conductive thin film [48]. These films are typically an epoxy that is doped with conductive metal particles [48]. While classical electrical resistance theory falls short for accurately predicting the contact resistance with an insulating thin film, a model is proposed by Jackson and Kogut to consider elastic-plastic behavior of the thin film and large deformations of the conductive particles [48]. While previous anisotropic conductive film models have under predicted electrical contact resistance, a conjecture is established that the difference may be accounted for by the quantum effect of electron tunneling that takes place through the energy barrier imposed by the thin film [48]. This "tunneling" resistance is higher than the constriction resistance [48]. Using empirical models, mechanical and electrical material constants were held constant and the radius of the conductive particles in the film were varied [48]. The results revealed that particle size influenced contact resistance and that the larger radius provided a lower resistance [48].

When it comes to contact resistance modeling, contact material deformation and the effective contact area radius are the two primary considerations [14]. An assumption that individual *a*-spots are sufficiently close and that a single effective area model is typically made to determine specific electron transport regions by comparing the effective radius and mean free path of an electron [14]. As seen by area models to characterize the surface topology, describing the appropriate effective area for modeling is difficult. From the modeling of the surface using statistical, deterministic, or fractal means to the models of contact resistance based on all the deformation modes, the development of a thin film will widen the variance between simulated and actual results. Contact materials also have an integral role in determining the performance and reliability of micro-switches. Hardness as well as conductivity and other material properties influence the contact resistance. Gold, palladium, and platinum are commonly used [49]. Due to the fact that these materials are very soft and wear easily, other materials such as ruthenium and combination of materials have been examined for their effectiveness at lengthening the lifecycle and the performance of the contact. For example, materials such as Au, Ru, Rh, Ni, were compared in mixed configurations to try and increase reliability [50].

12.3 Contact Materials for Performance and Reliability

The earlier discussion of micro-contact resistance modeling showed how the material properties of the contact impact the contact resistance. The intrinsic properties of the materials chosen for the contact are important for increasing the lifecycle of the contact. For instance, due to its low electrical resistivity and low sensitivity to oxidation [51], gold is widely employed as a contact material in MEMS [52]. In general, contacts are desired to have excellent electrical conductivity for low loss, high melting point to handle the heat dissipated from power loss, appropriate hardness to avoid material transfer and chemical inertness to avoid oxidation [53]. As will be discussed in the failure modes and reliability section, material transfer can take place less easily with harder materials.

Material hardness is an important property as the surface of the contact will change with actuations over time. As the surface changes, changes to contact resistance occur

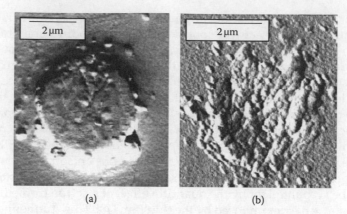

FIGURE 12.7
Au on Au contact SEM images of contact surfaces (a) is top electrode (b) is bottom electrode. (From Yang Z, et al., *IEEE Journal of Microelectromechanical Systems*, vol. 18, no. 2, pp. 287–295, 2009 [3].)

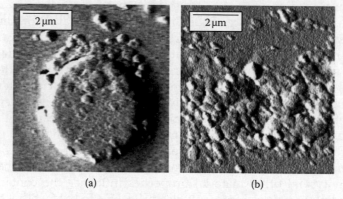

FIGURE 12.8
Au–Ni alloy contacts. (From Yang Z, et al., *IEEE Journal of Microelectromechanical Systems*, vol. 18, no. 2, pp. 287–295, 2009 [3].)

simultaneously. The surface change can be seen in Figure 12.7 which shows SEM images of Au–Au contacts after a lifecycle test [3]. Alloys are often created in order to take advantage of material properties to try and minimize the effect of material transfer [3]. As can be seen in by comparing Figure 12.8 to Figure 12.7, Zang et al. showed Au–Ni alloy contacts resist material transfer better than Au–Au contacts.

McGruer et al. showed that ruthenium (Ru), platinum (Pt), and rhodium (Rh) were susceptible to contamination and the contact resistance increased after a characteristic number of cycles, while gold alloys with a high gold percentage showed no contact resistance degradation under the same test conditions [3,54]. Similarly, Coutu et al. showed that alloying gold with palladium (Pd) or Pt extended the micro-switch lifetimes with a small increase in contact resistance [18]. Failure is typically defined as an increase in contact resistance beyond a given tolerance set by the circuit designer. As is shown in Figure 12.9, contact resistance tends to increase towards the end of a micro-switches lifetime.

As will be discussed later, frictional polymers are carbon based insulating films which develop over time and increase contact resistance [8]. Despite carbon being a core component to frictional polymers, Yaglioglu et al. examined the electrical contact properties of carbon nanotube (CNT) coated surfaces [56]. The high Young's Modulus and potential

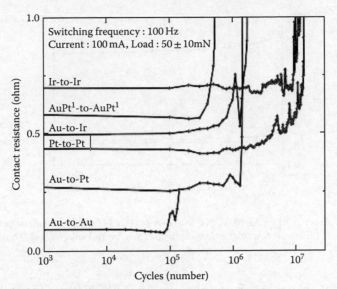

FIGURE 12.9
Evolution of contact resistance for hot switching 100 mA at a contact force of 50 mN. (From Kwon H, et al., *IEEE 20th International Conference on Micro Electro Mechanical Systems*, 2007 [55].)

FIGURE 12.10
2–4 μm of Au coating on MWNT. (From Yunus E, McBride J, Spearing S, *IEEE Transactions on Components and Packaging Technologies*, vol. 32, no. 3, pp. 650–657, 2009 [58].)

for low resistance of CNTs makes them worthy candidates for micro-switch contacts. For instance, Au contacts with a substrate coated with tangled single-walled CNTs were shown to have a resistivity between 1×10^{-4} and 1.8×10^{-4} Ωm [56]. CNTs have been reported to have an elastic modulus of approximately 1 TPa, which is comparable to diamond's elastic modulus of 1.2 Tpa [57]. Yunus et al. explored two contact pairs with carbon nanotubes: Au to multiwall carbon nanotubes (MWNTs), where one electrode is Au and the other is MWNTs, and Au to Au/MWNT composite, where the contact interface is Au on [58] jan. Figure 12.10 shows an SEM image of the Au/MWNT composite.

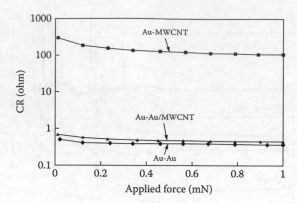

FIGURE 12.11

Comparison contact resistance (CR) of Au-MWCNT to Au–Au/MWCNT and Au–Au contacts. (From Yunus E, McBride J, Spearing S, *IEEE Transactions on Components and Packaging Technologies*, vol. 32, no. 3, pp. 650–657, 2009 [58].)

As shown in Figure 12.11, it was found that the Au/MWCNT was the better performer than Au-MWCNT in terms of contact resistance [58]. The data was collected with a nanoindentor apparatus which cycled for ten repeated operations with a maximum applied load of 1 mN [58]. The hardness of each material is also dramatically different, approximately 1 TPa for CNT and 1 GPa for Au [58]. The CNT structure supporting the Au film acts to allow the Au film to deform elastically under the applied load. In this study, a hard Au coated steel ball is making contact with the softer Au/MWCNT surface. The latter surface deforms to the shape of the steel ball, increasing the apparent contact area. With the Au coated steel ball in contact with the MWCNT surface the conduction path is through the lateral connection of the vertically aligned CNTs; leading to a higher contact resistance, as shown in Figure 12.11. A disadvantage to the mechanical design of the switch was discovered to be excessive bouncing on closure; that is, the contact takes time to settle in the closed position.

A study was conducted by Choi et al. to explore the current density capability of a CNT array with an average CNT diameter of 1.2 nm, site density of 2CNT/μm, and the number of CNTs for devices with 1 μm channel width ranged from one to three [59]. It was reported that a high current density of 330 A/cm^2 at 10 V bias was successfully transmitted through the contact without any noticeable degradation or failure [59]. A reliability test, as seen in Figure 12.12, with an input current of 1 mA showed repeatable and consistent contact characteristics over a million cycles of operation [59].

It is reported in literature that the small contact area between carbon nanotubes and a metal electrode makes electrical coupling between them extremely difficult [60–63]. An experiment was performed by Chai et al. to verify if a graphite interfacial layer would increase the electrical contact to the CNTs [60]. Graphite was chosen due to its close material properties to the CNTs namely, metal-like resistivity and similar chemical bonding [60]. A common technique for carbon deposition to the CNT contact region is to use the electron beam inside of a scanning electron microscope (SEM) to induce carbon deposition [60]. The technique is reported to successfully form low resistance electrical contact to multiwalled CNTs [64,65]. Chai's experiment validated that the graphitic carbon interfacial layer did reduce the contact resistance due to the increase in contact area to the CNT [60].

Another metal-coated-contact switch design was explored by Ke et al. which used ruthenium (Ru) on the Gold (Au) contact surface [66]. In an example of engineering a switch

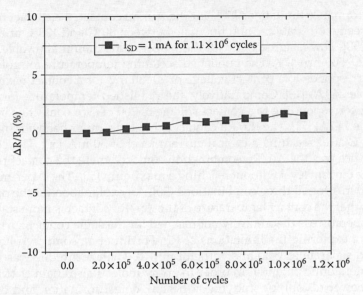

FIGURE 12.12
Reliability test results shown the resistance change over 1.1×10^6 cycles at 1 mA in non-hermetic environment. (From Choi J, et al., *IEEE 24th International Conference on Micro Electro Mechanical Systems (MEMS)*, 2011 [59].)

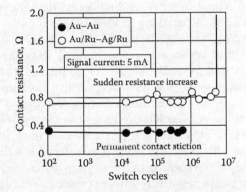

FIGURE 12.13
Hot switching life cycle tests for Au–Au switch compared to Au/Ru switch. (From Ke F, Miao J, Oberhammer J, *Journal of Microelectromechanical Systems*, vol. 17, no. 6, pp. 1447–1459, 2008 [66].)

for increased lifetime, the idea was to coat the Au contact with Ru, a harder material with low resistivity [66]. The contact resistance and life time of the Ru-layered Au switch were compared to the common Au–Au micro-contact switches [66]. The switches demonstrated a lifetime enhancement of over 10 times as measured in a non-hermetic environment as compared to pure, soft gold contacts as shown in Figure 12.13 [66]. Au on the other hand, any alloying of Au with other metals will result in increased hardness, but also an increased resistivity [67]. Atomic-level simulations and experimental observations have shown that the separation of gold contacts leads to considerable material transfer from one side of the contact to the other [68–70].

Broue et al. characterized Au/Au, Au/Ru, and Ru/Ru (upper and lower contact materials respectively) ohmic contacts by examining the temperature of the contact in the on-state to determine its performance limitations [50]. For the Au/Au contact, the contact temperature was linear until seemed to stabilize and fluctuate between 80°C and 120°C after the application of 40 mA [50]. This is agreeable with the reported maximum allowable current for gold contacts of 20–500 mA [71]. The published softening temperature for gold contact is ~100°C, which corresponds to a contact voltage of 70–80 mV for a contact near room temperature (see Table 24.1A) [13]. Comparatively, the published temperature for ruthenium contact is ~430°C, corresponding to a contact voltage of 200 mV for contact near room temperature (see Table 24.1A) [71]. The Ru/Ru contact exhibited similar behavior in that it fluctuated about 400°C after reaching a critical current level of 30 mA [50]. The contact with the best performance was the Au/Ru combination contact, where the contact temperature increased with the current level without reaching a maximum [50]. The experiment went as far as to apply 100 mA for all three combinations [50]. An explanation for the difference in performance was that the contact temperature of the Au/Ru contact is more stable because the softening temperature is theoretically not reached for the same contact current [50].

In the interest of exploring the limitations of Ru, Fortini et al. compared how asperity contacts form and separate in gold and ruthenium [72]. Their technique was to establish an appropriate interatomic potential in order to apply molecular dynamics (MD) simulation, which is a powerful tool for studying adhesion, defect formation and deformation on the nano-scale level [72]. The MD technique enabled the team to see the formation and separation of nanoscale asperity contacts by simulating the movement of the atoms [73]. The simulations showed that Ru was ductile at 600 K and more brittle at 300 K, where it separated by a combination of fracture and plasticity [72]. Gold exhibited ductile behavior at both 150 and 300 K [72,74]. The difference in ductile/brittle behavior of the Au and Ru contacts has consistent with FEM calculations in literature [75].

Other researchers have explored using Tungsten (W) as the contact material. Tungsten was chosen for its hardness and resistance to mechanical stress and physical deformation [76]. Kam et al. verified that W is beneficial for improved resistance to wear and micro-welding [77]. A disadvantage of W is its susceptibility to chemically react and form oxides on the surface [76]. It is reported that the oxidation of exposed W electrode surfaces occurs if there is any ambient oxygen, which increases the rate of oxidation exponentially with increasing temperature [78]. This was verified by Spencer et al., who studied the oxide layers as they became thicker with exposure time; they offered the theory that the exacerbated rate of oxidation is due to the widening of the oxygen diffusion path as the oxide gets thicker [79]. This thin film of oxide negatively impacts the contact resistance and requires higher contact loading to break through the film and obtain low resistance. The experiments of Chen et al. show that W electrodes show an undesirable increase in the on-state resistance over the lifetime of the device [76]. The oxidation of W was sped up by the amount of current flowing through the contact. Energy losses in the form of heat increased the opportunity and rate of thermal oxidation. They offered two solutions to the oxidation problem of W electrodes: either use another material or minimize device exposure to oxygen with a wafer-level encapsulation process [76,80].

Yamashita et al. investigated the use of an anti-stiction coating for ohmic micro-contacts under low loads (0–70 µN) [81]. The contacts were coated with thiophenol and 2-naphthalenethiol. The coatings successfully prevented the formation of the liquid meniscus, eliminate or reduce the capillary force better than the bare Au surface, and reduce the van der Waals forces [81]. They also noted that increased surface roughness could prevent stiction exponentially by reducing the effective contact area [81]. Increasing surface roughness would trade performance in terms of lower contact resistance which relies on large effective area of contact for anti-stiction properties. With the coatings applied, contact resistance decreased after 16 μN but required at least 4 μN of contact force for current to begin to flow [81]. It was found in the study that the contact resistances of the samples deposited with a 100 nm-thick Au layer were slightly smaller than those with a 20 nm-thick Au layer despite the larger resistivity value of the thicker layer due to the relationship between surface roughness and resistivity [81,82]. An answer was offered by Yamashita et al. that the contact area between the electrodes was larger for the thick layer because the electrodes made contact with large crystal grains [81]. Consistent with literature and classical theory, the results showed that contact resistance decreased proportionally with increasing contact force for all samples [81]. Because of increased contact forces, the contact resistance drops with an increase in asperity deformation, which provides a greater contact area as the micro-geometry changes [83,84].

As expressed earlier, increase in contact area decreases electrical contact resistance. One method to increase contact area was examined by Baek et al. using compliant nickel nanowire arrays [85]. The concept is to guarantee an approximate number of contact points for current to flow when the electrode surfaces mate instead of relying on rough approximations for asperity micro-geometry. Since the nanowires are compliant, the effective contact area would increase as contact force is increased overall decreasing contact resistance. The array was employed to achieve a minimum contact resistance of 73 mΩ for a contact area of 0.45 mm² using an array of compliant nickel nanowire [85]. The wires were fabricated by electrodeposition and porous filters in order to achieve a maximum aspect ratio of 300:1 (60 μm × 0.2 μm) [85]. Images of the nanowire arrays are shown in Figure 12.14.

Regarding the reduction in electrical contact resistance for tin contacts, Myers et al. proposed a new contact design in order to lower contact resistance which is not limited to tin [86]. The fundamental principles behind the concept was that when the classical Hertzian

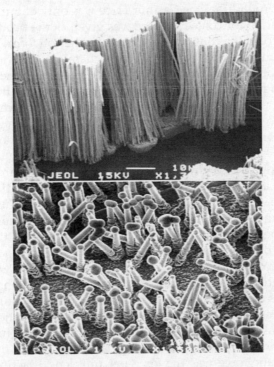

FIGURE 12.14
SEM images of nickel nanowires (0.2 × 60 μm on top and 2 × 20 μm nickel nanowires with sphere tips). (From Baek S, Fearing R, *IEEE Transactions on Components and Packaging Technologies*, vol. 31, no. 4, pp. 859–868, 2008 [85].)

surface makes contact, the mechanical load is carried by asperities in the center of the contact while the electrical load is distributed by the asperities along the outer rim. As oxides and other surface contaminants may appear, the team suggested designing the contract so that the outer rim asperities were the only asperities that would make contact i.e., bearing the mechanical and electrical load; this would allow the asperities to break through any developed contaminants as well as reduce electrical contact resistance by appropriately applying force along the conducting asperities. This novel concept was simulated to verify that the contact resistance of an outer rim maximum load and current density asperity contact interface design can be significantly smaller than a similarly finished Hertzian style contact interface. Based on simulation, the final results revealed that the greatest contact resistance reduction (up to a factor of 2) occurred for a mated tin finished surface.

12.4 Failure Modes and Reliability

Contact bouncing can greatly impact the lifetime and performance of electrical contacts. To that end, Peschot et al. performed experiments to better explain contact bouncing at the nanometer scale [87]. Using an AFM and a nano-indenter, the researchers controlled micro-contact make and break operations at low values of electrode velocity (few tens of nm/s). They discovered that the electrostatic force overcame the mechanical restoring force of the mobile contact near 10 nm. The team analytically ruled out the Casimir force by examining the effective distance for which the quantum electrodynamic force would have effect. It was found that the Casimir force was only dominant in the last few nanometers. The explanation for contact bounce was given as the product of competition between the restoring force of the contact beam and the adhesion force. The adhesion force is considered as contact interactions such as capillary, chemical, and van der Waals forces [88]. As the contact is made and the voltage between contacts is near zero the competition begins between adhesion force and restoring force. Upon opening of the contact, a potential difference is created and is the electrostatic force; which influences the contact to be made again. The outcome revealed that for the given mechanical design of the shaft, the velocity of the contact beam has to be higher than 1 μm/s in order to avoid bouncing due to the electrostatic force.

To analyze contact bounce, McCarthy et al. tried modeling the dynamic behavior of two different electrostatically actuated micro-switch configurations with time-transient finite difference analysis. The two configurations included one switch of uniform width and the other switch of nonuniform width. The model used dynamic Euler-Bernoulli beam theory for cantilevered beams, includes the electrostatic force from the gate, takes into account the squeeze-film damping between the switch and substrate, and includes a simple spring model of the contact tips [89]. The uniform width switch was modeled using Euler-Bernoulli beam theory with a constant cross-sectional area along the length of the beam. Once the equations of motion were found, a finite difference numerical solution to the resultant differential equation was obtained. The analytical solution was considered impractical due to the non-linearities in the electrostatic force and in the squeeze-film damping. The model expressed the applied voltage as a fraction of the static threshold voltage and the electrostatic force was the force per unit length acting on the beam in the region immediately above the gate. Squeeze-film damping pressure, due to the air film between the beam and the substrate, was determined using the simplified form of the Navier-Stokes equation known as the Reynolds equation [89]. The Reynolds equation assumes that the viscous and pressure force in the fluid film dominate the inertial terms [89]. The model of the contact tip is comprised of a simple spring at the free end of the beam. The tip acts as a constraint on the free end of the beam. Once the contact is made, the beam is considered no longer cantilevered, but fixed at one end and spring-supported at the other end [89]. To verify the effectiveness of the model, McCarthy et al. fabricated the contacts and tested against the simulation parameters. They chose to neglect bounces greater than 5 nm. The impact of contact bounces less than 5 nm is dependent on mechanical beam design. The researchers found that by applying higher voltages the threshold actuation voltages reduced the quantity of contact bounces. There was excellent agreement between theory and experiment with respect to the initial closing time and duration of the first bounce.

"Cold" switching is generally known to be actuating the switch repeatedly without applying RF or DC power during actuations, limiting the switch lifetime of mechanical failures such as structural fatigue, memory effect, stiction of the actuators, etc. [6]. Simply put, "cold" switching is powering the circuit off, then actuating the switch off then on, then powering the circuit back on. To model "cold" switching, the circuit elements would not contain stored energy at the time the switch closes and all energy would dissipate between actuations. This limits the types of failures of micro-switches to purely mechanical failure modes and extends the reliability of the micro-switch. "Hot" switching is considered to be actuating the switch repeatedly while applying RF or DC power during actuations [6]. Zavracky et al. reported over 2×10^9 cycles as the lifetime for Au sputtered contacts that were packaged in nitrogen [90]; a considerable difference compared to the 5×10 cycles Zavracky reported for "hot-switched" contacts. Majumder et al. reports greater than 10^7 "hot-switched" cycles and approximately 10^9 "cold-switched" cycles for micro-switches with a "platinum group" contact metal [91]. Newman et al. also performed lifetime measurements on high-reliability contacts and reported average lifetimes of cold switched contacts at 430×10^9 cycles [92]. In comparison, the "hot-switched" at 4 V, 20 mA, Au coated MWCNT surface exhibited 7×10^7 cycles in initial studies [92,93].

Toler et al. characterized the impact on reliability of external resistive, inductive, and capacitive loads for micro-switches [94]. Certain configurations of loads were determined to enhance micro-switch reliability. Specifically, that an external resistive load in series acts as a current limiter for both "hot" and "cold" switching conditions and reduces the chance of an electrical failure mode thereby enhancing the reliability of the micro-switch. In addition, there is a possibility of increasing the reliability of the switch by using a higher resistance contact metal with a matching external resistive load. The current limiting effect would restrict temperature and increased hardness of the higher resistance contact metal would most likely extend the reliability of the micro-switch further than a low resistance contact metal. Alternatively, it was found that certain configurations of resistive, inductive, and capacitive loads promote early failure via increased material transfer and current density. An external capacitive load in parallel was determined to be detrimental to micro-switch reliability under "hot" switching conditions since it compounded the current during discharge and raised the probability for increased current density, temperature, and material transfer. For "cold" switching conditions, the outpouring of the capacitor essentially continues to supply current through the contact after the signal has stopped transmitting and before the switch opens; effectively turning a "cold" switching condition into a "hot" switching condition and reducing reliability with the increased probability of electrical failure. Lastly, the external inductive load for DC conditions reduced susceptibility of failure via increased current density and temperature by limiting the current at the moment of initial contact in "hot" switching conditions. "Cold" switching conditions for external inductive loads have negligible effect to contact resistance and micro-switch reliability.

As mentioned earlier, stiction or adhesion is a failure mode which is commonly caused by capillary, electrostatic, chemical, and van der Waals forces [88]. The surface of contacts in air can become hydrophilic due to oxidation and the formation of a liquid meniscus by water vapor causes stiction [81]. Many researchers have proposed reducing the surface adhesion force by novel switch design, contact materials, and sealing the micro-contacts in inert gases [3,81,95–98]. Adhesion can be described by Hertz, JKR, or DMT theories [99]. Hertz theory, mentioned in the contact resistance modeling section, is traditionally used for modeling elastic adhesion between non-deformable surfaces [99]. For deformable surfaces, JKR or DMT theory is utilized. The JKR theory takes into account the surface energy of the contacting interfaces. Comparatively, DMT theory emphasizes the cohesive forces at the contact periphery [99]. The JKR model is valid for "soft" elastic materials with higher surface energy while the DMT model is applicable for "hard" stiff solids with low surface energy [99].

A multiscaled approach was developed by Wu et al. in order to predict stiction due to van der Waals forces [100]. For micro-scale calculations, the unloading adhesive contact-distance curves of two interacting rough surfaces were established from a combination of an asperity model and the Maugis transition theory [100]. The computed unloading distance curves were dependent on the material and surface properties such as roughness discussed earlier in this chapter [100]. The model was then integrated into a macro-model for the ease of finite element analysis [100]. The parameters for the FEM in terms of surface topography and micro-geometry were evaluated from theoretical models, surface energy measurements, or AFM measurements [100]. The key advantage of the model is its ability to account for a wide variety of micro-scale parameters such as surface topography, surface cleanliness, etc. while still enabling the complete modeling of the larger MEMS structure using FEM [100]. The disadvantage of this approach is the absence of the effect of capillary forces [100].

Fretting is a form of structural fatigue which is defined as accelerated surface damage occurring at the interface of contacting materials subjected to small oscillatory movements [8]. Braunovic [8] states that the lack of published information about failures due to fretting is because fretting is a "time-related process causing an appreciable effect only after a long period of time as a result of the accumulation of wear debris and oxides in the contact zone", see Chapters 5 and 7, However, contact force has significant influence on the contact resistance in fretting conditions. As the force applied on the contact is increased, the contact resistance declines until there is a significant amount of wear debris and oxide to form an insulating layer. As the insulating layer develops, the resistance increases despite larger applications of force. Fretting is a rate dependent phenomenon and the frequency of oscillations will affect the contact resistance.

Another "cold" switch mechanical failure cause is pitting. Pitting and hardening occur when two metals make contact repeatedly at the same location [1]. The repeated actuations create cavities on the surface and are confined to a point or small area [8]. The fields are described as being irregularly shaped and are filled with corrosion products over time [8]. The buildup of corrosion products in conjunction with pitting reduces the area available for current flow and will cause high temperatures in those areas while the switch is closed. The result will be a localized high temperature failure mode as seen in "hot" switching conditions.

According to Kim, the lifetime of a switch is more restricted by "hot"-switching than by "cold"-switching because most of the signals that are transmitted through the switch have high power loads [6]. Electrical failure mechanisms, like temperature, current density, and material transfer are all factors in reliability under "hot" switching [1]. With an emphasis on no arcing, the transfer of material between electrical contacts in MEMS devices below the minimum arcing voltage is known as "fine transfer" [93]. A major consideration in "hot" switching is a large temperature rise which occurs in the contact region due to the small contact area on the a-spots [1]. With a small contact region comes a large contact resistance, which in the case of "hot"-switching will result in large heat dissipation in that area at the time the switch closes. Increased temperature at these localized points may soften the contact metal and lead to bridge transfer. A problem with bridge transfer is that the internal stresses cause the contact metal to shrink and crack [8]. Oxidation then leads to a reduced number of electrical conducting paths thereby leading to overheating and ultimately failure [8].

An increase in current density raises the temperature for the contact areas on the cathode and anode. Concerning the topology of the contact surface, which has asperities, a higher current density will cause high temperature spots at asperities. The relationship between the temperature in the contact and voltage drop across the contact is described in Chapter 1 as:

$$V_c^2 = 4L(T_c^2 - T_o^2)$$
(12.20)

where V_c is the voltage drop across the contact, L is the Lorenz constant, $T_c(K)$ is the temperature in the contact, and $T_o(K)$ is the bulk temperature. It is important to note that the relationship between voltage and temperature above does not consider the size effects of the asperities in contact [101]. Examining Equation 12.20, an increase in current would result in an increase in temperature due to I^2R loss. The resistance is expected to increase because of the metal's positive temperature coefficient of resistance, α. The equation for resistance R_{ct}, at the new temperature T_c is then:

$$R_{ct} = R_{co}\left[1 + \frac{2}{3}\alpha(T_c - T_o)\right]$$
(12.21)

but Equation 12.21 only holds true until a temperature is reached that softening of the metal begins to occur, see Table 24.1A. When the contact metals are softening, the asperities collapse, increasing their areas to facilitate cooling. The collapsing of asperities increases the effective contact area and results in a decrease of the contact resistance. The plastic deformation of the asperities during the contact formation proceeds more rapidly when the softening temperature is reached [72]. This is caught by contact resistance as a function of the domain:

$$R_c = \frac{\rho}{2}\sqrt{\frac{1}{2R\alpha}}$$
(12.22)

and R is asperity peak radius of curvature and α is asperity vertical deformation [18]. Immediately following initial asperity deformation, contact asperities are susceptible to creep under compressive strain [40]. Creep deformation has been reported by Gregori et al. as well as Budakian et al. at micro-Newton level contact forces and low current levels [102,103]. With creep, the contact material deforms and reduces the contact force per unit area, resulting in increased contact resistance [8].

The softening of the metal at the asperities of the contact reduces the strain hardening of the a-spots and could accelerate the aging of the contract by the activation of thermal failure mechanisms such as bridge transfer [53]. Bridge transfer is only observed for circuit currents <20 mA. Above that value the metal transfer is from the arc that forms (see Chapter 10, Section 10.5.4). High temperature for the small volumes of material changes the softness of the contrast material and promotes bridge transfer. Holm noted that material transfer of very small volumes of material was known originally as fine transfer and said the phenomena is usually called bridge transfer. Bridge transfer is a form of material transfer which reduces the effective area of the asperities and increases the contact resistance [13]. Also, increased temperature decreases the mobility of electrons in a metal, resulting in increased resistivity. If the choice of contact materials is not appropriate, the materials may not be able to conduct away the resistive heat generated by currents passing through surface asperities, the large local temperature increases and will further the probability of a bridge transfer [83,104]. Changes to the surface topology are detrimental to contact resistance. When the contact opens, a newly ruptured bridge can provide better conditions for field emission when the electrodes are in close proximity and a voltage exists across them; see Section 9.3.2. Temperature is an important consideration for contact design. Increased contact temperatures can sometimes activate diffusion and oxidation processes that are driven by elevated temperatures, which ultimately reduces surface conductivity and contact resistance will increase [105,106].

Dickrell et al. simulated a Au–Pt micro-contact using a nanoindentor in order to test and examine the performance of Au–Pt contacts [24]. The experiment showed that the contact experienced a dramatic increase in contact resistance, by orders of magnitude, when hot-switched in both ambient and inert nitrogen environments [24]. The results indicated that arc formation at the time of opening or closing was the cause of increased resistance [24]. Arcing resulted in a decomposition of the surface contaminants and the creation of an insulating surface layer [24], see Section 10.6 and Chapter 19.

Considering DC, electromigration is another form of material transfer which causes micro-switch failure [8]. Electromigration is defined as "the forced movement of metal ions under the influence of an electric field" [8]. Atomic flux (J) is given by:

$$J = \frac{D}{kT} J\rho e Z^* \tag{12.23}$$

$$D = D_o e^{-\frac{Q}{kT}} \tag{12.24}$$

where D is the diffusion coefficient, J is the current density, ρ is the electrical resistivity and eZ^* is the effective charge, k is the Boltzmann constant, T is the absolute temperature, D_o and Q are the diffusivity constant and activation energy for diffusion, respectively [8]. As expressed by Equation 12.23, atomic flux is directly proportional to the current density. Voids form as a result of electro-migration and ultimately cause device failure [8]. Braunovic states that an increase in current density in the a-spots can be substantial and create the right conditions for electromigration to occur [8].

Distinct from electromigration, field emission is also responsible for material transfer phenomena [107]. Field emission is the transfer or emission of electrons induced by an electrostatic field. Literature in this area is limited, however, Poulain et al. conducted an investigation into the phenomena using a modified atomic force microscope [107]. The results showed a current increase when the contact gap became smaller than a few tens of nanometers [107]. At that range, the team deduced that the emission of electrons from the cathode follow the Fowler-Nordheim theory and lead to damage on the opposite contact member [107]. The damage to the opposite contact member consists of evaporated anode material caused by impact heating (electrons leaving the anode heat the material and cause evaporation of anode contact material to the cathode interface) [107]. The reported transfer of material due to field emission occurred with an open-circuit voltage across the two contact members at 5 V and the test current limited to 1 mA when the contact is closed.

For complete integration with CMOS processes, micro-switches need to withstand temperatures of about 400°C without a change in performance [108]. At high temperatures, cantilever beams normally begin to deflect due to intrinsic stresses in the layered materials making up the beam. Klein et al. designed an electrostatically actuated micro-switch based on a tungsten-titanium alloy to reduce the possibility of failure due to temperature and stiction [108]. Klein et al. chose tungsten for its high melting point of 3370°C, which is a good indicator of stability for temperatures a tenth of the melting point value [108]. The tungsten-titanium alloy switches were evaluated to temperatures up to 500°C and the results indicate that the design is stable with beam deflections of only 8% [108]. Insertion loss was reported to be slightly higher than compared to more conductive switches but isolation was comparable [108].

No discussion of failure modes is complete without referencing the development of frictional polymers. Metals most susceptible to the development of frictional polymers are the platinum group metals and any other "catalytically active metal" [8]. Holm points out that thin films, like oxides, develop over time on the contact surface and act as insulators, greatly increasing contact resistance [14]. The same is true for micro-switches. Though much smaller than the contacts studied by Holm, the effects of the films which develop on a micro-contact are orders of magnitude greater than those with macro scale contracts. Films on micro-contacts can render the contact useless and disabled. A particularly damaging film is the development of a frictional polymer [8]. Frictional polymers are organic films, sometimes referred to as deposits, that develop on commonly used contact materials when there are low levels of organic vapors or compounds evident in the operating environment of the contact [8]. Crossland et al., however, were able to show that the addition of a non-catalytically active metal, like silver, can significantly reduce the effects of frictional polymerization [109]. Though silver is not considered suitable for MEMS due to tarnishing, their experiment showed that silver must make up 36% or more of the contact materials in order to witness a significant reduction [109].

12.5 Conclusion

This review provides insight into the properties and concepts necessary for designing micro-electric contacts for DC and RF MEMS switches. The basic theories behind the aspects of design, contact resistance modeling, contact materials, and failure modes are discussed and explored. A survey of the challenges for these areas in ohmic contacts is provided. Complete models of contact resistance for various electron transport modes and deformation modes are shown. The decision for contact materials is investigated by examining the impact of material properties on the characterization of the contact.

References

1. Rebeiz G, *RF MEMS, Theory, Design, and Technology*, Hoboken: John Wiley and Sons, 2004.
2. Van Caekenberghe K, "RF MEMS on the radar," *IEEE Microwave Magazine*, vol. 10, no. 6, pp. 99–116, 2009.

3. Yang Z, Lichtenwalner D, Morris A, Krim J, Kingon A, "Comparison of Au and Au–Ni alloys as contact materials for MEMS switches," *IEEE Journal of Microelectromechanical Systems*, vol. 18, no. 2, pp. 287–295, 2009.

4. Dennard R, Cai J, Kumar A, "A perspective on today's scaling challenges and possible future directions," *Solid State Electronics*, vol. 51, no. 4, pp. 518–525, 2007.

5. Kaynak M, Ehwald K, Sholz R, Korndorfer F, Wipf C, Sun Y, Tillack B, Zihir S, Gurbuz Y, "Characterization of an embedded RF-MEMS switch," in *Topical Meeting on Silicon Monolithic Integrated Circuits in RF Systems (SiRF)*, IEEE Mircrowave Theory and Techniques Society, New Orleans, LA, pp. 144–147, 2010.

6. Kim J, Lee S, Baek C, Kwon Y, Kim Y, "Cold and hot switching lifetime characterizations of ohmic contact RF MEMS switches," *IEICE Electronics Express*, vol. 5, no. 11, pp. 418–423, 2008.

7. Majumder S, McGruer N, Adams G, Zavracky P, Morrison R, Krim J, "Study of contacts in an electrostatically actuated microswitch," *Sensors Actuators*, vol. 93, no. 1, pp. 19–26, 2001.

8. Braunovic M, Konchits V, Myshkin N, *Electrical Contacts—Fundamentals, Applications, and Technology*, New York: CRC Press, 2007.

9. Kaajakari V, *Practical MEMS*, Las Vegas, NV: Small Gear Publishing, 2009.

10. Kovacs G, *Micromachined Transducers Sourcebook*, New York: McGraw-Hill, 1998.

11. Comtois J, Michalicek M, Barron C, "Fabricating Micro-Instruments in Surface-Micromachined Polycrystalline Silicon," in *43rd Annual International Instrumentation Symposium of the Aerospace Industries and the Test Measurement Divisions of the Instrument Society of America*, Orlando, FL, 4–8 May 1997.

12. Lucyszyn S, "Review of radio frequency microelectromechanical systems technology," in *Science, Measurement and Technology, IEE Proceedings*, London, 2004.

13. Holm R, *Electric Contacts: Theory and Applications*, Reprint of the 4th Edition, Berlin: Springer, 2000.

14. Coutu R, McBride J, Starman L, "Improved Micro-Contact Resistance Model that considers Material Deformation, Electron Transport and Thin Film Characteristics," in *Proceedings of the 55th IEEE Holm Conference on Electrical Contacts*, Vancouver, British Columbia, Canada, pp. 296–299, 2009.

15. Greenwood J, Williamson J, "Contact of nominally flat surfaces," *Proceedings of the Royal Society: Series A*, vol. 295, pp. 300–319, 1966.

16. Pitney K, *Ney Contact Manual*, Bloomfield: The J. M. Ney Company, 1973.

17. Chang W, "An elastic-plastic model for a rough surface with an ion-plated soft metallic coating," *Journal of Wear*, vol. 212, pp. 229–237, 1997.

18. Coutu R, Reid J, Cortez R, Strawser R, Kladitis P, "Microswitches with sputtered Au, AuPd, Au-on-AuPt, and AuPtCu alloy electrical contacts," *IEEE Transactions on Components and Packaging Technologies*, vol. 29, no. 2, pp. 341–349, 2006.

19. Firestone F, Abbot E, "Specifying surface quantity—a method based on the accurate measurement and comparison," *ASME Mechanical Engineering* vol. 55, p. 569, 1933.

20. Edelmann T, Coutu R, Starman L, "Novel test fixture for collecting microswitch reliability data," in *Proceeding of SPIE, Photonics West: Reliability, Packaging, Testing and Characterization of MEMS/MOEMS and Nanodevices IX*, San Diego, CA, 2010.

21. Wexler G, "The size effect and the non-local Boltzmann transport equation in orifice and disk geometry," *Proceedings of the Physical Society*, vol. 89, pp. 927–941, 1966.

22. Timsit, R, "Constriction resistance of thin film contacts," *IEEE Transactions on Components and Packaging Technologies*, vol. 33, no. 3, pp. 636–642, 2010.

23. Karmalkar S, Mohan P, Kumar B, "A unified compact model of electrical and thermal 3-D spreading resistance between eccentric rectangular and circular contacts," *IEEE Electron Device Letter*, vol. 26, no. 12, pp. 909–912, 2005.

24. Dickrell D, Dugger M, "Electrical contact resistance degradation of a hot-switched simulated metal MEMS contact," *IEEE Transactions on Components and Packaging Technologies*, vol. 30, no. 1, pp. 75–80, 2007.

25. Patton S, Eapen K, Zabinski J, "Effects of adsorbed water and sample aging in are on uN level adhesion force between Si(100) and silicon nitride," *Tribology International*, vol. 34, pp. 481–491, 2001.

26. Lumbantobing A, Kogut L, Komvopoulos K, "Electrical contact resistance as a diagnostic tool for MEMS contact interfaces," *Journal of Microelectromechanical Systems*, vol. 13, no. 6, pp. 977–987, 2004.
27. Sawada S, Tsukiji S, et al., "Current density analysis of thin film effect in contact area on LED wafer," in *Proceedings of the 58th IEEE Holm Conference on Electrical Contacts*, pp. 242–247, 2012.
28. Norberg G, Dejanovic S, HesselBom H, "Contact resistance of thin metal film contacts," *IEEE Transactions Components Packaging Technology*, vol. 29, no. 2, pp. 371–378, 2006.
29. Timsit R, "Electrical conduction through small contact spots," *IEEE Transactions on Components and Packaging Technologies*, vol. 29, no. 4, pp. 727–734, 2006.
30. Agrait N, Yeyati A, Ruitenbeek J, "Quantum properties of atom-sized conductors," *Physics Reports*, vol. 377, pp. 81–279, 2003.
31. Jackson R, Crandall E, et al, "An analysis of scale dependent and quantum effects on electrical contact resistance between rough surfaces," in *Proceedings of the 58th IEEE Holm Conference on Electrical Contacts*, pp. 1–8, 2012.
32. Poulain C, Jourdan G, Peschot A, Mandrillon V, "Contact conductance quantization in a MEMS switch," in *Proceedings of the 56th IEEE Holm Conference on Electrical Contacts*, pp. 1–7, 2010.
33. Pennec F, Peyrou D, Leray D, Pons P, Plana R, Courtade F, "Impact of the surface roughness description on the electrical contact resistance of ohmic," *IEEE Transactions on Components, Packaging and Manufacturing Technology*, vol. 2, no. 1, pp. 85–94, 2012.
34. McCool J, "Comparison of models for the contact of rough surfaces," *Wear*, vol. 107, no. 1, pp. 37–60, 1986.
35. Majumdar A, Bhushan B, "Role of fractal geometry in roughness characterization and contact mechanics of surfaces," *ASME Journal of Tribology*, vol. 112, no. 2, pp. 205–216, 1990.
36. Majumdar A, Tien C, "Fractal characterization and simulation of rough surfaces," *Wear*, vol. 136, no. 2, pp. 313–327, 1990.
37. Kogut L, Jackson R, "A comparison of contact modeling utilizing statistical and fractal approaches," *ASME Journal of Tribology*, vol. 128, no. 1, pp. 213–217, 2005.
38. Kogut L, Etsion I, "A finite element based elastic-plastic model for the contact of rough surfaces," *Tribology Transactions*, vol. 46, no. 3, pp. 383–390, 2003.
39. Greenwood JA, "A unified theory of surface roughness," *Proceedings of The Royal Society A*, vol. 393, pp. 133–157, 1984.
40. Rezvanian O, Zikry M, Brown C, Krim J, "Surface roughness, asperity contact and gold RF MEMS switch behavior," *Journal of Micromechanics and Microengineering*, vol. 17, pp. 2006–2015, 2007.
41. Borri-Brunetto M, Chiaia B, Ciavarella M, "Incipient sliding of rough surfaces in contact: a multiscale numerical analysis," *Computer Methods in Applied Mechanics and Engineering*, vol. 190, no. 46/47, pp. 6053–6073, 2001.
42. Komvopoulos K, Ye N, "Three-dimensional contact analysis of elastic-plastic layered media with fractal surface topographies," *ASME Journal of Tribology*, vol. 123, no. 3, pp. 632–640, 2001.
43. Persson B, Bucher F, Chiaia B, "Elastic contact between randomly rough surfaces: comparison of theory with numerical results," *Physical Review B*, vol. 65, pp. 184106–184111, 2002.
44. Persson B, "Elastoplastic contact between randomly rough surfaces," *Physical Review Letters*, vol. 87, no. 11, p. 116101, 2001.
45. Jackson R, Streator J, "A multi-scale model for contact between rough surfaces," *Wear*, vol. 261, pp. 1337–1347, 2006.
46. Wilson W, Angadi S, Jackson R, "Electrical Contact Resistance Considering Multi-Scale Roughness," in *Proceedings of the 54th IEEE Holm Conference on Electrical Contacts*, Orlando, FL, pp. 190–197, 2008.
47. Jackson R, "A model for the adhesion of multiscale rough surfaces in MEMS," in *System Theory (SSST), 2011 IEEE 43rd Southeastern Symposium on*, 2011.
48. Jackson R, Kogut L, "Electrical contact resistance theory for anistropic conductive films considering electron tunneling and particle flattening," *IEEE Transactions on Components and Packaging Technologies*, vol. 30, no. 1, pp. 59–66, 2007.

49. Coutu R, Kladitis P, Leedy K, Crane R, "Selecting metal allow electric contact materials for MEMS switches," *Journal for Micromechanical Microengineering*, vol. 14, no. 8, pp. 1157–1164, 2004.

50. Broue A, Dhennin J, Charvet P, Pons P, Jemaa N, Heeb P, Coccetti F, Plana R, "Multi-physical characterization of micro-contact materials for MEMS switches," in *Proceedings of the 56th IEEE Holm Conference on Electrical Contacts*, pp. 363–372, 2010.

51. American Society for Metals, *Metals Handbook*, vol. 2, Materials Park, OH: ASM International, 1990.

52. Arrazat B, Duvivier P, Mandrillon V, Inal K, "Discrete analysis of gold surface apserities deformation under spherical nano-indentation towards electrical contact resistance calculation," in *Proceedings of the IEEE 57th Holm Conference on Electrical Contacts*, pp. 167–174, 2011.

53. Broue A, Dhennin J, Courtade F, Dieppedale C, Pons P, Lafontan X, Plana R, "Characterization of Au/Au, Au/Ru and Ru/Ru ohmic contacts in MEMS switches improved by a novel methodology," *Journal of Micro/Nanolithography, MEMS, and MOEMS*, vol. 9, no. 4, pp. 041102-1–041102-8, 2010.

54. McGruer N, Adams G, Chen L, Guo Z, Du Y, "Mechanical, thermal, and material influences on ohmic-contact-type MEMS switch operation," in *Proceedings of the 19th MEMS*, Istanbul, Turkey, 2009.

55. Kwon H, Choi D, Park J, Lee H, Kim Y, Nam H, Joo Y, Bu J, "Contact materials and reliability for high power RF-MEMS switches," in *IEEE 20th International Conference on Micro Electro Mechanical Systems*, 2007. DOI. 10.1109/MEMSYS.2007.4433055.

56. Yaglioglu O, Hart A, Martens R, Slocum A, "Method of characterizing electrical contact properties of carbon nanotube coated surfaces," *Review of the Scientific Instruments*, vol. 77, pp. 095105/1–095105/3, 2006.

57. Thostenson E, Ren Z, Chou T, "Advances in the science and technology of carbon nanotubes and their composites: a review," *Composites Science and Technology*, vol. 61, pp. 1899–1912, 2001.

58. Yunus E, McBride J, Spearing S, "The relationship between contact resistance and contact force on Au-coated carbon nanotube surfaces under low force conditions," *IEEE Transactions on Components and Packaging Technologies*, vol. 32, no. 3, pp. 650–657, 2009.

59. Choi J, Lee J, Eun Y, Kim M, Kim J, "Microswitch with self-assembled carbon nanotube arrays for high current density and reliable contact," in *IEEE 24th International Conference on Micro Electro Mechanical Systems (MEMS)*, 2011. DOI 10.1109/MEMSYS.2011.5734368.

60. Chai Y, Hazeghi A, Takei K, Chen H, Chan P, Javey A, Wong H, "Low-resistance electrical contact to carbon nanotubes with graphitic interfacial layer," *IEEE Transactions on Electron Devices*, vol. 59, no. 1, pp. 12–19, 2012.

61. Gambino J, Colgan E, "Silicides and ohmic contacts," *Materials Chemistry and Physics*, vol. 52, no. 2, pp. 99–146, 1998.

62. Kim W, Javey A, Tu R, Cao J, Wang Q, Dai H, "Electrical contacts to carbon nanotubes down to 1nm in diameter," *Applied Physics Letters*, vol. 87, no. 17, pp. 101–173, 2005.

63. Chai Y, Hazeghi A, Takei K, Chen Y, Chan P, Javey A, Wong H, "Graphitic interfacial layer to carbon nanotube for low electrical contact resistance," *IEDM Technology Digest*, IEDM10-210, pp. 210–213, 2010.

64. Chen Q, Wang S, Peng L, "Establishing Ohmic contacts for in situ current-voltage characteristic measurements on a carbon nanotube inside the scanning electron microscope," *Nanotechnology*, vol. 17, no. 4, pp. 1087–1098, 2006.

65. Rykaczewski K, Henry M, Kim S, Fedorov A, Kulkarni D, Singamaneni S, Tsukruk V, "The effect of geometry and material properties of a carbon joint produced by electron beam induced deposition on the elctrical resistance of a multiwalled carbon nanotube-to-metal contact interface," *Nanotechnology*, vol. 21, no. 3, pp. 1–12, 2010.

66. Ke F, Miao J, Oberhammer J, "A ruthenium-based multimetal-contact RF MEMS switch with a corrugated diaphragm," *Journal of Microelectromechanical Systems*, vol. 17, no. 6, pp. 1447–1459, 2008.

67. Lee H, Coutu R, Mall S, Leedy K, "Characterization of metal and metal alloy films as contact materials for MEMS switches," *Journal of Micromechanical and Microengineering*, vol. 16, pp. 557–563, 2006.
68. Cha P, Srolovitz D, Vanderlick T, "Molecular dynamics simulation of single asperity contact," *Acta Materialia*, vol. 52, pp. 3983–3996, 2004.
69. Song J, Srolovitz D, "Atomistic simulation of multi-cycle asperity contact," *Acta Materialia*, vol. 55, p. 14, 2007.
70. Kuipers L, Frenken J, "Jump to contact, neck formation, and surface melting in the scanning tunneling microscope," *Physcial Review Letters*, vol. 70, no. 25, pp. 3907–3910, 1993.
71. Patton S, Zabinski J, "Fundamental studies of Au contacts in MEMS RF switches," *Tribology Letters*, vol. 18, pp. 215–230, 2005.
72. Fortini A, Mendelev M, Buldyrev S, Srolovitz D, "Asperity contacts at the nanoscale: comparison of Ru and Au," *Journal of Applied Physics*, vol. 104, no. 7, pp. 074320.1–074320.8, 2008.
73. Frenkel D, Smit B, *Understanidng Molecular Simulation*, 2nd Edition, vol. 1 of Computational Science Series, New York: Academic Press, 2002.
74. Sorensen M, Brandbyge M, Jacobsen K, "Mechanical deformation of atomic-scale metallic contacts: Structure and mechanisms," *Physics Review B: Condensed Matter and Materials Physics*, vol. 57, no. 6, pp. 3283–3294, 1998.
75. Du Y, Chen L, McGruer N, Adams G, Etsion I, "A finite element model of loading and unloading of an asperity contact with adhesion and plasticity," *Journal of Colloid and Interface Science*, vol. 312, no. 2, pp. 522–528, 2007.
76. Chen Y, Nathanael R, Jeon J, Yaung J, Hutin L, Liu T, "Characterization of contact resistance stability in MEM relays with tungsten electrodes," *Journal of Microelectromechanical Systems*, vol. 21, no. 3, pp. 1–3, 2012.
77. Kam H, Pott V, Nathanael R, Jeon J, Alon E, Liu T, "Design and reliability of a micro-relay technology for zero-standby-power digital logic applications," *IEDM Technical Digest*, IEDM09–809, pp. 809–811, 2009.
78. Deal B, Grove A, "General relationship for the thermal oxidation of silicon," *Journal of Applied Physics*, vol. 36, no. 12, pp. 3770–3778, 1965.
79. Spencer M, Chen F, Wang C, Nathanael, Fariborzi H, Gupta A, Kam H, Pott V, Jeon J, Liu T, Markovic D, Alon E, Stojanovic V, "Demonstration of integrated micro-electro-mechanical relay circuits for VLSI applications," *IEEE Journal of Solid-State Circuits*, vol. 46, no. 1, pp. 308–320, 2011.
80. Chandler R, Park W, Li H, Yama G, Partridge A, Lutz M, Kenny T, "Single wafer encapsulation of MEMS devices," *IEEE Transactions Advance Packaging*, vol. 26, no. 3, pp. 227–232, 2003.
81. Yamashita T, Itoh T, Suga T, "Investigation of anti-stiction coating for ohmic contact MEMS switches with thiophenol and 2-naphtalenethiol self-assembled monolayer," *Sensors and Actuators A: Physical*, vol. 172, no. 2, pp. 455–461, 2011.
82. Tang W, Xu K, Wang P, Li X, "Surface roughness and resistivity of Au film on Si-(111) substrate," *Microelectronics Engineering*, vol. 66, pp. 445–450, 2003.
83. Hyman D, Mehregany M, "Contact physics of gold microcontacts for MEMS switches," *IEEE Transactions on Components and Packaging Technologies*, vol. 22, no. 3, pp. 357–564, 1999.
84. Kataoka K, Itoh T, Suga T, "Characterization of fritting phenomena on Al electrode for low contact force probe card," *IEEE Transactions on Components and Packaging and Technology*, vol. 26, pp. 382–387, 2003.
85. Baek S, Fearing R, "Reducing contact resistance using compliant nickel nanowire arrays," *IEEE Transactions on Components and Packaging Technologies*, vol. 31, no. 4, pp. 859–868, 2008.
86. Myers M, Leidner M, et al, "Contact resistance reduction by matching current and mechanical load carrying asperity junctions," in *Proceedings of the 58th IEEE Holm Conference on Electrical Contacts*, pp. 248–251, 2012.
87. Peschot A, Poulain C, et al, "Contact Bounce Phenomena in a MEM Switch," in *Proceedings of the 58th IEEE Holm Conference on Electrical Contacts*, pp. 49–55, 2012.

88. Maboudian R, Howe R, "Critical review: adhesion in surface micromechanical structures," *Journal of Vacuum Science & Technology B*, vol. 15, no. 1, pp. 1–20, 1997.
89. McCarthy B, Adams G, McGruer N, Potter D, "A dynamic model, including contact bounce, of an electrostatically actuated microswitch," *Journal of Microelectromechanical Systems*, vol. 11, no. 3, pp. 276–283, 2002.
90. Zavracky P, Majumber S, McGruer N, "Micromechanical switches fabricated using nickel surface micromachining," *Journal of Micromechanical Systems*, vol. 6, no. 1, pp. 3–9, 1997.
91. Majumder S, Lampen J, Morrison R, Maciel J, "MEMS switches," *IEEE Instrumentation and Measurement Magazine*, vol. 6, no. 1, pp. 12–15, 2003.
92. Newman H, Ebel J, Judy D, Maciel J, "Lifetime measurements on a high-reliability RF-MEMS contact switch," *IEEE Microwave and Wireless Components Letters*, vol. 2, pp. 100–102, 2008.
93. McBride J, "The wear processes of gold coated multi-walled carbon nanotube surfaces used as electrical contact for micro-mechanical switching," *Nanoscience and Nanotechnology Letters*, vol. 4, no. 2, pp. 357–361, 2010.
94. Toler B, Coutu R, "Characterizing external resistive, inductive and capacitive loads for microswitches," *SEM Proceedings on MEMS and Nanotechnology*, vol. 6, vol. 42, pp. 11–18, 2013.
95. Mercado L, Kuo S, Lee T, Liu L, "Mechanism-based solutions to RF MEMS switch stiction problem," *IEEE Transactions Components and Packaging Technologies*, vol. 27, pp. 560–570, 2004.
96. Peroulis D, Pacheco S, Sarabandi K, Katehi L, "Electromechanical considerations in developing low-voltage RF MEMS switches," *IEEE Transactions on Microwave Theory Technologies*, vol. 51, pp. 259–270, 2003.
97. Park J, Shim E, Choi W, Kim Y, Kwon Y, Cho D, "A non-contact-type RF MEMS switch for 24-GHz radar applications," *Journal of Microelectromechanical Systems*, vol. 18, pp. 163–173, 2009.
98. Brown C, Morris A, Kingon A, Krim J, "Cyrogenic performance of RF MEMS switch contacts," *Journal of Microelectromechanical Systems*, vol. 17, pp. 1460–1467, 2008.
99. Prokopovich P, Perni S, "Comparison of JKR and DMT based multi-asperity adhesion model: Theory and experiment," *Colloids and Surfaces A: Physicochemical and Engineering Aspects*, vol. 383, no. 1–3, pp. 95–101, 2011.
100. Wu L, Noels L, Rochus V, Pustan M, Golinval J, "A Micro-Macro approach to predict stiction due to surface contact in microelectromechanical systems," *Journal of Microelectromechanical Systems*, vol. 20, no. 4, pp. 976–990, 2011.
101. Nikolic B, Allen P, "Electron transport through a circular constriction," *Physical Review B*, vol. 60, pp. 3963–3969, 1999.
102. Gregori G, Clarke D, "The interrelation between adhesion, contact creep, and roughness on the life of gold contacts in radio-frequency microswitches," *Journal of Applied Physics*, vol. 100, no. 9, p. 094904, 2006.
103. Budakian R, Putterman S, "Time scales for cold welding and the origins of stick-slip friction," *Physical Review B*, vol. 65, p. 235429, 2002.
104. Kruglick E, Pister K, "Lateral MEMS microcontact considerations," *Journal of Micromechanical Systems*, vol. 8, no. 3, pp. 264–271, 1999.
105. Sun M, Pecht M, Natishan M, Martens R, "Lifetime resistance model of bare metal electrical contacts," *IEEE Transactions on Advanced Packaging*, vol. 22, no. 1, pp. 60–67, 1999.
106. Takano E, Mano K, "Theoretical lifetime of static contacts," *IEEE Transactions on Parts, Materials, and Packaging*, vol. 22, no. 1, pp. 60–67, 1967.
107. Poulain C, Peschot A, Vincent M, Bonifaci N, "A nano-scale investigation of material transfer phenomena at make in a MEMS switch," in *2011 IEEE 57th Holm Conference on Electrical Contacts*, Grenoble, France, 2011.
108. Klein S, Thilmont S, Ziegler V, Prechtel U, Schmid U, Seidel H, "High temperature stable RF MEMS switch based on tungsten-titanium," in *Solid-State Sensors, Actuators and Microsystems Conference, 2009. TRANSDUCERS 2009. International*, 2009.
109. Crossland W, Murphy P, "The formation of insulating organic films on palladium-silver contact alloys," *IEEE Transactions on Parts, Hybrids and Packaging*, vol. 10, no. 1, pp. 64–69, 1974.

13

Low Current Switching

John W. McBride

And oftentimes excusing the fault
Doth make the fault the worse by excuse.

King John, **William Shakespeare**

CONTENTS

13.1 Introduction and Device Classification 733
13.2 Device Types.. 734
    13.2.1 Hand-Operated Switches.. 734
        13.2.1.1 The Rocker Switch Mechanism.. 734
        13.2.1.2 Lever Switches 737
        13.2.1.3 Slide Switches 738
        13.2.1.4 Rotary Switches 738
        13.2.1.5 Push-Button Switches 738
        13.2.1.6 Switching Devices Used below 0.5 A... 739
    13.2.2 Actuated Switches 739
        13.2.2.1 Limit Switches 739
        13.2.2.2 Thermostatic Controls.. 740
        13.2.2.3 Electro-Mechanical Relay 741
13.3 Design Parameters for Static Switching Contacts... 741
    13.3.1 Small-Amplitude Sliding Motion 742
    13.3.2 Contact Force and Contact Materials... 742
        13.3.2.1 Contacts at Current Levels below 1 A.. 742
        13.3.2.2 Contacts at Current Levels between 1 and 30 A............................ 742
        13.3.2.3 Contact Force 743
13.4 Mechanical Design Parameters 743
    13.4.1 Case Study (1): Hand-Operated Rocker-Switch Mechanism....................... 744
        13.4.1.1 Moving-Contact Dynamics of a Rocker-Switch Mechanism........... 745
        13.4.1.2 Design Optimization of a Rocker-Switch Mechanism..................... 746
    13.4.2 The Opening Characteristics of Switching Devices 746
        13.4.2.1 Moving Contact Dynamics at Opening.. 747
    13.4.3 The Make Operation.. 747
        13.4.3.1 Impact Mechanics 748
        13.4.3.2 The Coefficient of Restitution.. 749

13.4.3.3 Impact Mechanics for a Pivoting Mechanism 751
13.4.3.4 The Velocity of Impact..752
13.4.3.5 Bounce Times..752
13.4.3.6 Total Bounce Times..754
13.4.3.7 Impact Times 755
13.4.3.8 Design Parameters for the Reduction of Contact Bounce.................755
13.5 The Measurement of Contact Wear and Contact Dynamics.......................................755
13.5.1 The Measurement of Contact Surfaces 756
13.5.2 Three Dimensional (3-D) Surface Measurement Systems 757
13.5.2.1 Contact Systems..757
13.5.2.2 Non-Contact Systems 758
13.5.3 Case Study (2): Example of Volumetric Erosion 758
13.5.4 The Measurement of Arc Motion and Contact Dynamics...........................759
13.6 Electrical Characteristics of Low-Current Switching Devices at Opening 761
13.6.1 Low-Current DC Arcs..761
13.6.1.1 Arc Voltage Characteristics...762
13.6.1.2 Voltage Steps below 7 A 762
13.6.1.3 Case Study (3): Arc Voltage, Current and Length under
Quasi-Static Conditions for Ag/CdO Contacts................................763
13.6.1.4 Opening Speed and Arc Length...764
13.6.1.5 Case Study (4): Automotive Systems 765
13.6.2 DC Erosion...766
13.6.2.1 Ag and Ag/MeO Contact Erosion/Deposition 766
13.6.3 Low-Current AC Arcs 768
13.6.3.1 Typical Waveforms and Arc Energy 769
13.6.4 AC Erosion...770
13.6.4.1 Point-on-Wave (POW) Studies Using Ag/CdO
Contact Materials...771
13.7 Electrical Characteristics of Low Current Switching Devices at Closure 773
13.7.1 Contact Welding on Make...774
13.7.2 Reducing Contact Bounce...775
13.7.3 Pre-Impact Arcing 775
13.7.4 Influence of Velocity during the First Bounce...777
13.7.4.1 The First Bounce...778
13.7.5 Bounces after the First..778
13.7.6 Summary of Contact Bounce 779
13.8 Summary...779
13.8.1 Switch Design...779
13.8.2 Break Operation 779
13.8.2.1 DC Operation..780
13.8.2.2 AC Operation..780
13.8.3 Make Operation 780
13.8.3.1 Design Parameters 780
13.8.3.2 Reducing Contact Bounce...781
13.8.3.3 Arcing during the Bounce Process...781
Acknowledgments...781
References..781

13.1 Introduction and Device Classification

In this chapter, consideration is given to the events occurring when switching low current, typically up to 30 A, at voltages below 300 V. Switching devices used in this range are normally mass produced and are generally optimized for a low number of component parts. It has been shown in the previous chapters that the phenomena associated with switching contacts and stationary contacts can be both complex and difficult to predict. The weakness and potential for failure of the electrical contact interface stems from the extreme operating conditions implicit with the interface. Some of the major mechanisms leading to failure are chemical corrosion owing to the environment, arc erosion, material transfer in the plasma developed between the contacts at opening and closing, and mechanical wear owing to sliding, rolling, and fretting motions. Often the overall wear can be a complex interaction of these processes, the study of which has necessitated an interdisciplinary approach. In addition to the electrical contact phenomena, a switching device will also be influenced by the mechanical action of the mechanism used. The mechanism used in a low-current device can often appear to be simple; however, in-depth studies reveal complex mechanical dynamic characteristics during switching.

Low-current switching devices are used in a range of products, for example; vehicles, domestic appliances, and consumer electronics. There are many variants of the low-current switch; however, in general there are two main groups, hand-operated devices and actuated devices. The former group includes devices such as rocker and push-button switches while the latter includes limit switches, thermostatic controls and electromagnetic relays. Devices used in the automotive sector operate at dc voltages between 12 and 300V, while in the consumer market the supply is usually the local distribution voltage. Other consumer products, such as keyboards and electronic devices, generally operate below 12V dc.

The automotive sector has seen rapid changes from the traditional 12–14V dc power supply. There was a period between 1990 and 2000 where a 42V replacement was considered. This was found to be limiting and is discussed here. Innovations that are more recent have seen the introduction of battery voltages for hybrid systems of 144V, and 288V; and of 345V for pure electric vehicles. Some systems use a convertor/inverter with an output of 650V ac. The area of automotive power systems is a rapidly evolving area; the fact remains that dc voltage levels are increasing above the traditional 12–14V level.

The evolution of low-current switch design has resulted in smaller devices with smaller operating forces and in the optimization of contact materials for particular applications. However, the reduction in size at a given current loading is limited by ergonomics; a hand-operated switch cannot be reduced in size to the point where the hand is unable to effectively operate the device. The size of a switch used to isolate a distribution voltage supply is also dominated by the need to maintain isolation distances between the conductors (see Section 9.3) and also the need to switch both the live and neutral conductor simultaneously. The isolation requirement has a direct influence on the contact gap used in a switch, and as will be seen in Section 13.4, the gap will also have a direct effect on the switching dynamics as well as the arc erosion characteristic.

The arc drawn when the contacts open causes contact erosion, but often the most severe operating condition is experienced when switching onto loads with a high inrush current. A current several times the rated operating current can flow as the switch contacts close to make the circuit. Contact bounce with high current can lead to welding of the contacts as well as severe arc erosion. We therefore need to understand the dynamics of the contact mechanism both at opening and closing (referred to here as "make" and "break" operation).

This chapter starts (Section 13.2) by describing various types of device that are used to switch low current. The aim here is to describe the general operation of a device and not to describe the design of specific mechanisms or devices. Static contact characteristics are considered in Section 13.3, with a subsection discussing the selection of contact force and the importance of small micro-motions on the switch contacts. Section 13.4, considers the closing operation, referred to as the "make" and the opening operation, referred to as the "break." Section 13.5, considers new approaches to the measurement of the mechanical characteristics of surfaces and to mechanism dynamics. The final sections (Sections 13.6 and 13.7), consider the electrical perspective to consider the contact erosion and wear resulting from arcing. To elaborate on the issues discussed, there are a number of numbered case studies within the sections.

13.2 Device Types

The majority of switches are designed such that the motion of the moving contact is ideally independent of the actuation method. With a hand-operated switch, the hand is used to move the switch to a position where a snap-action mechanism takes over the moving contact motion. This is also the case with a limit switch, although the actuation will be produced by a system-dependent movement, for example a thermostatic control switch will operate on a very small displacement generated by an expansion chamber, or other temperature-sensitive transducer. The ideal independence between the moving contact and the actuator is difficult to achieve and in many devices, the moving-contact dynamics is often affected by the actuation. In an electromechanical relay, the moving-contact dynamics is often a result of the interaction of the moving-contact system and a solenoid.

The following subsections give a brief overview of some typical types of devices used in low-current switching. These are considered in two categories, firstly, hand-operated mechanisms and secondly, actuated mechanisms.

13.2.1 Hand-Operated Switches

In this section, a range of hand-operated switching devices are described. Most have a wide range of variants, including single-pole (a single switch), double-pole (two switches in parallel controlled by a single actuator), change-over, and on-off configurations. In addition, momentary actions (a switch, which is held closed by a hand) and lamps (to indicate a condition) can be added. In switching a domestic supply voltage, the single-pole configuration is used to switch a live conductor only and is, therefore, normally used in appliances which are mobile and can be disconnected from the supply. The single pole configuration is also commonly used in wall-socket and lighting switches. A double-pole configuration is used to switch both a live and neutral conductor and is normally used in appliances, which remain connected to a supply. The change-over configuration is generally used to switch between circuits.

13.2.1.1 The Rocker Switch Mechanism

The rocker switch is the most widely used hand-operated switching device utilized in the low-current range (1–16 A); it is also used widely in applications below 1 A. It is commonly used in consumer products, and for general-purpose switching, for example, lighting circuits and computer power-supply switches. The switch mechanism is designed to have a snap action. A typical design is shown in Figure 13.1 Rocker switches can be designed in two basic configurations, *make-up* and *make-down*. In the make-down arrangement, the main fixed contact is mated with a moving contact which is moved downwards to close the switch. The design illustrated in Figure 13.1 has a make-down mechanism, which is shown schematically in Figure 13.2. Figure 13.1 can be used to identify the component parts.

A snap-action is used such that the moving blade contact moves independently of the hand forces after the toggle position has been reached. The toggle position corresponds to the center-line of the switch, shown in Figure 13.2. The switch is shown in a change-over configuration, with the main fixed contacts on the base of the switch; the current path is therefore through either of the main contact pairs depending on the rocker position and the central pivot contact. The contact force is provided by the spring-loaded plunger, which is also used as part of the snap-action mechanism, such that when the plunger is over the central pivot the moving contact can move under the force provided by the spring. The switch can be modified for the on-off configuration, shown in Figure 13.3, where in this case the main contact on the right is removed and replaced with a stop.

Figure 13.4 shows a make-up change-over arrangement. In the make-up arrangement, the contact force applied to the moving contact acts upwards. In this case, the fixed main contacts come through the base of the switch and lie above the moving contact. The change-over configuration can also be obtained using a combined make-up and make-down configuration, shown in Figure 13.5. A further design of rocker-switch mechanism

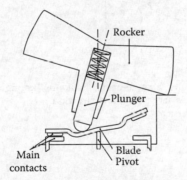

FIGURE 13.1
Schematic of a "make-down," "change-over" component parts identified. The blade is the moving contact.

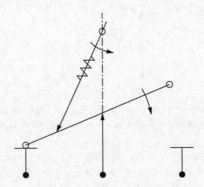

FIGURE 13.2
Rocker-switch layout corresponding to the "change-over" mechanism shown in Figure 13.1, showing the spring and direction of movement when the switch is actuated in one direction.

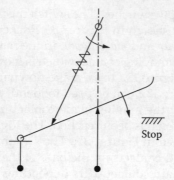

FIGURE 13.3
Rocker-switch layout for an on-off configuration, showing the main contact on the left. The blade or moving contact would be shaped to provide the contact gap required.

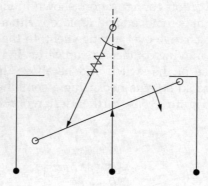

FIGURE 13.4
Rocker-switch layout corresponding to the "change-over," "make-up" configuration, showing the spring and direction of movement when the switch is actuated in one direction.

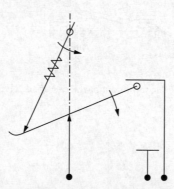

FIGURE 13.5
Rocker-switch layout corresponding to the "change-over," "make-up" and "make-down" configuration, showing the spring and direction of movement when the switch is actuated in one direction.

is shown in Figure 13.6. This shows a "hair-spring" moving contact, which can be used both with and without a spring-loaded plunger. In the case of the "hair-spring" device the contact force is provided by the mechanism, and no additional spring is necessary.

The simplicity of the mechanism and the small number of parts in the assembly, eight in the case of the single-pole switch shown in Figure 13.1, is a major factor in the production of such switches. The rocker-switch design dominates the hand-operated switch market because of its inherent flexibility.

The rocker-switch design has been in use for many years and has evolved through empirical developments. Mathematical models of the switch dynamics have been developed and resulted in improvements in some of the operational characteristics. A model is presented in [1–3]. In common with most hand-operated devices, the switch has two areas of electrical contact, the main switch contacts, and the interface between the moving contact and support, referred to as the pivot interface in Figure 13.1. It is common for arc erosion to occur at both these interfaces.

13.2.1.2 Lever Switches

A typical lever switch arrangement is shown in Figure 13.7; it is similar to the rocker switch, although the hand operation is applied to a lever rather than the rocker actuator, shown in Figure 13.1. The internal mechanism of the device is often similar to a rocker switch.

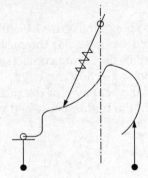

FIGURE 13.6
Rocker-switch layout for an on-off "hair spring" configuration, showing the main contact on the left. The blade in this case is flexible, and spring such that when the contacts are made the contact force is supplied from the blade deflection. The pivot point is offset to the right of the rocker pivot point.

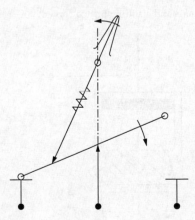

FIGURE 13.7
Rocker-switch layout corresponding to the "change-over" mechanism shown in Figure 13.1, showing the spring and direction of movement when the switch is actuated in one direction. In this case the rocker shown in Figure 13.1 is replaced with a lever actuator.

13.2.1.3 Slide Switches

The slide switch is often used to switch between two conductors and is generally used at the lower end of the low current range; usually for a voltage, or functional selection applications switching electronic signals typically 10^{-2} A, 5 V. It is not often used as an isolating switch, and is referred to as a micro-gap switch since the separation between live parts is less than that required for isolating a supply voltage. Figure 13.8 shows a typical schematic diagram of a switch mechanism. The principal switching action of the device is a sliding or wiping action. Arcing can occur between the contacts, which are usually lubricated.

13.2.1.4 Rotary Switches

The rotary switch is a similar design to the slide switch and is again often used for functional selection, or voltage selection applications. The device can be designed for isolation requirements.

13.2.1.5 Push-Button Switches

A push-button switch arrangement is shown in Figure 13.9. In this case, Figure 13.9 shows a simple arrangement where the contacts are mated as the push-button is pressed; the latching action is provided by the elasticity of the spring contacts, and the contacts are released by pulling the contacts apart. The latching action can be made more effective using a cam design, where a follower moves with the actuation of the push-button and latches into the closed position. To release the switch, the button is pushed which then releases the latch,

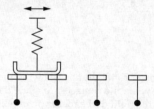

FIGURE 13.8
Slide-switch layout showing four terminals and a sliding bridge contact.

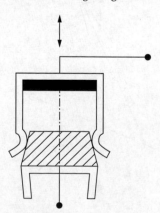

FIGURE 13.9
A simple push-button device with two terminals. The device makes contact when the upper contact is pushed into contact with the lower conductor.

allowing the contacts to open. These switches are generally used for on-off operations, and require two sets of main contacts in an arrangement similar to that used in contactors, with a bridge moving contact. The double-pole configuration can be accommodated but the change-over configuration is not generally used. Push-button switches are generally larger and have a greater number of parts than the equivalent rocker switch. The use of the cam type follower is not so robust as the rocker-switch design and is more prone to mechanical failure. This type of switch is often used where the push-button action is more desirable than the rocker-switch action. Its selection is thus often related to ergonomic design considerations.

13.2.1.6 Switching Devices Used below 0.5 A

There are a large number of devices used to switch current less than 0.5 A. At the low-current and low-voltage levels associated with switching electronic circuits, the applications can range from mobile phones, computer keyboards, electronic piano keys to membrane switches; [4,5]. In general at the low-current level typical of electronic switching there is no arcing, although a molten metal bridge may be formed [6], see Section 9.4.2. At these current and voltage levels, gold and other precious metal alloy contact materials can be used and with a robust mechanical design, devices are able to operate over millions of switching cycles [7]. As the current level increases above the minimum arc current level, silver-based contact materials are used.

A typical device for electronic switching is the membrane switch as shown in Figure 13.10. Upon hand actuation, the two parallel conductors are forced together. This type of switch offers an ultra-thin profile; both tactile and non-tactile soft touch operations can be designed into the arrangement. The contact material is often a carbon film while the lower surface can be a PCB coated with a gold layer [5].

13.2.2 Actuated Switches

Actuated switches can be separated into two categories. They can be triggered by small movements, for example, limit switches and thermostatic controls; or by an actuator, for example the electromechanical relay, where the contact motion is controlled by a solenoid. Developments in piezo-ceramics and memory metals have resulted in new possibilities [8].

13.2.2.1 Limit Switches

Two typical arrangements are shown in Figures 13.11 and 13.12. For such switches to operate reliably, it is necessary that the external device operates with an accurately controlled

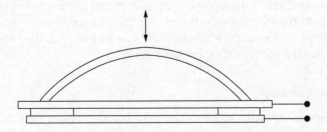

FIGURE 13.10
The membrane switch, showing a section through a device. The upper dome can be plastic or rubber which when pushed will force the upper contact surface to deform and make contact with the lower surface.

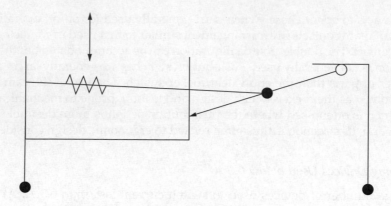

FIGURE 13.11

A limit switch layout in an on-off configuration. As the sprung-loaded element is forced down by an actuator the element will toggle the moving contact about the pivot opening the circuit.

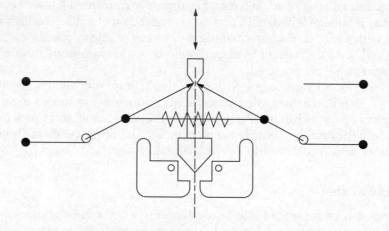

FIGURE 13.12

A limit-switch mechanism with a safety latch. As the central actuator is forced down the pair of moving contacts will pivot to the new position, and the actuator will latch requiring an upward force to release the latch.

movement and force. Although such switches may have mechanical and electrical lives in excess of 1 million operations, there are certain areas where this characteristic by itself is not a sufficient guarantee of certainty of operation. One such application is the guarding of machine tools. If the springs break in the device, there is no certainty that the elements of the switch will not produce a conductive path. In some systems, it is desirable that the guard should be able to start and stop the machine rather than merely act as a safety interlock; for this function some manufacturers use the "forced break" switch shown in Figure 13.12. This category of switching device, along with many of the others, can suffer from wear at both the main and pivoting contacts [9,10].

13.2.2.2 Thermostatic Controls

These mechanisms are essentially mechanical bi-stable devices which when heated or cooled provide movement. There are a number of methods which can be used, for example a bimetal strip can be used which when heated deforms to either make or break a circuit, another method uses a thermal transducer to actuate a bi-stable sprung mechanism. In the

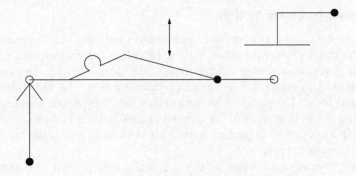

FIGURE 13.13
A thermostatic control switch, where the actuator action acts down upon the self-sprung cantilever; as the force is applied the element changes shape to move the contact upwards to make the circuit.

latter example the mechanism produces instability in the neutral plane, hence the mechanism will not remain in a position where the force holding the contacts together is zero. The non-linear velocity characteristics produced by this toggle mechanism have been analyzed using high-speed photography and show velocities of 0.16 m s^{-1} [11–13]. The design of switches for use in thermostatic controls is usually influenced by the requirement to operate from a slowly moving, thermally actuated prime mover. To obtain adequate speeds for producing well defined points of closure and interruption, the switch contacts are often arranged in a snap-action bi-stable type of mechanism shown in Figure 13.13.

13.2.2.3 Electro-Mechanical Relay

The electromechanical relay is usually operated by an electromagnetic solenoid. A microprocessor can be used with a solid-state device to operate a relay coil which switches the relay contacts. The underlying principle also applies to contactors, which normally operate at a switching current above the range considered in this section. Current-limiting circuit breakers also use a solenoid action to operate a mechanism at high-speed to enable circuit protection (see Chapter 14).

There are many designs of relay configuration [14,15]. It was often suggested that the onset of power electronics would remove the requirements for electro-mechanical relays; experience has shown this to be far from the case. The electro-mechanical relay exhibits two critical characteristic, which are not possible to replicate in the faster power electronic devices; the open circuit breakdown voltage is very high, with no current leakage; and the closed circuit losses are small, while the contact resistance remains low. The electromechanical relay is still a critical component of many systems, and will remain so.

13.3 Design Parameters for Static Switching Contacts

In the closed circuit condition, the main requirement is for a low contact resistance path to minimize the voltage drop across the interface and thus reduce the heat dissipation in the contact. When switching contacts are closed the contact interface can be described by the equations which govern all electrical contacts (see Chapter 1); and from these equations, estimations can be made of the force required to maintain a stable contact resistance and to prevent the contacts separating unintentionally [16].

13.3.1 Small-Amplitude Sliding Motion

A sliding action can occur at a contact interface when there is a small degree of motion either before or after a switching operation. Sliding can also occur when contacts are closed as a result of external vibrations, or temperature fluctuations. The influence on the contact interface is considered in Section 1.5. When the contacts are closed the sliding motion is normally considered to act tangentially to the surface, and results in a change of contact area, which can lead to fretting wear of the contacts (see Chapter 7). Motion vertical to the surface will result in a reduction of contact force, and can usually be accommodated by the selection of a suitable contact force.

The motions, which can occur prior to the contact opening, fall into the same category as with the fretting motions, and are sometimes used to produce a lateral force to break any contact welds, which may have occurred. Wiping motion during the closing phase of a switching process will result in the two contact surfaces rubbing together at the instant the arc is extinguished. The surfaces will thus be at a high temperature, and small areas of molten metal may be present, which can result in an increased wearing of the contact surfaces. This effect can also help produce lower weld forces, and rupture contaminating films.

13.3.2 Contact Force and Contact Materials

13.3.2.1 Contacts at Current Levels below 1 A

The main consideration in the selection and design of contacts for use in light-duty applications (less than 1 A) is that the contact resistance should be kept as low as possible. At low-current levels overheating of the contacts seldom occurs, making the electrical conductivity of the material of less importance. Arcing effects can also be less significant. The mechanical force, which can be applied, is usually restricted by the physical size of the device. To ensure that the contact pressure is as high as possible the correct design of the contact shape is important, flat contacts are not normally used. Flat contact surfaces can trap dust particles between the surfaces. The use of a hemispherical or domed contact against a flat contact is a common arrangement. The materials used for light-duty applications are selected primarily for their resistance to tarnishing under a range of conditions. Alloys of platinum, palladium, and gold are used.

13.3.2.2 Contacts at Current Levels between 1 and 30 A

In the range between 1 and 30 A, contact materials are selected which match a good current-carrying capacity to a reasonable resistance to tarnish and arcing. The shape of the contacts used is of importance in ensuring low contact resistance and adequate dissipation of heat from the contact surface; and as with the low current, flat contacts are usually not satisfactory. The spherical radius of a domed contact is often selected as it gives the best compromise between the high contact pressure obtained with a small radius and the superior heat dissipation from the contact area that can be achieved with a large radius. Contact materials used in this range are covered in Section IV of this book, and are generally silver-based alloys. Silver-nickel (80/20) and (90/10) is used extensively in the range, but in the case where there is the possibility of a high inrush current as the contacts make, and contact bounce occurs, silver cadmium oxide was used, but its application has been limited in some jurisdictions for health related issues and been replaced by silver tin oxide.

13.3.2.3 Contact Force

To estimate the contact force requirement for a given contact system, consider the contact resistance. This has been shown in Equation 9.39 to be given by;

$$R_c = \frac{\rho}{2}\sqrt{\frac{\pi H}{F}} \tag{13.1}$$

The normal force (F) is shown to have a significant influence on the contact resistance. To estimate a static contact force, assume the contact spot temperature T to be the main parameter. If it is also assumed that the rate of corrosion is temperature-dependent, and with current passing through the contacts an elevated temperature will develop in the area of the constriction, it has been shown that the temperature at the contact interface can be approximated to [17]:

$$T \approx 3200 V_c \tag{13.2}$$

Where, T is the temperature (in degrees Kelvin) at the center of the constriction, V_c is the potential drop across the constriction at a room temperature of 22°C, and the contact volt drop is given by

$$V_c = I R_c \tag{13.3}$$

Let us assume that for every 10°C rise in the contact spot temperature there is an approximate doubling of chemical reaction rates, which will greatly affect the reliability of the contact; then if 10 mV is taken as a critical volt drop across the contacts it can be shown that this will produce a 10°C rise in the contact region. Using Equations 13.2 and 13.3, the minimum contact force can be calculated for a given current rating. The force calculated is the minimum force for the assumptions made and does not account for external vibration, which can also lead to the contacts separating; thus, if a contact system is to be used in an environment with external vibrations, acceleration effects will need to be taken into account in evaluating a suitable contact force. At current levels outside those considered in this chapter additional contact force can be required to overcome electromagnetic repulsion forces (Section 10.4).

13.4 Mechanical Design Parameters

Mechanism dynamics are important in determining the reliability and endurance characteristics of a device. The dynamic response will have a direct relationship to the arc characteristic, affecting the duration of the arc, and contact erosion. The electrical contact materials used are also important as they provide the stable electrical contact interface when the contacts are closed and provide protection against arcing during both "make" and "break" operations. The contacts also offer protection against contact welding when contacts close. The selection of contact materials will be discussed in Section IV. The circuit parameters are also important, for example, an inrush of current during switch closure can have a large influence on the contact performance. In this section consideration is given to the mechanical characteristics of devices and the influence on make and break contact dynamics. The mechanical action can be split into two categories, the opening and the closing operation, but with many of the devices covered the actuation controls both actions.

13.4.1 Case Study (1): Hand-Operated Rocker-Switch Mechanism

The opening velocity profile of a hand-operated switch can be related to the nature of the hand actuation of the device. Even snap-action devices, which are designed to provide a fast opening independent from a hand operation, are affected by the nature of the operation. The hand operation, and how this affects contact velocity, can depend upon the way in which a switch is operated. The force required to switch the device depends upon the mechanism, the spring stiffness and the angle of operation; however, the force applied can be greater than the force required to switch the device. The operating force can be categorized into three types of action, described as soft, normal, and hard. These terms are defined as follows: the soft action requires control of the switch by operating it as slowly as possible. The hard operation requires a sudden abrupt action. Normal operation can be described as an application of force where the operator reacts to the stiffness of the mechanism and adjusts to toggle the switch. These criteria are applicable to all types of hand-operated snap-action devices. It can be assumed that in normal use, switches are operated to fall within the category of normal operation.

It has been shown that the hand operation can have a direct effect on the dynamics of a switch [1–3]. Using a test circuit it is possible to identify the dynamic events in a switch operation, [2]. This can be useful in identifying the mechanical characteristics of a device prior to electrical testing. Figure 13.14 shows a response using the circuit with a mechanism prone to bounce during hand operation. The figure applies to a standard change-over switch with a configuration similar to that shown in Figure 13.2, operated by hand. It shows the separation of the main contacts and the following impact of the second contact pair, between A and B. The time of this motion is defined as the change-over time (c/o). Immediately after the impact, a separation occurs at the pivot interface shown in the lower trace at C, and this is followed by a period during which both the main contact and pivot

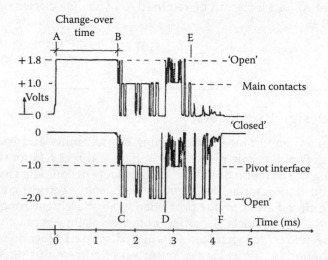

FIGURE 13.14
A typical switching event in a hand-operated rocker switch. The upper trace shows the events occurring at the main contacts, and the lower trace shows the events at the pivot interface.

contacts bounce. Using this method the mean normal hand-operated change-over time can be evaluated, and the average velocity obtained; this has been used in an investigation of the average velocities between 0.5 and 1.0 ms^{-1}. The pivot bounce time is the total time of separation between C and F. The method demonstrates a simple means of obtaining important information on the dynamic events in a switching system. Much of the details shown would not normally be observable using high-speed photographic techniques.

In general, the opening characteristics of hand-operated devices are dominated by the spring forces that need to be overcome to operate the device, thus with a large spring force the hand operation will need to be more forceful and the mechanism will therefore move faster.

13.4.1.1 Moving-Contact Dynamics of a Rocker-Switch Mechanism

In order to understand and improve the design of a switch it is possible to develop mathematical models of the mechanisms. Once a model of a mechanism has been developed and verified it can then be used to optimize the performance, for a given parameter. A full analysis of a rocker switch is given in [2], using the geometry of the rocker switch mechanism shown in Figures 13.1 and 13.2. The model defines the interaction of the moving rocker assembly α (the angle between the vertical and the spring-loaded plunger position) with the rotation of the moving contact β (the angle between the horizontal and the moving contact position). The interaction of the two parts is complex and depends upon the frictional characteristics of the spring-loaded plunger and the interaction with the surface of the moving contact, identified in Figure 13.1 as the blade. The mathematical model of the mechanism consists of six nonlinear simultaneous differential equations, which are solved numerically.

Figure 13.15 shows the response of the moving contact with the angular velocity of the rocker as a parameter $\dot{\alpha}$. The faster the rocker motion the greater the value $\dot{\alpha}$, therefore, the

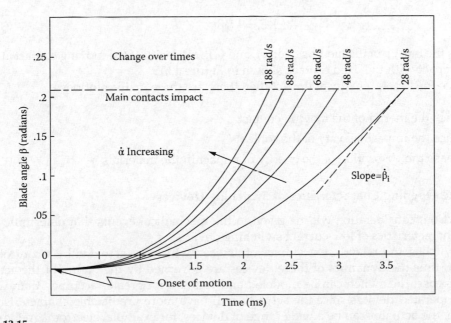

FIGURE 13.15
The response of a mathematical model of a rocker switch, showing the variation of the moving contact position with time, with the velocity of the hand-operated rocker actuator as a parameter.

faster the change-over time of the switch. The change-over time can be identified as the time between the onset of motion and the impact of the main contacts. With a rocker velocity of 100 rad s⁻¹ the change-over time is shown to be 2.25 ms. The moving contact (blade) accelerates from rest where the blade starts at a small negative angle, to the point of impact where the blade angle is defined by the dotted line at 0.21 radians. The slope of the line at impact gives the velocity of the contacts at impact, which increases with the rocker velocity.

13.4.1.2 Design Optimization of a Rocker-Switch Mechanism

It has been identified that when the make-down configuration rocker switch is operated by hand there is a complex mechanical event, where both the main and pivot contacts can bounce, as shown in Figure 13.14. By consideration of the mechanism shown in Figure 13.2, it can be identified that when the main contacts make, the inertia of the moving contact will allow it to continue moving in the same direction; the moving contact will then separate from the pivot. At the same time the restitutional effects at the main contacts will cause the main contact to bounce. This will be particularly the case in the change-over configuration, where the moving contact inertia will be higher. In the make-down, on-off configuration shown in Figure 13.3, the pivot will have less tendency to separate, because of the lower inertia. In the make-up configuration shown in Figure 13.4, the main contact impacts with the fixed contact. In this case, the moving contact is unable to continue moving in the same direction at the pivot interface, therefore the bounce that occurs at the main contact is the main dynamic event. The parameter that best describes the likelihood of bounce other than the geometry of the device is the kinetic energy in the moving contact, since it is the kinetic energy that needs to be dissipated in the switching action to bring the system to rest. The bounce of electrical contacts will be discussed in detail in the next section, but it can be observed here that the bounce of impacting contacts will be related to the magnitude of the kinetic energy at impact KE_i given by

$$KE_i = \tfrac{1}{2} I_b \beta_i^2 \tag{13.4}$$

where I_b is the inertia of the moving contact, and β_i the velocity of the moving contact at the point of impact with the fixed contact, shown in Figure 13.15.

To reduce the kinetic energy at impact then there are three main solutions:

1. Reduce the inertia of the moving contact
2. Reduce the impact velocity of the contacts
3. Increase the absorption of the energy of impact in the materials

13.4.2 The Opening Characteristics of Switching Devices

In this section, consideration will be given to the mechanical events that determine the opening characteristics of low-current switching devices.

Limit switches and relays are examples of devices, which are actuated by a controlled movement, thus the dynamics of these devices are governed by the nature of the actuator. In this case, the switch can be modeled and tested with greater accuracy than with the hand-operated devices since the actuator will have more predictable characteristics. In practice, the actuator can be a wide range of devices, for example, electromagnetic circuits in relays and residual current devices and thermal displacement devices in control switches.

13.4.2.1 Moving Contact Dynamics at Opening

The nature of an opening velocity profile depends upon the system considered. A simple mathematical model of moving contact dynamics can be given by assuming that there is an actuating device, which provides a constant force to a moving contact, which is pivoting about a fixed point A, in Figure 13.16. The figure shows the actuator force F_e acting to open the contacts, and the force F_0 from a spring acting against this to provide the static contact force when the contacts are in the closed condition. The moving contact is shown to have a position α relative to the horizontal, and the moving contact velocity is thus given by $\dot{\alpha}$. The dynamics of the mechanism can be given by

$$J\ddot{\alpha} = F_e h_e - F_0 h_0 \tag{13.5}$$

where

J = moving contact inertia

$\ddot{\alpha}$ = moving contact position

F_e = force of actuator

F_0 = static contact force

h_e, h_0 = the distance of the force F_e and F_0 from the pivot point

To simplify the dynamics, it can be assumed that the static contact force is small compared to the other forces, then Equation 13.5 can be simplified to assume no static force and a constant applied force, F_e;

$$J\ddot{\alpha} = F_e h_e \tag{13.6}$$

From Equation 13.6, the contact displacement as a function of time can be determined, and will show a linear acceleration, with the contact velocity increasing from zero at rest when the contacts are closed, to a maximum value as the moving contact hits an end stop.

13.4.3 The Make Operation

In this section, consideration is given to the mechanical events that determine the closing or make characteristics of devices used for low-current switching. The effect of current on the process will be considered in Section 13.7.

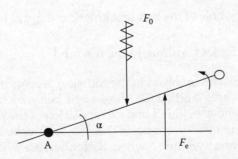

FIGURE 13.16
The opening dynamics of a pivoting switch mechanism, with the actuator force F_e and the static spring force F_0.

Electrical contact bounce occurs in most switching contacts. In the general case, when two solid objects collide, there is inevitably a rebound, which in the case of electrical contacts, if the circuit conditions permit, results in the generation of arcs. With arcing occurring during contact bounce, each bounce becomes a fast break-and-make operation, during which the arc is not usually extinguished, unless an ac zero occurs. The events occurring during the impact of electrical contacts thus reflect some of the events at the opening of the contacts. Often there are a number of bounces following the initial impact which, with associated arcing, can lead to severe erosion of the contacts. In most circuits, the current rise at impact is delayed by the effect of circuit inductance, thus reducing the energy dissipated by the arc and thus the arc erosion; conversely during the break operation transients occur which cause arcing. On switching a capacitive circuit, however, a high inrush of current can occur. Other devices can also give a high inrush, for example incandescent lamps, and electric motors. The high inrush of current can cause current as high as ten times the normal circuit current to pass through the interface during the bounce and arcing stages. The impact of electrical contacts in the closing process can be considered from the mechanical events which occur in the impact of solids. This can then be taken further to understand the effects of current on the process; and this is covered in Section 13.7.

13.4.3.1 Impact Mechanics

When two solids collide, there is a complex interaction of forces. To understand the events occurring in the electrical contact interface during impact, it is important to provide an understanding of the events, which govern mechanical impact. Mechanical impact is a general area applicable to many areas of mechanics. The study of impact between colliding solid bodies received its first theoretical treatment in 1867, by St Venant, who suggested that the total period of collision was determined by the time taken for an elastic compression wave to move through the solid and be reflected back. Experimental results show this theory not to be perfectly true and the evidence suggests that for small bodies the collision process is dependent upon the deformations occurring in the area of contact. Consequently for small bodies of the order of a few centimeters in length the problem of the compressive wave can be ignored. The physical explanation of impact and rebound can then be applied to the events occurring at the points of contact. Using these assumptions, the events occurring during impact were dealt with by Bowden and Tabor [18], who separated the impact process into four main stages.

1. The elastic deformation in the areas of contact
2. The partial plastic deformation when the local pressure exceeds 1.1 times the yield point of the materials
3. Full plasticity until the whole of the incident kinetic energy at impact is consumed by the collision
4. Release of the elastic stresses resulting in the rebound

A full analysis of the four phases involved in the collision process is complicated and difficult. Early research into this area studied the quasi-static behavior of contacts. Goldsmith and Yew [19], and Tabor [20] and concluded that extrapolation of the results to the dynamic impact process could be carried out on the basis that the difference between the dynamic and static yield stress and static dynamic energy dissipation are often within 25–50% for a number of common metals. These early tests encouraged the belief that a useful if only qualitative insight into bounce could be gained by quasi-static load tests [21].

Experimental investigations in the field of impact have been performed for a variety of purposes. Many have been conducted with the object of assessing the validity of a proposed theory or the accuracy of an assumed model of material behavior, for example, proving the Hertz law of contact or the study of viscoelastic wave propagation. Other efforts have been directed toward the collection of empirical information in the areas of cratering and penetration [22].

In the impact process applied to electrical contacts, one contact would normally be stationary. The kinetic energy of the moving contact is then absorbed at the interface, or in support structure. During the release stage, the moving contact is given back a degree of the energy, which results in a separation. With a spring acting on the moving contact, the contacts will again impact until all of the incident kinetic energy of the first impact is dissipated. The application of current to the interface during the impact process and the following bounces adds a further dimension of difficulty to the mechanical considerations.

13.4.3.2 The Coefficient of Restitution

To give analytical understanding of the events, a simple mathematical model can be developed to represent contact mechanics, based upon a system of masses, springs, and dampers, representing the elasticity of the contact surfaces. Consideration can be given to the coefficient of restitution e, and the conservation of linear momentum. With reference to Figure 13.17, the following definitions can be made. The velocity of the moving contact is identified in the figure as u_i and the velocity of the contact after the impact process as u_1 where "1" is used to indicate the first impact.

The coefficient is defined as the ratio of the impact velocity with the separation velocity, where the separation velocity can be defined as:

$$u_1 = e u_i \tag{13.7}$$

or for the more general case where there are a number of impacts during a bounce process;

$$u_n = e_n u_{n-1} \tag{13.8}$$

Figure 13.18 shows the terms used in defining the bounce and impact events, which occur as electrical contact make, where there are multiple impacts and bounce events.

In a simplified analysis, e is often taken as a constant; however, it can be shown that e is a function of the impact velocity:

$$e = f(u_i) \tag{13.9}$$

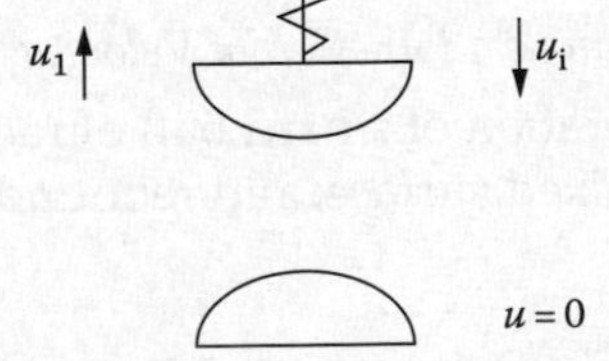

FIGURE 13.17
The closing event of contacts with an impact velocity u_i and a separation velocity u_1 after the impact event.

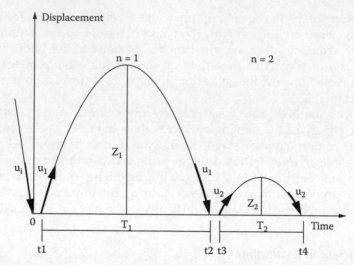

FIGURE 13.18
The notation used to describe the contact bounce process, in this case with two bounces after the first impact.

Often the coefficient of restitution is considered a material constant, and as such can be used to produce some useful relationships in describing contact bounce. However, in reality, the concept of restitution is used because of its attractive simplicity, and it would be more conceptually satisfactory to summarize the results in terms of energy dissipation. The coefficient of restitution has been shown to be a function of impact velocity and to some extent is affected by surface hardening, resulting from the work hardening of the contact surface [23]. In the case of electrical contact the event is also affected by arcing between the contacts.

The value for e can be determined from the bounce times as shown in the following analysis, with reference to Figures 13.17 and 13.18, which give the notation used; the mechanics of bounce without the passage of current can be described as follows, assuming e to be a constant:

$$e = \frac{u_1}{u_i} = \frac{u_2}{u_1} = \cdots = \frac{u_n}{u_{n-1}} \tag{13.10}$$

Assuming negligible gravitational effects, and a constant applied force,

$$Z_n = T_n \frac{u_i e^n}{4} \tag{13.11}$$

where Z_n and T_n are the maximum height and duration of the nth bounce.

13.4.3.2.1 The Coefficient of Restitution as a Function of Velocity

It has been shown from the consideration of a hard ball of radius r_1 mass m, falling from a height of h_1 with velocity u_i on to a fixed surface, and rebounding to a height h_2 with velocity u_1 [18], that

$$u_1 = k \left(u_i^2 - \frac{3}{8} u_1^2 \right)^{3/8} \tag{13.12}$$

where

$$k = \left(\frac{436 r_1^3}{mE} \right)^{1/8} p^{5/8} \tag{13.13}$$

where E is Young's modulus, taken as the same value for both sets of contacts, and p is the dynamic yield pressure.

Combining Equations 13.8, 13.12, and 13.13, it can be shown [23] that the impact event can be described by

$$u_i^2 e^8 + k^8 (0.053 e^6 - 0.43 e^4 + 1.13 e^2 - 1.0) = 0 \tag{13.14}$$

Equation 13.14 can only be solved using numerical methods; however, simplifications can be made to render the function more usable for design purposes. To a first approximation, since e is always <1, and if k is small, ignoring the powers of e in the brackets as they are much less than one, [23], then Equation 13.14 can be simplified to give

$$e \approx \left(\frac{k^8}{u_i^2} \right)^{1/8} \tag{13.15}$$

Using these equations, it is possible to relate the amplitude of a specific bounce to the duration of the bounce. For a given impact, velocity and assuming e to be a material constant there exists a unique relation between the time of bounce and the maximum displacement. These equations do not, however, allow for the vibrations in the fixed contact or the presence of a current or an arc between the contacts. They also do not take into account possible changes in the contact surfaces that can result from the arcing between bounces (see Section 10.6).

13.4.3.3 Impact Mechanics for a Pivoting Mechanism

Figure 13.19 shows the geometry of a pivoting contact system, with the terms used to define a bounce characteristic given in Figure 13.18. In this case the angle defining the position of

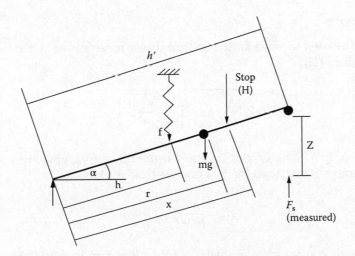

FIGURE 13.19
The closing dynamics of a pivoting mechanism, where H is the stop position corresponding to the maximum contact gap, f is the spring force bringing the contact together, and F_s is the static contact force after the event.

the moving contact is α, and h is the position of the spring, while h' is the position of the contacts from the pivot point.

The dynamics of the impact events are on the basis of the following assumptions:

1. There are no frictional losses.
2. There are no elastic vibrations in the contact support or contact arm.
3. There are no electromagnetic delaying forces.
4. The impact is dominated by surface effects and not by stress wave propagation.
5. The contacts are released from the stop position (H), and fall under the influence of the spring force f, with a spring stiffness of K.
6. The contact force when the mechanism is stationary is F_s.

13.4.3.4 The Velocity of Impact

If the angle (α) is small, then

$$\ddot{\alpha} + \frac{Kh^2}{J}\alpha + \frac{h'F_s}{J} = 0 \tag{13.16}$$

which can be solved (24) to give

$$u_i^2 = h'^2\left(\frac{h^2KH^2}{Jx^2} + \frac{2Hh'F_s}{Jx}\right) \tag{13.17}$$

Therefore, the velocity of impact therefore increases with both the initial setting H and F_s. Thus the greater the contact force used to hold the contacts in the closed position, and the greater the contact gap used in the device, the greater the velocity of impact. The analysis can be taken further to consider the impact and bounce events.

13.4.3.5 Bounce Times

Equation 13.16 can be used to solve for the bounce times, for example T_1, the bounce duration of the first bounce [24]:

$$T_1 = 2\sqrt{\frac{J}{h^2K}}\tan^{-1}\left(\frac{u_1\sqrt{h^2KJ}}{F_sh'^2}\right) \tag{13.18}$$

A simplification of this equation can be made if the bounce heights are small and the spring force remains constant; then the first bounce time is given by

$$T_1 = \frac{2u_1J}{(h')^2F_s} \tag{13.19}$$

With a known static contact force, F_s, inertia J, and pivot arm length, h', the duration of the first bounce can be predicted, if the velocity at which the contacts open after the first impact is known.

In the case of the pivoting mechanism used in this analysis the constant k given in Equation 13.13 needs to be modified [24], and is given by

$$k = 2.784 \left(\frac{p^5 r^3 (h')^2}{JE^4} \right)^{3/8}$$

(13.20)

In the case of both the pivoting contact system, and the system described in Equation 13.13, k will increase with an increase in p and the radius r. Both Equations 13.13 and 13.20 show that the surface profile of the impact area is significant and that the radius r needs to be investigated further since on a worn electrical contact the radius would be expected to be dominated by the worn contact surface and not the initial radius of the contacts. Equation 13.14 can be solved numerically to give e as a function of changes in velocity of impact for the pivoting mechanism, with k as a parameter, as shown in Figure 13.20 for a range of velocities. The figure shows that at low velocities $e = 1$, but as the velocity of impact increases e reduces. Also e is shown to be a major function of k, and reduces for a given impact velocity as k decreases. A reduction in k could be interpreted as a reduction in p, the dynamic yield pressure, or r, the radius of the contact surface. At high values of impact velocity e could be taken as a constant, and also at the velocities shown in the figure if k is small.

The use of Equation 13.8 to describe the bounce process involves a discontinuity for the last impact. This is irrespective of e being taken as a constant or not, since the equation will fail, because u_n cannot go to zero; indeed as the velocity, u_{n-1}, decreases e tends to unity, suggesting that the bounce will go on to infinity since $u_u = u_{n-1}$. Hence to model the process more precisely, energy equations should be used [21], giving

$$u_{n-1}^2 = u_n^2 + \frac{2 \in}{m}$$

(13.21)

for a moving contact of mass m, where $\in$ is the energy lost during the impact process.

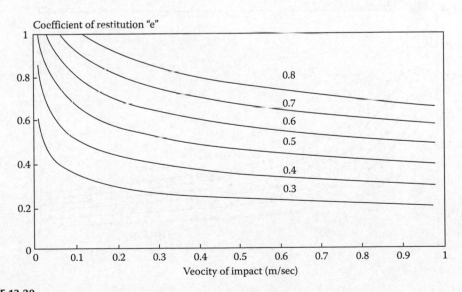

FIGURE 13.20
The coefficient of restitution as a function of impact velocity, based upon Equation 13.14, with k (0.3–0.8) as a parameter.

13.4.3.6 *Total Bounce Times*

It can be shown, using Equation 13.19 that if $e = $ constant, then

$$T_n = \frac{2u_i e^n J}{h'^2 F_s} \tag{13.22}$$

from which the total bounce time can be approximated using the binomial expansion, to give

$$T_n = \frac{2u_i j}{h'^2 F_s}\left(\frac{e}{1-e}\right) \tag{13.23}$$

If, however, $e = f(u_i)$, then this equation can be rewritten as

$$T_n = \frac{2u_i J}{h'^2 F_s}(e_1 + e_1 e_2 + e_1 e_2 e_3 + \cdots) \tag{13.24}$$

Hence, the total bounce times can be evaluated from Equations 13.23 and 13.24. Figure 13.21 shows an example characteristics, when $e = 0.4$, and $k = 0.3$. This shows that the total bounce time decreases as the static force increases. An increase in the static contact force could therefore be proposed as a method of reducing contact bounce, but the increase of the force will have a secondary affect in the increased impact forces as the contacts close, which wil result in increased deformation of the contact surface. The figure also shows that the contact gap from which the mechanism is released, H, will increases the bounce duration.

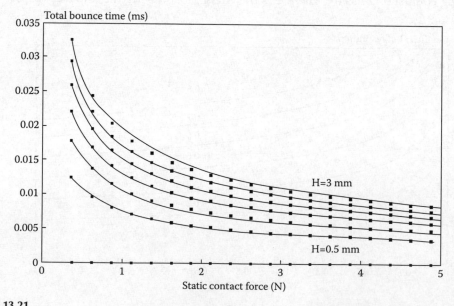

FIGURE 13.21
The total bounce time versus the static contact force, based upon Equation 13.23, with the drop height H (0.5, 1, 1.5, 2, 2.5 and 3mm) as a parameter.

13.4.3.7 Impact Times

For two contacts of radius r, and mass m, where one contact is stationary, the impact times can be given (24) as

$$T_i = \frac{\pi}{2}\sqrt{\frac{m}{p\pi r}} \tag{13.25}$$

Hence the impact time is shown to be independent of the velocity of impact. A re-evaluation of the impact time for the pivoting mechanism gives

$$T_i = \frac{1}{\omega}\tan^{-1}\left(\frac{\omega^2}{C\,u_i}\right) \tag{13.26}$$

where

$$\omega = \sqrt{\frac{h'^2 p\pi r}{J}} \tag{13.27}$$

and

$$C = -\frac{h'^2 F_s}{2J} \tag{13.28}$$

Hence, in the form of Equation 13.26 the impact time is shown to be a function of the impact velocity.

13.4.3.8 Design Parameters for the Reduction of Contact Bounce

It has been shown in Section 13.4.1 that contact bounce can be reduced by the control of a number of parameters, for example, reduction of impact velocity, reduction of moving contact inertia, energy absorption at the contacts or in the contact structure, and by increasing the contact force in a given mechanism. The selection of the parameter which best suits a given situation or device varies. For example, the control of the electromagnetic actuator can be used to reduce the impact velocity and thus reduce the contact bounce, but this would be a costly addition to a rocker-switch mechanism, and most other low-current switches. In general to reduce contact bounce in low-current hand-operated devices the choice of mechanism is important, as discussed in Section 13.2, followed by an understanding of the switching dynamics.

13.5 The Measurement of Contact Wear and Contact Dynamics

The previous sections have considered the mechanical operation of switching devices; consideration is now given to the electrical performance in terms of the interaction of arcing on the performance of these devices. As discussed in Chapters 9 and 10, the arcing process has a direct effect on the switching surfaces, in that the plasma interaction with the contacts results in material transfer. In addition to this, the plasma interaction with the surrounding gases also has an important effect of the surfaces of the electrodes. In many of the studies undertaken the mass change of the electrodes are often used to determine the wear processes. In this section the volumetric change of the electrode surfaces is introduced. It will be shown that this new method is a powerful tool for determining the evolution of wear on electrical contact surfaces. The methods discussed are equally applicable to connector wear, but the focus in this section is on the application to arcing surfaces.

13.5.1 The Measurement of Contact Surfaces

A study of wear and erosion performance of switching devices is essential in the prediction of switching device life. It is also fundamental in the development of new and improved switching contact materials. Much of the published research literature in the field of electrical contact studies is focused on these issues. To constrain the scope the main focus here is on devices types typically used in domestic and light industrial applications. Testing of commercial devices is normally defined by international and national testing specifications which are numerous. The testing regime is unique to a particular classification of device, and it is not possible to generalize these test. In studying the fundamental phenomena underlying the commercial devices and testing methods, a number of investigation have been reported. In general these investigations depend on the methods of reducing the number of experimental variables and designing a test method to reflect the condition to be investigated. In these studies, a controlled experimental condition is used, for example constant contact opening velocity. The contact materials are mounted in the test apparatus and evaluated for erosion after a fixed number of switching cycles. In the majority of the application the contacts are removed from the test system and evaluated for erosion by consideration of the mass change of the contacts. This method does not give an indication of the surface condition. For example, a contact surface can exhibit a zero mass change when the surface has been severely eroded; also the mass change will not reflect that some mass will be lost to the surrounding contact holders. This has in the past been known to cause errors when comparing different sets of result from different authors. To overcome the limitations of the mass change method a number of recent papers have focused on the evaluation of the contact surface using a range of contact and non-contact methods. These methods are referred to as Volumetric Methods, [25–33].

Surface profile can be measured in both 2-D and 3-D. Using a 3-D technique, a measurement of the change in volume above and below a datum, can be obtained, generating data on the distribution of mass on a contact surface. Figure 13.22, shows a schematic diagram of a typical eroded switching contact where mass has accumulated above the

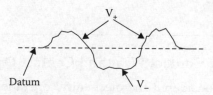

FIGURE 13.22
2-D Erosion profile of an electrical contact

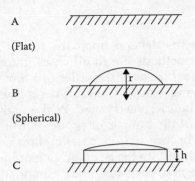

A (Flat)

B (Spherical)

C

FIGURE 13.23
Typical low voltage electrical contact shapes where h = 0.5mm.

new (datum) surface in some areas and below the surface in other areas. The figure identifies the volume (V_+), above the datum and (V_-), below the datum. The net volume change of the surface is then (ΔV),

$$\Delta V = |(V_+) - (V_-)| \qquad (13.29)$$

In addition to this a volume calculation (V) can be made relative to a fixed datum surface such as the material on which the contact surface is mounted. In Figure 13.23 surfaces B and C are examples of surfaces where the latter measurement can be made.

13.5.2 Three Dimensional (3-D) Surface Measurement Systems

Stylus and Laser-based metrology instruments have been used to investigate the surface profiles of eroded contacts. In early work on this method, [11,29], multiple 2-D stylus scans were used to investigate the erosion of contact surfaces and comparisons were made between the mass change and volume change. In [11], a study of the erosion of Ag-CdO contacts in low current switching, below 16 Amps, was presented. It was shown that the volume could be calculated above and below a curved contact surface.

Surface profiling using non-contact laser probes have been used to study contact erosion, [25–28]. In [26,27], a measurement method was developed on the basis of the non-contact measurement of an eroded surface using an auto-focus laser. This showed that an electrical contact welded to a substrate could be analysed and it was demonstrated that the near vertical surfaces of a typical electrical contact used in low voltage relays, type C in Figure 13.23, could be measured and the resultant data array of the surface used for the evaluation of the volumetric erosion (V) relative to the base. There follows a brief description of generic measurement systems.

13.5.2.1 Contact Systems

A stylus profilometer operates by accurately monitoring the vertical movement (Z axis) of a stylus of known radius (e.g. 2 µm), as it moves across a surface, (X axis). The stylus system is very slow in collecting data, as the contact nature of the measurement requires the stylus has to move slowly to maintain contact. Typical measurement times are between 1–2 hours.

13.5.2.2 Non-Contact Systems

Non-contact measurement systems depend upon the use of light incident upon the surface to be measured. These methods are in all cases much faster than the contact systems with typical measurement times of the order of a few minutes. These methods can be separated into the following groups, Confocal Laser Scanning Microscopy, Laser Ranging Non-Contact probes, Interferometry, and White Light probes. Laser ranging systems require the movement of the sample, or probe. With the scanning microscope, height data is gathered as series of equi-spaced, parallel slices, with each slice containing evenly spaced measurement points. With the laser ranging probes there are two common measurement techniques, using auto-focusing, and triangulation. With the auto-focus laser system a 1–2 μm spot is projected onto a surface, and the quality of the focus of the reflected beam assessed by a photo-diode array. An auto-focus controller drives a servo-system attached to the objective lens, adjusting its position until a best focus is achieved. By monitoring the movement of the focusing servomotor the displacement of the surface from the probe can be determined. Typical resolution is 10 nm with a range of 0.6 mm. With a triangulation probe a laser is directed at a surface, and the position of the reflected beam on a photo-diode array determines the displacement between the probe and the sample. The spot size will vary with the height, and is normally 20–30 μm, limiting the ability to measure surface detail. With interferometry a monochromatic light source (typically a laser) is passed through a beam splitter causing light to be directed onto a sample surface and reflect back onto a reference optical flat. The two beams recombine and the interference pattern formed is incident on a CCD array, connected to a computer with image processing facilities. White Light probes offer a similar resolution and spot size to the auto-focus systems but with a lower gauge range of 0.3 mm. The probe uses a White Light source which is focused on the surface though an optical fibre/lens system. The reflected light is the processed using a spectrometer, allowing the surface position to be detected. A review of the types of process measurement process is provided in [28].

13.5.3 Case Study (2): Example of Volumetric Erosion

To demonstrate the volumetric erosion measurement, an experiment is conducted using conventional Ag/Ni automotive relay contacts in a relay test fixture. The samples are cycled for 3000 switching cycles at 14.8 V, 10.2 A, in a resistive circuit. Figure 13.24 shows the resultant cathode surface and Figure 13.25 the anode surface. In both cases the underlying surface form is removed to provide the datum for the evaluation of the volumetric wear. To increase the accuracy of the volumetric wear process it is important to remove the roughness of the surrounding surface as this will generate significant errors in the evaluation. There are a number of methods for achieving this, [30,31]; although the most accurate is to use a data removal method, based upon the known 3D surface roughness of the datum surface, [34]. Resultant data is shown in Table 13.1. The net change in volume of both cathode ($1.355 \times 10^{-3}\,mm^2$) and anode surface ($1.367 \times 10^{-3}\,mm^2$) are similar. It is well know that the volume calculation should be adjusted for density changes in the material transfer for are relationship to mass, however a number of researchers have shown that for most electrical contact applications the volumetric method is sufficient, [11,35]. Figure 13.26 shows the interaction of the two surfaces using selected 2D data lines, identified in Figures 13.24c and 13.25c. Figure 13.26 shows the cross section through the pip formation on the cathode, superimposed on crater formation section from the anode surface. The difference in the two surfaces arises from the deposition on the anode surface shown Figure 13.25b. A 3D method has been developed using a replication methods, [36].

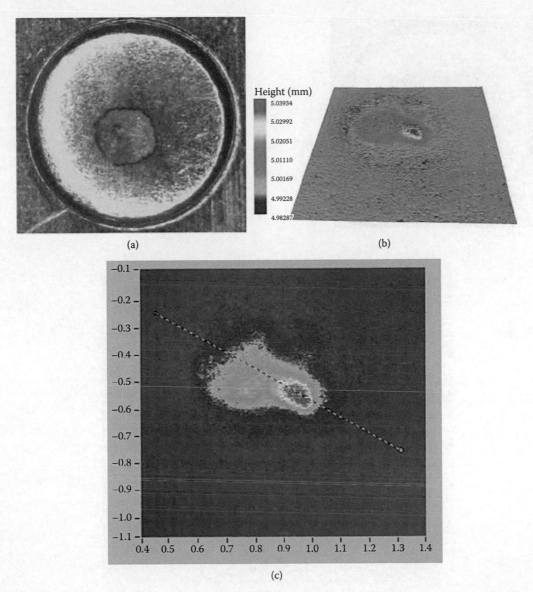

FIGURE 13.24
(a), Micrograph of the Cathode surface, (b) 3D surface data using TaiCaan XYRIS 4000CL, 90,000 data, over 1mm by 1mm, with the underlying spherical surface removed; (c) plan view of the 3D data with a cross section selected.

13.5.4 The Measurement of Arc Motion and Contact Dynamics

In order to understand the influence of the moving contact on the reliability of contacts, it is necessary to monitor moving-contact dynamics. High speed video has the ability to frame at high speeds, allowing detailed studies of the arc, [37], and the associated mechanism dynamics. Optical fiber arrays have also been developed and used in the study of mechanism motion. The optical fiber method is a low-resolution solid-state imaging system, where the images can be stored in computer memory; using this method images can be formed at the rate of 1,000,000 frames per second [38,39].

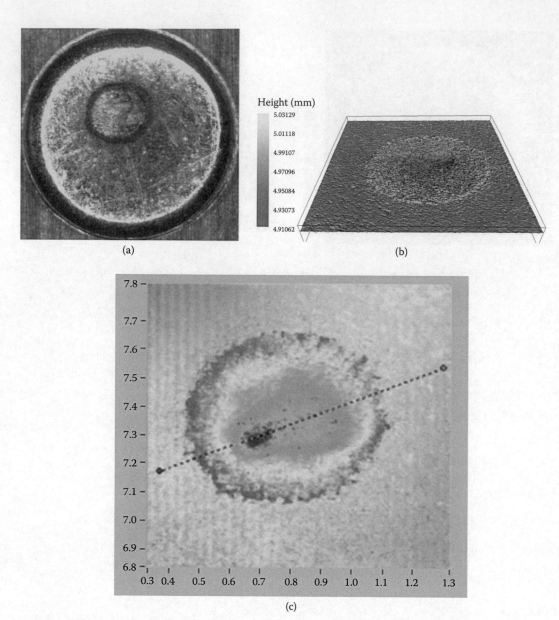

FIGURE 13.25

(a), Micrograph of the Anode surface, (b) 3D surface data using TaiCaan XYRIS 4000CL, 90,000 data, over 1mm by 1mm, with the underlying spherical surface removed; (c) plan view of the 3D data with a cross section selected.

TABLE 13.1

Determining Volumetric Erosion

	Cathode Surface × 10^{-3} mm²	Anode Surface × 10^{-3} mm²
Volume Above Datum (V_+)	1.526	0.833
Volume Below Datum (V_-)	0.171	2.200
Net Volume Change (ΔV)	1.355	1.367

(For the cathode from Figure 13.24b and anode from Figure 13.25b)

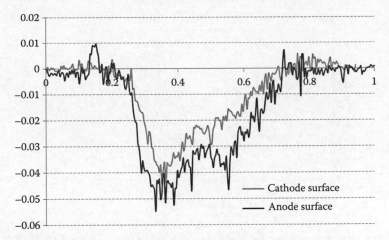

FIGURE 13.26
2D surface data using the cross sections shown in Figure 13.24c and 13.25c; aligned to show the interaction of the anode and cathode surfaces.

13.6 Electrical Characteristics of Low-Current Switching Devices at Opening

Having considered the types of switching devices in the low-current range and the mechanical characteristics of switching systems, consideration is now given to the electrical characteristics: this will include discussions of arc characteristics, arc energy, contact wear and erosion (see also Chapters 9 and 10). In this section, consideration is given to the arc characteristics at opening and in Section 13.7 consideration will be given to the events at switch closure.

13.6.1 Low-Current DC Arcs

In dc circuits where the supply current is well below the limiting arc current I_{lim} (0.01–0.06 A see Section 9.5.4) a showering arc can be formed and is described in Section 10.2.4. Above the minimum arc current in a dc circuit, arc voltage increases with an increase in the contact gap until the current is reduced to a current at which the arc can no longer be sustained, see Section 9.7.2; the arc will then extinguish. For a slowly moving contact the arc voltage can be determined at a given contact gap by use of the "Holm graphical method" [17]. In this method quasi-static curves of arc voltage and current for a particular contact gap of the type shown in Figures 9.37 through 9.39, can be used to define the contact gap at which an arc will extinguish in a given circuit. The arc length (d) can be approximated by the following empirical equation [40,41]:

$$d = A(V_{arc} - V_{min})^m \cdot I_{arc}^n \tag{13.30}$$

Where,

d = the contact gap (mm)

A = a constant

m = 1 to 2

n = 0.5–1

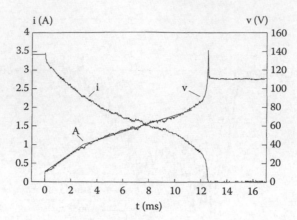

FIGURE 13.27

Typical arc voltage and current waveforms, for an opening velocity of 0.4 m s^{-1}, I = 3.3A, V = 110 V dc, and L/R = 1 μs.

The constants can be obtained experimentally, and vary with arc length and contact material. For a resistive dc circuit the contact gap (d) can be defined in terms of the supply conditions:

$$d = A(V_{arc} - V_{min})^m \left(1 - \frac{V_{arc}}{V} \right)^n I^n \tag{13.31}$$

Where, V and I are the supply voltage and current. Equations 13.30 and 13.31 depend upon the quasi-static nature of the opening contacts, and are not accurate in normal switching conditions where the arc is a transient event. In addition to this, in most opening events the contact velocity u_0 is neither constant, nor quasi-static (see Section 13.4.2). The empirical relationships defined in Equations 13.30 and 13.31, are a useful tool for modeling the switching performance of low voltage systems.

13.6.1.1 Arc Voltage Characteristics

Typical arc voltage and current waveforms are shown in Figure 13.27, for a break operation with a constant opening velocity, u_0 = 0.4 m/s, I = 3.3 A, V = 110 V dc [41]. The figure shows that the arc extinguishes after 12.6 ms; with a linear velocity this corresponds to an arc length of 5.04 mm.

13.6.1.2 Voltage Steps below 7 A

At a current below 7 A, arc voltage waveforms can be observed to exhibit step rises as the contacts are opened. Figure 13.28 shows a typical arc voltage waveform, [42]. At higher current (>7 A) the arc voltage increases smoothly. Early investigations [43–45], revealed the presence of only two steps. The first step corresponds to the minimum arc voltage. Boddy and Utsumi [46] proposed that the second step is owing to a transition from a metallic discharge medium to an arc burning in the ambient gas (see Section 10.3.5). The critical distance at which the metallic to the gaseous phase transition takes place is determined by the condition that the concentration of metal vapour falls below the value necessary for the maintenance of the discharge because of the diffusion of the ambient gas into the metal

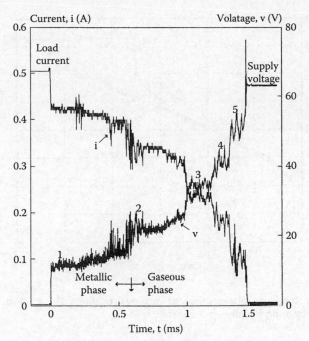

FIGURE 13.28

Arc voltage and current showing a case with five voltage steps, $I = 3$ A, $V = 65$ V, and opening velocity 0.5 m s^{-1}.

vapor. The critical distance r_c, [46], depends on the circuit current I and the pressure p of the ambient gas as well as the ionization potential V_i of the gas as follows:

$$r_c \propto p^{-\alpha} I^{\beta} V_i \qquad (13.32)$$

Where, $\alpha = 0.4$–0.5 and $\beta = 1.0$–1.1 for small molecular weight gases like air, N$_2$, Ne, He, etc. However, for large molecular gases, like SF$_6$ and CCl$_2$F$_2$, $\beta = 0.5$, which is may be owing to their electronegativity and to the ability of these gases to undergo numerous chemical reactions in the arc leading to a varying arc atmosphere in both time and space. The critical distance was found by Boddy and Utsumi to be independent of opening velocity of the contacts [46].

13.6.1.3 Case Study (3): Arc Voltage, Current and Length under Quasi-Static Conditions for Ag/CdO Contacts

To provide an example of the arc voltage current characteristics for silver cadmium oxide contacts, [41], Figure 9.39 shows the variation of the instantaneous arc length with the instantaneous arc current and voltage. These curves are obtained by compiling a large number of waveforms, for a range of break current. The contact speed was kept constant at 0.05 m/s to simulate a quasi-static operation. An empirical formula is derived from the waveforms relating the arc length to the arc voltage and current, on the basis of Equation 13.30.

$$d = A(V_{arc} - V_{min})^{m_1} I_{arc}^{n_1} \qquad (13.33)$$

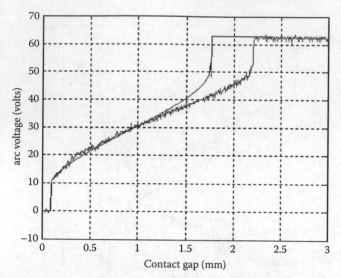

FIGURE 13.29
Model of the arc voltage on the basis of the quasistatic empirical relations. For a 64V dc 9A, resistive load. Experimental data contacts opening velocity of 0.1 mm/sec.

The values of m_1, n_1, and A depend on the length of the arc. For a short arc, $d < 1.5$ mm, the values are as follows, for AgCdO contacts:

$$m_1 = 1.57. n_1 = 0.49. A = 3.85 \times 10^{-3}$$

The 1.5 mm break point has been selected for the purpose of drawing comparisons, but is otherwise arbitrary. The break point can be identified in Figure 13.27, point A, as the point where the voltage waveform changes slope. For an arc longer than 1.5 mm and up to 7 mm the following values have been obtained:

$$m_2 = 2.12, m_2 = 0.76, A = 6.36 \times 10^{-4}$$

The empirical relationships defined in Equations 13.30 and 13.31, and modified here in Equation 13.33 are a useful tool for modeling the switching performance of low voltage system. Figure 13.29 shows an example where the coefficients are used to generate a model of the arc voltage, using a computer based numerical method to calculate the arc voltage for a given supply condition and contact gap, using the Newton-Raphson method. On the basis of this analysis, Figure 13.29 shows the resulting predicted arc voltage for the quasi-static model of the arc. It shows that for the 64 V, 9A resistive dc. Load that the arc model is able to model the initial stages of the arc voltage. However as the gap increases the simulation is unable to predict the resulting arc voltage for a contact velocity of 0.1 m/sec. The contact gap at arc extinction is 2.2 mm for the experimental data at 0.1 m/sec opening velocity, while the quasi-static model predicts a gap of 1.7 mm.

13.6.1.4 Opening Speed and Arc Length

Figure 13.30 shows the variation of arc current and voltage with the contact gap for three different opening speeds, for silver cadmium oxide contacts [41]. The contact gap at which the arc extinguishes, the total arc length d_x, increases with the opening speed. Increasing

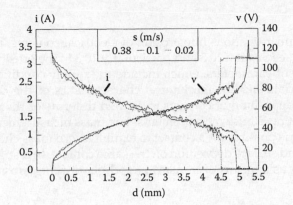

FIGURE 13.30
Instantaneous arc voltage and current versus contact gap for different opening velocities, $V = 110$ V, $I = 3.3$ A, and $L/R = 1$ μs.

the current increases the effect of the opening velocity. This demonstrates the limitations of the quasi-static arc characteristics. The quasi-static curve should only be used as an indication of the arc characteristics in dc circuits. The model used to determine the data in Figure 13.29 is used here to predict a contact gap at arc extinction of 5.13 mm. In Figure 13.30 the quasi-static value (0.02 m/sec) is 4.5 mm. Interestingly the extinction gap for (0.1 m/sec) is 5.0 mm; thus, the overestimation of the gap using the model brings the gap closer to the gap when the contacts are opened at a more realistic speed.

13.6.1.5 Case Study (4): Automotive Systems

The automotive sector has seen rapid changes from the traditional 12–14V dc power supply to a period in the 1990s-2000 where a 42V replacement was considered, but was found to be limiting. The key driver for the increase in the supply voltage is the increase in power requirements. This trend shows no sign of abating as the automotive sector moves more and more toward electrically driven sub systems, such as ABS and fuel pumps; and also to hybrid and fully electrical drive systems. In both a European forum and the American consortium of car manufacturers and their suppliers, the members agreed to a new standard of a new power network of 42 VDC, in 1998–99, [47,48]. The increase in a supply voltage from 12–14V dc to 42V dc has a number of consequences in terms of the switching systems, foremost is the fact that the supply voltage of 12–14V dc is close to the minimum arcing voltage for a resistive load for the materials used, thus minimizing the consequences of arcing. In simple terms opening a circuit with a 12–14V supply will cause minimal arcing. Opening a circuit with a 42V dc will produce arcing and as a consequence increase reliability issues and cause a significant fire risk if live conductors are damaged during an impact event. After many years of research, of which a few representative paper are provided here, [32,47–52], the outcome was that at the time of writing this in 2012 the majority of automotive systems are still dominated by the 12–14V dc power supply. The conclusion to this episode is that engineers need to consider the consequences of switching dc power systems as a priority.

Recent innovations have seen the introduction of battery voltages for hybrid systems of 144V, 288V, and of 345V for pure electric vehicles; some systems use a convertor/invertor, with an output of 650V ac. This is a rapidly advancing area and the fact remains that the dc voltage levels are increasing above the traditional levels, [53].

13.6.2 DC Erosion

There have been many investigations into erosion of electrical contacts, and as discussed elsewhere in this book (see Section 10.3 and particularly 10.3.5) the erosion characteristics depend upon a number of parameters, which include the supply conditions, the environment, the contact materials and the mechanical characteristics of the device [54–56]. In general, the erosion or deposition of contacts is defined in terms of the mass changes of the contacts, and contact resistance measurements. The mass measurement must however be treated carefully as it is possible for a contact to exhibit a zero mass change where there has been both erosion and surface deposition on the same contact. Investigations have considered surface analysis as a means of measuring and quantifying erosion, as discussed in section 13.5.

13.6.2.1 Ag and Ag/MeO Contact Erosion/Deposition

The dc erosion characteristics of low-current switching is shown in Figure 13.31, for a controlled test apparatus with silver cadmium oxide contacts where the contacts were opened at a constant velocity [41]. The erosion has been defined in terms of mass change at the anode and cathode; these are shown to be linear functions of the number of operations (see Section 10.6). The erosion can therefore be specified as an erosion rate. Figure 13.32 shows the erosion per operation of both anode and cathode contacts as a function of the circuit current. For the conditions stated in the figure the contact erosion rate is reduced as the contact velocity increases. Thus, in dc circuits there is an advantage in opening the contacts at high velocity. There are however consequences of a fast mechanism in that on making the contact there is likely to be more contact bounce, as a result of the increased kinetic energy in the system (Equation 13.4).

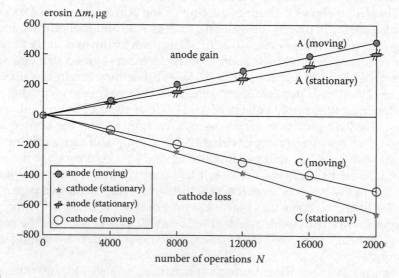

FIGURE 13.31

The dc erosion of Ag/CdO contacts with the number of operations, for a range of opening velocities with moving anode and fixed cathode cathode, and moving cathode and fixed anode, for $V = 62.5$ V dc, $I = 4$ A and opening velocity 0.1 m s^{-1}.

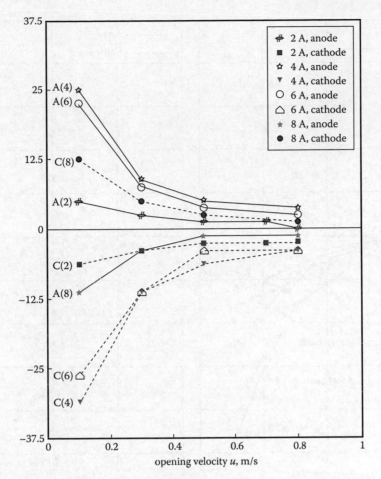

FIGURE 13.32
Erosion per operation, with opening velocity, for $V = 62.5$ V, moving anode in all cases.

In low-current dc switching with silver contacts below 4 A [57], it has been shown that at a current below 0.4 A material transfers from the anode to cathode and at higher current, above 0.4 A, the material transfers is reversed. In the current range 1–100 A there have been many investigations [40,42,55,58–60]. Figure 13.33 shows results of mass transfer per operation for a range of silver-based contact materials [40,60]. In Figure 13.33a, the materials tested all show material transfer increasing with current, and the direction of the transfer from the cathode to the anode, in Figure 13.33b silver cadmium oxide results show that at lower direct current (1–8 A) material transferred from cathode to anode, whereas at higher current (8–40 A) material transferred from anode to cathode.

It has been reported that the polarity of erosion between anode and cathode can vary with a large number of parameters [54] and that a change in one parameter can often lead to conflicting results.

In the automotive application, dc erosion is particularly important [61], although the majority of studies have concentrated on the erosion studies of the switching contacts in operating devices; where often the make operation is dominant in terms of reliability, and where both the de-gassing of the plastic housing and silicon contamination can have an important influence.

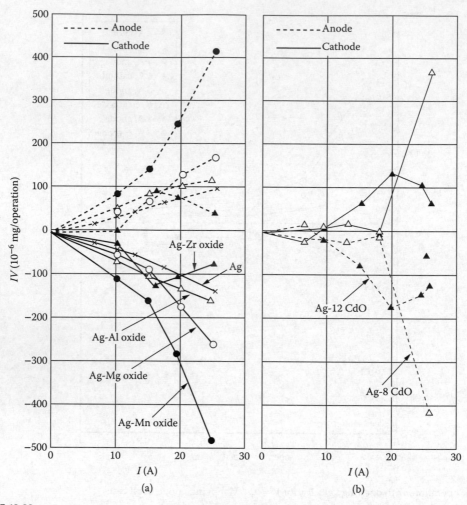

FIGURE 13.33
Erosion per operation for (a) a range of contact materials and (b) Ag/CdO [40].

To summarize, the following applies to dc arcs:

1. Very short arcs (less than a few microns depending on the contact material, activation, oxide layers etc) which occur when breaking very low current, typically below 0.4 A results in anodic erosion (from anode to cathode).

2. For arcs longer than a few microns where (1) above does not apply the erosion is cathodic (from anode to cathode) even for relatively short arcs. Increasing the arc length and/or current can change the arc transfer in some materials from cathodic to anodic.

13.6.3 Low-Current AC Arcs

On switching an ac circuit, the polarity of the contact will change with the supply frequency, the events controlling the arc erosion will, therefore, change with the changing the polarity of the contacts. The point at which the contacts are open will normally be a random event, and the arc will usually extinguish at the current zero, see Chapter 9.

13.6.3.1 Typical Waveforms and Arc Energy

To compare the arc voltage and current characteristics with the dc case, Figure 13.34 shows arc voltage with contacts opening at the same point of the ac waveform [56], defined as the point-on-wave, POW, for Ag/CdO, with a test current of 5.6 A, resistive load. The figure shows the contacts opening 1 ms after the current zero ($a = 1$ ms), producing an arc duration of nearly 9 ms for a 50 Hz ac supply. The figure shows that as the contact velocity increases the arc voltage increases at a given time, because the arc length will be greater; the arc is also shown to extinguish slightly earlier at the higher velocities. Similar waveforms have been obtained at different values of point-on-wave, POW, and with a range of current. The current always extinguished just before the first current zero after the ignition of the arc and no re-ignitions are observed in the range of reported velocities (0.1–0.8 m/s). Arc re-ignitions are observed at the much lower velocities of 0.01 m/s. Since the current waveform varies slightly with the contact opening velocity, the increase in arc voltage with contact opening velocity results in an increase of the arc energy. This is illustrated in Figure 13.35, which shows the variation of the arc energy per operation with the moment of contact opening a and contact velocity. The results shown in Figures 13.34 and 13.35 are for the ideal case with a linear opening velocity, and a variable maximum contact gap. The gap at which the arc extinguishes is governed by the velocity of the contacts and the POW at which the contacts start to open. In a commercial device the contact gap would be fixed. In all cases the arc energy is shown to reduce with the period of arcing, in Figure 13.35. Curves 1, 2 and 4 show the effect of current on the arc energy at a velocity of 0.3 m^{-1} s. This shows the arc energy increasing with the current. More significantly curves 3, 4, 5, and 6 show the effect of the velocity. As the opening velocity increases for a given current and arc duration the arc energy increases.

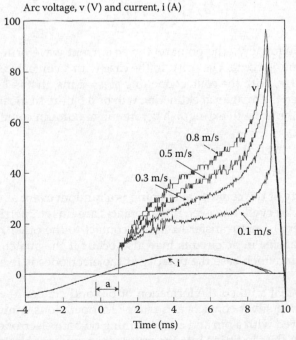

FIGURE 13.34
The influence of opening velocity and arc voltage and current, $V = 240$ V, 50Hz ac, $I = 5.6$ A, $a = 1$ ms POW.

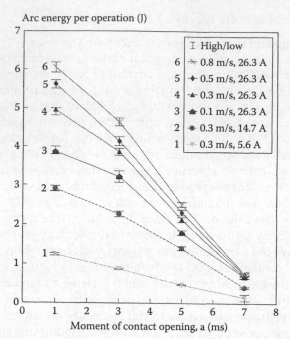

FIGURE 13.35
The influence of POW, opening velocity and current on arc energy, for $V = 240$ V, 50 Hz.

The contact gap at which an ac arc will extinguish, d_x, assuming there are no restrikes can be evaluated from the following equation:

$$d_x = \left(\frac{1}{2f} - a \right) u_0 \tag{13.34}$$

where f is the supply frequency, a the point on the ac current waveform the contact open at, and u_0 the linear opening contact velocity. In the case with a contact velocity of 0.5 m^{-1} s and an ac supply of 50 Hz, with the contact opening at $a = 1$ ms, that is 1 ms after the preceding current zero, then the arc gap at extinction will be 4.5 mm. In most cases in the current range considered, the arc will extinguish before the maximum contact gap is reached in the device.

13.6.4 AC Erosion

In ac circuits the polarity of the contacts at opening is a random event and in general little net transfer is usually observed, however, if the contacts are unsymmetrical (made of two different types of materials) net transfer from one contact to the other may be observed [62,63]. Net material transfer in ac circuits may also occur if the switch had a preferred point-on-wave (POW), in which case the polarity of the electrodes is fixed, this would be similar to dc erosion.

Holmes and Slade [64,65] studied the formation of pip and craters in ac circuits (220 V, 5.1 A, resistive load) using silver contacts at a rate of 1.5 operations/minute. A distinctive erosion pattern is observed with a pip and crater forming on both electrodes (each electrode has a crater beside a pip), the pip fitting into the crater of the opposite electrodes. It is shown that the rate of build-up of these erosion structures is a function of the average arcing time per switch operation. To explain these phenomena of double pip and crater formation, studies have been carried out using rectified ac. In this case, there is always a pip on the anode and a corresponding crater on the cathode. Based upon the rectified ac results Slade and Holmes explain the double pip and crater formation to be as a result of a similar mechanism to that of rectified ac transfer of material from the cathode to anode. However, in the ac case as the electrodes change polarity, the arc roots also change their position on the surface of the contacts when the polarity changes. This results in the double pip and crater formation.

13.6.4.1 Point-on-Wave (POW) Studies Using Ag/CdO Contact Materials

The aim here is to identify the influence of switching parameters on the ac arc, the variables considered are the POW, the point on the ac current waveform where the contacts open relative to the preceding current zero, the current and the velocity of the moving contact.

Figure 13.36 shows results of erosion studies [59], for a laboratory-based experiment using Ag/CdO contacts. The purpose here is to demonstrate how erosion polarity can be affected by some of the key parameters. The key parameters in ac erosion are:

1. The supply
2. The environmental conditions
3. The point at which the contacts are switched and the frequency of switching
4. The contact materials

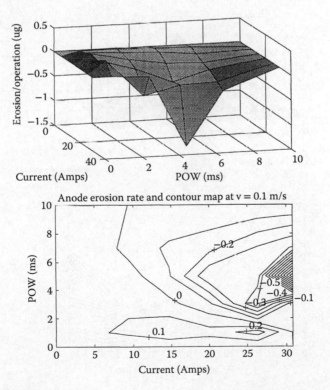

FIGURE 13.36
Anode erosion per operation for Ag/CdO contacts, with current and POW for $V = 240$ V, 50 Hz, and an opening velocity of 0.1 m s^{-1}. Upper figure shows the erosion vertically as a surface, lower figure shows the same data as a contour map.

Of these parameters the experiments show variation in the POW and the supply current. The contacts are opening at 0.1 m⁻¹ s and have a POW between 0 and 10 ms in a 50 Hz supply, thus the contact will always be the same polarity. Figure 13.36 shows a three-dimensional surface representing the erosion and deposition of the anode contact, for a range of current and POW. The erosion rate is plotted vertically in the upper graph. The figure shows that under most conditions shown the anode will lose material, indicated by the negative erosion. The exception to this is at a low POW (1 ms), where the anode gains material for the range of current investigated. The erosion at 0 and 10 ms point on wave is zero. The erosion rate peaks at 5 ms POW for the highest current in the range tested. In the lower graph the same data is presented as a contour map of the surface. This is particularly useful in identifying the point at which the erosion changes from positive to negative as a function of current. This is shown as the line (0) in the figure; as the supply current increases the cross-over point occurs at a lower POW. Thus, for low current the anode will gain material while at higher current the anode losses material.

For the same experimental conditions, Figure 13.37 shows the cathode erosion. The upper graph shows that under most conditions the cathode gains material. The exception is again at 1 ms POW, where the cathode losses material for the current range investigated. The contour map in the lower graph is again useful in identifying the cross-over from cathode gain to cathode loss. The figure is almost a mirror image of Figure 13.36, and shows that at low current the erosion is a cathode loss while at high current the erosion will be dominated by cathode gain. The degree of erosion varies with POW; with a commercial device the POW

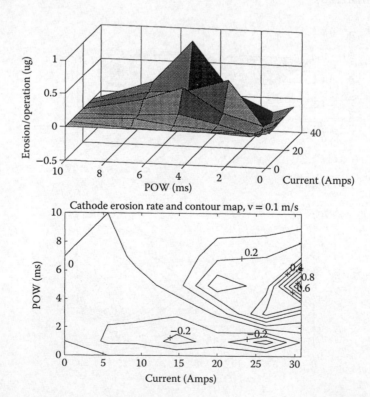

FIGURE 13.37
Cathode erosion per operation for Ag/CdO contacts, with current and POW for V = 240 V, 50 Hz, and an opening velocity of 0.1 m s⁻¹. Upper figure shows the erosion vertically as a surface, lower figure shows the same data as a contour map.

is generally a random process. The degree of mass transfer can be seen from the figures to increase with the current, and at the low current considered here the change of a contact from an anode to a cathode generally results in an effective elimination of any net transfer from one contact to the other. Both contacts, however, suffer from a change in the contact surface as a result of the arcing, which may lead to parts of the contact surface suffering more than other parts of the surface. In addition to this as the current increases there is more likelihood of mass being lost to the surrounding environment as droplets of molten metal are expelled from the arc roots. erosion at higher current will be considered in the next chapter.

13.7 Electrical Characteristics of Low Current Switching Devices at Closure

The mechanics of closing switch contacts have been considered in Section 13.4.3; the purpose of this section is to consider the effect of arcing on the bounce mechanics. For this purpose the "make" operation of low current contacts can be considered to occur in three distinct phases, [23,24]:

1. The pre-impact event
2. The first bounce with arcing
3. The bounces after the first bounce

When contacts open during a bounce, an arc may be initiated if the circuit conditions permit. This arc is similar to drawn arcs which occur during contact opening (see Chapters 10 through 12 and Section 13.5). Arcing changes the mechanical bounce characteristics, primarily as a result of heating of the contact surface. When a contact closes, a pre-impact arc is often formed for a short period prior to the first impact; this has small influence but the first impact is essentially a mechanical impact process. After the first impact period the contacts start to separate and the first bounce occurs. The first bounce produces an arc if the circuit conditions permit. An example voltage waveform is shown in Figure 13.38 for a dc supply, in the figure the voltage characteristic is shown off-set from the displacement data, the voltage shows the contacts closing at $t = 0$, the first impact occurs and the voltage rises to the minimum arc voltage, then as the contacts separate and come back together, the arc voltage is shown to rise, and to fall back to the minimum as the contacts come back together for the second impact. In the case of the dc supply the arc voltage reaches only a low voltage since the contact gaps during the bounce process are often small. Figure 13.38 shows the bounce characteristics of the same system used in Figure 13.39, but with a low voltage and a signal current supply, 2 V, 5 mA. This allows for a comparison between the mechanical bounce and the bounce with current for the same system conditions. The pre-impact arc is not observed since the duration is small compared to the overall event. A comparison of Figures 13.38 and 13.39 shows that the first bounce duration in both cases is similar; however, the bounce with arcing has a characteristic arc voltage. After the first bounce with arcing the surface of the contacts is heated such that when the contacts close for the second impact they will close on a molten metal surface; this will significantly influence the impact mechanics and allow more energy to be absorbed by the surface. The result of this process is that a bounce occurring after the first bounce will have a reduced duration and the whole process is over earlier than it would be with pure mechanical impact.

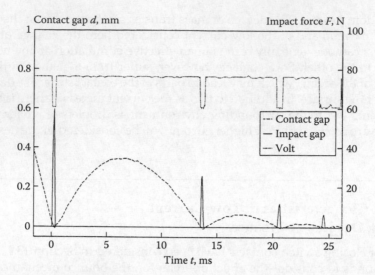

FIGURE 13.38
Mechanical make operation, showing the contact gap, contact force, and contact voltage as a function of time. The voltage is 2 V high and 0 V low.

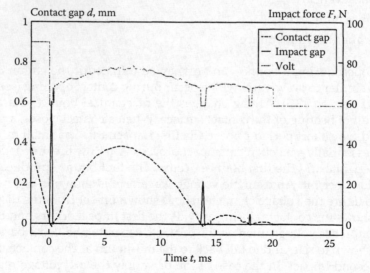

FIGURE 13.39
Make operation with arcing, showing the contact gap, contact force, and arc voltage as a function of time.

13.7.1 Contact Welding on Make

When contacts close, if the current is sufficient the arc roots are heated to their melting point, which can cause welding when the contacts re-close after a bounce. This dynamic welding is a danger to switching devices especially if the contacts close with a high inrush of current. The fact that the contacts have welded may only be discovered much later when the contacts are required to open. There are two solutions to this problem: the first is to reduce the contact bounce and the second is to use contact materials which are able to reduce the tendency to weld. Aggregate materials, for example Ag/C, Ag/CdO and

Ag/SnO_2 are used to overcome dynamic welding problems because their welding tendency is much less than pure silver, owing to a weaker structure.

13.7.2 Reducing Contact Bounce

The reduction of contact bounce can often be achieved by design as shown in Section 13.4.1, where a rocker switch was optimized to reduce moving-contact kinetic energy. In low-current devices cost is often a major design factor, thus no other elaborate bounce reduction methods can be used as these would increase the cost of the device, thus simple solutions are often sought. With rocker switch mechanisms the impact bounce can be reduced as already discussed by the reduction of the moving contact kinetic energy through a reduction in the moving-contact inertia, also by reduction of the impact velocity or by selecting a preferred switch arrangement (Section 13.4.1). In many low-current systems a combination of both approaches would appear to be the most satisfactory. In rocker-switch mechanisms the hair-spring system and the make-up arrangement are both beneficial in the reduction of contact bounce. In larger systems, for example contactors, control of the contact velocity can be shown to provide significant benefits to both the closing and opening characteristics.

13.7.3 Pre-Impact Arcing

When switching contacts close there is often a short-duration arc which occurs prior to the actual impact of the contacts; this arc is referred to as the pre-impact arc. The pre-impact arc can be seen to occur at voltages as small as 50 V, which is much lower than the minimum sparking voltage defined by Paschen's law (see Section 9.3.2). This suggests that low-voltage, pre-impact arc is initiated by field emission.

A typical arc voltage characteristic is shown in Figure 13.40; this shows that the pre-impact arc is full arc which is initiated and continues to burn in metal vapor from the contacts themselves. The duration of the arc is shown to be in micro-seconds and is thus much shorter than the typical bounce duration which is often of milli-second duration. The duration of

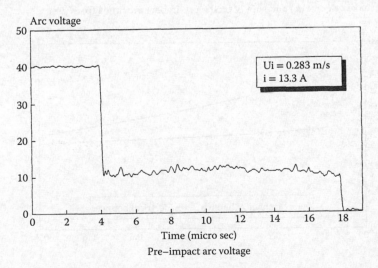

FIGURE 13.40

Pre-impact arc voltage for a system closing at 0.283 m/sec with a current of 13.3 A. The pre-impact arc time, T_p, is 14 μsec.

the pre-impact arc is not very repeatable, but experiments over a large number of impact events allow an understanding of the influence of contact velocity and supply voltage. Figure 13.41 shows the influence of impact velocity on the pre-impact arc time T_p, and shows that as the velocity of impact increases, the duration of the pre-impact arc reduces. This is shown to be affected by the supply, where a higher voltage gives a longer duration for the same impact velocity. The data presented here has been averaged over a large number of test conditions, [23]. As the contacts come together the greater the velocity the smaller the pre-impact arc, then the smaller the influence on the first impact. This is confirmed in Figure 13.42, which shows the evaluation of the coefficient of restitution for the first impact, e_u for a range of impact velocities, with and without current. The curves presented are regression fits taken over a large number of data samples. At the low impact velocities (eg 0.1 m^{-1} s) it is shown that the pre-impact arcing sufficiently softens the surface to reduce e_1 from 0.5 to 0.32 a reduction by 64%. At the higher impact velocities the reduction becomes marginal.

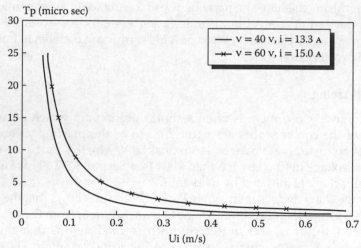

FIGURE 13.41
The effect of contact make velocity (u_i) and supply on the pre-impact arc time (Tp) in μsec.

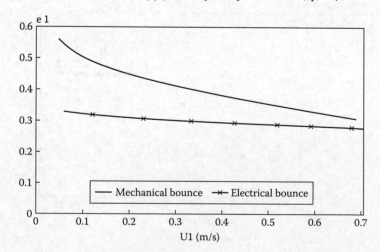

FIGURE 13.42
The effect of the pre-impact arc on the coefficient of restitution for the first impact event (e₁) as a function of impact velocity (u_i) and a comparison with a purely mechanical impact.

13.7.4 Influence of Velocity during the First Bounce

For the purpose of this section results are discussed from an experimental study of a pivoting contact mechanism shown in Figure 13.19. The values shown are thus specific to that mechanism; however, these results can be extrapolated to the more general case with arcing during the make process in manufactured devices. With the pre-impact arc occurring as contacts close there will be a small effect on the first impact process; this has been investigated in terms of the influence on the coefficient of restitution. Figure 13.43 shows a characteristic which is representative of the impact dynamics for both mechanical and electrical bounce during one contact closure. The mechanical bounce is defined as a bounce without arcing while the electrical bounce is defined as a bounce with arcing. The top trace shows four points corresponding to the evaluation of the coefficient of restitution e, during each impact for a single mechanical closure; with four impacts during the contact bounce process. The initial impact of $0.4 \, \text{m}^{-1} \, \text{s}$ gives a coefficient of $e_1 = 0.45$, which then increases with the subsequent impacts. The increase in the coefficient of restitution between consecutive impacts shows that the value is a function of the impact velocity; the curve closely follows Figure 13.20, but in this case the surface will change between impacts, leading to work hardening which will tend to alter k in Equation 13.13. The curve shows that as the velocity of impact decreases e increases, producing more elastic behavior.

The lower curve shows the changes in e under the same initial conditions with the same contacts, but with a 15 A dc current passing between the contacts. The initial impact velocity of $0.4 \, \text{m}^{-1} \, \text{s}$ produces a lower coefficient of restitution for the first impact e_1, than with the mechanical impact, as a result of the pre-impact arc discussed in Section 13.6.1. The following two impacts show the value of e decreasing. There are three impacts, compared to the four in the results without arcing. The relationship between the coefficient and the impact velocity can be defined by an empirical relationship:

$$e = 0.235 + 0.371u \tag{13.35}$$

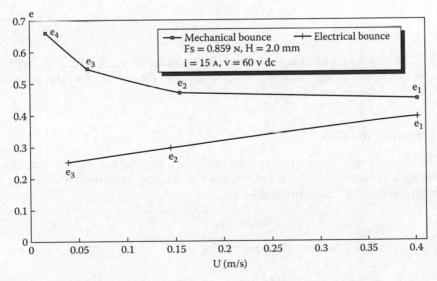

FIGURE 13.43
The variation of the coefficient of restitution during a single make event, with and without current flow.

This relationship gives *e* as a function of velocity for the current of 15 A, for the particular contact conditions used in the experiment. The reduction in *e* can be explained with reference to Equations 13.13 and 13.20, since the second impact will occur on a softened surface. The degree of softening will be a function of the arc energy dissipation, but the impact mechanics will be primarily affected by the solid surfaces rather than the liquid metal interface, hence reducing *p*, the dynamic yield pressure. The third impact will occur on a surface heated further, leading to a further reduction in *p* and *k*, as defined in Equation 13.13.

13.7.4.1 The First Bounce

Although the first impact event is influenced by pre-impact arcing it can be argued that, with reference to Figure 13.43, the reduction in e_x is small compared to the effects after the first impact and bounce. The first bounce time can then be given with a reasonable approximation by Equation 13.24, in the form

$$T_1 = \frac{2u_i e_1 J}{h'^2 F_s}$$
(13.36)

The evaluation of *k* can be established for a new contact from materials data; however, with a worn contact the value would be expected to change, as a result of changes in the surface roughness affecting the radius used in Equation 13.13. However, this only assumes that *r* changes and does not account for the heating effects on the contact materials. With contacts tested from new, it has been shown that the first bounce actually increases with wear, [24], which suggests that the changes in *r* are not important, and the impact mechanics depends more on the macro-surface characteristics rather than the micro-surface. Hence to a first approximation we can assume the *k* value used for the new contacts. Then in Equation 13.14, we can substitute for *k*, in doing so only the first and last terms become significant. In a further approximation, we can get a solution for *e*, shown in Equation 13.15, this can be substituted into Equation 13.35 to give the first bounce time T_1

$$T_1 \approx \frac{2kJ}{h'^2 F_s}$$
(13.37)

In most situations, the first electrical bounce will account for most of the arc erosion. If the arc voltage can be assumed constant, for a small contact gap during the bounce, the design process should focus on the reduction of T_1.

13.7.5 Bounces after the First

To account for the subsequent bounces, we can make two further assumptions: firstly that there are usually less than three electrical bounces, and secondly that the values of *e* can be expressed in the form of Equation 13.35, e.g.

$$e = a + bu$$
(13.38)

Then, by substitution into Equation 13.24 and 13.36,

$$T_N \approx \frac{2kJ}{h'^2 F_s} u_i^{3/4}(1 + a + bku_i^{3/4})$$
(13.39)

13.7.6 Summary of Contact Bounce

In this section, the phenomenon of contact bounce at make has been considered with respect to the low-current range. It has been shown that mathematical models can be developed for the analysis of the impact process and that this can be taken further from experimental observations to include the effect of current on the process. It has been shown that if the contact bounce follows the pattern where the first bounce is the largest and the subsequent bounces are smaller, then the event will be dominated by the first bounce. In many commercial systems the bounce process may not follow this pattern as the bounce may be affected by nonlinearities, such as the elastic properties of the main contact support structure or the interaction of a contact bridge arrangement. In low-current hand-operated switching devices the bounce may be made complex by the nature of the moving contact; for example, the characteristic shown in Figure 13.14 is highly nonlinear as a result of vibration affects.

13.8 Summary

In this section, the key elements from this chapter are summarized, and based upon the description of switching devices given, some design recommendations are offered. The design points are only general, as specific recommendations by definition would be unique to the device under consideration.

13.8.1 Switch Design

From the consideration of rocker-switch mechanisms, the impact of the moving contact during a make operation is very important to the successful operation of a switch.

In the case of the make-down configuration, it has been shown that a moving contact with a relatively high inertia will result in a separation at the pivoting contact as the main contact closes. The change-over configuration shown in Figure 13.2 will be particularly susceptible to this. In the make-down, on-off configuration shown in Figure 13.3 the pivot will have less tendency to separate, because of the lower inertia. In the make-up configuration shown in Figure 13.4, when the main contact impacts the moving contact is unable to continue moving in the same direction at the pivot interface, therefore the bounce that occurs at the main contact will be the main dynamic event. The parameter that best describes the likelihood of bounce other than the geometry of the device is the kinetic energy in the moving contact, since it is the kinetic energy that needs to be dissipated in the switching action to bring the system to rest. With an objective to reduce the kinetic energy at impact there are three main solutions:

1. Reduce the inertia of the moving contact
2. Reduce the impact velocity of the contacts
3. Increase the absorption of the energy of impact in the materials

13.8.2 Break Operation

In most devices, the opening velocity will be dominated by a nonlinear acceleration profile, which will depend upon the nature of the actuating method. In the case of an electromechanical relay, for example, it is the pull-in characteristic of the solenoid, or the release of the solenoid under spring forces.

13.8.2.1 DC Operation

In general, the erosion or deposition of contacts is defined in terms of the mass changes of the contacts and contact resistance measurements. The mass measurement must however be treated carefully as it is possible for a contact to exhibit a zero mass change where there has been both erosion and surface deposition on the same contact.

For dc conditions the contact erosion rate is reduced as the contacts are opened more quickly. There are, however, consequences of a fast mechanism in that on making the contact there is likely to be more contact bounce, as a result of the increased kinetic energy in the system.

To summarize, the following applies to dc arcs. Very short arcs (less than a few microns depending on the contact material, activation, oxide layers etc.) occur when breaking very small current results in anodic erosion (from anode to cathode). For arcs longer than a few microns the erosion is cathodic for relatively short arcs. Increasing the arc length and/or current can change the arc transfer in some materials from cathodic to anodic.

13.8.2.2 AC Operation

In ac circuits, the polarity of the contacts at opening is a random event and in general at low current very little transfer is usually observed; however, if the contacts are unsymmetrical (made of two different types of materials) net transfer from one contact to the other may be observed. Net material transfer in ac circuits may also occur if the switch operation has a preferred point-on-wave (POW), in which case the polarity of the electrodes is fixed; this would be similar to dc erosion. Both contacts will suffer from a change in the contact surface as a result of the arcing, which may lead to parts of the contact surface suffering more than other parts of the surface. In addition to this, as the current increases there is more likelihood of mass being lost to the surrounding environment as droplets of molten metal are expelled from the arc roots.

13.8.3 Make Operation

13.8.3.1 Design Parameters

It has been shown that contact bounce can be reduced by control of a number of parameters: for example, reduction of impact velocity, reduction of moving contact inertia, energy absorption at the contacts or in the contact structure, and increasing the contact force in a given mechanism. The selection of the parameter which will best suit a given situation or device varies. For example, a damped actuator could be used on an electromagnetic actuator to reduce the impact velocity and thus reduce the contact bounce, but this would be a costly addition to most low-current switches.

The "make" operation of low-current contacts can be considered in three distinct phases:

1. The pre-impact events
2. The first bounce with arcing
3. The bounces after the first bounce

With arcing occurring with a high in-rush of current during the bounce process here is significant heating of the contact surface, resulting in the possibility of welding. There are two solutions to this problem: the first is to reduce contact bounce and the second is to use contact materials which are able to reduce the tendency to weld. Aggregate materials like Ag/CdO and Ag/SnO$_2$ are used to overcome the dynamic welding problems because their welding tendency is much less than pure silver, owing to their weaker structure.

13.8.3.2 Reducing Contact Bounce

The reduction of contact bounce can often be achieved by design. In low-current devices cost is often a major design factor, thus no other elaborate bounce reduction methods can be used as these will increase the cost of the device, thus simple design solutions are often sought. With rocker-switch mechanisms the impact bounce can be reduced as already discussed by reduction of the moving-contact kinetic energy through a reduction in the moving-contact inertia, also by reduction of the impact velocity or by selecting a preferred switch arrangement. In many low-current systems a combination of both approaches would appear to be the most satisfactory. In larger systems, for example contactors, control of contact velocity can be shown to provide significant benefits to both closing and opening characteristics.

13.8.3.3 Arcing during the Bounce Process

It has been shown that mathematical models can be developed for the analysis of the impact process and that this process can be taken further by the consideration of experimental observations, to include the effect of current on the process. It has been shown that if the contact bounce follows the pattern where the first bounce is the largest and subsequent bounces are smaller, then the event will be dominated by the first bounce, since the subsequent bounces will be reduced by the influence of arc heating the surface and dissipating subsequent bouncing. In many commercial systems the bounce process may not follow this pattern as the bounce may be affected by nonlinearities such as the elastic properties of the main contact support structure or by the interaction of a contact bridge arrangement.

Acknowledgments

The author of would like to acknowledge a team of researchers with whom I have supervised and directed over the years on research in the field of electrical contacts; a real honor and privilege; these include Dr's Jonathan Swingler, Paul Weaver, Suliman Abu-Sharkh, Peter Jeffery, Kersorn Pechrach, Esa Yunnus, Alan Sumption, Christian Maul, and Wu Leung Ng. I would also thank my family and friends.

References

1. JW McBride, D Mapps, PJ White. Design optimisation of rocker-switch dynamics to reduce pivot bounce and the effect of load current in modifying bounce characteristics. *IEEE Trans Components Hybrids and Manuf. Technol.* CHMT9 (3): pp258–264, 1986.
2. JW McBride, D Mapps, PJ White. The influence of rocker switch dynamics on contact erosion. *IEEE Trans Components Hybrids and Manuf. Technol.* CHMT11 (1): pp137–144, 1988.

3. JW McBride. The accelerated testing of hand operated switches. *Proceedings of the 14th International Conference on Electrical Contacts*, Paris, pp61–67, 1988.

4. R. Hienonen, T. Saarinen. Reliability test on Membrane switches. *Proceedings of the IEEE Holm Conference on Electrical Contacts*, pp561–571, 1984.

5. J Praquin, C Gautherot, J Rivenc, N Benjemaa, R Elabdi, JBA. Mitchell, E Carvou. Influence of contact resistance in carbon-gold contacts in automotive keypad switches. *Proceedings of the IEEE Holm Conference on Electrical Contacts*, pp323–328, 2010.

6. Y Hayashi, M Baba, T Hara. Study on bridge phenomena of Sn-Rh and Rh-Sn-Rh contacts for dry reed switch. *Proceedings of the Holm Conference on Electrical Contacts*, pp541–547, 1978.

7. H Grossman, M Huck, G Schaudt, FJ Wagner. Make and break properties of electrodeposits. *Proceedings of the IEEE Holm Conference on Electrical Contacts*, pp551–560, 1984.

8. PM Weaver, K Pechrach, JW McBride, Arc root mobility on piezo-electrically actuated contacts in miniature circuit breakers. *IEEE Trans Components Packaging Technol*, 28, (4), pp734–740, 2005.

9. TJ Schopf, WF Rieder. The reliability of bearings in the current path of switches. *Proceedings of the 17th International Conference on Electrical Contacts*, Nagoya Japan, pp725–733, 1994.

10. M Maeda, M Kinoshita, T Yoshimura, K. Genma. Effects of joints on contact reliability in miniature precision snap-action switches. *Proceedings of the 36th Holm Conference on Electrical Contacts (15th ICEC)*, pp358–369, 1990

11. DJ Mapps, PJ White. Performance characteristics of snap-action switches with silver-cadmium-oxide contacts. *IEEE Trans. on Components, Hybrids and Manuf. Technol*. CHMT 4 (1): pp237–242, 1981.

12. DJ Mapps, PJ White. Investigation of available toggle switches for thermostatic control systems. *Proc. Inst. Electrical Engineers* part B 128(5): pp237–242, 1981.

13. PJ White, DJ Mapps. Zero current synchronization of opening switch contacts. *IEEE Trans Components Hybrids and Manuf. Technol* CHM-9 (3): pp252–257, 1986.

14. ML Gayford. *Modern Relay Techniques*. STC Monograph No 1. London: Butterworth, 1969.

15. RL Peek, HN Wagar. *Switching Relay Design*. New York: Van Nostrand, 1955.

16. W Ren, H Liang, Y. Chen, L Cui, L Wang, Y Kang. Mechanical and electrical characteristics of the electromagnetic relay under vibration and shock environment. *Proceedings of the IEEE Holm Conference on Electrical Contacts*, pp1–47, 2009.

17. R Holm. *Electric Contacts*. Berlin: Springer-Verlag, 4th Edition, Reprinted 2000.

18. FP Bowden, D Tabor. *The Friction and Lubrication of Solids*. Oxford: Clarendon Press, 1964.

19. W Goldsmith, CH Yew. Stress distribution in Soft Metals owing to static and dynamic loading by a steel sphere. *J Appl. Mech*. 31: pp635–646, 1964.

20. D Tabor. *The Hardness of Metals*. Oxford: Clarendon Press, 1951.

21. P Barkan. A study of the contact bounce phenomena. *IEEE Trans on Power App and Systems* PAS-86 (2): pp231–240, 1967.

22. W Goldsmith. *Impact*. London: Edward Arnold, 1960.

23. JW McBride. Electrical contact bounce in medium duty contacts. *IEEE Trans. Components, Hybrids, Manuf. Technol* 12 (1): pp82–90, 1989.

24. JW McBride, SM Sharkh. Electrical contact phenomena during impact. *IEEE Trans. Components, Hybrids, Manuf. Technol* 15 (2): pp184–192, 1992.

25. JW McBride, KJ Cross, SM Sharkh. The evaluation of arc erosion on electrical contacts using 3D surface profiles". *IEEE Trans. Components, Hybrids, Manuf. Technol* 19 (1): pp87–97, 1996.

26. JW McBride. The volumetric erosion of electrical contacts. *IEEE Trans. CPT* Vol. 23 (2): pp211–221, 2000.

27. JW McBride. A review of volumetric erosion studies in low voltage electrical contacts. *Technical Report of IEICE2001-45*, pp1–8, 2001.

28. JW McBride. A review of surface erosion measurements in low voltage switching devices. *International Conference on Electrical Contacts*, Zurich, Switzerland, pp462–470, 2002.

29. J Wolf. Erosion and material transfer measurement with the aid of a surface analyser. *Holm Conference on Electrical Contacts*, pp69–74, 1979.

30. D Zhang, M Hill, JW McBride. Evaluation of the volumetric erosion of spherical electrical contacts using the defect removal method. *IEEE Trans. Components. Packag. Manuf. Technol* 29 (4): pp711–717, 2006.

31. JW McBride, M Hill, D Zhang. A feature extraction method for the assessment of the form parameters of surfaces with localised erosion. *Wear* 256 (3–4): pp243–251, 2004.

32. J Swingler, A Sumption, JW McBride. The evolution of contact erosion during an opening operation at 42V. *Proceedings of the Fifty-First IEEE Holm Conference on Electrical Contacts*, pp346–351, 2005.

33. JW McBride, AP Sumption, J Swingler. On the evaluation of low level contact erosion in Electrical Contacts. *Proceedings of the 50ᵗʰ IEEE Holm Conference on Electrical Contacts and the 22ⁿᵈ International Conference on Electrical Contacts*, pp370–377, 2004.

34. BODDIES software analysis tool, TaiCaan Technologies Ltd, 2012. Available at http://www .taicaan.com/software_eng.html.

35. M Hasegawa, K Izumi, Y Kamada. New algorithm for volumetric analysis of contact damages with laser microscope data. *International Conference on Electrical Contacts*, pp337–342, 2006.

36. M Reichart, C Schrank, G Vorlaufer. Investigation of material transfer by 3D precision alignment of topographical data of electrical contact surfaces. *IEEE Holm Conference on Electrical Contacts*, pp313–318, 2008.

37. JJ Shea. High Current AC Break Arc Contact Erosion. *IEEE Holm Conference on Electrical Contacts*, pp22–47, 2008.

38. JW McBride, PM Weaver. Review of arcing phenomena in low voltage current limiting circuit breakers. *IEE Proc. - Sci. Meas. Technol.* 148 (1): pp1–7, 2001.

39. JW McBride, A Balestrero, L Ghezzi, G Tribulato, KJ Cross. Optical fiber imaging for high speed plasma motion diagnostics: applied to low voltage circuit breakers. *Rev. Sci. Instrum.* 81 (5): 055109-[6pp], 2010.

40. M Sato. Studies on the silver base electrical contact materials. *Trans Nat Res Inst. Metals* 18 (2): pp65–83, 1976.

41. JW McBride, SM Sharkh. The influence of contact opening velocity on arc erosion. *Proceedings of the 16ᵗʰ International Conference on Electrical Contacts*, UK, pp395–100, 1992.

42. JW McBride, SM Sharkh. Arc voltage fluctuations at low current. *Proceedings of the Electrical Contacts ISECTA*, Alma Alta, pp216–220, 1993.

43. EW Gray. Voltage fluctuations in low-current atmospheric arcs. *J Appl. Phys.* 43 (11): pp43–46, 1972.

44. EW Gray. Some spectroscopic observations of the two regions (metallic vapour and gaseous) in break arcs. *IEEE Trans Plasma Sci.* PS-1 (1): pp30–33, 1973.

45. DJ Dickson. A von Engel. Resolving the electrode fall spaces of electric arcs. *Proc. Roy Soc.* A300: pp316–325, 1967.

46. PJ Boddy, T Utsumi. Fluctuation of arc potential caused by metal-vapour diffusion in arcs in air. *J Appl. Phys.* 42 (9): pp3369–3373, 1971.

47. T J Schöpf, WF Rieder. Consequences for automotive relays of a 42VDC power network in vehicles. *IEEE Trans. Components, Packaging Technol.* 23: pp177–182, 2000.

48. TJ Schoepf. Electrical contacts in the automotive 42 VDC PowerNet. *International Conference on Electrical Contacts*, Switzerland, September, 2002.

49. N Ben Jemaa, L Doublet, L Morin, D Jeannot. Break arc study for the new electrical level of 42 V in automotive applications". *Proc. 47ᵗʰ IEEE Holm Conf. Montreal*, pp50–55, 2001.

50. E Carvou, N Ben Jemaa. Electrical arc study in the range of 14-112VDC for automotive power contacts. *International Conference on Electrical Contacts*, pp28–33, 2006.

51. J Swingler, JW McBride. Micro-arcing and arc erosion minimization using a 42 volt DC hybrid switching device. *IEEE Trans. Components, Packaging Technol.* 31 (2): pp425–430, 2008.

52. J Sekikawa, T Kubono, Motion of break arcs driven by external magnetic field in a 42V resistive circuit. *IEICE Trans. Electron* E91-C (8): pp1255–1259, 2008.

53. A Vassa, E Carvou, S Rivoirard, L Doublet, C Bourda, D Jeannot, P Ramini, N Ben Jemaa, D Givord. Magnetic blowing of break arcs up to 360 VDC. *International Conference on Electrical Contacts*, pp96–100, 2010.

54. AE Guile. Arc-electrode phenomena. *Proc. IEEE Rev* 18 (9R): pp1131–1154, 1971.

55. LH Germer. Physical processes in Contact Erosion. *J Appl. Phys.* 29 (7): pp1067–1082, 1958.

56. JW McBride, SM Sharkh. The effect of contact opening velocity and the moment of contact opening on the ac erosion of AgCdO contacts. *IEEE Trans Components, Packaging Manuf. Technol* 17 (1): pp2–7, 1994.

57. ZK Chen, H Mizukoshi, K Sawa. Contact resistance characteristics of Ag material in breaking low-load DC arcs. *IEEE Trans Components, Packing, and Manuf. Technol* Part A 17 (1): pp113–120, 1994.

58. DJ Mapps, PJ White. Performance characteristics of snap-action switches with silver-cadmium-oxide contacts. *Proceedings of the IEEE 26th Holm Conference on Electrical Contacts*, pp177–183, 1980.

59. SM Sharkh, JW McBride. Arc erosion in AC and DC circuits. *Proceedings of the 17th International Conference on Electrical Contacts (ICEC'94)*, Nagoya, Japan, pp569–576, 1994.

60. M Sato, M Hijikata, I Morimoto. Influence of oxides on materials transfer behaviour of silver base contacts containing various metal oxides. *Trans. Japan Inst. Metals* 15 (6): pp399–407, 1974.

61. CH Leung, A Lee. Electric contact materials and their erosion in automotive DC relays. *IEEE Trans Components, Packing, and Manuf. Technol*, Part A 15 (2): pp46–153, 1992.

62. M Lindmayer, KH Schroder. The effect of unsymmetrical material combination on contact and switching behaviour. *IEEE Holm Conference on Electrical Contacts*, pp265–271, 1978.

63. I Neveri. Proper pairing of different contact materials for increasing the reliability of low-voltage switches. *Proceedings of the International Conference on Electrical Contacts*, Japan, pp301–307, 1976.

64. FA Holmes, PG Slade. The erosion characteristics of Ag contacts and the effect of adding a small percentage of W. *IEEE Trans Parts, Hybrids and Packaging* PHP-13 (1): pp23–30, 1977.

65. FA Holmes, PG Slade. Suppression of pip and crater formation during interruption of AC circuits. *IEEE Trans Components Hybrids and Manuf. Technol.* CHMT-1 (1): p59–65, 1978.

14

Medium to High Current Switching: Low Voltage Contactors and Circuit Breakers, and Vacuum Interrupters

Manfred Lindmayer

Every fault seeming monstrous till his fellow fault came to match it.

As You Like It, William Shakespeare

CONTENTS

14.1 General Aspects of Switching in Air 787
    14.1.1 Arc Chutes 787
    14.1.2 Magnetic Blast Field 788
    14.1.3 Arc Dwell Time on the Contacts 789
    14.1.4 Sticking and Back-Commutation of the Arc 790
14.2 Contacts for Switching in Air 791
14.3 Low-Voltage Contactors 792
    14.3.1 Principle/Requirements 794
    14.3.2 Mechanical Arrangement 794
    14.3.3 Quenching Principle and Contact and Arc Chute
           Design 798
    14.3.4 Contact Materials 802
    14.3.5 Trends 803
           14.3.5.1 Contactors versus Electronics 804
           14.3.5.2 Vacuum Contactors 805
           14.3.5.3 Hybrid Contactors 805
           14.3.5.4 Integration with Electronic Systems 805
14.4 Low-Voltage Circuit-Breakers and Miniature Circuit-Breakers 806
    14.4.1 Principle/Requirements 806
    14.4.2 General Arrangement 807
    14.4.3 Quenching Principle and Design of Arc Chute and Contact
           System 808
           14.4.3.1 Quenching Principles 808
           14.4.3.2 Arc Chute and Contact Arrangement 810
    14.4.4 Trip System 814
    14.4.5 Examples of Miniature Circuit-Breakers 815

14.4.6 Contact Materials...816
14.4.7 Special Requirements for DC Switching...820
14.4.8 Current Limitation by Principles Other than Deion
       Arc Chutes...820
       14.4.8.1 Arcs Squeezed in Narrow Insulating Slots.......................................821
       14.4.8.2 Reversible Phase Changes of Liquid or
                Low-Melting Metal...821
       14.4.8.3 Temperature-Dependent Ceramics or Polymers...............................821
       14.4.8.4 Contact Resistance between Powder Grains......................................821
       14.4.8.5 Superconductors..821
14.5 Simulations of Low-Voltage Switching Devices...822
     14.5.1 Simulation of Low-Voltage Arcs..822
            14.5.1.1 General Principle of Simulation...822
            14.5.1.2 Arc Roots on Cathode and Anode..824
            14.5.1.3 Radiation...825
            14.5.1.4 Interaction between Arc and Electrode
                     or Wall Material (Ablation)..826
            14.5.1.5 Plasma Properties...826
            14.5.1.6 Simplification by Porous Media..827
     14.5.2 Further Simulations of Contact and Switching
            Device Behavior...827
14.6 Vacuum Interrupters...828
     14.6.1 Principle/Applications...828
     14.6.2 Design...829
     14.6.3 Recovery and the Influence of the Design...831
     14.6.4 Contact Materials for Vacuum Interrupters and Their Influence on
            Switching...835
            14.6.4.1 Requirements..835
            14.6.4.2 Arc Interruption..836
            14.6.4.3 Interruption of High Frequency Transients.................................837
            14.6.4.4 Current Chopping..838
     14.6.5 Simulation of Arcs in Vacuum Interrupters..839
References..840

This chapter emphasizes switching devices for low-voltage applications, that is, ≤1000 V. Medium- and high-voltage apparatus (SF_6 or oil circuit-breakers) have been treated elsewhere, for example, [1]. They will be disregarded here. Medium-voltage vacuum interrupters, however, are also discussed, because there is little difference between their physical phenomena and design aspects from those of low-voltage vacuum interrupters.

Low-voltage air contactors and air circuit-breakers are two categories of switching devices that have to switch regular loads. The contactor is used specifically to switch motor loads, the circuit-breaker additionally has to switch short-circuit currents. Both use arcs in air that have to be influenced to interrupt the current. The common aspects of these devices will be treated first, followed by specific questions of contactors and circuit-breakers in air. Finally, the use of vacuum interrupters for these switching duties will be presented.

14.1 General Aspects of Switching in Air

14.1.1 Arc Chutes

Above a certain voltage and current level, it is necessary to influence the arc established on contact opening in so-called arc chutes or arc chambers or quenching systems to lose its conductance (see also Section 9.7). For this, the arc is generally moved off the contacts by magnetic fields via arc runners or arc horns to an arc chute. Regardless of the switching principle—dc or ac interruption, current-limiting or not, whether the recovery of the plasma column or of electrode regions (see Section 9.7) is utilized—this always means that energy has to be removed from the arc, either by increasing its dimensions, by materials with high thermal diffusivity or latent heat, by gas flow from the walls, or by several of these methods at the same time. Figure 14.1 summarizes some of the characteristic methods. Figure 14.1a achieves arc elongation by V-shaped runners. Figure 14.1b is an example of meander-shaped arc by barriers of insulating material, where the effective length and the area in contact with cooling walls are increased. In Figure 14.1c, the arc is additionally confined between insulating walls, and its cross-section is reduced. Additional gas flow from thermally decomposing insulating walls (Figure 14.1d) may enhance cooling. All these methods cause an increase of arc voltage during the high-current arcing period and are suited for dc switching (Section 9.7.2), as well as current-limiting ac switching, where the arc voltage has to be raised quickly above the momentary system voltage (Section 9.7.4 and Figure 14.21, Section 14.4.3). They reduce the time constant of arc column cooling around current zero (see Section 9.7.1); this makes them also applicable to current zero switching devices. Both principles—current limitation and current zero switching—cannot be strictly separated because the arc voltage always helps to reduce

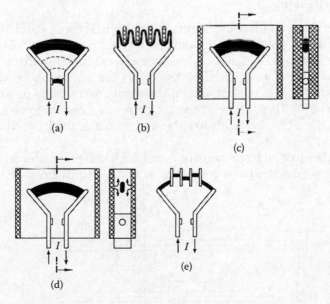

FIGURE 14.1
Methods to influence the arc on current interruption: (a) elongation on V-shaped runners; (b) meander-shaped insulating barriers; (c) elongation and confinement between insulating walls; (d) arc chute with gassing wall material; (e) partition between splitter plates.

the maximum current against the prospective current. In Figure 14.1e, the arc is directed into a stack of usually ferromagnetic "splitter plates" or "deion plates" (also "blades") one or several millimeters apart and isolated from each other. The denomination "deion arc chute," "deion chamber," or "deion grid" for this type of arrangement is historical [3] and says nothing about the physical principle. All sorts of arc chambers serve to deionize the arc. The function of the splitter plates is to split the arc up into several series arcs. By the formation of new anode and cathode fall regions with their minimum arc voltages (Sections 9.5.2–9.5.4), the voltage per unit total length during the high current period rises higher than in most other arrangements without arc splitting. Depending on current and geometry, 25–35 V are reached per partial arc [4]. Ferromagnetic material (mostly iron) is superior because it attracts the arc (Figure 14.3), and once the arc is split within the system there are magnetic forces that try to keep it there. This mechanism is used for dc interruption and current-limiting ac interruption, where the arc voltage must be raised. Deion plates are also suited for current-zero ac interruption. Then the "instantaneous" or "immediate recovery" effect of the new cathode sheath (Section 9.7.1), also called "self-extinction," is multiplied. Furthermore, the columns of the partial arcs are cooled by the metal plates. To make the situation even more complex, successful arc splitting as in Figure 14.1e cannot always be achieved. Depending on the actual current and the design, the arc may only split up partially or not at all. In this case, the metal plates act as coolants of the arc column only. A good example of such a behavior is given in [5].

Compared with the other arc chute principles, the deion arc chute system is mostly superior, and therefore it predominates in contactors, circuit-breakers, and similar switching devices for higher ratings. A rather global argument is that a stack of metal plates is able to consume more energy per unit volume than insulating material parts or gaseous matter.

14.1.2 Magnetic Blast Field

The effect of a current loop on the arc is shown in its simplest form, parallel current rails, in Figure 14.2. The current path generates a magnetic field B perpendicular to the drawing plane. Under the assumption that the current flows within a thin thread in the center of the rails and assuming $l_1 \gg k$, the flux density in the arc center between the rails ($y_1 = y_2 = k/2$) becomes $B = \mu_0 \cdot i/(4\pi k)$. Examples of magnetic fields for less simplified structures are given in [6]. Together with the arc current a Lorentz force F is generated, which is directed to the right and tends to enlarge the area of the loop, thus making the arc move.

Figure 14.3 shows the effect of ferromagnetic materials like deion plates on the arc, which is a current-carrying conductor. For a conductor at distance a in front of a semi-infinite

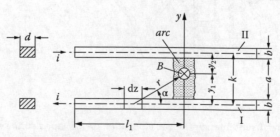

FIGURE 14.2
Generation of magnetic blast field by parallel rails.

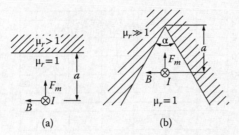

FIGURE 14.3
Principle of arc attraction by ferromagnetic plates: (a) straight front side; (b) V-shaped front side.

space with $\mu_r > 1$ (Figure 14.3a) the ferromagnetic distortion of the self-field of the arc generates a blast field.

$$B = \frac{\mu_0}{4\pi} \frac{\mu_r - 1}{\mu_r + 1} \frac{I}{a} \tag{14.1}$$

in its axis, which attracts the arc toward the ferromagnetic space. For a V-shaped ferromagnetic material with $\mu_r \to \infty$ (Figure 14.3b) it is increased by a factor of $(n - 1)$, where $n = 360°/\alpha$.

For this reason, the sides of deion plates facing the undivided arc are often V-shaped or serrated to facilitate the arc splitting. Some examples are given in Sections 14.3 and 14.4. These measures enhance the attractive magnetic field, at least locally, and by squeezing portions of the arc they additionally increase the local arc voltage drop adjacent to the metal plates. It is necessary to increase the voltage of the yet undivided arc above the minimum arcing voltage, before the formation of new anode and cathode spots splits it up.

Apart from deion plates, the ferromagnetic attraction of the arc or of parts of it is used in various forms, for example, as a ferromagnetic layer on arc electrodes to move the arc in the direction of the ferromagnetic material. A different application of ferromagnetic material is to concentrate the self-field of current-carrying conductors on certain areas, such as the space between the contacts. Detailed descriptions of such designs would require much more space. A very general qualitative formulation to assess the direction of forces by ferromagnetic parts is that the field lines tend to flow through the areas of lowest magnetic resistance. Of course, the magnetic fields and forces of rather complicated structures, including the effect of eddy currents, are nowadays accessible to numerical computation [7]. Typical values of the self-field in the arc axis lie in the order of several tens of mT per kA arc current [8,9].

14.1.3 Arc Dwell Time on the Contacts

When an arc is established at contact opening, it does not move immediately under the influence of the magnetic blow-out field [8,10,11]. Its roots first dwell at the location of arc establishment, while parts of the plasma are already deflected in the Lorentz force direction in the form of plasma jets. Only after a certain time—or when a certain minimum contact separation is reached—does the arc move off the contacts onto the arc horns. This process is accompanied by a steep voltage rise as the arc is lengthened. Figure 14.4 shows a typical current and voltage oscillogram of such a process. Dwell times typically lie between a fraction of a millisecond (Figure 14.32) and several

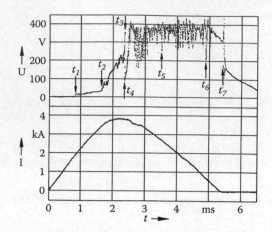

FIGURE 14.4

Oscillogram of an arc moving off the contacts into an arc chute under magnetic field influence (current-limiting miniature circuit-breaker), t_1, the contacts part and the arc forms; t_2, arc moves off contacts; $t_2 - t_1$, is the dwell time; t_3, the arc is elongated and splits up within arc chute; t_4, t_5, t_6, are examples of arc back-commutation; t_7, the current is zero and the arc extinguishes.

milliseconds. A more detailed consideration can further distinguish between different phases of dwelling [11], but for practical application, the time when the arc leaves the contacts is essential. The reason for dwelling is that the conditions for charge carrier production at first only exist at the original arc spots. For the arc to move, the preconditions for new arc spots must be created in the forward direction. This happens by magnetically deflected plasma jets that emanate from the old spots and that heat the electrodes to form new spots. For sufficient deflection of the plasma, a certain contact distance is necessary. Strongly vaporizing contact material produces rather stiff jets that are more difficult to deflect [12]. Therefore, the dwell time and minimum contact separation, respectively, depend on the material, current, opening speed, and magnetic blast field. A comparison of dwell times under conditions of miniature circuit-breakers is given in Figure 14.32, Section 14.4.6.

The start of arc motion as well as the subsequent motion is often not a continuous process but rather a sequence of forward-commutations [13]. Jumping across a gap in the course of arc motion, for example, from the moving contact part to the fixed arc runner in Figure 14.22b, is a similar process and can also slow down the arc motion [14,15].

14.1.4 Sticking and Back-Commutation of the Arc

At high currents, it may become a problem that the arc fails to completely elongate along the diverging arc horns but sticks at the edge of the parallel rails or remains across the whole length of the horns [16,17]. This can be explained by an increasing voltage demand of the elongated arc versus the well-conducting short old channel. The current limit where this may occur depends among other things upon the width of the arc chute and the angle of the horns. Figures 14.5 [17] and 10.22 give examples of such limits. Rather similar processes are back-commutations after the arc has already reached the chute and has split up into series arcs. Such a behavior, which is associated with voltage breakdown, can be seen in Figure 14.4. The arc voltage, which sharply rises as the arc elongates and splits, acts as a source for reheating and reignition of the still hot region between the arc horns or contacts

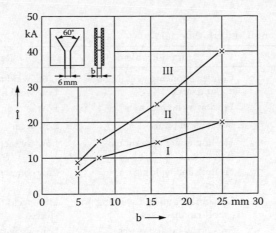

FIGURE 14.5
Current limits of unhindered arc elongation: I, arc moves to tips of arc runners; II, intermediate region; III, arc lasts as a broad band between arc runners.

that has been left behind by the arc [18]. These undesirable processes can be shifted toward higher currents by appropriate design, for example, vents in the walls, and magnetic fields.

14.2 Contacts for Switching in Air

Contact materials, their shapes and manufacturing technologies are treated in Chapters 16 and 17. Details about contact performance and testing are treated in Chapter 18. The contact parts in devices for medium- and high-current switching have to withstand a variety of different stresses [2,6,19], which are summarized, together with the requirements for contact materials, in Table 14.1. As several of these requirements are contradictory to each other, no contact material can fulfill all of them equally well. The development and selection of materials for certain applications is therefore always a compromise. Table 14.2 (according to [6,19]) is a summary of the sequence of different contact materials with respect to different stresses. It is of course only very crude, as the behavior depends on many design and circuit parameters, and the contact manufacturing details are important as well.

Table 14.3 finally is a summary of the application of contact materials in different low-voltage switching devices working in air [19,20]. The general trend is to use either pure silver or silver alloys for low duty, while for higher currents special erosion- and weld-resistant compound materials prevail. Some of those materials usually have some other drawbacks that must be overcome by the device design, like higher contact resistance, longer arc dwell time, or lower reignition voltage. As can be seen, the use of asymmetrical contact pairs, that is, different contact materials on both contacts, is widespread. Especially with AgC against either pure copper, AgNi, or materials from the WAg–WCAg group, the advantages of both partners are combined in circuit-breakers. At arcing the graphite content prevents the formation of tarnish films (contact resistance rise), and it reduces the weld tendency. On the other hand, the drawbacks of AgC—high erosion and long dwell time—are diminished by combination with materials like AgNi or Cu.

TABLE 14.1

Stresses and Requirements for Contact Materials

Operation	Strain/Problem	Requirement
Making	Contact bouncing, welding by bounce arcs	Little weld tendency Low weld forces
	Erosion by bounce arcs	Low and uniform make erosion
	Mechanical wear	Low wear
Closed contacts	Heating under operational conditions	Low contact resistance
	Heating and welding on short-circuit	Low contact resistance
	Dynamic lift-off and welding on short-circuit	Little weld tendency, low weld forces
Breaking	Erosion by breaking arcs	Low break erosion
	Immobility of arc roots	Short immobility time
	Arc extinction	Fast recovery
	Reaction with surrounding medium on arcing, contact resistance increase	Constant low contact resistance
	Reduction of insulating level by metal deposition	Nonconducting deposits
Open contacts	Electric stress	Sufficient insulation
	Tarnish formation, contact resistance increase	No formation of detrimental films

TABLE 14.2

Sequence of Silver-Containing Contact Materials with Respect to Different Stresses

	Welding of Closed Contacts	Welding on Make	Erosion by Arcing	Dwell Time	Contact Resistance after Arcing	Reignition Voltage
Inferior ↑	WAg, WCAg[a]	Ag, AgCu	Ag	WAg, WCAg	WAg, WCAg	WAg, WCAg
	AgNi	AgNi	AgCu	AgC	AgMeO	AgC
	AgMeO	AgMeO	AgC	AgMeO	AgNi	AgCu
	AgCu, Ag	AgC	AgNi	AgNi	AgCu	Ag
	AgC	WAg, WCAg	AgMeO	Ag, AgCu	AgC	AgNi
↓ Better		WCAg, WAg			Ag	AgMeO

[a] Early welding, but weld forces only weak.
AgMeO, silver-metal oxide ($AgCdO$, $AgSnO_2$).

14.3 Low-Voltage Contactors

The principal function and mechanical design aspects are treated first. The relevant physical mechanism of arc quenching leads to different typical arc chute designs, depending on the ratings of the contactor. The choice of contact materials is governed by the requirements for safe operation and long service life. While in most

TABLE 14.3

Application of Contact Materials in Different Low-Voltage Switching Devices

Application	Continuous Current Rating	Interrupting Capacity	Contact Material
Relays and auxiliary contacts	≤10 A	≤100 A	Ag, AgCu (3–10% Cu) AgCdO (10–15% CdO) AgNi (10–20% Ni)
Contactors	≤10 A	≤150 A	AgNi (10–20% Ni) AgCdO (10–15% CdO) $AgSnO_2$ (8–12% SnO_2)
	>10 A	>150 A–10 kA	AgCdO (10–15% CdO) $AgSnO_2$ (10–12% $AgSnO_2$)
Residential circuit-breakers, US type	≤125 A	≥10 kA	MoAg (25–50% Ag) WAg (50% Ag)
Switching duty residential breakers	≤30 A	≤10 kA	WAg (25–50% Ag) MoAg (Ag-enriched surface) AgCdO (10–15% CdO) AgZnO (8–10% ZnO) $AgSnO_2$ (8–10% SnO_2)
Residential circuit-breakers, European type	≤63 A	≤10 kA	AgCdO (10–15% CdO) $AgSnO_2$ (10–12% SnO_2) AgC (3–5% C) + Cu[a]
		>10 kA	AgC (3–5% C) + AgNi (40–50% Ni)[a] AgZnO (8% ZnO) MoAg (25–50% Ag), WAg
Industrial type circuit-breakers without extra arcing contacts	≤400 A	≤25 kA	AgC (3–5% C) + AgNi (40–50% Ni)[a] AgC (3–5% C) + WAg (25–50% Ag)[a]
	≤800 A	≤100 kA	WAg (25–50% Ag) WCAg (35–50% Ag) MoAg (30–50% Ag)
Circuit-breakers with main and arcing contacts	>400 A	<150 kA	Main contacts: AgNi (20–40% Ni) AgCdO (10–15% CdO) MoAg (50% Ag), AgW (25–50% W) WCAg (35–50% Ag) Arcing contacts: WAg (20–35% Ag) WCu (30–50% Cu) WCAg (30–40% Ag)

[a] Asymmetrical contact pair.

IEC contactors for low-current ratings AgNi contacts are predominant, silver–metal oxide compound materials prevail for larger sizes. For many years, intensive work has been undertaken to replace AgCdO by materials with less toxic metal oxides, especially $AgSnO_2$. Characteristic differences in their behavior and the present knowledge about the reasons are summarized. Finally, new trends in contactor development are mentioned.

14.3.1 Principle/Requirements

Contactors are remote control switches actuated electromagnetically or pneumatically. Pneumatic contactors are disregarded here as they play only a minor role. Most contactors are self-resetting, that is, they move into one position (in main circuits generally the ON-position) when their magnetic actuator is energized and they return to the original state when its excitation is switched off. They are used in a wide field of applications, such as switching small currents in auxiliary circuits, ohmic loads in heaters, capacitive currents for power factor compensation, or motor loads up to many hundred kilowatts or even megawatts. Standards distinguish between contactors in auxiliary circuits and in main circuits of electrical power engineering. I will mainly concentrate on motor switching, so-called motor starters, as this is one of the most common duties of contactors. There are other switching devices for motor load switching, for example, hand-operated motor starters. Apart from the actuator mechanism, their main parts like contacts and arc chutes are very similar to those of contactors.

Contactors provide a high operational life, for example, 10 million mechanical operations for smaller contactors and between tens of thousands and over a million switching operations under electric load, depending on the load conditions. A considerable amount of all the contact material produced worldwide goes into contactors, and many papers on the contact performance refer to contactors [21–26].

Mainly low-voltage (≤1000 V) air-break contactors are treated here. Vacuum contactors that also play a role, especially in the region of higher current and voltage ratings, are treated in Section 14.6.

As for other technical products, the duties of contactors, as well as other switching devices like circuit breakers are fixed in "Standards". Worldwide two main regions have influenced the technology and standardization: North America and Europe. The first is represented through NEMA, UL, and CSA and their standards [27–30], the second through IEC and VDE, today harmonized in the European Community as EN/IEC standards [31–33]. They represent two different philosophies in standardization as well as different habits in the selection and application of contactors. NEMA defines several sizes of contactors and prescribes ratings for different duties, depending on the size. IEC differentiates primarily between various utilization categories, representing the typical loads. As a very rough generalization, the NEMA standard requires a larger device for a certain switching purpose, while the IEC standard needs a more detailed consideration of the circumstances of application [34]. Successive harmonization between IEC and UL standards has been started, however, and will take place in the future.

Table 14.4 summarizes the major utilization categories and test duties of contactors for ac switching according to IEC [31]. The most frequently discussed categories are AC-3 and AC-4. AC-3 requires making currents of $6I_e$ and breaking currents $1I_e$ (I_e = rated operational current), representing starting of squirrel-cage motors and switching them off during running. AC-4 defines $6I_e$ on make and break. These conditions exist when the high motor inrush currents are switched off again before a drive has really started to move, or when inching, plugging or reversing is employed. Additionally, the required make and break capacities for occasional operations lie even higher.

14.3.2 Mechanical Arrangement

Figure 14.6 shows several principal arrangements of the magnetic actuator in relation to the contact system [35,36]. In all cases, the three or more current paths are arranged one behind the other in the viewing direction and actuated by the same magnet. Figure 14.6a

TABLE 14.4

Conditions for Life Test and Make-and-Break Test of Contactors According to IEC 60947-4-1 (Simplified)

Category	Value of the Rated Operational Current	Life Tests						Making and Breaking Capacity					
		Make			Break			Make			Break		
		I/I_e	U/U_e	cos φ	I/I_e	U/U_e	cos φ	I/I_e	U/U_e	cos φ	I/I_e	U/U_e	cos φ
AC-1	(All values)	1	1	0.95	1	1	0.95	1.5	1.1	0.95	1.5	1.1	0.95
AC-2	(All values)	2.5	1	0.65	2.5	1	0.65	4	1.1	0.65	4	1.1	0.65
AC-3	$I_e \leq 17$ A	6	1	0.65	1	0.17	0.65	10	1.1	0.65	8	1.1	0.65
	17 A $< I_e \leq 100$ A	6	1	0.35	1	0.17	0.35	10	1.1	0.35	8	1.1	0.35
	$I_e > 100$ A	6	1	0.35	1	0.17	0.35	8	1.1	0.35	6	1.1	0.35
AC-4	$I_e \leq 17$ A	6	1	0.65	6	1	0.65	12	1.1	0.65	10	1.1	0.65
	17 A $< I_e \leq 100$ A	6	1	0.35	6	1	0.35	12	1.1	0.35	10	1.1	0.35
	$I_e > 100$ A	6	1	0.35	6	1	0.35	10	1.1	0.35	8	1.1	0.35

represents the old system of clapper contactor, still in use for some special high-capacity contactors and in medium voltage vacuum contactors. The moving parts of the magnet and of the contact system are fixed on a rotating bar. By appropriate design of the lever arms of the magnet and the contacts, the available magnet force can be adjusted to the contact force characteristics. This type usually uses single break contacts. The space requirement of clapper contactors is relatively high. Figure 14.6b shows a variant of this principle with an additional toggle lever. By its means the closing speed can be diminished to reduce bouncing, and the contact force is increased. On the other hand, this measure also reduces the opening speed. The principle is mechanically complicated and therefore seldom used in contactors. In contrast to these two schemes, the two following ones, using magnets with linear motion, are more frequently found in modern contactors. Figure 14.6c is based on a rotating lever, coupling the magnet and the contact system. This implies an additional degree of freedom by choosing the gear ratio between the magnet and the contacts, and has some advantages with respect to the accessibility of both the contact system and the actuator magnet. As the magnet is arranged side-by-side with the contact system, and the lever is an additional part, the space requirement is somewhat higher than for the following arrangement. Finally, Figure 14.6d represents the mostly used principle, where the actuator and the contact system are arranged within one block one on top of the other and coupled directly. The lacking possibility of mechanically fitting the magnet characteristics to the contact system can be overcome by appropriate electrical design of the magnet and its coil. This system enables compact and cost-effective devices.

While Figure 14.6a is a single break, Figure 14.6b through d show double-break arrangements as used in most contactors. The consequences for the electrical behavior are discussed later. For the actuator, a double-break means roughly that it has to provide double the contact force.

Figure 14.7 is a schematic view of the mostly used contactor type with direct actuation, including the contact and opening springs [37]. The contacts are kept in the open position

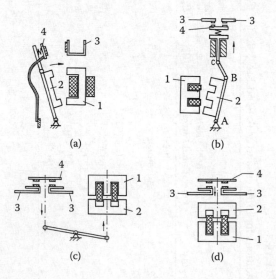

FIGURE 14.6
Arrangement of magnetic actuator in contactors: (a) clapper contactor; (b) clapper arrangement with toggle lever; (c) actuator with lever coupling; (d) direct actuator. 1, fixed magnet part; 2, moving magnet part; 3, fixed contact part; 4, moving contact part.

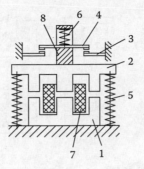

FIGURE 14.7
Arrangement of actuator, contact system, and springs: 1, fixed magnet part; 2, moving magnet part; 3, fixed contact part; 4, moving contact bridge; 5, opening springs; 6, contact spring; 7, magnet coil; 8, actuator element.

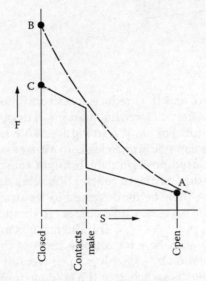

FIGURE 14.8
Scheme of static forces for a contactor magnet.

by the opening springs 5. When the coil is energized, the contacts are driven down together with the moving actuator element 8, until the contacts make, and the force of the contact spring suddenly becomes effective between the fixed and moving contacts. When the excitation ceases, the loaded spring 5 accelerates the moving parts toward the open position, at first supported additionally by the contact springs. The static force requirement for the actuator is schematized in Figure 14.8. It must be more than compensated, at least dynamically, by the magnetic force (dashed line).

Though it would require an extra monograph to cover the essentials of magnetic actuators, some main features shall be mentioned here. Figure 14.9 summarizes different typical shapes of magnets [35,37]. In addition to the shape, the main differentiation lies between ac and dc magnets. Many manufacturers offer these systems optionally and with different voltage ratings. To avoid ac losses, the iron cores of ac magnets have to be laminated. Additionally, they need shading coils, that is, shorted one-turn windings that surround partial areas of the magnetic poles (see Figure 14.9g). Their effect is to generate a partial flux phase-shifted against the remaining flux. The superposition of both fluxes prevents the force

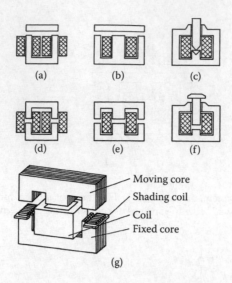

FIGURE 14.9
Typical shapes of magnets.

from becoming temporarily zero, and thus reduces noise and mechanical wear [35,37,38]. dc magnets have no problems with such force fluctuations of double line frequency.

All magnets must exceed the force of point A in Figure 14.8 when starting to close. The dependence of the force on the magnetic air gap leads to an excess of the available (B) over the necessary force (C) in the closed position, which might cause an unnecessarily high speed and kinetic energy at closure. Stronger contact bouncing and mechanical wear are the consequences. DC magnets can be better designed to feature lower excess forces and softer closing. For all magnets the necessary hold current in the closed state is only a small fraction of the pull-in current at point A. As their inductance is much smaller in the open than in the attracted state, ac magnets automatically meet both requirements, without the necessity to change the excitation by switching. In conventional dc-driven contactors, either the number of coil windings is switched or the hold current is reduced by inserting an additional resistor, both by means of auxiliary contacts. The latter method is associated with additional ohmic losses. New electronic solutions are increasingly gaining interest here (see Section 14.3.5). Depending on the contactor size, the contact stroke lies between a few millimeters and 1 cm per break. The average contact opening and closing speeds lie typically between 0.3 and 1.0 ms⁻¹.

14.3.3 Quenching Principle and Contact and Arc Chute Design

Contactors are generally devices quenching at current zero. In ohmic-inductive circuits, after the current has reached zero, the line voltage reappears across the switching gap in the form of the transient recovery voltage (TRV). Its frequency depends on the load and lies between several kilohertz and well above 200 kHz [31]. The arc that has been initiated by contact separation several milliseconds earlier must withstand this TRV within a microsecond frame. In the majority of contactors, the physical principle of immediate recovery is effective [3,39,40] (Section 9.7.1). Even without any additional arc chute, an ac arc between two electrodes is able to withstand a certain voltage immediately after current zero without restrike. Only when this voltage is exceeded does reignition and subsequent current flow occur (Section 9.7.1). This voltage is referred to as the instantaneous recovery voltage.

Besides the recovery of the cathodic space charge sheath, cooling of the arc column, especially when gases from decomposing insulating walls are involved, may additionally support arc extinction [41,42], but wall erosion also limits the operational life.

Figure 14.10 is a plot of measured instantaneous recovery voltage (average) for different contact materials at different currents. It clearly depends on the thermal and electrical properties of the electrode material, as well as the thermal arc stress on the contacts. The general tendency is a decrease with increasing arc current (see Figure 9.55), but contact materials with components of low boiling or sublimation temperatures, such as AgCdO, stay remarkably constant up to high currents [8,40]. Additives or impurities may also exert a strong influence [43]. While in small installation switches up to 20 A and 230 V one contact gap is sufficient to withstand the maximum TRV, higher currents and r.m.s. voltages of 400 or 690 V in three-phase systems already reach the limit of self-extinction. By connecting two gaps in series the breakdown voltage is nearly doubled, and thus, this principle extended to higher ratings. Figure 14.11 demonstrates this for several contact materials [44]. Most contactor relays for auxiliary circuits and small motor contactors up to $I_e = 20$ A use such double-break contacts without an additional system to influence the arc.

Another advantage of the usual double-break contact systems with two fixed contacts and one moving bridge (Figure 14.6b through d, Figure 14.12a through d) is that no flexible connections between the moving contacts and their terminals are needed. They might constitute a serious mechanical problem to contactors with their high number of operations. In contrast to circuit-breakers, the currents in contactors are not extremely high, so the drawbacks of the double break, namely higher contact resistance and higher total contact force, are acceptable.

There have been contactor designs in the market that even connected two double-breaks in series to form a fourfold break [6]. However, this arrangement is costly, and there are four sources of heating. When the plain double-break system is not sufficient any longer—typically at rated currents beyond 20 A—the next step is to move the arc on each side off the contacts by magnetic forces and to split it up into two arcs by means of an additional isolated metal electrode—often as a U-shaped iron sheet lining the arc chamber similar to that shown in Figure 14.12b. This results in a total of four immediately recovering gaps.

At ratings above $I_e = 50$ A, most contactor designs split up the arcs in even more partial arcs by means of iron splitter plates ("deion arc chute"). Figure 14.12c,d shows examples for such arrangements in contactors.

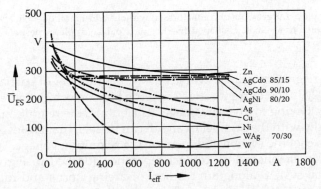

FIGURE 14.10
Instantaneous recovery voltage (average) for different contact materials vs. r.m.s. current.

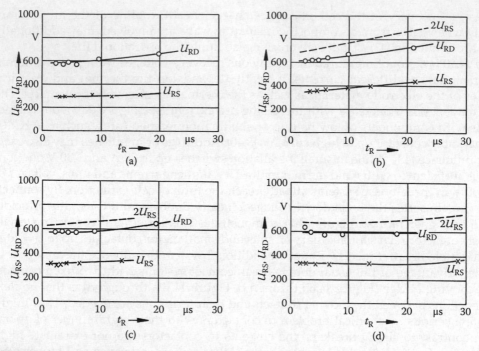

FIGURE 14.11
Reignition voltage vs. time after current zero for different contact materials on single and double break contacts:
(a) Ag; (b) AgNi 90/10; (c) AgCdO 90/10; (d) AgSnO$_2$ 88/12. $U = 750$V, $I = 103$ A, $f_{TRV} = 10$ kHz, U_{RD}, double break,
U_{RS}, single break.

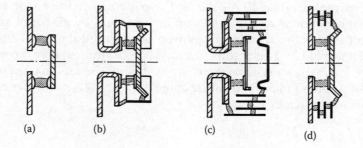

FIGURE 14.12
Typical contact and arc chute arrangements of contactors: (a) without additional quench system; (b) insulated
steel lining; (c) plates parallel to arc axis, U-shaped fixed contact parts; (d) plates perpendicular to arc axis,
straight fixed contact parts.

Figure 14.13 shows reignition voltages versus the actual number of partial arcs in series
between iron plates [2,8]. The current range concerns big contactors and smaller circuit-
breakers, respectively. Owing to uneven voltage distribution, the reignition voltage grows
less than proportionally with the number of arcs. It can be seen that gas-evolving polymer
wall material helps to cool the system and increase the reignition voltage.

In any case, the arcs established on contact separation (dashed in Figure 14.12) must
be moved to the deion system along arc runners or arcing horns and must create new
anode and cathode spots at the steel plates. This is facilitated by U-shaped current paths
of the fixed contacts (Figure 14.12b,c) that form a loop together with the current path in

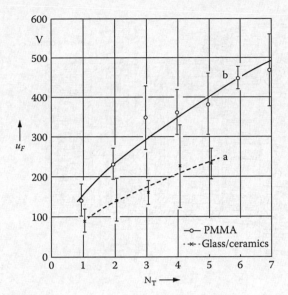

FIGURE 14.13
Reignition voltage (mean and standard deviation) vs. number of partial arcs in series: (a) nongassing wall material; (b) gassing wall material. $I = 5000$ A, $U = 500$ V, $f_{TRV} = 600$ kHz, steel splitter plates.

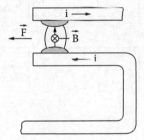

FIGURE 14.14
Typical current path arrangement for self blast field in contractors.

the bridge, thus generating a magnetic blowout field. This current path is drawn separately in Figure 14.14. As a drawback, the loop increases the magnetic blow-off forces that tend to open the contacts when high inrush currents are switched in. Furthermore, it is space-consuming. Some designs do not use U-loops but iron parts along the arcing horns to concentrate the magnetic self-field onto the arc area; others use both features. There is a wide variety of details with regard to the arrangement of arc runners, splitter plates, and additional iron parts to direct magnetic fields [2,6,45,46]. As an example, Figure 14.15 shows a complete cross-section of a larger contactor [45].

Figure 14.16 gives a summary of different shapes of deion splitter plates of larger contactors [6]. Their sides facing the undivided arc are often V-shaped or serrated to facilitate the arc splitting. They increase the attractive magnetic field, at least locally (compare Figure 14.3b), and by squeezing portions of the arc they increase the local voltage drop. It is necessary to increase the voltage of the yet undivided arc above the minimum arcing voltage before splitting occurs. Figure 14.12c,d also show that different orientations of the plates relative to the arc axis are usual. Figure 14.12c additionally is an example of how

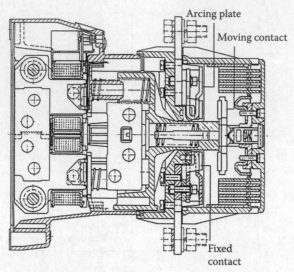

FIGURE 14.15
Cross-section of an ac contactor with 400 A rated operational current.

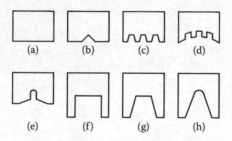

FIGURE 14.16
Shapes of deion splitter plates in contactors.

the moving bridge can be relieved from the arc stress by commutating the arc roots onto a fixed conductor connecting both arc chutes.

14.3.4 Contact Materials

When looking at the contact materials used in European contactors, two main groups can be distinguished:

Contactor relays and small contactors with operational currents of a few tens of amperes (switching currents to 200 A) use pure silver, hard silver with small alloying additions to increase the mechanical and thermal strength of silver (e.g., Ag with 3% Cu), or in most cases silver-nickel. Silver-nickel is a compound material produced by sintering processes from the metal powders. Usual compositions for contactors lie between 90/10 and 80/20 per cent by weight.

For contactors with higher ratings, silver-metal oxide (AgMeO) compound materials have been the standard worldwide for several decades, and they are still the first choice. Detailed summaries of their properties and their development are published in [23,24], see also Chapter 16. They are either manufactured by sintering processes or by internal oxidation of silver-metal alloys. It is a well-known fact that the characteristics of these contact materials may widely differ, depending on the composition, the additives and impurities, and the manufacturing process and its parameters. Usual compositions are Ag with 8–15 wt% of oxide. AgCdO materials used to be the optimal choice for a long time, but other oxides like CuO, ZnO, or SnO_2 have also been under consideration. Starting in the 1970s, the development work was intensified to replace the toxic CdO by less harmful components. The proceedings of all contact conferences since then reflects this development. $AgSnO_2$ was found to show remarkably low erosion at high loads in comparison with AgCdO, as well as high resistivity against welding. It was soon found, however, that $AgSnO_2$ led to higher contact resistances which, in turn, led to higher heating, as a result of its more stable oxide. After intensive research work in many places with additional small amounts of further components, this behavior could be considerably improved [22]. As far as it is understood now, such additives—for example, oxides of refractory metals [47,48]—modify the morphology of the melt at the contact surface.

$AgSnO_2$ materials have already replaced AgCdO in many new contactor designs. Especially in conjunction with the introduction of $AgSnO_2$ (10–12% SnO_2 by weight) it is a widely accepted truth that it is not optimal to just replace an existing contact material by a new one, but to optimize the switching device together with a contact material. While at first the longer service life of big contactors with these $AgSnO_2$ materials under AC-4 conditions (high make and break currents, Figure 14.17a [49]) was the outstanding feature, it was soon found that under AC-3 load $AgSnO_2$ may lead to a distinctively higher wear than AgCdO [49–51]. There are also strong dependences of the ranking on the design and size of the contactor, as can be seen from a comparison between Figure 14.17b and Figure 14.17c [49]. Measurements under make-only and break-only conditions show that the make erosion of $AgSnO_2$ may be two-to-three times higher than that of AgCdO, whereas the break erosion is smaller [46]. Owing to the much higher make current, the make erosion dominates on AC-3. Present interpretations suppose that the differences also lie in the properties of the contact material melt, and in the behavior under the impact when the contacts close (bouncing) [25,26,50]. Work is going on to improve this behavior by minor additives of third of fourth components.

It is often observed that especially the make erosion may widely differ between different designs, between different specimens of the same design, and between the three phases of one specimen. The latter is easily explained by a synchronizing effect of ac-fed actuator coils. Investigations where the actual bounce arc pattern and the arc energy on make have been evaluated [52] show that this energy strongly varies. In a three-phase circuit with floating star point a mechanical contact lift-off does not necessarily mean an arc. Depending on the mechanical properties of the design and of the individual specimen as well, the bounce pattern varies and consequently the time and current magnitude of the bounce arcs that are actually formed are subject to strong differences. Figure 14.18 gives an example of the measured lowest (1) and highest life (2) of a contactor type, as well as estimations (3) derived from the bounce patterns [52]. Variations of 3:1 or even 4:1 may occur.

14.3.5 Trends

Besides the general trend to develop better, even more reliable, and more compact contactors, and to further improve their handling and mounting, some specific trends should be mentioned briefly, as far as contactors in main circuits are concerned.

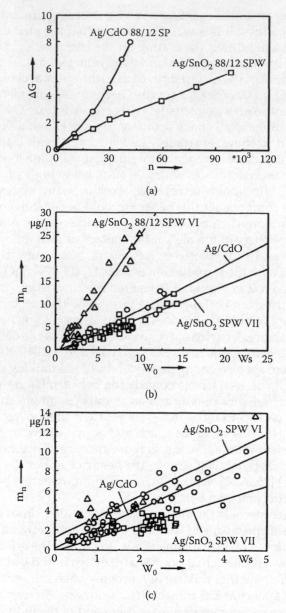

FIGURE 14.17
Erosion of AgCdO and of different variants of AgSnO₂: (a) AC-4, phase with strongest erosion, erosion vs. no. of operation; (b) AC-3, contactor type A, average erosion vs. arc energy; (c) AC-3, contactor type B, average erosion vs. arc energy.

14.3.5.1 Contactors versus Electronics

It is often discussed whether electromechanical devices like contactors will have chances in the future in comparison with fully electronic solutions [53,54]. The answer should be differentiated between main circuits and auxiliary circuits. Unless completely new electronic elements are discovered, the contactor will essentially keep its position as a switch for main circuits because of its advantages in view of much smaller losses, much

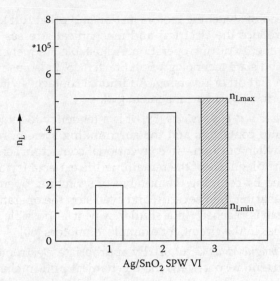

FIGURE 14.18
Measured and expected AC-3 life for AgSnO$_2$ 88/12 SPW VI in a contactor. 1, lowest; 2, highest measured life; 3, estimated spread of life (shaded).

smaller volume, smaller costs, and its isolating properties. Control circuits with contactors are being increasingly replaced by programmable controllers. But on the other hand, the number of contactor relays used as output interfaces between electronic controllers and actuators is rising, and the manufacturers still register constant or slightly growing numbers.

14.3.5.2 Vacuum Contactors

The low-voltage vacuum contactor (see Section 14.6) has already gained some importance, especially at higher current ratings (150 A–1000 A) and for special applications like mining. Its advantages are the completely sealed contact system (no contact resistance problems, no environmental problems), the high switching capacity, and a long operational life. It seems to be mainly a question of manufacturing costs whether the vacuum principle will be able to penetrate more into the lower-current regions.

14.3.5.3 Hybrid Contactors

Hybrid contactors are a combination of metallic contacts to carry the current with power electronics elements for the switching operation [50]. This reduces contact erosion and increases contact life considerably. Though technically excellent, this principle is costly for switching alone, and it is still limited to special cases.

14.3.5.4 Integration with Electronic Systems

There will be an increasing integration of electromechanical components into the systems of automation technology, and vice versa. Though solid state devices will partly replace traditional mechanical switching in various applications, the overall use of electromechanical contacts will remain and even be extended [55,56].

- *Motor Control.* Instead of switching motors just on and off at full load, more intelligence is used to reduce the electrical and mechanical stresses on closing and opening, or to save energy during operation. Adjustable Frequency Drive Control is fully based on solid state technology. Soft Starters [57] also use solid state control at switching on and off or reversing. Additional contacts switched parallel to reduce the losses form a hybrid arrangement [58].

- *Integration of contactors into bus systems* [59]. It is a tendency to integrate more and more intelligence into contactors and the surrounding system, from the simple monitoring of the switch position—in conventional contactors achieved by auxiliary contacts—to the diagnosis of the remaining life expectancy of the contactor. This can be achieved by counting the number of switching operations or better by additionally measuring their severity, and even stop the operation when a predefined contact wear is reached. In a similar way, it is possible to monitor the aging of other devices in the circuit, for example, of motors [60].

- *Electronic control of magnet coils* [33,61,62]. By appropriate electronic control of the coil voltage and current, respectively, the mechanical pull-in characteristics can be influenced much better than by the conventional technique. It is possible to minimize bouncing at contact closure, an appropriate measure to improve the performance of $AgSnO_2$ under AC-3 conditions [63]. Other features are that the operating voltage range can be increased and that the number of coil variants for different frequency and voltage ratings can be drastically reduced. Also the size of the magnet and coil can be minimized, especially with dc actuators.

14.4 Low-Voltage Circuit-Breakers and Miniature Circuit-Breakers

14.4.1 Principle/Requirements

Significant differences for example, between circuit breakers per IEC and UL/CSA are larger clearance and creepage distances and higher requirement of individual pole short-circuit breaking capacity with UL/CSA. Besides normal operational currents and overload currents, circuit-breakers must be able to switch over-load currents and short-circuit currents on and off. The IEC standards [64] differentiate between two utilization categories: Category A comprises circuit-breakers that are not specially designed for selectivity, and category B circuit-breakers that are designed for selectivity under short-circuit conditions. Selectivity of two circuit-breakers in series means that only the circuit-breaker on the load side interrupts, while the one on the feeder side remains inactive. This is usually achieved by short-time delay short-circuit release. The circuit-breaker must then be able to withstand the stress during the delay time. Some standardized data concerning the making and breaking capacities of circuit-breakers are summarized in Table 14.5 [64]. While the breaking capacity is defined as r.m.s. value, the making capacity is defined as the maximum peak value of the prospective short-circuit current. The required short-circuit switching operations are shown in Table 14.6. In any case, the circuit-breaker has to make and break its full rated short-circuit current only a few times. Depending on the rated current, the required number of switching cycles without current (mechanical) lies around several thousand, the number of closing-opening operations under normal load between 500 and 1500 [64].

TABLE 14.5

Standard Relationship Between Short-Circuit Making and Breaking Capacities, and Related Power Factor, for AC Circuit-Breakers (IEC 60947-2)

Short-Circuit Making Capacity (kA r.m.s)	Power Factor	$n = \dfrac{\text{Short-Circuit Making Capacity}}{\text{Short-Circuit Breaking Capacity}}$
$4.5 < I < 6$	0.7	1.5
$6 < I < 10$	0.5	1.7
$10 < I < 20$	0.3	2.0
$20 < I < 50$	0.25	2.1
$50 < I$	0.2	2.2

TABLE 14.6

Short-Circuit Test Sequences for Circuit-Breakers (IEC 60947-2)

Denomination	Test Sequence	Tests to Pass after Short-Circuit Sequence
Test sequence II: Rated service short-circuit breaking capacity	$O - t - CO - t - CO^a$	Insulation voltage Heating under load Overload tripping
Test sequence III: Rated ultimate short-circuit breaking capacity	$O - t - CO^a$	Insulation voltage Overload tripping

[a] O opening operation (by release) C closing operation.
 CO closing-opening operation t pause (3 min).

There are separate standards on miniature circuit-breakers for residential and industrial installations, for example, [65,66], but there are no basic differences to other circuit-breakers in function and general design of the contact and arcing systems.

14.4.2 General Arrangement

Low-voltage circuit-breakers are operated by latched springs which are charged either manually or by electric motors. As far as their operational principle is concerned, two variants can be differentiated:

- Circuit-breakers with high current-carrying capability, able to withstand the short-circuit current for some time, for example, several cycles to one second ("dynamically and thermally rigid" breakers). They are suitable for time-delay tripping (utilization category B), that is, the contacts must stay latched until the breaker is being tripped. They usually have higher rated continuous currents (630–4000 A) and are often built as open devices where the actuator mechanism, the trip system, the contacts, and the arc chutes are mounted together on a steel frame (example in Figure 14.19 [2]). Other designs of this type are built as molded-case circuit-breakers (MCCB). This type is more used in the upper hierarchy of low-voltage power distribution, for example, as feeder circuit-breakers for busbars. Owing to the time delay they do not limit the short-circuit current, and it makes no sense to construct their contact systems and arc chutes specially for current limitation. Nevertheless, the impedance of the tripping device and the arc voltage often cause a certain current reduction against the prospective current.

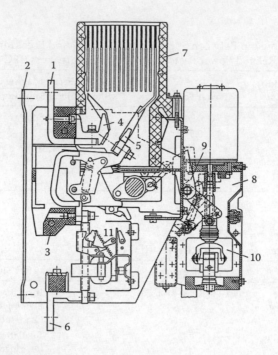

FIGURE 14.19
Non-current-limiting 1250 A circuit-breaker in open construction: 1, upper terminal; 2, steel frame; 3, insulating plate; 4, fixed contact piece; 5, moving contact piece; 6, lower terminal; 7, arc chute; 8, motor drive; 9, cam disk; 10, centrifugal weight; 11, overcurrent trip unit.

- Circuit-breakers provided with fast-acting trip- and contact-opening mechanisms, where the arc voltage is rapidly increased on short-circuit to limit the actual current to values well below the prospective currents. These breakers with rated continuous currents between 16 A and several 100 A are generally built as molded-case circuit-breakers [5,29]. The molded case, typically of reinforced polymer, provides both the insulation and the mechanical support structure for mounting all other components. Figure 14.20 [2] is an example of a current-limiting MCCB. Their main application lies more downstream in the power distribution system.

Combined principles are also realized, such as breakers behaving dynamically rigid up to a certain current and current-limiting beyond.

14.4.3 Quenching Principle and Design of Arc Chute and Contact System

14.4.3.1 Quenching Principles

Circuit-breakers may use two different physical principles of arc interruption. The first is *ac interruption at current zero*, where the arc has to lose its conductance quickly when the ac current passes through zero. It has already been treated in Section 9.7.1. These breakers are not current-limiting. The second principle is the principle of dc interruption as treated in Section 9.7.2, where the arc voltage must be increased above the system voltage. It is utilized for switching in dc circuits, which is presently still relatively rare, but becoming more important, and far more for *current-limiting switching in ac circuits* (see also Section 9.7.4). Figure 14.21 shows the principal current and voltage evolutions on current-limiting

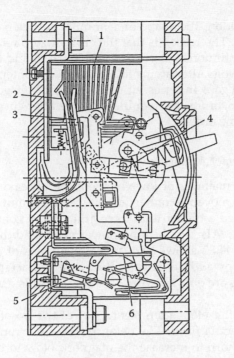

FIGURE 14.20
Current-limiting molded-case circuit-breaker (MCCB) for 225 A rated continuous current: 1, arc chute; 2, fixed contact piece; 3, moving contact piece; 4, operating mechanism; 5, undelayed trip device; 6, overload trip device.

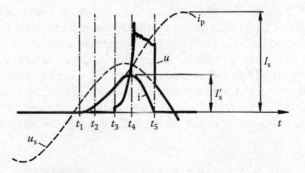

FIGURE 14.21
Principle of current limitation in ac circuits: u, arc voltage; u_s, system voltage; i, current; i_p, prospective current; I_s, prospective peak current; I'_s, let-through current; t_1, beginning of short-circuit; t_2, tripping; t_3, contact opening; t_4, arc voltage reaches system voltage, current is limited; t_5, current interruption.

switching. For an effective limitation, the arc voltage must be raised above the momentary system voltage well before the prospective short-circuit current has reached its first peak, that is, within a few milliseconds. By this the dynamic stress, which is proportional to the square of the peak let-through current $(I'_s)^2$, as well as the thermal stress, which grows with $\int i^2 \, dt$ of the system under protection, are reduced considerably. The current limitation works only under short-circuit conditions where the trip time is short enough. Under nominal load current or over-load current there is no limitation.

Nearly all low-voltage air circuit-breakers, if they are current-limiting or not, use arc chambers with steel splitter plates for quenching. Though, as explained in Section 14.1.1, the physical principles are quite different between current-limiting breakers and breakers using current zero interruption, the shape and number of deion plates may not differ substantially between them. The reason is that the current-zero reignition voltage of arcs between steel electrodes continuously decreases with current and finally approaches the arc voltage of a few tens of volts after arcing at high currents of tens of kiloamperes. The necessary number of partial gaps in series, either for the arc voltage to exceed the system voltage for current limitation, or for the reignition voltage after current zero to exceed the TRV, lies therefore in the same order.

14.4.3.2 Arc Chute and Contact Arrangement

Contrary to contactors the majority of low-voltage circuit-breakers use one break per phase, which necessitates a movable connection between the moving contact piece and its terminal. Two breaks need a higher total force and a more complicated construction of the operating mechanism. Also the heating power generated during continuous current load is double. Figure 14.22 shows two typical examples of contact and arc chute arrangements; there are many additional variants. Some characteristic arc positions are also drawn. The deion plates are only schematic, their number is usually higher (compare Figures 14.19 and 14.20).

In Figure 14.22a the arc after elongation has to stay at the tip of the moving arc horn on which the moving contact is fixed. The deion plates are arranged fan-like and surround the contacts and arc horn to increase the magnetic blow force. The steel plates are extended by insulating plates in this example. This provision which can be often found in open circuit-breakers for high currents (Figure 14.20) prevents the arc from being shorted behind the plates.

Figure 14.22b represents a system where the arc has to commutate from the moving contact onto a separate fixed arc runner that is connected to it by a flexible conductor. The moving contact piece is thus relieved from the arc stress, and the arc runners can be better fitted to the shape of the stack of deion plates. On the other hand the necessary commutation may constitute an additional source of delay [15]. There are also solutions where the fixed runner is not connected with the moving contact, but where this connection has to be made and maintained by an additional arc. The example of Figure 14.22b further has parallel plates, and has vents in the upper wall to relieve the pressure that builds up on arcing.

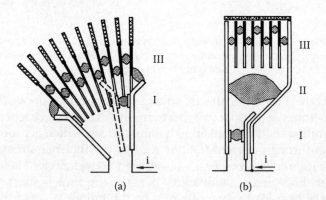

(a)   (b)

FIGURE 14.22
Schematic circuit-breaker arangements: (a) without commutation to a stationary arc runner; (b) with commutation to arc runner connected with the moving contact I, arc between contacts; II, arc in intermediate position; III, arc split up between deion plates.

Figure 14.23 [6] shows some typical shapes of circuit-breaker deion plates. They often surround the arc horns in a U- or V-shape, Figure 14.23a. This increases the magnetic attractive force. Additional slots in the center (Figure 14.23b) or staggered (Figure 14.23c) help to improve splitting at smaller currents.

The number of deion plates in circuit-breakers cannot be increased indefinitely. As explained in Section 14.1.4, sticking and back-commutations associated with voltage limitations may occur, when certain voltage and current levels are exceeded. Then it makes more sense to leave the principle of single break and to arrange two arc chutes in series, despite the other drawbacks. Some basic arrangements, which are increasingly used for quick-acting current-limiting breakers of lower continuous currents, are shown in Figure 14.24. Additionally, arrangements as in contactors (Figure 14.12) are possible.

The differences between rigid nonlimiting and current-limiting breakers lie especially in the contact and mechanical systems. Closed contacts have to withstand two different stresses when a short-circuit current flows across them (see also Sections 10.4 and 10.5):

- Ohmic heating of the constriction resistance with the danger of welding when the melting point is exceeded. The contact force necessary to avoid welding is

$$F_C \geq K_1 \hat{I}^2 \tag{14.2}$$

(a)   (b)   (c)

FIGURE 14.23
Deion plates for low-voltage circuit-breakers: (a) U- or V-shaped recess; (b) with central slot; (c) with staggered slots.

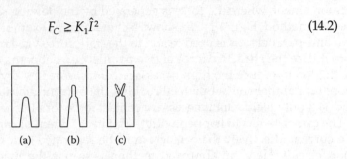

(a)

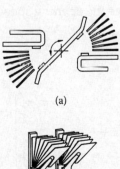

(b)

FIGURE 14.24
Arrangements of double-break systems in circuit-breakers: (a) rotating contact lever with opposite arc chutes; (b) double contact with side-by-side arc chutes.

where $\hat{I}$ is the peak current and K_1 is a constant that contains the electrical and thermal data and the hardness of the contact material [6,67], see also Section 10.5.1.

- Dynamic blow-off owing to the current path in the contact constriction, see Section 10.4. The contact force necessary to counteract the blow-off force is also roughly proportional to the square of the peak current,

$$F_C \geq K_2 \hat{I}^2 \tag{14.3}$$

where K_2 contains the geometry and also material properties.

As K_1 and K_2 are rather similar, it depends on the details of the design and the contact material, and one can hardly predict which of both mechanisms limits the short-circuit withstand capability of a closed circuit-breaker. In any case, the force necessary is a quadratic function of the peak current.

Rigid noncurrent-limiting breakers therefore need high contact forces. The simplest way is to use contact springs with the necessary static contact forces for the highest peak current. This requires, however, a heavy construction for the whole mechanism. Dynamic contact reinforcement, where the force is generated by the flowing short-circuit current, is a more elegant method. Figure 14.25a shows a characteristic example where the repulsion force of two anti-parallel rails is used to add to the static force via a pivot point [3]. A different way is to utilize the attractive force of two parallel current paths for dynamic inrease. Another method is to reduce the high necessary contact force by dividing the current into two or more (n) parallel contact paths. Provided the current distribution is even, Equations 14.2 and 14.3 only yield a total necessary contact force for all n paths $F_{C_n} = K\hat{I}^2/n$.

The opposite has to happen with the contact system of *current-limiting breakers*. To limit the current effectively the contacts must be separated quickly and with little delay upon short circuit. Here, an approved method is to use the magnetic repulsion of two anti-parallel and closely spaced current paths (principle in Figure 14.25b, see also Figure 14.21). This force can be considerably increased by concentrating the self-field of the current path to the moving contact arm where the force is needed. Figure 14.26 shows such a "slot motor" [5], a U-shaped laminated iron core that surrounds the anti-parallel contact arms of each phase. Another method to increase the opening speed that is often used in European

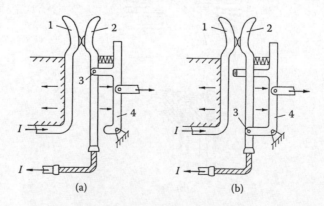

(a) (b)

FIGURE 14.25
Dynamic intensification of (a) contact force and (b) blow-off force: 1, fixed contact piece; 2, moving contact piece; 3, pivot point of moving contact piece; 4, switching lever.

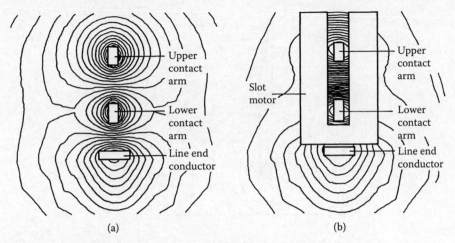

FIGURE 14.26
Function of a slot motor to increase the blow-off force: magnetic field lines (a) without and (b) with slot motor.

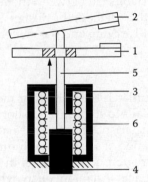

FIGURE 14.27
Principle of solenoid kicker for fast contact opening: 1, stationary contact piece; 2, moving contact piece; 3, stationary magnet yoke; 4, movable plunger; 5, nonmagnetic rod; 6, coil.

current-limiting miniature circuit-breakers uses a solenoid fed by the main current ("solenoid kicker," Figure 14.27 [5,7]). At short circuit its plunger directly acts on the moving contact part by hitting and pushing or pulling it. With such mechanisms speeds over 10 ms^{-1} are attainable at short circuit, while the speed at smaller currents is considerably lower, for example, 1–2 ms^{-1}. At the same time the actuation spring is unlatched by the plunger.

All air circuit-breakers use the magnetic self-field generated by current loops to move the arc off the contacts toward the arc chute (Figures 14.2, 14.14, 14.25). It can be increased by additional ferromagnetic flux concentrators, such as side plates isolated from the arc chamber. As an example, Figure 14.28 gives results of a two-dimensional numerical field calculation for such an arrangement [7]. Only one half of the symmetrical geometry is represented. As the field lines tend to flow within areas of minimal magnetic resistance, a force acts on the arc that moves it upward.

Some circuit-breaker designs of high interrupting capacity use two separate parallel contact paths, one to mainly carry the current during continuous load ("main contacts") and the other to withstand the arcing stress ("arcing contacts"). For this end the main contacts open first on breaking. The current commutates to the arcing contacts. When those open

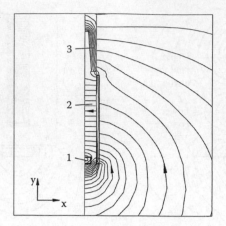

FIGURE 14.28
Increase of magnetic blowout force by ferromagnetic side plates: 1, arc; 2, ferromagnetic side plates; 3, ferromagnetic deion plates.

after a short delay the arc is being established between them. The opposite time sequence applies to the closing operation. By this method, the design and materials selection can be optimized separately for the current-carrying and the switching processes.

14.4.4 Trip System

The trip (= release) system has to unlatch the operating mechanism in the case of prolonged overload (e.g., when a motor fails to start up) and short-circuit current, respectively. A typical current-time tripping characteristic of circuit-breakers is shown in Figure 14.29 [2]. The classical method uses a combination of a directly or indirectly heated bimetal element for the current-dependent overload region, and a magnet in the main current path (or fed through a current transformer) that acts instantaneously when its force exceeds a spring counter-force. As mentioned before, the latter can be combined with direct action on the moving contact in current-limiting breakers. Rigid noncurrent-limiting breakers may be provided with an additional adjustable time delay (marked 2″ in Figure 14.29) of several half cycles or even much longer for the short-circuit protection coordination with other circuit-breakers.

While in miniature circuit-breakers the above-mentioned classical trip devices will prevail for the nearer future, digital electronic solutions have been gaining more importance for larger circuit-breakers. Additional and more complex tripping criteria can be incorporated in the protection characteristic, free from the physical limits of the device, such as the heating time constant of the bimetal. The adaptation to different ratings and overload characteristics does not need different designs, the setting is simpler, and there are no mechanical and thermal tolerances that have to be compensated by individual adjustment of each device.

Electronic thermal overload trip solutions are widely spread in circuit breakers (MCCB) of large frame sizes and high current ratings. Meanwhile also smaller sized Coordinated Switching and Protective devices (CPS) [68] and Motor Protection Circuit Breakers (MPCB) use electronic motor protection circuits down to 4 A [69]. In addition, reduction of power loss and energy efficiency considerations lead to additional functions like power metering and communication. Also in conjunction with fast-acting arc-free current limiters (see 14.4.8), digital rapid fault and short circuit detection is of interest. In addition to the

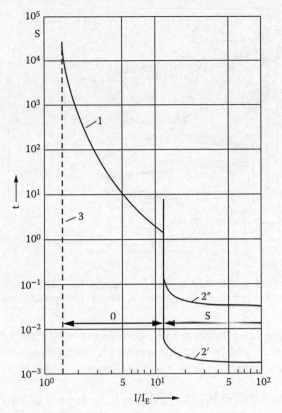

FIGURE 14.29
Trip characteristic for overload and short-circuit protection: 1, overload tripping; 2′, instantaneous short-circuit tripping; 2″, short-circuit tripping with short-time delay; 3, tripping limit for overload; O, overload region; S, short-circuit region.

momentary current, its first time derivative [70], or even both the first and second derivatives [71,72] can form the tripping algorithm. Electronic trip systems are being introduced to detect low current arc faults, which would not trip the old style trip units. This is discussed in Chapter 15.

14.4.5 Examples of Miniature Circuit-Breakers

Miniature circuit-breakers feature the basic attributes treated in Sections 14.4.1–14.4.4. They are used as single-phase units in large quantities in domestic installations, but also for cable or motor protection in industry. The following two examples represent two different philosophies.

The US types (example in Figure 14.30) are current zero switches from their general design, although there is always a current-limiting effect at 120 V system voltage. There is no need for extra-fast tripping and contact opening performance. The arc roots often stay on the contact tips that have to be erosion-resistant (typically WAg), the arc is elongated by contact opening and moderate magnetic blow-out within a simple arc chute. In the example shown, there is only one U-shaped floating iron lining surrounding the contact pieces (see also Section 14.3.3, Figure 14.12b). In other cases only two or three small deion splitter plates are used. Appropriate contact forces and wiping action on make and break help

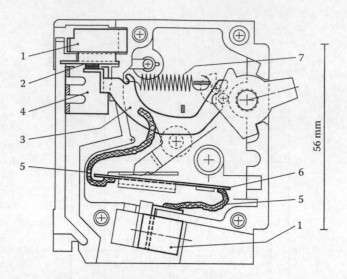

FIGURE 14.30
US type miniature circuit-breaker for 120 V, 10 kA (contacts closed): 1, terminals; 2, fixed contact piece; 3, moving contact piece; 4, U-shaped iron lining; 5, flexible connection; 6, bimetal trip system; 7, magnetic trip system.

to remove the unavoidable tarnish films on W-containing materials. In this example, the spring acts as contact spring and as actuator spring for contact opening as well. Tripping on overcurrent occurs by a bimetal release and the tripping on short-circuit current occurs by a magnetic trip.

The European miniature circuit-breakers (example in Figure 14.31) [73] act as current-limiters at short-circuit currents. The arc is quickly moved off the contacts by a magnetic field generated by a current loop and/or iron flux concentrators, commutated to arc runners, and directed to the arc chute. For this principle, the contact separation must occur extremely fast at current values still sufficiently low to ensure the arc motion necessary for voltage increase (see Section 14.1.4, Figure 14.5), and to enable limitation before the prospective current peak. The arc chute consists of a stack of around a dozen steel plates typically 1mm thick and 1mm apart. Their arc voltage lies at 350 V or higher and therefore exceeds the peak voltage in 230 V single-phase systems. The stack is often arranged so that the arc is subject to a 90° rotation during its motion. This helps to reduce back-commutations between the contacts. The contact points usually consist of AgC, mostly on the fixed contact, and of copper (often with a silver flash, mainly for tarnish protection) on the counter-contact. The graphite content (3–5% by weight) reduces weld forces and also helps to keep the contact resistance low after arcing. The trip system consists on the one hand of a solenoid that simultaneously acts as a kicker (Figure 14.27) to accelerate opening. On short circuit, tripping times below 1 ms and opening speeds up to 10 m/s can be reached by this. Total clearing times may lie well below 5 ms [15]. On the other hand, a directly heated bimetal element trips in the overcurrent range.

14.4.6 Contact Materials

The contact materials used in circuit-breakers are summarized in the lower part of Table 14.3. Their ranking with respect to different stresses can be seen in Table 14.2. In the following discussion, only some special features of asymmetrical combinations that are used in circuit-breakers will be highlighted.

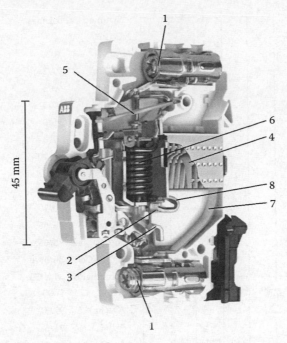

FIGURE 14.31
European type current-limiting miniature circuit-breaker for 230 V, 20 A, 10 kA [73]: 1, terminals; 2, fixed contact piece; 3, moving contact piece; 4, deion arc chute; 5, bimetal release; 6, magnetic release (solenoid kicker); 7, arc runner with flexible connection to moving contact; 8, arc runner connected to fixed contact.

As mentioned before, AgC is often used in European circuit-breakers because of its good anti-weld properties. Its main drawback, especially for current-limiting breakers, is the long dwell time. The following results of basic experiments show how nonsymmetrical combinations with other materials (AgNi 60/40, copper, or WAg) improve the situation. Figure 14.32 shows dwell times under conditions of miniature circuit-breakers [17]. The homogeneous metals Ag, Cu, AgCu own the shortest times; their anti-weld properties are however inferior. The heterogeneous compound material AgC shows by far the longest dwell times, silver/metal oxide compound materials and refractory metals (W, Mo) with silver lie in the middle. Nonsymmetrical combinations between Cu (Ag is similar) and the compound materials shorten the dwell time in all cases. In the same way, the dwell time of AgC is reduced by combination with any other material, whereby Ag and AgCu are best.

As already shown in Chapter 10, contact materials containing tungsten may have a tendency to high contact resistance. The same is true for copper in air. Figure 14.33 demonstrates the effect of AgC with results from a contact test machine, where always the same polarity was switched off [74,75]. Compared to the symmetrical arrangements, the 99.5% resistance values are considerably lower owing to the reducing atmosphere formed by the carbon content during arcing. A clear polarity influence can also be seen. In the case of random polarity the difference would even lie higher.

The effect of AgC in asymmetrical combination on the weld forces after making is demonstrated in Figure 14.34 [74,75]. The 99.5% value of copper lies nearly three times higher than that of either symmetrical AgC or AgC on only one side. The reason for the weld force reduction by the graphite content is believed to lie in the formation of a weak and spongy surface by gaseous reaction products between C and N_2 as well as O_2 from the air [76,77].

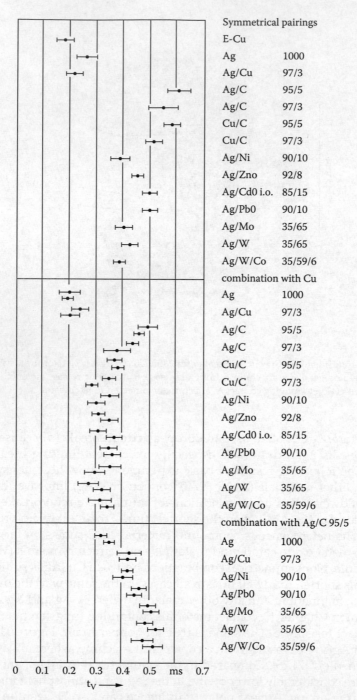

FIGURE 14.32
Dwell time of different contact material combinations under conditions of miniature circuit-breakers (current on contact separation 4 kA).

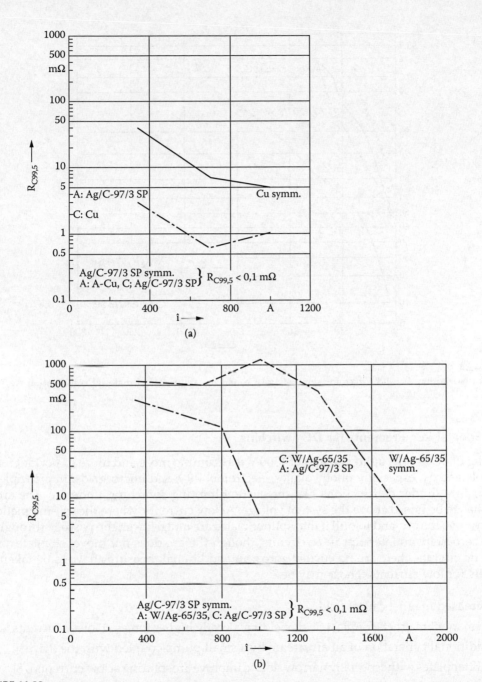

FIGURE 14.33
99.5% contact resistance vs. arc current for symmetrical and nonsymmetrical material combinations: (a) Ag/C against Cu; (b) Ag/C against W/Ag. Opposite polarity: 99.5%-resistance below 0.1 mΩ.

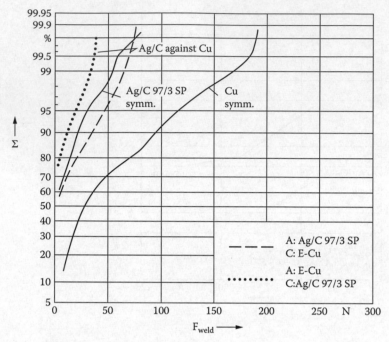

FIGURE 14.34
Statistical distribution of weld forces for symmetrical Cu and nonsymmetrical material combination Ag/C against Cu. $I = 1300$ A, 10 ms.

14.4.7 Special Requirements for DC Switching

Switching of dc circuits up to voltages of 1500 V is becoming more and more important for renewable energy, especially photovoltaics, electromobility, and micro-grids. In principle, arc quenching under dc functions like the current-limiting switching under ac: The arc voltage has to be raised above the system voltage. At low currents, where the self-magnetic forces to move the arc, and to pull it into splitter plates are much lower, the passage through current zero helps to interrupt in ac circuits, though the arc does not move, elongate, or split. In dc circuits, there are no current zeros, and additional measures have to be taken, especially for low currents. Those may be:

- Increased magnetic blast field
- Permanent magnets for blast field, yielding already higher forces at lower currents
- Additional generation of an air stream by a small pump coupled with the drive
- Splitter plates with especially narrow slots to improve arc splitting at low currents [78]
- Double break (see Figure 10.22)
- Hybrid arrangement of metallic contacts and electronic switching element [79]

14.4.8 Current Limitation by Principles Other than Deion Arc Chutes

There are many other principles than the most commonly used deion arc chutes that have been proposed and realized for the interruption of short-circuit currents. The classical ones are included in Figure 14.1a through d). Additionally some alternative methods, all current-limiting, have been investigated or are still under discussion. A few examples are given in the following subsections. Some of them drop out for environmental reasons.

14.4.8.1 Arcs Squeezed in Narrow Insulating Slots

The method of squeezing and elongating arcs between closely spaced insulating walls for voltage increase (Figure 14.1c) has been known for a long time. It needs high magnetic fields. A different way is to force the arc by a moving insulating screen into a narrow insulating slot [80,81]. Though the principle works well and the performance can be easily precalculated [82], it has not been used much so far. Problems lie in the erosion of insulating material, high pressure on the device's casing, insulation failure from arc erosion and metal deposit, and strong plasma jet formation.

14.4.8.2 Reversible Phase Changes of Liquid or Low-Melting Metal

The so-called permanent power fuse uses a fuse conductor of sodium in a special ceramic enclosure, parallel to a resistor [20,83,84]. The resistance increase on melting, vaporizing, and plasma formation is used to limit the current. The circuit is finally opened by a circuit-breaker in series. After cooling down the fuse element is ready for use again. A commercial version was on the market, but could not establish itself. A liquid metal alloy GaInSn within a special enclosure with constrictions acts in a similar way, additionally supported by magnetic pinch forces in the constrictions [85]. Another idea suggested was to use the strong resistivity growth of highly pretensioned Hg (2000 bar), when energy from the short circuit current is fed into it [86].

14.4.8.3 Temperature-Dependent Ceramics or Polymers

Instead of metallic contacts and arcs temperature-dependent elements have been developed that limit the current by strongly increasing their resistivity when a certain temperature is exceeded [87–92]. Such materials are barium titanate and vanadium oxide ceramics, or polymers made conducting with carbon or metal fillers. In the last case, additionally to the volume nonlinearity, the contact interface plays an important role in the limiting effect at high currents [91,92]. In power circuits they must be used in conjunction with parallel resistors to consume the energy stored in the circuit inductance. Additionally, a mechanical switch is needed to interrupt the residual current. Polymer current limiters with Ampere-ratings or lower are widely used in electronic circuits, however, this principle has not found a break-through yet in power circuit protection.

14.4.8.4 Contact Resistance between Powder Grains

Since the contact resistance (constriction resistance) between grains of conducting powders depends on the force pressing them together, a setup has been studied with compressed TiB_2 powder. Fast mechanical load release by magnetic or piezoelectric action can yield fast current-limiting switching [93].

14.4.8.5 Superconductors

The use of the transition from the superconducting state to normal conduction has been suggested and realized long ago. The discovery of high-temperature superconductors that do not need liquid helium but only liquid nitrogen for cooling has initiated new work on this subject worldwide [90]. The emphasis lies on current limiters in medium- and high-voltage systems. Owing to the high expenditure the application for low-voltage systems is unlikely, unless room-temperature superconductors are discovered.

14.5 Simulations of Low-Voltage Switching Devices

The rapid progress in computer technology on the hardware as well as on the software side has enabled more and more numerical simulations for principal studies and for design purposes as well, reducing gradually the necessary amount of experimental studies.

14.5.1 Simulation of Low-Voltage Arcs

14.5.1.1 General, Principle of Simulation

Arcs in low-voltage switching devices in air at atmospheric pressure are complex phenomena, where numerous plasma-physical and electromagnetic processes are closely coupled with each other. See, for example, [1] in Chapter 5, [45,46] in Chapter 9 and [94–101] in this Chapter. Figure 14.35 [102] represents a scheme of the processes in the arc column. Similarly, different coupled processes exist at the interface between the column and the electrodes.

The current density in the plasma generates ohmic heating and, together with the self-magnetic field from the arc plus the field generated from the external current path, brings forth magnetic forces. Both act on the gas flow and energy transport, and thus determine the temperature and pressure distribution. In turn, there results the local distribution of the electrical conductivity, as well as the other material parameters determining the flux flow or radiation behavior.

The numerical arc simulation, which should be in 3D owing to typical low-voltage arc chamber geometries, consists of the following components, which can be formulated as a system of 2nd order partial differential equations:

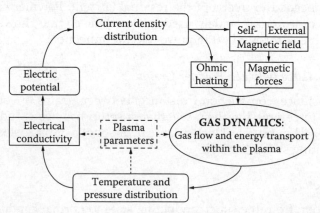

FIGURE 14.35
Scheme of coupled processes in the arc column.

Fluid dynamics of compressible gases (= plasma), so-called Navier-Stokes equations

The plasma is mostly treated in good approximation as a single continuum (fluid) in thermal equilibrium, to which the laws of fluid dynamics apply. Its transport properties depend on the local temperature and pressure. The Navier-Stokes equations consist of

- Mass balance
- Momentum balance
- Power balance (including radiation)
- Further transport equations, when further species such as metal vapor from the electrodes or gaseous decomposition products from the walls are to be modeled.

Maxwell equations (current flow, magnetic field)

It is mostly sufficient to treat both fields as time-stationary, leaving only the diffusive term in equation 14.4.

Further balance equations

For more complex radiation models, one or several additional transport equations may be necessary.

Though quite different in detail, all these equations follow the pattern of a general transport equation, which describes a physical quantity Φ transported at a flow speed $\vec{v}$:

$$\underbrace{\frac{\partial(\rho\Phi)}{\partial t}}_{\substack{\text{Transient}\\\text{Term}}} + \underbrace{\operatorname{div}(\rho\vec{v}\Phi)}_{\substack{\text{Convective}\\\text{Term}}} - \underbrace{\operatorname{div}(\Gamma_\Phi\operatorname{grad}\Phi)}_{\substack{\text{Diffusive}\\\text{Term}}} = \underbrace{S_\Phi}_{\substack{\text{Source}\\\text{Term}}}$$

(14.4)

ρ density, Γ_Φ diffusion coefficient, S_Φ source term.

The transported physical quantity Φ may stand for the speed components, the enthalpy, the electric potential, or the magnetic vector potential components.

Finite Element (FEM) and/or Finite Volume (FVM) software is used to solve this multiphysics problem, nowadays often in a combination of a standard FVM software package for fluid dynamics (e.g., FLUENT, CFX) with an electromagnetics FEM package (e.g., ANSYS).

The simulation of arcs in low-voltage switchgear requires special treatment of details, which can often be simplified in otherwise similar simulations of high-voltage switching arcs (in SF_6) or welding arcs:

- The geometry of arc chutes requires 3D modeling.
- Since arc movement, that is, the permanent new-formation of arc spots in the forward direction along rails or similar electrodes is a necessary prerequisite for functioning of most switching devices, it must be included in the modeling method.
- The electrode regions (cathode and anode) play an important role in low-voltage switching devices. They strongly influence arc movement, formation of new arcs (especially arc splitting between metal plates), the arc voltage, and the power dissipated in the roots.

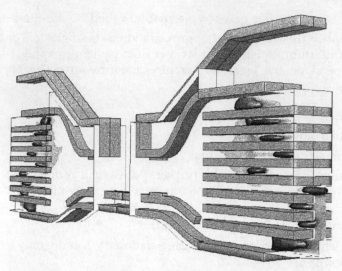

FIGURE 14.36
Example of simulated arcs in a stack of splitter plates [103].

Figure 14.36 depicts an example of simulated arcs in an arc chute with splitter plates from a motor protection circuit breaker [103]. Following are discussed some typical questions and challenges in modeling of low-voltage switching arcs.

14.5.1.2 Arc Roots on Cathode and Anode

As summarized in Sections 9.5.2 and 9.5.3 the processes in the cathode and anode falls, supplying the necessary charge carriers to the arc column and generating a considerable part of the arc voltage, are completely different from those in the column. They take place within a micrometer-scale rather than the millimeter or centimeter range of the arc chamber geometry. The conditions of charge carrier motion are partly free fall rather than random thermal movement, space charges exert a governing influence. There is no thermal equilibrium, but the temperatures of the electrons and of the heavy particles (ions and atoms) are quite different. To incorporate the cathode and anode spots directly into numerical arc simulations is therefore difficult and needs considerable simplifications. The following approaches have been taken:

- No arc root model, direct electrical and thermal contact between metal and plasma. Consequently the arc roots are completely neglected.
- Intermediate grid (3D or simplified as 1D) between metal and plasma in micrometer scale, solving for the balance equations of electron and ion currents, and the power balance in the fall regions [104–106]. This method is very laborious for an arrangement with a dozen or more splitter plates.
- Transfer or transition functions as boundary conditions between different zones, derived from more detailed theoretical considerations [105,107,108]
- A frequently used simple model is a thin intermediate layer or a transfer function on the boundary between metal and arc, modeling the essentially constant voltage

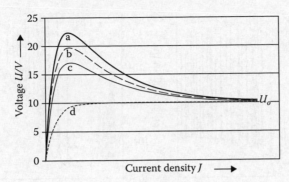

FIGURE 14.37
Variants of $U - J$ charactersistics of cathode or anode fall [110,111].

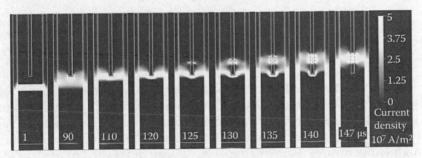

FIGURE 14.38
Simulated arc splitting process at a single metal plate. Characteristic b, current 1 kA RMS [110].

drop across the cathode and anode falls in a static current density–voltage characteristic or an equivalent conductivity characteristic. Dissipated power density as the product between fall voltage and current density has to be split between the electrode and the plasma according to physically reasonable considerations. Figure 14.37 shows several variants of a characteristic, enabling some tuning. Such characteristics are able to model the continuous transition of the arc current from zero to 100%, when the arc is split up in series arcs on splitter plates [109–111]. An example of a simulated arc splitting process on a single metal plate is shown in Figure 14.38 [110]. It agrees in principle with experimental results, the arc bending around first, and the newly formed arc channel overlapping in time with the old one.

14.5.1.3 Radiation

In high current arcs, a major part of the power generated in the arc is removed by radiation. Radiation is a very complex process which can hardly be incorporated into arc simulations in every detail. As a typical example, Figure 14.39 shows the absorption spectrum of air plasma which consists of numerous discrete lines and continuum as well [112]. A simplification often used is the net emission coefficient [113], which describes the difference between emitted and absorbed radiant power in a volume element. It is gained from data like Figure 14.39 by integration over the whole spectrum and over a characteristic arc volume, for example, a ball of 1-cm diameter. It can simply be used as a negative source term in the arc power balance, but is does not take

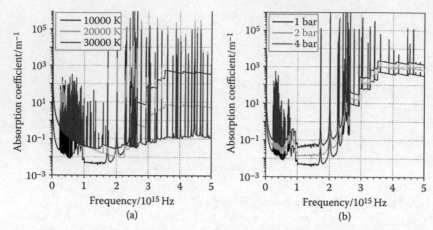

FIGURE 14.39
Absorption spectrum of air plasma [112, taken from 99] a) Variation of temperature, $p = 1$ bar b) Variation of pressure, $T = 10,000$ K.

into account that heat radiated in hotter regions is partly reabsorbed in colder regions, which is a diffusive process like thermal conduction. A different approach is to simplify the radiation completely by a diffusion coefficient [114,115]. This method neglects completely that a considerable part of the radiation leaves the volume unabsorbed and it yields too high temperatures. Many variants have been developed to model radiation better, but simplify the necessary processes of frequency and space integration. They are discussed elsewhere. In comparative simulations the so-called P1 models, which use additional transport equations either for the whole ("gray") or several individual frequency bands ("nongray") proved to show a good compromise between efficiency and precision [116,99].

14.5.1.4 Interaction between Arc and Electrode or Wall Material (Ablation)

The arc heats up the electrodes and the insulating chamber walls. They are partly vaporized and thermally decomposed. The gaseous constituents act back on the plasma properties and hence on the arc behavior. To model these processes correctly, one has to reproduce the heat balance in the solid, and the process of melting and/or vaporization in detail. For this the current density and, hence, the power density acting on the arc roots and radiating to the walls must be known, as well as details of the vaporization or decomposition process [117–120]. Since there are many unknowns, a simplified approach is often used, taking for example, an erosion rate depending on the local power density. It is also common to assume simple organic components such as H_2, PA 66 or POM gas [95,99,121,122] as decomposition products from organic wall materials.

14.5.1.5 Plasma Properties

The strongly nonlinear temperature- and pressure-dependent data for plasma of any known composition can in principle be derived theoretically. For simulations they are usually taken as tabulated functions [123–126]. In comparison with pure air, metal vapors or organic vapors can modify the plasma properties and consequently the arc behavior. The mixing rules for different properties are different and not completely known. It is therefore common to simplify them in a reasonable way [99]. Following are some examples of the influence of admixtures:

- Metal vapor increases the electrical conductivity, especially shifts its onset to lower temperatures. This has a positive effect on the splitting into partial arcs [127]. On the other side the arc voltage is reduced, and the danger of unwanted back-commutations behind the arc is increased [99,128].
- Metal vapor increases the net emission coefficient of radiation [129,130].
- In agreement with experiments, outgassing plastics vapor (PA 66) cools and compresses the arc and speeds up arc motion and splitting. Furthermore the modified radiation properties increase emission from the hot arc core, as well as absorption near the cold walls [131].

Further data about the effect of metal admixtures on plasma data are given in [132,133].

14.5.1.6 Simplification by Porous Media

The last sections have shown that many very detailed physical processes have to be taken into account, if the simulation is required to be very close to reality. Still all of them need more or less simplifications. A different approach into the other direction is taken in [134,135]. Instead of modeling every single anode and cathode sheath in a chute with splitter plates, the whole stack is replaced by a continuum with anisotropic properties, where the electrical sheath characteristics are averaged. Also all other properties have to be averaged in a suitable way. It has been shown that this method is feasible in principle. Since this method is based on stronger simplifications, some adjustments with experiments are necessary.

14.5.2 Further simulations of contact and switching device behavior

Many more phases in the function of contacts and arcs in switching devices are nowadays simulated in the course of basic investigations or design studies. They are often combined to simulate a whole system or parts of it. Here, only a few keywords are summarized:

- Magnetic fields and forces [136].
- Heat balance of switchgear under nominal current and overcurrent load [137].
- Kinematics of actuators including impacts, magnetic forces, contact lift-off forces [137,138].
- Numerical models of contact morphology and contact resistance, utilization and advancement of Holm's contact models by digital means [139,140].
- Thermal models of contacts under arcing influence, arc erosion, thermally induced mechanical stresses [117–119].
- Switching device—mechanics, arc voltage—in interaction with the surrounding electric network [141].

14.6 Vacuum Interrupters

This is only a short overview of switching in vacuum. Special reference is given to the book "The Vacuum Interrupter: Theory, Design and Application" by Paul Slade [142].

This section first describes the design principle of vacuum interrupters and of low- and medium-voltage vacuum switching devices. The arc constriction at higher currents necessitates either radial magnetic fields to rotate the arc, or axial fields to keep it diffuse. Some switching results with these constellations are given. The contact material plays an outstanding role because the vacuum plasma is entirely fed from the contact metal. Its influence on current interruption under low-frequency and high-frequency conditions, as well as its influence on premature current cessation (chopping) are discussed.

14.6.1 Principle/Applications

The principle of interrupting ac currents by vacuum arcs (see Section 9.7.3) is mainly used in medium-voltage circuit-breakers, predominantly at voltage ratings to 40.5 kV. In this range the high electric strength of the vacuum at short gap distances shows most advantages over other quenching principles such as SF_6 or oil. Vacuum breakers for voltages of 72 kV–145 kV are gradually encroaching on SF6 breakers, because of environmental concerns. Designers seem to be overcoming the problem of the disproportionately long gap distance necessary to withstand the voltage stress. The vacuum principle is furthermore applied in various contactor types for low-voltage (≤ 1 kV) and medium-voltage applications, as well as in medium-voltage load switches. The application in low-voltage circuit-breakers could not be established on a broad basis yet.

The active interrupter units of these devices are vacuum vessels shown in principle in Figure 14.40. They consist of a fixed and a movable contact, the latter fed into the vacuum vessel by means of a metal bellows. The contacts are especially shaped to generate

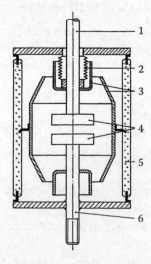

FIGURE 14.40
Principle of vacuum interrupter: 1, movable contact rod; 2, metal bellows; 3, vapor shield; 4, contacts; 5, glass or ceramic cylinder; 6, fixed contact rod.

magnetic fields (Section 14.6.3). One or more vapor shields act as condensation areas for the plasma and metal vapor produced during switching, and they serve to prevent the insulator from being coated with metal and to prevent other parts such as the bellows from being damaged [143,144].

14.6.2 Design

Figure 14.41 depicts some major design variants of vacuum interrupters, (a) and (b) differ by the arrangement of the necessary insulator gap and the central vapor shield; (c) is a simpler design used in interrupters for lower voltages, where the shield is directly connected to one of the poles. The high vacuum (typically lower than 10^{-5} mbar = 10^{-3} Pa after fabrication) over a lifetime of decades requires special materials and manufacturing processes [143].

A typical example of a low-voltage vacuum contactor is shown in Figure 14.42 [45]; medium-voltage contactors are similar. The actuator principle is equivalent to the scheme of Figure 14.6c with a lever between the magnet and the interrupter. Differently from air

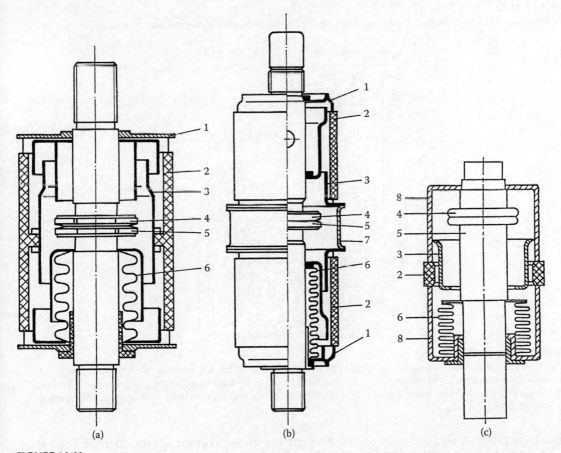

(a)   (b)   (c)

FIGURE 14.41
Different vacuum interrupter designs: (a) one insulator with floating shield inside; (b) two separate insulators with floating shield between; (c) LV contactor interrupter, shield connected to fixed contact. 1, metal cover; 2, insulator; 3, vapor shield; 4, fixed contact; 5, movable contact; 6, metal bellows; 7, metallic center ring; 8, housing.

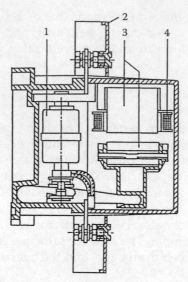

FIGURE 14.42
Low-voltage vacuum contactor: 1, vacuum interrupter; 2, terminal cover; 3, magnet core; 4, coil.

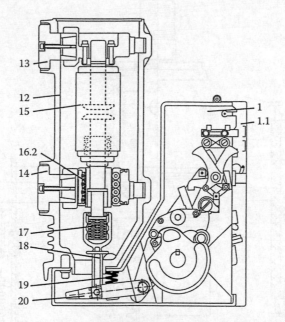

FIGURE 14.43
Medium-voltage vacuum circuit-breaker, poles mounted in insulator tube: 1, breaker mechanism housing; 1.1, front panel, removable; 12, insulating material pole tube; 13, upper breaker terminal; 14, lower breaker terminal; 15, vacuum interrupter; 16, roller contact; 17, contact pressure spring; 18, insulated coupling rod; 19, opening spring; 20 shift lever pair.

break contactors they need only one contact gap per phase to interrupt the current. Typical contact strokes lie at a few millimeters; the opening and closing speeds are $1~\text{ms}^{-1}$ or lower.

Figure 14.43 represents a medium-voltage vacuum circuit-breaker, whose actuator, as in low-voltage circuit-breakers, is spring-driven. In this special design the interrupter poles are arranged within tubes of epoxy resin [145]. Besides insulating these tubes take up the

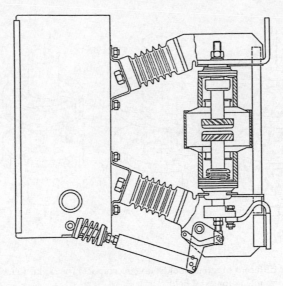

FIGURE 14.44
Medium-voltage vacuum circuit-breaker, poles mounted type insulators.

impact forces during switching and keep them off the interrupter vessel. A different variant is shown in Figure 14.44, where each interrupter pole is mounted on two post-type insulators [146]. In this case the forces are taken up by a separate insulating rod mounted parallel to the vacuum vessel. An alternative drive for vacuum circuit breakers uses a bi-stable electromagnet [147].

Depending on the rated voltage and breaking capacity, the final gap length usually lies at 1 cm and higher. For voltages over 50 kV several centimeters are necessary.

14.6.3 Recovery and the Influence of the Design

As discussed in Chapter 9, Section 9.7.3, the vacuum arc gap recovers rapidly if the arc has been staying diffuse or has gone diffuse again before current zero, and if the production of metal vapor that could be re-ionized under the influence of the transient recovery voltage (TRV) has ceased. On the other hand, vacuum arcs between plain electrodes become constricted and heavily melt large areas of the contact surface, when momentary currents of several kiloamperes are exceeded. The cooling time constants of these strongly vaporizing metal pools are large (millisecond range [148]) and may result in interruption failure after current zero.

There are two different methods used to overcome this problem. Both use magnetic fields generated by the current path through especially shaped contact structures:

- The first is to accept the arc constriction but to minimize its negative effect by forcing the arc to rotate instead of dwelling in one place. Thus, the arc energy is distributed evenly across the whole contact area. This is achieved by radial magnetic fields, by which, together with the axial current path, a magnet force $F = I \times B$ is exerted peripherally on the arc, see also Section 9.6.3. Figure 14.45a and b shows two typical examples of such structures [149]. Figure 14.45a represents so-called cup electrodes that have slanting slots in the outer cylinder, forming U- or V-shaped current paths, and a smooth contact ring without notches. The spiral electrodes, Figure 14.45b, are formed by appropriate slots in the contact

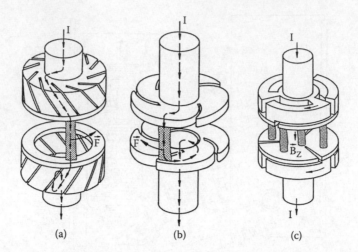

FIGURE 14.45

Vacuum contact structures for different magnetic fields: (a) cup contacts for radial field; (b) spiral contacts for radial field; (c) three thirds of a winding for axial field.

plate. Again a U-shaped current loop is formed, acting on the arc in the sense of increasing the loop (compare Figure 14.2). In this case the arc has to jump across the slot when it reaches the tip of a spiral. This is not a serious problem.

- The second way is to use axial magnetic fields that keep the arc diffuse to higher currents by forcing the charge carriers along the magnetic field lines (see Section 9.6.4). In addition to the even arc distribution, the arc voltage and thus the power transfer to the contacts and shield are lower [149–152]. A simple but space-consuming method to achieve this is an outside coil around the vacuum vessel [153]. Other solutions use more or less sophisticated constructions of the current path behind the contact plates [151] or they direct axial fields to the gap by ferromagnetic flux concentrators [153]. Figure 14.45c shows an example where the current path behind each electrode forms three thirds of a coil winding. In Figure 14.46, the distribution of the axial magnetic field component across the center plane of the contact gap, calculated by the three-dimensional finite element method (FEM), is shown [154]. The magnetic field is considerably enhanced by radial slots in the contact plate. On the one hand they improve the circular current flow necessary for the axial field, on the other they reduce eddy currents that cause a phase shift of the field against the current, with the consequence of a residual field at current zero [155,156]. This field reduces the effective mean free path of the charge carriers and may increase the reignition tendency. Figure 14.47 demonstrates the influence of radial slots by the FEM-simulated time-evolution of the center axial field without and with slots [154]. The higher field magnitude as well as the lower phase shift at current zero ($t = 10$ ms) becomes evident.

A comparison of interruption results with a demountable experimental vacuum chamber is given in Figure 14.48 for different contact configurations [152]. The "reignition voltage" plotted is either the breakdown voltage when the failure occurred in the first rising slope of the TRV, or the TRV peak in cases of successful interruptions or of late restrikes after the first peak. Because there is always a portion of reignitions after the first peak, the plot does not necessarily mean that the circuit-breaker will interrupt whenever the actual

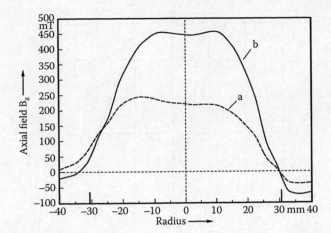

FIGURE 14.46
FEM-calculation of the local distribution of axial field at $t = 5$ ms for contacts like Figure 14.45c: (a) without slots; (b) with three radial slots in the contact plates. Diameter 62 mm, current shape: half sinewave 30 kA r.m.s.

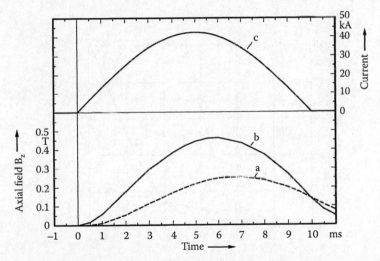

FIGURE 14.47
Simulated time evolution of axial field in the center for contacts like Figure 14.45c: (a) without slots; (b) with three radial slots in the contact plates; (c) current. Conditions as in Figure 14.46.

TRV peak stays below the curve of minimum reignition voltage, but it is a clear indication of the ranking of different configurations. While a sharp voltage decrease happens above 10 kA for the plain contacts and above 20 kA for the spiral contacts with moving constricted arc, the axial field contacts keep a higher voltage limit up to the highest current investigated. Results from postarc current measurements are shown in Figure 14.49 [152,154]. The lower postarc current of the axial field arrangement indicates less residual plasma around current zero at higher currents.

Of course the magnetic field configuration is not the only criterion determining the breaking capacity. Currently there are optimized interrupters of both principles on the market that do not substantially differ in size for the same high interruption capability [157]. Axial field contacts seem to have some advantages at the high-voltage end of the range.

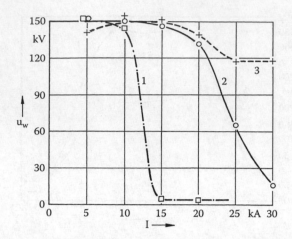

FIGURE 14.48
Lower boundary of reignition voltages for different contact arrangements vs. r.m.s. current. 50Hz, TRV frequency ≈ 20 kHz, TRV peak ≈ 160 kV; 1, plain cylinder; 2, spiral contacts; 3, axial field contacts; all contact diameters 60 mm, CuCr 75/25, gap distance at current zero 1 cm.

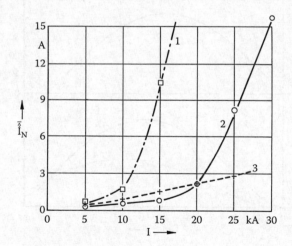

FIGURE 14.49
Postarc current maximum for different contact configurations vs. r.m.s. current: 1, plain cylinder; 2, spiral contacts; 3, axial field contacts; conditions similar to Figure 14.48.

Other factors influencing the current interruption are the same that determine the cold gap breakdown voltage, for example, smoothness of the contours of contacts and other parts stressed by the electric fields of the TRV, or cleanliness. The shape and size of the vapor shield also play an important role for the interruption behavior of vacuum breakers [158,159]. On the one hand a shield of larger diameter yields lower field strength of the partial gaps (contact–shield–contact). On the other hand the amount of residual plasma and therefore the intensity and duration of postarc current is increased by a wider shield. The shield potential follows the anode potential longer, which causes more uneven voltage distribution [159]. Also the heat capacity of the shield and its capability to take up the arc energy has an influence [158,159].

14.6.4 Contact Materials for Vacuum Interrupters and Their Influence on Switching

14.6.4.1 Requirements

Because the medium in which switching takes place is entirely formed by the contact material and because it acts back on the quality of the vacuum, the requirements for contact materials in vacuum interrupters are even more complex than for other principles [143,144,160]. The major ones are:

- Low gas content and gettering effect to keep the vacuum during the whole service life.
- Low erosion rate. Eroded contacts cannot be replaced during service life.
- Low welding tendency and weld forces. Welding may be especially critical because
  - There are no tarnish films, therefore tendency toward cold welding.
  - There is no wiping contact action and no torsion to break up welds.
  - There is a detrimental influence of welds on dielectric properties [161].
- Good interruption capability.
- Low chopping level.
- Low tendency to generate high-frequency transients.
- Low heat generation on current flow. As neither convection nor thermal conduction takes place within the vacuum, the heat can only leave the interrupter by thermal conduction through the contact rods.
- High electrical breakdown strength.

These requirements lead to consequences for some contact material properties:

- High electrical conductivity to reduce heat generation
- High thermal conductivity for good heat removal during continuous load and for fast electrode cooling after current zero
- Smooth surfaces after arcing
- Minimizing of particle formation on arcing which might lead to late restrikes [162,163]

The above requirements are partially contradictory, and the choice and development of contact materials is always a compromise.

The following groups of contact materials have been used for switching in vacuum [143,164–167]:

1. Materials produced by smelting:
   a. Single-component materials, for example, pure copper, are scarcely used; they hardly meet the requirements.
   b. Alloys such as copper with small additions of bismuth or tellurium for the reduction of weld tendency. These materials have been used in circuit-breakers by some manufacturers, but are now considered obsolete having been replaced almost universally by the Cu-Cr material.

2. Heterogeneous compound materials of two or more components with different melting and boiling points, mostly manufactured by sintering and/or infiltration techniques, such as

   a. Combination of refractory component (W, WC) with lower-melting component, especially WCu and WCAg, for switches with a high number of switching operations but moderate breaking capacity (load break switches and contactors). Similar compounds like WAg or MoCu have also been developed and investigated, but are less common.

   b. CuCr with compositions between 75/25% and 50/50% by weight. CuCr has proved optimum behavior in many respects for circuit-breakers: High breaking capacity, gettering effect, smooth surface after arcing with subsequent high electrical strength, moderate chopping current. To achieve special properties, further components such as high vapor pressure metals to influence the chopping and high frequency behavior (e.g., bismuth, antimony, zinc), are possible.

3. Other interesting suggestions concerned single-component materials or alloys with high vapor pressure to reduce the chopping current, for example, antimony, as solid depots in holes or grooves of erosion-resistant contact pieces (e.g., of molybdenum), so-called "low-chop" contacts [167–170]. Their interruption capacity is rather low, and they are not common.

In all cases, the manufacturing method and its parameters, size and distribution of the components, as well as minor additives or impurities, may decisively influence the performance of vacuum contact materials.

The following sections deal with the influence of contact material on the performance of vacuum arcs in the vicinity of current zero. Its influence on arc voltage, arc erosion, welding, contact resistance, and dielectric properties is treated elsewhere [168,171].

14.6.4.2 Arc Interruption

Figure 14.50 shows a simplified scheme of the coupled processes at cathode spots (see also Figure 9.46). Materials that own a high reluctance to re-establish these spots after current zero lead to a high breaking capacity. Therefore the electron emission and ionization

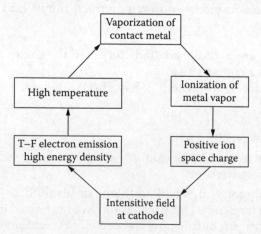

FIGURE 14.50
Simplified scheme of the processes in a vacuum cathode spot.

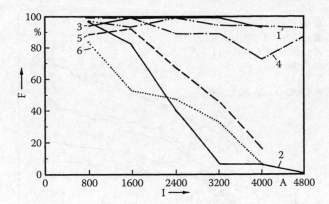

FIGURE 14.51
Frequency of successful interruptions vs. RMS current for different contact materials: 1, CuCr 75/25; 2, CuCr 33/67; 3, CuCrZn 65/24/11; 4, CuCrSb 68/24/8; 5, CuCrLi$_2$O 76/23.6/0.4; 6, WCu 70/30; contact diameter 15 mm, distance 1.5 mm; 50 Hz, f_{TRV} = 16–33 kHz, first TRV peak 35 kV.

parameters, the speed of cooling down, the readiness to vaporize or not, and the microstructure after arcing (smooth surface or sharp edges) play an important role.

Figure 14.51 summarizes results of model experiments with contacts of reduced size and different contact materials [172,173]. The percentage of successful interruptions is plotted versus the r.m.s. current. All results show the same pattern. While at low currents the interruption rate is equal or close to 100%, it decreases when a material dependent limit is exceeded. CuCr 75/25 without third components shows a nearly 100% interruption rate over the whole range investigated, whereas the approximately opposite CuCr composition (33/67) leads to a high number of failures even at low currents. The addition of the high vapor pressure metal zinc (it forms brass together with copper) has no deteriorating effect under the investigated conditions, while antimony, also with high vapor pressure, already reduces the ability to interrupt. Small additions of alkali or alkali oxide (Li$_2$O) result in considerably reduced breaking capacity, owing to the low work function and ionization potential. WCu also shows rather limited breaking capability. This is owing to agglomerates of tungsten that are formed by repeated arcing and that, owing to high melting and evaporation temperatures of tungsten, favor thermionic emission. Additionally it was found that the materials with the best performance had the smoothest surfaces after arcing, while the others showed sharp cracks and edges.

14.6.4.3 Interruption of High Frequency Transients

The fast recovery which enables high breaking capacity under line frequency conditons (50 or 60 Hz) can lead to unwanted voltage transients when the breaker successfully clears high frequency (HF) currents that are superimposed to the line frequency current under certain conditions ([142]; Section 5.3). This may occur when the gap distance at current zero is still too short to withstand the TRV. The reignition may initialize a HF current, and when the gap successfully clears in one of the HF current zero passages, a sequence of undesirable voltage escalations ("multiple reignitions," "virtual chopping") could start, depending on the circuit conditions. To avoid them, a poor interruption capability for HF currents would be ideal, while at the same time the high breaking capacity at line frequency should be unchanged [169,173–176].

Figure 14.52 summarizes statistical distributions of the HF reignition voltages of different contact materials [173,176]. For most materials at least two populations can be

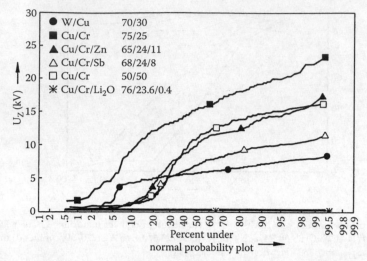

FIGURE 14.52
Statistical distribution of HF reignition voltages U_z for different contact materials after arcing with $I_{Peak} = 200$ A, $f = 200$ kHz. Gap distance 0.1 mm, $f_{TRV} = 1.8$ MHz.

distinguished, representing different reignition mechanisms. One approaches the cold gap breakdown, the other is governed by the residual plasma. CuCr 75/25 without additives exhibits the highest values. Additions of low boiling point metals (Zn, Sb) yield a reduction but cannot really avoid HF current interruptions. As under line frequency conditions a higher chromium content increases the reignition tendency. WCu again shows a rather low tendency to interrupt, and alkaline additions yield practically zero reignition voltage. A comparison of HF and line frequency data shows that the contact materials behave rather synchronously, and that the ideal behavior described above can hardly be reached.

In practice, today's contact materials do not require special precautions against HF transients in most load cases. For especially critical load situations, surge protection measures help to suppress overvoltages ([142]; Section 5.3).

14.6.4.4 Current Chopping

When the balance of charge carrier generation (Figure 14.50) is disturbed owing to lack of power input into the cathode region, the current may abruptly cease to flow ("chop") before the natural current zero [168,169,177,178,142; Sections 2.3.3 and 3.2.6]. The simplest equivalent circuit to explain the consequences is like Figure 9.53, but with a rigid voltage source on the left side. The current I_{ch} stored in the circuit inductance L at the moment of chopping commutates to the capacitance C' which is always present in a circuit (e.g., winding capacitance of a motor) and initiates an oscillation. The maximum voltage appearing across the load is roughly $U_{max} = I_{ch}Z$, where $Z = \sqrt{L/C}$ is the surge impedance. Depending on the circuit parameters it may constitute a direct overvoltage problem or favor the conditions for the generation of multiple reignitions or virtual chopping (see "interruption of high frequency transient"). These events may be critical on switching low inductive currents. The chopping current of vacuum contact materials, which in turn depends on the composition as well as on the circuit parameters, should not be too high. As explained in detail in ([142]; Section 3.2.6), much attention was given to the phenomenon of current chopping during

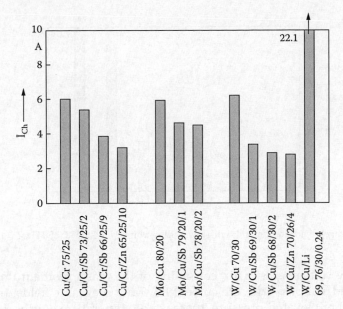

FIGURE 14.53
Chopping currents (average) of different heterogeneous contact materials. $U = 400$ V, $f = 50$ Hz, $I = 45$ A, $Z = 1000\ \Omega$.

the early stages of vacuum interrupter development, initiated by experiences with the old copper-bismuth, but is considered not critical with today's contact materials.

Chopping currents show different statistical distributions depending on composition, materials homogeneity, and circuit parameters [178,179]. Figure 14.53 compares average chopping currents of different compound materials [168,169]. First it is remarkable that each of them has a lower chopping level than their pure constituents. While copper lies around 15 A and tungsten as well as chromium may chop at even higher currents, their mixtures lie in the area of 6 A. At least for WCu this may be explained by strong vaporization of copper while the refractory tungsten skeleton essentially stays unaffected. The depletion of individual components may change the chopping level gradually during service life. Several per cent of a metal with low boiling point and high vapor pressure, respectively, such as antimony or zinc, clearly reduce the chopping level, but not drastically. Against first expectations a small amount of additive with low work function and low ionization potential (lithium) strongly increases the chopping level. WCAg 60/40, which is not shown, has a chopping level of only 2 A under the conditions of Figure 14.53.

Figure 14.54 shows the chopping levels of concentrated low boiling point metals [169]. They lie much lower, but their use is restricted to low currents because of their poor breaking capability. Again calcium, an earth alkali metal with low boiling point, low work function, and low ionization potential, is an exception with very high values. The influence of these and other material parameters on the chopping level may be explained by a model of the power and particle flux balance of the cathode region taking into account the processes of Figure 14.50 [169].

14.6.5 Simulation of Arcs in Vacuum Interrupters

Similar to switching arcs for low voltage in air or for medium and high voltage in SF_6, numerical simulations have gained more importance as study and design tools for vacuum interrupters. They have a great deal in common with those described in

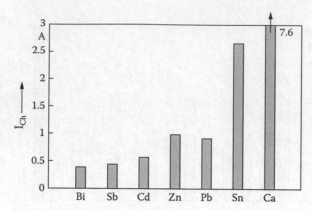

FIGURE 14.54
Chopping currents (average) of different metals with low boiling point. $U = 400$ V, $f = 50$ Hz, $I = 45$ A, $Z = 1000$ Ω.

Section 14.5 for low voltage arcs in air, consisting of the mass, momentum, and power balances of the fluid, coupled with the current flow and magnetic fields. In contrast to plasmas at ambient or elevated pressure, there is no temperature equilibrium between the species of the plasma, and they need appropriate formulations to take this into account. Often, the cathode and anode regions are more or less simplified. To determine the production of metal vapor at the electrodes, which strongly influences the arc shape and behavior, the heat balance at the electrodes must also be considered. Those simulations help to better understand the arc behavior, and to optimize designs, for example, the local distribution of the axial magnetic field. Without going into details, few references are given:

- Basics [180,181]
- Axial magnetic fields [180–183]
- Transverse magnetic fields [184]

References

1. TE Browne Jr. *Circuit Interruption, Theory and Techniques*. New York: Marcel Dekker, 1984.
2. M Lindmayer. *Schaltgeräte; Grundlagen, Aufbau, Wirkungsweise*. Berlin: Springer-Verlag, 1987.
3. J Slepian. Theory of the deion circuit-breaker. *Trans AIEE* 48:523–527, 1929.
4. G Burkhard. Über das Lichtbogenverhalten in Löschblechkammern und deren Bemessung. Thesis TH Ilmenau, 1962.
5. A Lee, PG Slade. Molded-case low-voltage circuit-breakers. In ThE Browne Jr (ed.) *Circuit Interruption. Theory and Techniques*. New York: Marcel Dekker, 1984, p 550.
6. A Erk, M Schmelzle. *Grundlagen der Schaltgerätetechnik*. Berlin: Springer-Verlag, 1974.
7. M Lindmayer, H Stammberger. Application of numerical field simulations for low-voltage switchgear. *IEEE Trans CPMT* 18(3):708–717, 1995.
8. M Lindmayer. Über die Vorgänge bei der Lichtbogenlöschung in kompakten Löschblechkammern bei Wechselströmen zwischen 2.5 und 8.5 kA. Thesis TU Braunschweig, 1972.
9. R Michal. Theoretical and experimental determination of the self-field of an arc. *Proceedings of the 26th Holm Conference on Electrical Contacts*, Chicago, 1980.

10. K-H Schroder. Das Abbrandverhalten offnender Kontaktstiicke bei Beanspruchung durch magnetisch abgelenkte Starkstromlichtbogen. Thesis TU Braunschweig, 1967.
11. W Rieder. Interaction between magnet-blast arcs and contacts. *Proceedings of the 28th Holm Conference on Electrical Contacts*, Chicago, 1982.
12. M Lindmayer. The influence of contact materials and of chamber wall materials on the migration and the splitting of the arc in extinction chambers. *IEEE Trans PHP* 9:45–49, 1973.
13. D Amft. Über die Lichtbogenwanderung im Bereich geringer Geschwindigkeiten. *5 Int. Tagung über Elektrische Kontakte*, Miinchen, 1970.
14. W Widmann. Arc commutation across a step or a gap in one of two parallel copper electrodes. *Proceedings of the 13th Holm Conference on Electric Contact Phenomena*, Chicago, 1984.
15. H Weichert. Lichtbogenwanderungs- und Kommutierungsvorgänge in strombegrenzenden Löschkammeranordnungen. Thesis TU Brauschweig, 1988.
16. H-W Sudhölter. Über den zwischen divergierenden Elekroden wandernden Lichtbogen bei Strömen bis 90 000 A. Thesis TU Braunschweig, 1975.
17. N Behrens. Lichtbogenwanderung in Leitungsschutzschaltern. Thesis TU Braunschweig, 1980.
18. M Lindmayer. Simulation of stationary current-voltage characteristics and of back-commutation in rectangular arc channels. *Proceedings of the 17th International Conference on Electrical Contacts*, Nagoya, 1994.
19. K-H Schröder. Kontaktverhalten und Schalten in der elektrischen Energietechnik-Grundlagen und Anwendungsbeispiele. 7. Seminar Kontaktverhalten und Schalten, Karlsruhe, 1983.
20. FW Kussy, JL Warren. *Design Fundamentals for Low-voltage Distribution and Control*. New York: Marcel Dekker, 1987.
21. M Poniatowski, K-H Schröder. Einfluß des Herstellverfahrens auf das Schaltverhalten von Kontaktwerkstoffen der Energietechnik. *7th International Conference on Electric Contacts*, Paris, 1974, pp 38–45, 477–483.
22. W. Böhm, N Behrens, M Lindmayer. The switching performance of an improved AgSnO$_2$ contact material. *Proceedings of the 27th Holm Conference on Electrical Contacts*, Chicago, 1981.
23. Y-S Shen, WD Cote, LJ Gould. A historic review of Ag-MeO materials. *Proceedings of the 32nd IEEE Holm Conference on Electric Contact Phenomena*, Boston, 1986.
24. W. Rieder. Silber/Metalloxyd-Werkstoffe für elektrische Kontakte. VDE-Fachbericht (VDE-Verlag Berlin) 42:65–81, 1991.
25. A Kraus, R Michal, KE Saeger. Schaltverhalten von Silber/Zinnoxid für Anwendungen im Bereich kleiner bis mittlerer Leistung. VDE-Fachbericht (VDE-Verlag Berlin) 42:83 87, 1991.
26. P Braumann, T Warwas. Analyse des Einschaltvorganges von Motorschaltern für die Entwicklung von Kontaktwerkstoffen. VDE-Fachbericht (VDE-Verlag Berlin) 42:47–57, 1991.
27. NEMA ICS 2-2000. Industrial control and systems controllers, contactors, and overload relays rated 600 Volts.
28. UL508, equivalent to CSA 22.2 No.14. Industrial control equipment.
29. UL489, equivalent to CSA 22.2 No. 5-09, Molded-case circuit breakers, molded-case switches, and circuit-breaker enclosures.
30. UL98, equivalent to CSA 22.2 No. 4. Enclosed and dead-front switches.
31. IEC 60947-4-1. Low-voltage switchgear and controlgear; Part 4-1: Electromechanical contactors and motor-starters. Equivalent: DIN VDE 0660. Teil 102, Niederspannungs-Schaltgeräte: Elektromechanische Schütze und Motorstarter.
32. Part -2: Circuit breakers
33. Part -3: Load switches
34. R Schnabel. Schütze nach NEMA und IEC: Philosophie der Realität. *Elektrische Energie-Technik* 31:8–10, 1986.
35. Sprecher + Schuh (Now Rockwell Automation). Various technical information brochures on contactors. Aarau, Switzerland.
36. M Lindmayer. Low voltage contactors: Design and materials selection considerations. *Proceedings of the 16th International Conference on Electrical Contacts*, Loughborough, 1992.
37. F Völker. Ein Verfahren zur Vorhersage des dynamischen Betriebsverhaltens von Schützen mit elektromagnetischem Antrieb. Thesis Universität/GH Duisburg, 1990.

38. AEG. ELSA, the contactor coil unit of the future. *Technical brochure*, Neumünster, 1990.
39. J Slepian. Extinction of an AC arc. *Trans AIEE* 47:1398–1407, 1928.
40. M Schmelzle. Grenzen der Selbstlöschung kurzer Lichtbogenstrecken bei Wechselstrombelastung. Thesis TH Braunschweig, 1968.
41. R Amsinck. Das Löschverhalten von Wechselstromlichtbögen in Schützen bei Ausnutzung der Säulenkühlung. Thesis TU Braunschweig, 1978.
42. N Reimann. Das Stromnulldurchgangs-Verhalten von Niederspannungs-Schaltlichtbögen unter Berücksichtigung von Kathodenschicht und dynamischer Lichtbogensäule. Thesis TU Braunschweig, 1986.
43. M Lindmayer. Effect of work function and ionization potential on the reignition voltage of Ag metal oxide contacts. *Proceedings of the 26th Holm Conference on Electrical Contacts*, Chicago, 1980.
44. A Takahashi, M Lindmayer. Reignition of arcs on double-break contacts. *IEEE Trans CHMT* 9:35–39, 1986.
45. Siemens. 3TF AC contactors. *Technical brochure*, Erlangen.
46. AEG. Low-voltage switchgear. Contactors and overload relays. *Technical Brochure*, Neumünster, 1988.
47. M Huck, A Kraus, R Michal, FJ Wagner. Silber/Zinnoxid in Relais und in kleinen Schaltgeräten. VDE-Fachbericht (VDE-Verlag, Berlin) 40:71–78, 1989.
48. P Braumann. Grenzen von Kontaktwerkstoffen der Energietechnik. VDE-Fachbericht, (VDE-Verlag, Berlin) 40:79–89, 1989.
49. P Braumann, J Lang. Kontaktverhalten von Ag-Metalloxiden für den Bereich höherer Ströme. VDE-Fachbericht (VDE-Verlag, Berlin) 42:89–94, 1991.
50. S Greitzke. Untersuchungen an Hybridschaltern. Thesis TU Braunschweig, 1988.
51. W Rieder, V Weichsler. Make erosion mechanism of Ag/CdO and Ag/SnO$_2$ contacts. *Proceedings of the 37th IEEE Holm Conference on Electrical Contacts*, Chicago, 1991.
52. P Braumann, A Koffler, K-H Schröder. Analysis of the interrelation between mechanical and electrical phenomena during making operations of contacts. *Proceedings of the 17th International Conference on Electrical Contacts*, Nagoya, 1994.
53. K-H Schröder. The future of electrical switching contacts under the influence of electronics. *34th IEEE Holm Conference on Electrical Contacts*, San Francisco, 1988, pp 3–11.
54. G Friedrich. *Hilfsschütze—eine überholte Technologie?* Bonn: Klöckner-Moeller, 1990.
55. Th Schoepf. Interplay between electromechanical and solid state switching technologies for meeting cost, sustainability, and safety demands of various applications. *54th IEEE Holm Conference on Electrical Contacts*, Orlando, 2008.
56. H Weichert. Halbleiter vs. Elektromechanik - Konkurrenz oder Ergänzung? *Electrosuisse Fachtagung ITG-Hardware-Technology, Kontakte und Schaltelemente*, Wintherthur, Switzerland 2010.
57. Siemens. 3RW Soft Starters. https://eb.automation.siemens.com/goos/catalog/Pages/ProductData.aspx?catalogRegion=WW&language=en&nodeid=10024029&tree=CatalogTree &regionUrl=/#topAnch&activetab=product& (04-03-2012)
58. Phoenix Contact. Contactron hybrid motor starters. http://www.phoenixcontact.de/produkte/54143_54157.htm (04-03-2012)
59. I Hackel. Mit dem Profibus gegen das Chaos. Siemens Drive & Control No 1:22–23, 1991.
60. D Amft, M Lindmayer. Schalt- und Schutzgeräte für Elektroantriebe. *12 Int. Fachtagung Industrielle Automatisierung—Automatisierte* Antriebe, Chemnitz, 1993.
61. V Glocke. Verfahren zur Bereichserweiterung der Betriebsspannung für ein elektromagnetisches Schaltgerät. German Patent Application DE 4003 179 A1, 1990.
62. JA Bauer, DA Mueller, RT Basnett, JC Engel. Electromagnetic contactor with algorithm controlled closing system. US Patent 4.833.565, May 23, 1989.
63. H Meyer. Sanfteinschaltung bei Schützen mit Kontaktstücken aus AgSnO$_2$. VDE-Fachbericht (VDE-Verlag, Berlin) 42:97, 1991.

64. IEC 60947–2: Low-voltage switchgear and controlgear; Part 2: circuit-breakers. Equivalent: DIN VDE 0660, Teil 101, Niederspannungs-Schaltgeräte; Teil 2: Leistungsschalter.

65. IEC 898: Circuit-breakers for overcurrent protection for household and similar installations. Equivalent: DIN VDE0641, Teil 11, Leitungsschutzschalter für den Haushalt und ähnliche Anwendungen.

66. EN 60934/VDE 0642, Geräteschutzschalter. Equivalent: IEC 934: Circuit-breakers for equipment.

67. R Holm. *Electric Contacts*. Berlin: Springer-Verlag, 1967.

68. IEC 60947-6-2. Multiple function equipment - Control and protective switching devices (or equipment)

69. Eaton. Motor protective circuit breakers PKE. http://www.eaton.eu/Europe/Electrical/ ProductsServices/CircuitProtection/MotorProtectiveCircuitBreakers/MotorProtective CircuitBreakerPKE/index.htm (04-03.2012).

70. M Stege. Kurzschluss-Erkennungsalgoritzmen zum strombegrenzenden Schalten. PhD thesis TU Braunschweig, 1992.

71. T Mützel, F Berger M Anheuser. Methods of Early Short-Circuit Detection for Low-Voltage Systems. *54th IEEE Holm Conf. on Electrical Contacts*, Orlando, 2008.

72. T Mützel. Verfahren der Kurzschlussfrüherkennung zur Verbesserung der strombegrenzenden Wirkung von Niederspannungs-Leistungsschaltern. PhD thesis TU Ilmenau, 2008.

73. ABB. Technische Daten. Sicherungsautomaten Baureihe S 200, S 200 M, S 200 P und S 280 UC. http://www05.abb.com/global/scot/scot209.nsf/veritydisplay/7368200c84b92309c12572eb0 031cc69/$file/2CDC002071D0102.pdf (04-05-2012).

74. N Behrens, M Lindmayer. Der Einffuß von Kontaktwerkstoffen auf Lichtbogenverharrzeit, Schweißkraft und Kontaktwiderstand unter Berücksichtigung unsymmetrischer Werkstoffpaarungen. Seminar Kontaktverhalten und Schalten, Karlsruhe, 1977.

75. M Lindmayer, K-H Schröder. The effect of unsymmetrical material combination on the contact and switching behavior. *Proceedings of the 9th International Conference on Electric Contact Phenomena*. Chicago, 1978.

76. R Michal, KE Saeger. Metallurgical aspects of silver-based contact materials for air-break switching devices for power engineering. *IEEE Trans CHMT* 12, 1989.

77. E Vinaricky. Das Abbrand und Schweißverhalten verschiedener Silber–Graphit–Kontaktwerkstoffe in unterschiedlichen Atmosphären. PhD thesis TU Wien, 1994.

78. V v.Döhlen. Lichtbogenverhalten beim Ausschalten von Gleichströmen. Experimente und Simulationen. PhD thesis TU Braunschweig, 1996.

79. M Wiersch. PV-Anlagen sicher trennen. *ETZ* S2: 26–29, 2011.

80. E Belbel, M Lauraire. A new breaking technology, independent from the contact material. *Proceedings of the 13th International Conference on Electrical Contacts*. Lausanne, 1986.

81. Z Huang, M Lindmayer. Current limiting switching by squeezing arcs into narrow insulating slots. *IEEE Trans CHMT* 15:160–165, 1992.

82. Z Huang, M Lindmayer. Switching arcs squeezed in rectangular insulating slots – scaling laws and comparison with measurements. *European Trans on Electric Power* 1:107–111, 1991.

83. T Ito, T Miyamoto, N Takano. Perminant power fuse self-recovering current limiting device. *IEEE Summer Power Meeting, Paper No. 70 CP 601-PWR*, July 1970.

84. T Itoh, T Miyamoto, Y Wada, T Mori, H Sasao. Design considerations on the permanent power fuse for a motor control center. *IEEE Trans PAS*, 92:1292–1297, 1973.

85. A Krätzschmer, F Berger, P Terhoven, S Rolle. Liquid metal current limiters. *20th Int. Conf. on Electrical Contacts*, Stockholm, 2000.

86. HCW. Gundlach. A self-restoring current-limiting device. *3rd Int. Symp. on Switching Arc Phenomena*, Lodz, Poland, 1977.

87. T Hansson, P-O Karlström. Device for motor and short-circuit protection. *Int. Pat. Publication WO 91/12643*, 1991.

88. FA Doljack. PolySwitch PTC devices—A new low-resistance conductive polymer-based PTC device for overcurrent protection. *IEEE Trans CHMT* 4:372–378, 1981.

89. Th Hansson. Polyäthylen-Stromwächter für den Kurzschlußschutz. ABB Technik, No 4: 35–38, 1992.

90. M Lindmayer, M Schubert. Current limitation by high temperature superconductors and by conducting polymers. *Proceedings of the 5th International Conference on Electrical Fuses and their Applications*, Ilmenau, 1995.

91. AR Duggal, LM Levinson. A novel high current density switching effect in electrically conductive polymer composite materials. *J.Appl. Phys.* 82: 5532–5539,1997.

92. A Wabner. Beitrag zur Kurzschlussstrombegrenzung mit leitfähigen Polymercompounds in der Niederspannungsebene. PhD thesis TU Chemnitz, 2001.

93. J Isberg, H Bernoff, L Liljestrand, L Jonsson, J Karlsson. A new powder based ac/dc switching technology. *20th Int. Conf. on Electrical Contacts*, Stockholm, 2000.

94. F Karetta. Dreidimensionale Simulation wandernder Schaltlichtbögen. PhD thesis TU Braunschweig, 1998.

95. LZ Schlitz. Simulation of gas dynamics and electromagnetic processes in high-current arc plasmas. PhD thesis, Univ. of Wisconsin, Milwaukee, 1998.

96. A Hauser, DW Branston. Numerical simulation of a moving arc in 3D. *17th Int. Conf. on Gas Discharges and their Applications*, Cardiff, 2008.

97. M Lindmayer. Lichtbogensimulation – Ein Überblick. VDE Fachbericht 65, VDE-Verlag Berlin, Offenbach, 2009.

98. M Anheuser, T Beckert, S Kosse. Electric arcs in switchgear – theory, numerical simulation and experiments. *19th Symp. on Physics of Switching Arc*, Brno, Czech Republic, 2011.

99. Ch Rümpler. Lichtbogensimulation für Niederspannungsschaltgeräte. PhD thesis TU Ilmenau, 2009.

100. R Gati, J Ostrowski, D Piva. Comparison of arc simulations with experiments using a simplified geometry. VDE Fachbericht 67, VDE-Verlag Berlin, Offenbach, 2011.

101. H Stammberger, Th Daube, C Dehning, M Anheuser. Arc simulations in realistic low-voltage arcing chambers. *21st Int. Conf. on Electrical Contacts*, Zürich, 2002.

102. F Karetta, M Lindmayer. Simulation of arc motion between divergent arc runners. *19th Int. Conf. on Electric Contact Phenomena*, Nürnberg, Germany 1998; VDE Fachbericht 51, VDE-Verlag Berlin, Offenbach, 1998.

103. Ch Rümpler, A Zacharias. Simulation des Schaltlichtbogens am Beispiel eines Motorschutzschalters. *24th CADFEM Users' Meeting*, Stuttgart, 2006.

104. P Zhu, JJ Lowke, R Morrow. A unified theory of free burning arcs, cathode sheaths and cathodes. *J. Phys. D: Appl. Phys* 25: 1221–1230, 1992.

105. A Kaddani. Modélisation 2D et 3D des arcs électriques dans l'argon à la pression atmosphérique avec la prise en compte du couplage thermique et électrique arc-électrodes et de l'influence des vapeurs métalliques. Thèse de doctorat de l'Université Pierre & Marie Curie Paris 6, 1995.

106. A Spille-Kohoff. Numerische Simulation des ChopArc Prozesses mit CFX-5. In: Final Report „ChopArc MSG-Lichtbogenschweißen für den Ultraleichtbau". Fraunhofer IRB Verlag, Stuttgart 2005. http://www.cfx-berlin.de/de/entwicklung/choparc/download/cfx_040916 .pdf (March 17, 2012).

107. JJ Lowke, R Morrow, J Haidar. A simplified unified theory of arcs and their electrodes. *J. Phys. D: Appl. Phys.* 30: 2033–2042, 1997.

108. J Wendelstorf. Ab Initio modeling of thermal plasma gas discharges (electric arcs). PhD thesis TU Braunschweig, 2000.

109. M Lindmayer, E Marzahn, A Mutzke, Th Rüther, M Springstubbe. The process of arc splitting between metal plates in low voltage arc chutes. *50th IEEE Holm Conf. on Electrical Contacts*, Seattle, 2004.

110. A Mutzke, Th Rüther, M Kurrat, M Lindmayer, E-D Wilkening. Modeling the arc splitting process in low-voltage arc chutes. *53rd IEEE Holm Conf. on Electrical Contacts*, Pittsburgh, 2007.

111. A Mutzke. Lichtbogen-Simulation unter besonderer Berücksichtigung der Fußpunkte. PhD thesis TU Braunschweig, 2009.

112. V Aubrecht, M Bartlova, O Coufal. Radiation transfer in air thermal plasmas. *16th Int. Conf. on Gas Discharges and Their Applications*, Xi'an, China, 2006.

113. RW Liebermann, JJ Lowke. Radiation Emission Coefficients for SF6 Arc Plasmas. *J. Quant. Spectrosc. Radiat. Transfer* 16: 253–264, 1976.

114. SV Dresvin (Ed.). *Physics and Technology of Low-Temperature Plasmas*. Ames, USA: Iowa State University Press, 1977.

115. R Siegel, JR Howell, J Lohrengel. *Wärmeübertragung durch Strahlung*, Teil 3. Berlin: Springer-Verlag, 1993.

116. C Lüders. *Vergleich von Strahlungs- und Turbulenzmodellen zur Modellierung von Lichtbögen in SF6-Selbstblasleistungsschaltern*. PhD thesis RWTH Aachen, 2005.

117. Ph Testé, T Leblanc, R Andlauer. An original method to assess the surface power density brought by an electric arc of short duration, and short electrode gap to the electrodes - Case of copper electrodes. *47th IEEE Holm Conference on Electrical Contacts*, Montreal, 2001.

118. F Pons, M Cherkaoui. An electrical arc erosion model valid for high current: vaporization and splash erosion. *54th IEEE Holm Conference on Electrical Contacts*, Orlando, 2008.

119. M Lindmayer. Simulation von Erwärmung, Abbrand und mechanischer Beanspruchung elektrischer Kontakte beim Schalten. VDE Fachbericht 65, VDE-Verlag Berlin, Offenbach, 2009.

120. M Anheuser, C Lüders. Numerical arc simulations for low voltage circuit breakers. *19th Symp. on Physics of Switching Arc*, Brno, Czech Republic, 2009.

121. Ch Rümpler, H Stammberger, A Zacharias. Low–voltage arc simulation with out–gassing polymers. *57th IEEE Holm Conference on Electrical Contacts*, Minneapolis, 2011.

122. Q Yang, M Rong, Y Wu. The simulation and experimental research on air arc plasma considering wall ablation in low-voltage circuit breakers. *23rd Int. Conf. on Electrical Contacts*, Sendai, Japan, 2006.

123. RN Gupta, K-P Lee, RA Thompson, JM Yos. Calculations and curve fits of thermodynamic and transport properties for equilibrium air to 30000 K, *Reference Publication 1260*, NASA Langley Research Center, Hampton, VA, 1991.

124. AB Murphy. Transport coefficients of air, argon-air, nitrogen-air, and oxygen-air plasmas. *Plasma Chemistry and Plasma Processing* 15, 1995.

125. M Capitelli, G Colonna, C Gorse, AD D'Angola. Transport properties of high temperature air in local thermodynamic equilibrium. *Eur. Phys. J. D* 11: 279–289, 2000.

126. S Selle. *Transportkoeffizienten ionisierter Spezies in reaktiven Strömungen*. PhD thesis Univ. Heidelberg, 2002.

127. F Yang, M Rong, Y Wu, AB Murphy, J Pei, L Wang, W Liqiang, L Zengchao, L Yiying. Numerical analysis of the influence of splitter-plate erosion on an air arc in the quenching chamber of a low-voltage circuit breaker. *J. Phys. D: Appl. Phys.* 43, 2010.

128. M Lindmayer, A Mutzke, Th Rüther. Messungen und Simulation von Schaltlichtbögen in Löschblechkammern. VDE-Fachbericht 61, VDE Verlag Berlin, Offenbach, 2005.

129. Y Cressault, R Hannachi, Ph Teulet, A Gleizes. Influence of metallic vapours on thermal plasmas properties: Application to Air - Fe/Cu/Ag mixtures. *XVI International Conference an Gas Discharges and their Applications*, Xi'an, China, 2006.

130. V Aubrecht, M Bartlova, O Coufal. Radiative emission from air thermal plasmas with vapour of Cu or W. *J. Phys. D: Appl. Phys.* 43, 2010.

131. Ch Rümpler, H Stammberger, A Zacharias. Berücksichtigung gasender Kunststoffe bei der Simulation des Schaltlichtbogens. VDE-Fachbericht 67. VDE-Verlag · Berlin, Offenbach 2011.

132. Y Cressault, A Gleizes. Calculation of diffusion coefficients in air–metal thermal plasmas. *J. Phys. D: Appl. Phys.* 43, 2010.

133. AB Murphy. The effect of metal vapor in arc welding. *J. Phys. D: Appl. Phys.* 43, 2010.

134. J Riß, M Lindmayer, M Kurrat. Considerations to simplify the numerical gas flow simulation of low voltage arcs. *19th Symp. on Physics of Switching Arc*, Brno, Czech Republic, 2011.

135. J Riß, M Lindmayer, M Kurrat. Simplification of the Arc Splitting Process in Numerical Gas Flow Simulations. *XXth Int. Conf. on Electrical Contacts*, Beijing, 2012.

136. H Stammberger. Magnetfeld- und Kraftberechnungen für strombegrenzende Niederspannungs-Schaltgeräte. PhD thesis TU Braunschweig, 1995.

137. F Barcikowski. Numerische Berechnungen zur Wärme- und Antriebsauslegung von Schaltgeräten. PhD thesis TU Braunschweig, 2003.

138. R Kalms. Simulation von Antrieben und Auslösern für Niederspannungsschaltgeräte. PhD thesis TU Braunschweig, 2000.

139. SV Angadi, WE Wilson, RL Jackson, GT Flowers, BI Rickett. A multi-physics Finite Element model of an electrical connector considering rough surface contact. *54th IEEE Holm Conf. on Electrical Contacts*, Orlando, 2008.

140. P Lindholm. Numerical study of asperity distribution in an electrical contact. *57th IEEE Holm Conf. on Electrical Contacts*, Minneapolis, 2011.

141. H Stammberger, H Pursch, A Zacharias, P Terhoeven. Simulation of the temporal behavior of circuit breakers and motor starters. *22nd Int. Conf. on Electrical Contacts*, Seattle, 2004.

142. PG Slade. *The Vacuum Interrupter - Theory, Design, and Application*. Boca Raton: Taylor & Francis Group, 2008.

143. M Lindmayer, L Schiweck. Vakuumschalterentwicklung und damit verbundene Werstoff-Fragen. ETG-Fachtagung "Mit Energie in die Zukunft," Nürnberg, 1986.

144. M Lindmayer. Das Vakuumschaltprinzip. *Vakuum-Technik* 36:101–108, 1987.

145. ABB Calor Emag. VD4 vacuum circuit-breaker. *Instruction manual*, Ratingen, 1993.

146. Siemens. Mittelspannung Vakuumschalttechnik, Leistungsschalter 3AF. *Technical Brochure*, Erlangen, 1991.

147. E Dullni. A vacuum circuit-breaker with permanent magnetic actuator for frequent operations. *18th Int. Symp. on Discharges and Electrical Insulation in Vacuum*, Eindhoven, the Netherlands, 1998.

148. G Frind, JJ Caroll, CP Goody, EJ Tuohy. Recovery times of vacuum interrupters. *IEEE Trans PAS* 104:283–288, 1987.

149. Fr-W Behrens. Über den Einfluß der Elektrodengeometrie auf das Ausschaltverhalten von Vakuumleistungsschaltern. PhD thesis TU Braunschweig, 1984.

150. WGJ Rondeel. The vacuum arc in an axial magnetic field. *J. Phys. D: Applied Phys.* 8:934–942, 1975.

151. S Yanabu, E Kaneko, E Okumura, T Aioshi. Novel electrode structure of vacuum interrupter and its practical application. *Proceedings of the IEEE PES Summer Meeting*, Minneapolis, 1980.

152. F Unger-Weber, Wiederverfestigung des Hochstrom-Vakuumbogens bei höheren Spannungen. PhD thesis TU Braunschweig, 1988.

153. H Schellekens, K Lenstra, J Hilderink, J ter Hennepe, J Kamans. Axial magnetic field type vacuum circuit-breakers based on exterior coils and horse shoes. *Proceedings of the 12th International Symposium on Discharges and Electrical Insulation in Vacuum*. Shoresh, Israel, 1986.

154. B Fenski, M Lindmayer. Vacuum interrupters with axial field contacts—3D Finite Element simulations and switching experiments. *Proceedings of the 17th International Symposium on Discharges and Electrical Insulation in Vacuum*, Berkeley, California, 1996.

155. T Sohme, H Toda, E Kaneko, T Tamagawa, S Yanabu. Basic characteristics of axial magnetic field electrode for vacuum circuit-breakers. *Proceedings of the International Symposium on Switching Arc Phenomena*, Lodz, Poland, 1980.

156. Y Kurosawa, S Sugawara, Y Kawakubo, N Abe, H Tsuda. Vacuum circuit-breaker electrode generating multi-pole axial magentic field. *IEEE PES Winter Meeting*, New York, 1980.

157. RK Smith. Tests show the ability of vacuum circuit-breaker to interrupt fast transient recovery voltage rates of rise of transformer secondary faults. *IEEE Trans PD* 10:266–273, 1995.

158. PG Slade, RE Voshall. Effects of arc shield proximity to the electric contacts on the current interruption capability of vacuum interrupters. *E&I* 107:138–140, 1990.

159. V Biewendt. Einfluß der Schirm- und Kontaktgeometrie auf das Löschverhalten von Vakuumleistungsschaltern. PhD thesis TU Braunschweig, 1993.

160. PG Slade. Vacuum interrupters, the new technology for switching and protecting distribution circuits. *IEEE Trans Industry Application*, 33(6): 1501–1511, Nov/Dec. 1997.

161. W Widl. Contact welding and field emission breakdown in vacuum. *ETZ-Archiv* 4:12–17, 1982.

162. L Cranberg. The initiation of electrical breakdowns in vacuum. *J. Applied. Phys.* 23:518–522, 1952.

163. S Anders, B Jüttner, M Lindmayer, C Rusteberg, H Pursch, F Unger-Weber. Vacuum breakdown with microsecond delay time. *IEEE Trans El* 28:461–467, 1993.

164. PG Slade. Contact materials for vacuum interrupters. *IEEE Trans PHP* 10:43–47, 1974.

165. PG Slade. Advances in material development for high power vacuum interrupter contacts. *IEEE Trans CPMT*, part A 17(1):96–106, 1994.

166. H Kippenberg, W Kuhl, W Schlenk. Kontaktmaterial für Vakuumschalter. *Siemens Energie und Automation* 7:76, 1986.

167. E Vinaricky (Ed.). *Elektrische Kontakte, Werkstoffe und ihre Anwendungen*. Berlin, Heidelberg: Springer-Verlag, 2002.

168. U Reininghaus. Schaltverhalten unterschiedlicher Kontaktwerkstoffe im Vakuum. PhD thesis TU Braunschweig, 1983.

169. L Czarnecki. Einfluß des Kontaktwerkstoffes auf Stromabriß und Löschung des Vakuumbogens. PhD thesis TU Braunschweig, 1986.

170. D Sämann. Schaltverhalten von Vakuumkontaktstücken mit Antimon-Einlagerungen, PhD thesis TU Braunschweig, 1984.

171. K Fröhlich, HC Kärner, D König, M Lindmayer, K Möller, W Rieder. Fundamental research on vacuum interrupters at Technical Universities in Germany and Austria. *IEEE Trans El* 28:592–606, 1993.

172. D Heyn. Untersuchungen Zum Einfluß des Kontaktwerkstoffes auf das Löschverhalten im Vakuum. PhD thesis TU Braunschweig, 1990.

173. D Heyn, M Lindmayer, E-D Wilkening. Effect of contact material on the extinction of vacuum arcs under line frequency and high frequency conditions. *IEEE Trans CHMT* 14:65–70, 1991.

174. J Panek, KG Fehrle. Overvoltage phenomena associated with virtual current chopping in three-phase circuits. *IEEE Trans PAS* 94:1317–1325, 1975.

175. J Helmer, M Lindmayer. Mathematical modeling of the high frequency behavior of vacuum interrupters and comparison with measured transients in power systems. *Proceedings of the 17th International Symposium on Discharges and Electrical Insulation in Vacuum*. Berkeley, 1996.

176. E-D Wilkening. Vorgänge im Hochfrequenz-Stromnulldurchgang von Vakuumstrecken. Thesis TU Braunschweig, 1991.

177. GA Farrall. Current zero phenomena. In *Vacuum Arcs, Theory and Application*. (Editor: J.M. Lafferty) New York: Wiley, 1980, pp. 184–227.

178. G Philip. Untersuchungen über das statistische Verhalten der Abreißströme von Vakuumschaltern. PhD thesis TU Wien, 1984.

179. L Czarnecki, M Lindmayer. Influence of contact material properties on the behavior of vacuum arcs around current zero. *Proceedings of the International Conference on Electrical Contacts, Electromechanical Components and their Applications*, Nagoya, 1986.

180. E Schade, DL Shmelev. Numerical simulation of high-current vacuum arcs with an external axial magnetic field. *IEEE Trans. Plasma Sci.* 31: 890–901, 2003.

181. DL Shmelev. Low current vacuum arcs in strong axial magnetic field. *21st Int. Symp. on Discharges and Electrical Insulation in Vacuum*, Yalta, Ukraine, 2004

182. W Hartmann, A Hauser, A Lawall, R Renz, N Wenzel. The 3D numerical simulation of a transient vacuum arc under realistic spatial AMF profiles. *14th Int. Symp. on Discharges and Electrical Insulation in Vacuum*, Braunschweig, Germany, 2010

183. S Jia, L Zhang, L Wang, B Chen, Z Shi, W Sun. Numerical simulation of high-current vacuum arcs under axial magnetic fields with consideration of current density distribution at cathode. *IEEE Trans. Plasma Sci.* 39: 3233–3243, 2011.

184. T Delachaux, O Fritz, D Gentsch, E Schade, DL Shmelev. Simulation of a high current vacuum arc in a transverse magnetic field. *IEEE Trans. Plasma Sci.* 37: 1386–1392, 2009.

15

Arc Faults and Electrical Safety

John J. Shea

...... should I escape thy flame,
   Thou wilt have helped my soul from Death:

The Dark Angel, **Lionel Johnson**

Stay alert, you can observe anything by watching,

Yogi (Lawrence Peter) Berra

CONTENTS

15.1 Introduction 849
15.2 Arc Fault Circuit Interrupters (AFCIs) 850
15.3 Arcing Faults 853
    15.3.1 Short-Circuit Arcing 854
    15.3.2 Series Arcing 855
15.4 Glowing Connections 857
15.5 Arcing Fault Properties 865
    15.5.1 Frequency 865
    15.5.2 Electrode Materials 866
    15.5.3 Arc Fault Current 867
    15.5.4 Cable Impedance and Cable Length Effects 868
15.6 Other Types of Arcing Faults 871
15.7 Conclusions 872
References 872

15.1 Introduction

Arcing faults in electrical wiring can occur for a wide variety of different reasons [1]. Many times arcing faults can lead to electrically initiated fires, especially in residential settings, where loss of life is more likely than in an office or industrial setting (75% of all fire deaths originating from electrical causes occur in a residential setting) [1]. 47% of electrical fires occur from "behind-the-wall" installed wiring owing to connections while 11% occur from "temporary" extension cords and plugs owing to misuse (overloaded, insulated) or abused (insulation damage, cut, crushed, pinched) wire, as seen in Figure 15.1 [2,3]. While other conditions exist, these two examples combine for the majority of causes of electrical fires and, as such, are areas for further investigation [2–4].

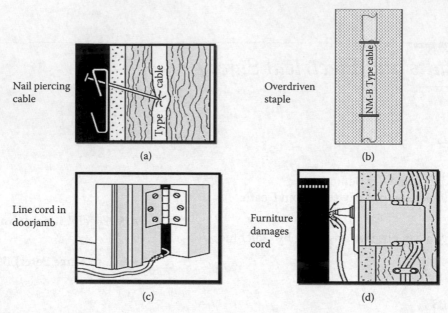

FIGURE 15.1
Examples of potentially hazardous for behind-the wall and exposed wiring (a) nail pierces behind-the-wall wire insulation, (b) over-driven wire staple, (c) cord crushed in door jamb, (d) cord pinched by furniture.

Since the late 1980s, research on mitigating residential electrical fires produced from electrical arcing faults has led to the development of the first commercially available Arc Fault Circuit Interrupter (AFCIs) in 1999 [5–8]. These circuit protection devices are designed to mitigate the effects of electrical arcing faults that are not detected by ordinary thermal-magnetic (T-M) circuit breakers or other types of circuit protection devices [5–10]. These new circuit protective devices have the potential to reduce property and equipment damage and minimize personnel injury by their ability to sense various types of arcing faults [5–10]. Today, AFCI technology is an actively researched field with many companies and universities pursuing advancements in detection methodology, testing standards, and basic arc physics to advance electrical safety [9, 11–17].

Because arcing fault properties are the key to the development of these protective devices, three types of fault conditions will be examined along with a basic description of how AFCI breakers operate and how wiring and loads can affect arc detection. Since residential applications are currently the only commercial application of this technology, the scope of this work covers residential applications on the basis of North American standards with some coverage on commercial and industrial arcing faults.

15.2 Arc Fault Circuit Interrupters (AFCIs)

The branch feeder (B/F) AFCI, commercially introduced in 1999, effectively lowered the magnetic instantaneous trip threshold of a thermal-magnetic (T-M) breaker from the typical range of $200A_{rms}$ to $250A_{rms}$ down to $75A_{rms}$ (as seen in Figure 15.2) while still retaining the standard thermal-magnetic overload trip features of a traditional breaker [5,8–10]. However, the B/F device had limitations in that it could not sense arcing faults below its detection limit of about $75A_{rms}$.

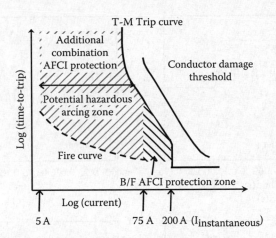

FIGURE 15.2
Additional protection zones provided by the B/F and combination breakers are indicated by the shaded areas. The fire curve is described in the following section.

The now outdated B/F device was replaced by the combination AFCI breaker which provides both branch feeder protection (i.e., behind-the-wall wiring having two wires with ground) and outlet protection (two wire cord), essentially combining both branch feeder and outlet protection, hence the name combination. A combination AFCI breaker has the ability to detect arcing faults without the need for ground fault, detecting arcing faults down to as low as $5A_{rms}$ as shown in Figure 15.2 [18–22].

One of the biggest challenges for designers of these devices was in the ability to discriminate between arcing faults and loads that normally arc during operation (e.g., motors, certain types of lighting, switches, etc.) and loads that have high di/dt [20]. Frequently, highly nonlinear currents with series arcing can be difficult to distinguish from the normal waveform without arcing. Figure 15.3 illustrates some examples of current waveforms from some loads which inherently produce nonsinusoidal current.

In designing AFCI breakers and other types of protective devices, the HF current from the arc is utilized [23] in algorithms, processed with microcontrollers, to identify arcing faults [18–22]. These algorithms generally use a combination of characteristics inherent in arcs including threshold signals obtained from broadband RF current generation produced from the arc and duration patterns. Since the focus of this section is on arcing fault phenomena, specific algorithms are not described in detail but rather can be found in the patent and nonpatent literature [18–23].

Commercially available combination AFCIs generally use a measure of the power frequency current and a bandpass filtered RF current signature to make a trip/no trip decision on the basis of proprietary algorithms [18–22]. A simplified block diagram of the RF sensing method is shown in Figure 15.4. Generally some means of bandpass filtering is used to only pass specific frequency range(s) of interest, consisting of one or more frequency bands. By measuring the amplitude and duration of the RF current at specific frequency band(s), a software algorithm can be used to determine if an arcing fault is present.

For currents below the instantaneous magnetic trip threshold of a breaker, decision times (i.e., time to trip) can be found in the latest UL1699 standard [24]. This standard has many requirements, including arcing tests used to determine if the AFCI device trips fast enough to meet minimum standards. One of the tests is named the carbonized path clearing time test because the test consists of creating a carbonized path or carbon bridge in SPT-2 lamp cord. This carbon path is intended to replicate the conditions formed when

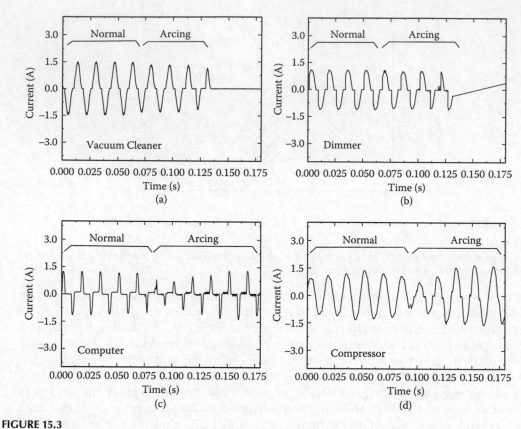

FIGURE 15.3
Illustrations showing how difficult it can be to detect series arcing faults for various nonsinusoidal loads; (a) Vacuum Cleaner; (b) Dimmer; (c) Computer; (d) Compressor.

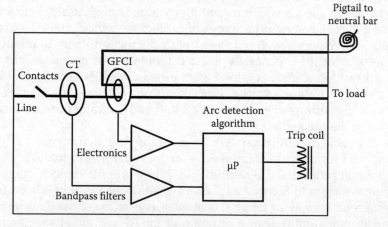

FIGURE 15.4
Block diagram showing RF current spectrum method of arc detection.

a broken conductor arcs while under load resulting in carbonized insulation [24]. Wire samples are prepared per the standard, by cutting the insulation across a lamp cord (SPT-2), exposing the copper wire strands and then wrapping the cut portion with electrical tape and cloth tape to replicate a repaired wire. HV is applied to the wire to create a carbonized path across the cut in the insulation, creating a wire with arcing damage. After the HV

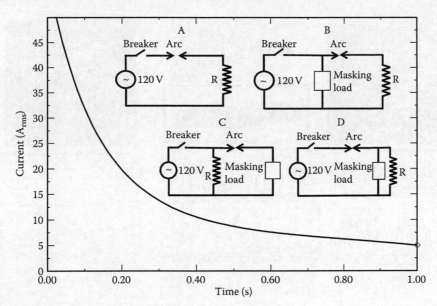

FIGURE 15.5
UL1699 "fire" curve. (Curve and circuits taken from *UL Standard for Safety for Arc-Fault Interrupters*, UL1699 rev. 4, Feb. 11, 2011.)

conditioning requirements have been meet, 120 Vac is applied to the wire sample at a specified current to initiate arcing. Continuous series arcing must occur for the test to be declared valid and the AFCI device must trip within a specified time per the standard [24].

The higher the current, the faster the breaker must trip to decrease the probability of starting a fire owing to the higher arc energy. It has been shown that the probability of igniting a fire under these conditions increases with increased arcing current [24]. On the basis of experimental data, a "fire" curve (Figure 15.5) was developed that indicates the maximum allowable time for AFCI device trip as a function of the power line current. This test requires that the arcing fault be located in various positions in the circuit (configurations A, B, C, D), representing the different locations an arcing fault can occur in a branch circuit [24].

Another test, called the opposing electrode test, consists of an arc generator (with copper and graphite electrodes), a resistive load R, and a masking load (configuration A, B, C, or D). There is no masking load used in configuration A. Like the carbonized path clearing time test, the AFCI device in this test must also trip within the allowed time indicated by the "fire" curve (i.e., below the fire curve).

15.3 Arcing Faults

It is generally considered that there are three classifications of arcing faults in residential circuits: short-circuits (i.e., parallel arcs), broken conductors/wire (i.e., series arcs), and loose overheated connections (i.e., glowing connections). Per UL1699, AFCI devices are intended to detect and trip within the required time on short-circuit and series arcs above $5A_{rms}$. They are not intended to detect glowing connections although it is recognized that a glowing connection may progress into a series arc which may be detected [25].

To illustrate the differences in faults, Figure 15.6 compares the three types of fault conditions that can occur in a residential electric circuit—short-circuit (overheated

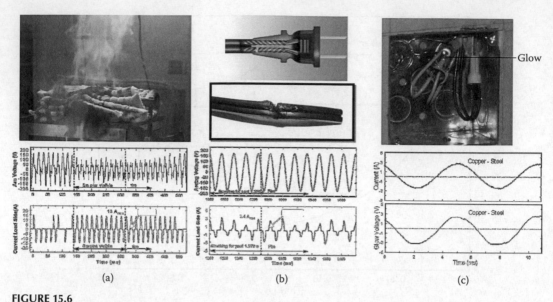

(a)   (b)   (c)

FIGURE 15.6
Photographs of three types of residential faults and associated waveforms; (a) Parallel arcing; (b) Series arcing; (c) Glowing connection [27,37,38].

extension cord under a rug), series arc (broken conductor inside an appliance cord), and glowing connection (loose wirenut) and corresponding waveform examples.

15.3.1 Short-Circuit Arcing

The stranded wire lamp cord (e.g., SPT-2) used in many extension cords, lamps, and appliances can become cut, crushed, pinched, over-heated, or deteriorated from old age exposing the metal conductors. Exposed live wires can create line-to-neutral or line-to-ground arcing faults, defined as a short-circuit arcing fault. The UL1699 standard refers to a guillotine test for measuring the high current arcing performance of an AFCI device [24].

In residential applications a short-circuit arcing fault is commonly referred to as a "parallel" arc fault. This "parallel" short-circuit arc generally produces a high current fault with a magnitude that depends on the available short-circuit current. This available short-circuit current is determined by available current at the load center and impedance of the circuit created by the fault location. Frequently, parallel faults can occur at the load which is connected by long lengths of lamp cord plus long lengths of "behind-the-wall" cabling (i.e., typically NM-B) cable length as illustrated in Figure 15.7. The longer these wires and the smaller their conductor size, the greater the impedance and the lower the available fault current becomes at the load.

It was recognized that a home fire could be initiated by a continuous short-circuit with a low available fault current [1,2]. It was also discovered that intermittent arcing can occur with a low available fault current, also called a sputtering arc [5,9]. Frequently, sputtering arcs are created by damaged stranded extension or lamp cords owing to the fine wire strands of the lamp (or appliance) cord (e.g., SPT-2) vaporizing [1,12,26]. The strands intermittently make and break creating sporadic arcing, carbonizing the wire insulation, which may not activate the thermal overload (i.e., bimetal) of a standard breaker to trip the breaker. This type of arcing fault can persist and create a hazardous condition, producing hot molten copper particles that get violently ejected from the fault along with ignited gasses and a hot arc, possibly igniting a fire. Consequently, a standard T-M breaker may

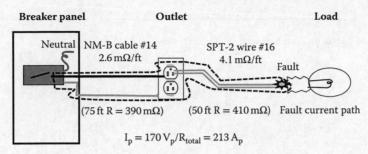

FIGURE 15.7
Drawing illustrates typical wiring setup in residential North American applications. This example shows how the available short-circuit current can be below the instantaneous pickup level of a traditional T-M breaker.

not trip or trip after significant arcing has occurred, greatly increasing the probability of igniting a fire in either the wire insulation or in the surrounding material [27].

Another form of parallel arcing can occur by "scintillation" [28–30]. Scintillations start as small, very low current, partial breakdown between two conductors at different potentials, across an insulating surface [26, 28–30]. Repeated breakdown can lead to carbonization of the surface (i.e., carbon bridge formation) and potentially a parallel arcing fault [28–30]. It is also known that old wiring will deteriorate and can be prone to arcing faults [31].

15.3.2 Series Arcing

Figure 15.6b shows how a broken wire in one conductor leg of a hair dryer cord created a series arc. A low current arcing fault, often referred to as a "series" arc fault because the arc fault is initially in series with a load, is an unintended arc created by either a break in the line side or neutral conductor. This section shows how continuous low current arcing can occur at residential voltage levels (i.e., $120V_{rms}$) and illustrates some examples of low current "series" arcing faults.

If one were to examine the Paschen breakdown curve for air at atmospheric pressure (see Figures 9.8 and 15.9) the lowest possible breakdown voltage. is around 327V [32–35]. This implies that for continuous arcing, the voltage must be greater than this value, otherwise the arc will extinguish at the first half-cycle zero crossing. It is possible for extremely small contact gaps that breakdown of the gap can occur for much lower voltages, see Section 9.3.2. Usually, however, when a $120V_{rms}$ circuit with two copper wires conducting current is pulled apart, even with a small gap, the arc extinguishes in ½ cycle. So, how does a series break in a conductor of a circuit operating at $120V_{rms}$ ($170V_p$) produce continuous "series" arcing at voltages below the Paschen minimum for air at atmospheric pressure? It turns out that in general, virtually all residential wiring and distribution equipment (i.e., outlets, fixtures, etc.) use some type of organic material (e.g., polyvinyl chloride [PVC], nylon or other types of thermoplastic materials) to insulate conductors. If this insulation is exposed to intermittent arcing or elevated temperatures, say from a loose connection, it can decompose and produce carbonaceous material (i.e., long chain hydrocarbons) [28–30,34]. This carbon is the reason why sustained "series" arcing is possible at $120V_{rms}$. Carbon is an excellent thermionic emitter (see Figure 15.9) with a very high vaporization temperature [31–35]. If a connection arcs intermittently or a broken wire occurs creating a series arc, the wire or nearby insulation can become carbonized, coating the copper wire surface and even creating a carbon bridge across the gap formed by the break in the conductor [29, 33]. Now rather than a pure copper wire, the wire acts like more like a carbon electrode and carbon electrodes, even at low currents, can sustain arcing since carbon is an excellent thermionic emitter with a negative temperature coefficient [32,33,35].

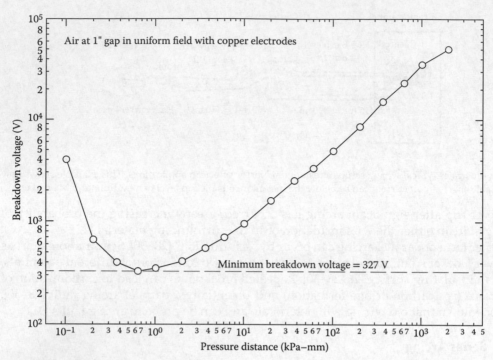

FIGURE 15.8
Paschen curve for air. Note minimum breakdown is 327V making $120V_{rms}$ ($170V_p$) seem too low for continuous arcing.

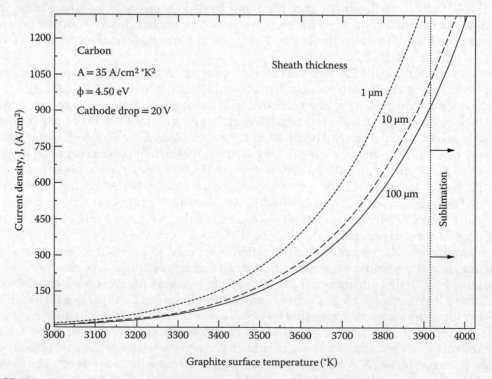

FIGURE 15.9
Thermionic emission for graphite [36].

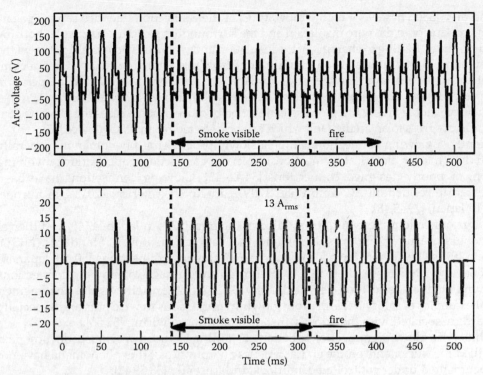

FIGURE 15.10
Arcing in SPT-2 cord at 13 A_{rms} initiated fire after 316 ms of arcing. Line-to-neutral arcing eventually occurred after the wire insulation burned through to both conductors at 6.7s [27].

Once the carbon becomes heated from the arc, its thermionic emission and negative temperature coefficient properties result in continuous "series" arcing faults. Once either a carbon bridge is formed or carbon coats the conductors in a broken connection, continuous arcing at $120V_{rms}$ becomes possible as shown in Figure 15.10 [27].

To ignite a fire, the arc needs to have a fuel source. Many times, the wire insulation itself becomes the fuel source, especially PVC insulation which can produce volatile gasses that can be ignited by the arc [37,27,28–30,34]. Gas ignition, not the arc itself, is generally observed during an arcing event since the chemical energy released from the burning gasses is far greater than the arc energy [37]. It is the ignited gases which can subsequently ignite surrounding materials such as structural wooden members, rugs, newspapers, or other materials and generally not the arc itself [37,27,28–30,34].

15.4 Glowing Connections

A third type of electrical fault that can occur in residential electrical systems is frequently referred to as a glowing connection (Figure 15.6c, see also Section 6.4.3.3) [38]. Glowing connections are molten metal and metal oxides that generally occur at a loose connection point in an electrical circuit [38,28]. The molten metal/oxide can reach temperatures around 1235°C (melting point of Cu_2O) [39–40]; well above the melting and vaporization temperature of polymeric insulation used in wiring [30,34,41]. Depending on current magnitude, 10s of watts of power can be dissipated in the loose joint causing overheating of electrical wiring and the surround area [38]. These very high temperatures at the connection interface can propagate down the conductor to the wire insulation and the surrounding structure (i.e., outlet housing or wood framing), leading to charring of the insulation and wood and potential ignition of the wire insulation, plastic outlets, wood framing, etc. [30]. While a glowing connection is technically not an arc, it does have characteristics similar to a plasma owing to its high temperature [42]. Its presence goes undetected by present day AFCIs [24]. However, it has been shown that fires can be prevented owing to a glowing connection because in some cases the glowing connection will progress to an arcing fault which can be detected by an AFCI device [25].

Frequently, a glowing connection can be a slowly progressing phenomenon, generally associated with loose electrical connections, worn out connections, applications involving vibration, or poorly designed connections [34,42–45]. Glowing connections have been documented in residential AC circuit (i.e., $120V_{rms}$ in North America, $220V_{rms}$ in Europe, $100V_{rms}$ in Japan) [29,45,48].

Glowing connections were initially studied in the mid 1970s in Japan [47,48] and then at the National Bureau of Standards (NBS) (now the National Institute of Standards (NIST)) in the late 1970s [49]. Others have also documented glowing connections [50,51]. Japanese researchers and a few researchers in the US showed some images of glowing connections and documented their effects on electrical safety [38,34,48]. Aronstein has also documented practical examples of safety hazards from glowing connections in wire nuts and especially the hazards associated with improperly installed aluminum wiring [52,53].

The ability of a glowing connection to be self-sustaining has been previously documented but detailed measurements on the characteristics and physics of this phenomena have not been documented under controlled conditions until recently [38,39,42].

Glowing connections can form at the interface of an electrical connection and are characterized by an incandescent orange glow of molten oxide [38,34,45]. Usually, the glowing connection would be the unintended result of a loose connection that intermittently arcs under load current. Depending on the connection material, it may take thousands of make/break arcing cycles and just the proper alignment of the wires to make a glow occur [38, 30]. In other materials, especially ferrous materials, the glow can occur with only a few or as little as one make/break operation under load [27]. Repeated make/break action of a loose connection is generally a precursor for the initiation of a glowing connection because the arc creates a metal oxide on the connection interface [38,30,39,40]. These semi-conductive oxides can form a resistive interface that can lead to glow formation [38,30,39,40]. Due to surface tension, it is suspected that the repeated make/break arcing action draws a thin strand of metal/metal oxide, heated to high temperatures owing to the high current density [38,39]. With sufficient oxide present on the surface of the interface, the glow will be initiated and sustained owing to the negative temperature coefficient of resistance of the metal oxide (e.g., Cu_2O, CuO). The glow can continue to consume the metal wire (i.e., convert metal to metal oxide) for extended periods of time (up to hours) [38,34]. The material type and current level also determine the type and stability of the glow.

The glowing phenomenon is not to be confused with I^2R resistive heating created from inherent connection resistance. Figure 15.11 illustrates how glowing connections can form in a loose or broken electrical connection. A "good" connection has a low resistance interface at the connection point along with a low mV drop and low and stable connection temperature. A loose connection can arc under load and create metal-oxide at the connection interface. The voltage drop across the connection will increase, intermittent short duration arcing may occur, along with an increase in the interface temperature. Even without arcing, millivolt drops may exceed softening temperatures and approach melting temperatures of the connection metals (see Lorenz voltage-temperature properties). A glowing connection can form

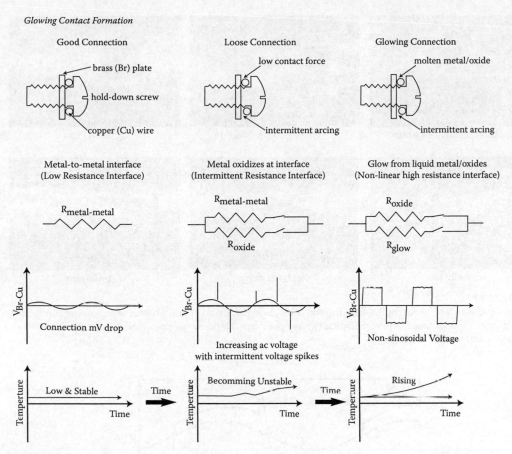

FIGURE 15.11
Conceptual sketches show the glowing connection formation process.

if the metal-oxide at the connection interface becomes heated to the molten state [39]. This heating can be caused by a high-current density created by a thin strand of metal oxide being drawn between the interface or if the joule heating of the metal oxide is high enough to melt the oxide then a glow will form [38,39]. The glow can be modeled as a molten metal-oxide in parallel with a solid metal-oxide with voltage drops of about 2 to $10V_{rms}$ (depending on current and materials) and very high and rising connection interface temperatures [38,39].

A device, consisting of a micrometer and a spring loaded wire holder, was designed to easily reproduce glowing connections to allow for controlled study of this phenomenon [34]. A series of photographs were taken, using this device that shows the progression of the glow over time (Figure 15.12).

An example of electrical waveforms, seen in Figure 15.13, illustrates the electrical characteristics of a copper-to-copper wire glowing connection at $2.9A_{rms}$. The voltage waveform is obviously distorted and high in magnitude ($\sim6V_{rms}$), about 1,000 times higher, as compared to a good electrical connection (typically for residential electrical connections 1 to 20mΩ)! Also, the glowing voltage is well above the reported values for softening (140mV) and melting voltages (430mV) for copper. The glow is actually modulated by the ac current as seen in Figure 15.13. The molten metal-oxide filament can be seen meandering across the solid oxide bridge surface of Figure 15.12 as it is modulated on-and-off by the ac current. At lower currents (i.e., about $<6A_{rms}$) the glow extinguishes near current

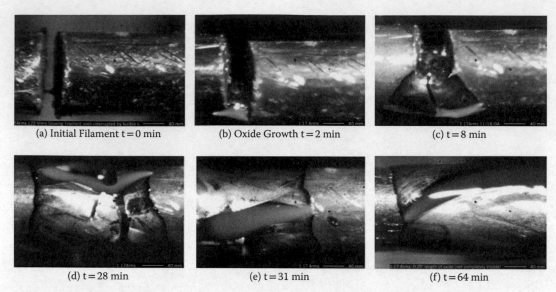

(a) Initial Filament t = 0 min   (b) Oxide Growth t = 2 min   (c) t = 8 min

(d) t = 28 min   (e) t = 31 min   (f) t = 64 min

FIGURE 15.12
A time lapse series of photographs showing the initial bridge formation and growth of a glowing connection (color images show orange filament) between two copper wires. Dark area (copper oxide) 1-mm diameter wires, 37x magnification at 1.17 A_{rms} [34].

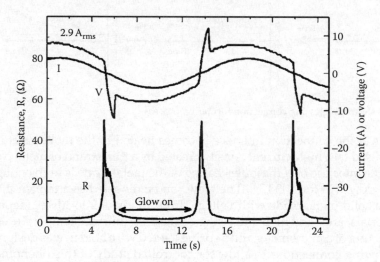

FIGURE 15.13
Dynamic resistance of glowing contact (2.9 A_{rms}, 1-mm diameter copper wire) shows large increase in resistance near current zero [38].

zero because the molten-metal oxide temperature drops. The current then flows through the solid oxide bridge. On the next half-cycle, as the current magnitude increases, the solid the glow reignites. At higher currents, the glow does not extinguish at current-zero because the molten-metal oxide temperature remains high enough to maintain the metal-oxide in a liquid state [38].

Glow voltage is a function of current. An average glowing contact voltage was measured at seven different current levels, at the time the glowing was initiated, to determine the

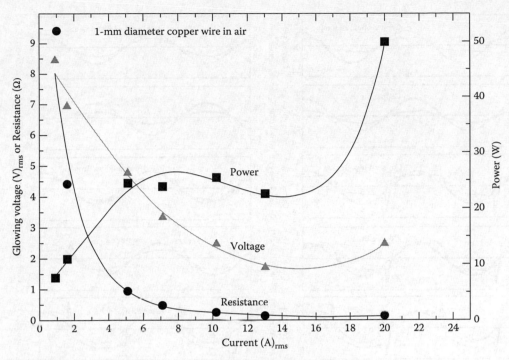

FIGURE 15.14

Typical average glowing contact voltage and power dissipated in contact for 1-mm diameter copper wire pair. Measurements taken just after glowing contact formation stabilized [38,39].

average power dissipated in the connection as a function of current (Figure 15.14). As the current is increased from 0.9 A_{rms} to 5 A_{rms}, power dissipated at the connection interface increased from 7W to 25W and remained fairly constant at 25W to about 16A_{rms}. At this point the wattage rapidly increased to 50W at 20A_{rms}. A voltage minimum is seen near 13A_{rms} and increases for both higher and lower currents. Thus, with increasing current, glowing voltage actually decreases but the power dissipated in the joint increases exponentially as shown [38]. The glow voltage is well below minimum arc voltage for copper (12V) indicating that the glow is not an arc [38,39].

Glowing connections are expected to occur in virtually any type of electrical circuit [38,39]. Glow initiation and sustainability depend on many factors including: current, wire interface shape, wire diameter, dc or ac, and wire material [38]. Glowing connections have been created with currents as low as 250mA_{rms} [38].

Glowing contacts have been studied for copper conductors but there are also other materials that may be present in electrical wiring that can glow. The metals in Figure 15.15 were chosen on the basis of their potential use in various types of electrical products, copper being the most common along with steel. Alloys of copper, brass, and phosphorous bronze are used in wall outlets and electrical switches. Steels, while not intended as a conductor path, can be found in outlet screws that secure the wires in wall receptacles, junction boxes, support structures, and conduit. Sometimes the steel screws are brass or nickel plated for polarity identification purposes. However, if arcing occurs, the plating can be removed exposing the base metal (often times steel). Wire nuts also frequently use tin plated steel springs to secure conductors and can result in glowing connections [38,28–31,39–53].

Research has shown that the material or combination of materials affects the glowing connection voltage wave-shape and thus the power dissipated in the glowing connection [38].

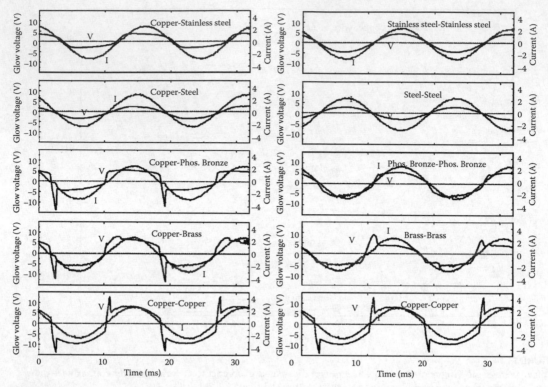

FIGURE 15.15
Current and voltage waveforms for the various material combinations all at 1.6 A_{rms}. All wires were 1.0 mm in diameter [38].

Figures 15.15 and 15.16 show the very dramatic difference in voltage waveform shape for various material combinations. And, for the same combinations, Figure 15.16 illustrates how the voltage wave-shape changes at a higher current. At higher current levels, the voltage distortion tends to decrease and become more sinusoidal owing to elevated temperatures (see prior explanation in Figure 15.13).

Figure 15.17 illustrates this material-current effect by showing the rms voltage across the glowing connection as a function of rms current for a number of different materials at 60Hz. Figure 15.18 shows the corresponding power dissipated.

The propensity of a material or material combination to begin glowing is an important measure since it reflects the probability of initiating a glowing connection which may lead to a fire. The relative ease to initiate glowing, for a given material pair, is indicated qualitatively in Table 15.1 (i.e., easy, very hard, etc.). This represents the number of make/break operations necessary to start a glowing connection. The lower the number of make/break to initiate a glow, in general, the easier it is to initiate a glowing connection.

The extinction current, in Table 15.1, shows the minimum current for existence of the glow in a steady state under the stated conditions. The minimum current to initiate glowing is considered to be about 0.25 A_{rms} for a brass-iron combination [38]. The outcome is a qualitative measure of how the oxide grows. Figure 15.12 shows how copper wires are consumed and form copper-oxides along the wire lengthwise. Slow growth, such as in case of an iron based wire, indicates that glowing occurs at the interface rather than over a distinct molten filament on the oxide surface as in the copper-to-copper case seen in Figure 15.12. Material properties of commonly used metals for electrical application are shown in Table 15.2.

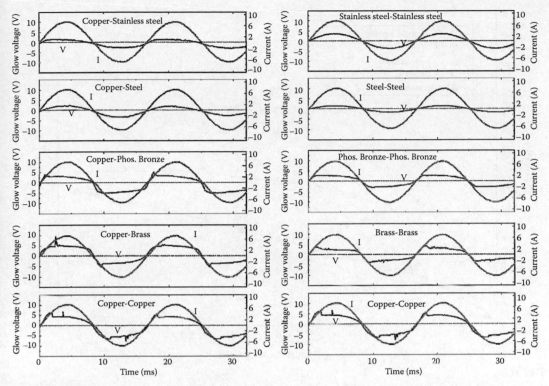

FIGURE 15.16
Current and voltage waveforms for the various material combinations all at 5 A$_{rms}$. All wires were 1.0 mm in diameter [38].

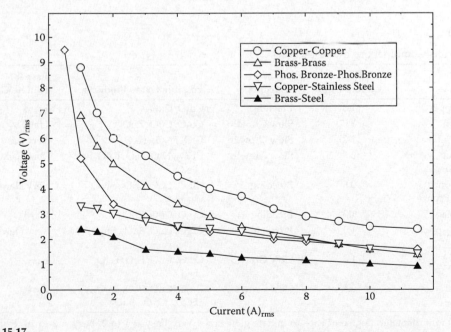

FIGURE 15.17
The true rms voltage as a function of current for the various combinations of wire materials at 60Hz. Except for the iron based materials there is an exponential decrease in voltage with increasing current [38].

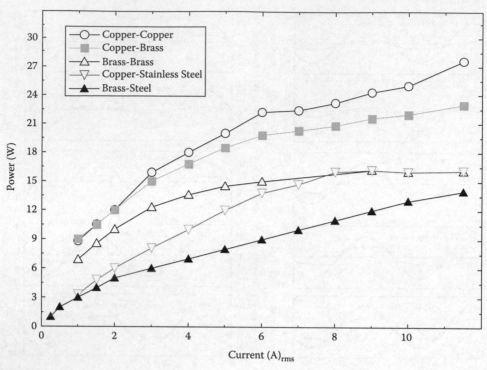

FIGURE 15.18
Power dissipated in the glowing connection as a function of current for various material combinations at 60Hz [38].

TABLE 15.1

Wire Materials Combinations and Outcomes

Material Type	Extinction Current (A_{rms})	Outcome	Possible Oxide Products	Ease to Initiate Glow
Copper - Copper	1.0	Breeding	Cu_2O, CuO	Hard
Brass - Brass	1.0	Slow Growth	Cu_2O, CuO, ZnO	Hard
Steel - Steel	1.0	Slow Growth	FeO, Fe_2O_3, Fe_3O_4	Very Easy
Stainless Steel - Stainless Steel	0.5	Slow Growth	Cr_2O_3, NiO, FeO, Fe_2O_3, Fe_3O_4	Very Easy
Phosphor Bronze - Phosphor Bronze	0.5	Breeding	Cu_2O, CuO, SnO_2, Sn_2O_4	Very Hard
Copper - Brass	1.0	Breeding	Cu_2O, CuO, ZnO	Hard
Copper - Phosphor Bronze	1.0	Breeding	Cu_2O, CuO, SnO_2, Sn_2O_4	Very Hard
Copper - Steel	1.0	Slow Growth	Cu_2O, CuO, FeO, Fe_2O_3, Fe_3O_4	Easy
Copper - Stainless Steel	0.5	Slow Growth	Cu_2O, CuO, FeO, Fe_2O_3, Fe_3O_4, Cr_2O_3, NiO	Easy

All wires 1.0-mm diameter. Degree of ease to initiate glow scale: Very Easy = 1 to 2, Easy = 2 to 10, Hard = hundreds, Very Hard = hundreds to thousands of make/break operations [38].

TABLE 15.2

Wire Material Properties

Material Type	Composition (wt%)	Metal MP (°C)	λ (W/m°K)	Melting Point (MP) of Various Stable Oxides (°C)
Copper	Cu (99.999)	1083	400	1235 (Cu_2O) and 1326 (CuO)
Brass 260	Cu/Zn (70/30)	915–955	109	1975 (ZnO) + copper oxides
Steel 1006	Fe/C/Mn (99.5/0.06/0.35)	1535 (Fe)	46	1457 (Fe_2O_3), 1597 (Fe_3O_4), 1360–1424 (FeO)
Stainless Stee 302	Fe/Cr/Ni/Mn (71./18/8/2)	1400–1420	16	2400 (Cr_2O_3) and 1984 (NiO) + steel oxides
Phosphor Bronze 510	Cu/Sn/P(94.8/5.0/0.2)	975–1060	84	540 (SnO_2) and 1100d. (Sn_2O_4) + copper oxides

All wires 1.0-mm diameter. Steel 1006 (Fe/C/Mn/P/S) (99.5/0.06/0.35/0.04/0.05). Stainless steel 302 (Fe/Cr/ Ni/Mn/Si/C/P/S) (71./18/8/2/0.75/0.15/0.04/0.03). Potential intermetallic oxides are not listed. (d. Decomposes, thermal conductivity, λ, given at 20°C) [38].

15.5 Arcing Fault Properties

AFCI breaker designers utilize the fact that arcing faults produce broadband RF currents [54–59]. Since many loads can produce RF current during normal operation, one of the greatest challenges in developing AFCI technology is the ability to distinguish between loads that produce RF currents and undesired arcing faults [20]. This is so difficult because many loads either arc or mimic arcing during normal operation. For example, arcing can occur in a light switch operated under load and, commutation brushes of a motor while high frequency noise, that mimics arcing, can occur in high intensity discharge lamps and compact fluorescent lamps. Various algorithms, using RF current magnitude and duration, pattern recognition, and frequency bandwidth, have been developed for distinguishing "good" intentional arcing (or electronics that create HF signals) from 'bad" or unintentional arcing [18–23]. Furthermore, there are many other factors that can influence the RF current signal generated by an arc fault (or HF produced from electronics). The main factors being the frequency range of interest, arcing current magnitude, cable and circuit influences, and arc gap materials. The following sections examine each of these factors to illustrate how they affect the magnitude of the RF current generated by the arc fault as seen by an AFCI protective device.

15.5.1 Frequency

It is well documented that broadband RF current can be generated by an arcing fault [54–59]. One example of the spectral output of a $5A_{rms}$ arcing fault over a 20MHz bandwidth is shown in Figure 15.19. Notice how the amplitude generally decreases across the frequency range shown. This data shows the level of detection circuit sensitivity needed to detect arcing in this particular circuit. There are situations where a load (e.g., inverters, compact fluorescent lights, or motors, etc.) may produce high harmonic content at a frequency used by the AFCI device [14,16,20]. Also, power-line carrier signals can operate in the 100kHz range and other smart grid devices can produce high frequency signals on a residential circuit that must be tolerated by the AFCI device. To prevent false tripping, AFCI devices generally utilize more than one frequency or operate over a broad range of frequencies to reduce false tripping [18–23]. Also, because there is generally a high

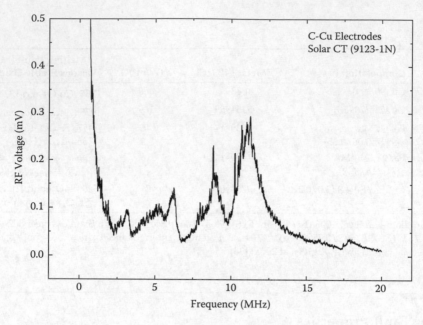

FIGURE 15.19
Spectrum analyzer output of a $5A_{rms}$ series arc illustrating the RF signal magnitude over DC to 20MHz bandwidth. Voltage input to spectrum analyzer shown (50 Ω load). Results not compensated for CT probe frequency response.

background noise in many ac power systems below 2.5kHz as seen by the dramatic rise in RF voltage shown in Figure 15.19, many designs tend to use a frequency range above this level [18–23]. The upper frequency limit is generally limited by transmission line effects combined with a diminishing signal and more costly electronics, making the useful detection range between about 2.5kHz to 20MHz [18–23,33].

15.5.2 Electrode Materials

The type of material that create the anode and cathode of an arcing fault gap also have an effect on the RF current generation. It has been shown that very little RF current is produced from carbon electrodes and high RF currents are generated from copper and other nonthermionic metals with RF current being produced from the cathode since it is an electron source [33,36]. The RF current generation is attributed to arc root instabilities on the cathode caused by an inherent high current density of the arc root on nonthermionic emitting material (i.e., copper) [36,60]. The arc voltage waveforms of Figure 15.20 illustrates an example of this instability in arc voltage for Cu-Cu (nonthermionic) electrodes as compared to a much smoother arc voltage for C-C (thermionic) electrodes. Figure 15.21, a real-time spectrum analyzer (RSA) output for a $2.5A_{rms}$ series arc created with the opposing electrode arc generator [24], with Cu-C electrodes, clearly illustrates this effect, with RF current generated from the cathode. The highest amount of RF current is produced when the cathode is made from a nonthermionic emitting material (e.g., copper). Very little RF current is produced from a graphite cathode except at very low currents [33,36]. Fortunately, virtually all residential wiring is made of nonthermionic materials (i.e., Cu), but significant amounts of carbon, produced from burned insulation, may reduce RF current generation or become intermittent and make arc detection more difficult in some situations.

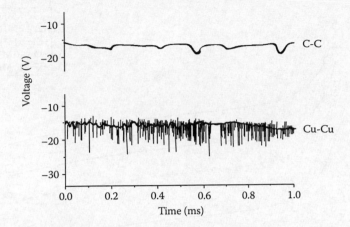

FIGURE 15.20
Expanded views of the arc voltage during a portion of a 60Hz half cycle that shows the difference in HF arc voltage activity between copper and carbon electrodes at $4.5A_{rms}$ [36].

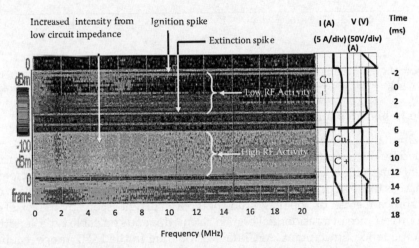

FIGURE 15.21
Real-time spectrum analyzer (RSA) output shows broadband RF current is predominantly produced when the cathode is made from a nonthermionic emitting material (copper) ($4.8A_{rms}$ at 120Vac copper-graphite electrodes in the opposing electrode arc generator) [33].

15.5.3 Arc Fault Current

The magnitude of the power frequency current (i.e., 60Hz) also has a significant effect on the magnitude of the RF current generated by an arc fault [33]. Generally, the lower the arcing current, the greater the RF current magnitude generated and the easier it is to detect with an AFCI device. Conversely, it was discovered that the amplitude of the RF current signal decreases with increasing arc current [33]. Above about $30A_{rms}$ there is a significant drop-off in the measured RF current as shown in Figure 15.22. This can be very significant effect for designers of AFCI protective devices since it may not be possible to use RF current sensing methods at high arcing currents [33]. Oftentimes, the only RF current is produced at the current zero crossings in ac systems.

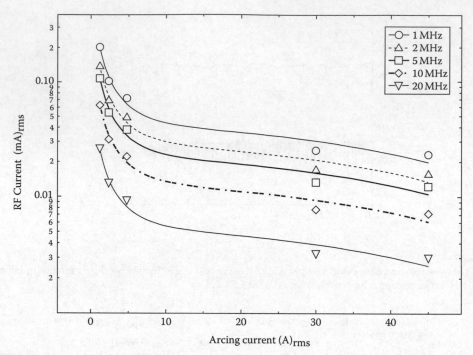

FIGURE 15.22

RF current produced by series arcing faults (copper-graphite opposing electrodes) at 120Vac for different arcing fault currents. RF current generally decreases in magnitude with increasing arcing current and increasing frequency.

15.5.4 Cable Impedance and Cable Length Effects

RF current, circulating through household wiring and through an AFCI device, can be greatly affected by the circuit impedance [14,33,61]. Wire capacitance and inductance affect the RF current magnitude, especially at higher frequencies. In particular, discrete shunt capacitance across the load or along the branch circuit, from loads in other upstream outlets, can have a significant effect on the magnitude of the RF current "seen" by the AFCI device.

The circuit impedance in a residential branch circuit depends on cable type, length, load, and source impedance. Since many AFCI devices operate in the MHz range, cabling can act like a transmission line—creating standing waves with varying levels of RF current magnitude along the cable's length causing the level of RF current at the AFCI device to depend on cable properties and length [33,61].

At the RF frequencies of interest, residential wiring can act like a transmission line with the impedance, Z_b, as seen by the protective device, can be represented by Equation 15.1.

$$Z_b = Z_o \frac{Z_L + jZ_oTan(\beta l)}{Z_o + jZ_LTan(\beta l)} \tag{15.1}$$

$$\beta = \frac{2\pi}{\lambda} \tag{15.2}$$

where Z_o is the wire's characteristic impedance given by

$$Z_o = \sqrt{\frac{R + j\omega L}{G + j\omega C}} \tag{15.3}$$

and λ the wavelength of the RF current and Z_L the magnitude of the load impedance.

Equation 15.3 can be simplified with a lossless case (i.e., R = 0 wire series resistance and G = 0 line-to-line wire conductance) resulting in the following:

$$Z_o = \sqrt{\frac{L}{C}} \qquad (15.4)$$

Typical values for commonly used residential wiring are shown in Tables 15.3 and 15.4.

Cable length can also affect the magnitude of the RF current seen at a protective device. With longer cable lengths (Figure 15.23), the cable can become multiple wavelengths long (λ = c/f), especially at higher frequencies, creating multiple impedance peaks and valleys (Equation 15.1) along the cable resulting in varying levels of RF current magnitude along the cable length.

The wave propagation speed, v, is a measure of how fast a signal travels through a media. A radio signal in free space travels at the speed of light, c, approximately 3×10^8 m/s, but a signal on a transmission line (e.g., NM-B cable at high frequencies) travels slower compared to the speed of light, effectively increasing the electrical line length. In twisted pair wires and NM-B type cable, the velocity of propagation may be 40% to 75% of the velocity in free space [61,62]. The relationship between propagation velocity v, and the speed of light is given by

$$v = cVF \qquad (15.5)$$

TABLE 15.3

Electrical Properties of Various Types of Commonly Used Residential Wire

Cable Type	AWG	R (mΩ/m)	C_{L-L} (pF/m)	C_{L-gnd} (pF/m)	L_{L-L} (μH/m)	Z_o (Ω)	v (%)
NM-B	12/2	12.1	43–59	62–69	0.66–1.64	150	44
NM-B	14/2	15.3	33–43	49–59	0.66–1.64	174	50
SPT-2	14/2	15.3	44	n/a	0.78	133	57
Armored (BX)	12/2	12.1	63	74	0.45	85	63
SOOW Cord	10/4	9.6	82	n/a	1.0	110	37

All values taken at 100kHz. Resistance is per wire conductor, Differential capacitance (line-line (L-L)), is measured between conductors, L is measured as loop inductance of the line-line conductors. Zo and VF are calculated using equations 15.4 and 15.7, respectively, using median values for CL-L and LL-L. NM-B has 2 conductors plus a bare ground wire. SPT-2 is 2 conductor PVC lamp/appliance cord. BX is 4 insulated conductors in a steel sheath.

TABLE 15.4

Dielectric Properties of Various Types of Insulation Used in Electrical Products

Insulation	ε_r	$VF = \dfrac{1}{\sqrt{\varepsilon_r}}$ (%)
PVC	4.0	50
Nylon	3.7	52
HDPE	2.7	60
Polypropylene	2.3	66
Polycarbonate	4.4*	49

(at 60Hz unless otherwise noted) *1kHz [61,62].

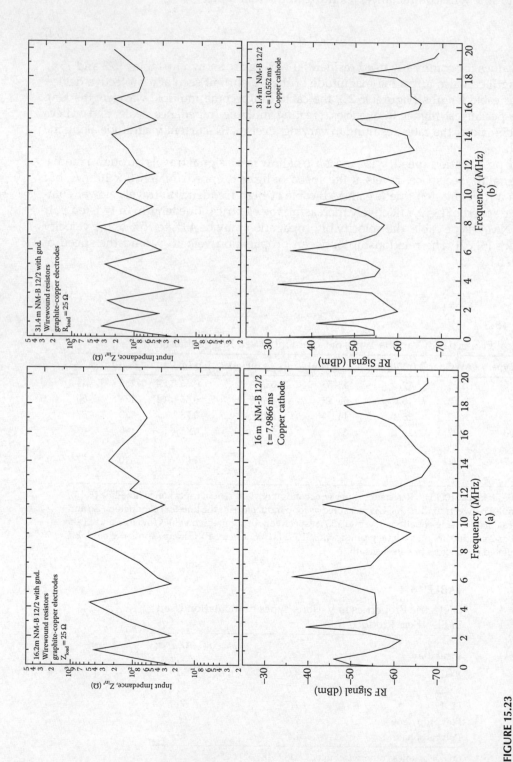

FIGURE 15.23

Change in cable length translates into a change in RF current magnitude versus frequency due to an approximate doubling of the wave propagation time from reflections from the load. Increased RF signal corresponds to low impedance (i.e. valley in input impedance). Velocity factor, VF, remains equal in both cases since the same cable was used in both cases. (NM-B 12/2 cable) (a) 16.2 m and (b) 31.4 m [33].

where VF is written as

$$VF = \frac{1}{\sqrt{\varepsilon_r}} \qquad (15.6)$$

or for the lossless case

$$VF = \frac{1}{c\sqrt{LC}} \qquad (15.7)$$

where ε_r is the relative permittivity of the cable insulation and L is wire inductance per unit length and C the wire capacitance per unit length. The velocity factor, VF, is a percentage of the speed of light, c, and for wire insulation made from PVC, the wave propagation velocity is approximately 50% that of the speed of light in free space. Meaning that for a given frequency, the wavelength is half as long as compared to the wavelength in free space making a given cable length effectively double as compared to the wave traveling at the speed of light. Knowing either the relative permittivity of the cable insulation or the L and C per unit length of the cable will allow for calculating VF.

Thus, for a given frequency, the magnitude of the RF current depends not only on the cable properties but also on the cable length and the location along the cable where the RF current is measured. The greater the circuit impedance at a given frequency, the lower the RF current and, conversely, the RF current magnitude will be higher at impedance valleys. This effect can be very significant since an AFCI protective device is expected to function properly for any cable length. There are many books and references on transmission line theory which the reader can refer for further study [61].

15.6 Other Types of Arcing Faults

While there is certainly much interest in residential arcing faults that may lead to fires, other applications such as commercial and industrial cover a wide range of circuits, currents, and voltages, especially 240Vac and higher [63–66]. Fault conditions that can occur in residential applications (i.e., series, parallel, and glowing) can also occur in commercial and industrial settings as well, but generally lead to arcing faults that can produce not only electrical fires but also produce significant arc flash which can cause serious injury or even death [67–73]. Accidents include series arcing that can progress to parallel arcing (line-to-line or line-to-ground), accidentally shorting a live conductor, animal shorting out live conductors, loose connections, degraded and/or contaminated insulation, especially at medium voltages [69]. Many of these problems can result in a large energy arc flash that can cause serious burns and even death to personnel near the arc flash zone and/or cause serious equipment damage [69]. A number of solutions used today to mitigate the effects of an arc flash include: protective clothing, crowbar devices that sense arcing faults and extinguish the arc flash in milliseconds, maintenance mode switches that reduce the trip time of a breaker when maintenance is being performed on a live circuit, zone selective interlocking (ZSI) (a method which allows two or more ground fault breakers to communicate with each other so that a short-circuit fault or ground fault will be cleared by the breaker closest to the fault in the minimum time), and fast shunt and arc flash energy diverters [66–69]. However, there is much room for improvement and new innovation to help further reduce the number and severity of injuries that occur each year.

15.7 Conclusions

Arcing faults can be effectively detected by AFCI protective devices. They effectively extend the protection level down to 5 Amps by responding to both sputtering arcing faults and low current "series" arcing faults. Both low current (series) and short-circuit (parallel) arcing faults can be reliably detected and discriminated from loads that inherently arc and loads that mimic arcing faults making AFCI technology a very reliable form of arc fault detection in residential applications. Many arcing fault properties have been discussed along with their effect on arc fault detection. AFCI technology, combined with continuous improvements in connection methods, insulation materials, installation standards, and insulation monitoring are all needed to further reduce electrical fires and enhance electrical safety. Continued study of arcing fault phenomena will also lead to new insights and developments.

References

1. L. Smith and D. McCoskrie, "Residential Electrical Distribution Fires," *US Consumer Product Safety Commission Report*, Dec. 1987.
2. J. Mah, L. Smith, and K. Ault, "1997 Residential Fire Loss Estimates," *CPSC Tech. Report*, 1997, Available at www.cpsc.gov/LIBRARY/fire97.pdf.
3. J.R. Hall, Jr., *Home Structure Fires Involving Electrical Distribution or Lighting Equipment*, NFPA, 1 Batterymarch park, Quincy, MA 02169-7471, March 2008.
4. Residential Building Electrical Fires, TFRS, *US Fire Administration National Fire Data Center*, Vol. 8, Issue 2, March 2008.
5. C. Kimblin, J. Engel, and R. Clarey, "Arc-Fault Circuit Interrupters," IAEI, Illinois July 2000.
6. *National Electric Code (NEC)*, National Fire Protection Association (NFPA) 70, 1 Batterymarch Park, Quincy MA, 02169-7471 1999.
7. P. Boden, R. Davidson, W. Skuggevig, R. Wagner, adn D. Dini "Technology for Detecting and Monitoring Conditions That Could Cause Electrical Wiring System Fires," Prepared for U.S. Consumer Product Safety Commission by Underwriters Laboratories Inc., Contract No. CPSC-C-94-1112, Sept. 1995.
8. "Report of Research on Arc-Fault Detection Circuit Breakers for National Electrical Manufacturers Association," UL Project Number 95NK6832, March 1996.
9. R.W. MacKenzie and J.C. Engel, "Electronic Circuit Breaker with Protection Against Sputtering Arc Faults and Ground Faults," US. Patent# 5,224,006 29 June 1993.
10. J.C. Engel and R.W. MacKenzie, "Low Cost Apparatus for Detecting Arcing Faults and Circuit Breaker Incorporating Same," US. Patent# 5,691,869 25 Nov. 1997.
11. J.A. Wafer, "The Evolution of Arc Fault Protection in Residential Systems," *Mort Antler Lecture at the 51st IEEE Holm Conference*, Chicago, IL, Sept. 2005, www.ewh.ieee.org/soc/cpmt/tc1/h2005wafer.pdf.
12. Underwriters Laboratories Inc. (UL). UL Standard for Safety for Arc-Fault Interrupters, UL 1699, Northbrook, IL 60062-2096 Feb. 26, 1999.
13. E. Carovou, J.B.A. Mitchell, N. Ben Jemma, S. Tian, and H. Belhaja, "AC Electrical Arcs with Graphite Electrodes," *57th IEEE Holm Conference*, Minneapolis, MN pp. 232–237, Sept. 2011.
14. P. Muller, S. Tenbohlen, R. Maier, and M. Anheuser, "Influence of Capacitive and Inductive Loads on the Detectability of Arc Faults," *57th IEEE Holm Conference*, Minneapolis, MN pp. 143–148, Sept. 2011.
15. J-M. Martel, M. Anheuser, and F. Berger, "A Study of Arcing Fault in the Low Voltage Electrical System," *25th International Conference on Electrical Contacts with 56th IEEE Holm Conference*, Charleston, SC pp. 199–209, Oct. 2010.

16. C. Strobel "Arc Fault Detection in AC and DC Systems," *Proc. 21ˢᵗ Albert-Keil-Kontaktseminar*, Karlsruhe, Germany, Sept. 2011.

17. J.M. Martel, M. Anheuser, and A. Hueber "Protection Against Parallel Arcing in Residential Installations," *Proc. 21ˢᵗ Albert-Keil-Kontaktseminar*, Karlsruhe, Germany, Sept. 2011.

18. K.L. Parker and T.J. Miller "Arc Fault Circuit Interrupter And Method Providing Improved Nuisance Trip Rejection," US Patent #8,089,737 Jan. 3, 2012.

19. X. Zhou, J.J. Shea, J.C. Engel, K.L. Parker, and T.J. Miller "Arc Fault Interrupter and Method of Parallel and Series Arc Fault Detection," US Patent #7,558,033 July 7, 2009.

20. C.E. Restrepo, "Arc Fault Detection and Discrimination Methods, *53ʳᵈ IEEE Holm Conference on Electrical Contacts*, Pittsburgh, PA Sept. 17–19, 2007.

21. C. Restrepo, P. Staley, A. Nayak, V. Mikani, and H. Kinsel "Systems and Methods for Arc Fault Detection," US Patent # 2008/0106832 A1 May 8, 2008.

22. S.J. Brooks et al, "Method of Detecting Arcing Faults in a Line Conductor," EP1657559, May 9, 2007.

23. F.K. Blades, "Method and Apparatus for Detecting Arcing in Electrical Connections by Monitoring High Frequency Noise," U.S. Patent # 5,223,795, 29 June 1993.

24. Underwriters Laboratories Inc. (UL). UL Standard for Safety for Arc-Fault Interrupters, UL 1699 rev. 4, Northbrook, IL 60062-2096 Feb. 11, 2011.

25. UL Report, "Special Services Investigation on Branch/Feeder Arc Fault Circuit Interrupter Incorporating Equipment Ground Fault Protection," File E45310, May 31, 2001.

26. "Household Extension Cords Can Cause Fires" CPSC Document# 5032 www.cpsc.gov/cpscpub/pubs/5032.pdf.

27. J.J. Shea, "Conditions that Can Cause Upper Thermal Limits on Residential Wiring to be Exceeded," Presentation at the Fire and Materials Conference, San Francisco, CA Jan 2007.

28. V. Babrauskas, *Ignition Handbook*. Issaquah, WA: Fire Science Publishers Society of Fire Protection Engineers, 2003, ch. 7, pp. 312–315, ch. 11, pp. 534–553, ch. 14, pp. 775–795.

29. V. Babrauskas, "How do electrical wiring faults lead to structure ignitions," *Proc. Fire Materials Conf.*, London, U.K, 2005, pp. 39–51.

30. V. Babrauskas, "Mechanisms and modes for ignition of low-voltage PVC wires, cables, and cords," *Fire and Materials Conference*, San Francisco, 2005, pp. 291–309.

31. D. Dini, "Some History of Residential Wiring Practices in the U.S.," Underwriters Laboratory," Presented at the NFPA sponsored Aging Wiring Conference, Chicago, IL, 2006.

32. P.G. Slade, *Electrical Contacts*. New York, NY: Marcel Dekker, 1999.

33. J.J Shea and X. Zhou, "RF Currents Produced from AC Arcs with Asymmetrical Electrodes," *Proc. 56th IEEE Holm Conference on Electrical Contacts*, Charleston, SC, pp.188–198, Oct 2010.

34. J.J. Shea, "Conditions for Series Arcing in PVC Wiring," *IEEE Transactions CPT* 2007; 30(3): 532–539.

35. H.F. Ivey, "Thermionic Electron Emission from Carbon," *Phys. Rev*, vol. 76, pp. 5567, July 8, 1949.

36. J.J Shea and J. Carrodus, "RF Currents Produced from Electrical Arcing," *Proc. 57th IEEE Holm Conference on Electrical Contacts*, Minneapolis, MN, Sept. 2011.

37. J.J. Shea, "Identifying causes for certain types of electrically initiated fires in residential circuits," *Fire and Materials, Fire Mater.* 2010, Published online in Wiley InterScience (www.interscience.wiley.com), DOI: 10.1002/fam.1033.

38. J.J. Shea and X. Zhou, "Material Effect on Glowing Contact Properties," *IEEE CPMT*, vol. 32, no.4, pp. 734–740, Dec. 2009.

39. J. J. Shea, "Glowing Contact Physics," in *Proc. 52ⁿᵈ IEEE Holm Conference on Electrical Contacts*, Montreal, QC, Sept. 2006, pp. 48–57.

40. J. Sletbak, R. Kristensen, H. Sundklakk, G. Navik, and M. Runde "Glowing Contact Areas in Loose Copper Wire Connections," *IEEE Trans. Comp., Hybrids, Manufact. Technol.*, vol. 15, no. 3, pp. 322–327, June 1992.

41. C.L. Beyler and M.M. Hirschler, *Thermal Decomposition of Polymers* (3rd ed.) 2002 Society of Fire Protection Engineers Handbook of Fire Protection Engineering, National Fire Protection Association, 1 Batterymarch Park, Quincy, MA 02169.

42. X. Zhou and J.J. Shea, "Characterization of Glowing Contacts Using Optical Emission Spectroscopy," *IEEE Holm Conference on Electrical Contacts*, Pittsburgh, PA Sept. 2007.

43. J. Urbas, "Glowing Connection Experiments with Alternating Currents below 1Arms," *Proc. 54th IEEE Holm Conference on Electrical Contacts*, Orlando, FL, pp. 212–217, Oct. 2008.

44. J. Wilson et al, "Glowing Electrical Connections," Tech. Rep., *Electrical Construction and Maintenance*, pp. 57–60, Feb. 1978.

45. Y. Hagimoto, K. Kinoshita, and T. Hagiwara, "Phenomenon of Glow at the Electrical Contacts of Copper Wires," *NRIPS (National Research Institute of Police Science) Reports*, Research on Forensic Science, vol. 41, No. 3, Aug. 1988.

46. B.V. Ettling, "Glowing Connections," *Fire Technology*, vol. 18, Issue 4, pp. 344–349 Nov. 1982.

47. T. Kawase (1975) "The Breeding Process of Cu2O," *IAEI News*, vol. 47: 24-25 July/Aug.; 1977, *second report*, vol. 49: 45–46, Nov./Dec.

48. H. Ohtani and K. Kawamura, "An Experimental Study on Thermal Deterioration of Electric Insulation of a PVC Attachment Plug," *J. Japan Society for Safety Engineering* vol. 42, no. 4, pp. 216–221, 2003.

49. W.J. Meese and R.W. Beausoliel, *Exploratory Study of Glowing Electrical Connections* (NBS Building Science Series 103). Washington, DC: U.S. Department of Commerce, National Bureau of Standards, Oct. 1977.

50. Glowing Electrical Connections. *Electrical Construction and Maintenance*, McGraw-Hill, vol. Feb. 1978, pp. 57–60, 1978.

51. D. Newbury and S. Greenwald, "Observations on the Mechanisms of High Resistance Junction Formation in Aluminum Wire Connections," *Journal of Research of the National Bureau of Standards (NBS)*, vol. 85, no. 6, p.429 Nov./Dec. 1980.

52. J. Aronstein, "Evaluation of receptacle connections and contacts," in *Proc. 39th IEEE Holm Conf. Elect. Contacts*, Sept. 1993, pp. 253–260.

53. J. Aronstein, "Environmental Deterioration of Aluminum-Aluminum Connections," in *Proc. 38th IEEE Holm Conf. Elect. Contacts*, Sept. 1992, pp. 105–111.

54. K. Uchimura, "Electromagnetic Interference from Discharge Phenomena of Electric Contacts," *IEEE Trans. Electromagnetic Compatibility*, vol. 32, no. 2, pp. 81–86, May 1990.

55. J.D. Cobine and J. Gallagher, "Noise and Oscillations in Hot Cathode Arcs," *J. Franklin Institute*, vol. 243, Issue 1, pp. 41–54, Jan. 1947.

56. T. Takakura, K. Baba, K. Nunogaki, and H. Mitani, " Radiation of Plasma Noise from Arc Discharge," *Journal Applied Physics*, vol. 26, no. 2, pp. 185–189, Feb. 1955.

57. M.I. Skolnik and H.R. Puckett, Jr., "Relaxation Oscillations and Noise from Low-Current Arc Discharges," *Journal Applied Physics*, vol. 26, no. 1, pp. 74–79, Jan. 1955.

58. I. Langmuir, "Oscillations in Ionized Gases," *Proceedings National Academy of Sciences* USA, vol. 14, Issue 8, pp. 627–637, 1928.

59. D. Klapas, R.H. Apperley, R. Hackam, and F.A. Benson, "Electromagnetic Interference From Electric Arcs in the Frequency Range 0.1-1000 MHz," *IEEE Trans. Electromagnetic Compatibility*, vol. EMC-20, no. 1, Feb. 1978, pp. 198–202.

60. A.E. Guile, "Arc-Electrode Phenomena," *Proc. IEE, IEE Reviews*, vol. 118, No. 9R, Sept. 1971.

61. P.C. Magnusson, G.C. Alexander, V.K. Tripathi, and A. Weisshaar, *Transmission Lines and Wave Propagation*, 4th ed. CRC Press, 2000 N.W. Corporate Blvd. Boca Raton, FL 33431 2001.

62. E. Oberg, Machinery's Handbook, 28 ed., New York, NY: Industrial Press, 2008, table also found at www.machinist-materials.com/comparison_table_for_plastics.htm

63. D. Kolker, S.Campolo, and N. DiSalvo, "A study of Time/Current Characteristics of the Ignition Processes in Cellulosic Material Caused by Electrical Arcing for Application in 240V Arc-Fault Circuit Interrupters," *53rd IEEE Holm Conference Proceedings on Electrical Contacts*, Pittsburgh, PA, pp. 105–114, Sept. 2007.

64. J.J. Shea, "Comparing 120Vac and 240Vac Series Arcing for Residential Wiring," *IEEE Holm Conference on Electrical Contacts*, Orlando, FL, 2007.

65. F. Berger, "Arcing Faults – An Overview," *Proc. 20th Albert-Keil-Kontaktseminar*, Karlsruhe, Germany, pp. 199–205, Oct. 2009.

66. J.M. Martel, "Series Arc Faults in Low Voltage Electrical Installations," *Proc. 20th Albert-Keil-Kontaktseminar*, Karlsruhe, Germany, pp.215–224, Oct. 2009.

67. G. Gregory and K. Lippert, "Applying Low-Voltage Circuit Breakers to Limit Arc Flash Energy," *IEEE Trans. Industry Applications*, vol. 48, no. 4, July/Aug. 2012, pp. 1225–1229.

68. C. Maroni, R. Cittadini, Y. Cadoux, and M. Serpinet, "Series Arc Detection in Low Voltage Distribution Switchboard Using Spectral Analysis," *International Symposium on Signal Processing and its Applications (ISSPA)*, Kuala Lumpur, Malaysia, I3–I6 Aug. 2001, pp.473–476.

69. J.W. Brown, S.P. Nowlen, and F.J. Wyant, "High Energy Arcing Fault Fires in Switchgear Equipment, A Literature Review," Sandia Report SAND2008-4820, Sandia National Labs, Albuquerque, NM, 87185, Feb. 2009.

70. A.D. Stokes and D.K. Sweeting, "Electric Arcing Burn Hazards," *IEEE Trans. Industry Applications*, vol. 42, no. 1, pp.134–141, Jan. 2006.

71. IEEE 1584, "Guide for Performing Arc-Flash Hazard Calculations," 445 Hoes Lane, Piscataway, NJ, 08855, 2002.

72. M.G. Drouet and F. Nadeau "Pressure Waves Due to Arcing Faults in a Substation," *IEEE Trans. Power Apparatus and Systems, (PAS)*, vol. PAS-98, no. 5, pp. 1632–1635, Sept/Oct. 1979.

73. T. Domitrovich and K. White, "Implementation of an Arc Flash Reduction Maintenance Switch – A Case Study," *IEEE Electrical Safety Workshop (ESW)*, ESW2012-08, Daytona Beach, FL, 2012.

Part IV

Arcing Contact Materials

16

Arcing Contact Materials

Gerald J. Witter

Inscrutable workmanship that reconciles Discordant elements, makes them cling together in one society

Lines composed a few miles above Tintern Abbey, William Wordsworth

CONTENTS

16.1 Introduction 880
16.2 Silver Metal Oxides 883
    16.2.1 Types 883
    16.2.2 Manufacturing Technology 884
        16.2.2.1 Internal Oxidation 884
        16.2.2.2 Post-Oxidized Internally Oxidized Parts (Process B 1.0) 885
        16.2.2.3 One-Sided Internally Oxidized Parts (Process B 2.01) 887
        16.2.2.4 Preoxidized Internally Oxidized Parts (Process B.2.02) 887
        16.2.2.5 Powder Metallurgical (PM) Silver Metal Oxides
                (Processes C and D) 887
    16.2.3 Electrical Performance Factors 890
        16.2.3.1 AC versus DC Testing 890
        16.2.3.2 High Current Inrush DC Automotive and AC Loads 890
        16.2.3.3 Inductive Loads 890
        16.2.3.4 Silver–Tin Oxide Type Materials and Additives 891
        16.2.3.5 Material Factor 895
        16.2.3.6 Interpreting Material Research, Example from Old Silver
                Cadmium Oxide Research 895
    16.2.4 Material Considerations Based on Electrical Switching Characteristics 899
        16.2.4.1 Erosion/Materials Transfer/Welding 899
    16.2.5 Transfer/Welding 902
    16.2.6 Erosion/Mechanisms/Cracking 904
    16.2.7 Erosion/Arc Mobility 906
    16.2.8 Interruption Characteristics 906
    16.2.9 Contact Resistance 906
        16.2.9.1 Summary Metal Oxides 908
16.3 Silver Refractory Metals 908
    16.3.1 Manufacturing Technology 909
        16.3.1.1 Manufacturing Technology/Press Sinter Repress (Process D 1.0) ...909
    16.3.2 Material Technology/Extruded Material 910
        16.3.2.1 Material Technology/Liquid Phase Sintering (Process D 2.0) 910
        16.3.2.2 Material Technology/Press Sinter Infiltration (Process D 3.0) 911

16.3.3   Metallurgical/Metallographic Methods..912
        16.3.3.1 Metallurgical/Metallographic Methods/Preparation.......................912
        16.3.3.2 Metallurgical/Metallography/Quantitative Analysis.......................912
16.3.4   Metallurgical/Structure/Strength and Toughness..914
16.3.5   Electrical Properties (EP) 918
        16.3.5.1 EP/Arc Erosion/Microstructure and Properties 919
        16.3.5.2 EP/Arc Erosion/Silver Refractory...921
        16.3.5.3 EP/Graphite Additions to Silver Tungsten and Silver Tungsten
                  Carbide..922
        16.3.5.4 EP/Copper Refractory Metals 922
        16.3.5.5 EP/Erosion/Summary..927
        16.3.5.6 EP/Composite Refractory Materials/Contact Resistance...............928
16.4   Vacuum Interrupter Materials 935
16.5   Tungsten Contacts...936
16.6   Non-Noble Silver Alloys 937
       16.6.1   Fine Silver..937
       16.6.2   Hard Silver and Silver–Copper Alloys..938
16.7   Silver–Nickel Contact Materials 939
16.8   Silver Alloys and Noble Metals 940
       16.8.1   Palladium and Silver–Palladium Alloys 940
       16.8.2   Platinum...942
16.9   Silver–Graphite Contact Materials 943
16.10 Conclusion 945
Acknowledgements 946
References..946

16.1 Introduction

The electrical contact field is mature yet material research continues in certain areas like silver metal oxides albeit at a lower level than twenty years ago. In the United States, Europe, and Japan, there are fewer university material research programs than in the past, but in China there are contact research activities at several universities bringing many new researchers to the contact field. In the last few years, the largest driving force for material research in the contact field has been the high price escalation and volatility of the noble and precious metal markets especially for silver and gold. In the next section on silver metal oxides another factor, RoHS (Restriction of Hazardous Substances) compliant materials, will be discussed as a factor for the direction of research. For some materials, it will be seen that little change has taken place while for others a wide variety of options are offered.

In this chapter, the various types of materials utilized for arcing contacts are described in terms of chemistry, general physical properties, and material structure. Some general information is given regarding the manufacturing technology that is utilized to make these materials. This is important to understand since some of the materials can be manufactured by several widely differing manufacturing processes and, as a result, materials with identical chemical compositions can have very different performance characteristics. This chapter also provides some performance information relating the materials to electrical loads and specific applications.

The materials are divided into different categories for this discussion on the basis of general similarities in chemical composition and metallurgical structure. Within each category the more common compositions are discussed. Many variations of material types may exist for each general composition as a result of process differences used by different manufacturers for making these contact compositions and also as a result of small differences in chemistry from the use of different minor additives in making these materials. For example, a common 90/10 wt% silver–tin oxide material may be made by using over five different processes without additives or with additives like In_2O_3, Bi_2O_3, WO_3, or more. From this, it is evident that over 20 different permutations of this general composition can exist.

As a result of the large variety of material types that are available, the switching device engineer may be overwhelmed by choices. In Section III of this book, it is shown that arcing contact switching involves many mechanical and electrical variables. With many numbers of choices being possible for matching device and material parameters, it is possible to give general guidelines for selection of materials for specific devices, but not absolute guarantees for material–device combination performance. For most applications, more than one material type can be used to perform the electrical switching functions successfully. The grouping of the materials into categories as shown in this chapter is aimed at simplifying the understanding of material choice differences and helping the engineer more quickly to narrow down the number of candidates that should be investigated. Once a specific material or range of materials is chosen, electrical tests can be made in the actual device or application to assess the performance of the material.

The word "material" as opposed to "alloy" is used in this section to describe the contact metal and metal oxide combinations. The word alloy is sometimes used in a general sense to describe these combinations, but many of the contacts are not alloys but instead are mixtures of metals or metals and ceramics that have either very little or no mutual solubility, thus they don't form true alloys. Since arcing contacts are made from many types of materials and material combinations, many different kinds of processes are involved in this technology. In order to better understand and discuss process technology, the various kinds of applicable manufacturing processes are divided into general process categories. The first major breakdown is the division into two major process differences, alloy and cast versus powder metallurgical. As shown in Figure 16.1, the processes can be divided further into four main process types designated as Processes A, B, C, and D. The subprocesses of these main processes can be subdivided into many variations much more than shown in the illustration. Although most contact material processes fit into one of the four processes shown there are some new technical methods being investigated such as "cold spray" for silver metal oxides that may have some future potential as reported by Rolland et al. [1].

Materials that are made by Process A, alloy and cast and form, in general show the least material variations among different manufacturing sources compared to materials made by processes B, C, or D. For example, Process A, a silver–copper alloy made to the same composition limits by two different sources, will have very similar metallurgical structures after casting, since alloying is on an atomistic basis and proper casting produces fully dense parts. In contrast, a material such as silver–nickel made by two different companies both using the same general powder metallurgical manufacturing methods, for example, Process C, is unlikely to be similar unless the two companies have identical starting powders, mixing processes, pressing parameters, and sintering methods. Variations in these process steps will cause structure variations in terms of silver and nickel distribution and retained porosity. If the two companies had made the silver–nickel materials using two

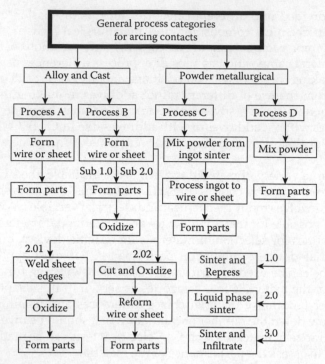

FIGURE 16.1
Processes used for making contact materials.

different general processes, process C versus process D, an even larger difference would be expected in the material properties. The materials made by process C would have a higher density and more grain directional properties than materials made by process D.

From the above examples the following deduction can be made. As opposed to materials such as steel and copper alloys where many generic classifications exist, many contact materials differ significantly although they are of the same general composition. As a result, it is not a recommended practice to substitute contact materials of the same composition made by different manufacturers without proper electrical testing of the material in the device for which its use is intended.

In the following sections of this chapter, some of the typical properties of the materials are listed in tables. A further listing of properties is given in Chapter 24. The properties listed in brochures of several manufacturers along with available standards were utilized to determine typical values for composition, density, hardness, and conductivity [2–5] For material hardness the Vickers scale is used although many companies list these values using the Rockwell scale. Most individual contacts are too small in size, however, for accurate measurement using the Rockwell scale. Vickers is therefore the better choice, because it allows comparison of hardness regardless of the contact size, and, more importantly, the Vickers values allow theoretical calculation of contact resistance using the following equation from Chapter 1 (see also Chapter 24, Section 24.2[f]):

$$R_c = \rho(\pi H/4F)^{1/2} \tag{16.1}$$

H is hardness in N (Newtons) mm^{-2}, F is force expressed as N, ρ is resistivity in Ω mm. Note: Since many contact catalogs list properties in terms of Vickers, kg mm^{-2}, you can convert these values to N mm^{-2} by multiplying the value by the factor 9.81, also most catalogs use conductivity (m Ω^{-1} mm^{-2}) that can be converted to resistivity in terms of Ω mm by taking the reciprocal of the value and multiplying it by 10^{-3}. For example if a silver cadmium oxide contact has hardness listed as 80 HV and a conductivity of 48 m Ω^{-1} mm^{-2}. The constriction resistance for a 0.5 N contact force is estimated as follows:

$$80 \times 9.81 = 784.8 \text{ N mm}^{-2} \text{ hardness} \quad (1/48) \times 10^{-3} = 20.810^{-6} \Omega \text{ mm}$$
$$R_c = 0.73 \times 10^{-3} \Omega$$

16.2 Silver Metal Oxides

16.2.1 Types

Silver metal oxides represent one of the most popular and important categories of arcing contact materials. There are many types of these materials in terms of both chemical composition and material structure. Table 16.1 shows some typical compositions with hardness and conductivity values for comparison. Shen et al. [6] give an historic review of silver metal oxide evolution since its beginning in the late 1930s and show examples of over 50 different chemical compositional systems patented or investigated. Silver metal oxide materials are composite materials consisting of a silver matrix containing a dispersion of fine particulate composed of single oxides or multiple oxides. Since the metal oxide phase has little or no solubility in the silver, the particles strengthen the contact structure without reducing the matrix conductivity.

Today, silver–tin oxide type material is the most popular contact of the metal oxide type, and it has replaced silver cadmium oxide to a large extent. There are several driving forces

TABLE 16.1

Some Typical Silver Metal Oxide Contact Compositions and Properties and Additives

Material Total Oxide	Process PM Powder Met IO internal Oxidization CP Chem. Precipitate B Blending	Additive	Particle Size Distribution C Coarse M Medium F Fine XF Extra Fine	Conductivity M Ω^{-1} mm^{-2}	Hardness HV 1
AgCdO -10	IO				
AgCdO-15	IO				
AgSnO$_2$-12	PM B	WO$_3$	MC	46	60–105
AgSnO$_2$-12	PM B	Bi2O$_3$ CuO	M	46	60–105
AgSnO$_2$-In$_2$O$_3$-12	IO	In$_2$O$_3$	F	44	95–120
AgSnO$_2$-12	PM CP	In$_2$O$_3$ (low)	XF	43	>100
AgSnO$_2$-In$_2$O$_3$-10	IO	In$_2$O$_3$	F	44	80–105
AgSnO$_2$-In$_2$O$_3$-15	IO	In$_2$O$_3$	F	37	100–130
AgZnO-8	PM B	Ag$_2$WO$_4$	MF	48	65–105
AgZnO-10	PM B	Ag$_2$WO$_4$	MF	46	65–110
AgMgONiO-0.6	IO	None	XF	44	105–125

See Table 24.5 for a broader listing of properties [1–4].
For the indium oxide, the IO materials have greater than 2.5% indium oxide and the PM grades can have any level, the (low) statement means less than 1%.

for the change to silver–tin oxide from silver cadmium oxide. Cadmium has been identified as a hazardous material and restrictions have been placed on it for many applications. The European Union for RoHS (Restrictions of Hazardous Substances) standards more recently allowed the use of CdO in electrical contacts as a temporary exception until the development of substitute materials is complete. This makes the future unclear about the use of silver cadmium oxide and certainly legal battles are possible. Another factor driving the conversion is improved electrical erosion resistance in many applications using silver–tin oxide type materials compared to silver–cadmium oxide, thus saving precious metal material. By the beginning of 2012, most all European device manufacturers and Japanese device manufacturers have converted to silver–tin oxide type materials yet in North America many devices still use silver–cadmium oxide. For automotive applications cadmium oxide contacts are banned worldwide. As a result of these factors, less emphasis will be put on silver cadmium oxide and more on the replacement materials for silver cadmium oxide which have been successful in most applications. Since there are many types of silver–tin oxide materials the generic term, silver–tin oxide will be used to describe the whole family of silver–tin oxide types. Therefore, in this section reference to silver–tin oxide will include $AgSnO_2$, $AgSnO_2/Bi_2O_3$, $AgSnO_2/In_2O_3$, $AgSnO_2/WO_3$ and other Ag-SnO_2/XX combinations.

16.2.2 Manufacturing Technology

Besides the differences in chemical compositions for silver metal oxides there are many differences in material structure owing to the many processing variations possible for manufacturing these materials. In order to have a good discussion of these materials it is important to understand the basics of the manufacturing technology. This section describes the general common processes used today. The processes for making silver metal oxide contacts can be first divided into two major categories: (1) internal oxidation (IO) (Process B), or (2) powder metallurgical (PM) (Processes C and D), see Figure 16.1.

Starting in the 1970s, different forms of silver–tin oxide materials were developed in Europe and Japan simultaneously. In Europe, most of the work was done using a PM approach. In Japan, all of the work was by IO as the manufacturing base. More recently, many variations of both processes are used around the world. Both processes have pros and cons with the PM approach being a little more flexible for additive additions.

16.2.2.1 Internal Oxidation

Internal oxidation consists of heating a silver alloy to a temperature below the melting point and allowing oxygen to diffuse into the alloy and react with solute atoms to form metal oxide particles. Following is an example of an equation for the oxidation of silver–cadmium alloy to become silver–cadmium oxide. This shows a parabolic relationship between oxidation depth and oxidation time and the oxidation of silver–tin oxide also has a similar parabolic relationship. Wagner [7] and Freudiger et al. [8] developed Equation 16.2 for oxidation of silver–cadmium alloys in the range 700–900°C. Equation 16.2 below is corrected for a typo in the original paper and also has the constant changed to yield mm versus cm:

$$X^2 = (2Ke^{-A/RT}(P)^{1/2}t)/N_{Cd} \tag{16.2}$$

where X = oxidation depth, mm, K = derived constant ($8.96 \ 10^{-4}$ cm^2 s^{-1}), A = activation constant (21,000 cal), R = gas constant (1.987 cal (°kg mole^{-1})), T = absolute temperature, P = partial pressure of O_2 (cm Hg) t = time, and N_{Cd} = mole fraction (atomic% Cd).

Example for Use of Equation 16.2: Compare the depth of oxidation for a silver–cadmium oxide alloy 10% by wt. CdO for oxidation at both 750°C and 850°C at atmospheric pressure for 10 h. An AgCdO 10 wt.% CdO needs a silver–cadmium alloy 9.55 wt.% Cd or by conversion 9.2 atomic% Cd. Use Equation 16.2 with the following values:

$P = (76 \times 0.21) = 15.96$ cm Hg,	$t = 36,000$ s,	$N_{Cd} = 0.092$,
$T = (750 + 273) = 1023$ K,	$T = (850 + 273) = 1173$ K	
$X_{750} = 0.30$ mm,	$X_{850} = 0.48$ mm	

Other silver metal oxide systems also follow the same general parabolic oxidation depth versus time relationship, but some systems like silver–tin oxide are more difficult to oxidize as a result of the formation of a tin oxide scale on the surface which passivates the surface to further oxidation. In order to solve this problem, silver–tin alloys are doped or alloyed with several different additives to aid in oxidation. Grosse et al. show the influence of additions of Bi, Cu, and In on the kinetics of oxidation of silver–tin alloys [9]. The amount added can equal or exceed the tin content and these materials are considered to be double oxides.

Additives are also used for altering and controlling the distribution of oxide particles for both silver tin oxide and silver cadmium oxide materials. Numerous patents exist for such additives [6]. Some of these additives serve as nucleation agents by providing sites for particle formation and refinement of the particle size distribution.

16.2.2.2 Post-Oxidized Internally Oxidized Parts (Process B 1.0)

In the early days of internal oxidation, the most popular process was to oxidize individual parts after they had been formed. This process is commonly called post-oxidation. For this type of oxidation, oxygen is diffusing from the outside toward the center of the part and the solute element diffuses in the opposite direction. This combination of opposite directional diffusion results in a depletion zone, void of metal oxide, in the center of the part. For AgCd, this zone is about 4% of the material thickness for oxidation at 800°C [8] see Figure 16.2. Silver–tin oxide post oxidized parts show similar depletion zones.

Fully oxidized AgCdO
showing Cd depleted center zone

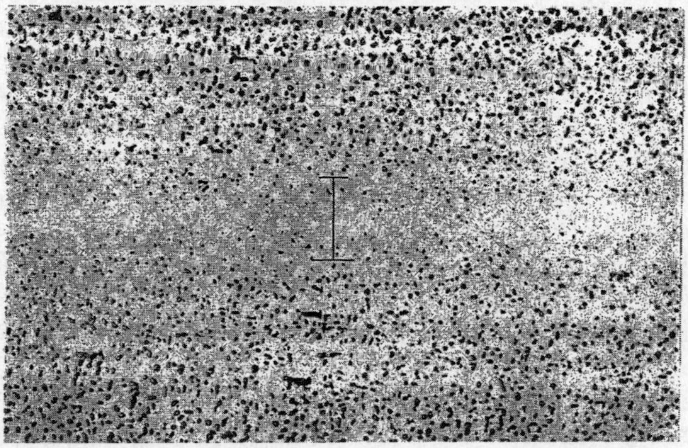

FIGURE 16.2
Post-oxidized internally oxidized silver–cadmium oxide cross section.

From Equation 16.2, it can be seen that the time of oxidation varies with the square of the depth of oxidation. As a result of this relationship, a particle size gradient is created during oxidation. As oxidation proceeds deeper into the part, particles become coarser since oxygen is arriving at the oxidation front at a slower rate relative to the solute atoms. Thick parts which require very long oxidation times can have excessive particle size growth if not oxidized under compensating conditions. From Equation 16.2, it also can be seen that the oxidation rate increases as a function of the partial pressure of oxygen, $P^{1/2}$. Jost and Santale describe a process of gradually increasing the oxygen pressure as the oxidation front progresses deeper into the parts in order to increase the oxidation rate and keep the particle size uniform and in the desired size range [10]. Also, as had been mentioned above, additives are used for controlling the particle size distribution.

Another characteristic of internal oxidation is the potential for defects called thermal arrest lines. These are lines parallel to the contact surface containing high concentrations of oxide particles. The lines are caused by abrupt changes in the oxidation rate, usually as a result of furnace malfunction such as a temporary temperature drop. The slowing of the oxidation rate allows solute atoms to penetrate beyond the oxygen front and temporarily reverse the direction of oxidation. After equilibrium is re-established and the oxidation again proceeds deeper into the part, the area in which the reversal occurred is left with a high concentration of oxide particles. Since the oxide particles are brittle, if the particles are concentrated and contiguous, the thermal arrest area will be prone to cracking or delaminating from thermal stress created by the heat from high current arcs.

Today, far fewer parts are made by post oxidation which produces an expensive monolithic silver metal oxide structure as opposed to a bimetal structure. Most of the background work on this process was done on silver–cadmium oxide yet most of this is applicable to silver tin oxide materials. The following shows why depletion zones are not as problematic as once thought. For contacts with depletion zones only 0.05 mm or less thick, it is questionable that the depletion zone has a detrimental effect. By the time the contact has eroded to the depletion zone, the surface of the contact has developed a new microstructure that bears little resemblance to the original microstructure and consists of a heat-affected layer with deposits of arc debris. Kim and Peters show cross sections of arced silver–cadmium oxide contacts in attempting to relate this surface melt layer, the bulk material microstructure, and the erosion process [11]. Their work shows that after arcing, the surface consists of oxide aggregates and platelets much coarser than the original oxide particles that are fed into the surface to form this layer. Also since the heat of the arc allows segregation of the CdO from the silver, large areas depleted of CdO exist at and just below the surface. Similar surface structures are seen in silver tin oxide materials after extensive endurance switching see Figure 16.3 [12] and also Figure 10.48 in Chapter 10. Certainly the chemistry of the arced surface layer is related to the microstructure that feeds into this surface, but the transition is more gradual than would be expected. The erosion process, as discussed in Chapter 10, is not like a machining operation, just cutting off surface material and leaving the structure below as the new surface. The erosion process is much more complicated and involves melting of the surface, arc deposits and transfer of material between anode and cathode, a heat-affected zone deeper into the surface, the under melt zone, and more. As discussed later in this chapter, the surface-layer characteristics are affected by both electrical load and material variables. The point of this discussion is that, as a result of thin depletion zones, it is unlikely that a surface completely depleted of metal oxides will be formed by the erosion process, but instead a surface with a reduced oxide content.

 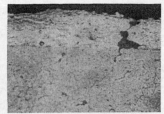

Anode   15% Oxide   Cathode   11.5% Oxide

FIGURE 16.3
An illustration of how the microstructure of the equilibrium melt layer established on the surface of a contact during endurance testing can differ from the mictrostructure of the bulk material. Cross section of silver metal oxide contact surfaces after 300,000 operations at 30 A 12 V dc inductive load.

16.2.2.3 One-Sided Internally Oxidized Parts (Process B 2.01)

This once very popular method is rarely used and consisted of welding two sheets of alloy together and oxidizing from one side. This eliminated the depletion zone in the middle of the part. Associated with this and internal oxidation in general is a problem of getting a very thin silver layer on the surface of the contact as a result of the oxidation process. This problem is discussed by Pedder for Ag–CdO and by Shen and Lima for Ag–SnO$_2$In$_2$O$_3$ [13,14]. This layer is more extensive for the silver–tin system. The extent of this problem also depends on the method of processing. The manufacturer of the contact must use special techniques to control this problem. Normally the cleaning process removes this layer.

16.2.2.4 Preoxidized Internally Oxidized Parts (Process B.2.02)

A very popular variation of the internally oxidized method for making silver metal oxides is to form small pieces of the alloy, internally oxidize the pieces, compact and extrude the pieces into wire or strip and form discrete parts from the wire or strip. Some manufacturers refer to this product as "preoxidized metal oxide parts," referring to the fact that the material is oxidized before the part is formed. There are many variations in methods used to form the small pieces of alloy for oxidation; for example cut wire, cut strip, atomization of melt to form shot, and more. The preoxidized process produces a product that is more heterogeneous in metallurgical structure throughout the contact body than post-oxidized parts. The extrusion of internally oxidized materials after oxidation produces material that is more ductile than material in the oxidized state before extrusion. This difference is very great for silver–tin oxide type materials. The metallurgical structure of the material is much different after extrusion and the electrical performance of the material can also be changed. This process has become the most popular type of process for making IO silver tin oxide wire and preoxidized strip since it offers materials with a uniform microstructure and good formability compared to other silver tin oxide processes. The use of large amount of machine made composite rivets which can use preoxidized wire has significantly increased the use of this IO process.

16.2.2.5 Powder Metallurgical (PM) Silver Metal Oxides (Processes C and D)

A powder metallurgical process was the first method used for making silver metal oxide contacts. Today many types of powder metallurgical processes exist for making silver metal oxide materials. Processing among companies can differ significantly in techniques for powder production, pressing, sintering, and forming. Some of the more common options in manufacturing technology for powder metallurgical silver metal oxide contacts are reviewed in this section.

16.2.2.5.1 PM/Powder

For powder metallurgical processes, the final properties of the materials are a product of many parameters including, powder size and characteristics, blending methods, pressing techniques, sintering processes, and methods of further consolidation [15–17]. For powder production, care must be taken to control contamination from retention of salt traces or chemicals used in production, as low levels of substances like alkali metals can cause electrical interruption problems, discussed later in this chapter.

16.2.2.5.2 PM/Blends of Elemental Silver and Metal Oxide

The simplest of the powder processes involves blending silver powder and metal oxide particles. The silver powder technology itself is complex and several different methods exist for making powder. Each technique, electrolytic, chemical precipitation produces powders with different particle size distributions and blending characteristics. Chemically precipitated powders in the low micron size ranges are common for this type of use. Metal oxide powders or compounds that break down during sintering to form metal oxides are used for blending with silver powders. For blending, there are sophisticated milling techniques for better intimate mixtures of the silver and metal oxides. Both the size and the degree of agglomeration of the powders are important for controlling the final microstructure in terms of metal oxide distribution in the silver matrix.

16.2.2.5.3 PM/Composite Powders

Today, many processes exist for producing composite silver tin oxide type powders. Most companies are very secretive about these types of processes. Chemical precipitation is one popular method where silver and metal oxides are precipitated from metal salts. These powders allow the formation of finer particle size distributions of the tin oxide and other oxides in the silver matrix. Another method that has been used is the internal oxidation of alloy powder (IOAP), for example fine atomized alloy particles containing silver, tin, and other additives. Pedder et al. used these types of processes for silver cadmium oxide powders [18]. Sometimes these methods produce particles so fine they are difficult to process and form without producing defects in the structure. The use of a process called Oswald ripening can coarsen the structure by heating pressed parts and allowing particle growth through fine particles dissolving and re-precipitating on larger particles [19].

16.2.2.5.4 PM/Processes/Press, Sinter, Repress (Process D 1.0)

This is one of the first basic metallurgical processes utilized for making silver metal oxide contacts. This process normally does not produce fully dense parts and is not used as much as the other PM processes discussed this section. Any of the above powder systems can be used with this process. The process consists of (1) compacting the powder in a die into the shape of the desired contact, (2) sintering the pressed compact to increase the strength, and (3) repressing the compact to increase the density. With this process significant porosity normally is left in the contact body after processing, with a range of 1–5% porosity being common.

If the pressing of the powders before sintering is done at too high a pressure, gas can be entrapped in the compact that affects the sintered structure. Gas can also be entrapped in the structures during repressing if there is a large change in density between sintering and repressing. On heating the structure after repressing it may expand if too much gas is entrapped. Electrical erosion tests were conducted by Lapinski for comparing silver–cadmium oxide contacts made both with and without excessive entrapped gas from processing variations [20]. The tests showed higher electrical erosion rates for the contacts with the entrapped gas.

16.2.2.5.5 PM/Wrought PM/Press Ingot Sinter, Extrude, Form (Process C)

This process is commonly referred to as wrought powder metallurgical for silver metal oxide material. As a result of the large amount of work put into the material the contacts made by this process are normally fully dense. The consolidation methods may vary with manufacturers, extrusion being the most common and other methods, such as swaging of ingots, also being used. The materials made by this process, depending on the characteristic of the starting powders and forming methods, can be anisotropic. Depending on the particle size and characteristics, the particles elongate and align parallel to the extrusion direction. Poniatowski and co-workers [21–23] studied anisotropic $AgSnO_2$ extruded material and found electrical erosion resistance improved for contacts made with the face perpendicular to the oxide particle elongation than for contacts with the particle elongation parallel to the face, see Figure 16.4. As a result of economics, however, most products made by the wrought PM process are made with the contact face parallel to the particle elongation or extruded into wire which is later formed into both monolithic and bimetal rivets or other forms.

16.2.2.5.6 Summary of Metal Processing Differences

The basic processes of IO and PM contact metal oxide production are quite different. The structures produced by IO post oxidized material and PM press, sinter, and repress materials are very different in appearance and microstructure particle distribution. On the other side the extruded IO and PM processes can produce very similar metallurgical structures. For silver–tin oxide type materials the IO process normally needs some additive in order to allow efficient oxidation to take place. The PM system is more flexible and only uses additives to enhance the properties of the contact material. Both processes need good controls to insure consistent product is made. Years ago the PM process resulted in coarser oxide dispersions than those made by the IO process even in extruded wire. Today most of the European PM manufacturers have developed wet chemical precipitation processes

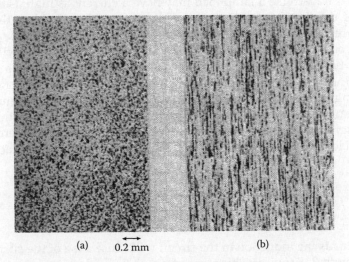

(a)   0.2 mm   (b)

FIGURE 16.4
Silver–tin oxide materials: cross section (a) perpendicular to tin oxide fibers and (b) parallel to tin oxide fibers.

for making composite powders, silver and metal oxides. These powders appear to be finer and more uniform than PM blended powders. As a result of this many PM materials now have particle size distributions similar to IO materials. It is important to understand the basic process that is being used to make a product and record the microstructure characteristics of any product that you are testing and evaluating. For electrical performance the chemistry is important but also the microstructure has a large influence.

16.2.3 Electrical Performance Factors

A. Electrical switching parameters and load considerations
Before going into a discussion of the effects of different additives and chemistries for silver tin oxide materials a short discussion will be made about some major differences in erosion characteristics with different electrical loads and switching conditions. When you look at the results of some research on contact performance you must consider the electrical parameters that were used to reach the results, since the results, may not apply to the parameters used in another device.

16.2.3.1 AC versus DC Testing

For a single switching operation for opening the contacts there is a transition in the direction of material transfer as a function of the arc erosion going from anodic to cathodic transfer, [24,25]. In the initial opening stage when the gap between the contacts is small, less than ~5–10 microns, short arc, the anode is eroding and there is some transfer of material to the cathode. As the gap between the contacts becomes larger the transfer of material during arc erosion reverses. For low voltage DC applications like automotive switching at 13VDC depending on the type of electrical load the accumulated material transfer can be significant. In AC, since the polarity changes with each operation if the timing of switching is random there is no cumulative material transfer effect (see Chapter 10, Section 10.3).

16.2.3.2 High Current Inrush DC Automotive and AC Loads

A high current inrush load like a lamp load may have a current as high as ten times the normal current for the first few milliseconds after contact closure. Most times there is some contact bounce that takes place after initial contact closure. This bounce will be associated with a high current arc and electrical erosion as a result of the arc will take place. For an automotive type DC load the arc will be short and anodic burning a crater in the anode and depositing material transfer onto the cathode contact. Since the contact opening part of the switching cycle will be at normal current the net result of endurance switching in this case will result in material gain on the cathode and material loss on the anode. For AC switching the same current with the polarity per operation random the erosion should result in a material loss for both contacts and no buildup of transfer on either contact (see also Section 10.3.5).

16.2.3.3 Inductive Loads

16.2.3.3.1 DC Automotive Inductive Loads

For these types of loads the induction in the circuit will cause a lag of the current buildup of several milliseconds, 3–8 typical. This means for switching closure under this load little erosion takes place even with moderate switching bounce taking place. For opening it is a different story since the inductance will prolong the arcing time and length of the arc. For a long arc on opening the arc will start out as being anodic and later make a transition to being cathodic with the material transfer changing from anode loss and cathode gain to cathode loss and anode gain. Chen and Witter [26,27] showed that the contact opening velocity and maximum contact gap in a device also have a significant effect on the erosion results under these conditions. They show the erosion of a silver–tin indium oxide to go from being anodic erosion to cathodic erosion as the opening speed changes from 0.47 m/s. to 1.28 m/s. It also should be noted that much of the fundamental testing is done using model switches which allow gathering of information not possible in actual commercial devices like sticking and welding force and actual weight loss. Sometimes these devices open and close the contacts at very slow speeds that are far slower than commercial devices and thus the result may not apply to the real world of switching.

16.2.3.3.2 Resistive Loads with no Inrush

The resistive loads have results that are mainly in between the examples shown for lamp and inductive loads.

B. Additive factors

From the above examples it should be apparent that electrical test results on materials and additives in materials will be subject to the electrical parameters and load type used in the testing and that loads differing from the tests performed may not agree with the results of a specific research.

16.2.3.4 Silver–Tin Oxide Type Materials and Additives

Silver–tin oxide type materials have become the most popular arcing materials for relays and contactors worldwide. Besides the many different variations in process for making the materials there is a large variety of additives used to adjust the properties of the materials. In this section we will discuss the most popular additives for silver–tin oxide materials: indium oxide, bismuth oxide, copper oxide, tellurium oxide, and tungsten oxide. The effect of the additives is a complex subject, since not only are the effects influenced by the manufacturing process and electrical application but also some of the additives interact with each other which change the effects. A complex study on combinations of CuO, Bi_2O_3, and WO_3 show the effects on ductility, arc erosion resistance, and contact resistance change significantly with different ratios combinations of these three additives, [28]. This means one must take into account the total package of additives being used.

16.2.3.4.1 Indium Oxide Additions and Tellurium Oxide

Indium oxide did not start out as an additive to silver tin oxide for the purpose of improving properties but instead as an aid to the oxidation process. The addition of indium to silver tin alloy prevents the formation of an imperious oxide wall from forming during the oxidation process. Silver tin indium oxides are still the most popular type of silver tin oxide type materials manufactured in Japan and China.

For AC contactors most manufacturers in Japan switched to silver tin indium oxide contacts in the late 1970s and 1980s. Since these contacts had a little higher contact resistance than silver cadmium oxide the companies increased the contact force in the contactors to compensate for the higher resistance. Besides for replacing cadmium as a black-listed element the companies were able to use smaller contacts and save on silver since the silver tin indium oxide material had a lower erosion rate. In 1992, Hetzmannseder and Rieder [29] did a study comparing erosion of silver tin indium oxide contacts and European silver tin oxide contacts that did not contain indium in a test device that simulated contactors operating in an AC-3 mode, high inrush, 6×, with normal break. The amount of erosion that take place in AC-3 testing is related to the amount of make bounce that takes place in closing of the contacts. Their testing controlled the bounce so a good comparison could be made for different materials. They showed a much lower erosion rate for the silver–tin indium contact than the European silver–tin oxide without indium and made by powder metallurgy. It should be noted that for bounce erosion the arc is short and the efficiency of material transfer between the contacts becomes a factor in determination of the actual erosion loss. This showed an advantage for the indium oxide addition for promoting a lower erosion rate in AC switch owing to good transfer characteristics. Much more recently Mützel and Niederreuther [30] did a comparison of some newer and older PM materials and included IO silver tin indium oxide. For erosion they did testing using an automotive inductive load and showed that the erosion rate of the silver tin indium oxide IO materials was much lower than the old PM material and similar to newer materials with other additives. Although this testing was DC testing the inductive load used had material transferring in both directions, from anode to cathode and as the contact gap widened from cathode to anode.

Although the indium oxide seemed to give an advantage for erosion on AC loads and certain DC loads it became somewhat problematic for high inrush DC loads. For lamp loads in DC automotive circuits the erosion of the contacts was mainly involved with contact bounce on closing since you had a large current on closing and normal current on opening. In this case, the material transfer that took place was always from the anode to the cathode and since the polarity stayed the same you had a buildup of material on the cathode. If the material had a high efficiency for transfer (i.e., good sticking of transfer material to the cathode), there was a higher probability of developing a condition called pip and crater erosion. In 1996, Witter and Polevoy [31] tested automotive relays with silver tin indium oxide contacts using a lamp load. Tests were run on the same batch of relays with some relay having a cover on enclosing the contacts and other relays with no covers just exposed to open air. These tests were done since some tests using a model switch were not duplicating the life results seen in the relays. The results of the testing were dramatic with regard to tests with and without relay covers. The relays with covers had long endurance life and those without covers failed very early as a result of severe pip and crater erosion. The relays with covers did not form pip and crater erosion as a result of organic outgassing from the plastic which activated the contact surface from carbon deposits of the organic gas and the arc. This phenomenon will be explained later in this chapter in the section on material transfer and in Chapter 19. The point learned from this work was that silver tin indium oxide had a high potential for forming pip and crater erosion which is very detrimental for DC high inrush loads.

Some companies in Japan that made silver tin indium oxide also discovered that a problem existed under certain applications with silver tin indium oxide. Work began on additives for improving this condition. In the early 1990s, several patents came out for adding tellurium to silver tin indium alloys for improving the performance of silver tin indium oxide materials. Some of the patents have expired and little has been published or explained about the actual phenomena. Witter and Chen [32] did an investigation using a model switch with a lamp load for comparing the performance of silver tin indium oxide materials with and without an additive of <0.5% Te which forms an oxide during the oxidation. The results showed much less pip and crater erosion for the silver tin indium oxide with the tellurium oxide for the 10.5 oxide materials and much less differences for higher level oxide materials, 14.3%, see Figure 16.5. Tellurium is an extremely toxic material and must be handled with care. The results for Te look similar to contact activation for moving the arc spot as discussed above for carbon deposits from organic vapors.

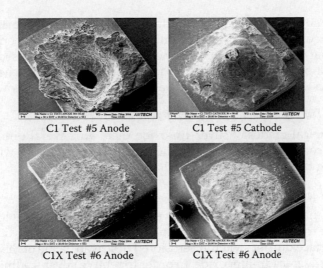

C1 Test #5 Anode   C1 Test #5 Cathode

C1X Test #6 Anode   C1X Test #6 Anode

FIGURE 16.5
Silver tin indium oxide 10.5% after lamp load switching, C1 without additive, CX with Te additive.

Chen and Witter [33] did further research on silver tin indium oxide materials by testing this material made by PM and made by IO. At the time the results were published in 2009, they were of the opinion that the PM and IO materials had the same level of indium oxide. It was found later by chemical analysis that the PM material had only 0.75% indium oxide compared to 3.5% in the IO material. The PM material was made by a separate company that made a composite powder by chemical precipitation. By this it was possible to make the PM material with a particle size distribution that was as fine as the IO particle size distribution. This is normally not possible by straight PM blending of components. A comparison of the IO and PM particle size distributions using automated scanning showed a mean average oxide particle size of 0.6 microns for both materials. Both materials were also fully dense being processed into wire by high pressure extrusion. The materials were tested using both a DC inductive load and a DC lamp load. The results for the inductive load showed similar results for both materials with a slightly less erosion rate for the PM material. For the lamp load testing, the IO material had over double the anode erosion and cathode material transfer than the PM material. At the time the paper was written, we concluded that more work had to be done to explain the lamp load results. After finding out the true difference in the indium oxide levels of the two materials, it is felt that the higher indium levels were the root cause for the higher level of material transfer. In this case, the higher indium level material had much more severe pip and crater erosion present. As the pip and crater formed the erosion rate was seen to increase dramatically from the concentration of the arc to a smaller area of the contact surfaces. Thus, in DC a material with less than 1% indium performs much better than one with over 3% indium. Under AC testing the materials were similar.

One more point found about silver tin indium oxide concerns contact structures that have the silver tin indium oxide attached directly to a copper backing. This is a common design for machine made composite rivets discuss in Chapter 17. In testing conducted by Chen and Witter, it was found that if the erosion level gets close to the interface level between the silver tin indium oxide and copper a phenomena called liquid metal embrittlement which can break up the contact structure [34]. This is from a low melting phase 110 to 120 C, recently found that form from copper and indium solders [35]. If the erosion actually reaches this level the contact's life is almost over. Up to this date there does not seem to be any other silver tin indium oxide failures by liquid phase embrittlement. From the above it can been seen that silver–tin indium oxide is complex and good or bad results can happen depending on the application and the level of indium oxide used. The use of Te will lessen the formation of pip and crater for high indium levels. It is believed that more work will continue on this material in the future since its erosion rate is low.

16.2.3.4.2 PM Silver–Tin Oxide Material with Different Additives

For powder metallurgical grades of silver–tin oxide there are several different additives being used and also combinations of these additives as described above which can result in interactions among the additives

16.2.3.4.3 Refractory Metal Oxide Additions

Both WO_3 and MoO_3 have been used as additives in PM silver tin oxide for many years. The main purpose of these additives is to minimize the buildup of tin oxide slags on the surface of the contact face during switching erosion. These were some of the first PM materials used to replace silver–cadmium oxide. In order to replace silver–cadmium oxide contacts as a retrofit in contactors the tungstate was added to the silver tin oxide to keep the temperature rise of the contacts lower during soak tests performed at intervals of endurance switching trials. These additives became popular mainly in Europe.

Today the PM metal processing is much more refined by use of newer powder processes like chemical precipitated powder which are as fine as or finer than IO materials. The tungstate additives are still used today for different grades of PM powders and many times with other additives. Mützel et al. [30] shows some recent results for DC testing with just tungstate additive alone compared to other materials. Kratzschmar et al. [36] shows results for tungstate additive with Bi for different levels of AC testing. The tungstate itself does not seem to have a large effect the erosion.

16.2.3.4.4 Bismuth as an Additive

Bismuth is an additive used in both PM and IO silver tin oxide materials. It is used most extensively in the PM materials in the form of Bi_2O_3. Kratzschmar et al. [36] show Bi additives to improve the erosion resistance when added alone or with tungsten additives for various levels of AC testing. Other testing [30] in DC shows it in combination with CuO and shows relatively low erosion and weld break force even with a coarse micro-structure. The problem is that there are too many combinations to get a good comparison on Bi. Hauner et al. [37] added 2% Bi_2O_3 to a number of chemically precipitated composite powders and show improved erosion and over temperature compared to conventionally blended silver tin oxide for AC endurance testing. In this case the effect of the microstructure and Bi are mixed together. Bismuth is compatible with both the IO and PM processes and more work is expected on using bismuth. The overall results indicate an improvement in erosion resistance and welding resistance with some addition of bismuth.

16.2.3.4.5 Copper Oxide Additions

Copper oxide, CuO, additions also have been used in PM silver tin oxide. Francisco et al. [38] tested a range, 0.25%–0.96%, of CuO additions in PM silver tin oxide and found a lower erosion rate in the range of 0.4% CuO. The testing was done using a 0.5 HP AC motor load. Most other testing was combined with other additives. For combined tests of Bi_2O_3, CuO, and WO_3 [28], it is indicated that CuO offers an improvement for this combination. The case for CuO improving erosion resistance seems to be true under certain conditions.

16.2.3.5 Material Factor

16.2.3.5.1 Silver–Cadmium Oxide Materials

As discussed earlier in this chapter, less emphasis is being put on this material since it has declined in use as a result of ecological concerns with cadmium. Although it is still used and manufactured in some areas of the world, pressure is present for eliminating it as an approved RoHS material by the European Union. As a metal oxide material silver cadmium oxide has some similar characteristics to silver tin oxide materials in types of processes used for manufacturing and physical structure. As a result of this some of the work done in development of silver cadmium oxide materials was also applicable to silver tin oxide.

16.2.3.5.2 Silver–Cadmium Oxide Versus Silver–Tin Oxide

Beginning in the late 1970s and through the 1990s a significant amount of work was done for development of new silver–tin oxide type materials. Many early papers were written comparing newly developed silver–tin oxide materials to silver–cadmium oxide and other contacts [21,23,39–41]. Most of the comparisons were for high current interruption duty and the $AgSnO_2$ materials showed low erosion compared to AgCdO. In 1984 Gengenbach et al. showed that for AC-4 testing, 320A make and 320A break, the $AgSnO_2$ showed much less erosion that the AgCdO, but for AC-3 testing, mainly bounce erosion, 660A make and 110A break, the AgCdO was better [42]. The testing was done on wrought powder metallurgical $AgSnO_2$. For a long period of time a general belief existed that $AgSnO_2$ was inferior to AgCdO for make arc erosion applications. In 1992, Hetzmannseder and Rieder presented testing done for make arc testing only in a device that simulated bounce in a contactor [29]. The work contained testing at different bounce times and at different currents. The testing included both wrought powder metallurgical type silver–tin oxide, wrought powder metallurgical AgCdO, and internally oxidized $AgSnO_2$ with additives. The results of this testing showed the internally oxidized $AgSnO_2$ materials with additives had significantly lower erosion than the AgCdO material and that the powder metallurgical $AgSnO_2$ remained higher in erosion as previous testing had shown.

As described in the previous section on silver tin oxide materials many types of silver tin oxides have been developed and refined in the last few decades. As a general statement it can be said that most silver tin oxides types have better erosion resistance and welding resistance than silver cadmium oxide materials. With regard to contact resistance after endurance testing some grades still have higher resistance and some materials have been developed that have similar resistance. Some device manufacturers have solved the higher resistance problem by using higher contact force. For the question about being able to replace silver cadmium oxide, the evidence of all the testing and success for silver tin oxide and other metal oxides that have replaced it, demonstrate that it would be rare to find an application for which there is no substitute. Besides the ecological incentive for replacing silver cadmium oxide the testing has shown that in most cases a smaller silver tin oxide contact can be used to replace a silver cadmium oxide contact as a result of the better erosion resistance.

16.2.3.6 Interpreting Material Research, Example from Old Silver Cadmium Oxide Research

The purpose of this section is to provide an interesting example that shows how many times the results of materials research is limited in scope as a result of how the material was manufactured, how the material was tested, and other variables involved in the research work. This example shows work by five independent groups doing similar studies on the effect of Li on the erosion characteristics of silver cadmium oxide. The main question was: Are Li additions to silver cadmium oxide beneficial or not.

For silver–cadmium oxide significant work on additives for powder metallurgical grades was done by several companies. Some of the results from the different companies are contradictory but the discussion of this serves as a good lesson for factors which must be considered in making material comparisons.

In the late 1970s two separate research groups reported significant improvements in erosion resistance for additions of Li to AgCdO [43]. Kim and Reid made the materials from a blend of silver powder and cadmium oxide powder and the contacts made by a press–sinter–repress process. Improved sintered properties were credited with a large improvement in erosion resistance. Brugner made materials with the IOAP powder technique [44] and a press–sinter–repress process. The densities of the material with and without the lithium were the same. IEC contactor AC-3 erosion tests [45], with six times normal inrush current and normal break current, showed about a 50% improvement for a 50 p.p.m. addition of lithium over no lithium. For greater or smaller amounts the erosion rate increased. Brugner attempted to explain the improvement in terms of interaction of the arc with lithium sites on the surface spreading the erosion more uniformly over the contact face, but admitted problems with support of this theory. Lindmayer and Bohm, a third team, conducted separate erosion tests on contacts with 0, 50, and 500 p.p.m. lithium [46]. The contact materials were made by wrought powder metallurgical process, hot extrusion, and, as a result, all materials were fully dense. No erosion resistance differences were seen for break arc tests run at 350, 700, 1000 and 1300 A for any of the materials. A fourth team, Jager et al. [47], investigated lithium additions for blend, press, sinter, and hot repressed AgCdO contacts. Testing was done in a contactor under IEC AC-testing [48], both six times rated current for make and break. The authors reported improved sintered densities with the lithium parts. Results showed lower erosion for lithium-containing contacts and improved arc mobility for up to 100 ppm. lithium and no improvement in arc mobility for lithium above 100 ppm.

Spectrographic analysis of the arc between AgCdO contacts by Kossowsky and Slade on materials made by internal oxidation showed a much higher cadmium content than silver, although the cadmium content in the bulk material is much less than silver content [49]. This dominance of cadmium in the arc is the result of sublimation of the CdO. They also observed that as the cathode surface became depleted in CdO the ratio of cadmium to silver in the arc decreased, as evident from alternating ratios of cadmium and silver deposits on the anode. Thus the arc chemistry is controlled by the chemistry of the melted surface layer which cycles in composition as a result of previous arc heating that causes both vaporization losses of material and gains in material from melting and diffusion of this layer with the sub-layer bulk material. Brecher and co-workers conducted similar spectrographic tests on powder metallurgical contacts with and without lithium [50,51]. The first tests they ran on new contact surfaces showed little difference between lithium-containing contacts and non-lithium-containing contacts. Further work on contacts taken from a contactor after 50,000 operations showed a higher ratio of cadmium to silver for non-lithium-containing contacts than contacts with lithium. The arc temperature was also calculated to be about 200–300 K cooler for lithium-containing materials. The arc chemistry alone would be difficult to use for supporting lithium having an effect on the arc since from the above work it was shown that the arc chemistry will vary with the contact surface chemistry. The arc temperature difference with the chemistry difference adds strength to this work. The results seem to support Brugner's speculation.

Several years later a fifth team, Witter and Lu, made samples with and without lithium using a multiple coining and resintering powder metallurgical process to obtain similar densities with 0 and 30 ppm. lithium [52]. Both make only and break only tests were done to measure electrical erosion. Samples were tested that had both 98% and 99% densities for all three lithium compositions. Break only tests showed lower erosion for higher densities but no differences for lithium content. Make only tests also showed lower erosion for higher densities but also significantly lower erosion for the lithium-containing materials. Arc retention time for break arc tests showed no effect for lithium.

From this example, the reader can obtain an idea of the complexity involved in predicting how materials made by different processes are going to compare when tested in different devices. For the above case we have five sets of testing, some supporting an improvement for lithium additions and some showing no effect. The five sets of materials were all made by different processing, some were close but none were the same. The testing was also done in five different devices and testing conditions varied. In addition to the testing, we have arc analysis data that suggests lithium has some effect on the arc. In order to get a better feeling for the facts Table 16.2 is put together to summarize the five tests.

For this example case, we can make the following comments on Table 16.2:

1. With four positive tests out of six we can believe that lithium is beneficial under some conditions for certain materials. This result is actually very typical of what is seen for materials in the contact field. Results are rarely universal because of the many variables involved with the material technology and differences in devices.

2. Since there are many variables, there may be more than one factor affecting the results and in this case this appears to be true.

3. For powder metallurgical products, the density effects should always be considered. Erosion is almost always increased from structural defects like porosity. In this case, there may be significant density effects for tests 1 and 4, so interpretation of lithium effects is difficult. For any testing investment in these types of contacts, the supplier should be willing to supply density information for the test record.

4. The kind of predominant arcing that the contacts will see should always be considered, lamp load versus inductive load, make arc versus break arc, dc versus ac, and more. The erosion mechanisms for make and break arcs are normally different since make arcs usually involve contact bounce, with mechanical splatter and short metallic arc erosion compared to a break arc which normally involves no mechanical splatter and a transition from a short metallic arc to a longer gaseous arc (see Section 10.3.5). The density effects for lithium can be eliminated from tests 3 and 5a which both show no effect for lithium. In this case it appears that the effect of lithium additives for break arcs is questionable.

TABLE 16.2

Testing Parameters and Results for Erosion Tests for Lithium in AgCdO

Test no.	Lithium Density Influence	Make Arc	Break Arc	Lithium Improvement
1	Yes	Yes	Yes	Yes
2	No	Large	Small	Yes
3	No	No	Yes	No
4	Yes	Yes	Yes	Yes
5a	No	No	Yes	No
5b	No	Yes	No	Yes

5. Two tests for make erosion can be separated from density effects, tests 2 and 5b. Both of these tests show large improvements for lithium additions. Since contact make involves only short arcs in duration and length, it is unlikely that arc mobility has much influence as it does on break arcs. The erosion primarily results from the short arc during contact bounce and also liquid metal splatter on re-closure of the contacts. For bounce arcs the wetting characteristics of the liquid metal melt on the contact surface will have an effect on the amount of splatter, therefore if lithium improves wetting the erosion will be less. This finding shows it is always important to look at both make and break separately to better understand a material's erosion for specific applications.

6. Arc analysis data supporting lithium influence on arc temperature and chemistry are not supported by the data for break arc erosion. A possible reason for this is that the spectrographic analysis was done for a long, 4 mm, fixed contact gap arc that does not represent the plasma composition generated during the initial stages of interruption of the arc in a device like a contactor. Much work remains to be done relating this type of data to practical results although contact technology is a mature science. There still is much to be learned regarding contact and erosion technology.

From this example, it can be seen that it would be difficult to generalize widely about the application of lithium additives for different silver cadmium oxide materials and electrical devices without more information. Even with more information it is unlikely that absolute predictions of performance can be made for any material in different contact devices. For this reason, extensive testing of contacts in the actual devices for which they are going to be used is always recommended before a specific contact is approved for use in a device. The variables tend to be more complicated than what we interpret.

16.2.3.6.1 *Silver Zinc Oxide Materials*

Silver zinc oxide has become the second most popular metal oxide for replacing silver cadmium oxide next to silver tin oxide. It is used in Europe for lower current switches, relays, and contactors. Schoepf et al. compared silver ZnO to other oxide systems with 92% silver and showed that silver zinc oxide without additives had a problem with high current inrush applications [53]. Chen and Witter also did testing using wall switches and a lamp load [54] and found strong welding for switches that had high bounce time, >0.5 ms. Work done that included the use of additives in the silver zinc oxide [53,55] showed a large improvement in service life of the silver zinc when silver tungstate, Ag_2WO_4 was added to the silver zinc oxide. The tests for high inrush loads showed 0.25% WO_4 gave the best results and for inductive loads a little higher level was best. This work showed a better performance with the silver zinc oxide with the additive than for silver cadmium oxide in these applications. Many of the tests comparing the silver zinc oxide contacts to the silver cadmium oxide contacts also showed a lower resistance after switching with silver zinc oxide.

Behrens et al. [55] also noted contact sticking for silver zinc oxide without additives and much less sticking with silver tungstate additions. In comparing the microstructures of the two silver zinc oxides it was noted that there was much less zinc oxide and silver segregation during switching endurance with the tungstate additive which probably explains the improvement in welding resistance with the additive. It was also reported that finer particle size distributions of the tungstate were more effective that coarse distributions.

Another additive, silver molybdate, has also shown improvements for silver zinc oxide but a little less improvement than the silver tungstate. The silver zinc oxide with silver tungstate additives has proven to be a good substitute for silver cadmium oxide in the lower current range especially when low contact resistance is a concern.

16.2.3.6.2 Other Metal Oxides

Besides the three metal oxide systems discussed above there are some other potential oxides. Rare earth metals have a few papers [56] but don't have much appeal since they pose a supply problem. There also does not seem to be much advantage over the current materials.

There is some small use of silver iron oxide, Fe_2O_3, but not much is published on this material. The material, AgMgONiO (0.3wt%MgO) (0.3wt%NiO), is used with some relay applications. This material is internally oxidized and has oxide particles that are extremely small. The small amount of oxide is effective in reducing transfer of silver in dc applications. This material shows very flat erosion even under high inrush DC since the oxides act as activation sites for new arcs and this keeps the erosion even.

16.2.4 Material Considerations Based on Electrical Switching Characteristics

16.2.4.1 Erosion/Materials Transfer/Welding

The material transfer that takes place with DC switching was explained earlier in the silver metal oxide sections of this chapter showing erosion transfer first going from the anode to the cathode and reversing as the contact gap increased, also see Chapter 10. A discussion of welding and sticking of contacts is also included in this section as related to a detrimental form of transfer called pip and crater erosion, also see Chapter 10.

Leung and Lee conducted work on silver alloys in automotive relays [57–59]. The relationship for anode and cathode erosion was shown for 0.5 mm gap switching at 12 V dc and resistive, lamp, and inductive loads. For short arcs like bounce arcs anodic erosion predominated and for interruption the erosion became more cathodic. They compared a combination of $AgSnO_2$ (10 and 12 wt.%) contact materials to a metal alloy contact material, AgCu (2 wt.% Cu) and a powder metallurgical material AgNi (20 wt.%). The results of comparisons were very dependent on the load type and current level. In general the silver–tin oxide material did much better compared to the other two materials for high inrush closure, lamp loads, since it exhibited much less transfer and contact welding [58]. The AgCu and AgNi materials welded early in life for the lamp load, 63 A peak current, with less than 40,000 operations compared to over 120,000 operations for the $AgSnO_2$ materials. For lower-current resistive and motor loads the results were different with severe pip and crater formation on erosion being a problem for the silver–tin oxide material. Cathodic pip and anodic crater formation is common for short arc dc loads and from this work it was shown that all three types of materials exhibited this type of erosion on low-current make. Further work done by Leung and Lee on silver–tin oxide materials showed that one of the advantages of a silver metal oxide material like silver–tin oxide is that the bonding of anodic material deposited on the cathode is weak [59]. They showed evidence of pips delaminating from the cathode and refilling the anodic crater, resulting in a low net transfer. It was also shown that for switching of asymmetric contact materials, silver–tin oxide mating with silver–copper, it was an advantage to use the silver–tin oxide material as an anode rather than a cathode. With the silver–tin oxide material as an anode a thin brittle melt layer containing silver and tin oxide material deposited on the cathode. This resulted in low transfer as a result of poor bonding to this surface by eroded anode material and good resistance to contact welding. For the opposite polarity, the silver–copper alloy transferred onto the silver–tin oxide cathode and resulted in a large amount of transfer forming a huge mound of silver copper over the silver–tin oxide original surface. This also resulted in poor contact welding resistance since the bonding of the silver–copper alloy to itself was strong.

From the above, it can be seen that the transfer characteristics vary with materials. As pointed out earlier in this chapter when discussing silver tin indium oxide material good wetting and thus efficient sticking characteristics of anode transfer to the cathode makes a material more prone to pip and crater erosion. Another important variable is the switching characteristics of the switching device in which the contacts are being used, in terms of contact bounce during contact closure (see Chapter 13). Witter and Polevoy studied material transfer for silver–tin–indium oxide materials in automotive relays [31]. They found that pip and crater type transfer was more severe as bounce frequency increased as opposed to increases in total bounce arc time. Figure 16.6 shows a comparison of pip

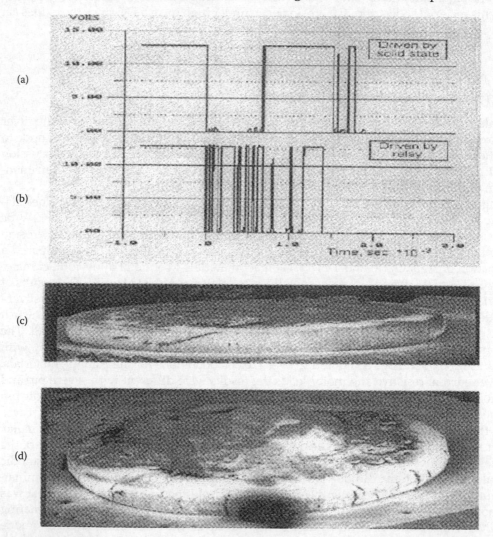

FIGURE 16.6
An illustration of transfer variation as a result of contact bounce frequency. Bounce trace (a) for contact (c) has a relatively low frequency compared to trace (b) for contact (d), but both have about the same bounce arc duration. Both silver–tin–indium oxide contacts saw the same lamp loads for 16,000 operations [31].

and crater formation for two different conditions of bounce. The reason for the increased pip and crater transfer with increasing bounce frequency can be rationalized as follows. As the bounce frequency increases two changes take place: (1) the number of bounces per operation increases thus the ratio of short anodic arcs increases compared to longer opening arcs per operation increasing net anodic transfer. (2) the amplitude of the bounce decreases, thus the contact gap during arcing is smaller which increases transfer efficiency from anode to cathode and reduces splatter For a device that both makes and breaks a dc circuit the ratio of the magnitude of anodic transfer on make to the magnitude of anodic and cathodic erosion that takes place on break has a major influence on the type of transfer that takes place. If only make and no break erosion takes place, pip and crater type erosion is normally present [57,58]. If the break erosion is much more severe than make, for example as with an inductive load, cathodic erosion that takes place during break is usually strong enough to prevent a pip build-up on the cathode. For the present discussion on the effects of increased bounce frequency it can be seen that as the number of closures per operation increases compared to the single break per operation, a point will be reached where the break erosion is not sufficient to prevent pip formation. This point, coupled with the reasons given above for increased concentration and efficiency of transfer, gives some of the reasons for seeing higher transfer with increased bounce frequency.

Another factor that influences material transfer and formation of pip and crater formation is contact activation, see Chapters 10 and 19. Witter and Polevoy showed that organic vapors given off by plastic components in the relay had a beneficial effect for preventing pip and crater formation [31]. Germer showed that as a result of contact activation by a substance such as carbon, the arc spot moves from one operation to another and pip and crater formation is prevented [60]. In addition to this it was shown by Germer et al. that contacts can be activated by non-organic materials such as minerals, silica, alumina, and others. A special silver metal oxide material that has been used for many years is AgMgO (0.3wt%)NiO(0.3wt%). The microstructure of this material is an extremely fine submicron random dispersion of oxide particles in the silver matrix. This microstructure continually produces an activated surface and as a result this small amount of oxide is effective in reducing transfer for dc switching.

From what has been discussed, it can be seen that many factors influence the transfer that results from a silver metal oxide material including circuit parameters, device parameters, and factors influencing contact activation. A material may show good transfer resistance at one level of current and high transfer at another as demonstrated above. The erosion characteristics and the sticking coefficient for anode transfer onto the cathode is important in influencing transfer build up. In testing relays with different silver–tin oxide type materials under severe bounce conditions, a material with a low arc erosion resistance formed no pip and crater type erosion but had very high anode material loss; another material with moderate erosion resistance had significant pip and crater formation under the same conditions, and yet another silver–tin oxide material with much better erosion resistance showed a low erosion rate with no pip and crater erosion [12]. Thus in this case a material with an erosion rate intermediate to two other materials, showed much more anodic transfer than materials above and below it in erosion rate. These results are not surprising considering the many factors that have been discussed that can influence transfer and pip and crater formation. For purpose of discussion and illustration let us focus on the fact that for this type of bounce erosion, the resulting short arc resulted in mainly metallic erosion. If a pip was prevented from forming on the cathode no pip and crater would form. This was a make and break application. We knew that if it had been only make erosion, pip and crater erosion would have been probable. Some of the factors that could have affected these results are as follows:

1. If the material with the high erosion rate had a low tendency for the anodic transfer material to stick onto the cathode surface and at the same time exhibited enough cathodic erosion, there would be a low probability of a pip forming.

2. Again for their material with the high erosion rate, owing to the large amount of erosion from the anode the position of the arcing will drift more from operation to operation, and that also will lower the probability for forming a pip.

3. For the medium erosion rate material, if the factors opposite to the above are true, good wetting of the anodic transfer material to the cathode and stability of the arc spot, a pip will probably form.

4. For the low erosion rate material there was a higher metal oxide content. This also decreased the anode transfer sticking efficiency and with some cathode erosion present there was no pip buildup.

5. If, for the low erosion rate material, there is some degree of activation from additives or external factors, the arc will tend to move from operation to operation as a result of the activation sites and this with even low cathode erosion rate will make pip and crater erosion a low probability. Activation can come from gases given off by the plastic components of a relay.

The above of course is only for illustrations of possible reasons for the transfer behavior. It should be kept in mind that regardless of the material, high frequency bounce will increase the probability of pip and crater formation.

16.2.5 Transfer/Welding

The welding resistance of silver metal oxide materials is affected by the surface microstructure and surface geometry developed and as a result of contact arcing and material transfer. There are two types of welding that should be considered, static welding and dynamic welding (see Chapter 10). Static welding involves welding which takes place with the contacts in a closed position under force. Static welding resistance relates to contact conductivity and surface resistance. Materials with lower contact resistance have a higher resistance to welding. For silver metal oxide types there is little difference among the various materials to resistance to this type of welding.

For dynamic welding, the condition of the contact surface after some switching duty determines the welding tendency. One of the largest causes of dynamic welding for dc devices is from formation of pip and crater geometry on the contact surface as discussed above. For this type of weld, the classic welding as a result of melting and solidification may or may not occur. Many times the pip and crater form mechanical latching that prevents the contacts from opening. Once pip and crater geometry is established the life of the contact is usually limited since all erosion is taking place in a limited area which increases the extent of melting and alteration of the material in that area, including oxide depletion, segregation, and surface roughness. In Chapter 10, there is a theoretical discussion of both static and dynamic welding with equations showing the relationship of maximum weld force, contact force, contact melting temperature, material hardness, switching current, constriction resistance and the total energy into the weld spot. It also has data on weld force measurements on several different materials.

In this section, some different kinds of dynamic welding will be discussed on the basis of experimentation using some model switches. By use of a special model switch that simulated NC (Normally closed) contact operation in a DC relay Chen and Witter studied contact welding characteristics of several materials under different conditions [61–63]. They found two types of dynamic welding or sticking. One was a very weak weld, cold adhesion that occurred after contacts were subjected to a switching operation that was mainly in the metallic arc state. Once the very clean arced surface had formed, this adhesion continued for a number of operations even without current present. Figure 16.7 [63] shows a switching sequence with a combination of no electrical load, resistive load, and inductive load. The resistive load promoted the adhesion by cleaning and annealing the surface and the inductive load left a thin layer of silver oxide or carbonate on the surface which reduced the adhesion. It was found that strong welds above the adhesion level strength, >35 cN, occurred infrequently. It was also found that strong welds were always associated with very short bounce or skip arcs, <50 μsec. These welds also could occur on contact make or break. For arcs with long bounce, strong welds were never seen. For very short bounce the contact gap is very small and is associated with a high concentration of arc heat and low impact closing which left more liquid metal to solidify between the contacts. Similar findings were made by Morin et al. [24] and Neuhaus et al. [64].

Considering the findings sited above on strong welds it should be kept in mind the important role contact force and over-travel of contacts on closure play in preventing contact welding. Figure 16.8 shows a typical weld strength distribution as a function of switching endurance life for a normally closed DC relay. In this case the beginning of the switching life has no strong welds but as erosion takes place and the contact over-travel decreases the conditions that favor the formation of strong welds.

There is another variation of dynamic welding that can occur if the current level is extremely high or the arcing time is very long. For example in doing some automotive DC testing of silver tin oxide composite rivets the test voltage was raised in increments to see the effect on arcing time. At 19VDC the tests showed normal erosion for about 15,000 operations and then a hard weld would occur that was completely across the face of the contacts.

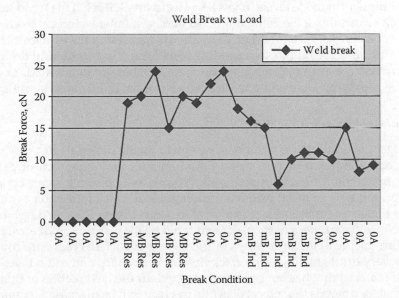

FIGURE 16.7
Weld break force with different sequential loads as follows: 0A current, 80A resistance, 0A current, 50A inductive, and 0A current.

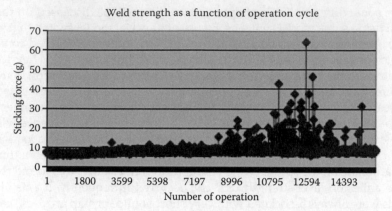

FIGURE 16.8
Typical weld strength distribution as a function of switching life [12].

What was happening in these cases was the surface of the rivet was curling up from stresses created from the arcing, see Chapter 17 for curling effects. The contact gap which was marginal to start with for this voltage became too small to allow the arc to extinguish and the arc time went from less than one msec. to hundreds of msec. The movable contact in this case was a thin contact spring that could not transfer the heat out fast enough which resulted in a large portion of the contact melting and forming a large weld area on contact closing. For this type of extreme heating almost all non-refractory contact materials will form a strong bond.

Silver metal oxide materials are known to be much better than non-oxide-containing silver alloys for being resistant to dynamic welding. The characteristics of the surface layers of the materials after switching determine the welding tendency. Erk and Finke showed threshold weld currents for silver metal oxides to be over three times higher than those for silver and silver alloys [65]. One important factor to keep in mind on silver metal oxides is the effect for the metal oxide level on welding resistance. Chen and Witter show a huge benefit for higher tin oxide levels in lowering weld break force. This trend holds true for almost all silver metal oxide combinations. So here you must weigh between welding resistance and contact resistance. Another alternative is increasing the contact force which will improve both contact resistance and welding resistance. The mechanical design of the switch is also critical. For example, impactive opening will break contact welds easier than a slow pull force. Applying an opening torque can also assist in breaking a contact weld.

16.2.6 Erosion/Mechanisms/Cracking

Like other contact materials, the erosion of silver–metal oxide materials involves evaporation of metal vapors and can involve metal liquid splatter and expulsion of particulate and chunks of materials. Unlike true silver alloys such as fine silver, silver copper, and silver palladium, some silver metal oxides are susceptible to crack formation as an erosion mechanism. At higher currents, hundreds of amperes, most materials can develop surface cracking and fissures from thermal stresses from rapid heating and arc root erosion on the surface. For more ductile materials the tips of the fissure of surface cracks are rounded or blunted off. For very brittle materials the crack tips are very sharp and extend well beyond the surface into material that has not been heat affected. In the next section of this chapter under silver refractory metals the subject of brittle fracture will be discussed in more detail. Normally brittle fracture mechanisms are considered impossible in ductile materials, that is materials which have face centered cubic structures. Although silver metal oxides are in the range of about 80% silver by volume, being composite materials, weak paths made up of contiguous brittle phases can exist. For example thermal arrest lines in internally oxidized materials have been reported to cause delamination of contacts. Here the oxide particles are so concentrated that a continuous plane of particles exists which allows the crack to propagate without going through the more ductile silver phase. For silver metal oxide materials, both internally oxidized grades and powder metallurgical grades, there is potential for creation of brittle paths in the materials by somewhat different mechanisms. For both types of material extruded material are less likely to present brittle cracking problems than material made without extrusion since the metal working will tend to break up the brittle phases.

For PM grades made by press sinter and repress some additives could create brittle paths through fine grain boundaries in these types of materials. An example of this type of defect is shown in (Figure 16.9) for Li additives for silver cadmium oxide [52]. For internally oxidized silver metal oxides brittle phases can form in the grain boundaries of the materials during oxidation as a result of certain additives. For some of the post-oxidized materials these brittle compounds in the grain boundaries limit the amount of plastic deformation that can be done on the materials after oxidation. Yet some of these materials withstand arcing without any major cracking. The grains in internally oxidized silver metal oxides are much larger than the grains of the brittle press, sinter and repress metal oxides. This may account for the better resistance to cracking from arcing thermal stress. Internally oxidized materials that are made from process 2.02, see Figure 16.1, wire or strip extruded after oxidation, have no problem since the metal working destroys the continuity of the brittle phases. Another variation of internally oxidized silver metal oxides that have appeared in the 1990s are ultra-fine materials made by high-pressure oxidation, they contain dispersions of oxide particles less than 100 nm in size [66]. These materials are over one order of magnitude finer than the general silver metal oxide grades. It is worth mentioning these materials in this section on cracking since as the particles become finer the

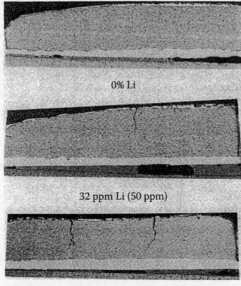

0% Li

32 ppm Li (50 ppm)

110 ppm Li (200 ppm)

FIGURE 16.9
Examples of brittle cracks going through an Ag/CdO 10wt% CdO with different levels of lithium as an additive, lithium levels shown both after sintering and before [52].

materials become more brittle from restriction of dislocation movement by the ultra-fine particles and thus more susceptible to cracking.

16.2.7 Erosion/Arc Mobility

For break erosion in devices designed to move the arc from the contact onto arc runners and arc chutes (see Chapters 14 and 15), arc mobility is a factor affecting the amount of erosion that takes place. Composite materials like silver metal oxides have lower arc mobility than pure silver and homogeneous silver alloys [67]. Little difference in arc mobility has been found among the various silver metal oxide compositions by Schroder [68,69]. He also has found that the manufacturing process has more effect than the chemistry with the internally oxidized grades showing faster arc mobility than those grades made by powder metallurgy. The effect is thought to be associated with differences in the silver grain boundaries between the two types of materials. Manhart et al. have shown that the arc running velocity is a function of contact gap and independent of metal oxide type, but that silver–tin oxides tend to begin movement at shorter gaps [70].

16.2.8 Interruption Characteristics

The ability to interrupt a circuit is an important characteristic for silver metal oxide materials, especially for use in devices like contactors. The tendency for arc retention after current zero in some ac switching devices owing to alkaline impurities in some new silver–cadmium oxide materials had been seen in the 1970s [71]. Some companies making silver–cadmium oxide materials by powder metallurgy at that time were experimenting with new processes for making silver powders and as a result unintentional increases in alkaline impurities, less than 100 p.p.m., ended up in the finished silver–cadmium oxide contacts. The example illustrates how careful companies must be in insuring that production processes for making contacts are not altered without extensive electrical testing of the contacts. At the same time some companies were experimenting with lithium as an additive to silver–cadmium oxide for improving the erosion life. The ability for a device to interrupt an ac circuit, reduction of reignition voltage, after current zero can be degraded by contacts containing small amounts of elements that have low work functions and low ionization potentials (see Chapter 9). As a result of this several separate groups ran tests and published the results for the effects of retained traces of sodium, potassium, and lithium in silver–cadmium oxide [71–73]. In general similar results were found by all groups. Lindmayer ran tests at 350 and 700 A and at 12 and 35 kHz. His conclusion was that potassium should be kept <20 ppm and sodium and lithium <50 ppm All three of these elements had a low work function and a low ionization potential, potassium being the worst. Braumann and Schroder ran comparisons between AgCdO and $AgSnO_2$ materials containing no alkali metals and found similar results for both, that is, good extinction properties of both materials in comparison to non-silver metal oxide materials like fine silver and silver refractory metals [67,74]. Gengenbach et al. compared interruption times and maximum arc voltage just before interruption for several types of silver–cadmium oxide and for one type of silver–tin oxide in a dc contactor at 200 A and 12 V [75]. No significant differences were found among the materials for arc interruption duration.

16.2.9 Contact Resistance

The contact resistance for silver metal oxide materials before switching duty is like other silver alloys dependent on the bulk resistivity, hardness of the material, surface geometry, and presence of oxide and/or corrosion films. As shown in this chapter's section on welding, contact arcing using a resistive load, metallic arc, cleans the contact surface and besides increasing adhesion between the contact pairs will lower the resistance. This is because silver oxide is not stable over 300 C. In general the initial contact resistance variations for clean un-arced contacts among the various types of silver metal oxides differ little and the values for resistance are low.

After contacts are subjected to switching duty the surfaces change mainly as a result of reactions between the contact surface and the electrical arc. As a result of arcing silver metal oxide contacts build up a surface layer that contains a mixture of metal and slag; that is, fused metal oxide aggregates. Like erosion characteristics, the microstructure of this layer will vary as a result of many factors including contact chemical composition, contact metallurgical characteristics, and switching load. The characteristics of this layer will determine many performance aspects of the contact and in particular both contact resistance and welding resistance. If the surface layer becomes depleted of oxide material the surface will react more like fine silver contacts, that is, low electrical resistance but high tendency for dynamic welding. If, on the other hand, the oxides build up to higher concentrations the contact will be more refractory in characteristic and the contact resistance will be high but the tendency for dynamic welding will be lower. This is an over-simplification for purpose of illustration.

Silver–tin oxide generally has higher contact resistance after switching duty than silver–cadmium oxide. In previous sections of this chapter it was shown that there are a large variety of silver tin oxide materials. Both the research on processes and additives has created materials more sophisticated than the materials developed in the past. Many times combinations of additive are used to reduce resistance and improve erosion resistance. Mutzel et al. [30] tested a variety of advanced materials which were both PM and IO with additives using a DC automotive load. Figure 16.10 shows a resistance comparison of six

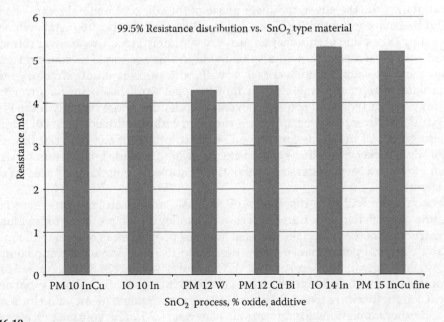

FIGURE 16.10
A comparison of contact resistance values of different silver tin oxide materials tested at 40 amperes inductive load at 13 VDC for 50,000 operations [30].

materials after endurance. The biggest difference in results is for higher resistance as the oxide level increases. The IO material is a little higher in resistance than the PM.

16.2.9.1 Summary Metal Oxides

Work will continue on silver tin oxide and silver zinc oxides systems with use of additives. Significant improvements have been made for both IO and PM grades. The use of composite powders with chemical precipitation has allowed the PM grades to improve significantly in erosion to compete with IO product. More so than ever, it can be said that there is no generic type of silver metal oxide material. This means it is very important that the user understand and test the material chosen for an application.

16.3 Silver Refractory Metals

Silver refractory metals are a special category of arcing contact materials for use in high fault current applications. These applications include residential circuit breakers, industrial breakers and high current switchgear used by utilities. A volume of silver-refractory metals are used in North America as a result of the high current level fault protection required by UL even for residential homes. It also should be noted that the requirements for fault current devices are different in different parts of the world so the material requirements also differ. The research level being done on these materials is continuing but not at the pace of work in the silver metal oxide field.

Silver refractory metals are mixtures or composite structures as opposed to true alloys since there is no or very limited solubility of the silver with the refractory metal. Since there is no alloying with the silver, the silver phase of the composite metal retains its high electrical and thermal conductivities. Since copper refractory composite metals which are discussed in the next section of this chapter are very similar in structure to silver refractory metals, points on those materials will also be made in this section. The silver refractory metal category includes combinations of silver with both the elemental refractory metals and also carbides of refractory metals. Much of the past work has been on silver tungsten, silver tungsten carbide, and some on silver molybdenum. More recently there is new work on mixtures of silver tungsten and silver tungsten carbide with additions of graphite. Some of these grades of materials are also paired with other contact material like silver graphite and silver nickel. Since these composite materials are made by different processes it is important to have a basic understanding of the technology in making choices on these types of materials.

In this section there will be a discussion of the basic manufacturing processes being used and some characteristics of those processes, metallographic methods for evaluation of the microstructure of the materials, physical material properties as related to the microstructure, and electrical performance properties. Table 16.3 lists some common compositions of W/Ag, WC/Ag, Mo/Ag, Mo/Ag, and W/Cu. Tables 24.4 and 24.8 in Chapter 24 have a more complete listing of different properties for these materials. The density, hardness, and electrical conductivity are listed as typical values, approximate mean value for several brochures and other tabulations for the various materials [2–5]. The volume percentage of the soft phase is also shown in Table 16.3 since there is a large density difference in silver,

TABLE 16.3

Typical Refractory Metal Contact Compositions Including Hardness and Conductivity Values

Material	Ag or Cu (wt.%)	Ag or Cu (Vol.%)	Hardness HV (kg mm^{-2})	Conductivity (m Ω^{-1} mm^{-2})
W/Ag-20	20	32	200	23
W/Ag-25	25	38	185	25
W/Ag-35	35	50	165	28
W/Ag-50	50	65	130	36
WC/Ag35	35	45	175	23
W/CAg-50	50	60	155	29
WCAg-60	60	69	145	32
Mo/Ag-35	35	34	155	24
Mo/Ag-50	50	49	140	29
W/Cu-25	25	42	200	23
W/Cu-50	50	68	125	29

Source: Adapted from G Rolland et al., *Proc. 26th Int'l Conference on Electrical Contacts*, 2012, pp 338–345 [1]; Chugai USA Brochure, Chugai USA Inc., Waukegan, Il, 1995 [3]; American Society for Testing and Materials. ASTM Standards, Section 3, *Metals Test Methods and Analytical Procedures*, Vol. 03.04, 1987; K Schroder. Dissertation, TH Braunschweig, 1969.

copper, and tungsten and the properties relate to volume percentage not weight percentage (see Chapter 24, Section 24.2[h]). The tungsten copper materials are for the infiltrated grades only and this will be explained in the following sections.

16.3.1 Manufacturing Technology

The silver refractory metal contact materials can be made by a wide variety of powder metallurgical techniques. Since the vast majority of the materials are made by one of three basic processes, the manufacturing technology discussion will be limited to the basic differences and characteristics for these three methods. The three methods are shown in Figure 16.1, all under Process D; (1.0) press sinter repress (PRS), (2.0) liquid phase sintering (LPS), and (3.0) infiltration. More recently a fourth process, hot isostatic pressing (HIP), is being used in combination with the basic three processes to increase density. For more detailed information on general powder metallurgy techniques including, HIP, liquid phase sintering, and infiltration, "Powder Metallurgical Science" by R. German may be useful [76]. It should be understood that saying that only three general processing techniques are used does not imply that the individual processes are the same for different companies. Although two companies, for example, have processes that fit the general description for infiltration, the process steps and controls for the individual steps may vary significantly. In order to make the discussion of the three processes easier, a description of the general processing for silver tungsten will be discussed for the three techniques.

16.3.1.1 Manufacturing Technology/Press Sinter Repress (Process D 1.0)

This, like other powder metallurgical processes, starts with powder manufacturing and blending. The starting powder particle size distributions will be the main factor for determining the possible finished microstructure and also have a large effect on pressing and sintering results. For silver tungsten, one or more silver powders and tungsten powders would be selected on the basis of particle size distribution. The powders are then blended with or without additives, pressed into the contact shape, sintered, and repressed to final density. The sintering is done in the solid state, under the melting point for silver, and in a reducing atmosphere such as hydrogen. The process is normally used only for grades with higher percentages of silver, >50% by weight, since coining is not possible when the percentage of tungsten in the material is large. This process is the main process for making silver tungsten with 75% or more silver as the other two processes will not work for high levels of silver since there is not enough tungsten in the contact body to hold it together for sintering above the melting point of silver. Contacts made by this process can have considerable porosity and entrapped gases. This process has become more popular for grades that add graphite powder to silver tungsten and silver tungsten carbide. Some of these grades have a high amount of silver and also a good volume of graphite so sintering and repressing is a way to increase the density.

16.3.2 Material Technology/Extruded Material

Some of the materials that use the PSR process can also be extruded. This will produce materials with higher densities and hardness. This does not mean that the material will have better electrical performance since at least hardness does not usually correlate with things like electrical erosion resistance.

16.3.2.1 Material Technology/Liquid Phase Sintering (Process D 2.0)

For this process powder is selected and blended to the final composition desired as was done with the press–sinter–repress method. The difference with this process is that the part is sintered to density and not dependent on coining or repressing to attain a high density. This means that the parts are pressed oversize, for example 5–15% linearly, and then shrunk to final size during sintering. The sintering is done at temperatures above the melting point of the silver in a reducing atmosphere, containing H_2, thus there is both a liquid and solid phase during the sintering operation. There are some important mechanisms that take place during the sintering with this process that influence the tungsten distribution which will be discussed later. The liquid-phase sintering process is composed of three stages: the first stage involves the flowing of the liquid phase into pores followed by rearrangement of the solid particles to form a denser packing; the second stage involves further densification and particle growth from liquid transport of solid phase through the liquid phase; and the third phase involves further densification by solid-state sintering of the solid phase [77,78]. Since silver and tungsten show no solubility even in the liquid state, for liquid-phase sintering to be effective in increasing the density of silver tungsten, additives, for example nickel or cobalt, are used for activation of the liquid-state sintering process [79]. Nickel is a very common additive since it has a very slight solubility in tungsten and silver in the 1000°C range, and nickel increases the self-diffusivity of tungsten by two orders of magnitude [80].

The liquid-phase sintering process can produce very high-density parts, 99%, for some silver or copper refractory systems. For metal carbides it is more difficult to obtain high-density parts with this process; therefore the process is limited in application. Some noteworthy characteristics of this process are as follows:

1. Additives for sintering are normally utilized in order to obtain density. Most additives have an effect on the conductivity of the tungsten phase. Nickel in the same composition range of these additives lowers the tungsten phase conductivity of heavy metals about 30%. This calculated out to be about a 10% decrease in conductivity for a 50 wt.% Ag and W composite that is about what is seen for composites with and without the Ni addition [81] should be noted that nickel does not decrease the conductivity of the silver phase of silver tungsten since it is not soluble in silver.

2. For the CuW system the use of additives not only lowers the conductivity of the tungsten phase but also lowers the copper phase conductivity, since copper forms a solid solution with nickel. Published data for 50 wt.% Cu/W show a reduction of 30% for liquid-phase sintered versus infiltrated grades without additives [81].

3. The first stage of liquid-phase sintering as described above involves the tungsten particle consolidation through lowering of surface free energy with the liquid phase. This phase of sintering has a major effect in increasing the contiguity of the tungsten particles, and it will be shown later in this section that, as the contiguity increases, the resistance of the material to crack propagation decreases.

This process requires less labor than the infiltration process that follows but the material toughness must be controlled to insure that the material resists cracking from the thermal shock of arc erosion.

16.3.2.2 Material Technology/Press Sinter Infiltration (Process D 3.0)

The infiltration process for making silver–tungsten material differs from the liquid-phase sintering process in that attainment of density is not dependent on shrinkage or consolidation of a pressed compact, but on filling the pores of a sintered tungsten skeleton with molten silver. The process basically consists of pressing a porous tungsten compact that contains a little silver or none at all, sintering the compact to weld the tungsten particles together, and then placing the tungsten skeleton in contact with silver and heating the combination in a reducing atmosphere to a temperature above the melting point of silver. The pores in the tungsten skeleton then suck the molten silver into them. Additives are not required as the process requires minimal tungsten sintering to attain density. There are many variations of this process both with and without additives and with different blending, pressing, sintering, and infiltration techniques. The strength of the tungsten skeleton can vary considerably with processing: for example, variations occur in the activation additives for sintering, sintering time and temperature, the starting powder mix, and both particle size and amount of silver in the powder blend. Some noteworthy characteristics of this process are as follows:

1. Good control of the infiltration process is required to insure a uniform and porous free contact structure. Several factors can cause sections of the tungsten skeleton to be void of the filler metal, such as the furnacing technique utilized and the matching of the silver weight needed to fill the tungsten skeleton. Too much silver may produce excess silver layers on the contact surface and too little will leave voids in the contact body. Metallography on cross sections of the contact and or fractures of samples can test for the quality of infiltration.

2. As a result of the sintering step for this process in which the refractory particles are welded together before the liquid phase is added, refractory particle alignment is prevented and therefore product made by the process has significantly lower contiguity of the refractory phase. This is a good property of this process which will be discussed later in the section.

3. The strength of the refractory skeleton can be varied significantly with processing and it, in turn, can affect performance results. This variable is difficult to check as an incoming property so controls in the manufacturing should monitor this property, for example checking for the extent of shrinkage (normally very small) that takes place during the sintering step.

16.3.3 Metallurgical/Metallographic Methods

The following sections relating microstructure to material properties for tungsten–silver also in principle will relate to tungsten copper, tungsten carbide–silver and other composite refractory metal systems. Since the metallographic structure plays a large role in electrical performance and metallographic preparation is unique for these materials, this section on metallographic techniques for preparation and measurements is included.

16.3.3.1 Metallurgical/Metallographic Methods/Preparation

This procedure is for tungsten–silver, but can be used for tungsten–copper and tungsten carbides–silver grades. Hand polishing of tungsten–silver alloys is difficult, especially in the case of very fine dispersions of silver and tungsten. The main problem which must be avoided is the development of relief between the hard phase and soft silver. If cloths with heavy nap are utilized the polishing grit erodes the silver at a higher rate than the tungsten. As the relief gets large the tungsten phase tends to smear over the silver phase and cover up porosity and grain structure detail. In order to avoid this, use of a no nap cloth, such as silk or some other commercial no nap cloth, is recommended, especially for the last stages of polishing. A good combination that works is the use of 0.25 alpha and 0.05 beta Al_2O_3 as the last two polishing stages. There may be some scratches with this technique but the structure will not have a distorted layer covering up true structure.

In order to see the tungsten grain boundaries an etchant must be used. Both modified and standard Murakami's reagent as listed in the *Metals Handbook* is too active [82]. Also, the ratio of $K_3Fe(CN)_6/Na(OH)$ is found to be important for attainment of preferential grain boundary etching versus total grain etching, see Figure 16.11. A dilution of 50/1 of the standard strength, a ratio of 8/1, and an etching time of 15–25 seconds has been found to provide good results [83]. The formulation is given below. The undiluted reagent over-etches the tungsten phase and makes rendition of the grain boundaries more difficult:

Na(OH) (g)	$K_3Fe(CN)_6$ (g)	H_2O (ml)
0.04	0.3	100

16.3.3.2 Metallurgical/Metallography/Quantitative Analysis

Currently there are many options in both software and hardware for doing image analysis. As long as some color or gray level exists between phases the systems can give you a multitude of statistics regarding size and distributions of the phases. This section will not cover the subject of modern image analysis, but will cover some fundamental formulas that have

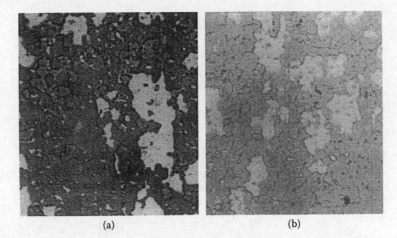

(a)　　　　　　　　　　　　　　　(b)

FIGURE 16.11
A comparison of polished tungsten silver etched with (a) modified Murakami's reagent and (b) diluted 8/1 Murakami's reagent [83].

been used for analyzing composite materials. Most modern image analysis equipment should be able to measure the parameters which will be discussed.

A very commonly used composite material metallographic parameter that can be measured even by hand is λ, the mean phase intercept size. This is defined by Underwood [84] as follows:

$$\lambda = V_V^A / N_L^A \tag{16.3}$$

where V_V^A is the volume fraction of the phase you want to measure and N_L^A refers to the number of areas of that phase intercepted per unit length.

Figure 16.12 illustrates the measurement of λ for both silver as a phase and tungsten as a grain. If you know the composition, you can calculate the volume fractions and use that number to get a rough idea of the λ value. Since grain size and phase size distributions are often widely varying throughout silver refractory contacts, accuracy depends on sampling technique and the number of measurements. Automatic systems will calculate the volume fraction of the phase of interest for each area measured.

The mean intercept technique provides a relative easy size measurement technique for comparing phases, grains, and particles in silver refractory metals, copper refractory metals, and even silver metal oxides. Below an example is provided for measuring tungsten grain size using the simulated drawing of a tungsten–silver microstructure in Figure 16.12.

Example: Measure the mean grain size of the tungsten particles, dark phase, in Figure 16.12 using Equation 16.3. Number of grains intercepted per unit length = $N_L^W = 18/40 = 0.45\ \mu m^{-1}$. The volume fraction of a phase is normally known from the chemistry or it can be calculated from the cross section using other quantitative techniques: in this case $V_V^W = 0.6$, Mean W grain size = $\lambda_W = 0.6/0.45 = 1.3\ \mu m$.

Another important metallographic measurement that can be made to silver and copper refractory metals is contiguity, C. The property of contiguity is defined by Gurland [85] as the ratio of grain boundary surface of a phase with itself to the total surface of that phase. The following formula, derived by Stjernberg [86] is convenient to use, and has been set up for W contiguity of AgW:

$$C = (N^{WG} - N^{Ag}) / N^{WG} \tag{16.4}$$

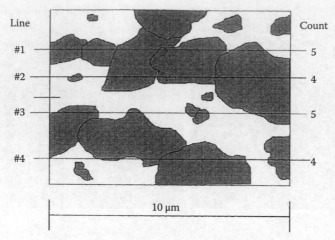

FIGURE 16.12
Illustration of the measurement of λ_w, mean W particle size, for a W/Ag 60 vol.%W composite material.

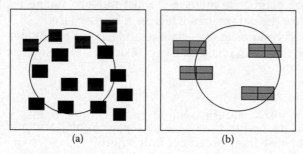

FIGURE 16.13
Illustration of contiguity for (a) particles having 0% contiguity and (b) particles having 50% contiguity.

The above formula allows contiguity to be measured without needing to know the size of any particle of the phase. A straight line or, better, a full circular line, can be used to count the particles intercepted, N^{WG}, tungsten grains and N^{Ag}, silver phase areas with the line. See Figure 16.13 for examples of measurement. Figure 16.13a is for a sample with 0% contiguity for the tungsten grains and Figure 16.13b has 50% contiguity for the tungsten grains. For the illustration a small sampling rate is used which can cause some error; it is best to use a particle count of 50 or more grains.

The contiguity parameter is mainly useful for the silver and copper refractory metal systems of contacts; it allows measurement of how continuous a phase or grain structure is with itself.

Example: Measure the contiguity of the particles in Figure 16.13 using Equation 16.4. Case a. N^{dg} = intercepts dark grains = 7, N^{Lp} = intercepts light phase = 7 C_{dg} = (7−7)/7 = 0 Case b. N^{dg} = 8, N^{Lp} = 4, C_{dg} = (8 − 4)/8 = 0.5.

16.3.4 Metallurgical/Structure/Strength and Toughness

For composite electrical arcing contacts, metallurgical structure along with chemical composition determines the performance capability of the contact. Metallurgical parameters like tensile strength, shear strength, fracture toughness, and hardness are useful for composite materials used as structural components or parts for mechanical wear. For the life of electrical contacts the interest is for electrical arc erosion resistance, contact resistance, welding resistance, material transfer and other arc-related wear phenomenon. The silver and copper refractory metal contact materials, as a result of the refractory metal content, have in general a high resistance to electrical erosion compared to other types of contact materials. As a result of this they are normally used in applications where they must switch high fault currents, thousands of amperes, and are subjected to high current density arcs and thermal stress. No simple direct relationship of mechanical properties like tensile strength or hardness to arc erosion wear or resistance has been found for composite contact materials. This is the result of the many parameters influencing testing results, both in terms of the material (structural, physical, and chemical); and electrical testing (contact size, current level, chamber, device, etc.) (see Chapter 10).

An important distinction of refractory metal composites compared to most metal alloys is that they are relatively more hard and brittle owing to their content of high melting refractory metals. An important attribute to measure on brittle materials is toughness which is the resistance the material has to crack propagation. Although the high melting point refractory metal component of these composites give them good arc erosion and welding resistance, they also lower the cracking resistance of these materials. Table 16.4 is shown to help illustrate the relationship of material toughness to three well know materials and also material strength and hardness.

The electrical erosion process for contact materials involves arc reactions with the contact surface that result in material melting, vaporization, ionization, materials transfer and rapid heating and cooling. The melting, re-solidification, rapid expansion on heating follow by cooling all leave residual stresses in the material. Most of the composite refractory metals are brittle and the degree varies depending on the composition, phase interface boundaries and particle size distributions. Depending on the degree of brittleness and the amount of stress generated from the electrical cycling small cracks can be generated that contribute to the erosion or large cracks can form that penetrate deep into the material. An important factor that must be considered in predicting electrical life of composite refractory metals is an understanding of the relationship of material structure and cracking resistance, toughness.

In the last section of this chapter it was shown how to measure metallographic parameters of the structure of composite refractory metal systems. In order to better understand and predict potential cracking problems with composite refractory metal materials like tungsten silver Witter and Warke [87] conducted a study on a matrix of tungsten silver materials that differed in composition and particle size distribution of the of the tungsten and silver phases. The matrix is shown in Figure 16.14. Besides having materials of different composition and particle size distribution the matrix includes a material made by a different process, infiltration.

In order to measure the brittleness of the material and potential for cracking a fracture mechanics technique was used to measure toughness of the material in terms of G_{IC} the

TABLE 16.4

Relative Relationship of Mechanical Parameters

Material	Strength	Hardness	Toughness
Copper	Low	Low	High
Tool steel	High	High	Medium
Ceramic	High	Very high	Low

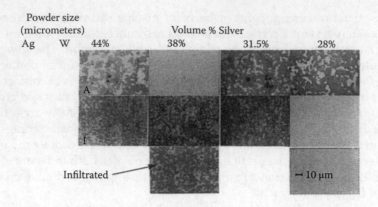

FIGURE 16.14
Microstructure of three coarse and three fine liquid phase sintered tungsten silver materials and one infiltrated material [87]. All materials were 97% of theoretical density or better.

critical energy release rate. This is the energy required to extend an existing defect or crack in the material under plain strain as shown for the following formula:

$$G_{IC} = ((P_c)^2/2t_n)dc/da \qquad (16.5)$$

where P_c is the critical load for crack growth, t_n is the material thickness at the crack area, c is compliance (deflection per unit load), and a is crack length [88]. For more details on the actual technique used and apparatus please see reference [87] and for more information on fracture mechanic techniques see references [88,89]. In Figure 16.15 the results of the G_{IC} measurements on the various materials are compared to results of transverse rupture strength (TRS) measurements and Vickers hardness measurements made on the same matrix of material. For the hardness measurements it shows a higher hardness for the finer materials and a higher hardness with an increase in tungsten content but there is little difference for infiltrated and liquid face sintered materials. The TRS measurements show slightly higher strength for finer materials but little difference with tungsten content. The G_I, critical energy release rate measurements show a clear large distinction for both phase size distribution and volume percentage of tungsten. It also shows a large increase in toughness for the infiltrated product compared to the liquid phase sintered product of the same composition. Arc erosion testing conducted at 8,000 amperes on the same matrix correlated well for prediction of cracking and showed the 69% tungsten fine grained material to have severe cracking.

Further work was done on this project to see if the measured metallographic parameters and the volume fraction parameters could be used to relate to the relative toughness of the materials in the matrix. A crack in one of the contacts was sectioned and studied and is shown in Figure 16.16. The cracks in these materials will try and follow the weakest phase interface in the structure. In this case there are tungsten to silver interfaces, silver to silver grain boundary interfaces, and tungsten to tungsten grain boundaries. The tungsten to tungsten grain boundaries are by far the weakest. So a crack that starts out in a fractured tungsten grain boundary will try to continue to grow by following and fracturing more tungsten grain boundaries which in this case is the path of least resistance. The obstacles to this path are the silver phase areas that separate the tungsten grain aggregates. So both the volume percent of silver and also the coarseness of the silver phase areas have an effect

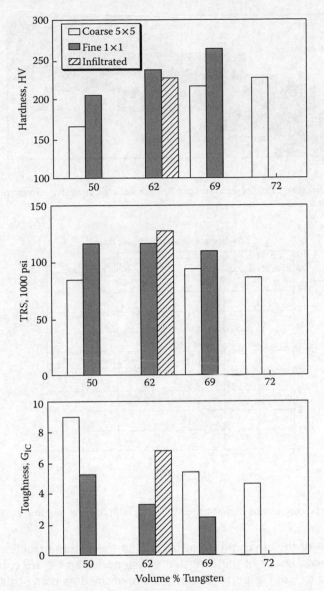

FIGURE 16.15
A comparison of strength, hardness, and toughness as a fundction of composition and phase-size distribution of tungsten-silver materials, including six liquid-phase sintered materials and one infiltrated material [83].

on the amount of energy it takes to either rupture and go through them or go around them. The other important factor is the contiguity of the tungsten phase as discussed in the metallographic section. The more contiguous the tungsten phase is the easier it is to go around the silver phase. From empirical work, Witter and Warke developed Equation 16.6 for the silver tungsten system relating metallographic and phase volume fraction data to fracture toughness:

$$G_{IC} = k \left(\lambda_{Ag} V_{Ag} / C_w \right) + \epsilon \tag{16.6}$$

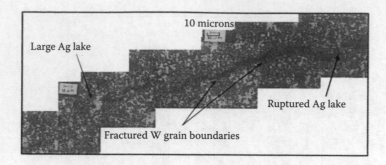

FIGURE 16.16
Cross section of a brittle fracture in an arc-eroded liquid-phase sintered sungsten–silver contact. The crack is perpendicular to the contact face.

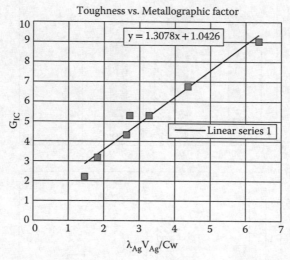

FIGURE 16.17
For tungsten–silver contacts: a correlation of the factor, $\lambda_{Ag}V_{Ag}/C_w$ with fracture toughness [83].

Where λ_{Ag} is the size of the silver phase areas, V_{Ag} is the volume fraction of the silver phase, and C_w is the contiguity of the tungsten phase and k and € are constants for the material system being tested. Figure 16.17 shows a plot of the data using this equation and constants.

16.3.5 Electrical Properties (EP)

Above the relationship of mechanical, compositional, and microstructural properties of refractory metal contacts was discussed mainly with data on the basis of the tungsten–silver system. It was seen that the relationships are complex and that the mechanical properties are significantly affected by the composition, particle size distributions of the phases, phase and grain interface properties and distribution, and method of fabrication. In this section if it will be seen that although the erosion properties of these materials are affected by the same structural variables as the mechanical properties, clear quantitative relationships do not exist between the mechanical properties and the erosion behavior of these types of materials. Another difficulty in studying the electrical erosion properties of these materials is that although a significant amount of work has been published on erosion of refractory metal electrical contacts, it is difficult to make comparisons since the results are also influenced by numerous device and testing variables as discussed in Chapter 10. In 1986 Slade [90] published a review of prior work for erosion of both silver and copper refractory metals which provides references to most of the significant work done at that time.

As mentioned earlier composite refractory metal contacts are mainly used in high current devices that provide fault current protection. Most of these materials are not suited for lower current repetitive switching devices since they develop high resistant films as a result of switching low and medium currents. This subject will be discussed in detail in a section later in this chapter. As a result of this the main interest in these materials is for switching currents of about 1000 A or more. It is generally agreed among researchers that at these higher currents the composite refractory metals offer superior erosion resistance compared to other silver-based alloys, including silver metal oxides. For electrical erosion the discussion will therefore be based on examples and work which involve only higher currents. This does not imply that these materials are not switched at low currents in some device applications but only that the device life is normally limited as a result of contact resistance. The discussion in this section will start with a discussion applicable to all of the composite refractory metal types looking at the correlation of erosion with mechanical and microstructural properties. The discussion will then be divided between specific work on the silver-based and copper-based materials. The copper-based materials are limited to devices offering oxidation protection such as oil interrupters, SF_6 devices, and vacuum switches.

16.3.5.1 FP/Arc Erosion/Microstructure and Properties

In Section 16.3.3, the relationship was discussed of mechanical properties and microstructure for composite refractory metal contacts using a matrix of tungsten–silver materials for demonstrating examples of various relationships. The same material matrix will now be used for discussing the relationship of mechanical and microstructural properties to arc erosion characteristics of these materials.

The same materials that were mechanically tested were subjected to arc erosion testing. The arc erosion testing was performed on 10.2 mm diameter by 2.0 mm thick disks brazed onto a copper carrier. The current source was a capacitor bank discharged through a transformer yielding a current of 8000 amperes peak of 60 Hz dc with a duration of 0.5 cycles. With a fixed contact gap of 3.2 mm, the arc was triggered by discharging the current through a very fine pure silver wire, 0.025 mm in diameter, placed between the contacts. Each contact pair was subjected to ten half-cycles of arcing. It should be noted that this testing involved no contact closure, so there was no contact bounce and no splatter from closing contacts coming together under arcing conditions, and also since the gap is fixed, the contacts do not see normal interruption conditions of a transition from a molten bridge to a metallic arc followed by a gaseous arc.

The erosion was measured in terms of total erosion from the contact pair in terms of volume of material lost versus weight loss that allows erosion comparisons for different material compositions. At high current levels the erosion process involves material evaporation, molten metal droplets being spattered from the contact surface, and chunks of composite material being broken off by thermal stress and oxidation and being ejected by the arc forces from the contact surface. Distortion of the contact can also take place without volume loss for softer materials, resulting in the contact material near the face surface increasing in diameter, mushrooming, compared to cooler layers of material under the surface. The materials picked for this study all had relatively high tungsten contents and thus were more prone to brittle behavior rather than distortion.

The results of the fault current level erosion versus mechanical properties for the material matrix are shown in Figure 16.18. The transverse rupture strength versus erosion shows very little correlation. The hardness plot versus erosion shows some trend with high hardness materials having high erosion rates. This would relate to brittleness of harder materials. The toughness plot versus erosion rate shows that materials with high toughness have much better

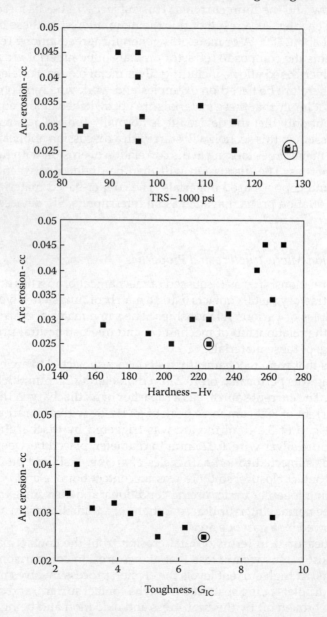

FIGURE 16.18

The arc erosion resistance of tungsten–silver materials as functions of mechanical properties, including 11 liquid-phase sintered materials and one infiltrated grade which is encircled [83].

erosion rates than those with low toughness readings. This clearly shows that for composite refractory metals there is a big influence in the erosion rate from thermal shock damage which results in cracking the surface and breaking off particles. The factor visible in these plots is the result of the infiltrated material compared to the liquid phase sintered materials. This material is a medium fine material and had the lowest erosion rate of all the materials. As a result of the low tungsten contiguity this material has a high resistance to cracking under the stress created by the arcing. This is a very significant factor for consideration of this manufacturing process.

16.3.5.2 EP/Arc Erosion/Silver Refractory

Walczuk tested tungsten-silver materials made to two compositions 50/50 and 65/35 by wt% W/Ag made by similar infiltration processes by four different suppliers [91]. He tested at 4 kA and 8 kA and found the erosion to be at least doubled at 8 kA. There were significant differences among the materials from the different sources despite the fact that similar processes were utilized. All of the 65/35 materials had more anodic than cathodic erosion (see Chapter 10 for definition), up to five times the amount. Walczuk attributed the high anodic erosion to the effects of the plasma jet of the cathode on the anode at the high currents. The 50/50 materials were mixed with regard to erosion being predominantly anodic or cathodic and were in general lower in total erosion than the 65/35 materials. No explanation was given but the data suggests some influence from the manufacturing method. SEM photographs of fractures for two of the materials studied show significant amounts of plastic deformation, indicating that these materials were much more ductile than the matrix of materials studied in the previous section. This would suggest more erosion from evaporation and droplet expulsion than brittle chunk expulsion.

Leung and Kim compared the erosion characteristics of infiltrated grades of Ag/W, AgWC, and Ag/Mo all with 50% by volume silver [92]. The materials were 98.5% dense or better. The testing of 4.7 mm diameter contacts was at 1600 A r.m.s. at 220 V ac with a long arc duration of two cycles at 60 Hz. The volume erosion rate differences found among the three types of materials were within a total spread of 7%. Although the erosion rates were very close significant differences were reported for the erosion mechanisms. The tungsten–silver eroded surface was reported to have many tungsten-rich cones and shows erosion of silver on the surface from capillaries or fissures which go below the surface followed by erosion of the tungsten-rich surface. The WC/Ag showed more even erosion with more composite chunk ejection from the surface as a result of the very weak WC/WC bonding and poor wetting of Ag to WC. The lower conductivity of the WC/Ag material also was thought to increase loss of Ag from the surface evaporation. The Mo/Ag showed an eroded surface with segregated silver and molybdenum droplets. There was less evidence of cracking but more evidence of silver being ejected as droplets as a result of poor wetting with the molybdenum.

Lindmayer and Roth also made comparisons of several types of tungsten–silver, both liquid-phase sintered and infiltrated, and also several types of tungsten carbide–silver, at currents of 350, 700, and 1000 A, and found that the erosion rates overlapped for the tungsten versus the tungsten carbide grades and that the tungsten–silver grades had a very wide spread at the 1000 A level [93]. They also ran tests on different sizes of tungsten–silver contacts and showed that at low currents the erosion increased linearly with current independent of size and, at higher current levels, depending on the size of the contact, the rate of erosion as a function of current level increases significantly. This bend in the erosion–current plot is stated to take place when the arc reacts with the major part of the surface area [94]. The bend point occurs at higher currents as the contact becomes larger.

Besides the refractory metal and silver, the chemistry of the composite materials can vary for various additives utilized during the material processing. Two commonly used additives for these types of materials are nickel and cobalt. Witter found that increasing the nickel content of liquid-phase sintered tungsten–silver increased the erosion rate significantly [95]. Kabayama et al. found the same thing true for cobalt additions to tungsten–silver [96]. Walczuk investigated the additions of 0.5, 1, 2, and 5% rhenium by wt. to tungsten–silver 50/50 [97]. He found that the erosion rate decreased for additions up to 2% and then increased. There is no explanation of the reason why the erosion rate drops and then increases, but the porosity level of the material increased as the rhenium level increased, which indicates an effect on the sintered structure.

16.3.5.3 EP/Graphite Additions to Silver Tungsten and Silver Tungsten Carbide

In some molded case circuit breaker, MCCB, applications asymmetrical contacts have been used. On the movable side a high erosion resistant material like silver tungsten carbide could be used and on the stationary side a low electrical resistance material like silver graphite may be used. Since the erosion rate for silver graphite is high, work is being done for replacing it with a material with more erosion resistance. Leung et al. [98] conducted comparisons of AgWC with graphite additions of 3% to AgC 5% made by different processes. Earlier work had been done by Allen et al. [99] on similar material with graphite additions that included both silver tungsten and silver tungsten carbide. The new results showed better erosion life with the AgWC + graphite made by the same process, press sinter repress. The anti-welding was not quite as good as AgC but the resistance to re-ignition was better. For these types of materials it was found that increasing the graphite amount increases the erosion rate. Decreasing the graphite particle size also increases the erosion rate a little but also improves the welding resistance. The materials in this work were made by press, sinter, and repress. The sintering must be done in the solid state. Some of these materials can be made by sintering and extrusion as an alternate process. The processing may be difficult because of the brittle nature of this type of material.

16.3.5.4 EP/Copper Refractory Metals

The structure of tungsten–copper composite contacts is very similar to tungsten–silver, a soft, ductile, high-conductivity metal phase mixed with a tungsten phase. The main difference between the two materials is that the oxides of silver are not stable at high temperatures and therefore is oxide free after arcing, as opposed to copper, which forms stable oxides at high temperatures. As a result, tungsten–copper can only be used in applications where it is protected from oxidation. These devices include, oil, SF_6, and vacuum interrupters, all of which are somewhat different and involve specialized technology. Some references for these types of devices are, "Circuit Interruption, Theory and Design," by Browne, and a review of material development for vacuum interrupters by Slade [100–101]. Another difference of tungsten–copper alloys from tungsten–silver is that some of the materials used as minor additives for activation and improving wetting characteristics of tungsten are not soluble in silver but are soluble in copper, and therefore significantly reduce the electrical conductivity of the copper. Figure 16.19 shows the effect of minor nickel additions on the conductivity of tungsten copper [81].

Several groups of researchers have investigated the erosion characteristics of tungsten–copper over the total composition range from pure copper to pure tungsten [102–104]. Both Haufe and Abdel-Asis show similar erosion rates in air and oil for tungsten–copper except at the copper-rich range where Haufe shows higher erosion in oil. In comparing erosion rate as a function of tungsten content, all three research groups show agreement that tungsten–copper composite mixtures have lower erosion rates than either pure copper or pure tungsten. Differences exist among the groups as to the compositions showing the lowest erosion rate, but this is expected since many differences exist, that is, contact size, processing, current density, and more. Zessack showed a linear relationship of erosion as a function of the number of operations for both 34% and 45% Cu by weight tungsten–copper. Figure 16.20 shows erosion rate results as a function of tungsten content in terms of volume loss per operation for switching in oil at six different current levels [104]. The 25 mm by 45 mm contacts were made by a laboratory non-infiltration process which produced a structure low in porosity and little contiguity of the tungsten particles. For this case the

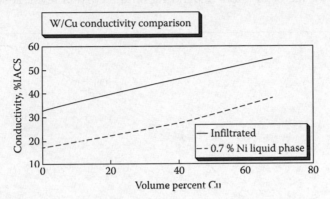

FIGURE 16.19
A comparison of conductivity for Cu/W materials with and without Ni additives.

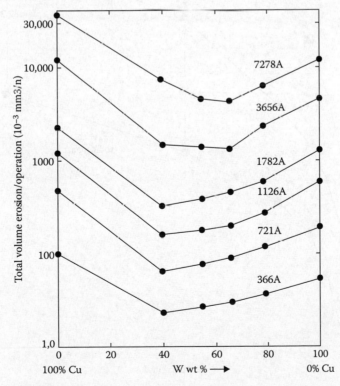

FIGURE 16.20
A comparison of erosion rate for various tungsten–copper compositions at different current levels in oil [104].

erosion results show that the optimum low erosion composition shifts to higher tungsten compositions as the current is raised. Zessack describes the erosion mechanism as taking place in three stages: stage 1 is mainly evaporation of copper from the surface, stage 2 involves the tungsten-rich surface with sub-layers of copper where the tungsten melts and erodes through evaporation and sputtering, and in stage 3 a stabilized surface is created with both copper and tungsten erosion.

Labrenz shows a comparison of erosion rates for tungsten–copper contacts in both oil and SF_6 breakers for currents up to 7.2 kA, see Figure 16.21 [105]. The erosion rate is much less in SF_6 than oil by a factor of 2 to 7 depending on the current level. Also, included in

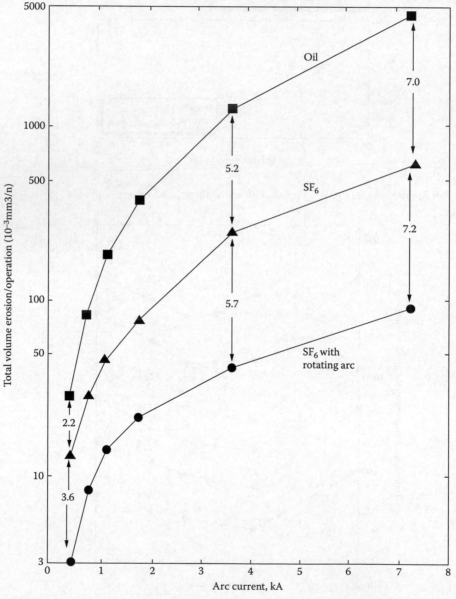

FIGURE 16.21
A comparison of tungsten–copper erosion in oil, SF_6, and SF_6 with a rotating arc ring [105].

Figure 16.21 is a comparison of erosion for ring-shaped electrodes in SF_6 for which the arc is magnetically forced to move around the circumference of the ring. By the rotary movement of the arc the erosion is further reduced by a factor of 3–7 depending on current level. Tungsten–copper ring contacts are stated to have about 10 times the erosion life of copper ring type contacts. Labrenz explains the erosion mechanism in SF_6 to be similar to oil for tungsten–copper with high copper evaporation occurring in early life followed by more stable lower erosion rates after several hundred operations.

Kaminski describes erosion of tungsten–copper and other contact nozzle electrode materials in SF_6 breakers in which the contacts are subjected to normal plasma erosion, heating and plasma jets, and also to mechanical erosion by high-pressure gas flow [106]. In this device the arc is moved off the contact electrodes in less than 2 ms by an SF_6 blast for currents below 10 kA for all materials tested. At higher currents a stable arc forms between the contacts which remains in place in spite of strong gas flow. The current level at which stable, immobile, arcs form varies with materials and is higher for homogeneous materials like copper and tungsten than heterogeneous materials like tungsten–copper [107]. The levels at which stable arcs formed were about 12 kA for the 81/19 wt.% tungsten–copper and 20 kA for pure copper and tungsten electrodes. Data taken from Kaminski's work was used to form Figure 16.22. It shows tungsten to have a slightly lower erosion than tungsten–copper at 10 kA, but the reverse to be true at 30 kA where both tungsten and tungsten–copper form stable arcs. In oil breakers even at 0.4 kA Figure 16.20, the tungsten–copper material showed lower erosion than tungsten contacts. The difference here is that the arc retention time on the tungsten or tungsten–copper contacts at 10 kA is about 2 ms compared to 10 ms for the oil breaker testing, thus less temperature effect on the contact surface. At 30 kA the arc retention time is long on the contacts and the tungsten–copper erodes at a lower rate than the tungsten as a result of surface cooling effects of evaporating copper and higher thermal conductivity of the composite structure. These examples illustrate that besides current level, arc duration is another important consideration when comparing materials having different thermal properties and melting points. Kaminski conducted another interesting material erosion experiment with the SF_6 nozzle type breaker [106]. A comparison was made for erosion under SF_6 for infiltrated tungsten–copper 81/19 material versus tungsten heavy metal material, W/Ni/Cu/Fe wt.% 94/4/2/1, both with and without high-force SF_6 flow. The test was conducted at 29 kA which results in stable arcs despite gas flow. The results in Figure 16.22b show that the SF_6 blast has almost no effect on the tungsten–copper but a large effect on the heavy metal. Another surprising result is that the heavy metal has only a slightly higher erosion rate than the tungsten–copper under the no-blast condition.

In order to discuss this result, first, the structure of heavy metal must be explained. Basically it consists of a tungsten phase and a binding phase which cements the tungsten particles together. The tungsten particles are normally much larger in size and more rounded in shape than what you would find in tungsten–copper as a result of significant grain growth through Oswald ripening during liquid-phase sintering [20]. The tungsten phase of such alloys normally has low contiguity of tungsten grains although the volume percent of tungsten in the structure is high. The binding phase is lower melting than the tungsten phase but slightly higher than the melting point of copper, and therefore in this respect is similar to tungsten–copper. Another significant difference for this material from tungsten–copper is that the nickel and iron alloy with the copper and tungsten and thus the composite heavy metal material has a very low conductivity, <10% of copper and about one-fifth the conductivity of the infiltrated tungsten–copper.

In this example, it can be seen that high temperatures from the arc will penetrate deeper into the heavy metal than for the tungsten–copper since the heavy metal has very low

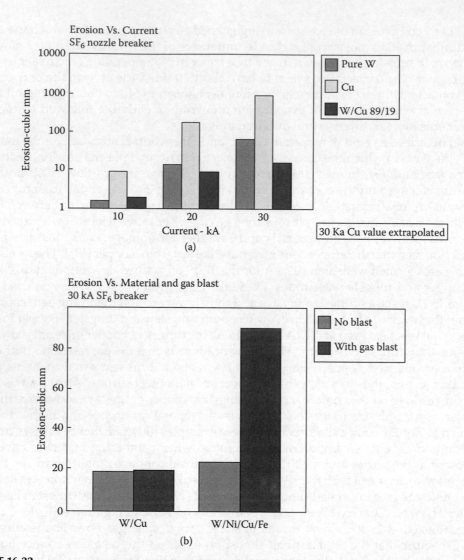

FIGURE 16.22
(a) Erosion of tungsten–copper, tungsten, and copper In SF$_6$ nozzle breaker at three current levels and (b) erosion in the same breaker with and without a gas blast [106].

conductivity and also a lower percentage of lower boiling point material for evaporating and cooling the contact surface. When the binding phase becomes molten the tungsten particles will be worn away by the gas blast since the tungsten particles have little contiguity. For the tungsten–copper material, layers just under the surface remain cooler and only molten copper right at the surface can be removed by the gas blast, since the tungsten on the surface remains bonded to the sub-layers. Also, the tungsten at the surface protects molten copper just below the surface from being wiped away by gas blast.

The relatively low erosion rate of the heavy metal, only slightly higher than the tungsten–copper, under the no-flow condition for this testing is more difficult to understand. As a result of the very low conductivity, you would think that at 30 kA current even without the gas blast the arc would splatter molten particles from the surface. Possibly the wetting of the cemented phase to tungsten phase is strong enough to resist the forces associated with arcing. The heavy metal material is also quite ductile compared to copper-tungsten materials in the same tungsten composition range, so erosion by brittle fracturing of small chunks will be low.

16.3.5.5 EP/Erosion/Summary

From the above discussion, it can be seen that the erosion process involves many factors for composite refractory contacts. In trying to summarize the various factors discussed above the following can be stated.

As the result of arc erosion on switching, the surface of the material changes. In most cases cited above the researchers for composite refractory metal materials state that an equilibrium type surface is established after some hundred arcing operations which yield fairly linear erosion as a function of switching life. The characteristics of this surface depend on both the characteristics of the subsurface from which the surface forms and the ambient conditions existing around the surface during arcing. There are many ambient variables, such as current level, type of electrical load, current density, oxidation protection, current duration, cathode jets, external gas blasts, magnetic forces, and gap size that influence the kind of surface that forms. Some of the material variables affecting the equilibrium surface are types of refractory, refractory content, particle size distribution, contiguity of particles, minor additives, material toughness, and processing.

As a new composite refractory metal contact is arced, the lower melting phase evaporates from the surface. As a result, the refractory content of the surface increases with the formation of a mixture of complex oxides of the refractory metal, slag containing both metal components, and molten metal of both the refractory phase and the lower melting phase. In the above examples we see that if the sub-layers have brittle interfaces, WC/WC bonds for example, chunk erosion is dominant and the surface layer contain little molten metal. If the bulk material is more ductile, like tungsten–silver, the surface layer is shown to become refractory rich with molten tungsten cones, tungsten oxides, and silver islands with many fissures going deep into the material. The erosion in this case is both from the surface and from evaporation of silver from the fissures. If the materials are very coarse, the layer may contain large areas of silver or copper, and droplet type erosion and evaporation may dominate erosion rate. As the bulk material becomes finer, chunk erosion will become more dominant. The oxide-rich surface layers only have small silver islands and most silver erosion is through the fissures in the surface. The brittle nature of the surface with the fissures results in small particles breaking off during erosion. A main point for understanding erosion of these materials is that for composite materials the surface characteristics of new contact surfaces only control erosion characteristics for early contact life and that stable erosion is dependent on an equilibrium surface that is established later in contact life.

Some general conclusions on erosion of composite refractory materials are as follows:

1. The composite refractory metals generally are found to have lower erosion rates than either of the components of which they are composed.
2. For the low refractory content range of these materials the erosion rate increases as the low melting phase increases and becomes less protected by the refractory material from erosion through expulsion of metal droplets and evaporation.
3. For high refractory content composite materials the metallographic property of contiguity of the refractory phase seems to correlate with erosion rate, thus as the brittle paths through the material become more continuous the erosion rate increases as pieces break off from thermal stress of the arc. Both increasing the refractory content and decreasing the refractory grain size increase contiguity.

4. The optimum refractory content for lowest erosion depends on many factors including current level, current duration, contact size, material structure and properties.

5. Most of the mechanical properties of these materials do not have a clear correlation with erosion properties, but materials with lower toughness, especially for high refractory contacts, seem to have higher erosion rates than high toughness materials.

6. The erosion rate is significantly affected by the relationship of the contact size to the current level and arc size. The erosion rate as a function of current rises steeply above the current level where the arc has reacted with the total contact face area.

7. A comparison of the relative erosion rates of different composite refractory metal systems by composition is difficult since processing of the materials has a major impact on properties.

8. Under dc the degree of anodic or cathodic erosion for contact materials depends on many factors such as gap, circuit voltage, current level and contact opening speed.

9. Erosion of tungsten–copper in air and oil seems to be similar for most ranges of composition.

10. High-pressure gas blasts associated with some high-current interruption devices have no significant detrimental erosion effect on tungsten–copper contacts but are detrimental to heavy metal materials that have little tungsten contiguity.

16.3.5.6 EP/Composite Refractory Materials/Contact Resistance

Even though silver-based refractory composites have very good erosion resistance properties, the use of these materials in switching devices is limited as a result of contact resistance problems associated with oxides that form on the surfaces of the contacts from arcing in air. As a result, these materials are mainly utilized in devices that need arc erosion resistance for interrupting high-current faults but do not require tens of thousands of switching duty endurance operations. These types of devices, circuit-breakers and interrupters, are normally designed to have high contact pressures and/or good mechanical wipe to help lower the resistance effects of the oxides. The instability of contact resistance with these materials has been recognized for a long period of time and has been the subject of many research projects [92,108–111]. The main goal of this type of research has been to better understand the mechanisms causing the contact resistance instability and to find ways to improve the resistance characteristics of these materials without detracting from the erosion and anti-welding characteristics of the materials.

The contact resistance that develops and is measured for arced silver refractory contacts depends both on the current level that is being switched and on the current used for measuring the resistance [93]. Lindmayer and Roth tested a variety of tungsten–silver and tungsten carbide–silver contacts at different current levels, 60–1800 A, and for several different contact sizes. At low arc currents, below 200 A, their data shows little relationship between contact size and contact resistance. Figure 16.23 is a plot of data taken from the data in the Lindmayer and Roth paper [93]. This Figure shows resistance versus switching current level for four contact sizes. It can be seen that as the contacts become larger for a fixed current the resistance increases and that also as the current level increases for a fixed contact size the resistance drops. Lindmayer and Roth have a qualitative model for explaining this relationship. After arcing they found a difference between the center and outer areas of the contacts. The center crater and splash ring surrounding the crater area they found to have

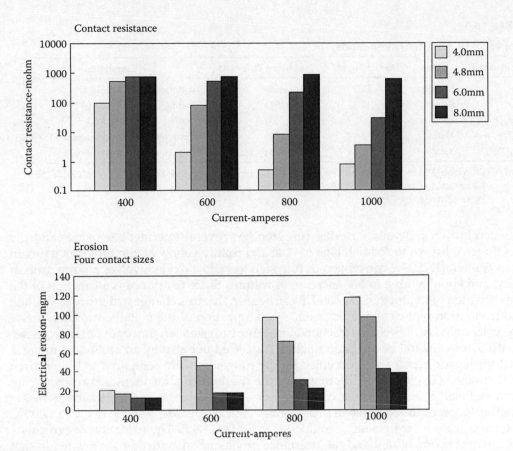

FIGURE 16.23
Contact resistance and electrical erosion versus switching current level for four different contact sizes at each current level [93].

metallic conduction when probed for voltage drop, and the area outside the ring was found to be mainly nonmetallic. As the current is increased, the crater and inner ring increase in diameter. They claim that when the metallic area reaches a diameter of about half the size of the contact diameter the resistance will remain low since the mating contacts will have overlapping metallic areas. In this example it can be seen that at 800 A the two smaller contacts, 4.0 mm and 4.8 mm diameters, have reached the critical current level to sustain low contact resistance.

The data in Figure 16.23 also shows that the erosion rate increases as the contact area decreases for a fixed current, hence as would be expected, an inverse relationship exists between erosion rate and contact resistance. The erosion rate for the two smaller contacts increases significantly at about the current range where the contact resistance drops off. For lower switching currents, under 100 A, such as the rated currents used by devices such as molded case circuit-breakers, the arc energy is not high enough to erode away the oxides that are building on the contacts with switching duty. For this reason much of the research has concentrated on studying resistance changes for lower current loads.

Since tungsten–silver and tungsten–carbide–silver are the most popular materials for these devices most of the work has focused on these systems. In order to better discuss results Table 16.5 is included to show some properties of the oxides of tungsten. Tungsten forms many oxide compounds with three of the most common shown in Table 16.5.

TABLE 16.5

Properties of Tungsten Compounds

Compound	Color	Density (g cm^{-3})	Melting Point (K)	Comments
W	Metallic	19.3	3683	Oxidation begins at 673 K
WO$_2$	Brown	11.4	1543	Heating in air converts to WO$_3$
WO$_{2.95}$	Blue			Heating converts to WO$_3$
WO$_3$	Yellow	6.47	1743	Vaporizes 1273 K, decomposes 1573 K
Ag$_2$WO$_4$	Yellow		823 forms	Silver tungstate, unstable 973 K

Source: Data from P Slade, *IEEE Trans. Components Packaging and Manuf Technol—Part A* 17 (1): 96–106, 1994; J Kaminski, *Proc. 9th Int'l Conference on Electrical Contacts*, p. 25, 1978; K Schroder. Thesis TU Braunschweig, 1967.

Between brown and yellow oxides, tungsten has several intermediate oxides and the colors go from brown to reddish blue to blue and finally yellow as the tungsten gains in oxygen content [112]. The conversion of tungsten to yellow oxide involves a large drop in density, and about a three to one increase in volume. Since resistance is a function of the volume fraction of conductors in a solid, it can be seen that this change in density will dilute the concentration of the conductors; thus, when a portion of the metallic tungsten phase converts to an oxide the volume fraction of metallic tungsten left for conduction decreases not only by the amount of tungsten fraction converted but also by an additional amount related to the additional space occupied by the tungsten oxide compared to the original metallic phase. This is much different than the mechanisms for increased resistance in silver metal oxides, since in those cases the oxides already exist and resistance increases through dilution of the conductive phase, silver. The WO$_3$ oxide is stable in air to 1,000°C where vaporization begins and decomposes at 1,300°C [112]. The oxidation of tungsten–silver is further complicated and the resistance problem is also further aggravated in that silver, which is normally free from oxides at elevated temperatures, in the presence of tungsten can combine with tungsten and form oxide compounds. This has been known for a long time and several researchers have identified and reported finding the compound silver tungstate, Ag$_2$WO$_4$, on arced tungsten–silver contacts [108,113,114]. Slade has shown the presence of crystals, cubic or orthorhombic in appearance, having compositions close to silver–tungsten on arced tungsten–silver, see Figure 16.24 [115]. Other researchers have reported the silver tungstate to have an appearance of a glass [116,117]. Silver also combines with tungsten to form amorphous silver tungstate, which may explain the difference in appearance and also the reason why some researchers using x-ray diffraction analysis on arced tungsten–silver contacts find only small amounts of silver tungstate, or none. Leung et al. have performed experiments on the oxidation of tungsten-silver materials and have found at 600°C a 30 μm layer of oxides formed which contained small isolated silver islands, WO$_3$, Ag$_2$WO$_4$, and a small amount of NiWO$_4$ [116]. Oxidation done at 700°C produced a much thicker oxide layer, 60–400 μm, and similar compounds except no Ag$_2$WO$_4$ is present. The layer is composed mainly of WO$_3$ and a large amount of segregated silver. The presence of this much silver compared to the silver in the oxide layer that formed at 600°C supports the non-existence of Ag$_2$WO$_4$ even in an amorphous state. It can be concluded from this that Ag$_2$WO$_4$ is not stable at 700°C. Silver also combines with molybdenum to form Ag$_2$MoO$_4$ but this compound has not been reported by researchers on arced molybdenum-silver, possibly since it can also exist as an amorphous compound [117]. Thus, it can be seen that, since silver also can be consumed as an oxide in the presence of tungsten, heating the tungsten–silver in air can theoretically convert the surface from a

FIGURE 16.24
Silver tungstate crystals found on arced tungsten–silver [115].

metallic to almost total oxide state. From the previous chapters it was explained that the temperature of the arc is very high and above temperatures where these oxides are stable. Therefore, the oxide layers form on the surface of the contacts just outside the arc roots during arcing and on the arc reacted surfaces after the arc is extinguished.

Leung and Kim have performed switching studies at 30 A for three refractory metal systems, tungsten–silver, molybdenum–silver, and tungsten carbide–silver all at 50% by volume silver [92]. The contact resistance measurements have been made at the same current as switching duty, 30 A and results are reported as voltage drop in millivolts (see Figure 16.25). Here, as a result of the lower switching current, the percentage of contact surface area melted by each arcing operation is much smaller than the 50% area cited by Lindmayer and Roth for maintaining low contact resistance, as discussed above. The voltage drop is seen to increase quickly for the first few hundred operations for tungsten–silver and at a slightly lower rate with molybdenum–silver and significantly slower with tungsten carbide–silver. At a switching life of 2000 operations the tungsten–silver is significantly higher in resistance than the tungsten carbide–silver but after 6000 operations all three materials are much closer although the tungsten–silver is still the highest in resistance. The erosion in this case did not follow an inverse relationship to the resistance, since they reported that the tungsten–silver contacts also had the highest erosion rate. In analysis of cross sections of the surface layer after testing, Leung and Kim found the tungsten carbide–silver material to have a layer with much more free silver, mainly segregated below the rest of the arc product layer, than was found for the layer of arc product on tungsten–silver. They also found a small amount of silver tungstate in the tungsten–silver layer and none for the tungsten carbide–silver material. These findings indicate that tungsten carbide being present versus just tungsten, may retard the oxidation of tungsten and formation of silver tungstate.

Witter and Abele have investigated the effect of composition, percent refractory metal, and also the size distribution of the microstructure on the contact resistance of tungsten–silver after switching [110]. This work was done at 20 amperes current for both switching and measurement on 4.8 mm contacts. Figure 16.26 shows the results for 4000 operations

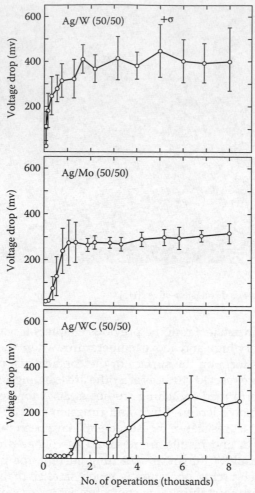

FIGURE 16.25
Average millivolt drop versus operating life for 50% by volume W/Ag, Mo/Ag, and WC/Ag switching 30 A at 100 V ac with a 4.7 mm diameter contact [92].

and it can be seen that as the volume percentage of tungsten increases in the tungsten–silver material the resistance increases. The fineness of the microstructure had a small influence with the coarser materials having slightly lower resistance. Cross sections made of all of the materials showed little difference among the materials for the oxide layers thickness. The oxide layers contained three types of inclusions, melted silver particles, and small chunks of tungsten–silver and melted tungsten particles. The amount and type of inclusions varied with composition and also fineness of structure. The high-silver compositions and the coarse materials had more and larger silver inclusions. The high-tungsten compositions had very few silver inclusions and many melted tungsten particles and cones. It also was reported that the voltage drop decreased significantly over the first few minutes of measurement. This was thought to be the result of softening and melting of the surface layer in the areas of constriction. The final resistance seemed to correlate with the amount of silver inclusions in the oxide layer which in turn was related to the amount of silver in the bulk material.

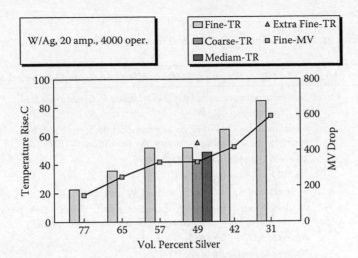

FIGURE 16.26
Temperature rise and resistance in terms of millivolt drop versus silver composition and microstructure for W/Ag, after switching 4.76 mm diameter contacts [110].

Slade did extensive studies of the voltage-drop characteristics for tungsten–silver 50 volume percent silver [115]. The work was done at 20 A on 4 mm diameter contacts and the voltage drop was continuously measured. The contacts were switched for 3000 operations and the contact resistance increased with contact switching life. Slade noted that the voltage trace varied considerably with the condition of the contact and described four typical types of voltage stability as follows (see Chapter 10, Figure 10.58):

- Type 1: A steady millivolt drop of less than 100 mV with no excursions. Typical for new and low switching life.
- Type 2: A plateau voltage from 200 to 600 mV with rapid excursions to lower voltages, indicating a semi-stable high-resistance contact surface.
- Type 3: Semi-stable mean voltage in the 400 mV range with continuous rapid oscillations of up to ±100 mV, indicating an unstable surface structure superimposed on a more stable high-resistance surface.
- Type 4: More violent type 3 oscillations with peaks as high as 1.3 mV leading to periods of lower stable voltages (200–300 mV), indicating an unstable high-resistance surface with periods of lower resistance.

Slade also developed Table 16.6 which uses the Kohlrausch voltage drop to temperature relationship to indicate possible reactions taking place at different millivolt readings. The voltage instabilities noted for tungsten–silver are in the range for the decomposition of silver tungstate (<300 mV) and tungsten oxides (390–490 mV). A significant conclusion from this work is that tungsten–silver contacts are self-limiting for contact resistance as a result of the oxide decomposition mechanisms. Thus, since all of the oxides possible on the surface of tungsten–silver decompose at or below 1300°C and tungsten melts at about 3400°C the time for millivolt drop excursions above the decomposition or melting temperature of tungsten oxides should be short and the upper limit for the excursions should be in the range of 1100 mV. Therefore, no thermal runaway conditions are expected for tungsten–silver contacts, although they may have high surface resistance and a resulting high temperature rise.

TABLE 16.6

The V_c Value for Temperatures at Which Physical and Chemical Effects Occur for the Ag, W, O_2, and C System

MV	Temperature (°C)	Oxidation, Softening, Melting Effects on Ag and W Contact Materials
110	180	Softening temperature of Ag, AgO_2 begins to decompose [114]
155	310	Ag_2O decomposes [114]
225	500	Initial oxidation of W begins as a compact protective film but as it becomes porous, the oxide film grows by oxygen diffusion [121]
235	550	$2Ag + WO_3 + \frac{1}{2}O_2 \rightarrow Ag_2WO_4$ [102]
255	580	$Ag_2WO_4 + CO \rightarrow AgWO_3 + CO_2$ [115]
260	600	Formation of Ag_2WO_4 on heated Ag-W contacts [107], Ag_2WO_4 melts [102]
295	700	Oxidation rate of heated Ag-W contact is twice that at 600°C with the formation of WO_3 and perhaps WO_2 but no Ag_2WO_4 [107]
330	800	$Ag_2WO_4 + CO \rightarrow 2AgWO_2 + 2CO_2$ [115]
380	960	Melting temperature of Ag [114]
390	1000	Softening temperature of W [114]. Above this temperature evaporation of W oxides increases, with the extent of evaporation being affected by the partial pressure of oxygen [121]. Conducting W_3O_8 can form with the reaction of W and WO_3 [102]
490	1300	W oxides evaporate as soon as they form. At high pressure a boundary layer of evaporated oxide limits the access of oxygen to the surface [121]
545	1470	WO_3 melts but can sublime at a lower temperature [121]
780	2210	Boiling temperature of Ag [114]
1140	3410	Melting temperature of W [101]

There also has been work done on other silver refractory systems and minor additives to the tungsten–silver and tungsten carbide–silver systems in search for materials with lower and more stable contact resistance on switching low-current loads. The system silver–titanium carbide and silver–tungsten carbide–titanium carbide were investigated and found to have improved contact resistance compared to tungsten–silver at low currents [111,118,119]. These materials never developed for several reasons: the difficulty for manufacturing, low silver content retained after higher-currency switching, and the formation of a slag on the contact surface from low-density TiO_2 which gave rise to high R_c [119]. The additions of nickel up to a few per cent by weight was investigated by Witter and found to lower the resistance of tungsten–silver materials after switching [95]. It was also found that the addition of nickel increased the erosion rate and lowered the dynamic welding resistance of the material. A similar effect was found by Kabayama for cobalt additions to tungsten–silver [96].

Slade et al. investigated the effect of additions of graphite, 0.5 to 1, to tungsten carbide–silver with high volume percentages of silver [120]. These materials exhibit considerable improvement for resistance performance as a result of the high silver content and graphite reducing the oxidation of the tungsten, but it must be kept in mind that the erosion and anti-welding must probably have also been compromised.

From the above, it can be seen that the resistance of silver refractory metal systems is a complex subject that does not have easy solutions. The surface of the materials changes after arcing and forms a surface that is a product of both the load it is switching and its own bulk properties. The characteristics of this layer will influence the performance of the material with respect to erosion, anti-welding and resistance. As seen above, adjustments in the material for improving resistance results in also changing the other properties. With regard to contact resistance, one point of distinction between silver refractory metals and silver metal oxides, which is important to understand, is that for silver metal oxides the change in contact resistance on switching is in general smaller than that seen for silver refractory metals. The reason for this is that for silver refractory metals the silver content is diluted beyond just the evaporation of silver from the surface, but by the growth of the refractory oxides and also consumption of silver in forming silver metal oxide compounds. For this reason, the use of these materials is usually in devices that employ high contact forces and wiping motion of the contacts on closing to mechanically disrupt surface oxides that have formed and create a low R_c region.

16.4 Vacuum Interrupter Materials

This is a very specialized area of contact materials. The subject of vacuum interrupters was introduced in Chapters 9, 14, and 15, and they also contain information on the contact materials that are used in them. A good reference on this subject is a review written in 1992 by Slade of the advances in materials for high-power vacuum interrupters which explains the unique material requirements for these materials, compares the different types of materials in use, and lists 68 references on this subject [121]. He updated this paper in the book published in 2008 [122]. Three main types of materials have dominated this market, copper-bismuth alloys, tungsten–copper type composite refractory materials, and chromium-copper contacts. At this point in time most of the copper bismuth materials have been replaced by chromium copper which is by far the dominant material for vacuum interrupters. The commercial type materials that are made and used for vacuum interrupters can be divided two groups today:

1. Tungsten–copper or tungsten carbide–silver

   This type of material is similar to these materials discussed earlier in this chapter, except that these materials have special processing to be suitable for vacuum application. As a result of interruption limitations these materials are usually limited to lower-current vacuum interrupter designs applied to load switches with currents less than about 2000 A [121]. Some work has been investigating problems with tungsten copper materials with high tungsten content, 90/10 W/Cu, Li el al. [123]. It was found that this material had cathodic type pip and crater erosion that left a pip on the anode and a crater on the cathode. The investigation showed that tungsten copper with a higher copper level, 30%, had smooth erosion compared to the 90/10 material. The cause of this problem was thought to be from that for the high tungsten material, the cathode spots (see Section 9.6) were slow to move from its bridge initiation area too long. Follow up work was done by Taylor [124] who made measurements with a high speed video camera. He found that there was about a 4 msec delay in the initial transition mode on contact opening that resulted in a high current density near the original bridge column. Some measurements he made on chrome coppers showed that the arc expansion during transition mode was 3.7 times faster than for tungsten copper. He also agreed that the erosion pattern for tungsten copper 70/30 was much improved over 90/10 W/Cu for providing a more even erosion and less with pip and crater formation. It should be also recalled from a previous section of this chapter that a higher erosion rate can also have an effect on reducing the probability of pip and crater erosion.

2. Chrome–copper

Chrome–copper is just as complicated as composite refractory metal contacts and maybe more. There are several different processes used for making this material which results in not only different properties but different metallographic structures. Hauner et al. [125] and Slade [122] provide some detail descriptions of the various processes and some general properties. A popular process is the press, sinter, repress, and annealing process. The sintering is done in the solid state and normally there is some level of porosity even after repressing and annealing. The microstructure is quite coarse compared to tungsten refractory metals. This is probably the most economical of the three main processes used.The infiltration process involves pressing a chrome or chrome rich skeleton, sintering it, and infiltrating the skeleton with copper. Since the infiltration is done at a high temperature above the copper melting point the bonding of the chrome and copper is very good. The chrome coarse similar to the press, sinter repress process. The process produces bars that are over 99% dense. After bars are formed with this material they can be cold extruded into smaller and longer bars. This results in an anisotropic structure with elongated chrome particles parallel to the rod. Melting and very fast freezing is the third process for making chrome copper. Induction melting of a chromium copper power mixture or arc melting chromium copper rods in an argon atmosphere is the first stage. The molten mixture is then very rapidly cooled in a water cooled crucible. This results in a copper matrix with very fine chromium inclusions. Large ingots can be produced that can be cold extruded into smaller and longer rods. The material is 100% dense and the structure is finer grained than the other two processes. Typical compositions for chrome copper are in the range of 25–50% chromium. In looking at the sophistication of the processes, it is believed that consistency of processing is very important. They are not generic and although two companies may use similar processes the material may be quite different. For any material that is evaluated it is important to record as much information as possible about the chemistry, microstructure and physical properties. In the literature there are some comments that infiltration process experiences less cracking on arcing than the press, sinter and repress materials. There is no experimental data to back those claims and it would be interesting to see some data on toughness measurements.

16.5 Tungsten Contacts

Tungsten as a result of its high melting and boiling points, highest of all the metals, and high electrical conductivity has been used for many years as a contact for special applications (see Chapter 24, Table 24.1). It is mainly used in applications which require high-frequency switching like automotive ignition systems and horn contacts. Tungsten contacts form oxide layers not considered protective on air exposure that reach about 50 Å at room temperature. By use of high contact force to break through most of the oxide layers, electron tunneling through the remaining layers of oxide can maintain reasonable conduction with tungsten contacts [126]. Forces of 5 N or more are common for devices using tungsten contacts. During arcing, as seen in the last section on composite materials, tungsten will be oxidized under the heat of the arc. Tungsten has many oxide compounds that are progressively oxidized to WO_3 and then sublimed. Recalling from Table 16.6, all tungsten oxides evaporate as soon as they are formed at or above 1300°C. Typical erosion of pure tungsten contacts shows WO_3 around the arc spot and a coating of thin sub-oxides in the center of the arc spots. Tungsten contacts were very popular for high-frequency switching operations, hundreds of operations per second, since it has low electrical erosion, but are not good for applications where contact force is not high, hundreds of grams, and the contacts must carry high currents for long periods in the closed position. High heating in the current constriction area can lead to oxide build-up and without mechanical wiping movement of the contacts high resistance can result with device overheating and or device failure [127]. This, of course, will depend on the current level and device contact forces. The usage of tungsten contacts began decreasing in the 1970s and 1980s as a result of replacement of electromechanical devices in which they were used, like automotive ignition systems, by solid-state devices.

16.6 Non-Noble Silver Alloys

Table 24.2 in Chapter 24 lists some of the common silver and silver copper alloys. Since these materials are made by melting and casting these alloys are much more consistent from supplier to supplier and batch to batch. For that reason there is no need to discuss variations as a result of processing. Fine silver and these alloys are some of the first contacts that were used yet very little is found in the contact literature discussing research or comparing electrical switching properties of these materials. The purity of these alloys can vary but is normally based on 99.95 wt.% silver as the alloying ingredient.

16.6.1 Fine Silver

Fine silver has the highest conductivity of all metals and as a result is useful in low-current applications, where resistance is critical owing to the low contact force. The oxides of silver are not stable at elevated temperatures and decompose at 200°C, and it is relatively stable to oxidation at room temperature except in the presence of pollution such as ozone [126]. The problem with silver with regards to corrosion is with the formation of sulfides and chlorides. This is discussed in detail in Chapter 2 and therefore will not be discussed here. For normal arcing applications the contact force, contact wiping, and circuit voltage and current are above values where corrosion of silver is a concern. Of the non-noble silver alloys, not containing noble metals such as palladium, fine silver offers the lowest resistance choice since Ag_2S forming on this alloy is soft and low in mechanical strength [128]. There is little data comparing electrical erosion characteristics of fine silver to other materials. For make bounce erosion, it is listed as having high erosion compared to silver–nickel and the silver metal oxides [67]. For break arcs some comparisons are published for long arcs where fine silver is shown to be very anodic and to yield lower erosion than palladium alloys [129]. Generally, fine silver is limited to low-current applications.

Some contact brochures suggest 1 A as an upper limit and others 10–20 A. The current is only one factor for consideration of a contact and other factors like load type are also important. Fine silver is a common contact for small snap-action thermostats which switch low currents and resistive loads.

16.6.2 Hard Silver and Silver–Copper Alloys

This term is used with silver alloys which have a small amount of nickel and or copper to increase the hardness. Again little research data is published on these materials. Nickel in small amounts increases the hardness of the silver without having much effect on the conductivity. Silver–nickel will be discussed in the next section as a separate category of contact material. Copper additions to silver as seen in Table 24.2 in Chapter 24 lower the melting point and decrease the electrical conductivity of silver–copper alloys. Silver and copper form eutectic type alloys with a eutectic melting point of 779°C [79]. Contact brochures indicated less electrical erosion wear of silver–copper alloys than fine silver but more erosion wear than silver–nickel. It is also indicated that dc transfer is less for alloys with 10% or more copper. Leung and Lee found that a silver copper 2 wt.% alloy performed well for automotive resistive and inductive loads in keeping low contact resistance, but that it welded and had high transfer for lamp loads compared to silver–tin oxide [58,59]. Under high make currents typical of lamp loads coupled with contact bounce, silver–copper alloys tend to erode at a very high rate compared to silver metal oxide contacts. Figure 16.27 compares erosion of silver–copper 2 wt.% to a silver–tin oxide 11% alloy for erosion under long bounce (>2 ms), with 100 A inrush current at 12 V dc as the copper content of silver copper alloys increases, the resistance of the material to corrosion decreases. For this reason, silver–copper alloys are normally limited to less than 30% copper. Silver copper alloys, similar to fine silver, are recommended for lower-current applications, with contact companies generally setting limits under 20 A.

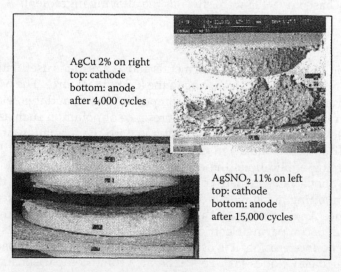

FIGURE 16.27
A comparison of silver–copper 2% alloy and silver–tin oxide 11% alloy after switching a 100 A inrush lamp load [12].

16.7 Silver–Nickel Contact Materials

A very popular type of contact material used worldwide, and especially in Europe, consists of silver–nickel contacts (see Chapter 24, Table 24.3). These are powder metallurgical contact materials since silver and nickel have virtually no mutual solubility. This makes them somewhat analogous to tungsten–silver except the nickel is used in much lower percentages similar in volume percentage to the oxide level in silver metal oxides. The sintering is done in the solid state and the process options for making these materials are similar to those used for powder metallurgical silver metal oxides. The conductivity varies with the volume percentage of silver since there is no decrease in the silver conductivity from alloying. Like powder metallurgical silver metal oxides silver–nickel contacts can be made by pressing, sintering, and coining individual parts or by wrought powder methods involving sintering a billet followed by extrusion or alternative forming methods. The wrought method produces fully dense materials, free of porosity. The particle size distribution of silver–nickel contacts can be varied greatly by the powder metallurgical processes used for making the powders and blending the powders. The nickel particle shape can also be varied for these materials by variances in starting powder combined with different wire and strip forming techniques.

A study of the particle size and shape effect on electrical erosion was done by Behrens et al. [130]. They tested materials which varied in particle size from submicron to over 100 μm). They also had materials which had nickel fibers perpendicular to the contact face and parallel to the face. The tests were conducted for break only at 115 A and 220 V ac AC-4 testing. The results showed no correlation for erosion with particle size. The orientation also only made a difference for the initial part of the testing until an equilibrium layer of silver nickel melt material had been established on the contact face. They concluded that silver–nickel is a unique material since it establishes a nickel particle distribution on the surface as a result of nickel melting and dissolving in the silver to a small extent during arcing and then re-precipitating on the surface. Therefore, regardless of the starting microstructure the surface melt layer microstructure is similar for a given arcing condition. It must be kept in mind that this study was limited to break erosion only and since the make erosion process is different these conclusions cannot be extrapolated to make and break results.

Balme et al. compared three silver nickel materials and both fine silver and silver metal oxide materials [131]. The testing was done in an automotive relay with relatively high contact force, 2 N, for lamp loads with over 100 A inrush current on closing. The results were reported in terms of contact resistance after endurance welding, and sticking that occurred during testing. For welding resistance all three silver nickel grades, 10–40% nickel, showed make welding resistance slightly better than fine silver but significantly inferior to the silver metal oxide grades. With regard to contact resistance, the 90/10 Ag/Ni material was just a little higher than fine silver but the 30% and 40% nickel grades were significantly higher and similar to the silver metal oxide materials. Leung and Lee tested silver–nickel in automotive relays and compared a 90/10 material to silver–tin oxide and silver–copper contacts. They found that the silver–nickel was intermediate for contact resistance with silver–copper and silver–tin oxide, but showed a higher amount of material transfer than either of the other two materials [58]. One of the reasons silver–nickel is very popular is that it can be directly welded onto copper substrates from wire. This lowers the contact assembly costs. This same advantage for fabrication is a disadvantage for limiting applications where the high currents cause contact welding in a device. Depending on the device and type of load silver–nickel materials work best in devices not exceeding currents of 50–100 A.

16.8 Silver Alloys and Noble Metals

16.8.1 Palladium and Silver–Palladium Alloys

The most common alloys of this category are silver–palladium alloys and palladium (see Chapter 24, Tables 24.2 and 24.7). These alloys have been used for many years in mainly telecommunication switching applications which require low electrical noise, therefore more stability in contact resistance. These alloys containing noble metals are generally more chemically stable and resistant to corrosion than non-noble silver alloys and silver metal oxide contact materials. The corrosion properties of the noble metal alloys are discussed in Chapters 3 and 8 and therefore will not be covered here. The telecommunication circuits that used relays with these materials operated at low currents near or below the minimum arcing current, so that normally the electrical erosion properties of these materials was not as important in these applications as the stability of these materials from forming resistive films. Arcing did frequently occur in these applications as a result of organic vapor contamination resulting in contact activation. Contact activation can lower the minimum current required for an arc to occur, activation is discussed in Chapters 10 and 19. These materials were thus used as arcing contacts but were mainly developed for low current or dry circuit switching use. After 1980 the switching function for these signal type circuits was largely converted from electromechanical relays to solid-state devices. Since noble metals are 10–100 times more expensive than common silver alloys, the use of these metals is limited to applications where there is an absolute need for their properties. These types of alloys are still used in some signal type circuits, but also in some higher current applications where corrosion resistance is important.

The silver–palladium alloys and palladium differ from silver alloys and metal oxides as a contact material in more ways than just being more chemically stable. The higher palladium-containing silver–palladium alloys are shown by Sawa et al. [132], to be more stable to changes in contact resistance as a result of electrical arcing than fine silver. They studied changes in contact resistance after 20,000 switching operations at 1.2 A for both make and break at 24 V dc. The largest difference was for break arcs several milliseconds in duration; palladium increased only 20% in contact resistance compared to a 60% increase for silver contacts. For protection against corrosion and oxidation for palladium silver alloys in many applications a thin gold overlay is put over the palladium silver alloy. Tamai [133] investigated the use of dopants in the palladium silver alloy to give it more resistance to oxidation on exposure to high temperatures. He found that the additions of 1% or less of magnesium to the alloy improved the alloys ability to retain a low contact resistance after being exposed to high temperatures. A thin diffused silver layer formed on the surface with the addition of magnesium. The mechanism is not clear for the formation of this layer but it resulted in less oxidation on the surface. It should be kept in mind that the gold will still be better for corrosion protection for non- arcing applications but if arcing is present the gold will be burned off.

With regard to electrical erosion, interruption studies done in the range of 5 A at 20–48 V dc show the erosion rate for pure palladium to be much higher than for fine silver [134,135]. Chen and Sawa compared the electrical erosion of silver contacts, palladium contacts and combination pairs of silver contacts mated with palladium contacts [134]. This study was done using an inductive load which produced relatively long arcs, 1–4 ms at 20 V dc. Comparisons were made in terms of anode and cathode losses or gains and net losses for both the anode and cathode combinations. At low currents of 1 A or less, the differences among the materials were small. A comparison of the erosion at 4 A is shown in Figure 16.28 from data taken from work by Chen and Sawa [134]. This figure serves well to illustrate the

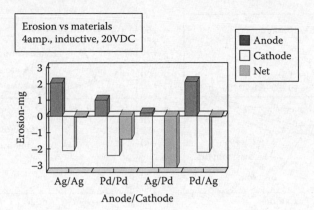

FIGURE 16.28
A comparison of electrical erosion for silver and palladium contacts after 60,000 operations.

differences in interruption erosion between palladium and silver. As a result of the long arc both the Ag–Ag and Pd–Pd contact pairs exhibit similar cathodic erosion. The net erosion, sum of cathode loss plus anode gain, is quite different, with silver showing very little loss compared to palladium. This difference is the result of a higher percentage of the material eroded from the cathode transferring to the anode for silver as compared to the percentage of eroded material transferring to the anode for palladium. The reason for this difference is not explained but some possible factors causing this will be discussed. From the data it can be seen that the gap at which the gaseous phase began for silver was shorter than where the gaseous phase began for palladium. A shorter gap should provide a higher ratio of material transfer versus eroded material being lost. The oxides of palladium are stable at higher temperatures than the oxides of silver. PdO forms on heating above 700°C and begins to decompose at 870°C. Some metastable oxides also form between 900°C and 1300°C [136]. The sticking coefficient for deposits of palladium on the anode may be affected by this. Figure 16.28 also shows some interesting results for the erosion of contact pairs containing both silver and palladium contacts. Those pairs that have the palladium contact as a cathode and a silver contact as an anode show erosion results similar to palladium pairs. The pairs that have silver as a cathode and palladium as an anode have results similar to silver pairs. Since the erosion is mainly cathodic, transfer from cathode to anode, for these long arcs the cathode material controls the erosion process. For example, for the silver cathode and palladium anode pair, the transfer of the silver from the cathode material onto the anode soon results in two contacts with silver surfaces.

Sone et al. show the erosion characteristics as a function of composition for a full range of silver and palladium alloys (see Figure 16.29 [135]). It can be seen that the erosion generally becomes more cathodic as the palladium content of the silver–palladium alloy increases. A large change in the erosion rate between pure silver and silver–palladium 10% is not explained. Silver and palladium form a solid solution alloy with complete solubility [79]. This results in a very sharp drop in conductivity as palladium is added to silver; see Figure 16.30. Possibly this might be a factor for influencing this type of shift in erosion. In comparing the results for pure silver and pure palladium in Figure 16.29 to Figure 16.28, it must be noted that the results in Figure 16.25 are expressed in erosion per number of switching cycles and Figure 16.29 is expressed in terms of erosion per arc second duration. For the switching represented by Figure 16.29, resistive load, the arc duration for the

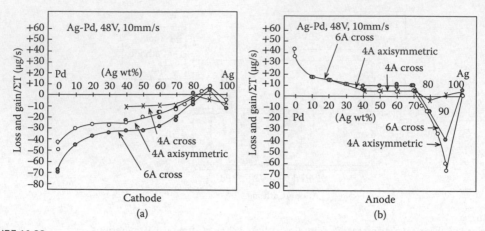

FIGURE 16.29

Electrical erosion of silver palladium alloys as a function of composition for interruption testing at 48 V dc resistive load, with a slow opening speed of 10 mm s^{-1}, for cross bar and cylindrical contacts, and erosion expressed as µg per second of arc duration [120].

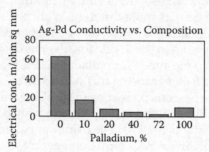

FIGURE 16.30

The electrical conductivity of silver palladium as as function of composition.

palladium contacts is reported as much shorter, a quarter to a tenth of the arc duration for the silver contacts. Taking this into account the results are comparable.

For erosion characteristics for contact make and shorter duration break arcs little published data was found. Palladium contacts are used in some automotive applications for special relays which are used to control lamps as hazard flashers. This relatively expensive material, as compared to silver-based contacts, is used since it erodes slowly in these applications and offers long switching life. This is mainly owing to the shorter arc duration found for switching resistive loads with palladium versus silver-based contacts as shown for palladium ruthenium contacts versus silver tin oxide contacts in Figure 16.31 [137]. From the above it can be seen that palladium alloys offer only lower resistance as an advantage for interrupting circuits which produce long arcs. For make and break dc applications such as automotive circuits, as a result of the good interruption properties of palladium these alloys offer longer switching duty life.

16.8.2 Platinum

Another metal from this group that is occasionally used as an arcing contact is platinum. It is similar to palladium in electrical conductivity but has a higher melting point, 1773°C. Platinum also has a very high density, 21.45 g cm^{-3}, which, coupled with its

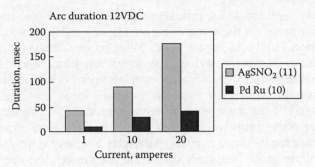

FIGURE 16.31
A comparison of arc duration in μs for clean AgSnO$_2$ 89/11 and Pd/Ru 90/10 contacts interrupting resistive loads with an opening velocity of 1.5 m s at 12 V dc.

very high price per ounce, make it a very expensive material to use as a contact. As an arcing contact material little has been published about platinum. Platinum can form some oxides but these oxides are only stable in a narrow temperature range, 900–1200°C [79,126]. Thus, oxidation is not a problem for using platinum as an arcing contact. Platinum has been used, but to a much lesser extent, in applications similar to where tungsten contacts are used such as distributor contacts. In these applications the contacts are subjected to millions of rapid and repetitive switching operations and a low duty cycle for carrying current. The platinum offers good erosion resistance plus more stable contact resistance than tungsten. In the 1980s solid-state devices, to a large extent, have replaced electromechanical distributors. This has cut down further the use of platinum as an arcing contact.

16.9 Silver–Graphite Contact Materials

Silver–graphite is a popular brush material (see Chapter 24, Tables 24.3 and 24.12) as a result of its anti-welding and electrical resistance properties (see Chapters 20–22). In this section, the discussion is limited to these materials used for arcing contact applications. For these applications the composition range of 2–5% graphite by weight, 9–20% by volume, is the most popular. Since graphite is very low in density compared to silver the volume percentage range compared to the weight percentage range is quite different. Silver–graphite is a composite material like previously discussed silver metal oxides. It is also manufactured by powder metallurgical methods since silver and graphite do not form any alloys. It is manufactured by a variety of powder metallurgical methods, including pressing, sintering, and extrusion of ingots to form wrought silver–graphite strip and wire and also the conventional press, sinter, and repress method. One of the differences in processing of silver–graphite from silver metal oxides is that it must be sintered in a protective atmosphere to prevent oxidation of the graphite.

Both the particle size and shape of the graphite phase can vary to a large extent for these materials [138,139]. The particle size can vary from submicron sizes to over 100 μm and the shape of the particles can be flakes, fibers, or spherical. Silver–graphite materials are noted for good anti-dynamic welding properties and stable and low-contact resistance [67,138,140]. The disadvantages of these materials are high electrical erosion rates and poor arc mobility (see Section 14.1.3 and Figure 14.32). Wingert et al. investigated the effect of particle size on the electrical erosion, dynamic welding resistance, and contact resistance of Ag/C, 95/5 wt.% materials [138]. They made materials with both 4 μm and 20 μm graphite and found that using the coarser graphite, the materials form very large silver mounds on the surface compared to the finer grade of silver–graphite. The results show that the coarser materials have poorer anti-welding resistance than the finer grades. The electrical erosion rate is higher for the finer particle materials. As the graphite particles become smaller the silver to silver grain boundary bounding is lowered, less contiguity of silver grains. This weakening of the contact structure by the decrease in silver to silver bounding results in higher erosion rates.

In devices such as current-limiting circuit-breakers, the silver–graphite contacts are frequently used as one contact component in unsymmetrical contact pairs. Turner and Turner investigated the use of silver–graphite contacts paired with fine silver contacts in railroad switches [141]. They found that there was a polarity effect with regard to the electrical erosion. This work was done at 250 V dc at 50 A for make arcs formed by opening and re-closing after 4 ms of arc time. In this study the silver–graphite showed lower erosion when used as the cathode than when it was used as an anode. The contact resistance was also lower when the fine silver was the anode and the silver–graphite was the cathode.

Lindmayer and Schroder found different results in studying unsymmetrical silver–graphite contacts paired with both copper and tungsten–silver contacts [140]. Their erosion test was done in a test fixture for interrupting a 220 V half-cycle current from 350 to 1000 A. Their tests show that the erosion resistance of silver–graphite paired with either material was much better when the silver–graphite was used as the anode than when it was used as a cathode. This is explained from erosion results of symmetrical pairs of silver–graphite under similar conditions. For symmetrical pairs of silver–graphite contacts erosion over an entire current range of 350–1000 A shows the erosion to be primarily cathodic with the ratio of the total erosion contributed by the anode ranging from nothing to a maximum of 20%. The difference in this work from that done by the Turners is that the Turners' work used shorter arcs with contact make which would tend to be more anodic in terms of erosion than tests done for only interruption. Lindmayer and Schroder also investigated dynamic welding resistance of the unsymmetrical contact pairs [140]. They conducted these tests in a fixture which created a contact bounce of about 3–4 ms. They show the weld strength to be lower when the silver–graphite was the cathode rather than the anode for both mating with tungsten–silver and copper. The difference between the two unsymmetrical pairs owing to polarity were much smaller and insignificant compared to the differences of these pairs to the much higher welding strengths generated by copper or tungsten–silver symmetrical pairs.

For contact resistance, tests were conducted in conjunction with the interruption cycling tests. They found that the resistance was also significantly affected by the polarity. If the silver–graphite was the cathode the resistance was less than 0.1 mΩ for both mating tungsten–silver and copper. For opposite polarity the resistance was much higher and in some cases over 100 mΩ. An explanation of this is that for the polarity of silver–graphite as the cathode the erosion rate was very high and resulted in a reducing atmosphere of C and CO which reduced the metal oxides and the contact electrodes. For the opposite polarity, recalling from above, the erosion rate of the silver–graphite anode was very low and, therefore, did not generate as much of a reducing atmosphere.

This work also shows that the poor arc mobility of silver–graphite can be improved on by using unsymmetrical contacts with the silver–graphite as the anode. With this configuration the arc mobility is shown to be intermediate between the mating cathode material and silver–graphite.

Vinaricky and Behrens [142] did a very thorough study of silver graphite that covered the questions of the importance of graphite particle orientation and also studied the erosion characteristics under different gasses and electrical loads. The question on the importance of fiber orientation, parallel or perpendicular to the contact surface, was that the processes were different for making the two different orientations. The extrusion process for making the perpendicular products produced fully dense material and the rolling process for parallel fibers had some porosity. This investigation the made samples of both materials using extruded silver graphite and machining it in different directions to get parallel and perpendicular samples. The other part of the study involved testing the materials in air, nitrogen, oxygen, and argon using the long arc. For this material the testing for long and short arcs is quite different as shown above from other research. First, for the short arc, the perpendicular fibers had lower erosion but this changed for the long arc where it didn't seem to matter. Many of the applications are for long not short arcs. The other big finding was that under argon the erosion was almost nothing compared to the rest of the gasses. The results for air, oxygen, and nitrogen were very similar. The conclusions were that even under nitrogen there are chemical reactions taking place during arcing that increase the erosion rate for silver graphite. This is an interesting finding showing chemical reactions during arcing are responsible for the high erosion of silver graphite. This is also an interesting finding especially for devices that have a protective atmosphere.

16.10 Conclusion

Arcing contact applications cover a very wide range of applications in terms of current levels being switched, types of electrical loads, voltages, ac versus dc ambient conditions, and mechanical opening and closing forces and velocities. Table 16.7 shows some of the common types of arcing contact applications matched with typical materials used at different current levels for those applications. As a result of the many differences in devices and electrical loads, certainly many other combinations of contacts versus applications exist. In this chapter it has been stressed that for composite materials like silver metal oxides and silver refractory metal materials that generic classification by chemistry is inadequate for comparing and allowing substitution of materials for specific applications. The chapter discussed how variations in processing and or small amounts of additive could significantly change performance results of materials. For the purposes of the table above the material designations are meant to be very general or generic, thus $AgSnO_2$, is meant for example to represent all types of silver tin oxide, $AgSnO_2$, $AgSnO_2 + Bi$, $AgSnO_2IN_2O_3$, etc. Thus, this table gives you a general idea of where different materials are applied, but the chapter shows that a much more detailed understanding of the materials being evaluated is required in making decisions on arcing contacts.

TABLE 16.7

Some Typical Applications for Arcing Contact Materials

Applications	Low Current (under 10 A)	Medium Currents (10–100 A)	Med–High Currents (100–1000 A)	High Currents (>1000 A)
AC switches, relays, and contactors	Ag Fine–grain Ag AgCu Ag/Ni	Ag/Ni Ag/CdO Ag/SnO$_2$	Ag/CdO Ag/SnO$_2$	
DC automotive relays and switches	Ag/Cu Ag/Ni Ag/SnO$_2$ Ag/NiO/MgO	AgCu Ag/Ni AgSnO$_2$ AgNiOMgO Pd		
Circuit-breakers USA		WAg WC/Ag	W/Ag WC/Ag Ag/SnO$_2$ Ag/W– Ag/CdO	
Circuit-breakers Europe	Ag/CdO Ag/ZnO Ag/SnO$_2$	Ag/CdO Ag/ZnO Ag/SnO$_2$	Ag/CdO Ag/SnO$_2$ Ag/C–Cu Ag/C–Ag/Ni	
Thermostats	Ag Fine Grain Ag AgCu Ag/Ni	Ag/CdO Ag/SnO$_2$		
Power interrupters			W/Cu vacuum	W/Ag WC/Ag W/Cu SF$_6$ or oil Cu/Cr vacuum

Acknowledgements

The author is very grateful to Dr. Zhuan-ke Chen, Chugai USA for technical discussions and Dr. Paul Slade discussions and suggestions. He also wants to thank the staff at Chugai USA: Sarah Bauer, Donna Witter, and Chris Anderson for their help in proofreading and editing.

References

1. G Rolland, Y Zerallil, V Guipoint, M Jeandin, S Hardy, L Doulet, C Bourda. Lifetime of cold-sprayed electrical contacts. *Proc. 26th Int'l Conference on Electrical Contacts*, 2012, pp 338–345.
2. *Doduco Data Book of Electrical Contacts*, Revised Edition, www.doduco.net, 2012.
3. Chugai USA Brochure, Chugai USA Inc., Waukegan, IL, 1995.

4. Umicore Brochure, Umicore, Brussels, Belgium.

5. American Society for Testing and Materials. ASTM Standards, Section 3, *Metals Test Methods and Analytical Procedures*, Vol. 03.04, 1987.

6. M Shen, W Cote, L Gould. A historic review of Ag–MeO materials. *Proc. 32nd Holm Conference on Electrical Contacts*, 1986, pp 71–76.

7. C Wagner. *Zeitschrift fur Elektrochemie* 63, 772–790, 1959.

8. E Freudiger, E Jost. Kinetics and thermodynamics of the internal oxidation of silver cadmium. *Proc. 1st Int'l Research Symposium on Electrical Contact Phenomena*, University of Maine, November 1961, pp 177–197.

9. J Grosse, T Moser, B Rothkegel. Internal oxidation of silver metallic oxide materials. *Proc. 13th Int'l Conference on Electrical Contacts*, 1986, pp 211–215.

10. E Jost, T Santala. Process and metallurgical variations affecting electrical performance of internally oxidized, homogeneous, monolithic silver cadmium oxide. *Proc. 33rd IEEE-Holm Conference on Electrical Contacts*, 1987, pp 175–179.

11. H Kim, T Peters. A metallurgical model of arc-erosion in Ag–CdO contacts. *Proc. 22nd Holm Conference on Electrical Contacts*, 1976, pp 98–97.

12. G Witter, Z Chen. Private research, Chugai Inc., USA, 1997.

13. D Pedder. Volatilisation of cadmium oxide and early welding in internally oxidized silver cadmium alloys. *Proc. 23rd Holm Conference on Electrical Contacts*, 1977, pp 69–76.

14. Shen Yuan-Shou, A Lima. Pure silver surface layer on the oxidized Ag–Sn–In material. *Proc. 34th IEEE Holm Conference on Electrical Contacts*, 1988, pp 13–15.

15. R German, *Powder Metallurgaical Science*, 2nd ed. Princeton, NJ: Metal Powder Industries Foundation, 1994

16. ASM. *Powder Metallurgy, Metals Handbook*, Vol. 7, 9th ed. American Society of Metals, 1984.

17. T Allen. *Particle Size Measurement*, London: Chapman and Hall, 1968.

18. D Pedder, P Douglas, J McCarthy, F Brugner. I.O.A.P.— a novel process for the manufacturing of long-life silver–cadmium oxide AC contact materials. *Proc. 22nd Holm on Electrical Contacts*, 1976, pp 109–120.

19. M Fischmeister, G Grimvall. Oswald ripening—a survey, sintering and related phenomena. *Material Science Research* 6: 119, 1972.

20. A Lapinski. Investigations on the effect of technological parameters on the properties of fine grained Ag CdO sinter. *Proc. 24th Holm Conference on Electrical Contacts*, 1978, pp 201–208.

21. M Poniatowski. The influence of production method on the properties of contact materials for high current applications. *Proc. 7th Int'l Conference on Electrical Contacts*, 1974, pp 477–484.

22. M Poniatowski, K Schroder, E Schulz. Behavior at closing or opening of silver metallic oxide materials in electrical power engineering. *Proc. 8th Int'l Conference on Electrical Contacts*, 1976, pp 353–358.

23. M Poniatowski, E Schulz, A Wirths. The replacement of silver–cadmium oxide by silver–tin oxide low voltage switching devices. *Proc. 8th Int'l Conference on Electrical Contacts*, 1976, pp 359–364.

24. L Morin, N Ben Jemaa, D Jeannott, H Sone. Transition from the anodic arc phase to the cathodic metallic arc phase in vacuum at low DC electrical level. *Proc. 47th IEEE Holm Conference* 2001, pp 88–93.

25. N Ben Jemaa, L Morin, Transition from the anodic arc phase to the cathodic arc phase in vacuum at low DC electrical level, *IEEE CPMT Trans*, V25 Issue 4, Dec. 2002, pp 651–655.

26. G Witter, Z Chen. A comparison for the effects of opening speed, contact gap and material type on electrical erosion for relays interrupting inductive automotive loads. *Proc. 23rd Int'l Conference on Electrical Contacts* 2006, pp 17–21.

27. Z Chen, G Witter. A comparison of contact erosion for opening velocity variations for 13 volt circuits. *Proc. 52nd IEEE Holm Conference* 2006, pp 15–20.

28. F Heringhaus, P Braumann, A Koffler, E Susnik, R Wolmer, R Chaussee. Quantitative correlation of additive use and properties of Ag-SnO$_2$-based contact materials. *Proc. 21st Int'l Conference on Electrical Contacts* 2002, pp 443–446.

29. E Hetzmannseder, W Rieder. Make erosion: Ag/MeO contact materials investigated in a contactor and in a new testing equipment. *Proc. 16th Int'l Conference on Electrical Contacts*, 1992, pp 371–377.

30. T Mützel, R Niederreuther. Advanced silver-tin oxide contact materials for relay application. *Proc. 26th Intern'l Conference on Electrical Contacts*, 2012, pp 194–199.

31. G Witter, I Polevoy. Contact erosion and material transfer for contacts in automotive relays. *Proc. 18th Int'l Conference and 42nd IEEE Holm Conference on Electrical Contacts*, 1996, pp 223–228.

32. G Witter, Z Chen. A comparison of silver tin indium oxide contact materials using a new model switch that simulates operation of automotive relays. I. *Proc.22nd Int'l Conference on Electrical Contacts*, 2004, pp 382–387.

33. Z Chen, G Witter. Comparison in performance for silver-tin-indium oxide materials made by internal oxidation and powder metallurgy. *Proc. 55th IEEE Holm Conference on Electrical Contacts*, 2009, pp 180–186.

34. Z Chen, G Witter. A study of contact endurance switching life as a function of contact bond quality, contact electrical load and residual stresses for silver tin indium oxide composite rivet contacts. *Proc. 58th IEEE Holm Conference on Electrical Contacts*, 2012, pp 172–177.

35. D Susan et al., Very long term aging of 51 In-48 Sn (at.%) solder joints on Cu-plated stainless steel substrates, *J Matter Sci* 2009 44:545–555.

36. A Krätzchmar, R Herbst, T Mützel, R Niederreuther, P Braumann. Basic investigations on the behavior of advanced Ag/SnO_2 materials for contactor applications. *Proc. 56th IEEE Holm Conference on Electrical Contacts*, 2010, pp 127–133.

37. F Hauner, D Jeannot, K McNeilly, Advanced $AgSnO_2$ contact materaisls with high total oxide content. *Proc. 21st Int'l Conference on Electrical Contacts*, 2002 Zurich, pp 452–456.

38. H Francisco, M Myers. Optimization of silver tin oxide chemistry to enhance electrical performance in AC application. *Proc. 44th IEEE Holm Conference on Electrical Contacts*, 1998, pp 193–201.

39. A Shibata. Silver–metal oxide contact materials by internal oxidation process. *Proc. 7th Int'l Conference on Electrical Contacts*, 1974, pp 214–220.

40. B Gengenbach, K Jager, U Mayer, K Saeger. Mechanism of arc erosion on silver–tin oxide contact materials. *Proc. 11th Int'l Conference on Electrical Contacts*, 1982, pp 208–210.

41. N Behrens, W Bohm. Switching performance of different Ag/SnO_2 contact materials made by powder metallurgy. *Proc. 11th Int'l Conference on Electrical Contacts*, 1982, pp 203–207.

42. B Gengenbach, U Mayer, R Michal, K Saeger. Investigations on the switching behavior of $AgSnO_2$ materials in a commercial contactor. *Proc. 12th Int'l Conference and 30th Holm Conference on Electrical Contacts*, 1984, pp 243–247.

43. H Kim, F Reid. Development of improved Ag–CdO contact material by a reactive sintering process. *Proc. 10th Int'l Conference on Electrical Contacts*, 1980, pp 775–785.

44. F Brugner. An erosion resistant, lithium additive modified Ag–CdO. *Proc. 9th Int'l Conference and 24th Holm Conference on Electrical Contacts*, 1978, pp 191–195.

45. International Electrochemical Commission, Publication 158–1.

46. M Lindmayer, W Bohm. Effect of alkali on the switching behavior of Ag/CdO. *Proc. 10th Int'l Conference on Electrical Contacts*, 1980, pp 849–862.

47. K Jager, U Mayer, K Saeger. On the influence of small alkali additions on the switching behavior and materials properties of silver–cadmium oxide materials. *Proc. 10th Int'l Conference on Electrical Contacts*, 1980, pp 765–774.

48. International Electrotechnical Commission. Low-voltage switchgear and controlgear. Pub. CEI 947–4-1, First Edition, Geneva, Suisse, 1990.

49. R Kossowsky, P Slade. Effect of arcing on the microstructure and morphology of Ag–CdO Contacts. *Proc. 6th Int'l Conference and Proc. 18th Holm Conference on Electrical Contacts*, 1972, pp 117–127.

50. C Brecher, J Gustafson, T Peters. Spectroscopic study of erosion in Ag–CdO materials. *Proc. 10th Int'l Conference on Electrical Contacts*, 1980, pp 377–386.

51. C Brecher, J Gustafson. The effect of lithium upon the erosion kinetics of Ag–CdO contacts. *Proc. 27th Holm Conference on Electrical Contacts*, 1981, pp 83–88.

52. G Witter, M Lu. The effect of lithium additions and contact density for silver–cadmium oxide contacts for make and break arcs. *Proc. 12th Int'l Conference and 30th Holm Conference on Electrical Contacts*, 1984, pp 229–236.

53. J Schoepf, V Behrens, T Honig, A Kraus. Development of silver zinc oxide for general-purpose relays. *Proc. 20th Int'l Conference on Electrical Contacts*, 2000, pp 187–192.

54. Z Chen, G Witter. Development of cadmium-free contact material for AC wall switch applications. *Proc. 23rd Int'l Conference on Electrical Contacts*, 2006, pp 549–553.

55. V Behrens, T Honig, O Lutz, D Späth. Substitute contact material for silver/cadmium oxide in AC applications in the low current range. *Proc. 23rd Int'l Conference on Electrical Contacts*, 2006, pp 294–299.

56. D Sallais et al., New silver-oxide composites to reduce break arc duration and its subsequent damages. *Proc. 23rd Int'l Conference on Electrical Contacts*, 2006, pp 278–283.

57. C Leung, A Lee. Contact erosion in automotive DC relay. *Proc. 15th Int'l Conference and 36th IEEE Holm Conference on Electrical Contacts*, 1990, pp 85–93.

58. C Leung, A Lee. Electric contact materials and their erosion in automotive DC relays. *Proc. 37th IEEE Holm Conference on Electrical Contacts*, 1991, pp 114–121.

59. C Leung, A Lee. Silver tin oxide contact erosion in automotive relays. *Proc. 39th IEEE-Holm Conference on Electrical Contacts*, 1993, pp 61–67.

60. L Germer. Physical process in contact erosion. *J Applied Phys* 29: 1067–1082, 1958.

61. G Witter, Z Chen. Dynamic welding resistance comparisons of silver and silver metal oxides. *Proc. 19th Int'l Conference on Electrical Contacts*, 1998, pp 355–359.

62. Z Chen, G Witter. Dynamic welding of silver contacts under different mechanical bounce conditions. *Proc. 45th IEEE Holm Conference on Electrical Contacts*, 1999, pp 1–8.

63. Z Chen, G Witter. A study of dynamic welding of electrical contacts with emphasis on the effects of oxide content for silver tin indium oxide contacts. *Proc. 56th IEEE Holm Conference on Electrical Contacts*, 2010, pp 121–126.

64. A Neuhaus, C Schrank, M Reichart, and W Rieder. Influence of the break arc conditioning effect on make-arc welding and material transfer. *Proc. 25th Int'l Conference on Electrical Contacts*, 2006, pp 137.

65. A Erk, H Finke. Uber die mechanischen Vorgange wahrend des Preliens einschaltender Kontakte. ETZ-A 86 (1964b): 231–241, 1964.

66. M Ohta, S Matsuda, K Ogawa, A Shibata. New contact materials of silver–metal oxide system. *Proc. 16th Int'l Conference on Electrical Contacts*, 1992, pp 421–426.

67. K Schroder. Development of contact materials for power engineering in Europe. *Proc. 33rd IEEE-Holm Conference on Electrical Contacts*, 1987, pp 163–173.

68. K Schroder. Das Abbrandverhalten offnender Kontaktstucke bei Beanspruchung durch Magnetisch. Dissertation, TH Braunschweig, 1969.

69. K Schroder. Silver-metaloxides as contact materials. *Proc. 13th International Conference on Electrical Contacts*, 1986, pp 186–194.

70. H Manhart, W Rieder, C Veit. Arc mobility on new and eroded Ag/CdO and Ag/SnO$_2$ contacts. *Proc. 34th IEEE-Holm Conference on Electrical Contacts*, 1988, pp 47–56.

71. F Brugner. The influence of Na and K impurities in Ag–CdO on the tendency for arc reignition during 600 VAC interruption by motor control gear. *Proc. 23rd Holm Conference on Electrical Contacts*, 1977, pp 77–79.

72. G Chen. The effect of Na, K, and Li on the arc interrupting capability of AgCdO contacts. *Proc. 25th Holm Conference on Electrical Contacts*, 1979, pp 199–204.

73. M Lindmayer. Effect of alkali additives on the reignition voltage of Ag/CdO. *Proc. 25th Holm Conference on Electrical Contacts*, 1979, pp 205–208.

74. P Braumann, K Schroder. Recovery of contact gaps with silver–metal-oxide contacts. *Proc. 13th Int'l Conference on Electrical Contacts*, 1986, pp 195–199.

75. B Gengenbach, U Mayer, G Horn. Erosion characteristics of silver based contact materials in a DC contactor. *Proc. 12th Int'l Conference on Electrical Contacts*, 1984, pp 201–207.

76. R German. *Powder Metallurgical Science*, 2nd ed. Princeton, NJ: Metal Powder Industries Foundation, 1994.

77. R Nelson, D Milner. Densification processes in the tungsten carbide–cobalt system. *Powder Metallurgy* 15 (30): 3, 1972.

78. G Gessinger, H Fischmeister, H Lukas. The influence of a partial wetting second phase on the sintering of solid particles. Powder Metallurgy 16 (31): 19, 1973.

79. M Hansen. *Constitution of Binary Alloys*, New York: McGraw-Hill, 1958.

80. Y Naidick, I Lavrinko, V. Eremenka. The role of capillary phenomena in the densification process during sintering in the presence of the liquid phase. *Powder Metallurgy* 16 (31): 19, 1973.

81. Fansteel Inc., *Make and Break*. Contact brochure, Fansteel Inc., North Chicago, IL, 1971.

82. American Society for Metals. *Metallography, Structures and Phase Diagrams*, Vol. 8. Metals Park, OH: ASM, 1973.

83. G Witter. A correlation of material toughness, thermal shock resistance, and microstructure of high tungsten, tungsten–silver composite materials. Master of Science Dissertation, Illinois Institute of Technology, Chicago, 1975.

84. E Underwood. *Quantitative Stereology*. Reading, MA: Addison-Wesley, 1970.

85. J Gurland. The fracture strength of sintered tungsten carbide–cobalt alloy in relationship to composition and particle spacing. *Trans. Metall. Soc.* AIME 227: 1146, 1963.

86. K Stjernberg. Some relationships between the structure and mechanical properties of WC–Co alloys. *Powder Metallurgy* 13 (25): 1, 1970.

87. G Witter, W Warke. A correlation of material toughness, thermal shock resistance, and microstructure of high structure composite materials, *Holm Conference on Electrical Contacts*, IIT Chicago, IL, 1974, pp 238–249.

88. American Society For Testing Materials. *Plane Strain Crack Toughness Testing Of High Strength Metallic Materials*, STP No. 410, Philadelphia, PA: ASTM, 1966.

89. A. Evans. *Fracture Mechanics of Ceramics*, Vol. 1, Edited by Bradt R, et al., Plenum Publishing Co., New York, 1974.

90. P Slade. Arc erosion of tungsten based contact materials: a review. *Refractory and Hard Metals* 4(4): 208–214, 1986.

91. E Walczuk. Investigations of technical properties of Ag/W contact materials for the moulded case circuit breaker. *Proc. 11th Int'l Conference on Electrical Contacts*, 1982, pp 180–188.

92. C Leung, H Kim. A comparison of Ag/W, Ag/WC, and Ag/Mo electrical contacts. *IEEE Trans. CHMT* 7 (1): 69–75, 1984.

93. M Lindmayer, M Roth. Contact resistance and arc erosion of W/Ag and WC/Ag. *Proc. 9th Int'l Conference on Electrical Contacts*, 1978, pp 421–30.

94. H Turner, C Turner. Discontinuous contact erosion. *Proc. 3rd Int'l Conference on Electrical Contacts*, 1966, pp 309–320.

95. G Witter. The effect of nickel additions on the performance of tungsten silver materials. *Proc. 11th Int'l Conference on Electrical Contacts*, 1982, pp 351–355.

96. S Kabayama, M Koyama, M Kume. Silver–tungsten alloys with improved contact resistance. *Powder Metallurgy International* 5 (3): 122–126, 1973.

97. E Walczuk. Experimental study of Ag–W–Re composite materials under high current conditions. *Proc. 32nd IEEE Holm Conference on Electrical Contacts*, 1986, pp 77–83.

98. C Leung, D Harman, L Doublet, C Bourda, Y Cui, L Hu. Electrical and mechanical lives of Ag/C and Ag/WC/C contacts. *Proc. 26th Int'l Conference on Electrical Contacts*, 2012, pp 233–239.

99. S Allen, E Streicher, C Leung. Electrical performance of Ag-W-C and Ag-WC-C contacts in switching tests. *Proc. 20th Int'l Conference on Electrical Contacts*, 2000, pp 109–114.

100. T Browne, ed., *Circuit Interruption, Theory and Design*, New York: Marcel Dekker, 1984.

101. P Slade. Advances in material development for high power vacuum interrupter contacts. *IEEE Trans. Components Packaging and Manuf Technol—Part A* 17 (1): 96–106, 1994.

102. W Haufe, W Reichel, H Schreiner. Abbrand Verschiedener Wcu-Sinter-Trankwerkastoffe an Luft bei Hohen Stromen. *A. Mettalde* 63 (10): 651–654, 1972.

103. A Abdel-Asis. Neue Entwicklungen und Forschungen auf dem Gebiet der Schaltgeratetechnik. E.T.Z. *Archiv* 3 (9): 301–308, 1981.

104. I Zessak. Burn up behaviour of sintered contact materials tested repetitively at currents ranging from 400 A to 7000 A in oil. *Proc. 9th Int'l Conference on Electrical Contacts*, 1978.

105. F Labrenz. Erosion of switching electrodes by rotating arcs in SF6. *Proc. 11th Int'l Conference on Electrical Contacts*, 1982, pp 326–330.

106. J Kaminski. Burn-up behavior of nozzle electrodes with strong SF6 flow. *Proc. 9th Int'l Conference on Electrical Contacts*, 1978, p 25.

107. K Schroder. Das Abbrandverhalten offnender Kontaktstücke bei Beanspruchung durch Magnetisch. Abgelenkte Starkstromlichtbogen. Thesis, TU Braunschweig, 1967.

108. A Keil, C Meyer. Uber die Bildung von isolierenden Deckschichten auf Kontakten aus Verbundmetallen. E.T.Z. 73 (2): 31–34, 1952.

109. P Slade, R Kossowsky. Effect of surface and contact resistance of a function of operating life in silver–tungsten contacts. *Proc. 7th Int'l Conference on Electrical Contacts*, 1974, pp 200–206.

110. G Witter, V Abele. The change in surface resistance of W–Ag contacts as a function of composition, microstructure, and environment. *Proc. 8th Int'l Conference on Electrical Contacts*, 1976, pp 445–451.

111. J Kosco. The effects of electrical conductivity and oxidation resistance on temperature rise of circuit breaker contact materials. *Proc. 14th Holm Conference on Electrical Contacts*, 1968, pp 55–66.

112. G Samsonov. *The Oxide Handbook*. New York: IFI/Plenum, 1973.

113. P Slade. Effect of the electric arc and the ambient air on the contact resistance of Ag, W, and W–Ag contacts. *J Appl Phys* 47 (8): 3438–3443, 1976.

114. H Kuhn, W Rieder. DerEinfluss natürlicher Hautschichten auf Konta. Stromen. Z. Metallde. 63 (10): 651–654, 1972.

115. P Slade. Variations in contact resistance resulting from oxide formation and decomposition in Ag–W and Ag–WC–C contacts passing steady currents for long times periods. *IEEE Trans. Components, Hybrids and Manuf Technol CHMT-9* (1): 316, 1986.

116. C Leung, R Bevington, P Wingert, H Kim. Effects of processing methods on the contact performance parameters for silver–tungsten composite materials. *IEEE Trans CHMT-5* (1): 23–31, 1982.

117. Hayashi et al. Surface-enhanced Raman scattering from layers of gas-evaporated silver small particles. *J Applied Phys* Part 1 28 (8): 1444–1449, 1989.

118. P Slade, C Lin, A Pebler. Titanium–tungsten carbide–silver, a new electric contact material. *IEEE Trans. CHMT 4* (1): 76–84, 1981.

119. P Slade, C Lin, A Pebler, N Hoyer. The switching and short circuit performance of (Ti, W)C–Co–Ag and (Ti, W)C–Ag contacts. *Proc. 11th Int'l Conference on Electrical Contacts*, 1982, pp 189–193.

120. P Slade, Y Chien, J Bindas. Contact resistance variations in high silver content, silver-refractory carbide contacts. *Proc. 13th Int'l Conference on Electrical Contacts*, 1986, pp 216–220.

121. P Slade. High current contacts: A review and tutorial, *21st ICEC*, Zurich, 2001, pp 413–424.

122. P Slade. *The Vacuum Interrupter: Theory, Design, and Application*, Boca Raton, FL: CRC Press, 2007.

123. W Li, P Slade, L Loud, R Haskins. Effect of Cu content on the electrical erosion of tungsten copper contacts switching load-current in vacuum. *Proc. 48th Holm Conference on Electrical Contacts*, 2002, pp 95–102.

124. E Taylor. Cathode spot behavior on tungsten-copper contacts in vacuum and the effect on erosion. *Proc. 51st Holm Conference on Electrical Contacts*, 2005, pp 135–138I.

125. F Hauner, G Tiefel, R Müller. CuCr for vacuum interrupters, properties and application. *Proc. 24th Int'l Conference on Electrical Contacts*, 2008, 61–68.

126. R Holm, E Holm. *Electric Contacts Theory and Application*. New York: Springer Verlag, 2000.

127. G Witter. Private Research.

128. H Wagar. *Bell Telephone Laboratories. Physical Design Of Electronic Systems*, Englewood Cliffs, NJ: Prentice-Hall, 1971, pp 439–594.

129. H Sone, H Ishida, T Takagi. Electrode temperature dependency on arc and erosion in electric contacts. *Proc. 15th Int'l Conference on Electrical Contacts*, 1990, pp 33–36.

130. V Behrens, R Michal, J Minkenberg, K Saeger, B Liang, W Zhang. Erosion mechanisms of different Ag/Ni 90/10 materials. *Proc. 14th Int'l Conference on Electrical Contacts*, 1988, pp 417–422.

131. W Balme, P Braumann, K Schroder. Welding tendency, contact resistance and erosion of contact materials in motor vehicle relays. *Proc. 15th Int'l Conference on Electrical Contacts*, 1990, pp 73–78.

132. K Sawa, M Hasegawa, K Miyachi. Contact characteristics of Ag and Pd contacts under various discharge conditions. *Proc. 33rd IEEE Holm Conference on Electrical Contacts*, 1987, pp 203–211.

133. T Tamai. Mechanisms of the low contact resistance properties for Ag-Pd-Mg contacts. *Proc. 46th IEEE Holm Conference on Electrical Contacts*, 2000, 94–101.

134. Z Chen, K Sawa. Polarity effect of unsymmetrical material combination on the arc erosion and contact resistance behaviour. *Proc. 40th IEEE Holm Conference on Electrical Contacts*, 1994, pp 79–88.

135. H Sone, I Hiroyuki, T Takagi. Electrode temperature dependency on arc and erosion in electric contacts. *Proc. 15th IEEE Holm Conference on Electrical Contacts*, 1990, pp 33–36.

136. C Hampel, *Rare Metals Handbook 2nd Edition*, New York: Reinhold, 1967.

137. Z Chen. Private Research, Chugai USA, Inc., Waukegan, IL, 1997.

138. P Wingert, S Allen, R Bevington. The effects of graphite particle size and processing on the performance of silver–graphite contacts. *Proc. 37th IEEE Holm Conference on Electrical Contacts*, 1991, pp 38–43.

139. K Muller, D Stockel. A new method for manufacturing of complex graphite-containing contact materials, *Proc. 11th ICEC*, 1982, pp 346–350.

140. M Lindmayer, K Schroder. The effect of unsymmetrical material combinations on the contact and switching behavior. *Proc. 9th IEEE Holm Conference on Electrical Contacts*, 1978, pp 265–271.

141. H Turner, C Turner. Polarity effects of silver/silver graphite contact combinations. *Proc. 6th IEEE Holm Conference on Electrical Contacts*, 1972, pp 351–356.

142. E Vinaricky, V Behrens. Switching behavior of silver/graphite contact material in different atmospheres in regard to contact erosion. *Proc. 44th IEEE Holm Conference on Electrical Contacts*, 1998, pp 292–300.

17

Contact Design and Attachment

Gerald J. Witter and Guenther Horn

I wish I did not have to write the instruction manual on the uses of a new metal.

The Instruction Manual, **John Ashbery**

CONTENTS

17.1 Introduction..954
   17.1.1 Arc-Induced Contact Stresses and Interface Bond Quality............954
17.2 Staked Contact Assembly Designs...955
   17.2.1 Contact Rivets...955
      17.2.1.1 Solid Rivets 955
      17.2.1.2 Machine-Made Composite Rivets.................................956
      17.2.1.3 Brazed Composite Rivets 959
      17.2.1.4 Rivet Staking...960
17.3 Welded Contact Assembly Designs 960
   17.3.1 Resistance Welding...962
      17.3.1.1 Button Welding...962
      17.3.1.2 Wire-Welding..963
      17.3.1.3 Contact Tape Welding 963
   17.3.2 Special Welding Methods..965
      17.3.2.1 Percussion Welding 965
      17.3.2.2 Ultrasonic Welding of Contacts 965
      17.3.2.3 Friction Welding of Contacts..966
17.4 Brazed Contact Assembly Designs 966
   17.4.1 Methods for Brazing Individual Parts..966
      17.4.1.1 Torch Brazing..967
      17.4.1.2 Induction Brazing 967
      17.4.1.3 Direct and Indirect Resistance Brazing.......................967
      17.4.1.4 Furnace Brazing 968
      17.4.1.5 Continuous Laminated Strip Brazing, "Toplay" 968
      17.4.1.6 Brazed Assembly Quality Control Methods................968
17.5 Clad Metals, Inlay, and Edge Lay 969
17.6 Contact Alloys for Non-Arcing Separable Contacts 970
   17.6.1 Gold and Gold Alloys...970
   17.6.2 Manufacturing Technology...970
   17.6.3 Physical and Chemical Properties..970

17.6.4 Metallurgical Properties 971
17.6.5 Contact Applications and Performance 972
Acknowledgments 972
References 972

17.1 Introduction

This chapter provides information for arcing contact design and information on methods used for attachment of contacts. It is the intent of this chapter to provide technical information useful for the contact user and not to discuss detailed methods used by various companies to make these contact designs. The chapter also contains a section on materials and contact designs used for non-arcing separable, dry circuit, and contact applications.

17.1.1 Arc-Induced Contact Stresses and Interface Bond Quality

In the following sections of this chapter, several different contact design options are discussed. Today most contact designs are composite metal structures having precious metal faces attached to a copper type base as opposed to a monolithic structure of precious metal since the cost of precious metals is very expensive. All discrete contacts have an interface between the contact and the device carrier blade or terminal. All composite contact structures have an additional interface between the precious metal contact face and the base metal part of the contact. Understanding the importance of the contact interface bond quality is a property that is sometimes ignored even though in some cases it can have a major effect on the endurance life of a switching device.

Compared to the number of publications made on contact materials very little has been written on contact interface bond quality. Spaeth et al. [1] and Janitzki et al. [2] have written similar papers showing that for high power devices with brazed contacts, the erosion rate of the contacts increases as the void level in the interface increases, especially when the void area exceeds 50% of the joint interface. They also show that voids near the edge of the contacts are more detrimental than voids in the middle of the interface. Shen et al. [3] did a similar study on high current devices and describes contact bond failure in terms of interface peeling from compressive stresses created as the contacts cool after surface melting from the switching arc.

Chen and Witter [4,5] have performed a study on the effect of bond quality for medium current DC devices using bimetal rivet contacts. They show a very large difference in contact switching endurance life, greater than 5/1, between contacts with good interface bonding, (i.e., +90%), versus contacts with poor interface bonds (i.e., 40%). They also state that the level of stress created from surface melting and re-solidification between the surface of a contact and an under layer is related to the temperature of the under-layer at the time the arc melted surface re-solidifies. This can be related as follows:

$$\text{Stress} = K \times L \times (\Delta T) \tag{17.1}$$

Where K is a factor that is dependent on geometry, extent of melting, and other things, L is the linear coefficient of expansion of silver and (ΔT) the temperature difference between the under-layer and the surface at the time of solidification. This stress causes the rivet head of a monolithic structure to curl as shown in Figure 17.1 during endurance switching. It also causes contacts with poor interface bond strength to delaminate and fail from base metal being exposed to the contact face. This type of curling can occur for all silver alloys subjected to arcing [6].

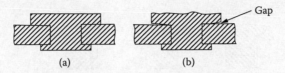

FIGURE 17.1
Typical distortion of rivet head after electrical testing: (a) before testing and (b) after endurance testing.

The stresses created in a contact from electrical endurance testing are more complex than just the stress described from the melting and re-solidification of the surface and include other stresses, such as thermal shock from rapid heating from the arc, annealing of sub-layers of material, and mechanical stresses and fatigue from the pounding of the contacts together. Regardless of the contact silver alloy or the specific discrete contact design some degree of curling is normally present from the melting and solidification cycling from endurance testing and this will vary with contact electrical load level. As the arc energy increases as a function of contact current or arcing time, the surface melting will become more severe and the curling effect will increase. Keeping these factors in mind will help in understanding the advantages and disadvantages of the various contact designs. For example by designing a slight radius for the surface of the contacts, you will not only initially bring the arcing spot and heat dissipation area closer to the center of the contact but also provide some delay in the curling effect reaching the edges of the contact.

17.2 Staked Contact Assembly Designs

17.2.1 Contact Rivets

Rivets are still one the most popular forms of contact construction for small contacts 2–10 mm in diameter. Figure 17.2 shows typical designs of contact rivets. Rivet constructions can be categorized in several different ways: (1) monolithic versus composite metal construction, (2) bimetal versus trimetal composites, and (3) machine made composites and (4) brazed composites. The bimetal construction uses the precious metal for the contact face and base metal, mainly copper, as the rest of the construction. The trimetal rivets have a precious metal end of the rivet shank plus the precious metal face. The composite constructions for rivets can be made using special heading machines or by forming the separate components and brazing the different materials together which results in significant labor and cost. The process that will be used for an application should be on the basis of both the quality of the contact required and the economics for manufacturing. To make any type of solid or machine made composite, the contact material must be headable, that is, able to be plastically deformed without cracking. Materials that are not headable such as refractory metals, refractory composite materials, many silver graphite compositions and more brittle materials are normally made into rivets by brazing the materials to copper based alloys [7].

17.2.1.1 Solid Rivets

This is the simplest rivet design, since no bonding is required for the rivet fabrication. Material deformation without cracking is the only concern. If all contact rivets were made of very low cost materials this design would be used in all cases where the material has

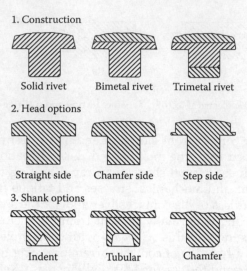

FIGURE 17.2
Different contact rivet designs.

sufficient ductility for heading. The main criterion to be concerned with for the design of solid rivets is the ratio of the head diameter to the shank diameter: this ratio depends on the choice of material. In general a ratio of 2/1 is good for many materials. A pure material like fine silver which has a low work-hardening rate can be made using higher ratios. Composite materials, such as some newer silver metal oxides with over 10% by weight metal oxide content, may be better suited for a lower ratio like 1.6/1.

From an economical standpoint it is tempting to design the shank diameter of a rivet to be as small as possible for staking a rivet into backing material. From a technical standpoint, this may or may not be wise, depending on the load and life a rivet will see in service. When a rivet is staked onto a backing or carrier, the most important contact area between the carrier and the rivet is the interface between the cylindrical shank circumference of the rivet and the inside diameter of the hole used for the rivet staking. Figure 17.1 shows an illustration of the typical distortion that takes place on a rivet head after the contact has been subjected to low-current switching duty of about 10 A for several thousand operations. Section 17.1.1 above describes the mechanism creating this curling phenomenon. It can be seen that because of this curling the bottom side of the rivet head and carrier interface contributes little or no heat conduction from the contact. It should be kept in mind that the main heat and current conduction path from the rivet to the carrier is through the interface between the rivet shank OD and the rivet hole ID. The use of a larger shank diameter, even though not economical, can minimize curling and can also provide more interface bond area for conducting heat away from the contact.

17.2.1.2 Machine-Made Composite Rivets

The most popular type of rivet used today is the headed bimetal composite. The heading process for making the rivet not only forms the wires into a rivet shape but also bonds the contact material to copper or in some cases a copper alloy. Many different machines and processes for making these rivets have been developed over the years, some involving different types of electrical welding and others using cold bonding and forming.

For example, ordinary resistance welding should not be used for direct bonding silver metal oxide materials to copper since oxide slags generated during the welding prevent good attachment. For this reason, silver metal oxide composite rivets are normally made by cold bonding and forming.

The ability to bond copper and other alloys in intimate contact by solid-state cold deformation is well known [8,9]. The cold bonding process for the attachment of a silver metal oxide to copper to form a contact rivet involves bonding mechanisms comparable to the better-known process of roll cladding. The cold bonding process consists of taking two cylinders, one cut from a copper wire and the other from a silver metal oxide wire, putting two of the flat faces of each metal cylinder in contact with each other and co-expanding each cylinder at the interface to create a bond. As the interface expands you have simultaneous plastic flow of the metals and fracturing of any surface films that may exist on the two components. The intimate metal to metal contact that takes place as a result of this co-expansion forms a bond between the two metals. The bonding process begins in the middle of the rivet and expands out to the circumference as the co-expansion progresses. Even for rivets with good quality interface bonds there is always a small amount of void area near the circumference. The process should also involve some post-forming heat treatment to remove stresses while at the same time producing some solid-state diffusion which further improves the bond quality.

The main concern for the contact user should be the quality of the bond that is created and not the exact process or machine that was used to make this bond. As a result of rivet geometry, it is difficult to measure the strength of this bond. Sometimes the rivet head is thick enough to perform shear testing but small variations in shear interface location cause very large errors in shear results. Many rivets, however, have an alloy layer that is too thin even to allow shearing and also work done by several researchers in an ASTM committee on contacts have found too much spread in shear strength results to be considered as a good test [10]. Two important quality factors for bonds are bond ductility and bond area. Brittle bonds result in the same kind of defect as voids in the bond interface since the bond interface separates under low stress. To evaluate bonds a simple test has been developed, called the squeeze test. An illustration of this squeeze test is shown in Figure 17.3a. In this case, the rivet head is compressed in a direction that is perpendicular to a void that exists in the copper precious metal interface bond. The head of the rivet is compressed from both sides of the diameter down to for example 50% of the original head diameter. Since there is no published standard for this ratio, it is somewhat arbitrary. The 50% ratio has been found by Witter [11] to be more effective than lesser compression for identification of voids and weak bonds in composite rivets. If the compression direction is as shown in this example the void will result in the rivet interface opening up in the rivet head area containing the void. For testing rivets unfortunately you do not know beforehand where a void will exist so you must squeeze test several rivets, (perhaps 5 to 10 from a batch) to insure you compress a void in a direction that will show a separation if significant voids do exist.

In Figure 17.3b, an illustration is shown that provide you a guide for the squeeze test results for different bond interface distributions. Good bonds, >85%, as depicted in (a) of Figure 17.3b will show little or no separation as compared to bonds of 70% or less. The bond level that is acceptable will depend on the application and an agreement between the vendor and the customer. With regard to the life to be expected with composite rivets that have good, >85%, bonds or poor bonds, ~40%, Chen and Witter [4,5] show a large difference in endurance life, see Figure 17.4. At 60 ampere load the good bond composite is comparable to a solid, monolithic, rivet for endurance life. The poor bond composite has less than 20% of the endurance life and fails by the silver metal oxide layer delaminating and

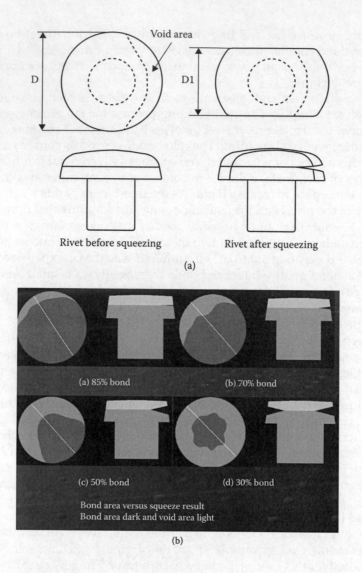

FIGURE 17.3

(a) Typical squeeze test for a composite rivet with weak bond and (b) Different bond distributions with squeeze shape for compression perpendicular to line in bond distribution.

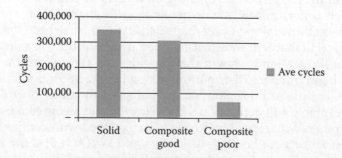

FIGURE 17.4

A comparison of endurance switching life for a solid rivet, good bond bimetal rivet, and a poor bond bimetal rivet.

falling off. Another test they have conducted for poor interface bonds shows thinner silver tin oxide layers last significantly longer before delamination than thicker silver layers. This unexpected result is supported by Equation 17.1 since the thinner layer would have a bond interface closer to the contact surface and have less of a ΔT difference in temperature. From a process control view point it is difficult to hold intermediate bond quality, therefore bonds of 70% or less would be considered marginal.

Other methods also exist for evaluation of bonds including metallographic examination and ultrasonic testing. Cross sectioning of the bond may not give complete information on the bond quality since the cross section represents only one line across the head diameter and it is possible for the bond interface to be very weak and still look good under the microscope. A better method is to squeeze or deform the rivet head to a small extent before making the cross section so that a weak bond will separate. Ultrasonic methods are expensive for checking small parts and also cannot distinguish between weak and strong bonds in intimate contact. Later in this chapter they will be shown to be very effective in evaluating brazed bonds.

Besides the cold bonding and resistance brazing discussed above, special welding processes exist that use an arc for fusing silver alloy materials and copper backings together. This type of welding can be used for attachment of composite materials like silver tin oxide to copper that normal resistance welding cannot accomplish.

17.2.1.3 Brazed Composite Rivets

Before the late 1960s and the development of machines for making composite contact rivets, almost all composite rivets were made by brazing. The brazing process for making composite rivets is much more expensive than the process for making machine composites since three components must be manufactured and then assembled together. The first step in the process consists of making a base metal rivet, contact material disk, and brazing alloy wafer. The components are then assembled together in a brazing fixture which is put through a furnace under a protective atmosphere to fuse the contact material disk to the base metal rivet. Beside cost, brazed composites have other disadvantages compared to machine composites. Firstly, the tolerances of the head diameter and thickness are wider since these dimensions are the product of two components and also the clearance in the brazing fixture must be taken into account. Another problem is the potential for getting brazing alloy running up the sides of the rivet head onto the contact face if the furnace temperature varies. Bond quality also varies with furnace conditions and cleanliness of the components.

The brazing process has the major advantage that brittle materials can be bonded by this method since no deformation is required for making the bond. Therefore, this process is used for making bimetal rivets of tungsten, silver-tungsten, silver-tungsten carbide, silver-graphite and brittle high oxide content silver oxide materials.

Another area where brazed composite rivet designs are used is for contact alloys that work harden rapidly on deformation. For example, silver copper alloys with 10% or more copper are more difficult to join by cold bonding, since co-expansion is difficult to achieve as a result of the large difference in work hardening rate between these alloys and copper. As a result of this, a brazed design may be selected over a machine composite design to obtain better bond quality, especially if the ratio of the head to shank diameter is large.

The bond area of a brazed rivet can be measured easily by cross sectioning the rivet and metallographic examination. The interface is normally 0.02 mm or more in thickness and therefore the voids are visible at low magnification. Since cross sectioning will show only one line through the diameter of a rivet head it is important however to section multiple samples of a given lot to get a representative result for bond quality. Ultrasonic testing and shearing are also effective for evaluating this type of bond.

17.2.1.4 Rivet Staking

A common method of rivet attachment is by a straight impact pressing action. The head of the rivet is placed in a die cavity which is shaped to conform to the head dimension and configuration. With the shank of the rivet sticking through a hole in the backing material to which it is being attached a punch presses on the end of the rivet shank to expand the rivet shank diameter to fill the rivet hole. The important factor to be considered for judging the quality of the rivet staking is the amount of contact there is between the outside diameter of the rivet shank and the inside diameter of the hole after staking. The contact area between the underside of the rivet head and the backing material is of little value since, as mentioned earlier, the head of the rivet will tend to curl upward as a result of switching cycling and produce a void in that area.

Orbital riveting is another popular method used for staking rivets. For orbital staking the punch that is supplying the staking force is not parallel to the end of the rivet shank but at a slight angle and rotating. This produces a swaging action that exerts both a radial and downward force on the rivet shank. By use of the orbital process the shank is expanded gradually and a high degree of contact can be obtained between the rivet shank and hole. The orbital process is also used for staking contacts with more brittle contact alloys.

Because staking of rivets is a simpler manufacturing process than welding, it is preferred by many manufacturers of relays and switches. Once the proper tooling and method of staking are developed the process runs very consistently, because there is little wear and the process parameters are easily controlled.

The bond quality from staking varies greatly depending on the tooling and technique utilized. Rivet to backing bonds of 10%–30% are common. Care must be taken, however, even though the rivet shank may appear by visual examination to be bonded well to the backing. Metallographic examination at higher power or mechanical twist tests are recommended to insure a good joint.

17.3 Welded Contact Assembly Designs

For the purpose of this chapter, welding will refer to processes where a contact is fused to a backing material by application of current through the contact and backing which either melts by resistance heating a portion of the contact and backing interface surface or, by some special processes described later in this section, forms an arc that melts the interface surfaces [12]. Figure 17.5 shows the basic components of a simple resistance welder. The current is applied to the contacts by use of two welding electrodes, one applied to the contact, the other to the base metal backing. An electronic weld control unit regulates the amplitude and duration of the current, the timing of the current, and the timing of the electrode force applied to the contacts. The devices for producing the force pushing the contact and backing together vary with different welding equipment but pneumatic air cylinders are widely used for this purpose. Other items used for this are springs,

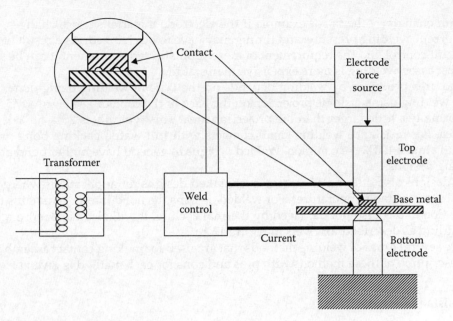

FIGURE 17.5
Illustration of a simple resistance welder.

torsion bars, cam drives, electromagnetic clamping devices and combinations of these various items. The welding equipment can vary from very simple on–off controls and transformer taps to equipment with very refined current waveform, frequency and timing, electrode force control, and voltage variation compensation.

Another area for clarification is the distinction between how the terms welding and brazing are differentiated in this chapter. Brazing processes are discussed in this chapter immediately following the sections on welding. The terms resistance welding, resistance brazing, spot welding, and spot brazing are often used interchangeably in the industry. A differentiation between welding and brazing is that brazing processes always employ the use of a filler metal which has a lower melting point than the components being joined. For the processes described in this section on welding, a filler metal may or may not be used depending on the contact material being attached. A resistance welding process as described in the following text can be technically described as a resistance brazing process when a filler metal is used. For the purpose of this chapter, the distinction between welding and brazing is based on process type rather than whether or not there is the use of a filler metal. For the processes where the term welding is used, the heating time is a fraction of a second and the attachment is for small to medium-size parts. The term brazing is used for processes for the attachment of medium to large parts that require longer heating times of several seconds to minutes for the liquefied brazing alloy to flow through the interface joint.

The welding of contacts onto backing material is more complicated than the riveting process. Resistance welding, which is the most popular welding method, involves controlling many parameters to obtain consistent weld bonds. The parameters for resistance welding include electrode materials, electrode shape, electrode surface condition, contact cleanliness, backing material cleanliness, electrode pressure, weld current control, and weld current timing. Many of these parameters are interrelated and changing one may require changing others. For example if the electrode material surface changes, the welding current, welding pressure, and timing may have to be changed. As a result of the more difficult control and the requirement of more sophisticated equipment it can be said that, in most cases, welding is more expensive than staking.

The bond area produced by welding depends on the type of welding being done. For resistance welding using a single projection on the back of the contact, see Figure 17.5, the bond is normally a little larger than the projection area. This typically is 25%–50% of the contact area. For resistance welding small contacts with full waffle backing bond areas from 70% to almost 100% are common. Welded designs in general have higher bond areas than staking designs.

Welded designs also offer an advantage over staked designs for applications where the device will see elevated temperatures. For welded designs the bond interface remains free from oxidation and is also not damaged by thermal cycling. It will also have more resistance to curling as described for rivets in Figure 17.2.

There are several types of welding methods that are used for making contact assemblies. A short description of these methods with pros and cons for each method is given below.

17.3.1 Resistance Welding

17.3.1.1 Button Welding

This type of welding refers to the attachment of individual contact disks, mostly called weld buttons, to a base metal carrier. Figure 17.6 shows some typical button designs. The buttons can be welded to a formed contact terminal or spring or the attachment can be made to the strip and the forming of the contact terminal or spring is done after attachment. The welding process for attachment of the contacts to the strip can be fully automated, for example the buttons can be fed from a feeder bowl onto a carrier strip which has pilot holes for locating the contacts. The strip is then fed from the welder onto a reel for later stamping and forming into a subassembly or into a punch press for immediate stamping and forming. This process has the advantage over clad strip in that the contact alloy is located only where it is needed.

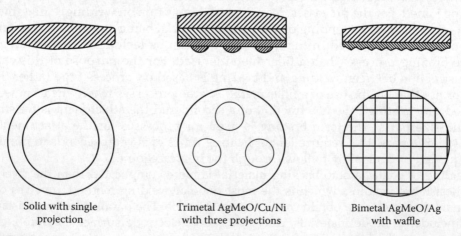

Solid with single
projection

Trimetal AgMeO/Cu/Ni
with three projections

Bimetal AgMeO/Ag
with waffle

FIGURE 17.6
Typical designs for button types of contacts.

The buttons can have one or several projections which will limit the current and bonding to specific areas. This provides easier attachment if bond areas of less than 50% can be tolerated for the specific application. This method has the disadvantage that the bonded area is in the center of the part and there is no bonding at the circumference. In applications where the electrical erosion is significant this type of bond normally does not work since the edges of the contact pull up. This may result in welding of the contacts or in excessive erosion from thermal stress.

Buttons made with a waffle pattern rather than a projection offer a more uniform distribution of bonding between the contact and the backing. For this type of welding a higher current is required, because there is less of a current restriction at the interface compared to projection welding. This results in faster electrode wear. Welds made by this method can be used in applications where a long switching life is required and even though the erosion may be significant the contact will not peel from the backing.

Buttons made for this type of welding can be of either composite or solid construction. For silver metal oxides and silver refractory contacts, a weld layer made of fine silver or a brazing alloy is required to obtain a strong attachment. For fine silver, silver-nickel, and most of the silver-copper alloys, no weld layer is required.

This method is applicable to contacts, 2.5–7 mm in diameter. Contacts smaller than 2.5 mm become difficult to feed as individual contacts and contacts larger than 7 mm are more easily bonded by brazing or special welding processes which are discussed in a later section of this chapter.

17.3.1.2 Wire-Welding

Wire welding, as a method of making contact assemblies, has been used extensively in Europe for many years. One of the reasons is the heavy use of silver–nickel in many low-current AC applications. Silver-nickel can be welded directly onto copper alloys very easily. Automated machines for cutting silver–nickel wire, welding the wire to a strip, and coining the welded wire into a contact shape were developed in Europe and are used to make large volumes of contact assemblies for wiring devices, relays and industrial or commercial controls. These machines are not limited to welding silver nickel but any silver alloy that can be directly welded to copper or copper alloys. The advantage of this type of process is cost. The disadvantage of this design is again that there is no bond at the circumference of the contact interface with the backing. Another possible negative consideration for this process is that once you have invested in this type of welding equipment, you are limited to using only contact materials that can be directly bonded, which eliminates the use of popular contact alloys like silver metal oxides. This is one reason why this process is not as popular in the North America and Japan, where half the voltage and twice the current are used for many high volume applications. Here for the same application silver metal oxide contacts have to be used.

17.3.1.3 Contact Tape Welding

Another major option used for arcing contact design is contact tape, also called contact profiles. Contact tape construction is a possible option for any material that is ductile enough to be rolled or swaged into strip or wire. Contact tape consists of a laminated metal strip material with two or more metal layers laminated together. Figure 17.7 shows some different laminated structures. For contact welding tape, the top layer is always the contact

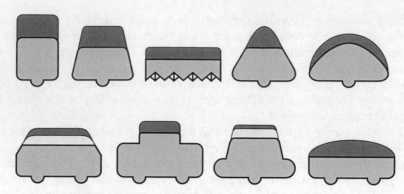

FIGURE 17.7
Typical designs for weld tape contacts.

material itself. The bottom layer is mostly a weldable alloy such as CuNi, Ni, Ag, CuAgP ("Silfos"), or other alloy that is directly weldable to a copper or copper alloy carrier. Besides having an alloy layer and weldable layer the tape can have a middle layer of copper. The purpose of this layer is to have a low cost, high conductivity filler that can be used to adjust the total thickness of the tape.

Tape welding has the advantage over wire welding in that the weld layer can be made of a higher resistance material than the contact alloy and thus provide an optimum contact design for attachment. The tape can be made by several different methods. A common method is to make a wide sheet of laminated contact material by roll cladding. In this method a contact alloy and a base metal material in sheet form are fed into a rolling mill together and bonded during deformation. The pressure from the rolls forces the interface of the two materials to expand and create clean metal-to-metal contact bonding similar to that described earlier in this section for cold-bonded composite rivets. After the cladding the bimetal or trimetal sheets are slit into narrower tapes of the desired width and then formed into the desired weld tape or profile shape.

Another way that tape is made is to clad the strip materials in a width that is close to the width of the final tape. This type of cladding is done under higher temperatures which create diffusion bonding of the laminated layers. The finished tapes are often referred to as mini or micro profiles, depending on their physical size and shape.

In making tapes for some material combinations that may have trouble for developing good quality bonds it is possible to place a silver layer between the silver contact alloy and the bottom weldable alloy layer. This is a common practice for many silver metal oxide tapes. The weld layer of the tape may have weld rails or a waffle pattern to increase the current density at the weld interfaces. Weld rails are popular since they are easy to form during tape manufacturing and provide a uniform weld profile throughout a coil of tape. For welding tape the area of bonding is normally larger than the area of the weld rails. The percentage of bond needed for tape welding should depend on the current and life cycling required for a contact application. As the percentage of contact substrate interface area that is bonded increases, the life of the welding electrodes for the tape welding decreases. The reason for this is that higher current densities for the electrode and contact interfaces are needed for increasing the percentage of the contact-to-substrate interface bond. This relates directly to the cost of welding.

In comparing the cost of using tape contacts to rivet contacts, in general the cost of rivets is lower since staking requires no current and electrodes and results in less tooling wear.

However, under certain conditions the cost advantage may favor tape welding. Some advantages for using tape are as follows:

1. If the contacts being used are very small, for example <2.2 mm in diameter, it becomes difficult to feed individual rivets with an automatic feeder bowl. This is not a problem for a continuous structure like tape for which machines have been developed for feeding and welding at speeds of over 200 parts per minutes.

2. For composite rivets the amount of contact alloy thickness that is used can only be decreased to a certain ratio of the contact head thickness, such as 40%–25%. For lower ratios beyond this limit the bonding between the silver based contact material and copper is decreased. For tape this ratio can be decreased much more resulting in lower contact material and silver use.

For bond quality, tape bonds can be evaluated by shear testing, ultrasonic scanning, and metallographic sectioning. Quite often a combination of methods is applied during initial equipment set-up and for process control purposes.

17.3.2 Special Welding Methods

17.3.2.1 Percussion Welding

Another type of welding that is used on a more limited basis is percussion welding [12]. This method can be used to directly weld materials like silver metal oxides onto a copper base metal with no intermediate layer. The method consists of using one or more small diameter projections, much smaller than used in projection welding, between the two components. The welding heating time is very short, typically one weld cycle at a very high current. The weld projection explodes and an arc is formed between the contact and backing. A high force is applied at the same time to the contact face, forcing the components together. During this short heating time there is no buildup of segregated metal oxide particles which will form when slower welding and brazing processes are used. The bonds, if the process is done correctly, have an area greater than 90% of the interface area. At the same time the very short heating cycle prevents the substrate materials from being softened by annealing.

The disadvantages of this process are the costs of the equipment, the necessary manual labor, and controlling weld flash that is expelled during the welding. The process requires special tooling for clamping the contact and backing during the welding. Even though the weld current application time is short, the total weld assembly cycle is long as a result of the time for clamping and unclamping. If the weld flash is not controlled well, additional expenses will result for removing the flash. This process is normally used for larger contacts of >8 mm diameter and for contact assembly designs that require high mechanical strength and stability. In general it is not competitive to other welding processes for smaller contacts which can more easily be automated or integrated into stamping operations for the contact supports.

17.3.2.2 Ultrasonic Welding of Contacts

Some experimental work has been done for ultrasonic welding of contacts [13]. Little information is available for this type of contact welding. The welding process can damage the contact material structure during the welding [13,14]. The process is also very sensitive to contact surface condition. This process can—in certain special instances and designs—produce good bond quality but has not developed as a popular production method for contacts since the process tends to damage the microstructure of many contact materials.

17.3.2.3 Friction Welding of Contacts

Friction welding is another process that is technically capable for welding some contact designs but has not been developed into a broader utilized manufacturing method. An example of how the process can be applied to a contact assembly is as follows: A silver metal oxide contact button is spun and then pushed against a copper contact carrier that is held stationary. The relative motion and force causes heating and plastic deformation at the weld interface that can result in a high-integrity solid-state bond [9]. The process has potential for minimizing both annealing of the carrier and formation of segregation of metal oxides which occurs when melting occurs during other welding processes. The process produces flash that must be controlled or removed. The economics and production practicality of this process for contact welding have remained limited to special low volume applications.

17.4 Brazed Contact Assembly Designs

For larger contact assemblies consisting of silver-based contact materials and predominantly copper alloy support carriers, brazing is employed as a traditional method of attachment [15]. The brazing processes discussed below all use silver based brazing alloys, sometimes also termed silver solders, as filler metals between the copper alloy carriers and the silver-based contact. The silver solders have a lower melting point than the contact or carrier, and this lowers the interface temperature required for fusion at the interface to take place. As discussed in Section 17.3, the brazing processes require much longer heating times than welding, seconds or minutes versus a fraction of a second, since time must be allowed for the brazing alloy to liquefy and flow through the whole joint.

The selection of the most suitable brazing alloys and methodology depends on the metallurgical and physical properties of the component materials, the geometry, bond requirements, hardness requirements, and economical considerations.

17.4.1 Methods for Brazing Individual Parts

Table 17.1 shows some of the common brazing methods used to attach medium- to large-size parts, 4 to over 80 mm in diameter. These processes mainly differ in the method used for application of the heat. For all of these processes the clearances between surfaces of the materials to be joined must closely match in order allow for capillary flow of the brazing alloy. Cleanliness of the surfaces is also important for all these processes. Parts are normally cleaned just before brazing. Except for furnace brazing the heating processes are normally done in air. Since all of these processes require long heating times compared to welding methods, oxidation can be significant and therefore some protective coatings such as fluxes are normally used. The flux can help in the wetting of the brazing alloy to the metals but it also generates gases that must be expelled from the braze joint. Some

TABLE 17.1

Brazing Methods for Electrical Contacts

Method	Heating Time	Annealing Amount	Productivity per Machine	Tooling Cost
Torch	Minutes	High	High	High
Induction	Many seconds	Medium	Medium	Medium
Direct Resistive	Few seconds	Low	Medium	Low
Indirect Resistive	Many seconds	Medium	Low	Low
Furnace	Many minutes	Complete	High	Medium

type of mechanical agitation of the contact being brazed is normally done during brazing to assist in gas removal. There are many choices for brazing alloys that can be utilized. Cadmium containing silver alloys were widely used in the past but have been replaced with Cd-free ones such as AgCuSn owing to the hazardous nature of cadmium. Some alloys that contain reducing components such as phosphorus can eliminate of the need for flux. For all of these processes the heating rate must be balanced in time and temperature for the attainment of good bonds without overheating the contact assembly, which can allow the brazing alloy to flow onto the contact face. For example, the alloys containing phosphorus wet the silver alloy so well that just a little overheating produces alloy flow onto the contact face. Brazing alloy on the contact face is a serious problem since it can cause mating contacts to stick or weld in service.

17.4.1.1 Torch Brazing

This process is normally automated into a continuous process that can have high productivity. Since the actual torches are relatively inexpensive, several can be used on an index table or in line in different stations to heat up the part in steps. The heating time for this continuous process can be long, and as a result significant annealing of the contact carrier takes place. Since it is automated and uses many stations, the tooling costs can be significant.

17.4.1.2 Induction Brazing

Induction brazing can be automated into a continuous process, but since the induction units are expensive normally a semi-automated non-continuous process is used. The heat-up time is normally shorter than that for torch brazing since for non-continuous operations there is no long preheating step. The part is instead heated up as fast as possible to obtain alloy flow, and then the heat is turned off. Since the heating time is less with this process a lesser degree of annealing of the carrier takes place.

17.4.1.3 Direct and Indirect Resistance Brazing

This method has some similarity to the welding process described in Section 17.3. The major difference is that the heating of the contact assembly is mainly from heat generated in the electrode which is made of a lower conductivity material like graphite. The terms "direct" and "indirect" are used, since most of the resistance heating is transferred from the electrode to the contact assembly and only a small amount of resistance heating takes place directly in the interface joint. Some manufacturers refer to the term "Direct" for this process if force is applied directly to the contact assembly during the heating process which aids in degassing of the interface and thinning the interface melt layer. The heat is mainly generated in the joint interface and some annealing occurs from contact with the electrodes. Only one part can be done at a time but tooling is inexpensive. When using the term "Indirect" they refer to heating the contact support backing close to the contact tip location and letting the heat travel to the brazing interface. This quite often allows also some manipulation of the contact tips to expel flux induced gases generated during the heating. Depending on the assembly geometry sometimes two tips—for moving contact bridge assemblies for example—can be attached in one heat-up cycle. This indirect resistance process is however only suited for assemblies that can tolerate significant annealing of the support material or require anyhow secondary forming operations which re-harden the substrate backings.

17.4.1.4 Furnace Brazing

For this type of brazing, a cover gas is normally used to prevent oxidation. For some alloys that do not contain metal oxides a reducing atmosphere can be used. A vacuum can also be used with proper controls and back filling with a cover gas to regulate the evaporation rate of the brazing alloy during melting. Fixtures made of graphite are commonly used to locate the braze components. The heating time is long, many minutes, so the parts are normally fully annealed after brazing. If many fixtures are made the productivity can be very high. Since a protective atmosphere is used no flux or only a light fluxing can be employed. Most of the contact structures used in vacuum interrupters are attached using this process.

17.4.1.5 Continuous Laminated Strip Brazing, "Toplay"

For larger quantity brazed contact assemblies, a continuous brazing process can be used, called "Toplay." This brazing process is popular for making contact assemblies for small contactors and definite purpose relays. In this process the contact material in strip form and the copper carrier strip are fed into a small tube furnace with or without a brazing filler material. The furnace is designed to apply induction or resistance heating to the strip sandwich and also applies a force pushing the laminations together as the metal interface melts while moving through the tube furnace. A protective atmosphere can also be used to prevent oxidation. For example, with this process it is possible to braze small width silver metal oxide strips which have a silver backing directly onto copper strip by heating and applying pressure to the interface joint, forming a low melting silver–copper eutectic alloy at the interface, and cooling the interface in the last zone of the furnace while it is still under pressure. This results in a high percentage bond area which is almost void free. After the bonding step of this process the toplay laminated strip can be profile rolled to adjust the assembly thickness and restore some hardness to the materials. The last step in the process is to form the individual contact assemblies by stamping them in a progressive forming die.

17.4.1.6 Brazed Assembly Quality Control Methods

Braze joint quality is usually defined as a minimum percentage of the interface area being bonded and often also by a qualifying definition of the non-bonded void areas by size and/or distribution. Typical bond quality requirements are in the range of 70%–80% of

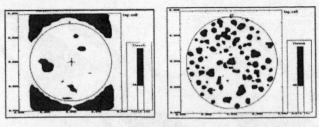

Ultrasonic C-scan pictures of BRAZE joint AgMeOx on
Copper backing – black: Voids [DODUCO Data Book]

(a)

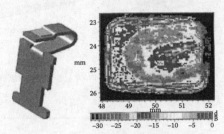

Ultrasonic C-scan pictures of WELD joint Ag-Graphite on
Copper backing – black: Voids [DODUCO Data Book]

(b)

FIGURE 17.8
(a) Ultrasonic C-scan pictures of BRAZE joint AgMeOx on Copper backing—black: Voids and (b) Ultrasonic C-scan pictures of WELD joint Ag-Graphite on Copper backing

the contact area above the interface even so for most arcing applications bond percentages above 50% do not influence the electrical life behavior. The actual braze area of brazed assemblies is much thicker than that for welded and cold formed bond designs. As a result it is easier to check for bond quality. Ultrasonic C-scans, X-ray, metallography, and shear testing can all be used to judge the quality and area of a bond.

As the example in Figure 17.8a shows, ultrasonic imaging can provide a clear indication of the true bond quality and the distribution of voids. Modern ultrasonic equipment with advanced software can also display the quality of weld joints as Figure 17.8b illustrates.

17.5 Clad Metals, Inlay, and Edge Lay

For some smaller low-current arcing contact applications and also non-arcing separable contacts clad metals in the form of inlay and edgelay are used (see Figure 17.9). Similar to the Toplay described in the previous section, the carrier strip has one or more contact alloy strips clad onto the carrier material [16]. These laminated structures can be made by roll cladding as described for tape above, seam welding, and special continuous brazing processes like Toplay. For smaller contact assemblies, e.g. 2–5 mm width contact alloy, this process offers an easy way for stamping companies to make a contact assembly. A problem with the method is that since the alloy is continuous on the carrier strip, often a large percentage of the alloy is wasted as skeleton scrap during the stamping process.

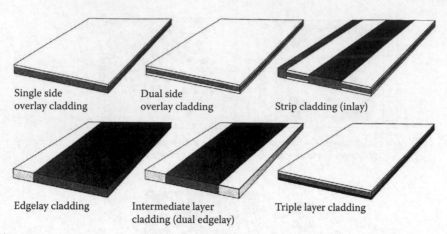

FIGURE 17.9

Typical configurations of clad contact strip materials. (From *Doduco Data Book.* Revised 3rd Edition, DODUCO GmbH–Stieglitz Verlag, Muehlacker, Germany, 2012 [16].)

17.6 Contact Alloys for Non-Arcing Separable Contacts

This section discusses gold contact alloys which are used in clad or electroplated composite structures for mainly non-arcing separable contact applications.

17.6.1 Gold and Gold Alloys

Historically gold is used for ornamental jewelry and coinage, because of its durability and resistance to environmental attack. For contact applications gold also combines high electrical conductivity with metallurgical properties that allow easy alloying with other metals. Its rarity and its link to the world's monetary systems, however, have in the past influenced its declining usage mostly owing to economic considerations [17,18].

17.6.2 Manufacturing Technology

Gold in its high-purity form is rather soft and prone to mechanical wear. Alloying with metals such as copper, nickel and other precious metals by melting will enhance the physical properties while maintaining most of its superior chemical resistance properties. Gold and gold alloys can easily be clad to copper base substrate materials to form overlay and inlay strip material or weldable contact tapes from which contact components can be economically manufactured. Owing to their high ductility very thin layers of the contact material on the surface of suitable carriers can be applied by these methods. Mostly used in electronic connections, pure gold and thin deposits with minor property-enhancing additives are also applied by electroplating (see Chapter 8) for use in dry circuit and low-level switching contacts.

17.6.3 Physical and Chemical Properties

Table 24.6 in Chapter 24 gives typical data for the most commonly used gold contact materials as measured and reported in the literature [19,20] and by various contact manufacturers in handbooks, technical brochures, and data sheets [21–23]. They are typically obtained from measurements on wrought material samples produced under manufacturing conditions and represent average values or ranges for these materials as experienced under practical fabrication and use conditions. The material compositions are given following standard conventions with the alloying additives expressed as weight percent and the first element as main constituent representing the balance to 100% in weight.

The almost total chemical resistance of gold against reactions with surrounding atmospheric constituents under conditions experienced during its use as a switching, sliding or connector contact is its outstanding property. Gold oxide decomposes at low temperatures above 200°C and is not retained in a solidified melt produced under normal air atmosphere. The only reported chemical reactions occur when gold comes into contact with mercury forming an amalgam which ultimately can lead at temperatures above 400°C to the gold almost completely dissolving in the mercury [24]. Another reaction of practical importance is the attack of liquid lead and tin, mostly in the form of tin-lead solders, onto gold surfaces, leading to alloying and the formation of brittle phases.

Pure gold and also gold alloys will however easily adsorb organic molecules, often present from various plastic sources surrounding electrical contacts in modern miniaturized switches and relays. Some of these adsorbed films can then further react with the gold surface or the surrounding air atmosphere to form contact-degrading reaction films [25,26].

17.6.4 Metallurgical Properties

Alloying gold with different other metallic elements is done to alter both its mechanical and its electrical switching properties and at the same time reduce the content of this high-cost metal. Preserving the superior resistance to chemical attack while strictly improving its mechanical strength and hardness requires the addition of other precious or noble metals, in most cases resulting in more costly but highly reliable contact alloys such as the AuPt and AuPtAg materials, Table 24.6 and [21].

Adding up to 30 wt-% of silver to gold does little to increase its hardness and mechanical strength, but does lower the cost with little lowering of the alloy's resistance to chemical attack. The two elements form a continuous series of solid solutions with the tendency of silver to the formation of silver sulfide showing a strong increase at higher silver contents. As a result, the alloy Au/Ag8 wt-% has gained widespread use as a contact material for dry circuit applications. To increase the mechanical stability the addition of copper to gold has practical limitations owing to the order–disorder transformations at typical annealing temperatures, which limit the cold workability of the resulting systems [27]. The binary alloy systems of gold-nickel and gold-cobalt exhibit a steep increase in mechanical strength and hardness at the low 2%–10% addition range which does not noticeably change the chemical resistance of gold. Cobalt is soluble in gold at higher temperatures for up to 8 wt-%, while at temperatures below 400°C the solubility drops to lower than 0.1%. This allows the precipitation hardening of higher percentage alloys which simultaneously in this heterogeneous condition have a substantially higher electrical conductivity, see Table 24.6 in Chapter 24. Tertiary alloy systems of gold with silver and copper or nickel have a lower electrical and thermal conductivity with low potential for surface oxidation and much higher values for hardness and mechanical strength. During the time of rapidly increasing commodity prices and fears of a reduced supply of gold in the early 1980s, other multi-component alloys on the basis of AuPd were developed and increasingly applied for telecommunication contacts.

17.6.5 Contact Applications and Performance

Pure gold finds only limited applications in electrical contacts owing to its softness and the resulting high mechanical wear and tendency to cold welding under even low mechanical forces. As a thin electroplated surface layer over other precious silver or palladium-based metal alloys pure gold is applied as a protective layer on high-reliability electronic connector components [28] (for examples see Chapter 8). Before the advent of electronic switching in telecommunication equipment, thin "gold flash" layers of typically 0.1–0.2 μm used to be the standard on silver contacts in small telephone relays, but this resulted in unexpected corrosion problems and thus is not a recommended practice. Continued usage for pure electroplated gold layers is still found today in high-reliability miniature relays by diffusing an electroplated gold layer into fine silver or silver alloy contacts with a post-plating heat treatment. A similar contact construction employing diffused gold is employed in reed relays with NiFe contact blades where the contacting ends of the plates are gold plated and undergo a heat treatment for controlled partial diffusion [20,29].

Gold with less than 1% alloying additives such as nickel and cobalt is widely used as an electroplated hard gold deposit in electronic connectors. Antler and Slade discuss these applications in more detail in Chapter 8, Materials, Coatings and Plating.

Acknowledgments

The authors thank Dr. Zhuan-Ke Chen, Chugai USA for his help with the figures. We thank Dr. Paul Slade for his discussions and suggestions. We also are grateful for the DODUCO GmbH to allow us to use some graphics from the latest edition of their Data Book on Electrical Contacts.

References

1. D Spaeth et al. "The Influence of the Bonding Area of Welded Contact Tips on Contact Erosion", *Proceedings of the 52nd IEEE Holm Conference on Electrical Contacts*, 2006, pp. 181–287.
2. AS Janitzki and B Schaefer. "The Influence of the Quality of Brazing on the Erosion of Contacts", *Proceedings of the ICEC*, IIT Chicago, IL, 1978, pp. 389–394.
3. Y Shen et al. "Peeling: A failure Mode of Arcing Contacts", *Proceedings of the 34th IEEE Holm Conference on Electrical Contacts*, 1990, pp. 538–542.
4. ZK Chen and GJ Witter. "A Correlation of silver tin indium oxide-Copper Composite Rivet Interface Bond Quality and Switching Endurance Life in DC Relays", *Proceedings ICEC 2012*, Beijing, China, May 2012, pp. 174–179.
5. ZK Chen and GJ Witter. " A Study of The Contact Endurance Switching Life as a Function of Contact Bond Quality, Electrical Load, and Residual Stresses for Composite Rivet Silver Tin Indium Oxide Contacts," *58th IEEE Holm Conference on Electrical Contacts*, Portland, OR, September 2012.
6. G Witter. Chugai USA, Inc., Waukegan, IL, Private Research.
7. *Doduco Data Book*. Revised 3rd Edition, DODUCO GmbH–Stieglitz Verlag, Muehlacker, Germany, 2012, pp. 131–138, 154.

8. R Nichting, D Olson, and G Edwards. "Low-temperature solid state bonding of copper". *J. Materials Engineering and Performance* 1(1): 35–44, 1992.
9. American Society For Metals, ASM Handbook. *Welding Brazing, and Soldering* Vol 6. ASM Publications, Metals Park, OH, ISBN 0-87170-382-3, 1993.
10. American Society for Testing Materials. Unpublished Work of ASTM Committee B4.04 - Round Robin Testing, 1985.
11. G Witter. Chugai USA Inc. private observations.
12. *Doduco Data Book*, Op Cit, pp. 158–66.
13. D Stoeckel. "Ultrasonic Welding of Silver-Metal Oxide Contact Materials" (in German). *Proceedings International Conference on Electrical Contacts*, Tokyo, Japan, August 1976, pp. 321–326.
14. G Witter. Fansteel Inc., Waukegan,IL, private development work, 1983.
15. *Doduco Data Book*, Op Cit, pp. 155–157, 167–169, 177–180.
16. *Doduco Data Book*, Op Cit, pp. 146–153.
17. L Vigdor. The gold bullion market in the United States. Precious Metals 1988, Intl. Prec. Metal Institute, 1988, pp. 535–362.
18. CA Waine and PMA Sollars. A comparison of gold and alternative low cost finishes for connector applications. *Proceedings 9th International Conference on Electrical Contact Phenomena*, IIT Chicago, IL, 1978, pp. 159–171.
19. A Keil, WA Merl, and E Vinaricky. *Elektrische Kontakte und ihre Werkstoffe*. Berlin, Heidelberg, New York, Tokyo: Springer-Verlag, 1984.
20. D Stoeckel, et al. *Werkstoffe fuer Elektrische Kontakte*. Grafenau, Germany: Expert Verlag, 1980, p. 45.
21. *Doduco Data Book*, Op Cit, pp. 21–36.
22. K Pitney. *Ney Contact Manual*. Bloomfield, CT: J. M. Ney Company, 1973, pp. 68–79.
23. Electrical Contacts for Switching Applications, Advanced Metallurgy, Inc. Export.
24. RP Elliot. *Constitution of Binary Alloys, First Supplement*. New York: McGraw-Hill, 1965, p. 89.
25. VA Lavrenko et al. "The corrosion of gold and silver coated copper by the thermal degradation products of chloroprene and silicone rubber". *Corrosion Science* 18: pp. 809–818, 1978.
26. G Horn. "The Influence of Vapors from Organic Insulating Materials on the Contact Resistance of Gold and Silver Alloys" (in German). *Proceedings of the 7th International Conference on Electrical Contacts*, Paris, 1974, pp. 72–79.
27. M Hansen. *Constitution of Binary Alloys*. New York: McGraw-Hill, 1958, pp. 198–203.
28. E Guancial et al. "Qualifications of connectors manufactured with diffused gold R156 inlay contacts". *Proceedings of 28th Holm Conference on Electrical Contacts*, IIT Chicago, IL, 1982, pp. 43–52.
29. CA Haque. "Diffusion effects of the heat treatment of Au-Ag on Fe-Ni dry sealed reeds". *Proceedings of the 9th International Conference on Electrical Contact Phenomena*. IIT, Chicago, IL, 1978, pp. 605–609.

18

Electrical Contact Material Testing Design and Measurement

Gerald J. Witter and Werner Rieder

Facts do not cease to exist because they are ignored.

Aldous Huxley

CONTENTS

18.1 Objectives.. 976
18.2 Device Testing and Model Switch Testing 976
    18.2.1   Device Testing 976
    18.2.2   Model Switch Testing 977
18.3 Electrical Contact Testing Variables... 978
    18.3.1   AC versus DC Testing 978
    18.3.2   Switching Load Type... 979
    18.3.3   Opening and Closing Velocity Effects.. 980
    18.3.4   Contact Bounce... 980
    18.3.5   Contact Carrier Mass and Conductivity 981
    18.3.6   Contact Closing Force and Over Travel.. 981
    18.3.7   Enclosed and Open Contact Devices 981
    18.3.8   Testing at Different Ambient Temperatures 982
    18.3.9   Erosion Measurement 982
    18.3.10 Summary Electrical Contact Testing Variables... 982
18.4 Electrical Testing Result Types and Measurement Methods 982
    18.4.1   Contact Resistance 982
           18.4.1.1 Model Testing.. 983
           18.4.1.2 Evaluation and Presentation of Results..................................... 985
    18.4.2   Contact Bounce Measurement 986
           18.4.2.1 Model Testing.. 987
           18.4.2.2 Evaluation... 988
    18.4.3   Contact Welding Measurement 990
           18.4.3.1 Weld Strength Measured... 991
    18.4.4   Contact Erosion Measurements 992
           18.4.4.1 Accelerated and Model Testing.. 992
           18.4.4.2 Extrapolation at Rated Stress... 993
           18.4.4.3 Increase of the Switching Frequency.. 993
           18.4.4.4 Testing at Increased Electrical Load 993

18.4.4.5 Fixed-Gap Models 994
18.4.4.6 Moving Contact Models 994
18.4.4.7 Evaluation and Presentation of Results..995
18.4.5 AC Arc Reignition Measurement 995
18.4.6 Arc Motion Measurements..995
18.4.6.1 Measurement..995
18.4.6.2 Electronic Optical 996
18.4.6.3 Model Switch Arc Motion Control..997
18.4.6.4 Evaluation and Presentation of Results..998
18.4.7 Arc-Wall Interaction Measurements 998
References...999

18.1 Objectives

The purpose of this chapter is to aid engineers in establishing testing methods for electrical contacts and testing of devices depending on the life and reliability of arcing type electrical contacts. The testing of electrical contacts involves many variables which make evaluations and comparisons very difficult. This chapter will summarize the important variables to consider and provide some guidance for designing test methods for your objectives.

18.2 Device Testing and Model Switch Testing

If you are an electrical device manufacturer, testing a new material in your device is mandatory before you can feel safe using it in your design. Testing in a model switch has the advantage of providing some specific information on electrical contact life that is difficult to measure in a commercial device. Two examples are: more precise electrical erosion data and quantitative welding measurements. The problem with model switches is that they don't duplicate the mechanics, heat sinking, thermal conductivity, atmosphere, and other variables of the actual commercial devices. Since these variables can have significant effects on the erosion process, contact welding, and material transfer, results are not always reliable for extrapolation from one device to another. So for one major point in understanding results of contact testing, do not think you can use information acquired on a separate device to exclude the need for actual commercial device testing. This section will discuss ways to make model switches better so the information is more useful in commercial devices.

18.2.1 Device Testing

Even though actual device testing is mandatory, it does not normally give you all the information you would like to know about one contact material as compared to another. Many companies will evaluate a new contact design, new material or contact size, by just running the standard UL or other standard test with their commercial device to see if the new design passes or fails. The problem with this approach, besides only obtaining limited information, is that you test not just the contact but all the other components that make up the commercial device plus the consistency of your manufacturing process. Any testing that is done on commercial devices without some statistical measurement controls to serve as a barometer to compare against, has limited value and only tells you that in this spin of the wheel that you have passed. Of course, if you substantially pass or fail you can get some idea of what the result means.

If you put a little more effort into the design of device testing you can achieve a much higher confidence in the results. One way of doing this is to use a DOE (Design of Experiment) approach for running your evaluation of a new contact design. This type of approach is good for all types of testing but especially for contacts where there are many variables. Today, there is a lot of very good software available for designing and doing DOE work. Although it is available, software companies have found that less than 10% of companies use DOE statistical approaches. An important part of using this type of approach is to make sure your preparation allows you to make a good comparison of a new contact to some known contact standard. An important step is to assemble both sets of contacts in your commercial device at the same period of time. This should eliminate both component and manufacturing variables from the test evaluation. DOE test methods also allow you to test as many sets of contacts and other variables at the same time with a minimum of additional work. When you get the final results you will have good data on the relative standings of the variables and a confidence level for judging the test result differences. Another advantage of the DOE approach is you will also get information on interactions between variables. If you keep a set of contacts made from your initial evaluation testing of a design you will also have a method of comparing your contacts and devices over a period of time. Although this approach seems basic, it is rare to find a company that has the option to go back and make such a comparison. There are many statistical methods for measuring how consistent your results are and where you stand with regard to meeting your specifications. For electrical contacts there are many variables and this is why a DOE approach is useful, since it allows you to find out which variables are the most important for controlling your process. The variables will be discussed in this chapter.

One more point in the discussion of actual device testing is how good your testing methods are for protecting you from field failures. Typical industrial standards for device life are normally set using maximum currents and loads. For contacts, there are circumstances where lower power loads may be more susceptible to contact resistance problems than higher current loads. This may be connected with contamination from films or vapors. Chapter 19 explains this type of problem especially the section on silicone contamination. The point here is that although you test at maximum loads it does not protect you from problems that can exist only at lower loads.

18.2.2 Model Switch Testing

This term involves both switches and relays. The advantage for model switch testing is that it makes getting quantitative data on electrical erosion and welding resistance possible. It also allows measuring contact resistance, contact force, temperature rise, and other parameters that can be measured with either commercial devices or model switches. Another advantage of using these devices is that they are normally made very precise and adjustable for parameters like opening and closing velocity, contact force, over travel, and other mechanical parameters so that testing can be more consistent over time than you would see for commercial devices made over a time span.

One of the problems with the use of data from model switches is that many times the model switch does not match the operating parameters of the commercial switches for which they are being used to supply supplemental information. For example, many very precise model switches used in research for comparing materials in erosion, material transfer and contact resistance as a function of operating life, have opening and closing speeds much different from most commercial devices which makes the results nonrepresentative or comparable to actual commercial devices. It is very important in designing model switches to make sure the operating range is within the parameters of the commercial devices for which you have interest. Sometimes this involves the size scale of the model switch and many times they are too big to adjust to the parameters of the smaller device. The second thing about these designs is to use similar mechanisms to what is used in the switch. For example, a model switch that is using an air cylinder to open and close the contacts will rarely mimic the opening and closing of a set of contacts that operates with an induction coil and armature as seen in relays. This may seem obvious but many times mechanical design engineers not familiar with these factors get carried away with elaborate and clever ideas for designs that never are able to recreate what happens in an actual relay or switch. It is not always easy to design the model switch to exactly duplicate the commercial switch in operation but you must try and come close to the same range of operation for the device of your interest.

In the following sections, some of the different electromechanical operating parameters will be discussed in addition to their effects on material performance such as erosion, material transfer, welding and sticking resistance and contact resistance.

18.3 Electrical Contact Testing Variables

This section will discuss the important parameters that need to be controlled for testing arcing contact materials. The parameters will be discussed in terms of how variations in these parameters influence the testing results for erosion, material transfer, welding, or contact resistance. This will be especially important if you are not just doing a fundamental study but are trying to use a model switch to duplicate results that will be seen in a commercial device.

Most of this discussion will concentrate on applications that are common for relays and switches used in residential homes or automotive circuits as opposed to more specialty devices used for protection for high voltage and current applications. Model switches for specialty items must almost duplicate the device.

18.3.1 AC versus DC Testing

For people working in this field, it is obvious that large differences exist between these two testing environments. In the automotive field, electromechanical relays are still the main devices used for switching the circuits on and off. For AC electromechanical relays, almost identical in design to the automotive relays, are used for switching industrial controls on and off. The mechanics of the two types of relays are very similar, but the contact materials for the two types of applications can be quite different. The reason that the AC and DC devices can be so similar in this case is that the DC voltage in automotive, 13VDC, is low enough to allow current interruption in somewhat the same time frame as for the AC relays. If the DC voltage is raised much above the 20 volt level the same designs would never work (see Sections 9.7.2 and 10.2.2). This was found out several years ago late in the game plan when an engineering study group sponsored by the auto industry was working on changing the automotive voltage to 42VDC. Although they were a highly technically refined group aimed on more efficient battery life, they failed to think about the interruption problems associated with increasing the automotive DC voltage.

Besides the difficulties for interrupting higher voltage DC, there are other differences as a result of AC changing polarity with each operation and DC staying with the same polarity.

To help understand this variable better a short explanation of the DC erosion process will be made. When the contacts open with the formation of a liquid bridge between the contacts that converts to an arc in metal vapor (see Section 9.4.2), the initial erosion is from the anode with some transfer to the cathode; anodic erosion (see Section 10.3.5). So the initial erosion when the contact opening gap is very small transfers material is from the anode to the cathode. As the contacts continue to open the arc transitions into one operating in the ambient gas (see Section 10.3.5). Here the material transfer direction reverses and the material net erosion is from the cathode and transfer to the anode; cathodic erosion. This happens at a gap of about 5–10 microns. As the gap becomes larger the efficiency of the transfer becomes lower and more material is lost to the environment. So for our DC automotive case depending when the arc interrupts and the load type we can have erosion of the anode and a buildup of material on the cathode or a little less likely but possible the reverse a buildup on the anode and more erosion on the cathode. This is a simplified explanation but gives you an idea of what is happening. For AC switching the material transfer variable is eliminated if the switching device operates random to the AC sine wave. Sometimes there is a problem with the switching not being random as a result of the mechanics or electronics controlling the opening and closing of the contacts in the model switch. The point here is checking to see that the AC testing is random with regard to the current sine wave and this is always important for any AC testing (See Section 10.3.6).

18.3.2 Switching Load Type

Contact load type is something that is specified for different relays and switches depending on the standard that the device manufacture wants to meet. The main point in this part of the discussion is that these different loads all have different effects on the results from different operating parameters of the switches and relays. We will discuss three different types of loads: lamp, resistive, and inductive loads. These loads also have different effects for AC and DC switching.

Lamp loads are mainly resistive loads with a high inrush current. For AC the typical closing current is 10 × the normal current and the break current just the normal rating. Because the closing current is so high, variations in the severity of contact bounce make a large difference in erosion results. Controlling contact bounce is a major factor for improving the contact life for AC lamp loads. For automotive DC lamp loads the ratio of inrush is about 6 × normal current, a little less than for AC. Minimizing bounce is also an important factor for DC lamp loads. For DC lamp loads, the variations in the severity of bounce also cause large differences in anode erosion but also result in large variations of material transfer and build up on the cathode. This will be discussed more under the section coming up on bounce.

Resistive loads, normal current levels, are a mild type of load that is the least problematic of the three types of loads. The erosion level for resistive loads is very mild compared to the lamp loads in both AC and DC switching.

In AC there are special specifications for motor load switching. An AC-3 test is for a typical motor switching with a high inrush make, 6 × normal current and normal break (See Section 14.3). There is also an overload testing AC-4 with both high inrush on make and high current on break. Contact bounce is again a major factor for the erosion because of the high inrush. For the AC-4 testing with the break being at high current the model switch has to have a special design that duplicates the commercial device for moving the arc off the contact.

For DC automotive testing, the inductive load presents an interesting contrast to the other loads. On contact make there is normally a large lag time in current buildup of 4–7 ms and then a prolonged arc on break (see Section 10.2.2). So for a normal automotive relay switching an inductive load almost all the erosion is on break and normally no erosion on make. This means bounce does not matter in this case. This means controlling the break operation is the most important factor for DC inductive loads.

18.3.3 Opening and Closing Velocity Effects

Chen and Witter [1,2] studied the effect of changes in opening speed on a model switch designed to simulate automotive relays. For an inductive load they showed the erosion material transfer to change from being anodic erosion to cathodic erosion when the opening speed changed from 0.47 m/sec to 1.28 m/sec. For slow speeds the contacts remain in the anodic erosion zone for a much longer time than at the high speed. The overall erosion is also much less at the higher opening speed (see Figure 10.33). This example illustrates the importance of designing the model switch to operate in a similar velocity range to the commercial device of your interest. Closing speed is also important for a different reason. High closing speed can increase contact bounce, therefore at higher velocities more force is needed for bounce control. Very slow closing speeds can be a problem for higher voltage AC switches which can result in a pre-arc to form just before closing. This is rarer than problems of closing too fast and having more bounce erosion.

18.3.4 Contact Bounce

As discussed above, contact bounce is the major factor for variations in electrical erosion especially for high inrush loads. Controlling and adjusting bounce is more difficult than controlling other operating parameters. Some factors that increase the difficulty for matching bounce in a model switch to a commercial device are size differences, different opening and closing mechanisms, and different vibration absorption materials. Besides making these items similar, adjustments in closing speed and closing force can be made. Higher closing force generally reduces bounce. Experimenting with absorption pads is also a way to make adjustments. There are, however, no easy answers for making these adjustments.

Bounce is a bigger problem for DC lamp loads than in AC since the transfer remains from anode to cathode and buildup of transfer accumulates with each operation. The bigger problem in DC is that the bounce can result in a serious type of erosion called pip and crater. Witter and Polevoy [3] studied bounce in DC and found that if the bounce frequency was high (i.e., many short duration bounces) pip and crater erosion was more likely to form than for longer less frequent bounce times, see Figure 16.6 in Chapter 16. There are three reasons for the increased build up with higher frequency: (1) Short bounce has less amplitude so material transfer is more efficient, (2) The low opening gap amplitude causes less splatter on closure and more material remains near the center of the arc, and (3) There is a higher ratio of make operations to break operations with higher bounce frequency so less transfer from cathode to anode takes place on break. Pip and crater formation causes several problems. Once the pip forms there is a higher concentration of erosion of the anode in the area opposite of the pip. The pip is prone to cause mechanical latching, sticking of the contacts, as it goes into the cavity on the anode. Adjusting the bounce out of the high frequency mode is a must for being able to get good contact life. If pip and crater erosion forms, and the bounce frequency is low, there is more likely to be a material related problem and your model switch is giving you good information.

18.3.5 Contact Carrier Mass and Conductivity

The contact erosion and stress build up in the contact from arcing is sensitive to the mass and thermal conductivity of the carrier for which the contacts are attached. For example, relays normally have a thin spring as the movable contact carrier and a thick terminal as the stationary contact carrier. The heat and melting of the contact is more severe for the spring side of the relay than the terminal side of the relay since the thickness ratio of the terminal/spring carrier is 4/1. For high inrush loads, the life of the contact is much longer if the thicker terminal is the anode than if the spring is the anode [4]. This means that if you want the model switch to simulate that relay or switch you should match the contact carrier mass ratio on your commercial device.

18.3.6 Contact Closing Force and Over Travel

Contact over travel refers to the amount of elastic compression put on the closed contacts from the point of initially mating to the final travel of the movable contact blade. In a relay this is applied by the spring being bent by the armature which is attached to the spring as it continues to travel after the contacts have mated until it bottoms on the relay coil. There are many ways to apply this over travel elastic force to the contacts. The important factor is that the contacts will remained closed even after some of the contact surface is removed by the electrical erosion of the contact by the arc. From experience, it is known that most failures even in good devices begin when over travel starts to be too small for the application, therefore, too little contact force. This can result in a cascading downturn. Some of the things happening as the force drops are as follows: (1) Constriction resistance goes up for the closed contacts and they go up in temperature, (2) The contact bounce increases and this increases erosion per operation and also contributes to increasing contact temperatures, and (3) Stress relaxation of the carrier spring can take place with the increased temperature and this further reduces contact force. The main point here is to show how important it is to have enough over travel and to make sure you can adjust the over travel for any model switch you design.

18.3.7 Enclosed and Open Contact Devices

Most model switches are made open to the atmosphere. This could be a problem if you are trying to duplicate a device that is sealed in a plastic housing. For relays some are vented and some are almost hermetically sealed. As devices like relays heat up from the electrical load many times there is out gassing from the plastic. These gasses can be adsorbed onto the contact surface and alter the contact erosion process and also have some effect on the contact resistance. In reference [3], a relay was tested both with and without the plastic cover. The relay with the cover had an endurance life of 300,000 operations. When the cover was taken off the relay would fail in less than 50,000 operations. The reason for this big difference in endurance life was related to contact surface activation (see Sections 10.2.3 and 19.2). The outgassing of the relay plastic produced carbon particles on the contact surface for the relay with the cover which made the arc move to new sites for each separate operation. The erosion under this condition was smooth. With the cover off, no carbon particles were created. The relay had high frequency bounce and the arc stayed in one area and created severe pip and crater erosion. Before you start designing a model switch for a device that is enclosed in plastic you should run some tests on the device with the cover off or with vents and compare to the results with the normal enclosure.

18.3.8 Testing at Different Ambient Temperatures

For some types of devices there is a requirement that some of the testing be done at different temperatures. Automotive normally requires a portion of the endurance testing to be done at three different temperatures: approximately −40°C, 25°C, and 125°C. This does make it more difficult for getting force transducers to work. The results of the testing can also be different for the different temperatures. Although the test involves combining the testing results for all the temperatures, it is recommended that the tests be run separate on each temperature to see how your device performs at each temperature. For relays the tests at the higher temperature are normally the hardest since the relay coil overheats and affects the temperature of the thin spring carrier.

18.3.9 Erosion Measurement

If you want to make periodic measurements of the erosion it is important to make removal easy and the ability to put the contacts back close to the same position. If the contacts are not close to the same position you will alter the erosion process especially if pip and crater erosion is possible. At the end of the testing it is good to measure the erosion and then clean the contacts with an ultrasonic bath to remove all the loose particles and then weigh the contact again so you get the true total erosion.

18.3.10 Summary Electrical Contact Testing Variables

Designing a model switch is complex and involves much more than the variables discussed in this section. The main purpose of this section is to make sure you have a better understanding of the variables and how they react with different loads so your design can better meet your objectives.

18.4 Electrical Testing Result Types and Measurement Methods

18.4.1 Contact Resistance

Contact resistance measurement is not problematic as a rule. The current and the electromotive force, e.m.f., of the test current circuit employed should not exceed the values of the minimum current and voltage, respectively, of the application device with regard to possible fritting processes (Section 1.3.6). The voltage measurement leads should be connected as close as possible to the contact spot. The reading must not be made before the dynamic contact phenomena level out; for example, only when three succeeding readings in properly chosen intervals are within a certain range of scatter, unless the dynamic contact resistance behavior is studied requiring automatic measurements with high time resolution [5,6].

If the contact resistance is measured under dry-circuit conditions, that is, excluding any A- or B-fritting owing to voltage and current applied for resistance measurement, respectively, the test voltage has to be limited to 80 mV [7]. Contact fritting can be studied, when first a minimum voltage not sufficient for A-fritting (e.g., 100 mV) is applied across the closed contacts while the current is at a certain minimum value (e.g., $I_{min} = 5$ mA). If high contact resistance limits the current to a value under that minimum, the voltage must be increased in small steps until the voltage breaks down owing to A-fritting and the minimum current value, I_{min}, is reached (Figure 18.1). The voltage values immediately before and after break-down are the actual values of the fritting voltage and the cessation voltage, respectively. As soon as I_{min} is flowing the open circuit voltage has to be increased to a value sufficient for B-fritting in the total current range. Then the test current is increased in small steps while the contact voltage is recorded. As soon as the latter does not raise proportional to the current but remains constant (at the value of the softening voltage of the contact material) B-fritting is indicated (Figure 18.2) [8,9]. Since fritting processes are not reliable because the resulting contact resistance value depends on voltage and current applied and since the absolute resistance values after fritting are comparatively high), contact surface films have to be destroyed mechanically in order to obtain reliable, well-defined and sufficiently low contact resistance values. Therefore, the influence of the contact force and more specifically the dynamics of closing in terms of impact and sliding components for force are more important than the influence of voltage and current. Systematic variation of the mechanical conditions (normal and tangential components of contact force, kinetic energy) require model switches, as a rule.

18.4.1.1 Model Testing

18.4.1.1.1 Crossed-Rod Arrangement

The classical model contact for contact resistance investigation is the crossed-rod arrangement (see Figure 2.10). A large number of contact measurements on virgin spots can be executed along each straight line of the specimen rod to investigate its surface as manufactured or after exposure to any atmosphere. Well-defined contact closing, contact force increasing, opening, and promotion to the next spot can be executed automatically; dynamic closure can be simulated realistically in a well-defined way by sliding one rod after closure or by application of a mechanical impact with a small hammer. Contact resistance versus contact load curves may yield informative insights into the nature and the structure of surface contamination [9–13].

18.4.1.1.2 Probe Measurements

Probe measurements employing a gold probe under very low mechanical and electrical load, (to avoid any mechanical or electrical damage of the surface layer), to scan the surface of a flat specimen yield detailed information about the cleanliness of the surface and perhaps even about the nature of the surface contamination [14–19]. The contact resistance

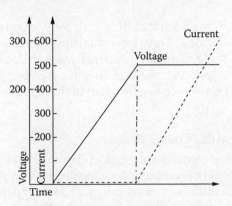

FIGURE 18.1
Contact resistance measurements by increase of voltage and current respectively, during a single test. (From Johler W. and Rieder W., *VDE-Fachbericht*, 38, 131–140, 1987.)

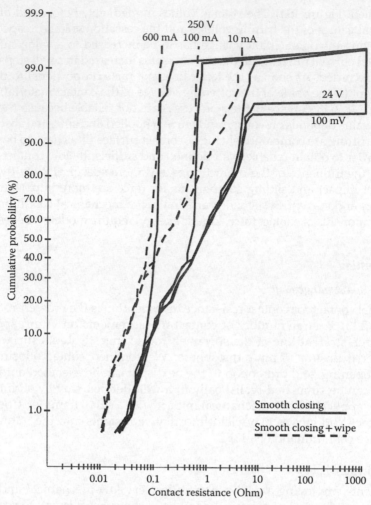

FIGURE 18.2
Contact resistance cumulative distribution curves of Ag/Ni 0.15 contacts after two months storage measured according the procedure shown in Figure 18.1.

measured with a probe on an intact surface film is usually much higher than would be seen for most practical applications. Therefore, this information may not be useful in predicting the resistance of these films after normal mechanical and electrical stresses are applied in typical device applications [20,21]. To solve this problem devices have been developed to measure contact resistance as a function of these types of stresses, force, wipe distance, and voltage [13,15].

18.4.1.1.3 *Individual Contact Quality Control Devices*

For quality control of individual contacts before assembly into devices, model switches are required which grip, test, and eject the contact automatically. This must be done without damaging the contact surface mechanically or electrically [17]. Most types of mineral contamination from deburring contacts cannot be found by using the above type of method, since the mineral particles are buried into the surface on the contact. In these cases, the contact must be arced with a few operations at low current to bring the contamination to the surface [22].

18.4.1.1.4 Organic Vapor Contamination Model Switches

Special model switches have been developed to measure the resistance of switching contacts exposed to the vapors emanating from one certain organic material only [23–28]. These sealed switches are made from glass and metal only (Figure 18.3). They cover an "offering cup" where any organic material can be heated and contacts switching a current in the critical range of tens to hundreds of milliamperes (ohmic load, e.m.f. $\leq$10 V) in order to crack and to carbonize the organic vapors adsorbed on the contact surfaces, thus contaminating the contacts with carbon which increases the contact resistance and/or "activates" the contacts (Section 19.2). Less generally applicable but more realistic are tests with modified relays [29].

18.4.1.2 Evaluation and Presentation of Results

18.4.1.2.1 Cumulative Distribution Curves

Corroded contact surfaces cause widely scattering resistance values that may yield useful information when properly evaluated. Statistical evaluations on the basis of log-normal distributions are adequate for items like contact resistance which has numerous independent parameters influencing the scatter of the results. This is especially true for contaminated contacts and these distributions are good since resistance values

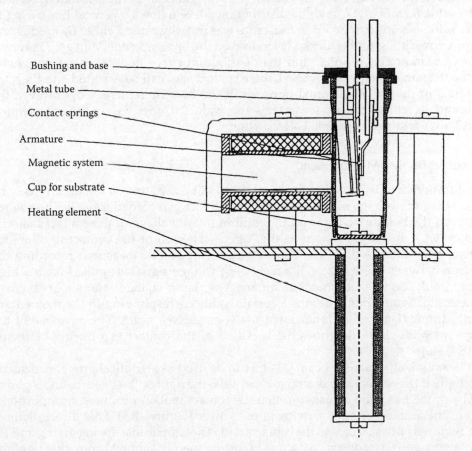

FIGURE 18.3
Model switch for testing the influence of individual organic materials on contact reliability. (From Neufeld C.N. and Rieder W.F., *IEEE Trans CPMT-A*, 18, 399–404, 1995.)

cannot become negative. The measured resistance values yield straight lines for correspondingly scaled coordinates. Mixed distributions characterized by a sudden change of the slope of the cumulative distribution curve at a certain numerical value indicate two sets of predominating parameters, material or inhomogeneous contamination of the surface (partly covered by skin layers or dust). The Weibull distribution, however, represents a "limit distribution" or "extreme value distribution" applicable to a chain of links with independent strength values rather than to the resistance values of contacts [30–33]. Cumulative probability curves of the resistance values obtained from measurements on corroded contacts employing rising voltage and current stresses as shown in (Figure 18.1) yield curves as shown in Figure 18.2. The resistance values obtained at minimum voltage and current yield normal distribution curves (straight lines) representing the contact resistance values caused by mechanical damage of the surface layer without or with wipe, respectively. There appears a sharp bend upwards, however, as soon as the product contact resistance times current exceeds the value 60 mV, corresponding to the softening temperature of the contact material. The contact spot which might have shown higher resistance values at the very moment of contact make immediately became deformed until the temperature of the contact restriction area fell below the softening point.

Therefore, at a certain current I_x the contact resistance cannot be higher than the critical resistance value $R_x = U_{soft} \times I_x$ and the distribution curve follows a vertical line owing to B-fritting, as long as any finite contact resistance was initially caused either by mechanical damage or by A-fritting of the surface layer. In case the layer was not damaged, however, the contact resistance is "infinite" and the distribution curve now follows a horizontal line. As the stressing voltage is increased, more contact sites can be A-fritted and the horizontal branch of the curve is shifted upwards. If the current is increased, however, the vertical branch is shifted toward lower resistance values. Under the conditions, yielding Figure 18.2 no A-fritting was needed after a wipe.

18.4.2 Contact Bounce Measurement

Both the duration of individual bounces and the total bounce time at make can easily be measured when the voltage across the bouncing contact is recorded [34–42]. The voltage limit between the boiling voltage and the minimum arc voltage of the contact material (e.g., Section 9.4.2) has to be chosen as the "open" criterion of the contact to allow for arcing during the bounce time. The bounce height can also be measured recording the voltage drop between the contacts, if a fast rising voltage signal is applied from a high resistance circuit too weak to maintain an arc. The rising voltage causes a breakdown across the contact gap when it attains a certain value corresponding to the momentary gap length. Immediately after breakdown the voltage rises again. The envelope of the resulting saw-tooth curve describes the variation of the contact gap during bouncing (Figure 18.4) [40].

A light beam [41,42] or a laser beam [43–45] can be used as an optical probe to measure the gap length, if the contact gap is arranged between the light or laser source and a photosensor. The probe has to be adjusted so that the contact motion produces a proportional variation in the amount of light arriving at the sensor (Figure 18.5). This direct method, however, requires optical access to the contacts and is not applicable during arcing and for the very narrow gaps of modern contactors bouncing merely about 0.1 mm, that is, of the order of surface irregularities inhibiting the light beam. Therefore, indirect optical methods are recommended, for example,

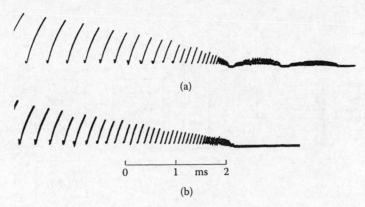

FIGURE 18.4
Contact bounce characteristics gained by no-load voltage oscillograms. (From Franken H., *ETZ-B*, 53–54, 1960.)

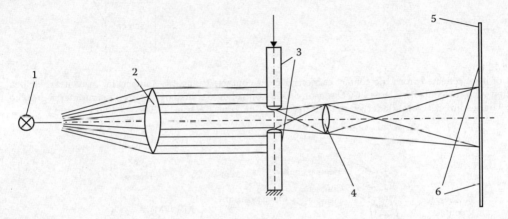

FIGURE 18.5
Principle of direct contact gap measurement with the aid of a light beam: 1, Sight source; 2, Sens; 3, contacts; 4, lens; 5, screen with scale; 6, shadow of the contacts. (From Erk A. and Finke H., *ETZ-A*, 86, 129–133, 1965.)

- Modulation of the intensity of a light beam by lightweight slot diaphragms connected with the contacts [42,46,47] or filters [48]
- Recording of the light reflected from white stripes on black papers fixed to each contact with the aid of a CCD camera (Figure 18.6) [34]

It has to be emphasized that it is not sufficient to record the motion of the moving contact only since even the fixed contact is usually subject to elastic oscillations caused by the impact at closure—not negligible when the electrode distance in moderately bouncing contacts has to be measured (see Section 13.4.3).

18.4.2.1 Model Testing

Mechanical models to investigate bouncing phenomena, for example, according to Figure 18.7, can be used to study impact phenomena or to determine the impact factor of a material [49,50]. A more sophisticated model also provides records of preimpact arcing [48]. Purely mechanical tests can be executed also employing model switches with controlled bouncing [37,42,51–53].

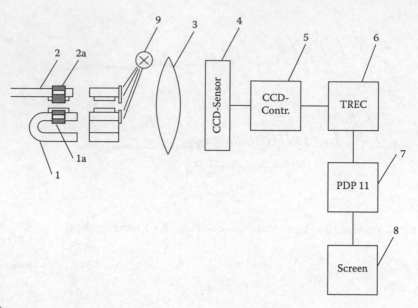

FIGURE 18.6
Record of light reflected from the contact carriers (1 fixed contact; 2 movable contact; 1a, 2a contrast stripes connected with the contacts; 3 lens; 4 CCD sensor; 5 CCD controller; 6 transient recorder; 7 computer; 8 graphic screen; 9 light source). (From Rieder W. and Weichsler V., *IEEE Trans CHMT*, 14, 298–303, 1991.)

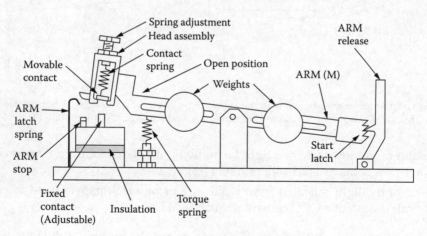

FIGURE 18.7
Bounce control mechanism according to Shaw. (From Shaw N.C., *Proc. 2nd ICEC*, Graz, 1964.)

18.4.2.2 Evaluation

A characteristic bounce diagram of a contact can be gained, when the "open" time of the contact recorded for each operation of a series (Figure 18.8a through d) is stored in a computer and finally the frequency of "open contact" is plotted versus time (Figures 18.8e and 18.9) [34,35]. Comparison of succeeding series shows the change of the bouncing behavior during the contact life [34].

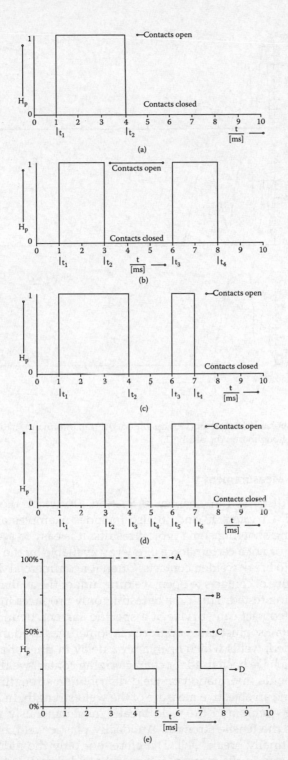

FIGURE 18.8
Statistical procedure for the representation of the bounce behavior: (a–d) Bounce patterns for four switching operations; (e) superposition of the four patterns (bounce diaigram); H_p: frequency of periods of open contacts. All four operations: $H_p = 100\%$ (A), three operations: $H_p = 75\%$ (B), two operations: $H_p = 50\%$ (C).

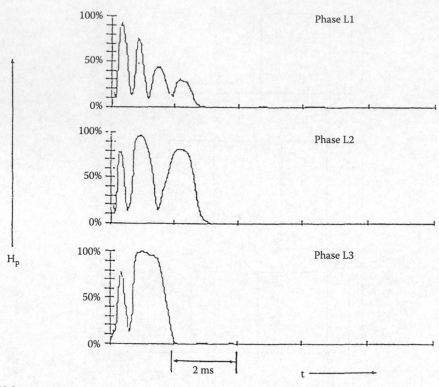

FIGURE 18.9
Bounce diagrams of the three poles of a contactor (1000 operations). (From Braumann P., Kolb H., Schroder K.H., and Weise W., *Proc.16th ICEC*, Loughborough, UK, 1992.)

18.4.3 Contact Welding Measurement

Contact welding is also discussed in Chapter 10, Section 10.5 where theoretical data and equations are discussed. Chapters 13 and 16 also includes examples and discussion of different kinds of welding showing some typical results. It is easy to see from voltage or current records whether or not a certain force or energy available by the drive of a certain switching device is able to break welded contacts. Often for commercial devices like relays life tests only record failure of contacts to open, welding, under the available opening force of the specific device being tested. This type of testing only produces information on the number of operations a contact can survive at a specific current, bounce condition, and opening force. A little more sophisticated approach is sometimes used in which partial or weaker welds are recorded, welds which open after a delay in time or after a few operations. These types of data have been used for characterizing contact welding [55,56]. These methods still do not provide information on weld distribution strength. It is more problematic, however, to define an objective measure of the weld strength. In an ideally brittle-elastic weld no yielding occurs and the weld breaks instantaneously when an applied increasing force exceeds the tensile strength. An ideally plastic weld, however, yields at constant force before it finally breaks [49]. Therefore, not only the momentary value of the opening force measured in the moment when real welds break but also the ranking of various materials may depend on the rate of rise of the force applied. Often the test force has been applied very slowly (10–100 N s^{-1}) as compared with the separating force in certain actual switching devices during an impact of accelerated masses (10^4–10^5N s^{-1}) [57]. Therefore, the fracture energy

$$W = \int (F(\delta))d\delta \qquad (18.1)$$

(where δ denotes strain) rather than the tensile strength, might be adequate to compare the weld strength of materials, yielding different rankings, for example, for various Ag/CdO materials [58,59]. The fracture energy depends very little on the strength velocity, if plastic deformation does not exceed velocity of sound in the material [60]. While the test force is usually applied perpendicular to the contact plane, the opening force in actual switching devices may provide a shearing component [61].

Freshly polished contacts yield reproducible results and the surface damage caused by an individual operation can be investigated in detail, thus offering some insight into the mechanisms of contact welding and erosion. Unfortunately, both welding and erosion of the first operation(s) are not necessarily representative of the contact behavior during the subsequent life.

Both welding and erosion are different whether or not

1. The direction of the test current is changed after each operation.
2. Each make operation is followed by a break operation as in service, depending on the break stress too (e.g., AC 3 or AC 4 duty conditions according to IEC-H 947-4).

Arc current and voltage have to be measured additionally if the weld strength is to be correlated with the make arc $\int dt$ or energy [55,62,63] (see Section 10.5.2).

Obviously the measured values of weld strength depend on a great number of uncontrolled design parameters and on the surface conditions generated by erosion and previous welds and last but not least the reproducibility of the opening force in test devices and actual switches. Therefore, it has to be emphasized that any kind weld strength measurements are yielding merely relative rather than absolute values.

18.4.3.1 Weld Strength Measured

If the weld strength is measured after each contact opening, usually its value is plotted versus current or $\int I \, dt$ of the arc preceding the welding process at make, other parameters being kept constant. Some authors have chosen the bounce arc energy $\int I U_{arc} \, dt$ as the independent variable but I or $\int I \, dt$ seems more meaningful for comparison of materials because the arc voltage cannot be chosen independently of the other characteristics of the material. The wide scatter of the weld strength requires a statistically relevant number of measurements; cumulative distribution curves for each set of parameters proved to be informative [64–70]. Often the first welds of each series yielded extremely high but well-reproducible values of the weld strength. Therefore, some authors tested only virgin contacts while others neglected a certain number of initial operations or tested only eroded contacts. Suppression of values obtained with new contacts, however, is not harmless since they may cause early failures in practical application. Since the surface of a contact material may change significantly with switching life owing to chemical and physical reactions, the welding resistance of the materials also changes with arcing life. Welding tests on materials should include weld data on both new and arced contact surfaces, Figure 18.10.

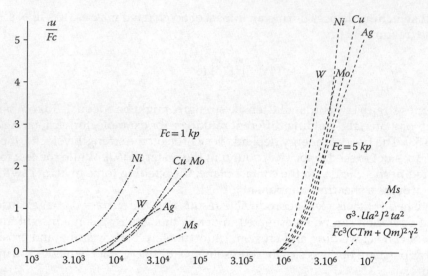

FIGURE 18.10
Dimensional analysis applied to welding of closing contacts. (From Rieder W. and Uranek J., Contact welding, *Proc. 2nd ICEC, Graz, 1964.*)

18.4.4 Contact Erosion Measurements

Weighing of contacts after a certain number of operations is not a problem, if the contacts are removable. In small contactors often only the contact bridge can be removed and both bridge contacts have to be weighed together. In most small devices the contacts are welded or staked into a copper alloy carrier. True erosion is difficult to measure since a portion of the erosion by splatter is collected on the carrier which can not be separated from the contact for weighing. If the contacts are replaced after weighing to continue the test series, their individual positions must be carefully re-established to avoid serious changes in the bouncing behavior.

Additional measurements are required if the amount of erosion is to be related to the stress data, for example, arc current, arc duration, arc energy or $\int I\,dt$. For that purpose, arc data have to be measured at each operation and properly processed by a computer [56,63,71–77]. Switching of small DC currents may cause contact material transfer from one contact to the other rather than absolute losses. In this case the amount of transferred matter may be less important than its shape since pip and crater formation may impede contact opening and cause failure (Chapters 10 and 13).

Mapping of contact surfaces can be done by use of mechanical or laser probes with scanning devices [78]. These methods give both quantitative volume loss information, material transfer information, and information on surface topography, such as pip and crater formation. These methods lend themselves to research applications rather than routine quality checks. SEM photographs can also give information on the surface characteristics of erosion [79,80] but do not give information on volume loss.

18.4.4.1 Accelerated and Model Testing

18.4.4.1.1 Accelerated Testing/Similarity Laws
There are various proposals to accelerate time consuming erosion tests by

1. Reducing the number of operations at rated load and extrapolation assuming a constant erosion rate per operation.

2. Increasing the switching frequency at rated electrical load assuming switching frequency does not influence the result.

3. Reducing the number of operations at increased arc stress per operation (current I_a or arcing time t_a) beyond the rated values and calculation of the erosion losses Δm at rated values assuming a purely empirical similarity law shown with no proof.

$$\Delta m = \text{const.} I_a^x \cdot t_a^y \tag{18.2}$$

18.4.4.2 Extrapolation at Rated Stress

This cannot be recommended because the erosion rates must not be assumed constant over the contact life as demonstrated in Section 18.2.5. Erosion curves might cross over and thus even qualitative ranking might depend on the number of operations, and also on the device. For example, in a relay, as the contact erodes, the over-travel of the contact decreases, which lowers the opening velocity and thus can change the erosion rate of the device.

18.4.4.3 Increase of the Switching Frequency

Increasing the switching rate at rated load is the only possibility to execute life tests in acceptable time. The contact system must not be thermally over-stressed, however. Otherwise welding may be favored and erosion losses may be unrealistically increased; in extreme cases the erosion mechanism may be changed (e.g., owing to cracking and peeling off according to Chapter 16) depending on both the design of the switching device (heat dissipation from the contacts, mechanical impact) and the contact material; thus even qualitative ranking might be changed, depending on the device. Accelerated testing is limited to much lower switching frequencies than those dictated by thermal overstressing if the critical effect, for example, fretting (Chapters 5 and 7), frictional polymerization (Chapter 7), corrosion (Chapters 2 through 4), pyrolysis of organic vapors in switching arcs (Section 19.2) depends on diffusion and/or adsorption phenomena occurring between two succeeding operations.

18.4.4.4 Testing at Increased Electrical Load

This may thermally over-stress the contact arrangement more severely, like increased switching frequency, and correspondingly falsify the results. Additionally deviations from the assumed similarity law have to be expected.

Similarity laws, however, are problematic at any rate since they are merely empirical and cannot be explained in terms of physics. Various authors gave values for the exponents in Equation 18.2 in the wide ranges $1 \leq x \leq 3.7$ and $0.6 \leq y \leq 3$. Obviously such laws, if valid at all, are restricted to comparatively narrow ranges and dependent at least on the contact size, the thermal, mechanical (bounce), and magnetic parameters of the test device, the circuit parameters and—last but not least—on the contact material itself. Since some of these items are interrelated to current, such as contact bounce and material transfer, testing at increased loads is unlikely to be reliable for use in predicting results at lower currents. The proof of their applicability in each special case might require much more effort than the initially wanted tests. It has to be emphasized that the erosion rate increases discontinuously about an order of magnitude as soon as the arc root covers the available contact surface [81] (see Sections 10.3.2 and 10.3.3).

It has been proposed [82] to avoid the above-mentioned problems by not changing the specified test conditions between two measurements of the contact in order to calculate the erosion rate. An then as a next step to insert a series of tests with increased current to accelerate the erosion process by overload switching simply in order to produce new conditions for the next series of operations at rated load between two measurements yielding another erosion rate valid for a more eroded state of the contacts, etc. The author did not state, however, how the validity of such results were proved and how cracking was avoided.

18.4.4.5 Fixed-Gap Models

Numerous authors simplified accelerated erosion testing employing fixed contact gaps and igniting the arc by a spark breakdown rather than by moving one of the contacts. Such procedures do not recognize that erosion is caused by both thermal (arc) stress and mechanical stress at closure. It has been shown that fixed-gap erosion may be just a fraction of actual make erosion of bouncing contacts with the same gap length and arc duration [83].

Fixed-gap tests providing only the thermal but not the mechanical stress on the contacts may enable certain insights into the erosion mechanism when compared with realistic tests but they are certainly not suited for development or quality control of contact materials.

18.4.4.6 Moving Contact Models

For make and break switching different erosion results depending on switching duty (See, for example, Figure 10.15) (ohmic, capacitive, resistive, lamp, motor load, etc.). In the special case of low DC switching, even the direction of material transfer may be different (Section 10.3.5). Since contacts in application may be used for:

- Make only
- Break only
- Various complete make-and-break cycles

There are contact erosion test machines applicable to each of those duties. Moreover, for research, separate investigations of make erosion and break erosion and even separation of make erosion at the site of contact parting (arc origin in the contact center) from that at the site of arc commutation (at the contact's edge) might yield very important information about the complex mechanisms such as material transfer and erosion rate and its dependence on various parameters of both material and design, for example, bouncing, magnetic blast field, opening velocity, shape of contact edge, etc. (See, for example, Chapter 14).

1. *Make-only erosion*: Bouncing test machines already discussed don't duplicate erosion in commercial devices but can give repeatable erosion results. Therefore, tests executed with such machines cannot replace development or acceptance tests of devices but they are needed for development and quality control of contact materials and for research. Bounce simulation models are also useful for erosion research [84]. Test machines developed to investigate contact welding at make can also be used for make erosion tests as a rule. Some of them are suited for break erosion tests too [64,65,85–89].

2. *Break-only erosion*: Test machines for investigation of erosion at break only need not consider bounce problems, but have to provide a magnetic blast field when magnet blast interrupter conditions are to be simulated because the erosion strongly depends on the time the arc remains on the contacts [90–93]. It has to be emphasized that the commutation delay causing edge erosion depends not only on the magnetic induction of the blast field but also on the opening velocity and on the geometry (height, rounding) of the contact edge at the site of commutation [94–96]. For fundamental investigations it may be important not only to separate make erosion from break erosion but also to distinguish between erosion at the site of arc origin (last point of mechanical contact before parting) and erosion at the site of arc commutation (at the contact's edge). The latter can be suppressed by crow-barring the arc when it approaches the contact edge [96].

3. *Make-and-break erosion*: It has to be emphasized that make-and-break erosion is not necessarily equal to the sum of make erosion plus break erosion (cf. Sections 10.3.1 and 14.3) especially not under AC-3 conditions according to IEC-H 947-4 [97] and when DC circuits are switched. Therefore, test machines executing the total make-and-break cycle are required [64,65,73]. It is convenient if such test machines are also able to measure the weld strength when opening and the contact resistance after closure.

18.4.4.7 Evaluation and Presentation of Results

Cumulative erosion versus number of operations may yield a straight line according to a constant erosion rate per operation. Initial deviations from the straight line may be caused by the structure of the virgin surface different from the equilibrium state attained after a certain number of operations, and often may be neglected. Changes in the erosion rate may be also caused by changes in contact force from erosion and the resulting changes in switching characteristics such as bounce and opening velocity. The end of the contact life is often announced by a final increase of the erosion rate. Systematic changes of the erosion rate in the main part of the erosion curve may be caused by various effects that change as a result of material microstructure or operating parameter as a result of contact erosion. They may yield important hints concerning design, material, or over-stressing.

18.4.5 AC Arc Reignition Measurement

The instantaneous reignition voltage of a contact gap is independent of the contact distance and may be investigated either in opening or in fixed gaps. During the first tens of microseconds after current it is nearly constant unless a blast field is applied. The values of the reignition voltage have wide scatter, with the minimum value being the most important (Figure 9.55 in Section 9.7 and Figures 14.10 and 14.11 in Section 14.3.3).

18.4.6 Arc Motion Measurements

18.4.6.1 Measurement

18.4.6.1.1 Optical Devices

1. *High-speed cinematography* is the most traditional method used to measure arc motion. There are two techniques available.

   (a) *Frame technique* yields two-dimensional pictures of the arc. Short exposure time and high movie frame rate are realized by a rotating shutter and a rotating prism yielding excellent information on arc structure at limited frame rate.

(b) *Streak technique* yields continuous but only one dimensional information since the second dimension is needed for the time.

Both techniques need free optical access; development, evaluation and storage of the exposed films is time consuming and expensive.

2. *Multiple exposed still photographs* require less effort than the above techniques and are rather convenient for measuring semi-continuous, not too fast arc motion in one direction.

A rotating screen with radial slits (rotor-core disk) positioned between a still camera with open shutter and the moving arc exposes the film to the arc only when a slit is passing the optical path between arc and camera. The resulting photograph (Figure 18.11) shows two-dimensional pictures of the arc in well-defined time intervals which can easily be evaluated [98].

18.4.6.2 Electronic Optical

Electronic optical devices do not yield immediate optical pictures like photographs but they are extremely fast (1000 images per millisecond) and convenient; the information gained can easily be stored and evaluated in various ways. Therefore, they are especially suited for series measurements.

1. *Analog devices.* A very fast, easily applicable and cheap type of arc position indicator yields reliable information of the position of the light emission center of an arc with the aid of a simple pinhole diaphragm. The photographic film is replaced by a large (5 cm × 5 cm) biaxial Au–Si–Schottky-semiconductor. The arriving light produces carriers at the site of incidence and finally an electrical signal exactly proportional to the x and y position of the center of (light emission's which can be fed directly into an oscilloscope or a computer [99]).

2. *Digital devices.* Both motion and structure of an arc can be recorded with the aid of a line of light guides arranged along the arc path, each light guide connected to a photodiode. Opaque walls between the arc path and the observer are not prohibitive because a very narrow slit (0.1 mm) in front of the receiving ends of the light guides yields sufficient access. Special resolution (distance between the axes of two neighboring light guides) down to 0.4 mm can be achieved easily [99,100].

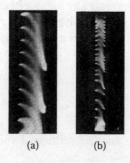

(a)   (b)

FIGURE 18.11
Arc motion (upwards) shown by multiple exposure of a still film (25 exposures per sec, cathode left). (From Eidinger A. and Rieder W., *Archiv für Elektrotechnik*, 43, 94–114, 1957.)

If two-dimensional information is needed, a number of light sensors (light guides) can be arranged, for example, in three lines along two arc runners and the middle line between them [101,102] or forming a two-dimensional matrix scanning the arc chamber of a circuit-breaker [103,104].

18.4.6.2.1 Magnetic Devices

Although an arrangement of light guides along the arc path requires just a 0.1 mm slit in the lateral wall of an arc chamber, there is also a possibility of avoiding any slit in an opaque wall when the moving arc is detected by magnetic probes arranged along the wall outside the arc chamber [105].

18.4.6.3 Model Switch Arc Motion Control

Self-Induced Fields: In most commercial magnet blast interrupters the switching arc is moved by the "self-field" induced by the current to be interrupted as it flows through the arc electrodes (contacts or runners) and their connecting leads. In this case the magnetic field strength is proportional to the arc current, but also depends on the geometry of the electrodes (including their distance!) and their connections. Therefore, the magnetic induction of such arrangements cannot be kept constant or arbitrarily varied independently of the arc current, the arc length, and the electrode geometry.

External Fields: Consequently in numerous experiments the effective magnetic field has been externally induced by a Helmholtz coil, independent of the arc current and of the electrode geometry, while any self-field has been avoided by symmetrical feeding of the current into the electrodes. Also there has been some use of permanent magnets in the development of dc switches for circuit voltages >100 V.

Figure 18.12a shows how identical electrodes can be employed for either self-blast or external field experiments. Of course, both fields can be superimposed. Thus, it has been shown [106,107] that there is always a certain equivalent value of the homogeneous external field which causes the same arc behavior as the self-field induced by any given electrode geometry. This equivalent external field value corresponds to the mean value of the self-field inside the volume of the arc.

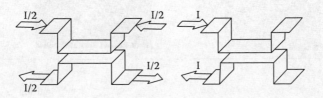

FIGURE 18.12
Electrode configuration used for experiments (a) without and (b) with magnetic self-field. (From Michal R., *IEEE Trans CHMT*, 4, 109–114, 1981.)

18.4.6.4 Evaluation and Presentation of Results

The electrical signals obtained from the light sensors may be evaluated with the aid of sophisticated computer programs. Results of this approach have been published by a number of researchers investigating arc motion and structure [99,100,108]. In Figure 18.13a, the marks along the abscissa correspond to the position of 30 light sensors along the arc path, and the light intensity received by each sensor is plotted along the ordinate. The curve obtained shows the momentary structure of the arc column in the direction of arc motion. The next curve of this type obtained a certain time interval (e.g., 1 μs) and later is shifted somewhat in the direction of the ordinate etc. The total diagram shows how the arc column proceeds (from left to right) and how it changes its shape. Figure 18.13b shows the corresponding smear photograph. Figure 18.14a shows a similar but three-dimensional view: time is plotted in the y-direction and light intensity in the z-direction (light–mountain chain). In Figure 18.14b is a contour plot of light intensity in terms of time and distance. It corresponds to the three-dimensional light intensity, time, and distance plot. It can be immediately compared with a smear photograph but gives more detailed information about light-intensity distribution.

18.4.7 Arc-Wall Interaction Measurements

Severe interaction of the switching arc with the insulating wall next to the contacts may not only damage the contacts but also the wall surfaces owing to either metallization of any material or carbonization of organic materials. The break-down effect can be characterized by measurement of either the surface resistance (leakage current after each interruption

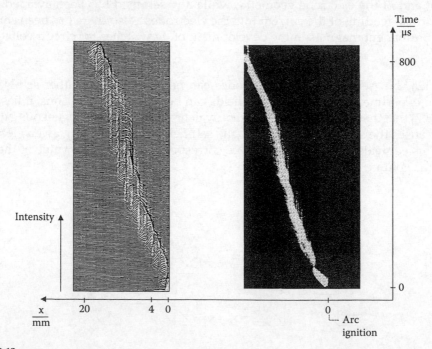

FIGURE 18.13
(a) Light-intensity signals indicated by 30 light sensors arranged along the abscissa. The individual lines (from down to top) have been plotted in 1-μs intervals showing the behavior of an arc moving from right to left, (b) Corresponding smear photograph.

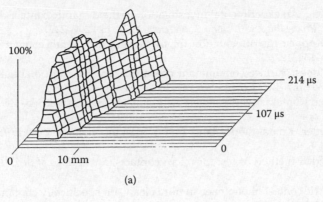

(a)

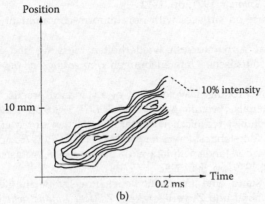

(b)

FIGURE 18.14
(a) Three-dimensional plot of light-intensity curves of an arc moving from left to right (time in y-direction), (b) Lines of constant light intensity in the time–position plane. (Contour lines of the light-mountain chain shown in [a]). (From Michal R., *IEEE Trans CHMT*, 4, 109–114, 1981.)

[109] or the surface flashover voltage between two probe wires inserted into the wall next to the contacts and energized a certain time (e.g., 1 ms) after interruption or arc commutation from the contacts [108]. What is recorded is dependent on what is important for your commercial device.

References

1. G Witter, Z Chen. A comparison for the effects of opening speed, contact gap, and material type on electrical erosion for relays interrupting inductive automotive loads. *ICECP*: 17–21, 2006.
2. Z Chen, G Witter. A comparison of contact erosion for opening velocity variations for 13 volt circuits. *IEEE Holm Conference* 2006, pp. 15–20.
3. G Witter, I Polevoy. Contact erosion and material transfer for contacts in automotive relays. *Proceedings of the ICECP and IEEE Holm Conference on Electrical Contacts*, Chicago, IL, 1996, pp. 223–228.
4. Z Chen, G Witter. A study of contact endurance switching life as a function of contact bond quality, contact electrical load and residual stresses for silver tin indium oxide composite rivet contacts. *IEEE Holm Conference* 2012, pp. 172–177.
5. F Borst, C. Neufeld, W. Rieder, Moderne Versuchstechnik. *VDE-Fachbericht* 42: 39–45, 1991.

6. L Jing, Z Guansheng. An experimental investigation of the dynamic contact characteristics on the relay contacts. *Proceedings 38th Holm Conference*, 1992, pp. 209–212.
7. L Jedynack, CH Kopper. Instrumentation for measuring dry-circuit contact resistance. *IEEE Trans PHP-11*: 130–134, 1975.
8. RE Cuthrell, LK Jones. Surface contaminant characterization using potential-current curves. *IEEE Trans CHMT-1*: 167–171, 1978.
9. W Johler, W Rieder. Kontaktzuverlässigkeit bei Relais im Kleinlastbetrieb. *VDE-Fachbericht* 38: 131–140, 1987.
10. K Millian, W Rieder. Kontaktwiderstand und Kontaktoberfläche. *Zeitschrift für angewandte Physik* 8: 28–31, 1956.
11. KE Pitney. The ASTM method of test for micro contacts. *Proceedings of the 11th Holm Seminar*, 1965, pp. 293–311.
12. DV Keller, Jr. Electric contact phenomena in ultra clean and specifically contaminated systems. *Proceedings of the 17th Holm Seminar*, 1971, pp. 1–12.
13. GR Crane. Contact resistance on surfaces with non-uniform contaminant films. *IEEE Trans CHMT-4*: 5–9, 1986.
14. W Merl, M Mittmann. Eine Apparatur zur wiederholten Messung und Registrierung des Kontaktwiderstandes für statistische Untersuchungen. *Proceedings of the 5th ICECP*, 1970, pp. 320–323.
16. M Antler. Contact resistance probing: Development of a standard practice by the American Society for Testing and Materials. *Proceedings of the 10th ICECP*, 1980, pp. 13–21.
17. GL Horn. Nachweis der Reinheit technischer Kontaktoberflächen durch eine automatisierte Meßmethode des Kontaktwiderstandes. *Proceedings of the 8th ICECP*, 1976, pp. 499–503.
18. M Füger, J Horn. Kontaktwiderstandsmeßplatz zum punktweisen Abtasten ebener Proben. *Radio fernsehen Elektronik* 28, 1979, pp. 596–599.
19. M Huck. Der Kontaktwiderstand, eine statistische Beurteilungsgröße zur Gewährleistung der Kontaktzuverlässigkeit, Qualität und Zuverlässigkeit. *Zeitschr f industrielle Qualitätssicherung* Nr. 10: 1980.
20. M Antler. New developments in the surface science of electric contacts. *Plating* 53: 1431–1439, 1966.
21. W Chow, E Stepke. An evaluation of the tarnish resistance of electroplatings and contact materials. *Proceedings of the 15th Holm Seminar*, 1969, pp. 99–110.
22. G Witter, R Leiper. A study of contamination levels measurement technique, testing methods, and switching results for silicon compounds on silver arcing contacts. *Proceedings of the IEEE-Holm Conference on Electrical Contacts*, 1992, pp. 173–180.
23. JL Queffelec, N BenJemaa, D Travers. Activation de contacts de relais électro-mécaniques par leurs propres produits organiques de dégazage en atmosphére confinée, en faction de la témperature. *Rev Gén El* 89: 577–586, 1980.
24. T Ide, T Sakurai, T Sasamoto. Contact reliability evaluation of plastic sealed relay. *Proceedings of the 31st Ann Nat Relay Conf (NARM)*, 1983, pp. 7-1–7-7.
25. K Yoshida, Y Aoyama, M Isagowa, K Shirakura. Evaluation of contact failure by vaporized organic gases. *Proceedings of the 31st Holm Conference*, 1985, pp. 191–199.
26. B Göttert. Eine Methode zur Auswahl von Kunststoffen bezüglich der Kontak-freundlichkeit in Miniaturrelais. *VDE-Fachbericht* 38: 79–84, 1987.
27. B Göttert, U Rauterberg. Evaluation of the influence of organic vapors on relay contacts. *Proceedings of the 36th Ann Nat Relay Conf (NARM)*, 1988, pp. 15-1–15-9.
28. CN Neufeld, WF Rieder. Carbon contamination of contacts due to organic vapors. *IEEE Trans CPMT-A18*: 399–404, 1995.
29. B Göttert. Kunststoffausgasungen von Klebefolien beeinflussen Kontaktmaterialien. *ETZ* 112: 1216–1221, 1991.
30. E Tittes. Über die Anwendung statistischer Methoden auf die Auswirkung von Versuchen mit elektrischen Kontakten. *Proceedings of the 7th ICECP*, 1974, pp. 327–332.

31. D Lange. Statistische Auswertung von Kontaktwiderstandsmessungen. *Elektrie* 34: 423–426, 1980.
32. D Lange. Der Zusammenhang zwischen einer Fremdschicht auf Kontaktoberflächen und der logarithmischen Normalverteilung. *Elektrie* 37: 371–374, 1983.
33. EA Santer. Mathematical statistics in electrical contact engineering. *Proceedings of the 11th ICECP,* 1982, pp. 68–70.
34. W Rieder, V Weichsler. Make erosion on Ag/SnO$_2$ and Ag/CdO contacts in commercial contactors. *IEEE Trans CHMT-14:* 298–303, 1991.
35. P Braumann, H Kolb, KH Schröder, W Weise. Optimization of Ag/SnO$_2$ contact material for use in contactors. *Proceedings of the 16th ICECP,* 1992, pp. 163–173.
36. R Jin. Relay contact bounce measurement. Electronics (USA): 28(8): 137–139, 1955.
37. NC Shaw. Investigations of contact bounce and contact welding. *Proceedings of the 2nd ICECP,* 1964, pp. 286–299.
38. F Rating, A Walter. Digitales Meßgerät zum Erfassen der Prelldauer. *ETZ-B* 17: 557–559, 1965.
39. G Kovács. Laboratory investigations and applicational testing of DIL relays for telecommunication. *Proceedings of the 16th ICECP,* 1992, pp. 315–321.
40. H Franken. Pie Messung von Kontaktprellungen bei Schaltgeräten. *ETZ-B:* 53–54, 1960.
41. KK Kennedy. Relay vibration studies. *Bell Lab Rec* 31: 141–146, 1953.
42. A Erk, H Finke. Über die mechanischen Vorgänge während des Prellens einschaltender Kontakte. *ETZ-A* 86: 129–133, 1965.
43. A Yonezawa, K Mano. Some investigations of measuring method of contact gap and contact motion. *Proceedings of 21st Holm Seminar,* 1975, pp. 93–98.
44. BZ Sandler, A Slonim, M Slonim, A Tslaf, M Weinberg. A method of measuring small displacements. *Israel J Technology* 16: 234–237, 1978.
45. AA Slonim. Analytical and experimental analysis of nonlinear bouncing contacts. *ICECP* 12: 531–537, 1984.
46. MR Swinehart. Instrumentation for analysis of contact wear. *El Eng* 70: 414, 1951.
47. G Cießow, F Koppelmann. Untersuchung des Prellens eines periodisch schaltenden Federkontaktes mit Hilfe eines optischen Schwingblendenverfahrens. ETZ-A 82: 111–114, 1961.
48. JW McBride, SM Sharkh. Electrical contact phenomena during impact. *IEEE Trans CHMT-15:* 184–192, 1992.
49. P Barkan. A study of the contact bounce phenomenon. *IEEE Trans PAS-86:* 231–240, 1967.
50. S Dzierzbicki, E Walczuk. Einfluß der Veränderlichkeit der Stoßzahl auf die Berechnungsergebnisse der Kontaktabhebungen. 3. Tagung "Kontakte der Elektrotechnik, 1967, pp. 161–166.
51. W Schaffer. Über das Schweißverhalten einschaltender Starkstromkontakte. *Int Wiss Koll Ilmenau* 13: 113–120, 1968.
52. A Wollenek. Die Schweißneigung prellender Starkstromkontaktstücke. *VDI-Zeitschr* 102: 1053–1088, 1960.
53. A Erk. Über die thermische Beanspruchung von Starkstromkontaktstücken bei Kurzzeitbelastung mit hohen Strömen. *ETZ-A* 85: 226–231, 1964.
54. W Rieder, J Uranek. Contact Welding. *Proceedings of the 2nd ICECP,* 1964, pp. 193–195.
55. T Ito, H Iwase, K Mano. Experimental study of contact arc discharge and contact welding phenomena. *Proceedings of the 27th Holm Conference,* 1981, pp. 35–44.
56. W Balme, P Braumann, K-H Schröder. Welding tendency, contact resistance and erosion of contact materials in motor vehicle relays. *Proceedings of the 15th ICECP,* 1990, pp. 79–84.
57. W Schaffer. Untersuchung des dynamischen Trennens verschweißter Starkstromkontakte. *Proceedings of the 5th ICECP* Pt 2, 1970, pp. 52–56.
58. W Schaffer. Untersuchung der Trennenergie verschweißter Starkstromkontakte. *Elektrie* 25: 305–306, 1971.
59. E Geldner, W Haufe, W Reichel, H Schreiner. Schweißkraft verschiedener Kontaktwerkstoffe beim dynamischen und statischen Öffnen der Kontaktstücke. *Bull SEV* 65: 236–240, 1974.
60. J Peters. Ein Verfahren zur Prüfung von Kontaktwerkstoffen unter den Bedingungen des dynamischen Verschweißens. *Elektrie* 29: 550–553, 1975.

61. E Geldner, W Haufe, W Reichel, H Schreiner. Ursachen der Schweißbrückenbildung und Einflüsse auf die Schweißkraft elektrischer Kontaktstücke in der Energietechnik. *ETZ-A* 93: 305–306, 1972.

62. A Carballeira, J Galand. Accelerated testing for determining erosion and welding behavior of silver based contact materials. *Proceedings of the 9th ICECP*, 1978, pp. 329–334.

63. K Mano, H Iwase. Automated measuring methods for discharge durations and energy in electrical contacts. *IEEE Trans CHMT*-2: 65–70, 1979.

64. H Schreiner, W Haufe. Messung der Schweißkraft, des Abbrandes und des Kontaktwiderstandes von Kontaktwerkstoffen für die Energietechnik. *Zeitschr f Werkstoffkunde/J of Materials Technol* 7: 381–389, 1976.

65. H Schreiner, W Haufe. The properties of p/m electrical contact materials. *Int J Powder Metallurgy & Powder Technology* 12: 219–228, 1976.

66. E Geldner, W Reichel, E Schreiner. Prüfschalter zur Messung der Schweißkraft von Kontaktwerkstoffen für Starkstromtechnik. *ETZ-A* 92: 637–642, 1971.

67. H Schreiner. Schweißkraft elektrischer Kontaktwerkstoffe und deren Prüfmethode. *Proceedings of the 4th ICECP*, 1968, pp. 187–190.

68. W Haufe, W Reichel, H Schreiner, R Tusche. Einfluß der Schaltzahl und Polarität des Prüfstromes auf die Statistik der Schweißkraftwerte von Reinsilber bei synchronem und unsynchronem Schließen der Kontaktstücke. *Bull SEV* 63: 461–467, 1972.

69. G Ludwar, W Rieder. Contact welding in vacuum at make. *Proceedings of the 13th ICECP*, 1986, pp. 156–160.

70. J Muniesa. Silver-tin oxide materials used in low voltage switching devices. *Proceedings of the 15th ICECP*, 1990, pp. 139–142.

71. JR Pharney, FS Gowaty. A technique for measuring net charge through transient arcs which occur across a contact breaking an inductive load. *Proceedings of the 20th Holm Seminar*, 1974, pp. 75–80.

72. MAK Pramanik, T. Takagi, K Mano. A new method for the measurements and integration of arc durations in electrical contacts. *IEEE Trans IM*-24: 188–193, 1975.

73. HJ Desmet. Testing and evaluating contact materials for erosion characteristics. *Proceedings of the 22nd Holm Seminar*, 1976, pp. 149–154.

74. E Needham. Computer controlled testing and evaluation of contact materials in 3-phase ac devices. *Proceedings of the 28th Holm Conference*, 1982, pp. 87–94.

75. K Suhara. Break arcs in inductive circuits and the minimum arc current. *Proceedings of the 15th ICECP*, 1990, pp. 94–101.

76. K Mano. New proposal for test methods of electrical contacts and their characteristics indications. *Proceedings of the 16th ICECP*, 1992, pp. 29–34.

77. H Fang, J Jiang, R Mou. The study of the relationship between arc duration and arc energy of relay and microswitch contacts. *Proceedings of the 38th Holm Conference*, 1992, pp. 167–172.

78. JW McBride, KJ Cross, SM Abu Sharkh. The evaluation of arc erosion on electrical contacts using three-dimensional surface profiles. *IEEE Trans CPMT*-A19: 87–97, 1996.

79. CH Leung, A Lee. Electric contact materials and their erosion in automotive dc relays. *Proceedings of the 37th Holm Conference*, 1991, pp. 114–121.

80. CH Leung, A Lee. Contact erosion in automotive dc relays. *Proceedings of the 15th ICECP*, 1990, pp. 85–93.

81. HW and C Turner. Discontinuous contact erosion. *Proceedings of the 3rd ICECP*, 1966, pp. 311–320.

82. D Dimitrov, P Parvanov, S Stefanov, M Tronkova. Accelerated test of the electrical endurance of contact members of commutation low-voltage apparatus by means of increasing the value of the test current. *Proceedings of the 8th ICECP*, 1976, pp. 627–631.

83. W Rieder, V Weichsler. Make erosion mechanism of Ag/CdO and Ag/SnO$_2$ contact. *IEEE Trans CHMT*-15: 332–338, 1992.

84. E Hetzmannseder, W Rieder. The influence of the bounce parameters on the make erosion of silver/metal-oxide contact materials. *IEEE Trans CPMT*-A17: 8–16, 1994.

85. H Schreiner. Prüfstrategie und Simultanprüfung der funktionswichtigen Eigenschaften von Kontaktwerkstoffen der Energietechnik mit einem Prüfschalter. *ETZ Archiv* 2: 307–310, 1980.
86. AM Suggs. An electrical contact testing machine. ASTM Bull No. 119: 25–30, 1942.
87. W Friedemann, P Leis. Modellprüfanlage zur Ermittlung des Einschaltabbrandes und des dynamischen Schweißverhaltens. *Proceedings of the 7th Kontakttagung der DDR*, 1983, pp. 158–162.
88. JP Guerlet, H Ladenise, C Lambert. A simple testing machine for the evaluation of the main electrical properties of contact materials. *ICECP* 12: 525–530, 1984.
89. A Carballeira, J Galand. A new equipment for evaluating welding and erosion tendencies of electrical contacts. *Proceedings of the 8th ICECP*, 1976, pp. 633–638.
90. KH Schröder, ED Schulz. Der Einfluß des Trennaugenblickes auf den Lichtobogenabbrand öffnender Kontaktstücke beim Wechselstromschalter. *Metall* 28 (5): 1974.
91. H Hauer. Über den Einfluß des magnetischen Blasfeldes auf den Ausschaltabbrand elektrischer Kontakte. *e & i* 107: 63–71, 1990.
92. Walczuk. Experimental study of Ag–W–Re composite materials under high current conditions. *IEEE Trans CHMP*-10: 283–289, 1987.
93. M Lindmayer, E-D Schulz. Prüfverfahren für ausschaltende Kontaktstücke. *ETZ-A* 98: 142–146, 1977.
94. W Widmann. Arc commutation across a step or a gap in one of two parallel copper electrodes. *Proceedings of the 12th ICECP*, 1984, pp. 329–339.
95. H Manhart, W Rieder, C Veit. Arc mobility on new and eroded Ag/CdO and Ag/SnO$_2$ contacts. *IEEE Trans CHMT*-12: 48–57, 1989.
96. H Manhart, W Rieder. Erosion behavior and "erodibility" of Ag/CdO and Ag.SnO$_2$ contacts under AC3 and AC4 test conditions. *IEEE Trans CHMT*-13: 56–64, 1990.
97. E Hetzmsnndrfrt, W Rieder. Make and break erosion of Ag/MeO contact materials. *IEEE Transactions CPMT*-A19: 397–403, 1996.
98. A Eidinger, W. Rieder. Das Verhalten des Lichtbogens im transversalen Magnetfeld. *Archiv für Elektrotechnik* 43: 94–114, 1957.
99. R. Michal, J Wassermann. Optoelectronical position indicators applied to arc-motion research. *IEEE Trans CHMT*-6: 92–99, 1983.
100. J Wassermann, W Ziegler. Quantitative recording of arc motion and structure through opaque walls. *Proceedings of the 14th ICEC*, 1988, pp. 69–74.
101. A Lee, YK Chien, PP Koren, PG Slade. High current arc movement in a narrow insulating channel. *IEEE Trans CHMT*-5: 51–55, 1982.
102. E Gauster, W Rieder. Arc–wall interaction phenomena immediately after contact separation in magnet-blast interrupters. *Proceedings of the 41st Holm Conference*, 1995, pp. 355–363.
103. JJ Shea, D Boles, YK Chien, R Zeigler. Computer animated digital arc diagnostic system. *IEEE Trans CHMT*-A17: 47–52, 1994.
104. PH Wearer, JM McBride. Magnetic and gas dynamic effects on arc motion in miniature circuit breakers. *IEEE Trans CHMT*-A17: 39–45, 1994.
105. J Wassermann, T Strof, W Rieder. Measurements of arc motion on electrical contacts through unperforated opaque walls. *Proceedings of the 14th ICECP*, 1988, pp. 69–74.
106. R Michal. Theoretical and experimental determination of the magnetic blast field due to the current flow in the electrodes. *IEEE Trans CHMT*-4: 109–114, 1981.
107. H Manhart, R Michal. Arc motion on AgMeO with special regard to the utilization categories of contactors. *Metall* 41(1): 49–53, 1987.
108. W Rieder, C Veit, E Gauster. Interaction of magnetically blown break arcs with insulating walls in the contact region of interrupters. *IEEE Trans CHMT*-15: 1123–1137, 1992.
109. HW Turner, C Turner, RC Melbourne. Testing and evaluation of contacts and insulating materials in switching devices with computerized control. *Proceedings of the 10th ICECP*, 1980, pp. 1069–1078.

19

Arc Interactions with Contaminants

Gerald J. Witter and Werner Rieder

What is this life if, full of care,
We have no time to stand and stare.

Leisure, **W. H. Davies**

CONTENTS

19.1 Introduction 1005
19.2 Organic Contamination and Activation 1006
    19.2.1 The Phenomena 1006
    19.2.2 Sources of Organic Vapors 1007
    19.2.3 Processes of Contact Activation 1007
    19.2.4 Activation Effects 1011
    19.2.5 Activation and Contact Resistance Problems 1012
    19.2.6 Methods for Detecting Carbon Contamination 1014
19.3 Mineral Particulate Contamination of Arcing Contacts 1015
19.4 Silicone Contamination of Arcing Contacts 1018
    19.4.1 Contamination from Silicone Vapors 1019
    19.4.2 Contamination from Silicone Migration 1023
    19.4.3 Summary of Silicone Contamination Mechanisms 1029
19.5 Lubricants with Refractory Fillers 1030
19.6 Oxidation of Contact Materials 1031
19.7 Resistance Effects from Long Arcs 1032
Acknowledgments 1034
References 1034

19.1 Introduction

This chapter will deal with contamination of arcing contacts. The emphasis will be on the effects of contact contamination as a result of arc interactions with contaminants and the contact material in contrast to contaminant and contact interactions from corrosion mechanisms discussed in Chapters 2, 3, 4, and 8. This is a broad subject and with many potential combinations of contaminants, contact materials, and electrical loads, so only the more common problems that have been studied and published will be discussed. The chapter begins by discussing organic contamination and contact activation, and includes references to a large long-term study mainly done by Bell Labs from 1950s through 1970s for telecommunication relays. This study uncovered many contact fundamentals that were useful in other contact applications. Another section discusses silicone contamination of contacts that continues to be a problem, although warnings about silicone compounds have existed for many years. Some of the basics for silicone vapor deposition and migration are discussed including relationships to electrical loads. It should be kept in mind when reading about specific examples of arc–contamination interactions given in this chapter that the results may be quite different when other electrical and mechanical load conditions from those cited in the specific example are used.

19.2 Organic Contamination and Activation

19.2.1 The Phenomena

Often the ambient air of switching contacts is contaminated by organic vapors either from outside a switch or produced by components inside the switch. Such organic contaminants may be adsorbed on the contact surfaces and build polymers, especially under the influence of friction between closed or dry switching contacts, as discussed in Chapter 7. In this section, the influences of arcing on the organic contaminants already introduced in Chapter 10 are discussed in more detail.

For applications where the current level is low, in the range of 1 A or less, adsorbed organic vapors may greatly reduce the switching life of noble and silver-based contact materials. For example, transient arc discharges appearing at loads below the minimum values of current and/or voltages required for steady arcs (Sections 9.5.4 and 10.2.4) may be just strong enough to crack such adsorbed organic molecules and convert them by pyrolysis into carbonaceous particles that build up on the surface with continued switching operations. Such produced deposits of carbon and heavily cracked molecules may enhance arcing under the same electrical load conditions for which transient discharges are hardly detectable on the initially clean contacts. This process, which extends the duration of the arc and thus increases the arc erosion, is called "contact activation" and has been studied by numerous researchers [1–6]. Germer, one of the key pioneers for researching this subject, first defined activation as a condition in the contact surfaces that increases the arcing when the contacts are made or separated as compared to that which occurs when the contacts are clean [7].

Most of the work done for studying activation was in connection with research on signal relays for telecommunications. In these applications the current level was low, normally less than 1 A, and the switching life requirements for these relays was very long, many millions of operations. At these low currents, contact activation could lower the minimum current required for an arc to occur and thus could extend the arc duration by an order of magnitude and this resulted in a similar decrease in relay life. Carbonaceous deposits produced by pyrolysis of organic compounds could also increase the low resistance between clean metal contacts (milliohm range) to values typical of carbon contacts (ohm range). These problems are not limited to signal relays but other switching devices operating in the low-current range such as load relays, microswitches, and thermostats. For higher-current devices the relative effects of activation on contact erosion are less than for lower-current applications and also the build-up of carbon is less pronounced. Later in this chapter it will be shown that activation can have a different effect on higher-current switching.

19.2.2 Sources of Organic Vapors

The sources for organic vapors are ubiquitous. Contacts in open switches may be exposed to vapors transported by the surrounding air from outside the building, car exhaust, industrial gases, refinery gases, or from indoor sources, electronic devices, electrical apparatus, etc. [8–12]. For example, more than 300 organic vapors have been identified in Skylab 4 [13] while 40 kinds of vapors have been detected emanating from two control devices at 65°C [14]. More problematic are those vapors that cannot be eliminated by sealing the housing of a device because the sources are from components of the device itself [5,15–19] that emit organic vapors particularly during internal or external heating of the device. An example of this for relays is the insulation on the relay coil which often contains the plasticizer diethyl phthalate, DEP. The internal heating of the coil during use can accelerate vaporization of DEP from the insulation. Levels of DEP as low as 3 p.p.m. have been shown to produce significant activation [5]. Researcher have also identified other organic vapor sources such as insulating materials with additives and greases and oils [12,20–23]. To prevent activation by internal components, work has also been done on baking components. The effectiveness of baking for reducing emanation of gas has been found to depend on the material [12,21,23,24].

19.2.3 Processes of Contact Activation

The activation process for electrical contacts involves four steps as follows:

1. Adsorption of organic vapors onto the contact surface
2. Pyrolysis of the organic films because of arcing, which produces deposits of carbonaceous particulate on the contact surface
3. Extended arc duration as a result of the carbonized particulate
4. Increased erosion owing to the increased arc duration [6]

The activation process involves many variables including, type of organic, type of metal, organic vapor pressure, type of carrier atmosphere, contact surface cleanliness, surface roughness, circuit load and energy type, rate of switching, and more that form many permutations that may or may not result in activation [2,7,25]. From the above it can be seen that the activation process does not just involve exposure to organic materials but usually requires more than 100 and sometimes thousands of switching operations before extended arc duration is detected [7]. Augis reported that insulating adsorbed monomolecular layers of water or organic, alone, were able to extend arcing [26]. This work involved very short, nanosecond range, anodic transient arcing conditions and such insulation layer activation effects are not the typical conditions detected for operating devices [5,25].

Initial research on activation led researchers to believe that activation is not produced by all organic gases. In later work it was found that, depending on the organic gas and metal combination, critical concentration levels required to produce activation can vary by several orders of magnitude. For example Koidl et al. show a concentration of 100 volume parts per million (v.p.m.) of toluene can activate palladium, while for the combination n-hexane and silver a concentration of 10,000 v.p.m. is needed [27]. It is now thought that activation may be possible for most organic gas and contact alloy combinations, but the concentrations and conditions for activation may be quite different [28]. Germer found, and Koidl later confirmed, that vapors of unsaturated ring hydrocarbons seem to favor activation over saturated ring compounds or straight chain compounds [2,7,27], but exceptions exist. Work by Takenaka et al., done 40 years later, reached a similar conclusion [29]. The specific combinations of metal type and organic compound type are critically important for determination of the potential for activation [30]. Table 19.1 shows a list of compounds tested by Germer for silver contacts and the results for either activation or no activation [2]. The condition of the surface is important to the activation process. When compounds are firmly adsorbed on the contact surface, they may decompose rather than vaporize without decomposition [2]. For example, benzene is not adsorbed firmly enough on clean silver to activate but will if the silver is contaminated by an oil film such as a fingerprint. Styrene, on the contrary, is very effective for activation of clean silver surfaces [2]. For certain metals, such as silver, the adsorption of some organic vapors is enhanced by already adhering carbon atoms from greases or CO_2 decomposition, while others such as platinum can absorb compounds such as benzene without prior contamination [2,31]. For gold, the critical concentration of organic vapors was found to be intermediate between silver and palladium, one order of magnitude lower than for silver and an order higher than for palladium [27].

TABLE 19.1

Substances and Compounds Which Activated or Did Not Activate Silver Contacts

Activate	Did Not Activate
Pump oil	Ethyl alcohol
Turpentine	Amyl alcohol
Penetrating oil	Ether
Kerosene	Carbon tetrachloride
Butyl alcohol	Trichlorethylene
Benzene	Hydrogen chloride
Toluene	Methyl methacrylate
Xylene	n-Butyl acetate
Pseudocumene	Amyl acetate
Styrene	Ethyl acetate
Ethyl styrene	Acetone
Phenyl acetylene	Oleic acid
Cyclohexanol	Propionic acid
Cyclohexanone	o-Dibromobenzene
Cyclopentanone	
d-Limonene	
l-Menthone	
Pinene	
Terpineol	
Ethyl silicate	
l-Octene	
l-Dodecene	
Octylene	
Crotonic acid	

Source: Data from LH Germer, *J Appl Phys* 22(7): 955–964, 1951[2].

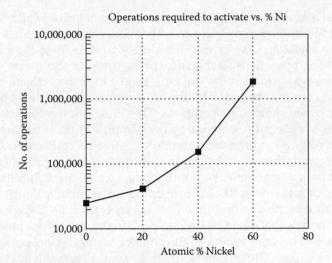

FIGURE 19.1
Resistance to activation for palladium as a function of nickel content.

In general, base metals are very difficult to activate by organic vapors. Palladium when alloyed with nickel becomes more difficult to activate as can be seen in Figure 19.1. Germer was not able to activate pure nickel at all [7]. It can be seen that the activation process is complex and also variable depending on the metal organic combination. To better understand the process a more generalized and simplified look at the process will be taken. The extent of activation, increase in arc duration, is directly related to the amount of carbon particulate that has been generated on the contact surface. During a single arcing operation the carbon and or adsorbed organic material in the area covered by the arc is burned away and new carbon is deposited in an annular ring surrounding the arc area [7]. In air, higher-energy arcs are less efficient at producing activation than low-energy arcs. On the other hand, extremely low arc energies are ineffective in producing carbon. Arc energies on the order of 100 ergs (10 μJ) seem to produce the largest difference between the arc area and annular ring of arc deposition area, and thus more rapidly activate contact surfaces [7]. The amount of carbon generated per pyrolysis operation is also dependent on the amount of organic material deposited per area of contact surface, density of adsorbed organic, carbon atoms per molecule, and surface roughness which affects the true surface area [30–32]. The amount of organic material that is accumulated on the surface before an arcing operation is dependent on the concentration of the organic vapor, exposure time, and sticking coefficient of the contact metal and vapor combination. During a series of switching operations, a dynamic equilibrium may be established in carbon balance among the various portions of the switching cycle as follows [7,31–33]:

- *Contacts in open position with no arcing.* During this period, there is a gain in carbon from organic material being deposited onto the contact surface.
- *Arcing during opening and closure.* During arcing there is a net loss of carbon as a result of several parallel processes taking place, dissociation of the organic material, sublimation and or oxidation of carbon particles, and formation and deposition of carbon particles.
- *Contacts in closed position.* Assumes no change and no reactions.

For a specific metal and organic compound combination, at a fixed vapor concentration, it would appear that the amount of carbon particulate concentration that is achieved on the contact surface for an established switching schedule would relate to the amount of open time between switching operations, if the adsorption time for saturation of the organic on the contact surface was less than the opening time. Since the activation arc duration is proportional to the carbon concentration, the arc duration should then be related to the contact open time between switching operations. This illustration is for helping in understanding the dynamics of the process and is an oversimplification of the processes.

The effect of operating rate on the arc duration of activated contacts has been studied by several groups of researchers and several models have been proposed [25,33,34]. Gray and Uhrig found for palladium contacts operating in a 5 p.p.m. diethylphthalate (DEP), air environment, the arc duration was 100 μs for opening times greater than 1 s and proportional to the square root of the opening time for times less than 1 s [34]. This work shows and discusses three models that could be used to explain the critical opening time, including classical diffusion of organic vapors to the contact surface as proposed by Schubert [33]. A broader study by Gray, Uhrig, and Hohnstreiter was done with 30 combinations of six different metals and seven different activates [35]. They developed a model for activation level as a function of exposure which correlated well with their actual experimental results. Exposure, η, was defined as the product of the contact open time and the partial pressure, and stated as pascal seconds. A typical activation curve generated for a single combination is shown in Figure 19.2. The model shows three regimes of contact activation varying with exposure level. For low levels of exposures there is the ion-drift-limited region in which arc duration is independent of exposure. For the intermediate exposure range the activation level is diffusion-limited and proportional to the square root of the opening time. At high exposure levels, the activation level is adsorption-site-limited and again independent of exposure. A unique activation curve occurs for each metal-organic pair. At low exposures, organic ions control the activation process and at higher levels of exposure diffusion of neutral organic molecules dominates the activation process. For the

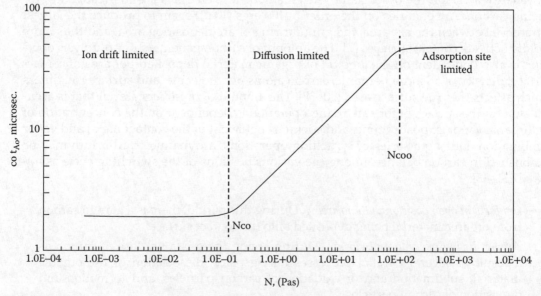

FIGURE 19.2
Activation characteristic curve, critical opening time, t_{Ao}, versus exposure, η, in pascal seconds.

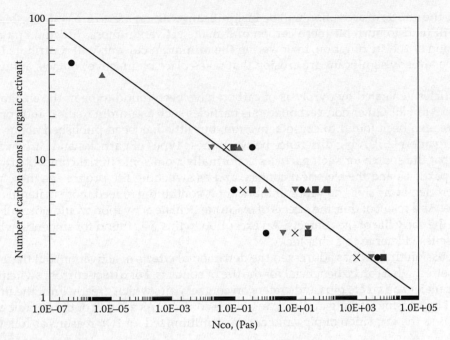

FIGURE 19.3
Critical exposure lower limit as a function of the number of carbon atoms in the organic molecule. (From Gray E.W., *IEEE Trans CHMT*, 1, 107–111, 1978.)

ion-drift-limited region no significant differences are noted among the metals or organic materials. The lower critical exposure level, the level at which the diffusion-limited range begins, varies with the different organic compounds as a function of the number of carbon atoms in the molecule (see Figure 19.3). The variation for organic compounds is greater than for metals in the diffusion-limited range. The adsorption-site-limited regime shows indistinguishable organic effects and significant differences among the metals [35]. From Figure 19.3, it can be predicted that organic materials with a large number of carbon atoms per molecule will activate at very low critical exposure levels. Some organic lubricants which contain molecules with a high number of carbon atoms have been observed to cause activation at a few parts per trillion [36]. On the other end of the scale, small molecules, CH_4, have not been observed to cause activation [36]. Koidl et al. measured the critical concentrations of various organic vapors in combination with three different contact materials and found a strong correlation between contact compatibility and the molecular properties of the organic vapor, e.g. number of carbon atoms, molecular structure, and volatility [27].

19.2.4 Activation Effects

The activation process lowers the erosion rate for a given discharge energy level, since some of the energy is used to dissociate organic molecules and to evaporate the carbon [37]. The overall erosion for low-current applications, however, is increased by activation owing to the much lower minimum arc current of carbon and consequent increase in arc duration. Noble metal contacts that are activated by organic vapors first see some short sparks that become short arcs that cause severe erosion and loss of switching life compared to non-arcing contacts [2,4,7]. At the other extreme, for contacts switching higher currents, above 2 A, the effects of activation will be insignificant with regard to erosion. The higher current will tend to burn off more carbon and make activation more difficult, and also the extension of the arc duration, by lowering the minimum current, will contribute little extra to the already significant arc erosion that takes place on inactive contacts at higher currents.

The particles generated by pyrolysis of carbon have been found to be in the size range of 10 nm to 3 μm [6]. Other non-carbonaceous particles such as alumina, silica, and copper oxide, have also been found to cause activation, but little has been published about this form of activation [2,6]. A big difference between these types of particles and carbon activation is that the source for such particles is normally from some initial contact between the oxide particles and the electrical contact, and as switching life progresses these particles are eroded away since no vapor or lubricant is continuing to feed more material onto the surface. As a result of this, the effects of oxide particulate activation would normally be limited to the early life of the contacts. An exception to this was found for silicones, which will be discussed later in this chapter.

The discussion thus far has addressed the detrimental effects of activation, but there are also some effects that can be beneficial for the life of contacts. For direct-current switching, material transfer can create pips and craters on contact pairs which greatly limit the life of the contact by causing sticking and welding of the contacts. As a result of the organic activation process the formation of pips and craters is minimized for two reasons as follows:

- *Arc movement.* For activated arcs carbon is being created at the periphery of the arc and the center of the arc is left clean. As a result of this the arc tends to move from one operation to the next seeking out a new carbon strike point. This movement from point to point results in erosion that spreads over the surface in a more uniform manner than for a non-active contact surface for which the arc can remain in the same place for many operations [7].
- For highly activated contacts the metal transfer from one contact to the opposite one has been observed to be inhibited by the metal not sticking to that contact, but mixing with carbon powder in loose deposits on its surface [7].

Above it was stated that the organic activation process does not have a significant effect on contact operation life for higher-current applications. For a relay operating at high-current, 90 A inrush, with severe make bounce, Witter and Polevoy showed that organic activation of the contacts could greatly reduce pip and crater type erosion [38]. The relay had pronounced pit and crater erosion at 25°C, but relatively flat erosion when operated under the same load at 125°C. By pumping preheated organic free air into the relay enclosure during the 125°C operation, they were able to duplicate the same pit and crater erosion seen for the relay operating at 25°C (see Figure 19.4). This indicates that controlled activation may have the potential for helping prevent pit and crater erosion for devices switching high inrush currents.

19.2.5 Activation and Contact Resistance Problems

The main focus on contact activation in the last section was on carbon deposits decreasing the minimum arcing current and resulting increased electrical erosion and noise for telecommunication circuits. Contact resistance is another problem associated with contact activation. Insulating layers on contacts, organic or inorganic, have to be mechanically

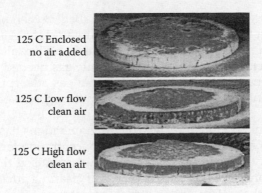

125 C Enclosed
no air added

125 C Low flow
clean air

125 C High flow
clean air

FIGURE 19.4
Relay testing in clean air versus atmosphere inside relay cover.

destroyed to gain small electrically conducting sites inside the area mechanically touching between two contacts. Such a condition yields widely scattered contact resistance values, from very low to open-circuit values, depending on whether and how successfully the layers were perforated, see Chapter 1. If, however, an organic compound is highly carbonized on the contact surface, the resulting carbon layer behaves as a well-defined semiconductor yielding a narrow range of scatter on the order of 1–10 Ω as a result of the electrical conductivity of carbon and the actual contact force used. This has been observed by numerous researchers [12,16,19,39–41] properly identified by Neufeld [42–44].

As stated in Section 19.2.3, in air higher-energy arcs are less efficient at producing activation than lower energy arcs. This is the result mainly from the chemistry of the arc. Low energy ohmic arcs are composed mainly of the elements from the surface of the contact at the area where the arc originates and therefore mainly remain metallic during the initial contact separation containing only the contact metal ions and material adsorbed on the contact surface (see Section 10.3.5). For higher energy arcs there is conversion from the metallic to a gaseous state as the contacts separate further with ions from the surrounding gas atmosphere. Vinaricky and Behrens show the erosion rate of graphite (carbon) increases by about 10 when the arc contains either oxygen or nitrogen as a result of the chemical reactions with carbon. [45]. Therefore, increases in contact resistance owing to carbonization are observed only when interrupting ohmic currents in the range of about 20 mA to 1 A for circuit voltages <24 Vdc where the arcing times are relatively short [46].

Contact resistance measurements under activating conditions for both gold and gold alloy contacts were done by two different research groups with similar results. For one of the studies switching 250 mA current in air with high concentrations of benzene, toluene, and xylene produced similar resistance increases in the range of 500–3000 mΩ [41]. The other study of a gold alloy at 100 mA current showed similar results, 500–2000 mΩ resistance for three organic vapors made from molecules containing six carbon atoms. There was no evidence of activation or resistance from 1,4-butanediol, a four-atom molecule [29].

Koidl, Rieder, and Salzmann [47] studied parameters that influence contact carbon resistance build up in telecommunication relays. Using a model switch operating at ohmic conditions, 10 Vdc and 100 mA, they investigated several parameters. Keep in mind that this is mainly a non-gaseous state. They found that it took lower concentrations of organic gasses to cause activation in nitrogen than in air. They concluded that this must be owing to lack of oxygen to decompose the carbon, but this is questionable for this type of weak arc. Maybe there is enough adsorbed oxygen on the surface of the contact when air is present to react with the adsorbed organic gas. Although this result is very repeatable it needs more work to understand the effect. They also found that higher humidity increased the critical concentration of organic gas needed for activation. This may be an effect on decreasing the amount of organic gas adsorption from the water vapor adsorbed or reaction of the adsorbed water with the organic gas. Again this effect was very repeatable.

The bounce frequency that occurs in relay switching operations has also been found to influence carbon build up by increasing the critical concentration level of organic gases. One thought about this is that increased bounce frequency raised the contact spot temperature and results in more carbon vaporization, a self-cleaning effect.

The overall data show that relays and switches operating in the presence of organic gases in these ohmic current ranges and low contact forces, 10 cN or less, are subject to carbon build up and large increases in contact resistance in the range of several ohms. Devices that operate at higher energy levels that support gaseous arcs normally don't have these problems with organic vapors since even though carbon particles are generated they are burned off and don't build up. In this case the carbon may create some benefit from moving the arc spot between operations and prevent pip and crater erosion as pointed out earlier in this chapter.

19.2.6 Methods for Detecting Carbon Contamination

Compatibility of certain organic materials with certain contact materials can be investigated with the aid of special model switches (see Chapter 18).

Carbonization can be detected as follows:

- The contact resistance increases compared to the characteristic contact resistance of carbon contacts (Figure 18.17d).

- For measurements across an opening contact pair, a characteristic plateau occurs in the voltage vs. time curve at a voltage level of 4 V corresponding to the sublimation temperature of carbon, 3650°C, as illustrated in Figure 19.5 [17,42–44].

- Minimum arc voltage increases to well above the minimum arc voltage for the noble metal contact and at a current below the minimum arc current of the contact metal [9,36].

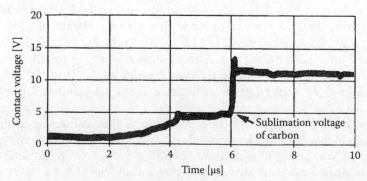

FIGURE 19.5
Contact voltage versus time for contact interruption with carbon present on the contact surface.

19.3 Mineral Particulate Contamination of Arcing Contacts

Mineral particles are present almost everywhere on our planet and finding some mineral particulate on electrical contacts should be no surprise. The effects of dust particulate have been discussed in Chapters 2 and 4 are not the focus of this chapter. Williamson et al. describe the effect of mineral particles on contact surfaces in terms of surface finish, particulate size, particulate concentration, and contact force [48]. They show that as particle size decreases and contact surface roughness increases to the point where the ratio of the particulate size to mean plane height of contact surface irregularities is unity, the probability of high resistance resulting from these particles is low. Also as the contact load is increased beyond what is required to produce a contact area equal to the cross-sectional area of particulate on the contacts, the probability of the particulate creating high resistance decreases rapidly. These same principles apply to arcing contacts in the initial unarced condition but become more difficult to interpret after arcing of the contact surface, owing to immense changes that occur in the particulate distribution and surface topography as a result of the arcing [49].

This section will discuss the effects of mineral particles on arcing contacts in terms of contact resistance rather than activation. Germer has shown that mineral particulate also caused activation [2,7], see Section 19.2.4. For mineral particles in contrast to carbon particles, since sources for building up or maintaining the concentration of the mineral on the contact surface during switching life normally do not exist, mineral particles have little effect on contact life.

With exception to very dusty environments and cases where there is no protective housing for the contacts, the effects of atmospheric mineral deposits on arcing contacts are low. A very small concentration of loose particles, which can be present on contacts from normal atmosphere having low dust content, will be eroded away by the arcing in most power devices. The main concern with mineral particulate is for cases where the concentration is much higher. A common source of mineral particulate contamination in contacts is from media used for removing burrs from fabricated contacts. In 1976, Trächslin [50] reported fine grit embedded in the contact surface for all parts he examined, which included contact rivets from seven different European manufacturers. For deburring and polishing of metal parts a wide variety of mineral compounds is used. Some common materials for deburring are silicon carbide, alumina, silica, and silicates in both particulate and stone form. By tumbling contacts in such media the burrs are worn away but also a portion of the media is left embedded in the contact surface. These refractory materials are not easily removed from the contact since they are chemically very stable. To remove such deposits, the contact surface layers must be etched away beyond penetration by these particles.

From the above cited mineral compounds, it can be seen that most contain the element silicon as a component. As a result of this, many times there is misinterpretation for the cause of resistance problems between mineral particles and silicone materials, organic compounds containing silicon. The large differences between them will become clear on reading this section and the next section of this chapter on silicones.

Witter and Leiper performed several studies for examining the effects of contamination from mineral particulate on resistance [36,51]. For Ag/CdO, 90/10 contacts contaminated with different concentrations of SiC they showed the contact resistance would increase rapidly for the first few dozen operations at 0.5 A and then gradually decrease on further switching (see Figure 19.6). Note that the initial resistance for the mineral-contaminated contact materials was very low and identical to the non-contaminated material. After a single switching operation, the contact resistance increased dramatically, one and two orders of magnitude respectively, for the 3% and 10% by volume SiC-contaminated materials. Under the same conditions the non-contaminated contacts decreased in resistance slightly owing to some cleaning effect of the arc. The reason for the low initial resistance readings and the large change after arcing is explained as follows: the silicon carbide materials, 1–2 μm in size, were embedded into the contact surfaces during the tumbling operation that was used to simulate contamination experienced by deburring. This is what would be expected for tumbling soft materials such as silver alloys with a hard refractory material. Using the principles developed by Williamson et al. [48], no effect of the contamination would be seen if the surface roughness of the contact were greater than the size of the portion of the embedded particle sticking out of the contact surface. In this case although the contamination level was high, the fraction of a micron of particulate that stuck out of the contact surface was insufficient to affect the resistance. When the contact surface was arced, the particles in the arced area were heated and reacted in the arc plasma to form aggregates composed of mixtures of silver and silicon oxides which were fused as deposits over the arced area.

Figure 19.7 is an SEM micrograph of the edge of an area from a single arc. SiC particles, triangular crystalline particles, can be seen embedded in the surface just outside the arced

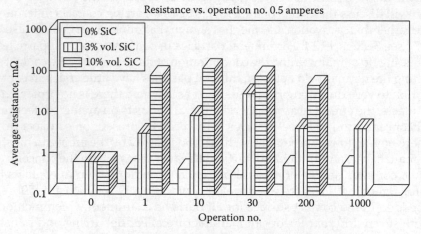

FIGURE 19.6
A comparison of contact resistance for silver cadmium oxide contacts contaminated by different levels of SiC after switching 0.5 A of current at 120 V ac.

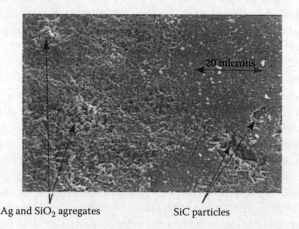

FIGURE 19.7
Edge of arced area for a contact contaminated with SiC. (From Witter G.J. and Leiper R.A., *Proc. 38th IEEE Holm Conference on Electrical Contacts*, pp 173–180, 1992.)

area and the silver-silicon oxide aggregates made of round and amorphous-shaped particles can be seen in the arced area. The arced area is very rough compared to the unarced surface and the individual particles making up the aggregates vary in composition with regard to the silicon to silver ratio. Although the surface is rough, there is no benefit as predicted by Williamson's work [48], since the aggregates form heaps that are significantly elevated above the unarced surface and the constriction areas are complex with current paths going through a series of many particles and interfaces making up the aggregate heaps. As a result of this, the contact resistance remains elevated but significantly variant as a function of the path required from the contact spots through this thick upraised resistive layer. The contact resistance initially increases with switching operations owing to increased refractory build up on the surface. Depending on the current level and the contamination concentration the resistance begins to decrease at some operation point as the refractory material is gradually eroded away through arc erosion.

The effects of contaminant type, refractory chemistry, and particle size were also investigated by Witter and Leiper [51]. Figure 19.8 shows a comparison of silica, a silicate, and silicon carbide. The relative concentrations in terms of ED AX analysis and resistance measurements are compared for the various mineral-contaminated contacts that all switched a 0.5 resistive load at 120 V for 200 operations. For these sets of samples, the particle size of the contaminant was fine, less than 3 μm. Although some differences existed among the results for the materials, they all seem to produce the same order of effect. In Figure 19.9 a comparison is made of silica contamination differing in particle size, 2 versus 8 μm, [51]. The finer material has a much larger effect on the elevation of resistance, which is opposite to the effect seen for particle size on non-arcing contacts. It is speculated that the finer material is more reactive in the arc and mixes more quickly with the silver on the surface. The material with the coarser silica showed a higher resistance level after 2000 operations compared to 200 operations. The current level was also found to have a large effect on the results for mineral contamination [51]. Tests at 0.1 A exhibited no arcing and showed no change in resistance, remained less than 1 mΩ, for short cycle switching up to 2000 operations. At currents greater than 2 A at 120 V ac, the build-up of resistance was very small owing to the cleaning effect of the arc. For work at lower voltage, 12 Vdc, the

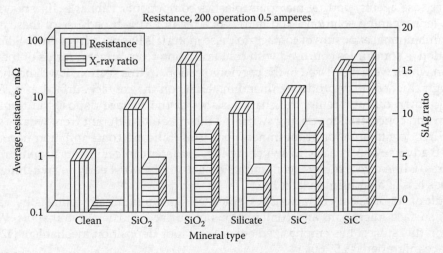

FIGURE 19.8
A comparison of various types of minerals for effects on contact resistance after switching 0.5 A current for 200 operations.

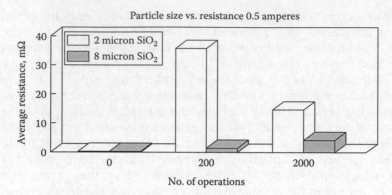

FIGURE 19.9
A comparison for particle size effect for SiO_2 contamination at the same concentration level.

effects of mineral particles on resistance are more severe and higher currents are less effective in reducing the effects.

In summary, the effects of mineral particulate contamination on contact resistance are mainly seen in the initial portion of the contact life. The severity of the effects will vary as a function of switching current and voltage and also concentration and particle size of the mineral. In many cases the particles are embedded or partially embedded in the surface and do not produce elevated resistance reading until some arcing takes place on the contact surface.

19.4 Silicone Contamination of Arcing Contacts

One of the most troublesome types of contaminants for arcing contacts consists of silicone compounds. Chapter 10 introduced a brief discussion of this phenomenon. Many types of silicones have been developed for numerous applications including lubricating oils, mold release agents, potting materials, plastics components, rubber seals, sprays for insulation, lens cleaning solutions, and more. As a result of the broad use of these compounds, unintentional exposure of contacts to these materials happens frequently. Silicone contamination is sometimes confused with contamination from mineral forms of the element silicon which were discussed in the previous section. In this section, it will be shown that although silicones may produce mineral materials on the contact surface, as a result of breaking down during arcing, the characteristics of the surface deposits in terms of distribution and concentrations during switching life are quite different from what is seen owing to direct contamination from mineral particles. Silicone contamination normally produces no adverse effects early in switching life, but later can cause catastrophic resistance failures, while direct mineral contamination is usually worst early in switching life and becomes less of a problem in later life.

As a result of its physical characteristics, silicones can contaminate contacts by vapor deposition, surface migration, and combinations of surface and vapor transport. Work published on this subject has emphasized either the vapor-deposition mechanism [52–58] or the surface migration [49,59,60].

Common silicone fluids are dimethyl silicones, which have the following typical structure, where n is the degree of polymerization:

$$CH_3 - \overset{\overset{\displaystyle CH_3}{|}}{\underset{\underset{\displaystyle CH_3}{|}}{Si}} - O - \overset{\overset{\displaystyle CH_3}{|}}{\underset{\underset{\displaystyle CH_3}{|}}{Si}} - CH_3$$

A typical silicone product is composed of a mixture of molecules having a broad molecular weight distribution. Some of the attributes of silicones that make it good for use as a lubricant are the stability of silicones over a wide temperature range and their low surface tensions, 17–22 dynes cm^{-1} [59] which allows good wetting of surfaces. These same attributes which make silicones good as a lubricant are harmful for applications using arcing contacts, since silicones will wet and spread, migrate, over nearly all types of materials. The molecular weight distribution of a silicone product will have an effect on the rate and ability for the silicones to spread. Lower molecular weight fractions of silicone have lower viscosity, which gives them faster surface migration rates than higher molecular weight fractions of silicone [59]. With regard to vapor transport, the vapor pressure of silicone exponentially becomes larger as the molecular weight decreases, and for low molecular fractions, evaporation takes place even at room temperature [54]. The low molecular weight fractions of silicones are clearly the most troublesome types of silicone contamination for contacts.

Although silicone transfer normally involves both vapor and migration the discussion will be divided into two sections: one where vapor transport is the main mechanism and the other where migration is the main factor for transport and contamination. In following this discussion, it should be kept in mind that factors connected with an electrical switching application such as electrical load type, current level, voltage, contact force, erosion rate, and switching rate can have a large effect on a specific result.

19.4.1 Contamination from Silicone Vapors

The work done on silicone contamination by vapor phase will be discussed first. This process can be considered very similar to the processes discussed in Section 19.2 on contact activation by organic vapors. The organic vapor activation process was mainly concerned with switching at low currents, less than 1 A. and the effects of activation on contact erosion more than the effects of contact resistance. For silicones, concerns are mainly with the effects of contact resistance. The effects also extend to much higher current ranges compared to resistance associated with organic activation. Automotive relays operating at 20 A or more have been found to fail as a result of silicone vapor and migration contamination [61]. The effects of silicone vapors on arcing contacts do not appear immediately on exposure to the silicone but normally after hundreds or thousands of switching operations, similar in timing to organic vapor activation.

Tamai studied the effects on contact resistance from decomposition of silicone vapor onto a heated gold surface [55]. He found that the decomposition and formation of silicon dioxide began at about 300°C and that films of sufficient thickness to cause high resistance readings for 10 cN of contact force started above 600°C. This work also showed that the decomposition at these temperatures left no carbon residue and that the amorphous glass deposits were stoichiometric silicon dioxide, SiO_2. Figure 19.10 shows the relationship of the contact resistance to the thickness of the silicon dioxide that was formed on the gold contact and indicates a sharp increase in resistance as the thickness approaches 1000 Å [0.1 µm] [54]. The contact forces utilized in this work were relatively low compared to forces used in typical arcing devices, but the results indicate that SiO_2 films thicker than 1000 Å will cause resistance problems even for devices having higher contact forces.

The effects of silicone vapors on arcing contacts will depend on many interrelated factors including vapor concentration, operating temperature, frequency of switching, type of silicone and more. The relationship of molecular weight in terms of degree of polymerization and temperature on vapor pressure of silicones is shown in Figure 19.11 [55]. It can be seen that silicones with degrees of polymerization less than 8 have vapor pressures of 10 p.p.m. or more at room temperature. For applications at elevated temperatures such as automotive relays, 125°C, the degree of polymerization must be greater than 12 to keep vapor pressures less than 10 p.p.m. The reason that 10 p.p.m. is used for this illustration is that the authors, on the basis of work they had performed, thought that below this level resistance failures would not occur. However, later work showed that this is not true [56,57].

From the published work on silicone vapor contamination, the effect of switching rate (operations per unit time) on failures from resistance may not seem clear. Ishino and Mitani [50] showed, for testing enclosed relays with silicone sleeves at 24 V dc and about 0.1–0.2 A, failures at 1000 operations for a switching rate of two operations per hour and failures at 150,000 operations for a switching rate of 30 operations per minute. Eskes and Groenendijk testing under similar electrical loads found that the number of operations to failure was independent of switching rate for the rate range of seven operations per second to only a few operations an hour [53]. This work was done in an artificial environment controlled at a rather high silicone concentration of 500 p.p.m. Tamai and Aramata tested relays at a single relatively fast switching rate of 10 operations per second under silicone concentrations from 0.1 to 300 p.p.m. and found that the time to failure was inversely proportional to the silicone concentration except for the low concentrations, for which no failure occurred even after prolonged life testing [55].

For the above three examples, using some of the same criteria discussed in Section 19.2 for organic vapor activation, an explanation can be made to fit each case as follows:

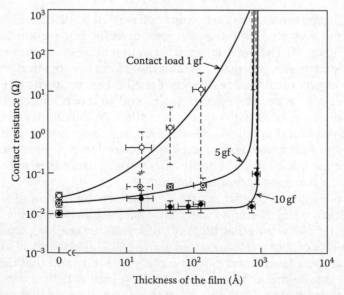

FIGURE 19.10
Contact resistance as a function of SiO_2 thickness.

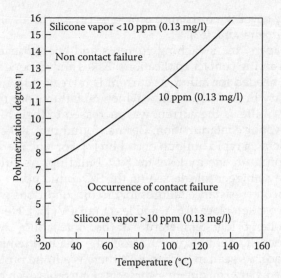

FIGURE 19.11
Relationship of temperature, vapor pressure, and degree of polymerization.

1. The values of the silicone vapor exposure, η, used in the work by Ishino and Mitani were in a diffusion-limited range for silicone adsorption, therefore increasing opening time increased the amount of SiO_2 generated per operation [52].

2. The silicone vapor concentration used by Eskes and Groenendijk in their work was very high and possibly the silicone exposures, η, they used were above the diffusion-limited range. Recall from Section 19.2.3 that the adsorption site limited range is independent of exposure. This would explain why no dependence on switching rate was found for this work [53].

3. The work by Tamai and Aramata for silicone vapor concentrations less than 10 p.p.m. was done at a very fast switching rate which greatly limited diffusion of silicone vapors onto the contact surface between switching operations and prevented a build-up of SiO_2. Later work by Tamai for the same concentration but at a slower switching rate of 1 Hz versus the previous 10 Hz, resulted in failures for concentrations at 7 p.p.m. and indicated that the former results were limited by diffusion time [56,57].

As mentioned earlier, the low molecular weight silicones are the most problematic. A method of predicting the presence of these harmful weight fractions is to heat the material and measure the weight loss. Eskes and Groenendijk measured the weight loss for a number of silicone materials after heating them at 175°C for 24 hours and found that resistance problems in general increased as the weight loss became greater than 0.3% [53]. Care must be taken to condition the silicone before testing to remove adsorbed water vapor by holding the silicone material at a low relative humidity for 24 hours. Removal of the lower molecular weight fractions of silicones heating the silicone components of a device at elevated temperatures before assembly into the device was shown by Ishino and Mitani to reduce the effects from silicones [52].

Another way for lessening the effects for vapor deposition of silicones is by ventilation. Ishino and Mitani did their work on an enclosed relay with a silicone sleeve. When they removed the plastic relay cover the resistance problems disappeared. Similar results were reported by Eskes and Groenendijk [53].

For the above-cited papers, the switching currents and energy levels were relatively low compared to typical arcing contact applications. Eskes and Groenendijk show a lesser number of operations is needed for failure as current is increased from 5 mA to 500 mA. Tamai predicts a boundary line of 1.6 W below which no failures occur and above which failures occurred earlier in life as the current was increased to their highest test current, 2 A [56,57]. For organic vapor contamination, Germer found that 100 ergs of energy was near optimum for producing a rapid buildup of carbon [7]. For silicone vapor deposits one would also expect an optimum energy level for SiO_2 build up. As current increases the electrical erosion will play a bigger role, lessening the SiO_2 build up.

Tamai also showed silicone resistance failures for circuit conditions where he suspected that no arcing would have taken place [0.6 A, and less than 4 V] [57]. Here the heating level could result from the molten bridge temperature of the contact alloy, which is more than sufficient to crack silicones. It could also result from the residual temperature of the arc roots on the contact surface. As seen in Section 9.5.4 there is a finite probability that a very short duration arc can form between opening contacts at currents even below 0.6 A. Tamai et al. [62] also did a study of a contact surface exposed to a D4 silicone, siloxane group, vapor environment. This work shows some interesting decomposition mechanisms not discussed before. The D4 exists as both a ring and a chain structure, 0.015–0.085 (ring/chain) equilibrium ratio, see Figure 19.12. A saturation film thickness of one monolayer was formed, 1 nm, in 1–100 hours with the time depending on the vapor concentration. A second layer can also form as a chain structure in the liquid state. For this material a reaction with oxygen in the air over time results in decomposition of the polymer room temperature into a vulcanized structure with higher oxygen content (See also Section 10.6.3).

This study included electrical testing at different currents and voltages using a small relay with less than 10 cN contact force and a resistive dc load. A resistance failure border was found at a switching power level of 1.6W below which failure rates were lower and less severe, see Figure 19.13. The contact surface for failed contacts had deposits of both silica and carbon.

Silicone such as other hydrocarbons will break down and form carbon as discussed in earlier parts of this chapter. In this case, a problem exists for interpretation of carbon

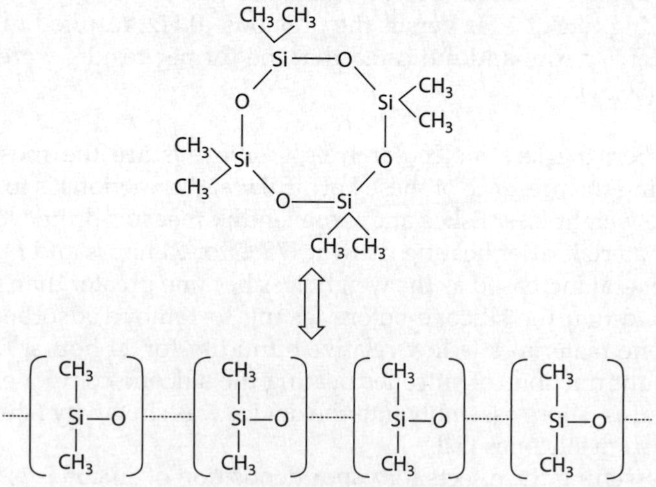

FIGURE 19.12
Silicone D4 which exists as both a ring and a chain structure.

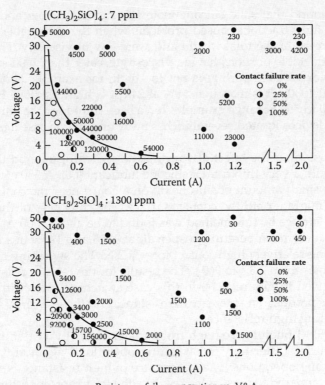

Resistance failure operation vs. V&A

FIGURE 19.13
Results of electrical testing showing a border for contact failure at 1.6 W.

versus silicone resistance failures. As seen in previous sections of this chapter, carbon build up alone can result in high resistance readings. For voltages 15 volts and above silica will easily form from the silicone. At the lower voltages the silica may still form from the added oxygen from the decomposition of the D4, but the high carbon build up at that power level may be the dominant cause for the high resistance in this range of voltage and current. This work clearly shows that in the low power range both silica and carbon are present from the decomposition of silicones.

From the above it can be seen that silicone vapors can cause device resistance problems over a wide range of electrical loads and for very low vapor concentrations. It also can be seen that the silicone vapor exposures follow some of the same rules developed from the long term work done by Bell Labs on carbon activation.

19.4.2 Contamination from Silicone Migration

The contamination of contacts through silicone surface migration is less restrictive than by vapor phase deposition since the contacts don't have to be encapsulated.

As stated earlier, silicones wet nearly all surfaces and will spread over the surfaces at a rate dependent on several factors including temperature, surface roughness, and silicone viscosity [59]. Similar to the case for silicone vapors, the low molecular weight fractions of silicone materials are most problematic since they have lower viscosity and will spread on surfaces ahead of higher-viscosity fractions of the same material. The rate at which the silicone spreads was shown by Kitchen and Russell to be related to the reciprocal of the viscosity [59]. They found that solid silicone materials such as rubber and sealants with uncured extractable fluid fractions caused problems when the extractable portions were more than a few percent of the total weight and were low in viscosity. They considered compounds with extractable portions having viscosity greater than 1000 centistokes, cSt, relatively safe if the extractable portion is not too high, for example 2.8% at 1200 cSt is relatively safe, 14% at 1030 cSt is marginal, 19% at 24 cSt is high risk [59]. For low viscosity silicones in the 20 to 50 cSt range examples have been found for the silicones to travel over wires easily to devices located several inches away and in one case items several feet away [59].

Witter and Leiper studied the effect of varying the switching rate on the failure rate for contacts coated with a very thin film of silicone lubricant [60]. A 350 cSt lubricant was applied by putting a small amount of silicone oil on a board near the contact and allowing the lubricant to migrate onto the contact surface. The thickness of the silicone film covering the contact surface by this method was found to be about 1000 Å. For this study switching was done at room temperature in open air at 0.5 A and 120 V in an experimental switching device using relatively high contact force, 1.2 N. The switching rate was varied with contact open times from 2.3 s to 960 s. The results for the resistance levels after only 200 operations are given in Figure 19.14. In this case switching rates with contact open times of less than 30 s produced no increase in resistance and those with 60 s or more open time resulted in creating high resistance.

Figure 19.15 is an SEM photomicrograph from a later study by Witter and Leiper [49] showing a contact surface after only 50 switching operations, again at 0.5 A and with 180 s between switching operations. The appearance of high resistance is for a relatively short switching life compared to most of the results cited for vapor-deposited silicone. The thickness of the glassy silicon-rich deposits are much greater than the 1000 Å deposits described by Tamai and Aramata [55] for silica from heated silicone-vapor deposits, and this explains why the resistance remains high even at high contact pressures.

The results show a switching rate dependence for silicone migration, similar to the findings for silicone-vapor depositions in the diffusion-limited exposure range. The mechanism for this build-up is termed "arc-driven migration" [49]. An explanation of this process and the surface chemistry distribution that would be seen as a result of this process is described as follows. Silica, SiO_2, will be used to describe the glassy silicon-rich deposits

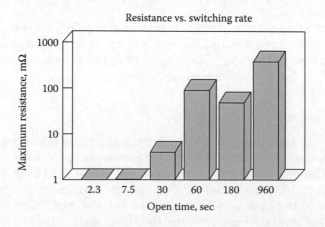

FIGURE 19.14
Resistance as a function of time between switching operations for contacts covered with a thin film of silicone lubricant and switched for 200 operations at 0.5 Å.

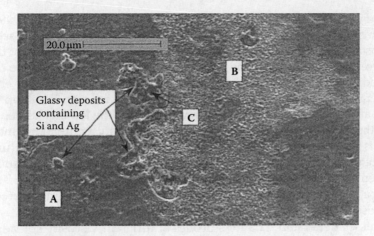

FIGURE 19.15
Arced area after 50 operations showing silicone glass deposits near edge of arced area. (From Rieder W. and Strof T., *IEEE Trans CHMT*, 15(2), 166–171, 1992.)

created by the reaction of the arc and the silicone, although these amorphous deposits may be a more complex glass or just a mixture of silica and silver.

- If a silicone fluid source is near a contact and has a surface path connecting to the contact, the silicone material will, given the time needed, migrate to the contact and cover the entire contact surface as a coating relatively uniform in thickness. The coatings are normally too thin, 1000 Å in the above examples, to be detected by conventional SEM and ED AX and thicker than what would be expected for vapor deposits. Before switching duty these coatings do not cause any contact resistance problems since metal-to-metal contact is made through the film.

- During a single switching operation the silicone film directly in and near the area of the arc is dissociated into silicone glass compounds that are deposited around the arced area. Since low-molecular silicones evaporate in the range of 200°C an area larger than the arc spot would have silicone removed by both evaporation and dissociation. The outer portions of the heat-affected area will have silica deposits from the arc reaction but the silicone film will be removed. The void or film thickness gradient in the silicone created by the arc is the driving force for the silicone material to migrate back into the hole. The migration begins immediately after the void in the film is created.

- The time between successive switching operations is an important factor. For example if the next operation is done immediately there will be insufficient time for silicone to be re-established in the void area and if the arc occurs in the same area there will be little or no silicon glass generated. After this operation as a result of electrical erosion the surface may become cleaner with regard to the glass deposits than it was after the first switching operation. On the other hand, if there is a long delay between switching operations, there will be sufficient time for the void to be filled and a second arc in this same region will again remove the silicone and deposit more silica. A sequence of successive arcing operations under these conditions will, in effect, cause silicone material to be drawn into the arced region and be converted to a build-up of silica, thus the "arc driven migration process."

Figure 19.16 shows EDXA spectra taken from different regions of Figure 19.15. These results for silicon chemistry spatial distributions are also typical of what is seen in field service failures as a result of silicone migration. Spectrum (a) is for an area of the contact away from the arced area, no silicone line is present since the silicone film is too thin in this area to be detected. Spectrum (b) gives the surface composition average for the entire arced area, after 50 operations, and shows silicon to be clearly detectable and built up compared to the non-arced area. Spectrum (c) gives the composition of the large glass deposit in the arced area: both the silver and silicon lines are strong, with only a minor amount of carbon present. From this analysis it is not known if this amorphous deposit is a mixture of silica and silver or a complex glass containing silver and silicon. This point is not so important since the glass deposit is very resistive in either case. The important characteristics of this analysis is the spatial distribution of silicon. If silicon is present only in the arced area as thick glassy deposits, this is indicative of silicone migration.

Kitchen and Russell generated data for silicones of different viscosity spreading over a phosphor bronze substrate [59]. They presented the following equation for approximating the spreading rate of silicones from knowing the spreading coefficient of the lubricant, σ:

$$\Delta A = \sigma t^{1/2} \tag{19.1}$$

Where ΔA is the incremental increase in area, cm^2, σ is the spreading rate coefficient, cm^2 h$^{-1/2}$ and t is the time in hours. The following equation was developed for relating the spreading coefficient of different silicones to their viscosity by using the data that Kitchen and Russell had generated for phophor-bronze substrates.

$$\ln \sigma = 0.65 \ln(1000/v) + 3.22 \tag{19.2}$$

In this equation v is the viscosity of the silicone in centistokes, cSt.

Using Equations 19.1 and 19.2 the estimated times for silicones to cover different diameter circular areas is plotted in Figure 19.17. It must be kept in mind that the calculations are not exact since factors such as surface roughness, wetting characteristics of the metal silicone combination, temperature, and others can change the results.

The plots in Figure 19.17 can be used to estimate the contact open time needed between switching operations to allow a maximum build-up of silica on the contacts. For example, in the above described work by Witter and Leiper for switching contacts at 0.5 A with

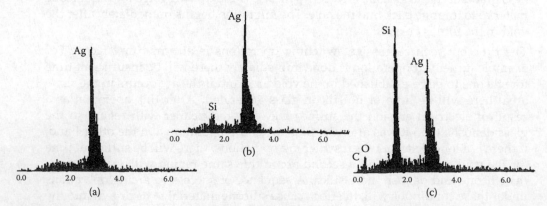

FIGURE 19.16
EDXA spectra for areas A, B, and spot C in Figure 19.15.

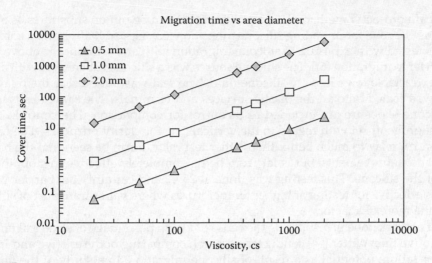

FIGURE 19.17
Estimated time required for silicone migration to cover different diameter areas as a function of silicone viscosity.

a 350 cSt silicone compound on the surface, the results in Figure 19.12 show that an open time of greater than 30 s, close to 60 s, is needed to generate a large build-up of silica. For this case the arc spot for the 0.5 A load would be about 0.5 mm in diameter. It is known that the silicone depleted area having been affected by a heat of 200°C or more was greater in size than the arc spot. For this example the size was estimated to be twice as large or 1 mm in diameter. Using Figure 19.15 the time estimate for refilling the hole is 35 s, which is lower but in the ball park for 60 s found from the experimental work.

There has been very little published on the effect of current on the results for silicone migration. For example, as the current increases from a level of 2 A to a much higher level such as 50 A, the erosion rate increases by a factor of 2 magnitudes. Cases have been cited where silicone contamination has caused problems for relays operating inrush currents in the 50–100 A range [61]. In such cases, depending on the contact material and the electro-mechanical device characteristics, the electrical erosion can be quite variant, and as a result of this effects from silicone migration may vary. For example, in one case a relay which was designed to operate with silver copper contact material showed no effect from silicone contamination while operating at 50 A, since the surface of the contact eroded away very rapidly. For the same high current load a relay with silver metal oxide contacts that operated with less than one third of the erosion rate failed after 50,000 operations from overheating owing to silica build-up on the contact surface. Further tests conducted on both relays for operating at 12 A with several minutes open time between operations produced failures on both relays in just a few thousand operations. This example shows that qualification testing for devices should include lower current testing with ample time between operations if silicones are at all suspected as a risk. Standard accelerated testing is normally not adequate for detecting problems from silicones since these tests use higher than normal operating current levels, which results in heavier than normal erosion, and also switching rates are too fast to allow replenishment of silicones to the arc spot by migration.

One method that has been recommended for prevention of failures by silicone migration is to coat the area around the contact with a substance with lower surface energy than silicone that will impede silicone migration. Bernett and Zisman found that poly-IH

IH-pentadecafluoro-octyl methacrylate, a polymer that can be put on substrates as a very thin film, was effective for impeding silicone migration [63]. Figure 19.18 shows a comparison of relays tested with exposure to silicone migration with and without the above polymer as a barrier to migration [61]. The silicone source was a silicone rubber seal with a high amount of extractable low-viscosity silicone fluid. The seal was put outside the relay and the barrier was coated onto all the plastic surfaces and terminals. The same relays tested with no silicone exposure are included as a control for comparison. The control results are shown slightly off set with regard to the vertical axis for clarity purposes. The results show the control relay to much better than either test relay. It can be seen that the barrier produces some improvements, but is far from being completely effective in stopping the migration of the silicone. This testing was done at 25°C, and certainly the barrier would be much less effective for testing at higher temperatures where vapor transport of silicone would become more of a factor.

As a result of work done investigating the cause of a high percentage of electrical failures in an automotive line, Witter [64] conducted research comparing both resistive and inductive loads for failure potential as a result of silicone migration. In a study of the failures in the actual cars it was found that the wiring harnesses were contaminated with silicone film and each harness housed about 20 automotive relays. The actual failure reports from the service dealerships showed that failures only occurred for the relays running high inductive loads and no failures were reported for resistive loads or mild inductive loads. All the relays had about the same level of contamination. To verify the findings similar tests were run in the laboratory comparing the two types of loads. Figure 19.19 shows the results comparing two inductive and two resistive loads. The testing was done with an open time of 360 seconds between operations and from 1 to 4 amperes for the inductive loads. The testing on the resistive loads included a range from 3 to 20 amperes and none of the tests showed any changes in electrical resistance as would be expected for contacts exposed to silicone migration.

For resistive and low inductance, automotive loads running at 13 V dc the relays normally interrupt in the metallic state and therefore there is no oxygen to react with the silicone in the arc (see Section 10.3.5). Popular silicones such as the dimethyl silicone shown in Figure 19.4 have too high a ratio of carbon and hydrogen to oxygen to allow the formation of silica without the arc reaching a gaseous state. This explains results seen in this study

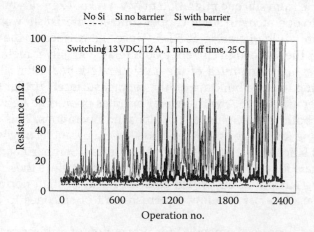

FIGURE 19.18
A comparison of resistance results for relays exposed to external silicone migration both with and without silicone migration barriers.

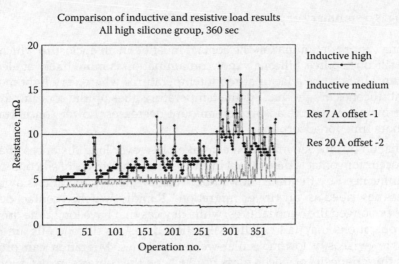

FIGURE 19.19
Comparing the effects of inductive and resistive loads for testing silicone contaminated contacts, bottom lines resistive loads off set and upper plots inductive loads.

where there were thousands of cars involved and a failure rate of 7% for inductive loads and 0% for other loads.

19.4.3 Summary of Silicone Contamination Mechanisms

Silicone contact contamination has been discussed in terms of both vapor transport and migration mechanisms. Actually in all cases both mechanisms will be active to some extent.

In both types of transport, the time between switching operations has an effect on the results. For vapor transport you must have enough time for the adsorption of the silicone vapor to take place unless you have a concentration above the diffusion limited range. In migration you need to provide enough time for silicone to migrate over the arc spot and recover the area that reacted with the arc. Most of the work on studying vapor transport used low currents and low device contact force. For devices using higher forces the results would be much less severe for the same contamination levels. For migration the silicone film can be much thicker than for adsorbed vapors and the resultant silica build up much thicker.

In either case of transport, the detection of silicone films before arc build up is not possible using SEM and EDX analysis and more sensitive surface analysis equipment such as auger and XPS must be used. Once the parts have been exposed to many electrical operations EDX is good for detecting the silica build up.

Different attempts have been made to counteract effects of silicone. The uses of lower surfaced energy coating such as fluorides as described in the previous section are only marginally successful. Another marginal solution is to use contact materials that have a high erosion rate so there is no significant built up of silica. This doesn't work for low current applications. Work by Schrank [65] looked at what had to be done for sealing relays against silicone contamination when put in silicone rich environments. This work showed that water proof seals are not effective and only true hermetic sealing worked. The other solutions being looked at are use of non-silicone type acryl-based polymeric products in the vicinity of the relays, Hasegawa [66], and this is may be the most successful solution.

The following is a summary of points relating to silicone contamination.

1. For silicone vapor contamination the contact must be in an enclosure which contains the silicone source. Silicone vapor contamination is more likely at elevated temperatures but can be a factor at room temperatures where very light molecular weight silicones are present. As the temperature goes up the concentration of the vapor increases but the adsorption amount decreases. Lower concentrations require more time for adsorption.

2. Contamination by silicone migration can take place over long distances and failures can occur for contacts that are not in enclosures. The effects of silicone migration are influenced by the time allowed between switching operations owing to the process described as "arc driven migration." Rapidly switched contact devices may have prolonged life or no failures, while devices that have long time duration between operations may fail rapidly. Migration times for silicone oils are much faster for low-viscosity low-molecular-weight silicones. Migration can produce relatively thick deposits of silicon glass products on the contacts in just hundreds of operations. Even under good contact force these deposits cause high resistance.

3. Qualification testing for electromechanical switching devices is normally done at accelerated switching rates that do not provide enough time between operations to see the effects of silicon migration, they are also normally done at maximum current which increases erosion of the silica, and therefore this type of testing is not effective for screening for silicone contamination. Silicone screening tests should be done with lower currents, inductive loads or voltages at 15 amperes or more, and a long delay between switching operations, 1 to 5 minutes.

4. Effects from silicone contamination do not show up until devices have been in operation for some time and if the contamination of a device is connected with the manufacturing process many parts can be affected before the problem surfaces. This is why screening tests are important.

5. Many different types of silicones have been developed and the decomposition processes can be quite different and involved. They all can cause failures and low molecular weight versions are the most troublesome.

6. Low power weak arcs involve breaking down silicones into both silica and carbon deposits and both contribute to the high resistance.

19.5 Lubricants with Refractory Fillers

Another type of contamination that is often confused with silicones consists of lubricants that contain mineral particles as fillers. The mineral fillers serve as thickeners for increasing the temperature stability of the lubricants. Lubricants of this type are sometimes used with base-metal contacts that have a sliding action for make and break. Little has been published about problems encountered with this type of lubricants for arcing contact applications. Some of the fillers used are silica, alumina, silicates, and magnesium oxide. Klungtvedt reported on resistance problems associated with the use of silicate-thickened lubricants used on the contacts of an automotive sliding switch that caused switch overheating and failure [67]. For this application the contact forces are over 1 N and the current is 20–50 A. The characteristics of these materials for causing resistance problems for arcing contacts can be thought of as a mixture of the mechanisms described in the preceding sections on mineral contaminants and silicones.

The minerals are suspended in the lubricants as fine particulate. Like silicone and dry mineral contaminants, before arcing these lubricants normally will not create a resistance problem if the particulate is very fine in size; they have a uniform distribution throughout the lubricant and the contacts have contact forces of about 0.5 N or more (see Figure 3.33). Failures of arcing devices owing to high resistance and overheating have occurred owing to these types of lubricants. On examining the contacts from such devices it was found that the mineral particulate is no longer uniform in distribution and that deposits are present which contain high concentrations of the minerals mixed with the contact base metal and carbon.

Although there is little research on this subject, we can make some assumptions as to what happens during the arcing of contacts with these lubricants. The heat from the arc will evaporate and also crack some of the lubricant and leave higher concentrations of the mineral materials around the arced area. The arc will also melt the mineral materials and form aggregates composed of the mineral particulate and base metal similar to that described in Section 19.3 on mineral contamination. The contact resistance that develops depends on the size and concentration of the aggregates that form. Different from the case of solid mineral contamination, where the supply of mineral particles is limited, the migration of the lubricant on the contact will form a continuous supply of mineral material to the arced region.

Another possible detrimental mechanism for this type of lubricant can take place from constriction heating. If a high-resistance spot is made on the contact surface from an arc deposit, heating from the constriction can result in evaporation of the oil near the contact spot. This evaporation will create a migration of replacement oil to the constriction area. The oil will carry more mineral material to the area and continuing evaporation will result in a concentration of mineral material.

Since little is known about this subject, care must be taken in using these types of lubricants where arcing will take place. If using this type of product switching tests should be run at several different current levels combined with periodic soak tests, periods with the contact closed at normal current. Switching should be done with enough time between operations to allow for migration of the oil.

19.6 Oxidation of Contact Materials

Silver alloys are the most popular of the arcing contact materials since they do not oxidize during normal switching operations. The oxides of silver are not stable at elevated temperatures. Silver is alloyed with up to 25% copper without any adverse effects for resistance owing to oxidation during arcing. Almost all of the contact materials that are operated in air are composed of silver or noble metal as a result of the need for oxidation resistance.

One of the exceptions for a class of materials that are operated in air and are not resistance to oxidation consists of the silver refractory composite contact materials, silver–tungsten, silver–tungsten carbide, and silver–molybdenum. These materials form complex oxide compounds with silver during arcing and as a result have limited switching life and must be used with high contact forces. A detailed explanation of this problem is described in Chapters 10 and 16.

Copper is another material that is used in some arcing contact applications but is very limited because it forms high resistance oxides during arcing in air. Most designers are aware of the problems with copper and avoid its use except under protective atmospheres such as SF_6 and oil. It has also been used extensively as a contact material in vacuum (see Section 16.4). Little has been published on the problems in using copper as an arcing contact. When copper is used for an arcing application in an air atmosphere, the following conditions are normally employed. A large contact force is used with good contact wiping during make and break. Also, oil coatings are sometimes used to inhibit the oxidation. Sometimes silver plate is used over the copper, but this is of little use if any significant arcing takes place.

Tungsten has been a popular contact material but only for limited applications since it oxidizes during arcing. Tungsten is used only in combination with very high contact forces. It is also used only for rapidly operated devices such as ignition sets and horns where it does not have to carry a continuous current load.

19.7 Resistance Effects from Long Arcs

For fine silver and other silver-based contact materials, it has been found that the contact resistance after the interruption of a long arc, a gaseous arc of approximately 1 ms or longer, will vary significantly and be much higher than the contact resistance after interruption of a short arc, a metallic arc about 0.5 ms duration or shorter. Several independent researchers have reported similar findings [68–70]. Witter and Polevoy found the contact resistance of an automotive relay switching an inductive load to be high and erratic regardless of the silver contact material tested which included silver tin oxide, silver cadmium oxide and fine silver [64]. Further tests were conducted on fine silver contacts in a model switch with a contact force of 1 N and 12 V dc and about 30 A current, for both an inductive load and a resistive load. The resistive load arc duration was about 300 μs and the inductive load was about 3000 μs. Figure 19.20 shows the contact resistance measurements made after each switching operation, starting with the inductive load and after 35 operations changing to the resistive load. For clean contacts these two very different contact resistance distributions for the long and short arcs are very repeatable. Sone et al. [68] conducted tests with a pure resistive load at 48 V dc and about 3 A with a similar contact force of 1 N. For this work a very slow opening speed was used (35 mm s^{-1}). As a result of this and the circuit, the arc duration was long and variable (2–8 ms). Sone et al. found a tendency for higher resistance readings after several longer arcs (>10 ms), and contact resistance dropping as arc duration decreased. They explained the difference in resistance between the long and short arcs in terms of surface roughness [71]. It was explained that as the arc transferred from the metallic phase to the gaseous phase the surface became rougher owing to the gaseous arc that resulted in higher resistance. Chen et al. [69] conducted switching tests at 20 V dc and 0.5 and 1.0 A for an inductive load. The resulting arcs were gaseous with a duration of about 1.5 ms. They tested with different mixtures of oxygen and nitrogen (0%, 5%, and 50% O_2). The results are shown in Table 19.2. The results show a clear correlation of oxygen level and resistance. Auger analysis of the anode shows a significant amount of oxide is present. The conclusion of these researchers is that silver oxide films are responsible for the higher resistance associated with long arc duration and not surface roughness.

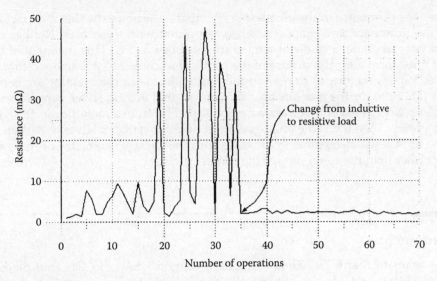

FIGURE 19.20
Contact resistance as a function of switching life for initial inductive load testing, long arc, followed by resistive load testing, short arc.

Witter and Polevoy also conducted surface analysis on the high contact resistance long arc samples they had tested [70]. The work was done using electron spectroscopy for chemical analysis (ESCA), which not only identifies elemental monolayer but also provides information on the valence of the element. This analysis showed that silver, carbon, and oxygen were present on the surface but also the valence state of the carbon and oxygen corresponded to that which matches the compound silver carbonate. From the analysis it was also determined that almost all of the silver on the surface was tied up as this compound.

A switching test conducted by Witter et al. was conducted in ordinary air, which contains 350 p.p.m. CO_2 and a small amount of hydrocarbon chemicals from pollution. Chen et al. had done their work with synthetic air which contained no significant amount of carbon compounds. Witter and Polevoy repeated their work in a test chamber with high purity synthetic air. The results for the resistance came out exactly the same as the previous work in ordinary air. The ESCA analysis of these samples showed a distinct shift in the oxygen spectrum to match the valence corresponding to silver oxide. This indicates that in the presence of carbon silver carbonate will form for these long arcs and in the absence of carbon silver oxide will form.

The reasons for these results are not well understood at this time. Some speculations can be made from what is known about the differences in the chemistries of the metallic and

TABLE 19.2

Average Contact Resistance for Fine Silver Contacts After Interrupting a 1.5 ms Arc Under Different Mixtures of Nitrogen and Oxygen

Oxygen (%)	Average Contact Resistance (mΩ)
0	20
5	60
50	400

Source: Data from GJ Witter, *Proc. 51st IEEE Holm Conference Proceedings*, pp. 1–9, 2—5. [63].

gaseous arcs. The difference in the arc plasma chemistries seems to be the main factor for explaining the difference in contact resistance associated with these arcs. Neither silver oxide nor silver carbonate is stable at temperatures above 300°C. This means that these compounds must form after the surface of the contact has cooled to this temperature. The ratio of silver vapor and ions to oxygen in the dying plasma of the metallic arc is much higher than that ratio for the gaseous arc. This means that a fresh silver deposit is more likely to pick up oxygen in cooling from a gaseous arc than from a metallic arc. If a source of carbon is present then the same analogy would hold for ratios of silver to carbon. The cause for the high resistance seems to correlate with the surface chemistry of the contact after arcing rather than the roughness of the contact surface.

Acknowledgments

The authors want to thank Dr. Zhuan-ke Chen, Chugai USA, and Dr. Paul Slade for their help in editing and encouragement. We thank Prof. Terutaka Tamai, Consultant in Japan.

References

1. LH Germer, FE Hawarth. Erosion of electrical contacts on make. *J Appl Phys* 20: 1085–1108, 1949.
2. LH Germer. Arcing at electrical contacts on closure. Part I. Dependence upon surface conditions and circuit parameters. *J Appl Phys* 22(7): 955–964, 1951.
3. LH Germer, WS Boyle. Two distinct types of short arcs. *J Appl Phys* 27(1): 32–39, 1956.
4. EW Gray. Voltage fluctuations in low-current atmospheric arcs. *J Appl Phys* 43(11): 4573–4575, 1972.
5. EW Gray, LG McKnight. Possible mechanism for deposition of trace organics on electrode surfaces of arcing relays. *Int J Electronics* 37(2): 257–260, 1974.
6. EW Gray, LG McKnight. A survey of possible mechanisms of activation and erosion of relay contacts. *IEEE Trans CHMT* PHP-11(2): 121–124, 1975.
7. LH Germer, JE Smith. Activation of electrical contacts by organic vapors. *Bell Syst Tech J* 36: 769, 1957.
8. LP Solos. Organic adsorption on precious metals and its effect on static constriction resistance. Proc. *16th Holm Conference on Electrical Contacts*, 1970, pp 27–36.
9. EW Gray. Effects of vapors from phenol boards and connector lubricants on activation of wire spring relay contacts. *IEEE Trans CHMT* 1: 107–111, 1978.
10. D Oblas, D Dugger, S Lieberman. The determination of organic species in the telephone central office ambient. Proc. *25th Holm Conference on Electrical Contacts*, 1979, pp 35–39.
11. T Ide, T Sakurai, T Sasamoto. Contact reliability evaluation of plastic sealed relay. Proc. *31st Annual National Relay Conference (NARM)*, 1983, pp 7–17.
12. K Yoshida, Y Aoyama, M Isagowa, K Shirakura. Evaluation of contact failure by vaporized organic gases. Proc. *31st Holm Conference on Electrical Contacts*, 1985, pp 191–199.
13. W Bertsch, A Zlatkis, H Liebich, H Schneider. Concentration and analysis of organic volatiles in skylab 4. *J Chromatography* 99: 673–687, 1974.
14. NE Lewis. A comparative study of the performance of electrical sliding contact lubricants in a chemically active atmosphere. *IEEE Trans PHP* 12(1): 45–50, 1976.

15. JL Queffelec, N Ben Jemaa, D Travers. Activation de contacts de relais électromécaniques par leurs propres produits organiques de dégazage en atmosphere confinée, en foction de la temperature. *Rev Générate de l'Electricité* 89(9): 557–586, 1980.

16. N Ben Jemaa, JL Queffelec, D Travers, P Guenot, A Monpert. Effect of evolved organic vapors from electromechanical relays on their contact behavior: erosion and contact resistance. *Proc. 11th Int'l Conferences on Electrical Contacts*, Berlin, 1982, pp 246–250.

17. A Motoyama, K Yoshida, M Isagawa, Y Miyazaki. Evaluation and selection of organic materials used for communication relays. *Proc. 36th Annual National Relay Conference*, 1988, pp 16.1–16.8.

18. B Göttert, U Rauterberg. Evaluation of the influence of organic vapors on relay contacts. *Proc. 36th Annual National Relay Conference (NARM)*, 1988, pp 15-1–15-9.

19. B Göttert. Kunststoffausgasungen von Klebefolien beeinflussen Kontaktmaterialien. *ETZ* 112: 1216, 1991.

20. M Tamura, K Hanada. Minimizing early and random failures of reed contact units by Wiedemann resonance testing. *Proc. 8th Int'l Conference on Electrical Contacts*, 1976, pp 419–424.

21. RS Barton, RP Gavier. A mass spectrometric study of the outgassing of some elastomers and plastic. *J Vacuum Sci Technol* 2: 113–122, 1965.

22. K Muhlethaler. Kunststoffausduenstungen und ihre Auswirkungen. *Bull Schweiz Elektrotecvh Verein* 64: 1374–1378, 1973.

23. WA Crossland, PM Murphy, IA Murdoch. The evolution of organic vapors from plastic materials associated with electromechanical relays. *IEEE Trans PHP* 10: 60–64, 1974.

24. SW Chaikin. Mechanics of electrical contact failure caused by surface contamination. *Electro Technol* 68: 70–75, 1961.

25. EW Gray. The role of carbonaceous particles in low current arc duration enhancement I. Break arcs. *IEEE Trans Plasma Sci* PS-3: 151–162, 1975.

26. JA Augis. Influence of adsorbed layers on the duration of short arcs on make. *Proc. 7th Int'l Conference on Electrical Contacts*, Paris, 1974, pp 87–93.

27. HP Koidl, WF Rieder, QR Salzmann. Contact compatibility of organic vapours and their physical/chemical properties. *Proc. 43rd IEEE Holm Conference on Electrical Contacts*, 1997, pp 328–332.

28. W. Reider. Private communciation, Vienna University of Technology, Vienna, Austria, 1996.

29. Y Takenaka, K Funaki, S Shimada. Effects of organic gas components on contact resistance. *Proc. 41st IEEE Holm Conference on Electrical Contacts*, 1995, pp 260–266.

30. EW Gray. How organic vapors cause contact activation. *Insulation Circuits*, March 23–25, 1977.

31. T Uhrig. A model of contact activation and the long-term behavior of RC protected contacts. *J Appl Phys* 46(11): 4705–4711, 1975.

32. LH Germer. Arcing at electrical contacts at closure IV. Activation at contacts by organic vapor. *J Appl Phys* 25: 332–335, 1954.

33. R Schubert. Electrical contact activation: a phenomenological model. *J Appl Phys* 45: 2421–2427, 1974.

34. TA Uhrig, EW Gray. Effect of operating rate on the arc duration of activated telephone relay contacts. *J Appl Phys* 46(11): 4712–4717, 1975.

35. EW Gray, TA Uhrig, GF Hohnstreiter. Arc duration as a function of contact metal and exposure to organic contaminants. *Proc. 8th Holm Conference on Electrical Contacts*, 1976, pp 15–25.

36. EW Gray. On the electrode damage and current densities of carbon arcs. *IEEE Trans Plasma Science* PS-6(4): 384–386, 1978.

37. LH Germer. Arcs in opening contacts. *Proc. 1st Int' Conference on Electrical Contacts*, 1961, pp 237–257.

38. GJ Witter, I Polevoy. Contact erosion and material transfer for contacts in automotive relays. *Proc. 42nd IEEE Holm Conference on Electrical Contacts*, 1996, pp 223–228.

39. M Tamura, K Hanada. Contact failure caused by organic vapor in hermetically sealed relay. *Proc. 8th Int'l Conference on Electrical Contacts*, 1976, pp 413–418.

40. K Kimura, M Ishino, K Matsui, S Mitani. Resistance increase of gold-plated silver contact by carbon and its acceleration factor. *Proc. 27th Relay Conference*, 1979, pp 1.1–1.6.

41. BJ Wang, N Saka, E Rabinowicz. The failure mechanism of low voltage electrical relays. *Proc. 38th IEEE Holm Conference on Electrical Contacts*, 1992, pp 191–202.
42. C Neufeld, W Rieder. Prufverfahren zur Untersuchung der Gefahrdung der Kontakt-zuverlassigkeit durch organishche Dampfe. *VDE-Fachbericht* 44: 127–135, 1993.
43. C Neufeld, W. Rieder. Electrical characteristics of various contact contamination. *IEEE Trans CPMT* 18: 369–374, 1995.
44. C Neufeld, W. Rieder. Carbon contamination of contacts due to organic vapors. *IEEE Trans CPMT* 18(2): 399–404, 1995.
45. E Vinaricky, V Behrens. Switching Behavior of Silver/Graphite Contact Material in Different Atmospheres in Regard to Contact Erosion. *Proc. 44th IEEE Holm Conference on Electrical Contacts*, 1998, pp 292–300.
46. W Rieder, T Strof. Reliability of commercial relays during life tests at low electrical load. *IEEE Trans CHMT* 15(2): 166–171, 1992.
47. H. Koidl, W. Rieder, QR Salzmann. Parameters Influencing the Contact Compatibility of Organic Vapous in Telecommunication and Control Switches. *Proc. 44th IEEE Holm Conf. Proceedings*, 1998, pp 220–225.
48. JB Williamson, JA Greenwood. Influence of dust particles on the contact of solids. *Proc. Roy. Soc. A* 237: 560–573, 1956.
49. GJ Witter, RA Leiper. A study of contamination levels measurement techniques, testing methods, and switching results for silicon compounds on silver arcing contacts. *Proc. 38th IEEE Holm Conference on Electrical Contacts*, pp 173–180, 1992.
50. W Trächslin. Contact behavior and surface condition of commercially available contact rivets. *Proc. 8th Int'l Conference on Electrical Contacts*, 1976, p 131.
51. GJ Witter, RA Leiper. A comparison for the effects of various forms of silicon contamination on contact performance. *Proc. 24th Holm Conference on Electrical Contacts*, 1978, p 371.
52. M Ishino, S Mitani. On contact failure caused by silicones and accelerated life test method. *Proc. 23rd Holm Conference on Electrical Contacts*, 1977, p 207.
53. A Eskes, HA Groenendijk. The formation of insulating silicon compounds on switching contacts. *Proc. 33rd Holm Conference on Electrical Contacts*, 1987, p 187.
54. T Tamai. Formation of SiO_2 on contact surface and its effect on contact reliability. *IEEE Trans CPMT* 16: 437–441, 1993.
55. T Tamai, M Aramata. Safe levels of silicone contamination for electrical contacts. *Proc. 39th IEEE Holm Conference on Electrical Contacts*, 1993, pp 267–273.
56. T Tamai, M. Aramata. Effect of silicone vapor concentration and its polymerization degree on electrical contact failure. *IEICE Trans Electron* E97-C: 1996, 1137–1143.
57. T Tamai. Effect of Silicone vapor and humidity on contact reliability of micro-relay contacts. *IEEE Trans CPMT A* 19: 329–338, 1996.
58. T Tamai, K Miyagawa, M Furukawa. Effect of switching rate on contact failure from contact resistance of micro relay under environment containing silicone vapor. *Proc. 43rd IEEE Holm Conference on Electrical Contacts*, 1997, pp 333–339.
59. NM Kitchen, CA Russell. Silicone oils on electrical contacts effects, sources and countermeasures. *Proc. 21st Holm Conference on Electrical Contacts*, 1975, p 79.
60. GJ Witter, RA Leiper. A study of the effects of thin film silicone contamination and the rate of testing on contact performance. *Proc. 10th Int'l Conference on Electrical Contacts*, Budapest, Hungary, 1980, p 829.
61. GJ Witter. Chugai USA, Inc. Unpublished Research, 1994.
62. T. Tamai, S. Sawada, Y. Hattori. Manifold Decomposition Processes of Silicone Vapor and Electrical Contact Failure. *Proc. 26th Int'l Conference on Electrical Contacts*, 2012, pp 261–266.
63. MK Bernett, WA Zisman. *J Phys Chem* 66: 1207, 1962.
64. GJ Witter. Contact Contamination and Arc Interactions. *Proc. 51st IEEE Holm Conference Proceedings*, 2005, pp 1–9.
65. C. Schrank. Reliability of hermetically sealed miniature relays in silicone-rich Environments. *ICEC 2012*, Beijing, China, May 2012, pp 113–119.

66. M. Hasegawa, N. Kobayashi, Y. Kohno, H. Ando. Contact resistance characteristics of relays operated in vapors evaporated from non-silicone-type acryl-based polymeric products cured in various manners. I *Proc. 26th Int'l Conference on Electrical Contacts*, May 2012, pp. 104–108.

67. K Klungtvedt. A study of effects of silicate thickened lubricants on the performance of electrical contacts. *Proc. 42nd IEEE-Holm Conference on Electrical Contacts*, 1996, pp 262–268.

68. H Sone, H Ishida, H Sugimoto, T Takagi. Relationship of contact resistance to arc duration in Ag. *Proc. 40th Int'l Conference on Electrical Contacts*, 1994, pp 63–70.

69. ZK Chen, K Arai, K Sawa. The influence of oxygen concentration on contact resistance behaviours of Ag and Pd materials in DC breaking arcs. *Proc. 40th Int'l Conference on Electrical Contacts*, 1994, pp 79–88.

70. GJ Witter, I Polevoy. A study of contact resistance as a function of electrical load for silver based contacts. *Proc. 40th Int'l Conference on Electrical Contacts*, 1994, pp 563–568.

71. H Sugimoto, H Sone, T Takagi. Relationship between contact resistance and surface profile in arcing Ag contacts. *Proc. 40th Int'l Conference on Electrical Contacts*, 1994, pp 71–78.

Part V

Sliding Electrical Contacts

20

Sliding Electrical Contacts (Graphitic Type Lubrication)

Kiochiro Sawa and Erle I. Shobert II

Let the world slide, let the world go

Be Merry Friends, John Heywood

CONTENTS

20.1 Introduction... 1042
20.2 Mechanical Aspects.. 1044
    20.2.1 Hardness 1045
    20.2.2 Friction and Wear 1047
    20.2.3 Tunnel Resistance and Vibration.. 1048
20.3 Chemical Aspects 1051
    20.3.1 Oxidation... 1051
    20.3.2 Moisture Film... 1051
20.4 Electrical Effects.. 1055
    20.4.1 Constriction Resistance... 1055
    20.4.2 Film Resistance... 1055
    20.4.3 Fundamental Aspects of Commutation 1056
    20.4.4 Equivalent Commutation Circuit and DC Motor Driving Automotive
        Fuel Pump 1061
    20.4.5 Arc Duration and Residual Current.. 1061
20.5 Thermal Effects 1064
    20.5.1 Steady State... 1064
    20.5.2 Actual Temperature... 1067
    20.5.3 Thermal Mound.. 1068
20.6 Brush Wear 1068
    20.6.1 Holm's Wear Equation.. 1071
    20.6.2 Flashes and Smutting... 1072
    20.6.3 Polarities and Other Aspects.. 1073
20.7 Brush Materials and Abrasion 1073
    20.7.1 Electro- and Natural Graphite Brushes 1074
    20.7.2 Metal Graphite Brush and Others 1075
20.8 Summary... 1076
References... 1077

20.1 Introduction

Sliding contacts are widely used to transfer electric power and/or signals between station-ary devices and moving devices. Although these contacts move relative to each other, much of what has been discussed previously in this book on stationary contacts also applies to sliding contacts albeit with some modifications.

As shown in Figure 20.1, real contact area is much smaller than apparent contact area [1]. And conductive area is smaller than real contact area. Therefore, like stationary contacts a constriction resistance is one of the important factors in the contact resistance. But the conducting surfaces on sliding contacts are usually long narrow strips [2].

The oxidation of metal surfaces, discussed previously in Chapter 2, applies in principle to sliding contacts. The fritting or breakdown of insulating surface films, considered in Chapter 1, takes place in sliding contacts. The disturbance owing to arcing in Chapter 9 takes place on the edges of commutator bars and brushes and contributes to brush and commutator wear. Surface films owing to condensed organic materials, discussed in Chapter 7, are involved in certain sliding contact applications.

The calculation of the temperature rise owing to the electrical currents discussed in Chapter 1 can be adapted to sliding contacts with the modification for the fact that one side of the contact is moving [2]. Depending on the mechanical design, the conducting surfaces may move around on both the moving and stationary contacts. With special designs, the contact surfaces may be stabilized for short periods on the stationary contact surface. Thus, we see that there is a considerable amount of the theory of stationary contacts that applies to sliding contacts.

There are several aspects in which stationary and sliding contacts are fundamentally different. For example, with some exceptions for small contact forces on metal wires, any two clean metallic surfaces in vacuum will cold weld and seize so that the surfaces are destroyed during sliding (see Figure 20.2). This is true even for electrographite against electrographite. When the two surfaces approach each other to within atomic distances, the electron clouds of the two materials comingle and the surfaces act as though they were welded together. When this occurs in practice, the brushes disappear in a cloud of dust and are worn out in minutes (see Section 20.3).

Thus, the surfaces must be several atomic distances apart and conduction takes place across an insulating film, water in most applications, approximately two molecular layers thick by means of the tunnel effect. As discussed in Chapter 1, the tunnel effect permits elec-tronic conduction through this film as described by quantum theory. As will be pointed out

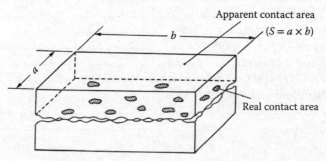

FIGURE 20.1
Apparent and real contact.

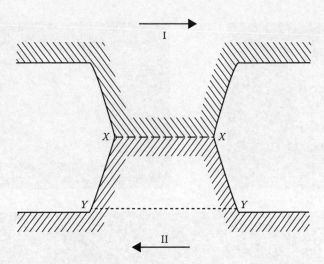

FIGURE 20.2
Rupture of a single contact.

in Section 20.3, Holm has discussed the thickness of these moisture films (see Section 20.3). The resistance of these films increases steeply with film thickness so that an increase of an additional layer or two of the film would cause a drastic increase in the voltage of the contact.

It is fortunate that water provides the material film for most brush applications at sea level. It is effective at dew points above about −10°C and is usually not harmful until liquid water becomes present and affects insulation. For electrographite brushes against copper, two molecular layers or water give a tunnel effect voltage, ΔV, of about 0.2–0.3 V. Three molecular layers of water add an additional 1 V drop. Since the normal total voltage drop on electrographite brushes is about 1 V, it is readily seen that the two surfaces must slide consistently within 5–10 Å (0.5–1 nm) of each other. Other filming agents can replace water when it is not available.

Another basic difference between stationary and sliding contacts is friction. Friction has two effects. First is the heat generated by sliding that adds to the heat generated electrically in determining the temperature. At high speeds this can be appreciable, and while it is generated over the mechanical contact surface, which is usually larger than the electrical conducting surface, it still must be dissipated in the brush collector system. The second friction effect is concerned with friction-excited vibration (see Section 20.2). Many ordinary machines have the possibility of having certain natural periods excited by friction, but it is possible to eliminate this by proper design.

Another important factor, which is not always recognized, is the fact that for long-term operation, the electrical situation must be the same on each commutation bar or there must not be synchronous effects of variable currents on rings. Since the collector films are influenced by the current, the friction will vary where the current varies. If this takes place consistently on any section of the collector, vibrations and eventual burning will take place, with disastrous results. This is discussed in Section 20.4.

Finally, there is a fundamental difference in the philosophy for successful brush operation compared with stationary contacts. Since the film is such an important component, and since it is complicated by moisture, oxides, organic compounds in the atmosphere, and brush materials, it is necessary to have materials that have a certain controlled amount of abrasive action. This prevents overfilming and provides consistent operation over a wide variety of circumstances (see Section 20.7). In practical applications, this is achieved with

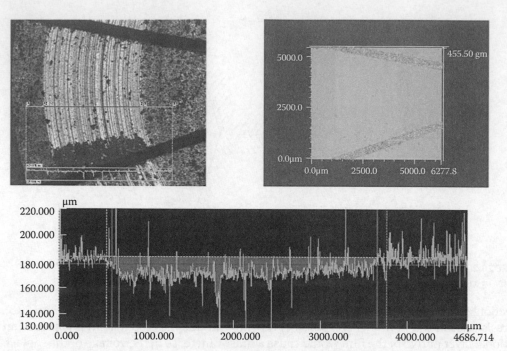

FIGURE 20.3
An example of laser microscopy analysis.

metal and graphite materials, with the addition of known abrasives, and with abrasive forms of carbon and graphite.

Recently, Finite Element Analysis and other simulation methods are used to examine phenomena of sliding contacts from electric, thermal and mechanical aspects [3]. In addition, the laser microscope and other analytical methods are used to evaluate friction wear and other parameters (see Figure 20.3) [4]. From fundamental viewpoints, the real area of contact between brush and copper ring has been measured, and the current density distribution was directly observed by using light emission diode wafer [5,6].

It should be noted that we are dealing in this section primarily with sliding contacts with graphite type lubrication. There are other types of sliding contacts, for example, fiber brushes and precious metal wires that operate on different fundamental bases: see Chapters 22 and 23. These are discussed in seperate sections below. With this background, we will discuss the mechanical, chemical, electrical, thermal, and material factors that work together in successful applications. Since we are dealing with phenomena, which are complicated combinations of mechanical, chemical and electrical effects, they are present at the same time, but they must be separated to understand the results and to solve the problems that arise.

20.2 Mechanical Aspects

There are reasons for considering the mechanical aspects of sliding electrical contacts first. The mechanical system determines several important factors which relate to operation, such as the radius of contact at the face of the brush, the possibility of friction-excited vibration, and the direction and characteristics of heat flow. Also, the mechanical system must ensure that the stationary and moving surfaces remain in contact at between 5 and 10 Å (0.5–1 nm) of each other owing to the reason mentioned later. This represents a distance of about two molecular layers of water.

20.2.1 Hardness

First, we consider the hardness.

Assuming that the deformation of two surfaces is elastic, the radius of their contact a is expressed by the following Hertz's formula [1].

$$a = \sqrt[3]{\frac{3}{4} P \left(\frac{1-\mu_1^2}{E_1} + \frac{1-\mu_2^2}{E_2} \right) \left(\frac{1}{r_1} + \frac{1}{r_2} \right)^{-1}} \tag{20.1}$$

where r_1, r_2: curvatures of two contacts, E_1, E_2: Young's modulus of elasticity of two contacts, μ_1, μ_2: Poisson ratios of two contacts, and P: contact force.

When two contacts are sphere, centers of the spheres approach each other by y

$$y = a^2 \left(\frac{1}{r_1} + \frac{1}{r_2} \right) \tag{20.2}$$

Actually, the surfaces have certain roughness and certain waviness. At contact make, asperities of one contact indent in the other contact. At small average pressure the indentations may be formed elastically. With increasing average pressure, indentations become plastically produced. Finally, almost all indentations could be deformed plastically, and the following equation is derived,

$$P = HA_b \tag{20.3}$$

where P: contact force, H: hardness, A_b: load bearing area.

However, actually, when some indentations deepen, other asperities obtain the opportunities to make contact. These initially generate shallow elastic indentations. So, the average pressure will be smaller than H, that is,

$$\bar{p} = \xi H \tag{20.4}$$

with $\xi < 1$. Hence,

$$P = \xi H A_b \tag{20.5}$$

Theoretically, the value of ξ is between 0 and 1, but according to measurements, the value is between 0.1 and 0.3 in many cases.

When two surfaces such as a ball and a plane are pressed together, the ball will indent into the plane until the elastic force supports the load. As the ball becomes larger for the same load, the surface pressure defined as the contact hardness becomes smaller as shown in Figure 20.4 [7]. Since we are dealing with relatively flat surfaces, the hardness that determines the surface area is the low point at specific depth 0. This is usually about two thirds of the value of the hardness at A', B', or C' that represents a small ball on a plane. Values of

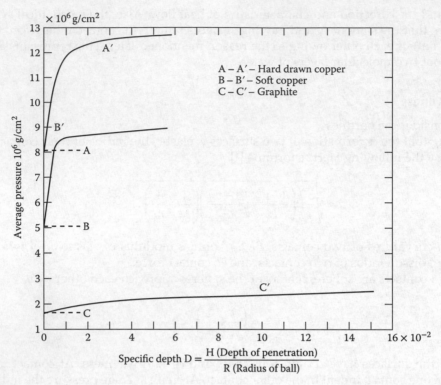

FIGURE 20.4
Contact hardness.

the contact hardness are given for various brush materials in Table 20.1. Because of certain small eccentricities in the rotating collector, the radius of the brush face is always slightly larger than the radius of the collector. Thus, we have two cylinders in compression. In addition to the eccentricity which makes the brush radius r_2 larger than that of the collector r_2, there is usually some possible side play of the brush that makes the face saddle shaped. Thus, we have two spheres one inside the other and the formula for the radius of the contact is obtained from (Equation 20.1) with $\mu = 0.3$ and $E_1 = E_2 = E$:

$$a' = 0.88 \sqrt[3]{\frac{P}{E}\left(\frac{1}{r_1} - \frac{1}{r_2}\right)^{-1}} \quad \text{(cm)} \tag{20.6}$$

TABLE 20.1

Contact Hardness

Material	Contact Hardness
40 Cu 60 C	1.0×10^2 N mm^{-2}
Resin-bonded graphite	2.6
Electrographite	2.9
Carbon graphite	3.5
Copper	4–9
Silver	3–7

where P is the force in grams and E is the elasticity of the softer brush materials. Going further, the elastic penetration, y, is given by

$$y = \frac{a'^2(r_2 - r_1)}{r_1 r_2} \quad \text{(cm)} \tag{20.7}$$

This elastic penetration is important since it provides the restoring force to permit the brush face to follow small variations on the collector surface such as low and high commutator bars. It is the only apparent mechanism that will permit the brush surface to follow known variations in collector surfaces.

The elastic recovery of the surface penetration should approach the speed of sound in the materials. This is faster than any commutator bar variation that is known in practice. Of course, the distance to be recovered must be within the limits of the penetration available at the brush face.

Thus, we have a means of providing that these surfaces stay within the 0.5–1.0 nm distance from each other that is required by the tunnel effect conduction.

20.2.2 Friction and Wear

The next mechanical effect is friction. The coefficient of friction is defined by the relation

$$\mu = \frac{F}{P} \tag{20.8}$$

where F is the tangential friction force and P is the normal force. As Holm [8] pointed out,

$$F = \psi A_b \tag{20.9}$$

where ψ is the shear strength of the film when normal sliding takes place and A_b is the load bearing area. Without a film ψ becomes the shear strength of the weaker of the two contact materials.

In the elastically supported, plastically deformed surfaces A_b can be obtained from Equation 20.5,

$$P = \xi H A_b \tag{20.10}$$

Thus,

$$\mu = \frac{F}{P} = \frac{\psi}{\xi H} \tag{20.11}$$

In this relation, ψ is determined by the complex film consisting of the moisture film, the oxide film, and any other materials or contaminants that might be in the atmosphere. The adjuvant added to the brushes provides this film in vacuum. H is a material constant of the brush which is generally much softer than the collector, and ξ is a factor that can vary with the machine and its operating condition. For example, high brush friction as a result of light load running may be caused by a decrease in this factor as well as the decrease in temperature.

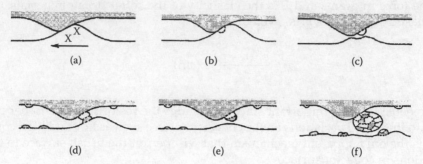

FIGURE 20.5
Transfer and growth wear model. (a) Contact; (b) Internal rupture; (c) Transfer element; (d) Attached transfer elements; (e) Transfer particle; (f) Grown transfer particle.

On the basis of the idea of adhesive wear, R. Holm proposed the following equation of wear,

$$V = Z \frac{P\ell}{p_m} \tag{20.12}$$

where V: wear volume, P: contact force, ℓ: sliding distance, p_m: hardness of softer contact and Z: a constant called wear factor by Holm. The equation is sometimes called Holm's law of wear, and regarded as one of fundamental laws of tribology together with Coulomb's law of friction. When the wear is expressed in volume V or mass M, wear per unit distance, V/ℓ or M/ℓ is defined as wear rate. Further, wear is proportional to friction distance and contact force according to (20.12) and then, wear per unit distance and contact force, $V/P\ell$ or $M/P\ell$ is useful. These are defined as specific wear rate or specific wear amount.

Dr. Sasada proposed another wear model called transfer and growth model [9]. According to the model, two small projections make contact on sliding surfaces (Figure 20.5a), and generally internal rupture occurs in a soft material (Figure 20.5b). A ruptured small particle is attached on a hard material, called transfer element (Figure 20.5c). During slide motion, another internal rupture occurs on the previous transfer element (Figure 20.5d) and transfer particle is formed (Figure 20.5e). The transfer particle grows to a large one as shown in (Figure 20.5f), and the large one drops off shortly as a wear particle. This model can explain how to produce various sizes of wear particles: see also Chapter 7, Section 7.2.

20.2.3 Tunnel Resistance and Vibration

As for molecular layers of water, a simple calculation is shown here. An electrographitic brush has a voltage drop of about 1 V with variations of about 0.2 V as shown on oscilloscope traces. This is demonstrated by a brush with a face area of 1 cm² and a force of 500 g with 20 A flowing at ordinary speed. About 0.2–0.3 V will be the drop in the water film. The specific resistance of the electrographitic materials is about 6000×10^{-6} Ω cm. Using the formulas for constriction resistance we find that the conducting surface area is 5×10^{-3} cm². Considering Figure 20.6 for $\varphi = 4$ V the tunnel resistance of this film at 6Å is 10^{-8} Ω cm². Thus the current conducting region of this film has a resistance of 2×10^{-4} Ω at 20 A. This represents a voltage of 4×10^{-5} V. The next two moisture layers if added to the system increase the tunnel resistance to 10^{-3} Ω cm², as can be seen from Figure 20.6 for

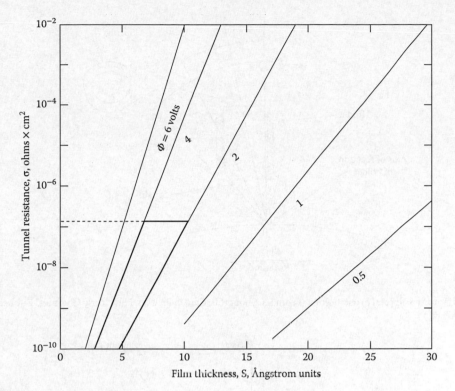

FIGURE 20.6
Tunnel resistance (φ is the work function) [1,2].

a film thickness of 12 Å and a work function of 4 V. The resistance of this layer would be 0.2 Ω, giving a voltage of 4 V at 20 A. Since this is not possible, we see that the brush does remain within a very close mechanical range.

The next mechanical consideration involves the generation of friction-excited vibrations. This was considered from the standpoint of brushes by Shobert [10]. A machine was built in which a brush could be rotated about the point 0 in Figure 20.7 [10]. The brush had its heel cut away so that the point of application of the friction force would not change as the brush was rotated. It was found that brushes could and would chatter and whistle within the angular range θ, but that outside this angle on either side, the brushes could not chatter. The reason for this is shown in Figure 20.8. At the angle θ, the friction force is equal to the component of P tangential to the surface. At a slightly greater angle the tangential component $P \sin \theta$ of the brush force P is greater than the friction force, $\mu P \cos \theta$, so there is no backward motion of the brush and therefore no chatter. Where the friction force $\mu P \cos \theta$ is greater than $P \sin \theta$, vibrations are excited because there is a component of the vibration normal to the surface that provides the periodicity. The friction force $\mu P \cos \theta$ has a component $\mu P \cos \theta \sin \theta$ opposite to the direction of the force P that causes the brush to shorten elastically. This shortening decreases the normal force $P \cos \theta$ that causes the friction force $\mu P \cos \theta$ to decrease. When $\mu P \cos \theta$ decreases, the brush force recovers elastically and the process is repeated. The brush then vibrates longitudinally at a frequency determining by its elasticity, length, and density.

At the limit where $\mu P \cos \theta = P \sin \theta$, we may write

$$\mu = \tan \theta \tag{20.13}$$

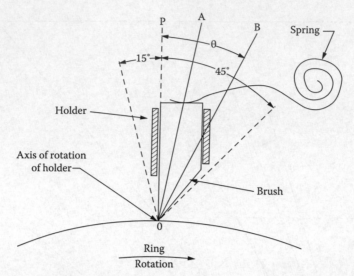

FIGURE 20.7
Schematic diagram of special brush holder. (From E.I. Shobert II, *Carbon Brushes*, New York: Chemical Publishing Co., p 62, 1965 [2].)

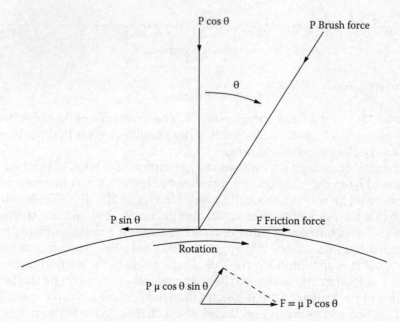

FIGURE 20.8
Chatter range.

and thus determine the maximum angular range θ within which brushes may chatter. If the force P is on the other side of the vertical, $P \sin \theta$ and the friction force F are in the same direction, forcing the brush against the holder, and there is no vibration. The experimental results in Reference [10] show this to be the case.

Radial brushes have a tendency to chatter depending on which part of the apparent surface is making contact. Chatter is inhibited and prevented by designs in which the brush is held against one or the other sides of the holder by various bevels [11].

As for other aspects, a ring surface has roughness, and its influence on contact voltage drop is reported [12], and the relation between contact voltage deviation and ring profile distortion is made clear [13]. As mentioned before, the friction is one of the key factors, and the measurement methods of the dynamic of friction are reported [14].

We cannot over-emphasize the importance of the mechanical design on successful brush operation. The actual requirements are very rigid. The presence of suitable lubricants, moisture, and adjuvants, and the properties of the brush face make successful applications possible. It is important to know why brushes work so that in designing a machine or working on a problem, we know what the problem really is and not waste effort on something that will not be productive.

20.3 Chemical Aspects

Under chemical effects, we will consider the oxidation of the conductor surface, the presence of the moisture layer, the effects of other environmental or ambient materials, and how these relate to friction and conduction.

20.3.1 Oxidation

The question of film formation on metals has been discussed in several places previously in this book (see Chapters 1 through 3). In sliding contacts, we are concerned primarily with collector materials of copper and copper-based alloys, silver and silver alloys, and several of the precious metals, gold, platinum, palladium, for example. The copper-based alloys oxidize by the diffusion of the metal ions through the film so that the rate of oxidation decreases as the film develops. The same is true of silver in an atmosphere containing sulfur, except that without sulfur the electrical conducting area can increase beyond the area determined by "fritting," and silver can act similarly to gold.

On the other hand, the rate of oxidation is influenced by the direction of the electric current. Since the oxidation proceeds by the migration of metal ions through the film, this will be enhanced under the cathode brush and impeded under the anode brush. This explains the fact that the voltage drop is usually higher for cathode brushes, although this may be reversed for metal graphite brushes.

When copper is machined, an oxide film several molecular layers thick forms quickly behind the cutting tool. Figure 20.9 [15] shows how copper oxidizes in relatively dry air. Figure 20.10 [16] shows the relative oxidation rates on different crystal faces of copper and Figure 20.11 [17] shows the effect of moisture in the air. Thus, we readily see that the oxide film alone is complicated. The addition of other atmospheric contaminants further complicates the result.

20.3.2 Moisture Film

As pointed out in the introduction to this chapter, the moisture film is the most important factor in the atmosphere affecting brush performance. The importance was first noted by Dobson [18] on rotary converters in the Chicago area where brush life was very short during the season of dry air in the winter. The problem was solved by releasing steam into the rooms when the humidity was low. The problem reappeared when high flying aircraft in World War II lost electric power when the brushes wore rapidly or "dusted." This problem

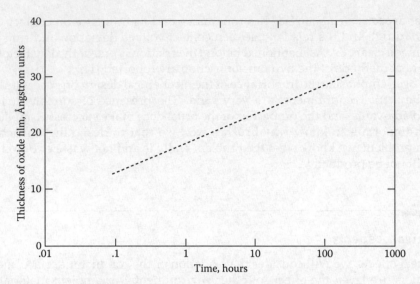

FIGURE 20.9
Oxidation of copper in air. (From U.R. Evans, *The Corrosion and Oxidation of Metals*. London: Edward Arnold, Table 1, p 53, 1960 [15].)

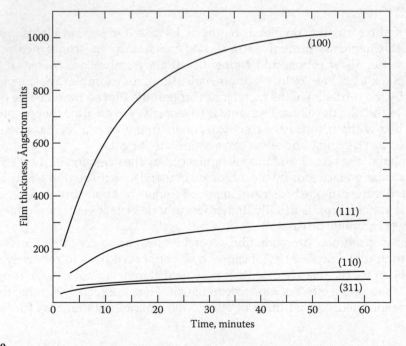

FIGURE 20.10
Oxidation of four faces of a copper single crystal at 178°C. H.C. Gatos, *The Surface Chemistry of Metals and Semiconductors*, New York: Wiley, pp 486 ff., 1959 [16].)

was solved by Elsey [19] who added metal halides to the brushes. PbI and BaF were the first to be considered. These had the drawback of requiring extensive run-in at sea level to develop a protecting film. MoS_2 solved this problem by providing a material that would protect immediately on running at altitude. It should be noted that many different materials have the same characteristic as MoS_2 to protect against this rapid wear, but that there

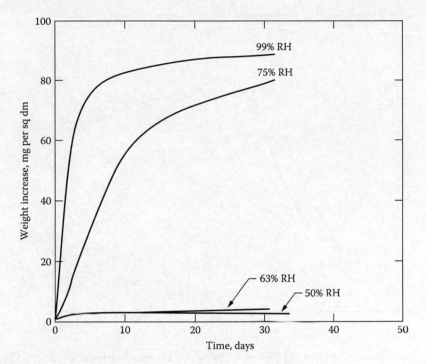

FIGURE 20.11
Corrosion of copper. (From W.H.J. Vernon, *Trans Faraday Soc.* 27:255, 582, 1931 [17].)

are other requirements for brushes. These include long-term storage with the brush in contact with the collector in high humidity. Many corrode the collector or the brush holder.

The above fact told us that moisture was the key, although it was later understood that other materials in the atmosphere could also prevent dusting. Oil vapors from the lubricants, smoke, organic vapors, any materials that could be quickly adsorbed on the collector could prevent the high wear. It was for these reasons that material tests had to be run under very clean conditions for consistent results.

The result, of course, is that brushes now operate in the vacuum of outer space for long periods. From a practical point of view the lower limit on moisture content can be taken as about −10°C dew point. While wear is somewhat variable around this point, catastrophic wear does not usually take place above this point.

We have seen that for friction, the friction force is given by (Equation 20.9),

$$F = \psi A_b \tag{20.14}$$

where ψ is the shear strength of the film and A_b is the mechanical load-bearing surface. ψ is the result of the chemistry of the system and A_b is dependent on the mechanics.

It would appear that the shear strength of the film is governed in part by an effect similar to an activated adsorption in which the rate of adsorption at the surface varies with temperature, as shown in Figure 20.12 [20]. Figure 20.13 shows the general nature of the way that friction varies with temperature in graphite-lubricated contacts. As will be shown later, temperature depends not only on the ambient conditions but also on the microscopic situation at the contact face. When sulfide or silicone vapor is included in air, contact failure may happen owing to increasing contact resistance. As for silicone contamination, safe

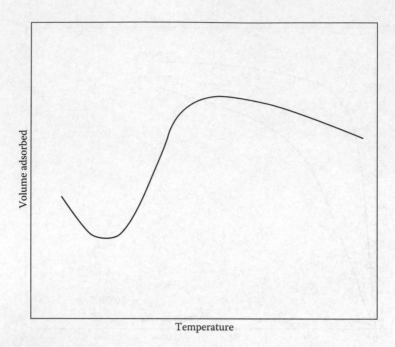

FIGURE 20.12
Activated adsorption. (From S. Glasstone, K.J. Laidler, H. Eyring. *The Theory of Rate Processes*, New York: McGraw Hill, pp 339 ff, 1941 [20].)

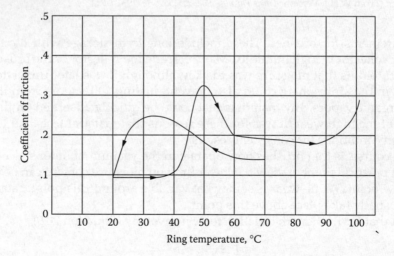

FIGURE 20.13
Coefficient of friction vs. ring temperature. Electrographite in air; dew point 14°C. (From E.I. Shobert II, *Carbon Brushes*, New York: Chemical Publishing Co., 1965 [2].)

level of silicone vapor was reported [21]. In addition, the idea of vapor lubrication was successfully proposed to protect contacts from surface contamination from atmosphere [22].

For special application of brushes brush-ring systems are sometimes used in gases other than air [23,24].

When the possibilities are considered it is easy to see why many brush tests are out of control. As previously stated, each part of the system is affected by and depends on various aspects of the other variables so that all must be taken together in consideration of a particular problem.

20.4 Electrical Effects

While the purpose of a sliding electrical contact is to conduct current between moving and stationary contacts, this cannot be done successfully unless the mechanical and chemical aspects are proper. When they are in order we can review the electrical effects and see how they relate to the other phenomena. We are concerned here with the contact voltage drop, the tunnel effect, the fritting or electrical breakdown of the oxide or non-conducting films, and the commutation phenomenon.

20.4.1 Constriction Resistance

One of main elements of the contact drop is the constriction resistance within the brush. As shown by Shobert [2] the conducting surfaces on the face of the collector are sometimes long narrow lines. The resistance of such considerations can be approximated by the relation

$$R = \frac{\rho}{2\pi l} \ln \frac{B}{b} \quad \text{(ohms)} \tag{20.15}$$

where

ρ is the resistivity of the brush;

l is the length of the conducting area;

B is a dimension similar to the width of the brush; and

b is the half-width of the conducting area.

This relation neglects the effects at the end of the conducting areas, but we are concerned with principles at this point. A further refinement would put a resistance of half of a "b" spot at each end of the strip or a resistance

$$R_b = \frac{\rho}{4b} \tag{20.16}$$

in parallel with that of Equation 20.15. As shown by Shobert [25] the conducting surface is influenced by the voltage direction and

$$R_c \sim \frac{1}{l} \tag{20.17}$$

under the cathode brush and

$$R_a \sim \ln \frac{1}{b} \tag{20.18}$$

for the anode brush. The other element of the contact drop or voltage drop in a sliding contact is the drop in the collector that is usually negligible except in the case of metal–graphite brushes with high-metal content.

20.4.2 Film Resistance

Then there is the drop in the moisture or other film. Since this film is insulating, (H_2O, CuO, MoS_2, etc.) as is required to prevent the electron clouds of the brush and collector from comingling, conduction takes place across this film by means of the tunnel effect. This is shown in Figure 20.14, in which it is seen that electrons with sufficient energy can "tunnel" through the potential barrier between the two surfaces. The effective resistance of this insulating film has been calculated by Holm as shown in Figure 20.6.

Figure 20.15 [25] shows the "B" fritting of a brush film on copper as probed with a gold wire. The test results in Figure 20.15 were taken on (a) an insulating region and (b) a conducting region of a carbon brush track with a gold wire. In Figure 20.15a "B" fritting is shown to occur around 0.3 V, while in Figure 20.15b, R-V characteristics are reversible. For comparison "B" fritting on an oxidized copper plate is shown in Figure 20.16 [1]. The temperatures were calculated on the same basis as those to be discussed in the next section.

In the case of a brush-ring sliding system, a voltage increase is observed just after the ring starts moving. As shown in Figure 20.17 [25] the increase value is relatively constant at about 0.3 V. Obviously, this additional voltage is caused by some film formed between moving ring and brush. The voltage between the brush and the ring at the edges of the tunnel effect film seems to be "B" fritting voltage, which is relatively constant and independent of the current for a copper collector and silver in the presence of sulfur. The voltage difference between stopped and running for electrographite brush on copper is about 0.3 V and the correlation is apparent. Thus the voltage drop can be summarized by Figure 20.18 [2].

We can now say that the conduction mechanism may be represented by Figure 20.18 where we have the constriction resistance in the collector, the voltage increase on sliding, and the constriction within the brush. Included in the equilibriums at low and high current is the effect of abrasives in the brush that will be discussed later. Important to note here is the fact that if the current is suddenly (60 Hz) reversed to zero and back as in the straight lines A and B, the resistance remains constant, indicating that the conducting area has not changed and that there is no semiconducting effect present.

20.4.3 Fundamental Aspects of Commutation

Large power dc motors used in steel plants, transportation vehicles, and other applications have been replaced by ac variable speed motors recently. However, small dc motors are widely used in various applications. Specially, many small dc motors are in

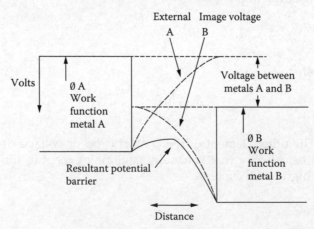

FIGURE 20.14
Schematic representation of the tunnel effect.

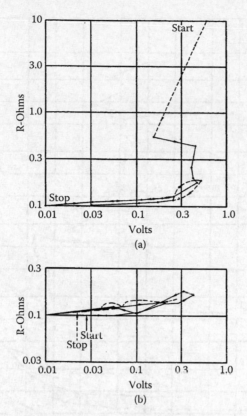

FIGURE 20.15
"B" fritting of copper oxide [25]: (a) insulating spot on the ring; (b) conducting spot on the ring.

automotive appliances. Approximately 30–40 motors are used in one automobile. Recently, the automotive electric power system is under consideration of changing from 14 V to 42 V power net. Supposing the 42 V power net, intensive research has been made [27–30].

Among automotive electromechanical devices, a dc motor has the feature that it is high in efficiency and easy to adjust its rotation speed by voltage of a power source. However, it has a sliding mechanism between commutator and brush, and arc discharge between brush and commutator segment is a big problem [31]. A commutation of the dc motor means that a coil current is reversed in its direction when the coil passes through neutral zone between magnetic poles.

At the termination of the commutation an arc discharge sometimes occurs, depending on commutation current, coil inductance and speed. Figure 20.19 shows a coil current distribution of the dc motor. In this case, a torque generates in counterclockwise direction. However, the armature coil rotates and the coil located around the magnetic neutral zone enters into the opposite magnetic pole zone. If the current direction of the coil remains unchanged, generated torque will be reversed in direction and the counterclockwise rotation cannot be continued. So, in order to keep counterclockwise rotation, the coil current should be changed in direction when it passes through the magnetic neutral zone. The current reverse is called commutation in a dc machines and Figure 20.20 shows how the commutation current changes its direction with time. The commutation current is expressed by Equation (20.19) during the commutation period.

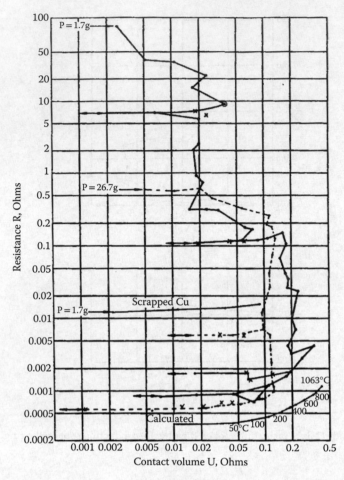

FIGURE 20.16
"B" fritting of copper oxide on copper rods. (From R. Holm, *Electric Contact Handbook*. Berlin: Springer-Verlag, pp 7–8, 1967 [1].)

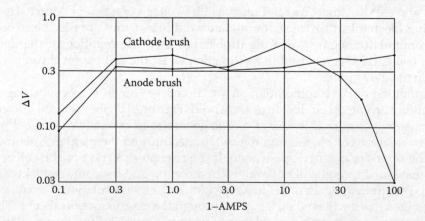

FIGURE 20.17
Voltage difference ΔV between stop and run as a function of current. (From E.I. Shobert II, Electrical resistance of carbon brushes on copper rings. *IEEE Trans. Power, Apparatus and Systems*, 13: 788–797, 1954 [25].)

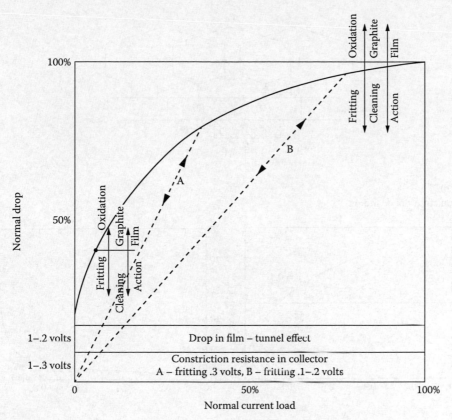

FIGURE 20.18
Schematic of brush voltage drop vs. current.

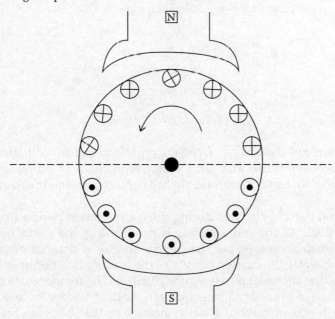

FIGURE 20.19
Coil current distribution of dc motor.

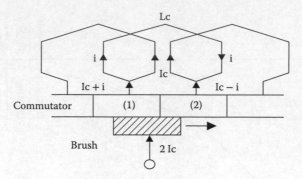

FIGURE 20.20
Commutation circuit of dc motor.

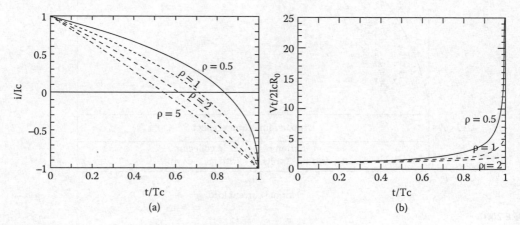

FIGURE 20.21
(a) Change of commutation current; (b) Change of trailing voltage ($\rho = L_c / R_0 T_c$).

$$L_c \frac{di}{dt} + R_1 (I_c + i) - R_2 (I_c - i) = 0$$

$$R_1 = \frac{R_0}{1 - t/T_c}, \quad R_2 = \frac{R_0}{t/T_c} \tag{20.19}$$

$$t = 0, i = I_c \text{ and } t = T_c, i = -I_c$$

where L_c: commutation coil inductance, i: coil current, I_c: armature coil current, R_1: trailing side resistance between brush and commutator segment, R_2: leading side resistance, T_c: commutation period, R_0: contact resistance at full contact between brush and commutator segment.

From (20.19), the coil current changes during the commutation period from I_c to $-I_c$ as shown in Figure 20.21(a). As the commutation is progressing, the contact surface of the trailing side of the brush decreases, but the current change is delayed owing to the coil inductance L_c. Consequently the current density of the trailing side becomes high, and the contact voltage drop v_t is increasing (see Figure 20.21(b)). The temperature of the contact area is rising, and reaches a threshold value (usually, boiling voltage in case of metal and sublimating voltage in case of carbon). Finally, metallic contact between brush and commutator segment is broken. At that time, the current still flows through the brush and segment that is called a residual current. If the residual current is larger than the minimum

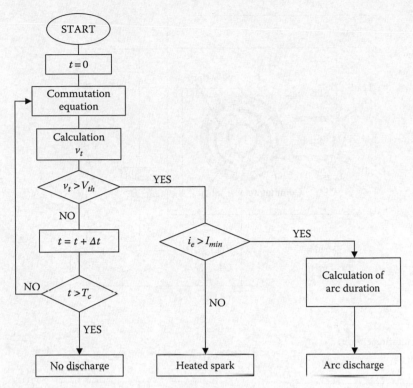

FIGURE 20.22
Calculation model of commutation arc.

arc current, an arc discharge will occur when the brush leaves the segment. This model is shown in Figure 20.22.

20.4.4 Equivalent Commutation Circuit and DC Motor Driving Automotive Fuel Pump

Generally an actual dc motor should be used to investigate its performance. However, in an actual dc motor, it is difficult to measure the duration of arc discharge and commutation current. So, an equivalent commutation circuit. was proposed. It represents only the commutation process of the dc motor in terms of an electrical circuit. Figure 20.23 shows an equivalent commutation circuit [32,33]. Among many dc motors used in a car, a motor driving a fuel pump has a unique feature that the commutation is carried out in gasoline. We have been much interested in the commutation in gasoline and its influence on commutator and brush. Figure 20.24 shows typical voltage and current waveforms during the commutation. The final stage of the commutation is enlarged, and the voltage between brush and commutator segment abruptly increases and an arc discharge is established. As a feature of arc voltage, the voltage value in gasoline is higher than in air [33].

20.4.5 Arc Duration and Residual Current

Figure 20.25 shows an equivalent circuit when an arc occurs between brush trailing edge and segment. The circuit equation is expressed as follows;

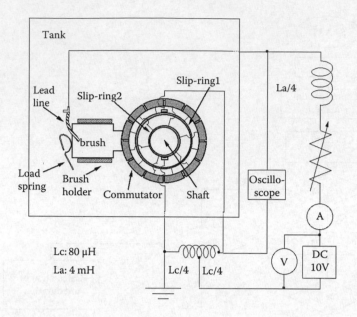

FIGURE 20.23
Equivalent commutation circuit.

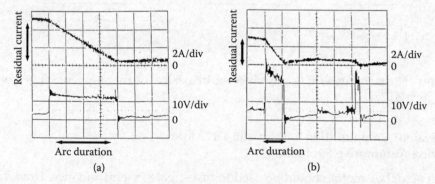

FIGURE 20.24
Voltage and current waveforms. (a) In air; (b) In gasoline.

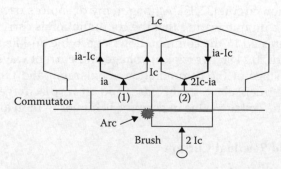

FIGURE 20.25
Commutation circuit during arc.

$$L_c \frac{d(I_c - i_a)}{dt} + R_c(I_c - i_a) + R_0(2I_c - i_a) - v_a = 0 \tag{20.20}$$

where R_c is a resistance of the commutation coil and i_a and v_a are arc current and arc voltage, respectively.

As the initial condition is $i_a = I_e$ (residual current) at $t = 0$, the solution of (20.20) is obtained as follows;

$$i_a = (I_e + K)\exp(-\frac{t}{T}) - K \tag{20.21}$$

where $K = \dfrac{v_a + (R_c - 2R_0)I_c}{R_c + R_0}$, $T = \dfrac{L_c}{R_c + R_0}$

As the arc distinguishes at $i_a = I_{min}$ (minimum arc current), the arc duration is expressed by;

$$\tau_a = T \ln \frac{I_e + K}{I_{min} + K} \tag{20.22}$$

If $\dfrac{\tau_a}{T} \ll 1$ and $K \simeq \dfrac{v_a}{R_c + R_0}$, the arc duration (20.22) is simply reduced to the followings;

$$\tau_a = \frac{L_c}{v_a}(I_e - I_{min}) \tag{20.23}$$

Figure 20.26 shows the relation between residual current and arc duration that are experimentally obtained in air and gasoline. From these results, the arc duration is almost proportional to the residual current as predicted by (20.23). On the other words, the Equation 20.23 is simple but useful [33].

In most machine designs, particularly in cases where the machines are made in large production quantities, the effort is made to use materials and processes to their economic limits. This results in problems that usually result in short brush life or related difficulties. To solve these problems it is necessary to understand all of the factors that are effective in successful operation and to see where deviations from the ideal take place. Again we are dealing with complicated situations where mechanical, chemical, electrical, thermal, and material problems are concerned.

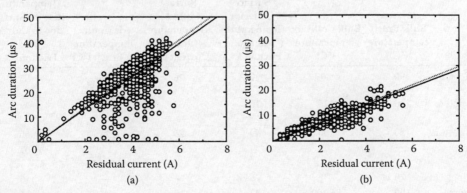

FIGURE 20.26
Arc duration vs. residual current Arc. (a) In air; (b) In gasoline.

20.5 Thermal Effects

Thermal effects in stationary contacts are discussed in Chapter 1. The basic principles for sliding contacts are similar, with the addition of friction heating which takes place over the mechanical sliding surface and transient phenomena which appear because the mechanical and electrical contact spots are moving to different positions on the collector and the conducting regions on both the brush and collector are elongated in the direction of motion.

From the standpoint of knowing how to analyze various problems of brushes, however, it is important to know the maximum temperature in the brush and the temperature at the surface of the collector. These have not been measured directly, but there are means to calculate them with sufficient accuracy for most purposes. These calculations involve, first, calculating hypothetical steady-state temperatures generated by both the friction and the electrical heat, and then applying a time-dependent factor that indicates what parts of these hypothetical temperatures are reached. Examples of these calculations and the details of the formulas are given in Shobert [2], in the chapter on brush temperatures. They are summarized here and several examples of the results are shown in Table 20.2.

Certain essential ideas of the contact theory provide the basis for these temperature calculations. These include the calculation of the mechanical sliding surface from the hardness of the brush materials, the calculation of the electrical contact surface from the voltage drop and the specific resistance of the brush materials, a preliminary calculation of the hypothetical steady state temperatures owing to friction and current—assuming that the contact surfaces have time to come to equilibrium—and then the calculation of the fraction of the steady-state temperature that is actually reached during the available time of contact while the contact spots are passing under the brush.

20.5.1 Steady State

The friction heat is generated over the mechanical contact surface, which is of the order of 1% or less of the apparent brush surface. Because the hottest isotherm is always located

TABLE 20.2

Brush Temperatures (Kohlrausch–Holm Method)

Application	Bulk Brush Temperature (°C)	Bulk Collector Temperature (°C)	Surface Collector Temperature due to Friction (°C)	Surface Collector Temperature due to Current (°C)	Maximum Temperature in Brush (°C)	Temperature Rise on Surface due to Current and Friction (°C)
dc power 25 kw exciter	90	80	29	26	78	55
Diesel locomotive motor	150	125	16	33	96	49
Aircraft generator	150	100	40	26	71	66
Automotive alternator	75	75	9	92	148	101
Silver slip rings	70	70	6	—	—	6
Fractional horse power (cleaner)	100	85-100	22	14	91	36

in the brush, it is assumed that all of the friction heat goes to the collector and causes a hypothetical steady-state temperature rise (ϑ_f) where

$$(\vartheta_f) = qWb_1 \quad (°C) \tag{20.24}$$

and where q is the friction heat current and W_{b_1} is the thermal resistance within the collector of the constriction pertaining to the mechanical contact area on the collector surface. For one circular contact area, A_b,

$$W_{b_1} = \frac{1}{4\lambda b_1} \quad (°C/W^{-1}) \tag{20.25}$$

where λ is the thermal conductivity of the collector and b_1 is the radius of A_b. This surface A_b is

$$A_b = \pi b_1^2 = \frac{P}{H} \quad (cm^2) \tag{20.26}$$

where P is the brush spring force and H is the contact hardness, about two-thirds the hardness as defined by ball indentation tests (see Figure 20.4).

Next, we calculate the hypothetical steady-state temperature rise of the collector surface and of the maximum temperature in the brush (see Figures 20.27 and 20.28). From the consideration of the unsymmetrical contact, we have the equations, which are derived in [2], but which are listed here for reference:

$$V^2 = Z\rho_1\lambda_1(\theta - \vartheta) \quad (V^2) \tag{20.27}$$

and

$$u^2 = 2\rho_1\lambda_1\theta \quad (V^2) \tag{20.28}$$

where now V is the voltage from the brush face to the point of maximum temperature in the brush, u is the voltage from the point of maximum temperature in the brush to the bulk of the brush, ρ_1 is the resistivity of the brush materials, and λ_1 is the thermal conductivity of the brush material. For simplicity, ρ_1 and λ_1 are considered in these calculations to be independent of the temperature. While involving an approximation, this is permissible since the steady-state super temperatures are fictitious quantities which never appear, and the values of ρ_1 and λ_1 near the bulk temperature of the brush and collector are suitable in the light of some of the other assumptions and approximations. ϑ is the steady-state surface super temperature of the collector, and θ is the steady-state maximum super temperature in the brush. ϑ and θ are related by the expression

$$\vartheta = \frac{4\lambda_1(\lambda_1 + \lambda)}{(2\lambda_1 + \lambda)}\theta \quad (°C) \tag{20.29}$$

We also find that

$$\Phi V = \rho\lambda\vartheta \quad (V^2) \tag{20.30}$$

where Φ is the voltage drop in the collector, which is usually small compared with $V + u$, the voltage drop in the brush.

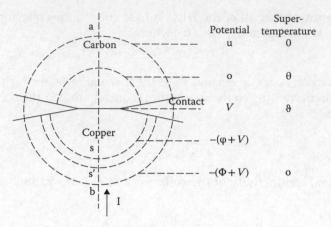

FIGURE 20.27
Relation of voltage and temperature in the asymmetric contact.

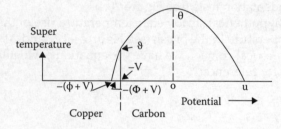

FIGURE 20.28
Temperature vs. voltage in the asymmetric contact.

The voltage will be distributed approximately proportionally to the average resistivities since the electrical conducting surface is the same for both the collector and the brush. We may therefore write

$$\frac{\Phi}{V+u} = \frac{\rho}{\rho_1} \tag{20.31}$$

if ρ and ρ_1 are averages within the temperature ranges.

The voltage $V + u$ represents, in most practical cases, the measured voltage brush drop. Thus from Equations 20.27, 20.28, 20.30, and 20.31, we can calculate ϑ and θ. Equation 20.28 can be used as a check on the calculations. It should also be noted that the tunnel resistance is considered a part *of $V + u$.*

These stationary temperatures are not reached because any contact spot is active for too short a time to reach thermal equilibrium. The general equation for the time rate of heating in a volume element is

$$\lambda \nabla^2 \vartheta + \rho J^2 = C \frac{d\vartheta}{dt} \quad \text{(W/cm}^{-3}\text{)} \tag{20.32}$$

where J is the current density and C is the heat capacity of the materials. Holm [1] gives solutions of this equation for circular contact spots. The solutions are represented by diagrams, in which the time is replaced by the dimensionless variable

$$z = \frac{\lambda}{cb^2} t \qquad (20.33)$$

The solutions we shall use are shown in Figure 20.29a, in which the fraction of the steady-state temperature that is reached is plotted as a function of z. In Figure 20.29, ϑ_f, ϑ_s, and θ_s represent the part of (ϑ_f), ϑ and θ that appears in the transient period. It is assumed that the temperature rise owing to the friction heat, ϑ_f, can be added to the other temperature rises to give the total temperature rise on the surface ϑ_v and the maximum temperature rise in the brush θ_v. Thus,

$$\vartheta_v = \vartheta_f + \vartheta_s \quad (°C) \qquad (20.34)$$

and

$$\theta_v = \vartheta_f + \theta_s \quad (°C) \qquad (20.35)$$

20.5.2 Actual Temperature

We now come to the crux of the problem, which involves the determination of the values of b and t under the circumstances involved in sliding brush contacts. We calculate t as t_v from the relation

$$t_v = \frac{\pi b_1}{2v} k \qquad (20.36)$$

where b_1 is chosen for the mechanical contact surface (assumed to be a single circular area), v is the velocity of the moving surfaces, and k was arbitrarily given the value 5. k represents an effective elongation of the surfaces in the direction of motion, which has been shown to exist [25], but the value 5 is purely an educated guess. Placing t_v from Equation 20.36 and b_1 from Equation 20.25 in Equation 20.32, the proportion of (ϑ_s), which appears at the collector

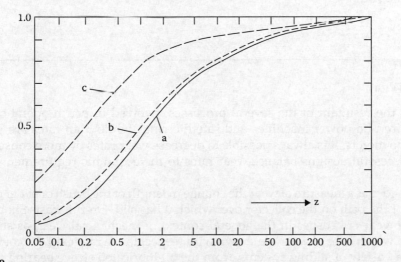

FIGURE 20.29
Correction for transient effects in brush and collector applications: (a) ϑ_s/ϑ for collector surface; (b) Θ_s/Θ for maximum temperature in brush; (c) $\vartheta_f/(\vartheta_f)$ for temperature rise at surface owing to friction.

surface, can be calculated from Figure 20.20a and we thus have ϑ_f, the temperature rise on the contact surface owing to friction.

The calculation of the value of b in Equation 20.33 for the case of electrical heating requires an assumption concerning the number n of electrically conducting contact spots operating in parallel. Here again, purely arbitrary choices are made.

The choices of $k = 5$ and $n = 10$, although arbitrary, have reasonable basis on other areas of general experience. However, in any particular problem the basis must be noted and the data considered in this light.

From a design standpoint, there are several factors that should be considered. Since the maximum temperature occurs back in the brush, the heat flow of about half of the electric heat is from the brush to the collector. Also the heat flow from the mechanical surface, owing to friction, is distributed between the brush and collector approximately in proportion to their thermal conductivities. This only emphasizes the importance of the collector in getting rid of heat. It also points out that where the limits are being stretched for increased currents and speeds, the overall heat flow and dissipation must be considered.

20.5.3 Thermal Mound

Another factor considered for applications, which are stretching the limits on current and speed is thermal mounding. This is a raised region that expands from both surfaces owing to the temperature rise in the current constriction and owing to friction. Particularly for high-current and high-speed application, this mounding can force the current to be transmitted in very few contact spots. The wear is concentrated on an individual spot until it wears away and the current and friction are transferred to another spot. The height of these mounds can be very small considering the film thickness of 10^{-7} for the water film. Bryant and others have calculated the temperatures and extensions of this mounding [34–37]. Practical observations on machines where selective action is taking place point to the fact that when brushes become overloaded, the number of contact spots under the brush decreases. This sometimes ends in glowing, that is, a bright glow on the trailing edge of a brush that moves slowly across the width.

20.6 Brush Wear

Brush wear is the resultant of the several processes involved in brush operation. Low rates are required on power machines and satellite rings while high rates are permissible on torpedo motors. It is always possible to decrease wear rates by more conservative design, but successful designs balance wear rates to the economic requirements of the application.

We shall consider as a measure of wear the change in length of the brush or slider divided by the length of the path on the collector over which it has slid. On this basis Shobert [38] has placed the wear of brushes in the general context of mechanical wear, as shown in Figure 20.30, in which rates of wear are shown with respect to the ranges of the coefficients of friction for a variety of sliding systems, from the oil-lubricated sleeve bearing to chalk on a chalkboard. The wear rates cover about 14 powers of ten, whereas the friction extends over 3 powers of ten. The reason for presenting this schematic diagram is to point out that

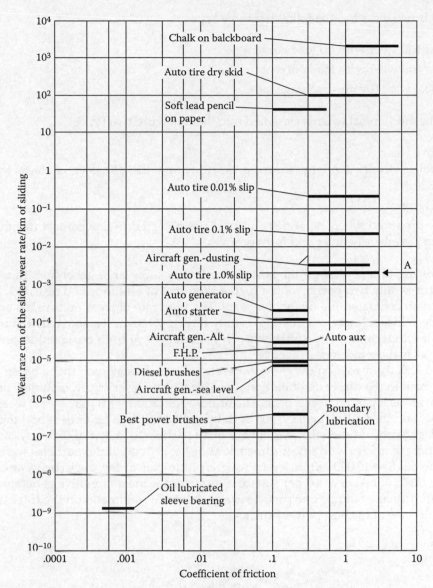

FIGURE 20.30
Schematic representation of friction and wear.

all the phenomena of wear cannot be lumped together and considered as being the result of the same physical effects. The wear of chalk on a blackboard results from the fact that the coarse asperities on the board are stronger than the chalk. Shear takes place within the chalk, leaving dust held to the board by weak cohesive forces that can be broken when the soft felt of an eraser rubs into these asperities. In fact, the wear and crumbling of chalk are so drastic that only a part of the dust sticks to the board.

By contrast, the wear in an oil-lubricated sleeve bearing is caused by the incidental touching of metallic asperities on the metal surfaces. These are held apart for the most part by pressure of the liquid that has been drawn into the load-bearing contact by the adhesion between the liquid and the two surfaces.

Generally brush wear has the following three types.

1. Mechanical wear with no load current
2. Mechanical wear with load current
3. Electrical wear (by arc discharge etc.)

Further, the mechanical wear is classified into the following two types;

1. Adhesive wear
   Material is dropped off by shearing or rupturing the adhesive contacts (see Figure 20.31).
2. Abrasive wear

   This is produced when asperities and particles attached to one contact member cut into the other member and form grooves.

It is interesting to note that on the best power brushes (on large generators in a steel mill), the wear is only two powers of ten higher than that of oil-lubricated bearings. Such brushes have an estimated life of about 4 years. The high rate of wear on brushes on aircraft or space vehicles, known as dusting, is only three powers of ten above that found in normal sea-level machines; and suitable brush materials have been designed to prevent this dusting as had been discussed above.

In Figure 20.30, the arrow on the right represents a rate of wear such that a brush with a face 1 cm thick in the direction of motion would leave one layer of carbon atoms on the collector surface as it passed over it once. The brush application with the highest wear has 1/10 this rate, and the lowest has 1/10,000 of it. In addition, it is to be understood that the wear particles are not individual atoms but are small particles of material that may contain some 10^{12} atoms (particles with linear dimensions of 1×10^{-4} cm). If the particles were this large, there would be 10,000 left on every square centimeter of the track for wear corresponding to the point A, or 1,000 per square centimeter for the automotive generator. On the same basis, there would be one particle every 10 square centimeter on the track of the collector under one of the best power brushes.

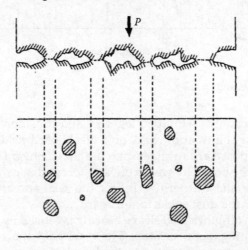

FIGURE 20.31
Contact surface model.

Of course, the wear debris is not as uniform as in the picture given in the previous paragraph. There is a wide spread in the size of the wear particles and, in addition, only a very small part of the total wear debris adheres to the collector track. In fact, some machines wear out many sets of brushes without building up excessive graphite films, indicating that an equilibrium must be reached very early on a new collector surface, in which the wear of the materials from the collector is just equal to the amount deposited, and no further accretion of carbon occurs.

Under normal conditions, then, we may consider the wear of brushes to be the result of the occasional contact between surface asperities.

The general subject of mechanical wear has been treated by Holm.

It has been pointed out that there can be a difference of three orders of magnitude in mechanical brush wear on commutators. The higher rates of wear could be reduced by improving the mechanical design, but in some cases such improvements would not be justified on economic grounds. Wear is usually increased when current is being carried, and as there is nothing about the passage of current through a normal contact which, by itself, should cause wear, the wear must be because current produces changes in the topography of the surfaces.

20.6.1 Holm's Wear Equation

The electrical wear owing to arcing and sparking, as such, is very low; but the resulting disturbance to the commutator and brush surfaces can result in high brush wear.

R. Holm proposed an equation of wear under the conditions of current and arc discharge [39].

$$W = P\left[W_o + C_1\{\tau(2) + 2\tau(5)\}I + g\sqrt{Q}\right] + \omega Q \ \ \text{cm}^3/\text{km} \tag{20.37}$$

where P: mechanical load on a brush, I: current per brush, Q: electric charge transported by arcs during 1km of sliding, ωQ: volume evaporated from the brush under the influence of the arcs, $\tau(2), \tau(5)$: fractions of the test time during which the voltage is over 2 and over 5 V respectively, the number 2 before $\tau(5)$: weight to the wear compared with $\tau(2)$.

Each term means that PW_o: wear without current, $PC_1\{\tau(2) + 2\tau(5)\}I$: wear owing to surface roughening by the flashes, $Pg\sqrt{Q}$: wear owing to surface roughening by arcing where the square root is motivated by earlier measurements and g is a coefficient that is determined so as to provide optimum agreement between formula and measurements.

In the above equation, the rate of brush wear is a function of the number and duration of arcs, defined as discharges in which the voltage drop of the cathode is 12–20 V; and the number and duration of so-called "flashes," in which the voltage does not reach 10 V. By placing an auxiliary brush on the commutator next to the normal brush and applying the brush voltage across a gated counter, the number of pulses of over 2 V, over 5 V, and over 10 V could be counted. From these numbers, and from the durations measured with an oscilloscope, the total time of the arcs as well as of the flashes could be calculated.

Figure 20.32 shows an oscillogram of such pulses; the lines show the voltages at which the counter was set. The wear caused by the short arcs was found to be smaller than the wear caused by the flashes. The relatively small influence of the short arcs is owing to their very short duration. The arcs usually form at the trailing edges of both brush and bar, and thus their duration is determined only by the residual inductive energy in the commutation coil. The flashes are usually generated under the brushes and can continue until the bar leaves the brush.

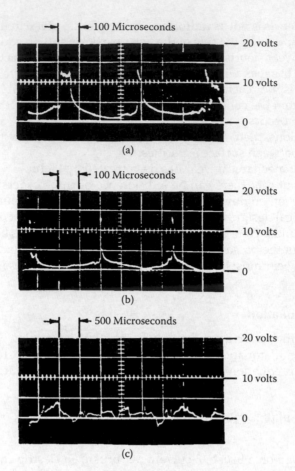

FIGURE 20.32
Oscillogorams of arcing, sparking, and flashes: (a) long arcs 20–60 ms (arcing); (b) short arcs 2–5 ms (sparking); (c) flashes 100–300 ms (over 2 V).

20.6.2 Flashes and Smutting

As the normal brush contact drop is of the order of 1–1.5 V, the higher voltage that appears during the flashes goes into heating of the region in both brush and commutator surfaces at and near the original conducting spots. These flashes are the result of mechanical circumstances, which cause enough loosening of the contact between the two surfaces to increase the resistance and therefore the voltage, but do not cause the complete separation that would lead to arcing. The temperatures in the conducting spots in both surfaces can be high enough to melt and even boil copper or, in longer flashes, to vaporize carbon. According to Holm [40] a flash of 4 V heats the copper surface to about 450°C. A flash of 6 V raises this temperature to about 900°C. In fact, this evaporation is very likely to be the cause of the "smutting" which appears on certain commutators when the machines are subject to excessive vibration or when the mechanical contact is disturbed for any reason.

This smutting, which is a heavy, dull black film, can appear irregularly over the face of the individual bars or over irregular area of the commutator and, in turn, causes a change in friction in the areas on which it appears. Occasional interruptions or flashes cause no trouble because the normal cleaning action of the brushes eliminates the excess carbon as fast as it appears. If the amount is excessive, the smut increases beyond the ability of the cleaning action to eliminate it. Smutting is dangerous because in most cases it leads to more serious commutator deterioration, with resulting high brush wear.

There is a certain limit of recovery from smutting for any given combination of brush and machine. Two equilibria are involved, in one of which fritting and oxidation oppose each other, and in the other filming and cleaning action oppose each other. Smutting adds to the effects of oxidation and filming, and when it occurs, cleaning action must be increased.

20.6.3 Polarities and Other Aspects

On slip rings on which brushes of different polarities slide on separate tracks, a difference between the tracks becomes apparent. Under the anode brush, the graphite film develops more uniformly; oxidation of the collector is inhibited by the direction of the current, and fritting of the film takes place at a lower voltage. The surface under the anode brush is thus not disturbed as much as the one under the cathode brush; as a result, the brush life may be several times higher.

Under the cathode brush, oxidation is enhanced, the oxide film tends to grow thicker, and there is less graphite deposited in it. Hence, as fritting must take place at a somewhat higher voltage, there is more disturbance of the areas near the conducting regions and therefore more brush wear. This effect of oxidation and fritting on wear is even more noticeable in the very long life of brushes running in hydrogen. It is interesting to note that in tests on metal graphite brushes reported by Hessler [41] the anode and cathode wear rates are reversed. The presence of metal in the brushes very probably accounts for the difference, because this metal can now become oxidized.

The brush wear is affected by various factors, and many papers have been reported [42]. Among them, there are papers concerning the effect of PV factor (product of contact pressure and peripheral speed) on the wear, relation between contact resistance and wear, the influence of vibration and surface waviness on the wear [43–46].

The atmosphere surrounding the brushes has an important influence on the rate of brush wear. Baker [47] showed that brushes running in H_2, N_2 and CO_2 had much lower rates of wear than brushes operating in O_2. This was particularly true for the negative brushes under which oxidation is enhanced. In the absence of O_2, the surface was less disturbed by oxidation and fritting and the wear rate was much lower. In addition the wear of copper fiber brush in carbon dioxide environment was reported [48]. Further, the wear of brush in various fuels, gasoline, ethanol and others has been examined recently [49–51].

20.7 BRUSH MATERIALS AND ABRASION

The experience of writing with a lead pencil or dusting a stubborn lock with graphite powder would lead one to believe that pure graphite should be the ideal brush material. But it is not. Graphite made by heating petroleum coke to temperatures about 2500°C, or pyrolytic graphite made by depositing carbon on active surfaces about 2500°C from carbon containing gases, and natural graphite, mined in Sri Lanka, Madagascar, and Mexico, show the largest crystals in their structures. However, none of these alone has provided a suitable brush.

Brushes that work on actual machines all contain certain abrasives. This leads to the idea that these abrasives are necessary to polish the collector surface to a fine enough finish so that the thin film of water (10^{-7} cm) can be effective.

20.7.1 Electro- and Natural Graphite Brushes

It is interesting to note that most electrographitic brushes are made using lampblack as a base. Figure 20.33 shows a schematic of the general process. In such a series of operations, it can readily be seen that there is a lot of art as well as science involved.

The important point is that with lampblack as the base, the crystal growth of the graphite is limited and the edges of many crystallites are available as abrasives to provide the mechanical strength to smooth down the roughness of the collector and to provide the electrical conductivity. Table 20.3 shows the physical properties of some of the generic types of brush materials. The physical properties are those required by different applications, as shown in Chapters. 21 and 22, but the different forms of carbon provide the proper sliding surface.

The natural graphite materials contain an appreciable amount of ash (SiO_2, CaO, etc.) and they are usually used in combination with metals (metal graphite) or resins.

From the viewpoint of crystallography, the carbon atoms of the graphite are held within the layer planesby homo-polar bonds, and in fact the bonding of a carbon atom in these planes is stronger than the bond in diamond (Fig. 20.34). On the other hand, the bond between the layer planes is due to van der Waals'forces which are much smaller. Further evidence of this effect is that the electric and thermal conductivities of pure single graphite crystals are very high in the direction of the layer planes, but very low perpendicular to them. The ratio had been shown to be as high as 100,000 to 1.

However, since carbon does not melt, during ordinary process of graphitization the crystal growth is limited by the way in which the original carbon was formed. The wide variation is illustrated by materials such as charcoal, which is different from different kinds of wood, lampblack, which comes from burning oil without enough oxygen, carbon black, the same from gas, and the limitless possibilities of charring various organic materials—resins, plastics, coconut shells, etc. This divergence in the end results is a

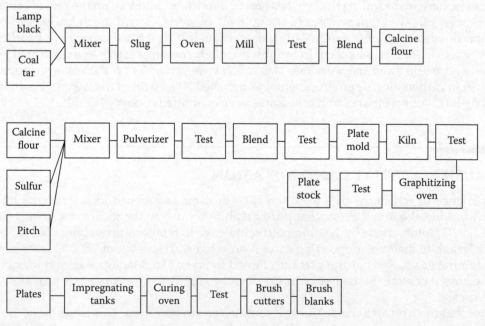

FIGURE 20.33
Typical process for the manufacture of electrographite brushes.

TABLE 20.3

Brush Materials

Material	Contact Hardness ($\times 10^2$ N mm^{-2})	Dynamic Elasticity (Mg cm^{-2}) ∥	Dynamic Elasticity (Mg cm^{-2}) ⊥	Density	Resistivity (μΩ cm)	Thermal Conductivity (W cm^{-1} $^{\circ}$C^{-1})	Heat Capacity (J cm^{-3} $^{\circ}$C^{-1})
40 Cu 60 C	1.0	45	155	2.75	75	—	1.79
Resin-bonded graphite	2.6	150	165	1.70	100,000	0.088	1.41
Electrographite	2.9	32	45	1.52	2,000	0.3	1.26
Carbon graphite	3.5	135	146	1.72	7,300	0.064	1.43
Silver	3–7	1200		10.5	1.63	4.18	2.5
Copper	4–7	700		8.89	1.75	3.8	3.4

∥ or ⊥ refers to the direction of the measurement with respect to the molding pressure.

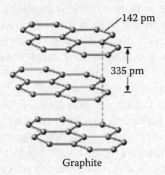

FIGURE 20.34
Layered structure of graphite atoms.

distinct advantage in the field of the brush applications, as a wide variety of combination of the various properties can be made possible [52].

20.7.2 Metal Graphite Brush and Others

Metal graphite brushes can be made by the ordinary powdered metal processes, or by processes using plastic molding techniques. These give another dimension to the applicability of brush materials.

Metal graphite brushes are suitable for a variety of applications because of their low resistivity. For examples, metal graphites are used on commutators of plating generators where low voltage and high brush current densities are encountered. They operate on rings of wound rotor induction motors where high brush current densities are also common. Metal graphites are used for grounding brushes because of their low contact drop.

For the metal-graphite brushes low contact drop is obtained for more content of metal, whereas the wear becomes large with high content of metal [53].

Recently, monolithic and metal/metal coated fiber brushes were developed for high velocity and high current applications [54,55,56,57]. In addition, from the environmental issues lead-free carbon brushes have been developed [58,59]. Further, in the railroad system copper-impregnated carbon fiber reinforced carbon composites are developed for pantograph contact strips [60].

The important factor which goes through all of the considerations is the effects and the necessity for abrasion. It is carried out by the edges of small carbon crystals or by metals or by added materials such as sand. With the proper amount of abrasive, a collector which may have many mechanical, electrical, chemical and thermal flaws can be kept operating for suitable periods.

20.8 Summary

Sliding mechanism has been used to transmit electric power or signal between stationary devices and moving devices for over 100 years. Recently, transmission mechanisms without mechanical contacts have been developed on the basis of transformer principle and others. However, brush–slip-ring systems are still widely used in the aerospace, commercial/industrial, wind turbine and other applications, because of high transmission efficiency and simple structure, while with mechanical contacts their reliability and lifetime are important issues. In order to realize high reliability and long lifetime, various components must be considered in working with sliding contacts. They are all—mechanical, chemical, electrical, thermal, abrasion, and wear—interconnected in any brush application. The followings are a summary of this chapter.

1. The progress of surface analysis methods can enable us to observe surface morphology in details (3D laser microscope) and to identify products on the contact surface (Auger electron spectroscopy, Atom force microscopy and others). In addition, electromagnetic analysis and linear or nonlinear stress analysis by FEM are also very powerful to know current distribution and micro deformation around contact surface. They are very useful tools for research and development.

2. Mechanically, the sliding contacts ride on a thin film—water usually—which is only some few molecular layers thick. This places the mechanical system under constraints to permit this to happen. Without this film, brushes wear catastrophically. Water may be replaced by laminar compounds such as MoS_2, WS_2 and others, for operation in dry atmospheres, at high altitude, or in space.

3. The mechanical contact surface is determined by the spring force and the contact hardness of the material. This may be divided into several regions scattered and moving around under the brush. At load bearing areas, elastic penetration of the collector into the face of the brush provides the ability of the face of the brush to accommodate small variations in the collector surface. This operates much faster than the whole brush that can move under the force of the spring, thus satisfying condition (2).

4. Under high current and high speed, thermal mounds are formed in the contact surface. In that case the current is concentrated in the thermal mounds and the temperature of the thermal mounds becomes very high, so that the wear rate becomes very high.

5. Under certain circumstances, brushes may squeal or chatter like chalk on a blackboard. This possibility can be eliminated by keeping the brush angle out of the range of possible chatter or by using suitable mechanical designs.

6. In dc motors, the commutation is one of the important factors to their performance and lifetime. The commutation is a complicated phenomenon including electrical, mechanical and material aspects, and commutation arc is predominant to the lifetime. Its duration time τ_a is obtained from residual current I_e, commutation inductance L_c and arc voltage v_a by the following equation; $\tau_a = \dfrac{L_c}{v_a}(I_e - I_{min})$.

7. Electric current differences built into the machine such as more than one coil per slot, differences in the resistance of armature windings, errors in armature coil turns, and eccentricities of the armature iron can cause exceptional currents at specific commutator bars. This usually leads to bar burning and the resulting short brush life.

8. Wear particles tend to abrade the collector and the face of the brush. There are many, but they are small, and they contribute to the fine smoothing of the collector. The strength of the homopolar bond in the layer planes of carbon provides the fine abrasive which keeps the surface smooth and permits a certain deviation from uniformity in current at different parts of the collector. Other materials may also be added to the brush to provide additional abrasion.

9. Fiber brushes have many distinct and measurable advantages over conventional slip ring contacts, that is, multiple points of ring contact per brush bundle and so low dynamic contact resistance (noise). Although much of the research have been on their use mainly for large current and high speed applications there is an interest in using these brushes at lower currents (see Chapter 23). Research does continue on carbon based sliding contact at high current for efficient railroad pantographs to carry current from overhead lines to high speed electric locomotives [60,61].

In a particular application, one or more of mechanical, chemical, electrical and thermal effects may dominate. It is then necessary to evaluate which is effective and work on the solution, considering materials. The fact that several may be operating together complicates the problem but proper analysis will usually permit a solution. Brush applications are usually a compromise in which cost, performance and life are balanced.

References

1. R. Holm, *Electric Contact Handbook*. Berlin: Springer-Verlag, pp 7 8, 1967.
2. E.I. Shobert II, *Carbon Brushes*, New York: Chemical Publishing Co., p.25, p.27, p.66, 1965.
3. M.F. Gomez, *Characterization and Modeling of Brush Contacts*, Ph.d Dissertation, University of the Federal Armed Forces Hamburg, 2005.
4. H. Tanaka, N. Morita, K. Sawa, T. Ueno, Carbon brush and flat commutator wear of DC motor, driving automotive fuel pump in various fuels, *Proc. of the 56th Holm Conf. on Electric Contacts*, 2010.
5. A.E. Bickerstaff, *The Real Area of Contact between a Carbon Brush and a Copper Collector*, Holm Seminar on Electric Contact Phenomena, 1969.
6. S. Tsukiji, S. Sawada, T. Tamai, Y. Hattori, K. Iida, Direct observation of current density distribution in contact area by using light emission diode wafer, *Proc. of the 57th Holm Conf. on Electric Contacts*, 2011.
7. R. Holm, E. Holm, E.I. Shobert II, Theory of hardness and measurements applicable to contact problems. *J Appl Phys* 20:310, 1949.

8. R. Holm, Friction force per unit area at true contact surface. Wissenschaftliche VeroffentLichungen a.d. Siemens-Werke 17(4):38, 1938.

9. S. Sasada, S. Norose, H. Mishina, The behavior of adhered fragments interposed between sliding surface and the formation process of wear particles, Trans. ASME, *J Lubrication Technology*, 103: 1981.

10. E.I. Shobert II, Carbon brush friction and chatter, *IEEE Trans Power Apparatus and Systems*, 30:268, 1957.

11. J.L. Johnson, *Brush Mechanical Support Characteristics in Inclined Holders*, Holm Seminar on Electric Contact Phenomena, 1970.

12. T. Ueno, K. Kadono, N. Morita, Influence of surface roughness on contact voltage drop of electrical sliding contacts, *Proc. of the 53th Holm Conf. on Electric Contacts*, 2007, IEEE.

13. M. Takanezawa, H. Yanagisawa, T. Ueno, N. Morita, T. Otaka, D. Hiramatsu, Brush sliding contact voltage deviation analysis based on the evaluation of displacement excited vibration caused by collector ring profile distortion, *Proc. of the 54th Holm Conf. on Electric Contacts*, 2008.

14. M. Milkovic, Test methods and measuring apparatus for measurement of the dynamic coefficient of friction of brushes, *Carbon*, 30(4): 587–592, 1992.

15. U.R. Evans, *The Corrosion and Oxidation of Metals*. London: Edward Arnold, 1960, Table 1, p 53.

16. H.C. Gatos, *The Surface Chemistry of Metals and Semi-conductors*, New York: Wiley, 1959, pp 486 ff.

17. W.H.J. Vernon, *Trans Faraday Soc.* 27:255, 582, 1931.

18. J.V. Dobson, The effect of humidity on brush operation. *Electric J*, 32:527, 1935.

19. H.M. Elsey, Treatment of high altitude brushes by application of metallic halides. *AIEE Paper No. TP45–108*, June, 1945.

20. S. Glasstone, K.J. Laidler, H. Eyring, *The Theory of Rate Processes*, New York: McGraw Hill, 1941, pp 339 ff.

21. T. Tamai, Safe levels of silicone contamination for electrical contacts, *Proc. of the 39th Holm Conf. on Electric Contacts*, 1993.

22. S. Uda, T. Higaki, H. Funami, K. Mano, The general idea of vapor lubrication system applied in small power motors, *Proc. of the 39th Holm Conf. on Electric Contacts*, 1993.

23. J.L. Johnson, J.L. McKinney, Electrical power brushes for dry inert-gas atmospheres, *Holm Seminar on Electric Contact Phenomena*, 1970.

24. T. Ueno, N. Morita, K. Sawa, Influence of surface roughness on voltage drop of sliding contacts under various gases environment, *Proc. of the 49th Holm Conf. on Electric Contacts*, 2003.

25. E.I. Shobert II, Electrical resistance of carbon brushes on copper rings. *IEEE Trans. Power, Apparatus and Systems*, 13: 788–797, 1954.

26. R. Holm. *Electric Contacts*, New York: Springer-Verlag, 1967, Fig. 27.13, p 147.

27. N. Ben Jemaa, L. Lehfaoui, L. Nedelec, Short aec duration laws and distribution at low current (<1A) and voltage (14-42VDC), *Proc. 20th Int. Conf. on Electrical Contacts*, pp. 379–383, 2000–6.

28. N. Ben Jemaa, L. Doublet, L. Morin, D. Jeannot, Break arc study for the new electrical level of 42 V in automotive applications, *Proc. 47th IEEE Holm Conf. Contacts*, pp 50–55, 2001–9.

29. T. Shoepf, G.Drew, Disengaging connectors under automotive 42 VDC loads, *Proc. 48th IEEE Holm Conf. on Electrical Contacts*, pp 120–127, 2002.

30. J.W. McBride, A review of volumetric erosion studies in low voltage electrical contacts, *IEICE Trans. Electron*, E86-C:908–914, 2003–2006.

31. K. Sawa, N. Shimoda, A study of commutation arcs of DC motors for automotive fuel pumps, *IEEE Trans. Components, Hybrids, & Manuf. Technol.*, 15:193–197, 1992.

32. F.D. Olney, Resistance commutation, *AIEE Trans.*, 69:1207–1218, 1950.

33. T. Yamamoto, K. Bekki, K. Sawa, A study on brush wear under commutation arc in gasoline, *IEEE Trans. Components, Hybrids, & Manuf. Technol.*, Part A. 19(3):376–383, 1996.

34. E.I. Shobert, Calculation of temperature transients in carbon brushes, *IEEE Trans. on Power Apparatus and Systems*, PAS-87(10): October 1968.

35. M.D. Bryant, J-W. Lin, Photoelastic visualization of contact phenomena between real tribological surfaces, with and without sliding, *Wear*, 170:267–279, 1993.

36. MD Bryant, YG Yune, Electrically and frictionally derived mound temperatures in carbon graphite brushes. *Proc. of the IEEE Holm Conference*, 1988, p 229.
37. C-T. Lu, M.D. Bryantb, Evaluation of subsurface stresses in a thermal mound with application to wear, *Wear*, 177:15–24, 1994.
38. E.I. Shobert II, *Carbon Brushes*, New York: Chemical Publishing Co., 1965, p 127.
39. R. Holm, *Electric Contact Handbook*. Berlin: Springer-Verlag, 1967, p 255.
40. R. Holm. Contribution to the theory of commutation on DC machines. *Power Apparatus and Systems*, 1124, December 1958.
41. V.P. Hessler. Abrasion-A factor in electrical brush wear. *Electrical Engineering (AIEE Trans)* 56(8):130, 1937.
42. R. Holm, D. Emmett, R. Bauer, W. Yohe. Brush wear during commutation. *IEEE Rotating Machinery Committee*, August 1963.
43. (W-2) H. Zhongliang, C. Zhenhua, X. Jintong, D. Guoyun, Effect of PV factor on the wear of carbon brushes for micromotors, *Wear* 265:336–340, 2008.
44. Dieter Beta, Relationship between contact resistance and wear in sliding contacts, *Holm Seminar on Electric Contact Phenomena*, 1973.
45. M.D. Bryant, A. Tewari, J-W. Lin, Wear rate reductions in carbon brushes conducting current and sliding against wavy copper surfaces, *Proc. of the 40th Holm Conf. on Electric Contacts*, 1994.
46. M.D. Bryant, A. Tewari, D, York, Effects of micro (rocking) vibrations and surface waviness on wear and wear debris, *Wear* 261:60-69, 2004.
47. R.M. Baker. Brush wear in hydrogen and in air. *Electric J*, 33(6):287, 1936.
48. N. Argibaya, J.A. Baresb, W.G. Sawyer, Asymmetric wear behavior of self-mated copper fiber brush and slip-ring sliding electrical contacts in a humid carbon dioxide environment, *Wear*, 268:455–463, 2010.
49. K. Nagakura, K. Sawa, Wear of carbon brushes under commutation arcs in gasoline, *Proc.43rd IEEE Holm Conf. on Electrical Contacts*, pp 264–271, 1997.
50. M. Takaoka, K. Sawa, An influence of commutation arc in gasoline on brush wear and commutator *Proc. 46th IEEE Holm conf. on Electrical Contacts*, pp 211–215, 2000.
51. T. Shigemori, K. Sawa, Characteristics of carbon and copper flat commutator on DC motor for automotive fuel pump, *50th IEEE Holm Conf. and 22nd ICEC*, pp 523–527, 2004.
52. J. Xia, Z. Hu, Z. Chen, G. Ding, Preparation of carbon brushes with thermosetting resin binder, *Trans. Nonferrous Met. Soc. China*, 17:1379–1384, 2007.
53. B. Emad, K. Sawa, N. Morita, T. Ueno, The effect of various atmospheric temperature on the contact resistance of sliding contact on silver coating slip ring and silver graphite brush, *Proc. of the 57th Holm Conf. on Electric Contacts*, 2011.
54. J.L. Johnson, Sliding monolithic brush systems for large currents, *Proc. of the 32nd Holm Conf. on Electric Contacts*, 1986.
55. R.A. Burton, R.G. Burton, Vitreous carbon matrix for low wear carbon/metal current collectors, *Proc. of the 34th Holm Conf. on Electric Contacts*, 1988.
56. J.A. Swift, S.J. Wallace, Key aspects of the materials and processes associated with the synthesis and use of distributed filament contacts from pan fiber based composites, *Proc. of the 43rd Holm Conf. on Electric Contacts*, 1997.
57. D. Kuhlmann-Wilsdorf, D. Alley, Commutation with metal fiber brushes, *Proc. of the 34th Holm Conf. on Electric Contacts*, 1988.
58. A. Uecker, Lead-free carbon brushes for automotive starters, *Wear*, 255:1286–1290, 2003.
59. R. Honbo, H. Wakabayashi, Y. Murakami, N. Inayoshi, K. Inukai, T. Shimoyama, K. Sawa, N. Morita, Development of the lead-free brush material for the high-load starter, *Proc. of the 51st Holm Conf. on Electric Contacts*, 2005.
60. Y. Kubota, S. Nagasaka, T. Miyauchi, Study on wear mechanisms of copper impregnated C/C composite under electrical current. *Proc. 26th Int'l Conf. on Electrical Contacts*, pp 73–77, 2012.
61. Z. Wang, F. Guo, Z. Chen, A. Tang, Z. Ren, Research on current-carrying wear characteristics of friction pair in pantograph catenary systems. *Proceedings IEEE Holm Conference on Electrical Contacts*, pp 92–96, 2013.

21

Illustrative Modern Brush Applications

Wilferd E. Yohe and William A. Nystrom

The heavens rejoice in motion, why should I Abjure my so much lov'd variety?

Elegies, **John Donne**

CONTENTS

21.1 Introduction 1081
21.2 Brush Materials 1082
    21.2.1 Electrographite 1082
    21.2.2 Carbon-Graphite 1083
    21.2.3 Graphite 1083
    21.2.4 Resin-Bonded 1083
    21.2.5 Metal-Graphite 1084
    21.2.6 Altitude-Treated Brushes 1084
21.3 Brush Applications 1085
    21.3.1 Minature Motors 1085
    21.3.2 Fractional Horsepower Motors 1085
        21.3.2.1 Wound Field/Permanent Magnet-Motor Characteristics 1086
    21.3.3 Automotive Brush Applications 1086
        21.3.3.1 Auxiliary Motors 1087
        21.3.3.2 Alternators 1088
        21.3.3.3 Starter Motors 1088
    21.3.4 Industrial Brushes 1089
    21.3.5 Diesel Electric Locomotive Brushes 1090
    21.3.6 Aircraft and Space Brushes 1090
    21.3.7 Brush Design 1091

21.1 Introduction

A brush is an electrical conductor that acts as a sliding contact to carry current to and from a rotating surface, the most common forms of which are known as a commutator or slip ring. The brush may be anodic or positive, where the current flow is from the brush to the collector (electron flow from the collector to the brush), and cathodic or negative, where the current flow is from the collector to the brush (electron flow from the brush to the collector).

The commutator is an assembly of bars or segmental sections, insulated from each other, to which the coil ends of an armature winding are attached. Commutation is the process of current reversal in the armature coils that occurs when they are shorted by the brushes contacting the segments to which the coil ends are attached.

Ideal commutation occurs when the current reversal takes place at a constant rate from full current at the instant the brush contacts the commutator bar to full current in the reversed direction at the instant the brush leaves the commutator bar. This is viewed as linear or straight line commutation. However, to attain linear commutation, we must assume that the resistance between the brush and commutator bar varies inversely as the area of contact, that the resistance of the armature coil is so small as to be considered negligible, that the coils have negligible self or mutual inductance, and that the coil undergoing commutation is cutting no flux during commutation. Since these assumptions do not hold true in practical applications, the attainment of strictly linear commutation is rarely possible. The presence of these variables results in either over-commutation or under-commutation, as discussed below.

Briefly, there is always a self-induced or reactance voltage build-up in the coil undergoing commutation. The coil reactance will tend to delay or oppose current reversal. Therefore, under-commutation is obtained. The current has not fully reversed when the contact is broken between the brush and the commutator bar, which results in light arcing. If the counter-reactance voltage is too great, the result is over-commutation; complete reversal of current has taken place before separation of brush and commutator bar and the current exceeds the normal current for a very short time. There is less chance of visible arcing on over-commutation than with under-commutation. However, with over-commutation, there is a possibility of slightly higher contact temperatures since the current is greater than normal; for a short time.

On large industrial-size machines, one can measure the commutation zone by a test, which is often referred to as a "buckboost" curve, whereby the current in the interpole winding is changed to determine the commutation zone for a machine and a specific grade of material. On small machines, one can check commutation by changing the brushholder position and selecting the one that gives the least amount of sparking.

The slip ring is a conducting, rotating ring to which a winding or circuit is connected. Brushes under this condition act only as sliding electrical contacts. Therefore, the major technical concerns are the current-carrying capacity or temperature rise effects, the stability of the coefficient of friction or brush rideability, and the filming properties of the brush material as it relates to the preceding and to electrical noise.

21.2 Brush Materials

Before beginning to discuss brush applications by motor type, it is desirable to briefly describe the general classes of material formulations that are in modern use for these applications.

21.2.1 Electrographite

The manufacture of electrographitic brush materials is an engineered process which is carried out with the technique, pride, and skill of expert craftsmen. Before beginning, representative samples of all raw materials are critically analyzed and approved before any lot of a specific material is accepted.

The lampblack, cokes, pitches, and tars, which are the major constituents for most electrographitic grades, are mixed, screened, and blended under controlled conditions to exacting specifications.

The "mix" is then molded in hydraulic presses into plates which are carefully checked by quality-control procedures before being baked to a typical temperature of 1000°C to convert the binder materials into amorphous carbon. After baking, the material is then converted into graphite by a special heat-treating process known as graphitization.

Graphitization is carried out in special furnaces designed to permit accurate control of temperatures of up to 3000°C. Engineers have developed systems of control which insure reasonably homogeneous material from lot to lot and from year to year.

The change from amorphous carbon to the graphite crystal structure takes place during the graphitization process. Electrographitic material is a form of carbon-approaching diamond in purity. As a brush material, the electrographitic grades of material generally have the best commutating characteristics and the lowest friction coefficients. However, in cases where very high current densities are required, or where high mechanical strength is a factor, it may be better to use another type of material.

21.2.2 Carbon-Graphite

The earliest carbon brushes were in the form of amorphous carbon or amorphous carbon plus natural graphite materials, There are many such grades active today although they have been greatly improved by modern manufacturing practices.

The raw materials are mixed and blended with the same care and under the same rigid process controls as those used for the electrographitic grades. They are molded into plates or pills which are baked to temperatures in the 1500–2800°F range to carbonize the binder. Carbon grades are stronger than the electrographitic materials and have a definite polishing action. There is a limit to the speeds at which they can be operated owing to their somewhat higher coefficients of friction and their current-carrying capacity is not as high as most electrographitic grades owing to their higher specific electrical resistivities. However, for applications where mechanical strength and adverse atmospheric conditions are a factor, this type of material can be very suitable.

21.2.3 Graphite

Natural and artificial or synthetic graphites are used in the manufacture of graphite-type brushes. Natural graphite deposits of varying structural types and qualities are found in many locations throughout the world. The selected graphite is usually ground to a very fine powder, controlled in particle size and ash content, and mixed with a tar, pitch, or resin binder. The material is molded into plates or pills which are then cured by heating to a temperature sufficient to set the binder.

The graphite brush is characterized by a low coefficient of friction and a cleaning action owing to the ash content inherent in the natural and artificial graphite. Strict quality control of the graphite raw material helps to insure a consistent degree of cleaning action for a given material.

21.2.4 Resin-Bonded

Resin-bonded brush materials are a special form of the graphite brush baked to a temperature in the range of 500°F. These materials are laminated in structure, and their

1084 Yohe and Nystrom

characteristics are such that the resistance across the laminations is often from five to eight times the resistance taken parallel to the laminations. This characteristic is effective in reducing short-circuit currents in the face of the brush. For this reason, resin-bonded brushes are used on machines with high commutating voltage. The current-carrying capacity of these brush materials is limited, however, because of their high electrical resistivities.

21.2.5 Metal-Graphite

Metal-graphite brushes are made from metal powders, natural and artificial graphite, and resins. The most commonly used metal is copper in percentages varying from 10% to 95% by weight although silver-graphite materials enjoy extensive use in speciality applications where their improved electrical performance is found to be cost effective. The low-metal-content brushes are generally mixed with a resin as a bonding agent. The medium-metal-content brushes may be bonded by either the sintering action of the metal or the use of a resin binder, depending on the specific grade. The high-metal-content brushes usually rely mainly on the sintering action of the metal for bonding.

Metal-graphite grades are susceptible to rapid wear when operated at absolute humidity levels below 1 grain per cubic foot or at a dewpoint of –10°C or lower. However, adjuvants which act as a film-forming agent can be added to the material to minimize the brush wear rate under these conditions (see Section 21.2.6 Altitude-Treated Brushes).

Metal-graphite materials are used where an exceptionally high current capacity is required and where the contact voltage drop must be kept low. In this regard, brushes of fine silver with graphite and other additives provide unusually low electrical noise levels, low and stable contact resistance, low friction, and high conductivity. As a result, silver-graphite brushes are suited for use on slip rings, commutators of low-voltage generators and motors, segmented rings, and many flat surfaces where the motion is reciprocating. Specific grades of this type of metal-graphite material have been developed for a wide variety of environmental conditions.

Silver-graphite brushes may be used against a variety of cooperating surfaces, although pure silver, coin silver, copper, bronze, and silver-plated copper and bronze are generally preferred. Radio interference noise levels—an increasingly important consideration in equipment design—show marked reductions in rotating equipment using silver-graphite brushes operating against coin silver slip rings.

21.2.6 Altitude-Treated Brushes

It is an interesting fact that in the absence of sufficient water vapor and/or hydrocarbon vapors, graphite can no longer function as a lubricant to provide the essential commutator film to enable the brush–collector system to function satisfactorily. In response to this problem, over the years a number of highly effective chemical compounds have been found to be effective in allowing a brush to function under low humidity conditions such as encountered in enclosed, dry atmosphere motor applications or at high altitudes where water vapor levels are very low. Typical examples of such materials include molybdenum disulfide and a number of halide salt formulations such as barium fluoride. Where it is known or found to be necessary for satisfactory brush performance, these materials can be incorporated by a variety of proprietary procedures into any of the material classes previously described.

21.3 Brush Applications

No two pieces of motorized equipment are ever exactly alike. Close, perhaps, but never exactly the same. As a result of this condition, the grade of brush material used for the particular application must be tolerant of handling these minor variations.

Optimum brush life, commutation, and freedom from commutator wear, burning, and noise are obtained only after careful tests have been made under the actual operating conditions. During these tests, in addition to the evaluation of the brush material for life, a study should be made of the overall brush system such as brush position, spring force, holder stability, commutator conditions, etc. Should problems arise which are unfamiliar and require assistance, engineering guidance should be sought from a competent brush supplier for their analysis and evaluation of the situation.

21.3.1 Minature Motors

Brushes for this class of motors are considered miniature since the cross-sectional area is less than 10.3 mm² (0.16 in²). We could just as well include them with the following fractional horsepower class, but since they are so small in size and their applications are so varied, it is important that a few separate comments be made. Also, most miniature motors are operated at applied voltages of 24 V or less, while most fractional horsepower motors operate using applied voltages of 110 or 220 V.

Miniature motors range from the very expensive, well-designed flea-power instrument and control motors to the relatively cheap motors used in toys.

Control motors are well constructed because the applications on which they are used required high reliability. Many of the brushes are similar in design to the fractional horsepower brush except that they are miniature in size. The grade of brush material which is used depends on the motor application so that all types of materials are used including altitude-treated grades, copper-graphite materials, and silver-graphite materials.

Toy motor requirements consist mostly of life and performance at the lowest possible price. Therefore, it is necessary to look at all grades and, specifically, at those that lend themselves to production at a very low cost

The crimped connection where the brush is crimped to a leaf spring is often used for this class motor. The load current is usually small; thus, the spring carries the current in addition to applying pressure to the brush. The spring material is normally phosphor bronze or beryllium copper.

21.3.2 Fractional Horsepower Motors

The small, but powerful, "universal" motor has earned a popular place in commercial and industrial applications. These motors are used in such applications as home appliances, sweepers, blenders, fans, sewing machines, to name a few, and in all types of power tools for home and industry. The range of applications is continually expanding as well as are the variations in brush and motor requirements.

Fractional horsepower motors are primarily universal-type motors which use alternating current. The motors operate at speeds up to 30,000 r.p.m. with relatively simple speed control since the motors are sensitive to voltage and flux changes. Universal motors may be either wound field or permanent magnet-type.

21.3.2.1 Wound Field/Permanent Magnet-Motor Characteristics

In the fractional horsepower dc permanent magnet (PM) motor, the magnet replaces the field (stator) winding in most cases, leaving the armature essentially the same as before. There are, however, several important differences that arise from this change:

- High peak efficiency
- Lower circuit inductance
- Generally lower resistance (higher stall current) for the same performance
- Generally longer brush life
- The rotational alignment of the armature to achieve the lowest reluctance path, known as "cogging in the PM"
- Less energy wasted as heat loss, but the permanent magnets do not conduct away this waste heat as well as the wound field

The PM motor has almost completely replaced the shunt motor and has replaced most of the series motors in the sub-horsepower size primarily because of its simplicity, low cost, high efficiency, and longer life. The series-wound field motor is still extensively used where high stall torque, high no-load speed (e.g., reel winders, winches), speed multiplicity and temperature extremes (e.g., where demagnetization effects must be considered) are the primary design requirements.

Because of design and economic limitations, sparkless commutation is rarely achieved. Sparking may often be traced to a number of causes such as:

- Rough or uneven commutator surface
- Commutator eccentricity
- Flats, high or low commutator bars
- Improperly cured phenolic resin on molded commutators
- High mica insulation on built-up commutators
- Incorrect spring pressure
- Incorrect positioning or spacing of brushes
- Excessive reactance voltage
- Selection of an unsuitable brush grade

In general, universal motor brushes should have a low coefficient of friction and medium-to-high contact drop to suppress the short-circuit currents induced between adjacent bars during the commutation process. They should also be capable of carrying the required load current without excessive heating owing to $I^2 R$ power losses within the brush material.

The resin-bonded grades or high-resistance graphite grades normally give very good performance on this type of motor. However, for low-voltage, high-current applications, a low-metal content, metal-graphite material may be desirable.

21.3.3 Automotive Brush Applications

Automotive applications can be classified into three broad groups; namely (1) auxiliary motors; (2) alternators; and (3) starter motors.

21.3.3.1 Auxiliary Motors

There are many auxiliary motors in use in automobiles today, and the list is ever increasing. Some of the more recent high-volume applications are fuel pump motors and electric fans for engine cooling. The number of auxiliary motors per automobile also depends to a large extent on the various options desired by the buyer.

Automotive motor requirements cover a wide range from the high-torque, intermittent duty demands of seat adjustors and window lifts, to the high-speed, low-noise continuous-duty requirements of the blower or wiper motors. Very long brush life, often at high operating temperatures, is required for most continuous-duty applications such as air-conditioner and engine-cooling fan motors. With the addition of legally mandated automotive controls and fuel economy requirements, many of the automotive motors must operate in even more severe, high-temperature environments. However, for intermittent automotive applications, such as seat movers and antenna lifts, motor torque is an important factor and brush life is usually not a serious problem because of the short duty cycle. Therefore, these applications normally use a high-metal-content grade to obtain low contact drop and high torque.

For continuous duty applications, both brush life and efficiency are very important. Therefore, a compromise must ordinarily be made resulting in the use of a lower-metal-content brush. In addition to life and efficiency, there is also a problem, of audible and/or electrical brush noise which can usually be broken down into three categories:

- Audible or electrical noise that develops as the brushes ride over the commutator segments
- Audible magnetic noise
- Audible noise from bearings or washers

The brush formulation and its physical properties have a considerable influence on the damping of the audible noise generated by the commutator-brush-slot bar frequency. The frequency excited in this manner can be transmitted or amplified through its mounting. Thus, it is important that this factor also be kept as low as possible. Electrical noise which arises from commutation effects can also be minimized by proper engineering design and brush material selection to achieve the best possible commutation at an affordable price. Also, in addition to material selection, it is often recommended to use a small angle on the face of the brush to provide for quicker brush seating to enhance brush stability with attendant better commutation (assuming proper spring and holder design).

Audible magnetic noise caused by an imbalance in the magnetic circuitry or flux can usually be minimized by using skewed armature slots. Audible bearing noise is usually attributed to defective bearings or bearing misalignment during assembly. Audible washer noise can usually be eliminated or reduced by the use of a combination of steel and fiber washers.

The use of electronic circuitry in automobiles has increased the importance of damping the radio-frequency interference (RFI) generated within the auxiliary motor because of commutation effects. The RFI may be dampened or controlled by the use of capacitors, ferrite beads, choke coils, or some combination of both. This RFI factor has always been present, but its influence on other apparatus is much more detrimental today because of the much more extensive usage of electronic circuitry in modern vehicles. Therefore, it must be controlled.

Auxiliary motors operate directly from the battery which is charged by the alternator; therefore, the need for load current continues to increase as additional apparatus is added to the vehicle.

The efficiency of the permanent magnet motor is higher than the wound field motor since it requires no field excitation, thus the energy required per motor is less. This is a major reason why the permanent magnet auxiliary motor has almost completely replaced the wound field motors in these applications.

21.3.3.2 Alternators

The conventional dc generator, a commutator-rectified generator, served for many years as the electrical power source for the automobile. However, its limitations resulted in its replacement by the diode-rectified generator, normally referred to as the alternator. Some of the advantages of the diode-rectified alternator over the dc generator are:

- The alternator has greater power at lower speed.
- The alternator is considerably lighter and smaller than the dc generator for the same power rating.
- The brushes in the alternator carry only excitation current to a rotating field and commutation is not involved. Thus, much longer life is normally achieved. The slip rings on which the brushes ride are usually made from copper, brass, or stainless steel.
- The overall service life of the alternator surpasses that of the dc generator with considerably higher reliability.
- The heavy current leads are in the stator, which does not rotate; and being near the outside surface, the heat generated owing to electrical resistance is more easily dissipated.
- The voltage regulator is much simpler since it does not require a current-limiting device nor a circuit breaker to prevent reverse current since the diodes conduct current in one direction only. Solid state rectifiers are very small in size and weight and, therefore, can be mounted inside the alternator.

The rating of alternators varies depending on the electrical output desired. Alternators are made in various sizes and are matched to the electrical output required for the vehicle on which they are used. Each manufacturer's alternator differs slightly in appearance and brush configurations. However the electrical and mechanical principles are the same in all cases. Brush materials in common use for alternators are either the electrographitic type or the low-metal-content, metal–graphite type.

21.3.3.3 Starter Motors

The invention of the electrical starter in the early 1920s was an important factor in the rapid growth and success of the automotive industry. The original concept invented by Kettering consisting of a high-torque motor used for short times at high speeds has remained unchanged in principle. However, the requirements of different applications have resulted in the development of a number of starter brush material types. The requirements of very high torque, high load current, and commutator speed for short periods of operation, in addition to the overspeed after the engine has started, must all be considered in the application of starter brush materials. Today, starter design trend is toward ever-decreasing engine cranking time under all weather conditions and environments. Starter grades generally contain moderate metal contents (50–70%) to give the low contact drop required and to carry high load currents with the graphite or lubricant content adjusted to achieve the required brush life.

Many starter motors require brushes with shunts that are quite large because of the heavy currents that must be carried. The voltage drop between the brush and the shunt must be kept low to keep electrical losses to a minimum. Since the cranking time is relatively short, normally a shunt is used at three to five times its normal current rating. The shunt connections are usually of the molded-in connection type. However, others can be used depending on the application.

Present automotive starter designs have changed from the wound field types used in the past to the almost universal use of permanent magnet field starters. Also, the small starter motors for outboard motors, lawn mowers, small garden tractors, and so forth are mostly permanent magnet starters. These differ in design, but they still have the same basic function and requirement. Many of these smaller permanent magnet starters use a faceplate commutator rather than a barrel commutator. Thus, a different brush configuration is used. These brushes are often wedge-shaped in design, conforming to the shape of the commutator bars.

21.3.4 Industrial Brushes

The field of industrial brushes is very broad in application and involves grades of all types which must function over a wide range of conditions. Many of the problems encountered with industrial motors can be attributed to maintenance, atmospheric conditions, or operational procedures.

Very large generators and motors, such as those used for the generation of power, are well designed, and preliminary testing has been conducted pertaining to machine parameters and operation. This background data is very helpful in the selection or development of a grade that will give excellent performance. The maintenance of these units is such that the problems that occur are spotted immediately, and corrective action can often be taken before they can become serious.

Normally, electrographitic or carbon–graphite grades are used on these motors and generators. Exceptionally long brush life is obtained on many of these machines.

Medium-sized motors and generators generally use the same grades of material as mentioned above. These machines also operate over a wide range of conditions and are involved in a large variety of applications. The number of motors and generators in this type of service greatly outnumber the volume of the larger units; thus more application problems are likely to be encountered in practice with these units.

Problems on these machines often arise because of the speed and load conditions, and as a result of atmospheric conditions. Many machines operate at full load (typical brush current densities of 80–90 A in^{-2}) and overload for short periods of time. Machine operation is often controlled by the workload and the nature of the application thus, the variation in its duty cycle can be extreme. Also, although it was more of a problem in the past before the principles of successful brush application were well known, light loading of brushes with current densities of 40 A in^{-2} (6.2 A cm^{-2}) or less can also cause severe problems owing to the inability of the graphite brush material to properly maintain a satisfactory commutator film.

Machines may also be operated in contaminated atmospheres, many of which are chemical in nature. In these instances, overfilming may occur which results in threading or grooving of the collector and, therefore, rapid brush and commutator wear. Special grades of brush material have been developed which exhibit polishing action to prevent a heavy film build-up in these atmospheres, which will then allow for excellent performance and brush life.

Motor maintenance is very important. However, it is often the case that the smaller the motor, the less care that it receives.

Many motors operate under very high ambient temperatures and are subject to the aforementioned wide range of load conditions. Thus, it is very important that the mechanical or stability aspects of the brushholder be considered and the other electrical aspects during machine design.

The turbo-generator requires slip ring brushes. No commutation is involved; therefore, the main concern is selective brush wear when one polarity wears more than the other. Turbo-generator brushes should have low, stable contact drop and exhibit low friction. Normally, carbon-graphite type materials which contain additional added lubricants are used for this application.

21.3.5 Diesel Electric Locomotive Brushes

The motors used on the diesel electric locomotive and similar type applications are referred to as traction motors, which are a special class of the industrial motor. Because of the number of motors involved, reliability, and type of service, they are treated as a separate entity.

Traction motors have undergone many changes over the years with continual increases in horsepower rating for the same relative size. These changes have been accompanied by basic motor design modifications, new insulating materials and brush changes in the way of design and material formulations. The scope of operation ranges from heavy load, low speed and low voltage to light load, high speed, and high voltage, which results in motor operation at both under- and over-compensated commutation conditions.

Multiple-wafer brushes have been in service for a number of years to improve the motor commutating characteristics. However, there are still traction motors in service today where a one-piece or single-wafer brush is used, but most modern applications call for a split- or multiple-wafer brush. A resilient pad is added to the top of these brushes which acts as a shock absorber, with the pad adsorbing vibrations so that the brush will have more intimate contact with the commutator. Each wafer of a multiple-wafer brush acts independently of the others, resulting in improved commutation and commutator conditions, which leads to vastly extended motor operation between commutator resurfacing or overhaul.

Normally, traction motors use electrographitic grades of material which are resin impregnated to give added strength to withstand vibration and to give longer brush life.

The modern diesel locomotive also has an alternator, driven by a diesel engine, to generate power for the motors. The brushes for the alternator carry only excitation current. The same basic considerations hold as previously described for the automotive alternators. Brush mounting and stability are very important. Because of the presence of vibration, it is very important to study the mechanical stability of the system to obtain good overall performance.

An impregnated electrographitic or carbon–graphite material is normally used for alternator brushes.

21.3.6 Aircraft and Space Brushes

The carbon industry has been involved in supplying brushes for the aircraft and space industries ever since it was recognized that the normal type of brush materials used on generators and motors at sea-level conditions gave exceptionally short life when operated above 20,000 feet (6,000 m) in altitude. Once the problem was defined as lack of sufficient moisture to maintain proper filming on the commutators or slip rings, it was necessary to find the correct additive or adjuvant which could be put into the brush, either by impregnation or during mixing of the material, to compensate for this lack of moisture. The adjuvants used had to be capable of providing suitable performance at sea level where moisture was present and at altitude where the dry conditions existed. Initially, a lead iodide additive was used which was soon followed by barium fluoride. These treatments served well; however, in many applications it was necessary to seat-in or run-in the brushes to establish a good film before operation at altitude.

Continued research and development resulted in the introduction of quick-filming brushes eliminating the need for preseating of brushes. The adjuvant for these brush materials is usually molybdenum disulfide, which minimizes threading of commutators and slip rings and provides higher altitude reliability without sacrificing the other desirable brush characteristics. Today there are a large number of material grades available which serve a large variety of applications from aircraft, missiles, and space satellites to sea-level applications such as totally enclosed motors where low humidity conditions can exist.

21.3.7 Brush Design

These applications cover a broad range of operating conditions since they include alternators, starters, inverters, dynamometers, and snychronous motors. Typical materials used for the various applications include carbon–graphites, electrographites, and metal–graphites (copper and silver). The state of the art of brush design and use continues to evolve. New uses repeatedly arise where brushes are required for novel purposes.

Some relatively recent advances include development of special brushes (see Chapter 23) for use at extremely high current densities, carbon-fiber brushes, and brushes with wear detectors added for special applications.

22

Sliding Contacts for Instrumentation and Control

Glenn Dorsey and Jax Glossbrenner

Curve: the loveliest distance between two points.

Mae West

CONTENTS

22.1 Introduction 1094
22.2 Sliding Contact—The Micro Perspective 1097
    22.2.1 Mechanical Aspects 1098
    22.2.2 Motion Initiation (Pre-Sliding) 1100
    22.2.3 Friction Forces 1100
    22.2.4 Motion Continuation 1102
    22.2.5 Adhesion 1102
    22.2.6 Adhesive Transfer 1103
    22.2.7 Plowing, or "Two-Body," Abrasion 1104
    22.2.8 Hard Particle, or "Three-Body," Abrasion 1105
    22.2.9 Motion Over Time 1105
22.3 Electrical Performance 1107
    22.3.1 Contact Resistance Variation (Noise) 1108
    22.3.2 Non-Ohmic Noise 1109
    22.3.3 Non-Linear Noise (Frequency Dependent) 1111
    22.3.4 Contact Impedance 1112
    22.3.4 Data Integrity 1114
22.4 Micro-Environment of Contact Region 1114
    22.4.1 Film Forming on the a-Spots 1115
    22.4.2 Unintentional Contamination 1117
        22.4.2.1 Particulates 1117
        22.4.2.2 Contamination or "Air Pollution" 1118
        22.4.2.3 Organic Off-Gasses 1118
        22.4.2.4 Friction Polymers 1119
    22.4.3 Lubrication (Intentional Contamination) 1119
    22.4.4 Lubrication Modes (Anaerobic and Aerobic) 1121
        22.4.4.1 Anaerobically Lubricated Contacts 1122
        22.4.4.2 Aerobically Lubricated Contacts 1123
        22.4.4.3 Temperature Extremes 1124
        22.4.4.4 Submerged in Flammable Fuels 1125
        22.4.4.5 Low-Pressure/Vacuum Operation 1125
        22.4.4.6 Vapor and Gas Lubrication 1125

22.5 Macro Sliding Contact.. 1126
    22.5.1 Counterface Configuration.. 1126
        22.5.1.1 Flat Surfaces 1127
        22.5.1.2 Cylindrical Surfaces.. 1127
        22.5.1.3 Counterface Contact Shapes... 1127
    22.5.2 Real versus Apparent Area of Contact 1128
    22.5.3 Brush Configurations 1128
        22.5.3.1 Cartridge Brush.. 1128
        22.5.3.2 Cantilever Composite Brush.. 1128
        22.5.3.3 Cantilever Metallic Finger 1129
        22.5.3.4 Cantilever Wire Brush... 1129
        22.5.3.5 Multifilament or Fiber Brush... 1130
        22.5.3.6 Benefits of Multiple Brushes... 1130
    22.5.4 Forces on the Brush 1131
22.6 Materials for Sliding Contacts... 1132
    22.6.1 Materials for Counterfaces 1132
    22.6.2 Solid Lubricated Composite Materials for Brushes 1133
    22.6.3 Wire Brush Materials Criteria... 1133
22.7 Friction and Wear Characteristics... 1135
    22.7.1 Friction.. 1135
    22.7.2 Wear 1137
22.8 Contact Parameters and Sliding-Contact Assemblies 1138
    22.8.1 Contact Noise 1138
    22.8.2 Slip Rings as Transmission Lines... 1139
    22.8.3 Results of Normal Operation.. 1140
22.9 Future 1141
22.10 Summary... 1142
Acknowledgments 1143
References.. 1143

22.1 Introduction

One important application for electrical contacts is the transfer of electrical signals and power between two members that are in relative motion to each other. This transfer is allowed through *sliding* electrical contacts. This chapter will discuss the theory of operation of sliding electrical contacts and the important performance parameters for their successful operation in common instrumentation and control applications. By far the most common of these applications are slip ring assemblies that transmit signals, data, and power across rotary platforms, and it is these rotary devices that will be the most common examples in this discussion.

The study of sliding electrical contacts has traditionally been divided into two areas of study: power capacity and signal quality. The study of power contacts is typically focused on an understanding of the impact on the sliding contact from power loading and the effect power (both current and voltage) have on the power rating and contact reliability. On the other hand, the study of sliding electrical contacts used for signal transfer is focused on the ability of the electrical contact to function within a signal transmission line with minimal impact on signal quality.

The phrase "instrumentation and control" is used to distinguish a large body of sliding contact applications that put a high level of emphasis on the integrity of electrical signal and power transfer where quality features of the signal and power transfer are key metrics. These applications are diverse and include wind turbine blade pitch control, rotating radar antennae, medical CT scanners, and armored vehicle turrets, to name just a few. Each of these applications requires the transfer of multiple signals and power across a rotating interface to ensure proper operation of the equipment, and the transfer quality is specified as part of the overall transmission line integrity. Although sliding motion most commonly occurs in a rotary interface, there are cases where linear motion requires electrical connection across the sliding interface.

Sliding electrical contacts are normally expected to perform their function while moving as well as in the static condition. The graphite lubricated sliding contacts discussed in Chapters 20 and 21 and the fiber brush contacts of Chapter 23 are commonly used in power transfer applications such as motor commutation, and these chapters discuss the important performance parameters required for successful power transfer. However, successful signal and data transfer impose other very important performance requirements on the sliding contacts focused on minimizing the impact of this sliding on signal quality, and it is these requirements that will receive our attention in this chapter.

Sliding is translation of one member over the other, with no intended separation between the surfaces.* Many sliding contacts are expected to undergo substantial motion distances and usually are subjected to repetitive motion over some or all the path areas of the larger of the two contact members. Sliding contact applications include slip ring and brush assemblies, motor commutators and brushes, rotary sliding switches, encoders, and potentiometers. The sliding contact consists of two members: the brush, or wiper, whose normal travel is measured as linear distance across the surface of the second member, or counterface, called the ring, segment, commutator, rail, or bar. The travel of the counterface member is usually counted in revolutions, passes, wipes, turns, and so on, under the wiper.

A distinction should be made between wiping and sliding contacts. Some tangential motion usually exists between surfaces in any "static" mechanical contact. This relative motion can occur because the members are not rigidly attached, are subject to thermal expansion resulting from temperature variations, or experience displacement from vibrations that exist in all structures. This small amplitude, rubbing motion can lead to a wear condition known as fretting which is discussed in detail in Chapters 5 through 7. Considerably more motion exists in the deliberate wipe of closing and opening of switches and relays and during connector insertion and withdrawal. If the motion is only during and incidental to closure, it is often referred to as "wiping" rather than sliding (see Chapter 7); these electrical contacts are not usually expected to conduct efficiently during the wiping motion. Wipe may also be used to refer to a single pass of a sliding contact over a point on the counter-face.

A sliding contact is not constrained in a fixed position relative to its counterface and is designed to perform well electrically while both moving and stationary. The sliding contacts of potentiometers, encoders, and slip rings are expected to conduct efficiently during intermittent or extended sliding as well as at rest. In most slip ring applications, 0.1–10 million cycles of partial or full rotation are required for successful operation.

---

* We will see in our investigations at the microscopic level that there is actually some separation between the surfaces over the majority of the apparent area of contact between the two surfaces. So in the case of our definition of sliding, it should be understood that "no separation" means that sufficient physical contact is maintained at the contact region to produce satisfactory electrical performance at all times during sliding or at rest.

For the past 50 years the critical parameter of signal quality in sliding electrical contacts has been the variation in contact resistance produced while sliding [1,2]. This variation in contact resistance produces signal noise as resultant voltage and/or current perturbations governed by electrical network laws. These changes in contact resistance with sliding are dependent on a number of key parameters including surface film changes [3,4], physical parameters such as contact force, sliding speed, surface roughness, vibration [2,5–8], or particulate contamination from wear debris or outside sources [1,9]. This activity occurs on the micro-level, i.e., in the region where the two elements of the electrical contacts actually have their interface.

The variation in ohmic contact resistance, or contact noise, is the figure of merit that has commonly been used to evaluate the suitability of sliding contacts to transmit analog and low speed digital data. This criterion is appropriate for the use of slip rings for low level analog signals where contact noise can be a significant component of signal noise. With the increased use of sliding contacts for transmitting high speed data, different signal-to-noise evaluation criteria are operative. Contact *impedance* that includes reactive effects of capacitance and inductance should be discussed. Intermittences are more descriptive than noise of the electrical events that cause problems with digital data, and noise bandwidth or frequency components are just as important as noise amplitude. In addition, the study of nano-devices is highlighting the presence of non-ohmic and non-linear contact phenomena that are important to consider for low force, high bandwidth sliding contacts.

To explain the critical parameters in the successful operation of sliding electrical contacts, it is necessary to first review the micro characteristics and environment of the sliding contacts. Much of this discussion will be an extension of the basic electrical contact information found in Chapters 1 and 12 with the added complexity of motion. The micro perspective will be followed by a macro look at the sliding contact—materials, design, forces, and tribological features of friction and wear. The final section will discuss the important performance parameters and design features of properly operating sliding electrical contacts.

There are some important characteristics or assumptions of sliding electrical contacts used in instrumentation and control that should be highlighted:

1. In most cases, it is a safe assumption that sliding electrical contacts for signal transfer are lightly loaded (i.e., the total normal force on the contact interface is low) to reduce wear. Brush force is typically in the range of 50–100 mN putting the contact well within the range that is considered elastic by traditional methods [10,11]. This assumption does not eliminate the possibility of plastic deformation some localized high asperities but does limit most contact to elastic.

2. The typical electrical contact surface is a noble metal or alloy, and gold or its alloy is the most prevalent of these. The noble properties of these materials are important for the electrical behavior, and their ductility has a significant impact on the friction and wear properties.

3. A typical sliding contact surface starts out with a surface roughness, R_f, of about 0.2 μm. The most common ring surface is hard gold plate with a hardness of about 2×10^3 N/mm². Using Bowden and Tabor's [12] classic formula $A = F/H$ where A is the estimated real area of contact, F is the brush force, and H is the hardness of the ring material, the estimated contact area is roughly 2.5×10^{-5} mm² or an equivalent contact spot size of roughly 5.6 μm in diameter.

4. The contact force is produced by a relatively compliant spring to produce a consistent low force under all conditions of mechanical tolerance. The low mass and high compliance of the spring results in an undamped and active response to all contact forces.

5. Most sliding contacts are lubricated and operate in a non-hermetically sealed environment so there are a wide variety of organic and inorganic sources of film producing agents available in the contact region.

6. The analog signals that are carried by instrumentation and control slip rings vary in frequency from DC to ~500 MHz. Digital signals vary from DC to 1.5 Gbps. The voltage and current on these signals is typically less than 30 volts and 1 amp.

7. Power is also required in these slip rings and the ability to integrate power and signals in the same assembly is typically a design challenge. Applications using small capsules (50 mm dia. x 150 mm long) normally require in the neighborhood of 500 W of power with a max voltage of 28 volts. Larger assemblies for applications such as radar, armored vehicles, and wind turbines can require as much as 25 KW of power or more with maximum voltage levels in the KV range.

22.2 Sliding Contact—The Micro Perspective

Consider the moving contact surface at an instant of time. The contact force which loads the brush (typically through a spring) is opposed at the interface by load-bearing areas made up of (h) metallic surface areas, (i) solid film covered asperity areas, (j) fluid-covered asperity areas or fluid pockets formed by the surface topography, and by any (k) particle supported areas that may intervene in the contact region (Figure 22.1). The total load-bearing areas (l) is equal to $\Sigma(h + i + j + k)$ areas, and each of these individual areas (h, i, and j) may be separate or portions of one or more individual asperity contacts. Mathematically the total force, F_t, is the sum of the forces exerted one each of these load bearing areas or,

$$F_l = \Sigma F_{mh} + \Sigma F_{oi} + \Sigma F_{fj} + \Sigma F_{pk} \tag{22.1}$$

These forces are all subject to dynamic conditions: the size and number of load-bearing asperities will change as the counter-face moves, the fluid pockets may not exist or may be

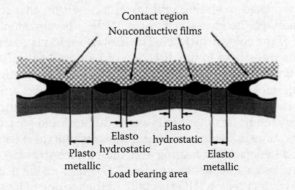

FIGURE 22.1
Load-bearing area of contacts coated with a thin film of fluid material. The figure illustrates the types of asperity distortions possible. Only two small areas of metal-to-metal contact are likely to conduct. The surfaces can be distorted by elastic and plastic loading either directly through metal-to-metal contact or through the film.

locally significant, and the number of particles and their sizes change. However, for each instant there are specific load-bearing asperities in the contact region that are supporting at least part of the load. Some of these can be partially covered by fluid or solid films that can support load and not conduct. Within the contact region, there may be pockets of trapped fluid materials that can also support part of the load, but not conduct. We shall see later that these solid and fluid films although not conductive do have the capacity to allow capacitive coupling as well as quantum conduction. If the contact does conduct, some of those asperities are likely to have some exposed metal from the destruction or penetration of the surface films. It is to this dynamic contact environment that we turn our attention by looking at the mechanical, electrical, and environmental characteristics of the contact region.

22.2.1 Mechanical Aspects

On surfaces sufficiently free of thick films, sliding is a series of asperity contact interactions while electrical contact is maintained through a-spot interaction [13]. We have an understanding from Chapter 1 of the static contact with its asperities and a-spots. What differences are imposed by sliding? Suitably effective sliding electrical contacts must be free to move to new locations with minimal change in electrical performance and manageable friction and wear of the contact surfaces. Just how well the contact performs as it slides depends on the micro-contact conditions and processes.

The primary interface interactions take place within the boundary of a *contact region.** This region, analogous to the apparent area of a static contact, includes the load-bearing area(s) together with the films, solid and fluid, and particles that support the load between one member and the counter-face. The word *region* is used to imply an inexact boundary enclosing features that are imprecisely located and constantly varying and moving. Surrounding the contact region may be a meniscus of any fluid present that may not support normal load, but does affect the environment around the contacting asperities or a-spots. During sliding this contact region will move on both the sliding and stationary member owing to surface topology and mechanical tolerances, so the contact region should be viewed as an instantaneous snapshot of the contact interface that will never be exactly reproduced over the life of the sliding contacts. The concept of the contact region is simply a construct that allows reference to the micro environment of the sliding interface. A representative unlubricated, particle-free contact region is illustrated in Figure 22.2 with parameters similar to Table 22.1.

All processes directly related to sliding will take place within this contact region: the friction, transfer of materials, the embedment of particles, the abrasion of the faces, loss of debris, and the wear of the contact surfaces. The contact region is most frequently ellipsoidal with the long axis transverse to the direction of motion. Most often it is a contiguous area on a single counterface, but the contact region can be noncontiguous if there are depressions in the counterface that can be spanned by the brush. For example, commutators and switches can have two contact regions as the brush spans two commutator bars. The contact region as represented in this manner is a two dimensional surface with infinitesimal thickness. There are some influences of sliding contacts that do extend beyond the 2-D contact region as defined here (such as heating and sub-surface stresses), but recent analyses and comparisons to physical data has shown this 2-D construct is a useful tool to understand the physics of the contact region [11] especially in the regime of low contact forces.

---

* In the earlier version of this chapter, this contact region was referred to as the *zone of closest approach* or ZCA. We have simplified and standardized the terminology to *contact region;* however, the meaning is the same.

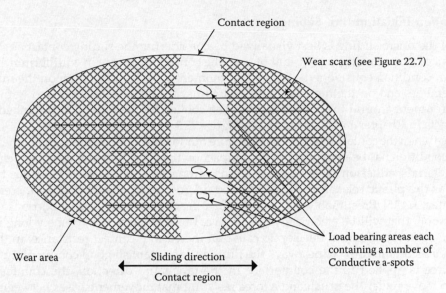

FIGURE 22.2
Contact region on worn face of a brush. The clear area in the shaded zone in the area of the wiped face is the contact region. The shape and size is determined by the relative contours of the two surfaces. Within the un-shaded contact region are three representative load-bearing areas, and within these load-bearing areas are still smaller conducting *a*-spots. The periphery of the shaded worn area is determined by the macro shape and size of the two mating contacts and the total wear. Ring travel has been in the horizontal direction and the horizontal lines represent longitudinal striations of the faces.

TABLE 22.1

Comments on Figure 22.2, Contact Region

Approx. hardness	2×10^3 N/mm^2 @100gf
Length × width	0.25 mm × 0.1 mm (0.025 mm2)
Light area	Contact region
Areas in contact region	Load-bearing asperities (LBA)
Est. load-bearing area	$0.03/2 \times 10^3 = 1.5 \times 10^{-5}$ mm^2 (total)
Number LBA/worn area	$0.025/1.5 \approx 1600$ fit in worn area
Max. LBA dia.	2×10^{-3} mm, 1 *a*-spot (0.5×10^{-3} mm, 19 *a*-spots)
a-spots	Within load-bearing areas, number is not predictable
Single cyclic motion	Contact region moves across entire shaded area
Constriction res.	3.5×10^{-3} Ω, @ρgold = 2.6×10^{-6} Ω cm

Shaded area of the figure represents the worn area of a typical crossed rod or wire brush contact.

The contact region will move about the face of the brush (the apparent area of contact) as it moves on the counterface. This movement will tend to be cyclic as the brush cycles through its full range of movement. Similarly, the contact region will move about on the counterface as the brush moves through the complete cycle (a revolution in the case of a ring). As motion occurs, the contact region will move along its transverse path but also side to side in response to mechanical tolerances of the assembly. All wear takes place in the contact region and the worn areas seen on the face of the brush or the wear track of the counterface show the extents of the contact region motion. Total movement of the contact region can be determined by observing the wear area on both the counterface and the brush.

22.2.2 Motion Initiation (Pre-Sliding)

The idea of the micro sliding is best visualized by considering the sliding contact as static at an instant of time. During this instant of time the contact forces are in equilibrium. This equilibrium condition considers the forces and moments, load, and friction on the macro contact members and on the metallic interfaces, films, fluids, particles involved with the microenvironment around the contact region. The contact region will be as described for static contacts in Chapter 1 and illustrated in Figure 22.1.

In the case when the contact has been at rest before motion initiation, the length of time that this static condition has been in effect has an impact on the condition of the contact region. In most materials, adhesion in the contact region strengthens with aging. This has been explained as the plastic relaxation of the stressed micro-contacts, leading to an increased area of contact [14,15]. Presumably, some of the load-bearing areas will have portions with metallic *a*-spots that will be electrically conductive. In some cases of extremely long rest, films in the contact region can develop to cause an increase in contact resistance. In these cases, motion is required to "wipe away" the films and re-establish good conductivity.

When force is applied to the counterface in the tangential direction, the counterface attempts to move under the brush and a force resisting that movement arises between the counterface and the brush. Two conditions contribute to that opposing force:

1. Areas that are adhesively joined are put under a shear stress as the transverse load is applied. Creep (in metals) and molecular re-ordering (in polymeric films) can occur to allow some micro-motion within the strain limits of the interface, but adhesive bonds must be broken to allow significant motion. It has been shown by experiment [16] and theoretical modeling [17] that pre-sliding displacement is on the order of 1/20 of the contact area diameter.

2. Any asperities of the slider, which are interleaved with those of the counterface in the direction of motion, will interfere and require force to either plow through or ride over one another.

22.2.3 Friction Forces

Both adhesion and asperity interlocking contribute to the tangential frictional force opposing the motion. The tangential forces act on the contact region to bring significant changes to the stress fields. The friction force combines with the normal force increasing the (resultant) total load on the asperities. Figure 22.3 shows the forces and moments acting on the counterface in the contact region [18]. The tangential force (Q_x) acts on asperities that are in adhesive, metal-to-metal contact causing the deformation and eventual rupture of the contact. The strength of the adhesive bond plays a significant role in the wear rate of the contacts (material transfer out of the contact region), as well as the friction coefficient [19]. The "stretching" or creep at the asperities tends to be velocity dependent [20] with a reduction in coefficient of friction with an increase in velocity.

Q_x also introduces a shear stress in the boundary film between contacts that can cause shear induced re-ordering in the thin film in the contact region [21]. This re-ordering shift can result in the friction "slip" events seen during sliding. The tangential force (Q_x) acts on asperities that are interleaved or interlocked and the interference of these asperities can also add to frictional counterforces.

So an electrical contact region in static equilibrium at the instant of motion initiation has a number of contact points that have supported the normal force to their elastic limit

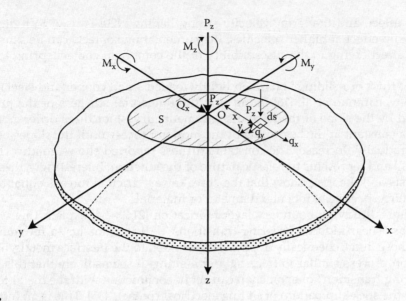

FIGURE 22.3
Typical forces and moments acting on the counterface by the brush contact. The principal force is the normal force holding the brush to the counterface. This is balanced by the force from the counterface spread on multiple asperities. A variable friction force is created as the counterface attempts to move under the brush, and this friction force is dependent on the local coefficient of friction in the contact region. The friction force is offset by the reaction force between the holder and the brush. The variability of the friction and its location results in variable moments about the heel-toe, side-side, and rotation axes. (Similar to diagram from K. L. Johnson, *Contact Mechanics*, Cambridge, U. K.: Cambridge University Press, 1985 [18].)

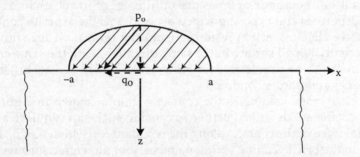

FIGURE 22.4
Semi-infinite elastically solid loaded by elliptically distributed load. The maximum normal and tangential stresses are p_o and q_o, respectively, and the figure shows the load distribution owing to a two dimensional cylindrical indenter with friction. (Redrawn from N. Suh, *Tribophysics*. Englewood Cliffs, New Jersey: Prentice-Hall, 1986 [37]. See the reference pp 114 ff for more detailed discussion and a derivation of the resultant stress distribution.)

with the likelihood of limited plastic deformation at the extreme asperity heights. The transverse force from friction begins to shift the stress field (Figure 22.4) that results in displacement of the brush without sliding until the interface "fails" suddenly. Sliding then occurs with a reduced transverse force, i.e., lower friction coefficient, until adhesive bonds are reformed, and then the process is repeated in a process known as "stick-slip." This asperity model of friction [22] is commonly used to explain the variation in static and dynamic friction coefficients. Typical sawtooth stick-slip curves [23–27] show this failure as a sudden event, but each event is actually a series of micro-failures of individual contact spots and is a time dependent process. A slow relative velocity produces a more dramatic stick and slip effect, and the term "velocity strengthening" [20] is used to indicate the reduced stick-slip effect at higher velocities. Instrumentation contacts can be sensitive to this stick-slip effect owing to the susceptibility of the compliant contact spring to vibration effects.

In a study of this "pre-sliding" phase on lightly loaded small copper and steel spheres on sapphire, Ovcharenko et al. [16] found that the tangential stiffness of the junctions (as represented by the slope of the force-displacement curves) for three different normal preloads was a constant at the beginning of the motion process until the slopes began to deviate and gradually decrease. The constant stiffness reported shows good correlation with theory [17] and represents the elastic nature of the junction. Interestingly, these same experimentalists and theorists show that the "breakaway" friction force is equal to ½ the area of contact irrespective of indenter diameter or material.

Nanotribologists studying boundary layer lubrication [21,28–34] explain this stick-slip phenomenon as "shear-induced ordering transitions" [21] in films just a few molecular layers thick. These films, chemically or physically bonded to the interface metals, undergo a crystallization process similar to freezing and melting. It seems likely that this "shear-induced ordering transition" phenomenon can act in conjunction with the metal-to-metal shear to affect the stick-slip pattern of sliding electrical contacts [35]. This is an intriguing area of research and provides useful explanations for temperature performance data of lubricated sliding contacts that have been a mystery for 40 years.

22.2.4 Motion Continuation

Motion continues as the contact region translates and encounters new surface conditions. The vacated a-spots will be left work hardened and with a somewhat cleaner surface. The process will repeat itself thousands of times per millimeter of travel, each time intersecting another set of asperities and exposing wiped asperities as the asperity contact passes. There is a cumulative cleaning action resulting from the process. If the sliding cycle is recurring over a recently wiped surface in the same direction, with the same contact force and attitude, then most of the counterface asperities are likely to contact brush asperities in a somewhat cleaner and harder condition.

The instantaneous average loading in the contact region is shown in Figure 22.4. The change in friction coefficient described earlier as causing stick-slip can have a significant effect on the normal force of the contacts along the resultant force line. As this force variation is transferred to the brush loading spring member, vibration effects from spring resonances can further exacerbate load variations in the contact region. Although Figure 22.4 shows these traction forces acting on the surface of the contact region, numerous studies have demonstrated that sliding causes the development of subsurface tensile stresses in the wake of the slider as a resultant of compressive brush and transverse frictional forces [36–39]. However, in the case of light loads, as is the case in instrumentation sliding contacts, the maximum strain occurs near the surface [36].

22.2.5 Adhesion

The adhesive bond at the interface is at the heart of the performance of the sliding contact since it is this bond that determines the quality of the electrical contact [40–43]. This contact quality is typically defined as low and consistent electrical contact resistance. Almost any discussion of performance issues with sliding contacts is, in the end, a discussion of the adhesive metal-to-metal adhesive bond where electrons actually flow from one surface to the other. But the term *adhesion* implies permanency, so what happens to this adhesive joint as motion is initiated and then continues? Ultimately, the adhesive bond at the interface is forced to fail as macro motion begins. The fracture of an *a*-spot will occur in one of two modes.

In the case of poor adhesion, the *a*-spot may merely shear at the original interface with no secondary effect other than the possible cold-work from the static load. This interface can be either elastic or plastic [44,45], so if there is plastic deformation cold work will likely result. One of the primary effects of lubrication and surface films is reduction of the adhesion at the junction. As a matter of fact, the perfect sliding contact would be one that has metal contact at the interface just good enough to provide low contact resistance, but adhesion poor enough to allow the junction to fracture without transferring metal.

In the second case, the adhesive forces are high resulting in subsurface fracture of the weaker of the two adhering asperities. Numerical modeling has demonstrated that the maximum Von Mises stress that result from normal compressive forces and tangential friction forces occurs slightly beneath the surface of the softer material. Suh [37] argues that dislocations "stacking up" at this high stress region during repeated cycling cause subsurface cracks and eventual material loss (or wear). With lightly loaded brushes and relatively low frictional forces (0.3–0.5 friction coefficient), the maximum forces are close to the contact points. It is reasonable to assume in the case of lightly loaded contacts that the adhesive force are not great enough to generate significant subsurface stress, but sufficiently high to cause some damage and material transfer at or near the original interface.

The physics of contact adhesion is advancing rapidly [44,45] due to the ability of researchers to both measure low adhesive forces between asperities using atomic and surface force measurement tools [82,83,142] and model asperities using sophisticated molecular dynamics simulation (MDS) models [141]. This research continues to support the seminal work done by Johnson, Kendall, and Roberts [143] on the influence of surface energy on contact adhesion. The need for precise adhesive responses of nano-contacts in a variety of nano-devices continues to drive research in this area.

22.2.6 Adhesive Transfer

The terminology and model developed by Antler [1,46–51] is still commonly used today to discuss the material transfer from counterface to brush. This transferred material tends to form a pad of laminar construction on the brush. This pad, or "prow", remains attached to the brush and the sliding contact interface is between the counterface and the prow. The prow is composed primarily of counterface material in an extremely cold-worked and hard condition. As sliding continues, small amounts of this prow material can back-transfer to the counterface and then back again to the brush, setting up an equilibrium condition of low wear or loss of material from the contact region. The wear debris lost from this system tends to be very small, fine particulates. Alternatively, entire prows can be dislodged from the system in situations of poor adhesion to the brush. These dislodged prows form larger, brighter wear particles. Figure 22.5 shows examples of these prows in various stages of their genesis, build-up, and separation. Chapter 7 contains a more detailed discussion of prow formation and separation. Chapter 23 provides a much more in depth discussion of adhesion, transfer, and sliding contact.

The counterface material must have a reasonable hardness or the *a*-spots will be so large and surface failure will be so deep into the counterface that there will be only gross adhesion, transfer of large particles, and major destruction of the counterface surface. This is sometimes referred to as "galling."

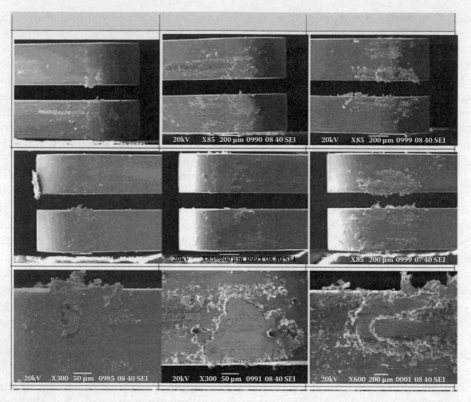

FIGURE 22.5
Wear on flat bi-furcated brush. Typical brush contact areas after several hundred hours of rotation at 30 RPM. These pictures illustrate ring material transferred to the brush to form a pad of material of laminar construction. After this wear pad is established, subsequent sliding occurs between the pad and ring. This prow material is quite hard with a measured hardness of roughly four times the hardness of the electro-deposited gold. The typical prow is about 6.5 μm thick and of various widths and lengths, although on average in the example slip ring, prows are about 200 μm in diameter.

22.2.7 Plowing, or "Two-Body," Abrasion

Some asperities may protrude below the peaks of asperities on the opposing member in the direction of slide and will collide. Depending on asperity size, normal force, and material properties, there are two basic responses at these interlocked asperities: either (1) the asperities "climb over each other" with a combination of elastic deformation and z translation, or (2) the harder asperity plows through the softer one. There is also the option of a combination of both of these responses (climbing and plowing) which along with straight plowing leads to plastic truncation of asperities. Yin et al. [19] report less than 2% of the contact area of sliding elastic contacts are these plastically truncated asperities.

Initial run-in periods of sliding contacts involve a significant amount of "asperity leveling" when the asperities of "as produced" contacts are plowed through or truncated (compare as-plated and post-run condition of the rings of Figure 22.6). Because the asperities on the smaller contact will likely have had more previous contacts per unit travel than those on the larger counterface, the asperities of the brush receive more cold-work per unit area and are more quickly hardened. These harder asperities then have a tendency to plow and force the brush profile onto the wear path surface of the counterface. In addition, the prow

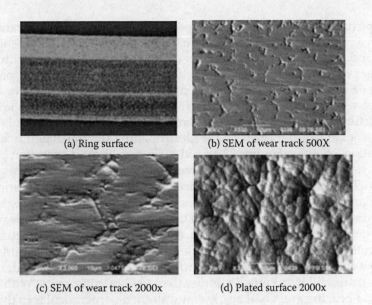

(a) Ring surface   (b) SEM of wear track 500X

(c) SEM of wear track 2000x   (d) Plated surface 2000x

FIGURE 22.6
Wear on the counterface. Ring surface after being wiped with brushes similar to those of Figure 22.5. The ring is becoming "burnished" during sliding (darker region of a) and the nodal finish formed by the electroplating (d) (Rc of 0.2 μm) is being worn smooth with rotation (compare (b and c) and (d)). Although the two halves of the bifurcated brush are identical, it can be seen that they make different widths of wear track, illustrating the variability that can occur with manufacturing tolerances. There are clearly plenty of "crevices" for lubricant to collect.

formation on the brush can also form a relatively hard, inelastic surface that can also serve to "plow through" any counterface asperities.

22.2.8 Hard Particle, or "Three-Body," Abrasion

If a hard particle, or third body, enters the contact region, it can act as the contact asperity and support load between the surfaces. Depending on the shape of the particle and the hardness of the materials, the particle may roll between the surfaces before becoming expelled from the contact region or embedded into one of the surfaces. If it embeds in one of the surfaces, the particle can begin to cut a groove of its protruding profile into the opposing surface material. Multiple particles will cut multiple grooves. If the entire load is supported by the hard particle(s), the conduction will depend on the conductivity of the particle(s), and most hard particles found on the contacts have much poorer conductivity than the contact surfaces, for instance, ordinary dust. If the particle is small enough that it supports only part of the load, some electrical contact is possible around the particle, but the contact resistance can become erratic and unreliable. One source of these particles is wear debris generated by the contacts during normal operation. End of life of sliding contacts most commonly occurs when the generation of wear debris accelerates to the point where the contacts can no longer operate successfully.

22.2.9 Motion over Time

Figures 22.5 and 22.6 illustrate ring and brush contact regions from a flat ring and a 2-fingered flat brush (bifurcated) after an initial run period. The photomicrographs in Figure 22.5 show typical brush contact areas after several hundred hours of rotation at

30 RPM. These pictures illustrate the earlier description of ring material transferring to the brush to form a prow of laminar construction. Subsequent sliding then occurs between the brush prow and ring. This prow material is quite hard with a measured hardness of roughly four times the hardness of the electro-deposited gold. The typical prow in this example is about 6.5 μm thick and of various widths and lengths.

Figure 22.6 shows typical ring surfaces (there is no relationship between the brush and rings of these pictures). While the prow is forming on the brush, the ring is becoming "burnished." The nodal finish formed by the electroplating (Rc of 0.2 μm) is being worn smooth with rotation (compare Figures 22.5c and d). In equilibrium conditions after prow formation, the system is composed of a hard prow sliding against a smooth, work hardened ring surface. This system is typically lubricated with a hydrocarbon lubricant that is applied in low quantities. In addition, other films can be found in the wear region that are products of environmental contamination. It is interesting to note the wide variability of the prow formation from brush to brush in Figure 22.6. Although we can discuss "typical" prows, it is difficult to exactly characterize these features. Some brushes build a distinct, well-defined prow and some seem to have a hard time getting a prow started.

At any instant, there will be a limited number (l) of load-bearing areas. Within the contact region the load-bearing areas will be generally aligned perpendicular to the direction of sliding, but they may be anywhere; peak, valley or side, on the transverse profile. The asperity interaction mechanisms of plowing and adhesion leave behind altered surfaces. Repetitive passes over the same area will leave both surfaces somewhat more level in the direction of motion. This leveling process is most pronounced during the initial passes of the slider over the counterface. Presuming that the slider is operating in a reasonably repetitive path over the counterface, leveling will reach a quasi-equilibrium after a number of passes depending on the materials and conditions. However, striations will develop parallel to the motion direction, making the surface transverse to the motion direction of both contact surfaces rougher than in the motion direction (Figure 22.7).

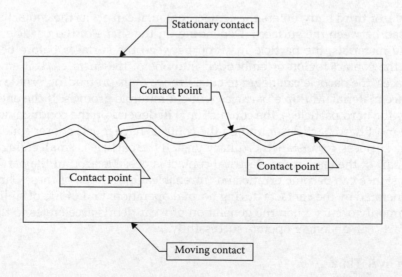

FIGURE 22.7
Transverse profile of ring and brush sliding contact. The lower member is moving into the page. Notice that the transverse profiles are almost identical, but the tongues will be slightly smaller than the grooves into which they fit. These striations of this pattern will be similar around the entire worn area of the faces. Illustrated are three load-bearing areas of points of contact. These points of contact are not always the most extreme asperity height.

22.3 Electrical Performance

As the sliding element of the electrical contact wipes across the stationary surface, there are number of events occurring at the micro-level in the contact region that affect the electrical performance:

1. Surface films are being continually wiped off, re-formed, and/or re-deposited. The presence/absence, thickness, and chemistry of these films have an effect on both the capacitance and the tunneling resistance.

2. The surfaces are being altered by the wiping action; surface roughness is changing; adhesive debris (prows) are being formed, broken, and re-formed; and the contact area is changing.

3. Dynamic effects (stick-slip, external vibration effects, surface run-out) are causing changes in the normal force pushing the contacts together.

4. Boundary lubrication parameters are being altered by surface physical and chemical changes, environmental conditions, and frictional and joule heating causing dramatic changes in the effect of lubricant [53,54].

5. Wear debris and contaminating particles (conducting, semi-conducting and insulating) are finding their way between the contacts and being embedded into the boundary layer.

The discussion of the electrical properties of sliding electrical contacts also starts out with consideration of the static contact (see Chapter 1). Early researchers [13,55] described the mechanics of electrical contact as multiple micro-asperities in contact. Electrical contact resistance resulting from a low number of contact points and low contact area constricts the electron flow near the asperities actually in contact [56]. This resistance is commonly called constriction resistance. Constriction resistance is defined as the increase in resistance owing to the replacement of a perfectly conducting interface by a constriction [57]. In the case of asperities that make contact through a thin film, current can tunnel through the film if a minimum tunneling voltage potential is exceeded [13], see Section 3.4. The model of constriction and tunneling resistance is shown in Figure 22.8. Much of the work on electrical contacts in the past 20 years has been focused on describing and modeling the contact asperities and their contact resistance using sophisticated numerical techniques [58–61]. This modeling has led to a better understanding of the thermal and electrical field distributions in the contact region and the bulk material adjoining this region.

It is the change in contact resistance with rotation that has been used as the critical parameter for sliding electrical contact performance. This change in contact resistance results in electrical noise on the signal or power channels being conducted through the interface. The methods of evaluation of sliding contacts is changing as the types of signals they transmit change, and an appreciation has been gained that more than DC resistance is at issue. There are non-linear and frequency dependent resistance effects [62,63] that can play a role in the use of sliding contacts in high frequency applications. Attention is also being drawn to non-ohmic "tunneling" conductance of electrical contacts, especially in micro and nano-sized contacts. As high speed data applications proliferate in the instrumentation slip ring applications (high definition digital video and high speed serial data buses), electronic systems designers are interested in the *impedance* variation and not just the DC resistance, so the reactive element of impedance becomes important. A number of

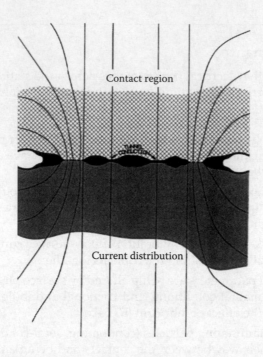

Contact region

Current distribution

FIGURE 22.8
Current-flow distribution in the *a*-spot region of Figure 22.1. Principal current-flow lines through the contact illustrated in Figure 22.1. The greatest current density passes through the metallic *a*-spots at right and left. Some small current may flow through very thin films by tunneling.

studies have expanded the understanding of contact impedance in static contacts to high speed digital data transmission [62,64–66], and it is equally important to recognize the importance of the reactive element of electrical contacts in sliding. A model of an electrical contact with all potential impedance elements, RLC, is found in Figure 22.9.

22.3.1 Contact Resistance Variation (Noise)

On relatively clean, conductive surfaces, load-bearing areas represent a fairly small ratio of the contact region. For this reason, the electric current density, more or less uniform in the bulk of the macro contact, is constricted to high densities in the small cluster of smaller spots that transfer from one contact to the other. This high current density passing through the small constrictions on each side of the metallic interfaces produce the constriction or "contact" resistance, R_c, as a simple relationship between the resistivity of the contacts, ρ, and the constriction radius. *a*, [13] (see Chapter 1).

$$R_c = \rho/2a \tag{22.2}$$

If we project a representative current distribution through the load-bearing area depicted in Figure 22.1, we see Figure 22.8. Here, we see higher current densities in the metallic *a*-spots and some lesser densities through thin film layers by tunnel conduction. Typically the contact area on gold signal contacts is ~25 μm² or ~5 μm diameter and the contact resistance is ~4 mΩ per contact, but with two contacts in parallel this contact resistance is ~2 mΩ. Electrical contact resistance is undoubtedly the most important element of sliding-contact systems, and performance of sliding electrical contacts have typically been

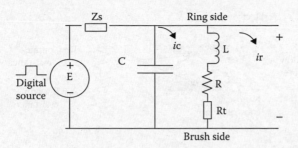

E = Digital data source	L = Contact inductance
Zs = Source impedance	C = Contact capacitance
Rt = Non-linear tunneling resistance	ic = Current through capacitor
P = Ohmic contact resistance	ir = Current through contact

FIGURE 22.9
The contact as a network. The diagram shows the sliding contact as with RLC circuit elements forming a contact impedance.

characterized by the variation in contact resistance (contact noise). This characterization has been in terms of linear (or ohmic) noise, but we will need to also look at non-linear resistance effects on the impedance.

Slip ring noise has always been considered a linear problem following Ohms Law. Low noise, precision slip rings were developed for the analog world. Frequently these slip rings were inserted in front of any system gain, and any noise introduced in the system by the slip ring was amplified and was seen as sensor error. A typical specification on slip ring noise for these slip rings was 20 mΩ that would equate to a few μV of noise when passing mA signals. It is important to note that these systems typically have a very low frequency response, so that any noise above a few KHz was filtered out and had no effect. And since the noise was measured on analog oscilloscopes, noise in the MHz range was not recognized. A thorough review of slip ring noise characteristics and properties is provided by Holzapfel in [7,67]. Over the years slip ring noise problems have generally been explained by contact film anomalies [68–70], but there is an increasing understanding that the vibration response of the brush to the mechanical perturbations of the contact zone are just as, and maybe more, important than film anomalies. However, it is true that film anomalies can be the primary agent behind the aforementioned mechanical perturbations.

22.3.2 Non-Ohmic Noise

This traditional analysis of contact resistance assumes that the spot size of the contact is significantly larger than the mean free path of electrons and that metal-to-metal contact is the conduction path. Holm [13,55] developed the concept of a non-ohmic, or quantum, component of conduction in electrical contacts due to the penetration of thin films between contacts by electrons. These semi-conducting films form potential barriers between metal contacts where the conductivity is dependent upon the voltage differential. In recent years mesoscopic [71] and nanoscopic [72] physics researchers have refined non-ohmic conductor theory where electrons traverse atomic-sized conductors "ballistically," and the resistance becomes independent of length [73]. When the contact size is reduced to a distance smaller than the mean free path of electrons, the Sharvin effect is the primary component of resistance [56]. As shown in Figure 22.10, this effect results in a transition of the conduction properties in the region near the contact spot from a continuum (Figure 22.10a) to a ballistic (Figure 22.10b) regime.

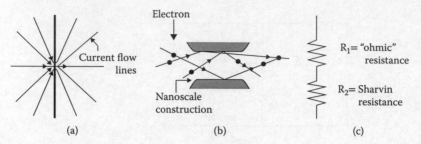

FIGURE 22.10

Maxwell (continuum) vs. Sharvin (ballistic) resistance. When the contact area is less than the mean free path of electrons, constriction resistance no longer follows continuum laws of Maxwell, and Sharvin calculations are required. At first approximation these two resistnace values can be treated as a series resistance. ([c] Redrawn from R. S. Timsit, *Components and Packaging Technologies, IEEE Transactions on*, vol. 29, pp. 727–734, 2006 [74].)

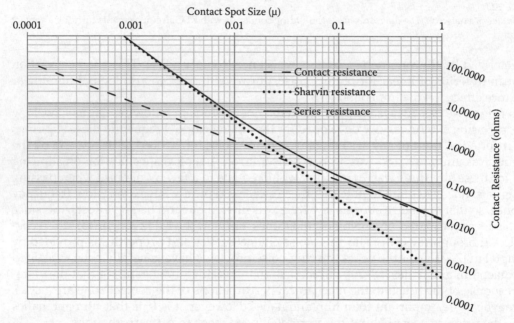

FIGURE 22.11

Resistance of a gold contact. When Maxwell and Sharvin resistance are treated as a series resistance [74] we see that Sharvin (ballistic, non-ohmic) resistance dominates with contact radius of less than 20 nm.

Timsit discusses this effect [74] and simplifies the traditional Sharvin equation to

$$R_s = C/a^2 \tag{22.3}$$

where C is a constant dependent only on the electronic properties of the material and a is the contact spot size. For gold, the value of this constant is 3.58×10^{-16} ohm m^2.

This Sharvin resistance component of resistance has significance when the contact spot is sub-micron. The standard ohmic resistance and Sharvin resistance can be evaluated as a series resistance [74] to get a quick idea of the relative effects, and the sum of these two resistance values for lightly loaded gold contacts are shown in Figure 22.11. This analysis shows the Sharvin resistance and continuum constriction resistance curves cross at about 0.03 μm. It is likely that the Sharvin resistance is actually a bigger component of the resistance in actual multi-spot contacts where the total contact area is divided among several contact spots. For example, we earlier estimated the contact spot size at about 5 μm in diameter, but that was estimated on the basis of only one asperity. If we assume 5–10 asperities (not an unreasonable assumptions [75]), some will clearly be submicron. It has been shown [75,76] that in the realm between Sharvin and pure ohmic contact (Maxwell contact), a formula developed by Wexler [77,78] for conductance, G_w, provides good fit to experimental data for gold contacts and can be used for a more precise estimate than that done for Figure 22.11. Wexler's formula is:

$$G_w = G_s \left[1 + \frac{3\pi}{8} \Gamma(K) \frac{a}{l} \right] \tag{22.4}$$

where G_s is the Sharvin conductance, a is the spot radius, l is the mean free path for gold, $\Gamma(K)$ is a slowly varying function of order unity, and $K = l/a$ is the Knudsen number.

Timsit argues that Sharvin resistance is the likely explanation of the apparent breakdown in the classical contact theory in the reporting of contact intermittences by Maul, et al. [79–81]. It is reasonable to assume that lightly loaded contacts of instrumentation slip rings are often within the mesoscopic range where ballistic effects can and often do apply. This is especially true as the metal junctions deform, stretch, and neck down before failure during sliding. Investigations with in situ transmission electron microscopy (TEM) and Scanning Tunneling Microscope (STM) [76,82,83] illustrate this contact necking. Adams in his review of the mechanics of adhesion [44,45] reviews the characteristics of ductile separation of the adhesive bond:

1. The contact radius decreases slowly before a sudden separation (typically at a significant reduction of the maximum contact radius);
2. The asperity if stretched significantly during separation owing to plastic deformation;
3. A neck is sometimes formed during ductile separation.

This research area of the existence and behavior of ballistic contacts as well as the tunneling effect of films in the contact region of sliding electrical contacts will continue to receive attention as MEMS devices continue to proliferate and contact areas become smaller and forces become lighter; see Chapter 12.

22.3.3 Non-Linear Noise (Frequency Dependent)

Electrical current is an operating parameter for any electric circuit or contact and the contact performance is affected by the current as a function of time. The load-bearing asperity area and metallic a-spot area are already stressed by the load to near the compressive yield of the softer material. As current passes through the metallic contact area, heat is generated in the interface and in the constriction adjacent to the a-spot area resulting in a temperature increase in the a-spot area. This "super-temperature" [13] causes:

1. The strength of solid films and the viscosity of fluid films adjacent to the a-spot to decrease
2. The evaporation of volatiles adjacent to the a-spot to increase
3. The yield strength of metal in the a-spot to decrease

These factors cause the a-spot area to increase and contact resistance to decrease. Most significantly, the latter occurs when the contact voltage (i.e., constriction resistance times the current) approaches the softening voltage described by Holm [13]. The resistance, and consequently the contact voltage, immediately falls to a lower value that suggests an increased metallic contact area resulting from more or larger *a*-spots.

Further increase of the current on the same contact to a new high will cause a repetition of the process described above. When the contact voltage drop reaches a level close to that described by Holm as the melting voltage, there will be melting of the asperities. In addition, with the higher current and heating, chemical reactions that can occur will occur at a more rapid rate; that is, the inorganic contaminant species, oxides, sulfides, chlorides, and so on, will react with base metals, and organic monomers that can be catalyzed to a polymeric state will be accelerated by the higher temperatures. When the current is reduced or removed, the *a*-spots and the surrounding contact will cool, the asperities will contract, more vapors can condense, and the viscosities of fluid, and polymer, will increase, but any cold welding and chemical reaction products will probably remain.

The response to these Joule heating effects have a very short time constant, and a resistance non-linearity can be generated in the contact region by a high frequency AC signal [57], i.e., variation of current/voltage with time. This non-linearity has been shown to be a function of the signal frequency as well as the higher order harmonics [43,84,85] owing to the cyclic variation in current and the resultant joule heating at the contact region. Primary studies have been performed on high frequency sine waves, but this non-linearity becomes more complicated in the case of digital data by the multiple frequency components of the square wave. This contact non-linearity is one source of Passive Intermodulation (PIM) interference on the transmission line carrying high frequency electrical signals [86] and is one potential source of signal jitter in the time domain of a high bandwidth square wave.

There is an additional non-linear component of contact resistance in contacts transmitting high frequency AC signals. An AC signal leads to skin effects in the constriction region [87] and produces a frequency dependent component of constriction resistance. There are actually two parts to constriction resistance that are lumped together in DC applications: the resistance at the constriction itself and the area on each side of the constriction where the current expands into the bulk material after passing through the constriction. Under DC and low frequency conditions, the current flow near the contact spot spreads widely and without constraint into the volume on both sides of the constriction. However, at high frequencies, skin effects limit the ability of the current to expand into the bulk material thereby increasing the effective resistance. There is some evidence to suggest that the resistance at the actual constriction decreases at increased frequencies [57], but the more significant increase in the bulk resistance skin effect component results in a net increase in constriction resistance as frequencies increase.

22.3.4 Contact Impedance

The reactive elements of contact impedance, capacitance and inductance, are important to consider with AC signals as discussed by Dorsey et al. [88]. The contact inductance is not an issue with low currents, but capacitance with high bandwidth signals is worthy of consideration. Estimates of capacitance of a typical sliding contact [88] show that making the assumptions of (1) separation between contacts is of the average surface roughness of the counterface, (2) the majority of this separation is filled with lubricant film, and (3) a coupling area of the apparent area of contact, yields the capacitive reactance $(Z_C = (2\pi f C)^{-1})$ of between 5 Ω and 50 Ω between 100 and 1,000 MHz (see Figure 22.12). Since this same study shows that 50 Ω is a reasonable threshold for intermittence failures

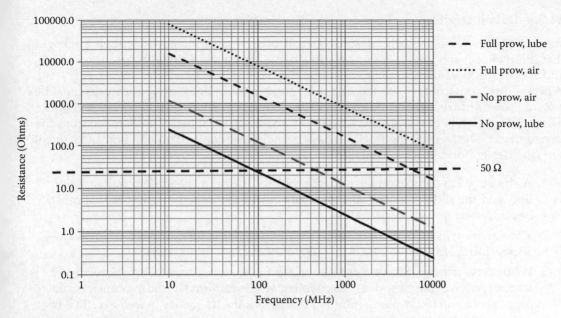

FIGURE 22.12
Estimated capacitive reactance expressed as equivalent resistance. These very approximate calculations were performed to show that capacitance will likely have minimal effect at data speeds less than 1 Gbps and between 1 and 10 Gbps the effect could be worthy of consideration for certain contacts.

of typical digital signals, this estimation shows that capacitance could be a transmission path for data with frequency components above 100 Hhz in the event of loss of metal-to-metal contact. However, the other curves of Figure 22.9 show more realistic assumptions on this same contact (not full lubricant film, prow to separate the bulk of the apparent area, etc.). Additionally, the assumption of a separation equal to approximately the surface roughness across the full apparent area is not practical with normal mechanical tolerances. The fact is that sliding contacts have small apparent areas of contact (capacitive coupling areas) and light brush pressures (capacitive separation), means that capacitive coupling at the contacts is not likely a significant factor in high speed data transmission with normal contact design.

This rough illustration of the possible effect of capacitance on a high speed sliding contact makes some simplifying assumptions about the size of the apparent area of the contact, the separation of this apparent area, and the relationship of the size of the apparent area to the real area of contact. Research on capacitive micro-electro-mechanical systems (MEMS) switches, as well as other nano devices, will serve to further the understanding of how capacitance can be understood and used to improve the transmission of high speed signals through sliding contacts.

A similar analysis can be performed on the inductive element of contact impedance. Although important for power contacts with high switching currents, inductive coupling is normally not a factor in instrumentation and control sliding contacts since the self-inductance values at the contacts is negligible. Inductance and capacitance both can play an important role in the overall interconnect structure of instrumentation and control slip rings and other devices, but the impact at the interface itself is relatively minor compared to resistance. Additional modeling and research is needed to further support these analyses.

22.3.4 Data Integrity

Passive circuit elements such as wires, circuit boards, and integrated circuit packages that interconnect active circuit elements in a digital system are called transmission lines, and most aspects of these transmission lines are well defined and characterized [89–94]. When sliding contacts are inserted into these transmission lines, design features must be selected to minimize signal propagation degradation (e.g., ringing, reflections, and phase delay), interaction between signals (crosstalk), and electromagnetic interference from the environment. The following guidelines have been defined for instrumentation and control slip ring design for high speed data [88]:

1. A slip ring is a complex interconnect problem in a high speed data transmission line, and the sliding contact model is only one of the impedance issues that must be addressed for proper operation.

2. The variation in AC impedance is the key parameter to be examined when evaluating sliding electrical contacts for suitability to carry digital data.

3. While capacitance can have a significant effect in stationary contacts with relatively large capacitive coupling area, this coupling area is minimal in sliding contacts and capacitance will likely play a minimal role unless the frequency is well over 1 GHz or the contact spot size is small relative to the capacitive coupling area. In the case of contact degradation, the increased dielectric value between the contacts could lead to higher capacitive reactance and shunting of higher frequency components.

4. Inductance in the contact region is low and in series with the resistance and should play no appreciable role in the variation in contact impedance with rotation in the case of signal and low power contacts.

5. Resistive effects continue to be the major parameter that should be used to characterize sliding contact performance. However, Sharvin resistance and non-linear frequency dependent resistance, in addition to "traditional" ohmic resistance, can be factors when sliding electrical contacts are transmitting digital data.

6. Intermittences are the primary issue with digital data transmission through electrical contacts. An obvious source of these intermittences is insufficient metal-to-metal contact leading to voltage drops in excess of the threshold value.

7. Experimental data shows that when the contact force is reduced, frequency intermodulation effects cause noise near the fundamental frequencies of a square wave leading to a rise in the noise floor.

The discussion regarding contact impedance dealt with the contacts only. The problem with instrumentation and control slip rings is that the contacts must be connected to the rest of the transmission line, and it is that connection process that imposes the greatest bandwidth limitation (see Section 22.8.3 and [88]).

22.4 Micro-Environment of Contact Region

The ability of the sliding electrical contact to transmit electrical signals reliably over the operating life is controlled by the micro-environment of the contact region. Figure 22.13 [95] illustrates representative contact surface with the materials to be expected on a gold or silver electroplated surface over a nickel layer on a copper or copper alloy substrate. A gold or

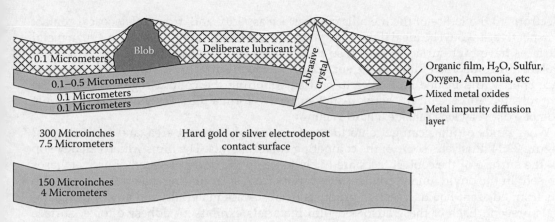

FIGURE 22.13
Representative schematic of a thick electroplated surface, normally "clean", after exposure in a typical environment. The copper alloy substrate has been plated with nickel and over-plated with either silver or gold. The surface of the plate is covered with base metals that have reached the surface by diffusion or "dragout." These have oxidized in air or reacted with other gases. Above that is a layer of all sorts of adsorbed gases and condensed organic vapors. Particulates and blobs of aerosols from the environment, together with lubricant deliberately added, complete the picture.

silver plate probably will have some small amount of some base metals co-deposited with it or "dragged out" with the plating solution. These contaminating base metals can diffuse to the metal surface where they can react with the atmosphere to produce an oxide layer, often mixed or overlaid with sulfides, chlorides, and nitrides generated by other reactive atmospheric pollutants. In addition, polymeric material can be co-deposited as part of the gold plating process [48,50]. On top of or within all this may lie abrasive particles and blobs of aerosols all surrounded by a film of any deliberately applied lubricant. Base metal surfaces have many of the same films except that the inorganic films tend to be thicker and more rigid.

Recognizing the need for some mechanical film disruption in establishing electrical contact, most connector, switch, and relay manufacturers build in some wipe during insertion to assure the mechanical breakthrough of existing films to obtain good contact. Recommendations for assembling many static connections include deliberate wipe under force to break up or move aside surface foreign materials. Sliding electrical contacts have the advantage of the sliding or wiping action as part of their function. However, sliding contacts typically have contact forces that are orders of magnitude lower than static contacts, and the chemisorbed and physisorbed materials can be tightly bound to the surface making them difficult to wipe away with light contact force. Additionally, as the contact moves forward, surface materials can become wedged into the interface of the contact region.

A tremendous amount of data is being produced by researchers working with atomic force microscope (AFM) instruments [139]. The tips of these instruments can be produced with a radius less than 10nm and they can resolve forces in the nN range. This allows testing of film dynamics on a contact the equivalent size of an asperity. The clarification being brought to boundary lubrication of micro- and nano-contacts is dramatic.

22.4.1 Film Forming on the *a*-Spots

In Section 22.2, we discussed the cleaning actions of *a*-spot interaction. These actions leave *a*-spots on both contact faces with clean, nascent metal spots where contact existed. The most efficient conduction occurs only where a metallic asperity of one surface transfers electrons to the lattice of the metallic counterface asperity and, from an electrical contact perspective, it would be ideal if these surfaces could be kept in this clean state. Clean metal surfaces have high surface energies, so they readily uptake gases from the environment [96] forming adsorbed films on the surface of the metal surface. Because of this high surface energy, the mobile molecules in and surrounding the contact regions rush to the unprotected and chemically active *a*-spot surface to form a film. Truly clean surfaces just do not exist very long outside a hard vacuum.

What kinds of films can get onto the surface, into the contact area, and to the *a*-spot? Figure 22.14 illustrates some of the competing contaminants. The films which can develop on the surface of the contact faces are likely to be, in the order of speed of arrival, gases present in the environment, condensable vapors from surrounding materials, surface fluids from adjacent liquids, reaction products from surface materials, and any of the chemicals present. Each of these adsorbed film materials exhibits an adhesive force (surface tension) to the surface metal. In general, the underlying films have lower surface energy and greater surface adhesion than those overlying them. Those of greater thickness have lower shear strength than the thinner ones.

An increase in time increases the effect of any of the aforementioned conditions: condensation, diffusion, chemical reaction, adhesive bonding, and film penetration. Some of these factors are short term while others are longer-term effects. Cycle time between repetitive events also has a significant effect on film formation and contact performance. Current flow causes heating of the contact and adjacent components and offgassing while the subsequent cooling allows condensation.

Since these time dependent factors play a critical role, one important consideration in the operation of a sliding contact is the sliding speed and operational duty cycle. Although not an exact science and difficult to predict, there is a surface speed at which a sliding

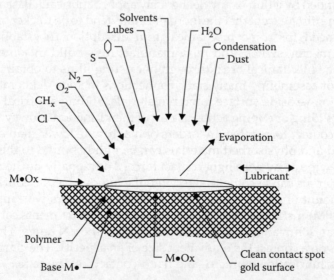

FIGURE 22.14
Competing materials rushing to cover a nascent *a*-spot just vacated by a contacting asperity. Gases and vapors in the vicinity and in the highest concentration will get to the surface first followed by fluids with the lowest viscosity. Base metals will diffuse to the surface as may plating polymers in the surface metal. The materials that react with the surface or have the lowest surface energy will tend to have the greatest adhesion to the nascent surface and exclude others.

contact with a certain lubrication strategy, design features, and material combination will no longer produce acceptable performance results owing to the inability of the surface films to repair themselves and provide sufficient protection of the metal contacts. This time factor of film repair is also the primary reason that accelerated wear tests often produce erroneous results—film formation and its effectiveness is dependent on the amount of time between surface wipes, i.e., rotations or linear cycles.

The presence of these surface films have both advantages and disadvantages. The advantages will be discussed in the lubrication section as we discuss the necessity of these films for reasonable friction and wear performance of sliding contacts. However, the presence of a film can be a disadvantage since the moving contact will have to push some of these aside or break through them mechanically to make some electrical contact. Hard, thick, and/or tenacious films can make contact difficult for low force and low power sliding contacts. Films that form on the unwiped portions of the contact are usually thick, but relatively benign since these relatively immobile films have less opportunity to interact with the surface materials. However, reactions between active and mobile lubricant and gaseous contaminants can be more problematic.

Mechanical action is the most effective method of maintaining metal contact by simply wiping the film away. Electrical current has the ability to tunnel through surface films at the asperities, so the films do not have to be completely removed, but they do have to be limited to one or two molecular layers for tunneling to create an acceptable path for conduction.

Lubricant spread on the contact surface has a high ratio of surface area compared to lubricant volume. This relatively large area facilitates the reaction of active chemical compounds with the lubricant independently or with the catalyzing effects of the nascent metal of the a-spots. This effect can produce an unintended surface film on the contacts. One striking illustration of the insidious nature of one off-gassing contaminant is reported in references [97–99]. The reaction of a halogenated lubricant with an off-gassed curing agent from a "cured" epoxy potting compound in a nearby component increased the resistance at the contacts to unacceptable levels.

Another example is the reported increase of viscosity of a lubricant on sliding contacts by vapors from a floor wax stripper being used near the lubricated slip ring assembly during installation into a critical application [100]. The contaminant reaction was found to increase the viscosity of the lubricant causing high circuit resistance at low temperatures and high rotational speeds. In both cases, the quantity of contaminant was quite small compared to the quantity of lubricant, but the effect was quite dramatic on the electrical performance.

22.4.2 Unintentional Contamination

"Environment" is a collective noun incorporating perhaps the most important, but least known and poorly controlled, factors affecting sliding contacts. Factors such as temperature, gases and vapors, moisture, particulates, fluids, and time all affect the contact in some form, but seldom are they measured and rarely controlled. The discussion on lubrication will review "adventitious" lubrication that can result from this uncontrolled contamination, but in many cases these environmental effects are deleterious.

22.4.2.1 Particulates

Particulates (dust, abrasives, organics, metallic debris, lint) can prevent contact closure or seriously affect the contact resistance, particularly if they are non-conducting. The effect depends on the material, location, sizes of the particles, the roughness of the surfaces, and contact radii. The contamination from particulate debris has been addressed by Williamson [101] and others [102,103] in their effect on static or closing contacts. Zhang et al. [104] made study of dust in Beijing, China and found three key elements: (1) inorganic compounds (abrasive), (2) organic compounds (film forming) and (3) water soluble salts (corrosion agents, see also Chapter 4.). The particle(s) can moderate the contact performance during wipe or slide. Depending on the hardness, the particle may smear out to make a film or impede and abrade the mating surface, causing wear and debris. Particles can agglomerate into clusters since they are often coated with fluid films. The clusters are usually nonconductive because of the insulating nature of the fluid that surrounds the particles. Particles can be drawn to a contact member because of electrostatic charges and caused to align with electrostatic fields, influencing both wear and electrical isolation of circuits [104].

Particles can be from external contamination caused by ingress of dust from the environment or they can be generated by the wear of the contacts themselves. One of the primary factors in the design of sliding contacts is the optimization of brush force, type and amount of lubricant, and materials to minimize the size and quantity of wear particles.

22.4.2.2 Contamination or "Air Pollution"

Typical macro-environment surrounding sliding contacts consist of oxygen, nitrogen, water vapor, and a myriad of "atmospheric contaminant" gases. The composition has been presented by several papers [105–107] as including S_8, Cl_2, NO_x, H_2S, SO_x, HCl, and others (see Chapters 2 through 4). These inorganic contaminants react with the moisture and the oxygen to produce various salts, hydroxides, oxides, chlorides, nitrates, sulfides, and sulfates, of the reacting metals which might be present on the contact surface, whether alloyed or not. The result is a complex mixture of chemicals, film materials, and particulates that will impede mechanical and electrical contact. Even with closed static contacts, the contaminants can intrude, react, and cause the contact to increase resistance.

Owing to the inevitable presence of reactive contaminants, most critical contacts, particularly those operating at low force and low voltage, use noble metals in the contact surface. These noble metals are not a cure-all for there is also diffusion of surface active elements from or through the surface layer to the surface even at relatively low temperatures. Chemical reactions of the diffused base-metal elements with the reactive agents on the surface can affect the contacts as if the contacts were made of the diffused metal. In addition, the active surfaces of most noble metals are reactive to some common contaminants.

Water molecules are inevitable in their presence and, in many cases, are quite beneficial. Chapter 23 provides a more detailed discussion of adsorbed water vapor and its beneficial effects on sliding contacts. Experiments have shown that even one layer of water molecules, approximately 0.25 nm thick, between two hydrophilic surfaces, can be sufficient to reduce the interfacial friction force significantly [29]. While metal contacts are not hydrophilic, water molecules in the "soup" in the contact region would provide similar lubricity. The disadvantage is that at cold temperatures this water mixture can become "slushy" or even solid and cause electrical resistance problems.

22.4.2.3 Organic Off-Gasses

Along with the inorganic contaminants present in the vicinity of contacts, there are usually organic vapors present that can condense on surfaces and react with or otherwise influence the behavior of the contacts. The most uncontrolled contaminants are the atmospheric contaminants and the gases from nearby or confined high-vapor-pressure materials and ingredients. These organic materials originate in construction plastics, their monomers, diluents, and plasticizers. The organics will condense on all exposed surfaces. On many surfaces, including precious metals, the organic monomers can undergo polymerization, becoming more viscous and hard to move [108–110]. They seldom are major problems for static contacts unless there is arcing present or the contact is subject to dithering or fretting motions. When they accumulate on sliding contacts, they can be difficult to penetrate mechanically or electrically.

22.4.2.4 Friction Polymers

Organic materials condensing on wiped surfaces can, with repeated wiping, convert from low-molecular-weight plastics (monomers) to long-chain polymers on precious metal surfaces [108,109]. Cyclic compounds like benzene were the first found that create friction polymers. These polymers form principally on palladium and platinum metal surfaces, but have been identified on gold surfaces with some reports on silver. Silicone vapors are notable because they will readily form polymers on gold sliding contacts. Other organic species observed to polymerize are polysulfides, terpenes, and some acrylics. The process is thought to be one of catalysis by the nascent metal *a*-spots in the organic vapor. The process requires many hundreds of thousands of passes before polymer can begin to be detected on gold used in miniature slip rings. The polymer is usually insoluble in all but the harshest acids.

22.4.3 Lubrication (Intentional Contamination)

Most sliding contacts are lubricated and these lubricants are selected from a general group of lubricants used in electrical contacts [111,112] bearing in mind the properties shown in Table 22.2. The most important property of these lubricants is their ability to bond with the contact surface. The bonding process can be a strong intermolecular force (a covalent or Coulomb bond) or a weak intermolecular force (most commonly known as Van der Waals force). In the discussion of contact films the terms chemisorbed (strong) and physisorbed (weak) are commonly used. The research activity related to the study of the bonding of thin films on solid surfaces is very robust and is one of the most promising for improvement in sliding electrical contact performance [116,140].

The lubrication of sliding contacts is best described by boundary layer language and is summarized by Singh et al. [53]:

(a) The normal load is supported by a few discrete asperities on the contacting surface.

(b) Adhesion occurs at the asperities.

TABLE 22.2

Ideal Properties of a Good Sliding-Contact Lubricant

Operates best if chemisorbed, or highly physisorbed, on the contact surface

Must not chemically degrade metals of the contact surfaces

Must not chemically react with insulation of the contact assembly

Any chemical reaction with gases of the operating environment should be benign

Must have low volatility at the operating temperature and pressure

Must have the proper viscosity for the velocity, temperature, and contact force

Must not form friction polymer of sufficient viscosity to separate contacts

Should have low surface energy to wet the contact surfaces

(c) When relative sliding occurs, resistance is offered by sheared adhered junctions.

(d) The presence of lubricant reduces the strength of asperity adhesion.

(e) Interposed boundary lubricating film is formed owing to physical adsorption and is homogenous throughout the interfacial surface.

(f) Adsorbed lubricant molecules orient themselves perpendicular to the contacting surface.

Figure 22.15 illustrates Singh's boundary layer adsorption description. Studies of these molecularly thin films [21,28–34,113–115] show that the film properties are dynamic in response to surface chemistry and physics and have a profound impact on the performance of the contact. These researchers have demonstrated that stick-slip can be a response to shear induced ordering transitions in these boundary films. A good summary of the boundary lubrication mode of sliding contacts is provided by Mate in *Chapter 10, Lubrication in Tight Spots* in Ref [116].

Another perspective on the lubrication scheme is provided again by the nanotribologists as they attempt to characterize contact films as elasto-hydrodynamic (EHL) or thin film. As summed up by Israelachvili, et al, in a forum discussion report [117], "one of the

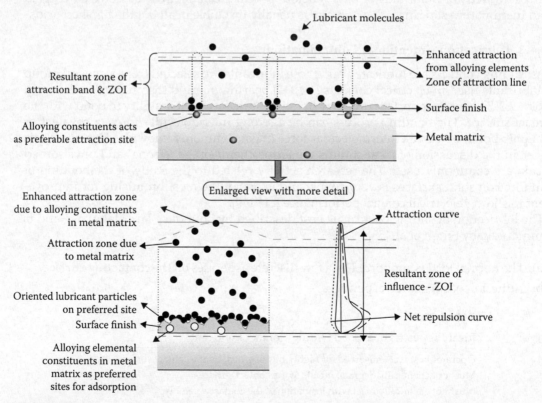

FIGURE 22.15
Boundary Lubrication Schematic representation of the active solid surface and the Zone of Influence (ZOI) where the surface forces are active for adsorption of lubricant. This research correlates friction and the adsorption energy at the surface. (From K. Singh et al, ASME Conference Proceedings, vol. 2008, pp. 229–231, 2008 [53].)

least understood types of classical macro-lubrication is the mixed lubrication regime. This occurs when the hydrodynamic film thickness is comparable to or less than the surface roughness, so that some of the load is supported by a thick (micro-scale) fluid film and some by asperity contacts." They go on further to report that once this partial or mixed lubrication regime is reached, the main contribution comes from the boundary layer regions (the nano-regime)—the EHD contribution (from the micro- and macro-regimes) being negligible.

One of the primary purposes of contact lubricant is to control the micro-environment with a material with known properties (see Reference [116], p. 265). A contact with a known adsorbed lubricant on its surface is less active to the "contaminating environment" surrounding it. But there are also lubricants known as adventitious lubricants [50] that have this "contaminating environment" as their source. Adventitious lubricants are many of the same materials discussed as contaminants. The most prevalent is water vapor. Oil vapors, organic vapors, sulfide vapors, silicone vapors, and so on, from the system, or the surrounding environment, will all act to lubricate any surface they reach as long as they, or their polymers, can remain on the surface in enough quantity to protect the *a*-spots. These adventitious contaminants can be undependable since their steady supply is usually unreliable.

22.4.4 Lubrication Modes (Anaerobic and Aerobic)

Two modes of lubrication have been defined based on the quantity in the contact region, anaerobic and aerobic. *Anaerobic* lubrication exists when the contact area is surrounded by a film or meniscus of lubricant and *aerobic* when the lubricant meniscus is not totally intact. Table 22.3 lists and compares the principal characteristics of these two modes of lubrication. Tribologists studying thin films discuss how a thin film of lubricant deposited on a surface redistributes when contact occurs. There are two driving forces for this

TABLE 22.3

Comparison of Lubrication Modes

Anaerobic Lubrication	Aerobic Lubrication
Contact region surrounded and sealed by lubricant meniscus	No lubricant meniscus "seal" around contact region
Meniscus reduces gas and vapor access to *a*-spots	Allows adsorption and reactions from reactive gasses at *a*-spots
Retards access of surface-active species	Allows condensation and polymerization from organic vapors
Contact behavior less time dependent	Contact behavior controlled by cyclic rate vs. condensation and reflow times
Loss of lubricant leads ultimately to an aerobic condition	Ultimate state of most contact surfaces
Lubricant suspends wear debris and can cause agglomeration	Wear debris less prone to agglomeration or clumping
Hydrodynamic lubrication considerations can apply	Boundary layer lubrication considerations apply
Wear debris color of counterface material and large, agglomerated	Wear debris fine black powder

redistribution: *capillary pressure* pulling the lubricant into the menisci around the contact points and the *disjoining pressure* drawing the liquid out of the menisci and onto the surface [116].

22.4.4.1 Anaerobically Lubricated Contacts

Figure 22.16 shows the anaerobic case diagrammatically where a lubricant meniscus is maintained to provide a continuous supply of lubricant to the contact region and protect the region from other contaminants. A low surface tension lubricant tends to reduce the creep of surface-active species back into the a-spot vicinity and prevent gaseous species from coating the contact surface. The maintenance of a meniscus may be difficult if both the viscosity of the lubricant and the surface energy are low, as it may migrate to adjacent surfaces.

While in the anaerobic condition, the lubricant controls the mechanisms at the a-spot; therefore, the lubricant must be optimum for the surface materials. Atmospheric contaminants can only reach the a-spot by absorption and diffusion through the lubricant [118]. This is particularly true if the lubricant is not miscible with these gases and vapors. These effects are all time-based functions: rate of absorption of contaminants into the lubricant, their diffusion rates through the lubricant to the nascent a-spot, and then the rates of reaction of the contaminant at the surface.

Sliding contacts that are intentionally anaerobically lubricated tend to be those that operate in a mode where film repair and reconstitution in the contact region is difficult. High speed operation is one example. Slip rings used in helicopter blade de-icing applications at 1000+ RPM are normally either lubricated with a solid lubricant or anaerobically lubricated. Lubrication reservoirs may be used to replenish the lubricant to insure that the anaerobic condition is maintained.

The contact surfaces depend on the contact forces to expel lubricant from asperities in the contact region and allow metallic a-spots to form. This requires the proper lubricant viscosity at the use and test temperatures to avoid hydrodynamic lift. The lubricant also helps to remove particles from the vicinity of the contact by partial suspension of the debris. Particles worn during the anaerobic period are typically the color of the metals of the contacts. This anaerobic lubrication mode is often a problem when intuitive thought processes lead to "over-lubrication" of sliding electrical contacts ("if a little is good, a lot

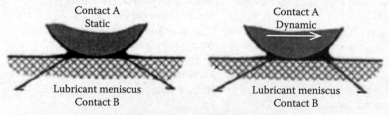

FIGURE 22.16
Illustration of an anaerobic lubricated contact region. A meniscus of lubricant surrounds the contact limiting access of air or vapors to the load-bearing areas. Capillary pressures cause the formation and maintenance of the meniscus, and these forces can increase friction.

will be even better!"). Many low to medium speed applications generate excessive wear debris when contacts are lubricated anaerobically, or at least excessive to the point that transition to aerobic is impossible or slow.

As the lubricant is lost from the contact area, the lubrication condition eventually changes to aerobic. Lubricant is lost from the meniscus area for a number of reasons:

- Surface spreading to adjacent unlubricated component surfaces
- Absorption by porous materials and crevices adjacent to contacts
- Evaporation from lubricated surfaces to the atmosphere and nearby surfaces
- Coating of wear debris as it forms and ejected from the contact region
- Reaction with chemical species to form reaction products that are non-lubricating or deleterious to lubrication
- Absorption in and around clusters of wear debris; higher surface energy of wear particles tends to attract lubricant and cause agglomeration

Only if the contact assembly is totally submerged and sealed in the selected contact lubricant will the anaerobic lubrication not degrade to the aerobic condition. The lifetime can be extended by use of lubricant reservoirs contained within the housing of the slip ring assembly. This technique has been effective for use in vacuum conditions and with high rotational speeds.

22.4.4.2 Aerobically Lubricated Contacts

Figure 22.17 shows a diagram of aerobic lubrication around a contact. There may be lubricant there, but it is so little that the meniscus is not contiguous and, therefore, atmospheric gases and vapors can reach the nascent a-spots without the shielding of the lubricant film. Even if there is no deliberate application of lubricant, there is almost always an adventitious lubricant which finds its way to the contact surface, providing an aerobic lubricant behavior to the a-spots.

In the case of aerobic lubrication, the determining factors in the behavior are the cyclic activity of the contact, the nature of the vapors, which can get to the nascent a-spot, the quantity that can be adsorbed in the period between successive encounters for the a-spot, and the reactions of the vapors on the a-spot. If the principal reactive gas arriving is oxygen

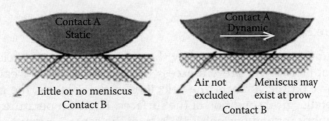

Aerobic creates nascent a-spots
subject to gas and vapor
absorption and reaction

FIGURE 22.17
Illustration of an aerobic lubricated contact region. There is no meniscus surrounding the load-bearing areas allowing easy access by gases and vapors. The disjoining forces exceed the capillary forces primarily owing to the limited availability of fluid.

and the surface is capable of oxidation, the result can be "fretting corrosion" of the surface. On the other hand, when the surface is platinum, palladium, or the like, and organic vapors are present cable of forming polymer products, the result will be the formation of so-called friction polymer providing lubrication, but leading ultimately to high contact resistance and noise. In these cases of reactive vapors, the lubricant serves the valuable role of protecting the surface.

Aerobic lubrication is preferred in cases where surface speeds are low to average and temperature extremes are great. The primary advantage of aerobic lubrication is the improved characteristics of the wear debris—fine, well-dispersed low-conductive particles. Gold contacts especially seem better lubricated with improved electrical characteristics when not sealed by a meniscus of lubricant and subjected to beneficial, adventitious lubricants. This is explained by Mate [116] as the very thin film having a better ability to flow and diffuse to repair displaced films on asperities than a thicker, closely packed chain of lubricant coating the surface.

Adventitious lubrication can provide beneficial results for limited periods of time on sliding contacts. By the nature of adventitious lubrication, its dependability is limited and subject to change on change of conditions at the contact. Aerobic lubricated wear debris tends to be dark in color and is normally non-conducting. Although these dark wear particles from aerobic lubricated contacts can cluster, they still tend to be much smaller and less problematic than the larger, contact-colored particles that result from anaerobic lubricated contacts.

If the cycle time is longer than the time for the gases or fluids to recoat the nascent a-spot, then the properties and effects of the film formed will control the behavior of the contact. If the cycle time is shorter than the film reformation time, the contact will operate as if unlubricated and severe wear will occur. As the aerobic lubricant becomes unavailable to the contact region, the friction begins to rise and the wear particles become larger and return to the color of the contact material, characteristic of lubricant starvation.

Some type of lubrication is required for metal on metal sliding contact, either as a solid, sacrificial inclusion in composite contact surfaces, as a supply of fluid lubricant to the contact face, or an environment that will provide contaminants capable of providing films with a lubricating quality. Lubrication is necessary to limit wear and to control the contact resistance variability. Table 22.2 summarizes the qualities of a good contact lubricant. There are a number of special conditions that require careful consideration for lubricant selection for sliding electrical contacts.

22.4.4.3 Temperature Extremes

High-temperature lubrication limits are set by the viscosity becoming so low that it will not provide sufficient lubricity to prevent excess friction or excessive wear. This limit is controlled by the force on the contacts, the velocities of the contacts, the material of the contacts, and the metallurgical condition of the surfaces. High temperature also increases the volatility of the lubricant and hastens its loss from the contact surface, limiting its life as an effective lubricant.

Low temperature becomes a limit on lubricant when its viscosity becomes so great that the contact force cannot penetrate the film, the brush begins to hydroplane, and contact cannot be sustained. As in the upper temperature limit, the temperature at which this occurs depends on the lubricant, its viscosity, contact velocity, and contact force. At still lower temperatures, the lubricant may become waxy or semisolid, allowing electrical contact to be established once the brush has pushed through the waxy buildup allowing electrical contact to be re-established. This "channeling" phenomenon can be seen in testing during temperature cycling as the contacts go through a transition temperature during both the ramp down to and up from the coldest temperature.

Temperature extremes have little influence on solid lubricated sliding contacts except in the presence of fluid contaminants although especially hot environments are often dry which could lead to accelerated wear of graphite lubricated brushes that depend on a minimum RH of ~15%.

22.4.4.4 Submerged in Flammable Fuels

It sounds hazardous, but contacts can slide submerged in gasoline, jet fuel, or other flammable fluids. Flammable liquids become flammable only when they have the proper fuel-to-air ratios, usually about 10–15% fuel-to-air. When totally submerged in the fluid, the gasoline or other fuel serves as an anaerobic lubrication system providing the lubricity required for metallic contacts. In this way electric contacts can be safely operated [119].

22.4.4.5 Low-Pressure/Vacuum Operation

The need for slip rings to operate in the vacuum of outer space has focused more attention on vacuum operation than any other unique condition for sliding contacts. As a result, slip rings and other sliding contact devices have been developed for satisfactory use in vacuum and space for extended periods of time. The solutions have boiled down to two: composites and fluid.

Graphite containing composites lose their lubricating capability at high-altitude, dry, low-pressure, or vacuum conditions because graphite requires a water vapor adsorption for its lubricity. Solid lubricated composite brushes with molybdenum disulfide solid lubricant embedded in a silver or copper matrix have been quite successful for both high and low current levels for over a decade in space where the pressure is well below 10^{-6} torr. Molybdenum disulfide (and other dichalcogenides) do not require moisture for lubrication and, therefore, can lubricate in vacuum and other low dew point conditions. In fact, moisture causes the molybdenum disulfide to over-film, react with the moisture, and break down into non-lubricating species.

Fluid-lubricated wire brushes in V-grooves have operated equally well for years when supplied with low vapor pressure oils in an essentially sealed capsule [120]. The lubricant must have a vapor pressure which is in the 10^{-3} to 10^{-4} torr range, and provision must be made for effective closure around the contacts and adequate reservoirs for lubricant supply throughout life.

22.4.4.6 Vapor and Gas Lubrication

Some researchers are utilizing the gases and vapors as lubrication schemes for sliding contacts. Among the most interesting is the use of polyhydric alcohols. Lubrication containing hexylene glycol has been found for small motor contacts to provide effective lubricity reducing friction, reducing contact resistance, and providing protection from contaminants such as silicone, H_2S, and organic gases [121–123].

Carbon dioxide and water vapor have been used with some types of fiber brushes (see Chapter 23) with an effective reduction of resistance and wear [124,125]. Other gases, such as 111-trichloroethane and alcohol, have been found to reduce contact noise and wear and friction of gold contacts in trace concentrations [126,127].

22.5 Macro Sliding Contact

Just as important as the micro perspective of sliding contacts is the macro perspective. In most cases, it is difficult to draw a clear line between the two but in general in the macro world lies the contact configuration, types or styles of contacts, contact materials, and operating parameters.

22.5.1 Counterface Configuration

Counterfaces take two forms: flat (pin on disk) (Figure 22.18a and cylindrical (crossed rods) (Figure 22.18b). These refer to the shape of the counterface surface; either a cylindrical surface rotating around its own axis or a disk-shape rotating around the center of the disk.

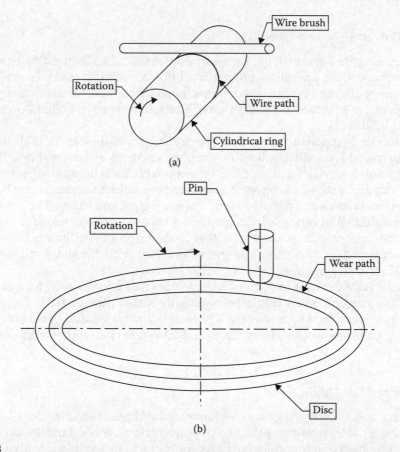

FIGURE 22.18
(a) Crossed-rod contact. This is a simple crossed-rod type of contact which is the basic concept behind all cantilever brushes on cylindrical ring designs. The larger rod (ring) rotates relative to the smaller rod (wire brush). The stiffness of the smaller rod and its mounting determine the stability of the brush. The brush is somewhat more stable when the ring rotation applies tension to the spring. (b) Pin-on-disk contact. The pin-on-disk type sliding contact is the basis for pancake type slip rings, switches, potentiometers, and so on. The structural support and spring system for the pin determine the stability of the brush.

The selection of the type of counterface configuration is typically made based on physical size constraints or producibility issues.

22.5.1.1 Flat Surfaces

Flat counterfaces are generally simple to produce in smaller sizes, but must be kept flat and mounted perpendicular to the axis of rotation. Potentiometers are most often flat surfaces, as are many switches. This is because the processes for producing precision resistor or complex conductor shapes are most amenable to flat patterns. Slip rings can be made in flat configurations, called pancakes, but require special techniques to control the flatness of the surface in large diameters. Large 1.5 m diameter platter slip rings are produced for medical imaging (CT) scanners.

22.5.1.2 Cylindrical Surfaces

Commutators for motors are most often cylindrical. Slip rings are, however, produced in either form, depending on the suitability for the use and the space available. The most compact form for use of materials, minimum cost, and maximum current capability for the available space is the cylindrical design, where the size is dictated only by the number and size of the electrical conductors passing through the rings.

22.5.1.3 Counterface Contact Shapes

Given that the counterface type has been selected, the counterface surface configuration may need to be shaped to maximize the performance of the brushes. Three general types of profiles are common (Figure 22.19):

1. *Flat.* The surface is essentially flat in cross section; therefore, there is no constraint on the brush position by the ring contour. This configuration is most common with composite brushes of all types.
2. *V-groove.* The cross section of the ring surface is in the form of a V. This is most frequently seen with single fibers on small slip rings. The groove provides the wire brush with lateral stabilization and two points of contact per wire brush that electrically act as contact resistances in parallel.
3. *U-groove.* This is similar to the cross section of the V-groove, but with a circular or oval bottom to the groove making the cross section into a U. This is often used with multiple wire brushes in cantilever operation. The groove confines the bundle of brushes aiding the collimation of the cluster.

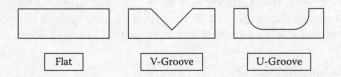

Flat   V-Groove   U-Groove

FIGURE 22.19
Surface profiles. Groove configurations for sliding contact counterfaces. The configuration is decided based on the type of brush and the lateral constraint expected from the ring. The grooved types are best for "side on" fibers, V for single fiber, and U for multifiber.

22.5.2 Real versus Apparent Area of Contact

A common belief is that after a small amount of wear, the brush face is entirely in contact with the counterface and it conforms to the ring surface. This is seldom, if ever, true. Motion is never without eccentricities, variations in attitude, and looseness or spring compliance in the members as time progresses. Most sliding contacts can be represented either as a wire brush, i.e., "crossed rods" (Figure 22.18a) or a cartridge brush, i.e., "pin on disk" (Figure 22.18b). Non-uniformities of motion result from wobble of the disk during rotation, eccentricity of the ring or commutator, and the clearances and elasticity of the supports and springs. All will cause the rubbing to sweep a worn area, or track, on the surface of each member. The wire brush on a cylindrical ring will develop a worn arc larger than the radius of the ring owing to the eccentricity of the ring and the freedom of movement of the brush. A pin running on a "flat" will correspondingly develop a convex face on the pin defined by the flexibility in the pin mount and the lack of flatness and perpendicularity of the disk. There will never be full face contact between the wearing face of the brush and the counterface under it. When the motion stops, reverses, or the friction changes, the attitude of the brush will change with a resultant change in the contact region.

And the above discussion is really about changing only the apparent area of contact, the run-in period as it is often called, when the wear track is established and the brush is contoured. This run-in period can be important for wear and electrical reasons, but this contouring of brush faces has negligible effect on the real area of contact except in the effect on the hardness of the contacts. The change is in the apparent area of contact only [128]. The real area of contact is dependent only on the force on the brush and the hardness of the softest material, and a bigger apparent area of contact does not mean a bigger real area of contact.

22.5.3 Brush Configurations

There are several types of brush structure: cartridge composite, cantilever composite, cantilever wire, and multi-fiber wire brushes.

22.5.3.1 Cartridge Brush

The traditional configuration is the brush one is accustomed to seeing on electric motors. This is a cartridge brush in which the brush member is confined tangentially and transversely by a channel in the brush holder while free to move axially against the spring force. Force is supplied by a spring to hold the brush against the counterface. The spring also provides some side load to stabilize the brush within its holder. Assuming the cartridge itself is stable, the stability of the brush is determined by the clearances, play, between the brush, and the cartridge containing it. Clearance between the brush and the cartridge must exist and it is usually limited, but there is little or no stability within the dimensions of that play. The material of the brush member is typically a composite material consisting of a conducting metal and a solid lubricant in a composite (Figure 22.20a).

22.5.3.2 Cantilever Composite Brush

The second type of brush is a composite material similar to the cartridge brush, rigidly attached to the end of a cantilever spring. The spring provides the contact force and restraint against any tangential or transverse deflection, and roll or pitching motions. The freedom will depend on the stiffness to the spring in each of these modes. There is no play

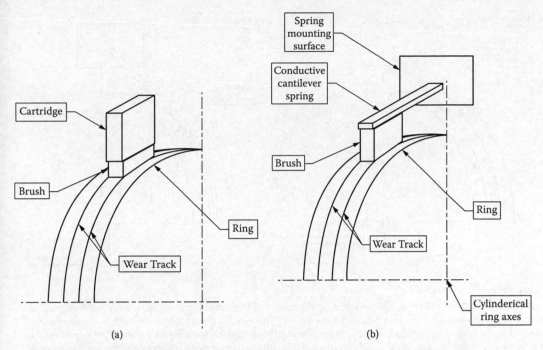

FIGURE 22.20
(a) Composite brush in a cartridge holder. The cartridge allows the composite block to move radially (force direction) with respect to the ring, with minimal play in the transverse or tangential directions. The clearances between the brush and the cartridge determine the stability of the brush. A spring will be mounted between the composite and the holder to apply force to the contact interface. The cartridge must be mounted rigidly to the stationary structure. (b) Cantilever flat spring with a composite brush attached. Attachment must be by a conductive means (soldering, welding, brazing, or clamping). Ring rotation may be either toward the spring or away from it, but the latter, with spring in tension, is a little more stable. The spring is rigidly held to the stationary structure by a nonconducting attachment. The thickness, width, and length of the spring determine the force and rigidity of the brush on the ring.

in any direction, but neither is there any rigid limit as in the cartridge. The cantilever type is most often used in spaces where the cartridge brush cannot be used because of the volume required by the cartridge (Figure 22.20b).

22.5.3.3 Cantilever Metallic Finger

Flat springs without a composite block on the end are used in slip ring and potentiometer applications. The manufacturing processes are economical, in that they can be stamped or etched to configuration. The contact points are often formed to have a dimple or an edge for a sharper point of contact, particularly for potentiometers and switches (Figure 22.21a).

22.5.3.4 Cantilever Wire Brush

For even smaller available circuit volumes, a cantilever round wire brush is simply a cantilever-mounted wire spring that uses the spring material itself as the contact material. Often called a wire brush, this type is a staple in the smallest sliding contact assemblies (Figure 22.21b).

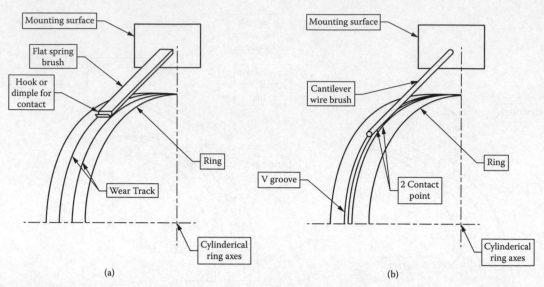

FIGURE 22.21

(a) Flat spring brush on a flat ring. The spring is rigidly mounted in a nonconducting manner to the structure. Its dimensions, stiffness and deflection determine the rigidity and force to the ring. A small hook is usually formed at the end of the spring to localize the contact area of the spring for contact to a flat ring. Ring direction is of little consequence, but the trailing brush is a little more stable. (b) Cantilever wire brush operating on a V-grooved ring. Similar to Figure 22.6, it is widely used in many slip-ring designs, especially miniature ones. The V-groove provides two contact areas to the brush, improving stability reliability.

22.5.3.5 Multifilament or Fiber Brush

The fourth type of brush structure is a multifiber variant of the wire brush. Many wires are operated either with the tips of the wires sweeping the surface (end on) of the counterface similar to a cartridge (Figure 22.22a) or with the sides of the cluster of wires as cantilevers (side on) operating against the counterface (Figure 22.22b). The multiple fibers act to provide multiple paths for the current, but can interfere mechanically with each other as individual fiber motion occurs if not totally collimated. Each conducting fiber has its own bulk resistance for its length and diameter from contact to conducting support, its own constriction resistance, and its own film resistance. See also Chapter 23 for more on the "end on" details.

22.5.3.6 Benefits of Multiple Brushes

Regardless of the brush type chosen for the contact, all sliding contact assemblies generally benefit from the use of multiple brushes on the same counterface. The resistance of a single brush is divided by the number of electrically parallel brushes in the assembly and the probability of an open circuit between the brush and the counterface is drastically reduced by the redundancy. Multiple brushes provide parallel paths, which divide the current between brushes, each of which normally has a lower current capability than the rings. This improves the current capability of the circuit. The lower force required on each contact can produce a reduced wear rate. The general effect of adding additional brush legs to a circuit is lowering of the RMS noise floor. Especially sensitive signal circuits can benefit from adding an extra pair of contacts (from one pair to two), but there are steeply diminishing returns for more than one pair of contacts.

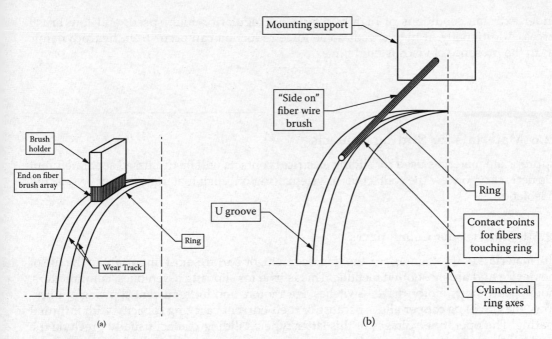

FIGURE 22.22
(a) Fiber brush operating "end-on" on a flat ring. The brush holder for this type of fiber brush must be mounted as a cartridge brush with a force-producing mechanism to maintain contact. Ring direction is typically unidirectional since the fibers are flexible and will align to trail the direction of motion of the ring. Capable of high currents per unit area. (b) Fiber brush operating "side-on" on a flat, V-grooved or U-grooved ring. The brush must be mounted in a nonconducting manner to the stationary structure. The U-groove ring configuration provides the maximum contacts to the fibers of the bundle. With properly defined fiber dimensions, the direction of rotation is of little consequence under moderate conditions. With high forces and surface speeds, longer life can be expected for brushes in the trailing direction. The force is controlled by the dimensions and deflection of the fibers of the bundle.

There are also disadvantages to multiple brushes; the friction of the assembly is proportional to the number of brushes in the assembly and wear debris generation is similarly increased. In switching or commuting brush systems, the precision of switching is impaired by the misalignment of the parallel brushes.

22.5.4 Forces on the Brush

The nominal forces on the contact members are relatively straightforward. There may be environmental forces on either or both of the members as the result of acceleration or vibration applied to the structure as a whole, but they will not be considered here (see Chapter 20).

The principal contact force is assumed to be normal to the counterface and applied to the brush by a spring in the direction of the contact. The counterface will be considered stiff in all directions. Either member may be the moving one, but for simplification it is normally assumed that the smaller one, the brush, is stationary and the larger counterface surface is moving relative to the brush.

Since the counterface may not be exactly a true cylinder or plane, existing irregularities will introduce transient inertial forces to the brush, and these forces can be significant at high surface or cyclic velocities. They will have the effect of varying the force at the interface of the brush, depending on the point in the cycle. Complete separation could occur under extreme conditions of high counterface irregularity, high speed, and low brush forces. Additionally, at high speeds some force reduction can occur from hydrodynamic lift in the presence of viscous fluid films.

22.6 Materials for Sliding Contacts

Appropriate materials used for sliding electrical contacts will be discussed separately, but it is necessary to view the contacts as a system. Ring or brush materials cannot be selected in isolation.

22.6.1 Materials for Counterfaces

The materials of the counterface, the ring, segment, or bar, are most important to the life of the device and are most often metallic. This is true for slip ring assemblies, commutators, motors, and many motor-driven switches. For motors and high-current slip ring applications, these will be copper alloys because of their current-carrying capacity with minimal heating. The operating voltages of this latter type of sliding contact usually preclude the need for corrosion-resistant metals.

Many sliding contact uses are required to carry substantial current at low voltages and with low contact resistance variation. A silver counterface or counterface surface can reasonable approach since silver is less susceptible to hard oxide or sulfide film formation than copper. An even more cost-effective approach is the use of a silver electroplate or clad counterface surface over copper or copper alloy.

When low voltages and currents (e.g. ≤ 1.0 V or ≤ 1.0 A) are expected with infrequent, intermittent motion, the counterface surface is frequently chosen to be gold or one of its alloys. This is by far the most common material selection. Gold has the distinction of being very noble, forming no oxides, sulfides, and so on, from reaction with atmospheric gases. This reduces the film-forming effect of these compounds and the consequent high resistances possible, particularly on long-term standing or slow operation. Gold has the additional benefit of having limited catalytic effect on organic materials compared to platinum and palladium group metals that cause copious formation of friction polymer in the presence of organic gasses and extended operation. Gold is most often used as a cladding or electroplate on a conductive substrate to minimize the economic impact of its selection, but its thickness must be consistent with the wear rate and desired life. Nickel or cobalt hardened electroplated gold is the most common material selected.

Most of the noble metals used in sliding contacts are relatively ductile, but counterface materials will have the longest wear lives if they are selected to be as low in ductility as reasonable. This lowest ductility reduces the metallic transfer to the brush and the tendency to produce large wear particles. If the selected metal cannot be hardened by heat treating, the metal should be cold-worked to the maximum practical hardness. Plated metals should be hardened as much as practical.

In applications where arcing may be expected, alloys should be chosen to be resistant to the high temperatures associated with the arc. However, the practice of using tungsten or metal oxide additions in the counterfaces has not been widely practiced for sliding contacts as it has for relays, breakers, and similar switches. These additions generally increase the abrasiveness of the material and, as a consequence, drastically reduce the wear life of the sliding contacts. Arc control in sliding contacts is better served by mechanical and electrical design than material selection.

The counter-faces for potentiometers present unique material challenges since the material must be selected to provide the required resistance gradient as the brush passes across it. Wire-wound potentiometers use fine precious metal wires of resistivity and sizes to obtain the correct resistance gradient. Film potentiometers similarly use a film of metal or graphite of the proper resistivity, thickness, and width to obtain the desired resistance distribution. In general, the potentiometer counterface metals are platinum group alloys with high resistivity. The graphite films are mixtures of graphite and other conductive materials to produce the selected resistivity. Potentiometers generally have to obtain fewer revolutions life than the sliding contacts in the motors, generators, and slip rings so that wear of conductive films, or development of polymer formation on wire wounds, may not be significant problems.

22.6.2 Solid Lubricated Composite Materials for Brushes

The most traditional types of brush for sliding contacts in motors and generators for over a century are those made of a composite of carbon and graphite (refer to Chapters 20 and 21 for a detailed discussion of metal graphite brushes). They are popular and effective for they contain graphite for lubrication and a conductivity-resistivity combination, that is, effective in commutation of motors. The graphite provides the lubrication and conductivity and the carbon from resin binder provides the strength and bulk resistance.

Composite materials for slip ring assemblies have different performance parameter from motor contacts. The principles still apply, but for most power and control slip ring applications there is greater interest in low noise and efficient power transfer. Solid lubricants, most commonly some type of graphite, are often combined with conductive powders of metal or carbon to produce monolithic blocks that are used for high current and long-running brush applications of slip rings. In general, higher speeds suggest greater percentages of lubricant while higher current requirements dictate greater volume fractions of metals. The metals used as matrix are usually the same copper, silver, and so on, that would be used for the environment in a metallic brush.

As an alternative to graphite as the lubricant, molybdenum disulfide, niobium diselenide, and other dichalcogenides have been added to or replaced graphite for vacuum or other environments where there is insufficient moisture to allow the graphite to form and maintain a lamellar film. Molybdenum disulfide has had the most effective usage as a replacement for graphite. There are also brush formulations that are formed by adding polymeric materials to metals in a "sintered" composite. Because of the relatively large minimum sizes required for practical composite brushes, the composite brush has found limited usage in low power and instrument slip ring assemblies.

22.6.3 Wire Brush Materials Criteria

On instrument slip rings, switches, and potentiometers, the brush or wiper is most frequently a metallic member, and the material used determines the performance of the contact. The principles involved are common among most materials and can be discussed in the general sense before looking at specifics. Materials for wire brushes have five characteristics that must be addressed: low electrical resistivity, low contact resistance, high corrosion resistance, good wear resistance, and appropriate spring properties. Metallic alloys appear to best meet these criteria. This discussion assumes that the wire brush is a single material that meets all the functional requirements. It is possible to create a brush that provides the separate functions as separate components of the brush—e.g., a good spring with a precious metal inlay for the contact surface and a copper braid for conductivity.

Electrical and thermal conductivity: Sliding contact metals are initially chosen for low electrical resistivity and high wear resistance. As usual there are trade-off decisions that must be made. Softer materials have lower contact resistance, but poorer wear resistance. Harder materials improve wear generally at the expense of higher resistivity.

Wear resistance: Good wear resistance for a sliding metallic contact generally means a material with a low ductility and high strength commensurate with other properties. "Low ductility and high strength" are relative terms with contact materials. Materials that satisfy the electrical properties (especially the noble metals) are not noted for low ductility and high strength. Alloying and hardening strategies are used to optimize these properties.

Spring properties: The spring characteristics of wire brushes are critical. The selection of metals for most wire brushes must incorporate the reasoning discussed above with the added consideration of high yield strength without sacrificing the conductivity. The spring requirement demands the elastic limit be sufficient to maintain the force on the contact despite the cyclic load variations imposed by millions of revolutions, the temperature rise from the joule-heating in the bulk of the wire spring, and the elevated temperature possible in many applications. The fatigue limit and the creep resistance of the spring alloy must also be considered in light of the mechanical stresses and expected temperatures.

Corrosion and filming properties: If the open circuit voltage is less than 100 mV, if there are infrequent passes over the surface, or the atmosphere is corrosive or laden with organic vapors, the material film properties need consideration. Gold and gold alloys are often chosen for their tarnish and oxidation resistance. This is particularly true where the contact surfaces must stand for extended periods of time between wiping events. Silver and its alloys can be a good alternative to gold although silver can form films that require some wiping to remove. Precious metal surfaces provide the best opportunity for film-free operation after extended exposure periods in air. The exposed surface of the precious metal is less likely to develop a corrosion film (see Chapter 2). Gold alloys can have condensation of organics, but this film should not be sufficiently thick to interfere with conduction.

The most common materials selected for low current instrumentation contacts are alloys of gold. Some of these are specified by ASTM standards and B477, B540, and B541 are widely used. There is a wide range of properties that can be achieved with these materials on the basis of metallurgical processes specified, and the common trade-off of mechanical and electrical properties must be made.

Fiber brushes, whether for end on or side contact, incorporate similar types of materials and considerations as wire brushes. In some instances, graphite fiber materials are being developed for use with this type of fiber brush [129]. Metal wipers, unless sliding against a solid lubricant counter-face, usually require some type of fluid lubricant, but the quantity may be quite small if the forces are very light [130]. Because of the multiplicity of fiber contacts and light forces on the fibers, fluid lubrication is not always necessary with fiber brushes, but CO_2 and water vapor have been found beneficial [131]. See also Chapter 23.

A final factor to be considered is that the material of the brush should, if possible, be of metals that are incompatible with the counterface material in the crystalline sense and solubility sense. Practically, this may not always be possible. However, it can provide a reduction in the wear from the metallic adhesion standpoint [130].

22.7 Friction and Wear Characteristics

Friction and wear are firmly linked by common causes, namely adhesion and plowing (abrasion). There are other wear modes identified in the literature (see Table 3.2 in [132]), but adhesion and abrasion are almost always involved in sliding wear issues. These principles were discussed on the micro-level in Section 22.2 and are also reviewed in Section 22.6. Adhesion at metal-to-metal junctions and shear stiffness in the fluid and film junctions serve to create resistance to sliding (friction) and cause metal transfer (wear). Asperity interlocking as well as third particle interference also creates resistance to sliding (friction) and can plow up slivers or abrade asperities in the softer surface (wear). Rather than try to provide a primer on friction and wear basics, the reader is pointed to standard tribology reference works [37,71,116,117,130,132–135], and this article will focus on the relationship of friction and wear to the performance of sliding electrical contacts.

22.7.1 Friction

It has been observed that with gold and silver alloy contacts stored in air, the surfaces become filmed with time. Some alloys and lubricants are less susceptible, but all show an increase in filming. This filming can show up in two ways: an increase in contact resistance variation (noise) and a decrease in friction. The friction decrease and noise increase owing to filming can be measured in some cases in as little as an overnight period. Operation of the assembly for several thousand revolutions causes an increase of friction and decrease of noise comparable to its overnight inverse. Brush pressure can be increased to avoid the contact resistance problems, but then the wear rate goes up. This noise/brush force quandary lurks behind all sliding electrical contact design.

The direct effect torque as a result of friction on the servo performance of modern systems with slip rings is normally minor given the contribution of slip ring torque to overall system torque, although the non-linearity of stick-slip friction can an annoyance to servo engineers. This torque insensitivity was not the case when small inertial guidance platforms were the primary slip ring applications (1950–1980), and torque of a few gm-cm (0.1 mN-m) was significant. The primary lesson from those years was that lubrication amount matters. This one parameter, lubrication amount, was the primary "torque adjuster" in these small slip ring capsules (~15 mm dia). And although counterintuitive, torque was lowered by *reducing* lubricant amount. It was this observation, years ago, that got this author interested in the science behind sliding contacts—lubricant is supposed to reduce friction!.

There are at least two theories explaining this increase of torque with increased lubrication amount. This increase is typically up to 25% and primarily static as opposed to dynamic (making the stick slip more pronounced). Theory 1 (which was the one espoused for many years) is that increased lubricant effectively "sealed" the contact zone by forming a meniscus around the contact. This contact sealing prohibited adventitious contaminants from healing film breaks at asperities resulting in increased metal-to-metal contact, greater adhesion, and higher torque values. A modern update to Theory 1 presented in Section 22.4.4 is that a thin film has faster creep and diffusion rates and is able to repair itself faster than a thicker one. Theory 2 is based on the AFM work being done by researchers showing the startling high forces that can be generated by the capillary forces of menisci surrounding a relatively small contact. Mate reports that meniscus forces from a thin film of lubricant can result in *excessive static friction* [116]. If the lubricant amount is sufficient to allow menisci to form around brush contacts, the resulting menisci forces serve to increase friction; this increase is primarily an increase in static friction since the menisci formation is time dependent. Theory 2 seems to fit observations better, but there is likely some contribution from Theory 1.

As discussed earlier, the friction forces in the interface are dynamic and have a stick-slip characteristic. The principal direction of this force is parallel to and in the direction of the counter-face motion. The spring character of the brush structure has compliance in the tangential direction. Friction causes the brush to move together with the ring (stick), increasing tangential stress in the spring until the increased force on the spring exceeds the static friction between the contact surface and the brush. When this breakaway occurs, the brush moves rapidly (slip) until the spring force is less than the kinetic friction and the brush is "recaptured" at a new position on the ring. There is little relative motion during the stick portion, but during the slip period the brush is moving rapidly through a short distance. Once at rest, the static friction may again take over and the cycle repeats.

Although the stick–slip force is tangential at the contact, the dynamics of the loading spring can convert these tangential forces into normal forces tending to reduce and increase the force to the contact region. If sufficiently high, these vibration forces can cause the contacts to separate from the counterface. The relationship of the fundamental frequency of the brush spring and the stick-slip disturbing force is important to understand as it relates to signal quality. Figure 22.23 shows the frequency response of a typical brush used in a slip ring assembly. There are typically a number of vibration modes in a brush spring that

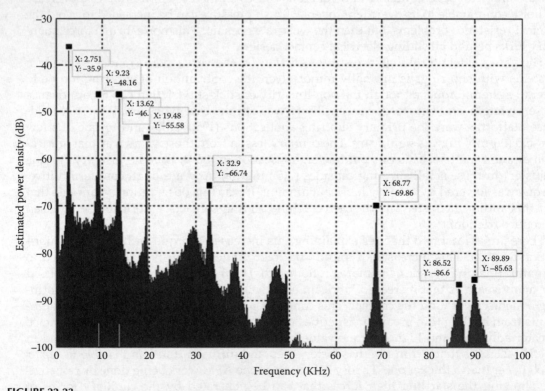

FIGURE 22.23
Laser vibrometer measurements taken at end of brush on a slip ring assembly; Rotation speed was 32 RPM. Multiple critical frequency values are shown.

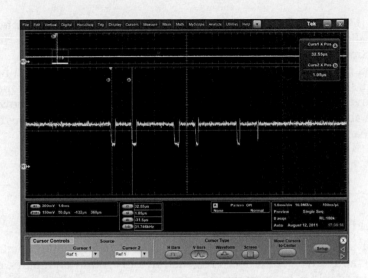

FIGURE 22.24
This oscilloscope trace of a signal on the brush illustrates the intermittent electrical contact that can result from brush vibration. When the time scale is expanded on bit error events, drop outs can be seen at ~30 kHz frequency that is one of the resonant frequencies. These events represent full open circuit events. Rotation at 30 RPM.

can be excited, and if excited at a resonant frequency or with an excess disturbing force, the contact force can be reduced enough to create an open circuit or contact noise.

Figure 22.24 shows the effect in the contact resistance in this same brush system from excitation at one of the resonant frequencies (in this case 30 kHz). Electrically open events occur at a 30 kHz frequency. The variation in friction coefficient (stick-slip) triggers a response at a critical vibration frequency of the brush. Stick–slip can be controlled for a contact design by choice of normal force, spring stiffness, and adjustment of the coefficient of friction by careful lubrication selection and control. The strategy to avoid this problem is to minimize the forcing function (stick-slip) and design the spring critical frequency out of the range of the expected frequency range of the stick-slip that is often associated with operating speed.

These charts demonstrate that the friction characteristics of the contact assembly have an important impact on their ability to reliably transmit electrical signals, especially digital data. The intermittences caused by brush vibration responses to stick-slip are one of the primary causes of bit errors in data transmission. All the factors that control this friction/vibration profile (brush force, lubricant, structure, brush design, surface roughness, to name just a few) must be controlled to eliminate this source of bit errors.

22.7.2 Wear

The wear process of metal brushes on metal rings is discussed in detail in Chapter 23, Section 23.2. The theory of wear of lightly loaded metal fibers on a metal ring is common between single fibers and bundles of fibers. Although theories of adhesive and abrasive wear are discussed, the basic concept is that lightly loaded, sliding metal electrical contacts have two wear regimes: (1) an equilibrium state that represents low wear, small wear particles, and low to moderate stick-slip and (2) an accelerated wear state with big chunky wear particles and high stick slip. Rabinowicz shows the difference on a well-known curve for gold contacts [130] shown in Chapter 23. The principles reviewed in Chapter 23 can be used to design a sliding contact that performs in the low wear regime and still performs its electrical function.

Wear studies on gold and silver alloy contacts loaded at a few grams, sliding at low to moderate (< 0.5 m/s) speeds, appropriately lubricated, carrying currents reasonable for the size of brushes, and protected from gross contamination have repeatedly shown counterface wear rates of about 10^{-12} in. (2.5×10^{-12} cm) per revolution (i.e., per wipe) when operating in this low wear equilibrium regime.

22.8 Contact Parameters and Sliding-Contact Assemblies

Up to now, we have focused on the conditions which exist at the single contact or wiper. Single wipers operating on a counterface are generally only found on motors, potentiometers, and switching devices. In these types of devices, wipers usually have the benefit of the full open circuit voltage of the circuit in which they operate to break down any films that may impede the current through the contact.

22.8.1 Contact Noise

The time variation of resistance in a sliding contact is referred to as contact noise. In one traverse of the slip ring surface, the brush will encounter both the highest and lowest resistance in the track. Because the contact interface cannot be accessed directly to determine the quality of the contact, contact resistance variation measurement is probably the best diagnostic tool for evaluating any sliding or moving contact. Contact resistance variation has been traditionally used in evaluating critical slip ring circuits and is one cause of noise or non-linearity in potentiometers. Section 22.3 discusses the contact noise issue on a micro-scale, but it is important to relate this micro-level noise discussion to a contact noise performance parameter (specification) of sliding contacts within an assembly performing a specific function. Within this macro-context there are two purposes for noise measurements. The first purpose is to assess the "health" of the contact system, and noise measurements are used to ensure proper contact force, contact alignment, materials, lubricant, and all the other factors discussed earlier that effect contact resistance. The second purpose for the noise specification is suitability for the specific application, i.e., what noise performance is required to satisfy the signal or power transmission quality level of the application.

For the purpose of contact quality control, the noise specification should be appropriate for the design. As previous discussions have shown, contact resistance can be effected by many different design parameters. A typical specification for circuit noise from sliding contacts on a signal slip ring is on the order of 20–80 mΩ per circuit (a circuit is typically at least two contacts in parallel as discussed in Section 22.5.3.6). Care must be taken to specify a "reasonable" noise specification since the design trade-offs to achieve a very low noise specification can result in high wear (e.g., high brush force and non-optimum lubrication). For purposes of ensuring contact health and quality, a noise specification of 80 mΩ is adequate. Low noise does not necessarily equate to slip ring quality. Consistent noise performance over the operating life of the contacts is the critical parameter.

The second factor of appropriate noise specification relates to the suitability for the application. As discussed in detail by Dorsey, et al. [88], in the case of signal circuits contact noise requirements depend on the nature of the signal. Very few applications remain for low level analog signals that can require exceptionally low noise levels (10–20 mΩ for example) to be replaced by digital signals. Noise levels on the order of 500 mΩ are required to rise above the noise floor of typical digital signals. Of increasing interest however are the frequency components of the noise [88], since digital signal processors (DSP) can recognize high bandwidth noise that normal analog signals will filter out. These types of contact noise events of fast rise times, short duration, and high amplitude are often called micro-interrupts or micro-cuts. In many cases bit error rates (BER) are more valuable to use to characterize noise on data circuits than contact noise.

The characterization of noise on power circuits must consider the fact that variation in contact resistance can represent a significant noise source when relatively high current levels are being transmitted. For example in the case of a 250 amp, 28 VDC power circuit, a noise level of 20 mΩ would represent a voltage ripple of 5 volts or 18%. For this reason noise specifications for power circuits must consider the allowable power ripple. Lower noise values can be achieved by providing multiple contacts (Section 22.5.3.6).

22.8.2 Slip Rings as Transmission Lines

The analysis above for circuit resistance is adequate for analog, power, and low bandwidth signals, but does not take into account the reactive elements of the circuit impedance in the case of high speed digital data.

When a slip ring is placed into a high speed data transmission line, there are a number of macro design features related to high data interconnect practices [94] that need to be put in place to ensure proper operation. These practices are related to how the contacts are terminated and "wired into" the transmission line. These practices include impedance control, crosstalk protection, and proper shielding and grounding. However, the basic question of the ability of the sliding contacts themselves to handle high speed data is an issue that inevitably arises and is worthy of consideration. Is there some maximum data speed past which sliding electrical contacts can no longer be expected to operate reliably?

Passive circuit elements such as wires, circuit boards, and integrated circuit packages that interconnect active circuit elements in a digital system are called transmission lines, and most aspects of these transmission lines are well defined and characterized [89–94]. With the continued increase in data speed and system complexity, the properties of these transmission lines must be selected to minimize signal propagation degradation (e.g., ringing, reflections, and phase delay), interaction between signals (crosstalk), and electromagnetic interference from the environment. In some high performance systems, these transmission lines require data to be moved from a rotating platform to a stationary platform, and the most common method of de-rotating the data is through the use of a slip ring.

A slip ring is a complex electro-mechanical assembly that inevitably introduces some degree of reactance and non-linearity into the transmission lines. This is not a contact problem, but rather an interconnect design problem. Good design practices can minimize these "lossy" and non-linear effects, but they cannot be eliminated. Figure 22.25a illustrates a two terminal transmission line with a slip ring included. For simplicity, we assume that the transmission line itself is lossless [136] to illustrate the slip ring's effect on the characteristics of the transmission line. This is equivalent to the common practice of wiring a slip ring with a controlled impedance twisted pair cable (twinax).

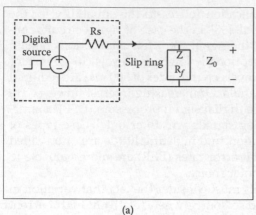
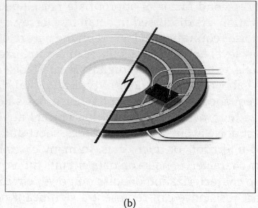

(a)   (b)

FIGURE 22.25
(a) Two terminal block diagram showing the sliding contacts in a lossless transmission line that would represent wiring a slip ring with controlled impedance pair of conductors. (b) Illustration of a platter slip ring with a single ring pair. As the brush rotates around the ring, the variation in length from the location of the ring lead termination to the brush lead termination in each direction can be a source of signal reflections.

A specific platter slip ring design is shown in Figure 22.25b. In a typical slip ring arrangement, each ring has two brushes per circuit contacting on the ring. In the case of our example, the brush itself is bifurcated, i.e., there are two distinct elements per brush. The bifurcation of each brush and the multiple brushes per ring provide redundancy and ameliorates the variations in contact parameters with rotation. Normally, connection to the ring is accomplished in one location with a permanent connection, for example a wire soldered to a ring. As rotation occurs, the change in the relative position between the brushes and the ring contact produces a variation in the electrical path length and a resistance variation [67]. Since rotational frequencies of slip rings are orders of magnitude slower than most data speeds and the amplitude of the variation is relatively small, this rotational variation in resistance can usually be ignored in high speed transmission line characterization.

However, this path length variation also produces two parallel paths from the brushes to the ring terminal that are of unequal and varying length, and this feature creates phase mismatching that can be significant on large diameters. Additional impedance mismatches can be created by wire and shield terminations and brush and ring impedance. A model of this slip ring configuration is illustrated in Figure 22.26 and traditional simulation tools can be utilized to simulate most of the circuit parameters [94, 137, 138]. Simulation of high speed interconnect circuits that contain "lossy" elements are discussed by researchers who are concerned with minimizing losses and signal distortion in electrical interconnects in the megahertz and gigahertz range and nanosecond and sub-nanosecond switching times.

There is still work that needs to be done to fully characterize the RLC properties of sliding electrical contacts.

22.8.3 Results of Normal Operation

Contacts slide in the environment surrounding them. Initially they operate in either an anaerobic or aerobic lubrication mode depending on the lubrication scheme and environmental conditions surrounding them. Contact resistance and noise remain generally low and constant. Wear is limited and appears as clean metal prows, mostly from the ring surface. The friction depends on the frequency of brush wipe over the contact point: typically

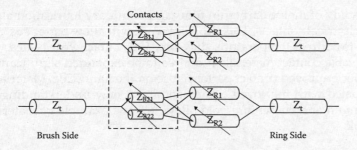

FIGURE 22.26
Electrical model of a slip ring assembly. An arrow through the impedance indicates a variable value. Wires to the brushes are assumed to be lossless (controlled impedance), but the contacts themselves (two pairs of bifurcated brushes) have a variable impedance and do not match Z_t (termination impedance) perfectly. And the brushes contact the ring at different spots, and the distance to the ring terminal varies with rotation causing additional impedance variation and mis-matching.

the higher the frequency, the higher the friction. As the operational life continues, the wear particles accumulate at a constant rate, each particle being surrounded by the lubricant and the debris matrix absorbing some of the fluid. At the same time, the lubricant evaporates and migrates from the contact interface as the disjoining pressure exceeds the capillary pressure and anaerobic lubrication transitions to aerobic.

In the case of aerobic lubrication (either as the initial condition or as a transition from anaerobic), the contacts continue to operate with about the same friction, contact resistance, and noise. The wear debris appears less metallic than contacts under anaerobic lubrication conditions and darker in color. The contaminants (or adventitious lubricants) from the surroundings can easily get to the contact region. This aerobic period can be quite a long period if the wipe rates are low and there is adequate adventitious lubrication. One the other hand, it may not last long if the frequency of wipe is high (i.e., the contact was designed to be anaerobic).

Suddenly contact resistance variation with rotation becomes high due primarily to the wear particulates interfering in the contact region causing both high noise and accelerated wear. This onset of the high resistance is usually taken as the effective end of electrical life for the contact. It is irreversible without effective cleaning and re-lubrication of the assembly.

If the contact continues to slide beyond the high noise period, the most likely scenario is that the noise suddenly becomes low and the wear debris again develops a clean metallic appearance as the wear rate skyrockets. This is the wear-out end of life because the contacts will soon be worn away, and cleaning and re-lubing is not a refurbishment option.

22.9 Future

The ongoing work with micro- and nanoscale diagnostic techniques such as scanning tunneling microscopy (STM) and atomic force microscopy (AFM) promises to provide a rich stream of information about the contact region of sliding electrical contacts. These improved techniques have allowed more accurate modeling and improved characterization on the macro-, micro-, and nano-scale of sliding contacts. Further characterization of the transition regions between continuum and particle mechanics as well as between continuum and ballistic electron flow will allow better understanding of friction, wear and electrical performance of this fascinating micro-world of sliding electrical contacts.

The associated study of molecularly thin films and boundary lubrication also promises to benefit sliding electrical contacts. This thin film work will allow better customization of contact lubricants to address temperature, stick-slip, and contact wear issues.

The list of acceptable contact materials promises to be expanded significantly as nano-tribologists push for improved contact performance on the nanoscale. Material properties are being re-evaluated with improved equipment and a new understanding of material properties. Ruthenium and ruthenium gold alloys are one example of materials that are getting a fresh look.

22.10 Summary

Sliding electrical contacts must perform a difficult tribological balancing act: metal-to-metal contact must be good enough to conduct electrical current with low contact resistance as well as resistance variation, but the wear must be low thereby limiting metal-to-metal contact. Surface films must be good enough to "lubricate" the contacts but not thick or strong enough to raise the resistance unacceptably. Contact force must be high enough to ensure good contact through full range of motion, but not high enough to cause high wear. This fundamental dichotomy of "good but not too good" caused by demands of good contact resistance and low wear rates effects every design decision surrounding sliding contacts. Table 22.4 summarizes the principle points that should be remembered about sliding contacts for instrumentation and control.

The electrical conduction through a contact, static or sliding, is dependent on the metallic a-spots. In most contacts there is a cyclic relative motion of the contact members, i.e., multiple passes over the same area. Most (but not all) sliding contacts are lubricated to control the environment in the contact region with a known material of chosen parameters. Environmental contaminating materials can and do reach the a-spots and modify their conductivity, their adhesion, and chemical nature. The effects of these contaminants are of greatest importance in controlling the performance and life of the contact.

Designers and users of sliding and static contacts should know and focus attention on micro environment at the contact a-spots as well as the construction of the contact assembly. The performance of any sliding contact device is dependent on its design and its suitability for the application and environment.

TABLE 22.4

Sliding Contact Summary

Low electrical noise depends on low resistance film or no film on the a-spot
Low friction depends on lubricated load bearing areas
Low wear rates depend on adequate lubrication of load bearing areas
Intended use is an important consideration for lubricant
a-spots have multicycle lives and repeatability
Sufficient brush force required to wipe away resistive films
Film formation is time dependent
Environment conditions can play a big role on contact performance
Contacts themselves do not limit digital data rate

There are no data rate limits imposed by sliding contacts themselves. There are, however, standard transmission line limitations imposed by the methods used to connect the contacts to the outside world that should be considered.

Acknowledgments

Jax Glossbrenner hired me as a slip ring engineer as my first job out of engineering school. It was in a small college town, and engineering opportunities were limited (before the high-tech boom hit). I was glad to have any job, and I'm sure my desperation showed during the interview—but Jax still sold hard. He told me that sliding electrical contacts was a technology you could never completely understand but you could have fun trying. He was right. Jax recently passed away, and it is a privilege to update this chapter that he originally authored. Special thanks to Becky Wills whose editorial and graphics help is appreciated.

References

1. M. Antler, "Wear, Friction, and Electrical Noise Phenomena in Severe Sliding Systems," *ASLE Transactions.*, vol. 5, pp. 297–307, 1962.
2. S. V. Prasad, P. Misra, and J. Nagaraju, "An Experimental Study to Show the Behavior of Electrical Contact Resistance and Coefficient of Friction at Low Current Sliding Electrical Interfaces," in *Electrical Contacts (Holm), 2011 IEEE 57th Holm Conference on*, 2011, pp. 1–7.
3. M. Antler, "Tribological Properties of Gold for Electric Contacts," *Parts, Hybrids, and Packaging, IEEE Transactions on*, vol. 9, pp. 4–14, 1973.
4. T. Ueno, N. Morita, and K. Sawa, "Influence of Surface Roughness on Voltage Drop of Sliding Contacts under Various Gases Environment," in *Electrical Contacts, 2003. Proceedings of the Forty-Ninth IEEE Holm Conference on*, 2003, pp. 59–64.
5. T. Ueno and N. Morita, "Influence of Surface Roughness on Contact Voltage Drop of Sliding Contacts," in *Electrical Contacts, 2005. Proceedings of the Fifty-First IEEE Holm Conference on*, 2005, pp. 324–328.
6. T. Ueno, K. Kadono, and N. Morita, "Influence of Surface Roughness on Contact Voltage Drop of Electrical Sliding Contacts," in *Electrical Contacts, 2007, the 53rd IEEE Holm Conference on*, 2007, pp. 200–204.
7. C. Holzapfel, P. Heinbuch, and S. Holl, "Sliding Electrical Contacts: Wear and Electrical Performance of Noble Metal Contacts," in *Electrical Contacts, 2010 Proceedings of the 56th IEEE Holm Conference on*, 2010, pp. 1–8.
8. D. Bansal and J. Streator, "Effect of Operating Conditions on Tribological Response of Al–Al Sliding Electrical Interface," *Tribology Letters*, vol. 43, pp. 43–54, 2011.
9. A. A. Conte Jr and V. S. Agarwala, "An Investigation of Gold Alloy Slip Ring Capsule Wear Failures," *Wear*, vol. 133, pp. 355–371, 1989.
10. R. L. Jackson and I. Green, "On the Modeling of Elastic Contact between Rough Surfaces," *Tribology Transactions*, vol. 54, pp. 300–314, 2011/01/10 2011.
11. M. Ciavarella, C. Murolo, and G. Demelio, "On the Elastic Contact of Rough Surfaces: Numerical Experiments and Comparisons with Recent Theories," *Wear*, vol. 261, pp. 1102–1113, 2006.
12. F. P. Bowden and D. Tabor, *Friction and Lubrication of Solids, Part 1 and Part 2.* Oxford, UK: Oxford Universty Press, 1950 Part 1, 1964 Part II.

13. R. Holm, *Electric Contacts, Theory and Applications*, 4th ed. Berlin, Heidelberg, New York: Springer-Verlag, 1967.

14. J. H. Dietrich and B. D. Kilgore, "Direct Observation of Frictional Contacts: New Insights for State-dependent Properties," *Pageoph*, vol. 143, pp. 283–302, 1994.

15. B. N. J. Persson, O. Albohr, F. Mancosu, V. Peveri, V. N. Samoilov, and I. M. Sivebaek, "On the Nature of the Static Friction, Kinetic Friction and Creep," *Wear*, vol. 254, pp. 835–851, 2003.

16. A. Ovcharenko, G. Halperin, and I. Etsion, "Pre-Sliding of a Spherical Elastic-Plastic Contact," *ASME Conference Proceedings*, vol. 2008, pp. 545–547, 2008.

17. V. Brizmer, Y. Kligerman, and I. Etsion, "Elastic–Plastic Spherical Contact under Combined Normal and Tangential Loading in Full Stick," *Tribology Letters*, vol. 25, pp. 61–70, 2007.

18. K. L. Johnson, *Contact Mechanics*. Cambridge, U. K.: Cambridge University Press, 1985.

19. X. Yin and K. Komvopoulos, "An Adhesive Wear Model of Fractal Surfaces in Normal Contact," *International Journal of Solids and Structures*, vol. 47, pp. 912–921, 2010.

20. T. Baumberger, "Dry Friction Dynamics at Low Velocities," in *Physics of Sliding Friction*, B. N. J. Persson and E. Tosatti, Eds., Kordrecht Germany: Kluwer Academic Publishers, 1996, p. 460.

21. B. Bhushan, J. Israelachvili, and U. Landman, "Nanotribology: Friction, Wear and Lubrication at the Atomic Scale," *Nature*, vol. 374, pp. 607–616, 1995.

22. K. De Moerlooze, F. Al-Bender, and H. Van Brussel, "A Generalised Asperity-Based Friction Model," *Tribology Letters*, vol. 40, pp. 113–130, 2010.

23. M. Leidner, H. Schmidt, and M. Myers, "A Numerical Method to Predict the Stick/Slip Zone of Contacting, Nonconforming, Layered Rough Surfaces Subjected to Shear Traction," in *Electrical Contacts, 2009 Proceedings of the 55th IEEE Holm Conference on*, 2009, pp. 36–41.

24. H.-Y. Cha, J. Choi, H. Ryu, and J. Choi, "Stick-slip Algorithm in a Tangential Contact Force Model for Multi-body System Dynamics," *Journal of Mechanical Science and Technology*, vol. 25, pp. 1687–1694, 2011.

25. J.-W. Liang and B. F. Feeny, "Wavelet Analysis of Stick-Slip Signals in Oscillators With Dry-Friction Contact," *Journal of Vibration and Acoustics*, vol. 127, pp. 139–143, 2005.

26. K. Hashiguchi and S. Ozaki, "Constitutive Equation for Friction with Transition from Static to Kinetic Friction and Recovery of Static Friction," *International Journal of Plasticity*, vol. 24, pp. 2102–2124, 2008.

27. S. Ozaki and K. Hashiguchi, "Numerical Analysis of Stick-Slip Instability by a Rate-Dependent Elastoplastic Formulation for Friction," *Tribology International*, vol. 43, pp. 2120–2133, 2010.

28. M. L. Gee, P. M. McGuiggan, J. N. Israelachvili, and A. M. Homola, "Liquid to Solidlike Transitions of Molecularly thin Films under Shear," *The Journal of Chemical Physics*, vol. 93, pp. 1895–1906, 1990.

29. A. M. Homola, J. N. Israelachvili, M. L. Gee, and P. M. McGuiggan, "Measurements of and Relation Between the Adhesion and Friction of Two Surfaces Separated by Molecularly Thin Liquid Films," *Journal of Tribology*, vol. 111, pp. 675–682, 1989.

30. A. M. Homola, J. N. Israelachvili, P. M. McGuiggan, and M. L. Gee, "Fundamental Experimental Studies in Tribology: The Transition from "interfacial" Friction of Undamaged Molecularly Smooth Surfaces to "normal" Friction with Wear," *Wear*, vol. 136, pp. 65–83, 1990.

31. J. Israelachvili, P. M. McGuiggan, M. Gee, A. M. Homola, M. O. Robbins, and P. A. Thompson, "Liquid Dynamics in Molecularly Thin Films," *Journal of Physics. Condensed Matter*, vol. 2, pp. 89–98, 1990.

32. J. N. Israelachvili, P. M. McGuiggan, and A. M. Homola, "Dynamic Properties of Molecularly Thin Liquid Films," *Science*, vol. 240, pp. 189–191, 1988.

33. P. M. McGuiggan, M. L. Gee, H. Yoshizawa, S. J. Hirz, and J. N. Israelachvili, "Friction Studies of Polymer Lubricated Surfaces†," *Macromolecules*, vol. 40, pp. 2126–2133, 2007/03/01 2007.

34. C. Yang, B. N. J. Persson, J. Israelachvili, and K. Rosenberg, "Contact Mechanics with Adhesion: Interfacial Separation and Contact Area," *EPL (Europhysics Letters)*, vol. 84, p. 46004, 2008.

35. B.N.J. Persson, "Contact Mechanics for Randomly Rough Surfaces," *Surface Science Reports*, vol. 61, pp. 201–227, 2006.

36. E. R. Kral and K. Komvopoulos, "Three-Dimensional Finite Element Analysis of Subsurface Stress and Strain Fields Due to Sliding Contact on an Elastic-Plastic Layered Medium," *Journal of Tribology*, vol. 119, pp. 332–341, 1997.

37. N. Suh, *Tribophysics*. Englewood Cliffs, New Jersey: Prentice-Hall, 1986.

38. A. Tangena and G. Hurkx, "Calculations of Mechanical Stresses in Electrical Contact Situations," *Components, Hybrids, and Manufacturing Technology, IEEE Transactions on*, vol. 8, pp. 13–20, 1985.

39. S. Liu and Q. Wang, "Studying Contact Stress Fields Caused by Surface Tractions with a Discrete Convolution and Fast Fourier Transform Algorithm," *Journal of Tribology*, vol. 124, pp. 36–45, 2002.

40. L. Kogut and K. Komvopoulos, "Electrical Contact Resistance Theory for Conductive Rough Surfaces," *Journal of Applied Physics*, vol. 94, pp. 3153–3162, 2003.

41. L. Kogut and K. Komvopoulos, "Analysis of Interfacial Adhesion Based on Electrical Contact Resistance Measurements," *Journal of Applied Physics*, vol. 94, pp. 6386–6390, 2003.

42. R. L. Jackson and L. Kogut, "Electrical Contact Resistance Theory for Anisotropic Conductive Films Considering Electron Tunneling and Particle Flattening," *Components and Packaging Technologies, IEEE Transactions on*, vol. 30, pp. 59–66, 2007.

43. L. Kogut, "Electrical Contact Resistance Theory for Conductive Rough Surfaces Separated by a Thin Insulating Film," *Journal of Applied Physics*, vol. 95, p. 576, 2004.

44. G. G. Adams, "The Mechanics of Adhesion: A Tutorial," *ASME Conference Proceedings*, vol. 2008, pp. 87–89, 2008.

45. G. G. Adams, "The Mechanics of Adhesion: Current and Future Research Trends," *ASME Conference Proceedings*, vol. 2008, pp. 91–93, 2008.

46. M. Antler, "Processes of Metal Transfer and Wear," in *Wear 7c*, ed, 1964, pp. 181–203.

47. M. Antler, "Processes of Metal Transfer and Wear," *Wear*, vol. 7, pp. 181–203, 1964.

48. M. Antler, "The Wear of Electrodeposited Gold," *A S L E Transactions*, vol. 11, pp. 248–260, 1968/01/01 1968.

49. M. Antler, "Stages in the Wear of a Prow-Forming Metal," *A S L E Transactions*, vol. 13, pp. 79–86, 1970/01/01 1970.

50. M. Antler, "Wear of Gold Plate: Effect of Surface Films and Polymer Codeposits," *Parts, Hybrids, and Packaging, IEEE Transactions on*, vol. 10, pp. 11–17, 1974.

51. M. Antler, "Sliding Wear of Metallic Contacts," *Components, Hybrids, and Manufacturing Technology, IEEE Transactions on*, vol. 4, pp. 15–29, 1981.

52. M. Antler, "Sliding Wear and Friction of Electroplated and Clad Connector Contact Materials: Effect of Surface Roughness," *Wear*, vol. 214, pp. 1–9, 1998.

53. K. Singh, S. Baghmar, J. Sharma, M. V. Khemchandani, and Q. J. Wang, "Boundary Lubrication and Its Stability," *ASME Conference Proceedings*, vol. 2008, pp. 229–231, 2008.

54. H. Zhang and L. Chang, "An Asperity-Based Mathematical Model for the Boundary Lubrication of Nominally Flat Metallic Contacts," *Lubrication Science*, vol. 20, pp. 1–19, 2008.

55. R. Holm, "The Electric Tunnel Effect across Thin Insulator Films in Contacts," *Journal of Applied Physics*, vol. 22, pp. 569–574, 1951.

56. A. Mikrajuddin, F. G. Shi, H. K. Kim, and K. Okuyama, "Size-Dependent Electrical Constriction Resistance for Contacts of Arbitrary Size: From Sharvin to Holm Limits," *Materials Science in Semiconductor Processing*, vol. 2, pp. 321–327, 1999.

57. R. S. Timsit, "High Speed Electronic Connectors: A Review of Electrical Contact Properties," *IEICE Transactions Electron.*, Vol. E88, No 8, pp. 1532–1545, August 2005.

58. C. Weißenfels and P. Wriggers, "Numerical Modeling of Electrical Contacts," *Computational Mechanics*, vol. 46, pp. 301–314, 2010.

59. R. Vijaywargiya and I. Green, "A Finite Element Study of the Deformations, Forces, Stress Formations, and Energy Losses in Sliding Cylindrical Contacts," *International Journal of Non-Linear Mechanics*, vol. 42, pp. 914–927, 2007.

60. J. J. Moody. (2007). *A finite element analysis of elastic-plastic sliding of hemispherical contacts*. Available: http://hdl.handle.net/1853/31992.

61. P. Põdra and S. Andersson, "Simulating Sliding Wear with Finite Element Method," *Tribology International*, vol. 32, pp. 71–81, 1999.
62. R. D. Malucci, "The Impact of Contact Resistance on High Speed Digital Signal Transmission," in *Electrical Contacts, 2002. Proceedings of the Forty-Eighth IEEE Holm Conference on*, 2002, pp. 212–220.
63. R. D. Malucci and A. P. Panella, "Wave Propagation and High Frequency Signal Transmission across Contact Interfaces," in *Electrical contacts - 2006, proceedings of the fifty-second ieee holm conference on*, 2006, pp. 199–206.
64. Y. Chen and S. Baisheng, "Investigate the Influence of Contact Impedance on Digital Signal Transmission by Computer Simulation Method," in *Electrical Contacts, 2003. Proceedings of the Forty-Ninth IEEE Holm Conference on*, 2003, pp. 29–37.
65. J. C. Gao and Z. Jian, "Effects of Electrical Contact Failure on High Speed Digital Signal Transmission," in *Electrical Contacts, 2008. Proceedings of the 54th IEEE Holm Conference on*, 2008, pp. 41–46.
66. S. B. Smith, V. Balasubramanian, D. Nardone, and S. S. Agili, "Effect of Nanosecond Electrical Discontinuities in High-Speed Digital Applications," in *Electrical Contacts, 2008. Proceedings of the 54th IEEE Holm Conference on*, 2008, pp. 47–52.
67. C. Holzapfel, "Selected Aspects of the Electrical Behavior in Sliding Electrical Contacts," in *Electrical Contacts, 2011 IEEE 57th Holm Conference on*, 2011, pp. 1–9.
68. N. Lewis and C. W. Reed, "Interaction of Electrical Sliding Contact Lubricants and Certain Aliphatic Amines," in *Holm Seminar on Electrical Contacts, 1974*, Chicago, IL, 1974, pp. 190–198.
69. K. R. North, "Characteristics of Slip-Ring Noise," *Third International Research Symposium on Electric Contact Phenomena 1966*, pp. 325–365, 1966.
70. E. W. Glossbrenner and J. Sun, K., "Effects of Parameters on Noise in Miniature Sliding Contacts," *Holm Conference on Electrical Contacts*, vol. 1963, 1963.
71. Y. Imry, *Introduction to Mesoscopic Physics*. Oxford, UK: Oxford University Press, 1997.
72. B. Bhushan, "Nanotribology and Nanomechanics," *Wear*, vol. 259, pp. 1507–1531, 2005.
73. N. Agrait, A. L. Yeyati, and J. M. van Ruitenbeek, "Quantum Properties of Atomic-Sized Conductors," *Physics Reports*, vol. 377, pp. 81–279, 2003.
74. R. S. Timsit, "Electrical Conduction Through Small Contact Spots," *Components and Packaging Technologies, IEEE Transactions on*, vol. 29, pp. 727–734, 2006.
75. R. D. Malucci, "Multi-Spot Model Showing the Effects of Nano-Spot Sizes," in *Electrical Contacts, 2005. Proceedings of the Fifty-First IEEE Holm Conference on*, 2005, pp. 291–297.
76. D. Erts, H. Olin, L. Ryen, E. Olsson, and A. Thölen, "Maxwell and Sharvin Conductance in Gold Point Contacts Investigated using TEM-STM," *Physical Review B*, vol. 61, pp. 12725–12727, 2000.
77. G. Wexler, "The Size Effect and the Non-Local Boltzmann Transport Equation in Orifice and Disk Geometry," *Proceedings of the Physical Society (1960)*, vol. 89, pp. 927–941, 1966.
78. G. Wexler, "The Non-Local Boltzmann Transport Equation in Orifice and Disk Geometry. II. The Absorbing Disk," *Proceedings of the Physical Society (1960)*, vol. 92, pp. 165–176, 1967.
79. C. Maul and J. W. McBride, "A Model to Describe Intermittency Phenomena in Electrical Connectors," in *Electrical Contacts, 2002. Proceedings of the Forty-Eighth IEEE Holm Conference on*, 2002, pp. 165–174.
80. C. Maul, J. W. McBride, and J. Swingler, "Measuring the Constriction Resistance in Low Power Applications Using Non-Linear Techniques," *ICEC*, vol. 1998, pp. 117–121, 1998.
81. C. Maul, J. W. McBride, and J. Swingler, "Intermittency Phenomena in Electrical Connectors," *IEEE Transactions on Components and Packaging Technologies*, vol. 24, pp. 370–377, 2001.
82. A. P. Merkle and L. D. Marks, "Liquid-Like Tribology of Gold Studied by in situ TEM," *Wear*, vol. 265, pp. 1864–1869, 2008.
83. D. Erts, A. Lõhmus, R. Lõhmus, H. Olin, A. V. Pokropivny, L. Ryen, and K. Svensson, "Force Interactions and Adhesion of Gold Contacts using a Combined Atomic Force Microscope and Transmission Electron Microscope," *Applied Surface Science*, vol. 188, pp. 460–466, 2002.

84. E. Rocas, C. Collado, N. D. Orloff, J. Mateu, A. Padilla, J. M. O'Callaghan, and J. C. Booth, "Passive Intermodulation Due to Self-Heating in Printed Transmission Lines," *Microwave Theory and Techniques, IEEE Transactions on,* vol. 59, pp. 311–322, 2011.

85. K. Hajek and J. Sikula, "Contact Voltage of Third Harmonic Distortion for Contact Reliability Investigation," *Components and Packaging Technologies, IEEE Transactions on,* vol. 28, pp. 717–720, 2005.

86. R. Kwiatkowski, M. Vladimirescu, and K. Engel, "Tunnel Conduction Consequences in High Frequency Microcontacts; Passive Intermodulation Effect," in *Electrical Contacts, 2004. Proceedings of the 50th IEEE Holm Conference on Electrical Contacts and the 22nd International Conference on Electrical Contacts,* 2004, pp. 152–159.

87. R. D. Malucci, "High Frequency Considerations for Multi-Point Contact Interfaces," in *Electrical Contacts, 2001. Proceedings of the Forty-Seventh IEEE Holm Conference on,* 2001, pp. 175–185.

88. G. Dorsey, D. Coleman, and B. Witherspoon, "High Speed Data Across Sliding Electrical Contacts," in *Proceedings 59th IEEE Holm Conference,* Portland Or., 2012.

89. H. W. Johnson and M. Graham, *High Speed Digital Design: A Handbook of Black Magic.* Upper Saddle River, NJ: Prentiss Hall, Inc., 1993.

90. H. W. Johnson and M. Graham, *High-Speed Signal Propagation.* Upper Saddle River, NJ: Prentice Hall, 2002.

91. K. Kaiser, *Transmission Lines, Matching, and Crosstalk.* Boca Raton, FL: CRC Press, 2006.

92. R. E. Matlick, *Transmission Lines for Digital and Communication Networks.* Piscataway, N. J.: IEEE Press, 1995.

93. B. J. Wadell, *Transmission Line Design Handbook.* Norwood, MA: Artech House, Inc., 1991.

94. R. Achar and M. S. Nakhla, "Simulation of High-Speed Interconnects," *Proceedings of the IEEE,* vol. 89, pp. 693–728, 2001.

95. E. W. Glossbrenner, "The Life and Times of an A-Spot: A Tutorial on Sliding Contacts," in *Electrical Contacts 1993, Proceedings of the Thirty-Ninth IEEE Holm Conference,* 1993, pp. 171–177.

96. S. Kim, "Environmental Effects in Tribology," in *Micro- and Nanoscale Phenomena in Triboloty,* Y. Chung, Ed., ed Boca Raton, FL: CRC Press, 2012, p. 210.

97. E. W. Glossbrenner and A. M. Jenney, "An Engineering Investigation of the Effect of the Environment in a Closed System on the Performance of Sliding Contacts," in *Electrical Contacts 1974, Proceedings of the Twentieth Annual Holm Seminar on Electrical Contacts,* 1974, pp. 185–189.

98. N. E. Lewis, C. W. Reed, J. J. DeCorpo, and J. R. Wyatt, "A Comparative Study of Performance of Electrical Sliding Contact Lubricants in a Chemically Active Atmosphere," in *Electrical Contacts 1975, Proceedings of the Twenty-First Annual Holm Seminar on Electrical Contacts,* 1975, pp. 155–163.

99. N. E. Lewis and C. W. Reed, "Interaction of Electrical Sliding Contact Lubricants and Certain Aliphatic Amines," in *Electrical Contacts 1974, Proceedings of the Twentieth Annual Holm Seminar on Electrical Contacts,* 1974, pp. 190–198.

100. D. Hively, "Interaction of Household Cleaning Products on the Lubricating Fluid and Cooling Fluid Used on Part DW4094/F (Litton Poly-Scientific Research and Development Test Report 892) (Internal Report)," Blacksburg 1992.

101. J. Williamson, "Influence of Dust on Lightly Loaded Electrical Contacts," presented at the Engineering Seminar on Electrical Contacts, Penn State University, 1958.

102. J. G. Zhang and X. M. Wei, "The Effect of Dust Contamination of Electric Contacts," in *Electric Contacts 1985, Proceedings of the Thirty-First IEEE Holm Conference,* 1985, pp. 175–179.

103. C. A. Haque, M. D. Richardson, and E. T. Ratliff, "Effects of Dust on Contact Resistance of Lubricated Connector Materials," in *Electric Contacts 1985, Proceedings of the Thirty-First IEEE Holm Conference,* 1985, pp. 141–146.

104. Z. Ji Gao, "Effect of Dust Contamination on Electrical Contact Failure," in *Electrical Contacts, 2007, the 53rd IEEE Holm Conference on,* 2007, pp. xxi–xxx.

105. W. Abbott and H. Odgen, "The Influence of Environment on Tarnishing Reactions," in *4th International Research Symposium on Electrical Contact Phenomena,* Swansea, UK, 1968, pp. 35–39.

106. W. Abbott, "Effects of Industrial Air Pollutants on Electrical Contact Components," in *Holm Conference on Electrical Contacts*, Chicago, IL, 1973, pp. 94–99.

107. T. Graedel, "The Atmospheric Environments Encountered by Electrical Components," in *Holm Conference on Electrical Contacts*, Chicago, IL, 1973, pp. 94–99.

108. H. W. Hermance and T. F. Egan, "Organic Deposits on Precious Metal Contacts," *Bell Systems Technical Journal 37*, p. 730, 1958.

109. H. W. Hermance, "Organic Deposits on Precious Metal Contacts," in *Proceedings of the Engineering Seminar on Electrical Contacts*, 1959, p. 127.

110. L. P. Solos, "Organic Adsorption on Precious Metals and Its Effects on Constriction Resistance," in *Electric Contacts 1970, Proceedings of the Sixteenth Annual Holm Seminar on Electrical Contact Phenomena*, 1970, p. 27.

111. M. Antler and T. Spalvins, "Lubrication with Thin Gold Films," *Gold Bulletin*, vol. 21, pp. 59–68, 1988.

112. B. H. Chudnovsky, "Lubrication of Electrical Contacts," in *Electrical Contacts, 2005. Proceedings of the Fifty-First IEEE Holm Conference on*, 2005, pp. 107–114.

113. M. Akbulut, N. Chen, N. Maeda, J. Israelachvili, T. Grunewald, and C. A. Helm, "Crystallization in Thin Liquid Films Induced by Shear," *The Journal of Physical Chemistry B*, vol. 109, pp. 12509–12514, 2005/06/01 2005.

114. H. Zeng, Y. Tian, B. Zhao, M. Tirrell, and J. Israelachvili, "Transient Surface Patterns and Instabilities at Adhesive Junctions of Viscoelastic Films," *Macromolecules*, vol. 40, pp. 8409–8422, 2007/11/01 2007.

115. B. N. J. Persson and E. Tosatti, *Physics of Sliding Friction* vol. Series E: Applied Sciences. Dordrecht/Boston/London: Kluwer Academic Publishers, 1995.

116. C. M. Mate, *Tribology on a Small Scale*. Oxford, UK: Oxford University Press, 2008.

117. B. Bhushan, *Fundamentals of Tribology and Bridging the Gap Between the Macro- and Mico/Nanoscales*. Dordrecht/Boston/London: Kluwer Academic Publishers, 2000.

118. N. E. Lewis, C. W. Reed, and B. K. Witherspoon, "Lubrication Quantity: Observations of the Effects on Gold Sliding Contact Wear Character and in situ Contamination Formation," *1979 IEEE Transactions on Components, Hybrids, and Manufacturing Technology*, pp. 145–149, 1979.

119. N. Shimoda, G. Mitsumatsu, and H. Kanoh, "A Study on Commutation Arc and Wear of Brush in Gasoline of DC Motor," in *Electrical Contacts 1993, Proceedings of the Thirty-Ninth Annual IEEE Holm Conference*, 1993, pp. 151–156.

120. C. J. Pentlicki and E. W. Glossbrenner, "The Testing of Contact Materials for Slip Rings and Brushes for Space Applications," in *Electrical Contacts 1971, Proceedings of the Seventeenth Annual Holm Seminar on Electric Contact Phenomena*, 1971, pp. 157–172.

121. S. Uda, H. Funami, S. Kondo, T. Higaki, and K. Mano, "A Proposal of Vapor Lubricants for Sliding Contacts," in *Proceedings of the Seventeenth International Conference on Electrical Contacts*, Nagoya, Japan, 1994, pp. 409–415.

122. S. Uda, T. Higaki, H. Takao, S. Kondo, K. Kawamura, K. Nobuta, H. Funami, and K. Mano, "The Development of Small Size Motor Characteristics Using Vapor Lubricants," in *Proceedings of the Seventeenth International Conference on Electrical Contacts*, Nagoya, Japan, 1994, pp. 419–426.

123. S. Uda, T. Higaki, H. Funami, and K. Mano, "The General Idea of Vapor Lubrication System Applied to Small Power Motors," in *Electrical Contacts 1993, Proceedings of the Thirty-Ninth Annual IEEE Holm Conference on*, 1993, pp. 141–145.

124. P. Reichner, "Metallic Brushes for Extreme High Current Applications," in *Electrical Contacts 1979, Proceedings of the Twenty-Fifth Anniversary Meeting of the Holm Conference on Electrical Contacts*, 1979, pp. 191–198.

125. P. Reichner, "High Current Tests of Metal Fiber Brushes," in *Electrical Contacts 1980, Proceedings of the Twenty-Sixth Annual Holm Conference on Electrical Contacts*, 1980, pp. 73–76.

126. D. Wright, "Personal Communication (Litton Poly-Scientific)," 1993.

127. E. W. Glossbrenner and J. K. Sun, "Effects of Parameters in Miniature Sliding Contacts (Paper 5)," in *Engineering Seminar on Electrical Contacts, University of Maine*, 1963.

128. G. Dorsey and R. Hayes, "The Effects of Apparent Contact Area on Thermal and Electrical Properties of Ag-MoS$_2$-C/sub Graphite/Brushes in Contact with Coin-Silver Slip Rings," in *Electrical Contacts, 1997., Proceedings of the Forty-Third IEEE Holm Conference on*, 1997, pp. 272–280.

129. Y. Wu, G. Zhang, and W. Tang, "The Research of Fiber/Graphite Composite," in *Proceedings of the Forty-First IEEE Holm Conference on Electrical Contacts*, 1995, p. 315.

130. E. Rabinowicz, *Friction and Wear of Materials*. New York: John Wiley & Sons, 1965.

131. C. M. Adkins and D. Kuhlmann-Wilsdorf, "Development of High-Performance Metal Fiber Brushes: I - Background and Manufacture," in *Electrical Contacts 1979, Twenty-Fifth Anniversary Meeting of the Holm Conference on Electrical Contacts*, Chicago, IL, 1979, pp. 165–170.

132. P. Blau, *Friction and Wear Transitions of Materials*. Park Ridge, NJ: Noyes Publications, 1989.

133. I. L. Singer and H. M. Pollock, *Fundamentals of friction : macroscopic and microscopic processes*. Dordrecht; Boston: Kluwer Academic, 1991.

134. D. Dowson, "Thinning films and tribological interfaces " in *Proceedings of the 19th Leeds-Lyon Symposium on Tribology*, University of Leeds, UK, 1992.

135. Y. Chung, *Micro- and Nanoscale Phenomena in Tribology*. Boca Raton, FL: CRC Press, 2012.

136. F. Gardiol, *Lossy Transmission Lines*. Norwood, MA: Artech House, 1987.

137. A. Dounavis, R. Achar, and M. S. Nakhla, "Efficient Sensitivity Analysis of Lossy Multiconductor Transmission Lines with Nonlinear Terminations," *Microwave Theory and Techniques, IEEE Transactions on*, vol. 49, pp. 2292–2299, 2001.

138. W. Jian, T. Min, M. Lizhuang, and M. Junfa, "FdSPICE: A Parallel Simulation Technique for Lossy and Dispersive Interconnect Networks," in *Electrical Design of Advanced Packaging & Systems Symposium (EDAPS), 2010 IEEE*, 2010, pp. 1–4.

139. H.-J. Butt, B. Cappella, and M. Kappl, "Force measurements with the atomic force microscope: Technique, interpretation and applications," *Surface Science Reports*, vol. 59, pp. 1–152, 2005.

140. J.N. Israelachvili, *Intermolecular and Surface Forces*. Amsterdam, Netherlands: Elsevier, 2011.

141. J. Song, and D. J. Srolovitz, "Adhesion effects in material transfer in mechanical contacts," *Acta Materialia*, vol. 54 pp. 5305–5312, 2006.

142. N. Alcantar, C. Park, J.-M. Pan and J.N. Israelachvili "Adhesion and coalescence of ductile metal surfaces and nanoparticles," *Acta Materialia*, vol. 51 pp. 31–47, 2003.

143. K. L. Johnson, K. Kendall and A.D. Roberts "Surface Energy and the Contact of Elastic Solids," *Proc. of the Royal Society of London. A. Mathematical and Physical Sciences*, vol. 324 pp. 301–313, 1971.

23

Metal Fiber Brushes

Glenn Dorsey and Doris Kuhlmann-Wilsdorf

Nature is always hinting at us. It hints over and over again.
And suddenly we take the hint

Comments, **Robert Frost**

CONTENTS

23.1 Introduction 1152
    23.1.1 Fiber Brushes for Power 1152
    23.1.2 Diversification of Applications 1154
    23.1.3 Outline of Chapter 1156
23.2 Sliding Wear of Multi-Fiber Brushes 1156
    23.2.1 Adhesive Wear 1157
    23.2.2 Holm-Archard Wear Equation 1158
    23.2.3 Low Wear Equilibrium 1159
    23.2.4 High Wear Regime 1160
    23.2.5 Plastic and Elastic Contact 1161
    23.2.6 Critical or Transition Brush Pressure 1163
    23.2.7 Wear of Fiber Brushes 1164
    23.2.8 Effects of Sliding Speed 1166
    23.2.9 Effect of Arcing and Bridge Transfer 1169
23.3 Surface Films, Friction, and Materials Properties 1170
    23.3.1 Thin Film Behavior 1170
    23.3.2 Water Molecules 1170
    23.3.3 Film Disruption 1172
    23.3.3 Lubrication 1172
23.4 Electrical Contact 1174
    23.4.1 Dependence of Electrical Resistance on Fiber Brush Construction 1174
23.5 Brush Dynamics 1177
    23.5.1 Speed Effect 1180
23.6 Future 1180
23.7 Summary 1181
Acknowledgments 1181
References 1187

23.1 Introduction

The term electrical "brushes," so evidently non-descriptive of the carbon and carbon–metal brushes discussed in Chapters 20 and 21, derives from the bundles of metal wires which "brushed" against the insulating, partly metal-covered electrostatic generator disks of two centuries ago. For more than one hundred years, such painter's style "brushes" and various modifications thereof in the form of assemblies of metal wires were universally used for sliding electrical contacts. They were displaced by "monolithic" graphite brushes only at the turn of the twentieth century. The compelling reason for the change-over was to decrease wear of both the brushes and the counter-surfaces.

Recent work has demonstrated that "bundled" fiber brushes can provide significant performance advantages over monolithic graphite or composite metal graphite brushes in certain applications. These advantages include higher current density, decreased contact resistance and contact resistance variation (noise), decreased wear debris generation, and improved insensitivity to environmental effects. These advantages are achieved by spreading the force of the electrical contact against its contacting counter-face across relatively lightly loaded contact spots of many independent metal fibers.

This chapter discussing fiber brushes is a natural extension of the discussion in Chapter 22 which discussed single metal fiber brushes. The notion of bundling a number of single fibers into a single brush to carry more current is an obvious one. The contribution made by recent research is to show that multiple contacts allow a significant reduction in brush force per individual fiber while maintaining low contact resistance. However, all the principles discussed in Chapter 22 apply to fiber brushes, and conversely the detailed discussion of wear and friction in Chapter 23 is relevant to the single brushes of Chapter 22.

23.1.1 Fiber Brushes for Power

Much of this fiber brush work was championed by Dr. Kuhlmann-Wilsdorf and her name appears on many of the references used in this chapter. The primary focus of her distinguished career was developing a high current density and low wear bundled fiber brush as shown simply in Figure 23.1 (a) taken from Dr. Kuhlmann-Wilsdorf's early patent [1]. This illustration shows a bundle of small, metal fibers collected in a holder and impinging on the contacting surface on the fiber tips. Dr. Kuhlmann-Wilsdorf recognized early in her work that one of the keys to successful and optimal operation of the fiber brush was in the control of the local environment of the contacts, and much of the data presented for fiber brush performance has been collected when operating in a controlled atmosphere, for example, moist argon and moist CO_2. This has been referred to as vapor phase lubrication [2].

Kuhlmann-Wilsdorf developed a very robust theoretical underpinning for the fiber brushes developed in her laboratory and reference [3] provides a very comprehensive outline of the theory behind multi-fiber electrical brushes. There are several important assumptions made about fiber brushes that overarch this theoretical work. These assumptions are not necessarily appropriate for all fiber brushes, but are inherent in the majority of Kuhlmann-Wilsdorf's work.

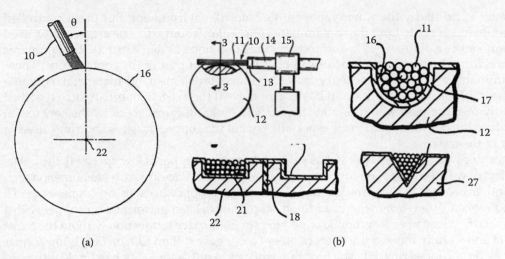

FIGURE 23.1

Fiber Brush Designs (a) This fiber brush illustration is from a 1982 patent of Kuhlmann-Wilsdorf [1]. The drawing shows the basic features of the multi-fiber sliding electrical contact that impinge on the counterface end-on. The slight cant or tilt of the brush allows the brush to be compliant, but does limit the brush to unidirectional rotation. (b) These graphics from the Lewis and Skiles patent of 1983 for a tangential fiber brush design. Unlike the end-on design of (a), this fiber brush design contacts the counterface tangentially allowing bi-directional rotation. The brush force is maintained by the cantilever bending of the fibers, so sufficient brush diameter is required. The brush can contact a flat ring or can be constrained in machined grooves as shown (U-groove, square groove, and V groove).

1. Fiber brushes are used for transmitting high current and the critical electrical parameters are related to conductivity.

2. Fiber brushes are relatively large bundles of very small (sub 100 μm. diameter), independently-acting fibers with their tips impinging on the counterface. The brush has been described as a "velvet-like mat" made of fibers.

3. The bulk brush length is short and the bulk resistance of the brush is insignificant compared to the contact resistance.

4. The environment and consequentially surface films are well controlled and the resistivity is roughly 10^{-12} Ω/m^2 or equal to 2 molecular layers of water. In most cases this requires a controlled atmosphere to achieve.

5. Fluid lubrication is never used.

6. Normal forces are equally distributed across individual fibers and the fibers are small enough that it is reasonable to assume that there is only one contact spot per fiber.

7. Individual contact spots can be evaluated using the Hertz criteria as elastic or plastic and a critical brush pressure can be determined where the contacts transition from elastic to plastic; low wear occurs under this transition pressure.

There will be a more detailed discussion later of point #4 requiring the strict control of the surface films, but the atmosphere control aspect of this work does highlight the key element of fiber brush design; unlike monolithic brushes (Chapter 20) that contain solid lubricant (usually graphite), metal fiber brushes depend on surface films from external sources for lubrication. As shown by controlled atmosphere tests, this external source of lubricant is primarily the water vapor in the "moist" environment. But these controlled environments serve to limit the contaminants of regular "room air." The argon or CO_2 used in many of the Kuhlmann-Wilsdorf tests ensures that none of the "normal" contaminants of standard industrial environments can play a significant part in the contact zone. These contaminants are one of the primary causes of contact problems in normal metal monofilament or multiple fiber brushes, and it is questionable if the extra light brush force specified in many of these tests would provide insufficient "wiping force" to clean the contacts in an industrial environment. This is especially true of the copper brushes and rings used in many of the tests.

A body of work by Argibay et al. has studied copper on copper at low speed [4], the polarity effect of wear on copper on copper (positive brush wore twice as fast, a phenomenon seen with other brushes to the same degree) [5], beryllium copper brushes on copper rings [2] all in a moist CO_2 environment. In addition, studies have been performed on copper with a third sacrificial member (as graphite wiper) to provide surface lubrication without the moist CO_2 [6] and a study showed the effect of other "cover gases" than CO_2 on beryllium copper brushes on a copper ring [7]. Much of this work by Argibay et al. as has Dr. Kuhlmann-Wilsdorf's, has been in support of high current sliding contacts for a high power homopolar motor application. The goal has been to provide brushes that can conduct a current density of $j = 2000$ A in^{-2} = 3.1 MA m^{-2} at velocity $v = 40$ m s^{-1} with a maximum total loss of $L_T = 0.25$ W A^{-1} and to be operated in a protective atmosphere of moist CO_2. (More information on this homopolar motor application, along with the High Speed Maglev Train Brushes—same high current with lower velocity—can be found in reference [3].)

The assumptions listed earlier are important to understand since the largest body of research work on fiber brushes is centered on *one specific approach* to fiber brush design, but this is not the only approach. There are successful fiber brush design approaches that contradict every one of these assumptions. The key elements of a fiber brush are: (1) multiple fibers to give multiple contact points, (2) some independence of action (even if limited) of the fibers, and (3) relatively light contact loading from each fiber in contact.

23.1.2 Diversification of Applications

It is unfortunate in many respects that the homopolar motor requirement has been so well funded since it has skewed much fiber brush research toward a very imposing high current requirement, and many of the advantages of the fiber brush technology—low debris generation, modestly high current density, and good reliability—have been pushed to the background. Widespread use of fiber brushes depends on operation in a standard industrial environment with environmental conditions from −50°C to +80°C, 0%–100% RH, sand and dust environments, and typical shock and vibration. Furthermore, they must operate with typical industrial tolerances, provide good reliability and maintainability, and be cost effective. A wind turbine nacelle in Denmark in mid-winter is a far cry from a moist argon lab environment.

Two application notes [8,9] report on an alternative fiber brush design that is used in industrial slip rings, although there is little published research on this approach. Figure 23.1b shows this tangential fiber brush as presented in the patent [10] for the design. The tangential fibers must be longer and larger in diameter than those reported by the Kuhlmann-Wilsdorf and Argibay teams (for structural integrity and contact force generation), and they impinge tangentially rather than on end. This tangential design has been used in medical imaging CT scanner applications (70–180 RPM on a 1.5m diameter, 100 amp, 20M revolution life), radar and wind turbine applications (40–100M revolution life), and military helicopter

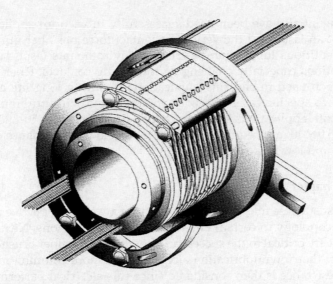

FIGURE 23.2
Illustrates a commericial slip ring product utilizing tangential fiber brushes for both power and signal transfer. This product line is used for a wide range of industrial and defense applications and has been in wide distribution since the late 1980s.

blade de-icing (harsh environment, high reliability). Figure 23.2 shows an example of one slip ring in a standard industrial slip ring product line used in packaging equipment, robotics, and industrial machinery. This entire product line utilizes tangential fiber brushes.

The important point to recognize is that the fiber brush concept of utilizing multiple fibers with lightly loaded contacts has a broad range of potential design approaches. Improving contact life, good maintainability, and high reliability are important design goals for many industrial, medical, and defense applications, and the whole range of lubrication schemes, contact design, and materials can be and should be utilized to approach each application. Sometimes (*but definitely not all the time*) a fiber brush approach is the best design approach. The generic approach of utilizing a bundle of multiple, lightly-loaded metal fibers to carry current in applications requiring long life and low maintenance has value in a number of important applications.

Almost all of the research on fiber brushes is centered on the advantages of this technology in power transfer requirements. However, instrumentation and control slip rings (see Chapter 22) have incorporated fiber brush technology to advantage primarily for the reduction of contact noise over long life. Fiber brushes are actually a collection or aggregation of monofilament wipers discussed in Chapter 22. In fact some researchers are starting to test single fibers as a method of gaining a better understanding of multi-fiber brushes [7] and "remove some of the complexities typically encountered in macroscopic scale experiments with multi-fiber metal brushes." This particular study is also interesting in that it evaluates the wear of the brush impinging on the ring tangentially rather that end on. The advantage of using a large number of metal brushes to carry current is obvious, but there are also advantages for transmitting signals. Multiple contact points can improve contact noise performance. Since contact noise is typically improved by the addition of multiple contact points, an entire fiber bundle can be useful in providing redundant signal paths and in reducing noise. It is important to understand that there is still a brush/pressure trade-off. If the contact pressure per fiber is insufficient to penetrate the surface film, having multiple contacts will still result in high noise.

Fiber brushes (tangential) have been used successfully in vacuum applications where surface films are not an issue and the very light contact force per fiber minimizes wear. The absence of adventitious lubrication in vacuum environments (especially moisture) makes these applications especially challenging for sliding contacts. Fiber brushes (tangential) have also been used in slip ring applications requiring long life and low wear debris generation. Two specific cases are radar antennae and wind turbine pitch control systems. These are both applications that require long life (frequently 50–100 million total revolutions or ~ 2.5-5.0 × 10⁷ m of equivalent linear travel), low maintenance intervals, and both efficient power and low noise signal transfer.

23.1.3 Outline of Chapter

This chapter will first address fiber brush contacts from a wear perspective to show wear advantages of the technology in certain environments or applications. Next, we consider the contact films that are critical to the successful operation of the fiber brushes since these brushes do not carry their own lubrication along like the more common metal graphite brushes. Electrical resistance is then considered since this electrical performance parameter is to most critical for successful implementation of fiber brushes. Finally we will consider some unique dynamic considerations for implementation of fiber brushes.

23.2 Sliding Wear of Multi-Fiber Brushes

Chapters 7, 20, and 22 provide an overview of sliding contact of common electrical contact materials. The discussion in this chapter will build on the ideas presented in these chapters to discuss the performance of multi-fiber groups of contacts impinging on a counterface. The load-bearing contact between a brush and its counterface occurs only at a number, n, of isolated areas called contact spots, or to the extent that they conduct current, called a-spots. In old-fashioned metal wire brushes, contact force was typically high to ensure good contact in the presence of contaminating films. This high contact force led to high wear. New design strategies with fiber brushes have significantly improved wear and electrical performance by "dialing back" on the contact force. Improved knowledge of contaminating films is being used to address the surface film problem that high force addressed previously.

Since the entire premise of utilizing multi-fiber brushes is based on the ability to control wear, the subject of sliding wear of electrical contact materials is worth exploring. The assumptions we make about the mechanical properties of potential contact materials are appropriate to highlight:

1. The materials that meet the conductivity and resistance to corrosion/filming requirements for sliding electrical contacts are the noble metals and their alloys (in the order of their nobility—Au, Pt, Ir, Pd, Os, Ag, Rh, Ru, and Cu*).

---

* Technically, copper should not be in the list of noble metals, but we will include it in our assumption since it is often used for sliding electrical contacts and it is reasonably close to Ru (the lowest of the noble group in standard electrical potential) in nobility. However, it is this poor nobility compared to gold and silver, for example, that prohibits the common use of copper in sensitive electrical applications; but copper is conductive, relatively cheap (compared to the others), and "reasonably" noble.

2. Most of these contact materials are ductile and the most common ones (Au, Ag, Pt, Pd, and Cu) are extremely so. They are sometimes alloyed to control hardness (and stiffness) which effects ductility, but in general these materials can be considered ductile.

3. These materials are relatively corrosion resistant, at least Au through Ag. Corrosion products have a significant effect on the free surface energy of wear particles and surfaces.

There are three common materials used for fiber brushes and each has its place in contact design. All of these materials can be used with or without lubricants depending on the environment and performance requirements.

1. *Gold and gold alloys*: Gold is used in applications that call for low contact resistance and resistance variation and environmental contamination cannot be controlled. Gold will allow low brush force with a lower risk of high resistance owing to surface contamination. Typically a hard gold counter surface is used, but other less noble surfaces can be used and the gold from the brush will transfer onto the counterface forming a gold wear track. This is clearly an expensive option for large diameters.

2. *Copper and copper alloys*: Copper is a good material for contacts owing to its high conductivity. As discussed above it is not one of the noble metals and corrosion products can cause high contact resistance. In the unalloyed state it is also quite soft and alloys are used to increase the hardness (BeCu is the most common alloy). To use copper and copper alloys the local environment must be controlled to permit low brush force without excessive insulating film development. The counterface is typically copper, but more noble materials can be used to control insulating film development.

3. *Silver and silver alloys*: Silver has the advantage of noble metal properties, high conductivity, and lower cost than gold. Counterfaces of gold, silver, or copper can be used depending on the electrical and wear properties desired. Typically silver alloys are used to alloy increased hardness without too great a loss in conductivity.

23.2.1 Adhesive Wear

Rabinowicz [11] discusses four types of wear: (1) adhesive, (2) abrasive, (3) corrosive, and (4) surface fatigue, and maintains that adhesive wear is the most common and difficult to avoid (see also Chapter 7). Blau [12] (pp. 58–59) compares wear classification schemes of 8 different researchers and Rabinowicz's list is common to most and certainly the most relevant to electrical contacts. It is very important to realize that these lists are research constructs to help explain physical phenomena, and nature is not as helpful in actual wear processes. Trying to identify wear modes is a tricky business.

Adhesive wear occurs through the formation of small wear particles generated by adhesion between asperities on the surfaces. The particles are generally produced from the softer side. Greater understanding of these adhesive connections has come from the advent of the atomic force microscope (AFM) and the surface forces apparatus (SFA) and the ability to study individual sliding junctions at the molecular level with these instruments. The reader is pointed to these works for a discussion of this fascinating topic of adhesive and surface dynamics at the molecular level [13–20]. Of particular interest to the area of sliding contacts is the discussion [13] of a contact discontinuity process called "jump in" in ductile materials (most electrical contact materials), wherein at a close proximity (nm), voids in a contact area become unstable and suddenly collapse and "coalesce" to become cold welded. This cold welding process has been discussed for years by tribologists, and the molecular perspective of these "nanotribologists" brought to this phenomenon has filled in a number of holes in the understanding of this process.

Bonds formed in sliding must in turn be broken as sliding continues. Adams, in his review of the mechanics of adhesion [21,22], reviews the characteristics of ductile separation of the adhesive bond:

1. The contact radius decreases slowly before a sudden separation (typically at a significant reduction of the maximum contact radius);
2. The asperity if stretched significantly during separation owing to plastic deformation;
3. A neck is sometimes formed during ductile separation.

Most sliding electrical contact materials are ductile in nature, although significant cold work in the contact region can reduce the ductility. In the case of alternative brittle separation (as is the case with Ru or heavily cold-worked Au or Cu), there is little or no plastic deformation on unloading.

23.2.2 Holm-Archard Wear Equation

Holm proposed a linear relationship for adhesive wear [23,24] that related the wear volume (V) to through a dimensionless wear parameter K_{HA}, the length of sliding path (L_s), force (P) and hardness (H):

$$V = K_{HA} P L_s / H \qquad (23.1)$$

K_{HA} is the probability that adhesive junctions when broken will form a wear particle. For a porous material and metal wire bundles, in which only the "packing fraction" f is occupied by solid and $(1 - f)$ is fraction of voids, this relationship may also be written in terms of dimensionless wear rate as

$$\Delta \ell / L_s = K_{HA} p / f H \qquad (23.2)$$

with $\Delta \ell$ the worn-away layer thickness, a is the average radius of the contact spot, and $p = P/\pi a^2$, and is the macroscopic normal pressure between the two sides. This relationship between wear and normal pressure is now commonly referred to as "Archard's law" [25], but should more properly be called the Holm–Archard wear law.

The problem is that the relationship between adhesive wear and force is much more complicated than can be expressed in a linear relationship. To quote Rabinowicz on the accuracy of the Holm-Archard law, "Many investigators have found that this equation is not always perfectly obeyed, but in almost all cases it represents the experimental data reasonably well." This is a "technical" way to say that wear data are usually ill-behaved primarily because the variables are so hard to control. In a study of three-dimensional fractal surfaces in normal contact, Yin and Komvopoulos [26] conclude that "the adhesive wear coefficient of rough surfaces in normal contact can vary by an order of magnitude depending on the material properties, surface compatibility, and environmental (lubrication) conditions."

23.2.3 Low Wear Equilibrium

In the case of adhesive wear, the first stage is simply the adhesive *transfer* of a wear particle from one surface to the other, and in the case of sliding electrical contacts, this transfer is almost always intended to be from the counterface to the brush. It is the next phase that is most important for brush design. In the case of a well-designed metal wire on metal counter-face contact system, this wear material is transferred back and forth from ring to brush forming an equilibrium condition for a long life system. This process is identified with ductile materials under light loads. The reader is referred to Chapters 7 and 22 for additional details of the adhesive wear process. Assisting in the long terms stability of a well-designed system is gradual wearing-in of the counterface surface eliminating the highest asperities.

In these equilibrium wear states, some wear particles are created as they "escape" from the system, but they tend to be agglomerations of very small wear particles (see Rabinowicz [11] pp. 142 ff for a detailed discussion of wear particle generation). Figure 23.3 shows a very good example of this transfer process in a noble metal fiber brush application in a standard atmosphere. The layered wear debris build-up on a single fiber with incipient wear particle generation is almost identical to those particles and wear patterns shown in Chapters 7 and 22, as it should be since it is essentially the same process.

It is clear in the derivation of the adhesive wear relationship of Equation 23.2 [25] that the adhesive wear refers to the transfer of the particles from the wearing surface to the rider. In the case of very ductile materials, the formation of a pad of wear material that is not really lost from the wear surface confuses both the understanding and the measurement of the

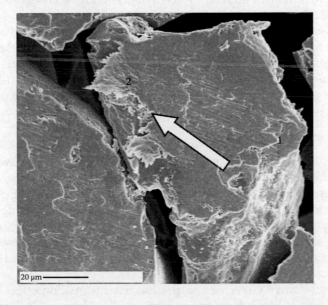

FIGURE 23.3
Photomicrograph from [42] showing transferred material on the fiber tip in a laminar construction. Wear particles can be seen about to separate from the brush. The arrow points in the direction of sliding. (From L. Brown, D. Kuhlmann-Wilsdorf, and W. Jesser, *Components and Packaging Technologies, IEEE Transactions on*, vol. 31, pp. 485–494, 2008 [42].)

wear volume. Yin et al. also assume that an asperity that is truncated from the counterface is directly transformed into a wear particle, and these analyses were performed on relatively brittle materials [26].

In this "low wear" state the wear process that actually transfers particles out of the wear region primarily occurs as shearing of interlocked asperities. Chang et al. describe the process observed in unlubricated copper multi-fiber brush sliding on a copper counterface as follows: wear particles are sheared off, principally from the softer side, where plastic contact spots momentarily "adhere" after traversing m^* times their own diameter on average [27]. This is a mechanical "interlocking" that develops as part of the adhesive transfer process. Chang observed that the average wear particle volume was in the order of $6a_{pl}^3$ and the number of wear particles was $nL_s/m \times 2a_{pl}$ where n = number of contact spots and a_{pl} = radius of average contact spot. Hence,

$$V = (nL_s/2m^* a_{pl})6a_{pl}^3 \approx PL_s/m^*H \qquad (23.3)$$

or Equation 23.2 with the constant K_{HA} in the Holm–Archard law equal to $1/m^*$. The effect of simultaneous current flow was to mildly increase the average wear chip size, presumably because of a decrease of hardness (H) through Joule heating.

This adhesion/abrasion process of ductile materials leads to some confusion in the study of sliding electrical contacts, since material *transfer* is occurring in an adhesive manner, but the final wear debris *generation* in the equilibrium condition is occurring as part of an abrasive process. If we look again at Figure 23.3, for example, we see laminar, heavily work-hardened material built up on the fiber tip. This is ring material that has become adhered to the brush either through an adhesive process or an asperity shearing and smearing process (or a combination of both). Wear particles that are expelled from the contact area are generated by abrading owing to asperity interlocking or just general instability of the wear particle or lack of sufficient adhesive force. Particles can be observed in Figure 23.3 ready to break off the edge of the wear pad.

Wear particles observed in gold-on-gold contacts in this wear mode are very small, dark in color, and much harder than either contact material (i.e., heavily cold-worked, very small, abraded gold particles). Wear measurements inevitably measure the particle generation rather than the transfer rate, and without a very long term test with the ability to capture average wear rates it is difficult to understand the wear of prow-forming, ductile materials.

23.2.4 High Wear Regime

Accelerated wear situations do not settle into this equilibrium, low wear condition. Rabinowicz argues that this is primarily the result of wear particles whose elastic energy exceeds the surface energy required to adhere to the wiper (internal elastic energy > surface energy or the condition of a prow-forming material). Obviously, the high wear condition can be caused by gross plowing of abrasive asperities or third party particles, but the most frequent cause of high wear with sliding electrical contacts is adhesion. The difference between contact adhesion that reaches a "low wear equilibrium" and adhesion that produces high wear can be frustratingly subtle.

Figure 23.4a is a well-known curve that shows the wear transition effect (wear particle generation) as a function of brush force for unlubricated gold-on-gold. This curve shows that below 5 grams (0.05N) the wear coefficient is low, and it is easy to get the perception (owing to the log x axis) that there is a *sharp* increase after 5 grams. Figure 23.4b shows the same data plotted without the log scale on the x axis and actually shows a reasonable

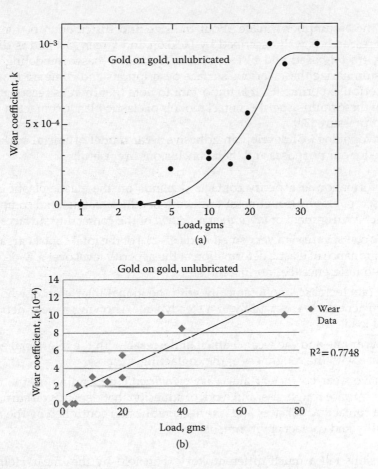

FIGURE 23.4

(a) A well-known curve redrawn from [11] p. 160. The log scale on the x-axis makes the data look more dramatic than it is. The curve redrawn on a linear scale shows a reasonably good fit for a linear trend. The (a) plot does give extreme emphasis to the three low wear points and low wear is notoriously difficult to measure consistently. Rabinowicz's conclusion that gold has a very low wear coefficient below 5 grams is still reasonable and widely accepted. (Redrawn from E. Rabinowicz, *Friction and Wear of Materials*. New York: John Wiley & Sons, 1965 [11].)

linear fit. The important point is not the exact relationship of the wear coefficient to force, but rather that this increase is in the *wear coefficient* that Holm-Archard specifies as a constant adhesion transfer probability.

23.2.5 Plastic and Elastic Contact

Molecular dynamics modeling has shown that in the case of nano-contacts of Au with a radius of curvature of 3 nm [28] there are three different separation modes for asperities joined by adhesion. These three modes in the order of increasing adhesive energy are: (1) elastic separation, (2) plastic interface separation, and (3) plastic non-interface separation (accompanied by material transfer). Adams points out [22] that these results are qualitatively similar to the continuum models of other researchers [29–31]. It is clear that if asperity contact can be controlled to minimize plastic non-interface separation, wear can be minimized.

The modeling of elastic-plastic contacts has received significant attention by recent researchers, and clear differentiation between elastic and plastic contact regimes is only as accurate as the assumptions made about the size and distribution of the asperities. This modeling research is well described by Jackson and Green [32] and is divided into statistics-based, fractal-based and FFT or spatial frequency–based modeling. Ciavarella et al. [33] also summarize these various surface descriptions and compare many of these models to numerical experiments. It is important to note that there is reasonable correlation between all these multi-asperity contact models of elastic-plastic contact in most of the range of contact pressure [34].

Yin and Komvopoulos [35] develop an adhesive wear model of fractal surfaces that is very instructive for our purposes and their conclusions are useful:

1. Plastic deformation at asperity contacts depends on the elastic-plastic material properties, topography (roughness), and work of adhesion (material compatibility and contact environment or *lubrication condition* of the contacting surfaces).

2. The plastic contact area is a very small part (<1–2%) of the total contact area, revealing the dominance of elastic deformation at the asperity level over a wide range of the global interference (i.e., normal force).

3. The wear rate increases monotonically with the global interference, whereas the wear coefficient decreases rapidly to a steady state, showing a wear dependence on normal load.

4. Both the wear rate and the wear coefficient decrease with the interfacial adhesion and increase with the roughness of the contacting surfaces.

5. The adhesive wear coefficient may vary *significantly* depending on the material properties, surface roughness, and work of adhesion that depends on surface energies of the contacting surfaces and interfacial adhesion controlled by the material compatibility and contact environment.

These conclusions tell a much different story than told by the linear Holm-Archard Equation 23.2. As put succinctly in a review article in *Nature* on the nonlinear nature of friction [20], "The multitude of asperities on two shearing surfaces are constantly coming into and out of contact, where the local pressure between them can fluctuate between ~ 1Pa (10^{-5} atmospheres of pressure) and GPa (10^4 atmospheres) within microseconds. These are extreme conditions that cannot always be treated by simple 'linear' theories."

In fact, the story on sliding contacts is coming into better focus from the aforementioned improved modeling on the macroscale and with the advent of the atomic force microscope (AFM) and the surface forces apparatus (SFA) and their ability to study individual sliding junctions at the molecular level. In the case of the sliding of two surfaces across each other where electrical contact is required, there is both reversible (elastic) deformation and irreversible (plastic or viscoelastic) deformation. The wear and friction properties of these surfaces depend on the adhesive properties of the contacting surfaces or the work of adhesion and these adhesive properties can be and often are non-linear. Quite often, friction and wear cannot be predicted strictly by force, hardness, and a constant.

Fiber brushes are able to maintain a low wear regime by utilizing multiple independent contact points with each contact point very lightly loaded. Kuhlmann-Wilsdorf et al. [3,36–38] define a critical brush force (transition from low wear to high wear) in terms of elastic and plastic asperity deformation, i.e., there is some critical brush pressure at which elastic contact transitions to plastic contact at each asperity. The argument is then made that this critical force (or pressure) is the transition from low wear to high wear. An equation is derived calculating this critical brush pressure [36,38]. This seems to be a reasonable approach for defining a design pressure for fiber brushes, but let's understand the assumptions of this approach better.

23.2.6 Critical or Transition Brush Pressure

If we turn our attention to the critical force or pressure that defines the transition between low and high wear, we do see repeated reference to a critical pressure. Depending on the model being used, there are a variety of expressions for this critical pressure. Wilson, et al. [39] express the critical force in terms of spatial frequency; Yan and Komvopoulos [40] express it in terms of fractal parameters. This critical force is typically normalized to the force required to achieve 100% contact in the apparent area. In light of the evidence that there is a critical brush force that represents the knee of a brush force-wear rate curve around the elastic/plastic transition pressure, it does seem desirable to find a simple equation that helps determine this critical force is for fiber brushes.

In the case of fiber brushes, the number of contact spots is high, each fiber (i.e., contact spot) is capable of acting independently (at least that is our assumption) and the applied force per fiber is as low as possible and still has good electrical contact. Kuhlmann-Wilsdorf [3,41] makes the simplifying assumptions of: (1) a fiber bundle that produces distributed multiple semi-circular "Hertzian" asperities that have a radius of curvature, r_c, approximately the same radius as the individual fibers, (2) a perfectly flat and rigid counterface, and (3) each of these asperities can be loaded up to, but not exceeding its elastic limit. A total critical or safe brush pressure is then defined in terms of this individual fiber elastic limit.

Each of these contact spots has the radius, a_{el}, of

$$a_{el} = 1.1(r_c P/nE)^{1/3} \tag{23.4a}$$

The average local pressure per elastic contact spot $_{el}p_b$ is found by

$$_{el}p_b = P/{n\pi a_{el}^2}, \text{ or by re-arranging} \tag{23.4b}$$

$$a_{el} = \left(P/{\pi n_{el}p_b} \right)^{1/2} \tag{23.4c}$$

To a useful first approximation we may write $H \approx 0.004E$ as shown by Table 24.1C. We may also define n^* as the number density of contact spots at the interface; therefore substituting (23.4c) into (23.4a) we can derive the average local pressure per elastic contact spot in terms of total brush force (P), elastic modulus (E) or alternatively in terms of hardness (H) and brush pressure ($p_b = P/A$):

$$_{el}p_c = P^{1/3}E^{2/3}/(n^{1/3}\pi 1.1^2 r_c^{2/3}) \approx 0.26(p_B E^2/n^* r_c^2)^{1/3} \approx 10.44(p_B H^2/n^* r_c^2)^{1/3} \tag{23.5}$$

The critical force at the transition between elastic and plastic contact spots, P_{trans}, is found by equating the average pressure of the Hertzian, elastic contact spot with H, that is,

$$P_{trans}/n\pi a_{el}^2 = H \tag{23.6}$$

which yields

$$P_{\text{trans}} = \pi^3 1.1^6 (H^3/E^2) n \, r_c^2 \approx 9 \times 10^{-4} H n r_c^2 \tag{23.7}$$

and for the macroscopic pressure at transition, p_{trans} and r_c = fiber radius,

$$p_{\text{trans}} \approx 9 \times 10^{-4} H n^* r_c^2 \approx 3.5 \times 10^{-6} E n^* r_c^2 \tag{23.8}$$

23.2.7 Wear of Fiber Brushes

Figure 23.5a from [36,38] shows the results of tests of silver-clad copper fibers on a copper interface (slip ring) in a moist CO_2 environment and data from tests performed with silver brushes on a gold-plated commutator [42] in a standard laboratory environment. The data show the wear vs. brush pressure relationship. The calculation of p_{trans} in this graph is based on two basic assumptions mentioned earlier: (1) each of the fibers acts independently and the force is equally distributed among the fibers, and (2) the radius of curvature of the contact spot equals the fiber diameter. The researchers report a power law relationship between wear (W) and normalized brush pressure (β, normalized to the transition pressure of Equation 23.8) proportional to $\beta^{4.6}$. These data suggest that under this very particular set of environmental conditions, by placing a reasonable factor of safety of two on the critical brush pressure (i.e., normalized force is 0.5), a dimensionless wear rate of $\sim 10^{-11}$ can be achieved on the slip ring and about 2×10^{-11} on the commutator. The paper's authors explain this difference as variation between a slip ring and commutator.

Figure 23.5b puts a different spin on the same data. There is an excellent linear fit on the slip ring data suggesting that in the narrow range of brush pressures below the transition pressure the wear rate follows the Holm-Archard law. The linear fit is not as good on the commutator data, but the best fit does match the slope of the slip ring data suggesting a similar wear coefficient. It is understandable that the data scatter on a commutator with slots would be greater than a slip ring. These data suggest that the wear coefficient of copper-on-copper fiber brush/slip ring pair in a controlled moist CO_2 environment is very similar to the silver-on-gold commutator in a laboratory environment after an initial run period. This correlation in slope is likely owing to similar material properties, but more importantly, similar film properties (controlled atmosphere vs. noble metal).

To put into perspective the values of $\Delta\ell/L_s$ in the 10^{-9} to 10^{-11} range for $\frac{1}{2}P_{\text{trans}}$ shown in Figure 23.5, they have been inserted into Figure 23.6 presented by Shobert [43], alongside the wear rates of a wide range of monolithic brushes and other substances. In order to gain an impression of expected wear lives, note that at 10 m s^{-1} sliding speed the sliding distance is $L_s \approx 3 \times 10^{10}$ cm in a year, so that at $\Delta\ell/L_s = 3 \times 10^{-11}$ the brush would wear $\Delta\ell \approx$ 1cm/year.

One of the most important advantages to fiber brushes is low quantity of wear debris generation compared to metal graphite brushes. As metal graphite brushes wear, significant amounts of conductive debris are generated as the graphite wears "sacrificially." Fiber brushes generate less debris as they wear which provides reliability and maintainability advantage as well as the potential for improved life. It is important to operate any fiber brush design at the minimum operating pressure that will provide adequate electrical performance. Equation 23.8 provides a reasonable starting point for brush force calculations and likely represents the minimum force for adequate electrical performance since the assumption is a minimum surface film thickness. As discussed later, operation of fiber brushes in standard industrial atmospheres might require lubrication or allowance for additional surface films which might require higher brush pressures. Tests will be required to ensure that wear rates are acceptable.

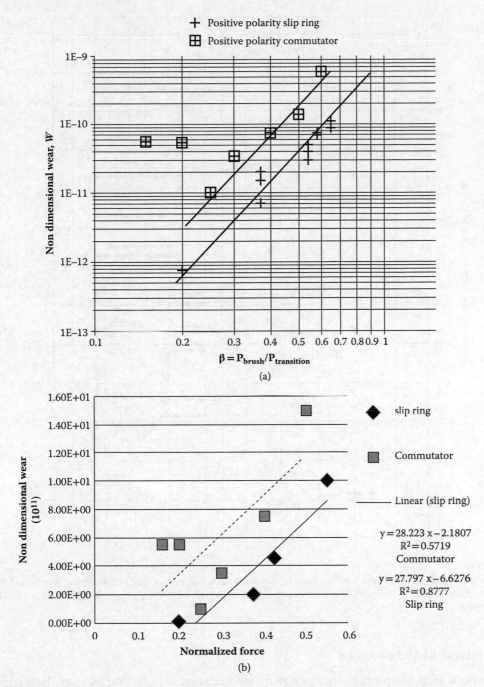

FIGURE 23.5
(a) Dimensionless wear rates for positive polarity brushes on a slip ring (+) and commutators □. The two trend lines show the wear rate dependence on brush force plotted on a log-log scale. Researchers report a power relationship of $\beta^{4.6}$ (b) Same data as (a) with a linear fit. Good fit of slip ring but less correlation for commutator. Greater scatter of wear data resulting from the commutator slots would be expected. (Data from L. Brown, D. Kuhlmann-Wilsdorf, and W. Jesser, *Components and Packaging Technologies, IEEE Transactions on*, vol. 31, pp. 485–494, 2008 [42].)

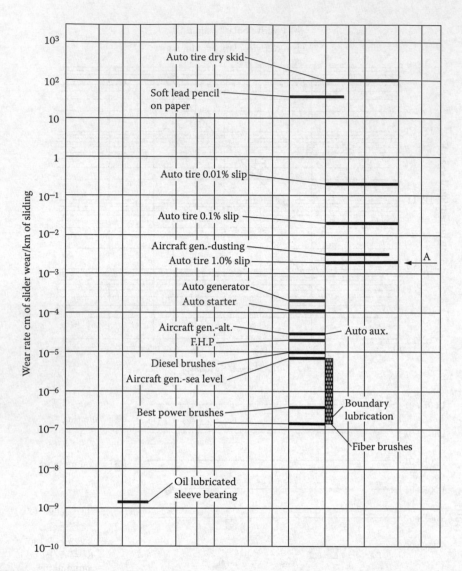

FIGURE 23.6
Dimensionless wear rates and friction coefficients of a wide range of monolithic brushes plus several other sliding situations owing to Shobert (Figure 6.1 of Ref. [42]), into which the data of Figure 23.5 have been added as "fiber brushes."

23.2.8 Effects of Sliding Speed

The transient local temperature peaks during electrical contact sliding as a result of friction are called the "flash temperature" [44], and this transient temperature has been recognized as a contributing cause to wear and deterioration of sliding electrical contacts. The temperature rise is attributable to both frictional heating and Joule heating. The actual asperity flash temperature can be an order of magnitude higher than the temperature in the apparent contact area [45]. The equations for computing the average temperature rise at contact spots above ambient are quite involved, especially when both friction and Joule heat must be considered. This is so even though the heating effects owing to the two contributions are simply additive, and to a good approximation Joule heating is independent of speed and friction heat is independent of current flow. The complexities arise because the rate of frictional heat generation increases with sliding speed while the rate of heat dissipation also increases as a result of an increased convection coefficient. The Joule heating effects of electrical contacts are discussed in detail in Chapter 1, so we will deal with the heating effect caused by sliding.

The pertinent mathematical theory of flash temperatures at contact spots was developed by Jaeger [46,47] and Blok [48]. Since that time notable research work on this subject has been done on flash temperatures of rubbing surfaces by Archard [49], Burton [50], and Barber [51] and, specifically, electrical contacts by Holm [24] and Rabinowicz [52]. Kalin presents a very thorough comparison of different theoretical models for flash temperature [53] and concludes that large discrepancies in the results can be obtained using any calculation technique because, typically contact conditions vary greatly in both time and place and all involve some very tenuous simplifying assumptions.

Research has been done using sophisticated mathematical models to simulate sliding frictional heating to address these problems. Vick and Furey [54], in a basic theoretical study on the temperature rise in sliding contacts with multiple contacts, found several important trends. Several of these points demonstrate why fiber brushes can be effective in ameliorating frictional heating:

1. Downstream contacts are hotter than upstream contacts owing to the convective effect.
2. The temperature rise decreases as the number of contacts increases.
3. The temperature rise decreases as the spacing between the contacts increases.
4. The shape of the contacts has only a small influence, except for very slender or long contacts.
5. Temperatures can be significantly overestimated using a single contact model.

Kuhlmann-Wilsdorf et al. [3,41,55–60] have developed a series of equations for using flash temperature calculations to predict speed and current density performance. These equations are included in Appendix A.1 for applications that closely follow the design constraints imposed by this design: negligible bulk brush resistance, optimum film formation (i.e., lowest possible film resistivity owing to controlled environment), and very small fibers, low packing factor). But it is truly hard to determine acceptable design criteria for flash temperatures, so the advantage of performing these calculations is unclear. Kuhlmann-Wilsdorf suggests that 200°C is the maximum acceptable flash temperature owing to damage to the lubricating moisture layer, but many researchers suggest that the melting point of the contact metal is the maximum. The reader is referred to the very complete derivations and discussions in the references above for more information. Beware however, Kalin [53], in his thorough review and comparisons of various flash temperature models, concludes that simplified, ready-to-use flash temperature calculations can be rather speculative and can lead to very different results on the basis of the model and input parameters.

As always, it is important to subject calculated values to the "engineering reasonableness" filter. Figure 23.7 shows the critical velocity of the Kuhlmann-Wilsdorf brush design using the assumptions inherent in the design and operational characteristics. If we pick one data point, $\beta = 0.5$ (safety factor of 2), fiber diameter of 50 Mm, the chart suggests a critical velocity of 400 m/s in a moist CO_2 environment which is faster than the speed of

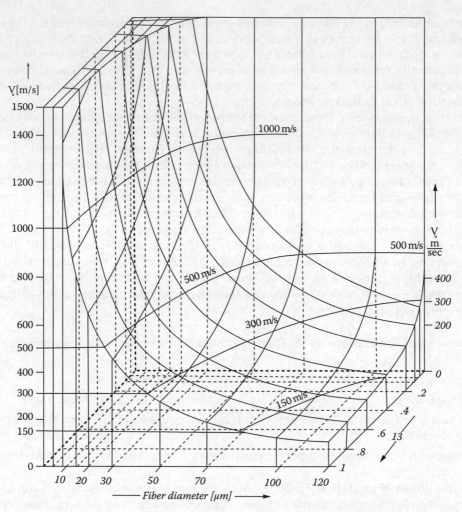

FIGURE 23.7
Critical velocities $(v_o^* = \kappa_{Cu}/r = 1.14 \times 10^{-4}/r$ [mks] units) with $r = \frac{1}{2}d\beta^{1/3}(3 \times 10^{-4})^{1/2}$ according to equations in Appendix A1, as a function of packing fraction (f) and fiber diameter (d) for copper. It is difficult to believe that velocities of 500+ m/s yield acceptable flash temperatures, but the graph suggests the capability.

sound. A 1.5 m diameter CT slip ring spinning at 240 RPM (4 rev/sec, 19 m/s) is a fast surface speed and challenges any brush technology. It is difficult to imagine brushes operating at 21 times that speed with enough force to maintain good electrical contact surviving more than a few minutes. Kuhlmann-Wilsdorf acknowledges the unrealistic appearance of these values [3] p. 370, but points to data that supports the values [41]. In fact, the data from these rail gun experiments referred to suggests a catastrophic contact failure.

Most researchers, starting with Blok and Jaeger, use the P_{eclet}, P_e, number to differentiate between low and high speed sliding contact. The P_e number is defined as Vd/4*χ, where V = velocity, d = contact diameter, and χ = thermal diffusivity. Researchers generally agree with Archard's summary of previous work [49] that contacts with P_e values of <0.1 are low speed and P_e values of >5.0 are high speed and >10 is very high speed. In the case of 50 µm. fiber brush design discussed above, the P_e number for 400 m/s for each fiber of the brush is

44. At issue is not necessarily Kuhlmann-Wilsdorf's method of estimating flash temperature, but the assumption that the very low brush forces typically specified by her designs will yield acceptable results with such high velocities (Figure 23.7).

Bansal, et al. have developed a very serviceable set of design curves for temperature rise in sliding elliptical contacts [61] using a rigorous regression based methodology [62,63]. Bansal finds that the Kuhlmann-Wilsdorf [44,55,56] analysis under-predicts the contact temperature (both average and maximum) by 15–40% for most practical cases. In cases where the brush design does not satisfy the design constraints of Kuhlmann-Wilsdorf, it would be better to use the more general and versatile design guidelines developed by Bansal et al. [61,63] to determine the maximum (flash) temperature and average temperature in the apparent contact area as a result of friction. These design curves have been developed to allow maximum variability for most design parameters.

Bansal et al. do report in their study of flash temperature models [63] that two very straightforward heuristic equations (one for plastic and one for elastic contact) developed by Tian and Kennedy [64] give very good correlation with Bansal's detailed models. These equations provide a very simple method for evaluating sliding electrical contacts for flash temperatures and related equations are presented for evaluation of average temperature rise in the contact zone.

Chen et al. [65] report on the development of a three dimensional thermo-elasto-plastic contact model on the basis of models derived in Ref. [66] and the algorithm presented in Ref. [67]. This model includes the mechanical, electrical, and thermal response at the sliding interface for both Joule and frictional heating. The analysis reveals a strong influence of transient heat transfer on plastic strain because of the dependence of plasticity on loading history; plastic strain is greater when predicted by the transient model. And changes in Joule loading has a much more pronounced influence on the time required for the inception of melting than does either normal load or friction coefficient. The value of steady state flash temperature calculations is questionable.

In some instances, fiber brushes do not have the thermal mass typical of a monolithic, metal graphite brush, so more care must be exercised to prevent bulk heating of the fiber bundle from I^2R losses as well as from frictional heating at higher speeds. Additional consideration must be given to the I^2R losses in the brush itself, as well as in the contact region. Once the thermal loads are established, the long term thermal stability of the system becomes a heat transfer problem. Experience has shown that care must be taken to provide an adequate heat sink at the brush end as well as the ring end of the contact to ensure heat can flow away from the brush. Contact performance is most effective in power transfer if both the brush and the ring transfer heat away from the contact region, so care must be exercised in the thermal design of the assembly. Extreme care must be exercised to ensure that a sharp heat gradient is established to pull heat away from the contact area.

23.2.9 Effect of Arcing and Bridge Transfer

It should also be noted that wear rates for power brushes of all types are typically about two times higher for positive polarity brushes than negative polarity brushes. Chapter 9 discusses the contact physics behind the tendency for material to transfer from the positive brush to the counterface when contact is broken for closely spaced contacts. The Chapter 9 discussion is centered on the macro-effect on electrical switching devices, but the same physics is in play at the asperity level. As contact is made and broken at the asperity level and as all the factors of energy balance at the contact surface come into play, there is a bias toward material loss at the positive brush that does not exist at the negative brush.

Repeated field experience and laboratory tests with metal composite brushes as well as metal fiber brushes show that a 2 × factor is typically demonstrated. Figure 4 of [42] illustrates this accelerated wear on positive brushes in a fiber brush test. In extreme cases microarcs can be observed causing extreme wear which normally leads to catastrophic failure. This condition is generally caused by intermittent full separation of the contacts owing to vibration or excessive particulate debris.

23.3 Surface Films, Friction, and Materials Properties

As in the case with all sliding electrical contacts, the films formed at the contact interface are critical for both the friction and wear characteristics, as well as the electrical performance. There are three primary issues to address: (1) these films can cause high film electrical resistance, (2) the lack of films can cause high wear, and (3) the films and their variation can cause friction perturbations that result in dynamic contact resistance problems.

23.3.1 Thin Film Behavior

Clean metal surfaces have very high surface energies, so they readily uptake gases from the environment [68] forming adsorbed films on the surface of the metal surface. The behavior of these molecularly thin films and their effect on friction and wear has been thoroughly reviewed by Israelachvili et al. [18], Homola et al. [69], and Gee et al. [15]. This research was summarized by Bhushan et al. [14] in a very useful review article in *Nature* in 1995. Persson and Tosatti [70] have collected a large number of useful articles on the subject of the effect of thin films on interfacial friction and wear during sliding. It is fair to say that sliding electrical contacts work in the boundary lubrication regime, and nanotribology studies listed above and the boundary lubrication studies in [71–74] have significantly contributed to the understanding of the impact of these surface films. This research has been primarily focused on the effect of the films on the mechanics of sliding, and we have to turn to Holm [24,75], Greenwood [76], and subsequent work [77–84] to understand the effect on electrical contact. A more thorough review of these films is contained in Chapter 22.

23.3.2 Water Molecules

One of the most prevalent of these environmental "contaminants" is water vapor and the presence of water molecules as a part of the film is inevitable except in a few unusual environments (a hard vacuum for example). On average, these films are much thicker than one or two monolayers. In humid conditions they make surfaces seem "sticky" and eventually, at the dew point, they increase to macroscopic sizes, i.e., give rise to sweating and dew formation. Adsorbed moisture films are very persistent indeed. They are present even in desert conditions, and the last monolayer of moisture desorbs from almost any surface only in a good vacuum (smaller than fractions of a mbar or a few Pa, whence the problem with graphitic brushes in space) or on heating to about 200°C.

In spite of the discussed wide variations of moisture film thickness, the measured film resistivity of "clean" metal fiber brushes is typically near $\rho_F \approx 10^{-12}$ Ω m², corresponding to a tunneling film thickness of ~ 5 Å as inferred from Figure 23.8. This tunneling film has been interpreted [36,38] as the result of a double molecular layer of water and suited to accommodate sliding as well as good current conduction. The previously referenced

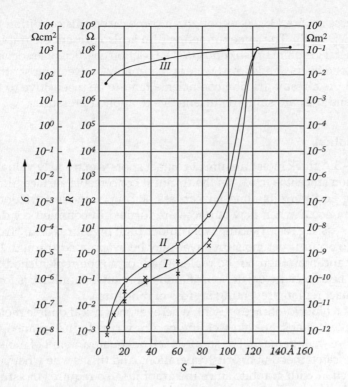

FIGURE 23.8
Tunnel resistivity for insulating "tunneling" films at metallic contact spots as a function of film thickness according to Holm (Figure. 26.11, p. 126 of Ref. [24]). Curve I, computed for very low voltages across the film, is most nearly applicable to the case of metal fiber brushes, while curve II represents average values up to 3 V across the films, and curve III is the classical resistivity for a $\rho = 10^7 \ \Omega$ m film material. Thus ordinary electrical conduction is negligible for insulating film thicknesses below 11 nm but becomes controlling above 13 nm.

works on molecularly thin films [14–19,69] explains this moisture film, as well as other thin films, as the source of many friction phenomena, including stick slip. As explained by Israelachvili [18], there can be long-range lateral order within the layer, and when this happens, the film becomes essentially solid-like and can sustain a finite shear stress in addition to a normal stress. Such solid-like films exhibit yield points or yield stresses where the film "melts" and begins to flow. These researchers point to this phenomenon as a cause of stick slip as the film layer melts and re-crystallizes.

Since much of fiber brush research has been conducted in intentionally moist environments, adsorbed moisture films have been studied in considerable detail using fiber bundles sliding painter's brush-style in the "hoop apparatus" [85–92]. It was concluded that the relative motion between the two sides of contact spots takes place between the two close-packed layers of water molecules, each absorbed to its respective surface. In the absence of deliberate boundary lubrication, adsorbed moisture (and perhaps similarly adsorbed oxygen, nitrogen, or other gases) serves as boundary lubricant. Even in adhesive wear with plastic contact spots, the energy dissipation through friction likely occurs within the adsorbed layers and not via subsurface plastic deformation.

Fortunately in protective atmospheres containing water vapor, management of these surface tunneling films for successful metal fiber brush operation is typically easy to the point of being automatic. In almost all previous research on unlubricated metal fiber brushes in protective atmospheres, tunneling films have been double layers of adsorbed moisture of thickness $s \sim 5$ Å [36,38], although, in some cases, this condition has not been acknowledged (e.g., [93–97]). The same tunneling film is also present on graphitic lubricating films and on metal graphite brush contacts. The friction coefficient associated with this sliding between monolayers of adsorbed water presumably between most other monolayers is near $M = 0.3$. As a result, friction coefficients tend to be insensitive to the choice of sliding materials, and they are similarly insensitive to applied pressure.

23.3.3 Film Disruption

As discussed, the ~ 5 Å thick moisture film at contact spots is extremely valuable for metal fiber brush operation since it is (close to) the thinnest conceivable surface film which prevents cold-welding and permits almost wear-less sliding at modest friction. A disruption of this film can occur when new surfaces are formed in confined conditions which inhibit moisture access, e.g., as in sliding wear under high pressures or in "fretting" wear. Disruption can also occur if the moisture layer may be evaporated through local heating near contact spots and may then not be replenished, or not be replenished fast enough. This depletion can be especially deleterious if ambient relative humidity is low or if a film is covering the contact region preventing ingress of moisture.

Unless other contact lubricants are present, wherever the critical double molecular moisture layer is disrupted or absent, cold-welding occurs virtually instantaneously with the corresponding severe wear damage. The supply of an adequate amount of moisture is, thus, essential for unlubricated metal fiber brush operation. And this is true whether or not they are used in conjunction with graphite since the graphite also requires moisture if it is not to "dust." The limits of current density, sliding speed, and surface coverage with brushes which may be achieved while yet supplying enough moisture still need to be established.

Real world conditions usually do not allow particularly good control over the atmosphere surrounding the contacts. Ideally, one would like to preserve the "standard" two adsorbed monolayers of water, with $\sigma_F = 10^{-12} \Omega m^2$, without any other surface coverage. However, in practice, virtually all films which form spontaneously from operation in standard environments on metals are insulating, their resistivity rises extremely rapidly with thickness in line with Figure 23.8 (compare also Chapter 1 and Ref. [98], for example), and they in turn are overlaid by adsorbed moisture. In fact, when the moisture is not firmly adsorbed in the metal matrix, it can create problems at temperatures below freezing by creating an insulating "slush." Therefore, even films much too thin to be visible can give rise to brush resistances orders of magnitude higher than our "ideal" layer. The challenge in metal fiber brush construction and operation is how to maintain a reasonable surface film resistivity value, σ_F, in the presence of contaminants that make the formation of insulating solid surface films highly likely. This filming issue is the foremost problem for the use of fiber brushes in the open atmosphere.

23.3.3 Lubrication

Non-tarnishing metals such as aluminum, stainless steel, chromium, and nickel are not useful for low-resistance fiber brushes (although they may have important applications). They do not visibly oxidize because they are protected by an oxide film of 30 Å thickness or more, with the correspondingly high film resistivity. As to other base metals, their oxidation rates can vary greatly even if they are of similar chemical composition. One useful screening parameter for candidate fiber brush or counterface materials is the time dependence of film resistivity of the materials [98].

One solution is the use of noble metals. In particular gold, metals of the platinum group, and a large variety of their alloys are obvious selections although rather expensive. Proper alloy selection and the use of plating instead of solid noble metals can ameliorate this cost disadvantage. This use of noble metals is not a foolproof solution, and sometimes a contact lubricant is required. There are certain speed and electrical power profiles that require some lubrication strategy.

The use of fluid contact lubricants are sometimes required for some fiber brush solutions. As discussed in Chapter 22, one of the primary advantages of fluid lubricants is the establishment of a known film in the contact region. Kuhlmann-Wilsdorf points to two problems with fluid lubricants: (1) over the long sliding distances involved, they do not protect well enough against oxidation, and (2) since fluid lubricants are more viscous than water, they form thicker layers and, thereby, raise the film resistivity substantially. Both of these problems do exist, but they do not represent prohibitive problems.

Most contact lubricants will not protect copper and copper alloys from oxidation over time. This is the problem with using copper in applications that are sensitive to contact resistance in standard atmospheres with metal brushes. It is also true that there is normally a slightly higher contact resistivity with lubricated contacts than with unlubricated ones; however, in most practical cases this slightly higher value is acceptable. The true advantage of fiber brushes in most real world applications is the reduced wear debris and improved life and reliability. Even with a slight increase in film resistivity, the current density is higher than metal graphite monolithic brushes. Boundary lubrication with contact lubricants extends the usefulness of fiber brushes into industrial applications where moist argon and CO_2 environments are not practical.

There is a fairly significant range of applications operating in standard industrial or military environments that utilize the advantages of fiber brushes benefiting from the use of noble metal contacts and specialty contact lubricants.

There are also options involving lubrication with solid lubricants. In principle a variety of different surface films could be considered, including MoS_2. However, by far the most widespread electrically conductive protective layer to inhibit insulating film formation and facilitate electrical contacts is graphite. It is graphite that permits operation of monolithic brushes in the open atmosphere. Its advantages include freedom from environmental problems (except in relative humidity conditions of less than 15% or greater than 85%), nontoxicity to plants, animals, and humans, resistance to atmospheric attack, and being inexpensive and widely available. There has been some promising work on the in situ lubrication of brushes with graphite for low speed and low current densities [6,99]. Most of these tests were performed on solid metal brushes, and the testing that was done on a fiber brush was performed in moist CO_2 atmosphere. Preliminary data suggests that the film resistivity is "high" resulting in low current density. It should be appreciated that "low current density" is relative. The low current density is the result of the higher resistivity of the graphite film which also exists with the next best option, metal graphite brushes, so, in fact, the "low current density" is relative to an ideal standard that is unattainable in standard environment [99].

The important point to understand in regards to surface films and lubrication is that moving fiber brushes into real world problem solutions involves making sound engineering trade-offs. In environments where contamination exists (i.e., most industrial environments), it is unlikely that the optimum 10^{-12} Ω m^2 surface resistivity can be achieved. So the engineering issue is one of the combinations of materials, lubricants, and design features can be brought to bear to solve a real-world sliding electrical contact problem with a good solution—maybe not a perfect solution, but a good solution.

23.4 Electrical Contact

In general, the electrical resistance of brushes, R_B, consists of three parts,

$$R_B = R_0 + R_F + R_C \qquad (23.9)$$

or the sum of the resistance of the brush body, R_0, the surface film resistance, R_F, and the "constriction resistance," R_C. All three values are significant and need to be considered in any brush design. The constriction resistance, R_C, is owing to the fact that all of the current has to pass through the contact spots. The current flow lines, therefore, have to constrict as they pass through the contact spots (as derived in considerable detail in Chapter 1). As a result, independent of brush type, indeed for any two objects which touch, the electrical effect of the current constriction through mechanical contact spots is much the same as if the contact spots were thin wires of a length equal to the contact spot diameter, $2a$. For multiple contact spots starting with the standard Tabor equation $A = F/H$, the constriction resistance is approximately,

$$R_C \approx \rho_a (A_c/n)^{1/2} / A_c = \rho_a / (A_c n)^{1/2} \approx \rho_a (H/Pn)^{1/2} \qquad (23.10)$$

Here, ρ_a is the averaged electrical resistivity of the two sides of the contact spots. On account of large n and small ρ_a for metal fiber brushes, R_c is negligible even if their running surface, A_B, should be quite small. For example, for $A_B = 0.1$ cm^2 = 0.015 in^2 with $d = 50$ Mm fibers occupying $f = 10\%$ of the volume, $n \sim 500$.

Film resistance cannot be avoided. Some nonmetallic film is necessary to prevent cold-welding and, thus, catastrophic gouging wear. If insulating, the current conduction through the thin films takes place by electron tunneling, and the dependence of constriction resistance on film thickness is very steep indeed, as shown in Figure 23.8 owing to Holm (Figure 26.11, p. 126 of Ref. [24]), but it is independent of the chemical nature of the film.

Kuhlmann-Wilsdorf discusses annular tunneling about contact spots as an advantage of fiber brushes [1,100]. Namely, the load-bearing part of a Hertzian contact spot is surrounded by an annular gap within which the separation between the two sides gradually increases. Current conduction is, therefore, not strictly limited to the load-bearing area, but occurs with rapidly diminishing intensity also via tunneling through the annular zone. In first approximation it may extend to a distance at which by purely geometrical construction the gap width, s, has increased to $s^* \approx 5$ Å [36,38]. Holm (p. 132 of Ref. [24]) had considered the effect theoretically, but concluded that it would always be insignificant. In the case of fiber brushes where the number of contact points is increased significantly above the number of contacting asperities in a monolithic brush, it could be expected that the effect could be moderately useful in increasing current capacity. References [36,38] can be used in the numerical evaluation.

23.4.1 Dependence of Electrical Resistance on Fiber Brush Construction

The brush resistance, R_B, of a fiber brush of macroscopic area, A_B, with elastic contact spots, subject to macroscopic pressure, p_B, is found in a straightforward manner using [36, 38] as:

$$R_b = \left(\rho_F \middle/ A_b K^2\right) \left\{ \left[\left(E \middle/ p_b\right)^2 \left(d \middle/ r_c\right)^2 \middle/ 70f\right] \right\}^{\frac{1}{3}}$$

(23.11)

The factor K^2 is used to correct the film resistivity, ρ_F, for peripheral tunneling and a detailed description can be found in the references.

The assumption of this relationship is that the bulk resistance of the brush and the constriction resistance are negligible and that the brush resistance is equal to the film resistance, or, $R_B = R_F$. Not unexpectedly, except for the complication of the factor K^2, the fiber brush resistance is proportional to ρ_F/A_b. Figure 23.9 presents Equation 23.11 for three different fiber diameters, namely 50 μm., 20 μm., and 10 μm., respectively. At still higher fiber diameters, there is almost no further difference in the diagrams since K^2 is close to unity. Actual measurements are in accord with theory. Experimental data [36,38] are presented in Figure 23.10, part of which (light symbols) pertain to the same brushes and in the same tests used for Figure 23.5 (slip ring). There is good correlation between theoretical values of Figure 23.9 and experimental values of Figure 23.10. In real world situations where R_c and R_o are not

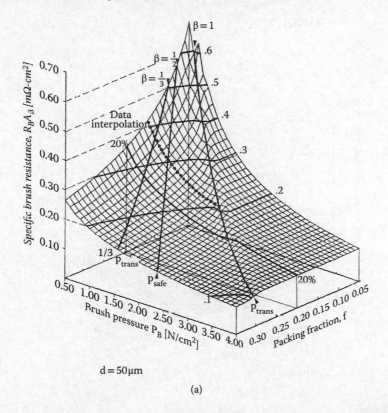

$d = 50\mu m$

(a)

FIGURE 23.9
Theoretical specific brush resistances ($R_B A_B$) as a function of pressure ($p_B = P/A_B$) and packing fraction (f) for fiber diameters of (a) $d = 50$ Mm, (b) $d = 20$ Mm, and (c) $d = 10$ Mm, according to Equation 23.11, for copper ($E = 1.2 \times 10^{11}$ N m^{-2}). The highlighted packing fraction of $f = 20\%$ and brush pressures between $p_{safe} = p_{trans}/2$ and $p_{trans}/3$, that is, $\frac{1}{2} \geq \beta \leq \frac{1}{3}$, are expected to be typical of best performance. The dotted line in (a) is the data interpolation line in Figure 23.6. Since brush resistance depends on Young's modulus only as $E^{1/3}$, this figure is fairly representative also for other common fiber materials, for example, silver and gold.

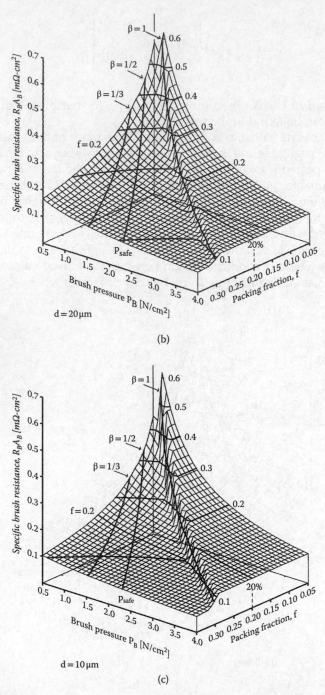

(b)

(c)

FIGURE 23.9 (Continued)

Theoretical specific brush resistances ($R_B A_B$) as a function of pressure ($p_B = P/A_B$) and packing fraction (f) for fiber diameters of (a) $d = 50$ μm, (b) $d = 20$ μm, and (c) $d = 10$ μm, according to Equation 23.11, for copper ($E = 1.2 \times 10^{11}$ N m^{-2}). The highlighted packing fraction of $f = 20\%$ and brush pressures between $p_{safe} = p_{trans}/2$ and $p_{trans}/3$, that is, $\frac{1}{3} \geq \beta \leq \frac{1}{2}$, are expected to be typical of best performance. The dotted line in (a) is the data interpolation line in Figure 23.6. Since brush resistance depends on Young's modulus only as $E^{1/3}$, this figure is fairly representative also for other common fiber materials, for example, silver and gold.

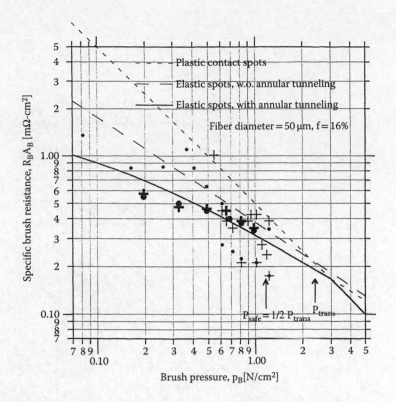

FIGURE 23.10

Specific resistances of four similar $d = 50\ \mu m$, $f \approx 15\%$ brushes in humid CO_2. Each point represents the slope of a voltage–current curve drawn through measured data, taken in short succession at fixed speed and pressure with the current varied between about 10 and 90 Å. Heavy and light symbols refer to two different brush pairs in separate, uncooled testers with somewhat different loading springs and accuracies of force measurements. The heavy symbols pertain to the same brushes as Figure 23.2. Each pair consisted of one brush made of bare copper fibers, and one made of silver-plated copper fibers; one brush pair was tested on a bare copper rotor, the other on a gold-plated rotor. None of these minor permutations appears to have affected the brush resistance. The interpolation curves represent theoretically expected values for plastic contact spots with $H = 5 \times 10^8\ N\ m^{-2}$ (- - -), and for elastic spots without (- -) and with (——) peripheral tunneling according to Equation 23.27, that is, assuming $r_c = d/2$ and $\sigma_F = 10^{-12}\ \Omega\ m^2$ The latter is shown as the dotted line in Figure. 23.5(a). Dots (• and ●) indicate negative polarity, plus signs (+ and +) positive polarity. Ohm's law was well obeyed.

negligible (longer bundles with fewer fibers), the bulk resistance of the brush and constriction resistance will have to be added to the film resistance to obtain the total brush resistance.

23.5 Brush Dynamics

Slip ring engineers are recognizing the importance of brush dynamics on the electrical and wear performance of slip rings and other sliding electrical contacts. Sliding contacts are spring loaded against their counterface. In the case of fiber brushes, this spring loading accommodates brush length changes as well as mechanical tolerances resulting in run-out (lack of perfect concentricity and irregular surface). The spring needs to ensure that the contact has the required brush pressure at the limit of brush wear and through the full extent of run-out tolerance. There is some spring loading from the fibers themselves as they are deflected onto the counterface. Figure 23.11 shows a diagram of a single fiber

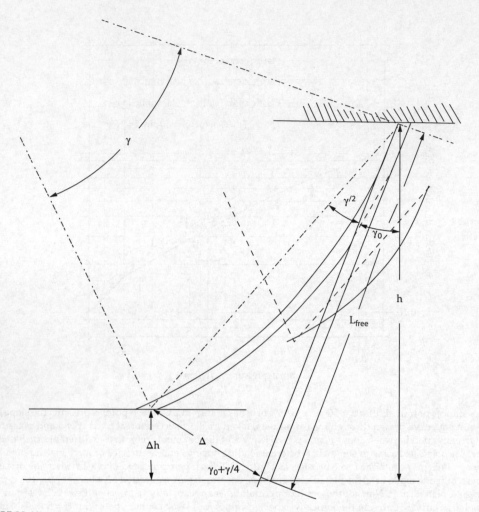

FIGURE 23.11
Geometry of free fiber length in a brush inclined by angle γ_0 the contact surface and elastically bent through angle γ by its share of the applied brush force.

deflected and the force and dynamic properties of the spring can be calculated from standard beam deflection equations. As a matter of fact, most of the fiber brush work in the literature uses brushes that have no other spring mechanism. A certain force is exerted on the brush, and the brush is then locked into position. This is a useful approach as a research technique since precise force can be established, and it's reasonable to assume a low run-out on the counterface.

Incorporating fiber brushes into diverse applications, however, must involve a method of producing a reasonably constant force on the contact points through a wider range of tolerances and wear than this "hard mount" method can accommodate. The loading technique commonly used is though a spring. The loading spring can be a cantilever (see Figure 23.12b or a plunger style (Figure 23.12a). The point of this design is that the fibers themselves do not need to accommodate the total travel required during the life of the contact. This removes a design constraint that would make the design of a fiber brush almost impossible for certain applications. For example, the fiber brush similar to that shown in

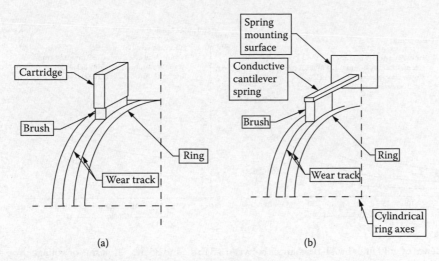

FIGURE 23.12

(a) and (b) show spring assemblies for a fiber brush as both a cartridge style or a cantilever style brush. The pictures are the same for either a composite brush or a fiber brush with the difference being the actual construction of the brush. The spring loading strategy is the same. Illustrations are from Chapter 22.

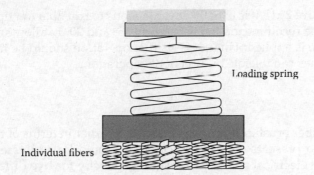

FIGURE 23.13

This diagram shows the spring model for a fiber brush that is independently loaded with a secondary spring. The secondary spring removes the design constraint that the compliance of the individual brush springs needs to accommodate all wear, mechanical alignment, and inertial forces.

Figure 23.12b operates on a 1.5 m diameter slip ring used on a CT machine with the potential of over 1.0 mm of run-out. Designing a fiber brush without a primary loading spring would be very difficult.

The addition of the second loading spring does make the dynamic modeling more difficult. Figure 23.13 new brush spring model gives the schematic of this approach showing small parallel brush springs in series with the larger loading spring. The larger spring compliance needs to be higher than the compliance of the total of the parallel compliance of the small springs. Constant force springs can be used to provide a constant force through the complete range of travel, although the most common approach is to use a linear spring with sufficient compliance (inverse of spring rate) to provide an acceptable range of forces through the total travel. In the case of high speed operation, analysis should be performed to ensure that dynamic effects do not reduce the brush pressure below the minimum level of acceptable performance.

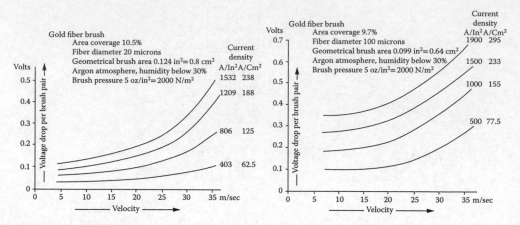

FIGURE 23.14

Speed dependence of gold fiber brush resistances between 3.5 m s⁻¹ and 36 m s⁻¹ in terms of voltage drop across different brush pairs at the indicated current densities in humid argon. (From C. Adkins and D. Kuhlmann-Wilsdorf, "Development of High Performance Metal Fiber Brushes II—Testing and Properties," presented at the Holm Conference on Electrical Contacts—1979, Chicago IL, 1979 Figure 6 [36]. With permission.)

As suggested in Figure 23.11, it is usually advantageous to run fiber (on tip) brushes with a trailing angle and the recommendation is between 15° and 20° to allow smooth bending of the fiber when force is applied. This suggests that operation should be in one direction only. Tangential brushes can tolerate bi-directional operation.

23.5.1 Speed Effect

The sliding speed of fiber brush contacts was discussed earlier in terms of thermal effects. But sliding speed also presents dynamic issues that should be addressed. Tests showing the dependence of electrical resistance on sliding velocity for two different gold-fiber brush pairs are shown in Figure 23.14. These brushes were tested under the same conditions, that is, in moist argon at $p_B = 2 \times 10^3$ N m⁻² (corresponding to $\beta = 0.2$), and were of similar construction with $f \approx 10\%$, but had different fiber diameters of $d = 20$ Mm and 100 Mm, respectively. It is obvious at first glance that the resistance under the conditions of Figure 23.8 rose substantially with speed, just about quadrupling between 5 and 35 ms⁻¹ for the thinner fibers on the left and doubling for the 100 Mm fibers on the right. The likely explanation is hydrodynamic lift from surface films [36,38].

23.6 Future

There are a number of very promising future directions for fiber brushes and most of them involve research into characterizing the performance of fiber brushes that can meet the requirements of standard industrial environments. This should extend the material range, lubrication options, design parameters (e.g., force, spring compliance, number, and length of fibers), and operational parameters of fiber brushes [100]. Careful attention should be given to the same developments listed in Chapter 22 from the nanotribologists that should provide support for lubrication, material, and design improvements to metallic sliding contacts.

23.7 Summary

Fiber brush technology can provide a solution for sliding contact applications that require low wear debris generation, long life without maintenance, and good electrical current density by utilizing multiple fibers impinging on the counterface with light contact force. Although much of the formal research has been with controlled environments, there are industrial and defense applications that support a more general application.

Acknowledgments

Dr. Kuhlmann-Wilsdorf started her fiber brush development about the same time I started designing slip ring assemblies. Dr. Norris Lewis and other colleagues in Blacksburg, VA, were familiar with her work, and we started integrating fiber brush techniques into our slip ring designs although our approach diverged significantly from her guidelines. Dr. Kuhlmann-Wilsdorf recently passed away, and I hope in this update to her chapter to show the broader application of her theories to general slip ring applications than she indicated in the original chapter. To do this, I have de-emphasized the considerable body of funded work done for the Navy in support of very high current requirements. As I said in the introduction, this very high current work pushed many of the advantages of the fiber brush approach into the background. But the general application of the fiber brush basics do require "push back" on many of Dr. Kuhlmann-Wilsdorf's main assumptions. The extensive equation derivation in the original chapter were for the very specific Kuhlmann-Wilsdorf fiber brush design and have very limited application to the problem of designing a fiber brush for general use. I have significantly reduced the number of equations and instead provided references for those interested. I have tried to show how Dr. Kuhlmann-Wildsdorf's basic principles have a much broader scope and application than she acknowledged. Those readers interested in fiber brushes for homopolar motors will probably be disappointed; readers interested in a modestly high current density, high reliability, low maintenance, long life brush for general industrial environments should be encouraged. Thanks to Becky Wills for editorial and graphic assistance.

Appendix A.1: Collection of Important Equations

The equations found in this appendix are a comprehensive collection of relationships derived by Kuhlmann-Wilsdorf for a specific fiber brush design [101]. The assumptions made for this brush design are listed in the Introduction but are included here for ease of reference. References are included that provide a more thorough discussion of the equations and their derivations. Appendix A.2 provides the description of the symbols used.

1. Fiber brushes are used for transmitting high current and the critical electrical parameters is related to conductivity.

2. Fiber brushes are relatively large bundles of very small (sub 100 μm. diameter), independently-acting fibers with their tips impinging on the counterface. The brush has been described as a "velvet-like mat" made of fibers.

3. The bulk brush length is short and the bulk resistance of the brush is insignificant compared to the contact resistance.

4. The environment and consequentially surface films are well controlled and the resistivity is roughly 10^{-12} ohms/m^2 or equal to 2 molecular layers of water. In most cases this requires a controlled atmosphere to achieve.

5. Fluid lubrication is never used.

6. Normal forces are equally distributed across individual fibers and the fibers are small enough that there is only one contact spot per fiber.

7. Contact spots can be evaluated using the Hertz criteria as elastic or plastic and a critical brush pressure can be determined where the contacts transition from elastic to plastic; low wear occurs under this transition pressure.

All Fibers Touch the Substrate, One Elastic Contact Spot per Fiber

Basic Properties of Contact Spots Ref. [3, 36, 38, 41]

Number of contact spots per unit area: $n^* = f / (\frac{1}{4}\pi d^2)$

Contact spot radius:

$a_{el} = 1.1(r_c p_B / n^* E)^{1/3} = 1.1d(\pi \beta 1.2 \times 10^{-6}/8)^{1/3} \approx 8.6 \times 10^{-3}\beta^{1/3}d$

"Transition" and "safe" pressures: $p_{trans} = 2p_{safe} \approx 3 \times 10^{-4} f H \approx 1.2 \times 10^{-6} f E$

Total contact spot area: $A_C = n^* A_B a_{el}^2 \approx 3 \times 10^{-4} \beta^{2/3} f A_B$

Pressure at contact spot: $_{el}pc(\beta) = \beta^{1/3}H \approx 0.04\beta^{1/3}E$

Force per fiber-end: $F_{fiber} = \beta \pi d^2 p_{trans}/4f = 1.2 \times 10^{-6} \beta \pi d^2 E/4$

Electrical Resistances Ref. [3, 36, 38, 41]

Brush resistance = $R_0 + R_C + R_B \approx R_B$

Brush body resistance: $R_0 = \rho_B L_{fiber}/f A_B$

Brush constriction resistance: $R_C = \rho_a/(nA_C)^{1/2} = 51\rho_a d/(\beta^{1/3} f A_B)$

Film resistance: $R_B = (\sigma_F/A_B K^2)\{(E/\rho_B)^2(d/r_c)^2/(70f)\}^{1/3}$

R_B for the typical case that $r_c = d/2$ and $\sigma_F = 10^{-12} \Omega$ m^2, using $p_B = \beta p_{trans}$

$R_B \approx (\sigma_F/A_B K^2) 3.4 \times 10^3/f \ \beta^{2/3} \approx 0.034 \, [\text{m}\Omega\text{cm}^2]/(K^2 f \ \beta^{2/3}$

Film resistivity for $r_c = d/2$ and $K^2 = 1.15$:

$\sigma_F = 3.4 \times 10^{-4} f R_B[\text{m}\Omega]A_B[\text{cm}^2]\beta^{2/3}$ from (23.27)

Equivalent length (length of a solid piece of metal of same electrical resistance):

$R_{equiv} = \rho_B L_{equiv}/A_B$

Equivalent length of body resistance: $iL_{O\,/equiv} = L_{fiber}/f = 5 L_{fiber}/\phi$

Equivalent length of constriction resistance, in general and specifically for

$\rho_a = \rho_B$ with $\beta = \frac{1}{2}$:

$iL_{C,equiv} = 51 \ d(\rho_a/\rho_B)(\beta^{1/3} f) = 0.016(\delta/\phi)[m]_{(\rho a=\rho b, \beta=\frac{1}{2}}} = 1.6[cm]_{standard}$

As above, but for the film resistance in general and with

$\rho_B = 1.65 \times 10^{-8} \Omega m$ and $\beta = 1/2$:

$iL_{F,equiv} = 3.4 \times 10^3 (\sigma_0/\rho_B K^2)(\beta^{2/3} f) = 1.03s/(\beta^{2/3}\phi)[m] = 1.64[m]_{standard}$

Wear [3, 36, 38, 41]

$w_o = (\Delta \ell/L_s)_o \approx 10^{-9} \beta^{4.6}$

Losses, Flash Temperatures, and Bulk Heating Through the Brush Body **Ref.** [3, 41, 44, 55–60]

Mechanical loss per unit of current: $L_M = \mu v P/i = \mu v p_B/j = W_M/j$

Electrical loss per unit of current: $L_E = R_B i = R_B A_B j = W_E/j$

Total power loss per unit area: $W = W_M + W_E = \mu p_B v + R_B A_B j^2$

$= 3 \times 10^{-4} \beta f \ H \mu v + (\sigma_F/K^2) 3.4 \times 10^3 j^2/(f\beta^{2/3})$

Flash temperature:

$\Delta T = [F(Z,S)/\lambda_B]\{6.8 \times 10^{-3} \beta^{2/3} H \mu v d + (\sigma_F/K^2)(7.71 \times 10^4 d \ j^2/f^2 \beta)\}$

Relative pressure for which $L_M = L_E$, at which W as well as ΔT are close to their minimum

$\beta \min \approx 1.71 \times 10^4 (\sigma_F/K^2)j^2/(H\mu v f^2)\}^{3/5}$

Flash temperature near β_{min}:

$\Delta T_{min} = [F(Z,S)/\lambda_B]9.02\{\sigma_F^2 H^3 \mu^3/f^2 K^2\}^{1/5} dv^{3/5} j^{4/5}$

As above, but for "standard brush" at slow speed: $\Delta T_{min}^* = 3.01 \times 10^{-6} j^{4/5} v^{3/5}$

Total equivalent voltage loss for "standard brush" near β_{min}:

$L_{T,min}^* = 1.76 \times 10^{-3} (jv^2)^{1/5}$

Relative contribution to flash temperature through brush bulk heating:

$\Delta T_{ave}/\Delta T_{spot} = (v_{ave}/v_{spot})0.78[cm^{-1}]\phi\beta^{1/3} D_{\parallel}/\delta$

Heat conduction through brush body:

$W_{cond} = f\lambda_B \Delta T_{cond}/L_{fiber} = 72\phi\ell(\Delta T_{cond}/L_{fiber})$

Equivalent length in regard to heat transfer through fiber brush material:

$_h L_{F,equiv} = 3.4 \times 10^3 (t_{Film}\lambda_B/\lambda_F K^2)/(\beta^{2/3} f) = (1/\phi)1.7 \times 10^7 \times t_{Film}[m]_{standard}$

Brush Body Construction: Only the Fraction ξ of Fibers "Track" Ref. [101]

Body brush resistance: $R_O(\xi) = R_O/\xi$

Constriction resistance: $R_C(\xi) = R_C/\xi^{2/3}$

Film brush resistance: $R_B(\xi) = R_B/\xi^{1/3}$

Equivalent lengths of

body resistance: $_i L_{o,equiv}(\xi) = 5L_{fiber}/\xi\phi = \{5L_{fiber}/\xi\}_{standard}$

constriction resistance:

$_i L_{C,equiv}(\xi) = 0.016\delta/(\phi\xi^{2/3})[m]_{(\rho_a=\rho_b, \beta=1/2)} = \{0.016/\xi^{2/3}\}[m]_{standard}$

film resistance: $iL_{F,equiv}(\xi) = 1.03s/(\beta^{2/3}\phi\xi^{1/3}[m] = \{1.64/\xi^{1/3}\}[m]_{standard}$

Value of ξ below which the body resistance becomes larger than the film

resistance: $\xi_{o=F} \approx \{5 \ L_{fiber}[m]/1.64\phi[m]\}^{3/2}$

Wear: $(\Delta \ell/L_s)_\xi \approx (\Delta \ell/L_s)o/\xi^{4.6} \approx 10^{-9}(\beta/\xi)^{4.6}$

Variation of wear rate with small variations of local pressure:

$\delta(\Delta \ell/L_s)/(\Delta \ell/L_s) = (2 \times 4.6 \times 5.6)\delta\beta/\beta = 51.5\delta\beta/\beta$

Brush Body Construction: Limitations Through Elastic Fiber Flexing **Ref.** [101]

Difference of gap width through surface undulation, accommodated through

fiber flexing of: $\Delta h = \Delta \sin(\gamma_0 + \gamma/4) \approx \Delta\sqrt{2}$

Average fiber flexing owing to brush pressure ($A = L_{free}/d$ = aspect ratio):

$\Delta = 1.95 \times 10^{-6} \beta L_{free}^3/d^2 = 1.95 \times 10^{-6} \beta A^2 L_{free} = 1.95 \times 10^{-6} \beta A^3 d$

Maximum amplitude of substrate undulations compatible with full "tracking" (maximum elastic strain
$2 \times 10^{-4} \leq \epsilon_{max} \leq 5 \times 10^{-3}$):
$\Psi_{max}[\text{rad}] = \Delta h / v\tau = 312\, \epsilon_{max} / v = 3.6°/v[\text{ms}^{-1}] \leq \Psi°_{max} \leq 89.4°/v[\text{ms}^{-1}]$
Maximum speed with full tracking in the "fundamental" oscillation mode:
$3.6°[\text{ms}^{-1}]/\Psi° \leq v_{max} \leq 89.4°[\text{ms}^{-1}]/\psi°$

A.2 Symbols

a_{el}	Radius of average elastic contact spot
a_{pl}	Radius of average plastic contact spot
A	Aspect ratio of free fiber length, i.e. L_{free}/d
A_B	Area of macroscopic brush/substrate interface
A_C	Total area of contact spots; P/H or Equation 23.45 in the plastic or elastic case
A_{pl}	Total area of plastic contact spots (equal to P/H)
B	Factor to correct flash temperature for overall brush heating
C	Specific heat
D	Fiber diameter
D	Mechanical density of brush or substrate material
$D_{||} \approx (4A_B/\pi)^{1/2}$	Brush diameter in sliding direction
E	Ellipticity of contact spot (ratio of long to short axis)
E	Averaged Young's modulus of brush and substrate materials
F	Packing fraction, percentage of metal in brush volume at interface
F_{fiber}	Force applied to a fiber end, deflecting it by Δ normal to its axis
$F(Z, S)$	Factor (≤ 1) by which flash temperature is reduced compared to ΔT_0, i.e., of an asperity on an insulating substrate at zero velocity
H	Hardness of brush material in units of $5 \times 10^8 \text{N/m}^2$, the average hardness of copper
H	Meyer hardness of brush or substrate, whichever is the smaller
I	Current conducted through brush
$j = i/A_B$	Current density through brush
K^*	Dimensionless constant in the Holm-Archard wear law
K_{HA}	K^* when expressed in terms of dimensionless wear
K^2	Factor by which σ_F must be divided to correct for peripheral tunneling
K^2_{safe}	Probable value of K^2 at the safe pressure, p_{safe}
ℓ	The thermal conductivity of the brush material in units of λ_{Cu}
L_E	Electrical brush loss per ampere conducted
L_E^*	L_E for the "standard case," that is, $s = h = m = \varphi = \ell = \delta = v = 1$
L_{fiber}	Fiber length from contact spot to brush current connection
L_{free}	Free fiber length between the interface and the nearest fixed point
$hL_{j,equiv}$	Same as $iL_{j,equiv}$ but in regard to heat transfer resistance
$iL_{j,equiv}$	Length of solid fiber brush material of same electrical resistance (subscripts j = O, C, or F for body, constriction and film resistance, respectively)
L_M	Mechanical brush loss per ampere conducted
L_M^*	L_M for the "standard case," i.e., $s = h = m = \varphi = \ell = \upsilon = v = 1$
L_{min}	Minimum loss per ampere conducted, obtained by adjusting pressure to β_{min} so that $L_E = L_M$
L_s	Sliding distance
L_{soft}	Loss per ampere conducted equal to Holm's "softening voltage" (equal to 0.12V for copper)
$L_T = L_E + L_M$	Total brush loss per ampere conducted

$L_{T,min}$	The total loss per ampere conducted at β_{min}, i.e. when the pressure is adjusted to obtain $L_M = L_E$
M	The coefficient of friction in units of 0.35, its conservative value
m^*	Number of spot diameters through which, on average, a contact spot slides before it forms a wear particle
N	Number of contact spots at brush/substrate interface
n^*	Number density of contact spots at brush/substrate interface
P	Macroscopic pressure at interface between sliding solids
p_B	Brush pressure: P/A_B
$p_{Bmin} = \beta_{min}p_{trans}$	Brush pressure at which the total heat evolution is (nearly) minimized
$p_{contact}$	Average local pressure at contact spots
$_{el}p_c = \beta^{1/3}H$	Average local pressure at elastic contact spots
$_{pl}p_C = H$	Average local pressure at plastic contact spots
$p_{safe} = \frac{1}{2}p_{trans}$	Brush pressure at which expected dimensionless wear is in the 10^{-10} range
p_{trans}	Value of p_B (namely $\approx 3 \times 10^{-4} fH$) at the transition between elastic and plastic contact spots
P	Normal force between brush and substrate
P_{trans}	Value of P at p_{trans} i.e., the transition between elastic and plastic contact spots
Q	Local heat evolution at contact spots per unit area and time
q_{ave}	Average heat evolution at brush-substrate interface per unit area and time
R	Characteristic dimension of contact spot (i.e., normally its radius)
r_c	Radius of the asperity forming an elastic, "Hertzian" contact spot (in fiber brushes, normally $r_c \approx d/2$)
R_B	Electrical resistance of a brush, in a well-constructed metal fiber brush equal to R_F
R_{Bmin}	Brush resistance at condition of minimum loss, i.e., at $p_B = \beta_{min}P_{trans}$
R_C	Constriction resistance of a brush
R_F	Film resistance of a brush
R_O	Electrical resistance of the brush body
S	σ_F/K^2 in units of 10^{-12} Ω m^2, the most common value of σ_F
s^*	Gap width through which effective tunneling can take place (~ 0.5 nm)
$S(c,v)$	Function by which the flash temperature is modified on account of contact spot ellipticity
t_{Film}	Thickness of surface film
V	Sliding velocity of brush on substrate
v_{crit}	Highest sliding velocity at which $L_E = L_M$ can be achieved for $V < V_{soft}$
$v_0 = \kappa_B/r$	Characteristic velocity
$v_r = v/v_0$	Relative velocity (Peclet number)
v_{rB}	Relative velocity of contact spots on substrate (almost always $= v_r$)
v_{rS}	Relative velocity of contacts spots on brush (almost always $= 0$)
V	Wear volume
V_{melt}	"Melting voltage" as defined by Holm, see Table 23.2
V_{soft}	"Softening voltage" as defined by Holm, see Table 23.2
W	Heat developed per unit area and time at the brush/substrate interface
W_{cond}	Heat conducted per unit area and time through fiber brush from interface to cooled fiber-end
W_E	Rate of electrical heat evolution, i.e., the part of W owing to Joule heat
W_M	Rate of mechanical heat evolution, i.e., the part of W owing to friction
Z_0	Function describing the velocity dependence of contact spots for $e = 1$ and $\lambda r = 0$
*(Asterisk)	Pertaining to the "standard brush" with $s = h = m = \varphi = \ell = \delta = v = 1$
A	Number of contact spots per fiber end
B	Ratio p_B/p_{trans}
β_{min}	The value of β yielding equal friction and Joule heating, approximating the condition of minimum flash temperature as well as minimum heat loss

β_{safe}	The β-value (namely $\frac{1}{2}$) believed to confer wear rates $\Delta\ell/L_s < 10^{-10}$
Γ	Average bending angle of the free fiber lengths on account of applied force
γ_0	Angle of inclination of brush body relative to the sliding interface
Δ	The fiber diameter in units of 50 Mm, its most typical value
$\Delta\beta$	Incidental local change of β at fiber-ends on account of surface undulations
Δ	Elastic deflection of fiber end at interface, normal to L_{free}
Δh	Decrease of distance between the interface and the nearest fixed fiber point on account of elastic flexing of fiber length L_{free}
$\Delta\ell$	Reduction of brush length through wear
$\Delta\ell/L_s$	Dimensionless wear rate
ΔT	"Flash temperature": Rise of contact spot temperature above ambient
ΔT_{ave}	Temperature rise at fiber brush as a whole
ΔT_{cond}	Temperature drop owing to heat conduction from interface to cooled end of fiber brush
$\Delta T_{cor} = B\Delta T$	Flash temperature corrected for superimposed overall brush heating
ΔT_E	Contribution to flash temperature through Joule heat
ΔT_E^*	ΔT_E for the "standard case" $of\, s = h = m = \varphi = I = \ell = v = 1$
ΔT_M	Contribution to flash temperature through friction heat
ΔT_M^*	ΔT_M for the "standard case" $of\, s = h = m = \varphi = \ell = \delta = v = 1$
ΔT_{min}	Near-minimum flash temperature, obtained at $W_M = W_E$ or equivalents $L_M = L_E$, at which also W is nearly minimized
ΔT_{min}^*	ΔT_{min} for the "standard case" of $s = h = m = \varphi = \ell = \delta = v = 1$
$\Delta T_0 = \pi q r / 4\lambda_B$	Flash temperature at zero velocity on a thermally insulating substrate
$_M\Delta T_{ave}$	Temperature rise at monolithic brush as a whole
$_M\Delta T_{spot}$	Flash temperature at contact spot of monolithic brush
$\Delta\rho_a$	Averaged electrical resistivity of brush and substrate materials
$\kappa = \lambda/Dc$	Thermal diffusivity
$\kappa_B = \lambda_B/D_B c_B$	Thermal diffusivity of brush material
Λ	Thermal conductivity
λ_B	Thermal conductivity of brush material
λ_F	Thermal conductivity of surface film material
$\lambda_r = \lambda_S/\lambda_B$	Relative thermal conductivity of substrate material relative to brush material
λ_S	Thermal conductivity of substrate material
M	Coefficient of friction
V	$F(Z, S)$ in units of $\frac{1}{2}$, its value for copper slowly sliding on copper
v_{ave}	v for the brush as a whole
v_{spot}	v for the average individual contact spot
Ξ	Fraction of fiber ends which actually touch the substrate
ρ_a	Average electrical resistivity of substrate and fiber material
ρ_B	Electrical resistivity of fiber material
σ_F	Specific resistivity of surface film at contact spots, in units of flm^2
σ_F/K^2	Surface film resistivity as modified by peripheral tunneling
τ	Period of mechanical oscillation of fiber ends
Φ	The packing fraction, f, in units of 0.2, its "standard" value
Ψ	Angular local deviation from planarity of the interface

References

1. D. Wilsdorf, "Versatile Electrical Fiber Brush and Method of Making," USA Patent 4,358,699, Nov. 9, 1982, 1982.
2. N. Argibay, J. A. Bares, J. H. Keith, G. R. Bourne, and W. G. Sawyer, "Copper–beryllium metal fiber brushes in high current density sliding electrical contacts," *Wear*, vol. 268, pp. 1230–1236, 2010.
3. D. Kuhlmann-Wilsdorf, "Electrical fiber brushes-theory and observations," *Components, Packaging, and Manufacturing Technology, Part A, IEEE Transactions on*, vol. 19, pp. 360–375, 1996.
4. J. A. Bares, N. Argibay, N. Mauntler, G. J. Dudder, S. S. Perry, G. R. Bourne, and W. G. Sawyer, "High current density copper-on-copper sliding electrical contacts at low sliding velocities," *Wear*, vol. 267, pp. 417–424, 2009.
5. N. Argibay, J. A. Bares, and W. G. Sawyer, "Asymmetric wear behavior of self-mated copper fiber brush and slip-ring sliding electrical contacts in a humid carbon dioxide environment," *Wear*, vol. 268, pp. 455–463, 2010.
6. J. A. Bares, "Lubrication of High Current Density Metallic Sliding Electrical Contacts," PhD, University of Florida, 2009.
7. N. Argibay and W. G. Sawyer, "Low wear metal sliding electrical contacts at high current density," *Wear*, vol. 274–275, pp. 229–237, 2012.
8. G. Dorsey, Application Note 207: Fiber Brushes: The Low Maintenance, Long Life, High Power Slip Ring Contact. Blacksburg, VA, Moog Components Group, 2005.
9. G. Dorsey, Application Note 203: Fiber Brushes: The Maintenance-Free Wind Turbine Slip Ring Contact Material. Blacksburg, VA, Moog Components Group, 2005.
10. N. Lewis and J. A. Skiles, "Fiber Brush Slip Ring Assembly," USA Patent 4398113, Aug 9, 1983, 1983.
11. E. Rabinowicz, *Friction and Wear of Materials*. New York: John Wiley & Sons, 1965.
12. P. Blau, *Friction and Wear Transitions of Materials*. Park Ridge, NJ: Noyes Publications, 1989.
13. N. A. Alcantar, C. Park, J.-M. Pan, and J. N. Israelachvili, "Adhesion and coalescence of ductile metal surfaces and nanoparticles," *Acta Materialia*, vol. 51, pp. 31–47, 2003.
14. B. Bhushan, J. Israelachvili, and U. Landman, "Nanotribology: friction, wear and lubrication at the atomic scale," *Nature*, vol. 374, pp. 607–616, 1995.
15. M. L. Gee, P. M. McGuiggan, J. N. Israelachvili, and A. M. Homola, "Liquid to solidlike transitions of molecularly thin films under shear," *The Journal of Chemical Physics*, vol. 93, pp. 1895–1906, 1990.
16. A. M. Homola, J. N. Israelachvili, P. M. McGuiggan, and M. L. Gee, "Fundamental experimental studies in tribology: The transition from "interfacial" friction of undamaged molecularly smooth surfaces to "normal" friction with wear," *Wear*, vol. 136, pp. 65–83, 1990.
17. J. Israelachvili, S. Giasson, T. Kuhl, C. Drummond, A. Berman, G. Luengo, J. M. Pan, M. Heuberger, W. Ducker, and N. Alcantar, "Some fundamental differences in the adhesion and friction of rough versus smooth surfaces," in *Tribology Series*. vol. 38, D. Dowson, Ed., Elsevier, 2000, pp. 3–12.
18. J. Israelachvili, P. M. McGuiggan, M. Gee, A. M. Homola, M. O. Robbins, and P. A. Thompson, "Liquid dynamics in molecularly thin films," *Journal of Physics. Condensed Matter*, vol. 2, pp. 89–98, 1990.
19. J. N. Israelachvili, P. M. McGuiggan, and A. M. Homola, "Dynamic properties of molecularly thin liquid films," *Science*, vol. 240, pp. 189–191, 1988.
20. M. Urbakh, J. Klafter, D. Gourdon, and J. Israelachvili, "The nonlinear nature of friction," *Nature*, vol. 430, pp. 525–528, 2004.
21. G. G. Adams, "The Mechanics of Adhesion: A Tutorial," *ASME Conference Proceedings*, vol. 2008, pp. 87–89, 2008.
22. G. G. Adams, "The mechanics of adhesion: current and future research trends," *ASME Conference Proceedings*, vol. 2008, pp. 91–93, 2008.
23. R. Holm, "Über Metallische Kontaktwiderstände. Wiss Veröff.," Siemens Werke 7/2: 217,1929.

24. R. Holm, *Electric Contacts, Theory and Applications*, 4th ed. Berlin, Heidelberg, New York: Springer-Verlag, 1967.
25. J. F. Archard, "Contact and rubbing of flat surfaces," *Journal of Applied Physics*, vol. 24, pp. 981–988, 1953.
26. X. Yin and K. Komvopoulos, "An adhesive wear model of fractal surfaces in normal contact," *ASME Conference Proceedings*, vol. 2008, pp. 529–531, 2008.
27. Y. J. Chang and D. Kuhlmann-Wilsdorf, "A case of wear particle formation through shearing-off at contact spots interlocked through micro-roughness in "adhesive" wear," *Wear*, pp. 175–197, 1987.
28. J. Song and D. Srolovitz, "Material effect in material transfer in mechanical contacts," *Acta Materialia*, vol. 54, pp. 5305–5312, 2006.
29. Y. Du, L. Chen, N. E. McGruer, G. G. Adams, and I. Etsion, "A finite element model of loading and unloading of an asperity contact with adhesion and plasticity," *Journal of Colloid and Interface Science*, vol. 312, pp. 522–528, 2007.
30. Y. Du, G. G. Adams, N. E. McGruer, and I. Etsion, "A parameter study of separation modes of adhering microcontacts," *Journal of Applied Physics*, vol. 103, pp. 064902–064909, 2008.
31. D. Maugis and H. M. Pollock, "Surface forces, deformation and adherence at metal microcontacts," *Acta Metallurgica*, vol. 32, pp. 1323–1334, 1984.
32. R. L. Jackson and I. Green, "On the modeling of elastic contact between rough surfaces," *Tribology Transactions*, vol. 54, pp. 300–314, 2011/01/10 2011.
33. M. Ciavarella, C. Murolo, and G. Demelio, "On the elastic contact of rough surfaces: Numerical experiments and comparisons with recent theories," *Wear*, vol. 261, pp. 1102–1113, 2006.
34. G. Carbone and F. Bottiglione, "Asperity contact theories: Do they predict linearity between contact area and load?," *Journal of the Mechanics and Physics of Solids*, vol. 56, pp. 2555–2572, 2008.
35. X. Yin and K. Komvopoulos, "An adhesive wear model of fractal surfaces in normal contact," *International Journal of Solids and Structures*, vol. 47, pp. 912–921, 2010.
36. C. Adkins and D. Kuhlmann-Wilsdorf, "Development of High Performance Metal Fiber Brushes II—Testing and Properties," *Presented at the Holm Conference on Electrical Contacts—1979*, Chicago IL, 1979.
37. C. Adkins and D. Kuhlmann-Wilsdorf, "Development of High Performance Metal Fiber Brushes 1—Background and Manufacture," *Holm Conference on Electrical Contacts*, pp. 165–170, 1979.
38. C. Adkins and D. Kuhlmann-Wilsdorf, "Development of High Performance Metal Fiber BrushesIII—Further Tests and Theoretical Evaluation," *Holm Conference on Electrical Contacts*, pp. 67–72, 1980.
39. W. E. Wilson, S. V. Angadi, and R. L. Jackson, "Surface separation and contact resistance considering sinusoidal elastic–plastic multi-scale rough surface contact," *Wear*, vol. 268, pp. 190–201, 2010.
40. W. Yan and K. Komvopoulos, "Contact analysis of elastic-plastic fractal surfaces," *Journal of Applied Physics*, vol. 84, pp. 3617–3624, 1998.
41. D. Kuhlmann-Wilsdorf, "Uses of theory in the design of sliding electrical contacts," in *Electrical Contacts, 1991. Proceedings of the Thirty-Seventh IEEE Holm Conference on*, pp. 1–24, 1991.
42. L. Brown, D. Kuhlmann-Wilsdorf, and W. Jesser, "Testing and evaluation of metal fiber brush operation on slip rings and commutators," *Components and Packaging Technologies, IEEE Transactions on*, vol. 31, pp. 485–494, 2008.
43. E. I. Shobert, *Carbon Brushes: The Physics and Chemistry of Sliding Contacts*. New York: Chemical Publishing Company, Inc., 1965.
44. D. Kuhlmann-Wilsdorf, "Temperatures at interfacial contact spots: Dependence on velocity and on role reversal of two materials in sliding contact," *Journal of Tribology*, vol. 109, pp. 321–329, 1987.
45. M. Kalin, "Influence of flash temperatures on the tribological behaviour in low-speed sliding: a review," *Materials Science and Engineering: A*, vol. 374, pp. 390–397, 2004.
46. J. C. Jaeger, "Moving sources of heat and the temperature at sliding contacts," in *Proc Roy Soc*, New South Wales 56, 1942, pp. 203–224.

47. H. S. Carslaw, *The Conduction of Heat in Solids*. London: Oxford University Press, 1959.
48. H. Blok, "Theoretical study of temperature rise at surfaces of actual contact under oiliness lubricating condition." *Proceedings of the Institution of Mechanical Engineers* 2: 222–235, 1937.
49. J. F. Archard, "The temperature of rubbing surfaces," *Wear*, vol. 2, pp. 438–455, 1959.
50. R. A. Burton, "Thermal deformation in frictionally heated contact," *Wear*, vol. 59, pp. 1–20, 1980.
51. J. R. Barber, "The conduction of heat from sliding solids," *International Journal of Heat and Mass Transfer*, vol. 13, pp. 857–869, 1970.
52. E. Rabinowicz, "The temperature rise at sliding electrical contacts," *Wear*, vol. 78, pp. 29–37, 1982.
53. M. Kalin and J. Vižintin, "Comparison of different theoretical models for flash temperature calculation under fretting conditions," *Tribology International*, vol. 34, pp. 831–839, 2001.
54. B. Vick and M. J. Furey, "A basic theoretical study of the temperature rise in sliding contact with multiple contacts," *Tribology International*, vol. 34, pp. 823–829, 2001.
55. D. Kuhlmann-Wilsdorf, "Demystifying flash temperatures I. Analytical expressions based on a simple model," *Materials Science and Engineering*, vol. 93, pp. 107–118, 1987.
56. D. Kuhlmann-Wilsdorf, "Demystifying flash temperatures II. First-order approximation for plastic contact spots," *Materials Science and Engineering*, vol. 93, pp. 119–133, 1987.
57. D. Kuhlmann-Wilsdorf, "Flash temperatures due to friction and Joule heat at asperity contacts," *Wear*, vol. 105, pp. 187–198, 1985.
58. D. Kuhlmann-Wilsdorf, D. D. Makel, and N. A. Sondergaad, Refinement of flash temperature calculations, in F. A. Smidt and P. J. Blau (eds.), *Engineered Materials for Advanced Friction and Wear Applications*, ASM International, Metals Park, OH, pp. 23–32, 1988.
59. D. Kuhlmann-Wilsdorf, "Sample calculations of flash temperatures at a silver-graphite electric contact sliding on copper," *Wear*, vol. 107, pp. 71–90, 1986.
60. D. Kuhlmann–Wilsdorf, "Theoretical speed and current density limits for different types of electrical brushes," *Magnetics, IEEE Transactions on*, vol. 20, pp. 340–343, 1984.
61. D. G. Bansal and J. L. Streator, "Design curves for temperature rise in sliding elliptical contacts," *Tribology International*, vol. 42, pp. 1638–1650, 2009.
62. D. G. Bansal and J. L. Streator, "A method for obtaining the temperature distribution at the interface of sliding bodies," *Wear*, vol. 266, pp. 721–732, 2009.
63. D. G. Bansal and J. L. Streator, "On estimations of maximum and average interfacial temperature rise in sliding elliptical contacts," *Wear*, vol. 278–279, pp. 18–27, 2012.
64. X. Tian and F. E. Kennedy, "Maximum and average flash temperatures in sliding contacts," *Journal of Tribology*, vol. 116, pp. 167–174, 1994.
65. W. W. Chen, Q. J. Wang, and W. Kim, "Transient thermomechanical analysis of sliding electrical contacts of elastoplastic bodies, thermal softening, and melting inception," *Journal of Tribology*, vol. 131, pp. 021406–021410, 2009.
66. S. Liu and Q. Wang, "Studying contact stress fields caused by surface tractions with a discrete convolution and fast Fourier Transform algorithm," *Journal of Tribology*, vol. 124, pp. 36–45, 2002.
67. C. Jacq, D. Nelias, G. Lormand, and D. Girodin, "Development of a three-dimensional semi-analytical elastic-plastic contact code," *Journal of Tribology*, vol. 124, pp. 653–667, 2002.
68. S. Kim, "Environmental Effects in Tribology," in *Micro- and Nanoscale Phenomena in Triboloty*, Y. Chung, Ed., Boca Raton, FL: CRC Press, 2012, p. 210.
69. A. M. Homola, J. N. Israelachvili, M. L. Gee, and P. M. McGuiggan, "Measurements of and relation between the adhesion and friction of two surfaces separated by molecularly thin liquid films," *Journal of Tribology*, vol. 111, pp. 675–682, 1989.
70. B. N. J. Persson and E. Tosatti, *Physics of Sliding Friction* vol. Series E: Applied Sciences. Dordrecht/Boston/London: Kluwer Academic Publishers, 1995.
71. W. Campbell, "The lubrication of electrical contacts," *Components, Hybrids, and Manufacturing Technology, IEEE Transactions on*, vol. 1, pp. 4–16, 1978.
72. K. Komvopoulos, N. Saka, and N. P. Suh, "The mechanism of friction in boundary lubrication," *Journal of Tribology*, vol. 107, pp. 452–462, 1985.

73. S. Andersson, A. Söderberg, and S. Björklund, "Friction models for sliding dry, boundary and mixed lubricated contacts," *Tribology International*, vol. 40, pp. 580–587, 2007.

74. K. Singh, S. Baghmar, J. Sharma, M. V. Khemchandani, and Q. J. Wang, "Boundary lubrication and its stability," *ASME Conference Proceedings*, vol. 2008, pp. 229–231, 2008.

75. R. Holm, "The electric tunnel effect across thin insulator films in contacts," *Journal of Applied Physics*, vol. 22, pp. 569–574, 1951.

76. J. A. Greenwood, "Constriction resistance and the real area of contact," *British journal of applied physics*, vol. 17, pp. 1621–1632, 1966.

77. L. Boyer, "Contact resistance calculations: generalizations of Greenwood's formula including interface films," *Components and Packaging Technologies, IEEE Transactions on*, vol. 24, pp. 50–58, 2001.

78. S. Lee, H. Cho, and J. Yong Hoon, "Multiscale electrical contact resistance in clustered contact distribution," *Journal of Physics D: Applied Physics*, vol. 42, pp. 165302–165308, 2009.

79. L. Kogut and K. Komvopoulos, "Analytical current-voltage relationships for electron tunneling across rough interfaces," *Journal of Applied Physics* **97**, pp. 73701–1/73701–5, 2005.

80. L. Kogut, "Electrical performance of contaminated rough surfaces in contact." *Journal of Applied Physics* **97**(10): pp. 10372–1/10372–5, 2005.

81. G. Norberg, S. Dejanovic, and H. Hesselbom, "Contact resistance of thin metal film contacts," *Components and Packaging Technologies, IEEE Transactions on*, vol. 29, pp. 371–378, 2006.

82. Z. Peng, Y. Y. Lau, W. Tang, M. R. Gomez, D. M. French, J. C. Zier, and R. M. Gilgenbach, "Contact resistance with dissimilar materials: bulk contacts and thin film contacts," in *Electrical Contacts (Holm), 2011 IEEE 57th Holm Conference on*, 2011, pp. 1–6.

83. C. Pradille, F. Bay, and K. Mocellin, "An experimental study to determine electrical contact resistance," in *Electrical Contacts (HOLM), 2010 Proceedings of the 56th IEEE Holm Conference on*, 2010, pp. 1–5.

84. P. Zhang, Y. Y. Lau, and R. M. Gilgenbach, "Thin film contact resistance with dissimilar materials," *Journal of Applied Physics*, vol. 109, pp. 124910–1–12490–10, 2011.

85. C. Gao and D. Kuhlmann-Wilsdorf, "Observations on the effects of surface morphology on friction and sliding modes," in *Tribology of Composite Materials*, ed Metals Park, Ohio: Am. Soc. for Metals Intl., 1990, pp. 195–201.

86. C. Gao and D. Kuhlmann-Wilsdorf, "On stick-slip and the velocity dependence of friction at low speeds," *Trans. ASME J. Tribology*, vol. 112, pp. 354–360, 1990.

87. C. Gao and D. Kuhlmann-Wilsdorf, "Experiments on, and a Two-component Model for, the Behavior of Water Nano-films on Metals," in *Thin Films: Stresses and Mechanical Properties II*, M. F. Doerner, W. C. Oliver, G. M. Pharr, and F. R. Brotzen, eds., Pittsburgh, PA: Materials Research Society, 1990, pp. 237–242.

88. C. Gao and D. Kuhlmann-Wilsdorf, "Adsorption films, humidity, stick-slip and resistance of sliding contacts," in *Electrical Contacts 1990, Proceedings of the Thirty-Sixth IEEE Holm Conference on Electrical Contacts*, Piscataway, NJ, 1990, pp. 292–300.

89. C. Gao and D. Kuhlmann-Wilsdorf, "Adsorption films, humidity, stick-slip and resistance of sliding contacts," in *IEEE Transactions on Components, Hybrids, and Manufacturing Technology*, pp. 37–44, 1991.

90. C. Gao and D. Kuhlmann-Wilsdorf, "On the tribological behavior of adsorbed layers, especially moisture," *Wear*, vol. 149, pp. 297–312, 1991.

91. C. Gao, D. Kuhlmann-Wilsdorf, and D. D. Makel, "Moisture effects including stiction resulting from adsorbed water films," *Trans. ASME J. Tribology*, vol. 114, pp. 174–180, 1992.

92. C. Gao, D. Kuhlmann-Wilsdorf, and D. D. Makel, "Fundamentals of stick-slip," *Wear*, pp. 162–164; 1139–1149, 1993.

93. I. R. McNab and P. Reichner, "Environment and Brushes for High-current Rotating Electrical Machinery," U.S. Patent 4,277,708, 1981.

94. P. Reichner, "Metallic brushes for extreme high current applications," in *Electrical Contacts 1979, Proceedings of the Twenty-Fifth Anniversary Meeting of the Holm Conference on Electrical Contacts*, pp. 191–198, 1979.

95. P. Reichner, "High current tests of metal fiber brushes," in *Electrical Contacts 1980, Proceedings of the Twenty-Sixth Annual Holm Conference on Electrical Contacts*, pp. 73–76, 1980.

96. L. Boyer, J. P. Chabrerie, and J. Saint-Michel, "Low wear metallic fiber brushes," *Wear*, vol. 78, pp. 59–68, 1982.

97. L. Boyer and J. P. Chabrerie, "Current collection in a homopolar machine using metallic fiber brushes," in *Proceedings of the 1983 International Current Collector Conference*, Annapolis, 1983.

98. D. Kuhlmann-Wilsdorf and A. M. Rijke, "Surface film resistivity of cu and cu-alloy crosswires," in *Electrical Contacts 1996, Proceedings of the Forty-Second IEEE Holm Conference on Electrical Contacts*, Chicago, IL, 1996, pp. 291–302.

99. J. A. Bares, N. Argibay, P. L. Dickrell, G. R. Bourne, D. L. Burris, J. C. Ziegert, and W. G. Sawyer, "In situ graphite lubrication of metallic sliding electrical contacts," *Wear*, vol. 267, pp. 1462–1469, 2009.

100. D. kuhlmann-Wilsdorf, "Theoretical performance liomits for different types of electrical brushes," in *Electrical Contacts 1983, Proceedings of the Twenty-ninth IEEE Holm Conference on Electrical Contacts*, Chicago, 1983, pp. 21–30.

101. D. Kuhlmann-Wilsdorf, "Metal Fiber Brushes, Chapter 20," in *Electrical Contacts: Principles and Applications*, P. Slade, Ed., New York: Marcel Dekker, Inc., 1999.

Part VI

Contact Data

24

Useful Electric Contact Information

Paul G. Slade

Nature never deceives us; it is always we who deceive ourselves.

The Social Contract, **Jean Jacques Rousseau**

CONTENTS

24.1 Introduction 1195
24.2 Notes to Tables 1196
References 1210

24.1 Introduction

In a book of this scope, many sources were consulted to prepare the tables of data presented in this chapter. Most of the data, however, were based primarily on the information drawn from the *Doduco Data Book* [1]. I am extremely grateful to the Doduco GmBH for permission to freely use the data in this book. Other sources of data were used when that information was not available in the Doduco book. In addition, some of these sources were also used to check the values presented in the Doduco book: examples include, Holm's book [2], and other materials reference books [3–7]. Other sources of contact data used to check and to add to these tables were information gathered from brochures and books published by a wide range of contact manufacturers [8–13] as well as information from the technical literature [14,15] and the internet. It was interesting to note that for some data there seemed to be a considerable variability between some of the values from the different sources, while for others there seemed to be close agreement. Where there was disagreement, I have endeavored to use consensus values as much as possible in these tables. I have also purposely left blank spaces where I could not find values to place there. I would expect readers, as they use this book, to fill in these blanks as the information becomes available.

24.2 Notes to the Tables

a. For kg m⁻³, multiply by 10^3.

b. Using Equation (1.34), $V_m = [4L(T_m^2 - T_1^2)]^{1/2}$

Where V_m is the melting voltage,

L is the Lorenz constant, 2.45×10^{-8} V² K⁻²

T_m is the melting temperature, K

T_1 is the bulk temperature of the contacts, K.

c. For cal g⁻¹, multiply by 0.239.

d. For torr (mm of Hg), multiply by 7.5×10^{-3}.

e. The range is from annealed metal to approximately 40% work hardened.

f. Hardness is usually denoted by a name. The three most common hardness designations are:

Vickers hardness (VH) ≈ Knoop hardness (KH) ≈ Brinell hardness (BH)

These are written "The Vickers hardness is VH80 or HV80"

that is, 80 kgf mm⁻² etc.

Note: 1 kg mm⁻² = 9.81 N mm⁻²

It measures an area left in a surface after an indenter has impinged on it with a given force. This number can be used to calculate the true area of electric contact given in Equation (1.10).

Rockwell hardness measures the depth of the indentation and thus *cannot be used in Equation (1.10).*

Note: Hardness ≈ 3× tensile strength.

g. Softening temperature is ≈ 1/3× melting temperature.

h. AgCd (10) means 90% by weight Ag and 10% by weight Cd.

If a contact is made from two materials A and B, with densities δ_A and δ_B respectively, and the weight fraction for material A is m_A, then *the volume fraction v_A for material A is*

$$v_A = \frac{m_A \delta_B}{m_A \delta_B + (1 - m_A)\delta_A}$$

i. These materials are distinct mixtures, their manufacture is discussed in Chapter 16.

j. The melting point value given is that for the lower melting point material in the mixture.

k. These are approximate values.

l. These values critically depend upon the treatment (e.g., cold working, tempering, heat treatment etc.).

m. The aluminum association number.

n. Good for the temperature range 0–100°C, but for most metals their resistivity (ρ) follows the linear equation:

$$\rho_T = \rho_0 \left[1 + \alpha(T - T_0) \right]$$

where α is the temperature coefficient of resistance, to within about 20% up to the metal's melting point.

o. Carbon sublimes above 3500°C.

p. Viscosity (dynamic) η, 1 N s m^{-2} = 10 poise (1P = 1 dynes cm^{-2}) = 0.1cP.

q. These are dynamic values for rapidly opening contacts [16]. For cases where the "thermal runaway" effect is valid, see Section 1.4.6, then the melting voltage for $Fe \cong 0.19$V and for $Ni \cong 0.16$V.

r. See Section 9.5.4 for a complete discussion of I_{min} and what the values given here mean.

s. This type of material is a possible replacement for Cu-Be spring materials

TABLE 24.1A

Physical Properties of the Most Important Contact Metals

Material	Chemical Symbol	Atomic Number	Atomic Weight	Density[a] (g cm^{-3})	Softening Temperature[g] (°C)	Softening Voltage (measured) (V)	Melting Point (°C)	Melting Voltage (measured) (V)	Melting Voltage (calculated) [b] (V)
Aluminum	Al	13	26.98	2.70	150	0.1	660	0.3	0.29
Antimony	Sb	51	121.75	6.7		0.2	630		0.28
Beryllium	Be	4	9.01	1.65			1,277		0.48
Bismuth	Bi	83	208.98	9.80			271		0.15
Cadmium	Cd	48	112.40	8.65			321		0.17
Carbon	C	6	12.01	2.3			(0)		
Chromium	Cr	24	52.00	7.19			1,875		0.67
Cobalt	Co	27	58.93	8.65			1,490		0.54
Copper	Cu	29	63.54	8.95	190	0.12	1,083	0.43	0.42
Gallium	Ga	31	69.72	5.91			30		0.04
Gold	Au	79	196.97	19.32	100	0.08	1,063	0.43	0.42
Indium	In	49	114.82	7.31			156		0.11
Iridium	Ir	77	192.20	22.5			2,450		0.86
Iron	Fe	26	55.85	7.87	500	0.19	1,537	0.6[a]	0.57
Lead	Pb	82	207.19	11.36	200	0.12	327	0.19	0.17
Magnesium	Mg	12	24.31	1.74			650		0.28
Manganese	Mn	25	54.94	7.43			1,245		0.47
Mercury	Hg	80	200.59	13.55			−39		
Molybdenum	Mo	42	95.94	10.21	900	0.3	2,610	0.75	0.91
Nickel	Ni	28	58.71	8.90	520	0.16	1,453	0.65[a]	0.54
Niobium	Nb	41	92.91	8.57			2,469		0.78
Osmium	Os	76	190.20	22.60			3,050		1.04
Palladium	Pd	46	106.40	12.02	540	0.25	1,552	0.57	0.57
Platinum	Pt	78	195.09	21.45			1,769	0.71	0.64
Rhenium	Re	75	186.20	21.04			3,180		1.09
Rhodium	Rh	45	102.91	12.41			1,966		0.70
Ruthenium	Ru	44	101.07	12.30	430	0.2	2,350		0.81
Silver	Ag	47	107.87	10.49	180	0.09	961	0.37	0.38
Tantalum	Ta	73	180.95	16.60	850	0.3	2,996		1.03
Tin	Sn	50	118.69	7.30	100	0.07	232	0.13	0.14
Titanium	Ti	22	47.90	4.51			1,668		0.61
Tungsten	W	74	183.85	19.32	1,000	0.4	3,410	1.1[a]	1.16
Vanadium	V	23	50.94	6.10			1,900		0.68
Zinc	Zn	30	63.37	7.13	170	0.1	420	0.17	0.20
Zirconium	Zr	40	91.22	6.49			1,852	0.67	0.67

TABLE 24.1B

Physical Properties of the Most Important Contact Metals

Material	Heat of Fusion[c] (Jg⁻¹)	Vapor Pressure at Melting Point[d] (Nm⁻²)	Boiling Point (°C)	Heat of Vaporization[c] (Jg⁻¹)	Thermal Conductivity @ 20°C (Wm⁻¹K⁻¹)	Coeff. of Thermal Expansion (×10⁻⁶K⁻¹)	Volume Change at Solidification (%)	Specific Heat[c] @ 20°C (Jg⁻¹K⁻¹)
Aluminum	395	2.5×10^{-6}	2,450	10,470	222	23.6	−6.5	0.896
Antimony	160	24	1,380	1,970	21	9.5	+0.95	0.205
Beryllium	1,090	4.3	2,770	24,700	147	11.6		1.892
Bismuth	52.3	6.5×10^{-4}	1,560	1,425	9	13.3	−0.33	0.121
Cadmium	55.3	16	765	879	92	29.8	−4.0	0.230
Carbon					50–300	0.6–4.3		0.670
Chromium	282	1030	2,480	5,860	67	6.2		0.461
Cobalt	245	190	2,900	6,660	71	13.8		0.415
Copper	212	5.2×10^{-2}	2,595	4,770	394	16.5	−4.2	0.385
Gallium	80.4	9.6×10^{-36}	2,237	3,895	33	18.0	+3.0	0.331
Gold	67.4	2.4×10^{-3}	2,966	1,550	297	14.2	−5.1	0.130
Indium	28.5	1.5×10^{-17}	2,000	1,970	25	33.0	−2.5	0.239
Iridium	144	1.5	4,600	3,310	59	6.8		0.130
Iron	274	7.3	2,750	6,365	75	11.8	−3.0	0.461
Lead	26.4	4.4×10^{-7}	1,750	858	33	29.3	−3.5	0.130
Magnesium	369	370	1,120	5,440	155	27.1	−4.1	1.026
Manganese	267	125	2,095	4,100	50	22.0	−1.7	0.482
Mercury	11.7	3.1×10^{-4}	357	292	9	60.8	−3.7	0.138
Molybdenum	292	3.6	5,560	5,610	142	4.9		0.276
Nickel	309	240	2,800	6,450	92	13.3	−2.5	0.439
Niobium	289	7.9×10^{-2}	4,900	7,790	54	7.3		0.272
Osmium	141	2.7	5,500	3,610	88	4.6		0.130
Palladium	143	1.3	3,170	3,475	72	11.8	−5.5	0.243
Platinum	113	3.2×10^{-2}	3,850	2,615	72	8.9	−6.0	0.130
Rhenium	178	3.5	5,900	3,420	72	6.7		0.138
Rhodium	211	6.5×10^{-1}	3,900	5,190	88	8.3	−10.8	0.247
Ruthenium	252	1.5	4,900	6,615	105	9.1		0.239
Silver	105	3.6×10^{-1}	2,212	2,387	419	19.7	−3.8	0.234
Tantalum	157	8×10^{-1}	5,425	4,315	54	6.5		0.142
Tin	60.7	6×10^{-21}	2,270	1,945	63	23.0	−2.8	0.226
Titanium	403	5.1×10^{-1}	3,280	8,790	17	8.4		0.519
Tungsten	193	4.4	5,930	3,980	167	4.6		0.138
Vanadium	330	3.2	3,400	10,260	29	8.3		0.498
Zinc	102	20	907	1,760	113	39.7	−4.7	0.385
Zirconium	224	1.7×10^{-3}	3,580	4,600	21	5.9		0.281

TABLE 24.1C

Physical Properties of the Most Important Contact Metals

Material	Modulus of Elasticity (kN mm⁻²)	Shear Modulus (kN mm⁻²)	Hardness[e][f] (×10² Nmm⁻²)	Resistivity @ 20°C (μΩ cm)	Temperature Coeff. of Electric Resistance[o](× 10⁻³K⁻¹)	Electrical Conductivity @ 20°C (m Ω⁻¹ mm⁻², MS·m⁻¹)
Aluminum	65	27	1.8–4	2.65	4.6	37.7
Antimony	56	20.4		38.6	5.4	2.6
Beryllium	298	150		4.0	10.0	25.0
Bismuth	33	13		106.8	4.5	0.94
Cadmium	57.5	29	≈3.5	6.83	4.3	14.7
Carbon	5		2.6–3.5	6500		0.015
Chromium	160		7–13	14.95	3.0	6.7
Cobalt	216			6.22	6.6	16.1
Copper	115	48	4–9	1.65	4.3	59.9
Gallium	9.6			43.2	4.0	2.3
Gold	80	29	2–7	2.19	4.0	45.7
Indium	11			8.33	4.9	12.0
Iridium	548	225	≈27	5.3	4.1	18.9
Iron	208	83	10–19	9.72	6.6	10.3
Lead	14.5	6	≈0.5	20.7	4.2	4.9
Magnesium	46	18		4.46	4.2	22.4
Manganese	165	77		185.0	0.5	0.54
Mercury				96.0	1.0	1.04
Molybdenum	347	122	15–26	5.15	4.7	19.4
Nickel	216	83	8–18	6.84	6.8	14.6
Niobium	113	39		13.1	3.4	7.6
Osmium	580	220	≈40	9.5	4.2	10.5
Palladium	117	50	4–9	10.8	3.8	9.3
Platinum	154	57	4–8.5	10.6	3.9	9.5
Rhenium	480	215	25–35	19.3	4.6	5.2
Rhodium	386	153	12–30	4.51	4.4	22.2
Ruthenium	430	193	≈25	6.71	4.6	14.9
Silver	79	29	3–7	1.59	4.1	63.0
Tantalum	188	70	10–30	12.4	3.5	8.1
Tin	47	18	0.45–0.6	11.6	4.6	8.7
Titanium	120	43	≈11	41.6	5.5	2.4
Tungsten	360	158	12–40	5.55	4.8	18.0
Vanadium	136	32		26.0	3.9	3.8
Zinc	96	36	3–6	5.92	4.2	16.9
Zirconium	98	36		43.5	4.4	2.3

TABLE 24.1D

Physical Properties of the Most Important Contact Metals

Material	Minimum Arc Voltage (V)	Minimum Arc Current (A)[a]	Work Function Range (eV)	Ionization Potential (eV)
Aluminum	11.2	0.4	4.0–4.4	5.98
Antimony	10.5		4.0–4.1	8.64
Beryllium			3.2–3.9	9.32
Bismuth			4.1–4.5	8.0
Cadmium	12	0.1	3.7–4.1	8.99
Carbon	20	0.01–0.02	4.0–4.8	11.27
Chromium	16	0.4	4.4–4.7	6.76
Cobalt			4.4–4.6	7.86
Copper	13	0.4	4.5	7.72
Gallium			3.8–4.1	6.0
Gold	11.5	0.35	4.0–4.9	9.22
Indium			4.0–4.1	5.79
Iridium	11.5		4.6–5.3	9.1
Iron	11.5	0.45	4.1–4.8	7.9
Lead	11.5	0.1	4.0–4.1	7.42
Magnesium			3.4–3.8	7.64
Manganese			3.8–4.1	7.43
Mercury			4.5	10.44
Molybdenum	12	0.75	4.1–4.5	7.18
Nickel	13.5	0.5	4.7–5.2	7.63
Niobium			4.0	6.77
Osmium			4.5	8.7
Palladium	14	0.8	4.5–5.0	8.33
Platinum	14	0.9	4.1–5.5	8.96
Rhenium			4.7–5.0	7.8
Rhodium	14	0.35	4.6–4.9	7.7
Ruthenium			4.5	7.5
Silver	12	0.4	4.2–4.5	7.57
Tantalum	12		4.0–4.2	7.89
Tin	11		3.6–4.1	7.33
Titanium	12		4.0–4.4	6.83
Tungsten	13	0.8–1.2	4.3–5.0	7.98
Vanadium			3.8–4.2	6.71
Zinc	11	0.1	3.1–4.3	9.39
Zirconium	12.5		3.7–4.3	6.92

TABLE 24.2

Silver-Based Alloys

Material[h]	Density (g cm^{-3})	Melting Range (°C)	Resistivity @ 20°C (μΩ cm)	Temp. Coeff. Elec. Resistance (× 10^{-3}K^{-1})	Thermal Conductivity @ 20°C (W m^{-1}K^{-1})	Modulus of Elasticity (kN mm^{-2})	Hardness[e] (× 10^2 Nmm^{-2})
Ag	10.5	960	1.59	4.1	419	79	3–7
AgCd (10)	10.3	910–925	4.3	1.4	150	60	4.5–10
AgCd (15)	10.1	850–875	4.8	2.0	109	60	5–11.5
AgCu (3)	10.4	900–938	1.92	3.2	385	85	4.5–11.5
AgCu (5)	10.4	905–940	1.96	3.0	380	85	6–13.5
AgCu (10)	10.3	779–900	2.08	2.8	335	85	6.5–15
AgCu (20)	10.2	779–860	2.17	2.7	335	85	8–16
AgCu (28)	10.0	779–820	2.0	2.7	325	92	10–17.5
AgNi (0.15)	10.5	960	1.7	4.0	414	85	4–10
AgCu (2) Ni	10.4	940	1.92	3.5	385	85	5–11
AgPd (30)	10.9	1150–1220	14.7	0.4	60	116	7–14
AgPd (40)	11.1	1225–1285	20.8	0.36	46	134	7.5–15
AgPd (50)	11.2	1290–1340	32.3	0.23	33.5	137	8–16
AgPd (60)	11.4	1330–1385	41.7	0.12	29.3		8.5–19.5
AgPd (30) Cu (5)	10.8	1120–1165	15.4	0.37	28.0	108	9–17

TABLE 24.3

Silver-Based Materials

Material[h][i]	Density (g cm^{-3})	Melting Temperature[j] (°C)	Resistivity @ 20°C (μΩ cm)	Temp. Coeff. Elec. Resistance[n] (× 10^{-3} K^{-1})	Thermal Conductivity @ 20°C (Wm^{-1} K^{-1})	Modulus of Elasticity (kN mm^{-2})	Hardness[e] (× 10^2 Nmm^{-2})
AgNi (10)	10.2	960	1.9	3.5	310	84	5–11
AgNi(15)	10.1	960	2.0	3.5	290	90	7–11.5
AgNi (20)	10.0	960	2.1	3.5	270	98	8–12
AgNi (30)	9.8	960	2.4	3.4	240	115	8.5–13.5
AgNi (40)	9.7	960	2.7	2.9	210	129	9–15
AgNi (50)	9.6	960	3.1		185	145	9–16
AgNi (60)	9.4	960	3.7		155	160	9–18
AgNi (70)	9.3	960	4.0		140	170	9–18
AgC (3)	9.1	960	2.0	3.5	325		4.2–4.2[k]
AgC (5)	8.5	960	2.2	3.3	318		4.0–4.2
AgC (10)	7.4	960	2.9				3.1
AgC (15)	6.5	960	4.5				2.6
AgFe(8.4) Re(0.4)		960	1.9				≈6.3

TABLE 24.4

Silver–Refractory Materials

Material[h][i]	Volume% Ag	Density (g cm^{-3})	Melting Temperature[j] (°C)	Resistivity[k] @ 20°C (μΩ cm)	Temp. Coeff.[k][n] Elec. Resistance (× 10^{-3} K^{-1})	Thermal Conductivity[k] @ 20°C (W m^{-1} K^{-1})	Hardness[k] (× 10^2 Nmm^{-2})
AgW (30)	81	11.9	960	3.1	1.9	326	11–13
AgW (50)	65	13.5	960	3.85		292	12–14
AgW (60)	55	14.2	960	4.17		276	14–16
AgW (65)	50	14.7	960	4.55		265	15–18
AgW (70)	44	15.2	960	5.0		257	16–19
AgW (75)	38	15.7	960	5.36		248	17–20
AgW (80)	32	16.3	960	5.56		239	18–22
AgWC (40)	69	11.9	960	4.17		255	13–16
AgWC (50)	60	12.4	960	4.55			14–17
AgWC (65)	45	13.2	960	5.0			16–19
AgWC (80)	27	13.5	960	5.3			≈40
AgMo (50)	49	10.2	960	5.56	3.9	234	12–16
AgMo (65)	34	10.3	960	5.7	4.1	281	14–17

TABLE 24.5

Silver–Metal Oxide Materials

Material[h][i]	Density (g cm^{-3})	Melting Temperature[j] (°C)	Resistivity @ 20°C (μΩ cm)	Temp. Coeff. Elec. Resistance[n] (× 10^{-3} K^{-1})	Thermal Conductivity @ 20°C (W m^{-1} K^{-1})	Modulus of Elasticity (kN mm^{-2})	Hardness[e] (× 10^2 Nmm^{-2})
AgCdO(10)	10.2	960	2.1	3.6	330		6–11
AgCdO(12)	10.2	960	2.2	3.6	325		7–12
AgCdO(15)	10.1	960	2.3	3.6	315		8–12.5
AgSnO$_2$(8)	10	960	2.0		335		5.8–9.5
AgSnO$_2$(10)	9.9	960	2.1		330		6.4–10
AgSnO$_2$(12)	9.8	960	2.4		315		7–12
AgSnO$_2$CdO(total oxide, 15)	10	960	2.6		315		≈9.8
AgSnO$_2$(11.5)WO$_3$(0.5)	9.75	960	2.3		315		6.4–10
AgSnO$_2$(11.5)MO$_3$(0.5)	9.68	960	2.4		315		7.7–12
AgSnO$_2$(11.5)BiO$_2$(0.5)	9.7	960	2.4		315		≈8.5
AgSnO$_2$In$_2$O$_3$(total oxide, 6)	10.2	960	2.3		335		≈8.2
AgSnO$_2$In$_2$O$_3$(total oxide, 10)	10	960	2.4		330		≈8.4
AgSnO$_2$In$_2$O$_2$(total oxide, 11.5)	9.96	960	2.8		325		≈9.6
AgSnO$_2$In$_2$O$_3$ (total oxide, 15.5)	9.76	960	2.9		310		≈10.6
AgZnO(8)	9.9	960	2.2		330		6–9.5
AgZnO(8)WO$_3$(0.5)	9.7	960	2.0		325		5.5–10
AgMgO(0.3)NiO(0.3)	10.4	960	2.3		355		≈10.1
AgFe$_2$O$_3$(5)ZrO$_2$(1)	10.0	960	2.08		285		≈7.5

TABLE 24.6

Gold-Based Alloys

Material[h]	Density (g cm⁻³)	Melting Range (°C)	Resistivity @ 20°C (μΩ cm)	Temp. Coeff. Elec. Resistance[n] (× 10⁻³ K⁻¹)	Thermal Conductivity @ 20°C (W m⁻¹ K⁻¹)	Modulus of Elasticity (kN mm⁻²)	Hardness[e] (× 10² Nmm⁻²)
Au	19.3	1063	2.19	4.0	317	80	2–7
AuAg(8)	18.1	1060	6.1	1.25	147	82	4–8
AuAg(10)	17.8	1058	6.3	1.25	147	82	4–8.5
AuAg(20)	16.4	1036–1040	10	0.86	75	89	4–11.5
AuAg(30)	15.4	1025–1030	10.2	0.7			4.5–11
AuNi(5)	18.3	995–1010	13.3	0.71	52	83	11.5–19
AuCo(5)	18.2	1010–1015	6.2–55.5	0.68		88	9.5–15
AuPt(10)	19.5	1150–1190	12.2	0.98	66	95	8–10.5
AuAg(25) Cu(5)	15.2	980	12.2	0.75		89	9–18.5
AuAg(20) Cu(10)	15.1	865–895	13.7	0.52	66	87	12–23
AuAg(26) Ni(3)	15.4	990–1020	11.4	0.88	59	114	9–15.5
AuAg(25) Pt(6)	16.1	1060	15.9	0.54	46	93	6–12.5
AuCu(14) Pt(9) Ag(4)	16.0	955	14.3–25				19–27

TABLE 24.7

Platinum- and Palladium-Based Alloys

Material[h]	Density (g cm⁻³)	Melting Range (°C)	Resistivity @ 20°C (μΩ cm)	Temp. Coeff. Elec. Resistance[n] (× 10⁻³ K⁻¹)	Thermal Conductivity @ 20°C (W m⁻¹K⁻¹)	Modulus of Elasticity (kN mm⁻²)	Hardness[e] (× 10² Nmm⁻²)
Pt	21.4	1769	10.6	3.9	72	173	4–8.5
PtIr(5)	21.5	1774–1776	22		42	190	8–15
PtIr(10)	21.6	1780–1785	17.9	2.0	29	220	10.5–17
PtRu(10)	21.6	«1800	33.3	0.83		235	19–30
PtNi(8)	20.6	1670–1710	30	1.5		180	20–31
PtW(5)	19.2	1830–1850	43.4	0.7		185	15–27
Pd	12.02	1552	10.8	3.8	72	117	4–9
PdCu(15)	11.6	1370–1410	38.5	0.49	17	175	9–22
PdCu(40)	10.9	1200–1230	33.3	0.28	38	175	12–26
PdNi(5)	11.0	1455–1485	16.9	2.47		175	9.5–20

TABLE 24.8

Copper–Tungsten Materials

Material[h][i]	Volume% Cu	Density (g cm⁻³)	Melting Temperature[j] (°C)	Resistivity[k] @ 20°C (μΩ cm)	Temp. Coeff. Elec. Resistance[k] (× 10⁻³ K⁻¹)	Thermal Conductivity[k] @ 20°C (W m⁻¹ K⁻¹)	Hardness[k] (× 10² Nmm⁻²)
CuW (50)	68	12.0	1083	4.2		200	11–16
CuW (60)	59	13.0	1083	4.5		195	14–20
CuW (70)	48	14.1	1083	5.0		175	16–22
CuW (75)	42	14.9	1083	5.3		160	17–24
CuW (80)	35	15.4	1083	5.6		150	20–26

TABLE 24.9

Copper-Based Spring Alloy Materials

Material[h]	Density (g cm⁻³)	Melting Range (°C)	Resistivity @ 20°C (μΩ cm)	Temp, Coeff Electrical Resistance[n] (× 10⁻³ K⁻¹)	Thermal Conductivity @ 20°C (W m⁻¹ K⁻¹)	Coeff. of Linear Expansion (× 10⁻⁶ K⁻¹)	Modulus of Elasticity (kN mm⁻²)	Max. Spring Bending Strength[l] (N mm⁻²)
Cu	8.96	1083	1.65	4.3	394	16.5	115	230
CuSn(6)P(0.2)	8.93	910–1040	11.1	0.7	75	18.5	118	370–430
CuSn(8)P(0.2)	8.93	875–1025	13.3	0.7	67	18.5	115	390 480
CuSn(6)Zn(6)	8.86	900–1015	10.5	0.8	79	18.4	114	500
CuNi(9)Sn(2)	8.93	1060–1120	16		48	16.0	132	500
CuNi(2)Si(1)[s]	8.82		4.2	9.6	200	13.0	130	450–830
CuNi(18)Zn(20)	8.73	1055–1105	23	3	33	17.7	132	390–550
CuBe(1.7)Co(0.2)	8.4	890–1000	8–11	1	110	17.0	130	780–1000
CuBe(2.0)Co(0.2)	8.3	870–980	8–11	1	110	17.0		850–1050
CuBe(0.5)Co(2.5)	8.8	1030–1070	3.6	1.5	210	18.0	132	370–650

TABLE 24.10

Copper-Based Carrier Alloy Materials

Material[h]	Density (g cm⁻³)	Melting Range (°C)	Resistivity @ 20°C (μΩ cm)	Temp, Coeff Electrical Resistance[n] (× 10⁻³ K⁻¹)	Thermal Conductivity @ 20°C (W m⁻¹ K⁻¹)	Coeff. of Linear Expansion (× 10⁻⁶ K⁻¹)	Modulus of Elasticity (kN mm⁻²)	Hardness[e] (× 10² Nmm⁻²)
Cu	8.96	1083	1.65	4.3	394	16.5	115	4–9
CuZn(5)	8.87	1055–1065	3.0	2	243	18	125	4.5–9
CuZn(10)	8.79	1030–1043	4.0	1.8	184	18.2	125	6–11
CuZn(15)	8.75	1005–1025	4.8	1.6	159	18.5	122	6–15
CuZn(20)	8.67	980–1000	5.3	1.5	142	18.8	120	6.5–15.5
CuZn(30)	8.53	910–940	6.3	1.5	124	19.8	114	7–13
CuZn(40)	8.41	895–900	6.7	1.7	117	20.3	103	8–14
CuNi(25)	8.94	1150–1210	32	0.2	29	15.5	147	85
CuAg(0.1)	8.89	1082	1.8	3.9	380	11	110	5–9
CuAg(2)	9.0	1050–1075	2.0	3.0	330	17.5	123	5–13
CuCd(1)	8.94	1040–1080	2.2	3.4	320	17	124	9.5–14
CuCr(1)	8.89	980–1080	3.5	2.5	240	17	112	8–15
CuAg(2) Cd(1.5)	9.0	970–1055	2.3	2.4	260	17.8	121	6–15
CuAg(5) Cd(1.8)	9.1	920–1040	2.6	2.4	240	17.8	120	7–13.5
CuTe(0.5)P	8.93	1050–1075	1.8	3.7	356	18	118	6–10

TABLE 24.11

Other Types of Spring and Carrier Materials

Material[h]	Density (g cm⁻³)	Melting Range (°C)	Resistivity @ 20°C (μΩ cm)	Thermal Conductivity @ 20°C (Wm⁻¹ K⁻¹)	Coeff. of Linear Expansion (× 10⁻⁶ K⁻¹)	Modulus of Elasticity (kN mm⁻²)	Hardness[e] (× 10⁻² Nmm⁻²)
301 stainless steel	7.9	≈ 1400	71	16	17	190	≈16
302 stainless steel	7.9		72	16	17.5	190	
304 stainless steel	7.9		70	16	17.5	190	
305 stainless steel	8.0		72	16	17.5	190	
430 stainless steel	7.7		60	26	10.5	196	
1100[m] Al (99)	2.71	660	2.9	222	23.6	65	2.3–4.4
5052 Al (Al, Mg, other)	2.68	660	4.9	131	23.8	65	4.6–7.7
3003 Al (Al, Mn, other)	2.73	660	4.2	193	23.3	65	2.8–5.5
Inconel (Ni, Cr, Fe alloy)	8.5		98	26	12.9		≈9
Constantan (Cu, Ni(≈43) alloy)	8.8	≈1190	50	23	14.0	170	8–20
Monel (Ni, Cu(≈30) alloy)	8.8	≈1300	48.2	25	14.0		≈14

TABLE 24.12

Selection of Sliding-Contact Materials

Material[h]	Density (g cm⁻³)	Melting Range (°C)	Resistivity[k] @ 20°C (μΩ cm)	Hardness[k] (× 10² Nmm⁻²)
C	2.3	sublimes @ ≈3675	87	
AgC(5)	8.5	960	2.9	4.0
AgC(10)	7.4	960	3.9	3.1
AgC(25)	5.1	960	14.5	
AgC(50)	3.2	960	23	
AgC(70)	2.6	960	53	
AgC(90)	2.14	960	87	
AgCu(5)C(10)	6.8	905–940	4.7	
AgCu(48.5)C(3)	8.3	779–875	4.0	5.2
AgCu(47.5)C(7)	7.4	779–875	8.0	
AgCu(35)C(30)	4.2	779–815	63.6	
AgCu(69.75)C(5)	8.2	779–945	3.2	5.2
AgCu(72)C(3)	8.3	779–960	5.3	7.1

TABLE 24.13

SI Units and Other Commonly Used Units

	SI Units			Other Commonly Used Units	
Length	meter	m		angstrom	Å
Mass	kilogram	kg		bar	bar
Time	second	s		British Thermal Unit	Btu
Electric current	ampere	A		calorie	cal
Thermodynamic	degree	K		cycles per second	c/s
Temperature	kelvin		w	day	D
Amount of Substance	mole 1			degree	°
				degree Celsius	C
Frequency	hertz	Hz	s^{-1}	degree fahrenheit	F
Force	newton	N	$M \cdot kg/s^2$	electron volt	eV
Pressure	pascal	Pa	N/m^2	foot	Ft
Energy, work	joule	J	N.m	gauss	G
Quantity of heat				gram	g, gm
Power	watt	W	J/s	horse power	hp
Radiant flux				hour	h, hr
Quantity electricity	coulomb	C	$\Lambda \cdot s$	inch	in
Electric charge				liter	l
Electric potential	volt	V	W/A	minute of time	min
Capacitance	farad	F	C/V	ounce	oz
Electric resistance	ohm	Ω	V/A	pond	p
Magnetic flux	weber	W	V-s	relative humidity	rh
Magnetic flux	tesla	T	Wb/m^2	revolution	rev
Density				standard atmosphere	atm
Magnetic field strength	amperes/ meter		A/m	standard temperature and pressure @273K & 1 atm.	STP
Magnetomotive	ampere		A	torr	torr
Force				year	yr
Inductance	henry	H	V·s/A	Poise	P

TABLE 24.14

Miscellaneous Conversion Table

Length	m	meter	
	cm	centimeter	100 cm = 1 m
	mm	millimeter	1000 mm = 1 m
	μm	micrometer(micron)	10^6 μm = 1 m
	in	inch	1 inch = 25.4 mm
	μin	microinch	1 μin = 0.0254 μm
	Å	angstrom	1 Å = 10^{-10} m
Mass	g	gram	1 g = 10^{-3} kg
	kg	kilogram	
	lb	pound	1 lb = 0.454 kg
	oz	ounce (avdp)	1 oz = 28.35 g
	oz	ounce (troy) [precious metals]	1 oz = 31.1g
Force	N	newton	9.81N = 1kgf
	P	pond (gram force, gF)	1 kp = 9.81 N
	lbf	pound force	1 lbf = 4.45 N
Pressure	Pa	pascal	1 Pa = 1 N m^{-2}
	bar	barr	1 bar = 10^5 Pa
	atm	atmosphere	1 atm = 1.013 bar
	torr	torr (mm of Hg)	1 torr = 1 atm/760
	lbf/in^2	pounds/square inch	1 lbf in^{-2} = 6.9 × 10^3Pa
Energy	J	joule	1 J = 1 Nm
	kWh	kilowatt hours	1 kWh = 3.6 × 10^6J
	ft lb	foot pounds	1 ft lb = 1.36 J
	eV	electron volts	1 eV = 1.6 × 10^{-19}J
		erg	1 erg = 10^{-7} J
Heat	cal	calories	1 cal = 4.189J
	Btu	british thermal units	1 Btu = 1055 J
Power	W	watt	1 W = 1Js
	hp	horse power	1 hp = 746 W
			1 hp = 550 ft lb s^{-1}
	ft lb/s	foot pound/second	1 ft lbs^{-1} = 1.36 W
Charge	C	coulomb	1 C = 1 As
		1 electron charge	1.6 × 10^{-19} C
Temperature	°C	degrees Celsius	°C = K − 273
	F	degrees Fahrenheit	F = (9°C/5) + 32
Volume	l	liter	1 liter = 10^3 cm^3

TABLE 24.15

Some Fundamental Constants

Speed of light in vacuum	c	3 × 10^8 ms^{-1}
Elementary charge	e	1.6 × 10^{-19} C
Electron rest mass	m_e	9.11 × 10^{-31} kg
Proton rest mass	m_p	1.67 × 10^{-27} kg
Neutron rest mass	m_n	1.68 × 10^{-27} kg
Permittivity constant	ε_o	8.85 × 10^{-12} Fm^{-1}
Permeability constant	μ_o	1.26 × 10^{-6} Hm^{-1}
Plank's constant	h	6.63 × 10^{-34} Js
Molar gas constant	R	8.31Jmol^{-1} K^{-1}
Avogadro's constant	N_A	6.02 × 10^{23} mol^{-1}
Boltzman's constant	k	1.38 × 10^{-23} JK^{-1}
Molar volume of an ideal gas at STP	V_M	2.24 × 10^{-2} m^3 mol^{-1}

TABLE 24.16

Water Vapor in Saturated Air at 1 Atmosphere Pressure as a Function of Temperature

Temp. (°C)		0	1	2	3	4	5	6	7	8	9
0	Pa (Nm^{-2})	611	656	705	757	812	871	934	1,001	1,071	1,146
	(gm^{-3})	4.84	5.18	5.54	5.92	6.33	6.74	7.22	7.70	8.21	8.76
10	Pa (Nm^{-2})	1,226	1,311	1,400	1,495	1,596	1,703	1,815	1,935	2,061	2,194
	(gm^{-3})	9.33	9.93	10.57	11.25	11.96	12.71	13.50	14.34	15.22	16.14
20	Pa (Nm^{-2})	2,335	2,483	2,640	2,805	2,980	3,164	3,357	3,561	3,775	4,001
	(gm^{-3})	17.12	18.14	19.22	20.35	21.54	22.80	24.11	25.49	26.93	28.45
30	Pa (Nm^{-2})	4,239	4,488	4,750	5,026	5,314	5,618	5,936	6,270	6,620	6,987
	(gm^{-3})	30.04	31.70	33.45	35.27	37.18	39.18	41.3	43.5	45.8	48.2

$$\text{Relative humidity}(\%) = \frac{\text{pressure of water vapor present at a temperature}}{\text{saturation pressure at the same temperature}} \times 100$$

$$\approx \frac{\text{weight of unit volume water vapor present at a temperature}}{\text{saturation weight in unit volume at the same temperature}} \times 100$$

TABLE 24.17

Physical Properties of Some Contact Metals at their Melting Points

Material	Melting Point(°C)	Density[a] (g cm^{-3})	Resistivity (μΩ cm)	Thermal Conductivity (W m^{-1} K^{-1})	Viscosity[p] (Dynamic) (× 10^{-3} Nsm^{-2})	Surface Tension (J m^{-2})
Aluminum	660	2.39	20.0	92.7	4.5	0.86
Bismuth	271	10.06	128.1	11.0	1.68	0.39
Cadmium	321	8.02	33.7	44.0	1.4	0.67
Chromium	1,875	6.46		25.1	0.68	1.59
Copper	1,083	7.96	21.3	49.4	3.34	1.29
Gallium	30	6.2	26	33.5	2.04	0.74
Gold	1,063	17.32	30.8			0.75
Indium	156	7.03	33.1	41.9	1.69	0.56
Iridium	2,450	20.0				2.25
Iron	1,537	7.15	139		2.2	1.68
Lead	327	10.68	95.0	16.3	2.63	0.47
Magnesium	650	1.59	27.4	13.9	1.24	0.56
Mercury	values @ 20C	13.55	96	9	1.55	0.47
Molybdenum	2,610	9.34				2.25
Nickel	1,453	7.9	109			1.76
Palladium	1,552	10.7				1.5
Platinum	1,769	19.7				1.74
Rhodium	1,966	10.65				2.0
Silver	961	9.33	17.2		3.9	0.93
Tantalum	2,996	15.0				2.15
Tin	232	6.97	48.0	33.5	1.97	0.58
Titanium	1,668	4.13				1.51
Tungsten	3,410	17.6				2.31
Zinc	420	6.64	37.4	60.3	3.93	0.82

TABLE 24.18

Contact Materials Used in Vacuum Interrupters

Material[h][i]	Volume% Cu or Ag	Density[k] (g cm^{-3})	Melting Temperature[j] (°C)	Resistivity[k] @ 20°C (μΩ cm)	Chopping Current (99% value) (A)	Hardness[k] (× 10^2 Nmm^{-2})
CuCr (50)	45	7.1–7.7	1083	4.5–6.3	5	10
CuCr (40)	55	7.3–7.8	1083	3.8–5.0	4.5	9.5
CuCr (25)	71	7.9–8.3	1083	3.1–3.5	4.5	9
CuCr (20)	76	8.0–8.4	1083	3.0–3.3	5	8.5
CuCr (25) Bi (5)	65	7.9–8.4	271	3.9	1.5	8
CuW (80)	35	15.2–15.8	1083	5.0	5	25
CuW (85)	28	16.1–16.5	1083	5.6	6.0	27
CuW (69) Sb (1)	48	13.8–14.5	630	6.3	5.5	22
CuMo (75)	28	9.8	1083	4.2	5.5	19
CuMo (55)	53	9.6	1083	3.4	6.5	16
AgWC (60)	50	12.7–13.0	960	3.0–5.9	1.5	18–22
AgWC (50)	60	12.2	960	3.0–4.3	1.8	14–18
CuFe (50)	47	8.3	1083	8.3	5	16
CuBi (0.5)	99.5	8.95	271	1.8	14	8

References

1. E Vinaricky, G Horn, V Behrens (Eds.). *Doduco Data Book of Electrical Contacts*, New Edition. Pforzheim: Doduco GmbH, (pub. Stielitz Verlag, Muehlacker) 2012.
2. R Holm. *Electric Contacts, Theory and Application*. New York: Springer-Verlag, 2000, pp 436–438.
3. *Metals Handbook*, desk edition. (Ed. J Davis) Chicago, IL: American Society of Metals, 1998.
4. GWC Kaye, TH Laby. *Tables of Physical and Chemical Constants*, 13th ed. New York: Wiley, 1966.
5. *CRC Handbook of Chemistry and Physics*. (Ed. in Chief, W Haynes) CRC Press, Boca Raton, FL, 2013.
6. JF Shackleford, W Alexander, JS Park. *CRC Materials Science and Engineering Handbook*. Boca Raton, FL: CRC Press, 1994.
7. M Bauccio (ed.). *ASM Metals Reference Book*. Chicago, IL: American Society of Metal, 1993.
8. *Elektrische Kontakte, Werkstoffe und Technologie*. Pforzheim: Rau GMBH, 1984.
9. Elektrotechnik, Hanau: Degussa AG.
10. *Handbook for Clad Metals*. Lincoln, RI: Technical Materials Inc., 1974.
11. *General Contact Guide*. Waukegan, IL: Chugai Inc.
12. *Electrical Contacts*. St. Marys, PA: Stackpole Carbon Company, 1960.
13. *Electric Contacts and Assemblies*. Export, PA: AMI Inc.
14. PG Slade. Advances in material development for high power, vacuum contacts. *IEEE Trans Components, Packaging and Manuf Technol* – Part A 17(1): 96–106, 1994.
15. F Heitzinger, H Kippenberg, K Saeger, K Schroder. Contact materials for vacuum switching devices. *Proc. of the 15th International Symposium on Discharges and Electrical Insulation in Vacuum*. VDE-Verlag, Berlin, September 1992, pp 373–378.
16. PG Slade, MD Nahemow. Initial separation of electric contacts carrying high currents, *J. Appl. Phys.* 42(9): 3290–3297, August 1972.
17. PG Slade. *The Vacuum Interrupter, Theory, Design and Application*. Boca Raton, FL: CRC Press, 2008.

Author Index

A

Abbasi, M., 28, 33, 34, 35
Abbot, E., 706
Abbott, W. H., 52, 116, 117, 128, 131, 133, 134, 141, 148, 153, 154, 155, 156, 157, 163, 164, 167, 175, 177, 179, 180, 187, 334, 346, 394, 395, 422, 439, 442, 445, 447, 448, 449, 452, 458, 466, 475, 488, 506, 1118
Abdul Assis, A. M., 633, 922
Abe, N., 832
Abele, V. R., 661, 662, 928, 931, 933
Aberg, M., 126, 129
Abys, J. A., 458, 459, 524
Achar, R., 1114, 1140
Adams, G. G., 704, 705, 706, 708, 712, 714, 718, 720, 721, 1103, 1111, 1158, 1161
Adams, J. B., 489
Adamson, A. W., 502
Adkins, C., 1162, 1164, 1170, 1172, 1174, 1175, 1180
Adkins, C. M., 1134
Agarwala, V. S., 1096
Agili, S. S., 1108
Agrait, N., 710, 1109
Aichi, H., 9
Aiosh, T., 832
Aitchison, I., 449
Akbulut, M., 1120
Alamarguy, D., 272, 466, 479, 502, 503, 505, 540
Al-Bender, F., 1101
Albohr, 1100
Alcantar, N. A., 1103, 1158
Alessandrini, E. I., 51
Alexander, G. C., 868, 869
Alexander, W., 1195
Alexandrov, N., 28, 31, 32, 40, 42, 290
Allen, P., 723
Allen, S., 922, 943, 944
Allen, T., 888
Alley, D., 1076
Alon, E., 718
Alyabev, A. Y., 269
Ambier, J., 287, 456, 455
Amft, D., 631, 654, 790, 806
Amimoto, T., 125
Amin, A. N., 439
Amsinck, R., 799

Anders, A., 594
Anders, S., 835
Anderson, J. J., xxxiii
Anderson, J. S., 47
Andersson, G., 126, 127
Andersson, S., 1107, 1170
Andlauer, R., 636, 649, 826, 827
Ando, H., 1038
Angadi, S. V., 79, 712, 827
Anheuser, A., 815, 822, 826
Anheuser, M., 850, 868
Anli, H., 525
Ansel, G. S., 73
Antler, M., 23, 28, 131, 134, 144, 153, 154, 163, 175, 177, 179, 274, 283, 334, 396, 419, 420, 422, 423, 426, 427, 428, 429, 430, 431, 432, 433, 434, 436, 437, 439, 442, 447, 449, 450, 451, 452, 456, 458, 459, 460, 462, 464, 465, 466, 467, 468, 469, 470, 472, 473, 476, 479, 482, 483, 484, 485, 486, 487, 489, 490, 492, 493, 495, 496, 497, 498, 499, 500, 504, 505, 506, 525, 530, 531, 533, 535, 536, 537, 543, 544, 983, 984, 1096, 1103, 1115, 1119, 1121
Aoki, T., 392
Aoyama, Y., 985, 1007, 1013
Apperley, R. H., 865
Arai, K., 1032
Aramata, M., 1018, 1019, 1020, 1024
Archard, J. F., 13, 17, 75, 271, 415, 418, 1158, 1159, 1167, 1168
Argibay, N., 1152, 1154, 1155, 1173
Argibaya, N., 1073
Ario, A., 659, 660
Arnell, S., 126, 127
Aronstein, J., 296, 333, 385, 489, 858, 861
Arrazat, B., 713
Arushanov, K., 689
Asar, M. P., 27
Ashby, M., 355
Ashby, M. F., 73
Ashcroft, N. W., 81, 89, 96, 97, 99, 100
Aspden, R., 645
Asthner, B., 427, 432
Atermo, R., 300
Aubrecht, V., 825, 827
Augis, J. A., 1007

Aukland, N., 174, 180, 442, 449, 458, 467, 473,
    476, 486, 487, 504, 521
Ault, K., 849, 854
Axon, R. T., 317, 326

B

Baba, K., 865
Baba, M., 737
Babrauskas, V., 855, 857, 858, 861
Bader, F. E., 458, 521, 674
Baek, C., 704, 723
Baek, S., 720
Baghmar, S., 1107, 1119, 1170
Baily, F. G., xxxvi
Baisheng, S., 1108
Bajard, M. T., 115, 116
Baker, R. G., 23, 486, 525
Baker, R. M., 1073
Balasubramanian, V., 1108
Balestrero, A., 759
Balme, W., 939, 989, 993
Balog, M., 505
Bansal, D., 1096, 1169
Barber, J. R., 79, 1167
Barber, S. A., 421
Barbetta, M., 423
Barcikowski, F., 827
Bardolle, J., 115, 116
Bare, J. P., 458, 469, 486
Bares, J. A., 1152, 1154, 1173
Baresb, J. A., 1073
Barkan, P., 15, 646, 648, 650, 748,
    988, 990
Barron, C., 704
Bartlova, M., 825, 827
Barton, R. S., 1007
Basnett, R. T., 806
Batzill, M., 49
Bauccio, M., 1195
Bauer, C. L., 28, 35, 292, 332
Bauer, E. J., 125, 126
Bauer, J. A., 806
Bauer, R., 1073
Baumberger, T., 1100, 1102
Bay, N., 79
Bay, F., 1170
Bayer, R. G., 134, 428, 431
Bazaykin, V. I., 300
Beausoliel, R. W., 858, 861
Beckert, T., 822
Behrens, F. W., 831, 832
Behrens, N., 790, 796, 803, 817, 895

Behrens, V., 198, 520, 541, 542, 898, 939, 945,
    1013, 1195
Beilis, I., 594
Bekki, K., 1058, 1063
Belbel, E., 821
Belhaja, H., 850
Bell, J. A. E., 525
Benedetto, A., 505
Benhenda, S., 525
Benitez, G., 502
Ben-Jemaa, N., 585, 586, 587, 588, 621, 636, 637,
    641, 648, 713, 718, 737, 765, 850, 890,
    903, 985, 1007, 1013, 1057
Benson, F. A., 865
Bentounes, H. A., 599
Berg, H., 41
Berger, F., 649, 815, 821, 850, 871
Berger, H. H., 22
Berman, A., 1158, 1171
Bernett, M. K., 1028
Bernoff, H., 821
Bertsch, W., 1007
Bet, M., 653
Bevington, R. C., 661, 662, 930, 943, 944
Beyler, C. L., 857, 861
Bhattacharya, A., 501
Bhavnani, S. H., 79
Bhushan, B., 711, 1100, 1102, 1109, 1120, 1135,
    1158, 1170, 1171
Bickerstaff, A. E., 1044
Biewendt, V., 834
Bindas, B., 934
Bindas, J. A., 630
Binder, L., xxxv
Biswas, A. B., 47
Biunno, N., 423
Bizjak, M., 649
Björklund, S., 1170
Bjorkman, C., 133
Blades, F. K., 851, 865, 866
Blake, B. E., 143
Blanchette, Y., 341
Blanks, H. S., 448, 449
Blau, P., 1135, 1157
Blee, J., 524
Blessington, D. R., 422
Blok, H., 1167
Bochkarev, M., 576
Bock, E. M., 274, 441, 449, 453, 464, 488, 492
Boczowski, D., 655, 656
Boddy, P. J., 640, 762, 763
Boden, P., 850
Bodin, C., 466

Boguslavsky, I., 458, 525
Bohland, J., 23
Bohlin, A., 254, 307
Böhm, W., 794, 803, 895, 896
Bohr, N., 556
Boissel, J., 501
Boisvert, S., 341
Boiziau, C., 501
Boles, D., 997
Bond, N., 278, 331
Bonifaci, N., 724, 725
Bonwitt, W. F., 331
Booth, H. C., xxxv, 308
Booth, J. C., 1112
Borkowski, P., 630, 633
Borri-Brunetto, M., 712
Borst, F., 982
Bortel, M., 355
Bottiglione, F., 1162
Bourda, C., 621, 637, 665, 765, 881, 922
Bourne, G. R., 1152, 1154, 1173
Bouzera, A., 272
Bowden, F. P., 14, 15, 69, 415, 425, 481, 482, 748,
        750, 1096
Boxman, R. L., 593, 594
Boychenko, V. I., 309, 312
Boyer, L., 123, 466, 501, 502, 1170, 1172
Boyle, W. S., 1006
Bozack, M. J., 456, 466, 541
Bradford, K. F., 23, 27, 28, 458, 674
Brailovski, V., 328, 331
Braithwaite, E. R., 503
Brandbyge, M., 718
Branston, D. W., 822
Bratton, R., 645
Braumann, P., 794, 803, 891, 894, 906, 939, 986,
        988, 989, 993
Braunovic, M., xxxiv, 23, 28, 31, 32, 40, 42, 175,
        176, 179, 249, 259, 266, 274, 275, 278, 283,
        285, 286, 287, 290, 302, 307, 311, 315,
        322, 328, 330, 331, 333, 334, 335, 343,
        345, 349, 351, 355, 385, 393, 449, 456,
        464, 479, 539, 704, 705, 706, 722, 723,
        724, 725
Breach, C. D., 41
Brecher, C., 896
Breck, G. F., 458, 459
Bresgen, H., 440, 441
Brinen, J. S., 693
Brizmer, V., 1100, 1102
Brockman, I. H., 417, 506
Bron, O. B., 313
Brooks, S. J., 851, 865, 866

Broue, A., 713, 718, 724
Brown, C., 712, 722
Brown, J. W., 871
Brown, L., 1159, 1164, 1165
Browne, T. E., xxxiv, 556, 557, 578, 579, 597, 598,
        603, 604, 650, 786, 788, 822, 922
Bruckner, R., 419
Bruel, J. F., 333, 474
Brugner, F., 888, 896, 906
Brusse, J., 541
Brussel, H., van 1101
Bryant, M. D., 274, 447, 1068, 1073
Bu, J., 715
Buchenauer, C. J., 571
Bucher, F., 712
Buckley, D. H., 421
Buehler, W. J., 351
Bugaev, A., 594
Buldyrev, S., 718, 723
Bureau, C., 501
Burkhard, G., 788
Burris, D. L., 1173
Burton, R. A., 1076, 1167
Burton, R. G., 1076
Burwell, J. T., 418
Bushby, S. J., 53
Butt, H. J., 1115
Byon, E., 594

C

Cabrera, N., 53
Cadoux, Y., 871
Caekenberghe, K. van, 703
Cai, J., 704
Callen, B. W., 53, 346
Calvin, H. A., xxxiii
Campbell, W. E., 47, 52, 129, 142, 144, 169,
        258, 269, 334, 445, 452, 458, 474, 488,
        491, 1170
Campel, R., 380, 392
Campolo, S., 871
Cao, J., 716
Cao, W. D., 301
Capitelli, M., 826
Capp, P., 458
Cappell, R. J., 439
Cappella, B., 1115
Caprile, C., 295, 301
Carballeira, A., 117, 131, 333, 456, 457, 466,
        474, 991
Carballeira, M., 117, 131
Carbone, G., 1162

Caroll, J. J., 831
Carrob, P., 502
Carrodus, J., 856, 866, 867
Carslaw, H. S., 57, 1167
Carter, R. O., 492
Carvou, E., 272, 621, 637, 648, 737, 765, 850
Cassie, A. M., 604
Castell, Ph., 456, 457, 466
Castner, D. G., 693
Celis, J. P., 422
Cha, H. Y., 1101
Cha, P., 717
Chabrerie, J. P., 1172
Chabrier, J. P., 583, 649
Chai, Y., 716
Chaikin, S. W., 458, 474, 1007
Chalandon, P., 449
Chan, P., 716
Chana, M. S., 129
Chance, A. B., 331
Chandler, R., 718
Chang, L., 1107
Chang, W., 706, 707
Chang, Y. J., 1160
Chao, J. L., 128, 129, 133
Charitidis, C., 51
Charlier, J., 501
Charpentier, J. P., 123
Charvet, P., 713, 718
Chaussee, R., 891
Chauvin, W. J., 449
Chen, B., 840
Chen, C., 180, 396
Chen, C. L., 594, 630
Chen, D., 608, 609
Chen, F., 718
Chen, G., 906
Chen, H., 716
Chen, L., 714, 718, 1161
Chen, N., 1120
Chen, Q., 716
Chen, W., 203, 204, 506
Chen, W. W., 1169
Chen, Y., 716, 718, 741, 1108
Chen, Z., 1075
Chen, Z. K., 198, 459, 640, 643, 644,
        659, 660, 767, 891, 892, 893,
        898, 903, 940, 954, 957,
        981, 1032
Cheng, Y. T., 539
Cherkaoui, M., 826, 827
Chiaia, B., 712
Chianrabutra, C., 572, 573, 575, 639

Chiarenzelli, R. V., 52, 144, 154, 439, 506
Chidsey, C. E. D., 501
Chien, Y. K., 934, 997
Cho, D., 722
Cho, H., 1170
Choi, D., 715
Choi, J., 716, 717, 1101
Choi, W., 722
Choo, S. C., 22
Chou, T., 715
Chow, E. M., 449
Chow, L. W., 87
Chow, W., 984
Chrétien, P., 466, 479
Christenson, T., 698
Christofel, L., 126
Chua, C. L., 449
Chudnovsky, B. H., 42, 179, 524, 542, 1119
Chung, Y., 1135
Ciavarella, M., 79, 712, 1096, 1098, 1162
Cießow, G., 988
Cifaldi, N. A., 541, 542
Cittadin, R., 871
Clarey, R., 850
Clarke, E. G., 47
Clarke, M., 527
Cleghorne, W. S. H., xxxvi
Cobine, J. D., 584, 865
Coccetti, F., 713, 718
Cocks, M., 419, 420
Cohen, J. B., 428
Cohen, M., 51
Colchen, L., 648
Colclaser, R. G., 592
Coleman, D., 1112, 1114, 1139
Colgan, E., 716
Collado, C., 1112
Colonna, G., 826
Compton, K. J., 541
Comtois, J., 704
Conrad, H., 301
Conte, A. A., 1096
Cook, J., 657, 658
Corby, I. H., 384
Cordes, F., 423
Corman, N. E., 488, 492
Cornwell, L. R., 300
Correia, S., 272, 540
Cortez, R., 706, 708, 726
Cosack, U., 126, 163
Cote, W. D., 794, 802, 883, 885
Coufal, O., 827
Courtade, F., 711, 712, 724

Coutu, R., 417, 705, 706, 707, 708, 713, 716, 717, 721, 725
Cox, R. H., 22
Craggs, J. D., 555
Cranberg, L., 835
Crandall, E., 541, 711
Crane, G. R., 132, 486, 983, 984
Crane, R. L., 417, 713
Crawford, F. W., 599, 602
Cressault, Y., 827
Cross, K. J., 756, 757, 759, 993
Crossland, W. A., 163, 164, 449, 458, 474, 725, 1007
Crutcher, R., 125
Cui, L., 741
Cui, Y., 665, 922
Cunningham, F. L., 267
Cupp, G. M., 343
Curren, W. J., 169
Cuthrell, R. E., 15, 983
Czanderna, A. W., 47
Czarnecki, L., 836, 837, 838, 839

D

D'Heurle, F. M., 28
Daalder, J. E., 593, 594
Dai, H., 716
Daijima, H., 637, 638
Dalrymple, E., 388
Dang, C., 275, 315
Daube, T., 822
Davidson, R., 850
Davies, D. K., 594, 630
Davies, W. D., 594
Davis, G. J., 28, 35, 288
Davy, H., 556
Day, S., 698
De Moerlooze, K., 1101, 1102
Deal, B., 718
DeCorpo, J. J., 1117
Dehning, C., 822
Dejanovic, S., 20, 710, 1170
Delachaux, T., 840
Delhalle, J., 501
Demelio, G., 79, 1096, 1098, 1162
Denhoff, M. W., 22
Deniau, G., 501
Dennard, R., 704
Dervos, C. T., 52, 524
Desmet, H. J., 993, 995
Devautou, J., 649
Dhennin, J., 713, 718, 724

Di Meo, A., 502, 505
Dickrell, D., 709, 724
Dickrell, P. L., 1173
Dickson, D. J., 762
Diebold, U., 49
Diehl, R. P., 541, 542
Dieppedale, C., 724
Diesselhorst, H., xxxiii, 58
Dieter, B., 1073
Dietrich, J. H., 1100
Dijk, P., van, 442
Dimitrov, D., 994
Ding, G., 1075
Dini, D., 850, 855, 861
DiSalvo, N., 871
Dittes, M., 439
Dixon, C. R., 288
Djakov, B. E., 592
Do, T. K., 41
Dobner, R., 49
Dobson, J. V., 1051
Döhlen, V., von 820
Dol'nikov, S. S., 290
Doljack, F. A., 821
Domitrovich, T., 871
Don, J., 428
Donati, E., 309
Doremieux, J. L., 572, 574, 576, 662
Dorsey, G., 1112, 1114, 1128, 1139, 1154
Doublet, L., 621, 636, 637, 665, 765, 922, 1057
Douglas, P., 888
Doulet, L., 881
Dounavis, A., 1140
Dovzhenko, V. A., 37, 288
Downey, R., 388
Dowson, D., 1135
Dresvin, S. V., 826
Drew, G., 380, 1057
Driel, H. M., van, 53
Drouet, M. G., 871
Drozdowicz, E. S., 274
Drozdowicz, M. H., 23, 428, 429, 430, 451, 452, 462, 463, 464, 465, 467, 468, 476, 496, 497, 537
Drummond, C., 1158, 1171
Du, C. X., 187
Du, Y., 714, 718, 1161
Duboy, G., 131
Ducker, W., 1158, 1171
Dudder, G. J., 1154
Duggal, A. R., 821
Dugger, D., 1007

Dugger, M., 709, 724
Dullni, E., 831
Dumbauld, L. D., 674
Duquette, D. J., 271
Duvivier, P., 713
Dzektser, N. N., 309, 312
Dzierbicki, S., xxxiv, 988

E

Eapen, K., 709
Early, J. G., 73
Ebel, J., 721
Ebling, W., 576
Echlin, P., 467
Edelmann, T., 707
Edels, H., 592, 599, 602
Eden, E. M., 267
Edwards, G., 957
Egan, T. F., 23, 122, 125, 126, 154, 441, 449, 451,
    457, 458, 469, 474, 690, 1119
Ehwald, K., 704
Eidinger, A., 992, 996
Eilers, J., 382, 386
Elabdi, R., 737
Elenbaas, W., 556, 580
El-Hadachi, M., 174
Eliezer, Z., 428
Elliot, R. P., 971
Elsey, H. M., 1052
Emad, B., 1075
Emmett, D., 1073
Engel, A., von 556, 762
Engel, J. C., 806, 850, 851, 865, 866
Engel, K., 1112
Engel, P. A., 543
Enos, D. G., 121, 123, 528
Epstein, A., 458
Erben, H. F. T., xxxvi
Eremenka, V., 910
Eriksson, P., 394
Erk, A., 633, 788, 791, 799, 801, 811, 812, 904, 986,
    988, 989
Erts, D., 1103, 1111, 1112
Eskes, A., 665, 1018, 1020, 1021, 1022
Estroff, L. A., 502
Etsion, I., 711, 718, 1100, 1102, 1161
Ettling, B. V., 858, 861
Eun, Y., 716, 717
Evans, A., 355, 916
Evans, J., 572, 574
Evans, U. R., 1051, 1052
Eyring, H., 1053, 1054

F

Fagerstrom, W. B., 441, 447
Fairweather, A., 15, 274, 441, 448, 449
Fang, H., 993
Fantini, F., 295, 301
Fariborzi, H., 718
Farrall, G. A., 838
Farrell, T., 254, 278, 300, 318
Fasaki, I., 51
Faure, F., 636
Fearing, R., 720
Féchant, L., xxxiv
Feder, M., 23, 482, 483, 484, 485, 486
Feeny, B. F., 1101
Fehrl, K. G., 837
Fei, X., 456, 466
Feigenbaum, H., 380, 392
Feller, H. G., 420
Feng, C. F., 197, 198, 211, 222
Feng, I. M., 270, 416, 447, 448
Fenner, A. J., 274
Fenski, B., 608, 832, 833
Ferguson, T. P., 79
Ferris, C., 115, 116
Fiacco, V. N., 492
Fiaud, C., 52, 439
Finke, H., 904, 986, 988, 989
Finkelnburg, W., 556, 579, 582
Firestone, F., 706
Fischmeister, H., 47, 910
Fischmeister, M., 888
Fisher, R. M., 541
Fitzgerald, D., 657, 658
Fleck, N. A., 355
Fleischbein, H., von, xxxvi
Flowers, G. T., 180, 456, 466, 541, 827
Fluehmann, W. F., 431
Flurscheim, C. H., xxxiv
Fork, D. K., 449
Forsell, K. A., 278, 441, 448, 449, 464
Fortes, R. A. F. O., 302
Fortini, A., 718, 723
Foster, A., 576
Fouvry, S., 442, 443, 444, 445, 446, 449
Fowler, R. R., 567
Francisco, H., 894
Franken, H., 986
Frankenthal, R. P., 51
Franz, S., 520, 541, 542
Freeman, A. H., xxxvi
Freeman, G. H., 592
Freeman, I., 555

Freitag, W. O., 334, 449, 453
French, D. M., 1170
Frenkel, D., 718
Frenken, J., 717
Freudiger, E., 884, 885
Friedemann, W., 995
Friedrich, G., 804
Frind, G., 831
Fritz, O., 840
Fröhlich, K., 836
Frost, L. S., 604
Fry, H., 420
Fu, R., 180
Füger, M., 983
Fujita, Y., 125
Funaki, K., 624, 1008, 1013
Funami, H., 1054, 1125
Furey, M. J., 1167
Furukawa, M., 1018
Fuzier, G., 662
Fyodorov, V. N., 313

G

Gagnon, D., 283, 285, 355, 393
Gal, J. Y., 117
Galand, J., 333, 991
Gallagher, J., 865
Galyon, G. T., 541
Gambino, J., 716
Gan, H., 45
Gao, C., 1171
Gao, J., 490
Gao, J. C., 193, 197, 198, 222, 1108
Gardiol, F., 1139
Garte, S. M., 179, 431, 449, 479, 527, 529
Garzon, R., 618
Gashler, R. J., 674
Gast, A. P., 502
Gati, R., 822
Gatos, C., 1051, 1052
Gaul, D. J., 271
Gauster, E., 621, 997
Gautherot, C., 737
Gauyacq, J. P., 583
Gayford, M. L., 741
Geckle, R. J., 75, 132
Gee, M. L., 1102, 1118, 1120, 1158, 1170, 1171
Geldner, E., 653, 991
Gellman, A. J., 489
Gelmont, B., 22
Gendre, P., 540
Gengenbach, B., 895, 906

Genma, K., 740
Gentsch, D., 840
Gerdinand, F., 637
German, R., 888, 909
Germer, L. H., 622, 636, 638, 766, 767, 901, 1006,
        1007, 1008, 1009, 1011, 1012, 1015
Gervais, P., 274
Gessinger, G., 910
Gharpurey, M. K., 47
Ghezzi, L., 759
Giasson, S., 1158, 1171
Gibson, L. J., 355
Gilfrich, J. V., 351
Gilgenbach, R. M., 20, 21, 22, 1170
Ginsberg, G. L., xxxiv
Ginsburg, R., 134
Girodin, D., 1169
Givord, D., 765
Gjosten, N. A., 290
Glashorster, J. H. A., 274, 466
Glasstone, S., 1053, 1054
Glauner, C. S., 121, 123, 528
Gleizes, A., 827
Glinkowski, M., 608
Glocke, V., 806
Glossbrenner, E. W., 417, 1109, 1114, 1117, 1125
Godet, M., 271
Godfrey, D., 270
Goldsmith, W., 748, 749
Goldstein, J. I., 216, 467
Golego, N. L., 269
Golinval, J., 722
Gomez, M. F., 1044
Gomez, M. R., 1170
Goodier, J. N., 70, 76
Goody, C. P., 831
Gore, R. R., 128, 129, 133, 163
Gorman, J. G., 592, 595, 621
Gorse, C., 826
Gosele, U., 29
Göttert, B., 985, 1007, 1013
Gould, L. J., 794, 802, 883, 885
Gourdon, D., 1158, 1162
Gowaty, F. S., 993
Graddick, W. F., 28, 274
Graedel, T. E., 126, 439, 1118
Graham, A. H., 458, 469, 486
Graham, M., 1114
Graham, M. J., 51
Gray, E. W., 622, 623, 640, 762, 1006, 1007, 1008,
        1009, 1010, 1011, 1012, 1014, 1015
Green, I., 253, 1096, 1107, 1162
Greenwald, S., 858, 861

Greenwood, A., xxxiv, 583, 608, 597, 618
Greenwood, J. A., 4, 11, 13, 14, 17, 22, 25, 26, 58, 60, 63, 64, 66, 76, 77, 176, 422, 706, 712, 1015, 1016, 1017, 1170
Greenwood, N. N., 47
Gregory, G., 871
Greitzke, S., 803, 805
Griffiths, K., 53
Grimvall, G., 888
Groenendijk, H. A., 665, 1018, 1020, 1021, 1022
Gromov, V. E., 300
Grosjean, A., 540
Grosse, J., 885
Grossman, H., 737
Grossman, S., 252, 254, 267
Grove, A., 718
Grunewald, T., 1120
Guan, R., 609
Guancial, E., 972
Guansheng, Z., 982
Guenot, P., 1007, 1013
Guerlet, J. P., 995
Guerry, J. P., 649
Guery, J., 123
Guile, A. E., 583, 766, 866
Guinement, J., 52, 439
Guipoint, V., 881
Gumley, R. H., 441
Gundlach, H. C. W., 821
Gunther, O. E., xxxv
Guo, Z., 714
Guoyun, D., 1073
Gupta, A., 718
Gupta, D., 31
Gupta, R. N., 826
Gurbuz, Y., 704
Gurland, J., 913
Gurov, K. P., 290
Gusak, A. M., 290
Gustafson, J., 896
Guttmann, K., 639
Gwathmey, A. T., 47
Gyurina, D., 449, 456

H

Hackam, R., 865
Hackel, I., 806
Hagimoto, Y., 858, 861
Hagiwara, T., 858, 861
Haidar, J., 824
Hajek, K., 1112
Halkola, V., 211

Hall, J. R., 849
Hall, P. M., 21, 22
Halperin, G., 1100, 1102
Hamada, K., 698
Hamer, A. D., 693
Hammam, T., 438, 466, 476
Hamon, J., 53
Hampel, C., 941
Hanada, K., 1007, 1013
Hanada, R. P., 1007
Hanlon, J. T., 180
Hannach, R., 827
Hannel, S., 442, 443, 444, 445, 446, 449
Hansen, M., 910, 941, 943, 971
Hanson, D., 125
Hansson, T., 821
Hantschel, T., 449
Hantzsche, E., 593, 594
Hapase, M. G., 47
Haque, C. A., 23, 135, 144, 153, 163, 187, 395, 458, 525, 972, 1118
Hara, T., 737
Hardee, H. C., 442, 449, 458, 467, 472, 476, 486, 487, 504, 521
Hardy, S., 881
Hare, T. K., 333
Harlow, J., 125
Harman, D., 665, 922
Harmsen, H., 458
Harris, J., 186
Harris, J. H., 503
Harris, L. P., 593, 594
Hart, A., 714, 715
Hart, R. K., 74
Hartmann, W., 840
Hasegawa, M., 585, 588, 633, 657, 758, 940, 1038
Hashiguchi, K., 1101
Hashimoto, Y., 689
Haskins, R. E., 647, 935
Hatanaka, M., 382, 386, 459
Hattori, Y., 22, 49, 50, 259, 665, 666, 1022, 1044
Hauer, H., 995
Haufe, W., 653, 922, 991, 995
Haug, R., 572, 574, 576
Hauner, F., 894, 936
Hauser, A., 822, 840
Hauser, F., 636
Hausner, H. H., 73
Hawarth, F. E., 1006
Hayashi, Y., 737, 930
Hayes, R., 1128
Hazeghi, A., 716
He, A. Q., 539, 540

He, Y., 609
Heaton, C. E., 479
Hebb, M. H., 53
Heber, J., 520, 541, 542
Heberlein, J. V. R., 579, 583, 585, 592, 594, 595, 596, 597
Hector, L. G., 489
Heeb, P., 713, 718
Heilmann, P., 428
Heinbuch, P., 1096, 1109
Heinonen, R., 126, 129
Heitzinger, F., 1195
Heller, G., 580
Helm, C. A., 1120
Helmer, J., 837
Henmi, Z., 431, 463, 465, 481
Hennepe, J., ter, 832
Henriksen, J. F., 126
Henry, B. C., 506
Herbst, R., 649, 894
Heringhaus, F., 891
Herklotz, G., 525
Hermance, H. W., 23, 125, 126, 441, 449, 451, 457, 458, 469, 474, 690, 1119
Herring, C., 73
Hesselbom, H., 20, 710, 1170
Hessler, V. P., 1073
Hetzmannseder, E., 631, 891, 895, 995
Heuberger, M., 1158, 1171
Heyn, D., 837
Hienonen, R., 737
Higaki, S., 1125
Higaki, T., 1054
Hijikata, M., 767
Hilderink, J., 832
Hill, M., 756, 758
Hiltmann, K., 639
Hinohara, K., 674, 686, 688, 693, 695, 697
Hiramatsu, D., 1051
Hirschler, M. M., 857, 861
Hirst, W., 418
Hirth, J. P., 73, 74
Hirz, S. J., 1102, 1120
Hively, D., 1117
Ho, J. W., 428
Ho, P. S., 28, 31
Hodne, E., 42, 295, 301
Hoft, H., 641
Hohnstreiter, G. F., 622, 623, 1010, 1011
Hoidis, M., 252, 254, 267
Holden, C. A., 132, 431, 486
Holl, S., 1096, 1109
Holm, E., 557, 372, 588, 639, 646, 937, 943, 1045

Holm, R., xxxiv, xxxv, 5, 6, 8, 9, 15, 19, 23, 51, 53, 56, 57, 58, 59, 60, 62, 63, 87, 88, 89, 130, 131, 141, 176, 199, 249, 253, 258, 415, 425, 487, 557, 572, 588, 639, 646, 674, 705, 706, 707, 709, 710, 718, 724, 743, 812, 937, 943, 1042, 1045, 1047, 1057, 1058, 1066, 1071, 1072, 1073, 1098, 1107, 1108, 1109, 1111, 1112, 1158, 1167, 1170, 1174, 1195
Holmbom, G., 458, 459
Holmes, F. A., 643, 770
Holmes, R., 592
Holtgreven, L., 579
Holzapfel, C., 1096, 1109, 1140
Homma, H., 572
Homola, A. M., 1102, 1118, 1120, 1158, 1170, 1171
Honbo, R., 1076
Honda, F., 458, 460
Honig, T., 898
Honjo, G., 47, 48
Honma, N., 186
Hooghan, K. N., 439
Hooyer, J. M., 274
Hope, D., 525
Hopkins, M. R., 572, 574, 640
Höpp, W., xxxv
Horn, G. L., 27, 198, 432, 906, 971, 983, 984, 1195
Horn, J., 983
Horn, R. H., 201
Hornig, C. F., 23
Hosokawa, N., 125
Hotkrvich, V., 300
Houze, F., 123, 466, 501, 502, 525
Hovgaard, O. M., 674
Howe, R., 720
Howell, J. R., 826
Hoyaux, M. F., 557, 572
Hoyer, N., 934
Hu, L., 665, 922
Hu, Z., 1075
Huang, K., 87
Huang, Z., 821
Hubbard, D. C., 331
Huber, A., 334
Huber, B. F., 649
Huck, M., 535, 737, 803
Hueber, A., 850
Huemoeller, R., 423
Humbeek, J., van, 422
Humiec, F., 458, 459
Hunt, R. T., 77, 78, 79
Huntington, H. B., 42, 294, 295, 301
Hurkx, G., 1102
Hurricks, P. L., 269

Hutchings, I. M., 415, 416, 425, 481, 482, 488, 491
Hutchinson, J. W., 355
Hutin, L., 718
Hyman, D., 720, 724

I

Ide, T., 985, 1007
Iida, K., 1044
Ike, H., 79
Imada, Y., 458, 460
Imrell, T., 133, 274, 439
Imry, Y., 1109, 1135
Inal, K., 713
Inayoshi, N., 1076
Ingvarson, F., 22
Inoue, H., 572
Inukai, K., 1076
Ioyama, K., 635
Ireland, H., 396
Isagowa, M., 985, 1007, 1013, 1014
Isberg, J., 821
Ishida, H., 572, 693, 937, 1032
Ishino, M., 1013, 1018, 1021
Ishitani, A., 693
Israelachvili, J. N., 69, 1100, 1102, 1103, 1118,
        1119, 1120, 1158, 1162, 1170, 1171
Ito, T., 396, 821, 989, 991
Itoh, T., 718, 719, 720, 722
Ittner, W. B., 619
Ivanova, V. S., 271
Ivey, D. G., 539, 540
Ivey, H. F., 855
Iwase, H., 989, 991, 993
Izmailov, V. V., 349, 350
Izumi, K., 758

J

Jackson, R. L., 79, 253, 258, 275, 277, 307, 333,
        541, 711, 712, 713, 827, 1096, 1102, 1162
Jacobsen, K., 718
Jacobson, S., 274, 283, 439, 440, 450, 451
Jacq, C., 1169
Jaeger, J. C., 57, 1167
Jager, K., 895, 896
Jan, H., 126
Jang, Y. H., 79
Janitzki, A. S., 954
Jansenr, A. G. M., 80, 81
Janz, S., 53
Javey, A., 716
Jeandin, M., 881

Jeannot, D., 621, 636, 765, 890, 894, 903, 1057
Jedrzejczyk, P., 449
Jedynack, L., 982
Jenney, A. M., 1117
Jensen, B. D., 87
Jensen, E., 380, 392
Jeon, J., 718
Jeppson, K. O., 22
Jerz, J., 355
Jesser, W., 1159, 1164, 1165
Jesvold, H., 315
Jia, S., 840
Jian, W., 1140
Jian, Z., 1108
Jiang, L., 572, 573, 575, 639, 993
Jin, R., 986
Jing, L., 982
Jintong, X., 1073
Johannet, P., 53, 54
Johansson, L., 132, 133
Johler, W., 609, 610, 611, 983
Johnson, B., 346
Johnson, H. W., 1114
Johnson, J. L., 276, 441, 451, 464, 1050, 1054, 1076
Johnson, K. L., 71, 1100, 1101, 1103
Jondhal, D. W., 343
Jones, C. R., 640
Jones, G. R., 592
Jones, H., 71
Jones, L. K., 983
Jones, R., 572, 574
Jonsson, L., 821
Joo, Y., 715
Joos, G., 555
Jorgen, S., 382
Jost, E., 884, 885, 886
Jourdan, G., 711
Joy, D. C., 467
Judy, D., 721
Junfa, M., 1140
Juttner, B., 593, 835

K

Kaajakari, V., 704
Kabayama, S., 661, 922, 934
Kaddani, A., 824
Kadija, I., 524
Kadish, A., 571
Kadono, K., 1051, 1096
Kagawa, H., 301
Kaiser, K., 1114
Kalin, M., 1166, 1167

Kalms, R., 827
Kaltenborn, U., 252, 254, 267
Kam, H., 718
Kamada, K., 758
Kamada, Y., 585, 588
Kamans, J., 832
Kaminski, J., 925, 926, 930
Kaneko, E., 832
Kang, Y., 741
Kanno, M., 13
Kano, M., 177
Kanoh, H., 1125
Kanter, H., 81
Kapin, A. T., 594
Kappl, M., 1115
Kapsa, Ph., 442, 443, 444, 445, 446, 449
Karabanov, S., 689
Karasawa, K., 459
Karetta, F., 822
Karin, S., 649, 654
Karlsson, J., 821
Karlström, P. O., 821
Karmalkar, S., 709, 710
Kärner, H. C., 836
Kassman-Rudolphi, A., 133, 274, 283, 332, 439,
          440, 442, 450, 451, 466, 475, 476
Kataoka, K., 720
Katehi, L., 722
Kato, K., 674
Kawakita, C., 674, 686, 688, 693, 695, 697
Kawakubo, Y., 832
Kawamura, K., 858, 861, 1125
Kawarai, H., 125
Kawase, T., 858, 861
Kawase, Y., 650
Kay, P. A., 539
Kay, P. J., 41, 438, 467
Kaye, G. W. C., 1195
Kaynak, M., 704
Ke, F., 716, 717
Keefer, J. H., 441
Keil, A., xxxiv, 431, 661, 928, 930, 970
Keilien, S., xxxv
Keith, J. H., 1152, 1154
Keller, D. V., 15, 983
Kelsall, G. H., 129
Kendall, K., 71, 1103
Kennedy, F. E., 1169
Kennedy, K. K., 986
Kenny, T., 718
Kesaev, I. G., 583, 593
Key, P. L., 541
Khanna, V. D., 428

Kharin, S., 631
Khemchandan, M. V., 1107, 1119, 1170
Khlomov, V. S., 290
Khudyakov, K. I., 290
Kilgore, B. D., 1100
Kim, H., 52, 886, 896, 921, 928, 930, 931, 932
Kim, H. J., 661, 662
Kim, H. K., 1107, 1109
Kim, J., 704, 723, 716, 717
Kim, M., 716, 717
Kim, S., 1170
Kim, W., 716, 1169
Kim, Y., 704, 715, 722, 723
Kim, Z., 126
Kimblin, C. W., 579, 593, 594, 596, 602, 621, 629,
          850
Kimura, K., 1013
King, J., 28
King, L. A., 590, 591
King, P., 346
Kingon, A., 703, 714, 722
Kingsbury, E. F., xxxvi
Kinoshita, K., 858, 861
Kinoshita, M., 740
Kinsel, H., 851, 865, 866
Kippenberg, H., 1195
Kirby, J. R., 129, 133
Kitchen, N. M., 665, 1018, 1019, 1022, 1023,
          1024, 1026
Kladitis, P. E., 417, 706, 708, 713, 726
Klafter, J., 1158, 1162
Klapas, D., 865
Klein, K., 449
Klein, S., 725
Klenke, C. J., 496
Kligerman, Y., 1100, 1102
Klimov, K. M., 300, 301
Klonowski, T., 636
Klungtvedt, K., 504, 1032
Klypin, A. A., 300, 305
Ko, J. S., 489
Kobayashi, N., 1038
Kobayashi, T., 674, 686, 688, 689, 697
Koch, F. B., 422, 525
Koffler, A., 803, 891
Koga, N., 79
Kogut, L., 709, 711, 712, 713, 1102, 1112, 1170
Kohlrausch, F., xxxiii, 58, 255
Kohno, Y., 1038
Koidl, H. P., 1007, 1008, 1011, 1013
Kolb, H., 986, 988, 989
Kolker, D., 871
Kompitsas, M., 51

Komvopoulos, K., 709, 712, 1100, 1102, 1104,
        1158, 1160, 1162, 1163, 1170
Konchits, V., 704, 705, 706, 722, 723, 724, 725
Konchits, V. K., xxxiv, 177, 249, 349, 355
Kondo, K., 288, 296
Kondo, S., 1125
Kongsjorden, H., 274, 449, 456
König, D., 836
Koppelmann, F., 988
Kopper, C. H., 982
Koren, P. P., 653, 572, 575, 997
Korndorfer, F., 704
Kosco, J., 928, 934
Kosse, S., 822
Kossowsky, R., 645, 657, 896, 928
Kouakou, T., 572, 574, 576
Koutoulaki, A., 51
Kovacik, J., 355
Kovacs, G., 704
Koysma, M., 661, 922, 934
Kral, E. R., 1102
Krätzcschmar, A., 649, 821, 894
Kraus, A., 794, 898
Kraus, A. Q., 803
Kravchenko, V. Y., 300, 301
Kriebel, J. K., 502
Krim, J., 703, 704, 705, 706, 708, 712, 714, 722
Krinberg, I., 594
Kristensen, R., 857, 858, 861
Kristiansson, S., 22
Kruglick, E., 724
Krumbein, S. J., 120, 124, 274, 465, 466, 506, 525
Kubo, S., 355
Kubono, T., 588, 637, 765
Kubota, Y., 1076
Kuczynski, G. C., 73
Kudrak, E. J., 458, 459, 524, 539
Kuhl, T., 1158, 1171
Kuhlmann-Wilsdorf, D., 1076, 1134, 1152, 1154,
        1159, 1160, 1162, 1163, 1164, 1165, 1166,
        1167, 1168, 1169, 1170, 1171, 1172, 1174,
        1175, 1180, 1181
Kuhn, A. T., 129
Kuhn, H., 930, 934
Kuhn, H. D., 661
Kuipers, L., 717
Kulinovich, V., 649
Kulpa, S. H., 51
Kulsetas, J., 274, 449, 456
Kumar, A., 704
Kumar, B., 709, 710
Kume, M., 661, 922, 934
Kunkle, R. W., 331

Kuo, S., 721, 722
Kurabayashi, K., 87
Kurosawa, Y., 832
Kurrat, M., 637, 825, 827
Kurz, R., 125
Kussy, F. W., 791
Kutzner, J., 594
Kuxmann, U., 49
Kuznetsov, V. A., 300
Kuznetsov, V. D., 420
Kwiatkowski, R., 1112
Kwon, H., 715
Kwon, Y., 704, 722, 723

L

Labrecque, C., 328, 331
Labrenz, F., 924
Laby, T. H., 1195
Ladenise, H., 995
Lafferty, J. M., 557, 584
Lafontan, X., 713
Laidler, K. J., 1053, 1054
Lall, P., 541
Lambert, C., 995
Lampen, J., 721
Landman, U., 490, 1100, 1102, 1120, 1158,
        1170, 1171
Landman, W. R., 490
Lang, J. H., 22
Lang, W., 639
Lange, D., 986
Langeron, J. P., 662
Langmuir, I., 865
Lanouette, C., 341
Lapham, T., 419
Lapinski, A., 889, 925
Laskey, R. C., 134
Latham, R. V., 566, 567
Lau, Y. Y., 20, 21, 22, 1170
Lauraire, M., 821
Laurat, P., 272
Lavers, J. D., 90, 91, 92, 93, 95
Lavrenko, V. A., 971
Lavrinko, L., 910
Law, H. H., 132, 486, 501
Lawall, A., 840
Lawless, K. R., 47
Lazenby, F., 274, 441, 448, 449
Lebedev, V. P., 300
Leblanc, T., 636, 649, 826, 827
Lécaudé, N., 466, 479, 503, 540
Lécayon, G., 501, 502

Leclercq, M., 123
Lee, A., 274, 396, 402, 434, 435, 445, 475, 476, 478,
        479, 504, 579, 603, 608, 621, 638, 642,
        767, 788, 808, 812, 813, 899, 901, 938, 939,
        993, 997
Lee, H., 715, 717
Lee, J., 716, 717
Lee, J. H., 49
Lee, K. P., 826
Lee, K. Y., 449, 453, 454, 455, 466, 467, 472, 475,
        476, 479
Lee, R. E., 458, 474
Lee, S., 704, 723, 1170
Lee, T., 721, 722
Lee, T. H., xxxiv, 557, 583
Leech, P. W., 525
Leedy, J., 125
Leedy, K., 713, 717
Leedy, K. D., 417
Lees, P., 442, 458, 521
Lees, P. W., 539
Lefebre, J., 333
Legrady, J., 306
Lehfaoui, L., 585, 586, 587, 588, 1057
Leibel, J. M., 422
Leidecker, H., 541
Leidner, M., 720, 1101
Leiper, R. A., 488, 665, 984, 1015, 1016, 1017, 1018,
        1014, 1024
Leis, C., 995
Lemp, E., 639
Lencl, F. V., 73
Lenstra, K., 832
Leong, M. S., 22
Leray, D., 711, 712
Leslie, I., 458
Lessman, G. G., 28, 35, 332
Leung, C. H., 198, 396, 402, 434, 435, 445, 504,
        638, 642, 657, 658, 661, 662, 665, 767,
        899, 901, 921, 922, 928, 930, 931, 932,
        938, 939, 993
Levinson, L. M., 821
Levinson, Y., 83
Lewand, L., 125
Lewis, N. E., 1007, 1109, 1117, 1122, 1154
Leygraf, C., 126, 135
Li, D., 404
Li, H., 718
Li, W. P., 935
Li, X., 720
Li, Z., 637, 657
Liang, B., 939
Liang, J. W., 1101

Liang, Y. N., 192
Lichtenwalner, D., 703, 714, 722
Lieberman, S., 1007
Liebermann, R. W., 825
Liebich, 1007
Likhtman, V. I., 300, 301
Liljestrand, L. G., 427, 432
Lima, A., 887
Lin, C., 934
Lin, C. Y., 653
Lin, J. W., 1073
Lin, L. B., 220
Lin, X. Y., 193, 203, 208, 209, 531
Lindborg, U., 540
Lindholm, P., 827
Lindmayer, M., 608, 661, 662, 663, 665, 770, 787,
        789, 790, 791, 794, 799, 800, 801, 803,
        806, 807, 808, 813, 814, 817, 821, 822, 825,
        826, 827, 829, 832, 833, 835, 836, 837,
        839, 896, 906, 921, 928, 929, 944, 995
Ling, F. F., 75
Liou, M. J., 479
Lippert, K., 871
Liqiang, W., 827
Liu, B., 582
Liu, H., 608, 609
Liu, H. D., 200
Liu, J. J., 192
Liu, L., 721, 722
Liu, Q., 539, 540, 608
Liu, S., 1102, 1169
Liu, T., 718
Liu, Y., 637
Lizhuang, M., 1140
Llewellyn, Jones, F., xxxiv, 562, 639, 640
Lobl, H., 252, 254, 267
Lõhmus, A., 1103, 1111
Lõhmus, R., 1103, 1111
Lohrengel, J., 826
Long, T. R., 23, 458
Longo, T. A., 131
Lorant, S., 114
Lormand, G., 1169
Loud, L., 630, 935
Love, J. C., 502
Lowke, J. J., 580, 581, 590, 591, 824, 825
Lozovskaya, A. V., 37, 288
Lu, C. T., 1068
Lu, J. G., 219
Lu, M., 897, 905
Lu, X. P., 301
Lu, Y., 135, 200, 220
Lucyszyn, S., 704

Ludwar, G., 991
Lüders, C., 826
Luengo, G., 1158, 1171
Lui, L., 135
Lukas, H., 910
Luke, G. E., xxxv
Lumbantobing, A., 709
Lund, A., 41
Lundstrom, P., 466, 476
Luo, G. P., 211, 219
Lutz, M., 718
Lutz, O., 520, 541, 542, 898
Lyman, C. E., 467
Lythall, R. T., xxxiv

M

Ma, A. L., 200, 203
Ma, J. T., 186
Maboudian, R., 720
Maciel, J., 721
Mackay, C. A., 28, 41, 438, 467, 539
MacKenzie, R. W., 850
Maddox, H. L., 27
Maecker, H., 556, 579, 582
Maeda, M., 740, 1120
Magie, T., 252, 254, 267
Magnusson, P. C., 868, 869
Mah, J., 849, 854
Mahle, E., 431
Maier, R., 850, 868
Maier, W. B., 571
Maitournam, M. H., 123
Maizels, R., 689
Majumber, S., 721
Majumdar, A., 711
Majumder, S., 704, 705, 706, 708, 712
Makel, D. D., 1167, 1171
Makinouchi, A., 79
Mall, S., 717
Malucci, R. D., 79, 249, 274, 396, 447, 456, 466,
　　　　475, 650, 1107, 1108, 1111, 1112
Mamrick, M., 274, 396, 402, 475, 476, 478, 479
Mancosu, F., 1100
Mandrillon, V., 711, 713
Manhart, H., 906, 995, 995
Mann, J. Y., 274
Mannan, S. L., 301
Mannheimer, W. A., 302
Mano, K., 186, 351, 724, 986, 989, 991, 993,
　　　　1054, 1125
Manukyan, N., 355
Mao, A., 475, 479

Mapps, D., 737, 741, 744, 745, 757, 758, 767
Marinkovic, Z., 28
Marjanov, M., 266
Marks, L. D., 1103, 1111
Maroni, C., 871
Marsolais, R. M., 333
Martel, J. M., 404, 850, 871
Martens, R., 22, 714, 715, 724
Martin, J. L., 458
Martin, K. M., 179
Martin, P. J., 593, 594
Marzahn, E., 825
Masui, S., 396
Mate, C. M., 1119, 1120, 1121, 1122, 1124, 1135
Mateu, J., 1112
Matlick, R. E., 1114
Matsuda, S., 905
Matsui, Y., 431, 463, 465, 481
Mattoe, C. A., 432
Maugis, D., 1161
Maul, C., 88, 89, 449, 1111
Mauntler, N., 1154
Mausli, P. A., 431
Mayer, J. M., 28
Mayer, U., 895, 896, 906
Mayr, O., 604
McBride, J. W., 88, 89, 396, 449, 572, 573, 575, 588,
　　　　590, 605, 639, 640, 641, 642, 644, 645,
　　　　705, 707, 713, 715, 716, 721, 723, 725,
　　　　737, 739, 744, 745, 750, 751, 753, 756, 757,
　　　　758, 759, 761, 762, 764, 765, 766, 767, 769,
　　　　771, 773, 776, 778, 997, 988, 993, 997,
　　　　1057, 1111
McCarthy, B., 720, 721
McCarthy, J., 888
McCarthy, S. L., 479, 492
McCool, J., 711
McCoskrie, D., 849, 854
McDonald, S., 41
McGeary, F. L., 331
McGruer, N., 704, 705, 706, 708, 712, 714, 718,
　　　　720, 721, 1161
McGuiggan, P. M., 1102, 1118, 1120, 1158,
　　　　1170, 1171
McKinney, J. L., 1054
McKnight, L. G., 1006, 1007, 1012
Mclntyre, N. S., 449
McNab, I. R., 1172
McNeilly, K., 894
Meaden, G. T., 81, 89
Meckler, P., 637
Meek, J. M., 555
Meese, W. J., 858, 861

Mehregany, M., 720, 724
Mei, C. H., 506
Meijl, F., van, 442
Melera, A., 693
Melsom, S. W., xxxv, 308
Mendelev, M., 718, 723
Mendizza, A., 122, 154
Menet, A., 456, 457, 466
Mentel, J., 585
Mercado, L., 721, 722
Merkle, A. P., 1103, 1111
Merl, W. A., xxxiv, 432, 970, 983
Mermin, N. D., 81, 89, 96, 97, 99, 100
Messina, F. D., 496
Meyer, C., 928, 930
Meyer, C. L., 661
Meyer, H., 806
Meyer, R., 502, 505, 636
Meyerhofer, D., 54
Mezec, J. L., 186
Miao, J., 716, 717
Michael, J. R., 467
Michal, R., 789, 796, 803, 817, 895, 939, 992, 997, 998
Michalicek, M., 704
Miedzinski, B., 689
Mikani, V., 851, 865, 866
Miki, N., 610
Mikrajuddin, A., 1107, 1109
Miksic, B. A., 180
Milievskii, R. A., 290
Milkovic, M., 1051
Miller, H. C., 594, 585
Miller, T. J., 851, 865, 866
Miller, Y. G., 290
Millian, K., 983
Mills, G. W., 441, 453, 626
Min, T., 1140
Mindlin, R. D., 270
Minkenberg, J., 939
Minowa, I., 13, 24
Miroschnikov, I. P., 313
Mishina, H., 1048
Misra, B. N., 501
Misra, P., 1096
Mitani, H., 865
Mitani, S., 1013, 1018, 1021
Mitchell, B., 648
Mitchell, G. R., 594
Mitchell, J. B. A., 737, 850
Mitra, N. K., 386
Mitsumatsu, G., 1125
Mittmann, M., 983

Miyachi, K., 459, 940
Miyagawa, K., 1018
Miyajima, K., 572, 573, 627
Miyamoto, T., 821
Miyauchi, T., 1076
Miyazaki, Y., 1007, 1014
Mizuguchi, Y., 674, 697
Mizukoshi, H., 640, 767
Mizusawa, Y., 383
Moberly, L. E., 276, 441, 451, 464, 665
Mocellin, K., 1170
Moffat, H., 528
Moffat, H. K., 121, 123
Mohan, P., 709, 710
Molanin, K. I., 420
Mollen, J. C., 144
Möller, K., 836
Mollimard, D., 115, 116
Monpert, 1007, 1013
Moody, J. J., 1107
Moore, A. J. W., 77
Moos, F., 689
Mori, H., 650
Mori, S., 452, 458
Mori, T., 821
Morimoto, I., 767
Morin, L., 765, 890, 903, 1057
Morita, N., 1044, 1051, 1054, 1075, 1076, 1096
Morris, A., 703, 714, 722
Morrison, R., 704, 705, 706, 708, 712, 721
Morrow, R., 824
Morse, C. A., 458, 476, 486, 487
Moser, T., 885
Motoyama, A., 1007, 1014
Mott, N. F., 53
Mottine, J. J., 201, 274, 441, 448, 453, 463
Mou, R., 993
Mouanda, B., 501
Mroczkowski, R. S., xxxiv, 75, 132, 249, 253, 391, 392, 417, 506
Mueller, D. A., 806
Mueller, F. M., 80, 81
Müller, R., 936
Mützel, T., 815, 892, 894, 907
Mugele, F., 490, 491
Muhlethaler, K., 1007
Muller, K., 943
Muller, P., 850, 868
Muniesa, J., 599, 991
Murakami, Y., 1076
Murakawa, K., 431, 463, 465, 481
Murakawa, M., 79
Murdoch, I. A., 1007

Murolo, C., 1096, 1098, 1162
Murphy, A. B., 826, 827
Murphy, P., 725
Murphy, P. M., 1007
Murphy, P. M. K., 449, 458, 474
Mutoh, A., 572, 573, 627
Mutzke, A., 825, 827
Myasnikova, M. G., 313
Myers, M., 53, 440, 442, 524, 720, 894, 1101
Myshkin, N., 704, 705, 706, 722, 723, 724, 725
Myshkin, N. K., xxxiv, 249, 349, 355

N

Na, S. J., 17
Nabeta, Y., 49, 259
Nadeau, F., 871
Nagae, A., 492
Nagai, A., 693, 695
Nagakura, K., 1073
Nagao, E., 125
Nagao, M., 674
Nagaraju, J., 1096
Nagasaka, S., 1076
Nahemow, M. D., 572, 574, 575, 577
Naidick, Y., 910
Naillat, M., 525
Nakahara, S., 422, 525
Nakajima, K., 458, 460
Nakamura, A., 393
Nakamura, M. M., 10, 13, 24, 288, 296
Nakamura, S., 132
Nakanishi, T., 288, 296
Nakhla, M. S., 1114, 1140
Napp, D.T., 460
Nardone, D., 1108
Nathanael, R., 718
Natishan, M., 724
Navik, G., 857, 858, 861
Nayak, A., 851, 865, 866
Nayak, P. R., 75
Naybour, R. D., 254, 278, 300, 318
Nedelec, L., 585, 586, 587, 588, 1057
Needham, E., 993
Neijzen, J. H. M., 274, 466
Nelias, D., 1169
Nelson, F. G., 288
Nelson, R., 910
Neri, B., 295, 301
Neufeld, C., 1013, 1014
Neufeld, C. N., 624, 982, 985
Neuhaus, A., 903
Neveri, I., 770

Newbury, D., 858, 861
Newbury, D. E., 467
Newman, H., 721
Newton, P., 501, 502
Nguyên-Duy, P., 341
Nichting, R., 957
Nicolet, M. A., 41
Nicotera, E. T., 441, 447
Niederreuther, R., 892, 894, 907
Nikolic, B., 723
Nilsson, S., 328, 329
Nitta, S., 572, 573, 627
Nobel, F. I., 422, 458, 486
Nobuta, K., 1125
Noël, S., 180, 272, 393, 466, 479, 501, 502, 503, 505, 540
Noels, L., 722
Nogita, K., 41
Nomura, Y., 380
Norberg, G., 20, 710, 1170
Nordheim, L. W., 567
Norose, S., 1048
North, K. R., 1109
Norton, P. R., 53
Nothnagel, F., 649
Nottingham, M. C., 431
Nouri, H., 631, 649, 654
Novakovic, J., 524
Novikov, N. I., 300, 301
Novoselova, M. V., 349
Nowlen, S. P., 871
Noyan, C., 428
Nunogaki, K., 865
Nussbaum, A., 22
Nuzzo, R. G., 502

O

Ober, A., 393
Oberg, A., 254, 307, 328, 329
Oberg, E., 869
Oberhammer, J., 716, 717
Oberndorff, P., 439
Oblas, D., 1007
Obrist, 53
O'Callaghan, J. M., 1112
Ochs, S. A., 54
Odgen, H., 1118
Oding, I. A., 271
Ogawa, K., 905
Ogden, H. R., 141
Ohmae, N., 421
Ohno, A., 383

Ohta, M., 905
Ohtani, H., 858, 861
Okada, M., 431, 463, 465, 481
Okamoto, H., 186
Okazaki, K., 301
Okinaka, Y., 525
Oks, E., 594
Okumura, E., 832
Okuyama, K., 1107, 1109
Olin, H., 1103, 1111
Olney, F. D., 1058
Olsen, D., 41, 957
Olsson, E., 1111
Olsson, K. E., 254, 307, 385, 393
Ono, K., 674
Oosterhout, H., van 52
Opila, R. L., 486
Orloff, N. D., 1112
Osenbach, J. W., 439
Osias, R. J., 45, 465, 463
Ostendorf, F., 393
Ostrowski, J., 822
Otaka, T., 1051
Ovcharenko, A., 1100, 1102
Ozaki, S., 1101

P

Pacheco, S., 722
Padilla, A., 1112
Palacin, S., 501
Palumbo, T. A., 486, 525
Pamies-Teixeira, J. J., 427
Pan, J. M., 1103, 1158
Panashchenko, L., 541
Panek, J., 837
Panella, A. P., 1107
Park, C., 1103, 1158
Park, J., 722, 715
Park, J. S., 1195
Park, S. J., 49
Park, S. W., 17
Park, W., 718
Park, Y. W., 449, 453, 454, 455, 466, 467, 472, 475,
    476, 479
Parker, A., 274, 441, 448, 449
Parker, K. L., 851, 865, 866
Partridge, A., 718
Parvanov, P., 994
Paschen, F., 564
Patton, S., 709, 718
Pavageau, V., 42
Pebler, A., 934

Pechrach, K., 737
Pecht, M., 524, 724
Peck, R. I., xxxiv
Pedder, D., 887, 888
Pedersen, K., 53
Peek, R. L., 674, 741
Peekstok, K., 274
Pei, J., 827
Peng, L., 716
Peng, Z., 1170
Pennec, F., 711, 712
Pentlicki, C. J., 1125
Perdigon, P., 287, 456, 466
Peretyatko, V. N., 300
Perni, S., 722
Peroulis, D., 722
Perreault, G. E., 674
Perrin, C., 394
Perry, S. S., 1154
Persson, B., 712
Persson, B. N. J., 79, 1100, 1102, 1120, 1170
Peschot, A., 711, 724, 720, 725
Peters, J., 991
Peters, P., 886
Peters, T., 896
Peterson, A. S., 79
Peterson, P., 439
Pethica, J. B., 15
Petit, L., 439
Peveri, V., 1100
Peyrou, D., 711, 712
Pfender, E., 583, 585
Pharney, J. R., 993
Philip, G., 838, 839
Philips, C. F. T., 439
Phipps, P. B. P., 439
Pike-Biegunski, M. J., 458
Pimenov, V. N., 290
Pinnel, M. R., 27, 28, 478, 479, 533, 537
Pister, K., 724
Pitney, K., 706, 970, 983
Piva, D., 822
Plana, R., 711, 712, 713, 718, 724
Plyutto, A. A., 594
Poate, J. M., 28
Põdra, P., 1107
Pokropivny, A. V., 1103, 1111
Polchow, J. R., 79
Polevoy, I., 659, 660, 980, 981, 892, 900, 901,
    1012, 1032
Pollock, H. M., 1135, 1161
Poniatowski, M., 794, 889, 895
Pons, F., 630, 826, 827

Pons, P., 711, 712, 713, 718, 724
Porte, M., 637
Porter, J., 621
Porto, D. R., 594
Porto, M. F. S., 302
Potinecke, J., 163, 164
Pott, V., 718
Potteiger, B. D., 439
Potter, D., 720, 721
Poulain, C., 123, 711, 720, 724, 725
Pradille, C., 1170
Pramanik, M. A. K., 993
Praquin, J., 737
Prasad, S. V., 1096
Prechtel, U., 725
Preston, P. F., 23
Probert, S. D., 11
Prokopovich, P., 722
Puchkarev, V., 576
Puckett, H. R., 865
Pullen, J., 77, 79
Pursch, H., 827, 835
Putvinski, T. M., 501

Q

Qi, L., 582
Queffelec, J. L., 585, 586, 587, 588, 641, 985,
        1007, 1013

R

Rabinowicz, E., 271, 274, 415, 421, 425, 428,
        431, 452, 1013, 1134, 1135, 1157, 1159,
        1161, 1167
Rabkin, D. M., 37, 288
Radtke, R., 576
Rakkolainen, J., 126, 129
Ramini, P., 765
Ramoni, P., 621
Rapeaux, M., 42
Rating, F., 986
Ratliff, E. T., 135, 187, 1118
Rauterberg, U., 1007
Rayne, J. A., 28, 292
Razafiarivelo, J., 637
Read, M. B., 22
Reagor, B. T., 186, 201, 274, 441, 448, 453, 459,
        463, 474
Rebeiz, G., 703, 704, 723
Reed, C. W., 1109, 1117, 1122
Reichart, M., 758, 903
Reichel, W., 653, 922, 991

Reichner, P., 1125, 1172
Reid, F., 896
Reid, F. H., 431, 460
Reid, J., 706, 726
Reider, W., 588, 589, 590, 591
Reimann, N., 799
Reininghaus, U., 836, 838, 839
Ren, W., 741
Ren, Z., 715
Renz, R., 840
Restrepo, C. E., 851, 865, 866
Revay, L. B., 427, 432, 537, 538
Rezvanian, O., 712
Rice, D. W., 144, 153, 163, 164, 439
Richardson, M. D., 135, 187, 1118
Rickett, B. I., 827
Rieder, W. F., 621, 624, 631, 632, 636, 661, 740,
        765, 789, 790, 794, 802, 803, 836, 891,
        895, 903, 906, 930, 82, 983, 985, 986, 988,
        989, 990, 991, 992, 994, 995, 996, 999,
        1007, 1008, 1011, 1013, 1014, 1025
Rigby, E. B., 439
Rightmire, B. G., 270, 447, 448
Rigney, D. A., 428
Rijke, A. M., 1172
Ring, K., 380, 381
Riß, J., 827
Risto, H., 126
Rivenc, J., 737
Rivene, J., 648
Rivoirard, S., 621, 637, 765
Robbins, M. O., 1102, 1120, 1158, 1170, 1171
Robbins, R. C., 187
Roberts, A. D., 71, 1103
Roberts, M. W., 54
Robiscoe, R. T., 571
Rocas, E., 1112
Rochus, V., 722
Rockfield, M. L., 343
Rockland, J. G. R., 73
Rodekohr, C., 541
Rokni, M., 83
Rolland, G., 881, 909
Rolle, S., 821
Rondeel, W. G. J., 832
Rong, M., 826, 827
Ronnquist, A., 47
Roos, J., 422
Roschupkin, A. M., 301
Rose, W. N., 267
Rosen, G., 525
Rosenberg, K., 1102, 1120
Rosenfeld, A. M., 7, 89, 90

Rosenfield, A. R., 428
Roth, M., 661, 662, 663, 921, 928, 929
Rothkegel, B., 885
Roullier, L., 278
Royce, B., 343
Rozno, A. G., 300
Rümpler, C., 609, 822, 824, 826, 827
Ruengsinsub, P., 439
Rüther, T., 825, 827
Ruitenbeek, J. M., van, 710, 1109
Runde, M., 42, 290, 295, 301, 315, 857, 858, 861
Ruppert, C., 42
Russell, C. A., 125, 126, 186, 665, 1018, 1019,
        1022, 1023, 1024, 1026
Russ, G. J., 465
Russo, R. A., 458
Rusteberg, C., 835
Ryabov, V. R., 37, 288
Ryen, L., 1103, 1111
Ryu, H., 1101
Ryzhkov, V. G., 300
Ryzhkov, V. N., 594

S

Saarinen, T., 126, 129, 737
Saba, C. S., 496
Saeger, K., 895, 896, 939, 1195
Saeger, K. E., 794, 817
Saeid, Z., 126
Saha, M. M., 579
Sainsot, Ph., 123
Saint-Michel, J., 1172
Saitoh, Y., 49
Saka, N., 274, 427, 428, 479, 1013
Sakura, T., 985, 1007
Salehi, M. T., 28, 33, 34, 35
Sallais, D., 899
Salmeron, M., 490, 491
Salvarezza, R. C., 502
Salzmann, Q. R., 1007, 1008, 1011, 1013
Sämann, D., 836
Samoilov, V. N., 1100
Samsonov, G., 930
Samsonov, G. V., 47, 51
Sanders, D. M., 593, 594
Sandler, B. Z., 986
Sandmaier, H., 639
Sankara Narayanan, T. S. N., 449, 453, 454, 455,
        466, 467, 472, 475, 476, 479
Sano, Y., 17, 18
Santala, T., 886
Santer, E. A., 986

Sapjeta, J., 132, 501
Sarabandi, K., 722
Sasada, S., 1048
Sasamoto, T., 985, 1007
Sasao, S., 821
Sato, M., 761, 767, 768
Sato, T., 431, 453, 465, 481
Savio, L., 125
Sawa, K., 459, 610, 637, 638, 640, 643, 659, 660,
        767, 940, 1032, 1044, 1054, 1057, 1061,
        1063, 1073, 1075, 1076, 1096
Sawada, S., 22, 49, 50, 259, 665, 666, 709, 1022,
        1044
Sawyer, E., 467
Sawyer, L., 467
Sawyer, W. G., 1073, 1152, 1154, 1155, 1173
Schade, E., 840
Schaefer, B., 954
Schaffer, W., 988, 990, 991
Schaudt, G., 737
Schellekens, H., 832
Schetky, L. M., 328, 331
Schetky, McD., 328, 351
Schey, J. A., 75
Schiff, K. L., 431
Schiff, K. L., 163, 164
Schindler, J. J., 317, 326
Schiweck, L., 829, 835
Schlegel, S., 252, 254, 267
Schlitz, L. Z., 822, 826
Schmeizle, M., 633, 788, 791, 798, 799, 801, 811, 812
Schmidt, H., 1101
Schmitt, W., 520, 541, 542
Schnabel, R., 794
Schneegans, O., 502, 503, 505
Schneider, H., 588, 590, 1007
Schoepf, T., xxxiii, xxxiv, 636, 805, 898, 1057
Schopf, T. J., 740, 765
Schrank, C., 758
Schrank, C., 903, 1038
Schreiber, C., 380, 392, 448, 449, 466
Schreiber, F., 502
Schreier-Alt, T., 380, 381
Schreiner, H., 653, 922, 991, 995
Schroder, D. K., 22
Schroder, K. H., 665, 770, 789, 791, 794, 803, 804,
        817, 889, 906, 909, 925, 930, 937, 939, 944,
        986, 988, 989, 993, 995, 1195
Schubert, M., 821, 822
Schubert, R., 447, 1009, 1010
Schulman, M. B., 592, 595, 596, 597, 630
Schultz, L. K., 492
Schulz, E., 889, 895

Schulz, E. D., 995
Schulz-Gulde, E., 579
Schumacher, A., 639
Schuylenbergh, K., van, 449
Sears, F. W., 100
Seeger, E. W., xxxvi
Seelikson, B., 131
Seibles, L., 458, 453, 459, 474
Seidel, H., 725
Sekikawa, J., 765
Selle, S., 826
Serpinet, M., 871
Shackleford, J. F., 1195
Shao, C. B., 203
Sharkh, S. M. S., 588, 590, 605, 640, 641, 642, 645,
        753, 756, 757, 761, 762, 764, 766, 767, 769,
        771, 773, 778, 988, 993
Sharma, J., 1107, 1170, 1119
Sharma, S. P., 144, 153, 163
Sharvin, Y. V., 81
Shaw, N. C., 986, 988, 990
Shea, J. J., 404, 645, 649, 759, 851, 855, 856, 857,
        858, 859, 860, 861, 862, 863, 864, 865,
        866, 867, 868, 870, 871, 997
Shearer, M. P., 28
Sheeler, F. G., 27
Shen, M., 883, 885
Shen, Y. S., 794, 802, 887
Sheu, S., 79
Shevelya, V. V., 269
Shi, F. G., 1107, 1109
Shi, Z., 840
Shibata, A., 895, 905
Shigemori, T., 1073
Shim, E., 722
Shimada, S., 22, 624, 1008, 1013
Shimizu, Y., 698
Shimoda, N., 1057, 1125
Shimoyama, T., 1076
Shinchi, A., 458, 460
Shirakura, K., 985, 1007, 1013
Shitara, Y., 452, 458
Shklyarevskii, O. I., 81
Shmelev, S. D., 840
Shnyrev, V. D., 300, 301
Shobert, E. I., 15, 1042, 1045, 1049, 1055, 1056,
        1057, 1058, 1064, 1065, 1067, 1068, 1164
Sholz, R., 704
Shook, R. L., 439
Shrank, T., 637
Shugg, W. T., xxxiv
Shur, M., 22
Sieber, C. S., 417, 506

Siegel, D. J., 489
Siegel, R., 826
Sikula, J., 1112
Silveira, V. L. A., 302
Simancik, P., 355
Simic, V., 28
Simmons, J. G., 54
Simon, D., 115, 116
Simon, S., 394
Singer, I. L., 1135
Singh, K., 1107, 1119, 1170
Sivebaek, L. M., 1100
Sjogren, L., 427, 432
Sjovall, R., 133, 439
Skiles, J. A., 1154
Skolnik, M. I., 865
Skuggevig, W., 850
Slade, H. C., xxxiii
Slade, P. G., xxxiii, xxxiv, 125, 187, 564, 566, 568,
        572, 574, 575, 576, 577, 579, 592, 593, 594,
        596, 597, 598, 603, 607, 608, 609, 618, 620,
        621, 630, 639, 640, 643, 645, 647, 649,
        650, 652, 653, 657, 659, 661, 662, 664,
        770, 788, 808, 812, 813, 828, 843, 835,
        838, 896, 919, 922, 928, 930, 931, 933,
        934, 935, 936, 997, 1195
Slepian, J., 599, 788, 798, 812
Sletbak, J., 274, 449, 456, 857, 858, 861
Slocum, A., 714, 715
Slocum, A. H., 22
Slonim, A. A., 986
Slonim, M., 986
Smirou, P., 474
Smit, B., 718
Smith, E., 396
Smith, E. F., 449, 456
Smith, J. E., 1006, 1007, 1008
Smith, J. L., 622
Smith, L., 849, 854
Smith, R. K., 833
Smith, S. B., 1108
Smith, T., 417
Smythe, W. R., 6, 90
Snowdon, A. C., 646
Söderberg, A., 1170
Soderberg, S., 442, 443, 444, 445, 446
Sohme, T., 832
Sollars, P. M. A., 384, 970
Solomon, A. J., 422
Solos, L. P., 15, 16, 1007, 1119
Sommerville, J. M., 556
Somorjai, G. A., 693
Sondergaad, N. A., 1167

Sone, H., 890, 903, 937, 941, 1032
Song, J., 717, 1103, 1161
Sorensen, M., 718
Sorzoni, A., 295, 301
Souma, S., 596
Souques, G., 653
Souter, J. W., 274, 431
Spaeth, D., 954
Spalvins, T., 482, 1119
Spatami, C., 583
Späth, D., 898
Spearing, S., 715, 716
Spencer, M., 718
Spille-Kohoff, A., 824
Spitsyn, V. I., 300, 301
Sprackling, M. T., 71
Sprecher, A. F., 301
Sprecher, J., 343
Springstubbe, M., 825
Sproles, E. S., 132, 163, 164, 271, 274, 441, 447,
        440, 470, 501
Srolovitz, D., 717, 718, 723, 1103, 1161
Staley, P., 851, 865, 866
Stammberger, H., 609, 789, 822, 826, 827
Starman, L., 705, 707, 713, 725
Stashenko, V. I., 300, 301
Staunton, W., 274
Steadly, H., 179
Steenstrup, R. V., 492
Stefanov, S., 994
Stege, M., 815
Steinemann, S. G., 431
Stennett, N. A., 476
Stepke, E., 984
Stjernberg, K., 913
Stoeckel, D., 943, 965, 970, 972
Stojanovic, V., 718
Stokes, A. D., 871
Stowers, L. F., 271
Strack, H., 22
Straschil, H. K., 524
Strawser, R., 706, 708, 726
Streator, J., 712, 1096, 1169
Streator, R. L., 79
Streicher, E., 657, 658, 922
Strobel, C., 850
Strof, T., 997, 1013, 1025
Sudhölter, H. W., 790
Suga, T., 718, 719, 720, 722
Sugawara, S., 832
Suggs, A. M., 995
Sugimoto, H., 1032
Sugimura, K., 492

Sugiura, T., 637
Suh, N., 1101, 1103, 1135
Suh, N. P., 271, 425, 427, 479
Suhara, K., 588, 993
Suhling, J., 541
Suits, C. G., 579
Sumption, A., 756, 765
Sun, A., 528
Sun, A. C., 121, 123
Sun, B. S., 192
Sun, H., 608
Sun, J., 1109, 1125
Sun, M., 724
Sun, T. C., 428
Sun, W., 840
Sun, Y., 704
Sundberg, R., 466
Sundklakk, H., 857, 858, 861
Suratkar, P., 439
Susan, D., 893
Susnik, E., 891
Suzuki, K., 637, 638, 674
Svedlung, O. A., 132, 133
Svensson, K., 1103, 1112
Sweeting, D. K., 871
Swift, J. A., 1076
Swinehart, M. R., 988
Swingler, J., 88, 89, 476, 642, 644, 756, 765, 1111

T

Tabor, D., 14, 15, 69, 415, 422, 425, 481, 482,
        748, 1096
Tahara, N., 9
Taheri, A. K., 28, 33, 34, 35
Takagi, T., 937, 941, 993, 1032
Takago, T., 572
Takahashi, A., 799
Takahashi, M., 674
Takakura, T., 865
Takanezawa, M., 1051
Takano, E., 351, 724
Takano, N., 821
Takao, H., 1125
Takaoka, M., 1073
Takei, K., 716
Takenaka, Y., 624, 1008, 1013
Tamagawa, T., 596, 832
Tamai, T., 22, 49, 50, 176, 177, 259, 393, 665, 666,
        940, 1018, 1019, 1020, 1021, 1022, 1024,
        1044, 1054
Tamaki, Y., 585, 588
Tamashita, S., 596

Tamimura, J., 125
Tamura, M., 1007, 1013
Tan, L. S., 22
Tanaka, H., 1044
Tang, L., 203
Tang, W., 720, 1134, 1170
Tangena, A., 1102
Taniguch, M., 572
Tanii, T., 674
Tarquin, D., 504
Taschke, H., 382
Taylor, D. E., 271
Taylor, E. D., 568, 647, 935
Tegehall, P. E., 41
Teillet-Billy, D., 583
Tenbohlen, S., 850, 868
Terhoven, P., 821, 827
Terriault, T., 328, 331
Terriault, Y. Y., 328, 331
Terrien, J., 53
Testé, Ph., 466, 479, 583, 636, 649, 826, 827
Teufet, P., 827
Tewari, A., 1073
Thiede, H., 458
Thiesen, P. J., 278, 441, 448, 449, 464, 649
Thilmont, S., 725
Thölén, A., 1111
Thomas, K., 54
Thomas, T. R., 4, 11, 17
Thomas, U. B., 47, 129, 142, 144, 169
Thompkins, H. G., 163
Thompson, D. H., 143
Thompson, J. J., 556
Thompson, P. A., 1102, 1120, 1158, 1170, 1171
Thompson, R. A., 826
Thomson, D. W., 422
Thostenson, E., 715
Tian, H., 274, 428
Tian, S., 850
Tian, X., 1169
Tian, Y., 1120
Tiefel, G., 936
Tien, C., 711
Tierney, V., 122
Tillack, B., 704
Timoshenko, S., 70, 76
Timsit, R. S., 7, 8, 19, 20, 21, 22, 28, 29, 30, 31, 32,
　　　　33, 34, 35, 36, 37, 38, 39, 42, 43, 45, 46,
　　　　47, 53, 56, 62, 66, 67, 68, 69, 70, 71, 72,
　　　　73, 74, 75, 81, 82, 83, 84, 85, 86, 87, 89,
　　　　90, 91, 92, 93, 95, 290, 302, 317, 326, 346,
　　　　385, 391, 438, 488, 492, 639, 709, 710,
　　　　1107, 1110, 1112

Tipping, D. W., 15
Tirrell, M., 1120
Tisley, D. G., 693
Tittes, E., 986
Toben, M. P., 458
Toda, H., 832
Toler, B., 721
Tomlison, G. A., 267
Tompkins, H. G., 28, 274, 478, 479, 537
Tosatti, E., 1100, 1120, 1170
Tøtdal, B., 42, 295, 301
Townsend, J. S., 556
Toyama, S., 125
Trachslin, W., 665, 1015
Travers, D., 585, 586, 587, 588, 641, 985,
　　　　1007, 1013
Trcka, M. J., 334
Treadwell, W. D., 53
Tremoureaux, R., 439
Tribulato, G., 759
Tripathi, V. K., 868, 869
Tripp, J. H., 45, 463, 465
Tristani, J. I., 466
Tristani, L., 502, 503, 505
Trivedi, H. K., 496
Trochu, F., 328, 331
Troitskii, O. A., 300, 301
Tronkova, M., 994
Trzeciak, M., 144
Tsuda, H., 832
Tsukiji, S., 22, 709, 1044
Tsutsum, T., 596
Tu, R., 716
Tu, T. N., 28, 29, 45
Tuma, D. T., 594, 630
Tuohy, E. J., 15, 831
Turner, C., 630, 632, 656, 657, 921, 944, 994, 999
Turner, H. W., 630, 632, 656, 657, 921, 944, 994,
　　　　999
Tusche, R., 991
Tveite, G. I., 315
Tylecote, R. F., 259, 415

U

Uchimura, K., 865
Uda, S., 1054, 1125
Uecker, A., 1076
Ueno, T., 1044, 1051, 1054, 1075, 1096
Uhlig, H. H., 270
Uhrig, T., 1008, 1009, 1010, 1011
Uhrig, T. A., 622, 623
Umemura, S., 392

Umezaki, S., 698
Underwood, E., 913
Unger-Weber, F., 832, 833, 835
Unsworth, D. A., 28, 539
Updergraff, S. W., 458
Uranek, J., 989, 992
Urbakh, M., 1158, 1162
Urbas, J., 858, 861
Urquhart, A. W., 28
Utsumi, T., 572, 640, 762, 763

V

Vaccaro, B. T., 439
Vanderlick, T., 717
Vannerberg, N., 132, 133
Varma, S. K., 300
Vassa, A., 621, 637, 765
Vassiliou, P., 52, 524
Vatchiantz, S., 355
Veit, C., 906, 999
Vela, M. E., 502
Vericat, C., 502
Vernon, W. II. J., 143, 1051, 1053
Vick, B., 1167
Viel, P., 501, 502, 505
Vigdor, L., 970
Vijaywargiya, R., 1107
Villien, P., 126
Vinaricky, E., xxxiv, 817, 835, 836, 945, 970, 1195
Vincent, L., 442, 443, 444, 445, 446, 449
Vincent, M., 724, 725
Vingsbo, O., 442, 443, 444, 445, 446
Vithayathil, J., 621
Vižintin, J., 1167
Vladimirescu, M., 1112
Vogel, V., 641
Völker, F., 796, 797, 798
Vorlaufer, G., 758
Voshall, R. E., 596, 621, 834

W

Wabner, A., 821
Wada, Y., 821
Wadell, B. J., 1114
Wadley, H. N. G., 355
Wadlow, H. V., 125, 126
Wadwalkar, S., 253
Wafer, J. A., 850
Wagar, H. N., xxxiv, 544, 545, 557, 568, 569, 570,
    588, 624, 625, 626, 627, 628, 674, 741, 937
Wagner, C., 884

Wagner, F. J., 737, 803
Wagner, M., 126
Wagner, R., 850
Waine, C. A., 970
Wakabayashi, H., 1076
Wakatsuki, N., 572, 697
Walcuk, E., xxxiv, 633, 634, 652, 653, 655, 656,
    921, 922, 988, 995
Wallace, S. J., 1076
Wallach, E. R., 28, 35, 288
Wallinder, I. O., 394
Walsh, M., 423
Walter, A., 986
Walton, K. A., 693
Wan, J. W., 193
Wan, N., 192, 193
Wang, B. J., 1013
Wang, C., 718
Wang, D., 395
Wang, H., 200, 220
Wang, K., 637
Wang, L., 741, 827, 840
Wang, P., 720
Wang, Q., 716, 1102, 1169
Wang, Q. J., 1107, 1119, 1169, 1170
Wang, R., 637
Wang, S., 716
Wang, W. M., 539, 540
Wanheim, T., 79
Waram, T., 328
Warke, W., 915, 916
Warren, J. L., 791
Warwas, T., 794
Warwick, M., 439
Wassermann, J., 997, 999
Wassink, R. J. K., 41
Watanabe, M., 637, 638, 674
Watanabe, Y., 15, 16, 572
Waterhouse, R. B., 269, 271
Waymouth, J. F., 603
Weaver, P. H., 997
Weaver, P. M., 737, 759
Webber, S. W., 452
Weber, W. H., 492
Weertman, J., 73
Wehr, A., 442
Wehr-Aukland, A., 521
Wei, I.-Y., 465, 466
Wei, X. M., 1118
Weichert, H., 790, 805, 810, 816
Weichsler, V., 631, 632, 803, 986, 988, 989,
    990, 994
Weilley, K. C., 351

Weißenfels, C., 1107, 1120
Weisshaar, A., 868, 869
Wen, X. M., 199, 263, 506
Wendelstorf, J., 824
Wenzel, N., 840
Wexler, G., 80, 81, 708, 1111
White, J., 301
White, K., 871
White, L., 187, 195
White, P. J., 737, 741, 744, 745, 757, 758, 767
Whitehouse, D. J., 75
Whitesides, G. M., 502
Whitlaw, K. J., 431, 458
Whitley, J. H., 133, 274, 334, 441, 448, 449, 453,
    464, 465, 466, 506
Widl, W., 835
Widmann, W., 790, 995
Wien, R. E., 621
Wiersch, M., 820
Wilkening, E. D., 825, 837
Wilkening, W.-D., 637
Williams, D. W., 122
Williams, D. W. M., 486
Williamson, J. B. P., 4, 14, 17, 22, 25, 26, 58, 60,
    63, 64, 66, 75, 76, 77, 78, 79, 186, 208,
    219, 249, 251, 253, 255, 256, 284, 464,
    706, 1015, 1016, 1017, 1118
Wilson, J., 858, 861
Wilson, J. T., 41
Wilson, W., 712
Wilson, W. E., 827, 1163
Wilson, W. R. D., 79
Wiltshire, B., 144
Windred, G., xxxvi
Wingert, P. C., 661, 662, 930, 943, 944
Wipf, C., 704
Wirths, A., 889
Witherspoon, B. K., 1112, 1114, 1122, 1139
Witska, R., 129, 133
Witter, G. J., 198, 488, 643, 644, 659, 660, 661, 665,
    891, 892, 893, 897, 898, 900, 901, 903,
    905, 912, 913, 915, 916, 917, 918, 920, 922,
    928, 931, 933, 934, 954, 957, 980, 981,
    984, 1012, 1015, 1016, 1017, 1018, 1024,
    1018, 1028, 1032
Wojcik, D., 656
Wolf, J., 756, 757
Wollenek, A., 988
Wolmer, R., 891
Wolowodiuk, W., 422
Wong, H., 716
Wong, L., 449
Wriggers, P., 1107, 1120

Wright, D., 1125
Wright, I. S., 431
Wright, K. H., 274
Wright, R., 41
Wu, J., 524
Wu, L., 722
Wu, Y., 826, 827, 1134
Wulff, F. W., 41
Wyant, F. J., 871
Wyatt, J. R., 1117
Wyder, P., 80, 81

X

Xia, J., 1075
Xu, K., 720
Xu, L. J., 222, 689

Y

Yaglioglu, O., 714, 715
Yakowitz, H., 216
Yama, G., 718
Yamada, N., 125
Yamaguchi, M., 186
Yamamoto, A., 697
Yamamoto, T., 1058, 1063
Yamashita, T., 718, 719, 722
Yamazaki, S., 674
Yan, W., 1163
Yanabu, S., 596, 832
Yanabu, T. S., 596
Yanagisawa, H., 1051
Yang, C., 1102, 1120
Yang, F., 827
Yang, Q., 826
Yang, S., 524
Yang, Z., 703, 714, 722
Yano, T., 674, 686, 688, 693, 695, 697
Yanobe, T., 635
Yanson, I. K., 81
Yasuda, K. T., 392
Yasuda, M., 674, 686, 697
Yaung, J., 718
Ye, N., 712
Yew, C. H., 748
Yeyati, A., 710
Yeyati, A. L., 1109
Yin, X., 1100, 1104, 1158, 1160, 1162
Yiying, L., 827
Yokokawa, T., 688, 691, 693, 695
Yokoyama, K., 689
Yonezawa, A., 986

Yonezawa, Y., 288, 296
Yong, H. J., 1170
Yoon, K. H., 621
York, D., 1073
Yos, J. M., 826
Yoshida, K., 637, 638, 985, 1007, 1013, 1014
Yoshimura, T., 740
Yoshioda, K., 177
Yoshioka, Y., 650
Yoshizawa, H., 1102, 1120
Yu, P., 135, 195
Yune, Y. G., 1068
Yunus, E., 715, 716
Yushkov, G., 594

Z

Zabinski, J., 709, 718
Zacharias, A., 609, 824, 826, 827
Zachariassen, K., 126
Zakipour, S., 126, 135
Zalucki, Z., 594
Zavracky, P., 704, 705, 706, 708, 712, 721
Zeigler, R., 997
Zeltser, I., 689
Zeng, G., 41
Zeng, H., 1120
Zengchao, L., 827
Zerallil, Y., 881
Zessak, I., 922, 923
Zhai, G., 582
Zhang, D., 756, 758

Zhang, G., 1134
Zhang, H., 657, 1107
Zhang, J. G., 135, 175, 187, 190, 191, 192, 193, 195, 197, 198, 199, 201, 203, 204, 206, 208, 209, 210, 211, 219, 222, 263, 492, 506, 531, 608, 689, 1118
Zhang, L., 840
Zhang, P., 20, 21, 22, 1170
Zhang, W., 939
Zhao, B., 1120
Zhao, L. Z., 657
Zhenhua, C., 1073
Zhong, H., 489
Zhong, J., 489
Zhongliang, H., 1073
Zhou, K. D., 187
Zhou, X., 582, 583, 649, 851, 855, 857, 858, 859, 860, 861, 862, 863, 864, 865, 866, 867, 868, 870
Zhou, Y. L., 135, 175, 200, 208, 209, 220, 492, 531
Zhu, P., 824
Zhuravlev, V. A., 75
Ziegert, J. C., 1173
Ziegler, V., 725
Ziegler, W., 997, 999
Zier, J. C., 1170
Zihir, S., 704
Zikry, M., 712
Zindibe, E. M., 272
Zindibe, L., 272
Zindine, E. M., 466
Zisman, W. A., 1028
Zlatkis, A., 1007

Subject Index

A

Abnormal glow, 570
Abrasion, 424–426, 1073–1076, 1104–1105
Abrasive particles, 425
Abrasive wear, 424–426, 1070
Absolute cathode gain, 639
Accelerated current-cycling tests, 345, 363–365
Accelerated testing, contact erosion measurements, 993
Acceleration factor (corrosion), definition of, 158–159
Accelerator Pedal Position (APP) sensor, 541
Account vaporization of metal, 630
ACSR conductors, 327
AC-1, -2, -3, -4, testing, 794–795
AC testing vs. DC testing, 978–979
Activated adsorption sliding contacts, 1053, 1054
Activation of switching contacts, 526, 588, 622–624, 643, 1006–1014, See also Contact activation
  absorption ion site limitation, 622
  activation curve, 622, 1010–1011
  activation process, 1007
  arc duration, 622
  arc formation, 1007
  arc motion, 1012
  closed contacts, 1009
  contact resistance, 1012–1014
  contacts after arcing, 1009
  contacts no arcing, 1009
  critical exposure, 1011
  detection, 1014
  effects, 1011–1013
  ranking metals and organic activants, 623
Actuated switches, 739–741
Adhesion, 416–417, 1102–1103
Adhesive bond, 1102–1103, 1111
Adhesive forces, 1103
Adhesive transfer, 1103, 1104
Adhesive wear, 439, 481, 1070, 1157–1158
  in gold electrodeposits, 422
  mechanism, 415
Adsorbed moisture films, 1170
Adventitious lubrication, 1121, 1124

Aerobic lubrication, 1141
  anaerobic vs., 1121
  contacts, 1123–1124
  wear debris, 1124
AES, See Auger electron spectroscopy
AFCIs, See Arc Fault Circuit Interrupters
AFM, See Atomic force microscope
A-fritting, See Fritting
Ag–CdO contacts, 605, 645, 655, 771–773, 884–906
Air blown cleaning, 225
Air-break contactors, 794
Air contaminants, expected maximum average value for, 126
Aircraft brushes, 1090–1091
Air cylinder switch, 699
Air pollution, 126–128, 190, 1118
Alternating current arcs
  low-current, 768–770
  reignition measurement, 996
Alternating current circuits, arc interruption in, 597–600
  arc plasma and dielectric reignition, decay of, 602–603
  arc quenching, 798–802, 808–814
  current limiting technique, 608–609, 820–822
  instantaneous dielectric recovery, 600–602, 799–800
    effect of current level, 599, 800
    effect of arc chamber material, 801
  low frequency circuits, 609
  mega and giga hertz circuits, 609–612
  reignition
    dielectric, 602
    thermal, 603
    thermionic, 603
  stages, 599–602, 800
  thermal reignition, 603–604
  transient recovery voltage, 598, 798, 831
  vacuum, 607–608
Alternating current erosion, 770–773
Alternating current operation, 780
Alternators, brush applications, 1088
Altitude-treated brushes, 1084
Aluminium–copper, mechanical properties of, 294

Aluminum, 385
  and aluminum alloys, 243–244
  coating of, 331
  foam materials, 355–358
  oxidation kinetics of, 258
  wiring connections, 406–408
Aluminum-aluminum contacts
  current–voltage characteristics from, 56
  Sharvin resistance role in, 86–87
  voltage-temperature relationship for, 61
Aluminum-base connectors, 23
Aluminum–brass interfaces
  equivalent contact radius in, 39
  high resistivity intermetallic layer in, 40
  and intermetallic compounds, 30–33
  intermetallic growth at, 42–45
Aluminum conductors
  fretting damage, 285
  properties of, 241
  stress relaxation of, 304
Aluminum connections, fretting in,
      283–285
Aluminum contacts, breakdown of classical
      electrical contact theory in, 84–87
Aluminum-copper bolted joints, relative
      motion effect in, 278
Aluminum–copper interfaces, 32, 33
Aluminum oxide growth rate and electrical
      resistivity, 53–56
Ambient effects, on arcing contact surface,
      665–667
Ambient temperatures, testing at, 982
Ameliorating frictional heating, effective
      in, 1167
American Society of Testing and
      Measurement (ASTM), 377
Anaerobically lubricated contacts,
      1122–1123
Anaerobic *vs.* aerobic lubrication, 1121
Analog devices, arc motion measurements,
      996–997
Anode of the arc, 584–585
  active or passive mode, 584
  fall, 578
  gain for gold contacts, 641
  spot formation, 594
Anodic arc, 638
Anodic contact material, 638
Apparent contact, 1042
Arc
  anode region of, 584–585
  atmospheric pressure, 578
  cathode region of, 581–584

column, 579–582, 822
column decay, 602–603
contact materials, *See* Arcing contact
      materials
and electrode interaction, 826
Elenbaas and Heller's equation, 580
electron emission at cathode, 583
energy, 769–770
energy balance at cathode/anode, 581–585,
      628–630
energy transfer at cathode/anode, 584
ferromagnetic attraction, 789
floating, 649
gaseous, 642, 762–763
interruption, *See* Arc interruption
local thermodynamic equilibrium, 578
long-gap gas breakdown, 558–565
metallic, 641, 762–763
metal vapor in, 579
minimum arc current at initial contact
      parting, 585–588
minimum arc current, DC interruption, 606
minimum arc voltage, 585–588
motion, 596, 637, 759, 782–791, 800–802,
      816–820, 992, 998
  dwell time, 789–790, 818
  measurement, 759, 996
  self induced magnetic fields, 999
parameters to sustain, 577
plasma, decay of, 602–603
radius as f(current), 582
Saha's equation, 579
short gap breakdown, 566
showering, 624–628
simulation of, 822–827, 839–840
squeezing and elongating of, 821
temperature of free burning, 579
time constant, 592
vacuum breakdown, 566
vacuum, *See* Vacuum arc
voltage characteristics, 762
voltage as f(contact gap), 589–591, 605
volt–ampere characteristics, 588–592
volt–current characteristics of separated
      contacts, 569–570
Arc behavior, 611
Arc chambers, 787–788
Arc chutes, 787–788
  deion, 820–822
  design, 798–802
  low-voltage circuit-breakers,
      810–814
  V-shaped runners, 787

Arc column, 578–581, 822, *See also* Arc
    cooling, 582, 602
    electron mobility, 579
    electron temperature, 578
    Elenbaas-Heller equation
        arc boundary cooling effect, 581
        high currents, 580
        low currents, 580
        thermal loses, 581
    gas temperature, 578–579
    ionization potential, 561
    ion mobility, 579
    ion temperature, 578–579
    local thermodynamic equilibrium, 578
    radiation loses, 581
    radius as f(current), 580–582
    Saha's equation, 579
    voltage as f(current), 588–592
Arc control in sliding contacts, 1133
Arc current, effect of, 631–633
Arc driven migration process, 1024,
        1025, 1030
Arc duration
    activated contacts, 622, 1011
    function of circuit inductance, 621
    for gold contacts, 620
    and residual current, 1061–1063
Arc dwell time, 789–790, 818
Arced tungsten–silver, 659–665, 930, 931
Arc erosion of electrical contacts, 628–630,
        See also Contact erosion
    activated contacts, 1012
    arc chamber material effect, 707
    arc current effect, 631–633
    contact size effect, 633–635
    contact surface as result of arcing, *See*
        Contact surface as result of arcing
    electric load effect, 994
    in low-current AC circuits, 644–645,
        770 773
    in low-current DC circuits, 636–644,
        766–770, 990–996
    on make and erosion on break, 631, 995
    measurement, 982
    model testing, 983–985
    movement on contact surface, effect
        of, 1012
    parameters, 628–629
    pip and crater formation, 638, 757, 899,
        901–902, 980
    point on wave, 771–773
    radioactive tracer measurement of, 639–640
    resistive loads, 979

silver- metal oxide contacts, 899–906
    switching frequency effect, 993
    W–Ag contacts and effect, 661, 662,
        918–927
Arc Fault Circuit Interrupters (AFCIs),
        850–853, *See also* Arcing faults
Arc fault, low current, 867, 868
Arc formation
    during contact closing, 570–571
        time lag, 571
    during contact opening, 571–577
        molten metal bridge formation, 571–577
        molten metal bridge rupture, 573–577
        pseudo arc, 576
        metallic arc 577
Arc simulation
    arc roots, cathode and anode, 824–825
    plasma properties, 826–827
    principle of, 822–824
    radiation, 825–826
Archard/Holm law, 1158
Archard's adhesion model for wear, 271
Arcing contact
    lubricants with refractory filler, 1030
    mineral particles, 1015–1018
    silicone contamination of, *See* Silicone
        contamination, arcing contacts
Arcing contact design
    arc-induced contact stresses and interface
        bond quality, 954–955
    brazed contact assembly designs, 966–969
    clad metals, inlay, and edge lay, 969–970
    contact alloys, *See* Contact alloys, non-
        arcing separable contacts
    determination of contact size in AC
        operation, 635
    staked contact assembly designs, 955–960
    welded contact assembly designs, 960–966
Arcing contact materials, *See also* Contact
        materials, switching
    applications, 945, 946
    manufacturing processes, 881
    noble metals, *See* Noble metals
    non-noble, *See* Non-noble silver alloys
    silver cadmium oxide, *See* Silver metal
        oxide switching contacts
    silver graphite contact materials, 939
    silver metal oxides, *See* Silver metal oxide
        switching contacts
    silver–nickel, *See* Silver nickel contact
        materials
    silver refractory metals, *See* Silver-
        refractory metals

silver tin oxide, *See* Silver metal oxides
tungsten contacts, 689, 718, 936–937
vacuum interrupter materials, *See* Vacuum
 interrupter contact materials
Arcing faults, low current, 849–872
 detection of, 872
 other types, 871
 properties, 865–871
 series arcing, 855–857
 short-circuit arcing, 854–855
Arcing region, 577
Arcing time, 642
 in AC circuits, 618
 activated contacts, 622–624, 1012
 in DC circuits, 618–621, 624–628, 761–767
 in very low current inductive circuits, *See*
  Showering arcs
Arc interactions, contaminants
 contact materials, oxidation of, 1031–1032
 long arcs, resistance effects,
  1032–1034
 mineral particles, 1015–1018
 organic contamination and activation,
  1006–1014
 refractory fillers, lubricants, 1030
 silicone contamination, 665–667, 1018
Arc interruption, 597–612, 808, 836–837
 in alternating current circuits, *See*
  Alternating current circuits, arc
  interruption in
 in direct current circuits, *See*
  Direct current circuits, arc
  interruption in
 of low frequency circuits, 609
 of megahertz and gigahertz electronic
  circuits, 609–612
 showering arc, 624–628
 vacuum, in alternating circuits, 607–608,
  836–838
Arc motion, measurement of, 586, 637, 759,
  788–791, 816–820, 992, 996–999
Arc plasma, decay of, 602–603, 800–803, *See
 also* Arc interruption
Arc root
 on cathode and anode, 824–825
 erosion at, 635
Arc, computer simulation of, low voltage
  circuit breakers and low voltage
  contactors
 arc roots cathode and anode, 824–825
 plasma properties, 826–827
 principles of 822–824
 radiation, 825–826

Arc splitting process, 825
Arc voltage characteristics, 633
Arc-wall interaction measurements, 999
Arrhenius equation, 116, 532
Asperities
 center line average, 421
  deformation of, 4, 11
 elastic deformation, 253, 706–707
 plastic deformation of, 13
 plasticity index, 76–77
 work-hardening of, 16
Asperity model, 447, 1101
a-spots, 4, 1111–1112, *See also* Constriction
  resistance, Contact resistance
 boundary of, 1098
 circular, 5–7
 conduction in small, 79–83
 constriction resistance of, 6–23
 electrical conduction in small, 79–83
 film forming on, 1115–1117
 Joule heat flow through, 82–83
 micro-environment of, *See* Micro-
  environment of contact region
 multiple, 11–17
 non-circular, 7–11
 regular array of, 13
 ring, 7–11
 rough interfaces, 75–79
 sintering effect, 72–75, 85
 smooth interfaces, 69–75
 surrounded by oxide film, 55
 temperature, 57
 temperature distribution in, 63–64
 voltage-temperature relation, *See* Voltage–
  temperature relation
ASTM, *See* American Society of Testing and
  Measurement
Asymmetric contact, temperature *vs.* voltage
  in, 1066
Atmospheric gases, 126–128, 144, 190, 1122
Atmospheric corrosion, 260–261
Atmospheric air particle concentration,
  function of temperature, 557
Atmospheric gas concentrations, outdoor
  and indoor environments, 126–128,
  394–395
Atmospheric pressure, arc in air
 anode region, 584–585
 arc column, 578–581
 arc volt–ampere characteristics, 588–592
 cathode region, 581–584
 minimum arc current and voltage,
  585–588

Atomic force microscope (AFM), 712, 1115, 1141, 1157

Attractive magnetic-force curves, 678, 797

AuCuCd platings, 539

Audible bearing noise, 1087

Audible magnetic noise, 1087

Audible washer noise, 1087

Auger analysis, 130

Auger electron spectroscopy (AES), 693, 695

Automatic clamping, 327

Automatic splices, 326–327

Automobile applications, DC relays for, 636

Automotive brush applications
  alternators, 1088
  auxiliary motors, 1086–1088
  starter motors, 1088–1089

Automotive connector contacts, 405

Automotive electrical circuits, 621

Automotive electric power system, 1057

Automotive generator, 1070

Automotive position sensor connector, 402

Automotive sector, 733

Automotive systems, 765

Auxiliary motors, automotive, 1086–1088

Axial magnetic field, 596

B

Back-transfer prows, 421

Base metals, 456 457
  cladding of, 246
  contacts, 441

Bathtub life curve for connectors, 345

Belleville washers, *See* Disc-spring washers

BER, *See* Bit error rates

Bessel function, 20

B/F, *See* Branch feeder

B-fritting, *See* Fritting

Biasing spring, 354

Bimetallic insertion, 336–337

Bimetallic system, galvanic corrosion in, 118, 263–264

Bi-metals in electrical contact, 64–65
  maximum temperature plane in, 100–104
  operation of, 96

Bismuth oxide, 894

Bit error rates (BER), 1139

Blade-box terminal, 386–387

Blade-leaf terminal, 387

Blow-off force, 646–650, 811–812
  arc effect on, 649
  minimum contact force to balance it, 647, 648

Blow-out field, magnetic, 788–792, 995

Bohr model of atom, 556

Boiling voltage *vs.* break voltage for metals, 575

Bolted connections, 236, 278–283

Bolt torque, 318–321

Boltzmann's constant, 97

Bounce, *See* Contact bounce

Boundary lubrication, 488, 1107, 1120, 1173

Branch feeder (B/F), 850–851

Brass–indium interface, 36

Brass–tin interface, 37

Brass–zinc interface, 35

Brazed composite rivets, 959–960

Brazed contact assembly designs
  braze joint quality, 968–969
  direct and indirect resistance brazing, 967–968
  methods of, 966–967
  toplay, 968
  torch and induction brazing, 967

Breakaway friction force, 1102

Breakdown, *See* Long-gap gas breakdown; Short-gap breakdown; Vacuum breakdown
  breakdown voltage, 563–568, *See also* Reignition voltage

Break-only contact pairs, 631

Break-only erosion, 631, 995

Break operation, 779–780

Bridge, molten metal, *See* Arc formation during contact opening; Molten metal bridge
  bridge transfer, 723–724
  effect of, 1169–1170
  erosion, 639
  problem with, 723

Bridging contact, 637

Brinell hardness (BH), 1196

Brittle fracture, 426

Brittle interface, 41

Brittle materials, 915

Brush applications
  aircraft and space brushes, 1090–1091
  automotive, 1086–1089
  brush design, 1091
  diesel electric locomotive, 1090
  fractional horsepower motors, 1085–1086
  industrial brushes, 1089–1090
  instrument and control, 1093
  minature motors, 1085
  power, 1152–1154

Brush configurations, *See also* Sliding
contacts for instrument control
cantilever composite brush, 1128–1129
cantilever metallic finger, 1129
cantilever wire brush, 1129–1130
cartridge brush, 1128
multiple brushes, benefits of, 1130–1131
Brushes
design, 1091
dynamics, 1177–1180
forces on, 1131–1132
multiple, benefits of, 1130–1131
solid lubricated composite materials
for, 1133
Brush materials, 1073, 1082–1084, 1206
electro- and natural graphite, 1074–1075
metal fiber, 1152–1154
metal graphite, 1075–1076
Buckboost curve, 1082
Bump-flat terminal, 389
Bus joint, 307–313, 361–363
Bus-stab contacts, fretting effect on, 276–278
Butt contacts blow-off, 647–650
Button welding, 962–963

C

Cable impedance, 868–871
Cable length effects, 868–871
Cable termination, 381, 389–392
Capacitance, 1114
Capillary pressure, 1122
Carbon, 855
Carbon dioxide, 1125
Carbon-graphite brushes, 1083
Carbonized path clearing time test,
851–853
Carbon nanotubes (CNT), contact properties
of, 714–716
Carrier materials, 1205–1206
Cartridge brush, 1128
Casimir force, 720
Cathode
arc, 638
brush, 1073
erosion, 772–773
fall, 570, 578
field emission, 567
gain for gold contacts, 641
region, 581–584
spots, 593–594
T–F electron emission, 583
thermionic electron emission, 566, 583

Cathodic arc, 638
Cathodic contact material, 638
Cathodic polarization, 501
Cathodic protection, 261
Cathodic reduction method, 129
Center line average (CLA), 421
Chattering, contact, 657
Chemical oxidation reaction, 119
Chemical vapor deposition (CVD), 689
Chlorine, 117
in atmosphere, 537
single-gas corrosion, 163–164
Chop current, 587, 838–840
Chrome–copper materials, 593, 836, 936, 1210
Circuit-breakers, *See also* Low-voltage
circuit-breakers
high interrupting capacity designs,
806–815
medium-voltage vacuum, 828, 830–831
miniature, 815–820
European type, 817
US type, 816
molded case, 807–809
Circuit breaker terminal contacts, fretting in,
287–288
Circular *a*-spots, 5–7, 1066
Circular contact spots, 1066
CLA, *See* Center line average
Clad contact assemblies, 969–970
Cladding, 246, 435–437, 520, 525
material, 521, 526
requirement, in contact terminal, 393
Clad palladium–gold–silver alloys, 458
Classical electrical contact theory, 5–17
in aluminum contacts, breakdown
observations of, 84–87
electrical conduction in *a*-spots, 79–83
in gold contacts, breakdown observations
of, 87
in small contact spots, breakdown of, 79
in tin contacts, breakdown observations of,
88–89
Clay-thickened greases, 503–504
Closed contact, 811
spring force on, 649
welding of, 651
Clusters, 1118
Coatings, *See also* Platings
of aluminum/copper, 331
fretting in, 286–287
on metal finishes, 520–526
power connectors, palliative measures,
331–333

properties related to porosity, 526–532
requirements of, 520
terminology, 521
Cobalt–gold coating, 485
Cobalt–gold electrodeposit, 429
Cobalt–gold electroplates, 422
Cobalt-hardened gold, variation of, 434
Co-deposited particulate material, 527
Coefficient of friction *vs.* ring temperature, 1053, 1054
Coil current distribution of dc motor, 1059
Coil reactance, 1082
Cold cathode, electron emission, 582, 583
Cold resistance of contact, 67
Cold-working, 651
Collisions electron with air, 555
Columnar vacuum arc, 595
Common galvanic cell, 118
Commutation, 1056–1061, 1082
Commutator, 1081–1082, 1127
Complex electro-mechanical assembly, 1139
Compliant pin connector, 380–381
Composite refractory materials, 928–935
Compression connectors, 274–275
Compression sleeve connector, 314–316, 344–345
Compression-type connectors, 313, 317
Conducting coatings, electrically, 23–28
Conductive plating, 253
Conductivity, effects on 237
    deformation dislocations, 239
    grain boundaries, 239
    Hall effect, 240
    lattice imperfections, 238
    longitudinal magneto resistance, 239–240
    magneto-resistance, 239
    Matthiessens' rule, 238
    skin effect, 91, 240
    temperature, 238
    vacancies, 239
Conductor materials
    aluminum and aluminum alloys, 243–244
    conductivity, 237
    copper and copper alloys, 240–243
    grain boundaries, 239
    Hall effect, 240
    imperfections, 238
    longitudinal magneto resistance, 239
    Matthiessen's rule, 238
    properties of, 237
    skin effect, 240
    temperature effect, 238
    transverse vacancies, 239

Conductors
    aluminum, *See* Aluminum conductors
    copper, properties of, 241, 1198–1201
    electric field between, 124
    factors affecting conductivity, 237–240
    solid metallic, classification of, 237
Connection resistance, constriction *vs.*, 94–95
Connections, *See also* Connection systems, electrical
    aluminum wiring, 406–408
    for high-vibration environment, 408
    welded, 339–342
Connection systems, electrical
    application parameters of, 376
    contact, degradation of, 693
Connector contacts, fluid lubricants for, 491
Connector lock, 383
Connector position assurance (CPA), 383
Connectors
    bolted joints, 361
    compression, 274–275
    contact, degradation of, 393
    contacts, automotive, 405
    dead-end, 327
    degradation mechanism, 456
    electronic, *See* Low power and electronic connectors
    flexible, 297, 298
    functional requirements, 377–378
    generic, 234, 235, 342
    high power for electric and hybrid vehicles, 405–406
    low power, *See* Low power and electronic connectors
    mechanical, 324, 381–383
    position sensor, 402
    power, *See* Power connectors
    stepped deep indentation, 315, 316, 344–345
    types of, 378–381
    wedge, 325–326
Connector seal, 382
Connector systems, 237, 244–246
Consistent electrical contact resistance, 1102
Constant stiffness, 1102
Constriction resistance, 4–5, 57, 251, 543, 1055, 1107, *See also* a-spots; Contact resistance
    asperity shape of, 17–18
    circular a-spots, 5–7
    vs. connection resistance, 94–95
    contact forces, 14, 719
    elastic deformation, 14, 253, 706–707
    for copper-copper contact, 82

at high frequencies, 89–95
Holm radius, 12–13
load relationship, 14, 15
multiple contact spots, 11–17
non-circular and ring *a*-spots, 7–11
plastic deformation, 14, 239, 252
surface films effect on, 18–28
surface profile, 69–79
values of, 56
Contact activation, 622–624, 665–667, *See also*
    Activation of switching contacts
    contact resistance problems, 1012–1014
    definition, 1006
    detecting carbon contamination, 1014
    effects, 1011–1013
    exposure level, 1010–1011
    organic vapors, sources of, 1007
    palladium, 1008, 1009
Contact ageing process, 350
Contact aid compounds, 332, 334–336
Contact alloys, non-arcing separable contacts
    contact applications and
        performance, 972
    manufacturing technology, 970
    metallurgical properties, 971
    physical and chemical properties,
        970–971
Contact area, *See also a*-spots
    effective, Hertz's model for, 706
    electrical, 11–15, 250
    mechanical, 249–250
    method to increase, 719
    palliative measures, 307–313
    performance factors, 249–252
Contact asperity on constriction resistance,
    17–18
Contact blow-off force, 646–650, 652
Contact bounce, 631, 720, 745–756, 980
    arcing during bouncing, 631, 781
    design parameters, for reduction of, 755
    measurement, 986–989
    reduction of, 775
Contact chattering, 657
Contact cleaners, 622
Contact closing force, 15–16, 646, 652, 981
Contact closure, schematic diagrams of, 631,
    773–778
Contact degradation, rate of, 472
Contact deterioration, economical
    consequences of, 346
Contact switching test devices, 876–978,
    981–982
Contact dynamics, measurement of, 759, 996

Contact erosion, 633, 634, 641, *See also* Arc
    erosion of electrical contacts
    effect of, 661, 663
    in low-current AC circuits, 644–645,
        770–773
    in low-current DC circuits, 636–644,
        766–770, 990–996
    measurements, 755–761, 991, 993–996
Contact film resistance, 56, 251
Contact failures
    electrical, 402–405
    electromechanical components, 189
    preliminary attachment and final
        attachment, 223
    single particle and ideal model, 222–223
Contact finishes, 539
    contact resistance of, 543–544
    hardness, 543
    produced by non-chemical methods, 525
    requirements, 520
    topography, 545–546
Contact force, 15–16, 214–216, 313, 392,
        404–405, 572, 743, 1097, 1098
Contact geometry (MEMS), 479
Contact impedance, 1096, 1112–1113
Contact inductance, 1112
Contact lubricants, *See* Lubricants
Contact material, switching
    device testing and model switch testing,
        976–978
    testing variables, 978–982
Contact materials, switching, 742, 880–946,
        954–972
    Ag, 937, 1202
    Ag-C, 943, 1202
    Ag-CdO, 883, 1203
    AgCu, 938, 1202
    Ag-Mo, 908, 1203
    Ag-Ni, 939, 1202
    Ag-W, 908, 1203
    Ag-WC, 908, 1203
    Cu-W, 1025, 1203
    Pd, 940, 1204
    Pt, 940, 1204
    dwell time of, 818
    low-voltage contactors, 802–805
    manufacturing processes, 884–890,
        909–912
    in MEMS, 713
    switching in air, 791–793, 883–945
    used in vacuum interrupters,
        834–839, 1210
    wear-out, 480

Contact metals, physical properties of, 1198–1205, 1209–1210

Contact noise, 1138–1139

Contactors, *See* Low-voltage contactors

Contact over travel, 981

Contact region, 5–17, 249–250, *See also* a-spots, Contact area

Contact resistance (CR), 5, 384, 662, 882, 982–983, 1140, *See also* Constriction resistance, a-spots

  activated contacts, 1012

  Ag_2WO_4, 661, 930

  in aluminum/aluminum contact, 46

  arced tungsten–silver, 659–665, 930, 931

  "a" spot temperature effect of, 703

  behavior, 469

  bolted joint configuration, 311–313

  breakdown of classical theory, 84–89

  characteristic, 259

  composite refractory materials and, 928–935

  and contact force, 719

  contact load on, 14–17, 76, 77, 255, 313, 317, 318

  dust effects, 212–214, 219–223

  effects of fretting corrosion, 457

  electrical conduction in small a-spots, 79–83

  and electrical erosion *vs.* switching current level, 929

  electrically conducting films on, 23–28

  electrically weakly conducting and insulating films, 45–57

  evaluation and presentation of results, 985–986

  of finishes, 543–544

  vs. fretting cycles, 497, 500

  function of load, 15, 16

  gold probe, 983

  at high frequencies, 89–95

  intermetallic layer growth on, 28–45, 464

  low energy and higher energy arcs, 1013

  lubricant effect on, 175–177

  magnitude, 7

  measurement, 27, 130–131, 364, 982

  models, 708–710, 713

  model testing, 983–985

  organic gas, 1014

  oxide layers, 932

  between powder grains, 821

  of pure gold platings, 537

  plating factor, 25

  rate of rise of, 533

  silicone effect of, 665–667, 1018–1030

  silver metal oxides, 907

  silver refractory metal systems, 934–935

  skin effect, 89–95

  stability, categorization of, 449

  static *vs.*, dynamic, 445

  surface films effect on, 18, 45, 50, 53, 657–665

  vs. switching current level, 819, 928

  vs. switching current duration, 1032

  temperature development, *See* Voltage–temperature relation

  thermal stability of, 534, 535

  thin films and spreading resistance, 18

  of tin on various substrates, 540

  tungsten compounds, properties of, 929, 930

  variation (noise), 1108–1109, 1138

Contact rivets

  brazed composite rivets, 959–960

  cold bonding, 956–957

  composite, 956

  machine-made composite rivets, 956–959

  rivet staking, 960

  squeeze test, 958

  solid rivets, 955–956

  ultra-sonic testing, 959

Contacts

  arc dwell time on, 789–790, 818

  bounce, measurement, 777–781, 986–989

  carrier mass and conductivity, 981

  closed, 811–812

  at current levels, 742

  mechanical area of, 14

  under mineral oil, 125–126

  nominal/true area of, 13

  size in AC operation, 635

    effect of, 633–635

  switching device simulations, 827

  welding, measurement, 989–991

Contact spots, *See* a-spots

Contacts switching, 618–628, 636–644, 674–725, 703–725, 733–781, 787–839, 890–898, 918–934

Contact super-temperature, 59

Contact surfaces, measurement of, 756–757

Contact surface as result of arcing, 657–665, 906, 930–934, *See also* Arc erosion of electrical contacts

  ambient effects on, 665–667, 1006–1034

  silver-based contacts, 659, 889–900

  silver refractory metal contacts, 659–665, 930–934

Contact tape welding, 963–965
Contact welding, *See also* Welding
  of closed contacts, 651–654
  during contact closure, 654–657, 774, 902
  as contacts open, 657
  on make operation, 654–657, 774–775,
    902–903
  measurement, 989–991
Contactor- air, *See* Low-voltage contactors
Control motors, 1085
Conventional catalysis reactions, poisons
    for, 459
Conventional d.c. generator, 1088
Conversion table for; Imperial, cgs and SI
    units, 1208
Coordinated Switching and Protective devices
    (CPS), 814
Copper
  coating of, 331
  in contact coating, 392–393
  and copper alloys, 240–243
  corrosion mechanisms, 151–152
  corrosion rates, 145, 146
  degradation of, 128
  diffusion rates, 532
  electrolytic refining of, 240
  foam materials, 358–360
  intermetallics on, 539
  metallographic sections of gold plated
    on, 530
  stress relaxation of, 303
Copper-based carrier alloy materials, 1205
Copper-based spring alloy materials, 1205
Copper conductors, properties of, 241
Copper–copper contacts, 469
Copper–copper power connections, 133
Copper–copper systems, 462
Copper–graphite contacts, 943, 1202
Copper oxide
  additions, 894
  fritting of, 1057
  growth rate and electrical resistivity, 47–49,
    54, 141
Copper refractory metals
  high erosion rate, 925
  low erosion rate, 926
  tungsten–copper composite contacts,
    922–923, 1205
Copper rods, copper oxide on, 1058
Copper–tin systems, 296–299
Copper-to-copper wire glowing connections,
    859–860
Copper–tungsten materials, 922–927, 1205

Corrosion, 114–134, 140–151, 260–263,
    383–405
  atmospheric environment, 126–128,
    143–144
  contacts under mineral oil, 125–126
  copper, 47–49, 118, 151, 1053
  creep, 122–123, 154–157, 262–263
  crevice, 261
  dry, 117–118
  dust, 134–135, 205–211, 263, *See also* Dust
    contamination
  electrolytic, 404
  electro-migration, 123
    temperature effect, 124
  electronic connectors, 131–133
  environment factors, 144
  film, kinetic growth of, 116
  fretting, 282, 283, *See also* Fretting
  galvanic, 118–120, 263–264, 394, 404, 407
  history, 140, 159
  inhibitors, 179
  laboratory gas effects on, 128, 159, 160
  localized, 261
  lubrication and inhibition of, 174–180
  measurement, 129–131, 140–143
  mechanisms, 151–157
  metallic electromigration, 123–124
  mineral oil effect of, 125–126
  nickel, 152
  pitting, 262
  pore, 120–122, 153, 262, 527–532, *See also*
    Porosity
  pore corrosion product, 114
  porous gold, 153
  power connectors, 133–134
  properties, 1134
  protection, 179
  reactivity distributions, 148–151
  shielding effects, 135, 147
  silver, 151
  stress corrosion cracking, 124–125
  thermal expansion, 264
  tin, 152
  vibration effect of, 180
  visual inspection, 129
Corrosion inhibition, *See* Lubrication
Corrosion rates, 114–116, 145–181
  copper and silver, 145, 146
  film effects, 146–147
  other metals, 145
  oxide growth rate and electrical resistivity,
    47–56
  reactivity distribution, 148–149

Corrosion stains, dust contamination
corrosion products *vs.* dust, 209
fretting experiments, 209–211
structure of, 208
thin gold-plated surface, 206, 207
Corrosive gases, single, 117, 159–164
Corrosive gases, mixed flowing, 167–171
humidity, 117–118, 124, 127, 144, 164
Coulomb's law of friction, 1048
Counterface, sliding contacts
configuration, 1126–1127
material, 1103, 1132–1133
CPA, *See* Connector position assurance
CP-AFM technique, 272
CPS, *See* Coordinated Switching and
Protective devices
CR, *See* Contact resistance
Crater erosion, *See* Pip and crater formation
Creep, metal, 299, 305–306, 1100
evolution, 305–306
logarithmic law, 305
stress relaxation *vs.*, 299–300
Crimped connections, 274, 315, 390–392
Critical vacuum breakdown field for metals, 568
Crossed-rod contact, 1126
Crossed rod contact resistance measurement,
130–131, 983
Crystal structure, graphite, 1075
CSA, Canadian Standards Association, 794, 806
CTF, *See* Cycles to failure
Cumulative contact resistance, 984
Cumulative probability, 652–653, 984
Cumulative weld strength, 653
Cu radiation, 576
Current-cycling tests for fixed connections
accelerated, 345, 363–365
results of, 315, 316, 333
Current distribution in parallel
contacts, 650, 813–814
Current-induced diffusion, *See* Intermetallics
Current limiting
in a.c. circuits, principle of, 608–609,
808, 809
breakers, 812
other techniques, 821
contact resistance between powder
grains, 821
liquid or low-melting metals, 821
narrow insulating slots, 821
reversible phase change materials, 821
superconductors, 821–822
temperature-dependent ceramics or
polymers, 821

CVD, *See* Chemical vapor deposition
Cyanide nickel-hardened gold plates,
polymers in, 422
Cycles to failure (CTF), 401, *See also* Fretting

D

"Dark discharge," 563
DC circuit, *See* Direct current arcs, Direct
current circuit, arc interruption in
DC motor driving automotive fuel
pump, 1061
DC testing *vs.* AC testing, 978–979
Dead-end connectors, 327
Debris-filled interface, 447, *See also* Fretting
Debris generation, rate of, 415, *See also*
Fretting
Deformation
of connector sleeve, 314
elastic, 253, 706
plastic, 13, 75–79, 239, 252, 1161–1163
yield stress, 250, 651
Deion arc chutes, 820–822
Deion plates, 621, 788–789
Delamination wear, 480, *See also* Wear
mechanism of, 271
and subsurface wear, 427–428
DEP, *See* Diethylphthalate
Design of Experiment (DOE) approach, 977
Device testing, 976
DG R-156, 482, 521, 539
definition, 526
Dielectric reignition, arc plasma decay of,
602–603
Diesel electric locomotive brushes, 1090
Diethylphthalate (DEP), 1010
Differential electrochemical cell, 118
Diffused gold 60Pd40Ag, 521
Diffuse vacuum arc, 592–595
Digital devices, arc motion measurements,
759, 997
Digital signal processors (DSP), 1139
Diode-rectified generator, 1088
Direct current arcs, 604, 618, 761
current erosion, 766–768, *See also* Arc
erosion
direct current operation, 780
low-current, 761–762
opening speed and arc length, 764–765
quasi-static conditions, Ag/CdO contacts,
763–764
showering arc, 624–628
voltage below 7 V, 762–763

Direct current (DC) circuit arc interruption in, 604–607, 637, 820
  arcing time in, 618–621
  arc plasma and dielectric reignition, decay of, 602–603
  effect of current level, 599, 800
  instantaneous dielectric recovery, 600–607
  load type, 890, 979–780
  minimum arc current at interruption, 606
  restrike, 637
  self extinction, 788
  stages, 607
  voltage generation, 604, 618, 820
Discontinuous erosion, 632, 633
Disc-spring washers, 322–325, 329–331
Dissociation process, 559
DOE approach, *See* Design of Experiment approach
Double-break arrangements, contactors, 796
Double-break contact systems, 637, 799
Dry corrosion, 117–118
Dry tin-plated connection, systematic studies of, 280, *See also* Fretting
DSP, *See* Digital signal processors
Dust contamination
  adhesive effect, 220–221
  atmospheric environment, 143–144
  background, 134, 186–187
  chemical behavior, 205
  collection and shape, 191
  complexity problem, 187, 189
  composition, 191–193
  contact failure mechanism, 222–223
  contact interface, 219
  corrosion effects, 134–135
  corrosion mechanisms, 151–157, 263
  corrosion products, trapping effect of, 221, 222
  corrosion stain, 208
  deposition, 187
  dust particle, 187–188, 191
  dust problem, 225–226
  dust shield, 135
  dust test, 224–225
  dusty water solutions, 205–206
  electrical behavior
    electric charge, measurement of, 193–194
    electrostatic attracting force, 195–198
  elements atomic percentage *vs.* accelerating voltage, 216, 217
  environment, 190–191, 211
  fretting, 204
  fritting, 204
  high and erratic contact resistance, 212–213
  indoor exposure results, 206–208
  lubricants effect of, 201–204
  mechanical behavior, *See* Mechanical behavior
  micro motion, 211–212, 223–224
  organic and inorganic materials, 192
  pore creation, 205, 527
  SEM/XES, 216
  short life *vs.* longer life contacts, 218
  sodium lactate, 215, 216
  source, 190
  stress corrosion cracking, 125
  testing spots, 213–214
  test systems, 189
  water soluble salts, 192–193
Dwell time of arc on contacts, 789–790, 818
Dynamic blow-off, closed contacts, 646–650, 812
Dynamic welding, 654–657, 774, 903, 904, 989

E

Edge lay, 969–970
Edge-on type connector, 379–380
EDX, *See* Energy dispersive X-ray
EIA, *See* Electronics Industries Association
Elastic contact, 253, 702–707, 1161–1163, 1182–1184
Elastic deformation, 14, 253, 706–707
Elastic electron-gas collisions, 559
Electrical breakdown of gases, *See* Long-gap gas breakdown
Electrical constriction resistance, *See* Constriction resistance
Electrical current
  on metal creep, 305–306
  on stress relaxation, 302–305
Electrical interface, *See* a-spots; Constriction resistance; Contact resistance; Corrosion; Fretting; Fritting; Intermetallics
Electrically conductive coatings, 18–28, 253, *See also* Spreading resistance
Electrically conductive layers
  on conducting substrate, 22–23
  on insulated substrate, 18
  and thin contaminant films, 23–28
Electrically insulating films, 45–57
Electrical measurement methods
  AC arc reignition measurement, 996, *See also* Alternating circuit currents, arc interruption in
  arc motion measurements, 996–999

arc-wall interaction measurements, 999
contact bounce, 759, 986–989
contact erosion, 756–761, 991–996
contact resistance, 130, 982–985
contact welding, 989–991
Electrical performance in sliding contacts,
    1107–1108, *See also* Sliding contacts for
    instrument control
Electrical resistivity, 6–26, 1200
constants, 68, 1200
of metals and Wiedemann–Franz Law,
    96–100
at room temperature, 82
temperature coefficients of, 1200
Electrical safety, 849
Electric arc, 578–596, *See also* Arc
Electric arc formation, *See* Arc formation
Electric field, definition of, 558
Electrode materials, low current arcing fault
    properties, 866, 867
Electrodeposits, 522–525
contact resistance trends and, 533
hard golds, 422
microthrowing characteristics of, 545
porosity of, 530–531
Electrographite brushes, 1048, 1056,
    1074–1075, 1206
Electrographitic brush materials, 1082–1083
Electroless gold plating, 423
Electroless Nickel and Immersion Gold
    (ENIG) plating, 423, 522
Electroless nickel–phosphorus, 423
Electroless plating, 522–525
Electrolytically tough pitch (ETP)
    copper, 240
Electrolytic refining of copper, 240
Electromagnetic penetration depth, 90
Electromigration, 295–296, 639
corrosion, 123–125
on intermetallic growth rates, possible
    effect, 42–45
Electron avalanche, 561
Electron ballistic motion of, 80, 708
Electron bombardment, 640, *See also* Arc
    erosion
Electron-dislocation interaction, *See* Stress
    relaxation
Electron emission, 566–570, 583, 603
energy for at arc cathode, 583
field emission, 566–567
Fowler-Nordheim rquation, 567
potential energy near contact surface,
    566–567

Richardson-Dushman equation, 566
thermally enhanced field emission (T-F), 583
thermionic, 566, 583
Electron-gas collisions, 569
dissociation, 559
elastic, 559
excitation and relaxation, 559
ionization, 560
Electron mean free path in air, 559, 562
Electron volt, 558–559, 1208
Electronic connectors, 114, 131–133, 171–172,
    378–381, 388–389
contact platings in, 522
field studies of, 180, 441
finishes, 522
Electronic mean free path in metals, 82
Electronic optical devices, 996–997
Electronics Industries Association (EIA), 377
Electronic/switching systems, integration
    with, 805–806, 849–872
Electronic thermal overload trip solutions,
    814–815
Electron mean free path in air, 558–559
Electron microprobe, 130
Electron probe microanalysis (EPMA), 690
Electron spectroscopy for chemical analysis
    (ESCA), 130, 1033
Electron theory of metals, 96, 238
Electroplastic effect, 301, 302
Electroplated palladium–nickel alloy, 458
Electroplating, 226, 244–246, 249, 253–256,
    422–423, 522–526
intrinsic polymers in hard gold, 422
Electropolymerization, 458, 462
Electrostatic actuation (MEMS), 704
Electrostatic micro-switch, 704
Elenbaas–Heller equation, 581
Energy balance from arc at contacts, 581–585,
    628–630
Energy dispersive X-ray (EDX), 464, 1026
Energy transfer at arc cathode/anode, possible
    mechanism of, 584
ENIG plating, *See* Electroless Nickel and
    Immersion Gold plating
Environmental classes, reactivity
    distributions, 150
EPMA, *See* Electron probe microanalysis
Erosion of contacts, *See* Arc erosion of
    electrical contacts; Contact erosion
ESCA, *See* Electron spectroscopy for chemical
    analysis
European miniature circuit-breakers, 816, 817
Excitation process, 559, 560

F

Fabrication characteristics of metal
        plates, 526
Failed mobile phones, performance of, 218
Fermi–Dirac distribution function, 97–98
Ferromagnetic materials, conduction in, 240
Fiber brushes, *See* Metal fiber brushes
Field emission, 566–568
Field environments for electrical contacts,
        143–157
Film effects, 18, 146–147
Film resistance, 56, 251, 1055–1056
Film rupture, 255, *See also* Fretting; Fritting
Film thickness, growth of silicones,
        665–666
Filtered connectors, 383
Fine transfer, 723, 724, *See also* Molten metal
        bridge
Finite element analysis, 13, 1044
Finite Element Method (FEM), 91, 711,
        722, 823
Finite Volume (FVM) software, 823
"Fire" curve, 853
Fixed contact gap test devices, 994
Flash coatings, effect on porosity, 530–531
Flash gold coating, 174
Flash gold plating, 131, 486, 522
Flattening of rough surface, 79
Flexible connectors, 297, 298
Flexible printed circuit (FPC), 388
Floating arc, 649
Flowing mixed gas (FMG) environment, 132,
        167–171
Fluid contact lubrication, 1173
Fluid lubricants, *See* Lubricants
FMG environment, *See* Flowing mixed gas
        environment
Foam metals, 361
Force–displacement friction curve, 444
Formative time lag, gas breakdown, 571
Four-gas tests, 168
Four states of matter, 554
Fowler-Nordheim equation, 567
FPC, *See* Flexible printed circuit
Fractal model for rough surfaces, 79
Fractional horsepower motors,
        1085–1086
Free-burning arc column, 579–582,
        588–592
    decay of, 602–603
Frequency, low current arcing fault
        properties, 865–866

Fretting, 134, 267, 398–402, 414, 441–482, 722
    apparatus, materials studies, 449–450
    bolted connections, 278–283
    bus-stab contacts, 276–278, 287
    in circuit breaker terminals, 287–288
    in coatings, 286–287
    compression connectors, 274–275
    current effect of, 476
    cycle rate, 468–470
    cycles to failure (CTF), 401
    degradation, control of, 496–500
    displacement, 447–449
    dust effect, 204
    electrical current effect, 285–286,
        476–479
    environmental effects, 474–475
    examples, 274
    factors affecting, 268–270
    field and laboratory testing methodologies,
        447–449
    force, 472–474
    frictional polymer-forming metals,
        457–460
    gross slip, 444
    lubrication effect of, 179–180, 496–500
    maps, 445
    material transfer, wear, film formation, and
        contact resistance, 479–481
    mating contacts, dissimilar metals on,
        460–464
    mechanisms of, 270–274, 447–449
    mixed slip, 444
    no film-forming tendency, 450–452
    non-noble metals/fretting corrosion,
        452–457
    oxide debris, 271
    partial slip, 443
    plug-in connectors, 278
    power connection, 274
    regimes, 442–445
    static *vs.* dynamic contact resistance,
        445–447
    sticking, 442
    surface finish and contact geometry,
        479
    thermal effects, 475–476
    Type I (Unstable) contact resistance
        behavior, 449–452
    Type II (Intermediate) contact resistance
        behavior, 449–452
    Type III (Stable) contact resistance
        behavior, 449–452
    underplate effect of, 467

vibration, 180, 441
wear-out phenomena, 464–468
wipe distance, 470–472
Fretting corrosion, 282, 283, 396–398, 401–402, 445, 452–457
  characteristics, 399, 400
  debris filled interface, 447
  effect on, 499–500
  frictional polymers, 457–460
  material studies
    aluminum *vs.* aluminum, 283–285, 456
    copper *vs.* copper, 452
    gold *vs.* gold, 451, 461
    gold *vs.* copper, 464
    gold *vs.* nickel, 462
    gold *vs.* palladium, 460, 465
    gold *vs.* tin/lead, 180, 463, 456
    identical base metals, 456, 479
    identical non film forming metals, 479
    mono-metallic interfaces, 456
    nickel, 452, 461
    silver and silver alloys, 282–283, 450, 461
    palladium and palladium alloys, 458, 461
    tin and tin alloys, 179, 280, 283 285, 396–402, 453, 466
  oxide debris, 271
  substrate effect of, 467
  of surface, 1124
Friction, 444, 1170–1173
  adhesion theory of, 481
  characteristics, 1135–1137
  force displacement curve, 444
  forces, 1100–1102
  and wear, 1047–1048
Frictional polymer-forming metals, 457–460, 480
Frictional polymerization, 487
  effect on, 497–499
  mechanisms of, 458–460
Frictional polymers, 23, 149, 447, 458, 1124
  conventional catalysis reactions, poisons for, 459
  effects of, 480
  formation rates, 471
  yield, 469
Friction welding, 339–340, 966
Fritting, 56–57, 258
  A, 56–57, 482
  B, 56–57, 483, 1057–1058
Fuel injector connector, 403
Furnace brazing, 968
Fusion coatings, 522

G

Galling, 1103
Galvanic corrosion, 118–120, 263–264, 404, 407
Galvanic series, 120
Gas breakdown, *See* Long-gap gas breakdown
Gas chromatography/mass spectrometry (GC/MS), 192
Gas concentrations in atmosphere, 126–128
Gas corrosion, *See also* Corrosion
  field environments for electrical contacts, *See* Field environments for electrical contacts; Atmospheric gas concentrations
  laboratory accelerated testing, *See* Laboratory accelerated testing, corrosion
Gas exposure tests, 528
Gas flow effects, single-gas corrosion, 165–168
Gas, molecular mean free path in air, 562
Gas 3rd form of matter, 554
Generators, 1089
Gigahertz electronic circuits, interruption of, 609–612
Glassy silicon-rich deposits, 1024–1025
Glow discharge, 570
Glowing connections
  contact voltage, 403, 860–861
  copper-to-copper wire, 403, 859–860
  current and voltage waveforms, 861–863
  description, 857–858
  floating arc, 649
  formation process, 858–859
  wire materials, 864, 865
Glowing contacts, 403–404, 857
Glow voltage, 565, 570, 860–861
Gold, 245, 521, 970
  degradation of, 128
  and gold alloy contacts, 714
  vs. SnPb solder, 463–464
Gold alloys, 521, 970, 1134, 1204, *See also* Hard golds
Gold-based systems, 464–465
Gold contacts, 639–641, 1124
  arc duration for, 620
  classical electrical contact theory in, 87
  non-arcing, 970–972, 1204
Gold fiber brush resistances, speed dependence of, 1180
Gold-flashed palladium, 480, 487, 546
Gold-plated connectors, 532
Gold-plated contacts, 245, 431, 435, 523
  thickness, 522–523

Gold-plated copper system, 535
Gold-plated film on copper, 123
Gold plating process, 1115
Grafted lubricant layers, 500–503
Granular interface model, 447
Graphite, 1125, *See also* Activation of switching
    contacts
    atoms, layered structure of, 1074–1075
    thermionic emission for, 855, 856
Graphite-lubricated sliding contacts, 1053
Graphitization, 1083
Ground layer plating, 686–687
Growth wear model, 1048
Guillotine test, 854

H

Hall Effect, 240
Halogen-containing fluid, 495
Hard ductile underplates, 431
Hardener content of gold plates, 432–435
Hard golds, 523
    adverse effect of nickel underplate
      for, 537
    fabrication characteristics of, 526
    intrinsic polymers in, 525
Hard mount method, 1178
Hardness, 17, 428–431
    Brinnel, 1196
    Knoop, 341, 1196
    material, 713
    Mohs (dusts), 199
    platings, 523, 543–544
    Rockwell, 1196
    and tensile strength, 651
    Vickers, 1196
    vs. yield stress, 250, 261
Hard nickel underplate, beneficial role
    of, 428
Harmonic distortion, power quality
    issues, 348
Heavy-duty connectors, *See* Power connectors
Hertzian contacts, 392, 706
HF transients, *See* High frequency transients
High current arc, 581
High-current vacuum arc, 595, 596
High frequency current circuits
    interruption of, 609
    solid state *vs.* mechanical switches, 610
High frequency (HF) transients, interruption
    of, 837–838
High-speed cinematography, 996
High-temperature lubrication limits, 1124

High voltage interlock circuit (HVIL), 406
Holm-Archard wear law, 415, 1048, 1158
Holm graphical method, 761
Holm radius, 12, 13
Holm's wear equation, 1071, 1072
Hot-dip tinning process, 298
Hot wax dip, 261
Humidity, 117–118, 124, 127, 144, 164
    water vapor in saturated air, 1209
HVIL, *See* High voltage interlock circuit
Hybrid contactors, 805
Hydrogen sulfide, 117–118, 127, 142, 159,
    161–162
Hysteresis, 705

I

IACS, International annealed copper
    Standard, 241
IDC, *See* Insulation displacement
    connection
IEC, *See* International Electrotechnical
    Commission
IMCs, *See* Intermetallic compounds
Incompatible plating, 404–405
Indium oxide, *See also* Silver metal oxide
    switching contacts
    PM and IO materials, 893
    PM silver–tin oxide material, 894
Indoor industrial environments, 126, 128
Induction brazing, 967
Inductive-coupled plasma (ICP)
    spectrometer, 192
Industrial brushes, 1089–1090, 1152
Inelastic electron collisions, 559
Inlay, 969–970
Instantaneous dielectric recovery, 600–602,
    799–801, *See also* Arc interruption
Insulation displacement connection (IDC),
    381–382, 386
Insulation piercing connectors, 236, 382
Intentional contamination, 1119–1121
Inter-contact plasma, initial rapid adjustment
    of, 601, 607
Interdiffusion bands, growth of, 38
Interdiffusion in electrical interfaces, 41
Intermediate displacement amplitudes, 444
Intermetallic compounds (IMCs), 32, 288–298,
    532, 539–541
    in Al-Au systems, 132
    in Al-brass systems, 29–33, 39, 44
    in Al-Cu systems, 32, 289
    in In-brass systems, 32, 36, 37, 38

in In-bronze systems, 32, 37, 38
in Ni-Sn systems, 540
in Sn-brass systems, 32, 37, 38, 405
in Sn-Cu systems, 32, 296–299, 438, 540
in Sn-bronze systems, 32, 37, 38
in Zn-brass systems, 32, 35, 38
in Zn-bronze systems, 32, 37, 38
diffusion process, 28–45, 288–292
establishment and growth of, 290
failure examples, 298
formation of, 476
Intermetallic layers growth, 28
activation energy, 29–30
current induced diffusion, 295
electromigration on, 42–45
temperature, 33
Intermittences, 1096, 1114
International Electrotechnical Commission
(IEC), 315
International Standards Organization
(ISO), 377
Interruption, power quality issues, 348
Intrinsic polymers
degradation of, 532
in hard gold plates, 525
Intrinsic porosity in electrodeposited
coatings, 527
Ion chromatographic (IC) spectrometer, 192
Ion mean free path in air, 562
Ionization energy, 560
Ionization potential, values of, 561
Ionization process, 556, 559–560
Iridium, 689

J

Joule heat, 18, 43, 57, 280, 1166–1167
effects, 1112
flow, through *a*-spots, 82–83
non-locality, 82–88

K

Key switch, 698
Kinetic energy of molecules, 555, 556–577
Kinetic gas theory, 555, 556–577
Kinetic growth of corrosion film, 116
Kirkendall porosity, 532
Knoop hardness (KH), 341, 1196
Knoop microhardness of intermetallic layer, 34
Knudsen number, 1111
Kohler effect, 638–639
Kohlrausch–Holm Method, 1065–1066

L

Laboratory accelerated testing, corrosion
acceleration factor, 158–159
description, 157–158
mixed-gas environments, 167–171
test applications, 171–174
Lamp loads, effects from operation
parameters, 890, 979
Laplace's equation, 7
Laser microscopy analysis, 1044
Laser vibrometer measurements, 1136
Lattice imperfections, effect of, 238–239
Layered systems, characteristics of,
543–544
Lever switches, 737
Life test of contactors, conditions for, 795
Light-duty connectors, 235
Limit switches, 739–740, 746
Linear momentum, conservation of, 749
Linear tarnish kinetic equation, 115,
1197
Line insulation, 381
Liquid 2nd state of matter, 554
Load-bearing areas, 14, 249–250, 705–707,
1097, 1098, 1106
Load line, 678, 679
Local thermodynamic equilibrium, 578
Logarithmic tarnish kinetic equation, 115
Long duration arcs, effect on contact
resistance, 906–908, 1032
ESCA analysis, 1033
metallic and gaseous arcs, 1033–1034
switching tests, 1032
Long-gap gas breakdown, 558–565
dissociation, 559
excitation and relaxation, 559
ionization, 559
Paschen's law, 564
time lag, 571
Townsend avalanche, 561
Longitudinal magneto resistance, 240
Lorenz constant, 62, 100
Low contact force, 79, 404–405, 705–707
Low-current switching devices, 733–734,
See also Switches, low current
Low ductility, 1134
Low frequency power circuits, interruption
of, 609
Low power and electronic connectors
applications, 376
blade-box, 378, 385–387
blade-leaf, 385–387

bump contacts, 385, 389
connector position assurance (CPA),
      379, 393
compliant pin, 380
corrosion, 493–405
crimp terminals, 390
edge-on, 379
environments, 394–395
eye of needle, 380
failure examples, 402–405
flexible printed circuits (FPC), 380
fretting, 396–402, 407
insulation displacement (IDC),
      382, 386
lug-screw, 385
materials, 385
pin-hyperloid, 388
pin-sleave, 385, 387, 402
pin-socket, 417
plug and receptical, 379
rack and panel, 378
terminal position assurance (TPA),
      379, 383
terminals, 381, 384
terminal temperature, 393
terminal types, 384–392
terminal bulk resistance, 393
terminal-terminal, 378
wire-screw, 385
wire-wire twist, 384, 385
zero insertion force (ZIF), 383
Low-pressure, vacuum operation, 1125
Low surface tension lubricant, 1122
Low-temperature lubrication limits, 1124
Low-voltage circuit-breakers, *See also* Molded-
      case circuit-breakers; Miniature
      circuit-breakers
arc chute and contact arrangement,
      787–788, 810–814
arc dwell time, 789–790, 818
arc simulation, 822–827
arc sticking and back commutation, 637,
      790–791
arrangement, 807–808
contact materials, 791–793, 815–820
current limitation, deion arc chutes,
      608–609, 820–822
electronic control, 814
magnetic blast field, 788–789
principle/requirements,
      806–807
quenching principles, 808–810
trip system, 814–815

Low-voltage contactors, 792–806
air, 794
arc chute design, 798–802
arc simulation, 822–827
bus system integration, 806
contact materials, 802–805
electronic control of magnet, 806
vs. electronics, 804–805
mechanical arrangement, 794, 796–798
pneumatic, 794
principle/requirements, 794, 795
quenching principle and contact and arc
      chute design, 798–802
relays, 802
testing: AC-1,-2, -3 & -4, 794–795
vacuum, 805
Low-voltage switching devices computer
      simulation, 822–827
Lubricants
aerobic, 1121, 1124, 1141
chemical grafting, 500–503
durability, 504–505
effectiveness, 423, 497
fretting on, 179–180, 204–205, 487, 496–500
fundamental properties of, 488–491
halogen containing fluids, 495
liquid, grease and wax lubricants, 175–177,
      203–204, 334–336, 488–496, 503–505
lubricants coated, 175, 201–203
making contact with, 175–177
requirements, 177–178, 491–492
role of, 442
solid, 504, 1133
Lubrication, 134, 174–180, 334–336, 481–506, 1119
adventitious, 1121, 1124
aerobic, 1141
contact aid compounds, 175–176, 334–336
contact resistance effect on, 175–177
fluid lubricants, 488–500
grafted and self-assembled lubricant
      layers, 500–503
greases, 175–176, 503–504
liquids, 177–179
lubricant durability, 177, 504–505
metallic films, 481–487, 482
modes, 1121–1125
particle displacement, 201–204, 505–506
sliding contacts, 1121–1125, 1172–1173
solid lubricants, 504, 1133
studies, 179
surface films and, 1172–1173
vapor, 1125
Lug–screw terminal, 385–386

M

Machine-made composite rivets
  cold bonding process, 956–957
  squeeze test, 957, 958
  ultrasonic methods, 959
Macro sliding contact
  brush configurations, *See* Brush
    configurations
  counterface configuration, 1126–1127
  forces on brush, 1131–1132
  real *vs.* apparent area of contact, 1128
Magnet coils, electronic control of, 806
Magnet drive characteristics of reed switch,
    683–686
Magnetic actuator, 794, 796
Magnetic blast field, 788–789
Magnetic devices, arc motion
    measurements, 997
Magnetic driven reed switches, applications
    of, 698–701
Magnetic fields
  of coil, 674
  generation, 831–832
Magnetoresistance, 239–240
Magnets, shapes of, 797, 798
Make-and-break erosion, 631, 995
Make-and-break test, of contactors, 795
Martensitic transformation, 351–352
Mass transfer, between contacts, 636–644, 757,
    766–768, 899
Mated exposures, *vs.* unmated, 172–174
Matter, 4th state of, 554–558
Matthiessen's rule, 238
Maxwell–Boltzmann distribution
    function, 556
Maxwell (continuum) *vs.* Sharvin (ballistic)
    resistance, 1110
MCCB, *See* Molded-case circuit-breakers
MD simulations, *See* Molecular dynamics
    simulations
Mean free paths of molecules, electrons and
    ions in air, 562
Mechanical behavior
  fretting, *See* Fretting
  dust particles, 198–205
  lubricants, *See* Lubricants
  sliding contacts, 201
  stationary contacts, 199–201
  wax and liquid lubricants, 203–204
"Mechanochemical reaction," 460
Medium-duty connectors, *See* Low power and
    electronic connectors

Medium-sized motors, 1089
Megahertz electronic circuits, interruption of,
    609–612
Melting voltage, 60, 61, 89, 95–96, 562, 1198
  deviation of, 64–69
  for metals, 573, 1198
  nickel, 67–69
  nickel-aluminum system, 67–68
MEMS, *See* Microelectromechanical systems
Metal conductor classifications, 237
Metal dendrites, growth of, 124, 125
Metal fiber brushes, 1152–1181, *See also* Sliding
    wear, of multifiber brushes
  applications, diversification of,
    1154–1156
  brush construction, electrical resistance,
    1174–1177
  brush dynamics, 1177–1180
  clean metal fiber brushes, 1170
  electrical contact, 1174–1177
  electrical resistance on construction,
    1174–1177
  elements of, 1154
  fiber brush designs, 1152, 1153
  film disruption, 1172
  flash temperatures, 1166
    estimating, 1169
    pertinent mathematical theory of, 1167
  future directions for, 1180–1181
  kinetic tarnish equation, 1197
  Kuhlmann-Wilsdorf tests, 1154
  non-tarnishing metals, 1172
  materials, 1157
  metal coated, 1076
  pertinent mathematical theory of flash
    temperatures, 1167
  plastic and elastic contact, 1161–1163
  resin-bonded brush materials, 1083–1084
  sliding wear, multi-fiber brushes, *See*
    sliding wear, of multi-fiber brushes
  sliding speed of, 1180
  spring model for fiber brush, 1178–1179
  steady state flash temperature, 1169
  surface films, friction, and materials
    properties, 1170–1173
  thin film behavior, 1170
  water surface films, 1170–1172
Metal foam, materials of, 355–361
Metal-in-elastomer materials, 525–526
Metallic *a*-spots, 56
Metallic electromigration, 123–125
Metallic film lubrication, principles of,
    481–482, 485

Metallic finishes, coatings on, 520
  electrodeposits and electroless deposits,
      522–525
  metal-in-elastomer materials, 525–526
  overview, 526
  thickness, 529–530
  wrought metals, 521–522
Metallic phase arc, 642–643, 762–763
Metallographic methods, 912–918
Metallurgical methods, 912–918
Metals
  boiling voltage *vs.* break voltage for, 575
  in contact with non-metal, 65
  critical vacuum breakdown field for, 568
  electrical resistivity and thermal
      conductivity, 96–100
  fabrication characteristics of, 526
  melting voltages for, 573, 1198
  properties, 1198–1201
  properties at melting point, 1209
Metal–semiconductor contacts, 22
Metal surfaces, 624, 1042
Metal-to-metal *a*-spots, 69
Metal-to-metal contact, 27, 55, 118
Metal transfer, *See* Arc erosion
Methods B735, B741B799, B809 (ASTM), 528
MFG, *See* Mixed flowing gas
Micro-contact resistance modeling, 705–713
  contact resistance in, *See* Contact
      resistance
  elastic modeling, 706–707
  spreading resistance in, 709
Microcrystalline wax, 504
Microelectromechanical systems (MEMS), 77,
      610, 703–725, 1113
  activation methods, 704–705
  contact materials for performance and
      reliability, 713–720
  failure modes and reliability, 720
  micro-contact resistance modeling, *See*
      Micro-contact resistance modeling
Micro-environment of contact region,
      1114–1115
Micro motion, 211–212
Micro sliding, 1100
Microtopography
  evolution of, 79
  of surfaces, 7
MIL Specs, Military Specifications, 179, 377
Minature motors, brushes for, 1085
Mindlin model, 273
Mineral oil, contacts under, 125–126
Mineral particles, arcing contact, 1015–1018

Mineral particulate contamination, arcing
      contacts
  contaminant type, 1017
  finer material, 1017–1018
  SiC-contaminated materials, 1015–1016
Miniature circuit-breakers, 815–816, *See also*
      Low-voltage circuit-breakers
  contact materials, 816–820
  dwell time, 818
  examples, 815
Minimum arc current, 585–588
Minimum arc voltage, 585–588
Miscellaneous conversion table, 1208
Mixed flowing gas (MFG)
  corrosion testing, 167–171
  laboratory testing, 131–133
Mixed-gas sulfur environments, 164
Mixed slip, 444
Model test switch arc motion control, 998–999
Model test switch testing, 977–978
  contact erosion, 993
Moisture, 124
Molded-case circuit-breakers (MCCB),
      807–809, *See also* Low voltage circuit
      breakers
MOLE, *See* Molecular optical laser examiner
Molecular attrition, 270
Molecular dynamics (MD) simulations, 490,
      718, 1103
Molecular optical laser examiner (MOLE),
      693, 695
Molecular re-ordering in polymeric films,
      1100
Molten metal bridge, 572–577, 619, *See also* Arc
      formation during contact opening
  boiling voltage, 575–576
  break/rupture voltage, 575–576
  bridge formation, 571–577
  bridge rupture, 573–577
  for Cu–Cr contacts opening, 593
  erosion
      Buhl effect, 639
      electro migration, 639
      fine transfer, 639, 723–724
      Kohler effect, 639
      Thompson effect, 639
  melting voltage, 574–575
  stages of, 574–577
  voltage characteristics of, 573
Molybdenum disulfide, 1125, 1133
Monte Carlo techniques, 13
Motion continuation, 1102
Motion initiation (pre-sliding), 1100

Motion of high current columnar vacuum
        arc, 596
Motion over time, 1105–1106
Motor control, 806
Motor Protection Circuit Breakers (MPCB), 814
Moving-contact dynamics, 745–747
Moving contact test devices, 994–995
MPCB, *See* Motor Protection Circuit Breakers
Multi-fiber brushes, sliding wear of, *See*
        Sliding wear, of multi-fiber brushes
Multifilament brush, 1130, 1131
Multilam contact elements, 338–339
Multiple exposed still photographs
        technique, 996
Multiple-wafer brushes, 1090
Multiscale model for rough surfaces, 79
Multiwall carbon nanotubes (MWNTs), 715

N

Natural graphite, 1075
NEC, National Electric Code, 377
NEMA, National Electrical Manufacturers
        Association, 794
Net cathode gain for gold contacts, 641
Nickel, 245–246
    vs. copper contacts, 452
    corrosion mechanisms, 152
    corrosion rates, 146–147
    degradation of, 128
    melting voltage, 67–69
Nickel–aluminum system melting voltage, 67, 68
Nickel-hardened gold, 427
Nickel nano-wires, 719
Nickel–nickel systems, 462
Nickel oxide, 141–142, 215
    growth rate and electrical resistivity,
        51–52, 54
Nickel plating, 245, 524
    thickness, 523
Nickel underplate, 524, 536
    adverse effect of, 537
    gold-flashed palladium on, 546
    noble metal layer on, 532
Nitrogen dioxide (NO_2), 117, 163
Nitrogen ion, acceleration of, 558
Noble metals
    palladium and silver–palladium alloys,
        940–942
    platinum, 942–943
    in sliding contacts, 1132
Nodular plating finish, 546
No film-forming tendency, 450–452

Noise, 1108–1109
Nominal plating thicknesses, 523
Non-catalytic elements, 459
Non-chemical methods, contact finishes
        produced by, 525
Non-circular *a*-spots, 7–11
Non-contact surface profiling systems,
        758
Non-corrosive film formation, 493
Non-linear noise, 1111–1112
Non-metal, metal in contact with, 65
Non-noble metals, 240–245, 415, 437, 452–457,
        524–525
Non-noble silver alloys
    fine silver, 937–938
    hard silver and silver–copper alloys, 938
    silver alloys, 244, 439, 524
Non-ohmic noise, 1109–1111
Non-ohmic "tunneling" conductance, 1107
Nut factor, 319

O

OFHC, *See* Oxygen-free high-conductivity
        copper
Oleic acid, lubrication with, 177
Optical devices, arc motion measurements,
        759, 996
Organic contamination and activation, *See*
        Activation; Contact activation
Outdoor industrial environments, 126
Oxidation, 117, 118, 151–152, 256–260,
        1051, 1052
Oxidation-inhibiting additives, 492
Oxide debris, *See* Fretting corrosion
Oxide films, 47–57, 140–143, 256–260
    fracture, 255, 464
    growth rate, 47–56, 557
    properties of, 256–260
Oxide-free aluminum surface, 69, 70
Oxides of contact materials, growth rate and
        electrical resistivity, 47–56
Oxidizing gas, 117
Oxygen-free high-conductivity copper
        (OFHC), 241, 434–436
Ozone, 117

P

Palladium, 245, 441, 521, 524, 537–538
    contact resistance behavior of, 458
    gold properties and, 686, 687
Palladium-based alloys, 1204

Palladium-based plated contacts, contact resistance of, 473
Palladium-based systems, 465–466
Palladium–nickel alloys, 460, 521, 524
Palladium–palladium contacts, 469, 471
Palladium plating, 245, 522
  thickness, 523
Palladium rider sliding, 419
Palladium–silver alloys, 458
Palladium switching contacts, 940–942, 1204
  palladium silver, 941
Palladium *vs.* gold alloys, 460–462
Palliative measures, power connectors, *See* Power connectors, palliative measures
Panel type connector, 378
Paper electrography, 528
Parabolic tarnish kinetic equation, 115
Parallel arc fault, 854
Paschen curves for air and SF_6, 564, 565, 855–856
Paschen's law, 564
Passive Intermodulation (PIM) interference, 1112
PCBs, *See* Printed circuit boards
Percussion welding, 654–657, 903–904, 965
Perfluorinated polyether (PFPE), 180, 488
Perfluoroalkyl polyether fluid, 495–496
Permanent magnet (PM) motor, 1086, 1088
Permanent power fuse, 821
Permittivity *vs.* dust deposition density, 202, 203
Peroxides, 117
Petrolatum grease, 503
PFPE, *See* Perfluorinated polyether
Photoelectric emission, 566
Photovoltaic solar power systems, 637, 647–648
Piezoelectric actuation in MEMS, 705
PIM interference, *See* Passive Intermodulation interference
Pip and crater formation, 638, 757, 889, 901–902, 980
Pipe diffusion, 290, 296
Planck's constant, 54
Plasma state, 554, 556, 826, *See also* Arc column
Plastic contact, 1161–1163
Plastic deformation, 239, 252
  of asperities, 13, 77
  creep and, 305
Plasticity index, 76
Plated contacts, 23, 131–132, 253–256, 414–505, 520–545

Plating-factor, 25
Platings
  clad, 521
  classification of, 244–246, 523–526, 688–689
    copper, 245, 688
    ENIG, electroless Ni and immersion Au, 423
    gold, 245, 523, 714–715
    hard gold, 523, 714
    intrinsic polymers on hard gold, 525
    iridium, 689
    nickel, 245, 524
    nickel alloys, 523–524, 689
    other deposits, 525
    palladium and palladium alloys, 245, 523–524, 714
    platinum, 714
    rhodium, 245, 714
    ruthenium, 523, 714, 717
    silver, 244, 524
    tin, 525
    tungsten, 689
  contact resistance, 533, 543
  electroless, 522
  electrolysis, 522
  fretting in, 286–287, 441–479
  gold over carbon nano-tubes, 714–717
  hardness, 523, 543–544
  intermetallic, 539–541
  Kerkendall porosity, 532
  MEMS contacts, 713–720
  porosity, *See* Porosity
  power connectors, 246, 253, 331–333
  reed relay contacts, 686–689
  thermal effects, 532–539
  substrate diffusion, 532
  thickness of, 522–523
  whiskers Sn and Ag, 541–542
Platinum-based alloys, 1204
Platinum plated contacts, 245
  thickness, 523
Platinum switching contacts, 942–943, 1204
Plots of formative time lag *vs.* fractional over voltage, 571
Plowing, 419–421, 1104–1105
Pneumatic contactors, 794
Point-on-wave (POW), using Ag/CdO contact materials, 771–773
Polyglycol diproxamine D157, lubrication with, 177
Polymer-forming materials, 457
Polymers, 422, 451–452

Pore corrosion, 120–122, 153, 262, 526, *See also* Corrosion and Porosity
  corrosion product, 114
Porosity, 131–133, 526–532
  classic test, 423
  corrision product, 114
  origins of, 527
  methods B735, B741B799, B809 (ASTM), 528
  porous gold, 153, 161, 162, 423
  properties related to, 526
  tests of, 527–529
  thickness of finish, and substrate roughness, 529–530
  underplatings, flash coatings, and strikes on, 530–532
Position sensor connector, automotive, 402
Possible mechanism of energy transfer at cathode and anode, 584
Potential energy *vs.* distance for electron, 566
POW, *See* Point-on-wave
Powder metallurgical (PM), 887–890
Power connections, installation, 361–363
Power connectors, 133–134
  automatic splices, 326–327
  bolted, 236, 278–283
  compression, 315–316, 344–345
  dead end, 327
  fired wedge, 343–344
  insulation piercing, 236
  intermetallics formation in, 298–299
  physical model, 351
  plug and socket, 236
  prognostic models, 349–351
  shape memory (SMA), 327–331, 351–355
  Souriau, 354
  stepped deep indentation, 315–316, 344–345
  thermite welding, 339
  types of, 234–236
  wedge, 325
  welded connections, 339–342
  wire, 236
Power connectors, palliative measures, 264, 306
  bimetallic inserts, 336–337
  coating, 331–333
  connector design, 342–345
  contact area, 307–313
  contact pressure, 313–320
  disc-spring washers, 322–325
  lubrication, 334–336
  mechanical contact device, 320–322
  multilam contact elements, 338–339
  transition washers, 337–338

Power connectors, performance factors, 233, 247
  aluminum and aluminum alloys, 243–244
  contact area, 249–252
  copper and copper alloys, 240–243
  corrosion, *See* Corrosion
  creep, *See* Creep, metal
  current cycling tests, 315–316, 333, 345, 363–365
  degradation, 345–348
  elastic deformation, 253
  electric current effect, 301–302
  factors affecting reliability of, 247–249
  fretting, *See* Fretting
  installation, 361–363
  intermetallic compounds, *See* Intermetallic compounds
  lubrication, 334–346
  nut factor, 319
  other metals, 244–246
  oxidation, 256–260
  plastic deformation, 252
  plated contacts, 253–256, 331
  power quality, 346–348
  prognostic models, 349–351
  reliability, 247–250
  stress relaxation, *See* Stress relaxation
  thermal expansion, 264–267, 321–322
  thermoelastic ratcheting effect in bolted aluminum joints, 266
  tightening torque, 318–320
  testing, 363
Power quality, 346–348
  disturbances, 347
  harmonic distortions, 348
  interruption, 348
  sag, 348
  swell, 348
  transients, 347
Pre-impact arcing, 570, 631, 654, 775–776
Preoxidized metal oxide parts, 887
Pre-sliding phase, 1100, 1102
Pre-strike arc, 570, 631, 654, 775–776
Primary cathode spot parameters, 594
Primary interface interactions, 1098
Primary operating conditions for electromigration, 124
Printed circuit boards (PCBs), 423
Probability of anode spot formation, 594
Probe contacts, 543
Prognostic models for power connectors, 349–351
Progressive subsurface plastic deformation, 428

Prow formation, 415, 419–421, 437, 1103
Pure palladium, 537–538
Purple plague, 132

Q

Quantum-mechanical tunneling
mechanism, 567
Quenching principles (arc)
contactors, 798–802
low-voltage circuit-breakers, 808–810
Quenching systems (arc), 787–788

R

Race theory of dielectric recovery, 599
Rack type connector, 378
Radial brushes, 1050
Radial stresses, 73
Radiation, low-voltage arcs simulation,
825–826
Radioactive tracer contact erosion
measurement, 639, 640
Radio-frequency interference (RFI), 1087
Radio frequency microelectromechanical
system (RF MEMS), 703
Real contact area, 15, 249–251, 1042
vs. apparent contact area, 15, 249, 1128
Reasonable noise specification, 1138
Recovery, stages of, 599–600
Reed relays and switches, 674–698
application examples, 698–701
contact plating, 686–689
contact adhesion prevention, 695–696
contact surface treatments 690–693
design principles, 674–685
drop-out characteristics, 682–683
ground plating layer, 687
magnet drive characteristics, 683–686
magnetic driven, 698–701
neutral position detection switch, 700
oxygen, 693
polymerization, 690
pull-in characteristics, 674–682
surface deactivation, 690–695
thermal, 701
Reflection high-energy electron diffraction
(RHEED), 693
Reignition voltage, 599, 637, 800, 801,
832–834
arc chamber wall material effect, 799–800
Relative humidity (RH), 117–118, 126, 127
Residential low current arcing faults, 871

Residential wiring, 869
Residual resistivity, 238
Resistance brazing, 341–342
Resistance welding, 341
button welding, 962–963
contact tape welding, 963–965
wire-welding, 963
Restrike, 637, 790
Reynolds equation, 721
RFI, *See* Radio-frequency interference
RF MEMS, *See* Radio frequency
microelectromechanical system
RF sensing method, 851, 852
RH, *See* Relative humidity
RHEED, *See* Reflection high-energy electron
diffraction
Rhenium, 686
Rhodium, 245, 687, 690
Rhodium oxide (Rh$_2$O$_3$), 693–695
Rhodium-plating solution, 687
Richard-Dushman equation, 566
Rigid noncurrent-limiting breakers, 812
Ring *a*-spots, 7–11
Ring temperature, coefficient of friction *vs.*,
1053, 1054
Rivets, *See* Contact rivets
Rockwell hardness, 1196
Roughness, surface, 431
Round dots, 521
Ruthenium, 687–688

S

SAE, Society of Automotive Engineers, 377
Saha's equation, 579, 603
SAMs, *See* Self-assembled monolayers
Scanning electron microscope (SEM), 29,
187, 716
Scanning electron microscopy with energy-
dispersiveX-ray spectroscopy (SEM/
EDAX), 129
Scanning Tunneling Microscope (STM),
1111, 1141
SCC, *See* Stress corrosion cracking
Scintillations, 855
Second Townsend coefficient, 563
Self-assembled lubricant layers, 500–503
Self-assembled monolayers (SAMs), 502
Self-healing, 504
of electrical contacts, 75, 96
process, 256
Self-induced fields, switch arc motion
control, 998

SEM, *See* Scanning electron microscope
Series arcing, 855–857
Series-wound field motor, 1086
Servomotors, 758
SFA, *See* Surface forces apparatus
Shape-memory alloy (SMA), 351–355
    connector devices, 327–331
    and disc-spring washers, 329–331
Shape-memory effect (SME), 327
    electronic application, 329
    one-way, 352–353
    origin of, 351–353
    two-way, 353
Sharp contacts, 543
Sharvin resistance, 81, 83, 210, 708,
        1110–1111
    heating effect, 82
    role in small aluminum-aluminum
        contacts, 86–87
Shear-induced ordering transitions, 1102
Sheet clad metal, 521
Shielding effects, 147
    corrosion, 134
Short-circuit arcing, 854–855
Short-gap breakdown, 566–569
    critical fields, 568
    electron emission, 566
    enhancement factor, 568
Showering arcs, 624–628
SI, International System of Units, 1207–1208
Signal quality, 1094
Silicone contamination, arcing contacts,
        665–667, 1018–1030
    film growth, 665–666
    lower molecular weight fractions, 1019
    lubricant, 1024
    migration, contamination from, 666–667,
        1023–1029
    molecules, 665, 1022
    SEM and EDX analysis, 1029
    silicone migration, 665–667,
        1023–1029
    silicone vapors, 665–667, 1019–1023
    types of, 1018
    vapors, contamination from, 665–667,
        1019–1023
Silver, 524, 1202
    corrosion mechanisms, 151
    corrosion rates, 145, 146
    drawbacks of, 244–245
    foam materials, 360–361
    and silver alloys, 282–283
    sulfide resistivity, 54, 141

Silver alloys, 1202
    noble metals, *See* Noble metals
    non-noble, *See* Non-noble silver alloys
    silver and, 282–283
Silver-coated contacts, stability of, 287
Silver coatings, contact properties of, 339
Silver contacts, 638, 1202
    voltage drop across, 642
    silver-based alloy switching contacts,
        937–938, 1202
    silver-based switching contacts, 939, 1202
        Ag-C, 943–945
        Ag-Cu, 938
        Ag-Ni, 939
        Ag-Pd, 940
        Ag-Pt, 942
Silver counterface, 1132
Silver-graphite brushes, 1084
Silver graphite switching contacts
    anode and cathode, 944–945
    dynamic welding, 944
    erosion rate, 922
Silver metal oxide switching contacts, 802,
        880–908, 1203
    AC *vs.* DC testing, 890
    contact resistance, 906–908
    cracking, 904–906
    erosion/arc, 632, 644, 896–902, 904–906
    erosion/arc mobility, 906
    erosion/materials transfer/welding,
        899–902
    erosion/mechanisms/cracking,
        904–906
    inductive loads, 890–891
    internal oxidation, 884–889
    interpreting Ag-CdO material research,
        895–899
    interruption characteristics, 800, 906
    manufacturing technology, 884–895
    materials
        Ag-CdO, 883, 887
        Ag-Fe$_2$O$_3$, 899
        Ag-MgO-NiO, 883
        Ag-SnO$_2$, 632, 644, 883, 889,
        891–894
        Ag-ZnO, 883, 898–899
    silver–cadmium oxide *vs.* silver–tin
        oxide, 895
    thermal arrest lines, 886
    transfer/welding, 902–904
    types, 883–884
    welding, 902–904
Silver migration, 124

Silver nickel contact materials
    electrical erosion, 939
    manufacturing methods, 881
    use, 793, 802
Silver-plating, 244, 332, 439, 524
    thickness, 522–523
Silver-refractory metals, 908–935,
        1203, 1205
    arc erosion, 919
    contacts, 659–664
    contact resistance, 663, 928–935
    electrical properties (EP), 918
    graphite additions, 922
    manufacturing technology/press sinter
        repress, 909–910
    materials
        Ag-Mo, 921, 1203
        Ag-W, 908–922, 1203
        Ag-WC, 921, 1203
        Cu-W, 922, 1205
    material properties, 908, 1203
    material technology, 909–912
    metallurgical/metallographic methods,
        912–914
    metallurgical/structure/strength and
        toughness, 914–918
    oxides and tungstates, 663, 928–935
Silver–silver contacts, 450, 661
Silver sulfide, 141–142, 542
    growth rate and electrical resistivity,
        52–54, 142–143
Silver tungstates, 663, 930–935
Silver whisker, 542
Single controlled bounce, 656
Single-gas corrosion effects
    chlorine, 163–164
    gas flow effects, 165–168
    hydrogen sulfide, 159, 161–162
    temperature, 164–165
Single gas mixtures, 131
Single-wafer brush, 1090
Six VDC circuit, 639–640
    mass transfer between contacts, 644
Skin effect/depth, 240
    and constriction resistance, 89–95
    electromagnetic penetration depth, 90
Sliding contacts for instrument control,
        1094–1143
    adhesion, 1102–1103
    adhesive transfer, 1103–1104
    adventitious lubrication, 1121, 1124
    anaerobically lubricated contacts,
        1122–1123

    anaerobic *vs.* aerobic lubrication, 1121
    assemblies, 1138–1141
    cantilever composite brush, 1128–1129
    cantilever metallic finger, 1129
    cantilever wire brush, 1129–1130
    channeling, 1124–1125
    characteristics of, 1096–1097
    conductivity, 1134
    contact impedance, 1096, 1112–1113
    contact resistance variation (noise),
        1108–1109
    contaminating environment, 1121
    cylindrical surface, 1127
    data integrity, 1114
    direct effect torque, 1135
    disjoining pressure, 1122
    dynamic effects, 1107
    electrical performance in, 1107–1108
    excessive static friction, 1135
    film formation on *a*-spots, 1115–1117
    filming properties, 1134
    film potentiometers, 1133
    flammable fuels, submerged in, 1125
    flat counterfaces, 1127
    fluid-lubricated wire brushes in
        V-grooves, 1125
    fretting, 1124
    frictional polymers, 1124
    friction and wear characteristics, 1135–1138
    friction forces, 1100–1102
    gas lubrication, 1125
    groove configurations, 1127
    guidelines, 1114
    hard particle/three-body, 1105
    inductance in contact region, 1114
    low pressure and vacuum operation, 1125
    lubrication, 1119–1121
    lubrication modes, 1121–1125
    macro, *See* Macro sliding contact
    materials, 1132–1134, 1206
    mechanical action, 1117
    mechanical aspects, 1098–1099
    micro-cuts, 1139
    micro-environment of contact region,
        1114–1115
    micro-interrupts, 1139
    molybdenum disulfide, 1125, 1133
    motion continuation, 1102
    motion initiation (pre-sliding), 1100
    motion over time, 1105–1106
    non-linear noise (frequency dependent),
        1111–1112
    non-ohmic noise, 1109–1111

numerical modeling, 1103
organic off-gasses, 1118–1119
over-lubrication of sliding electrical
    contacts, 1122
particulates, 1117
passive circuit elements, 1114, 1139
pin-on-disk contact, 1126
platter slip ring, 1140
plowing/two-body, 1104–1105
polymeric films, molecular re-ordering
    in, 1100
potentiometers, 1127
pressure, 1135
resistive effects, 1114
ring sliding system, 1056
sawtooth stick-slip curves, 1101
spring force, 1128
spring properties, 1134
stick-slip force, 1136, 1137
stick-slip process, 1101
temperature extremes, 1124–1125
tunnel resistance and vibration, 1048–1051
types of, 1044
unintentional contamination, 1117–1119
voltage drop, 1059
water molecules, 1118
Wexler's formula, 1111
wire brush, 1129–1130
    materials for, 1133–1134
Sliding contacts, metal fiber brushes, *See*
    Metal fiber brushes
adhesive wear, 1157–1158
arcing and bridge transfer, effect of,
    1169–1170
critical/transition brush pressure,
    1163–1164
film disruption, 1172
high wear regime, 160–161
Holm-Archard wear equation, 1158–1159
low wear equilibrium, 1159–1160
sliding speed effects, 1166–1169
steady state flash temperature, 1169
wear *vs.* brush pressure, 1164
Sliding graphite based contacts, 1042–1077
adhesive wear, 1048
atoms, layered structure of, 1074–1075
brush materials and abrasion, 1067,
    1073–1076
brush wear, 1068–1073
chatter range, 1049, 1050
chemical aspects, 1051–1054
commutation, 1061–1063
contact area, 1042

contact resistance, 1055–1061
elastic penetration, 1047
electrical effects, 1055–1063
electrical noise, 1087
film resistance, 1055–1056
flashes, 1071–1073
friction-excited vibrations, 1049
graphite crystal structure, 1075
graphite, natural, 1075, 1083
Hertz's formula, 1045
Holm's law of wear, 1048
Holm's wear equation, 1071, 1072
homo-polar bonds, 1074
ideal commutation, 1082
Kohlrausch-Holm Method, 1065–1066
mechanical aspects, 1044–1051
mechanical contact surface, 1064–1065
mechanical wear, 1070–1071
metal/graphite brushes, 1075–1076
moisture film, 1051–1054
monolithic fiber brush, 1076
occasional interruptions, 1072–1073
oil-lubricated sleeve bearing, 1068–1069
optimum brush, 1085
over-commutation, 1082
oxidation, 1051
particle transfer, 1048
residual current, arc duration and,
    1061–1063
single contact, rupture of, 1042–1043
steady-state super temperatures, 1065
temperature effects, 1064–1068
thermal effects, 1064–1068
transfer and growth model, 1048
transfer wear model, 1048
transient effects in brush and collector
    applications, 1067
tunnel resistance, 1048, 1057
under-commutation, 1082
vibration, 1048
wear, 1068–1073
wear debris, 1107, 1141
wear resistance, 1134
Sliding motion, small-amplitude, 742
Sliding speed, effects of, 1166–1169
Sliding wear, 415
abrasion wear, 424–426
adhesion, 416–417
brittle fracture, 426–427
clad metals, 435–437
delamination and subsurface wear,
    427–428
early studies, 415–416

electroless gold plating, 423
electroless nickel plating, 423
fretting, 441–448, *See also* Fretting
gold platings, 422–423
lubrication, 488–503
of multifiber brushes, 1156–1170
mild and severe, 418–419
prow formation, 419–421
rider wear, 421
silver, 439–441
tin and tin-lead alloys, 437–439
underplate and substrate, effect of,
    428–431
underplate hardness, relationship of wear
    to, 431–435
unlubricated adhesive wear, 432
    electrographic wear indexes from, 429
unlubricated palladium-palladium
    contacts, 487
wiping contaminant from contact surfaces,
    417–418
Slip events, 1100
Slip rings, 1082, 1122, 1127
    composite materials for, 1133
    electrical model of, 1141
    noise, 1109
    as transmission lines, 1139–1140
SMA, *See* Shape-memory alloy
SME, *See* Shape-memory effect
Smutting, 1072–1073
Soft ductile metals, 419
Softening voltage, 95–96, 1198
Soft magnetic nickel-iron alloys, 246
Soft sticking, 695
Solid lubricants, 504
    composite materials for brushes, 1133
Solid state *vs.* mechanical switches, high
    frequency circuits, 609–610
Solid 1st state of matter, 554
Space brushes, 1090–1091
Spark, 563
Sparkless commutation, 1086
Special brush holder, 1049, 1050
Speedometer switch, 700
Splitter plates, 603, 788, 801, 802, 824
Spreading resistance
    calculation in thin film, 18–20, 709
    remarks on calculation of, 20–22
Spring materials, 526, 1205–1206
Square *a*-spots, spreading resistance of, 10, 11
Stannic oxide (SnO_2), 49
Stannous oxide (SnO), 49
Starter motors, automotive, 1088–1089

Static *vs.* dynamic contact resistance, *See*
    Contact resistance
Static friction coefficient, 489
Static mechanical contact, 1095
Stationary anode spots, 608
Statistical time lag, 571
Steady state temperature distribution, closed
    contacts, 65, 103–104
STM, *See* Scanning Tunneling Microscope
Streak technique, 996
Streamline effect, 307–309
Stress corrosion cracking (SCC), 124–125
Stress relaxation, 299–305
    electric current effect on, 302–305
    and metal creep, 305–306
    static *vs.* dynamic, 267
Strikes effect on porosity, 530–531
Stylus profilometer, 757
Substrate
    effect of underplate and, 428–431
    role of underplate and, 467–468
    roughness, porosity and, 529–531
Subsurface wear, 427–428
Sulfur dioxide (SO_2), 117, 131, 163
Superconductors, 821–822
Superficial initial oxides, mechanical
    disruption of, 452
Supplementary organic lubricant, 485
Surface deactivation treatment, 690–691
    life test of samples, 691–695
Surface film effects, 141
    on constriction and contact resistance, 18
    electrically insulating/weakly conducting,
        45–47
Surface films, 393–396, 1107, 1170–1173
Surface forces apparatus (SFA), 1157
Surface micro-roughness, 4
Surface penetration, 1047
Surface pressure, 1045
Surface profile
    measurement of, 756–757
    using non-contact laser probe, 757
Switch arc motion control, 998–999
Switches, low current, 733–781
    AC current erosion, 720–723
    actuated, 739–741
    bounce times, 752–754
    DC current erosion, 755–753
    design, 779
    change-over configuration, 735–737
    change-over time, 744–746
    classification, 733–734
    coefficient of restitution, 749–751

contact bounce, 779
contact force, 742–743
contact welding on make, 774–775
"hair spring" configuration, 736, 737
hand-operated, 734–739
hand-operated rocker-switch mechanism,
    744–746
impact mechanics, 748–749, 752
impact times, 755
lever, 737
low-current alternating current arcs,
    768–770
low-current direct current arcs, 761–765
low current, electrical characteristics on
    closing, 773–774
make-down configurations, 735, 736
make operation, 747–748, 780–781
make-up configurations, 735, 736
membrane switches, 739
normal operation, rocker-switch,
    mechanism, 744
on-off configuration, 735, 736
opening characteristics, switching devices,
    746–747
pivoting mechanism, impact mechanics
    for, 751–752
pre-impact arcing, 775–776
push-button switches, 738–739
remote control switches, 794
rocker-switch mechanism, 734–737
rotary switches, 738
slide switches, 738
snap action, 735
soft action, rocker type, 744
static design parameters, 741–743
switching devices, 739
thermostatic control switch, 740–741
velocity influence, during first bounce,
    777–778, *See also* Point-on-wave (POW)
Switching of d.c. circuits, *See* Arc interruption
Synthetic graphites, 1083

T

Tabor equation, 1174
Tangential brushes, 1180
Tangential fiber brush, 1154–1156
Tangential forces, 1100
Tarnish films, 129, 256–260
    observation and analysis of, 117–118, 129,
        140–143
    reaction, 114
    tarnish rates, 114–116, 145

Teflon-thickened greases, 503–504
Tellurium oxide, 891–892
TEM, *See* Transmission electron microscopy
Temperature-dependent ceramics/
    polymers, 821
Temperature-dependent electrical resistivity,
    voltage-temperature relation with,
    60–62
Temperature distribution in *a*-spot vicinity,
    63–64
    plane of maximum value, 100–104
Temperature *vs.* voltage, in asymmetric
    contact, 64, 1066
Ten-finger parallel contacts, 649
Tensile strength, hardness and, 651
    cold working, 651
Terminal position assurance (TPA), 383
Testing switching contacts
    AC *vs.* DC, 978
    bounce, 980
    contact resistance, 982
    device *vs.* model, 976, 988
    erosion, 981
    load type, 974
    motion, 996
    welding, 989
Tests of porosity, 527–529
Test systems, mixed-gas environments, 169
T–F electron emission, *See* Electron emission
Thermal conductivity, 1198
    constants, 68
    of metals and Wiedemann–Franz Law,
        96–100
    voltage-temperature relation with, 60–62
Thermal diffusion in platings, 532–539
Thermal expansion
    coefficients, 321–322, 1199
    power connectors, 264–267
Thermal-magnetic (T-M) breaker, 850–851
Thermal reignition, 603–604
    power balance equations, 604
Thermal runaway effect, 65–69
Thermionic emission, 566, 600
Thermionic reignition, 603
Thermoelastic ratcheting effect in bolted
    aluminum joints, 266
Thin contaminant films, electrically
    conducting layers and, 23–28
Thompson effect, 638–639
Three-body abrasion, 424, 425
Three dimensional (3–D) surface
    measurement systems, 757–758
Three-gas tests, 168

Tin
    coatings, 525, 539
    corrosion mechanisms, 152
    oxide and resistivity, 54, 141
    plating thickness, 521
    properties, 1198–1201
    as terminal coating material, 393
    and tin alloys, 280–282, 437–439
    whiskers, 541–542
Tin and tin–lead alloy systems, 466–467
Tin-base coatings *vs.* tin-base surfaces,
        453–456
Tin-based contacts, degradation of, 439
Tin Commandments, 133–134
Tin contacts, classical electrical contact theory
        in, 88–89
Tin–copper interfaces, intermetallic growth
        in, 40–41
Tin–lead alloy, 420, 539–540
Tin–lead-coated checkerboard surface, 479
Tin oxide, growth rate and electrical
        resistivity, 49–50
Tin-plated contact regions, surface
        morphology of, 454–455
Tin-plated copper contacts, surface
        morphology of, 466–467
Tin-plating, 331, 525
    thickness, 522–523
Tin–tin contacts, 134
    degradation of, 478
T-M breaker, *See* Thermal-magnetic breaker
Torch brazing, 967
Torque, 318–320
    nut factor, 319
Townsend avalanche, 561
Townsend breakdown, 563
TPA, *See* Terminal position assurance
Transient recovery voltage (TRV), 598, 798, 831
Transients, power quality issues, 347
Transition washers, 337–338
Transmission electron microscopy,
        (TEM), 1111
Transverse magnetic field, 596
Transverse magneto resistance, 240
Transverse rupture strength (TRS), 916
Trip system, 814–815
TRS, *See* Transverse rupture strength
TRV, *See* Transient recovery voltage
Tungsten, 245
Tungsten contacts, 689, 718, 936–937
Tungsten–silver contacts, 659–665, 908–928
Tungsten–silver microstructure, 913
Tunnel effect, 566–567, 1056

Tunnel resistance, 713
    and vibration, 1048–1051
Turbo-generator brushes, 1090
Two-body abrasion, 424, 427, 433
Two-body abrasive wear, electrographic wear,
        indexes from, 430
Type I (Unstable) contact resistance, behavior,
        449–452
Type II (Intermediate) contact resistance
        behavior, 449–452
Type III (Stable) contact resistance behavior,
        449–452

U

Underwriters Laboratory, (UL), 327, 794, 806
Ultrasonic cleaning, 225
Underplate
    effect of, 530–531
    nickel, *See* Nickel underplate
    reduction in chemical reactivity of finishes
        by, 531–532
Unintentional contamination, 1117–1119
Universal motors, 1085–1086
Unmated exposures, mated *vs.*, 172–174
US type miniature circuit-breaker,
        815–816

V

Vacuum arc, 578, 828
    anode spot, 594
    axial magnetic field, 596–597, 832
    cathode spots, 593–594
    columnar, 595
    diffuse, 592–595
    formation, 592–593
    interruption in alternating circuits,
        607–608
    modes, 595
    and short-gap breakdown, 566–569
    transverse magnetic field, 596, 832
Vacuum breakdown, 566–569, 602
    critical fields, 568
    electron emission, 566
    enhancement factor, 568
    metal vapor, 567–568
Vacuum contactors, 805, 830
Vacuum circuit breakers, 831
Vacuum interrupter contact materials
    chop current, 836, 838–389
    chrome–copper, 836, 936, 1210
    other suggestions, 836

requirements, 835
tungsten–copper/tungsten carbide–silver,
    836, 935–936, 1210
Vacuum interrupters
    design, 829–831
    principle/applications, 828–829
    recovery and influence of design, 607,
        831–839
    simulation of arcs in, 839–840
Van der Waals force, 1119
Vapor deposition, 393
Vapor phase corrosion inhibitors
    (VCIs), 180
Verband der Elektrotechnik (VDE), 794
Velocity effects
    opening and closing, 980
    switches, make operation, 752
Velocity factor (VF), 871
Velocity strengthening, 1102
Vertical free burning arc, 580
VF, *See* Velocity factor
Vibration, corrosion effect, 180
Vickers hardness (VH), 1196
Visual inspection of corrosion, 129
Voltage characteristics of molten metal
    bridge, 573
Voltage–current characteristics
    for free burning, 589, 590
    function of contact gap, 591
    for gas breakdown, 570
    of separated contacts, 569–570
Voltage regulator, 1088
Voltage–temperature relation, 5, 58 60, 96,
    572, 723
    deviation in assymetric contact, 64–65
    with temperature-dependent electrical
        resistivity and thermal conductivity,
        60–62
    validity of, 84
Volumetric erosion, 758–761

W

Water-sealing products, 343
Water soluble salts, 192–193
Water vapor, 1125, *See also* Humidity
    in saturated air, 1209
Weakly conducting films, 45–47
Wear, 415, 1047–1048, 1071, *See also* Sliding
    wear; Fretting
    abrasion, 424–426
    adhesion, 416–417
    clad metals, 435–437

delamination and subsurface wear,
    427–428
delamination mechanism of, 271, 481
equation, 415, 1048
gold platings, 422–423
mild and severe, 418–419
prow formation, 419–421
rate, reduction in, 496–497
rider wear, 421
stages in, 438
tin and tin–lead alloys, 437–439
underplate and substrate, 428–431
to underplate hardness, relationship of,
    431–435
unlubricated adhesive wear, 432
    electrographic wear indexes from, 429
unlubricated palladium–palladium
    contacts, 487
Wear characteristics, 1137–1138
Wear debris, 1107, 1141
Wear resistance, 1134
Wedge-flow mechanism, 437
Weibull distribution, 986
Weight gain measurements of corrosion, 129
Welded connections, 339–342
Welded contact assembly designs
    components, 960, 961
    friction welding, 966
    parameters of, 961–962
    percussion welding, 965
    resistance welding, *See* Resistance welding
    special welding methods, 965–966
    ultrasonic welding of, 965–966
Weld force, 654–657
Welding
    of closed contacts, 651–654
        current level, 652, 811
    as contacts close, 654–657
    as contacts open, 657
    dynamic welding, 654–657, 902–903
    measurement, 989–981, 992
    silver metal oxide materials, 904
    static welding resistance, 902
    switching contacts, *See* Contacts switching
Wexler's formula, 1111
Whiskers
    silver, 542
    tin, 541–542
White plague, 132
Wiedemann–Franz law, 62–63, 96–100, 477
    electrical resistivity and thermal
        conductivity, 96–100
Wipe cleaning motion, 417

Wiping, 459, 1095
    contacts, sliding and, 482–486
    motion, 742
Wire connectors, 236
Wire materials, 864, 865
Wire–screw terminal, 385–386
Wire-welding, 963
Wire–wire twist terminal, 384–385
Wire-wound potentiometers, 1133
Work hardening
    first bounce, 777
    switches, make operation, 750
Worst-case office air condition, 145
Wound field magnet motor, 1086, 1088
Wrought metals, 521–522
Wrought noble contact metals, 426

X

XES, *See* X-ray energy spectroscopy
XPS, *See* X-ray photoelectron spectroscopy

X-ray diffraction analysis of fretting
    debris, 280
X-ray energy spectroscopy (XES), 188
X-ray photoelectron spectroscopy (XPS), 129,
    272, 693, 695

Y

Yielding, *See* Deformation
Yield stress *vs.* hardness, 250, 651
Young's modulus, 681
    of elasticity, 1045

Z

ZCA, *See* Zone of closest approach
Zero-insertion-force (ZIF), 383
ZOI, *See* Zone of influence
Zone of closest approach (ZCA), 1098
Zone of influence (ZOI), 1120
Zone selective interlocking (ZSI), 871